THE STATISTICAL MECHANICS OF IRREVERSIBLE PHENOMENA

This book provides a comprehensive and self-contained overview of recent progress in nonequilibrium statistical mechanics, in particular, the discovery of fluctuation relations and other time-reversal symmetry relations. The significance of these advances is that nonequilibrium statistical physics is no longer restricted to the linear regimes close to equilibrium but extends to fully nonlinear regimes. These important new results have inspired the development of a unifying framework for describing both the microscopic dynamics of collections of particles, and the macroscopic hydrodynamics and thermodynamics of matter itself. The book discusses the significance of this theoretical framework in relation to a broad range of nonequilibrium processes, from the nanoscale to the macroscale, and is essential reading for researchers and graduate students in statistical physics, theoretical chemistry, and biological physics.

PIERRE GASPARD is a professor in physics at the Université Libre de Bruxelles and co-director of the Interdisciplinary Center for Nonlinear Phenomena and Complex Systems. He is the author of the book, *Chaos, Scattering and Statistical Mechanics* (Cambridge University Press, 1998), and he has published more than 200 related papers in the fields of statistical physics, nonlinear physics, and chemical physics.

THE STATISTICAL MECHANICS OF IRREVERSIBLE PHENOMENA

PIERRE GASPARD
Université Libre de Bruxelles

CAMBRIDGE
UNIVERSITY PRESS

University Printing House, Cambridge CB2 8BS, United Kingdom

One Liberty Plaza, 20th Floor, New York, NY 10006, USA

477 Williamstown Road, Port Melbourne, VIC 3207, Australia

314–321, 3rd Floor, Plot 3, Splendor Forum, Jasola District Centre, New Delhi – 110025, India

103 Penang Road, #05–06/07, Visioncrest Commercial, Singapore 238467

Cambridge University Press is part of the University of Cambridge.

It furthers the University's mission by disseminating knowledge in the pursuit of education, learning, and research at the highest international levels of excellence.

www.cambridge.org
Information on this title: www.cambridge.org/9781108473729
DOI: 10.1017/9781108563055

First published 2022

A catalogue record for this publication is available from the British Library.

Library of Congress Cataloging-in-Publication Data
Names: Gaspard, Pierre, 1959- author.
Title: The statistical mechanics of irreversible phenomena / Pierre Gaspard, Université Libre de Bruxelles.
Description: First edition. | Cambridge, United Kingdom; New York, NY: Cambridge University Press, 2022. | Includes bibliographical references and index. Identifiers: LCCN 2021061928 (print) | LCCN 2021061929 (ebook) |
ISBN 9781108473729 (hardback) | ISBN 9781108563055 (epub)
Subjects: LCSH: Nonequilibrium statistical mechanics.
Classification: LCC QC174.86.N65 G37 2022 (print) |
LCC QC174.86.N65 (ebook) | DDC 530.13/2–dc23/eng20220314
LC record available at https://lccn.loc.gov/2021061928
LC ebook record available at https://lccn.loc.gov/2021061929

ISBN 978-1-108-47372-9 Hardback

In memory of Grégoire Nicolis

Contents

Preface

Time asymmetry is observed in many phenomena, which are referred to as irreversible. At the macroscale, the arrow of time is expressed by the second law of thermodynamics in terms of the so-called entropy. Yet, at the microscale, the laws of electrodynamics and mechanics are symmetric under time reversal. Irreversibility and microreversibility are often opposed, and we may wonder how such contrasted aspects may be compatible with each other.

Recent advances in statistical mechanics are shedding a new light on this issue. Since pioneering work by Maxwell and Boltzmann, statistical mechanics has been building a bridge between the motion of atoms and molecules composing matter and its macroscopic properties. During recent decades, discoveries in the nanosciences have revealed the existence of diverse molecular structures and processes on all the scales intermediate between the size of atoms and the macroscopic world, this latter being usually characterized by the Avogadro number equal to $N_{\mathrm{A}} = 6.02214076 \times 10^{23}$ particles per mole.[1] With covalent bonds, atoms can form large molecules such as fullerenes and carbon nanotubes, as well as arbitrarily long macromolecules. Nanoclusters with a few hundred atoms may undergo liquid–solid transitions (Haberland et al., 2005). Chemical nanoclocks can manifest themselves in heterogeneous catalytic reactions, as observed with field ion microscopy (McEwen et al., 2009). Single-electron transport and irreversibility are measured in submicrometric semiconducting quantum dots (Küng et al., 2012). In biological cells, linear and rotary molecular motors made of proteins can perform mechanical power developing about 10^{-18} W (Alberts et al., 1998). More generally, the metabolism and the self-reproduction of biological cells are driven by nanometric enzymes dissipating energy.

At the nanoscale, thermal and molecular fluctuations are prevailing due to the atomic structure of matter, so that statistical mechanics plays a fundamental role in the description of such small systems. Statistical mechanics supplements the laws of mechanics by making assumptions on the initial and boundary conditions for externally prepared or controlled systems, and by methods to predict the properties of the systems from these assumptions. Indeed, the laws of mechanics formulated by Newton, Hamilton, Schrödinger, and others are based on ordinary or partial differential equations, leaving unspecified the initial and

[1] By its historical definition as an SI unit, this number refers to artefacts such as the kilogram and, thus, probably more to human muscular strength than to any property of the inanimate world of atoms.

boundary conditions. In this regard, the symmetry under time reversal can be considered either for the equations of motion, which define microreversibility, or for the initial conditions, which concern the statistical level of description. This key point is used, in particular, to establish the so-called fluctuation relations, which constitute a major advance in statistical mechanics, allowing us to understand today the properties of irreversible phenomena on the basis of the reversible microscopic dynamics of atoms and electrons.

The aim of this book is to provide a comprehensive overview of these advances in statistical mechanics. For this purpose, the successive chapters explain how statistical mechanics can make predictions by linking together different theories, including thermodynamics, hydrodynamics, and the theory of stochastic processes.

Chapter 1 presents thermodynamics, where the equilibrium and nonequilibrium properties can be identified using the second law of thermodynamics.

Chapter 2 is devoted to statistical mechanics, where the concepts of statistical ensembles and probability distributions are introduced on the basis of classical mechanics. In this framework, the distinction between equilibrium and nonequilibrium statistical ensembles can be made by considering their symmetry under time reversal. The concept of entropy is associated with the probability distributions describing the system. Moreover, linear response theory and the fluctuation-dissipation theorem are formulated in the classical setting and the projection-operator methods are summarized.

The deduction of hydrodynamics from the underlying microscopic dynamics is carried out in Chapter 3 by considering local equilibrium distributions. The method is shown to extend to the phases of matter with broken continuous symmetries such as crystals and liquid crystals.

In Chapter 4, the theory of stochastic processes is elaborated for physicochemical systems described as Markovian processes. In this context, the rate of entropy production is deduced and the Hill–Schnakenberg network theory is explained. The case of Brownian motion is used to illustrate how the probabilistic description can be inferred from statistical mechanics.

The fluctuation relations are presented in Chapter 5, first for the nonequilibrium work and then for the energy and particle fluxes across open systems in contact with several reservoirs. Their deduction is performed on the basis of microreversibility in the framework of classical statistical mechanics, leading to exact fluctuation relations and the connection to entropy production. Furthermore, the fluctuation relations are shown to have fundamental consequences about the linear and nonlinear response properties. The Onsager–Casimir reciprocal relations are found for the linear response properties. Their generalizations up to arbitrarily high orders are obtained for the nonlinear response properties. Moreover, the multivariate fluctuation relation for the currents is also established within the theory of stochastic processes using the Hill–Schnakenberg network theory.

In Chapter 6, path probabilities and temporal disorder are defined and their properties under time reversal are inferred, showing that the rate of entropy production can be related to time asymmetry in temporal disorder. The analogy with other symmetry-breaking phenomena is discussed.

Chapters 7–13 apply the previous results to different types of nonequilibrium processes.

Chapter 7 deals with driven Brownian particles and analogous electric circuits, as well as to related stochastic processes.

The case of effusion processes is presented in Chapter 8, for which the fluctuation relation and the connection between the entropy production and the time asymmetry in temporal disorder can be directly proved from mechanics.

The processes ruled by Boltzmann's kinetic equation in dilute and rarefied gases are studied in Chapter 9, where the fluctuation relation for the energy and particle fluxes is obtained from the fluctuating Boltzmann equation.

Chapter 10 presents several processes where fluctuation relations can be obtained from fluctuating chemohydrodynamics: transport by diffusion, diffusion-influenced surface reactions, ion transport, diodes, transistors, and Brownian motion described by non-Markovian generalized Langevin processes deduced from fluctuating hydrodynamics.

The stochastic approach to reactive systems is developed in Chapter 11, where fluctuation relations are obtained for chemical reactions.

Chapter 12 considers several cases of active processes: transmembrane ion transport, molecular motors, and chemically propelled Janus particles. In these active processes, energy transduction is ruled by a fluctuation relation for the fluxes that are coupled together.

In Chapter 13, transport is studied using Hamiltonian dynamical models. The periodic Lorentz gases and the multibaker map are used to investigate deterministic diffusion and to mathematically construct the diffusive modes on the basis of the microscopic dynamics. Fourier's law for heat conduction is shown to hold in many-particle billiard models. Furthermore, the importance of the nonlinear response properties is illustrated with models for mechanothermal coupling.

The last two chapters are concerned with quantum systems. Quantum statistical mechanics is summarized in Chapter 14, showing how quantum master equations and stochastic Schrödinger equations can be deduced with methods similar to those used in Brownian motion theory. Finally, transport in open quantum systems is presented in Chapter 15, where the fluctuation relation for the energy and particle fluxes is established within the framework of quantum mechanics. Systems with interacting and noninteracting particles are considered. The scattering approach is developed for the full counting statistics of noninteracting particles and for their temporal disorder. The Onsager–Casimir reciprocal relations and their generalizations beyond the linear regime are shown to hold in quantum as well as classical systems. The transport properties are described in particular for fermions, bosons, and electrons in mesoscopic devices such as quantum dots, quantum point contacts, and single-electron transistors.

Several appendices provide complements on thermodynamics, dynamical systems theory, statistical mechanics, hydrodynamics, stochastic processes, and fluctuation relations.

I wish here to express my gratitude to my students, postdoctoral associates, coworkers, and colleagues. This research was supported by the "Université libre de Bruxelles (ULB)," the "Fonds de la Recherche Scientifique – FNRS," and the Belgian Federal Government under the Interuniversity Attraction Pole programme.

1
Thermodynamics

1.1 Generalities

Thermodynamics aims to describe many-particle systems on the macroscale, i.e., on spatial scales larger than the distances between the particles and temporal scales longer than the corresponding time intervals. Thermodynamics enunciates general principles governing the balance of physical quantities characterizing such macroscopic systems. These physical quantities are the state variables, also called macrovariables, that are defined by observing the system on the macroscale. The state variables include mechanical variables such as the energy E and the particle numbers N_k, which are defined in the framework of the underlying microscopic mechanics, as well as the nonmechanical variable called entropy S. This latter was introduced by Clausius (1865), who established its existence at the macroscale in addition to the mechanical properties, in particular, using the study of Carnot (1824) on the behavior of gases in idealized steam engines.

Basically, the system is delimited by a boundary and has a volume V. The system can be an engine, a device, a machine, a motor, or part of a larger system, such as a volume element in a continuous medium like a fluid or a solid.

The time evolution of the system may result from internal transformations and also from exchanges with its environment, as schematically represented in Figure 1.1. During the evolution of any kind (i.e., spontaneous time evolution or evolution under some external drive), some state variable X changes by some infinitesimal amount dX at every infinitesimal step of the evolution. Mathematically speaking, dX is the differential of X. This differential may have two contributions

$$dX = d_{\rm e}X + d_{\rm i}X. \tag{1.1}$$

The contribution $d_{\rm e}X$ is due to the exchanges of X with the exterior of the system (i.e., its environment) and the contribution $d_{\rm i}X$ is caused by the transformations inside the system (Prigogine, 1967). The symbols $d_{\rm e}X$ and $d_{\rm i}X$ denote contributions that are not given by the differential of some function. The notation $đX$ is also often used for such nondifferential contributions. If there is no environment, we have that $d_{\rm e}X = 0$ for any quantity X and the system is said to be *isolated*.

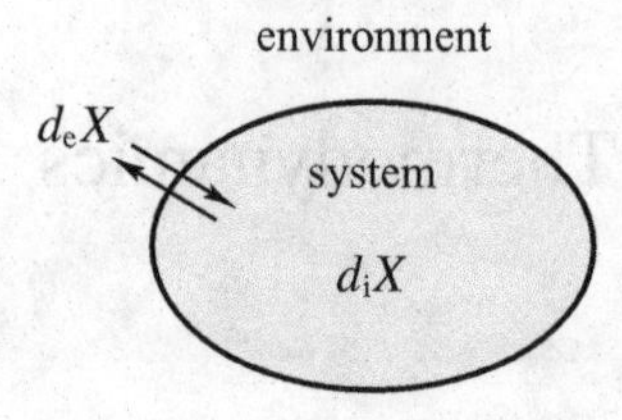

Figure 1.1 Schematic representation of a system in contact with its environment. Changes of the macrovariable X with the *exterior* of the system are denoted by $d_e X$ and $d_i X$ denotes those occurring within its *interior*.

Thermodynamics is formulated as follows with three laws, specifically concerning energy and entropy.

1.2 Energy and Other Conserved Quantities

The first law of thermodynamics is a principle of conservation.

First law: *There exists a state variable called the energy E that is conserved in every internal transformation of the system, i.e.,*

$$dE = d_e E + d_i E \qquad \textit{with} \qquad d_i E = 0. \tag{1.2}$$

Energy is measured in joules (SI unit), calories, or electron-Volts (eV), depending on the context. The first law of thermodynamics expresses the conservation of energy in any form. The energy is the sum of all the forms of energy: kinetic, potential, electric, magnetic, thermal, chemical, nuclear, gravitational, etc. The first law is justified in all the mechanical theories of physics as resulting from the symmetry of the equations of motion under time translations,[1] which implies the conservation of a quantity identified as energy by the theorem of Noether (1918). We note that the first law defines energy up to a constant value that remains arbitrary.

Beside energy, there exist other quantities that are also conserved as the result of fundamental symmetries:

- linear momentum (by symmetry under spatial translations[1]);
- angular momentum (by symmetry under rotations[1]);
- electric charge (by local gauge symmetry[2]);
- leptonic number (by global symmetry[3]);
- baryonic number (by global symmetry[3]);

[1] These fundamental symmetries of Minkowski's spacetime belong to the Poincaré group, also called the inhomogeneous Lorentz group. This group reduces to the Galilean group in the nonrelativistic limit (Weinberg, 1995).

[2] This fundamental symmetry holds at every spacetime point for the quantum fields associated with electrically charged particles (Weinberg, 1996).

[3] This other fundamental symmetry is independent of spacetime and holds for the quantum fields associated with leptonic or baryonic particles (Weinberg, 1996).

according to experimental observations (Weinberg, 1995, 1996). Every one of these quantities obeys equation (1.1) with $d_{\mathrm{i}}X = 0$, as expressed for energy by the first law.

Among the state variables, we also have the numbers N_k of the particles of different species $k = 1, 2, \ldots, c$. The particles are supposed to be identical objects that should be considered in the description of the system, such as photons, leptons, baryons, nuclei, atoms, molecules, and supramolecular entities. If some particles undergo reactions, their numbers are not conserved so that

$$dN_k = d_{\mathrm{e}}N_k + d_{\mathrm{i}}N_k \tag{1.3}$$

with $d_{\mathrm{i}}N_k \neq 0$, depending on the reaction rates and the stoichiometric coefficients of the species k in the reactions. However, if there is no reaction and the species k is conserved, we again have that $d_{\mathrm{i}}N_k = 0$ and the particle number N_k goes along the other conserved quantities.

It is also possible that the particle numbers $\{N_k\}_{k=1}^{c}$ are not conserved, but that some linear combination of them, $L_j = \sum_{k=1}^{c} \ell_{jk} N_k$, is nevertheless conserved, so that $d_{\mathrm{i}}L_j = 0$, which defines an effective conservation law. The existence of the conserved quantities L_j depends on the energy scale of the reactions taking place inside the system. For low collision energies, in the absence of chemical reactions the molecules are preserved so that $d_{\mathrm{i}}N_k = 0$, where k denotes a molecular species. At higher collision energies, though still below the energy of the strongest chemical bonds, some parts of molecules called moieties (Nelson and Cox, 2017) may be preserved by the reactions, in which case the numbers L_j of these moieties are conserved. At collision energies higher than the energy of the chemical bonds, the molecules break up into atoms so that only the numbers A_j of atoms are conserved. If ionization occurs, the numbers of electrons and ions become the relevant state variables, as in electrolytes or plasmas. Moreover, different isotopes may be distinguished by their mass m_j. The numbers of isotopes are conserved as long as there is no radioactivity. Within the nonrelativistic description, the law of mass conservation holds, which is expressed as $dM = d_{\mathrm{e}}M + d_{\mathrm{i}}M$ with $d_{\mathrm{i}}M = 0$, where $M = \sum_j m_j A_j$ is the total mass of the system. For still higher energies at the scale of MeV or higher, radioactivity and nuclear reactions break the conservation laws of the mass and the numbers of atomic nuclei, so that systems should be described in terms of nucleons and possibly other particles such as photons, electrons, positrons, and neutrinos. At energies above about 100 MeV, further particles should be included in the description (Weinberg, 1995, 1996).

We note that entities much larger than atoms or molecules may also be counted, such as atomic or molecular clusters, colloidal particles, crystalline particles, or biological entities such as viruses, organelles, or cells. In every case, an issue is to assess the relevance of the thermodynamic description adopted.

A system is said to be *closed* if only energy is exchanged with its environment, i.e., if $d_{\mathrm{e}}E \neq 0$ but $d_{\mathrm{e}}N_k = 0$. A system is said to be *open* if energy and matter are exchanged with its environment, i.e., if $d_{\mathrm{e}}E \neq 0$ and $d_{\mathrm{e}}N_k \neq 0$.

The environment is often supposed to be much larger than the system, in which case it plays the role of energy or particle reservoir. The environment may also be composed of several such reservoirs in contact with the system.

1.3 Entropy

In addition to the mechanical state variables, there is a nonmechanical variable that obeys the **Second law:** *There exists a state variable called entropy S such that*

$$dS = d_e S + d_i S \quad \textit{with} \quad d_i S \geq 0. \tag{1.4}$$

The entropy production $d_i S$ is thus always nonnegative. The evolution or transformation undergone by the system is said to be *reversible* if $d_i S = 0$ and *irreversible* if $d_i S > 0$. The system remains at *thermodynamic equilibrium* if $d_i S = 0$ and it is *out of equilibrium* if $d_i S > 0$. In this latter case, there is a time asymmetry in the macroscopic description of the system. We note that $d_e S$ may be positive, negative, or zero, depending on the exchanges between the system and its environment.

1.3.1 Equilibrium Macrostates

If the system is at equilibrium, i.e., if $d_i S/dt = 0$, its (absolute) *temperature* is defined by differentiating the energy with respect to the entropy,

$$T \equiv \left(\frac{\partial E}{\partial S}\right)_{V,\{N_k\}_{k=1}^c}, \tag{1.5}$$

where all the other variables remain constant. The SI unit of temperature is the kelvin (K), which is related to the SI unit of energy by Boltzmann's constant $k_B = 1.380649 \times 10^{-23}$ J/K. Accordingly, the entropy has the units of joule per kelvin (J/K). At equilibrium again, the (hydrostatic) *pressure* is defined as

$$p \equiv -\left(\frac{\partial E}{\partial V}\right)_{S,\{N_k\}_{k=1}^c}, \tag{1.6}$$

and the *chemical potential* of species k as

$$\mu_k \equiv \left(\frac{\partial E}{\partial N_k}\right)_{V,S,\{N_j\}_{j(\neq k)=1}^c}. \tag{1.7}$$

As a consequence, the energy of an equilibrium macrostate varies according to the *Gibbs relation*

$$dE = T\,dS - p\,dV + \sum_{k=1}^{c} \mu_k\, dN_k, \tag{1.8}$$

when changing its entropy, its volume, and particle numbers. In equation (1.8), $đQ = T dS$ corresponds to the change of heat under the transformation. We note that other contributions may be included for instance from electromagnetism

$$dE\big|_{\text{em}} = \int_V (\boldsymbol{\mathcal{E}} \cdot d\boldsymbol{\mathcal{D}} + \boldsymbol{\mathcal{H}} \cdot d\boldsymbol{\mathcal{B}})\, d^3r, \tag{1.9}$$

where $\boldsymbol{\mathcal{E}}$ is the electric field, $\boldsymbol{\mathcal{D}}$ the electric displacement, $\boldsymbol{\mathcal{H}}$ the magnetizing field, $\boldsymbol{\mathcal{B}}$ the magnetic field, and d^3r the volume element (Landau and Lifshitz, 1984); or from the interface between two bulk phases

$$dE\Big|_{\text{surf}} = \gamma \, d\Sigma, \tag{1.10}$$

where γ is the surface tension and $d\Sigma$ some change of the interfacial surface area Σ. Between three bulk phases, a further contribution from line tension should be added (Rowlinson and Widom, 1989).

Since energy, entropy, and particle numbers are *extensive variables* proportional to the volume, the thermodynamically conjugated variables, which are temperature, pressure, and chemical potentials, are *intensive variables* independent of the volume at the macroscale. Further intensive variables can be defined by dividing the extensive variables, for instance, with the volume to get the *densities*.

An important consequence of the second law is that the entropy should be maximal at equilibrium. In turn, the Gibbs relation (1.8) implies that the temperature, the pressure, and the chemical potentials must be uniform across an equilibrium system, as shown in Appendix A. This fundamental property of equilibrium macrostates does not preclude the existence of equilibrium spatial structures since thermodynamically conjugated variables, i.e., the entropy, mass, and particle densities, are left unconstrained. In particular, crystals are equilibrium spatially periodic structures classified by the 230 space groups in three dimensions (Ashcroft and Mermin, 1976). Vortex lattices in type-II superconductivity are other examples of equilibrium spatial structures. In any case, equilibrium macrostates are stationary at the macroscale (although dynamical at the microscale).

Since the second law is formulated in terms of a differential, the entropy is only defined up to a constant, as in the case of energy. Nevertheless, the constant of entropy can be determined with the

Third law: *If the system has a unique microstate of minimal energy, the entropy vanishes at absolute zero temperature:*

$$\lim_{T \to 0} S = 0. \tag{1.11}$$

Accordingly, the absolute value of the entropy can be defined with the third law on the basis of an assumption about the microstates of minimal energy (Pauling, 1970).

Another consequence of the Gibbs relation (1.8) is that the energy E is a state variable that depends on the entropy S, the volume V, and the particle numbers $\{N_k\}_{k=1}^{c}$. The energy therefore plays the role of thermodynamic potential $E(S, V, \{N_k\}_{k=1}^{c})$ for a system with independently fixed values of these variables. However, another set of independent variables may be required if the entropy, the volume, and the particle numbers are not fixed in the system of interest. We are thus led to define other thermodynamic potentials by performing Legendre transforms, substituting one variable by the thermodynamically conjugated variable that is fixed, as explained in Appendix A. This leads to the definition of the enthalpy describing systems where the pressure is fixed instead of volume, the Helmholtz free energy for systems where the temperature is fixed instead of entropy, the Gibbs free energy (or free enthalpy) if the temperature and the pressure are fixed instead of entropy and volume, or the grand thermodynamic potential if the temperature and the chemical potentials are fixed instead of entropy and particle numbers. Various thermodynamic potentials can thus be introduced depending on the experimental conditions imposed on the system of interest. Moreover, inverting equation (1.8), we obtain an expression for the change of entropy

$$dS = \frac{1}{T}\,dE + \frac{p}{T}\,dV - \sum_{k=1}^{c} \frac{\mu_k}{T}\,dN_k, \tag{1.12}$$

showing that the entropy can also play the role of thermodynamic potential given by the function $S(E, V, \{N_k\}_{k=1}^{c})$.

The thermodynamic properties of chemical substances have been measured experimentally and they are known, in particular, under standard conditions ($T^0 = 298.15$ K, $p^0 = 100$ kPa). The values of the standard molar enthalpy and the Gibbs free energy of formation, as well as the standard molar entropy, are tabulated for many chemical substances (Lide, 2000). Since the values of the state variables do not depend on the pathway followed to reach some equilibrium macrostate, the thermodynamic properties can be determined in mixtures on the basis of their composition.

1.3.2 Nonequilibrium Macrostates

The system is out of equilibrium if entropy is produced inside the system, i.e., if $d_iS/dt > 0$.

Isolated Systems

If the system is isolated, there is no environment, which implies that $d_eS/dt = 0$. In this case, the time derivative of the entropy is only determined by the entropy production rate according to

$$\frac{dS}{dt} = \frac{d_iS}{dt} \geq 0. \tag{1.13}$$

Therefore, the entropy increases in the system up to its maximal value corresponding to the equilibrium macrostate, as shown in Figure 1.2(a).[4] The second law thus conveys the

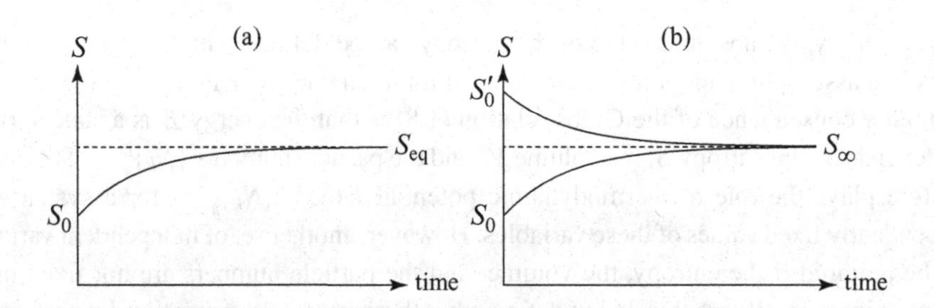

Figure 1.2 Possible time evolutions of the entropy towards an asymptotic stationary value in (a) an isolated system and (b) a nonisolated system.

[4] Clausius (1865) expressed the first and second laws for the universe. If the entropy state variable was known everywhere in the universe, the entropy of the universe could be decomposed as $S_{univ} = S_{sys} + S_{env}$, i.e., into the entropies of the system and its environment shown in Figure 1.1. Since the universe contains everything, it is isolated, so that $dS_{univ}/dt = d_iS_{univ}/dt \geq 0$ by equation (1.13), which is the statement of Clausius (1865). Neither the system nor its environment being isolated, the second law (1.4) gives $dS_{sys}/dt = d_eS_{sys}/dt + d_iS_{sys}/dt$ and $dS_{env}/dt = d_eS_{env}/dt + d_iS_{env}/dt$. Moreover, the amounts of entropy exchanged between the system and its environment and *vice versa* are opposite to each other: $d_eS_{sys}/dt = -d_eS_{env}/dt$. Therefore, the sum of the entropies produced inside the system and its environment is equal to the one produced in the universe: $d_iS_{univ}/dt = d_iS_{sys}/dt + d_iS_{env}/dt \geq 0$. We note that the system of interest is often significantly smaller than the universe, in which case $d_iS_{univ}/dt \gg d_iS_{sys}/dt \geq 0$.

observation that, during some nonequilibrium transients, an isolated macroscopic system undergoes a relaxation towards the macrostate of thermodynamic equilibrium and that energy is dissipated in the sense that the macroscopic movements in the system come to rest when the equilibrium macrostate is reached. This time asymmetry is a characteristic feature of nonequilibrium systems at the macroscale.

Systems in Contact with One Reservoir

If the system is not isolated, its environment may form an energy or particle reservoir. Now, the system is closed or open due to the exchanges of energy or particles with the environment, so that

$$\frac{dS}{dt} = \frac{d_{\rm e}S}{dt} + \frac{d_{\rm i}S}{dt} \tag{1.14}$$

with some entropy exchange rate $d_{\rm e}S/dt$. If these conditions hold, the system will undergo a relaxation towards a macrostate of global equilibrium with its environment. In the long-time limit, the entropy of the system will reach a stationary value, as schematically represented in Figure 1.2(b) with $\lim_{t\to\infty} S = S_\infty$. Since equilibrium is global in this stationary macrostate, there is no entropy exchange between the system and its environment, $\lim_{t\to\infty} d_{\rm e}S/dt = 0$. Therefore, the entropy production rate also vanishes in this limit, $\lim_{t\to\infty} d_{\rm i}S/dt = 0$, and the stationary macrostate reached after the nonequilibrium transients is the equilibrium macrostate corresponding to the temperature, pressure, and chemical potentials of the environment. Since the system is not isolated, the entropy does not have to increase with time and the initial value of the entropy may be smaller or larger than its asymptotic value, as shown in Figure 1.2(b). For instance, the system may initially be hotter than its environment, in which case there will be a heat flux outgoing the system during the nonequilibrium transients, the system will thus cool, and its entropy will decrease from S_0' to $S_\infty = S_{\rm eq}$.

Systems in Contact with Several Reservoirs

If the environment is composed of several energy or particle reservoirs at different fixed values of their temperature, pressure, and chemical potentials, the system in contact with these reservoirs cannot reach an equilibrium macrostate as long as these differences persist. Since arbitrarily large reservoirs keep their temperature, pressure, and chemical potentials, the system can be maintained in macrostates with persistent exchanges of energy or particles between the system and the different reservoirs. Remarkably, several types of macrostates are possible under nonequilibrium conditions. After some possible transitory relaxation, stationarity can be reached in the system. Again, since the system is not isolated, the initial value of the entropy may be smaller or larger than its asymptotic value S_∞, as depicted in Figure 1.2(b). In such a stationary macrostate, the entropy of the system remains stationary so that $dS/dt = 0$. Therefore, the second law and equation (1.14) imply that

$$\frac{d_{\rm i}S}{dt} = -\frac{d_{\rm e}S}{dt} > 0 \tag{1.15}$$

in the stationary macrostate. Consequently, the entropy produced inside the system is evacuated to the environment, thus keeping invariant the system entropy. Such macrostates are called *nonequilibrium steady states*. Since entropy is continuously produced, energy should be supplied in order to compensate energy dissipation and maintain these steady states.

In addition, there also exist macrostates that are nonstationary even after transients. Such macrostates may be oscillatory with periodic, quasiperiodic, or chaotic dynamical behavior. The latter is nonperiodic, manifesting sensitivity to initial conditions and dynamical randomness over long timescales, as in turbulence. In these regimes, the macrostates evolve with time towards attractors in the space of macrovariables because of dissipation. These attractors are limit cycles, tori, or fractals, whether the dynamical behavior is periodic, quasiperiodic, or chaotic (Bergé et al., 1984; Eckmann and Ruelle, 1985; Strogatz, 1994; Nicolis, 1995).

Systems with Time-Dependent Driving

Systems may also be driven out of equilibrium by time-dependent external forces. Examples are systems heated by electromagnetic waves or driven by the periodic motion of pistons. In such circumstances, the system cannot reach a stationary macrostate and its state variables remain time dependent.

1.4 Thermodynamics in Continuous Media

1.4.1 Balance Equations

In continuous media, the principles of thermodynamics are applied to every volume element d^3r of the macrosystem, which is here assumed to be nonrelativistic. This latter is described in terms of densities associated with the slowest observable quantities, which include the locally conserved quantities such as mass, energy, linear momentum, and possibly other variables such as particle numbers or order parameters. The set of these quantities depends on the continuous medium whether it is a fluid with one or several compounds, a liquid crystal, a crystal, a superfluid, a plasma, or something else. Since these systems differ by their compositions, the relevant variables will be different, but the breaking of continuous symmetries, for instance in liquid crystal, crystals, and superfluids, may introduce order parameters and extra slow modes called the Nambu–Goldstone modes that arise from the fast kinetic modes of normal fluids at phase transitions (Forster, 1975).

A continuous medium is described in terms of fields $x(\mathbf{r}, t)$ defined at any position $\mathbf{r} \in \mathbb{R}^3$ inside the system and any time $t \in \mathbb{R}$. The time evolution of some density x is ruled by the balance equation

$$\partial_t x + \nabla \cdot \mathbf{J}_x = \sigma_x, \tag{1.16}$$

where $\mathbf{J}_x$ is the associated current density and σ_x the corresponding production rate density. The current density has the units of the density x multiplied by a velocity or, equivalently, the units of the transported quantity X per unit surface and unit time.[5] Integrating the density x

[5] Current densities are also called flows (Balescu, 1975; de Groot and Mazur, 1984).

over some volume V that is assumed to be fixed in space, we obtain the amount of this quantity in this volume

$$X \equiv \int_V x\, d^3r. \tag{1.17}$$

Carrying out the same integration for the balance equation (1.16), we get the time derivative of this quantity as

$$\frac{dX}{dt} = \frac{d_e X}{dt} + \frac{d_i X}{dt}, \tag{1.18}$$

where

$$\frac{d_e X}{dt} = -\int_{\partial V} \boldsymbol{J}_x \cdot d\boldsymbol{\Sigma} \tag{1.19}$$

is the contribution due to the exchanges of the quantity X at the boundary ∂V of the system with the exterior ($d\boldsymbol{\Sigma}$ being the vector surface element) and where

$$\frac{d_i X}{dt} = \int_V \sigma_x\, d^3r \tag{1.20}$$

is the production rate of X inside the system. We thus recover the global form (1.1) at the basis of the formulation of thermodynamics.

If the quantity x is locally conserved, the production rate density is equal to zero, i.e., $\sigma_x = 0$.

In normal fluids, the fluid elements are advected by the motion of the fluid described by the velocity field $\mathbf{v}$. In every element of the fluid, the velocity is defined as the velocity of the center of mass of the element. Denoting $d\mathbf{P}$ to be the linear momentum in the fluid element of volume d^3r and mass dM, the velocity is thus defined as $\mathbf{v} \equiv d\mathbf{P}/dM$. Introducing the mass density $\rho \equiv dM/d^3r$ and the linear momentum density $\mathbf{g} \equiv d\mathbf{P}/d^3r$, the velocity is thus given by $\mathbf{v} = \mathbf{g}/\rho$. The advection contributes to the current density $\boldsymbol{J}_x$ associated with the density x according to

$$\boldsymbol{J}_x = x\mathbf{v} + \boldsymbol{\mathscr{J}}_x, \tag{1.21}$$

where $\boldsymbol{\mathscr{J}}_x$ is the rest of the current density due to the flow of x with respect to the center of mass of the fluid element, which is either identical or related to the corresponding diffusive or dissipative current density $\boldsymbol{\mathcal{J}}_x$.

Table 1.1 gives the different quantities that are relevant in normal fluids with chemical reactions (Prigogine, 1967; de Groot and Mazur, 1984). Every quantity with $\sigma_x = 0$ is locally conserved. This is the case in particular for mass, which thus obeys the well-known continuity equation. The local conservation of mass results from the balance equations of the different molecular species k because the diffusive current densities are defined with respect to the center of mass of every fluid element, so that $\sum_k m_k \boldsymbol{\mathcal{J}}_k = 0$, and because every chemical reaction conserves mass, $\sum_k m_k \nu_{kr} = 0$, where m_k is the mass of the molecules of species k and ν_{kr} the stoichiometric coefficient of species k in the reaction r of rate density w_r. We note that the local conservation of angular momentum implies that the

Table 1.1. *Normal fluids with chemical reactions: The relevant quantities, their density x, the rest* $\mathscr{J}_x$ *of the current density, and production rate density* σ_x. *Here,* $n_k \equiv dN_k/d^3r$ *denotes of the density of species k (also called concentration),* $\mathcal{J}_k$ *the corresponding diffusive current density,* ν_{kr} *the stoichiometric coefficient of species k in the reaction r of rate density* w_r, ρ *the mass density,* m_k *the mass of the particles of species k,* **g** *the linear momentum density,* **v** *the fluid velocity field,* **P** *the pressure tensor,* ϵ *the total energy density, e the internal energy density,* $\mathcal{J}_q$ *the heat current density, s the entropy density, and* $\mathcal{J}_s$ *the diffusive current density of entropy. The pressure tensor is composed of the hydrostatic pressure p multiplied by the* 3×3 *identity matrix* $\mathbf{1}$, *and its viscous part* $\mathbf{\Pi} \equiv \mathcal{J}_{\mathbf{g}}$.

Quantity	x	$\mathscr{J}_x \equiv J_x - x\mathbf{v}$	σ_x
Number of particles k	n_k	$\mathcal{J}_k$	$\sum_r \nu_{kr} w_r$
Mass	$\rho = \sum_k m_k n_k$	0	0
Momentum	$\mathbf{g} = \rho\mathbf{v}$	$\mathbf{P} = p\mathbf{1} + \mathbf{\Pi}$	0
Energy	$\epsilon = \frac{\rho}{2}\mathbf{v}^2 + e$	$\mathbf{P}\cdot\mathbf{v} + \mathcal{J}_q$	0
Entropy	s	$\mathcal{J}_s$	$\sigma_s \geq 0$

pressure tensor is symmetric $\mathbf{P} = \mathbf{P}^{\mathrm{T}}$, where the superscript T denotes the transpose.[6] In the presence of external force fields, the balance equations of linear momentum and energy have nonvanishing source terms σ_x describing the force and work exerted by the resulting external force on the fluid element (de Groot and Mazur, 1984).

At every time t, the macrostate of a normal fluid with c components is determined by their densities $\{n_k(\mathbf{r},t)\}_{k=1}^{c}$, the velocity field $\mathbf{v}(\mathbf{r},t)$, and the temperature field $T(\mathbf{r},t)$, at every point $\mathbf{r}$ of the system. An alternative set of fields is given by the mass density $\rho(\mathbf{r},t)$, the fluid velocity, the temperature, and the mass fractions of the solute species because the mass fraction of the solvent can be deduced from them and the mass density. Since the temperature determines the internal energy, the time evolution of the fluid macrostate is ruled by $c+4$ partial differential equations given by the balance equations for the particle densities, the linear momentum, and the energy. However, these balance equations do not yet form a closed set of partial differential equations because knowledge of the fluid properties is still missing.

1.4.2 Local Thermodynamic Equilibrium and Consequences

In order to determine the still missing properties in accordance with the second law, the hypothesis of local thermodynamic equilibrium is supposed to hold in every fluid element. Using the entropy density as thermodynamic potential, its variations satisfy the Gibbs relation

[6] Because of the local conservation of linear momentum $\partial_t \mathbf{g} + \nabla\cdot J_{\mathbf{g}} = 0$, the angular momentum density $\boldsymbol{\ell} = \mathbf{r}\times\mathbf{g}$ obeys the balance equation $\partial_t \boldsymbol{\ell} + \nabla\cdot J_{\boldsymbol{\ell}} = \sigma_{\boldsymbol{\ell}}$ with the angular momentum current density $J_{\boldsymbol{\ell}} = \mathbf{r}\times J_{\mathbf{g}}$ and the source density with components $(\sigma_{\boldsymbol{\ell}})_i = -\sum_{jk}\epsilon_{ijk}P_{jk}$ expressed in terms of the Levi-Civita totally antisymmetric tensor such that $\epsilon_{ijk} = \epsilon_{jki} = -\epsilon_{ikj}$ and $\epsilon_{xyz} = +1$. Accordingly, the source density is equal to zero if the pressure tensor is symmetric, $P_{jk} = P_{kj}$. The assumption here is that there is no intrinsic angular momentum (spin), which should otherwise be included in the balance equation, leading to a possible antisymmetric part for the pressure tensor (de Groot and Mazur, 1984).

Table 1.2. *The irreversible processes in normal fluids, their affinity $\mathcal{A}_\alpha$, their associated diffusive current density $\mathcal{J}_\alpha$, their space character, and the time-reversal parity of their affinity (Prigogine, 1967; Nicolis, 1979; de Groot and Mazur, 1984). The symmetrized gradient of the velocity field is denoted by $(\nabla\mathbf{v})^{\mathrm{S}} = (\nabla\mathbf{v} + \nabla\mathbf{v}^{\mathrm{T}})/2$ and $\mathring{\mathbf{\Pi}} = \mathbf{\Pi} - \Pi\,\mathbf{1}$ denotes the traceless part of the viscous pressure tensor $\mathbf{\Pi} = \mathbf{P} - p\,\mathbf{1}$ with $\Pi = (\mathrm{tr}\,\mathbf{\Pi})/3$, T the temperature, μ_k the chemical potential of the species k, and ν_{kr} the stoichiometric coefficient of the species k in the reaction r of rate w_r.*

Irreversible Process	$\mathcal{A}_\alpha$	$\mathcal{J}_\alpha$	Space	Time
Shear viscosity	$\mathring{\mathbf{A}}_{\mathbf{g}} = -\left[(\nabla\mathbf{v})^{\mathrm{S}} - \frac{1}{3}(\nabla\cdot\mathbf{v})\mathbf{1}\right]/T$	$\mathring{\mathbf{J}}_{\mathbf{g}} = \mathring{\mathbf{\Pi}}$	Tensor	Odd
Dilational viscosity	$\mathcal{A}_{\mathbf{g}} = -(\nabla\cdot\mathbf{v})/T$	$\mathcal{J}_{\mathbf{g}} = \Pi$	Scalar	Odd
Reaction r	$\mathcal{A}_r = -\sum_k \mu_k\,\nu_{kr}/T$	$\mathcal{J}_r = w_r$	Scalar	Even
Heat conductivity	$\boldsymbol{\mathcal{A}}_q = \nabla(1/T)$	$\boldsymbol{\mathcal{J}}_q$	Vector	Even
Diffusion of species k	$\boldsymbol{\mathcal{A}}_k = \nabla(-\mu_k/T)$	$\boldsymbol{\mathcal{J}}_k$	Vector	Even

$$ds = \frac{1}{T}\,de - \sum_{k=1}^{c} \frac{\mu_k}{T}\,dn_k, \tag{1.22}$$

as shown in Appendix A. Accordingly, the entropy density is given by the equilibrium function $s = s(e, \{n_k\}_{k=1}^{c})$ locally defined at every point and every time in the fluid. Therefore, the hypothesis of local thermodynamic equilibrium assumes that the entropy density depends on the densities of the quantities relevant to the continuous medium, $s(\{x\})$. However, this hypothesis may have to be extended to include the gradients of some densities, e.g., taking $s(\{x\}, \{\nabla x\})$, for inhomogeneous fluids (Penrose and Fife, 1990; Wang et al., 1993), or some systems with chemical reactions (Mátyás and Gaspard, 2005). The inclusion of gradients leads in particular to the Ginzburg–Landau theory of the free-energy functional density (Landau and Lifshitz, 1980a,b; Evans, 1979).

According to equation (1.22), which is based on the hypothesis of local thermodynamic equilibrium in normal fluids, the balance equations (1.16) for the particle and energy densities allow us to deduce the balance equation for the entropy density, as shown in Appendix A. This equation has the form of equation (1.16) with $x = s$, with the diffusive current density of entropy given by

$$\boldsymbol{\mathcal{J}}_s = \frac{1}{T}\,\boldsymbol{\mathcal{J}}_q - \sum_{k=1}^{c} \frac{\mu_k}{T}\,\boldsymbol{\mathcal{J}}_k \tag{1.23}$$

in terms of the heat current density and the diffusive current densities of the particles, and the entropy production rate density

$$\sigma_s = \sum_\alpha \mathcal{A}_\alpha\,\mathcal{J}_\alpha \geq 0 \tag{1.24}$$

expressed with the affinities $\mathcal{A}_\alpha$ and current densities $\mathcal{J}_\alpha$ of the irreversible processes taking place in the system. These latter quantities are given in Table 1.2 for normal fluids. The entropy production rate density (1.24) should always be nonnegative in accordance with the second law. Integrating the balance equation of the entropy density s recovers the global balance equation (1.14) with the entropy exchange and production rates given by equations (1.19) and (1.20) for $X = S$ and $x = s$. Entropy is conserved if the irreversible processes are negligible.

We note that the diffusive current densities of particle species should satisfy the mass conservation condition $\sum_{k=1}^{c} m_k \mathcal{J}_k = 0$. Consequently, one of these current densities can be related to the other ones:

$$\mathcal{J}_c = -\sum_{k=1}^{c-1} \frac{m_k}{m_c} \mathcal{J}_k. \tag{1.25}$$

The species c is often taken as the solvent and the other ones as the solutes. The total contribution of diffusion to the entropy production rate density (1.24) can thus be expressed in terms of the mutual diffusion of the solute species in the solvent as

$$\sum_{k=1}^{c} \mathcal{A}_k \mathcal{J}_k = \sum_{k=1}^{c-1} \mathcal{A}'_k \mathcal{J}_k, \tag{1.26}$$

by redefining the affinities according to

$$\mathcal{A}'_k \equiv \nabla\left(-\frac{\mu'_k}{T}\right) \qquad \text{with} \qquad \mu'_k \equiv \mu_k - \frac{m_k}{m_c}\,\mu_c. \tag{1.27}$$

These affinities are the thermodynamic forces of mutual diffusion of the solute species k with respect to the solvent c. Since these redefinitions are linear, they can be performed every time the constraint of mass conservation is met.

1.4.3 Equilibrium and Nonequilibrium Constitutive Relations

Closing the partial differential equations of the fluid first requires the knowledge of two equilibrium equations of state, one for the pressure $p(T, \{n_k\}_{k=1}^{c})$ and another one for the internal energy $e(T, \{n_k\}_{k=1}^{c})$. These equations of state are the equilibrium properties for the material composing the system.

Besides this, we also need the nonequilibrium properties given by some relations between the affinities and the current densities in Table 1.2. These nonequilibrium constitutive relations should satisfy the symmetries of the continuous medium. For this purpose, we use the Curie symmetry principle (Curie, 1894; Prigogine, 1967). Since normal fluids are isotropic, their properties are symmetric under continuous spatial rotations, so that the tensorial character of the affinities $\mathcal{A}_\alpha$ and current densities $\mathcal{J}_\alpha$ should be respected. In crystals, the space group of the lattice should determine the relations between the affinities and the current densities. Moreover, the underlying microscopic mechanics is symmetric under time reversal according to electrodynamics. Therefore, the relations between the affinities and the current densities should also satisfy the consequences of this discrete symmetry called *microreversibility*.

The nonequilibrium constitutive relations may be linear or nonlinear.

Linear Relations

Typically, the relations are linear if the gradients of the macrofields extend over distances $\|(\nabla x)/x\|^{-1}$ larger than the mean free path of the particles in the fluid. This is usually the case for transport properties such as viscosity, heat conduction, and particle diffusion. Under such circumstances, the current densities are linearly related to the affinities as

$$\mathcal{J}_\alpha = \sum_\beta \mathcal{L}_{\alpha,\beta}\, \mathcal{A}_\beta \tag{1.28}$$

with some linear response coefficients $\mathcal{L}_{\alpha,\beta}$ characterizing the nonequilibrium properties of the fluid. In order to satisfy the Curie symmetry principle, such relations may only exist between current densities and affinities of the same tensorial character.

Since there is only one quantity given by a tensor, the traceless part of the viscous pressure is related to the corresponding affinity by $\overset{\circ}{\mathbf{\Pi}} = \overset{\circ}{\mathcal{L}}_{\mathbf{g},\mathbf{g}} \overset{\circ}{\mathbf{A}}_\mathbf{g}$ where the linear response coefficient is proportional to the coefficient η of shear viscosity by $\overset{\circ}{\mathcal{L}}_{\mathbf{g},\mathbf{g}} = 2T\eta$. Among the scalar quantities, the direct linear relation $\Pi = \mathcal{L}_{\mathbf{g},\mathbf{g}} \mathcal{A}_\mathbf{g}$ holds in the absence of chemical reactions, which defines the dilational or bulk viscosity $\zeta \equiv \mathcal{L}_{\mathbf{g},\mathbf{g}}/T$. Consequently, the viscous part of the pressure tensor is given by

$$\mathbf{\Pi} = -\eta \left(\nabla \mathbf{v} + \nabla \mathbf{v}^{\mathrm{T}} - \frac{2}{3} \nabla \cdot \mathbf{v}\, \mathbf{1} \right) - \zeta\, \nabla \cdot \mathbf{v}\, \mathbf{1}. \tag{1.29}$$

For the vectorial quantities, the direct relations $\mathcal{J}_q = \mathcal{L}_{q,q} \mathcal{A}_q$ and $\mathcal{J}_k = \mathcal{L}_{k,k} \mathcal{A}_k$ give, respectively,

$$\textit{Fourier's law:} \qquad \mathcal{J}_q = -\kappa\, \nabla T \tag{1.30}$$

with the coefficient of heat conductivity $\kappa \equiv \mathcal{L}_{q,q}/T^2$, and

$$\textit{Fick's law:} \qquad \mathcal{J}_k = -\mathcal{D}_k \nabla n_k \tag{1.31}$$

with the coefficient of diffusion $\mathcal{D}_k \equiv (\mathcal{L}_{k,k}/T)(\partial \mu_k/\partial n_k)_T$.

Beyond the direct effects described by the coefficients $\mathcal{L}_{\alpha,\alpha}$, there also exists the possibility of thermodiffusive coupling between heat conduction and particle transport with the Soret effect relating $\mathcal{J}_k$ to $\mathcal{A}_q$ and the reciprocal Dufour effect relating $\mathcal{J}_q$ to $\mathcal{A}_k$, as well as possible cross-diffusion expressed by linear relations between $\mathcal{J}_k$ and $\mathcal{A}_l$ with $k \neq l$ (de Groot and Mazur, 1984; Haase, 1969). These effects are described by the coefficients $\mathcal{L}_{\alpha,\beta}$ with $\alpha \neq \beta$, coupling together the affinity and the current density of different processes. These couplings often play essential roles because they make possible the driving of a process by the thermodynamic force of another process, as in thermoelectric and mechanochemical effects. They may thus induce energy transduction of different kinds. For such couplings between different transport processes, microreversibility leads to the Onsager–Casimir reciprocal relations, as discussed in the following chapters.

With the linear relations (1.28), the entropy production rate density (1.24) is given by the quadratic form

$$\sigma_s = \sum_{\alpha\beta} \mathcal{L}^{\rm S}_{\alpha,\beta}\, \mathcal{A}_\alpha\, \mathcal{A}_\beta \geq 0, \qquad \text{where} \qquad \mathcal{L}^{\rm S}_{\alpha,\beta} \equiv \frac{1}{2}(\mathcal{L}_{\alpha,\beta} + \mathcal{L}_{\beta,\alpha}) \tag{1.32}$$

forms the symmetrized matrix of linear response coefficients. According to the second law, this matrix should be nonnegative, i.e., $(\mathcal{L}^{\rm S}_{\alpha,\beta}) \geq 0$. Therefore, the transport coefficients such as the shear and dilational viscosities, the heat conductivity, and the diffusion coefficients should be nonnegative: $\eta \geq 0$, $\zeta \geq 0$, $\kappa \geq 0$, and $\mathcal{D}_k \geq 0$. In the case of coupling between two processes $\alpha, \beta = 1, 2$, the symmetrized linear response coefficients should satisfy the condition $\mathcal{L}^{\rm S}_{1,1}\mathcal{L}^{\rm S}_{2,2} \geq (\mathcal{L}^{\rm S}_{1,2})^2$ in order for the entropy production to be always nonnegative.

Nonlinear Relations

However, if the macrofields rapidly vary over distances comparable to or smaller than the mean free path of the particles, the constitutive relations are nonlinear and of the general form

$$\mathcal{J}_\alpha = \sum_\beta \mathcal{L}_{\alpha,\beta}\, \mathcal{A}_\beta + \frac{1}{2}\sum_{\beta\gamma} \mathcal{M}_{\alpha,\beta\gamma}\, \mathcal{A}_\beta\, \mathcal{A}_\gamma + \frac{1}{6}\sum_{\beta\gamma\delta} \mathcal{N}_{\alpha,\beta\gamma\delta}\, \mathcal{A}_\beta\, \mathcal{A}_\gamma\, \mathcal{A}_\delta + \cdots \tag{1.33}$$

with nonlinear response coefficients $\mathcal{M}_{\alpha,\beta\gamma}$, $\mathcal{N}_{\alpha,\beta\gamma\delta}$, ... This is the case, in particular, for chemical reactions because their reactants and products are separated by molecular distances corresponding to the rearrangement of atoms in the reaction. Therefore, the relations between the reaction rates w_r and the associated affinities $\mathcal{A}_r$ are typically nonlinear. In dilute solutions, the rates of elementary chemical reactions[7] are proportional to the densities of all the species incoming the reactive events, which is the basis of the so-called *mass action law* (Pauling, 1970; Moore, 1972; Berry et al., 1980; Kondepudi and Prigogine, 1998). Accordingly, the elementary chemical reaction

$$\sum_{k=1}^{c} \nu^{(+)}_{kr}\, {\rm X}_k \underset{k_{-r}}{\overset{k_{+r}}{\rightleftharpoons}} \sum_{k=1}^{c} \nu^{(-)}_{kr}\, {\rm X}_k \tag{1.34}$$

between the molecular species $\{{\rm X}_k\}_{k=1}^{c}$ has the net rate density

$$w_r = w_{+r} - w_{-r} = k_{+r} \prod_{k=1}^{c} \left(\frac{n_k}{n^0}\right)^{\nu^{(+)}_{kr}} - k_{-r} \prod_{k=1}^{c} \left(\frac{n_k}{n^0}\right)^{\nu^{(-)}_{kr}} \tag{1.35}$$

expressed in terms of the rate constants $k_{\pm r}$ and the numbers $\nu^{(\pm)}_{kr}$ of molecules of species k respectively incoming the forward and reverse reactions, as well as the standard density n^0

[7] An elementary reaction is associated with a single barrier or transition state separating reactants from products, as opposed to a reaction that goes through several barriers or transition states. Every barrier or transition state forms a bottleneck where the process is slowed down and the transition rate is lower than the pace of dynamics in the potential wells of reactants and products. The crossing of the barrier may result from thermal activation or quantum tunneling. An overall reaction or reaction network should be decomposed into elementary reactions before evaluating the entropy production.

equal to one mole per liter. We note that the stoichiometric coefficient of the species k in the reaction (1.34) is given by $\nu_{kr} = \nu_{kr}^{(-)} - \nu_{kr}^{(+)}$. Since the chemical potential of a solute in a dilute solution is given by

$$\mu_k = \mu_k^0 + k_{\rm B} T \ln \frac{n_k}{n^0}, \tag{1.36}$$

where $k_{\rm B}$ is Boltzmann's constant, the net rate density depends on the affinity $\mathcal{A}_r = k_{\rm B} \ln(w_{+r}/w_{-r})$ of the reaction according to the nonlinear relations

$$w_r = w_{+r}\left(1 - {\rm e}^{-\mathcal{A}_r/k_{\rm B}}\right) = w_{-r}\left({\rm e}^{\mathcal{A}_r/k_{\rm B}} - 1\right). \tag{1.37}$$

The contribution of the elementary reaction r to the entropy production rate density is given by

$$\sigma_{s,r} = \mathcal{A}_r\, w_r = k_{\rm B}\,(w_{+r} - w_{-r}) \ln \frac{w_{+r}}{w_{-r}} \geq 0, \tag{1.38}$$

which is always nonnegative since the rate constants $k_{\pm r}$ and the rate densities $w_{\pm r}$ are nonnegative. At chemical equilibrium, the rate vanishes together with the affinity because there is *detailed balance* for every reaction r, resulting in equality between the rates of the forward and reversed reactions, $w_{+r} = w_{-r}$ (Wegscheider, 1901; Fowler, 1929). As a consequence, the net reaction rate (1.35) is equal to zero at equilibrium, which implies that the equilibrium densities of the reacting species should satisfy the Guldberg–Waage condition

$$\prod_{k=1}^{c} \left(\frac{n_k}{n^0}\right)_{\rm eq}^{\nu_{kr}} = K_r \tag{1.39}$$

with the equilibrium constant $K_r \equiv k_{+r}/k_{-r}$ (Guldberg and Waage, 1879; Moore, 1972). Since the affinity of the reaction is equal to zero at equilibrium $\mathcal{A}_r = 0$, the equilibrium values of the chemical potentials should obey the identity $\sum_{k=1}^{c} \mu_{k,{\rm eq}} \nu_{kr} = 0$. We note that, close to chemical equilibrium, the reaction rate is approximately proportional to the affinity, $w_r \simeq \mathcal{L}_{r,r}\mathcal{A}_r$, which defines the linear response coefficient $\mathcal{L}_{r,r} = w_{+r}/k_{\rm B}$. Beyond, i.e., for larger values of the affinity, we should include nonlinear terms, obtained with the Taylor expansion (1.33) for equation (1.37). Similar nonlinear relations also exist in diodes and transistors where electron and hole densities sharply vary across the junctions at the core of these nonlinear electric devices. We also note that, in virtue of Curie's symmetry principle, the divergence of the velocity field may contribute to the reaction rates and, reciprocally, the chemical reactions may contribute to the scalar component of the viscous pressure tensor (de Groot and Mazur, 1984; Haase, 1969).

Using these equilibrium and nonequilibrium constitutive relations, the balance equations can be closed and the fluid macrofields may be obtained by solving their partial differential equations with the boundary conditions applied to the system. The balance equation for linear momentum leads to the Navier–Stokes equations, while the balance equation for energy leads to the heat equation for the temperature. Several paradigmatic examples will be presented below. Beyond normal fluids, similar considerations are known for the different bulk phases of matter (Martin et al., 1972; Fleming and Cohen, 1976), as well as for active matter (Jülicher et al., 2018). Moreover, nonequilibrium thermodynamics can be extended to systems with two bulk phases separated by an interface, as shown in Section A.9.

1.5 Hydrodynamics and Chemohydrodynamics

1.5.1 Hydrodynamics in One-Component Fluids

In one-component fluids, the local conservation of mass leads to the continuity equation

$$\partial_t \rho + \nabla \cdot (\rho \mathbf{v}) = 0. \tag{1.40}$$

According to Tables 1.1 and 1.2, the balance equation (1.16) for linear momentum combined with the continuity equation (1.40) and the viscous pressure tensor (1.29) gives the Navier–Stokes equations for fluid mechanics,

$$\rho\,(\partial_t \mathbf{v} + \mathbf{v} \cdot \nabla \mathbf{v}) = -\nabla p + \eta \nabla^2 \mathbf{v} + \left(\zeta + \frac{\eta}{3}\right) \nabla(\nabla \cdot \mathbf{v}), \tag{1.41}$$

assuming that the viscosity coefficients do not depend on the densities so that $\nabla\eta = \nabla\zeta = 0$.

If the viscosity coefficients are equal to zero ($\eta = \zeta = 0$), we recover the Euler equations of hydrodynamics where fluid entropy is conserved and which thus describe a fluid without dissipation. If the fluid is incompressible, its mass density is constant in time and the continuity equation reduces to the constraint $\nabla \cdot \mathbf{v} = 0$, so that the property of dilational viscosity ζ does not exist in an incompressible fluid.

Furthermore, the continuity and Navier–Stokes equations should be coupled to the heat equation for the temperature field (see Appendix A).

1.5.2 Chemohydrodynamics in Multicomponent Fluids

Reactions may occur in multicomponent fluids. If the heat of reactions is negligible, the fluid may be supposed to be isothermal. Moreover, if the fluid is at rest, the velocity field $\mathbf{v}$ is equal to zero, the Navier–Stokes equations are satisfied, and the continuity equation implies that the mass density is constant in time. Therefore, the quantities evolving in such systems are the particle densities or concentrations of the different species reacting in the system. Since the different particle species $\{k\}$ composing the system may be transported by diffusion and transformed by the reactions $\{r\}$, the evolution is ruled by diffusion–reaction equations

$$\partial_t n_k = \mathcal{D}_k \nabla^2 n_k + \sum_r \nu_{kr}\, w_r\left(\{n_l\}_{l=1}^{c}\right) \tag{1.42}$$

under the assumptions that the diffusion coefficients have a negligible dependence on the densities such that $\nabla\mathcal{D}_k = 0$ and cross-diffusion can also be neglected.

As an example, we have the following diffusion–reaction equation for the density $n = n_{\mathrm{X}}$ of the species X generated by the autocatalytic reaction

$$\mathrm{X} \underset{k_-}{\overset{k_+}{\rightleftharpoons}} 2\mathrm{X}: \qquad \partial_t n = \mathcal{D}\,\nabla^2 n + k_+ n - k_- n^2, \tag{1.43}$$

which is nonlinear because of the autocatalytic character of the reaction. This equation describes the Verhulst model of population dynamics (Nicolis, 1995).

If the reactions are exothermic or endothermic, the diffusion–reaction equations should be coupled to the heat equation (see Appendix A).

1.6 Hydrodynamic Modes of Relaxation to Equilibrium

1.6.1 Hydrodynamic Modes in One-Component Fluids

As previously mentioned, the second law of thermodynamics predicts the relaxation towards equilibrium in an isolated system. This system can be considered as a one-component fluid of infinite extension in space. The equilibrium macrostate corresponds to a fluid at rest with zero velocity, uniform temperature T, uniform pressure p, and thus uniform mass density ρ. The hydrodynamic modes are the solutions of the fluid equations of motion corresponding to small deviations from equilibrium. Linearizing the continuity equation, the Navier–Stokes equations, and the balance equation of entropy around equilibrium given in Appendix A, we obtain the following set of partial differential equations for the deviations in the mass density $\delta\rho$, the entropy per unit mass $\delta\mathfrak{s} = \delta(s/\rho)$, and the fluid velocity $\delta\mathbf{v}$:

$$\partial_t \delta\rho = -\rho\,\nabla\cdot\delta\mathbf{v}, \tag{1.44}$$

$$\partial_t \delta\mathfrak{s} = \frac{\kappa}{\rho T}\,\nabla^2 \delta T, \tag{1.45}$$

$$\partial_t \delta\mathbf{v} = -\frac{1}{\rho}\,\nabla\delta p + \frac{\eta}{\rho}\,\nabla^2\delta\mathbf{v} + \frac{1}{\rho}\left(\zeta + \frac{\eta}{3}\right)\nabla(\nabla\cdot\delta\mathbf{v}). \tag{1.46}$$

The hydrodynamic modes of relaxation to equilibrium are spatially periodic solutions with an exponential dependence on time of the form $\delta\rho, \delta\mathfrak{s}, \delta\mathbf{v} \sim \exp(\imath\,\mathbf{q}\cdot\mathbf{r} + zt)$ with $\imath = \sqrt{-1}$. The dispersion relations of these modes give the dependence of their exponential rate z on their wave number $q = \|\mathbf{q}\|$ and they are obtained by solving equations (1.44)–(1.46) (Balescu, 1975; Résibois and De Leener, 1977). As shown in Appendix A, the dispersion relations of the hydrodynamic modes are given by

$$\text{2 shear modes:} \quad z = -\frac{\eta}{\rho}\,q^2, \tag{1.47}$$

$$\text{1 heat mode:} \quad z = -\frac{\kappa}{\rho \mathfrak{c}_p}\,q^2 + O(q^3), \tag{1.48}$$

$$\text{2 sound modes:} \quad z = \mp\imath\, v_{\mathrm{l}}\, q - \frac{1}{2\rho}\left[\zeta + \frac{4}{3}\eta + \kappa\left(\frac{1}{\mathfrak{c}_v} - \frac{1}{\mathfrak{c}_p}\right)\right] q^2 + O(q^3), \tag{1.49}$$

in terms of the sound velocity,

$$v_{\mathrm{l}} \equiv \sqrt{\left(\frac{\partial p}{\partial\rho}\right)_{\mathfrak{s}}}, \tag{1.50}$$

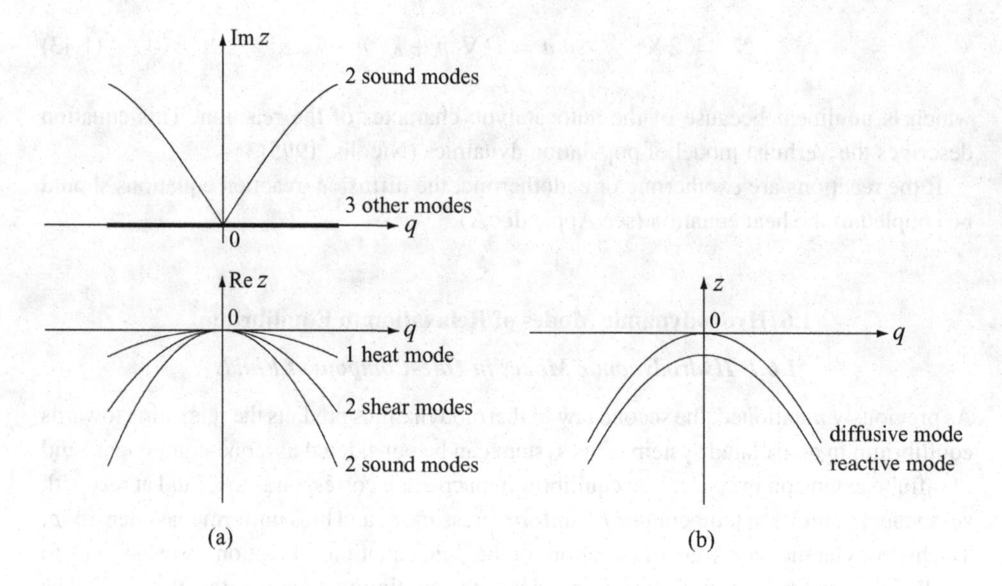

Figure 1.3 (a) Imaginary and real parts of the dispersion relations (1.47)–(1.49) for the five hydrodynamic modes in a normal fluid with one component. (b) Dispersion relations of the diffusive and reactive modes (1.54)–(1.55) for the reaction of isomerization A $\rightleftharpoons$ B taking place in an isothermal fluid at rest.

the heat capacities per unit mass at constant volume and pressure,

$$\mathfrak{c}_{\mathfrak{v}} \equiv T \left(\frac{\partial T}{\partial \mathfrak{s}} \right)_{\rho}^{-1} \qquad \text{and} \qquad \mathfrak{c}_{p} \equiv T \left(\frac{\partial T}{\partial \mathfrak{s}} \right)_{p}^{-1} , \tag{1.51}$$

and the transport coefficients η, ζ, and κ. The shear modes are transverse to the wave vector $\mathbf{q}$, while the heat and sound modes are longitudinal.[8]

The dispersion relations of the five modes are shown in Figure 1.3(a). All of them are vanishing with the wave number q since the five modes are associated with the five conserved quantities, namely, mass, energy, and the three components of linear momentum. The two shear modes and the heat mode are diffusive, while the two sound modes are propagating. They are all damped because the transport properties, i.e., the viscosities and the heat conductivity, dissipate energy. Consequently, the deviations given by the linear superpositions of these five modes undergo relaxation towards equilibrium. We note that this relaxation is here expressed in terms of the transport coefficients. If the latter are equal to zero, there is no energy dissipation and the fluid flow is isoentropic.

[8] These results hold if the wavelength, $\lambda = 2\pi/q$, of the deviations is larger than the mean free path, under which circumstances the particles are interacting. Nevertheless, the ideal gas law may still be used (after expressing the pressure in terms of the entropy per unit mass) as long as the mean free path is larger than the range of interaction between the particles. In this regard, the sound velocity (1.50) is the feature of interacting particles.

1.6.2 The Relaxation Modes in Diffusion–Reaction Systems

In fluids with several components but no reaction, there exist extra hydrodynamic modes associated with the mutual diffusion between the different species. There are as many extra modes as there are solutes in the solvent. The dispersion relations of these extra modes are also vanishing with the wave number q because of the conservation of species in the absence of chemical reactions.

However, this is no longer the case in the presence of reactions, as shown by the following example. We consider the isothermic reaction of isomerization, $\mathrm{A} \underset{k_-}{\overset{k_+}{\rightleftharpoons}} \mathrm{B}$, between two solute species A and B. The molecules A and B are diffusing in a solvent of inert molecules, the whole fluid being at rest. This system is a ternary mixture in which we expect five standard hydrodynamic modes plus two modes of mutual diffusion in the absence of reaction. For simplicity, the fluid is assumed to be at rest with a uniform temperature. If cross-diffusion is neglected, the system can be modeled by the two coupled diffusion–reaction equations:

$$\partial_t n_{\mathrm{A}} = \mathcal{D}_{\mathrm{A}} \nabla^2 n_{\mathrm{A}} - k_+ n_{\mathrm{A}} + k_- n_{\mathrm{B}}, \tag{1.52}$$

$$\partial_t n_{\mathrm{B}} = \mathcal{D}_{\mathrm{B}} \nabla^2 n_{\mathrm{B}} + k_+ n_{\mathrm{A}} - k_- n_{\mathrm{B}}. \tag{1.53}$$

Supposing that the deviations with respect to uniform densities behave as $\delta n_{\mathrm{A}}, \delta n_{\mathrm{B}} \sim \exp(\imath\, \mathbf{q} \cdot \mathbf{r} + zt)$, we obtain two dispersion relations:

$$\text{1 diffusive mode:} \quad z = -\mathcal{D} q^2 + O\big(q^4\big), \tag{1.54}$$

$$\text{1 reactive mode:} \quad z = -k_+ - k_- - \mathcal{D}_{\mathrm{r}} q^2 + O\big(q^4\big), \tag{1.55}$$

with the diffusion coefficients

$$\mathcal{D} \equiv \frac{k_+ \mathcal{D}_{\mathrm{B}} + k_- \mathcal{D}_{\mathrm{A}}}{k_+ + k_-} \quad \text{and} \quad \mathcal{D}_{\mathrm{r}} \equiv \frac{k_+ \mathcal{D}_{\mathrm{A}} + k_- \mathcal{D}_{\mathrm{B}}}{k_+ + k_-}. \tag{1.56}$$

These dispersion relations are depicted in Figure 1.3(b).

The dispersion relation of the diffusive mode is vanishing with the wave number q, because of the conservation of the total number of molecules A and B. However, the dispersion relation of the reactive mode is not vanishing because the reaction breaks the separate conservations of the molecule numbers of the two species. We notice that the dispersion relation (1.55) of the reactive mode satisfies $z(q = 0) = 0$ if the reaction rates vanish, i.e., $k_+ = k_- = 0$, in which case the ternary mixture has two diffusive modes in addition to its five standard hydrodynamic modes, as expected. Accordingly, we conclude that some hydrodynamic diffusive modes become kinetic modes with $z(q = 0) \neq 0$ in reacting systems. These reactive modes may evolve over timescales ranging from femtoseconds (for the fastest chemical reactions) to eons.

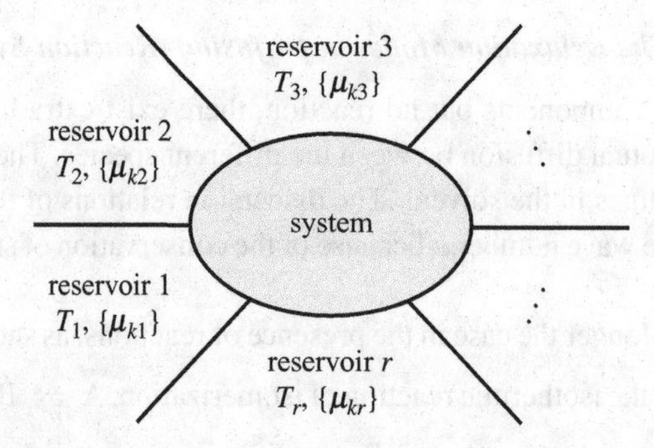

Figure 1.4 Schematic representation of a system in contact with r reservoirs at fixed temperatures T_j and chemical potentials $\{\mu_{kj}\}$.

1.7 Nonequilibrium Steady States

1.7.1 From Local to Global Affinities

Let us consider an open system formed by a solution at rest ($\mathbf{v} = 0$) and surrounded by an environment composed of several heat or particle reservoirs $j = 1, 2, \ldots, r$ at fixed temperatures T_j and chemical potentials $\{\mu_{kj}\}_{k=1}^{c}$, as shown in Figure 1.4.

We suppose that there is no reaction going on in the system and its environment. If the reservoirs have different temperatures and chemical potentials, the equilibrium conditions are not satisfied, so the system is out of equilibrium and crossed by fluxes of energy and/or matter. After some time, the system evolves towards a nonequilibrium steady state where the macrovariables take invariant values, in particular, $\partial_t T = 0$ and $\partial_t n_k = 0$. Accordingly, the balance equations for energy and particle numbers reduce to $\nabla \cdot \mathcal{J}_q = 0$ and $\nabla \cdot \mathcal{J}_k = 0$ in the nonequilibrium steady state. The entropy production rate is thus given by

$$\frac{d_i S}{dt} = \int_V \left(\mathcal{A}_q \cdot \mathcal{J}_q + \sum_{k=1}^{c} \mathcal{A}_k \cdot \mathcal{J}_k \right) d^3 r \geq 0, \tag{1.57}$$

in terms of the local affinities $\mathcal{A}_q = \nabla(1/T)$ and $\mathcal{A}_k = \nabla(-\mu_k/T)$. Substituting these expressions back into equation (1.57) and integrating by parts, the entropy production rate becomes

$$\frac{d_i S}{dt} = \int_V \left[\nabla \cdot \left(\frac{1}{T} \mathcal{J}_q \right) - \frac{1}{T} \underbrace{\nabla \cdot \mathcal{J}_q}_{=0} - \sum_{k=1}^{c} \nabla \cdot \left(\frac{\mu_k}{T} \mathcal{J}_k \right) + \sum_{k=1}^{c} \frac{\mu_k}{T} \underbrace{\nabla \cdot \mathcal{J}_k}_{=0} \right] d^3 r \geq 0. \tag{1.58}$$

Using the conditions of stationarity and the divergence theorem, the volume integral is reduced to a surface integral over the boundary ∂V of the system

$$\frac{d_i S}{dt} = \int_{\partial V} \left(\frac{1}{T} \mathcal{J}_q - \sum_{k=1}^{c} \frac{\mu_k}{T} \mathcal{J}_k \right) \cdot d\mathbf{\Sigma} \geq 0, \tag{1.59}$$

where $d\mathbf{\Sigma}$ is a vector surface element pointing towards the exterior of the volume V. Now, we notice that the boundary is composed of several parts, $\partial V = \cup_{j=1}^{r} \partial_j V$, each one in contact with one reservoir. On the part $\partial_j V$ of the boundary, the temperature and the chemical potentials are, respectively, fixed at the values T_j and $\{\mu_{kj}\}_{k=1}^{c}$ of the corresponding reservoir. Therefore, equation (1.59) can be written as

$$\frac{d_{\rm i}S}{dt} = \sum_{j=1}^{r} \left(-\frac{1}{T_j} J_{qj} + \sum_{k=1}^{c} \frac{\mu_{kj}}{T_j} J_{kj} \right) \geq 0 \tag{1.60}$$

in terms of the surface integrals of the heat and particle current densities that are *incoming* the system from the reservoir j:

$$J_{qj} \equiv - \int_{\partial_j V} \mathcal{J}_q \cdot d\mathbf{\Sigma} \qquad \text{and} \qquad J_{kj} \equiv - \int_{\partial_j V} \mathcal{J}_k \cdot d\mathbf{\Sigma}. \tag{1.61}$$

These quantities are the current intensities or, more simply, the currents. By conservation of energy and particle numbers (since there is no reaction), we have that

$$\sum_{j=1}^{r} J_{qj} = 0 \qquad \text{and} \qquad \sum_{j=1}^{r} J_{kj} = 0. \tag{1.62}$$

Accordingly, the current from one reservoir is determined by the currents from all the other reservoirs. Taking the reservoir $j = r$ as reference, the expression (1.60) finally becomes

$$\frac{1}{k_{\rm B}} \frac{d_{\rm i}S}{dt} = \sum_{j=1}^{r-1} \left(A_{qj} \, J_{qj} + \sum_{k=1}^{c} A_{kj} \, J_{kj} \right) \geq 0 \tag{1.63}$$

in terms of the global affinities respectively defined as

$$\text{thermal affinity:} \qquad A_{qj} \equiv \frac{1}{k_{\rm B} T_r} - \frac{1}{k_{\rm B} T_j}, \tag{1.64}$$

$$\text{chemical affinity:} \qquad A_{kj} \equiv \frac{\mu_{kj}}{k_{\rm B} T_j} - \frac{\mu_{kr}}{k_{\rm B} T_r}, \tag{1.65}$$

for the different reservoirs $j = 1, 2, \ldots, r-1$ (except the reference one) and the different particle species $k = 1, 2, \ldots, c$. The global affinities are here introduced by dividing the entropy production rate with Boltzmann's constant $k_{\rm B}$, so that the thermal affinities have the units of J^{-1} and the chemical affinities are dimensionless. These global affinities are the direct control parameters of the nonequilibrium drives because they are fixed in the reservoirs at the boundaries of the system, where the conditions are supposed to be under experimental control. If all these global affinities are equal to zero, the temperatures and chemical potentials are all equal in the reservoirs and the stationary macrostate becomes the equilibrium macrostate where the currents are equal to zero, together with the entropy production rate.

We note that similar considerations apply to nonequilibrium systems with fluid flow where the affinities are determined by the boundary conditions on the velocity field, as in Couette–Taylor or Poiseuille flows. Mechanical affinities may also be introduced in systems where external forces or torques perform work driving the system out of equilibrium, as ions driven by an electric field.

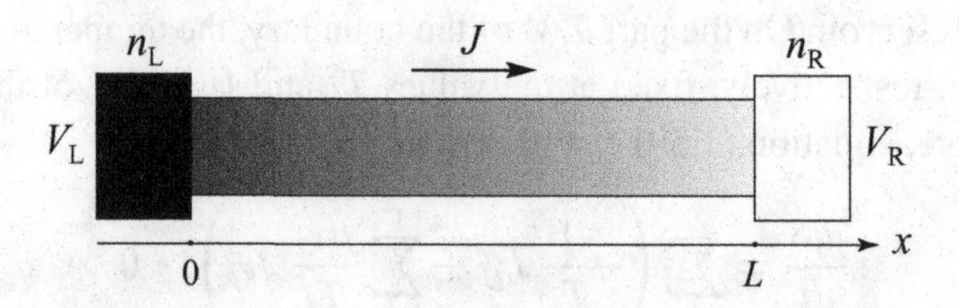

Figure 1.5 Schematic representation of a nonequilibrium steady state in the transport process of diffusion between two reservoirs with particle densities n_L and n_R and separated by the distance L.

1.7.2 Diffusion

As an example, let us consider the diffusion of a single solute species in an isothermal dilute solution in contact with two reservoirs located at $x = 0$ and $x = L$ and, respectively, having the densities n_L and n_R, as schematically depicted in Figure 1.5. The velocity of the solution is assumed to be equal to zero. This transport process is ruled by Fick's law $\jmath = \mathcal{J} = -\mathcal{D}(n)\nabla n$ with a diffusion coefficient $\mathcal{D}(n)$ *a priori* depending on the density n. The diffusion equation is thus given by

$$\partial_t n = \nabla \cdot [\mathcal{D}(n)\nabla n] \tag{1.66}$$

in the three-dimensional physical space (x, y, z). Taking the boundary conditions $n(0, y, z, t) = n_L$ and $n(L, y, z, t) = n_R$ for all times, the time evolution of the density will undergo a relaxation towards a stationary state $n(x)$ obeying

$$\frac{d}{dx}\left[\mathcal{D}(n)\frac{dn}{dx}\right] = 0. \tag{1.67}$$

Accordingly, the current density should take the uniform value $\jmath = -\mathcal{D}(n)dn/dx$ across the system, so that the density profile is given by solving the ordinary differential equation $dn/dx = -\jmath/\mathcal{D}(n)$, which can be integrated to obtain the current flowing across the sectional area Σ as

$$J \equiv \Sigma\, \jmath = \frac{\Sigma}{L}\int_{n_R}^{n_L} \mathcal{D}(n)\, dn. \tag{1.68}$$

If the diffusion coefficient is independent of the density, the profile is linear,

$$n(x) = n_L - \frac{n_L - n_R}{L}\, x, \tag{1.69}$$

and Fick's law is recovered at the global scale in the form,

$$J = \Sigma\, \frac{\mathcal{D}}{L}\,(n_L - n_R). \tag{1.70}$$

Otherwise, Fick's law only holds locally. The global affinity is here given by

$$A = \frac{1}{k_B T}\,(\mu_L - \mu_R) = \ln\frac{n_L}{n_R}. \tag{1.71}$$

Taking the left or right reservoir as reference, the current (1.70) can be expressed in terms of the global affinity, respectively, as

$$J(A) = \Sigma \frac{\mathcal{D}}{L} n_{\rm L}\left(1 - {\rm e}^{-A}\right) \qquad \text{or} \qquad J(A) = \Sigma \frac{\mathcal{D}}{L} n_{\rm R}\left({\rm e}^{A} - 1\right). \tag{1.72}$$

In general, the current (1.68) is a function $J(A)$ of the global affinity and the other properties of the macrostate. The entropy production rate is thus given by

$$\frac{1}{k_{\rm B}} \frac{d_{\rm i} S}{dt} = A\, J(A) \geq 0. \tag{1.73}$$

We note that, although we use a linear local relation between the current density and the local affinity with Fick's law, we find a nonlinear relation between the current and the global affinity for the nonequilibrium steady state driven by boundary conditions on the system.

If both reservoirs have the same density $n_{\rm L} = n_{\rm R}$, the global affinity together with the current and the entropy production rate (1.73) are equal to zero.

If we suppose that the reservoirs on the left and right sides of the conductive medium have finite volumes larger than the volume of the open system, i.e., $V_{\rm L}, V_{\rm R} \gg V_{\rm s} = \Sigma L$, there will be a slow evolution of their particle numbers $N_{\rm L} = V_{\rm L} n_{\rm L}$ and $N_{\rm R} = V_{\rm R} n_{\rm R}$ ruled by $dN_{\rm L}/dt \simeq -J$ and $dN_{\rm R}/dt \simeq +J$ with the current given by equation (1.68) or (1.70). In the simple diffusive case (1.70), there will thus be an exponential equilibration of the particle densities in the reservoirs according to

$$\frac{d}{dt}\left(n_{\rm L} - n_{\rm R}\right) \simeq -\Gamma\left(n_{\rm L} - n_{\rm R}\right) \qquad \text{with the rate} \qquad \Gamma = \frac{\mathcal{D}}{L^2}\left(\frac{V_{\rm s}}{V_{\rm L}} + \frac{V_{\rm s}}{V_{\rm R}}\right). \tag{1.74}$$

The equilibration time, $t_{\rm equil} = 1/\Gamma$, should be compared with the relaxation time, $t_{\rm relax} \sim L^2/\mathcal{D}$, taken to reach the nonequilibrium steady state in the diffusive medium of length L. We thus find that

$$\frac{t_{\rm relax}}{t_{\rm equil}} \sim \frac{V_{\rm s}}{V_{\rm L}} + \frac{V_{\rm s}}{V_{\rm R}} \ll 1, \tag{1.75}$$

i.e., the relaxation towards the nonequilibrium steady state is faster than the time taken by the reservoirs to reach a global equilibrium across the whole system if the reservoirs are much larger than the open system, i.e., $V_{\rm L}, V_{\rm R} \gg V_{\rm s}$. Accordingly, the reservoirs should be arbitrarily large in order to maintain a nonequilibrium steady state in an open system in contact with them.

Analoguous considerations apply to heat conduction ruled by Fourier's law in terms of the temperature instead of the particle density.

1.7.3 Ohm's Law for Electric Resistance

If we consider the transport of electric charges in a conductor, we need to include the effects of the electric field $\boldsymbol{\mathcal{E}} = -\boldsymbol{\nabla}\Phi$, or, equivalently, the electric potential Φ. For simplicity, we consider positive charge carriers moving in a conductor such as a resistor or an electrolytic solution. The electric charge density is thus given by $\rho_{\rm e} = e(n - n_0)$ where $e = |e|$ is the elementary electric charge, n the density of positively charged particles, and n_0 the uniform and invariant density of negatively charged particles forming a background, which is called the jellium model. The associated electric current density can be expressed as $j_e = e j = e\mathcal{J}$ in terms of the diffusive current density of the mobile particles

$$\mathcal{J} = -\mathcal{D}\nabla n + \beta e \mathcal{D} n \mathcal{E} = -\mathcal{D}\mathrm{e}^{-\beta e \Phi}\nabla\left(\mathrm{e}^{\beta e \Phi} n\right), \tag{1.76}$$

where $\mathcal{D}$ is their diffusion coefficient and $\beta = (k_B T)^{-1}$ is the inverse temperature, which is known as the Nernst–Planck equation (Probstein, 2003). Since the electric charge is locally conserved, the continuity equation

$$\partial_t \rho_e + \nabla \cdot j_e = 0 \tag{1.77}$$

is satisfied. Moreover, the electric field obeys Gauss' law

$$\nabla \cdot \mathcal{E} = \frac{\rho_e}{\epsilon}, \tag{1.78}$$

where ϵ is the dielectric coefficient of the material. The coupled equations (1.76), (1.77), and (1.78) define the so-called Nernst–Planck–Poisson problem.

We consider electric conduction in a piece of length L and cross-sectional area Σ in contact with two reservoirs, as shown in Figure 1.5. Here, the reservoirs have fixed values for the particle density and the electric potential: $n(0) = n_L$ with $\Phi(0) = \Phi_L$, and $n(L) = n_R$ with $\Phi(L) = \Phi_R$. In the presence of electric potential, the chemical potential should be replaced by the electrochemical potential, $\tilde{\mu} = \mu + e\Phi$. The global affinity (1.71) is thus given by $A = \beta(\tilde{\mu}_L - \tilde{\mu}_R) = \beta e V$, where

$$V = \Phi_L - \Phi_R + \frac{1}{\beta e}\ln\frac{n_L}{n_R} \tag{1.79}$$

is the applied voltage difference with respect to the Nernst potential with the assumption that the charge carriers are dilute in the conductor. The voltage (1.79) is equal to zero at equilibrium.

In the stationary macrostate, the current density $j = (j, 0, 0)$ is invariant and uniform because of the continuity equation (1.77) and the stationary condition $\partial_t n = 0$. Moreover, the electric field $\mathcal{E} = (\mathcal{E}, 0, 0)$ is determined by Gauss' law. Consequently, the stationary profiles of the particle density and the electric field are obtained by solving the coupled equations

$$\frac{dn}{dx} = \beta e n \mathcal{E} - \frac{j}{\mathcal{D}} \quad \text{and} \quad \frac{d\mathcal{E}}{dx} = \frac{e}{\epsilon}(n - n_0). \tag{1.80}$$

These latter admit the uniform solution with $n = n_0$ (electroneutrality) and $j = \beta e \mathcal{D} n_0 \mathcal{E}$, corresponding to *Ohm's law*, $j_e = \sigma\mathcal{E}$ with the electric conductivity $\sigma = \beta e^2 \mathcal{D} n_0$. In this case, the uniform charge density requires the boundary conditions $n_L = n_R = n_0$. Moreover, integrating the uniform electric field over the length L of the conductor, we obtain the potential difference $\mathcal{E} = (\Phi_L - \Phi_R)/L$, so that the voltage (1.79) is related to the electric field by $V = \mathcal{E}L$, as expected. In the stationary state, the electric current is given by $I = eJ = e\Sigma j$. Therefore, we find Ohm's law $V = RI$ with the resistance $R = L/(\sigma\Sigma)$.

For general boundary conditions, there are deviations with respect to electroneutrality near the contacts with the reservoirs and a uniform electric field. These deviations typically extend over a distance of the size of Debye's screening length

$$\ell_{\rm D} = \sqrt{\frac{\epsilon k_{\rm B} T}{e^2 n_0}} \tag{1.81}$$

if the electric field is moderate, i.e., $|\mathcal{E}| \ll \sqrt{k_{\rm B} T n_0/\epsilon}$. Since Debye's screening length is usually much smaller than the size of the conductor, $\ell_{\rm D} \ll L$, the assumption of uniform electric field and electroneutrality is well satisfied.

Otherwise, the x-component of the second expression of the current density in equation (1.76) gives

$$\frac{d}{dx}\left({\rm e}^{\beta e \Phi} n\right) = -\frac{\jmath}{\mathcal{D}}\,{\rm e}^{\beta e \Phi}, \tag{1.82}$$

which can be integrated from $x = 0$ to $x = L$ to obtain (Andrieux and Gaspard, 2009)

$$J = \Sigma\,\jmath = \Sigma\,\mathcal{D}\,\frac{n_{\rm L}\,{\rm e}^{\beta e \Phi_{\rm L}} - n_{\rm R}\,{\rm e}^{\beta e \Phi_{\rm R}}}{\int_0^L {\rm e}^{\beta e \Phi(x)} dx}. \tag{1.83}$$

This expression is equal to zero at equilibrium where the applied voltage (1.79) is equal to zero. Again, if the density is uniform with $n_{\rm L} = n_{\rm R} = n_0$ and the electric field uniform with $\Phi(x) = \Phi_{\rm L} - \mathcal{E}x$, the integral in the denominator can be performed and we recover the current density $\jmath = \beta e \mathcal{D} n_0 \mathcal{E}$, giving Ohm's law.

In the presence of an electric field, the entropy production rate is given by (de Groot and Mazur, 1984)

$$\frac{1}{k_{\rm B}}\frac{d_{\rm i}S}{dt} = \int_V \frac{\mathcal{D}}{n}\,(\nabla n - \beta e n \boldsymbol{\mathcal{E}})^2\, d^3r \geq 0. \tag{1.84}$$

If the charge and current densities are uniform so that Ohm's law holds, this entropy production rate becomes

$$\frac{1}{k_{\rm B}}\frac{d_{\rm i}S}{dt} = \frac{VI}{k_{\rm B}T} = \frac{P}{k_{\rm B}T} \geq 0, \tag{1.85}$$

where $P = VI = RI^2$ is the power dissipated by the electric current flowing in the resistor according to *Joule's law*.

1.7.4 Electric Circuits

Electric components can be wired together to form circuits. Figure 1.6 shows common examples of such components. Electric generators such as batteries are characterized by their electromotive force $\mathscr{E} = V$. Capacitors, inductors, and resistors are components with a linear relation between the voltage V and, respectively, the electric charges $\pm Q$ on the capacitor plates, the time derivative dI/dt of the current I, and the current itself. In this regard, these components are linear. However, there are also nonlinear components such as diodes and transistors. For instance, the current–voltage relation of diodes can be

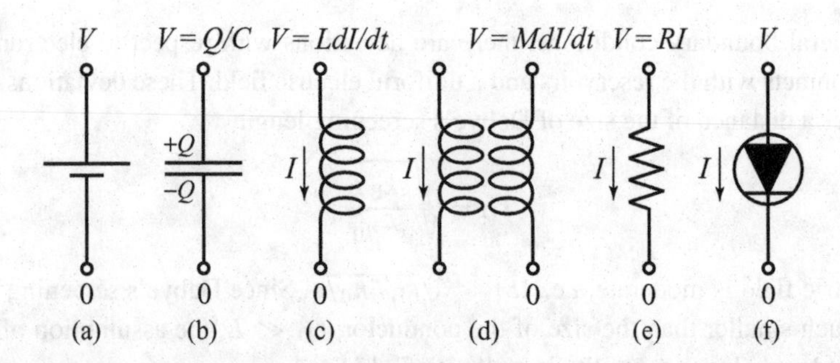

Figure 1.6 Various components of electric circuits: (a) battery of electromotive force $\mathscr{E} = V$, (b) capacitor of capacitance C, (c) inductor of inductance L, (d) two coupled inductors with mutual inductance M, (e) resistor of resistance R, (f) diode.

approximated by the expression $I = I_{\rm s}\left(e^{\beta e V} - 1\right)$ with the saturation current $I_{\rm s}$, as will be further discussed in Section 10.6.2.

Electric circuits are networks with nodes connected by edges. As long as the electromagnetic radiation of the circuit is negligible, the electric currents and potentials can be determined in the circuit using the current–voltage relations characterizing every component and the laws of Kirchhoff (1847):

1. *The sum of electric currents in all the edges arriving at any node in the circuit is zero.*
2. *The sum of the electric potential differences along any loop in the circuit is zero.*

Kirchhoff's first law, or current law, results from the local conservation of electric charge and the assumption that conduction is large enough so that electroneutrality is maintained in the wires connected together at any node. Kirchhoff's second law, or voltage law, is the consequence of Faraday's law of electromagnetism, provided that the magnetic field is localized inside the inductors (Reitz and Milford, 1967). Accordingly, the circulation of the electric field around any loop is equal to zero, i.e., $\oint_{\rm loop} \boldsymbol{\mathcal{E}} \cdot d\mathbf{r} = \sum_{i\to j} V_{i\to j} = 0$, where the sum extends over all the oriented edges $i \to j$ in the loop and $V_{i\to j} = \Phi_i - \Phi_j$ in terms of the electric potentials $\{\Phi_i\}$ at the nodes $\{i\}$. In Kirchhoff's second law, the contribution of every electromotive element is equal to minus its electromotive force, this latter driving the circuit out of equilibrium.

Energy is supplied by the electromotive forces of batteries. Capacitors and inductors conserve energy. Inside capacitors, energy is stored in the electric field between oppositely charged plates. Inside inductors, energy is stored in the magnetic field generated by the electric current. Other components dissipate energy and produce entropy, which is the case for resistors, diodes, and transistors. The entropy production rate in an electric circuit at temperature T can be evaluated as $d_{\rm i}S/dt = P_{\rm diss}/T \geq 0$ in terms of the power $P_{\rm diss}$ that is dissipated in all the components.

1.8 Reaction Networks

Reaction networks are envisaged in different fields of science. Nuclear reaction networks are considered for primordial or stellar nucleosynthesis to explain the abundance

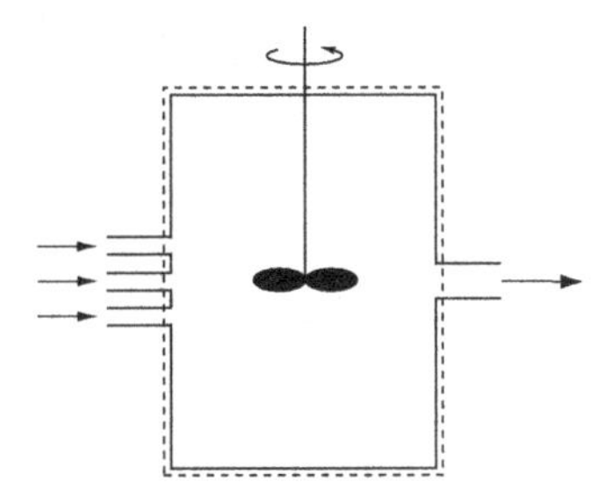

Figure 1.7 Schematic representation of a continous-flow stirred tank reactor (CSTR). The open system itself is delimited by the dashed line. Reactants are continuously pumped into the reactor by several pipes shown on its left-hand side. The outflow of products is carried out by the large pipe on its right-hand side. The solution inside the reactor is mechanically stirred by the rotating blades of an impeller.

of isotopes in the universe. Moreover, the chemical elements may combine to form millions of known chemical compounds (Pauling, 1970). These compounds are transformed in chemical or biochemical reaction networks (Nelson and Cox, 2017). There exist networks of different sizes depending on the number of relevant species included in the description. A famous example is the Belousov–Zhabotinsky reaction, which involves about fourteen species, but can be described by the Oregonator model with only three variables (Nicolis and Prigogine, 1977; Bergé et al., 1984; Scott, 1991; Nicolis, 1995; Epstein and Pojman, 1998). Complex reaction networks are considered in astrochemistry, atmospheric chemistry, petrochemistry, studies in prebiotic chemistry, and combustion theory. Biochemical reaction networks are also known in enzyme kinetics, metabolic pathways, signal transduction pathways, cellular rhythms, and gene regulation (Segel, 1975; Hill, 1989; Nicolis and Prigogine, 1977; Goldbeter, 1996; Qian and Beard, 2005; Michal and Schomburg, 2012; Wachtel et al., 2018).

Chemical or biochemical reactions can be controlled in reactors of different types. Batch reactors are closed systems at controlled temperature where reactants are initially poured in a stirred solution, yielding products until chemical equilibrium is reached. In contrast, continuous-flow stirred tank reactors are open systems continuously fed by reactants, the products exiting by an outflow. These reactors are equipped with a stirrer to guarantee the quasi-uniformity of the reacting mixture.

1.8.1 Flow Reactors

A flow reactor called a continuous-flow stirred tank reactor (CSTR) is schematically represented in Figure 1.7 with the inflow of reactants and the outflow of the solution in excess, also containing products (Aris, 1989; Nicolis, 1995; Epstein and Pojman, 1998; Blokhuis et al., 2018). The stirrer induces hydrodynamic mixing, so that the concentrations of reactants and products are made uniform inside the reactor.

Kinetics

In the flow reactor, the reactant and product densities $\{n_k\}_{k=1}^{c}$ are ruled by the balance equations

$$\partial_t n_k + \nabla \cdot (n_k \mathbf{v} + \mathcal{J}_k) = \sum_r \nu_{kr} w_r, \tag{1.86}$$

where $\mathbf{v}$ is the fluid velocity, $\mathcal{J}_k = -\mathcal{D}_k \nabla n_k$ the diffusive current density, and ν_{kr} the stoichiometric coefficient of species k in the reaction r of rate w_r.

The number of the molecules of species k inside the volume V delimited by the dashed line in Figure 1.7 is defined by $N_k = \int_V n_k \, d^3r$. This number evolves in time according to

$$\frac{dN_k}{dt} = \frac{d_e N_k}{dt} + \frac{d_i N_k}{dt}, \tag{1.87}$$

which has a form reminiscent of equation (1.3) with the contribution $d_e N_k/dt$ due to the exchanges of molecules at the boundaries of the volume where inflow and outflow are controlled, and the internal contribution $d_i N_k/dt$ due to the reactions taking place inside the system. By the divergence theorem, the exchanges with the exterior contribute according to

$$\frac{d_e N_k}{dt} = -\int_{\partial V} (n_k \mathbf{v} + \mathcal{J}_k) \cdot d\boldsymbol{\Sigma}, \tag{1.88}$$

where ∂V is the boundary of the volume V. Since the solution is well stirred by the mixer, the density n_k is practically uniform inside the reactor, so that $N_k \simeq V n_k$ and $\mathcal{J}_k = -\mathcal{D}_k \nabla n_k \simeq 0$. Moreover, exchanges only happen where the fluid velocity is not equal to zero in the direction transverse to the boundary ∂V, i.e., at the portions of the boundary, $\partial_{k,\text{in}} V$ and $\partial_{\text{out}} V$, respectively, corresponding to the inflow of species k and the outflow of the solution in excess. Consequently, we have that

$$\frac{d_e N_k}{dt} \simeq -\int_{\partial_{k,\text{in}} V} n_k \mathbf{v} \cdot d\boldsymbol{\Sigma} - \int_{\partial_{\text{out}} V} n_k \mathbf{v} \cdot d\boldsymbol{\Sigma} = \phi_{k,\text{in}} \, n_{k,\text{in}} - \phi_{\text{out}} \, n_k \tag{1.89}$$

in terms of the density $n_{k,\text{in}}$ inside the inlet pipe of species k, the ingoing flux $\phi_{k,\text{in}} = -\int_{\partial V_{k,\text{in}}} \mathbf{v} \cdot d\boldsymbol{\Sigma}$, and the outgoing flux $\phi_{\text{out}} = \int_{\partial V_{\text{out}}} \mathbf{v} \cdot d\boldsymbol{\Sigma}$. We note that $\phi_{\text{out}} = \sum_k \phi_{k,\text{in}}$ because of the fluid incompressibility, $\nabla \cdot \mathbf{v} = 0$. In addition, the reactions inside the system contribute to

$$\frac{d_i N_k}{dt} = \int_V \sum_r \nu_{kr} w_r \, d^3r \simeq V \sum_r \nu_{kr} w_r, \tag{1.90}$$

since the solution is well stirred and the densities are thus uniform inside the reactor. Substituting these results back into equation (1.87), we find that the density of species k is ruled by

$$\frac{dn_k}{dt} = \sum_r \nu_{kr} \, w_r + \frac{1}{\tau}(n_{k0} - n_k) \qquad \text{with} \qquad n_{k0} \equiv \frac{\phi_{k,\text{in}}}{\phi_{\text{out}}} n_{k,\text{in}}, \tag{1.91}$$

where $\tau \equiv V/\phi_{\text{out}}$ is the mean residence time of the species inside the reactor. The macrostate inside the open system is thus determined by the control parameters τ and n_{k0}. We note that the conservation of the total mass, $M = V \sum_k m_k n_k$, implies that $d_i M/dt = 0$, so that $dM/dt = d_e M/dt = (M_0 - M)/\tau$, and $M = M_0$ in steady regimes.

If the fluxes are equal to zero, the residence time is infinite ($\tau = \infty$), and the system is closed, corresponding to a batch reactor. However, the system is open and out of equilibrium if the residence time is finite and some reactants are injected inside the reactor. In the limit of a vanishingly small residence time ($\tau = 0$), the densities converge towards their injection values $\lim_{t\to\infty} n_k(t) = n_{k0}$.

Thermodynamics

The thermodynamics of the reactions can be investigated using the Gibbs free energy as thermodynamic potential if the temperature and the pressure are supposed to be uniform inside the system. The Gibbs free energy density is given by

$$g = \sum_k \mu_k \, n_k, \tag{1.92}$$

where μ_k is the chemical potential of species k. For this quantity, Gibbs' relation has the form

$$dg = -s\, dT + dp + \sum_k \mu_k \, dn_k. \tag{1.93}$$

Under isothermal and isobaric conditions, $dT = 0$ and $dp = 0$, the time evolution of the Gibbs free energy density is thus ruled by the changes of densities according to equation (1.91), so that

$$\frac{dg}{dt} = \sum_k \mu_k \frac{dn_k}{dt} = \sum_{kr} \mu_k \nu_{kr} w_r + \frac{1}{\tau} \sum_k \mu_k \,(n_{k0} - n_k), \tag{1.94}$$

which can be written in the equivalent form

$$\frac{dg}{dt} = -T\sigma_{s,\,\mathrm{react}} + \frac{1}{\tau}\,(g_0 - g), \tag{1.95}$$

in terms of $g_0 = \sum_k \mu_k \, n_{k0}$ and the entropy production rate density $\sigma_{s,\,\mathrm{react}}$ due to the reactions. If the solution is dilute and the kinetics obeys the mass action law, the chemical potentials of the reacting solute species are given by equation (1.36) and the entropy production rate density reads as

$$\sigma_{s,\,\mathrm{react}} = -\frac{1}{T} \sum_{kr} \mu_k \nu_{kr} w_r = k_\mathrm{B} \sum_r (w_{+r} - w_{-r}) \ln \frac{w_{+r}}{w_{-r}} \geq 0. \tag{1.96}$$

If the residence time is infinite ($\tau = \infty$), there is no exchange with the exterior of the reactor. Therefore, the system is closed and evolves from an initial macrostate with a high content of Gibbs free energy, towards the equilibrium macrostate where Gibbs free energy is minimal and the entropy maximal, as required. Accordingly, the solution undergoes relaxation towards chemical equilibrium where the conditions of detailed balance are satisfied for all the reactions, $w_{+r} = w_{-r}$, corresponding to the minimum of Gibbs free energy in the reactor. For dilute solutions, the Gibbs free energy (1.92) can be written as

$$g = g_\mathrm{eq} + k_\mathrm{B} T \sum_{k=1}^{c-1} \left[n_k \ln \frac{n_k}{n_{k,\mathrm{eq}}} - (n_k - n_{k,\mathrm{eq}}) \right], \tag{1.97}$$

where $n_{k,\text{eq}}$ are the equilibrium densities of the solute species in the closed reactor, corresponding to the free energy minimum given the constraints coming from the initial densities and the reactions. Since $g \geq g_{\text{eq}}$ and $dg/dt \leq 0$ in a closed reactor, the function (1.97) plays the role of Lyapunov function, implying that the equilibrium macrostate is the unique attractor in the closed reactor (Shear, 1967; Horn and Jackson, 1972; Rao and Esposito, 2016).

However, if the reactor is open the condition $dg/dt \leq 0$ is no longer always satisfied because equation (1.95) has the extra term $(g_0 - g)/\tau$ due to the exchanges of the reactor with the exterior. Accordingly, the function (1.97) is no longer a Lyapunov function and the existence and uniqueness of a stationary macrostate no longer hold. Actually, complex dynamics with multistability, as well as periodic, or chaotic oscillations become possible for flow reactors driven far from equilibrium, i.e., beyond some instability threshold for the stationary macrostate issued from equilibrium by increasing the nonequilibrium constraints (Bergé et al., 1984; Scott, 1991; Nicolis, 1995; Epstein and Pojman, 1998). In the limit where the residence time is vanishingly small ($\tau = 0$), the free energy density becomes equal to its injection value $g = g_0$.

We note that, if the system is closed (as in a batch reactor) and large enough pools of some species are maintained during long time intervals, these species are practically chemostatted and their densities can be assumed to remain invariant. During such lapses of time, the system can be maintained far enough from equilibrium to sustain dynamical behaviors similar to those observed in flow reactors.

1.8.2 Stoichiometric Analysis of Reaction Networks

The theory of chemical reaction networks has been developed since the 1960s on the basis of the stoichiometric matrix and in close relation with thermodynamics (Polettini and Esposito, 2014; Rao and Esposito, 2016; Feinberg, 2019).

The densities of all the species in the reaction network are ruled by the closed set of ordinary differential equations (1.91), defining a so-called *dynamical system*. These equations can be written in the vectorial form

$$\frac{d\mathbf{n}}{dt} = \boldsymbol{\nu} \cdot \mathbf{w} + \frac{1}{\tau}(\mathbf{n}_0 - \mathbf{n}), \tag{1.98}$$

ruling the time evolution for the c-dimensional vector $\mathbf{n} \in \mathbb{R}^c$ of the reactant and product densities. The control parameters are the components of the invariant vector $\mathbf{n}_0 \in \mathbb{R}^c$ and the residence time τ. The reaction rates form the m-dimensional vector $\mathbf{w} = (w_r)_{r=1}^m$ and the stoichiometric coefficients the $c \times m$ matrix $\boldsymbol{\nu} = (\nu_{kr})$ with $k = 1, 2, \ldots, c$ and $r = 1, 2, \ldots, m$. The time evolution generates trajectories $\mathbf{n}(t)$ in the *phase space* $\{\mathbf{n} \in \mathbb{R}^c\}$ of the dynamical system (1.98) (Bergé et al., 1984; Nicolis, 1995).

Closed Reactors

Let us first assume that the reactor is closed ($\tau = \infty$). On the one hand, the left null eigenvectors $\mathfrak{l} \in \mathbb{R}^c$ of the stoichiometric matrix are defined by

$$\mathfrak{l}^{\mathrm{T}} \cdot \boldsymbol{\nu} = 0. \tag{1.99}$$

Since the solution is uniform in the volume V of the system because of stirring, we may introduce the quantities

$$L \equiv V\,\mathfrak{l}^{\mathrm{T}} \cdot \mathbf{n} \tag{1.100}$$

that are conserved by the reaction network and are thus constants of motion for the set of ordinary differential equations, $dL/dt = 0$. These *conserved quantities* include, in particular, the total mass and the moieties that are preserved by the reaction network (Haraldsdóttir and Fleming, 2016). The number l of these conserved quantities is given by the dimension of the null space of the transpose of the stoichiometric matrix, also called the cokernel: $l = \dim \operatorname{coker} \boldsymbol{\nu}$. On the other hand, the right null eigenvectors $\mathbf{e} \in \mathbb{R}^m$ of the stoichiometric matrix such that

$$\boldsymbol{\nu} \cdot \mathbf{e} = 0 \tag{1.101}$$

define the so-called *stoichiometric cycles* of the network,[9] forming cyclic reaction pathways in the network. The number, o, of cycles is given by the dimension of the null space of the stoichiometric matrix, also called the kernel: $o = \dim \ker \boldsymbol{\nu}$. A general property of linear algebra shows that the rank of the stoichiometric matrix is given by

$$\operatorname{rank} \boldsymbol{\nu} = c - l = m - o. \tag{1.102}$$

Open Reactors

Next, the reactor is supposed to be open with a finite residence time ($\tau < \infty$). In this case, the contribution $d_e\mathbf{n}/dt = (\mathbf{n}_0 - \mathbf{n})/\tau$ due to the exchanges should be added to the internal contribution $d_i\mathbf{n}/dt = \boldsymbol{\nu} \cdot \mathbf{w}$. Consequently, the quantities L are no longer constants of motion, but instead they obey

$$\frac{dL}{dt} = \frac{1}{\tau}(L_0 - L) \tag{1.103}$$

with $L_0 \equiv V\,\mathfrak{l}^{\mathrm{T}} \cdot \mathbf{n}_0$, because of the inflow and outflow generating exchanges with the exterior. Consequently, the quantities L evolve in time according to

$$L(t) = L(0)\,\mathrm{e}^{-t/\tau} + L_0\left(1 - \mathrm{e}^{-t/\tau}\right). \tag{1.104}$$

Nevertheless, these quantities reach constant values L_0 fixed by the inflow over a timescale longer than the residence time: $L(t) \simeq L_0$ for $t \gg \tau$. We notice that in a closed reactor, these quantities take the constant values $L(t) = L(0)$ fixed by their initial conditions, $L(0)$, that may differ from the constant values L_0 fixed by the inflow in an open reactor. A general remark is that the quantities L continue to obey conservation laws if the system is open. Indeed, equation (1.103) can be written in the form $dL/dt = d_eL/dt + d_iL/dt$ with $d_eL/dt = (L_0 - L)/\tau$ and $d_iL/dt = 0$, showing that these quantities get their time dependence because of the exchanges with the exterior.

[9] The stoichiometric cycles of the network should not be confused with the limit cycles of the dynamical system, which are periodic trajectories for the dynamics.

In a flow reactor, the contributions $d_e\mathbf{n}/dt = (\mathbf{n}_0 - \mathbf{n})/\tau$ due to transport between the system and the exterior can be handled in the same way as for reactions, writing equation (1.98) in the equivalent form

$$\frac{d\mathbf{n}}{dt} = \boldsymbol{\nu}' \cdot \mathbf{w}' \qquad \text{with} \qquad \boldsymbol{\nu}' = (\boldsymbol{\nu}, \mathbf{1}), \tag{1.105}$$

by extending the stoichiometric matrix $\boldsymbol{\nu}$ with the $c \times c$ identity matrix $\mathbf{1}$, and the vector of reaction rates into $\mathbf{w}' = (\mathbf{w}, \tilde{\mathbf{w}})^{\mathrm{T}}$, where $\mathbf{w} \in \mathbb{R}^m$ is the previous one and $\tilde{\mathbf{w}} = (\mathbf{n}_0 - \mathbf{n})/\tau \in \mathbb{R}^c$. The new stoichiometric matrix $\boldsymbol{\nu}'$ thus has dimensions $c \times m'$ with $m' = m + c$. Instead of equation (1.102), here we have the relation, rank $\boldsymbol{\nu}' = m' - o' = c - l'$, with $l' = 0$, since there is no longer any constant of motion as a consequence of equation (1.103). Therefore, the number of cycles is equal to $o' = m' - c = m$ in the flow reactor, instead of $o = m - c + l \leq m$ in the closed one. There are thus new cycles in the open reactor due to the exchanges with the exterior, which are called external cycles. A general cycle can be split as $\mathbf{e}' = (\mathbf{e}, \tilde{\mathbf{e}})^{\mathrm{T}}$, such that $\boldsymbol{\nu}' \cdot \mathbf{e}' = \boldsymbol{\nu} \cdot \mathbf{e} + \tilde{\mathbf{e}} = 0$. Here, a distinction can be made between the previously identified internal cycles $\mathbf{e}'_\gamma$ with $\boldsymbol{\nu} \cdot \mathbf{e}_\gamma = 0$ and $\tilde{\mathbf{e}}_\gamma = 0$ for $\gamma = 1, 2, \dots, o$, and the external cycles $\mathbf{e}'_\alpha$ such that $\tilde{\mathbf{e}}_\alpha = -\boldsymbol{\nu} \cdot \mathbf{e}_\alpha \neq 0$ for $\alpha = 1, 2, \dots, m - o$ (Blokhuis et al., 2018). These external cycles have a pathway involving transport from or to the exterior of the flow reactor.

Thermodynamics in Open Reactors

Noting that Gibbs free energy density can be expressed as $g = \boldsymbol{\mu}^{\mathrm{T}} \cdot \mathbf{n}$ in terms of the c-dimensional vector of chemical potentials $\boldsymbol{\mu} = \{\mu_k\}_{k=1}^{c}$, equation (1.94) can be written as

$$\frac{dg}{dt} = \boldsymbol{\mu}^{\mathrm{T}} \cdot \boldsymbol{\nu}' \cdot \mathbf{w}'. \tag{1.106}$$

In a stationary macrostate, the relation $\boldsymbol{\nu}' \cdot \mathbf{w}' = 0$ holds, so that the vector of reaction rates can be decomposed in the basis of the right null eigenvectors as $\mathbf{w}' = \sum_{\gamma=1}^{o} w_\gamma \mathbf{e}'_\gamma + \sum_{\alpha=1}^{m-o} w_\alpha \mathbf{e}'_\alpha$. If these conditions are satisfied, the entropy production rate density (1.96) simplifies to

$$\sigma_{s,\,\mathrm{react}}\Big|_{\mathrm{st}} = -\frac{1}{T}\,\boldsymbol{\mu}^{\mathrm{T}} \cdot \boldsymbol{\nu} \cdot \mathbf{w} = +\frac{1}{T}\sum_{\alpha=1}^{m-o} w_\alpha\, \boldsymbol{\mu}^{\mathrm{T}} \cdot \tilde{\mathbf{e}}_\alpha \geq 0, \tag{1.107}$$

because $\boldsymbol{\nu} \cdot \mathbf{e}_\gamma = 0$ for internal cycles and $\boldsymbol{\nu} \cdot \mathbf{e}_\alpha = -\tilde{\mathbf{e}}_\alpha$ for external cycles. The entropy production rate can thus be written in the form

$$\left.\frac{d_{\mathrm{i}}S}{dt}\right|_{\mathrm{st}} = V\sigma_{s,\,\mathrm{react}}\Big|_{\mathrm{st}} = k_{\mathrm{B}}\sum_{\alpha=1}^{m-o} A_\alpha\, J_\alpha \geq 0 \tag{1.108}$$

in terms of the global affinities $A_\alpha \equiv \boldsymbol{\mu}^{\mathrm{T}} \cdot \tilde{\mathbf{e}}_\alpha/(k_{\mathrm{B}}T)$ and the currents $J_\alpha = V w_\alpha$ associated with the external cycles $\alpha = 1, 2, \dots, m - o$.[10] The consequence is that, in stationary

[10] We note that the currents associated with the reactions are proportional to the volume, although the currents (1.68) or (1.70) associated with transport between reservoirs are proportional to the surface area of the interfaces with the reservoirs. The reason is that the reaction currents – also called reaction fluxes – are microscopic, since the bottlenecks of the reactions have submolecular sizes (Moore, 1972; Berry et al., 1980).

macrostates, the entropy production is only determined by the external cycles able to drive the open system away from equilibrium (Blokhuis et al., 2018). The entropy produced inside the system is evacuated to the exterior by the flow term. At equilibrium where detailed balance is satisfied, the reaction rates are equal to zero, i.e., $w_r = w_{+r} - w_{-r} = 0$, together with the entropy production rate.

If the solution $\mathbf{n}(t)$ of equation (1.98) depends on time, we may consider the time average

$$\overline{X} \equiv \lim_{\mathscr{T}\to\infty} \frac{1}{\mathscr{T}} \int_0^{\mathscr{T}} X(t)\, dt \tag{1.109}$$

for any quantity X of interest. Taking the time average of equation (1.105), we find that $\overline{\mathbf{w}}'$ is still a right null eigenvector of the stoichiometric matrix introduced in equation (1.105), $\boldsymbol{\nu}' \cdot \overline{\mathbf{w}}' = 0$, which can again be decomposed in the basis of the right null eigenvectors to give

$$\overline{\mathbf{n}} = \mathbf{n}_0 + \tau\, \boldsymbol{\nu} \cdot \overline{\mathbf{w}} = \mathbf{n}_0 - \tau \sum_{\alpha=1}^{m-o} \overline{w}_\alpha\, \tilde{\mathbf{e}}_\alpha . \tag{1.110}$$

Therefore, the deviations of the mean densities with respect to the effective injected densities $\mathbf{n}_0$ are given in terms of the external cycles of the network. The time average can also be applied to the balance equation (1.95) for the Gibbs free energy. In this way, we obtain the time average of the entropy production rate

$$\overline{\frac{d_{\rm i}S}{dt}} = V\, \overline{\sigma}_{s,\,\rm react} = \frac{V}{T\tau}\left(\overline{g}_0 - \overline{g}\right) \geq 0, \tag{1.111}$$

giving the mean value of the Gibbs free energy density as

$$\overline{g} = \overline{g}_0 - T\tau\, \overline{\sigma}_{s,\,\rm react} \leq \overline{g}_0 . \tag{1.112}$$

This result shows that the mean value of the Gibbs free energy inside the reactor is always lower than or equal to its mean injection value $\overline{g}_0$.

1.9 Dissipative Dynamics and Structures

Systems are driven out of equilibrium by control parameters such as the global affinities or the reactant inflow rates for reactors. If these control parameters are switched on, some fluxes of energy or matter are generated inside the system, and the equilibrium macrostate of the undriven system turns into a nonequilibrium steady state for the driven system. Since the perturbations with respect to this macrostate are damped by dissipation, the nonequilibrium steady state is a stationary attractor for the time evolution of the system, as is the case at equilibrium. When the control parameters are increased, a critical threshold may be reached where the nonequilibrium steady state becomes unstable and the system undergoes a transition, called bifurcation, leading to the emergence of new attractors.

These emerging attractors may be stationary or dynamical. In the latter case, they can manifest periodic, quasiperiodic, or chaotic oscillations (see Appendix B). The attractors of periodic oscillations are called *limit cycles*, in reference to the convergence in the long-time limit towards a cyclic time evolution. Quasiperiodic oscillations have tori as attractors in the phase space of macrovariables. The dimension of the attractor is a quantitative characterization of the effective number of macrovariables that are dynamically active in the system. Steady states and limit cycles have their dimensions equal to zero and one, respectively. For tori, the dimension gives the number of incommensurable frequencies in the quasiperiodicity they represent. Chaotic oscillations are characterized by the property of sensitivity to initial conditions, generating aperiodicity over long timescales. The attractors of chaotic oscillations typically form fractals in the phase space (Bergé et al., 1984; Eckmann and Ruelle, 1985; Strogatz, 1994; Nicolis, 1995).

In spatially extended systems, the attractor may correspond to stationary patterns or spatiotemporal structures, referred to as *dissipative structures* (Prigogine, 1967; Glansdorff and Prigogine, 1971; Kondepudi and Prigogine, 1998). The formation of these macroscopic structures is possible because the system is open and the entropy produced by dissipation inside the system is evacuated to the exterior, allowing self-organization to happen far from equilibrium at the macroscale (Nicolis and Prigogine, 1977).

These nonequilibrium phenomena manifest themselves in different physicochemical systems.

In hydrodynamics, the Rayleigh–Bénard instability in a fluid layer subjected to gravity and a temperature gradient induces the formation of stationary convective rolls (Mareschal and Kestemont, 1987). This dissipative structure may undergo further instabilities, leading to turbulence, which is a chaotic behavior of high dimension in the phase space (Bergé et al., 1984; Nicolis, 1995).

Nonequilibrium phenomena are also a feature of reactions with nonlinear mechanisms caused by autocatalysis or cross-catalysis (Nicolis and Prigogine, 1977; Scott, 1991; Epstein and Pojman, 1998). The so-called chemical clocks are periodic oscillations observed in the Belousov–Zhabotinsky reaction and other reactions. Quasiperiodic and chaotic oscillations have also been observed in these reactions. Here, the time evolution takes place in the phase space of the chemical concentrations (i.e., the densities). In spatially extended systems where the reactions are coupled to the diffusion of the reacting species, stationary dissipative structures called *Turing patterns* may emerge, as well as spatiotemporal structures forming circular or spiral waves, which may become turbulent if the system is driven far enough from equilibrium.

In addition, similar phenomena are observed in lasers, nonlinear optics, electronics, and other areas (Haken, 1975; Lugatio and Lefever, 1987; Schöll, 2001).

Dissipative structures often emerge through symmetry breaking at the macroscale. For instance, the symmetry under temporal translations is broken at the onset of oscillations, while the symmetry under spatial translations is broken in the Rayleigh–Bénard instability or the formation of Turing patterns in reaction–diffusion systems (Prigogine and Nicolis, 1967; Prigogine and Lefever, 1968).

The transitions, called bifurcations, occurring between the different nonequilibrium regimes have been classified (Strogatz, 1994; Nicolis, 1995). They include the pitchfork bifurcation shown in Figure 1.8(a), where two new stable steady states emerge from

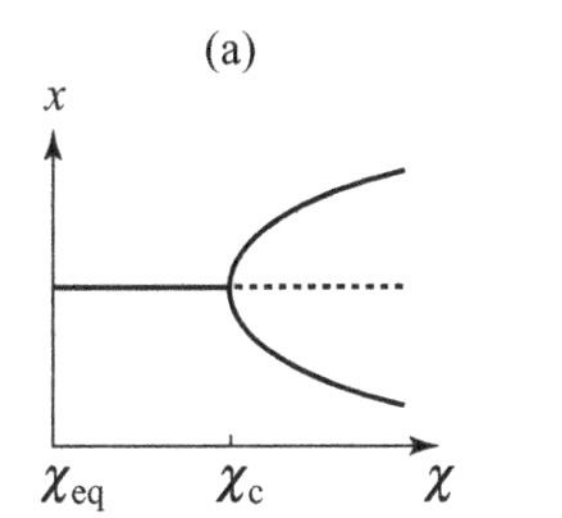

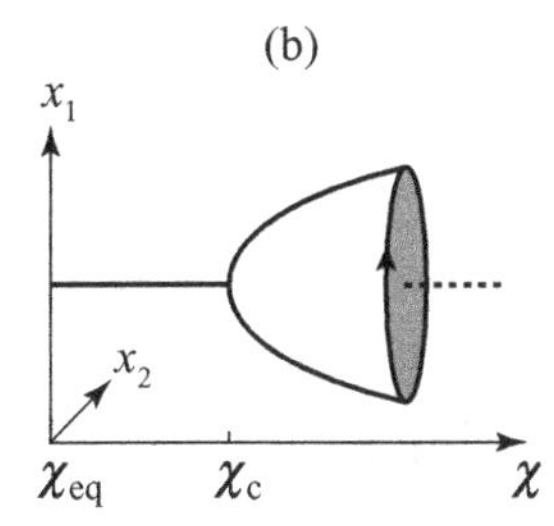

Figure 1.8 Schematic representation of bifurcations in the phase space of the dynamical variables versus the control parameter χ: (a) Pitchfork bifurcation with the emergence of two new stable steady states in the variable x; (b) Hopf bifurcation with the emergence of a limit cycle in the phase space of the variables (x_1, x_2). In both cases, the equilibrium macrostate is located at the value $\chi_{\rm eq}$ of the control parameter, and the bifurcation happens at the critical threshold $\chi_{\rm c}$. The solid lines depict the stable solutions and the dashed lines the unstable ones.

the instability of a previously existing steady state, leading to multistability. Another important nonequilibrium transition is the Hopf bifurcation shown in Figure 1.8(b), where a limit cycle emerges from the instability of a steady state, leading to rhythmic behavior. Successive bifurcations may lead to chaotic regimes, such as the cascade of period-doubling bifurcations.

In time-dependent regimes, the attractor can be characterized by the time average (1.109) for the different quantities of interest. Indeed, in dissipative dynamical systems, every attractor is typically surrounded by a basin of attraction where all the time evolutions converge towards the attractor. Therefore, the time averages (1.109) of any time evolution starting in the basin of attraction will have values associated with the same attractor (Eckmann and Ruelle, 1985).

1.10 Engines

Heat engines such as steam engines or internal combustion engines are mechanical devices that convert heat into work. More generally, engines and motors achieve the transduction of heat, chemical energy, or electric energy into work, i.e., mechanical energy. Many engines function in such a way that gases undergo a cycle of transformations, including compression and expansion, heating and cooling, and/or inflow and outflow. Engines may have an autonomous periodic motion, which can be represented as the limit cycle of a dissipative dynamical system. In general, engines should be described as piecewise continuous media in terms of hydrodynamics and transport theory. Otherwise, the cycle of an engine can be idealized as the succession of several transformations driven by time-dependent external forcing, as conceived by Carnot (1824) for heat engines. Moreover, there also exist engines that function under isothermal conditions using a difference of pressure or chemical potential to power their motion. Such idealized engines can be directly analyzed in terms of the first and second laws of thermodynamics in order to determine their efficiencies, as discussed below.

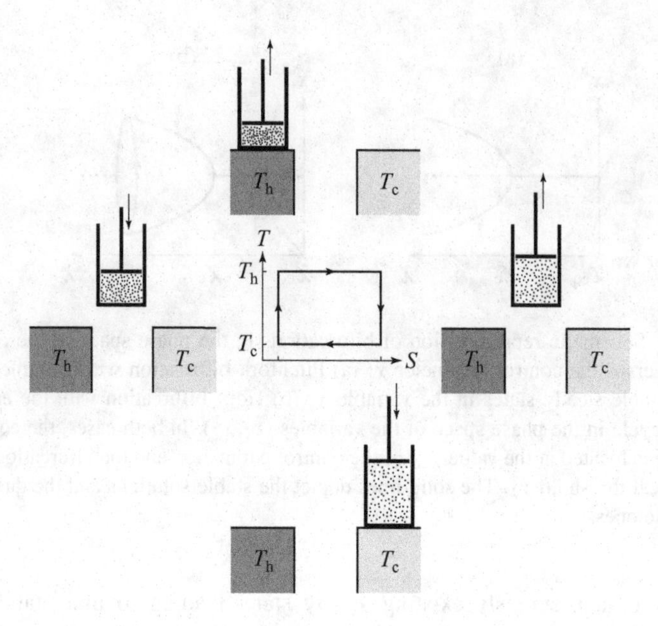

Figure 1.9 The cycle of the Carnot heat engine in the plane of the entropy S and the temperature T. The piston is filled with gas and alternately put in contact with the hot heat reservoir at the temperature T_h and the cold heat reservoir at the temperature T_c, or held isolated from them.

1.10.1 Carnot Heat Engine

This famous engine is composed of a piston containing a dilute gas and two heat reservoirs at the temperatures T_h and T_c. The piston forms a closed system where the number of molecules does not change.

The cycle of this engine is shown in Figure 1.9. First, the piston is put in contact with the hot reservoir, which generates the isothermal expansion of the gas and the heat transfer $Q_h > 0$ from the reservoir to the gas. Next, the piston is isolated from the reservoirs and the gas is subjected to an adiabatic (i.e., isoentropic) expansion, cooling the gas. When the temperature T_c is reached, the piston is placed in contact with the cold reservoir. Now, the isothermal compression of the gas can be carried out, releasing some heat $|Q_c|$ (with $Q_c = -|Q_c|$) towards that reservoir. Finally, the piston is again isolated from the reservoirs and the gas is compressed adiabatically, increasing its temperature back to the one of the hot reservoir. When this cycle is completed, the state variables of the gas have recovered their initial values, which is the case, in particular, for the energy E and the entropy S of the gas. Consequently, the integrals of their changes over the cycle are equal to zero: $\oint dE = 0$ and $\oint dS = 0$.

Integrating the first law (1.2) over the cycle and using the conservation of molecules in the closed system ($dN = 0$), we get

$$0 = \oint dE = \oint d_e E = \oint (đQ - p\, dV), \tag{1.113}$$

where

$$\oint đQ = |Q_{\rm h}| - |Q_{\rm c}| \qquad \text{and} \qquad \mathcal{W} \equiv \oint p\,dV \tag{1.114}$$

are, respectively, the heat exchanged with the reservoirs during the isothermal transformations and the work performed by the piston on the exterior during the cycle. Thus, energy conservation implies that

$$\mathcal{W} = |Q_{\rm h}| - |Q_{\rm c}|. \tag{1.115}$$

Next, integrating the second law (1.4) over the cycle gives

$$0 = \oint dS = \oint d_{\rm e}S + \oint d_{\rm i}S \tag{1.116}$$

with $\oint d_{\rm i}S \geq 0$. Since $d_{\rm e}S = đQ/T$, we have that

$$0 \geq \oint d_{\rm e}S = \oint \frac{đQ}{T} = \frac{|Q_{\rm h}|}{T_{\rm h}} - \frac{|Q_{\rm c}|}{T_{\rm c}}. \tag{1.117}$$

The efficiency is defined as the ratio between the work performed by the engine and the heat supplied by the hot source. According to the first and second laws, the efficiency is bounded as

$$\eta \equiv \frac{\mathcal{W}}{|Q_{\rm h}|} = 1 - \frac{|Q_{\rm c}|}{|Q_{\rm h}|} \leq \eta_{\rm C} = 1 - \frac{T_{\rm c}}{T_{\rm h}} \tag{1.118}$$

by the Carnot efficiency $\eta_{\rm C}$ reached in the absence of entropy production during the cycle, i.e., if $\oint d_{\rm i}S = 0$. In any case, the efficiency is equal to zero if both reservoirs are in equilibrium (i.e., if $T_{\rm h} = T_{\rm c}$).

Most often, engines are running at speeds that maximize their power, which has the effect of reducing efficiency, as compared to arbitrarily slow regimes aiming at the optimization of efficiency. Using linear relations between currents and global affinities, the efficiency at maximum power is estimated to reach the value

$$\eta_{\rm max\,power} = 1 - \sqrt{\frac{T_{\rm c}}{T_{\rm h}}}, \tag{1.119}$$

which is thus smaller than the Carnot efficiency of a reversible cycle (Curzon and Ahlborn, 1975; Van den Broeck, 2005; Esposito et al., 2010).

1.10.2 Isothermal Engines Working on Potential Differences

There also exist isothermal engines, which use a difference of pressure or chemical potential to perform work. An example is the pneumatic engine schematically depicted in Figure 1.10. This engine is composed of a piston connected with two reservoirs by two valves, which are successively open or closed, allowing some dilute gas to be transferred from each reservoir to the piston and vice versa. The piston thus forms an open system, where the number of molecules changes along the cycle. The gas pressure has the high value $p_{\rm h}$ in the left

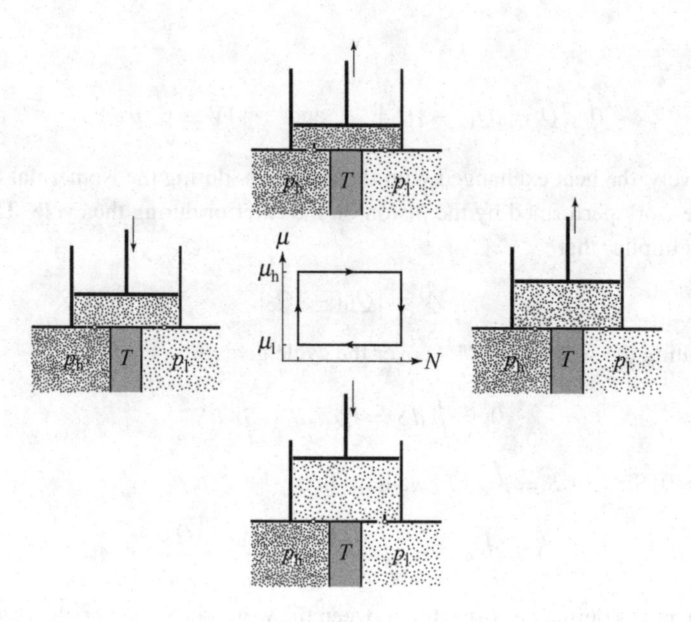

Figure 1.10 The cycle of the isothermal pneumatic engine in the plane of the particle number N and the chemical potential μ. The piston can be filled with gas or emptied via two valves opening to the left reservoir at the high pressure p_h and the right one at the low pressure p_l. The system is kept isothermal at the temperature T by the heat reservoir in contact with the piston and both gas reservoirs.

reservoir and the low value p_l in the right reservoir. Moreover, there is a heat reservoir in contact with the piston and the two gas reservoirs, keeping the temperature uniform at the value T. Since the gas is dilute, its chemical potential is related to its pressure by $\mu(p, T) = \mu^0(T) + k_B T \ln(p/p^0)$, where p^0 is the standard pressure.

The cycle of this engine starts when both valves are closed and the gas pressure in the piston has the high value p_h. The pressures in the piston and the high pressure reservoir being equal, the valve between them can be slowly opened without causing dissipation by viscosity or friction. The piston can thus move up with an inflow of ΔN gas molecules at the high pressure p_h. Next, the valve is closed and the gas in the piston undergoes an isothermal expansion, decreasing the pressure to the low value p_l. The valve with the low pressure reservoir may now be opened and the piston moved down, releasing ΔN gas molecules into that reservoir. After closing the valve on the right-hand side, the gas is isothermally compressed, increasing its pressure back to the high value p_h. In this cycle, the number N of molecules in the piston, the energy E, and the entropy S recover their initial values since they are state variables, so that $\oint dN = 0$, $\oint dE = 0$, and $\oint dS = 0$.

Here, the first law (1.2) gives

$$0 = \oint dE = \oint d_e E = \oint (đQ - p\,dV + \mu\,dN), \tag{1.120}$$

where $\oint \text{đ}Q$ is the heat exchanged with the reservoirs, $\mathcal{W} \equiv \oint p\,dV$ the work performed by the piston during the cycle, and

$$\Delta\mathcal{G} \equiv \oint \mu\,dN = (\mu_\text{h} - \mu_\text{l})\Delta N \tag{1.121}$$

the free energy used by the engine. Since $d_\text{e}S = \text{đ}Q/T$ and the cyclic process is isothermal, the second law (1.4) implies that

$$0 \geq \oint d_\text{e}S = \oint \frac{\text{đ}Q}{T} = \frac{1}{T}\Big(\underbrace{\oint dE}_{=0} + \underbrace{\oint p\,dV}_{=\mathcal{W}} - \underbrace{\oint \mu\,dN}_{=\Delta\mathcal{G}}\Big). \tag{1.122}$$

Consequently, the work $\mathcal{W} > 0$ performed by the engine satisfies $\mathcal{W} \leq \Delta\mathcal{G}$. Here, we may introduce the thermodynamic efficiency

$$\eta_\text{th} \equiv \frac{\mathcal{W}}{\Delta\mathcal{G}} \leq 1, \tag{1.123}$$

which may never exceed the unit value because of the second law. We note that the thermodynamic efficiency (1.123) is defined using the net free energy $\Delta\mathcal{G} > 0$ consumed to drive the engine, while the usual efficiency (1.118) involves the heat $|Q_\text{h}|$ supplied by one reservoir, which explains the difference between their upper bounds.

If we consider the reverse process where the work $W \equiv -\mathcal{W} > 0$ is performed by the exterior onto the system in order to store the free energy $\Delta G \equiv -\Delta\mathcal{G} > 0$ inside the system, we obtain the so-called *Clausius inequality*

$$W \geq \Delta G. \tag{1.124}$$

Therefore, the thermodynamic efficiency of free-energy storage should satisfy the following inequality

$$\eta'_\text{th} \equiv \frac{1}{\eta_\text{th}} = \frac{\Delta G}{W} \leq 1, \tag{1.125}$$

according to the second law. The same results hold for multicomponent mixtures with $\Delta\mathcal{G} = \oint \sum_{k=1}^{c} \mu_k dN_k = -\Delta G$.

Electric motors working with the electric potential difference supplied by a battery obey similar relations with the high and low electrochemical potentials $\mu_\text{h} = e\Phi_\text{h}$ and $\mu_\text{l} = e\Phi_\text{l}$, and the total electric charge $e\Delta N$ transferred during the cycle. If the cycle has the period $\mathscr{T}$, the mechanical power of the electric motor is given by $P_\text{mech} \equiv \mathcal{W}/\mathscr{T} \leq \Delta\mathcal{G}/\mathscr{T} = VI$ with the voltage $V = \Phi_\text{h} - \Phi_\text{l}$ and the electric current $I = e\Delta N/\mathscr{T}$. The difference $P_\text{diss} = VI - P_\text{mech}$ represents the power dissipated by the irreversible processes.

Isothermal engines working on potential differences constitute an important class of engines from the macroscale down to the nanoscale, e.g., for molecular motors running at ambient temperature using chemical energy.

1.11 Open Issues

This chapter has been devoted to thermodynamics, which provides a general framework to identify the equilibrium and nonequilibrium properties in relation to the conservation laws of energy and particle numbers within the macroscopic description of matter. In this regard, thermodynamics plays a fundamental role in the energetics of various dissipative phenomena.

As already understood by Maxwell (1871), the domain of validity of thermodynamics is the macroscopic world. Since the discovery of the atomic structure of matter and the development of the molecular kinetic theory of heat, thermodynamics has been confronted with the microscopic description of matter in terms of atoms and molecules moving according to the laws of classical or quantum mechanics on the basis of electrodynamics.

At the microscale, the particles undergo collisions that conserve energy and their dynamics are symmetric under time reversal, which is the property of microreversibility. This is in contrast with the time asymmetry observed at the macroscale in processes dissipating energy and producing entropy, in particular, during relaxation towards thermodynamic equilibrium. Such irreversible phenomena manifest a loss of memory of their initial conditions, which should be understood in terms of the microscopic dynamics of atoms and molecules.

If thermodynamics provides the framework to consistently formulate the equilibrium and nonequilibrium constitutive relations, these latter remain unknown in theory without using the microscopic dynamics. Furthermore, microreversibility has consequences for the coupling between the currents and affinities of different irreversible processes, and thus for the description of energy transduction down to the nanoscale.

These open issues will be addressed in the following chapters.

2
Statistical Mechanics

2.1 Introduction

At the Ångström scale (10^{-10} m), matter is observed to be composed of atoms and molecules. Their motion is ruled by classical or quantum mechanics, depending on whether their de Broglie wavelength is respectively smaller or larger than the mean distance between nearest neighboring particles. For positive temperatures, these particles undergo ceaseless movements, called thermal or molecular fluctuations. Yet, at the macroscale, these fluctuations are not observed and matter appears continuous. In this context, the issue is to understand the macroscopic properties of matter on the basis of the microscopic dynamics of atoms and molecules. The difference of spatial and temporal scales between the microscopic realm of atoms and the macroscopic world is such that statistical considerations are required to bridge the gap between these scales and to conceive the properties that emerge at the macroscale. In this endeavor, statistics should be considered not only for large numbers of particles but also for time intervals that are longer than the timescale of microscopic motion, or for many trajectories of the system, repeatedly observed under given macroscopic conditions. In this way, probability distributions are introduced in the framework of mechanics, laying down the foundations of statistical mechanics.

Statistical mechanics has a long and rich history going back to pioneering work on the thermal probability distribution of particle velocities (Maxwell, 1860), the dynamical theory of gases (Maxwell, 1867), the thermal probability distribution of energy (Boltzmann, 1871), the kinetic equation for dilute gases (Boltzmann, 1872), the statistical interpretation of entropy (Boltzmann, 1877, 1896, 1898), and the statistical ensembles describing equilibrium systems (Gibbs, 1902). If the first law of thermodynamics directly finds its origin in mechanics and its conservation laws resulting from continuous symmetries according to Noether's theorem, the second law of thermodynamics has statistical origins, the entropy being related to the probability distribution associated with some statistical ensemble, as already understood by Maxwell, Boltzmann, and Gibbs.

In 1900, Planck discovered the quantum character of blackbody radiation, thus playing a key role in the development of statistical mechanics (Planck, 1914). The reality of atoms was established by the efforts of Einstein (1905, 1926), von Smoluchowski (1906), Perrin (1910), and others to determine Avogadro's number, in particular, from the phenomenon of

Brownian motion. Since the rise of quantum mechanics, statistical mechanics has known fundamental advances, especially, with the discovery of Bose–Einstein and Fermi–Dirac statistics for bosons and fermions, which are essential to understand the condensed phases of matter at low temperature or high density.

The different phases of matter are characterized not only by their equilibrium properties such as the pressure and energy equations of state but also by their nonequilibrium properties, including the viscosities, the conductivities, the diffusivities, and the reaction rates. The goal of statistical mechanics is to determine these equilibrium and nonequilibrium properties on the basis of the microscopic dynamics of atoms and molecules, combined with the macroscopic conditions of temperature, pressure, and chemical potentials at the boundaries of the system.

The understanding of nonequilibrium properties is particularly challenging because the Gibbsian equilibrium statistical ensembles of microstates no longer apply to systems that are time dependent or driven away from equilibrium by external constraints. From the microscopic viewpoint, nonequilibrium statistical ensembles should thus be introduced for the trajectories or histories of the system, i.e., the possible time evolutions of its microstates, instead of the instantaneous microstates as in the equilibrium statistical ensembles. Nevertheless, great advances have been made since Boltzmann obtained his famous kinetic equation (Boltzmann, 1872, 1896). In the 1930s, Onsager discovered his reciprocal relations on the basis of microreversibility (Onsager, 1931a,b). In the 1950s, the fluctuation–dissipation theorem was established (Callen and Welton, 1951) and statistical-mechanical formulae were deduced for the transport coefficients by Green (1952b, 1954), Kubo (1957), and others, in the linear regime close to equilibrium.

Crucial progress has been made about the conceptual foundations of statistical mechanics with the development of ergodic theory for classical and quantum systems (Ehrenfest and Ehrenfest, 1911; Arnold and Avez, 1968; Cornfeld et al., 1982; Thirring, 1983), as well as the large-deviation theory in the framework of dynamical systems theory (Bowen, 1975; Ruelle, 1978; Eckmann and Ruelle, 1985). Progress in this area allows us to nowadays understand how microscopic dynamics can generate relaxation processes towards an equilibrium macrostate (Pollicott, 1985, 1986; Ruelle, 1986a,b; Jakšić and Pillet, 1996a,b; Gaspard, 1998).

More recently, fundamental results have been obtained for systems driven away from equilibrium beyond the linear regime, especially with the advent of time-reversal symmetry relations called fluctuation relations, allowing us, in particular, to generalize the Onsager reciprocal relations to the nonlinear response and transport properties (Bochkov and Kuzovlev, 1977, 1979, 1981a,b; Stratonovich, 1992, 1994; Evans et al., 1993; Gallavotti and Cohen, 1995; Gallavotti, 1996; Jarzynski, 1997; Kurchan, 1998; Crooks, 1999; Lebowitz and Spohn, 1999; Evans and Searles, 2002; Andrieux and Gaspard, 2004, 2007a; Derrida, 2007; Esposito et al., 2009; Campisi et al., 2011; Jarzynski, 2011; Seifert, 2012; Gaspard, 2013a). These results will be presented in Chapter 5. Furthermore, significant progress has been made in the foundations of nonequilibrium statistical mechanics towards understanding how time-reversal symmetry can hold for the microscopic dynamics of atoms and molecules, while being broken at the statistical level of description.[1]

The purpose of this chapter is to introduce the basic concepts of statistical mechanics.

2.2 Classical Mechanics

2.2.1 The Quantum Roots of Classical Mechanics

The fundamental theory for the microscopic dynamics of atoms and other particles composing matter is quantum mechanics. In this framework, the system of particles is described by a wave function depending on the particle positions, and the time evolution of this wave function is ruled by Schrödinger's equation. Often, the wave function has undulatory spatial variations, every particle position having a wavelength $\lambda_{\rm dB}$ related to the particle linear momentum p and Planck's constant $h = 6.62607015 \times 10^{-34}$ J/Hz, according to de Broglie's formula $\lambda_{\rm dB} = h/p$.

Quantum mechanics can be approximated by classical mechanics if de Broglie's wavelength is much smaller than the mean distance between nearest neighboring particles, $\mathcal{R} \sim (V/N)^{1/3}$, where N is the number of particles in the volume V.[2] Since nonrelativistic particles have velocities v much smaller than the speed of light $c = 299{,}792{,}458$ m/s, their linear momentum is given by $p = mv$ in terms of their mass m. In systems at low enough temperature $T \ll mc^2/k_{\rm B}$, the particle momentum can thus be estimated as $p \sim \sqrt{mk_{\rm B}T}$. Therefore, such particles have a classical behavior if the following condition holds:

$$\lambda_{\rm dB} \sim \frac{h}{\sqrt{mk_{\rm B}T}} \ll \mathcal{R} \sim \left(\frac{V}{N}\right)^{1/3}. \tag{2.1}$$

For condensed phases such as liquids or solids at room temperature, this condition is satisfied for nuclei, thus having classical dynamics. However, electrons being much lighter than nuclei manifest quantum behavior, in particular, forming orbitals in atoms and molecules. In this regard, the electronic orbitals determine the force fields between the nuclei according to the Born–Oppenheimer quantum theory of molecules (Moore, 1972; Berry et al., 1980).

2.2.2 The Hamiltonian Function

In the nonrelativistic limit, the mechanics of nuclei is described by the following Hamiltonian function:

$$H = \sum_{a=1}^{N} \left(\frac{\mathbf{p}_a^2}{2m_a} + u_a^{(\rm ext)} \right) + \sum_{1 \le a < b \le N} u_{ab}^{(2)} + \sum_{1 \le a < b < c \le N} u_{abc}^{(3)} + \cdots, \tag{2.2}$$

[1] In this book, we say that some symmetry is broken at the dynamical level of description if the equations of motion such as Hamilton's equations are not symmetric. We say that some symmetry is broken by the selection of initial conditions or at the statistical level of description if the equations of motion are symmetric, but their solutions are not, as discussed in Section 2.6.

[2] In a random distribution of particles with density $n = N/V$, the mean distance between nearest neighbors is given by $\mathcal{R} = \Gamma(4/3)\,(4\pi n/3)^{-1/3} \simeq 0.554\,n^{-1/3}$ in terms of the Gamma function $\Gamma(z)$ (Chandrasekhar, 1943).

where $\mathbf{r}_a = (r_{ax}, r_{ay}, r_{az}) = (x_a, y_a, z_a) \in \mathbb{R}^3$, $\mathbf{p}_a = (p_{ax}, p_{ay}, p_{az}) \in \mathbb{R}^3$, and $m_a > 0$ are, respectively, the position, the momentum, and the mass for every one of the N particles composing the system. The spin of the particles is here discarded. The function $u_a^{(\text{ext})} = u_a^{(\text{ext})}(\mathbf{r}_a)$ gives the potential energy of the particle a in an external force field, for instance, containing the particles in a vessel of volume V. The functions $u_{ab}^{(2)} = u_{ab}^{(2)}(r_{ab})$ and $u_{abc}^{(3)} = u_{abc}^{(3)}(r_{ab}, r_{bc}, r_{ca})$ are the potential energies of binary and ternary interactions between the particles, which are separated by the distances $r_{ab} = \|\mathbf{r}_a - \mathbf{r}_b\|$. These potential energies are determined within the Born–Oppenheimer theory (Moore, 1972; Berry et al., 1980). For rare gases, a good approximation consists in using only the potential energy $u_{ab}^{(2)}$ of binary interaction, neglecting further terms. For molecular gases, the mutual interactions between the nuclei lead to the formation of molecules, in which case the Hamiltonian function should be expressed in terms of the translational, rotational, and vibrational degrees of freedom of the molecules, as long as the temperature is low enough to avoid the dissociation of the molecules.

If the particles have electric charges $\{q_a\}_{a=1}^N$ and the system is subjected to electric and magnetic fields $\boldsymbol{\mathcal{E}}$ and $\boldsymbol{\mathcal{B}}$, the linear momenta $\mathbf{p}_a$ of the particles should be replaced by the quantities $\mathbf{p}_a - q_a\boldsymbol{\mathcal{A}}(\mathbf{r}_a, t)$, where $\boldsymbol{\mathcal{A}}$ is the electromagnetic vector potential, such that $\boldsymbol{\mathcal{B}} = \nabla \times \boldsymbol{\mathcal{A}}$. Moreover, the Hamiltonian function (2.2) should include the contribution $q_a\Phi(\mathbf{r}_a, t)$ from the electromagnetic scalar potential Φ. In general, the scalar and vector potentials and thus the Hamiltonian function may be time dependent, e.g., if the system is heated by an electromagnetic wave. In static and uniform electric and magnetic fields, these potentials can be chosen to be in the forms $\Phi(\mathbf{r}) = -\boldsymbol{\mathcal{E}} \cdot \mathbf{r}$ and $\boldsymbol{\mathcal{A}} = \boldsymbol{\mathcal{B}} \times \mathbf{r}/2$, in which case the Hamiltonian function remains independent of time.

All the particles of the same species are identical and they have the same mass. In this respect, their labels should be grouped together into sets, $a \in \mathcal{S}_k$, for all the particles of the same species ($k = 1, 2, \ldots, c$). The set $\mathcal{S}_k$ contains a number of labels equal to the number N_k of particles of the species k, and $N = \sum_{k=1}^{c} N_k$ is the total number of particles in the system. The Hamiltonian function (2.2) should thus be totally symmetric under the $N_k!$ permutations of the labels $a \in \mathcal{S}_k$, so that $m_a = m_k$ for all the labels $a \in \mathcal{S}_k$. Similarly, the potential energies should satisfy the identities $u_a^{(\text{ext})}(\mathbf{r}_a) = u_k^{(\text{ext})}(\mathbf{r}_a)$, $u_{ab}^{(2)}(r_{ab}) = u_{kl}^{(2)}(r_{ab}), \ldots$ for $a \in \mathcal{S}_k$, $b \in \mathcal{S}_l, \ldots$

2.2.3 Phase Space

The positions and momenta of the particles take their values

$$\boldsymbol{\Gamma} = (\mathbf{r}_1, \mathbf{p}_1, \mathbf{r}_2, \mathbf{p}_2, \ldots, \mathbf{r}_N, \mathbf{p}_N) \in \mathcal{M} \tag{2.3}$$

in the so-called *phase space* $\mathcal{M} \subset \mathbb{R}^{6N}$. Every degree of freedom corresponds to a pair of conjugate position-momentum variables, so that the Hamiltonian system has $f = 3N$ degrees of freedom and its phase-space dimension is equal to $d = 2f = 6N$. In this regard, every classical microstate of the system should be associated with a point in the phase space.

If all the particles are identical (i.e., $m_a = m$ for $a = 1, 2, \ldots, N$), an observer should assign labels to the particles according to some rule, because all the points $\mathbf{\Gamma}_P = \left(\mathbf{r}_{P(1)}, \mathbf{p}_{P(1)}, \mathbf{r}_{P(2)}, \mathbf{p}_{P(2)}, \ldots, \mathbf{r}_{P(N)}, \mathbf{p}_{P(N)}\right) \in \mathbb{R}^{6N}$ obtained by exchanging the labels with some permutation $P \in \operatorname{Sym} N$ are physically equivalent.[3] If the observer makes the choice of sorting the particles according to their position coordinate in the x-direction, the phase space is defined as

$$\mathcal{M} = \left\{\mathbf{\Gamma} = (\mathbf{r}_1, \mathbf{p}_1, \mathbf{r}_2, \mathbf{p}_2, \ldots, \mathbf{r}_N, \mathbf{p}_N) \in \mathbb{R}^{6N} \text{ such that } x_1 < x_2 < \cdots < x_N\right\}, \tag{2.4}$$

which contains all the physically distinct classical microstates of the system. Consequently, any phase-space integral can be evaluated as

$$\int_{\mathcal{M}} (\cdot)\, d\mathbf{\Gamma} = \frac{1}{N!} \int_{\mathbb{R}^{6N}} (\cdot)\, d\mathbf{\Gamma}, \tag{2.5}$$

where $d\mathbf{\Gamma} = d^{6N}\Gamma = \prod_{a=1}^{N} d^3 r_a d^3 p_a$ is the phase-space volume element.

If the system contains N_k particles of species $k = 1, 2, \ldots, c$, the phase space $\mathcal{M}$ of physically distinct classical microstates should be defined by sorting all the identical particles $a \in \mathcal{S}_k$, e.g., using their position coordinate in the x-direction, so that equation (2.5) should be written as

$$\int_{\mathcal{M}} (\cdot)\, d\mathbf{\Gamma} = \frac{1}{\prod_{k=1}^{c} N_k!} \int_{\mathbb{R}^{6N}} (\cdot)\, d\mathbf{\Gamma}. \tag{2.6}$$

We note that, taking the limit from the quantum to the classical description, every quantum microstate of the particle system (e.g., every eigenstate of the energy operator) corresponds to the elementary quantal volume

$$\Delta\mathbf{\Gamma} = \Delta^{6N}\Gamma = \prod_{a=1}^{N} \prod_{i=x,y,z} \Delta r_{ai}\, \Delta p_{ai} = h^{3N} \tag{2.7}$$

in the classical phase space $\mathcal{M}$. This $6N$-dimensional volume defines the finest grain to be considered in the classical phase space from the viewpoint of quantum mechanics. Finer grains may be classically considered, but they have no physical meaning since Heisenberg's uncertainty relations $\sigma_{r_{ai}} \sigma_{p_{ai}} \geq h/(4\pi)$ may not be satisfied for grains with a volume smaller than that given by (2.7).[4] The elementary quantal volume (2.7) is essential for counting quantum microstates using the classical approximation and, in particular, for the evaluation of the density of quantum microstates at the energy E, i.e., the density of quantum energy levels, according to

[3] Sym N denotes the discrete group containing the $N!$ permutations of N elements, called the *symmetric group*.

[4] Considering a function of Wigner (1932) equal to $p(\mathbf{\Gamma}) = 1/\Delta\mathbf{\Gamma}$ in some volume (2.7) with $\Delta r_{ai} \Delta p_{ai} = h$ and zero elsewhere, the standard deviations on positions and momenta are given by $\sigma_{r_{ai}} = \Delta r_{ai}/\sqrt{12}$ and $\sigma_{p_{ai}} = \Delta p_{ai}/\sqrt{12}$, so that $\sigma_{r_{ai}} \sigma_{p_{ai}} = h/12$, which closely satisfies Heisenberg's uncertainty relation. This latter arises from the impossibility of jointly defining the position and the momentum of a particle in quantum mechanics, where the microstates are described by wave functions depending on either position or momentum, but never on both.

$$\mathscr{D}(E) \simeq \frac{1}{\Delta\Gamma}\int_{\mathcal{M}} \delta\left[E - H(\Gamma)\right] d\Gamma = \frac{1}{h^{3N}N!}\int_{\mathbb{R}^{6N}} \delta\left[E - H(\Gamma)\right] d\Gamma, \tag{2.8}$$

where the second equality holds for systems with N identical particles of the same species. The classical approximation (2.8) for the density of states can be obtained by taking the transform of Wigner (1932), which justifies the expression (2.7) for the elementary quantal volume corresponding to a single quantum microstate in the classical phase space. This elementary quantal volume was unknown to Boltzmann and Gibbs, since it could only be identified with the advent of quantum mechanics.

2.2.4 Hamiltonian Dynamics

In classical mechanics, the positions and momenta of the particles obey Hamilton's equations

$$\begin{cases} \dfrac{d\mathbf{r}_a}{dt} = \dfrac{\partial H}{\partial \mathbf{p}_a}, \\[2mm] \dfrac{d\mathbf{p}_a}{dt} = -\dfrac{\partial H}{\partial \mathbf{r}_a}, \end{cases} \qquad \text{for} \quad a = 1, 2, \dots, N, \tag{2.9}$$

which forms a set of $6N$ coupled ordinary differential equations to be solved to determine the time evolution of the system.

If the Hamiltonian function has the nonrelativistic form (2.2), the Hamilton equations (2.9) are equivalent to Newton's equations

$$m_a \frac{d^2\mathbf{r}_a}{dt^2} = \mathbf{F}_a, \qquad \text{for} \quad a = 1, 2, \dots, N, \tag{2.10}$$

where

$$\mathbf{F}_a = \mathbf{F}_a^{(\text{ext})} + \sum_{b(\neq a)} \mathbf{F}_{ab} \tag{2.11}$$

is the force exerted on the particle a by the external force $\mathbf{F}_a^{(\text{ext})} = -\partial u_a^{(\text{ext})}/\partial \mathbf{r}_a$ and the other particles. In the case of binary interactions, the force exerted on the particle a by the particle b is given by

$$\mathbf{F}_{ab} = -\frac{\partial u_{ab}^{(2)}}{\partial \mathbf{r}_a} = -\frac{du_{ab}^{(2)}}{dr_{ab}} \frac{\mathbf{r}_{ab}}{r_{ab}} \tag{2.12}$$

with $\mathbf{r}_{ab} = \mathbf{r}_a - \mathbf{r}_b$. The *action-reaction principle*, $\mathbf{F}_{ab} = -\mathbf{F}_{ba}$, is satisfied as the potential energy $u_{ab}^{(2)}$ of binary interaction is assumed to depend only on the distance r_{ab} between the two particles.

Besides the Hamiltonian function giving the total energy of the system, there exist other observables such as the total linear momentum $\mathbf{P} = \sum_{a=1}^{N} \mathbf{p}_a$, the total angular momentum $\mathbf{L} = \sum_{a=1}^{N} \mathbf{r}_a \times \mathbf{p}_a$, the mass density $\hat{\rho}(\mathbf{r}) = \sum_{a=1}^{N} m_a \delta(\mathbf{r} - \mathbf{r}_a)$, the linear momentum density $\hat{\mathbf{g}}(\mathbf{r}) = \sum_{a=1}^{N} \mathbf{p}_a \delta(\mathbf{r} - \mathbf{r}_a)$, and the electric energy $\sum_{a=1}^{N} q_a \Phi(\mathbf{r}_a, t)$. All these observables are real functions or distributions of the phase-space variables with a possible explicit dependence on time: $A(\mathbf{\Gamma}, t)$. Supposing that the phase-space point evolves in time according to some solution $\mathbf{\Gamma}_t$ of the Hamilton equations (2.9), the total time derivative of this observable obeys

$$\frac{dA}{dt} = \{A, H\} + \frac{\partial A}{\partial t}, \tag{2.13}$$

where $\{\cdot, \cdot\}$ denotes the Poisson bracket (Goldstein, 1950; Arnold, 1989)

$$\{A, B\} \equiv \sum_{a=1}^{N} \left(\frac{\partial A}{\partial \mathbf{r}_a} \cdot \frac{\partial B}{\partial \mathbf{p}_a} - \frac{\partial A}{\partial \mathbf{p}_a} \cdot \frac{\partial B}{\partial \mathbf{r}_a} \right). \tag{2.14}$$

Equation (2.13) determines the time evolution of all the observables. In particular, Hamilton's equations can be written in the form $d\mathbf{\Gamma}/dt = \{\mathbf{\Gamma}, H\}$ since the positions and momenta are observables that do not explicitly depend on time ($\partial \mathbf{\Gamma}/\partial t = 0$). We note that the Poisson bracket (2.14) can be decomposed as $\{A, B\} = \sum_{a=1}^{N} \{A, B\}_a = \sum_{i=1}^{f} \{A, B\}_i$ in terms of the Poisson brackets associated with every particle a or degree of freedom i.

An important property of Hamiltonian dynamics is

Liouville's theorem: *The time evolution ruled by the Hamilton equations (2.9) preserves the phase-space volumes.*

This property results from the fact that the phase-space divergence of Hamilton's equations is equal to zero:

$$\text{div}_{\mathbb{R}^{6N}} \dot{\mathbf{\Gamma}} = \sum_{a=1}^{N} \left(\frac{\partial \dot{\mathbf{r}}_a}{\partial \mathbf{r}_a} + \frac{\partial \dot{\mathbf{p}}_a}{\partial \mathbf{p}_a} \right) = \sum_{a=1}^{N} \left(\frac{\partial^2 H}{\partial \mathbf{r}_a \partial \mathbf{p}_a} - \frac{\partial^2 H}{\partial \mathbf{p}_a \partial \mathbf{r}_a} \right) = 0, \tag{2.15}$$

where $\dot{\mathbf{X}} \equiv d\mathbf{X}/dt$. Since the elements of phase-space volumes expand if this divergence is positive and contract if it is negative, these elements keep their volume if it is equal to zero, so that the phase-space volumes are preserved by Hamiltonian dynamics (Goldstein, 1950; Arnold, 1989). Liouville's theorem finds its origin in the unitarity of underlying quantum mechanics.

2.2.5 Existence and Uniqueness Theorem

In classical mechanics, the time evolution of the particle system is determined by the phase-space trajectories that are the solutions of Hamilton's equations. In general, Hamilton's equations can be written in the form

$$\dot{\mathbf{\Gamma}} = \mathbf{V}(\mathbf{\Gamma}, t), \tag{2.16}$$

where $\mathbf{V}(\mathbf{\Gamma}, t)$ is a vector field defined in the phase space $\mathbb{R}^d$ of dimension $d = 6N$ by the right-hand side of the Hamilton equations (2.9), possibly varying with time $t \in \mathbb{R}$. Accordingly, the phase-space point $\mathbf{\Gamma}$ evolves in time, advancing in the direction of the vector $\mathbf{V}(\mathbf{\Gamma}, t)$ with the speed given by the magnitude of this vector.

The conditions for the existence and uniqueness of the solutions of the ordinary differential equations (2.16) are given by the

Existence and uniqueness theorem: *If the vector field* $\mathbf{V}(\mathbf{\Gamma},t)$ *is bounded as*

$$\sup_{(\mathbf{\Gamma},t)\in\mathcal{D}} \|\mathbf{V}(\mathbf{\Gamma},t)\| \leq M \tag{2.17}$$

in the domain $\mathcal{D} \subset \mathbb{R}^{d+1}$ *such that* $|t-t_0| < T$ *and* $\|\mathbf{\Gamma}-\mathbf{\Gamma}_0\| < R$ *with* $\|\mathbf{X}\| \equiv \sup_{i=1,2,\dots,d}|X_i|$ *and, moreover, if it satisfies the Lipschitz condition,*

$$\|\mathbf{V}(\mathbf{\Gamma},t) - \mathbf{V}(\mathbf{\Gamma}',t)\| \leq L\, \|\mathbf{\Gamma}-\mathbf{\Gamma}'\| \tag{2.18}$$

with some constant $L>0$, $\mathbf{\Gamma}\in\mathcal{D}$, *and* $\mathbf{\Gamma}'\in\mathcal{D}$, *then the solution* $\mathbf{\Gamma}(t)$ *of the equations (2.16), starting from the initial conditions* $\mathbf{\Gamma}_0 = \mathbf{\Gamma}(t_0)$ *at time* t_0, *exists and is unique in the time interval* $|t-t_0|<\tau$ *with* $\tau = \min(T, R/M)$. *This solution is continuous in the initial conditions because*

$$\|\tilde{\mathbf{\Gamma}}(t) - \mathbf{\Gamma}(t)\| \leq (\|\delta\mathbf{\Gamma}_0\| + M|\delta t_0|)\, \mathrm{e}^{L|t-t_0|}, \tag{2.19}$$

where $\tilde{\mathbf{\Gamma}}(t)$ *is the solution of initial conditions* $\mathbf{\Gamma}_0 + \delta\mathbf{\Gamma}_0$ *at time* $t_0 + \delta t_0$.

A proof of this theorem is given by Coddington and Levinson (1955). This theorem, which is also known as the Cauchy–Lipschitz or Picard–Lindelöf theorem, is sometimes referred to as the principle of determinism since the phase-space trajectory is uniquely determined by the initial conditions on the positions $[\mathbf{r}_a(t_0)]_{a=1}^N$ and the momenta $[\mathbf{p}_a(t_0)]_{a=1}^N$ of the particles. This confirms that every classical microstate is defined by the positions and momenta of the particles, i.e., as a point in the classical phase space (2.4). The $d=6N$ initial conditions $\mathbf{\Gamma}_0 = [\mathbf{r}_a(t_0), \mathbf{p}_a(t_0)]_{a=1}^N$ are often called constants of integration or integrals of motion (Goldstein, 1950; Landau and Lifshitz, 1976) because they are introduced by integrating the $d=6N$ ordinary differential equations (2.16) over time t.

If the system is autonomous, i.e., time independent, the vector field $\mathbf{V}(\mathbf{\Gamma})$ is fixed in phase space, so that there exists a *flow* $\mathbf{\Phi}^t$ defined as the set of $6N$ functions of time t and initial conditions $\mathbf{\Gamma}_0$, giving the phase-space point,

$$\mathbf{\Gamma} = \mathbf{\Phi}^t\mathbf{\Gamma}_0, \tag{2.20}$$

at time t for the trajectory issued from the initial conditions. This whole trajectory is the phase-space curve $\mathcal{T} = \{\mathbf{\Phi}^t\mathbf{\Gamma}_0 : t\in\mathbb{R}\}$ composed of all the points (2.20) as time varies from past to future. This curve is invariant under the phase-space dynamics, $\mathcal{T} = \mathbf{\Phi}^t\mathcal{T}$ for $t\in\mathbb{R}$. The phase-space flow defined by $\mathbf{\Phi}^t$ forms a one-parameter Lie group because

$$\mathbf{\Phi}^t\,\mathbf{\Phi}^{t'} = \mathbf{\Phi}^{t+t'}, \qquad \left(\mathbf{\Phi}^t\right)^{-1} = \mathbf{\Phi}^{-t}, \qquad \text{and} \qquad \mathbf{\Phi}^0 = \mathbf{1}, \tag{2.21}$$

where $\mathbf{1}$ denotes the identity map in phase space (Arnold, 1989). Time is the parameter of this continuous group.

If classical mechanics can, in principle, determine the time evolution of the system from the knowledge of its initial conditions, classical mechanics cannot predict the values of the initial conditions, which are inputs to be provided by other considerations such as the measurement or the preparation of initial conditions by devices external to the system described by Newton's equations. The initial conditions are not specified by the equations of mechanics, although the initial conditions determine the various behavior of systems (Wigner, 1963). In this regard, classical mechanics appears incomplete.

Furthermore, the phase space forms a continuum because the positions and the momenta of the particles are given by real numbers, $\mathbf{\Gamma} \in \mathbb{R}^{6N}$. Indeed, real numbers are defined by an infinite sequence of bits or digits, so that they form an uncountable set, namely a continuum. Since physical measurements are always limited to a finite accuracy, the knowledge of any real number is restricted to a finite number of bits or digits in the sequence defining the real number. There is thus always a dichotomy between the existence and the knowledge of a real number. Although the basic postulate of classical mechanics is to suppose that there exist $6N$ real numbers $\mathbf{\Gamma} \in \mathcal{M}$ representing the positions and momenta of the particles at some time t, the knowledge of these positions and momenta is always limited to some accuracy $\|\delta\mathbf{\Gamma}\| < \epsilon$. The lack of complete knowledge of the phase-space point $\mathbf{\Gamma}$ has the consequence that a probability distribution should be introduced to describe the possible values of $\mathbf{\Gamma}$ resulting from the measurement or the preparation of $\mathbf{\Gamma}$. This is the fundamental reason for extending mechanics into statistical mechanics.

2.2.6 Dynamical Stability or Instability

As shown by equation (2.19) of the existence and uniqueness theorem, trajectories starting from nearby initial conditions at most diverge exponentially in time over the finite time interval $|t - t_0| < \tau$. This result raises the question of the stability or instability of the solutions, in particular, with respect to arbitrarily small perturbations of the initial conditions. Infinitesimal perturbations $\delta\mathbf{\Gamma}$ obey the following set of linear equations,

$$\delta\dot{\mathbf{\Gamma}} = \frac{\partial \mathbf{V}}{\partial \mathbf{\Gamma}}(\mathbf{\Gamma}_t, t) \cdot \delta\mathbf{\Gamma}, \tag{2.22}$$

defined in terms of the Jacobian matrix $\partial\mathbf{V}/\partial\mathbf{\Gamma}$ once the solution $\mathbf{\Gamma}_t$ of Hamilton's equations is known. The infinitesimal perturbations $\delta\mathbf{\Gamma}_t$ may have different types of time dependence, which can be characterized by their rate of exponential growth in time, defining the (maximum) Lyapunov exponent

$$\lambda \equiv \lim_{t\to\infty} \frac{1}{t} \ln \frac{\|\delta\mathbf{\Gamma}_t\|}{\|\delta\mathbf{\Gamma}_0\|}. \tag{2.23}$$

In many systems, this Lyapunov exponent is positive ($\lambda > 0$), so that they manifest the property of *sensitivity to initial conditions*, also called dynamical instability. In such systems, the prediction of the future time evolution from the initial conditions is no longer possible beyond some time interval referred to as *Lyapunov's horizon*. If a prediction with accuracy $\|\delta\mathbf{\Gamma}_t\| < \varepsilon$ is required from initial conditions known with accuracy $\|\delta\mathbf{\Gamma}_0\| = \varepsilon_0$, prediction fails beyond Lyapunov's time

$$t > t_{\mathrm{L}} = \frac{1}{\lambda} \ln \frac{\varepsilon}{\varepsilon_0}, \tag{2.24}$$

which is of the order of magnitude of the inverse of the maximum Lyapunov exponent. Lyapunov's horizon can only be pushed to longer time at the cost of increasing exponentially the accuracy on the knowledge of initial conditions.

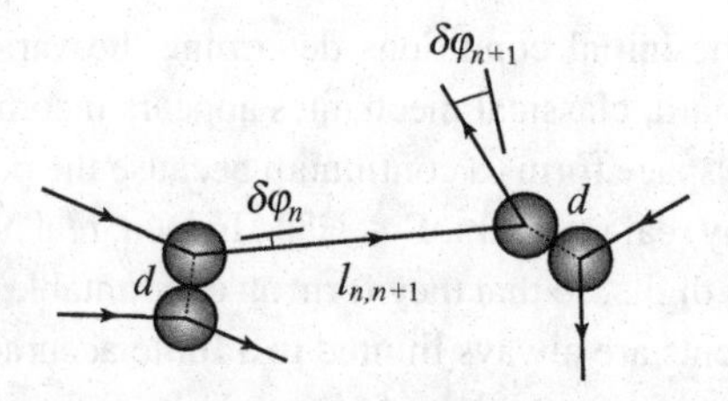

Figure 2.1 Mechanism of dynamical instability in a hard-sphere gas due to the exponential growth of a perturbation $\delta\varphi_n$ on the velocity angle for a particle of diameter d undergoing successive collisions, which are separated on average by the mean free path $\ell = \langle l_{n,n+1} \rangle$. Accordingly, the perturbation goes as $\delta\varphi_{n+1} \sim (l_{n,n+1}/d)\,\delta\varphi_n$ at the next collision and as $\delta\varphi_n \sim (\ell/d)^n \delta\varphi_0 \sim \exp(\lambda t)\,\delta\varphi_0$ after n collisions, which corresponds to the time lapse $t \simeq n\ell/\bar{v}$, where $\bar{v}$ is the mean particle velocity.

In dilute gases, the maximum Lyapunov exponent can be estimated to take the value

$$\lambda \sim \frac{\bar{v}}{\ell} \ln \frac{\ell}{d}, \tag{2.25}$$

where $\bar{v}$ is the mean thermal velocity of the particles, ℓ their mean free path, and d their diameter, as shown in Figure 2.1 (Krylov, 1979). For gases under standard conditions, this Lyapunov exponent takes the value $\lambda \sim 10^{10}$ digits/s, so that Lyapunov's horizon falls on the timescale of the mean intercollisional time $\tau_{\text{intercoll}} \simeq \ell/\bar{v}$. Therefore, the classical microscopic motion of interacting particles typically manifests an extreme sensitivity to initial conditions.

We note that a Lyapunov exponent can be defined with equation (2.22) in every phase-space direction, so that there exists a spectrum of Lyapunov exponents $\{\lambda_i\}_{i=1}^{6N}$ characterizing the dynamical instability of the phase-space trajectory (Eckmann and Ruelle, 1985; Nicolis, 1995). The calculation of these exponents is explained in Appendix B. In Hamiltonian systems, the sum of all these Lyapunov exponents is always equal to zero according to Liouville's theorem, $\sum_{i=1}^{6N} \lambda_i = 0$. A spectrum of Lyapunov exponents is shown in Figure 2.2 for a gas of hard spheres undergoing elastic collisions between free flights.

However, there also exist systems without this exponential sensitivity to initial conditions, such as the ideal gases and other fully integrable systems for which the maximum Lyapunov exponent is vanishing. In between the fully integrable and fully chaotic systems, there is a whole diversity of systems manifesting intermediate behavior with intertwined quasiperiodic and chaotic motions. Indeed, the famous theorem by Kolmogorov (1954), Arnold (1963), and Moser (1973) (the KAM theorem) predicts quasiperiodic motion in small perturbations $H = H_0 + U$ of integrable Hamiltonian systems H_0. Since quasiperiodic dynamics have several zero Lyapunov exponents, regular motion is thus predicted by the KAM theorem (Arnold and Avez, 1968; Ott, 1993). Nevertheless, as mentioned, the knowledge of initial conditions is always limited by finite accuracy in any kind of system.

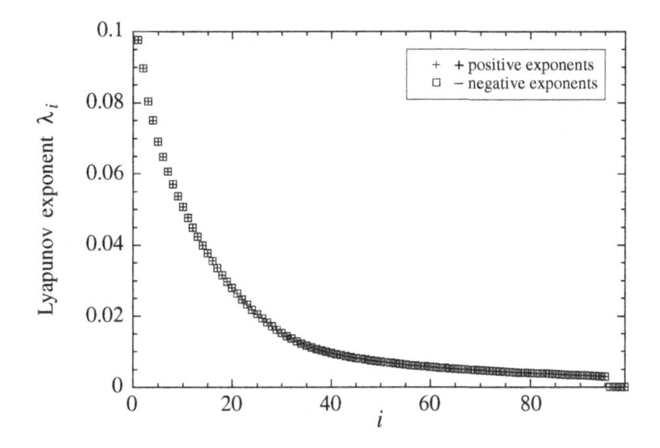

Figure 2.2 Spectrum of Lyapunov exponents of a system with $N = 33$ hard spheres of unit diameter and mass at unit temperature and with density 0.001, moving in a cubic domain with periodic boundary conditions. The Lyapunov exponents obey the pairing rule that the Lyapunov exponents come in pairs $\{\lambda_i, -\lambda_i\}$ because the dynamical system is symplectic. Eight Lyapunov exponents vanish because the system has four conserved quantities, namely, energy and the three components of momentum, and the pairing rule should hold. The total number of Lyapunov exponents is equal to $6N = 198$.

2.2.7 Symmetries of Hamiltonian Dynamics

The Hamilton equations (2.9) may be invariant under continuous or discrete phase-space transformations.

Continuous Symmetries and Conservation Laws

As a consequence of the theorem by Noether (1918), there exist conserved quantities that are given by regular functions of the phase-space variables and that are invariant under time evolution, thus, also called (analytic) constants of motion or first integrals $C(\mathbf{\Gamma})$, such that $dC/dt = \{C, H\} = 0$ everywhere in phase space. Consequently, these conserved quantities are constants of integration because they keep their initial values along trajectories: $C(\mathbf{\Gamma}_t) = C(\mathbf{\Gamma}_0)$. In fully integrable systems, the number of independent analytic constants of motion is equal to the number of degrees of freedom according to the Arnold–Liouville theorem (Arnold, 1989). However, in chaotic systems there exist strictly less independent analytic constants of motion than degrees of freedom (Moser, 1973).

If Hamilton's equations are invariant under time translations $t \to t + \tau$ for $\tau \in \mathbb{R}$, there exists a conserved quantity called *energy*. This is the case if the Hamiltonian function is independent of time, $\partial_t H = 0$, so that $dH/dt = \{H, H\} + \partial_t H = 0$ along the phase-space trajectories of the system. This result justifies the first law of thermodynamics, i.e., the principle of energy conservation, on the basis of the microscopic Hamiltonian dynamics.

If the system is subjected to time-dependent external forces, work is performed on the system and the energy of the system changes accordingly.

If Hamilton's equations are invariant under spatial translations $\mathbf{r}_a \to \mathbf{r}_a + \mathbf{R}$ for $\mathbf{R} \in \mathbb{R}^3$, the total linear momentum $\mathbf{P} \equiv \sum_{a=1}^N \mathbf{p}_a$ is also a conserved quantity. This condition is satisfied if there is no external force field, i.e., if $\mathbf{F}_a^{(\text{ext})} = -\partial u_a^{(\text{ext})}/\partial \mathbf{r}_a = 0$, and the interaction forces only depend on the relative positions $\mathbf{r}_a - \mathbf{r}_b$ between the particles, thus obeying the action-reaction principle $\mathbf{F}_{ab} = -\mathbf{F}_{ba}$. In contrast, an external force field confining the particles inside an infinitely heavy vessel breaks the conservation of total linear momentum.

Similarly, the total angular momentum $\mathbf{L} \equiv \sum_{a=1}^N \mathbf{r}_a \times \mathbf{p}_a$ is a conserved quantity if Hamilton's equations are invariant under spatial rotations $\mathbf{r}_a \to \mathsf{O} \cdot \mathbf{r}_a$ for $\mathsf{O} \in \text{SO}(3)$,[5] which is satisfied if there is no external force field and the interaction forces are central, i.e., they only depend on the distances $r_{ab} = \|\mathbf{r}_a - \mathbf{r}_b\|$ between the particles.

Moreover, the center of mass of the system $\mathbf{R}_{\text{cm}} \equiv \left(\sum_{a=1}^N m_a \mathbf{r}_a\right) / \left(\sum_{a=1}^N m_a\right)$ is also constant in time if there is no external force field.

These ten conserved quantities result from the symmetry of the system under the ten-dimensional continuous group of Galilean transformations, as a consequence of Noether's theorem. However, the microscopic motion of particles confined inside an infinitely heavy vessel of arbitrary shape often conserves only the total energy because of the presence of the external force field exerted by the walls of the vessel. If the dynamical system is considered on a three-dimensional torus with periodic boundary conditions for computational purposes, the total linear momentum is also conserved.

Beyond the conservation laws of fundamental origin, there may exist further conserved quantities, e.g., in fully integrable systems such as finite systems of harmonic oscillators or noninteracting particles, or one-dimensional chains of particles with special interaction (Arnold, 1989).

We note that the symmetry condition under some phase-space transformation $\mathbf{G}$ can be expressed in terms of the flow (2.20) as $\mathbf{G}\,\mathbf{\Phi}^t = \mathbf{\Phi}^t \mathbf{G}$ for all $t \in \mathbb{R}$.[6]

Discrete Symmetries

Beyond the continuous symmetries, Hamilton's equations may also be invariant under discrete symmetries, to which Noether's theorem does not apply.

The *spatial inversion* or *parity* is defined as the phase-space transformation

$$\mathbf{\Pi}: \qquad \mathbf{\Gamma} = (\mathbf{r}_a, \mathbf{p}_a)_{a=1}^N \to \mathbf{\Pi}\,\mathbf{\Gamma} = (-\mathbf{r}_a, -\mathbf{p}_a)_{a=1}^N. \tag{2.26}$$

The kinetic energy is invariant under parity. This is also the case for the interaction energy potentials since they only depend on the distances between the particles. Hamilton's equations are thus invariant under parity in the absence of external forces, or if $u_a^{(\text{ext})}(\mathbf{r}_a) = u_a^{(\text{ext})}(-\mathbf{r}_a)$. Parity is an involution because $\mathbf{\Pi}^2 = \mathbf{1}$, so that it generates the group $\mathbb{Z}_2 = \{\mathbf{1}, \mathbf{\Pi}\}$.

[5] SO(3) denotes the special orthogonal group of three-dimensional space.

[6] If $\mathbf{G}_1$ and $\mathbf{G}_2$ are two phase-space transformations or maps such that $\mathbf{\Gamma}' = \mathbf{G}_1(\mathbf{\Gamma})$ and $\mathbf{\Gamma}'' = \mathbf{G}_2(\mathbf{\Gamma}')$, we use the short notation $\mathbf{G}_2\mathbf{G}_1$ for the composed transformation such that $\mathbf{\Gamma}'' = \mathbf{G}_2[\mathbf{G}_1(\mathbf{\Gamma})] \equiv \mathbf{G}_2\mathbf{G}_1(\mathbf{\Gamma})$.

A very important discrete symmetry is the *time-reversal* transformation

$$\boldsymbol{\Theta}: \qquad \boldsymbol{\Gamma} = (\mathbf{r}_a, \mathbf{p}_a)_{a=1}^N \to \boldsymbol{\Theta}\,\boldsymbol{\Gamma} = (\mathbf{r}_a, -\mathbf{p}_a)_{a=1}^N \tag{2.27}$$

acting in phase space. Time reversal is also an involution, generating the group $\mathbb{Z}_2 = \{\mathbf{1}, \boldsymbol{\Theta}\}$, because $\boldsymbol{\Theta}^2 = \mathbf{1}$. Hamilton's equations are invariant under time reversal if the Hamiltonian function satisfies the condition

$$H(\boldsymbol{\Theta}\boldsymbol{\Gamma}) = H(\boldsymbol{\Gamma}), \tag{2.28}$$

and is independent of time. In this case, if the phase-space trajectory $\mathcal{T} = \{\boldsymbol{\Phi}^t\boldsymbol{\Gamma}_0,\, t \in \mathbb{R}\}$ is a solution of Hamilton's equations, then the time-reversed trajectory $\boldsymbol{\Theta}\mathcal{T}$ is also a solution of Hamilton's equations. This fundamental symmetry is called *microreversibility*. An equivalent statement of microreversibility is that $[\mathbf{r}_a(t), \mathbf{p}_a(t)]_{a=1}^N$ and $[\mathbf{r}_a(-t), -\mathbf{p}_a(-t)]_{a=1}^N$ are both solutions of Hamilton's equations. In terms of the flow (2.20), microreversibility is expressed as the condition

$$\boldsymbol{\Theta}\,\boldsymbol{\Phi}^t = \boldsymbol{\Phi}^{-t}\,\boldsymbol{\Theta} \tag{2.29}$$

for all $t \in \mathbb{R}$.

Microreversibility is satisfied if the Hamiltonian function has the form (2.2). In this case, we can indeed verify that $[\mathbf{r}_a(t)]_{a=1}^N$ and $[\mathbf{r}_a(-t)]_{a=1}^N$ are both solutions of the Newton equations (2.10). We note, however, that the trajectories $\mathcal{T}$ and $\boldsymbol{\Theta}\mathcal{T}$ do not necessarily coincide in phase space, as further discussed below. If the Hamiltonian function depends on time, the time dependence should also be reversed to establish the symmetry under time reversal.

If the particles have electric charges and they are subjected to an external constant magnetizing field $\mathcal{H}$, the time-reversal transformation has the effect of also reversing the magnetizing field, so that the symmetry should be generalized into

$$H(\boldsymbol{\Theta}\boldsymbol{\Gamma}; \mathcal{H}) = H(\boldsymbol{\Gamma}; -\mathcal{H}). \tag{2.30}$$

Indeed, electrodynamics being symmetric under time reversal, the currents in the solenoids generating the external magnetizing field should also be reversed, hence the reversal of the magnetizing field itself.[7] In such systems, the condition (2.30) for microreversibility means that, if the phase-space trajectory $\mathcal{T}_{\mathcal{H}} = \{\boldsymbol{\Phi}^t_{\mathcal{H}}\boldsymbol{\Gamma}_0,\, t \in \mathbb{R}\}$ is a solution of Hamilton's equations in the presence of the magnetizing field $\mathcal{H}$, then the time-reversed trajectory $\boldsymbol{\Theta}\mathcal{T}_{-\mathcal{H}}$ is also a solution of Hamilton's equations in the magnetizing field $-\mathcal{H}$. Therefore, we should have that

$$\boldsymbol{\Theta}\,\boldsymbol{\Phi}^t_{\mathcal{H}} = \boldsymbol{\Phi}^{-t}_{-\mathcal{H}}\,\boldsymbol{\Theta} \tag{2.31}$$

for all $t \in \mathbb{R}$.[8]

[7] The absence of symmetry $H(\boldsymbol{\Theta}\boldsymbol{\Gamma}; \mathcal{H}) \neq H(\boldsymbol{\Gamma}; \mathcal{H})$ for the system itself is often referred to as time-reversal symmetry breaking in the sense that it concerns the Hamiltonian dynamics in none other than the given magnetizing field $\mathcal{H}$. In this respect, Hamilton's equations are not invariant under the time-reversal transformation without considering the further operation $\mathcal{H} \to -\mathcal{H}$, since there exist two distinct Hamiltonian systems if $\mathcal{H} \neq 0$.

[8] In relativistic quantum field theory, the time-reversal transformation $\boldsymbol{\Theta}$ should be replaced by the transformation $\mathbf{C}\boldsymbol{\Pi}\boldsymbol{\Theta}$ combining charge conjugation $\mathbf{C}$, parity $\boldsymbol{\Pi}$, and time reversal $\boldsymbol{\Theta}$, because this combined symmetry is known to hold in general (Weinberg, 1995).

Furthermore, there is the aforementioned symmetry of all the observables including the Hamiltonian function under the permutations of the labels of identical particles.

2.3 Liouvillian Dynamics

2.3.1 Introduction to Statistical Ensembles

As mentioned, the initial conditions are free real variables that are not determined by the sole equations of motion. Therefore, the question arises about their determination during their preparation or their measurement.

For a ball thrown by hand and moving under Earth's gravity, the trajectory is a parabola according to Newton's equations (if friction on air is neglected). However, the initial conditions are not ruled by these equations, but free to be given some values by the biomechanics of the hand and the arm throwing the ball. This example illustrates that the preparation of initial conditions is, in general, performed by a system different to and interacting with the one ruled by the equations of immediate interest.[9] Moreover, the preparation of initial conditions is affected by inaccuracies that are specific to the preparing device (and not to the system itself). In this regard, successive preparations of initial conditions will lead to different values forming the following statistical ensembles of initial conditions, $\{\mathbf{\Gamma}_0^{(j)}\}_{j=1}^{\infty}$. These values are distributed according to some probability density

$$p_0(\mathbf{\Gamma}) = \lim_{\mathcal{N}\to\infty} \frac{1}{\mathcal{N}} \sum_{j=1}^{\mathcal{N}} \delta\left(\mathbf{\Gamma} - \mathbf{\Gamma}_0^{(j)}\right) \tag{2.32}$$

associated with the statistical ensemble and which can be determined empirically by repeating the preparation of the initial conditions.

The situation is similar for the preparation of initial conditions in experiments on gases. For instance, the particles can initially be confined on the left-hand side of a vessel of length L with $0 < x_a(t = 0) < l$ and $l < L$, as shown in Figure 2.3. The preparation is made

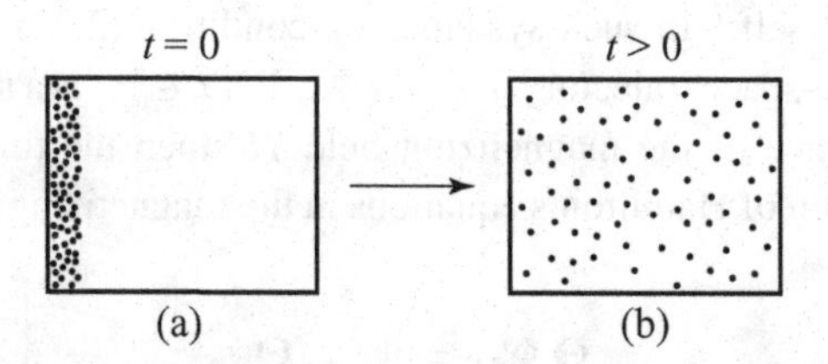

Figure 2.3 Isolated system of particles evolving under Hamiltonian microscopic dynamics and modeling a gas in a vessel: (a) Initial condition where all the particles are located in the left-hand side $0 < x < l$ at time $t = 0$. (b) Configuration after a relatively short time $t > 0$.

[9] This is the limitation of all physical theories based on ordinary or partial differential equations, which require initial or boundary conditions to be solved.

by first placing a wall at the position $x = l$ and then filling the compartment $0 < x < l$ with the gas, while the other compartment $l < x < L$ remains empty. The wall separating both compartments is quickly removed at time $t = 0$ and, thereafter, the gas expands into the whole vessel $0 < x < L$. In this experiment, the initial condition is only known by the number of particles initially contained in the compartment $0 < x < l$ and the initial temperature of the gas, which again defines a probability distribution such as (2.32), here with a poor knowledge of the initial conditions every time the experiment is repeated.

The control of initial conditions can be improved if the temperature is lowered and if the particles are initially trapped in small and tight potential wells, as performed for ultracold atoms. Nevertheless, lowering temperature is challenging, and the preparation of initial conditions remains described by some probability distribution (2.32).

Furthermore, measuring the positions and momenta of the particles is limited by some accuracy, as for the measurement of any real number, since an arbitrary real number is specified by an infinite sequence of bits or digits, thus requiring infinite knowledge. In this respect, statistical ensembles and their associated probability distribution should be introduced in the framework of mechanics.

Another motivation for the introduction of statistical ensembles is the

Poincaré recurrence theorem: *For autonomous Hamiltonian dynamics taking place inside a compact invariant domain $\mathcal{D}$ of phase space, every neighborhood $\mathcal{U}$ of some point $\mathbf{\Gamma}_0 \in \mathcal{D}$ contains a point $\mathbf{\Gamma} \in \mathcal{U}$ returning inside the neighborhood $\mathcal{U}$ after some recurrence time.*

This theorem is a consequence of Liouville's theorem and it applies to an isolated system of particles moving inside a bounded vessel, such as the one schematically depicted in Figure 2.3. In this example, the finite phase-space domain is defined by the set of points corresponding to a total energy in the range $[E, E + \Delta E]$:

$$\mathcal{D} = \{\mathbf{\Gamma} \in \mathcal{M} : E \leq H(\mathbf{\Gamma}) \leq E + \Delta E\}. \tag{2.33}$$

This domain is invariant because energy is conserved. According to Liouville's theorem, the phase-space volumes of the time-evolved subsets $\mathcal{U}_n = \mathbf{\Phi}^{n\Delta t}\mathcal{U} \subset \mathcal{D}$ remain constant in time for $n \in \mathbb{N}$ and $\Delta t > 0$. For any integer n_1, there should exist another integer $n_2 > n_1$ such that $\mathcal{U}_{n_1}$ and $\mathcal{U}_{n_2}$ intersect each other, otherwise the union of all the subsects $\mathcal{U}_n$ for $n \in \mathbb{N}$ would occupy an infinite phase-space volume in contradiction with the compactness of the domain $\mathcal{D}$. Consequently, there exists a point such that $\mathbf{\Gamma} \in \mathcal{U}$ and $\mathbf{\Phi}^{t_{\rm rec}}\mathbf{\Gamma} \in \mathcal{U}$ for the recurrence time $t_{\rm rec} = (n_2 - n_1)\Delta t$ (Arnold, 1989).

In order to evaluate the Poincaré recurrence time, we may consider the time evolution of the indicator function of the neighborhood $\mathcal{U}$, such that $I_{\mathcal{U}}(\mathbf{\Gamma}) = 1$ if $\mathbf{\Gamma} \in \mathcal{U}$ and zero otherwise. The time evolution $I_{\mathcal{U}}\big(\mathbf{\Gamma}_t^{(j)}\big) = I_{\mathcal{U}}\big(\mathbf{\Phi}^t\mathbf{\Gamma}_0^{(j)}\big)$ is schematically represented in Figure 2.4 for several initial conditions. All these functions of time start from the value $I_{\mathcal{U}}\big(\mathbf{\Gamma}_0^{(j)}\big) = 1$ since all the initial conditions are located inside the neighborhood $\mathcal{U}$ of the representative initial condition $\mathbf{\Gamma}_0$ shown in Figure 2.3(a). After some relatively short time interval, the functions $I_{\mathcal{U}}\big(\mathbf{\Gamma}_t^{(j)}\big)$ drop to zero as soon as one particle leaves the region $0 < x < l$. At their Poincaré recurrence time $t = t_{\rm rec}^{(j)}$, the functions $I_{\mathcal{U}}\big(\mathbf{\Gamma}_t^{(j)}\big)$ recover

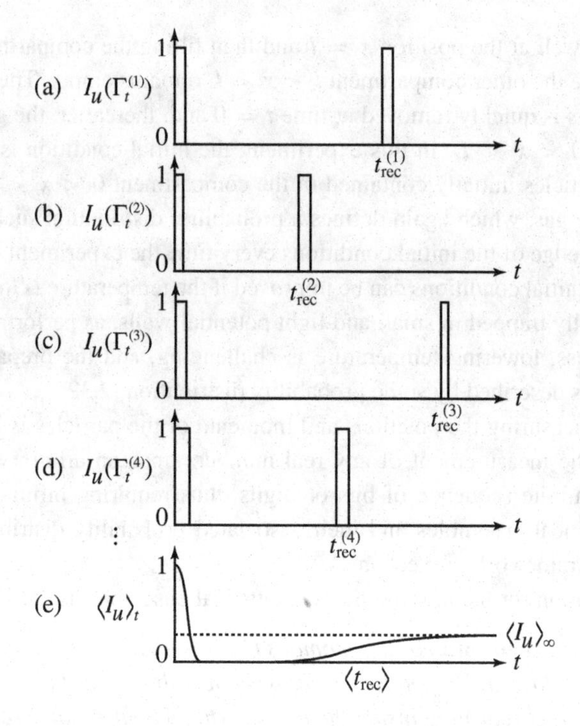

Figure 2.4 Illustration of Poincaré recurrences for the system of Figure 2.3, where $t_{\rm rec}^{(j)}$ denotes the Poincaré recurrence time for the jth trajectory. Panels (a)–(d) depict the time evolution of the indicator function $I_{\mathcal{U}}(\mathbf{\Gamma})$ of some neighborhood $\mathcal{U}$ of the configuration shown in Figure 2.3(a) for several trajectories $\{\mathbf{\Gamma}_t^{(j)}\}$ of the statistical ensemble. Panel (e) shows the statistical average (2.34) of all these functions.

the unit value, indicating the recurrence inside the neighborhood $\mathcal{U}$. Unless the dynamical system is periodic or quasiperiodic, the recurrence times $\{t_{\rm rec}^{(j)}\}$ take very different values, depending on the initial condition of the trajectories, as schematically depicted in Figure 2.4. Therefore, the statistical average,

$$\langle I_{\mathcal{U}} \rangle_t = \lim_{\mathcal{N}\to\infty} \frac{1}{\mathcal{N}} \sum_{j=1}^{\mathcal{N}} I_{\mathcal{U}}\left(\mathbf{\Gamma}_t^{(j)}\right), \tag{2.34}$$

has a shape similar to the one shown in Figure 2.4(e). Its initial value is equal to unity, $\langle I_{\mathcal{U}} \rangle_0 = 1$, since all the functions $I_{\mathcal{U}}\big(\mathbf{\Gamma}_t^{(j)}\big)$ start inside the neighborhood $\mathcal{U}$, but $\langle I_{\mathcal{U}} \rangle_t$ drops to zero after a relatively short time interval due to the gas expansion in the whole vessel. It is only on the timescale $\langle t_{\rm rec} \rangle$ of the Poincaré recurrences that the statistical average (2.34) increases, finally converging towards its asymptotic value $\langle I_{\mathcal{U}} \rangle_\infty$.[10] For the

[10] For a gas of N particles with the initial condition shown in Figure 2.3(a), the mean Poincaré recurrence time can be estimated as $\langle t_{\rm rec} \rangle \sim \bar{\tau}(L/l)^N$ where $\bar{\tau} \sim (V/N)^{1/3}/\bar{v}$ is the time for the first passage of a particle at the plane $x = l$, where $\bar{v}$ is the mean velocity of the particle. For $l = L/2$, the mean Poincaré recurrence time already exceeds the age of the universe for only $N \simeq 100$ particles.

statistical ensemble, this asymptotic value is not equal to unity for irregular enough dynamics. Accordingly, the Poincaré recurrence phenomenon does not concern statistical ensembles composed of infinitely many copies of the system, as described by the probability distribution (2.32).

We further note that, in chaotic dynamics, the sensitivity to initial conditions amplifies any error in the initial conditions, so that the statistical ensemble of trajectories issued from nearby initial conditions quickly spread in phase space, requiring us to set up a probabilistic description beyond the Lyapunov timescale (2.24).

2.3.2 Liouville's Equation

The next issue is to determine the time evolution of the statistical ensemble $\{\mathbf{\Gamma}^{(j)}\}_{j=1}^{\infty}$ of phase-space points $\mathbf{\Gamma}^{(j)} \equiv \mathbf{\Gamma}_t^{(j)} = \mathbf{\Phi}^t \mathbf{\Gamma}_0^{(j)} \in \mathcal{M}$. The probability density describing the distribution of this statistical ensemble is defined as in equation (2.32) by

$$p(\mathbf{\Gamma}) = \lim_{\mathcal{N}\to\infty} \frac{1}{\mathcal{N}} \sum_{j=1}^{\mathcal{N}} \delta\big(\mathbf{\Gamma} - \mathbf{\Gamma}^{(j)}\big) \qquad \text{with} \qquad \int_{\mathcal{M}} p(\mathbf{\Gamma})\, d\mathbf{\Gamma} = 1. \tag{2.35}$$

The mean value of some observable $A(\mathbf{\Gamma})$ is thus given by

$$\langle A \rangle = \lim_{\mathcal{N}\to\infty} \frac{1}{\mathcal{N}} \sum_{j=1}^{\mathcal{N}} A\big(\mathbf{\Gamma}^{(j)}\big) = \int_{\mathcal{M}} A(\mathbf{\Gamma})\, p(\mathbf{\Gamma})\, d\mathbf{\Gamma}. \tag{2.36}$$

Since probability is locally conserved in phase space, the probability density (2.35) obeys the continuity equation

$$\partial_t p + \text{div}\,\big(\dot{\mathbf{\Gamma}} p\big) = 0, \tag{2.37}$$

where $\dot{\mathbf{\Gamma}} = \mathbf{V}(\mathbf{\Gamma}, t)$ is the vector field of the dynamical system in phase space and "div" denotes the corresponding divergence. Equation (2.37) is known as the *generalized Liouville equation*, since it applies to any vector field $\mathbf{V}(\mathbf{\Gamma}, t)$, whether it preserves phase-space volumes or not.

For Hamiltonian dynamics, Liouville's theorem holds and $\text{div}\,\dot{\mathbf{\Gamma}} = 0$, so that equation (2.37) reads

$$\frac{dp}{dt} = \partial_t p + \dot{\mathbf{\Gamma}} \cdot \text{grad}\, p = 0, \tag{2.38}$$

where "grad" denotes the phase-space gradient. Accordingly, the probability density remains constant along trajectories because Hamiltonian flows are incompressible in phase space. Using Hamilton's equations, equation (2.38) leads to the *Liouville equation*,

$$\partial_t p = \{H, p\}, \tag{2.39}$$

in terms of the Poisson bracket (2.14), which defines the *Liouvillian operator*:

$$\hat{L} \equiv \{H, \cdot\}. \tag{2.40}$$

This operator is anti-Hermitian with respect to some stationary probability distribution of density $p_{\rm st}(\boldsymbol{\Gamma})$ defined in phase space with $\{H, p_{\rm st}\} = 0$ so that $\hat{L}^\dagger = -\hat{L}$. In this regard, the Hermitian Liouvillian operator

$$\hat{\mathcal{L}} \equiv \imath \{H, \cdot\} = \hat{\mathcal{L}}^\dagger \tag{2.41}$$

is often considered acting, for instance, in the space of square integrable functions with respect to the stationary probability density $p_{\rm st}$, so that the Liouville equation (2.39) can be written in the following form,

$$\imath\, \partial_t p = \hat{\mathcal{L}} p. \tag{2.42}$$

Liouville's equation is a partial differential equation of first order in all the variables. Moreover, it is linear so its solution can be formally written as

$$p_t = \mathrm{T}\, \mathrm{e}^{\int_0^t \hat{L}(t')\, dt'} p_0 = \mathrm{T}\, \mathrm{e}^{-\imath \int_0^t \hat{\mathcal{L}}(t')\, dt'} p_0, \tag{2.43}$$

if the system is time dependent, where T denotes time ordering.

2.3.3 Liouvillian Dynamics in Autonomous Systems

If the system is time independent, the solution (2.43) reads

$$p_t = \mathrm{e}^{\hat{L} t} p_0 = \mathrm{e}^{-\imath \hat{\mathcal{L}} t} p_0. \tag{2.44}$$

In this case, the solution of Liouville's equation can be written, equivalently, as

$$p_t(\boldsymbol{\Gamma}) = p_0\left(\boldsymbol{\Phi}^{-t}\boldsymbol{\Gamma}\right) \tag{2.45}$$

in terms of the flow (2.20) giving the solution of Hamilton's equations. Equation (2.45) shows that the value of the probability density at the phase-space point $\boldsymbol{\Gamma}$ is given by the initial value of the probability density at the initial condition $\boldsymbol{\Gamma}_0 = \boldsymbol{\Phi}^{-t}\boldsymbol{\Gamma}$. This confirms that the time evolution of the probability distribution is induced by the phase-space dynamics. Equation (2.45) also confirms that the time-dependent probability density remains constant in time along any phase-space trajectory, $dp_t(\boldsymbol{\Gamma}_t)/dt = 0$, because $p_t(\boldsymbol{\Gamma}_t) = p_t\left(\boldsymbol{\Phi}^t \boldsymbol{\Gamma}_0\right) = p_0(\boldsymbol{\Gamma}_0)$.

The adjoint $\hat{L}^\dagger$ of the Liouvillian operator (2.40) rules the time evolution of time-independent observables according to $A_t = \exp\left(\hat{L}^\dagger t\right) A_0 = \exp\left(-\hat{L} t\right) A_0$, which defines the so-called Koopman dynamics of observables (Koopman, 1931; Gaspard, 1998). Therefore, the mean value of some observable at time t can be equivalently expressed as

$$\langle A \rangle_t = \int_{\mathcal{M}} A_0(\boldsymbol{\Gamma})\, p_t(\boldsymbol{\Gamma})\, d\boldsymbol{\Gamma} = \int_{\mathcal{M}} A_t(\boldsymbol{\Gamma})\, p_0(\boldsymbol{\Gamma})\, d\boldsymbol{\Gamma}, \tag{2.46}$$

as a consequence of Liouville's theorem.

Autonomous dynamical systems can be characterized by the spectrum of the Hermitian Liouvillian operator (2.41). This spectrum may be composed of discrete eigenvalues, $\hat{\mathcal{L}}\psi = \omega\psi$, which define characteristic frequencies, ω, and a continuous spectrum. Every discrete eigenvalue is associated with a periodic time dependence $\exp(-\imath\omega t)$ and there exist several nonvanishing eigenfrequencies if the time evolution is quasiperiodic. In the presence of a continuous spectrum, the time evolution manifests relaxation processes.

The eigenfunctions ψ_0 corresponding to zero eigenvalues $(\hat{\mathcal{L}}\psi_0 = 0)$ define the stationary probability densities of the Liouvillian dynamics. This null subspace contains all the conserved quantities C such that $\{C, H\} = 0$, in particular, the Hamiltonian function. Accordingly, the null subspace includes the eigenfunctions $\psi_0 \doteq \eta[H(\boldsymbol{\Gamma})]$ where $\eta(E)$ is some function of the variable E. Such eigenfunctions can be used to describe the equilibrium macrostate of isolated systems since they are stationary.

2.3.4 Time-Reversal Symmetry of Liouville's Equation

If the underlying Hamiltonian system is symmetric under time reversal as a consequence of microreversibility, Liouville's equation is also symmetric under time reversal in the sense that, if $p_t(\boldsymbol{\Gamma})$ is a solution of Liouville's equation, then $p_{-t}(\boldsymbol{\Theta\Gamma})$ is also a solution. Indeed, the Poisson bracket (2.14) changes its sign under the time-reversal operation (2.27). If the Hamiltonian function has the symmetry (2.28), we thus have that

$$\hat{\Theta}\{H(\boldsymbol{\Gamma}), p_t(\boldsymbol{\Gamma})\} = -\{H(\boldsymbol{\Theta\Gamma}), p_t(\boldsymbol{\Theta\Gamma})\} = -\{H(\boldsymbol{\Gamma}), p_t(\boldsymbol{\Theta\Gamma})\} \tag{2.47}$$

for any time t, where $\hat{\Theta}\Psi(\boldsymbol{\Gamma}) \equiv \Psi(\boldsymbol{\Theta\Gamma})$ is the time-reversal operator acting on arbitrary phase-space functions Ψ. Therefore, we find

$$\partial_t p_{-t}(\boldsymbol{\Theta\Gamma}) = \{H(\boldsymbol{\Gamma}), p_{-t}(\boldsymbol{\Theta\Gamma})\}, \tag{2.48}$$

which establishes the symmetry.

In the presence of an external magnetizing field, the symmetry relation (2.30) holds for the Hamiltonian function. In this case, if $p_t(\boldsymbol{\Gamma}; \mathcal{H})$ is a solution of Liouville's equation, then $p_{-t}(\boldsymbol{\Theta\Gamma}; -\mathcal{H})$ is also a solution because

$$\hat{\Theta}\{H(\boldsymbol{\Gamma}; \mathcal{H}), p_t(\boldsymbol{\Gamma}; \mathcal{H})\} = -\{H(\boldsymbol{\Gamma}; -\mathcal{H}), p_t(\boldsymbol{\Theta\Gamma}; \mathcal{H})\}, \tag{2.49}$$

so that

$$\partial_t p_{-t}(\boldsymbol{\Theta\Gamma}; -\mathcal{H}) = \{H(\boldsymbol{\Gamma}; \mathcal{H}), p_{-t}(\boldsymbol{\Theta\Gamma}; -\mathcal{H})\}. \tag{2.50}$$

These results can also be obtained using equation (2.29) or (2.31) with equation (2.45). The time-reversal symmetry of Liouville's equation is thus the expression of microreversibility.

2.3.5 BBGKY Hierarchy

In the 1930s and 1940s, work by Bogoliubov (1946a,b), Born and Green (1946), Kirkwood (1946), and Yvon (1935) showed that Liouville's equation can be transformed into a hierarchy of equations for multiparticle distribution functions (Balescu, 1975).

The system is assumed to be composed of N identical particles, so that the phase-space probability density is totally symmetric under the permutations of the particle labels,

$$p(\mathbf{x}_1, \mathbf{x}_2, \ldots, \mathbf{x}_N) = p\left(\mathbf{x}_{P(1)}, \mathbf{x}_{P(2)}, \ldots, \mathbf{x}_{P(N)}\right) \qquad \forall\, P \in \text{Sym}N \tag{2.51}$$

with $\mathbf{x}_a \equiv (\mathbf{r}_a, \mathbf{p}_a)$, since this is the case for all the observables.

The one-particle distribution function can be defined as

$$f_1(\mathbf{r}, \mathbf{p}) \equiv \left\langle \sum_{a=1}^{N} \delta^3(\mathbf{r} - \mathbf{r}_a)\, \delta^3(\mathbf{p} - \mathbf{p}_a) \right\rangle, \tag{2.52}$$

which is normalized to the total number of particles in the system, $\int_{\mathbb{R}^6} f_1(\mathbf{x})\, d\mathbf{x} = N$. In general, s-particle distribution functions can be defined for $s = 1, 2, \ldots, N$. These functions are related to the phase-space probability density (2.51) by

$$f_s(\mathbf{x}_1, \ldots, \mathbf{x}_s) \equiv \frac{1}{(N-s)!} \int_{\mathbb{R}^{6(N-s)}} p(\mathbf{x}_1, \mathbf{x}_2, \ldots, \mathbf{x}_N)\, d\mathbf{x}_{s+1} \cdots d\mathbf{x}_N \tag{2.53}$$

with $d\mathbf{x}_a = d^3 r_a d^3 p_a$, and they are normalized according to

$$\int_{\mathbb{R}^{6s}} f_s(\mathbf{x}_1, \ldots, \mathbf{x}_s)\, d\mathbf{x}_1 \cdots d\mathbf{x}_s = \frac{N!}{(N-s)!}. \tag{2.54}$$

The introduction of these distribution functions is motivated by the observation that the physical observables are, in general, expressed in terms of functions involving a finite number of particles as

$$\begin{aligned} A(\mathbf{x}_1, \ldots, \mathbf{x}_N) = A^{(0)} &+ \sum_{1 \le a \le N} A^{(1)}(\mathbf{x}_a) + \sum_{1 \le a < b \le N} A^{(2)}(\mathbf{x}_a, \mathbf{x}_b) \\ &+ \sum_{1 \le a < b < c \le N} A^{(3)}(\mathbf{x}_a, \mathbf{x}_b, \mathbf{x}_c) + \cdots + A^{(N)}(\mathbf{x}_1, \ldots, \mathbf{x}_N). \end{aligned} \tag{2.55}$$

For instance, the total kinetic energy involves the one-particle observables $A^{(1)}(\mathbf{x}) = \mathbf{p}^2/(2m)$, while the interaction potential energy is expressed in terms of the two-particle functions $A^{(2)}(\mathbf{x}, \mathbf{x}') = u^{(2)}(\|\mathbf{r} - \mathbf{r}'\|)$. Consequently, the mean values of such observables can be written as

$$\langle A \rangle = \sum_{s=0}^{N} \frac{1}{s!} \int_{\mathbb{R}^{6s}} A^{(s)}(\mathbf{x}_1, \ldots, \mathbf{x}_s)\, f_s(\mathbf{x}_1, \ldots, \mathbf{x}_s)\, d\mathbf{x}_1 \cdots d\mathbf{x}_s \tag{2.56}$$

in terms of the multiparticle distribution functions (2.53) (Balescu, 1975).

If the particles are subjected to binary interactions, the Liouville equation (2.39) can be written as

$$\partial_t p = \sum_{1 \le a \le N} \hat{L}_a^{(1)} p + \sum_{1 \le a < b \le N} \hat{L}_{ab}^{(2)} p \tag{2.57}$$

with the one- and two-particle Liouvillian operators defined as

$$\hat{L}_a^{(1)} \equiv -\frac{\mathbf{p}_a}{m} \cdot \frac{\partial}{\partial \mathbf{r}_a} - \mathbf{F}_a^{(\text{ext})} \cdot \frac{\partial}{\partial \mathbf{p}_a}, \tag{2.58}$$

$$\hat{L}_{ab}^{(2)} \equiv -\mathbf{F}_{ab} \cdot \left(\frac{\partial}{\partial \mathbf{p}_a} - \frac{\partial}{\partial \mathbf{p}_b} \right), \tag{2.59}$$

in terms of the forces $\mathbf{F}_a^{(\text{ext})}$ and $\mathbf{F}_{ab}$ introduced in Section 2.2.4. These operators have the properties that

$$\int_{\mathbb{R}^6} \hat{L}_a^{(1)} p \, d\mathbf{x}_a = 0 \qquad \text{and} \qquad \int_{\mathbb{R}^{12}} \hat{L}_{ab}^{(2)} p \, d\mathbf{x}_a d\mathbf{x}_b = 0, \tag{2.60}$$

because the forces only depend on positions and the probability density p is supposed to vanish at the boundaries of the vessel, as well as for large values of momenta.

Now, if we integrate the Liouville equation (2.57) over the phase-space variables of the particles $a = s+1, s+2, \ldots, N$, we obtain the following equation for the s-particle distribution function (Balescu, 1975):

$$\partial_t f_s(\mathbf{x}_1, \ldots, \mathbf{x}_s) = \sum_{1 \leq a \leq s} \hat{L}_a^{(1)} f_s(\mathbf{x}_1, \ldots, \mathbf{x}_s) + \sum_{1 \leq a < b \leq s} \hat{L}_{ab}^{(2)} f_s(\mathbf{x}_1, \ldots, \mathbf{x}_s)$$
$$+ \sum_{1 \leq a \leq s} \int_{\mathbb{R}^6} \hat{L}_{a,s+1}^{(2)} f_{s+1}(\mathbf{x}_1, \ldots, \mathbf{x}_s, \mathbf{x}_{s+1}) \, d\mathbf{x}_{s+1}. \tag{2.61}$$

All these equations with $s = 1, 2, \ldots, N$ form the so-called *BBGKY hierarchy*. Explicitly, the equations for the one- and two-particle distribution functions can be written in the following form,

$$\frac{\partial f_1}{\partial t} = -\frac{\mathbf{p}_1}{m} \cdot \frac{\partial f_1}{\partial \mathbf{r}_1} - \frac{\partial}{\partial \mathbf{p}_1} \cdot \left(\mathbf{F}_1^{(\text{ext})} f_1 + \int_{\mathbb{R}^6} \mathbf{F}_{12} \, f_2 \, d\mathbf{x}_2 \right), \tag{2.62}$$

$$\frac{\partial f_2}{\partial t} = -\frac{\mathbf{p}_1}{m} \cdot \frac{\partial f_2}{\partial \mathbf{r}_1} - \frac{\partial}{\partial \mathbf{p}_1} \cdot \left(\mathbf{F}_1^{(\text{ext})} f_2 + \mathbf{F}_{12} \, f_2 + \int_{\mathbb{R}^6} \mathbf{F}_{13} \, f_3 \, d\mathbf{x}_3 \right)$$
$$-\frac{\mathbf{p}_2}{m} \cdot \frac{\partial f_2}{\partial \mathbf{r}_2} - \frac{\partial}{\partial \mathbf{p}_2} \cdot \left(\mathbf{F}_2^{(\text{ext})} f_2 + \mathbf{F}_{21} \, f_2 + \int_{\mathbb{R}^6} \mathbf{F}_{23} \, f_3 \, d\mathbf{x}_3 \right), \tag{2.63}$$

$$\vdots$$

We observe that every one of these equations depends on the distribution function with one more particle, so that all these equations are coupled together. Nevertheless, the BBGKY hierarchy is useful for specific situations.

If the particles are noninteracting ($\mathbf{F}_{ab} = 0$), all the equations become decoupled and they can be solved using the one-particle flow $\boldsymbol{\phi}^t$ as $f_s(\mathbf{x}_1, \ldots, \mathbf{x}_s; t) = f_s(\boldsymbol{\phi}^{-t}\mathbf{x}_1, \ldots, \boldsymbol{\phi}^{-t}\mathbf{x}_s; 0)$.

If the particles may be assumed to be statistically independent, i.e., uncorrelated, the s-particle distribution function factorizes as $f_s(\mathbf{x}_1, \ldots, \mathbf{x}_s) = \prod_{a=1}^{s} f_1(\mathbf{x}_a)$ in large enough systems, so that the equations of the hierarchy are also decoupled. The first equation (2.62) of the hierarchy becomes the so-called *Vlasov equation*:

$$\frac{\partial f_1}{\partial t} = -\frac{\mathbf{p}_1}{m} \cdot \frac{\partial f_1}{\partial \mathbf{r}_1} - \left[\mathbf{F}_1^{(\text{ext})} + \int_{\mathbb{R}^6} \mathbf{F}_{12}\, f_1(\mathbf{x}_2)\, d\mathbf{x}_2 \right] \cdot \frac{\partial f_1}{\partial \mathbf{p}_1}, \tag{2.64}$$

which is useful if the interactions are medium or long ranged.

It should be noticed that the s-particle distribution function goes as $f_s = O(n^s)$ with the particle density n as a consequence of their normalization. Accordingly, the term involving the distribution function f_{s+1} in the equation for f_s appears as a perturbation if the particle density is low enough. Therefore, the BBGKY hierarchy can be truncated to obtain a set of closed equations. In this way, the famous *Boltzmann kinetic equation* is obtained for dilute gases by keeping the first two equations (2.62) and (2.63), while supposing that f_3 is negligible in the equation for f_2. Consequently, equation (2.63) becomes Liouville's equation for the two-body problem describing the binary collisions, which can be solved in terms of the corresponding differential cross section. Moreover, in dilute gases every binary collision involves two incoming particles that have undergone previous collisions at remote distances, so that the incoming two-particle distribution function can be supposed to be factorized, as first assumed by Boltzmann (1872, 1896). These considerations lead to his famous kinetic equation (Balescu, 1975; Résibois and De Leener, 1977). The Boltzmann kinetic equation provides an excellent description of kinetic processes in dilute gases (without or with reactions).

The distribution functions may also be defined in the phase-space domain outside the interaction zones and their evolution equations form another hierarchy (Grad, 1958). In Grad's hierarchy, the equations are coupled together by gain and loss terms due to the particles that are either incoming or outgoing collisions, as for Boltzmann's kinetic equation obtained in the limit of dilute gases.

2.4 Ergodic Properties

2.4.1 Time Average

Since pioneering work by Boltzmann (1887), ergodic theory has been developed in order to justify the existence and uniqueness of the stationary probability distribution corresponding to the equilibrium macrostate and to understand the macroscopic processes of relaxation towards equilibrium. More generally, ergodic theory applies to abstract dynamical systems with the aim to construct and select the stationary probability distributions of any dynamical system.

Because of equation (2.45), the probability density remains constant in time along the phase-space trajectories, so that there is no possibility for a point-like convergence of the time-dependent probability density towards any smooth stationary probability distribution that would describe an equilibrium macrostate. Yet equilibrium is reached in the long-time limit after some transient behavior and it remains stationary if the conditions for its existence are maintained indefinitely. In this respect, Boltzmann (1887) proposed to construct a stationary probability distribution by considering the time average, thus founding ergodic theory. In the 1930s, Birkhoff (1931) and Khinchin (1932) showed that time averages can indeed be used for such a purpose according to the

Birkhoff–Khinchin ergodic theorem: *For almost all initial conditions* $\mathbf{\Gamma}_0$ *in phase space, the time average of some observable* $A(\mathbf{\Gamma})$ *converges towards the function*

$$\overline{A}(\mathbf{\Gamma}_0) \equiv \lim_{\mathscr{T}\to\infty} \frac{1}{\mathscr{T}} \int_0^{\mathscr{T}} A\left(\mathbf{\Phi}^t \mathbf{\Gamma}_0\right) dt \tag{2.65}$$

that is stationary under time evolution, $\overline{A}\left(\mathbf{\Phi}^t\mathbf{\Gamma}_0\right) = \overline{A}(\mathbf{\Gamma}_0)$.

A proof of this theorem is given by Cornfeld et al. (1982). The stationary probability distribution constructed by the time average takes the form

$$p_{\rm BK}(\mathbf{\Gamma}|\mathbf{\Gamma}_0) = \lim_{\mathscr{T}\to\infty} \frac{1}{\mathscr{T}} \int_0^{\mathscr{T}} \delta\left(\mathbf{\Gamma} - \mathbf{\Phi}^t\mathbf{\Gamma}_0\right) dt, \tag{2.66}$$

which indeed satisfies the stationarity condition, $p_{\rm BK}\left(\mathbf{\Gamma}|\mathbf{\Phi}^t\mathbf{\Gamma}_0\right) = p_{\rm BK}(\mathbf{\Gamma}|\mathbf{\Gamma}_0)$, because the time average is carried out over an arbitrarily long time interval. In general, the time average thus depends on the initial condition $\mathbf{\Gamma}_0$. Different initial conditions may lead to different values for the time average, $\overline{A}(\mathbf{\Gamma}_0) \neq \overline{A}(\mathbf{\Gamma}_0')$, if the initial condition $\mathbf{\Gamma}_0$ does not belong to the trajectory issued from the other initial condition $\mathbf{\Gamma}_0'$. For instance, some initial conditions may lead to quasiperiodic motion while others can generate trajectories meandering in chaotic zones (Ott, 1993). This phenomenon is general according to the KAM theorem, but the phase-space extension of quasiperiodicity tends to decrease compared to the chaotic zones as the perturbation or the total energy increases.

However, the Birkhoff–Khinchin ergodic theorem does not establish the uniqueness of the stationary probability distribution (2.66). Some assumption is needed to go farther. In any case, the basic assumption is that the system starts from initial conditions distributed according to some probability density (2.32), so that the stationary probability distribution corresponding to the equilibrium macrostate is, in general, given by

$$p_{\rm eq}(\mathbf{\Gamma}) = \int_{\mathcal{M}} p_{\rm BK}(\mathbf{\Gamma}|\mathbf{\Gamma}_0)\, p_0(\mathbf{\Gamma}_0)\, d\mathbf{\Gamma}_0 = \lim_{\mathscr{T}\to\infty} \frac{1}{\mathscr{T}} \int_0^{\mathscr{T}} p_0\left(\mathbf{\Phi}^{-t}\mathbf{\Gamma}\right) dt. \tag{2.67}$$

Moreover, the analytical construction of the Birkhoff–Khinchin stationary distribution requires detailed knowledge of the system trajectories. Since early work by Boltzmann (1887) and Gibbs (1902), simplifying assumptions have been formulated based on the idea that the stationary probability distribution should only depend on the fundamental constants of motion given by analytic functions such as the total energy. In this way, a stationary probability distribution is assumed and the issue is to verify if this distribution satisfies specific properties. For this purpose, it is convenient to define a dynamical system $\left(\mathcal{M}, \mathbf{\Phi}^t, P\right)$ by its phase space $\mathcal{M}$, its flow $\mathbf{\Phi}^t$, and one of its stationary probability distributions P, such that $P\left(\Phi^t\mathcal{A}\right) = P(\mathcal{A}) = \int_{\mathcal{A}} p(\mathbf{\Gamma})\, d\mathbf{\Gamma}$ for any phase-space subset $\mathcal{A} \subseteq \mathcal{M}$ and the corresponding stationary probability density $p(\mathbf{\Gamma})$ such that $\hat{L}p = \{H, p\} = 0$.

2.4.2 Ergodicity

The property of ergodicity was first introduced by Boltzmann (1887), discussed by Ehrenfest and Ehrenfest (1911), and defined on rigorous mathematical ground since the 1930s (Arnold and Avez, 1968; Cornfeld et al., 1982).

> **Definition:** *The dynamical system $(\mathcal{M}, \mathbf{\Phi}^t, P)$ is said to be* **ergodic** *if the time average $\overline{A}$ of some observable $A(\mathbf{\Gamma})$ is equal to the phase-space average $\langle A \rangle$ over the stationary probability distribution P of density p:*
>
> $$\overline{A} \equiv \lim_{\mathcal{T}\to\infty} \frac{1}{\mathcal{T}} \int_0^{\mathcal{T}} A(\mathbf{\Phi}^t \mathbf{\Gamma}_0)\, dt = \int A(\mathbf{\Gamma})\, p(\mathbf{\Gamma})\, d\mathbf{\Gamma} \equiv \langle A \rangle \tag{2.68}$$
>
> *for almost all initial conditions $\mathbf{\Gamma}_0$ chosen with the stationary probability distribution P.*

If energy is the only analytic constant of motion, the stationary probability distribution is given by

$$p(\mathbf{\Gamma}|E_0) = \frac{1}{\kappa(E_0)}\, \delta[E_0 - H(\mathbf{\Gamma})] \quad \text{with} \quad \kappa(E_0) = \int_{\mathcal{M}} \delta[E_0 - H(\mathbf{\Gamma})]\, d\mathbf{\Gamma}, \tag{2.69}$$

which is concentrated on the energy shell $H(\mathbf{\Gamma}) = E_0$ corresponding to the energy $E_0 = H(\mathbf{\Gamma}_0)$ of the initial condition $\mathbf{\Gamma}_0$. In this case, the property of ergodicity assumes that the time average does not depend on the initial condition $\mathbf{\Gamma}_0$ almost everywhere in the whole energy shell, so that the stationary probability distribution P is only specified by the initial energy E_0.

If the dynamical system is defined on the three-dimensional torus with periodic boundary conditions, the total linear momentum is also conserved, so that we should instead expect the property of ergodicity for the stationary probability density

$$p(\mathbf{\Gamma}|E_0, \mathbf{P}_0) = \frac{1}{\kappa(E_0, \mathbf{P}_0)}\, \delta[E_0 - H(\mathbf{\Gamma})]\, \delta^3[\mathbf{P}_0 - \mathbf{P}(\mathbf{\Gamma})] \tag{2.70}$$

with the appropriate constant $\kappa(E_0, \mathbf{P}_0)$ to satisfy the normalization condition $\int_{\mathcal{M}} p(\mathbf{\Gamma}|E_0, \mathbf{P}_0)\, d\mathbf{\Gamma} = 1$.

However, in the case where the total energy is the only conserved quantity, we should not expect the property of ergodicity for the stationary probability distribution P of density $p(\mathbf{\Gamma}) = \eta[H(\mathbf{\Gamma})]$ defined with some smooth function $\eta(E)$ of energy. Indeed, this stationary probability density is positive for initial conditions with different values of energy, so that the corresponding trajectories will thus remain on their respective energy shell and the equality (2.68) will not hold in general. In such systems, the phase space is decomposed into different ergodic components, every one given by a single energy shell. More generally, the different KAM quasiperiodic trajectories and chaotic zones constitute many ergodic components (Arnold and Avez, 1968; Moser, 1973; Ott, 1993).

The property of ergodicity can also be expressed in terms of the spectral properties of the Liouvillan operator (2.41) acting on square integrable functions with respect to the stationary probability distribution P (Arnold and Avez, 1968):

Theorem: *The dynamical system* $(\mathcal{M}, \mathbf{\Phi}^t, P)$ *is* ***ergodic*** *if and only if the discrete eigenvalue* $\omega = 0$ *is simply degenerate.*

In addition, the Liouvillian operator (2.41) of an ergodic system may have nonzero discrete eigenvalues and/or a continuous spectrum.

The property of ergodicity is established for several dynamical systems.

A simple example is given by the quasiperiodic flow $\mathbf{\Phi}^t(x_1,x_2)=(x_1+\omega_1 t, x_2+\omega_2 t)$ defined modulo one on the two-dimensional torus $\mathcal{M} = [0,1[^2$ with the stationary Lebesgue distribution dx_1dx_2 and with two positive frequencies $\omega_1,\omega_2 > 0$. This flow is ergodic if the ratio of its frequencies is incommensurate, i.e., if $\omega_1/\omega_2 \neq p_1/p_2$ for any positive integers p_1 and p_2. Because of the incommensurability, the trajectory starting from any initial condition can approach arbitrarily near any point of the torus. The eigenfrequencies of this system are given by $\omega = n_1\omega_1 + n_2\omega_2$ with $n_1, n_2 \in \mathbb{Z}$. If the frequencies ω_1 and ω_2 are incommensurate, we have that $\omega = 0$ if and only if $n_1 = n_2 = 0$, so that the zero eigenfrequency is simply degenerate and the system is indeed ergodic. In this example, the dynamical system does not manifest sensitivity to initial conditions because the Lyapunov exponents are vanishing.

Ergodicity has also been proved for chaotic systems such as the Sinai and Bunimovich billiards (Sinai, 1970; Bunimovich, 1979). In billiards, a point particle moves in free flight between elastic collisions undergone on the boundaries of the billiard table (see Appendix B). These dynamical systems have Hamiltonian character because they conserve energy and preserve phase-space volumes. Since they have two degrees of freedom, the phase space is four-dimensional and the energy shells are three-dimensional. In Sinai's billiard, the collisions happen on a hard disk fixed inside a square with periodic boundary conditions. In Bunimovich's billiard, the boundary has the shape of a stadium. Both billiards are ergodic with respect to the stationary probability distribution of density (2.69). Furthermore, they manifest the property of sensitivity to initial conditions with a positive Lyapunov exponent.

In the seventies, Sinai further proved the ergodicity of the gas composed of two hard disks moving on the torus and undergoing elastic collisions and this with respect to the stationary probability distribution of density (2.70) (Sinai, 1970). Indeed, this dynamical system reduces to the ergodic Sinai billiard in the frame moving on the torus with the center of mass of the two disks, as shown in Figure 2.5. Since then, ergodicity has also been proved for gases with hard spheres (Szász, 1996).

Ergodicity also holds for dynamical systems with infinitely many degrees of freedom such as the ideal gases of noninteracting particles having stationary probability distributions defined with the Poisson distribution for their positions and the Maxwell distribution for their velocities (Cornfeld et al., 1982), as well as the infinite harmonic crystals with the Boltzmann probability distribution (van Hemmen, 1980). Therefore, systems with infinitely many degrees of freedom may be ergodic even if the corresponding finite systems are not.

Since ergodicity is defined by infinitely long time averaging, no timescale is defined here. In particular, ergodicity does not provide any timescale that would characterize relaxation towards the stationary probability distribution. In this regard, further ergodic properties are needed.

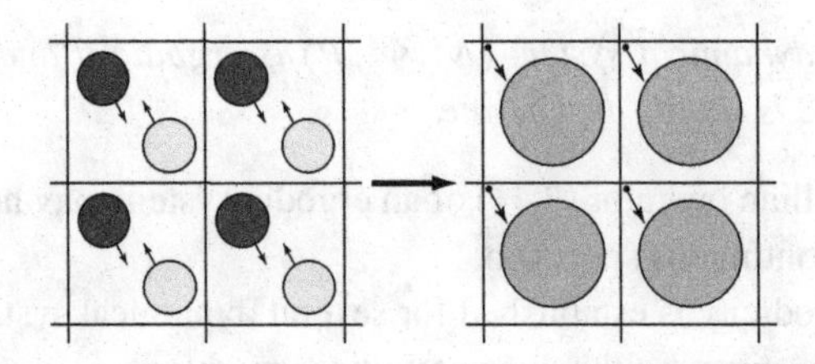

Figure 2.5 System of two hard disks of equal mass moving in a square with periodic boundary conditions and undergoing elastic collisions, and its reduction to Sinai's billiard in the space of the relative position $\mathbf{r} = \mathbf{r}_1 - \mathbf{r}_2$ after separating the center of mass $\mathbf{R}_{\rm cm} = (\mathbf{r}_1 + \mathbf{r}_2)/2$.

2.4.3 Dynamical Mixing

Because of Liouville's theorem and its consequence (2.45), the probability density keeps its initial value along every trajectory in phase space, so that the point-like convergence towards the equilibrium probability density is, in general, impossible, $p_t(\mathbf{\Gamma}) \nrightarrow p_{\rm eq}(\mathbf{\Gamma})$ for $t \to \infty$. In order to understand the observation that particle systems undergo relaxation processes towards equilibrium, Gibbs (1902) introduced the property of dynamical mixing, which can be formulated as follows.

> **Definition:** *The dynamical system* $(\mathcal{M}, \mathbf{\Phi}^t, P)$ *is said to be* ***mixing*** *if the phase-space average* $\langle A \rangle_t$ *over the time-dependent probability distribution of density* $p_t(\mathbf{\Gamma}) = p_0(\mathbf{\Phi}^t \mathbf{\Gamma})$ *converges in time towards the phase-space average* $\langle A \rangle$ *over the stationary probability distribution* P *of density* p*:*
>
> $$\lim_{t\to\infty} \langle A \rangle_t \equiv \lim_{t\to\infty} \int A(\mathbf{\Gamma})\, p_t(\mathbf{\Gamma})\, d\mathbf{\Gamma} = \int A(\mathbf{\Gamma})\, p(\mathbf{\Gamma})\, d\mathbf{\Gamma} \equiv \langle A \rangle \tag{2.71}$$
>
> *for any observable* $A(\mathbf{\Gamma})$ *and any initial probability density* p_0.

According to this mixing property, the relaxation towards equilibrium can be conceived as a weak convergence (i.e., after averaging the probability density with a smooth enough observable A) instead of a point-like convergence at every phase-space point.

The property (2.71) of dynamical mixing can be equivalently formulated by the condition

$$\lim_{t\to\infty} P\left(\mathcal{A} \cap \mathbf{\Phi}^t \mathcal{B}\right) = P(\mathcal{A})\, P(\mathcal{B}) \tag{2.72}$$

for any phase-space subsets $\mathcal{A}, \mathcal{B} \subseteq \mathcal{M}$. This condition expresses the statistical independence or memory loss between the initial event $\mathcal{A}$ and the event $\mathcal{B}$ evolved after some time $t \to \infty$. Indeed, taking the observable $A(\mathbf{\Gamma}) = I_{\mathcal{A}}(\mathbf{\Gamma})$ as the indicator function of the subset $\mathcal{A}$ and the initial probability density $p_0(\mathbf{\Gamma}) = I_{\mathcal{B}}(\mathbf{\Gamma}) p(\mathbf{\Gamma})/P(\mathcal{B})$ concentrated in the subset $\mathcal{B}$, we have that $\langle A \rangle_t = P\left(\mathcal{B} \cap \mathbf{\Phi}^{-t} \mathcal{A}\right)/P(\mathcal{B})$ and $\langle A \rangle = P(\mathcal{A})$, hence the definition (2.71) implies the condition (2.72) since P is stationary. Therefore, dynamical mixing conveys the idea of relaxation towards the stationary probability distribution over some timescale.

Dynamical mixing can also be expressed in terms of the spectral properties of Liouvillian operator (2.41) acting on square integrable functions with respect to the stationary probability distribution P (Arnold and Avez, 1968; Cornfeld et al., 1982):

> **Theorem:** *If the dynamical system* $(\mathcal{M}, \mathbf{\Phi}^t, P)$ *is* ***mixing****, the discrete eigenvalue* $\omega = 0$ *is simply degenerate and the rest of the frequency spectrum is continuous.*

Since dynamical mixing also requires a further condition on the Liouvillian spectrum in addition to ergodicity, we have the result that *dynamical mixing implies ergodicity*. Therefore, dynamical mixing is a stronger property than ergodicity itself.

Indeed, the converse statement does not hold since there exist systems that are ergodic without being mixing, such as the aforementioned quasiperiodic flow, which has countably many discrete eigenvalues.

However, the Sinai and Bunimovich billiards, as well as the aforementioned hard-disk and hard-sphere gases are mixing. In these systems, the property of dynamical mixing is induced by the stretching of phase-space volumes in the unstable directions with positive Lyapunov exponents and their contraction in the stable directions with negative Lyapunov exponents, leaving a unique eigenfunction and thus a unique discrete eigenvalue in the spectrum. The mechanism of dynamical mixing is reminiscent of the mixing of an ink droplet in the flow of an incompressible fluid, as envisaged by Gibbs (1902) for the relaxation towards equilibrium.

Moreover, the infinite ideal gases and harmonic crystals are also known to be mixing (van Hemmen, 1980; Cornfeld et al., 1982).

In the discussion carried out here above for the Poincaré recurrence theorem, the property of dynamical mixing would imply that the mean value $\langle I_{\mathcal{U}} \rangle_t$ converges over a long timescale towards its asymptotic value $\langle I_{\mathcal{U}} \rangle_\infty = P(\mathcal{U})$ equal to the probability to visit the neighborhood $\mathcal{U}$ of the initial condition $\mathbf{\Gamma}_0$. This probability is much lower than unity if the initial condition is prepared far enough from equilibrium.

2.4.4 Pollicott–Ruelle Resonances

If the Liouvillian operator has a continuous spectrum, the frequencies can be analytically continued towards complex values to obtain the so-called Pollicott–Ruelle resonances (Pollicott, 1985, 1986; Ruelle, 1986a,b), which can be expressed as the generalized eigenvalues $\omega = \imath z$ of the Liouvillian operator (2.39): $\hat{L}\psi = z\psi$. Typically, the eigenmodes ψ are not defined as functions, but as distributions in phase space. The time evolution associated with these eigenmodes goes as $\exp(zt) = \exp(\text{Re}\, z\, t + \imath\, \text{Im}\, z\, t)$, thus describing exponential relaxation over the decay time $\tau = -(\text{Re}\, z)^{-1}$ for Pollicott–Ruelle resonances with $\text{Re}\, z < 0$. For resonances such that $\text{Im}\, z \neq 0$, the time evolution may manifest damped oscillations with the period $T = 2\pi\, |\text{Im}\, z|^{-1}$, if $\tau > T$. Therefore, the corresponding eigenmodes ψ would evolve in time according to $\exp(zt) = \exp(-t/\tau) \exp(\pm\imath\, 2\pi t/T)$.

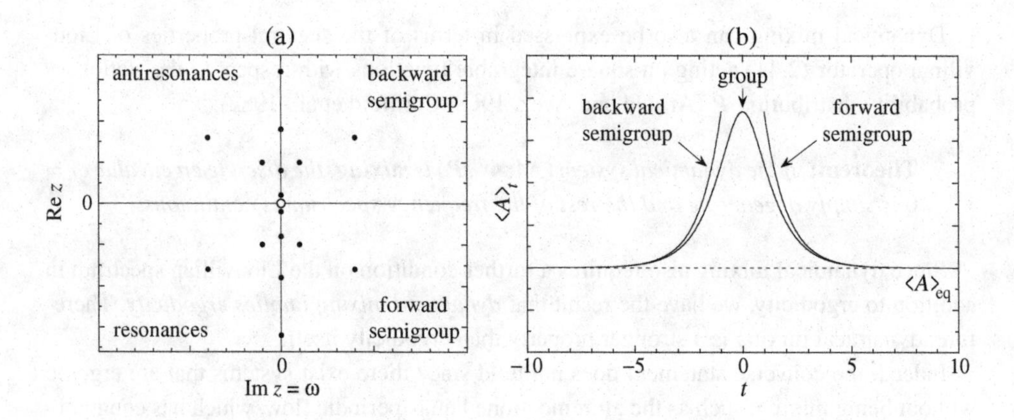

Figure 2.6 (a) Pollicott–Ruelle resonances and antiresonances in the complex plane of the variable z. (b) Time decay of the mean value of an observable towards its equilibrium value according to the group of time evolution, as well as the truncated asymptotic expansions of the forward and backward semigroups for $t > 0$ and $t < 0$, respectively.

The analytic continuation can be performed either towards the lower- or the upper-half plane of complex frequencies in order to pick up the contributions from several complex singularities, which can be poles, branch cuts, or else, as shown in Figure 2.6(a). In the lower half-plane, the poles are the Pollicott–Ruelle resonances, while those in the upper half-plane are the antiresonances. The sum of the contributions from the resonances in the lower half-plane gives us an expansion, which is valid for positive times $t > 0$ and which defines the forward semigroup:

$$\langle A \rangle_t = \int A(\boldsymbol{\Gamma}) \exp\left(\hat{L}t\right) p_0(\boldsymbol{\Gamma})\, d\boldsymbol{\Gamma} \simeq \sum_\alpha \langle A|\psi_\alpha\rangle \exp(z_\alpha t) \left\langle \tilde{\psi}_\alpha | p_0 \right\rangle + \cdots, \tag{2.73}$$

where the dots denote the contributions from singularities other than the simple poles. These contributions may include Jordan-block structures if a resonance has a multiplicity m_α higher than unity. In this case, the exponential decay is modified by a power-law dependence on time as $t^{m_\alpha - 1} \exp(z_\alpha t)$ (Gaspard, 1998). The coefficients of the expansion (2.73) are given by

$$\langle A|\psi_\alpha\rangle = \int A(\boldsymbol{\Gamma})\, \psi_\alpha(\boldsymbol{\Gamma})\, d\boldsymbol{\Gamma} \qquad \text{and} \qquad \left\langle \tilde{\psi}_\alpha | p_0 \right\rangle = \int \tilde{\psi}^*_\alpha(\boldsymbol{\Gamma})\, p_0(\boldsymbol{\Gamma})\, d\boldsymbol{\Gamma}, \tag{2.74}$$

in terms of the right- and left-eigenmodes of the Liouvillian operator:

$$\hat{L}\psi_\alpha = z_\alpha \psi_\alpha \qquad \text{and} \qquad \hat{L}^\dagger \tilde{\psi}_\alpha = z^*_\alpha \tilde{\psi}_\alpha. \tag{2.75}$$

The eigenmodes $\psi_\alpha(\boldsymbol{\Gamma})$ and $\tilde{\psi}_\alpha(\boldsymbol{\Gamma})$ define, in general, mathematical distributions of Gelfand–Schwartz type (Gaspard, 1998). Accordingly, the observable $A(\boldsymbol{\Gamma})$ and the initial probability density $p_0(\boldsymbol{\Gamma})$ must be sufficiently smooth functions for the integrals (2.74) to exist.

On the other hand, the analytic continuation to the upper half-plane gives a similar expansion valid for negative times $t < 0$, which defines the backward semigroup. The time-reversal symmetry implies that a singularity located at the complex value $-z_\alpha$ in the upper half-plane corresponds to every complex singularity z_α in the lower half-plane and the associated eigenmodes are related to each other by the time-reversal transformation $\boldsymbol{\Theta}$.

The analytic continuation has the effect of breaking the time-reversal symmetry at the statistical level of description because the semigroups are necessarily restricted to one of the two semiaxes of time, as schematically represented in Figure 2.6.

The spectrum of resonances may include the discrete spectrum of eigenvalues on the axis of real frequencies and, in particular, the eigenvalue $z_0 = 0$ if it exists. In this case, the property of ergodicity guarantees the existence of a unique stationary state ψ_0 associated with an eigenvalue $z_0 = 0$ of multiplicity $m_0 = 1$. The property of mixing implies that $z_0 = 0$ is the unique eigenvalue on the axis $\mathrm{Re}\, z = 0$.

If the system is open with escape, as in the disk scatterers (Gaspard, 1998), the leading Pollicott–Ruelle resonance has a nonvanishing real part $\mathrm{Re}\, z_0 < 0$. In these systems, the leading Pollicott–Ruelle resonance defines the escape rate $\gamma \equiv -z_0$ corresponding to the associated eigenmode, which defines a conditionally invariant probability distribution. Pollicott–Ruelle resonances can also be obtained in spatially extended dynamical systems defined on a lattice, as will be further discussed in Chapter 13.

Pollicott (1985, 1986) and Ruelle (1986a,b) proved the existence of these Liouvillian resonances for the so-called axiom-A systems, which are defined as dynamical systems having the properties that: (1) their nonwandering set Ω is hyperbolic and (2) their periodic orbits are dense in Ω. The nonwandering set Ω contains all the points for which any neighborhood $\mathcal{U}$ of these points has recurrent nonempty intersections $\mathcal{U} \cap \boldsymbol{\Phi}^t \mathcal{U}$ at arbitrarily long times t. This invariant subset is said to be hyperbolic if every one of its trajectories has unstable and stable directions, in which infinitesimal perturbations $\delta \boldsymbol{\Gamma}_t$ are exponentially growing or shrinking. Furthermore, the Pollicott–Ruelle resonances of an axiom-A system can be given in terms of its dense periodic orbits. The periodic orbit theory is based on a trace formula, giving the trace of the evolution operator $\exp\left(\hat{L}t\right)$ as a so-called zeta function, which is a product over all the unstable periodic orbits of factors involving only the period and the instability eigenvalues of each periodic orbit (Cvitanović and Eckhardt, 1991; Gaspard, 1998).

The Pollicott–Ruelle resonances are therefore providing the decay or relaxation rates as intrinsic properties for the deterministic dynamics. If it exists,[11] the leading nontrivial Pollicott–Ruelle resonance $z_\alpha = \mathrm{Re}\, z_\alpha + \imath\, \mathrm{Im}\, z_\alpha$ with $\mathrm{Re}\, z_\alpha < 0$ gives the timescale $\tau_\alpha = -(\mathrm{Re}\, z_\alpha)^{-1}$, over which the equilibrium stationary state ψ_0 is effectively reached, completing in this regard the program of ergodic theory.

2.5 Equilibrium Statistical Ensembles

2.5.1 Equilibrium Systems under Different Conditions

As discussed above, an equilibrium stationary probability distribution corresponding to some initial probability distribution can be defined by the time average (2.67). For an

[11] The leading nontrivial resonance is the one with the largest nonzero real part in finite systems. If the system is spatially extended, the leading nontrivial resonance should be defined for eigenmodes with a nonvanishing wave number, i.e., a finite wavelength.

isolated system of particles contained in an infinitely heavy vessel of arbitrary shape, the total linear and angular momenta are not conserved, in which case the probability density only depends on the total energy if there is no further analytic constant of motion, and the equilibrium stationary probability density can be written as

$$p_{\rm eq}(\mathbf{\Gamma}) = \eta[H(\mathbf{\Gamma})] \tag{2.76}$$

with some arbitrary function η (Balescu, 1975). More generally, they may be functions of conserved quantities such as the total energy, linear momentum, and angular momentum (Landau and Lifshitz, 1980a):

$$p_{\rm eq}(\mathbf{\Gamma}) = \eta[H(\mathbf{\Gamma}), \mathbf{P}(\mathbf{\Gamma}), \mathbf{L}(\mathbf{\Gamma})]. \tag{2.77}$$

They may also depend on the particle numbers $\{N_k\}_{k=1}^{c}$ or the volume V.[12]

Different statistical ensembles have been introduced to describe equilibrium systems for different conditions at their boundary (Gibbs, 1902; Tolman, 1938; Hill, 1956, 1960; Pathria, 1972; Huang, 1987; Diu et al., 1989; Balian, 1991). The microcanonical, canonical, grand canonical, isobaric-isothermal ensembles describe, respectively, isolated systems with given total energy and particle number, closed systems at given temperature and particle number, open systems exchanging particles with a chemostat, and systems at given pressure, temperature, and particle number. In the grand canonical and isobaric-isothermal ensembles, the equilibrium probability distribution depends not only on the phase-space variables $\mathbf{\Gamma}$ but also on the random number of particles inside the system, and on the volume of the system, respectively. Moreover, for systems containing several particle species, semigrand canonical ensembles have been considered that are open for some particle species but not others (Hill, 1956, 1960). These equilibrium statistical ensembles are presented in detail in Appendix C.[13]

For each ensemble, the given parameters determine the corresponding thermodynamic potential, while the equilibrium probability density is normalized to the unit value using the so-called partition function. The latter typically grows exponentially with the system size, although the thermodynamic potential is extensive. In this regard, the link between the statistical ensemble and thermodynamics is established by relating the thermodynamic potential to the logarithm of the partition function.

Table 2.1 summarizes the properties of the ensembles for systems with a single particle species. All these ensembles are equivalent from the viewpoint of thermodynamics in the limit where their volume is arbitrarily large, but they differ by the fluctuations of their macrovariables. Indeed, the novelty of statistical mechanics is that the macrovariables of energy, volume, and particle numbers may fluctuate, depending on the boundary conditions. Therefore, different statistical ensembles may have different values for the standard deviations of these macrovariables, $\sigma_E \equiv \sqrt{\langle E^2\rangle - \langle E\rangle^2}$, $\sigma_V \equiv \sqrt{\langle V^2\rangle - \langle V\rangle^2}$, and

[12] The equilibrium distributions (2.76) and (2.77) are stationary solutions for the Liouville equation (2.39), so that $p_{\rm eq} = p_t = p_0\ \forall t \in \mathbb{R}$. In this respect, the function η is fixed by the choice of the initial probability distribution (2.32).

[13] We note that every equilibrium probability distribution can be considered as the stationary solution of Liouville's equation for the Hamiltonian function of the corresponding isolated system. In this respect, the temperature and the other given parameters can be interpreted as the parameters of the probability distribution (2.32) for the initial conditions.

Table 2.1. *Equilibrium statistical ensembles for systems with one particle species. The ensembles are defined by their probability density p_{eq}, which is expressed in terms of the Hamiltonian function H and which may depend on the phase-space variables $\mathbf{\Gamma}_N = (\mathbf{r}_a, \mathbf{p}_a)_{a=1}^N$, the particle number N, and the volume V, as random variables; the energy E, the inverse temperature $\beta = (k_{\mathrm{B}}T)^{-1}$, the pressure p, and the chemical potential μ as parameters; as well as Planck's constant h and an elemental volume constant υ. For every ensemble, the given parameters are fixed by the choice of initial conditions if the system is considered to be isolated. The partition function of each ensemble (Ω, Z, Υ, or Ξ) is related to the corresponding thermodynamic potential, which is respectively the entropy S, the free energy F, the free enthalpy G, or the grand potential J. The standard deviations of energy σ_E, volume σ_V, and particle number σ_N are expressed in terms of the heat capacity at constant volume C_V, the isothermal compressibility κ_T, and the particle density $n = \langle N \rangle / V$. The Gibbs and Euler relations of the thermodynamic potentials are given in Table A.1 of Appendix A.*

Ensemble	Given Parameters	Random Variables	p_{eq}	Thermodynamic Potential	σ_E	σ_V	σ_N
Microcanonical	E, V, N	$\mathbf{\Gamma}_N$	$\frac{\Delta E}{h^{3N}\Omega}\delta(E-H)$	$S = k_{\mathrm{B}} \ln \Omega$	0	0	0
Canonical	T, V, N	$\mathbf{\Gamma}_N$	$\frac{1}{h^{3N}Z}\mathrm{e}^{-\beta H}$	$F = -k_{\mathrm{B}}T \ln Z$	$\sqrt{k_{\mathrm{B}}T^2C_V}$	0	0
Isobaric-Isothermal	T, p, N	$\mathbf{\Gamma}_N, V$	$\frac{1}{h^{3N}\upsilon\Upsilon}\mathrm{e}^{-\beta H-\beta pV}$	$G = -k_{\mathrm{B}}T \ln \Upsilon$	$\sqrt{k_{\mathrm{B}}T^2C_V + \left(\frac{\partial E}{\partial V}\right)_{T,N}^2 \sigma_V^2}$	$\sqrt{k_{\mathrm{B}}T\kappa_T\langle V \rangle}$	0
Grand Canonical	T, V, μ	$\mathbf{\Gamma}_N, N$	$\frac{1}{h^{3N}\Xi}\mathrm{e}^{-\beta H+\beta\mu N}$	$J = -k_{\mathrm{B}}T \ln \Xi$	$\sqrt{k_{\mathrm{B}}T^2C_V + \left(\frac{\partial E}{\partial N}\right)_{T,V}^2 \sigma_N^2}$	0	$n\sqrt{k_{\mathrm{B}}T\kappa_T V}$

$\sigma_N \equiv \sqrt{\langle N^2\rangle - \langle N\rangle^2}$, as well as for their covariances. We note that, away from phase transitions, the standard deviations of the extensive macrovariables behaves as $\sigma_E, \sigma_V, \sigma_N \sim V^{1/2}$ with the volume V, so that the relative fluctuations $\sigma_E/\langle E\rangle, \sigma_V/\langle V\rangle, \sigma_N/\langle N\rangle \sim V^{-1/2}$ become negligible in the limit $V \to \infty$ of an arbitrarily large system, confirming the thermodynamic equivalence between the statistical ensembles.

The statistical-mechanical expression for the thermodynamic entropy was first identified by Boltzmann (1896, 1898), as being proportional to the logarithm of the number Ω of quantum microstates compatible with the macroscopic conditions. The relation was completed by Planck (1914) into the well-known expression $S = k_{\rm B} \ln \Omega$ with Boltzmann's constant $k_{\rm B}$. Using Gibbs' relations for the thermodynamic potential associated with the different ensembles, all the quantities of interest can be deduced from their partition function and, in particular, the entropy as further discussed in Section 2.7.

In this way, the equations of states for pressure and energy can be deduced from the microscopic dynamics.

2.5.2 Detailed Balance

The principle of detailed balance was introduced in the context of chemical kinetics by Wegscheider (1901) and more generally formulated by Fowler (1929) and others. This principle holds if the whole system is at equilibrium described by an equilibrium probability distribution. As the previous discussion shows, the equilibrium distribution for a system of particles isolated in a vessel is expressed by equation (2.76) in terms of the Hamiltonian function (2.2). Since the Hamiltonian function is symmetric under time reversal according to equation (2.28), the equilibrium probability of any event $\mathcal{A}$,

$$P_{\rm eq}(\mathcal{A}) = \int_{\mathcal{A}} p_{\rm eq}(\mathbf{\Gamma})\, d\mathbf{\Gamma} = \int_{\mathcal{A}} \eta[H(\mathbf{\Gamma})]\, d\mathbf{\Gamma}, \tag{2.78}$$

is also symmetric under time reversal because $d\mathbf{\Theta\Gamma} = d\mathbf{\Gamma}$:

$$P_{\rm eq}(\mathcal{A}) = P_{\rm eq}(\mathbf{\Theta}\mathcal{A}). \tag{2.79}$$

Therefore, the equilibrium probability of any two consecutive events $\mathcal{A}$ and $\mathcal{B}$,

$$P_{\rm eq}\left(\{\mathbf{\Gamma} \in \mathcal{A}\} \cap \{\mathbf{\Phi}^t\mathbf{\Gamma} \in \mathcal{B}\}\right) = P_{\rm eq}\left(\mathcal{A} \cap \mathbf{\Phi}^{-t}\mathcal{B}\right), \tag{2.80}$$

can be transformed as follows using the symmetry (2.29) of the dynamics under microreversibility. Since the equilibrium probability distribution is stationary, i.e., invariant under time evolution, we first have that

$$P_{\rm eq}\left(\mathcal{A} \cap \mathbf{\Phi}^{-t}\mathcal{B}\right) = P_{\rm eq}\left(\mathbf{\Phi}^t\mathcal{A} \cap \mathcal{B}\right). \tag{2.81}$$

Next, the symmetry (2.79) of the equilibrium probability distribution and the time-reversal symmetry (2.29) imply that

$$P_{\rm eq}\left(\mathbf{\Phi}^t\mathcal{A} \cap \mathcal{B}\right) = P_{\rm eq}\left[\mathbf{\Theta}\left(\mathbf{\Phi}^t\mathcal{A} \cap \mathcal{B}\right)\right] = P_{\rm eq}\left(\mathbf{\Theta}\,\mathbf{\Phi}^t\mathcal{A} \cap \mathbf{\Theta}\mathcal{B}\right) = P_{\rm eq}\left(\mathbf{\Phi}^{-t}\,\mathbf{\Theta}\mathcal{A} \cap \mathbf{\Theta}\mathcal{B}\right). \tag{2.82}$$

Combining equations (2.81) and (2.82), we thus have that

$$P_{\text{eq}}\left(\mathcal{A} \cap \mathbf{\Phi}^{-t}\mathcal{B}\right) = P_{\text{eq}}\left(\mathbf{\Theta}\mathcal{B} \cap \mathbf{\Phi}^{-t}\mathbf{\Theta}\mathcal{A}\right), \tag{2.83}$$

which means that, at equilibrium, the probability of observing the event $\mathcal{A}$ at time zero and the event $\mathcal{B}$ at time t is equal to the probability of observing the time-reversed event $\mathbf{\Theta}\mathcal{B}$ at time zero and the time-reversed event $\mathbf{\Theta}\mathcal{A}$ at time t. If we define the conditional probability to observe these two consecutive events as

$$P_{\text{eq}}(\mathcal{B},t|\mathcal{A},0) \equiv \frac{P_{\text{eq}}\left(\mathcal{A} \cap \mathbf{\Phi}^{-t}\mathcal{B}\right)}{P_{\text{eq}}(\mathcal{A})}, \tag{2.84}$$

the relation (2.83) reads

$$P_{\text{eq}}(\mathcal{A})\, P_{\text{eq}}(\mathcal{B},t|\mathcal{A},0) = P_{\text{eq}}(\mathcal{B})\, P_{\text{eq}}(\mathbf{\Theta}\mathcal{A},t|\mathbf{\Theta}\mathcal{B},0), \tag{2.85}$$

which is the general expression of detailed balance.

If, moreover, the events are themselves symmetric under time reversal, i.e., $\mathcal{A} = \mathbf{\Theta}\mathcal{A}$ and $\mathcal{B} = \mathbf{\Theta}\mathcal{B}$, the relation (2.85) becomes

$$P_{\text{eq}}(\mathcal{A})\, P_{\text{eq}}(\mathcal{B},t|\mathcal{A},0) = P_{\text{eq}}(\mathcal{B})\, P_{\text{eq}}(\mathcal{A},t|\mathcal{B},0). \tag{2.86}$$

Therefore, the probability of observing the history formed by two consecutive self-reverse events is equal to the probability of observing the time-reversed history, as a consequence of microreversibility together with the assumption that the system is at equilibrium.

We may, furthermore, introduce the rate of the transition $\mathcal{A} \to \mathcal{B}$ over some time interval τ as

$$W(\mathcal{A} \to \mathcal{B}) \equiv \frac{1}{\tau}\, P_{\text{eq}}(\mathcal{B},\tau|\mathcal{A},0) \tag{2.87}$$

in the system at equilibrium (if $\mathcal{A}$ and $\mathcal{B}$ are disjoint). The expression (2.86) of detailed balance can thus be written in terms of the opposite transition rates as

$$P_{\text{eq}}(\mathcal{A})\, W(\mathcal{A} \to \mathcal{B}) = P_{\text{eq}}(\mathcal{B})\, W(\mathcal{B} \to \mathcal{A}) \tag{2.88}$$

for self-reverse events $\mathcal{A}$ and $\mathcal{B}$. The principle of detailed balance is thus a consequence of microreversibility and the further feature of equilibrium probability distributions to themselves be symmetric under time reversal.

We note that the transition $\mathcal{A} \to \mathcal{B}$ may result from several distinct paths $\mathcal{A} \stackrel{\rho}{\to} \mathcal{B}$ composed of trajectories issued from disjoint phase-space subsets $\{\mathcal{A}_\rho\}$ such that $\mathcal{A} = \cup_\rho \mathcal{A}_\rho$ and $\mathcal{A}_\rho \cap \mathcal{A}_{\rho'} = \emptyset$. Indeed, $\mathcal{A}$ and $\mathcal{B}$ may be envisaged as two valleys separated by a chain of mountains with several mountain passes going from $\mathcal{A}$ to $\mathcal{B}$, as illustrated in Figure 2.7. In this case, the transition rates $W\big(\mathcal{A} \stackrel{\rho}{\to} \mathcal{B}\big) \equiv (1/\tau) P_{\text{eq}}(\mathbf{\Phi}^\tau \mathcal{A}_\rho \cap \mathcal{B})/P_{\text{eq}}(\mathcal{A})$ may be associated with every possible path ρ and the total transition rate is given by $W(\mathcal{A} \to \mathcal{B}) = \sum_\rho W\big(\mathcal{A} \stackrel{\rho}{\to} \mathcal{B}\big)$. Under such circumstances, the detailed balance conditions read

$$P_{\text{eq}}(\mathcal{A})\, W\big(\mathcal{A} \stackrel{\rho}{\to} \mathcal{B}\big) = P_{\text{eq}}(\mathcal{B})\, W\big(\mathcal{B} \stackrel{-\rho}{\to} \mathcal{A}\big), \tag{2.89}$$

where $-\rho$ denotes the path ρ travelled in the reversed direction.

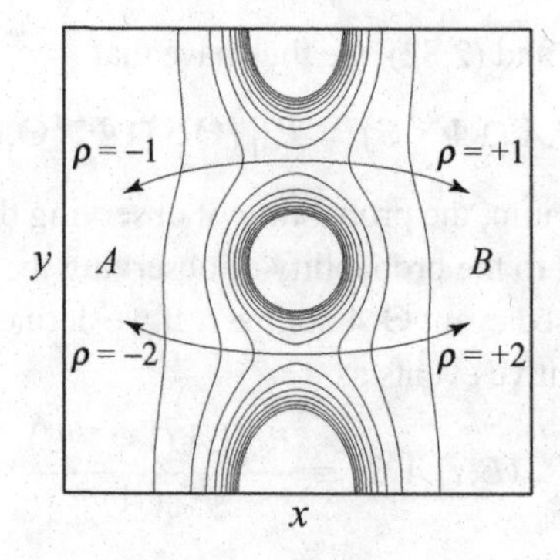

Figure 2.7 Schematic representation of two transition paths ($\rho = +1, +2$) and their reversals ($\rho = -1, -2$) between the potential wells A and B. These paths go through the two saddles separating the wells A and B in the depicted potential energy landscape $U(x, y)$, forming two bottlenecks for the transitions.

More generally, if we introduce the equilibrium probability to observe k successive events as

$$P_{\rm eq}(\mathcal{A}_1, t_1; \ldots; \mathcal{A}_k, t_k) \equiv P_{\rm eq}\left(\left\{\mathbf{\Phi}^{t_1}\mathbf{\Gamma} \in \mathcal{A}_1\right\} \cap \cdots \cap \left\{\mathbf{\Phi}^{t_k}\mathbf{\Gamma} \in \mathcal{A}_k\right\}\right), \tag{2.90}$$

the principle of detailed balance extends into the reversibility condition,

$$P_{\rm eq}(\mathcal{A}_1, t_1; \ldots; \mathcal{A}_k, t_k) = P_{\rm eq}(\mathbf{\Theta}\mathcal{A}_k, -t_k; \ldots; \mathbf{\Theta}\mathcal{A}_1, -t_1). \tag{2.91}$$

If the events are time-reversal symmetric, i.e., $\mathcal{A}_j = \mathbf{\Theta}\mathcal{A}_j$, equation (2.91) becomes

$$P_{\rm eq}(\mathcal{A}_1, t_1; \ldots; \mathcal{A}_k, t_k) = P_{\rm eq}(\mathcal{A}_k, \tau - t_k; \ldots; \mathcal{A}_1, \tau - t_1) \tag{2.92}$$

for any time $\tau \in \mathbb{R}$, given that the equilibrium probability distribution is stationary. Taking $t_j = j\Delta t$ $(1 \leq j \leq k)$ and $\tau = (k+1)\Delta t$, we get

$$P_{\rm eq}(\mathcal{A}_1, \Delta t; \ldots; \mathcal{A}_k, k\Delta t) = P_{\rm eq}(\mathcal{A}_k, \Delta t; \ldots; \mathcal{A}_1, k\Delta t), \tag{2.93}$$

showing that this event sequence and its time reversal have equal probabilities as a consequence of microreversibility under equilibrium conditions.

We emphasize that the detailed balance conditions hold for any event $\mathcal{A}_j$ no matter how small the phase-space domains corresponding to these events may be because the system is assumed to be described by a time-reversal symmetric equilibrium probability distribution.

2.6 Nonequilibrium Statistical Ensembles

The challenging issue is to determine what statistical ensembles should replace Gibbsian equilibrium ensembles under nonequilibrium conditions. Since the nonequilibrium properties are typically time dependent, the Gibbsian ensembles of phase-space points should be replaced by statistical ensembles of trajectories or histories. The trajectories of the system can be determined either exactly by solving Hamilton's equations to obtain the flow $\mathbf{\Phi}^t\mathbf{\Gamma}_0$

for every phase-space initial condition $\mathbf{\Gamma}_0$, or approximately by using stochastic methods on timescales longer than the collision timescale of deterministic motion.

Out of equilibrium, there are different time-dependent properties of interest.

In regard to dynamical mixing, there are the time-dependent statistical averages $\langle A\rangle_t = \int p_0(\mathbf{\Gamma}_0)\, A\left(\mathbf{\Phi}^t\mathbf{\Gamma}_0\right) d\mathbf{\Gamma}_0$ for some observable $A(\mathbf{\Gamma})$ evolving under the flow $\mathbf{\Phi}^t\mathbf{\Gamma}_0$ from initial conditions distributed according to the probability density $p_0(\mathbf{\Gamma}_0)$. Such averages are important to investigate the processes of relaxation towards equilibrium or other stationary macrostates.

There are also the time-dependent correlation functions $\langle A(t)B(0)\rangle_{\rm eq} = \int p_{\rm eq}(\mathbf{\Gamma}_0)\, A\left(\mathbf{\Phi}^t\mathbf{\Gamma}_0\right)\, B(\mathbf{\Gamma}_0)\, d\mathbf{\Gamma}_0$ between two observables separated by some time interval t for the whole system being in the equilibrium macrostate. Such correlation functions arise in linear response theory at first order in the amplitude of a time-dependent external perturbation applied to the system. In this case, the use of perturbation theory allows us to solve the problem in terms of the flow $\mathbf{\Phi}^t$ of the unperturbed system.

If the system is in contact with several reservoirs at different temperatures or chemical potentials, an important issue is to determine the stationary probability distribution describing the nonequilibrium steady macrostate of the system. In such circumstances the system is closed or open, and its dynamics involve the degrees of freedom of the reservoirs. If there are exchanges of particles with the reservoirs, the trajectories of particles incoming or outgoing the system of interest should be determined. Moreover, a question also arises about the relaxation of the probability distribution from some initial distribution towards such nonequilibrium stationary distributions.

In addition to these systems where the equations of microscopic motion are time independent, there also exist systems that are driven away from equilibrium by arbitrary time-dependent external forces. Solving such problems requires the determination of the trajectories for the full Hamiltonian dynamics including the time-dependent external forces.

In all these problems, the Hamiltonian trajectory starting from some initial condition should be known and a key issue is to understand the symmetries of these trajectories, especially under time reversal.

2.6.1 Symmetry Breaking by the Selection of Initial Conditions

Symmetry-breaking phenomena find their origin in the general result that the symmetry of the equations ruling some phenomenon does not imply that the solutions of these equations themselves have this symmetry.

In order to illustrate this fundamental result, let us consider the Hamiltonian dynamics of a particle moving in a double-well potential. For this system, the Hamiltonian function is given by

$$H = \frac{p^2}{2m} + U(x), \qquad \text{where} \qquad U(x) = -\frac{a}{2}x^2 + \frac{b}{4}x^4 \tag{2.94}$$

with $a > 0$ and $b > 0$, and the phase space is the plane $\mathcal{M} = \left\{\mathbf{\Gamma} = (x,p) \in \mathbb{R}^2\right\}$. The Hamiltonian function (2.94) is invariant under the parity transformation

$$\mathbf{\Pi}\,(x,p) = (-x,\,-p). \tag{2.95}$$

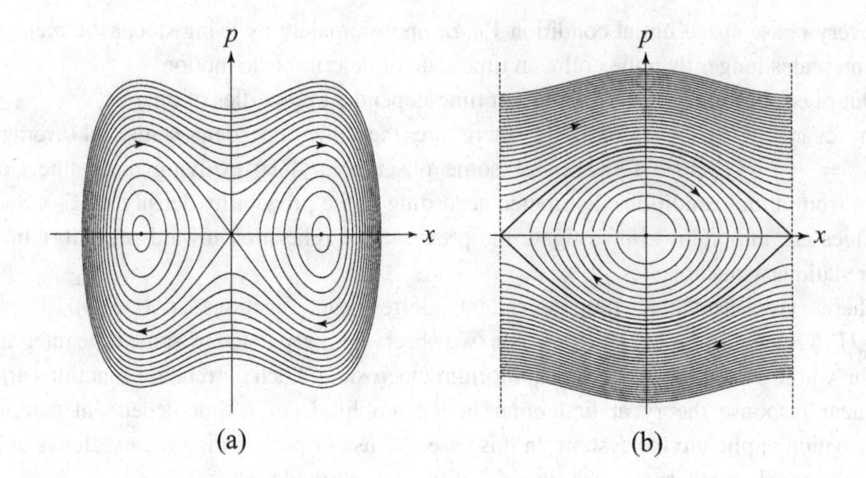

Figure 2.8 Phase portraits of the Hamiltonian flows for (a) the double-well potential (2.94) and (b) the simple pendulum (2.99).

Consequently, the Hamilton equations of motion are also invariant under parity. The phase portrait of this dynamical system is plotted in Figure 2.8(a). There are three stationary points: the origin $x = p = 0$, which is symmetric but unstable, and the two minima of the potential at $x = \pm\sqrt{a/b}$ and $p = 0$, which are stable at the minimum energy $U_{\rm min} = -a^2/(4b)$. At zero energy, there are two homoclinic loops connecting the unstable point to itself and playing the role of separatrices between the periodic orbits at negative and positive energies. Otherwise, all the other trajectories are periodic with some period $\mathscr{T}$ and every one of them forms a curve such as

$$\mathcal{T} = \left\{ \mathbf{\Gamma} = \mathbf{\Phi}^t \mathbf{\Gamma}_0 \ \text{for} \ 0 \le t < \mathscr{T} \right\}, \tag{2.96}$$

which is invariant under the phase-space Hamiltonian dynamics. To every positive value $E > 0$ of energy, there corresponds one and only one periodic orbit $\mathcal{T}_E$ that oscillates between the two wells. In contrast, there exist two periodic orbits $\mathcal{T}_E^{(\pm)}$ corresponding to every negative value of the energy $U_{\rm min} < E < 0$ and they oscillate in *either* the well on the right-hand side *or* the one on the left-hand side. If the periodic orbits at positive energies are symmetric under space inversion, this is no longer the case for those at negative energies:

$$\mathbf{\Pi}\,\mathcal{T}_E = \mathcal{T}_E \qquad \text{for} \qquad E > 0, \tag{2.97}$$

$$\mathbf{\Pi}\,\mathcal{T}_E^{(+)} = \mathcal{T}_E^{(-)} \neq \mathcal{T}_E^{(+)} \qquad \text{for} \qquad E < 0. \tag{2.98}$$

Therefore, the selection of one periodic orbit at negative energy breaks the symmetry under space inversion for the Hamiltonian dynamics. This phenomenon of symmetry breaking is well known in condensed matter physics. The selection of some initial condition may thus result in the selection of a trajectory that does not have the symmetry of the equations of motion.

Remarkably, the same phenomenon also concerns the time-reversal symmetry, as illustrated by the simple pendulum. For this system, the Hamiltonian function is given by

$$H = \frac{p^2}{2ml^2} - mgl\cos x, \tag{2.99}$$

where m is the mass of the body attached to a massless rod of length l and subjected to the gravitational acceleration g oriented towards the minimum $x = 0$ of potential energy. The angle x varies between $-\pi$ and $+\pi$, so that the phase space is the cylinder $\mathcal{M} = \{\mathbf{\Gamma} = (x, p) : -\pi < x \leq +\pi,\ p \in \mathbb{R}\}$. The Hamiltonian function and Hamilton's equations are symmetric under the time-reversal transformation

$$\mathbf{\Theta}(x, p) = (x, -p). \tag{2.100}$$

The phase portrait of the pendulum is depicted in Figure 2.8(b). The system has two stationary configurations: the stable one $x = p = 0$ at the minimum energy $U_{\text{min}} = -mgl$ and the unstable one $(x = \pm\pi, p = 0)$ at the energy $U_{\text{max}} = +mgl$. On the cylinder, the two unstable points $(x = \pm\pi, p = 0)$ are equivalent and they correspond to the same configuration for the pendulum. Here, there are two homoclinic orbits connecting the two points together at the energy $E = U_{\text{max}}$. They form two separatrices between the oscillating periodic orbits at lower energies $U_{\text{min}} < E < U_{\text{max}}$ and the rotating periodic orbits at higher energies $E > U_{\text{max}}$. To every value of the energy $U_{\text{min}} < E < U_{\text{max}}$, there corresponds one and only one oscillating periodic orbit $\mathcal{T}_E$, which is self-reverse because it is mapped onto itself by time reversal. In contrast, there exist two periodic orbits $\mathcal{T}_E^{(\pm)}$ rotating either anticlockwise or clockwise for every energy value $E > U_{\text{max}}$. None of them is self-reverse, so each one breaks the time-reversal symmetry. In analogy with equations (2.97)–(2.98), we here have that

$$\mathbf{\Theta}\mathcal{T}_E = \mathcal{T}_E \qquad \text{for} \qquad U_{\text{min}} < E < U_{\text{max}}, \tag{2.101}$$

$$\mathbf{\Theta}\mathcal{T}_E^{(+)} = \mathcal{T}_E^{(-)} \neq \mathcal{T}_E^{(+)} \qquad \text{for} \qquad E > U_{\text{max}}. \tag{2.102}$$

The fact is that the symmetry of the equations of motion does not imply the symmetry of all their solutions. Already, in this simple example, the breaking of time-reversal symmetry may arise by the selection of initial conditions.

This symmetry breaking by the selection of initial conditions is of direct consequence to the probability distribution describing the system behavior. Indeed, over timescales longer than the period of the trajectory selected by the initial condition, a stationary probability distribution can be defined by time averaging over the period $\mathscr{T}$ to get the stationary probability density

$$p_{\mathcal{T}}(\mathbf{\Gamma}) = \frac{1}{\mathscr{T}} \int_0^{\mathscr{T}} \delta\left(\mathbf{\Gamma} - \mathbf{\Phi}^t\mathbf{\Gamma}_0\right) dt \tag{2.103}$$

according to the Birkhoff–Khinchin ergodic theorem. The support of this stationary probability distribution is the curve (2.96) depicting the periodic orbit in the phase space. As a consequence of previous discussion, the stationary probability density (2.103) also breaks the time-reversal symmetry if it is supported by one of the periodic orbits $\mathcal{T}_E^{(\pm)}$:

$$p_{\mathcal{T}_E^{(+)}}(\boldsymbol{\Theta\Gamma}) = p_{\mathcal{T}_E^{(-)}}(\boldsymbol{\Gamma}) \neq p_{\mathcal{T}_E^{(+)}}(\boldsymbol{\Gamma}) \tag{2.104}$$

for $E > U_{\rm max}$. More generally, it is possible to define a continuous family of stationary densities supported by both periodic orbits according to

$$p_\lambda(\boldsymbol{\Gamma}) = \lambda\, p_{\mathcal{T}_E^{(+)}}(\boldsymbol{\Gamma}) + (1-\lambda)\, p_{\mathcal{T}_E^{(-)}}(\boldsymbol{\Gamma}) \tag{2.105}$$

with $0 \leq \lambda \leq 1$. They satisfy the symmetry relation

$$p_\lambda(\boldsymbol{\Theta\Gamma}) = p_{1-\lambda}(\boldsymbol{\Gamma}) \tag{2.106}$$

and they break the time-reversal symmetry unless $\lambda = 1/2$. In this latter case, we recover the microcanonical stationary density

$$p_E(\boldsymbol{\Gamma}) = \frac{\delta[E - H(\boldsymbol{\Gamma})]}{\int \delta[E - H(\boldsymbol{\Gamma}')]\, d\boldsymbol{\Gamma}'}, \tag{2.107}$$

which is always symmetric under time reversal since it is defined in terms of the symmetric Hamiltonian function $p_{\lambda=1/2}(\boldsymbol{\Gamma}) = p_E(\boldsymbol{\Gamma}) = p_E(\boldsymbol{\Theta\Gamma})$ (Gaspard, 2012a).

Although Newton's equation is time-reversal symmetric, its solutions do not necessarily have the symmetry. Therefore, the selection of a trajectory by the initial condition can break the time-reversal symmetry, as in other symmetry-breaking phenomena. This happens if the trajectory selected by the initial condition is physically distinct from its time reversal, as shown in Figure 2.9.

This time-reversal symmetry breaking does not manifest itself in every system. In particular, all the trajectories of Newton's equation $md^2x/dt^2 = -kx$ for the harmonic oscillator are the ellipses $E = p^2/(2m) + kx^2/2$ in the phase space (x, p). Every ellipse is mapped onto itself by time reversal (2.100). All the trajectories are thus self-reverse in the harmonic oscillator. In contrast, all the trajectories with a nonvanishing momentum are distinct from their time reversal in the case of the free particle, for which Newton's equation reads $md^2x/dt^2 = 0$ in some reference frame. Indeed, the trajectories are straight lines with $x(t) = x(0) + p(0)t/m$ and $p(t) = p(0)$, which are distinct from their reversal if the momentum is nonvanishing, $p(0) \neq 0$. Therefore, the selection of an initial condition may likewise break the time-reversal symmetry in this simple system (Gaspard, 2008). *A fortiori*, this breaking can also happen in a chaotic system with a spectrum of positive Lyapunov exponents indicating many stable and unstable directions in phase space. These directions are mapped onto each other by time reversal, but they are physically distinct, so that the time-reversal symmetry will be broken if one specific direction is selected by the initial condition (Gaspard, 2006, 2007b).

2.6.2 Systems Evolving from Nonequilibrium Initial Macrostates

Similar considerations apply to Liouville's equation. This equation is symmetric under time reversal, but its solutions do not need to coincide with its time reversal. This concerns the stationary probability distributions describing nonequilibrium steady macrostates, but also

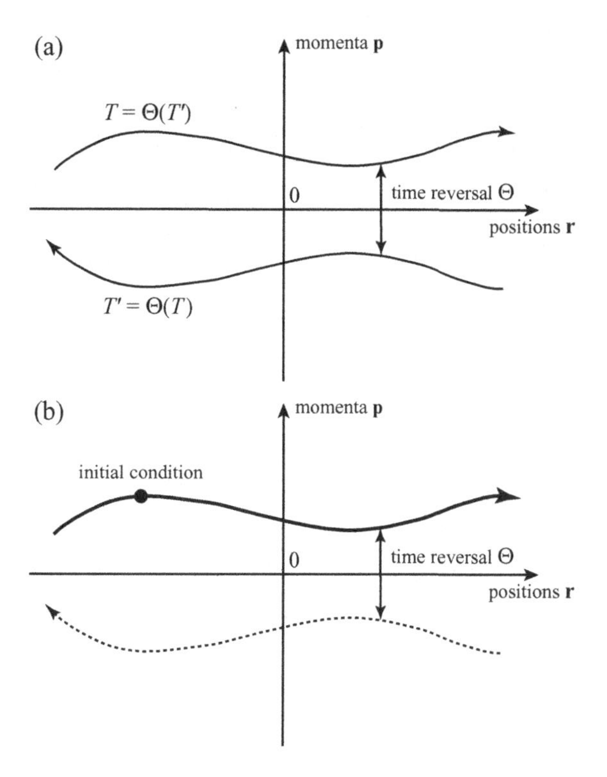

Figure 2.9 (a) If Newton's equation is time-reversal symmetric, the time reversal $\Theta\mathcal{T}$ of every solution $\mathcal{T}$ is also a solution, as shown here in the phase space of the positions and momenta of the system. (b) If the time reversal $\Theta\mathcal{T}$ is physically distinct from the trajectory $\mathcal{T}$, the selection of the trajectory $\mathcal{T}$ by the initial condition gives a unit probability to $\mathcal{T}$ and zero to all the other solutions including the time-reversal image $\Theta\mathcal{T}$, therefore breaking the time-reversal symmetry.

typical time-dependent solutions. According to equation (2.48), if $p_t(\boldsymbol{\Gamma}) = p_0(\boldsymbol{\Phi}^{-t}\boldsymbol{\Gamma})$ is a solution of Liouville's equation, then $p_t'(\boldsymbol{\Gamma}) = p_0'(\boldsymbol{\Phi}^{-t}\boldsymbol{\Gamma})$ with $p_0'(\boldsymbol{\Gamma}) = p_t(\boldsymbol{\Theta\Gamma})$ is also a solution. However, these two solutions do not coincide if $p_t(\boldsymbol{\Gamma}) \neq p_0(\boldsymbol{\Theta\Gamma})$, which is generally the case. For instance, it may be that the initial density is time-reversal symmetric, but the solution is not stationary $p_t(\boldsymbol{\Gamma}) \neq p_0(\boldsymbol{\Gamma})$; or the solution may be stationary but not symmetric under time reversal $p_{\rm st}(\boldsymbol{\Gamma}) \neq p_{\rm st}(\boldsymbol{\Theta\Gamma})$.

Therefore, time-reversal symmetry breaking is possible at the statistical level of description, although the microscopic equations of motion are always symmetric under time reversal. In this respect, the thermodynamic time asymmetry is compatible with microreversibility.

2.6.3 Systems in Nonequilibrium Steady Macrostates

In statistical mechanics, every initial condition – and thus every phase-space trajectory – is weighted with a probability giving the statistical frequency of its occurrence in a sequence of repeated experiments. This is, in particular, the case for a nonequilibrium system with

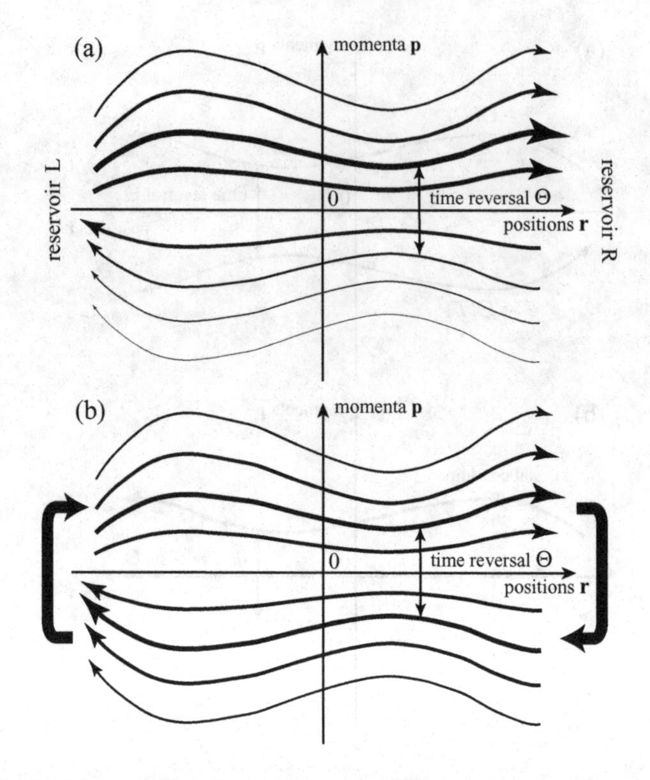

Figure 2.10 (a) Schematic phase portrait of a nonequilibrium system with particles moving between the high concentration reservoir L and the low concentration reservoir R. The trajectories from L to R typically have a higher probability weight than the trajectories from R to L in order that the mean current flows from L to R. The probability weight is indicated by the thickness of the trajectories, showing the time asymmetry of the probability distribution. (b) Schematic phase portrait at equilibrium after closing the contacts with the reservoirs so that the system goes to equilibrium, in which case the probability distribution is time-reversal symmetric and detailed balance is satisfied.

particles diffusing between two reservoirs at different concentrations. After some time, the system reaches a nonequilibrium steady macrostate, which can be described by a stationary probability distribution. Averaging over this distribution gives the mean current of particles diffusing from high to low concentrations. In this case, the phase-space trajectories coming from the high concentration reservoir typically have a larger probability weight than the trajectories from the low concentration reservoir, as schematically illustrated in Figure 2.10(a). Since the set of these latter trajectories contains the time reversal of the former ones, we conclude that the time-reversal symmetry is broken by this nonequilibrium stationary probability distribution

$$p_{\rm neq}(\boldsymbol{\Theta\Gamma}) \neq p_{\rm neq}(\boldsymbol{\Gamma}). \tag{2.108}$$

Of course, if the system is isolated by closing contacts with the reservoirs, the stationary probability distribution becomes the equilibrium one after relaxation, in which case detailed balance is satisfied and the time-reversal symmetry is restored:

Table 2.2. *Symmetries of equilibrium and nonequilibrium dynamical systems under time reversal* $\boldsymbol{\Theta}$. *The systems are denoted* $(\mathcal{M}, \boldsymbol{\Phi}^t, P)$, *where* $\mathcal{M}$ *is the phase space,* $\boldsymbol{\Phi}^t$ *the flow, and* P *the stationary probability distribution.* $\mathcal{A}$ *denotes some phase-space subset.*

Equilibrium System	Nonequilibrium System
$(\mathcal{M}, \boldsymbol{\Phi}^t, P_{\rm eq})$	$(\mathcal{M}, \boldsymbol{\Phi}^t, P_{\rm neq})$
$\boldsymbol{\Theta}\mathcal{M} = \mathcal{M}$	$\boldsymbol{\Theta}\mathcal{M} = \mathcal{M}$
$\boldsymbol{\Theta}\boldsymbol{\Phi}^t = \boldsymbol{\Phi}^{-t}\boldsymbol{\Theta}$	$\boldsymbol{\Theta}\boldsymbol{\Phi}^t = \boldsymbol{\Phi}^{-t}\boldsymbol{\Theta}$
$P_{\rm eq}(\boldsymbol{\Theta}\mathcal{A}) = P_{\rm eq}(\mathcal{A})$	$P_{\rm neq}(\boldsymbol{\Theta}\mathcal{A}) \neq P_{\rm neq}(\mathcal{A})$

$$p_{\rm eq}(\boldsymbol{\Theta}\boldsymbol{\Gamma}) = p_{\rm eq}(\boldsymbol{\Gamma}) \tag{2.109}$$

(see Figure 2.10(b)). Since the nonequilibrium stationary probability distributions are solutions for the Liouville equation (2.39), we here have a symmetry-breaking phenomenon similar to the one for Newton's equations. The nonequilibrium stationary density (2.108) is a solution of Liouville's equation with a lower symmetry than the equation itself. In this way, irreversible behavior can be described by weighting the trajectories $\mathcal{T}$ and their time-reversal images $\boldsymbol{\Theta}\mathcal{T}$ with different probabilities (Gaspard, 2008).

As summarized in Table 2.2, equilibrium and nonequilibrium dynamical systems are specified by their phase space, their flow, and their stationary probability distribution. The time-reversal symmetry of the system concerns every one of these specifications. Here, the systems are supposed to be symmetric under time reversal at the dynamical level of description because both their phase space and their flow are symmetric.[14] However, the symmetry of their stationary probability distribution depends on whether the system is at equilibrium or not. In nonequilibrium systems, the time-reversal symmetry is broken at the statistical level of description, although this symmetry remains satisfied at the dynamical level of description. Therefore, the detailed balance expression (2.88) or the reversibility condition (2.93) are no longer satisfied for nonequilibrium probability distributions $P_{\rm neq}$. This general mechanism will be illustrated in detail for effusion processes in Chapter 8.

2.7 Entropy

2.7.1 Coarse Graining

For any equilibrium statistical ensemble, the thermodynamic entropy can be obtained using the Gibbs relation for the thermodynamic potential defined in terms of the associated

[14] The time-reversal symmetry is broken at the dynamical level of description if $\boldsymbol{\Theta}\boldsymbol{\Phi}^t \neq \boldsymbol{\Phi}^{-t}\boldsymbol{\Theta}$, which is the case, in particular, for electrically charged particles in an external magnetizing field.

partition function. Since the quantum microstates correspond to phase-space cells of volume (2.7), the phase space should be partitioned into disjoint cells $\{\mathcal{C}_i\}$ such that

$$\cup_i \, \mathcal{C}_i = \mathcal{M} \qquad \text{and} \qquad \mathcal{C}_i \cap \mathcal{C}_j = \emptyset \quad \text{if} \quad i \neq j. \tag{2.110}$$

All the cells associated with quantum microstates are supposed to have the same volume, $\text{vol}(\mathcal{C}_i) = \Delta\mathbf{\Gamma} = h^{3N}$.

If the system is described by some probability density $p(\mathbf{\Gamma})$, the probability that some cell $\mathcal{C}_i$ is visited is given by

$$P_i \equiv P(\mathcal{C}_i) = \int_{\mathcal{C}_i} p(\mathbf{\Gamma})\, d\mathbf{\Gamma}, \tag{2.111}$$

which is normalized according to $\sum_i P_i = 1$.

In all the equilibrium statistical ensembles, the expression of thermodynamic entropy deduced from the Gibbs relation is compatible with the formula

$$S = -k_{\rm B} \sum_i P_i \ln P_i, \tag{2.112}$$

which is called the *coarse-grained entropy*.[15] In particular, in the microcanonical ensemble, every quantum microstate in the energy shell $E < H(\mathbf{\Gamma}) < E + \Delta E$ has the probability $P_i = 1/\Omega$, where Ω is the total number of microstates in the energy shell (see Appendix C). Therefore, Boltzmann's expression $S = k_{\rm B} \ln \Omega$ for entropy can be recovered from equation (2.112).

Moreover, the third law of thermodynamics is established, for instance, in the canonical ensemble. If the minimum energy of the system has the multiplicity g_0, all the probability is distributed over g_0 corresponding quantum microstates as the temperature reaches absolute zero, so that $P_i = 1/g_0$ for the microstates i of minimal energy, and $P_i = 0$ otherwise. According to equation (2.112), the entropy takes the value $S = k_{\rm B} \ln g_0$, which is vanishing if the microstate of minimum energy is unique ($g_0 = 1$), hence the third law (1.11). We note that the entropy density is also vanishing if the microstates of minimum energy have a finite multiplicity. In contrast, there exists a positive residual entropy density as the temperature reaches absolute zero if the multiplicity of the microstates of minimum energy grows exponentially with the number of particles in the system, as in the case for ice or other molecular crystals at very low temperature (Pauling, 1970).

If the phase-space cells are small enough, their probabilities can be approximately expressed in terms of the probability density at some phase-space point $\mathbf{\Gamma}_i \in \mathcal{C}_i$ as

$$P_i \simeq p(\mathbf{\Gamma}_i)\Delta\mathbf{\Gamma}, \tag{2.113}$$

so that the entropy (2.112) becomes

$$S = k_{\rm B} \ln \frac{1}{\Delta\mathbf{\Gamma}} - k_{\rm B} \int p(\mathbf{\Gamma}) \ln p(\mathbf{\Gamma})\, d\mathbf{\Gamma} + O(\Delta\mathbf{\Gamma}) \tag{2.114}$$

[15] This relation finds its origin in pioneering work by Boltzmann (1877, 1896, 1898) and has been known since the beginning of the twentieth century (Ehrenfest and Ehrenfest, 1911; Planck, 1914), long before the work by Shannon and Weaver (1949).

in the limit $\Delta\mathbf{\Gamma} \to 0$. In equation (2.114), the first term is essential to fix the problem of the entropy constant in order to obtain the absolute value of the entropy that is compatible with the counting of quantum microstates (Fermi, 1937). This term is also important to remain consistent with the units, since the probability is dimensionless while the probability density appearing in the logarithm in the second term has the units of $\Delta\mathbf{\Gamma}^{-1}$. The second term is the so-called differential entropy. The third term contains the corrections of order $\Delta\mathbf{\Gamma}$ in the approximation (2.113).

Now, for an ideal monatomic gas in the canonical ensemble (C.8), we recover the Sackur–Tetrode formula for the equilibrium value of the thermodynamic entropy with $\Delta\mathbf{\Gamma} = h^{3N}$:

$$S_{\rm eq} = Nk_{\rm B} \ln \frac{{\rm e}^{5/2}(2\pi mk_{\rm B}T)^{3/2}V}{h^3 N}, \tag{2.115}$$

showing the consistency of the general expression (2.112) in this case as well (Pathria, 1972; Huang, 1987). Since the kinetic energy of this gas has the mean value $\langle E\rangle_{\rm eq} = 3Nk_{\rm B}T/2$, the entropy in the microcanonical ensemble is given by the expression (2.115) with $k_{\rm B}T = 2E/(3N)$.

We note that there exist many different ways of coarse graining to define the entropy. The phase-space cells may be chosen in such a way that the coarse graining is achieved in terms of the one-particle distribution function (2.52) in dilute gases. In such systems, the N-particle probability density factorizes as

$$p(\mathbf{\Gamma}) = C_N\, f_1(\mathbf{x}_1)\, f_1(\mathbf{x}_2)\, \cdots\, f_1(\mathbf{x}_N) \tag{2.116}$$

into one-particle distribution functions (2.52), where the normalization constant should be equal to $C_N = N!/N^N \simeq \exp(-N)$. In such systems, the entropy (2.114) becomes

$$S = k_{\rm B} \int_{\mathbb{R}^6} d\mathbf{x}\, f_1(\mathbf{x}) \ln \frac{\rm e}{h^3 f_1(\mathbf{x})}, \tag{2.117}$$

which obeys the H-theorem of Boltzmann (1872, 1896) under the time evolution of his kinetic equation. In denser gases, where the two-particle correlations are important, the entropy should be modified according to

$$S = k_{\rm B} \int_{\mathbb{R}^6} d\mathbf{x}\, f_1(\mathbf{x}) \ln \frac{\rm e}{h^3 f_1(\mathbf{x})} + \frac{k_{\rm B}}{2} \int_{\mathbb{R}^{12}} d\mathbf{x}\, d\mathbf{x}'\, f_2(\mathbf{x},\mathbf{x}') \ln \frac{f_1(\mathbf{x})\, f_1(\mathbf{x}')}{f_2(\mathbf{x},\mathbf{x}')}, \tag{2.118}$$

where the one- and two-particle distribution functions are locally defined in the limit of an arbitrarily large system (Green, 1952a; Yvon, 1966; Wallace, 1987; Evans and Morriss, 1990). We note that the Boltzmann expression (2.117) for the entropy is recovered from equation (2.118) if the particles are uncorrelated, i.e., if $f_2(\mathbf{x},\mathbf{x}') = f_1(\mathbf{x})f_1(\mathbf{x}')$.

For an ideal monatomic gas of density $n = N/V$ and temperature T, the equilibrium one-particle distribution function is given by

$$f_{1,\rm eq}(\mathbf{x}) = n\, \frac{\exp\left(-\frac{\mathbf{p}^2}{2mk_{\rm B}T}\right)}{(2\pi mk_{\rm B}T)^{3/2}}, \tag{2.119}$$

in which case the Sackur–Tetrode formula (2.115) is recovered from equation (2.117).

2.7.2 Interpretation

According to equation (2.112), the entropy is directly associated with the probability distribution defined in phase space. In this regard, the entropy (2.112) can be interpreted as a measure of *disorder in phase space*. If all the probability is concentrated inside a single cell, i.e., $P_i = 1$ and $P_j = 0$ for $j \neq i$, the entropy (2.112) is equal to zero. Instead, if the probability distribution is equally distributed over all the microstates compatible with the macroscopic conditions, as in the microcanonical ensemble, the entropy (2.114) reaches its maximal value.

The entropy (2.112) is also a measure of the amount of data required to describe the statistical ensemble associated with the probability distribution $\{P_i\}$. If the statistical ensemble is composed of $\mathcal{N}$ copies of the system, the instantaneous state of the ensemble is given by the microstates of all these copies $\{i_j\}_{j=1}^{\mathcal{N}}$. The number of copies in the microstate i is approximately given by $\mathcal{N}_i \simeq P_i \mathcal{N}$ according to the law of large numbers and $\sum_i \mathcal{N}_i = \mathcal{N}$. The entropy (2.112) is recovered by taking

$$S = \lim_{\mathcal{N}\to\infty} \frac{k_{\rm B}}{\mathcal{N}} \ln \mathcal{W}_{\mathcal{N}}, \qquad \text{where} \qquad \mathcal{W}_{\mathcal{N}} = \frac{\mathcal{N}!}{\prod_i \mathcal{N}_i !} \tag{2.120}$$

is the number of possible configurations $\{i_j\}_{j=1}^{\mathcal{N}}$ of the ensemble, the copies being statistically independent. The number of digits required to specify the configuration of the $\mathcal{N}$ copies in the ensemble is given by $\log_{10} \mathcal{W}_{\mathcal{N}}$. Accordingly, the entropy is proportional to the size of computer memory storage required to specify the microstate of one copy if the system is described by the probability distribution $\{P_i\}$.

In this regard, the concept of entropy is extra-mechanical, being the macroscopic property emerging from the statistical description of the system in terms of some probability distribution.

We note that entropy can also be defined in the framework of large-deviation theory, which justifies the thermodynamic formalism in the limit of large systems or large statistical ensembles (Ellis, 1985; Touchette, 2009).

2.7.3 Coentropy

An interesting quantity to define is the *coentropy* associated with some symmetry transformation $\mathbf{G}$ applied to the probability distribution

$$S^{\rm G} \equiv -k_{\rm B} \sum_i P_i \ln P_i^{\rm G} \tag{2.121}$$

with the notation $P_i^{\rm G} \equiv P(\mathbf{G}\mathcal{C}_i) = \int_{\mathbf{G}\mathcal{C}_i} p(\mathbf{\Gamma})\, d\mathbf{\Gamma}$. The difference between the coentropy and the entropy is related to the Kullback–Leibler divergence:

$$S^{\rm G} - S = k_{\rm B} \sum_i P_i \ln \frac{P_i}{P_i^{\rm G}} \equiv k_{\rm B}\, D_{\rm KL}\big(P \| P^{\rm G}\big) \geq 0, \tag{2.122}$$

characterizing deviations between the two probability distributions (Cover and Thomas, 2006). This divergence is always nonnegative. It vanishes if the probability distribution is symmetric under the transformation $\mathbf{G}$, i.e., if $P_i = P_i^{\mathbf{G}}$. Otherwise, the divergence is positive and it measures how much the symmetry is broken.

2.7.4 Further Coarse Graining

An important issue is to express the entropy (2.112) if the phase-space partition is carried out with coarser cells

$$\mathcal{C}_\omega = \cup_{i\in\omega} \mathcal{C}_i \qquad \text{with the probabilities} \qquad P_\omega = \sum_{i\in\omega} P_i . \tag{2.123}$$

The entropy (2.112) can actually be written in the more general form

$$S = \sum_\omega P_\omega \big(S_\omega^0 - k_{\rm B} \ln P_\omega \big), \tag{2.124}$$

in terms of the entropy of the system found in the coarse-grained state ω,

$$S_\omega^0 \equiv -k_{\rm B} \sum_{i\in\omega} P_{i|\omega} \ln P_{i|\omega}, \tag{2.125}$$

which is expressed with the conditional probabilities

$$P_{i|\omega} \equiv \frac{P_i}{\sum_{i\in\omega} P_i} . \tag{2.126}$$

Indeed, substituting (2.125) into equation (2.124) and using equation (2.126), the expression (2.112) is recovered. Furthermore, equation (2.124) reduces to equation (2.112) if the coarser cells $\{\mathcal{C}_\omega\}$ are replaced by the finer ones $\{\mathcal{C}_i\}$, because $S_\omega^0 = 0$ if the cell $\mathcal{C}_\omega$ coincides with some cell $\mathcal{C}_i$. Therefore, equation (2.124) continues to give the thermodynamic entropy including the contributions of all the degrees of freedom whatever the level of coarse graining might be. For this reason, expression (2.124) can be generally considered as the coarse-grained entropy.

2.7.5 Dynamical Mixing and Entropy Time Evolution

As already understood by Gibbs (1902), the entropy (2.112) evolves in time towards its equilibrium value if the system has the property of dynamical mixing.[16] Indeed, if the property (2.71) holds, any probability to visit a phase-space cell at time t will converge towards its equilibrium value:

$$P_t(\mathcal{C}_i) \to_{t\to\infty} P_{\rm eq}(\mathcal{C}_i). \tag{2.127}$$

Therefore, we have that

$$S_t = -k_{\rm B} \sum_i P_t(\mathcal{C}_i) \ln P_t(\mathcal{C}_i) \to_{t\to\infty} S_{\rm eq} = -k_{\rm B} \sum_i P_{\rm eq}(\mathcal{C}_i) \ln P_{\rm eq}(\mathcal{C}_i). \tag{2.128}$$

[16] If the system is not ergodic, its dynamics should be decomposed into its different ergodic components according to its constants of motion.

As time increases, the probability is distributed over more and more cells in phase space because of the mixing property, so that the entropy tends to increase with time. After a long enough time interval, the entropy will thus reaches its equilibrium value, which is thus maximal. The increase of entropy according to the second law can thus be understood if the system is dynamically mixing and the entropy is defined by coarse graining.

If the approximate expression (2.114) is used, the first two terms remain constant in time as a consequence of Liouville's theorem, so that all the time dependence is in the third term $O(\Delta\boldsymbol{\Gamma})$, which confirms the need for coarse graining in order to get the time dependence of the entropy. In this regard, the coarse-grained entropy (2.112) can be extended from equilibrium to nonequilibrium systems.

2.7.6 Entropy Production

If we consider a large system composed of several parts such as an open system in contact with reservoirs, the coarse-grained entropy can be defined in some phase-space domain $\mathcal{D}$ corresponding to some part of the total system (Gaspard, 1997, 1998; Dorfman et al., 2002). This domain should thus be partitioned into small enough cells $\{\mathcal{C}_i\}$ to define the entropy as before

$$S_t(\mathcal{D}|\{\mathcal{C}_i\}) = -k_{\rm B} \sum_{\mathcal{C}_i \subset \mathcal{D}} P_t(\mathcal{C}_i) \ln P_t(\mathcal{C}_i) \tag{2.129}$$

for some time-dependent probability distribution P_t. This entropy evolves in time since

$$P_{t+\tau}(\mathcal{C}_i) = P_t(\boldsymbol{\Phi}^{-\tau}\mathcal{C}_i), \tag{2.130}$$

which implies

$$S_{t-\tau}(\mathcal{D}|\{\mathcal{C}_i\}) = S_t(\boldsymbol{\Phi}^{\tau}\mathcal{D}|\{\boldsymbol{\Phi}^{\tau}\mathcal{C}_i\}). \tag{2.131}$$

Now, the time variation of the entropy over some time interval τ is given by

$$\Delta^{\tau} S = S_t(\mathcal{D}|\{\mathcal{C}_i\}) - S_{t-\tau}(\mathcal{D}|\{\mathcal{C}_i\}). \tag{2.132}$$

Since the domain $\mathcal{D}$ is, in general, smaller than the whole phase space $\mathcal{M}$, the entropy (2.132) can be separated as

$$\Delta^{\tau} S = \Delta^{\tau}_{\rm e} S + \Delta^{\tau}_{\rm i} S \tag{2.133}$$

into the *entropy exchange* with the environment of the domain $\mathcal{D}$, $\Delta^{\tau}_{\rm e} S$, and the rest to be identified with the *entropy production*, $\Delta^{\tau}_{\rm i} S$. The entropy exchange can be naturally defined as the difference between the entropies of the subsets ingoing and outgoing the domain $\mathcal{D}$ during the time interval τ, respectively denoted $\mathcal{D}_{\rm in}$ and $\mathcal{D}_{\rm out}$:

$$\Delta^{\tau}_{\rm e} S \equiv S_{t-\tau}(\mathcal{D}_{\rm in}|\{\mathcal{C}_i\}) - S_{t-\tau}(\mathcal{D}_{\rm out}|\{\mathcal{C}_i\}). \tag{2.134}$$

Noting that the ingoing subset $\mathcal{D}_{\rm in}$ belongs to $\boldsymbol{\Phi}^{-\tau}\mathcal{D}$ and not to $\mathcal{D}$, while $\mathcal{D}_{\rm out}$ belongs to $\mathcal{D}$ and not to $\boldsymbol{\Phi}^{-\tau}\mathcal{D}$, the entropy exchange is thus given by

$$\Delta^{\tau}_{\rm e} S = S_{t-\tau}(\boldsymbol{\Phi}^{-\tau}\mathcal{D}|\{\mathcal{C}_i\}) - S_{t-\tau}(\mathcal{D}|\{\mathcal{C}_i\}). \tag{2.135}$$

Taking the difference with the entropy variation (2.132) and using the property (2.131), we obtain the following expression for the entropy production:

$$\Delta_{\rm i}^{\tau} S = \Delta^{\tau} S - \Delta_{\rm e}^{\tau} S = S_t(\mathcal{D}|\{\mathcal{C}_i\}) - S_t(\mathcal{D}|\{\boldsymbol{\Phi}^{\tau}\mathcal{C}_i\}). \tag{2.136}$$

If the system has the property of dynamical mixing in phase space, time-dependent probability distributions P_t tend to become smoother and smoother in the stretching directions, but more and more heterogeneous in the contracting directions. For the same reason, the cells $\{\boldsymbol{\Phi}^{\tau}\mathcal{C}_i\}$ of the time-evolved partition are longer in the stretching directions and thinner in the contracting directions, so that they form a finer partition with respect to the probability distribution P_t than the partition $\{\mathcal{C}_i\}$ itself. Under these circumstances, we should expect that the entropy $S_t(\mathcal{D}|\{\mathcal{C}_i\})$ can be larger than $S_t(\mathcal{D}|\{\boldsymbol{\Phi}^{\tau}\mathcal{C}_i\})$, which may explain that the entropy production turns out to be positive in nonequilibrium situations. This mechanism of entropy production is indeed confirmed for mixing systems manifesting deterministic diffusion (Gaspard, 1997, 1998; Dorfman et al., 2002).

For an isolated system at equilibrium, the domain $\mathcal{D}$ being taken as the whole phase space $\mathcal{M}$, the entropy production (2.136) is vanishing,

$$\Delta_{\rm i}^{\tau} S\big|_{\rm eq} = 0, \tag{2.137}$$

because $\boldsymbol{\Phi}^{-\tau}\mathcal{D} = \boldsymbol{\Phi}^{-\tau}\mathcal{M} = \mathcal{M}$ and $P_{\rm eq}(\boldsymbol{\Phi}^{\tau}\mathcal{C}_i) = P_{\rm eq}(\mathcal{C}_i)$.

For a system in a nonequilibrium stationary macrostate, the probability distribution is stationary, $P_t = P_{t-\tau} = P_{\rm st}$, so that the entropy variation (2.132) is vanishing $\Delta^{\tau} S = 0$, implying that the coarse-grained entropy production is directly related to the entropy exchange according to

$$\Delta_{\rm i}^{\tau} S\big|_{\rm st} = -\Delta_{\rm e}^{\tau} S\big|_{\rm st}. \tag{2.138}$$

The evaluation of this coarse-grained entropy production in specific systems shows that the macroscopic expressions are recovered (Gaspard, 1997, 1998; Dorfman et al., 2002).

2.8 Linear Response Theory

This theory was developed in the 1950s by Callen and Welton (1951), Green (1952b, 1954), Kubo (1957), and others.

2.8.1 Response Function and Dynamical Susceptibility

We consider a Hamiltonian system subjected to some time-dependent external forcing of small amplitude. For instance, the system is composed of N particles with electric charges $\{e_a\}_{a=1}^N$ interacting with an electromagnetic wave. In the dipolar approximation, the total Hamiltonian function of this system is given by

$$H_{\rm tot}(t) = H - \sum_{a=1}^{N} e_a \mathbf{r}_a \cdot \boldsymbol{\mathcal{E}}(t), \tag{2.139}$$

where H is the Hamiltonian function of the autonomous unperturbed system and $\mathcal{E}(t)$ is the time-dependent electric field. The system can also be some dielectric or magnetic material subjected to a time-dependent electric or magnetic field. In all these cases, the total Hamiltonian function takes the form

$$H_{\rm tot}(t) = H - A\,\mathcal{F}(t), \tag{2.140}$$

where $\mathcal{F}(t)$ is the time-dependent external forcing and $A(\boldsymbol{\Gamma})$ is the observable coupling the system to the external forcing.

The response of the forcing may manifest itself on the statistical averages of some observable B,

$$\langle B\rangle_t = \int B(\boldsymbol{\Gamma})\, p_t(\boldsymbol{\Gamma})\, d\boldsymbol{\Gamma}, \tag{2.141}$$

where p_t is the probability density at time t in the phase space. The time evolution of the latter is ruled by Liouville's equation

$$\partial_t p_t = \{H_{\rm tot}(t), p_t\} = \{H, p_t\} - \mathcal{F}(t)\,\{A, p_t\}. \tag{2.142}$$

In order to solve Liouville's equation, we may introduce the *interaction representation* by defining

$$p_{\rm I}(\boldsymbol{\Gamma},t) \equiv p_t\big(\boldsymbol{\Phi}^t\boldsymbol{\Gamma}\big) \qquad \text{and} \qquad A_{\rm I}(\boldsymbol{\Gamma},t) \equiv A\big(\boldsymbol{\Phi}^t\boldsymbol{\Gamma}\big), \tag{2.143}$$

where $\boldsymbol{\Phi}^t$ is the phase-space flow of the unperturbed dynamics corresponding to the Hamiltonian function H without external forcing. In this representation, the Liouville equation (2.142) becomes

$$\partial_t p_{\rm I}(\boldsymbol{\Gamma},t) = -\mathcal{F}(t)\,\{A_{\rm I}(\boldsymbol{\Gamma},t), p_{\rm I}(\boldsymbol{\Gamma},t)\}, \tag{2.144}$$

which can be solved by time integration as

$$p_{\rm I}(\boldsymbol{\Gamma},t) = p_{\rm I}(\boldsymbol{\Gamma},t_0) - \int_{t_0}^{t} dt'\,\mathcal{F}(t')\,\{A_{\rm I}(\boldsymbol{\Gamma},t'), p_{\rm I}(\boldsymbol{\Gamma},t')\}. \tag{2.145}$$

Iterating the integration up to second order in the external forcing, we get

$$\begin{aligned} p_{\rm I}(\boldsymbol{\Gamma},t) = {} & p_{\rm I}(\boldsymbol{\Gamma},t_0) - \int_{t_0}^{t} dt_1\,\mathcal{F}(t_1)\,\{A_{\rm I}(\boldsymbol{\Gamma},t_1), p_{\rm I}(\boldsymbol{\Gamma},t_0)\} \\ & + \int_{t_0}^{t} dt_1\,\mathcal{F}(t_1) \int_{t_0}^{t_1} dt_2\,\mathcal{F}(t_2)\,\{A_{\rm I}(\boldsymbol{\Gamma},t_1),\{A_{\rm I}(\boldsymbol{\Gamma},t_2), p_{\rm I}(\boldsymbol{\Gamma},t_0)\}\} + O\big(\mathcal{F}^3\big), \end{aligned} \tag{2.146}$$

which is expressed in terms of the initial distribution $p_{\rm I}(\boldsymbol{\Gamma},t_0)$.

We suppose that the system was unperturbed before the external forcing started at the initial time t_0. Accordingly, the initial distribution can be considered as the equilibrium stationary distribution, $p_{\rm I}(\boldsymbol{\Gamma},t_0) = p_{\rm eq}(\boldsymbol{\Gamma})$. Substituting this assumption into the expansion (2.146) limited to first order and using equation (2.143), we obtain the mean value (2.141) of the observable B as

$$\langle B\rangle_t = \langle B\rangle_{\rm eq} + \int_{t_0}^{t} \phi_{BA}(t-t')\,\mathcal{F}(t')\,dt' + O\big(\mathcal{F}^2\big) \tag{2.147}$$

in terms of the *response function*

$$\phi_{BA}(t) \equiv \int \{p_{\rm eq}(\mathbf{\Gamma}), A(\mathbf{\Gamma})\}\, B\left(\mathbf{\Phi}^t \mathbf{\Gamma}\right) d\mathbf{\Gamma} = \int p_{\rm eq}(\mathbf{\Gamma})\, \left\{A(\mathbf{\Gamma}), B\left(\mathbf{\Phi}^t \mathbf{\Gamma}\right)\right\} d\mathbf{\Gamma}. \tag{2.148}$$

The final expression in the response function (2.148) is obtained using the definition of the Poisson bracket (2.14) and integration by parts. Accordingly, the linear response of the system to the external forcing is expressed in terms of the equilibrium time-dependent correlation function between the coupling observable A and the observable B measuring the response. The response function has the properties

$$\phi_{BA}(t)^* = \phi_{BA}(t) \qquad \text{and} \qquad \phi_{BA}(-t) = -\phi_{AB}(t) \tag{2.149}$$

because the observables are real and the equilibrium distribution is stationary.

If the system has the property of dynamical mixing, this time-dependent correlation function is expected to vanish in the long-time limit, $\lim_{t\to\infty} \phi_{BA}(t) = 0$. In this case, the initial time can be taken in the remote past $t_0 \to -\infty$. After the change of time variable $t' = t - \tau$, the response at first order can be written as

$$\langle B\rangle_t = \langle B\rangle_{\rm eq} + \int_0^\infty \phi_{BA}(\tau)\, \mathcal{F}(t-\tau)\, d\tau + O\left(\mathcal{F}^2\right), \tag{2.150}$$

showing the progressive loss of memory of the forcing in the response of the system, as illustrated in Figure 2.11.

If the equilibrium distribution is canonical, $p_{\rm eq} = \exp(-\beta H)/\left(h^{3N} Z\right)$, the response function can be expressed as

$$\phi_{BA}(t) = \beta \int p_{\rm eq}(\mathbf{\Gamma})\, \{A(\mathbf{\Gamma}), H(\mathbf{\Gamma})\}\, B\left(\mathbf{\Phi}^t \mathbf{\Gamma}\right) d\mathbf{\Gamma} = \beta \int p_{\rm eq}(\mathbf{\Gamma})\, \dot{A}(\mathbf{\Gamma})\, B\left(\mathbf{\Phi}^t \mathbf{\Gamma}\right) d\mathbf{\Gamma} \tag{2.151}$$

in terms of the time derivative of the coupling observable, $\dot{A} = \{A, H\}$. Introducing the time correlation function

$$C_{BA}(t) \equiv \langle B(t) A(0)\rangle_{\rm eq} - \langle B\rangle_{\rm eq} \langle A\rangle_{\rm eq} \tag{2.152}$$

with respect to the equilibrium statistical average $\langle\cdot\rangle_{\rm eq} \equiv \int p_{\rm eq}(\mathbf{\Gamma})(\cdot)\, d\mathbf{\Gamma}$, the response function (2.151) is thus obtained as

$$\phi_{BA}(t) = \beta\, C_{B\dot{A}}(t) = -\beta\, \dot{C}_{BA}(t). \tag{2.153}$$

For a periodic forcing $\mathcal{F}(t) = {\rm Re}[\mathcal{F}_0 \exp(-\imath\omega t)]$ of frequency ω, the first-order response is given by

$$\langle B\rangle_t = \langle B\rangle_{\rm eq} + {\rm Re}\left[\mathcal{F}_0\, {\rm e}^{-\imath\omega t}\, \chi_{BA}(\omega)\right] + O\left(\mathcal{F}^2\right) \tag{2.154}$$

in terms of the *generalized susceptibility*

$$\chi_{BA}(\omega) \equiv \int_0^\infty \phi_{BA}(\tau)\, {\rm e}^{\imath\omega\tau}\, d\tau, \tag{2.155}$$

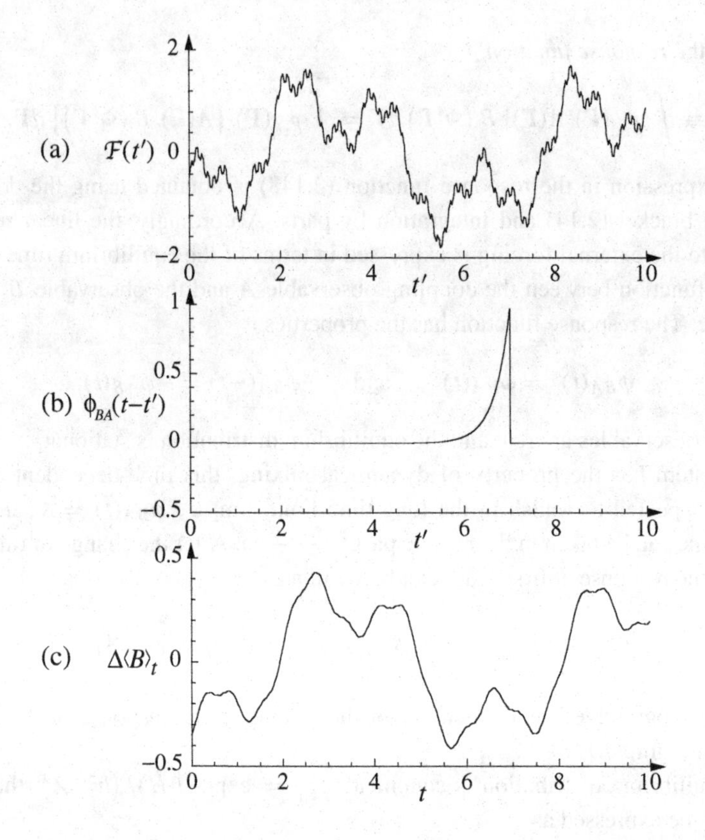

Figure 2.11 Example of linear response (2.147) in the limit $t_0 \to -\infty$: (a) Time-dependent external forcing $\mathcal{F}(t)$; (b) Response function (2.148); (c) Linear response of the observable B given by $\Delta\langle B\rangle_t = \langle B\rangle_t - \langle B\rangle_{\text{eq}}$. The response function $\phi_{BA}(t)$ filters time variations of the forcing that are faster than its damping time t_c, so that the response $\Delta\langle B\rangle_t$ is smoother than the forcing $\mathcal{F}(t)$.

which satisfies $\chi_{BA}(\omega)^* = \chi_{BA}(-\omega)$ since the response function is real (Landau and Lifshitz, 1980a; Balescu, 1975). The generalized susceptibility also satisfies the Kramers-Kronig relations as a consequence of causality in the response of the system to external forcing (Landau and Lifshitz, 1980a; Pottier, 2009).

Expanding $\mathcal{F}(t-\tau)$ in powers of τ, equation (2.150) becomes

$$\langle B\rangle_t = \langle B\rangle_{\text{eq}} + \sum_{n=0}^{\infty} c_{BA}^{(n)} \frac{(-1)^n}{n!} \frac{d^n \mathcal{F}(t)}{dt^n} + O\big(\mathcal{F}^2\big) \tag{2.156}$$

with the coefficients

$$c_{BA}^{(n)} \equiv \int_0^{\infty} \tau^n\, \phi_{BA}(\tau)\, d\tau = (-\imath)^n \frac{d^n \chi_{BA}(0)}{d\omega^n}, \tag{2.157}$$

which may exist if the response function $\phi_{BA}(\tau)$ decays fast enough. Denoting $\omega_{\mathcal{F}}$ to be the characteristic frequency of the time-dependent forcing $\mathcal{F}(t)$, and t_c to be the correlation time (i.e., the characteristic damping time of the response function $\phi_{BA}(\tau)$), equation (2.150) can be approximated as

$$\langle B\rangle_t = \langle B\rangle_{\rm eq} + \mathcal{F}(t)\int_0^\infty \phi_{BA}(\tau)\, d\tau + O(\mathcal{F}\omega_{\mathcal{F}}t_{\rm c}) + O\big(\mathcal{F}^2\big), \tag{2.158}$$

up to corrections that are negligible if $\omega_{\mathcal{F}}\, t_{\rm c} \ll 1$, i.e., if the forcing slowly varies with respect to the characteristic timescale of the response function.

We note that, if the coefficients (2.157) exist for arbitrarily large integers n, the response function could formally be written as

$$\phi_{BA}(\tau) = 2\sum_{n=0}^{\infty} c_{BA}^{(n)}\, \frac{(-1)^n}{n!}\, \frac{d^n\delta(\tau)}{d\tau^n} \tag{2.159}$$

in terms of the Dirac distribution $\delta(\tau)$ and its derivatives. Over timescales that are long with respect to the correlation time ($\tau \gg t_{\rm c}$), the response function can be approximated as $\phi_{BA}(\tau) \simeq 2\, c_{BA}^{(0)}\, \delta(\tau)$, which means that the memory loss of the response function is faster than the observational time under such circumstances.

2.8.2 Electric Conductivity

At the microscale, the electric current density is given by $\hat{\jmath}_{\rm e}(\mathbf{r}) \equiv \sum_{a=1}^{N} e_a \dot{\mathbf{r}}_a \delta^3(\mathbf{r}-\mathbf{r}_a)$. At the macroscale, the *electric conductivity* is defined by the tensor σ_{ij} giving the linear response of the mean electric current density $\langle \hat{\jmath}_{{\rm e}i}\rangle$ to some electric field $\mathcal{E}_j$.

On the one hand, the coupling of the electric charges to the electric field $\mathcal{E}_j$ in the direction $j = x, y, z$ is given according to equation (2.139) by

$$A = \sum_{a=1}^{N} e_a r_{aj}. \tag{2.160}$$

On the other hand, the observable quantity measuring conductivity in the direction $i = x, y, z$ can be defined by averaging the electric current density over the volume V of the system to get

$$B = \frac{1}{V}\int_V \hat{\jmath}_{{\rm e}i}(\mathbf{r})\, d\mathbf{r} = \frac{1}{V}\sum_{a=1}^{N} e_a \dot{r}_{ai}. \tag{2.161}$$

As a consequence of equation (2.151), here the response function takes the form,

$$\varphi_{ij}(t) \equiv \phi_{BA}(t) = \frac{1}{V k_{\rm B} T}\, \langle J_i(t)\, J_j(0)\rangle_{\rm eq}, \tag{2.162}$$

where $\langle\cdot\rangle_{\rm eq}$ denotes the statistical average with respect to the equilibrium canonical distribution of the unperturbed system and

$$J_i \equiv \sum_{a=1}^{N} e_a \dot{r}_{ai} \tag{2.163}$$

is the total electric current in the system, which vanishes at equilibrium because of $\langle J_i \rangle_{\rm eq} = 0$, so that $\langle B \rangle_{\rm eq} = 0$. Now, the response of the mean current density (2.161) to a periodic electric field $\mathcal{E}_j(t) = {\rm Re}[\mathcal{E}_{0j} \exp(-\imath\omega t)]$ is given according to equation (2.154) by

$$\langle \hat{j}_{\rm ei} \rangle = {\rm Re}\left[\sigma_{ij}(\omega)\, \mathcal{E}_{0j}\, {\rm e}^{-\imath\omega t}\right], \qquad \text{where} \qquad \sigma_{ij}(\omega) = \int_0^\infty \varphi_{ij}(t)\, {\rm e}^{\imath\omega t}\, dt \tag{2.164}$$

is the *ac* electric conductivity at frequency ω. The *dc* electric conductivity can thus be expressed at zero frequency $\omega = 0$ by the time integral of the electric current autocorrelation function,

$$\sigma_{ij} = \frac{1}{V k_{\rm B} T} \int_0^\infty \langle J_i(t)\, J_j(0) \rangle_{\rm eq}\, dt, \tag{2.165}$$

in the large-system limit where $N, V \to \infty$ for a given particle density $n = N/V$. Ohm's law $\langle \hat{j}_{\rm ei} \rangle = \sigma_{ij} \mathcal{E}_j$ is thus justified within the framework of linear response theory, providing a microscopic expression for the conductivity tensor $\{\sigma_{ij}\}$.

If the electric charges are identical and dilute enough (as ions of one species in a dilute electrolyte), their velocities $\mathbf{v}_a = \dot{\mathbf{r}}_a$ are statistically uncorrelated and their correlation functions satisfy the conditions, $\langle v_{ai}(t)\, v_{bj}(0) \rangle_{\rm eq} = \delta_{ab} \langle v_i(t)\, v_j(0) \rangle_{\rm eq}$, where $\mathbf{v}$ is the velocity of any representative charge. Under such circumstances, the electric conductivity is given by Einstein's relation,

$$\sigma_{ij} = \frac{e^2 n}{k_{\rm B} T} \mathcal{D}_{ij}, \tag{2.166}$$

where $n = N/V$ is the density of mobile electric charges and

$$\mathcal{D}_{ij} = \int_0^\infty \langle v_i(t)\, v_j(0) \rangle_{\rm eq}\, dt \tag{2.167}$$

is the tensor of their diffusivities in the directions $i, j = x, y, z$. The relation (2.167) is the so-called Green–Kubo formula, giving the diffusion coefficients in the framework of linear response theory.

2.8.3 Onsager–Casimir Reciprocal Relations

If we suppose that the system is also subjected to a uniform external magnetizing field $\mathcal{H}$ that is independent of time, the Hamiltonian function and flow obey the time-reversal symmetries (2.30) and (2.31) as a consequence of microreversibility. The observables may be even or odd under time reversal,

$$A(\boldsymbol{\Theta}\boldsymbol{\Gamma}) = \epsilon_A\, A(\boldsymbol{\Gamma}) \qquad \text{and} \qquad B(\boldsymbol{\Theta}\boldsymbol{\Gamma}) = \epsilon_B\, B(\boldsymbol{\Gamma}), \tag{2.168}$$

depending on whether $\epsilon_A, \epsilon_B = \pm 1$.[17] Consequently, the time reversal of the response function (2.151) is obtained by using the change of variable $\boldsymbol{\Gamma} \to \boldsymbol{\Theta}\boldsymbol{\Gamma}$, leaving invariant the phase space, to obtain

[17] We note that the associated forcing $\mathcal{F}_A$ or $\mathcal{F}_B$ in the total Hamiltonian function (2.140) should have the same parity under time reversal since the Hamiltonian has the symmetry (2.30).

$$\phi_{BA}(t;\mathcal{H}) = \frac{\beta}{h^{3N}Z(\mathcal{H})}\int e^{-\beta H(\Theta\Gamma;\mathcal{H})}\,\dot{A}(\Theta\Gamma)\,B(\Phi^t_{\mathcal{H}}\Theta\Gamma)\,d\Theta\Gamma$$
$$= -\epsilon_A\,\epsilon_B\,\frac{\beta}{h^{3N}Z(-\mathcal{H})}\int e^{-\beta H(\Gamma;-\mathcal{H})}\,\dot{A}(\Gamma)\,B\left(\Phi^{-t}_{-\mathcal{H}}\Gamma\right)d\Gamma, \tag{2.169}$$

by using equations (2.30) and (2.31), the symmetry $Z(\mathcal{H}) = Z(-\mathcal{H})$, and the result $d\Theta\Gamma = d\Gamma$. According to the second relation in equation (2.149), the response function and the generalized susceptibility thus satisfy the symmetry relations

$$\phi_{BA}(t;\mathcal{H}) = \epsilon_A\,\epsilon_B\,\phi_{AB}(t;-\mathcal{H}) \quad \text{and} \quad \chi_{BA}(\omega;\mathcal{H}) = \epsilon_A\,\epsilon_B\,\chi_{AB}(\omega;-\mathcal{H}), \tag{2.170}$$

as a consequence of microreversibility.

These relations have been discovered by Onsager (1931a,b) in the absence of magnetizing field, in which case they are called *Onsager reciprocal relations*. They have been generalized by Casimir (1945) to the presence of magnetizing field, so that they bear the name of *Onsager–Casimir reciprocal relations* in the general case.

Since $\epsilon_i = -1$ for the total electric current (2.163), the *ac* electric conductivity introduced in equation (2.164) obeys the following Onsager–Casimir reciprocal relation:

$$\sigma_{ij}(\omega;\mathcal{H}) = \sigma_{ji}(\omega;-\mathcal{H}). \tag{2.171}$$

In the absence of external magnetizing field ($\mathcal{H} = 0$), the *dc* electric conductivity (2.165) satisfies the Onsager reciprocal relation

$$\sigma_{ij} = \sigma_{ji}. \tag{2.172}$$

We can also check that the diffusivity tensor also has the Onsager reciprocal symmetry $\mathcal{D}_{ij} = \mathcal{D}_{ji}$, as a consequence of microreversibility.

Microreversibility also implies that the linear response coefficients introduced in equation (1.28) obey the Onsager–Casimir reciprocal relations

$$\mathcal{L}_{\alpha,\beta} = \epsilon_\alpha \epsilon_\beta \mathcal{L}_{\beta,\alpha}, \tag{2.173}$$

where $\epsilon_\alpha, \epsilon_\beta = \pm 1$ whether the associated thermodynamic forces $\mathcal{A}_\alpha$ and $\mathcal{A}_\beta$ are even or odd under time reversal.

The experimental verification of the Onsager reciprocal relations has been performed for several irreversible processes including thermoelectricity, electrokinetics, and cross-diffusion (Miller, 1960).

2.8.4 Fluctuation–Dissipation Theorem

If a system is driven by a time-dependent external forcing, it usually heats up. This response of the system can be evaluated by taking the observable B in equation (2.141) as the energy of the system given by the unperturbed Hamiltonian function H. However, equation (2.148) implies that the first response is vanishing in this case, because the Hamiltonian function H is a constant of motion for the flow Φ^t of the unperturbed system, so that $H(\Phi^t\Gamma) = H(\Gamma)$

and $\{H,H\} = 0$. We thus need to go to second order in the perturbation. Using equation (2.146), the definition (2.14) of a Poisson bracket, and integration by parts, we obtain that

$$\langle H\rangle_t = \langle H\rangle_{\rm eq} + \int_{t_0}^{t} dt_1\, \mathcal{F}(t_1)\int_{t_0}^{t_1} dt_2\, \mathcal{F}(t_2) \times \int d\boldsymbol{\Gamma}\, p_{\rm eq}(\boldsymbol{\Gamma})\left\{A\left(\boldsymbol{\Phi}^{t_1}\boldsymbol{\Gamma}\right), \left\{A\left(\boldsymbol{\Phi}^{t_2}\boldsymbol{\Gamma}\right), H(\boldsymbol{\Gamma})\right\}\right\} + O(\mathcal{F}^3). \tag{2.174}$$

By also using $\left\{A\left(\boldsymbol{\Phi}^{t_2}\boldsymbol{\Gamma}\right), H(\boldsymbol{\Gamma})\right\} = dA\left(\boldsymbol{\Phi}^{t_2}\boldsymbol{\Gamma}\right)/dt_2$ and the stationarity of the equilibrium distribution, the energy gain of the system during the time interval $[t_0 = 0, t]$ is thus given by

$$\Delta E_t = \langle H\rangle_t - \langle H\rangle_0 = \int_0^t dt_1\, \mathcal{F}(t_1)\int_0^{t_1} dt_2\, \mathcal{F}(t_2)\,\frac{d}{dt_2}\phi_{AA}(t_2 - t_1) + O\left(\mathcal{F}^3\right). \tag{2.175}$$

Since

$$\operatorname{Im}\chi_{AA}(\omega) = \int_0^\infty d\tau\, \phi_{AA}(\tau)\sin\omega\tau, \tag{2.176}$$

the power dissipated by the external forcing inside the system can be thus expressed as

$$P = \lim_{t\to\infty}\frac{1}{t}\,\Delta E_t = \frac{1}{2}\,\omega\operatorname{Im}\chi_{AA}(\omega)\,\mathcal{F}_0^2 + O\left(\mathcal{F}_0^3\right) \tag{2.177}$$

in terms of the imaginary part of the generalized susceptibility characterizing the first-order response of the observable to itself.

In order to characterize the fluctuations of the quantity $A(t)$, the following spectral density function is introduced,

$$\mathcal{S}_{AA}(\omega) \equiv \lim_{\mathcal{T}\to\infty}\frac{1}{\mathcal{T}}\left\langle\left|\int_0^{\mathcal{T}} e^{\imath\omega t}A(t)\,dt\right|^2\right\rangle_{\rm eq} = \int_{-\infty}^{+\infty} dt\, e^{\imath\omega t}\langle A(0)A(t)\rangle_{\rm eq}, \tag{2.178}$$

which is always nonnegative. Integrating the spectral density function over the whole frequency spectrum gives

$$\langle A^2\rangle_{\rm eq} = \frac{1}{2\pi}\int_{-\infty}^{+\infty}\mathcal{S}_{AA}(\omega)\,d\omega, \tag{2.179}$$

which is the second statistical moment of the observable A.

For the canonical equilibrium distribution of inverse temperature $\beta = (k_{\rm B}T)^{-1}$, equations (2.148) and (2.151) with $B = A$ imply the property

$$\langle\{A(0), A(t)\}\rangle_{\rm eq} = -\frac{1}{k_{\rm B}T}\left\langle A(0)\dot{A}(t)\right\rangle_{\rm eq}. \tag{2.180}$$

Substituting this result into equation (2.178) and integrating by parts, we obtain

$$\operatorname{Im}\chi_{AA}(\omega) = \frac{\omega}{2k_{\rm B}T}\,\mathcal{S}_{AA}(\omega), \tag{2.181}$$

giving the first-order response in terms of the spectral function for the fluctuations of the observable A. As a consequence, the dissipated power (2.177) is related to the equilibrium fluctuations of the observable A according to

$$P = \frac{\omega^2}{4k_{\rm B}T}\, \mathcal{S}_{AA}(\omega)\, \mathcal{F}_0^2 + O\big(\mathcal{F}_0^3\big). \tag{2.182}$$

Therefore, the dissipation of energy is determined by the spectral function characterizing the fluctuations of the observable A coupling the system to the external field. This fundamental result is called the *fluctuation–dissipation theorem* and was first obtained by Callen and Welton (1951). Since the spectral function is nonnegative, $\mathcal{S}_{AA}(\omega) \geq 0$, the second-order contribution to the power dissipated in the system is always nonnegative. The further contributions $O\left(\mathcal{F}_0^3\right)$ are negligible if the amplitude $\mathcal{F}_0$ is supposed to be small enough. For a system at temperature T, the entropy production rate due to the periodic external forcing is thus given by $d_{\rm i}S/dt = P/T$ in terms of the dissipated power (2.182). Therefore, linear response theory is consistent with the second law of thermodynamics as well as microreversibility. In this regard, the fluctuation–dissipation theorem only holds in the linear regime close to equilibrium.

In the case of electric conduction in a homogeneous system of volume V, the dissipated power given by equation (2.177) leads to

$$P = \frac{V}{2} \sum_{i,j} \sigma_{ij}(\omega)\, \mathcal{E}_{0i}\, \mathcal{E}_{0j} + O\big(\mathcal{E}_0^3\big). \tag{2.183}$$

The factor $\frac{1}{2}$ arises because the dissipated power is given by time averaging a quadratic expression of the electric field that is periodic in time, $\boldsymbol{\mathcal{E}}(t) = \boldsymbol{\mathcal{E}}_0 \sin \omega t$. The result (2.183) is the consequence of Ohm's law. Because of the Einstein relation (2.166) between the electric conductivity and the diffusivities if the electric charges are dilute, energy dissipation is thus related to the fluctuations of their random motion, which is the key idea of the fluctuation–dissipation theorem.

2.9 Projection-Operator Methods

Since the 1960s, projection-operator methods have been systematically developed for deducing kinetic and stochastic equations in nonequilibrium statistical mechanics, in particular, following the works by Zwanzig (1961) and Mori (1965).

2.9.1 Zwanzig Projection-Operator Method

In the method pioneered by Zwanzig (1961), the projection operators $\hat{P}$ and $\hat{Q} = 1 - \hat{P}$ acting on the probability density p are introduced. Since they are defined as projectors, they must satisfy the conditions

$$\hat{P}^2 = \hat{P}, \qquad \hat{Q}^2 = \hat{Q}, \qquad \text{and} \qquad \hat{P}\hat{Q} = \hat{Q}\hat{P} = 0. \tag{2.184}$$

These projectors are applied to the Liouville equation (2.39) to get

$$\partial_t \hat{P}p = \hat{P}\hat{L}\hat{P}p + \hat{P}\hat{L}\hat{Q}p, \tag{2.185}$$

$$\partial_t \hat{Q}p = \hat{Q}\hat{L}\hat{P}p + \hat{Q}\hat{L}\hat{Q}p, \tag{2.186}$$

which form a set of two coupled linear differential equations for the time evolution of $\hat{P}p$ and $\hat{Q}p$. Integrating equation (2.186), we obtain

$$\hat{Q}p_t = e^{\hat{Q}\hat{L}t}\,\hat{Q}p_0 + \int_0^t d\tau\, e^{\hat{Q}\hat{L}\tau}\,\hat{Q}\hat{L}\hat{P}p_{t-\tau}, \tag{2.187}$$

which can be substituted into equation (2.185) to find

$$\partial_t \hat{P}p_t = \hat{P}\hat{L}\hat{P}p_t + \hat{P}\hat{L}e^{\hat{Q}\hat{L}t}\,\hat{Q}p_0 + \int_0^t d\tau\, \hat{P}\hat{L}\,e^{\hat{Q}\hat{L}\tau}\,\hat{Q}\hat{L}\hat{P}p_{t-\tau}. \tag{2.188}$$

This evolution equation for $\hat{P}p_t$ may be supposed to be closed if the evolution operator $\exp(\hat{Q}\hat{L}t)$ is known, although it involves the initial condition p_0 and the knowledge of the solution over the time interval $0 < \tau < t$. In this regard, this equation describes a process with the memory of the initial condition $\hat{Q}p_0$ as a consequence of the projection onto the part $\hat{P}p_t$ of the probability distribution. Memory may be lost by the evolution operator $\exp(\hat{Q}\hat{L}t)$ after some characteristic time, in which case the second term of equation (2.188) is vanishingly small. The third term describes damping or relaxation of the part $\hat{P}p_t$ due to the time evolution $\exp(\hat{Q}\hat{L}t)$ of the other part. Nevertheless, equation (2.188) is exact.

This method is particularly useful for obtaining kinetic equations ruling the time evolution of systems with slow and fast degrees of freedom, such as Brownian motion, as will be discussed in Chapter 4. The method may also be applied to other systems (Zwanzig, 1961). The key result is that the processes deduced by such methods from the underlying microscopic dynamics have, in general, the memory of the past. Such memory typically decays over some correlation or relaxation times, which are characteristic of the dynamics.

2.9.2 Mori Projection-Operator Method

The method proposed by Mori (1965) uses projection operators acting on the observables instead of the probability density, the observables being defined as functions in the phase space of the system. The purpose of Mori's projection-operator method is to deduce stochastic equations, called generalized Langevin equations, for the time evolution of some observables $\mathbf{A} = (A_1, A_2, \ldots, A_n)$ that are supposed to be slower than all the others. This is in particular the case for the hydrodynamic modes associated with conserved quantities such as energy or linear momentum, which are slowly evolving on macroscopic timescales, as compared to the much faster microscopic degrees of freedom. This method is thus based on the separation of timescales that often exists in large systems. With this aim, Mori (1965) defines the projection operator

$$\hat{\mathcal{P}}B \equiv \langle B\,\mathbf{A}\rangle \cdot \langle \mathbf{A}\,\mathbf{A}\rangle^{-1} \cdot \mathbf{A} \tag{2.189}$$

acting on generic observables $B(\mathbf{\Gamma})$. In expression (2.189), the notation $\langle\cdot\rangle$ stands for the statistical average $\langle X\rangle = \int p_{\rm eq}(\mathbf{\Gamma})X(\mathbf{\Gamma})\, d\mathbf{\Gamma}$ over the equilibrium probability distribution.[18] The quantity $\langle\mathbf{A}\,\mathbf{A}\rangle$ is the correlation matrix with the elements $\langle A_i A_j\rangle$ between the observables $\mathbf{A}$ that are considered. The operator (2.189) and its complement, $\hat{\mathcal{Q}} = 1 - \hat{\mathcal{P}}$, are projection operators because they satisfy conditions similar to those in equation (2.184). Since $\hat{\mathcal{P}}\mathbf{A} = \mathbf{A}$, Mori's projector maps the subspace of the observables $\mathbf{A}$ onto itself.

The deduction of the generalized Langevin equations starts from the identity

$$\mathrm{e}^{-\hat{L}t} = \mathrm{e}^{-\hat{\mathcal{Q}}\hat{L}t} - \int_0^t dt'\, \mathrm{e}^{-\hat{L}(t-t')}\,\hat{\mathcal{P}}\hat{L}\,\mathrm{e}^{-\hat{\mathcal{Q}}\hat{L}t'}, \tag{2.190}$$

which can be checked by differentiation with respect to time t. Following Zwanzig (2001), the identity (2.190) can be multiplied on its right-hand side by $\hat{\mathcal{Q}}\hat{L}\mathbf{A}$ to get

$$\mathrm{e}^{-\hat{L}t}\hat{\mathcal{Q}}\hat{L}\mathbf{A} = -\mathbf{F}(t) + \int_0^t dt'\,\mathbf{K}(t')\cdot\mathbf{A}(t-t') \tag{2.191}$$

with the noises

$$\mathbf{F}(t) \equiv -\mathrm{e}^{-\hat{\mathcal{Q}}\hat{L}t}\hat{\mathcal{Q}}\hat{L}\mathbf{A} = \mathrm{e}^{-\hat{\mathcal{Q}}\hat{L}t}\hat{\mathcal{Q}}\dot{\mathbf{A}} \tag{2.192}$$

and the memory kernel

$$\mathbf{K}(t) \equiv \big\langle \hat{L}\,\mathbf{F}(t)\,\mathbf{A}\big\rangle\cdot\langle\mathbf{A}\,\mathbf{A}\rangle^{-1}. \tag{2.193}$$

The noises (2.192) satisfy the identity $\langle\mathbf{F}(t)\mathbf{A}\rangle = 0$ because they belong to the subspace orthogonal to the slow variables $\mathbf{A}$. Furthermore, they obey the generalized fluctuation–dissipation theorem

$$\langle\mathbf{F}(t)\,\mathbf{F}(t')\rangle = \mathbf{K}(t-t')\cdot\langle\mathbf{A}\,\mathbf{A}\rangle, \tag{2.194}$$

which can be proved as follows by using equation (2.192), the stationarity of the statistical average $\langle\cdot\rangle$, the anti-Hermiticity of $\hat{L}$, and equation (2.193):

$$\begin{aligned}\langle\mathbf{F}(t)\,\mathbf{F}(t')\rangle &= \Big\langle \mathrm{e}^{-\hat{\mathcal{Q}}\hat{L}t}\hat{\mathcal{Q}}\hat{L}\mathbf{A}\,\mathrm{e}^{-\hat{\mathcal{Q}}\hat{L}t'}\hat{\mathcal{Q}}\hat{L}\mathbf{A}\Big\rangle = \Big\langle \mathrm{e}^{-\hat{\mathcal{Q}}\hat{L}(t-t')}\hat{\mathcal{Q}}\hat{L}\mathbf{A}\,\hat{L}\mathbf{A}\Big\rangle \\ &= -\big\langle\mathbf{F}(t-t')\,\hat{L}\mathbf{A}\big\rangle = \mathbf{K}(t-t')\cdot\langle\mathbf{A}\,\mathbf{A}\rangle.\end{aligned} \tag{2.195}$$

Now, using the relation $\hat{\mathcal{Q}} = 1-\hat{\mathcal{P}}$, the left-hand side of equation (2.191) can be transformed according to

$$\mathrm{e}^{-\hat{L}t}\hat{\mathcal{Q}}\hat{L}\mathbf{A} = \mathrm{e}^{-\hat{L}t}\hat{L}\mathbf{A} - \mathrm{e}^{-\hat{L}t}\hat{\mathcal{P}}\hat{L}\mathbf{A} = -\frac{d}{dt}\mathbf{A}(t) - \big\langle\hat{L}\mathbf{A}\,\mathbf{A}\big\rangle\cdot\langle\mathbf{A}\,\mathbf{A}\rangle^{-1}\cdot\mathbf{A}(t), \tag{2.196}$$

because $\mathbf{A}(t) = \mathrm{e}^{-\hat{L}t}\mathbf{A}$. Since equations (2.191) and (2.196) are equal, we obtain the coupled generalized Langevin equations

$$\frac{d}{dt}\mathbf{A}(t) = -\mathsf{L}\cdot\mathbf{A}(t) - \int_0^t dt'\,\mathbf{K}(t')\cdot\mathbf{A}(t-t') + \mathbf{F}(t) \tag{2.197}$$

[18] The method can be generalized using statistical averages over time-dependent nonequilibrium probability distributions (Oppenheim and Levine, 1979; Robertson et al., 2020).

with the matrix

$$\mathsf{L} \equiv \langle \hat{L}\mathbf{A}\,\mathbf{A}\rangle \cdot \langle \mathbf{A}\,\mathbf{A}\rangle^{-1}. \tag{2.198}$$

The first term describes the autonomous dynamics of the observables **A**, the second term their damping or relaxation due to their coupling to the other observables, and the third term the noises (2.192), which are fluctuating in time with the correlations (2.194). Because of the presence of the noises $\mathbf{F}(t)$, the equations (2.197) are stochastic and they describe mesoscopic phenomena such as Brownian motion. Moreover, the so-defined stochastic process has a memory of the past due to the term with the memory kernel (2.193). This latter may decay over some correlation timescale, after which the process loses the memory of its history.

Since projection operators are linear, Mori's method leads to linear stochastic equations for the observables **A**. If these latter were nonlinear functions $\mathbf{A}(\mathbf{x})$ of other observables $\mathbf{x}$ or if the projection operators were depending on other observables **B** participating to the dynamics, nonlinear stochastic equations would be obtained for the time evolution of the system. An important issue is thus the selection of observables that are physically relevant for the system of interest. Mori's method has been considered for several problems of nonequilibrium statistical mechanics (Kapral, 1972; Desai and Kapral, 1972; Forster, 1975; Oppenheim and Levine, 1979; Szamel, 1997; Zwanzig, 2001; Mazo, 2002; Robertson et al., 2020).

The projection-operator methods show that kinetic and stochastic processes may be deduced from the underlying microscopic dynamics. The memory kernels and the related time-dependent correlation functions allow us to identify the different timescales in the system of interest. If there is a large enough separation between these timescales, approximations with memoryless stochastic processes, called Markovian processes, can possibly be justified, as further discussed in Chapter 4.

3
Hydrodynamics

3.1 Nonequilibrium Statistical Mechanics and Hydrodynamics

The spatiotemporal scales of such phenomena as laminar or turbulent flows, sound waves, or heat conduction that we observe around us in fluids or solids are much larger than the size and dynamics of their molecular constituents. At the macroscale in fluids or the other phases of matter, the observable quantities are thus the slowest modes. From the viewpoint of microscopic mechanics, the slowest modes are expected to be related to the locally conserved quantities, which are energy, momenta, and particle numbers. Therefore, the microscopic approach to hydrodynamics considers the densities of locally conserved quantities and aims at deducing their balance equations from Newton's equations for the constituent particles. With this purpose, the probability distribution ruled by Liouville's equation can be projected onto the subspace of Fourier modes with the largest wavelengths, corresponding to the slowest modes in the mechanical system of particles. In order to deduce a closed set of balance equations, the probability distribution can be assumed to have the form of a local equilibrium distribution depending on macrofields such as local temperature, chemical potentials, and velocity, which are conjugate to the densities. This approach has been systematically developed since the late 1950s for fluids (i.e., gases and liquids), liquid crystals, and crystals, as well as magnetic systems and quantum fluids (Mori, 1958; McLennan, 1960, 1961, 1963; Kadanoff and Martin, 1963; Zubarev, 1966; Robertson, 1966, 1967; Piccirelli, 1968; Desai and Kapral, 1972; Forster, 1975; Oppenheim and Levine, 1979; Akhiezer and Peletminskii, 1981; Brey et al., 1981; Kavassalis and Oppenheim, 1988; Spohn, 1991; Reichl, 1998; Mazenko, 2006; Sasa, 2014; Mabillard and Gaspard, 2020, 2021).

In normal fluids and the other phases of matter, the slowest modes can be decomposed into Fourier modes $\exp(\imath\mathbf{q}\cdot\mathbf{r})$ with a wave number $q = \|\mathbf{q}\|$ related to their wavelength λ according to $q = 2\pi/\lambda$. Their wavelength is typically much larger than the size of their molecular constituents, which often have the same order of magnitude as the range d of intermolecular forces. An important characteristic length scale is given by the mean free path $\ell \simeq (\sqrt{2}\,\sigma_{\rm tot} n)^{-1}$, where $\sigma_{\rm tot}$ is the total cross section of binary collisions and n is the particle density. In gases, the mean free path is larger than the range of intermolecular forces, $\ell \gg d$. In liquids, they are of the same order of magnitude, $\ell \sim d$. The wavelength of the slowest modes is supposed to be larger than the mean free path, so that

$$\lambda \gg \ell \gtrsim d \qquad \text{or, equivalently,} \qquad q = \frac{2\pi}{\lambda} \ll \frac{2\pi}{\ell} \lesssim \frac{2\pi}{d}. \tag{3.1}$$

As discussed for the framework of the macroscopic description in Chapter 1, the balance equations of the locally conserved quantities rule the process of relaxation towards the global thermodynamic equilibrium, which is characterized by uniform temperature, chemical potentials, and velocity. In this regime close to equilibrium, the balance equations can be linearized and their solutions may thus be decomposed into Fourier modes behaving as $\exp(zt + \imath \mathbf{q} \cdot \mathbf{r})$. The associated eigenvalues give the so-called *dispersion relations* $z = z_\alpha(\mathbf{q})$, linking their temporal dependence to their spatial periodicity. In this way, the different modes are characterized by specific dispersion relations of the general form

$$z = -\imath\, v_\alpha\, q - D_\alpha\, q^2 + o(q^2), \tag{3.2}$$

where v_α is the propagation velocity of the mode, D_α its diffusivity, and $o\left(q^2\right)$ denotes terms vanishing faster than the second power of the wave number q in the limit $q \to 0$. The propagation velocity is a conservative property since wave propagation typically conserves energy and thus preserves entropy. However, the diffusivity is responsible for the damping of the mode, the dissipation of energy, and thus entropy production. As shown in Section 1.6, the sound velocity is determined by the equation of state for pressure, which is a property deduced from equilibrium statistical mechanics, as explained in Section 2.5.1. In contrast, the diffusivities are associated with transport properties such as viscosity and heat conductivity, and their microscopic deduction requires using nonequilibrium statistical mechanics, since time evolution plays a key role in transport, energy dissipation, and entropy production. The main goal of this chapter will thus be the microscopic derivation of these transport properties within nonequilibrium statistical mechanics. The transport coefficients will be shown to be given by Green–Kubo formulas, as established for any linear response property in Section 2.8.

In normal fluids with one component there exist five slow modes called hydrodynamic modes, which are associated with the five locally conserved quantities that are mass, energy, and the three components of linear momentum. If there are c components but no reaction, these components may undergo mutual diffusion still conserving the particles of each species, so that the number of hydrodynamic modes is equal to $c + 4$.

In phases with continuous symmetry breaking, such as crystals where the translational symmetry prevailing in normal fluids is broken into a spatially periodic crystalline structure, long-ranged order emerges and the so-called Nambu–Goldstone modes having dispersion relations vanishing with the wave number combine with the hydrodynamic modes (Nambu, 1960; Goldstone, 1961). If there are b broken continuous symmetries in a system with c components, the number of slow modes having dispersion relations such that $\lim_{\mathbf{q}\to \mathbf{0}} z(\mathbf{q}) = 0$ is therefore equal to $b + c + 4$, according to the theorem by Goldstone (1961). The slow modes of the dynamics are often called collective modes because they are associated with collective movements for the constituent particles, such as sound waves or fluid flows forming vortices.

The techniques of neutron scattering (Van Hove, 1954) and light scattering (Berne and Pecora, 1976; Boon and Yip, 1980) provide the experimental observations of the dispersion relations for the collective modes down to spatial scales as short as the distance between the atoms in crystals (Svensson et al., 1967), liquids (Sköld et al., 1972; de Schepper and Cohen, 1980), magnetic systems (Lovesey, 1980), and superfluids (Leggett, 2006).

3.2 Multicomponent Normal Fluids

3.2.1 Microscopic Densities and Balance Equations

We assume that the system is composed of N particles of masses $\{m_a\}$, positions $\{\mathbf{r}_a\}$, and momenta $\{\mathbf{p}_a = m_a\dot{\mathbf{r}}_a\}$ with $1 \leq a \leq N$. The Hamiltonian of this system is thus given by equation (2.2) with short- or medium-ranged interactions. The particles may be supposed to evolve in a finite volume where $u_a^{(\text{ext})} = 0$. This is the case, for instance, if the volume is a parallelepipedon with periodic boundary conditions. Newton's equations for the particles are expressed by equation (2.10) in terms of the forces (2.11) and (2.12). The particles of species k have the labels $a \in \mathcal{S}_k$ and $N_k = |\mathcal{S}_k|$ is the number of these particles. The total number of all the particles is thus given by $N = \sum_{k=1}^{c} N_k$. Such systems can be simulated by molecular dynamics using the numerical integration of Newton's equations (Frenkel and Smit, 2002; Allen and Tildesley, 2017).

We introduce the microscopic densities:

$$\hat{n}_k(\mathbf{r};\boldsymbol{\Gamma}) = \sum_{a\in\mathcal{S}_k} \delta(\mathbf{r}-\mathbf{r}_a), \tag{3.3}$$

$$\hat{\rho}(\mathbf{r};\boldsymbol{\Gamma}) = \sum_{a=1}^{N} m_a\, \delta(\mathbf{r}-\mathbf{r}_a) = \sum_{k=1}^{c} m_k\, \hat{n}_k(\mathbf{r};\boldsymbol{\Gamma}), \tag{3.4}$$

$$\hat{\mathbf{g}}(\mathbf{r};\boldsymbol{\Gamma}) = \sum_{a=1}^{N} \mathbf{p}_a\, \delta(\mathbf{r}-\mathbf{r}_a), \tag{3.5}$$

$$\hat{\epsilon}(\mathbf{r};\boldsymbol{\Gamma}) = \sum_{a=1}^{N} \varepsilon_a(\boldsymbol{\Gamma})\, \delta(\mathbf{r}-\mathbf{r}_a), \tag{3.6}$$

with the energy of the ath particle given by

$$\varepsilon_a(\boldsymbol{\Gamma}) \equiv \frac{\mathbf{p}_a^2}{2m_a} + \frac{1}{2}\sum_{b(\neq a)} u_{ab} + \cdots, \tag{3.7}$$

where $u_{ab} = u_{ab}^{(2)}(r_{ab})$ is the potential energy of binary interaction between the particles a and b separated by the distance $r_{ab} = \|\mathbf{r}_a - \mathbf{r}_b\|$.

The first is the density of the particles of species k (where the sum extends over all the labels $a \in \mathcal{S}_k$ corresponding to the species k), the second the mass density, the third the linear momentum density, and the fourth the energy density.

Integrating these densities over space with the volume element $d\mathbf{r} \equiv d^3r$, we recover the total mass $M = \int_V \hat{\rho}\, d\mathbf{r} = \sum_{a=1}^{N} m_a = \sum_{k=1}^{c} m_k N_k$, the total linear momentum

$\mathbf{P} = \int_V \hat{\mathbf{g}}\, d\mathbf{r} = \sum_{a=1}^N \mathbf{p}_a$, and the total energy $E = \int_V \hat{\epsilon}\, d\mathbf{r} = \sum_{a=1}^N \varepsilon_a$, this latter being equal to the Hamiltonian function $H = E$ of the system.

According to the Newton equations (2.10) and the action-reaction principle $\mathbf{F}_{ab} = -\mathbf{F}_{ba}$, the densities obey the balance equations:

$$\partial_t \hat{n}_k + \nabla \cdot \hat{\jmath}_k = 0, \tag{3.8}$$

$$\partial_t \hat{\rho} + \nabla \cdot \hat{\mathbf{g}} = 0, \tag{3.9}$$

$$\partial_t \hat{\mathbf{g}} + \nabla \cdot \hat{\jmath}_{\mathbf{g}} = 0, \tag{3.10}$$

$$\partial_t \hat{\epsilon} + \nabla \cdot \hat{\jmath}_{\epsilon} = 0, \tag{3.11}$$

in terms of the current density of particle of species k

$$\hat{\jmath}_k(\mathbf{r};\boldsymbol{\Gamma}) = \sum_{a\in\mathcal{S}_k} \dot{\mathbf{r}}_a\, \delta(\mathbf{r} - \mathbf{r}_a), \tag{3.12}$$

the mass current density $\hat{\mathbf{g}} = \hat{\jmath}_\rho = \sum_{k=1}^c m_k\, \hat{\jmath}_k$, which is identical to the momentum density (3.5), the momentum current density tensor

$$\hat{\jmath}_{\mathbf{g}}(\mathbf{r};\boldsymbol{\Gamma}) = \sum_a \dot{\mathbf{r}}_a\, \mathbf{p}_a\, \delta(\mathbf{r} - \mathbf{r}_a) + \frac{1}{2} \sum_{a\neq b} \mathbf{r}_{ab}\, \mathbf{F}_{ab} \int_0^1 \delta[\mathbf{r} - \boldsymbol{R}_{ab}(\lambda)]\, d\lambda, \tag{3.13}$$

and the energy current density vector

$$\hat{\jmath}_\epsilon(\mathbf{r};\boldsymbol{\Gamma}) = \sum_a \dot{\mathbf{r}}_a\, \varepsilon_a\, \delta(\mathbf{r} - \mathbf{r}_a) + \frac{1}{4} \sum_{a\neq b} \mathbf{r}_{ab}\, (\dot{\mathbf{r}}_a + \dot{\mathbf{r}}_b) \cdot \mathbf{F}_{ab} \int_0^1 \delta[\mathbf{r} - \boldsymbol{R}_{ab}(\lambda)]\, d\lambda, \tag{3.14}$$

with $\mathbf{r}_{ab} \equiv \mathbf{r}_a - \mathbf{r}_b$ and $\boldsymbol{R}_{ab}(\lambda) = \lambda\, \mathbf{r}_a + (1-\lambda)\, \mathbf{r}_b$. Indeed, the divergence of a vector field can be formed using the identity

$$\delta(\mathbf{r} - \mathbf{r}_a) - \delta(\mathbf{r} - \mathbf{r}_b) = -\nabla \cdot \int_0^1 \frac{d\boldsymbol{R}_{ab}(\lambda)}{d\lambda}\, \delta[\mathbf{r} - \boldsymbol{R}_{ab}(\lambda)]\, d\lambda, \tag{3.15}$$

which holds for any smooth curve $\boldsymbol{R}_{ab}(\lambda)$ joining $\boldsymbol{R}_{ab}(0) = \mathbf{r}_b$ to $\boldsymbol{R}_{ab}(1) = \mathbf{r}_a$ and, in particular, the aforementioned linear one (Robertson, 1967; Piccirelli, 1968; Spohn, 1991; Sasa, 2014).

Since the balance equations (3.8)–(3.11) have a right-hand side equal to zero, they express the local conservation laws of mass, linear momentum, energy, and also particles because there is no reaction in this system. We note that further densities may also be considered such as the angular momentum density, which also obeys a local conservation law.

3.2.2 *Time Evolution*

Equations (3.8)–(3.11) for the locally conserved quantities can be written as

$$\partial_t \hat{\mathbf{c}}(\mathbf{r},t) + \nabla \cdot \hat{\jmath}(\mathbf{r},t) = 0, \tag{3.16}$$

where

$$\hat{\mathbf{c}} = (\hat{\epsilon}, \hat{n}_k, \hat{\mathbf{g}}) \quad \text{and} \quad \hat{\jmath} = (\hat{\jmath}_\epsilon, \hat{\jmath}_k, \hat{\jmath}_{\mathbf{g}}) \tag{3.17}$$

are, respectively, the densities and the corresponding current densities or flows. At time t, the densities are expressed in terms of the Liouvillian operator $\hat{L}$ or the phase-space trajectories $\mathbf{\Gamma}_t = \mathbf{\Phi}^t \mathbf{\Gamma}$ as

$$\hat{\mathbf{c}}(\mathbf{r}, t) \equiv \mathrm{e}^{-\hat{L}t}\, \hat{\mathbf{c}}(\mathbf{r}; \mathbf{\Gamma}) = \hat{\mathbf{c}}(\mathbf{r}; \mathbf{\Gamma}_t) \tag{3.18}$$

with similar expressions for the current densities.

Furthermore, any phase-space probability density $p_t(\mathbf{\Gamma})$ at time t is given by

$$p_t(\mathbf{\Gamma}) = \mathrm{e}^{\hat{L}t}\, p_0(\mathbf{\Gamma}) = p_0(\mathbf{\Gamma}_{-t}) \tag{3.19}$$

in terms of the initial probability density $p_0(\mathbf{\Gamma})$ and the reversed trajectory going back from the current phase-space point $\mathbf{\Gamma}$ to the initial point $\mathbf{\Gamma}_{-t} = \mathbf{\Phi}^{-t}\mathbf{\Gamma}$ along the trajectory according to equations (2.44) and (2.45). Consequently, the macroscopic densities can be obtained by taking the mean value of the time-independent densities over the time-evolved probability distribution $p_t(\mathbf{\Gamma})$ or, equivalently, the mean values of the time-dependent densities over the initial probability distribution $p_0(\mathbf{\Gamma})$, so that

$$\langle \hat{\mathbf{c}}(\mathbf{r}; \mathbf{\Gamma}) \rangle_t \equiv \int d\mathbf{\Gamma}\, p_t(\mathbf{\Gamma})\, \hat{\mathbf{c}}(\mathbf{r}; \mathbf{\Gamma}) = \int d\mathbf{\Gamma}_0\, p_0(\mathbf{\Gamma}_0)\, \hat{\mathbf{c}}(\mathbf{r}; \mathbf{\Gamma}_t), \tag{3.20}$$

because of Liouville's theorem $d\mathbf{\Gamma}_0 = d\mathbf{\Gamma}_t$, the mean value with respect to $p_t(\mathbf{\Gamma})$ being denoted as $\langle \cdot \rangle_t$. Similar results hold for the current densities and other fields.

3.2.3 Local Equilibrium Distribution

The key assumption of the formalism is the initial condition being the local equilibrium distribution

$$p_{\mathrm{leq}}(\mathbf{\Gamma}; \chi) = \frac{1}{\Delta\mathbf{\Gamma}} \exp\left[-\Omega(\chi) - \chi * \hat{\mathbf{c}}(\mathbf{\Gamma})\right], \tag{3.21}$$

where χ are inhomogeneous fields conjugate to the density fields $\hat{\mathbf{c}}$, and the asterisk $*$ corresponds to the following integration over space:

$$f * g \equiv \int f(\mathbf{r})\, g(\mathbf{r})\, d\mathbf{r}. \tag{3.22}$$

Accordingly, knowledge of the particle system is reduced to the information contained in the conjugate fields $\chi(\mathbf{r})$. This reduction of information on the distribution of particles constitutes a form of coarse graining, which keeps the large spatial scales in the description of the fluid. Since the fields $\chi(\mathbf{r})$ are conjugate to the densities obeying the local conservation laws (3.8)–(3.11), all these fields are expected to evolve on the longest timescales of the system. This coarse graining does not describe the same statistical correlations as the particle distribution functions of Section 2.3.5, in particular, because the conjugate fields only depend on position.

The normalization condition for the local equilibrium distribution (3.21) gives the functional

$$\Omega(\chi) = \ln \int \frac{d\mathbf{\Gamma}}{\Delta\mathbf{\Gamma}} \exp\left[-\chi * \hat{\mathbf{c}}(\mathbf{\Gamma})\right]. \tag{3.23}$$

The mean value with respect to the local equilibrium distribution (3.21) is denoted by $\langle\cdot\rangle_{\text{leq},\boldsymbol{\chi}}$. In this formalism, the mean values of the densities can be obtained by taking the functional derivative of (3.23) with respect to the conjugate fields as follows:

$$\mathbf{c}(\mathbf{r}) = -\frac{\delta\Omega(\boldsymbol{\chi})}{\delta\boldsymbol{\chi}(\mathbf{r})}, \qquad \text{where} \qquad \mathbf{c}(\mathbf{r}) \equiv \langle\hat{\mathbf{c}}(\mathbf{r};\boldsymbol{\Gamma})\rangle_{\text{leq},\boldsymbol{\chi}}. \tag{3.24}$$

In this regard, the mean densities $\mathbf{c}(\mathbf{r})$ and the conjugate fields $\boldsymbol{\chi}(\mathbf{r})$ both contribute to the description of the fluid in terms of the local equilibrium distribution (3.21).[1]

In addition, the entropy of the local equilibrium distribution (3.21) is defined as[2]

$$S \equiv -k_{\text{B}} \int p_{\text{leq}}(\boldsymbol{\Gamma}) \ln \left[p_{\text{leq}}(\boldsymbol{\Gamma})\, \Delta\boldsymbol{\Gamma}\right] d\boldsymbol{\Gamma}, \tag{3.25}$$

leading to the following entropy functional of the mean densities $\mathbf{c}$:

$$S(\mathbf{c}) = k_{\text{B}} \inf_{\boldsymbol{\chi}} \left[\Omega(\boldsymbol{\chi}) + \boldsymbol{\chi} * \mathbf{c}\right], \tag{3.26}$$

which is the Legendre transform of the previously introduced functional (3.23). The conjugate fields are thus given by the functional derivatives,

$$\boldsymbol{\chi}(\mathbf{r}) = \frac{1}{k_{\text{B}}} \frac{\delta S(\mathbf{c})}{\delta\mathbf{c}(\mathbf{r})}. \tag{3.27}$$

Similarly, the Legendre transform of the entropy functional (3.26) gives back the functional (3.23) according to

$$\Omega(\boldsymbol{\chi}) = \sup_{\mathbf{c}} \left[k_{\text{B}}^{-1} S(\mathbf{c}) - \boldsymbol{\chi} * \mathbf{c}\right]. \tag{3.28}$$

The second functional derivatives define the matrix of correlation functions

$$\mathsf{C}(\mathbf{r},\mathbf{r}') \equiv \frac{\delta^2\Omega(\boldsymbol{\chi})}{\delta\boldsymbol{\chi}(\mathbf{r})\delta\boldsymbol{\chi}(\mathbf{r}')} = \langle\delta\hat{\mathbf{c}}(\mathbf{r})\,\delta\hat{\mathbf{c}}(\mathbf{r}')\rangle_{\text{leq},\boldsymbol{\chi}} \qquad \text{with} \qquad \delta\hat{\mathbf{c}} \equiv \hat{\mathbf{c}} - \langle\hat{\mathbf{c}}\rangle_{\text{leq},\boldsymbol{\chi}}, \tag{3.29}$$

and its inverse

$$\mathsf{C}^{-1}(\mathbf{r},\mathbf{r}') \equiv -\frac{1}{k_{\text{B}}} \frac{\delta^2 S(\mathbf{c})}{\delta\mathbf{c}(\mathbf{r})\delta\mathbf{c}(\mathbf{r}')}, \tag{3.30}$$

such that

$$\int \mathsf{C}(\mathbf{r},\mathbf{r}') \cdot \mathsf{C}^{-1}(\mathbf{r}',\mathbf{r}'')\, d\mathbf{r}' = \delta(\mathbf{r}-\mathbf{r}'')\,\mathbf{1}. \tag{3.31}$$

The correlation functions connect infinitesimal perturbations of the mean densities and conjugate fields according to

$$\delta\mathbf{c}(\mathbf{r}) = -\int \mathsf{C}(\mathbf{r},\mathbf{r}') \cdot \delta\boldsymbol{\chi}(\mathbf{r}')\, d\mathbf{r}' \qquad \text{and} \qquad \delta\boldsymbol{\chi}(\mathbf{r}) = -\int \mathsf{C}^{-1}(\mathbf{r},\mathbf{r}') \cdot \delta\mathbf{c}(\mathbf{r}')\, d\mathbf{r}'. \tag{3.32}$$

[1] We note that further density fields may be included in the local equilibrium distribution (3.21), e.g., the density of angular momentum, which may be relevant in fluids with particles carrying an intrinsic angular momentum (spin), or local order parameters in phases with broken continuous symmetries, as explained in Section 3.3.

[2] Here, we use the expression (2.114) for entropy with the probability density $p(\boldsymbol{\Gamma})$ replaced by the local equilibrium one (3.21). Since this latter is coarse graining the general probability density $p(\boldsymbol{\Gamma})$, the entropy (3.25) is expected to change with time if the time dependence of the local equilibrium distribution is given by the conjugate fields $\boldsymbol{\chi}_t(\mathbf{r})$. In contrast, the entropy (3.25) would remain constant in time if the exact probability distribution (3.19) was used, as explained in Section 2.7.5.

At equilibrium, these correlation functions can be obtained as explained in Section C.2.[3]

Since the particle numbers $\mathbf{N} \equiv (N_k)_{k=1}^{c} \in \mathbb{N}^c$ and $N = \sum_{k=1}^{c} N_k$ are also random variables for the local equilibrium distribution (3.21), we note that $\boldsymbol{\Gamma}$ stands for $(\boldsymbol{\Gamma}_N, \mathbf{N})$ with $\boldsymbol{\Gamma}_N \in \mathcal{M}_{\mathbf{N}} = \bigotimes_{k=1}^{c} \mathcal{M}_{N_k}$, where $\mathcal{M}_{N_k}$ is the phase space (2.4) for the N_k identical particles of species k. Accordingly, the normalization of such distributions is based on the following combination of integrals over phase spaces and sums over particle numbers:

$$\int (\cdot)\, d\boldsymbol{\Gamma} \equiv \prod_{k=1}^{c} \sum_{N_k=0}^{\infty} \frac{1}{N_k!} \int_{\mathbb{R}^{6N_k}} (\cdot)\, d^{6N_k}\Gamma. \tag{3.33}$$

Furthermore, we notice that the grand canonical equilibrium distribution

$$p_{\text{eq}}(\boldsymbol{\Gamma}) = \frac{1}{\Xi\, \Delta\boldsymbol{\Gamma}} \exp\left[-\beta \left(H - \sum_{k=1}^{c} \mu_k N_k - \mathbf{v} \cdot \mathbf{P}\right)\right] \tag{3.34}$$

in the lab frame where the fluid is moving at the uniform velocity $\mathbf{v}$ is recovered if the conjugate fields χ are uniform with $\chi_\epsilon = \beta$, $\chi_{n_k} = -\beta\mu_k$, $\chi_{g_i} = -\beta v_i$, and $\Omega = \ln \Xi$. In equation (3.34), $\beta = (k_{\text{B}}T)^{-1}$ is the inverse temperature, $\mathbf{P}$ is the total momentum, μ_k is the chemical potential of species k, and $\Delta\boldsymbol{\Gamma} = h^{3N}$. The same distribution (3.34) may also be considered in the frame where the fluid is at rest. Indeed, by expressing equation (3.34) in terms of the particle momenta $\mathbf{p}_{a0} = \mathbf{p}_a - m_a\mathbf{v}$ in the frame moving with the fluid, the standard form of the grand canonical equilibrium distribution is recovered with $\mathbf{v} = 0$, the chemical potentials $\mu_{k0} \equiv \mu_k + m_k\mathbf{v}^2/2$, and the total energy $E = E_0 + M\mathbf{v}^2/2$ in the lab frame, where E_0 is the total energy in the frame where the fluid is at rest. Under such equilibrium conditions, the functional (3.23) reduces to the generalized Massieu function (Callen, 1985),

$$\Omega = \beta\left(TS - \langle E\rangle + \sum_{k=1}^{c} \mu_k \langle N_k\rangle + \mathbf{v} \cdot \langle \mathbf{P}\rangle\right) \tag{3.35}$$

in the lab frame where $\langle E\rangle = \langle E_0\rangle + M\mathbf{v}^2/2$ and $\langle \mathbf{P}\rangle = M\mathbf{v}$, and

$$\Omega = \beta\left(TS - \langle E_0\rangle + \sum_{k=1}^{c} \mu_{k0} \langle N_k\rangle\right) = \beta p V \tag{3.36}$$

in the frame where the fluid is at rest, p being the hydrostatic pressure and V the volume of the system. At equilibrium, the functional (3.23) is thus related to the thermodynamic grand potential (C.14) according to $\Omega = -\beta J$.

3.2.4 Time Evolution of the Local Equilibrium Distribution

The basic idea of the formalism is that the time-evolved probability density (3.19) should remain close to the local equilibrium distribution (3.21) with the conjugate fields χ_t considered at time t. In this regard, these latter should be determined by the conditions

$$\langle \hat{\mathbf{c}}(\mathbf{r}; \boldsymbol{\Gamma})\rangle_t = \langle \hat{\mathbf{c}}(\mathbf{r}; \boldsymbol{\Gamma})\rangle_{\text{leq},\chi_t} \equiv \mathbf{c}(\mathbf{r}, t), \tag{3.37}$$

[3] In normal fluids, such correlation functions typically decay as $C(r) \sim r^{-1} \exp(-r/\ell_c)$ with $r = \|\mathbf{r} - \mathbf{r}'\|$ and the correlation length ℓ_c, so that order is short ranged in normal fluids.

according to which the mean values (3.20) of the densities with respect to the probability distribution $p_t(\mathbf{\Gamma})$ are equal to their mean values with respect to the local equilibrium distribution with the conjugate fields χ_t at time t. The conditions (3.37) thus define the macroscopic densities $\mathbf{c}(\mathbf{r},t)$, which are given by the functional derivatives (3.24) with respect to the conjugate fields $\chi = \chi_t$ at time t.

We note that the time derivatives of the mean densities and the conjugate fields can be related to each other by equation (3.32) with δ replaced by ∂_t.

Now we consider the exact time evolution of the probability density starting from the initial condition given by the local equilibrium distribution (3.21) with $\chi = \chi_0$,

$$p_t(\mathbf{\Gamma}) = \mathrm{e}^{\hat{L}t} p_{\text{leq}}(\mathbf{\Gamma};\chi_0) = \frac{1}{\Delta\mathbf{\Gamma}} \exp\left[-\Omega(\chi_0) - \chi_0 * \hat{\mathbf{c}}(\mathbf{\Gamma}_{-t})\right]. \tag{3.38}$$

Since the normalization of this probability distribution should be preserved during the time evolution, we must have that $(d/dt)\int p_t(\mathbf{\Gamma})\, d\mathbf{\Gamma} = 0$ for the probability density (3.38), which gives the condition $\chi_0 * \langle \partial_t \hat{\mathbf{c}} \rangle_{\text{leq},\chi_0} = 0$. Using equations (3.16), (3.20), and (3.37), and integrating by parts leads to the following relation, which should hold for any conjugate field χ_0 that may thus be replaced by χ:

$$\nabla\chi * \langle \hat{\jmath} \rangle_{\text{leq},\chi} = 0 \tag{3.39}$$

in the absence of boundaries (Oppenheim and Levine, 1979; Sasa, 2014).

The key issue of the formalism is to relate the time-evolved probability distribution (3.38) depending on the initial conjugate fields χ_0 to the local equilibrium distribution (3.21) for the time-evolved conjugate fields χ_t. Remarkably, we have that

$$p_t(\mathbf{\Gamma}) = p_{\text{leq}}(\mathbf{\Gamma};\chi_t)\, \mathrm{e}^{\Sigma_t(\mathbf{\Gamma})} \tag{3.40}$$

with the quantity (McLennan, 1963; Sasa, 2014)

$$\Sigma_t(\mathbf{\Gamma}) \equiv \int_0^t d\tau\, \partial_\tau \left[\Omega(\chi_\tau) + \chi_\tau * \hat{\mathbf{c}}(\mathbf{\Gamma}_{\tau-t})\right]. \tag{3.41}$$

Indeed, performing the integration over time in equation (3.41), we find that

$$\Sigma_t(\mathbf{\Gamma}) = \left[\Omega(\chi_t) + \chi_t * \hat{\mathbf{c}}(\mathbf{\Gamma})\right] - \left[\Omega(\chi_0) + \chi_0 * \hat{\mathbf{c}}(\mathbf{\Gamma}_{-t})\right], \tag{3.42}$$

so that the arguments of the exponential functions in equation (3.21) with $\chi = \chi_t$ and equation (3.38) are interconnected, as required.

As a consequence of equation (3.40), the mean value (2.36) of any observable $A(\mathbf{\Gamma})$ with respect to the time-evolved probability distribution $p_t(\mathbf{\Gamma})$ is thus exactly given in terms of the mean value with respect to the local equilibrium distribution according to (Sasa, 2014)

$$\langle A(\mathbf{\Gamma})\rangle_t = \left\langle A(\mathbf{\Gamma})\, \mathrm{e}^{\Sigma_t(\mathbf{\Gamma})}\right\rangle_{\text{leq},\chi_t}. \tag{3.43}$$

In particular, the conditions (3.37) are equivalent to the relations,

$$\left\langle \hat{\mathbf{c}}(\mathbf{r};\mathbf{\Gamma})\left[\mathrm{e}^{\Sigma_t(\mathbf{\Gamma})} - 1\right]\right\rangle_{\text{leq},\chi_t} = 0. \tag{3.44}$$

3.2.5 Entropy Production and Dissipative Current Densities

Taking $A(\boldsymbol{\Gamma}) = \mathrm{e}^{-\Sigma_t(\boldsymbol{\Gamma})}$ in the identity (3.43), we find the integral fluctuation relation of Sasa (2014):

$$\left\langle \mathrm{e}^{-\Sigma_t(\boldsymbol{\Gamma})} \right\rangle_t = 1. \tag{3.45}$$

Using Jensen's inequality (Cover and Thomas, 2006),

$$\left\langle \mathrm{e}^{x} \right\rangle \geq \mathrm{e}^{\langle x \rangle}, \tag{3.46}$$

the integral fluctuation relation implies that

$$S(\mathbf{c}_t) - S(\mathbf{c}_0) = k_{\mathrm{B}} \langle \Sigma_t(\boldsymbol{\Gamma}) \rangle_t \geq 0, \tag{3.47}$$

because the statistical average of equation (3.42) gives the difference between the values of the entropy functional (3.26) at the times t and $t = 0$ as a consequence of the conditions (3.37) or, equivalently, the relations (3.44).[4]

In open systems, the entropy S changes in time due to the exchanges $d_{\mathrm{e}}S$ with the environment and its production $d_{\mathrm{i}}S$ inside the system: $dS = d_{\mathrm{e}}S + d_{\mathrm{i}}S$. Since the system is here isolated, there is no exchange with the environment, i.e., $d_{\mathrm{e}}S = 0$, so that the change in time of the entropy is equal to the entropy production $dS = d_{\mathrm{i}}S$. In this regard, the result (3.47) may be interpreted as the nonnegativity of the entropy production, in agreement with the second law of thermodynamics.

Using equations (3.23) and (3.24), we have that

$$\frac{d}{dt}\,\Omega(\chi_t) = -\partial_t\, \chi_t * \langle \hat{\mathbf{c}} \rangle_{\mathrm{leq},\chi_t} \tag{3.48}$$

and

$$\frac{d}{dt}\,\chi_t * \langle \hat{\mathbf{c}} \rangle_{\mathrm{leq},\chi_t} = \partial_t\, \chi_t * \langle \hat{\mathbf{c}} \rangle_{\mathrm{leq},\chi_t} + \chi_t * \partial_t \langle \hat{\mathbf{c}} \rangle_{\mathrm{leq},\chi_t}. \tag{3.49}$$

According to definition (3.26) for the entropy functional, the relation (3.37), equation (3.16), and integrations by parts, the entropy production rate is obtained as

$$\begin{aligned} \frac{1}{k_{\mathrm{B}}}\frac{d_{\mathrm{i}}S}{dt} &= \frac{1}{k_{\mathrm{B}}}\frac{dS}{dt} = \frac{d}{dt}\left[\Omega(\chi_t) + \chi_t * \langle \hat{\mathbf{c}} \rangle_{\mathrm{leq},\chi_t}\right] \\ &= \chi_t * \partial_t \langle \hat{\mathbf{c}} \rangle_t = -\chi_t * \nabla \cdot \langle \hat{\jmath} \rangle_t = \nabla \chi_t * \langle \hat{\jmath} \rangle_t, \end{aligned} \tag{3.50}$$

where $\langle \hat{\jmath} \rangle_t = \langle \hat{\jmath}(\mathbf{r};\boldsymbol{\Gamma}) \rangle_t$. Now, using the identity (3.43) with A taken as $\hat{\jmath}$, the mean values of the current densities with respect to the phase-space probability distribution (3.19) can be decomposed as (McLennan, 1963)

$$\jmath(\mathbf{r},t) \equiv \langle \hat{\jmath}(\mathbf{r};\boldsymbol{\Gamma}) \rangle_t = \left\langle \hat{\jmath}(\mathbf{r};\boldsymbol{\Gamma}) \left\{1 + \left[\mathrm{e}^{\Sigma_t(\boldsymbol{\Gamma})} - 1\right]\right\} \right\rangle_{\mathrm{leq},\chi_t} = \bar{\jmath}(\mathbf{r},t) + \mathcal{J}(\mathbf{r},t) \tag{3.51}$$

into the mean values of the current densities over the local equilibrium distribution

$$\bar{\jmath}(\mathbf{r},t) \equiv \langle \hat{\jmath}(\mathbf{r};\boldsymbol{\Gamma}) \rangle_{\mathrm{leq},\chi_t} \tag{3.52}$$

[4] We note that the integral fluctuation relation (3.45) is obtained without using the property of microreversibility.

and the rest

$$\mathcal{J}(\mathbf{r},t) \equiv \left\langle \hat{j}(\mathbf{r};\Gamma)\left[e^{\Sigma_t(\Gamma)} - 1\right]\right\rangle_{\text{leq},\chi_t} . \tag{3.53}$$

The entropy production rate can thus be expressed as

$$\frac{1}{k_{\text{B}}}\frac{d_{\text{i}}S}{dt} = \underbrace{\nabla\chi_t * \langle\hat{j}\rangle_{\text{leq},\chi_t}}_{=0} + \nabla\chi_t * \mathcal{J}(t), \tag{3.54}$$

where the first term vanishes because of the identity (3.39). This term thus expresses the conservation of entropy in adiabatic (isoentropic) processes induced by the dissipativeless current densities defined by equation (3.52).[5] The second term in equation (3.54) is, in general, nonvanishing and related to the production of entropy, leading to the definition of the dissipative current densities given by equation (3.53). Accordingly, the entropy production rate is given by

$$\frac{1}{k_{\text{B}}}\frac{d_{\text{i}}S}{dt} = \nabla\chi_t * \mathcal{J}(t) \geq 0 \tag{3.55}$$

in terms of the dissipative current densities (3.53) and the gradients of the conjugate fields, which play the role of thermodynamic forces, also called the affinities. The statistical-mechanical expression (3.55) for the entropy production rate is in accordance with macroscopic nonequilibrium thermodynamics (Prigogine, 1967; Haase, 1969; Nicolis, 1979; de Groot and Mazur, 1984; Callen, 1985).

As we will explicitly show below, the three identities (3.27), (3.39), and (3.43) allow us to deduce the macroscopic equations of hydrodynamics and identify the dissipative transport coefficients. The local conservation equations for the mean values (3.37) can indeed be obtained from the mean values of equation (3.16), giving

$$\partial_t\, \mathbf{c} + \nabla \cdot (\bar{\jmath} + \mathcal{J}) = 0 \tag{3.56}$$

in terms of the dissipativeless (3.52) and dissipative (3.53) current densities. We note that the expressions here above are exact and do not involve any approximation.

The method is carried out as follows using expansions in powers of the gradients:

- At leading order in the gradients, the local thermodynamic relations can be established between the mean densities and the conjugate fields, and the identity (3.27) can be used to obtain the conjugate fields χ.
- Next, the dissipativeless current densities can be computed by taking the mean values of the microscopic current densities over the local equilibrium distribution, according to equation (3.52).
- The dissipative current densities arise from equation (3.53) as a direct consequence of the identity (3.43).
- The Green–Kubo relations giving the linear response coefficients between the dissipative current densities and the gradients of the conjugate fields can thus be deduced.

[5] We note that discontinuities due to shock waves may already produce entropy within the framework of Eulerian hydrodynamics that rule perfect fluids where viscosity and heat conductivity are assumed to be vanishing (Landau and Lifshitz, 1987). Such discontinuities are not described if the conjugate fields are smooth functions.

3.2.6 Local Thermodynamics

The identity (3.39) shows that the dissipativeless current densities (3.52) keep the entropy constant in time. All the processes involved by the dissipativeless current densities may thus be considered as reversible (i.e., adiabatic or isoentropic). The approach of nonequilibrium statistical mechanics based on the local equilibrium distribution (3.21) therefore provides the local thermodynamic relations in every element of matter, as shown in the remainder of this section.

At the leading order of the expansion in the gradients, the functionals (3.23) and (3.26) may be supposed to be of the forms,

$$\Omega(\chi) = \int \omega(\chi)\, d\mathbf{r} + O(\nabla^2) \qquad \text{and} \qquad S(\mathbf{c}) = \int s(\mathbf{c})\, d\mathbf{r} + O(\nabla^2), \tag{3.57}$$

defined by introducing the densities $\omega(\chi)$ and $s(\mathbf{c})$, which are, respectively, functions of the conjugate fields χ and the mean densities $\mathbf{c}$.[6] Since both functionals are interrelated by Legendre transforms, we have that

$$\mathbf{c} = -\frac{\partial \omega}{\partial \chi} \qquad \text{and} \qquad \chi = \frac{1}{k_{\rm B}} \frac{\partial s}{\partial \mathbf{c}}, \tag{3.58}$$

giving the local relations

$$s = k_{\rm B}\,(\omega + \chi \cdot \mathbf{c}), \qquad ds = k_{\rm B}\, \chi \cdot d\mathbf{c}, \qquad d\omega = -\mathbf{c} \cdot d\chi, \tag{3.59}$$

up to terms of second order in the gradients. In equation (3.59), the first relation can be identified as the local Euler relation, the second as the local Gibbs relation for the entropy density, and the third as the corresponding Gibbs–Duhem relation.

At leading order in the gradients, these relations can be compared with the known local thermodynamic relations (A.1) in the frame moving with the fluid. We may introduce the energy density and the chemical potential of species k in the lab frame where the fluid moves at the velocity $\mathbf{v}$, respectively, as

$$\epsilon = \epsilon_0 + \rho \mathbf{v}^2/2 \qquad \text{and} \qquad \mu_k = \mu_{k0} - m_k \mathbf{v}^2/2 \tag{3.60}$$

in terms of the corresponding quantities in the frame moving with the fluid, denoted $e = \epsilon_0$ and $\mu_k = \mu_{k0}$ in Chapter 1. In the lab frame, we thus have that

$$s = \frac{\epsilon + p}{T} - \sum_k \frac{\mu_k}{T} n_k - \frac{\mathbf{v}}{T} \cdot \mathbf{g}, \qquad ds = \frac{1}{T} d\epsilon - \sum_k \frac{\mu_k}{T} dn_k - \frac{\mathbf{v}}{T} \cdot d\mathbf{g}, \tag{3.61}$$

up to terms of $O(\nabla^2)$. Therefore, the conjugate fields are given by

$$\chi_\epsilon(\mathbf{r}, t) \equiv \frac{1}{k_{\rm B}} \frac{\delta S(\mathbf{c})}{\delta \epsilon(\mathbf{r}, t)} = \beta(\mathbf{r}, t) + O(\nabla^2), \tag{3.62}$$

[6] The notation $O(\nabla^2)$ stands for terms that are quadratic (or of higher degrees) in the gradients $\nabla \mathbf{c}$ or $\nabla \chi$. We note that the inclusion of the terms of order $(\nabla \mathbf{c})^2$ leads to entropy functionals similar to the Ginzburg–Landau functional. Such terms may be important to describe inhomogeneous fluids, multicomponent fluids with intermediate characteristic lengths, or phase transitions (Landau and Lifshitz, 1980a,b; Evans, 1979; Penrose and Fife, 1990; Wang et al., 1993; Mátyás and Gaspard, 2005).

$$\chi_{n_k}(\mathbf{r},t) \equiv \frac{1}{k_\mathrm{B}} \frac{\delta S(\mathbf{c})}{\delta n_k(\mathbf{r},t)} = -\beta(\mathbf{r},t)\,\mu_k(\mathbf{r},t) + O\big(\nabla^2\big), \tag{3.63}$$

$$\chi_{g_i}(\mathbf{r},t) \equiv \frac{1}{k_\mathrm{B}} \frac{\delta S(\mathbf{c})}{\delta g_i(\mathbf{r},t)} = -\beta(\mathbf{r},t)\,v_i(\mathbf{r},t) + O\big(\nabla^2\big), \tag{3.64}$$

confirming results obtained in equations (3.34)–(3.36) at global equilibrium.

Furthermore, the comparison between Euler's thermodynamic relations in equations (3.59) and (3.61) shows that the Massieu density ω is related to the hydrostatic pressure p according to $\omega = \beta p$, which gives the Gibbs–Duhem relation,

$$d\omega = -\epsilon\, d\beta + \sum_k n_k\, d(\beta\mu_k) + \mathbf{g}\cdot d(\beta\mathbf{v}). \tag{3.65}$$

3.2.7 Dissipativeless Time Evolution

The mean values (3.52) of the current densities with respect to the local equilibrium distribution (3.21) can be obtained by considering the change of variables $\mathbf{p}_a = \mathbf{p}_{a0} + m_a\mathbf{v}$ from the lab frame to the frame moving with the fluid element. Under this change of variables, the quantities of interest are transformed according to

$$\hat{\epsilon} = \hat{\epsilon}_0 + \hat{\mathbf{g}}_0\cdot\mathbf{v} + \frac{1}{2}\hat{\rho}\,\mathbf{v}^2, \qquad \hat{\jmath}_k = \hat{\jmath}_{k0} + \hat{n}_k\,\mathbf{v}, \qquad \hat{\jmath}_\rho = \hat{\mathbf{g}} = \hat{\mathbf{g}}_0 + \hat{\rho}\,\mathbf{v}, \tag{3.66}$$

$$\hat{\jmath}_\mathbf{g} = \hat{\jmath}_{\mathbf{g}0} + \hat{\mathbf{g}}_0\,\mathbf{v} + \mathbf{v}\,\hat{\mathbf{g}}_0 + \hat{\rho}\,\mathbf{v}\,\mathbf{v}, \tag{3.67}$$

$$\hat{\jmath}_\epsilon = \hat{\jmath}_{\epsilon0} + \hat{\epsilon}_0\,\mathbf{v} + \hat{\jmath}_{\mathbf{g}0}\cdot\mathbf{v} + \hat{\mathbf{g}}_0\cdot\mathbf{v}\,\mathbf{v} + \frac{1}{2}\mathbf{v}^2(\hat{\mathbf{g}}_0 + \hat{\rho}\,\mathbf{v}) - \hat{\boldsymbol{\Delta}}, \tag{3.68}$$

where the subscript 0 denotes those with the momenta $\mathbf{p}_{a0}$ replacing $\mathbf{p}_a$, and the quantity (Sasa, 2014),

$$\hat{\boldsymbol{\Delta}} \equiv \frac{1}{2}\sum_{a\neq b} \mathbf{r}_{ab}\left[\mathbf{v}(\mathbf{r}) - \frac{\mathbf{v}(\mathbf{r}_a)+\mathbf{v}(\mathbf{r}_b)}{2}\right]\cdot\mathbf{F}_{ab}\int_0^1 \delta\left[\mathbf{r} - \boldsymbol{R}_{ab}(\lambda)\right] d\lambda. \tag{3.69}$$

This latter would vanish if the velocity field $\mathbf{v}$ was uniform. Furthermore, we note that the mean value of $\hat{\boldsymbol{\Delta}}$ goes as the square of gradients, in particular, the square of the gradient of the velocity field.

The quantities with odd powers for the momenta $\mathbf{p}_{a0}$ are equal to zero after averaging over local equilibrium, so that we get $\langle\hat{\mathbf{g}}_0\rangle_\mathrm{leq} = 0$, $\langle\hat{\jmath}_{k0}\rangle_\mathrm{leq} = 0$, $\langle\hat{\jmath}_{\epsilon0}\rangle_\mathrm{leq} = 0$, $\langle\hat{\rho}\rangle_\mathrm{leq} = \rho$, $\langle\hat{\epsilon}_0\rangle_\mathrm{leq} = \epsilon_0$, $\langle\hat{n}_k\rangle_\mathrm{leq} = n_k$, and $\langle\hat{\jmath}_{\mathbf{g}0}\rangle_\mathrm{leq} = p\,\mathbf{1}$. Moreover, $\langle\hat{\boldsymbol{\Delta}}\rangle_\mathrm{leq}$ can be neglected because it contributes to the square of the gradients. Thus, we find

$$\epsilon = \langle\hat{\epsilon}\rangle_\mathrm{leq} = \epsilon_0 + \frac{\rho}{2}\mathbf{v}^2, \qquad \bar{\jmath}_k = \langle\hat{\jmath}_k\rangle_\mathrm{leq} = n_k\mathbf{v}, \qquad \bar{\jmath}_\rho = \langle\hat{\mathbf{g}}\rangle_\mathrm{leq} = \rho\,\mathbf{v}, \tag{3.70}$$

$$\bar{\jmath}_\mathbf{g} = \langle\hat{\jmath}_\mathbf{g}\rangle_\mathrm{leq} = \rho\mathbf{v}\mathbf{v} + p\,\mathbf{1}, \tag{3.71}$$

$$\bar{\jmath}_\epsilon = \langle\hat{\jmath}_\epsilon\rangle_\mathrm{leq} = \left(\epsilon_0 + p + \frac{\rho}{2}\mathbf{v}^2\right)\mathbf{v} + O\big(\nabla^2\big), \tag{3.72}$$

locally in the fluid element centered around the position $\mathbf{r}$. Accordingly, the Galilean invariance is satisfied. We note that $\bar{\jmath}_\rho = \sum_k m_k\bar{\jmath}_k$ because $\rho = \sum_k m_k n_k$.

By substituting into the balance equations, the dissipativeless equations of fluid mechanics are obtained in their two equivalent Eulerian and Lagrangian forms using the streamline time derivative $df/dt = \partial_t f + \mathbf{v}\cdot\nabla f$ for any field f:

$$\partial_t n_k + \nabla\cdot(n_k\mathbf{v}) = 0, \qquad \frac{dn_k}{dt} = -n_k\nabla\cdot\mathbf{v}, \tag{3.73}$$

$$\partial_t \rho + \nabla\cdot(\rho\mathbf{v}) = 0, \qquad \frac{d\rho}{dt} = -\rho\,\nabla\cdot\mathbf{v}, \tag{3.74}$$

$$\partial_t(\rho\mathbf{v}) + \nabla\cdot(\rho\mathbf{v}\mathbf{v} + p\,\mathbf{1}) = 0, \qquad \rho\,\frac{d\mathbf{v}}{dt} = -\nabla p, \tag{3.75}$$

$$\partial_t\left(\epsilon_0 + \frac{\rho}{2}\mathbf{v}^2\right) + \nabla\cdot\left[\left(\epsilon_0 + p + \frac{\rho}{2}\mathbf{v}^2\right)\mathbf{v}\right] = 0, \qquad \frac{d\epsilon_0}{dt} = -(\epsilon_0 + p)\nabla\cdot\mathbf{v}. \tag{3.76}$$

These equations are known to preserve entropy, thus describing adiabatic (i.e., isoentropic) processes in normal fluids, as long as there are no shock wave discontinuities (Landau and Lifshitz, 1987). We note that these dissipativeless balance equations only involve the terms of first order in the gradients.

The dissipativeless equations for the time evolution of the conjugate fields can be deduced from equations (3.73)–(3.76) at the same level of approximation. The deduction is carried out in Appendix D, giving

$$\frac{d\beta}{dt} = (\partial_t + \mathbf{v}\cdot\nabla)\beta = \beta\left(\frac{\partial p}{\partial\epsilon_0}\right)_{\{n_k\}}\nabla\cdot\mathbf{v}, \tag{3.77}$$

$$\frac{d(\beta\mu_{k0})}{dt} = (\partial_t + \mathbf{v}\cdot\nabla)(\beta\mu_{k0}) = -\beta\left(\frac{\partial p}{\partial n_k}\right)_{\epsilon_0,\{n_j\}_{j\neq k}}\nabla\cdot\mathbf{v}, \tag{3.78}$$

$$\rho\,\frac{d\mathbf{v}}{dt} = \rho(\partial_t + \mathbf{v}\cdot\nabla)\mathbf{v} = \frac{\epsilon_0 + p}{\beta}\,\nabla\beta - \sum_k \frac{n_k}{\beta}\,\nabla(\beta\mu_{k0}). \tag{3.79}$$

3.2.8 Dissipative Time Evolution

We proceed with the expansion in powers of the gradients in order to deduce the dissipative terms that should be added to the Euler equations of hydrodynamics. These dissipative terms are of second order in the gradients, defining the linear transport properties. In this way, the Green–Kubo formulas will be obtained for the transport coefficients.

Entropy Production Rate

First, we should verify that the entropy production rate (3.55) has the form expected from nonequilibrium thermodynamics (de Groot and Mazur, 1984). Using the expressions (3.62)–(3.64) for the conjugate fields, the entropy production rate can be written in the form

$$\frac{1}{k_{\mathrm{B}}}\frac{d_{\mathrm{i}}S}{dt} = \int \nabla\chi : \mathcal{J}\,d\mathbf{r} \simeq \int\left[\nabla\beta\cdot\mathcal{J}_\epsilon - \sum_k \nabla(\beta\mu_k)\cdot\mathcal{J}_k - \nabla(\beta\mathbf{v}) : \mathcal{J}_{\mathbf{g}}\right]d\mathbf{r}, \tag{3.80}$$

up to higher order terms in the gradients. Expanding the gradient $\nabla(\beta\mathbf{v})$ and expressing the chemical potentials in the frame moving with the fluid element with equation (3.60), we obtain

$$\frac{1}{k_{\rm B}}\frac{d_{\rm i}S}{dt} \simeq \int \left[\nabla\beta \cdot (\mathcal{J}_\epsilon - \mathbf{v}\cdot\mathcal{J}_{\mathbf{g}}) - \sum_k \nabla(\beta\mu_{k0})\cdot\mathcal{J}_k - \beta\,\nabla\mathbf{v} : \mathcal{J}_{\mathbf{g}} \right] d\mathbf{r}, \tag{3.81}$$

since $\sum_k m_k \mathcal{J}_k = 0$ because of mass conservation. Accordingly, we can identify the dissipative part of the pressure tensor as

$$\mathbf{\Pi} \equiv \mathcal{J}_{\mathbf{g}}, \tag{3.82}$$

the heat current density as

$$\mathcal{J}_q \equiv \mathcal{J}_\epsilon - \mathbf{v}\cdot\mathcal{J}_{\mathbf{g}}, \tag{3.83}$$

and the diffusive current density of species k as $\mathcal{J}_k$. Decomposing the dissipative pressure tensor as $\mathbf{\Pi} = \overset{\circ}{\mathbf{\Pi}} + \Pi\,\mathbf{1}$ into its traceless part such that tr $\overset{\circ}{\mathbf{\Pi}} = 0$ and $\Pi = (\mathrm{tr}\,\mathbf{\Pi})/3$, the entropy production rate becomes

$$\frac{d_{\rm i}S}{dt} \simeq \int \Bigg[\nabla\left(\frac{1}{T}\right)\cdot\mathcal{J}_q - \sum_k \nabla\left(\frac{\mu_{k0}}{T}\right)\cdot\mathcal{J}_k - \frac{1}{2T}\left(\nabla\mathbf{v} + \nabla\mathbf{v}^{\rm T} - \frac{2}{3}\nabla\cdot\mathbf{v}\,\mathbf{1}\right) : \overset{\circ}{\mathbf{\Pi}} - \frac{1}{T}(\nabla\cdot\mathbf{v})\,\Pi \Bigg] d\mathbf{r}, \tag{3.84}$$

which corresponds to the expression from nonequilibrium thermodynamics for normal fluids without reaction according to Table 1.2 where $\mu_{k0} = \mu_k$.

Dissipative Current Densities

Now we go on with the deduction of the dissipative contribution (3.53) to the current densities at the next order in the gradients. For this purpose, equation (3.53) is expanded in powers of Σ_t, keeping the term of first order as

$$\mathcal{J}(\mathbf{r},t) \equiv \langle \hat{\jmath}(\mathbf{r};\mathbf{\Gamma})\Sigma_t(\mathbf{\Gamma})\rangle_{{\rm leq},\chi_t} + O\big(\Sigma_t^2\big). \tag{3.85}$$

In order to evaluate the dissipative current densities, the quantity (3.41) is transformed by taking the time derivative ∂_τ under the integral, and using equation (3.48) and the balance equations (3.16) for the microscopic densities (3.18), leading to

$$\begin{aligned} \Sigma_t(\mathbf{\Gamma}) &= \int_0^t d\tau \left[\partial_\tau\Omega(\chi_\tau) + \partial_\tau\chi_\tau * \hat{\mathbf{c}}(\mathbf{\Gamma}_{\tau-t}) + \chi_\tau * \partial_\tau\hat{\mathbf{c}}(\mathbf{\Gamma}_{\tau-t})\right] \\ &= \int_0^t d\tau \left[-\partial_\tau\chi_\tau * \langle\hat{\mathbf{c}}\rangle_{{\rm leq},\chi_\tau} + \partial_\tau\chi_\tau * \hat{\mathbf{c}}(\mathbf{\Gamma}_{\tau-t}) - \chi_\tau * \nabla\cdot\hat{\jmath}(\mathbf{\Gamma}_{\tau-t})\right]. \end{aligned} \tag{3.86}$$

After integrating the last term by parts and subtracting the identity (3.39) integrated over time, we obtain the following expression:

$$\Sigma_t(\mathbf{\Gamma}) = \int_0^t d\tau \left[\partial_\tau\chi_\tau * \delta\hat{\mathbf{c}}(\mathbf{\Gamma}_{\tau-t}) + \nabla\chi_\tau * \delta\hat{\jmath}(\mathbf{\Gamma}_{\tau-t})\right] \tag{3.87}$$

with

$$\delta\hat{\mathbf{c}}(\mathbf{r};\boldsymbol{\Gamma}_{\tau-t}) \equiv \hat{\mathbf{c}}(\mathbf{r};\boldsymbol{\Gamma}_{\tau-t}) - \langle\hat{\mathbf{c}}(\mathbf{r})\rangle_{\text{leq},\chi_\tau}, \tag{3.88}$$

$$\delta\hat{\jmath}(\mathbf{r};\boldsymbol{\Gamma}_{\tau-t}) \equiv \hat{\jmath}(\mathbf{r};\boldsymbol{\Gamma}_{\tau-t}) - \langle\hat{\jmath}(\mathbf{r})\rangle_{\text{leq},\chi_\tau}. \tag{3.89}$$

Next, the expressions for the conjugate fields, the densities, and the current densities are substituted into equation (3.87) and equations (3.77)–(3.79) are used for the time evolution of the conjugate fields. The detailed calculations are performed in Section D.1.3. At first order in the gradients, we find the following result in the frame moving with the fluid element where $\mathbf{v} = 0$:

$$\partial_\tau\chi\cdot\delta\hat{\mathbf{c}} + \nabla\chi : \delta\hat{\jmath} = \nabla\beta\cdot\delta\hat{\jmath}'_\epsilon - \beta\,\nabla\mathbf{v} : \delta\hat{\jmath}'_{\mathbf{g}} - \sum_k \nabla(\beta\mu_{k0})\cdot\delta\hat{\jmath}'_k \tag{3.90}$$

with the fluctuations of the current densities given by

$$\delta\hat{\jmath}'_\epsilon \equiv \delta\hat{\jmath}_\epsilon - \rho^{-1}(\epsilon_0 + p)\,\delta\hat{\mathbf{g}}, \tag{3.91}$$

$$\delta\hat{\jmath}'_{\mathbf{g}} \equiv \delta\hat{\jmath}_{\mathbf{g}} - \left[\left(\frac{\partial p}{\partial\epsilon_0}\right)_{\{n_k\}}\delta\hat{\epsilon} + \sum_k\left(\frac{\partial p}{\partial n_k}\right)_{\epsilon_0,\{n_j\}_{j\neq k}}\delta\hat{n}_k\right]\mathbf{1}, \tag{3.92}$$

$$\delta\hat{\jmath}'_k \equiv \delta\hat{\jmath}_k - \rho^{-1}n_k\,\delta\hat{\mathbf{g}}. \tag{3.93}$$

After substituting equation (3.90) into equation (3.87), the dissipative current densities (3.85) are given at leading order by

$$\begin{aligned}\mathcal{J}(\mathbf{r},t) = {} & \int_0^t d\tau\int d\mathbf{r}'\,\langle\delta\hat{\jmath}(\mathbf{r},0)\,\delta\hat{\jmath}'_\epsilon(\mathbf{r}',\tau-t)\rangle_{\text{leq},t}\cdot\nabla'\beta(\mathbf{r}',\tau) \\ & - \int_0^t d\tau\int d\mathbf{r}'\,\langle\delta\hat{\jmath}(\mathbf{r},0)\,\delta\hat{\jmath}'_{\mathbf{g}}(\mathbf{r}',\tau-t)\rangle_{\text{leq},t} : [\beta(\mathbf{r}',\tau)\,\nabla'\mathbf{v}(\mathbf{r}',\tau)] \\ & - \sum_k\int_0^t d\tau\int d\mathbf{r}'\,\langle\delta\hat{\jmath}(\mathbf{r},0)\,\delta\hat{\jmath}'_k(\mathbf{r}',\tau-t)\rangle_{\text{leq},t}\cdot\nabla'[\beta(\mathbf{r}',\tau)\mu_{k0}(\mathbf{r}',\tau)].\end{aligned} \tag{3.94}$$

Since the conjugate fields are macroscopic, their spatial and temporal scales are much larger than the correlation length and time of the current fluctuations (3.91)–(3.93), so that their gradients $\nabla'\chi(\mathbf{r}',\tau)$ can be replaced by $\nabla\chi(\mathbf{r},t)$, giving

$$\begin{aligned}\mathcal{J}(\mathbf{r},t) = {} & \nabla\beta(\mathbf{r},t)\cdot\int_0^t d\tau\int d\mathbf{r}'\,\langle\delta\hat{\jmath}(\mathbf{r},0)\,\delta\hat{\jmath}'_\epsilon(\mathbf{r}',\tau-t)\rangle_{\text{leq},t} \\ & - \beta(\mathbf{r},t)\,\nabla\mathbf{v}(\mathbf{r},t) : \int_0^t d\tau\int d\mathbf{r}'\,\langle\delta\hat{\jmath}(\mathbf{r},0)\,\delta\hat{\jmath}'_{\mathbf{g}}(\mathbf{r}',\tau-t)\rangle_{\text{leq},t} \\ & - \sum_k\nabla[\beta(\mathbf{r},t)\mu_{k0}(\mathbf{r},t)]\cdot\int_0^t d\tau\int d\mathbf{r}'\,\langle\delta\hat{\jmath}(\mathbf{r},0)\,\delta\hat{\jmath}'_k(\mathbf{r}',\tau-t)\rangle_{\text{leq},t},\end{aligned} \tag{3.95}$$

where the gradients in front of the integrals are contracted with the second current density in the time correlation functions.

Moreover, under the same assumptions, the local equilibrium distribution may be considered as the equilibrium distribution at the local values of the conjugate fields in the frame moving with the fluid element. Since the equilibrium distribution is stationary, the following property holds: $\langle \delta\hat{a}(\mathbf{r},0)\,\delta\hat{b}(\mathbf{r}',\tau-t)\rangle_{\rm eq} = \langle \delta\hat{a}(\mathbf{r},t-\tau)\,\delta\hat{b}(\mathbf{r}',0)\rangle_{\rm eq}$. The integral over time becomes $\int_0^t d\tau\,\langle \delta\hat{a}(\mathbf{r},0)\,\delta\hat{b}(\mathbf{r}',\tau-t)\rangle_{\rm eq} = \int_0^t d\tau\,\langle \delta\hat{a}(\mathbf{r},\tau)\,\delta\hat{b}(\mathbf{r}',0)\rangle_{\rm eq}$, after replacing $t-\tau$ by τ. Here, the limit $t\to\infty$ can be taken since the timescale t of the conjugate fields is longer than the correlation time of the current densities.

Over spatial scales much larger than the distance between the atoms, the material properties can be defined by averaging over space. In this regard, the microscopic total currents are introduced as

$$\delta\mathbf{J}(t) \equiv \int_V \delta\hat{\jmath}'(\mathbf{r},t)\,d\mathbf{r}. \tag{3.96}$$

Using equations (3.91)–(3.93), the microscopic total currents are given by

$$\delta\mathbf{J}_\epsilon = \int_V \delta\hat{\jmath}_\epsilon\,d\mathbf{r} - \frac{\epsilon_0+p}{\rho}\,\delta\mathbf{P}, \tag{3.97}$$

$$\delta\mathbf{J}_\mathbf{g} = \int_V \delta\hat{\jmath}_\mathbf{g}\,d\mathbf{r} - \left[\left(\frac{\partial p}{\partial\epsilon_0}\right)_{\{n_k\}}\delta E + \sum_k \left(\frac{\partial p}{\partial n_k}\right)_{\epsilon_0,\{n_j\}_{j\neq k}}\delta N_k\right]\mathbf{1}, \tag{3.98}$$

$$\delta\mathbf{J}_k = \int_V \delta\hat{\jmath}_k\,d\mathbf{r} - \frac{n_k}{\rho}\,\delta\mathbf{P}, \tag{3.99}$$

where the volume integrals of the microscopic current densities have the expressions,

$$\int_V \hat{\jmath}_\epsilon\,d\mathbf{r} = \sum_a \frac{\mathbf{p}_a}{m_a}\,\varepsilon_a + \frac{1}{4}\sum_{a\neq b}\mathbf{r}_{ab}\left(\frac{\mathbf{p}_a}{m_a}+\frac{\mathbf{p}_b}{m_b}\right)\cdot\mathbf{F}_{ab}, \tag{3.100}$$

$$\int_V \hat{\jmath}_\mathbf{g}\,d\mathbf{r} = \sum_a \frac{\mathbf{p}_a\mathbf{p}_a}{m_a} + \frac{1}{2}\sum_{a\neq b}\mathbf{r}_{ab}\,\mathbf{F}_{ab}, \tag{3.101}$$

$$\int_V \hat{\jmath}_k\,d\mathbf{r} = \sum_{a\in\mathcal{S}_k}\frac{\mathbf{p}_a}{m_a}, \tag{3.102}$$

where $\delta\mathbf{P} = \int \delta\hat{\mathbf{g}}\,d\mathbf{r}$, $\delta E = \int \delta\hat{\epsilon}\,d\mathbf{r}$, and $\delta N_k = \int \delta\hat{n}_k\,d\mathbf{r}$ are the fluctuations in the total momentum, energy, and particle numbers, and $\delta X \equiv X - \langle X\rangle_{\rm eq}$. Since the total momentum, energy, and particle numbers are constants of motion, they do not fluctuate in time, but they are still random variables because the initial conditions in phase space may give different values to these quantities. At equilibrium, their fluctuations may thus be added or subtracted in the time correlation functions since $\langle\delta\mathbf{J}\rangle_{\rm eq} = 0$. Consequently, the unprime quantities can be replaced by the prime ones in the time correlation functions. Accordingly, equation (3.95) becomes[7]

[7] As in equation (3.95), the gradients in front of the integrals are contracted with the microscopic currents $\delta\mathbf{J}_\alpha(0)$ in the time correlation functions because equation (3.103) is deduced from equation (3.94). Moreover, $\langle\cdot\rangle_{\rm eq}$ denotes the mean value with respect to the equilibrium distribution at the local values of the conjugate fields.

$$\mathcal{J}(\mathbf{r},t) = \frac{1}{V}\,\nabla\beta(\mathbf{r},t)\cdot\int_0^\infty d\tau\,\langle\delta\mathbf{J}(\tau)\,\delta\mathbf{J}_\epsilon(0)\rangle_{\text{eq}}$$
$$-\frac{1}{V}\,\beta(\mathbf{r},t)\,\nabla\mathbf{v}(\mathbf{r},t) : \int_0^\infty d\tau\,\langle\delta\mathbf{J}(\tau)\,\delta\mathbf{J}_{\mathbf{g}}(0)\rangle_{\text{eq}}$$
$$-\frac{1}{V}\sum_k \nabla[\beta(\mathbf{r},t)\mu_{k0}(\mathbf{r},t)]\cdot\int_0^\infty d\tau\,\langle\delta\mathbf{J}(\tau)\,\delta\mathbf{J}_k(0)\rangle_{\text{eq}}, \tag{3.103}$$

holding for the dissipative current densities $\mathcal{J} = (\mathcal{J}_\epsilon, \mathcal{J}_k, \mathcal{J}_{\mathbf{g}})$ in the frame where $\mathbf{v} = 0$ and $\mathcal{J}_\epsilon = \mathcal{J}_q$.

3.2.9 Green–Kubo Formulas for the Transport Coefficients

As a consequence of the previous considerations, the dissipative current densities associated with viscous pressure, heat, and diffusion are thus obtained as

$$\mathcal{J}_{\mathbf{g}} = \mathbf{\Pi} = -\eta\left(\nabla\mathbf{v} + \nabla\mathbf{v}^{\mathrm{T}} - \frac{2}{3}\,\nabla\cdot\mathbf{v}\,\mathbf{1}\right) - \zeta\,\nabla\cdot\mathbf{v}\,\mathbf{1}, \tag{3.104}$$

$$\mathcal{J}_q = \mathcal{L}_{q,q}\,\nabla\frac{1}{T} + \sum_k \mathcal{L}_{q,k}\,\nabla\left(-\frac{\mu_{k0}}{T}\right), \tag{3.105}$$

$$\mathcal{J}_k = \mathcal{L}_{k,q}\,\nabla\frac{1}{T} + \sum_l \mathcal{L}_{k,l}\,\nabla\left(-\frac{\mu_{l0}}{T}\right), \tag{3.106}$$

with the transport coefficients given by the Green–Kubo formulas (McLennan, 1963; Akhiezer and Peletminskii, 1981; Sasa, 2014),

$$\eta = \lim_{V\to\infty}\frac{1}{Vk_{\mathrm{B}}T}\int_0^\infty dt\,\langle\delta J_{xy}(t)\,\delta J_{xy}(0)\rangle_{\text{eq}}, \tag{3.107}$$

$$\zeta + \frac{4}{3}\,\eta = \lim_{V\to\infty}\frac{1}{Vk_{\mathrm{B}}T}\int_0^\infty dt\,\langle\delta J_{xx}(t)\,\delta J_{xx}(0)\rangle_{\text{eq}}, \tag{3.108}$$

$$\mathcal{L}_{q,q} = \lim_{V\to\infty}\frac{1}{Vk_{\mathrm{B}}}\int_0^\infty dt\,\langle\delta J_{\epsilon x}(t)\,\delta J_{\epsilon x}(0)\rangle_{\text{eq}}, \tag{3.109}$$

$$\mathcal{L}_{q,k} = \mathcal{L}_{k,q} = \lim_{V\to\infty}\frac{1}{Vk_{\mathrm{B}}}\int_0^\infty dt\,\langle\delta J_{kx}(t)\,\delta J_{\epsilon x}(0)\rangle_{\text{eq}}, \tag{3.110}$$

$$\mathcal{L}_{k,l} = \mathcal{L}_{l,k} = \lim_{V\to\infty}\frac{1}{Vk_{\mathrm{B}}}\int_0^\infty dt\,\langle\delta J_{kx}(t)\,\delta J_{lx}(0)\rangle_{\text{eq}}, \tag{3.111}$$

expressed in terms of the microscopic total currents (3.97)–(3.99) defined by the microscopic current densities (3.12)–(3.14) with $\delta J_{ij} = (\delta\mathbf{J}_{\mathbf{g}})_{ij}$, $\delta J_{\epsilon i} = (\delta\mathbf{J}_\epsilon)_i$, and $\delta J_{ki} = (\delta\mathbf{J}_k)_i$ for $i, j = x, y, z$. Here, the limit $V \to \infty$ is taken at constant values of the chemical potentials in order to define the bulk transport properties in arbitrarily large systems, in this way eliminating possible finite-size effects encountered in molecular dynamics simulations (Frenkel and Smit, 2002; Allen and Tildesley, 2017). The transport coefficients are intensive quantities because the correlation functions of the current densities in equation (3.95) are vanishingly small at large distance in thermal systems due to the absence of correlations between far enough spatial regions. Accordingly, the time correlation functions $\langle \delta J_\alpha(t)\, \delta J_\beta(0)\rangle_{\rm eq}$ of the microscopic total currents are extensive, so that the transport coefficients are intensive, as expected.

The Green–Kubo formula (3.107) gives the shear viscosity η, (3.108) the dilational viscosity ζ, (3.109) the heat conductivity $\kappa = \mathcal{L}_{q,q}/T^2$, (3.110) the coefficients of thermodiffusion, and (3.111) the diffusion coefficients. Because of mass conservation, the diffusion current densities should obey $\sum_{k=1}^{c} m_k \mathcal{J}_k = 0$, so that these coefficients are not all independent, but satisfy the conditions $\sum_{k=1}^{c} m_k \mathcal{L}_{k,q} = 0$ and $\sum_{k=1}^{c} m_k \mathcal{L}_{k,l} = 0$. They can thus be reduced to the coefficients describing mutual diffusion of solute into solvent species. We note that the Green–Kubo formula (2.165) for the electric conductivity is recovered because the total electric current (2.163) is given by the sum $\delta\mathbf{J} = \sum_k e_k \delta\mathbf{J}_k$, combining the microscopic total currents (3.99) with the electric charge e_k of species k. The extra term involving $\delta\mathbf{P}$ is required if the center of mass is mobile in the statistical ensemble considered.

Normal fluids being isotropic, the fourth-order tensor of viscosity coefficients has the form,

$$\eta_{ijkl} = \eta\left(\delta_{ik}\delta_{jl} + \delta_{il}\delta_{jk} - \frac{2}{3}\,\delta_{ij}\delta_{kl}\right) + \zeta\,\delta_{ij}\delta_{kl}, \tag{3.112}$$

so that we also have

$$\zeta - \frac{2}{3}\,\eta = \lim_{V\to\infty} \frac{1}{Vk_{\rm B}T} \int_0^\infty dt\, \langle \delta J_{xx}(t)\, \delta J_{yy}(0)\rangle_{\rm eq} \tag{3.113}$$

and similar formulas between δJ_{xx} and δJ_{zz}, or between δJ_{yy} and δJ_{zz}. Therefore, we recover the following alternative Green–Kubo formula for the dilational viscosity coefficient (Akhiezer and Peletminskii, 1981; Sasa, 2014):

$$\zeta = \lim_{V\to\infty} \frac{1}{Vk_{\rm B}T} \int_0^\infty dt\, \langle \delta J_\zeta(t)\, \delta J_\zeta(0)\rangle_{\rm eq} \quad \text{with} \quad \delta J_\zeta \equiv \frac{1}{3}\left(\delta J_{xx} + \delta J_{yy} + \delta J_{zz}\right). \tag{3.114}$$

Moreover, the isotropy of normal fluids also implies that tensorial transport properties cannot be coupled to vectorial ones according to the Curie symmetry principle. Therefore, the tensorial transport of linear momentum cannot be coupled to the vectorial transport of heat or particles in fluids. Nevertheless, the vectorial transport properties of heat conduction and particle diffusion can be coupled together by the coefficients of thermodiffusion (3.110).

Since the Hamiltonian function ruling the microscopic dynamics has the time-reversal symmetry (2.28), we have the Onsager–Casimir reciprocal relation (2.173), which here reads

$$\int_0^\infty dt\, \langle \delta J_\alpha(t)\, \delta J_\beta(0)\rangle_{\rm eq} = \epsilon_\alpha\, \epsilon_\beta \int_0^\infty dt\, \langle \delta J_\beta(t)\, \delta J_\alpha(0)\rangle_{\rm eq}, \tag{3.115}$$

where $\epsilon_\alpha = \pm 1$ if δJ_α is even or odd under time reversal. The currents $\mathbf{J}_\epsilon$ and $\mathbf{J}_k$ are odd, since $\hat{\epsilon}$ and $\hat{n}_k$ are even, while $\mathbf{J}_{\rm g}$ is even because $\hat{\mathbf{g}}$ is odd. As a consequence, the Onsager reciprocal relations given by equations (3.110) and (3.111) are satisfied.

With the Green–Kubo formulas, the transport coefficients can be computed using molecular dynamics (Frenkel and Smit, 2002; Allen and Tildesley, 2017). The time correlation functions of the microscopic currents (3.97)–(3.99) should be calculated by numerical simulation in some equilibrium statistical ensemble. In the grand canonical ensemble, the quantities $\delta\mathbf{P}$, δE, and δN_k should be included because they have nontrivial probability distributions. In the canonical ensemble, the particle numbers are fixed, so that $\delta N_k = 0$. The calculation can also be performed in the microcanonical ensemble (2.70) with a given energy E and zero total linear momentum $\mathbf{P} = 0$, in which case $\delta\mathbf{P} = 0$, $\delta E = 0$, and $\delta N_k = 0$. The results of the different statistical ensembles should be equivalent in the thermodynamic limit $V \to \infty$ with given values for the chemical potentials.

An alternative to the integration of the time correlation functions in the Green–Kubo formulas is given by introducing the so-called Helfand moments as

$$\delta G_\alpha(t) \equiv \int_0^t \delta J_\alpha(\tau)\, d\tau \tag{3.116}$$

for every microscopic total current J_α (Helfand, 1960). Therefore, the transport coefficients can be equivalently obtained using the Einstein–Helfand formula

$$\mathcal{L}_{\alpha,\beta} = \lim_{V\to\infty} \frac{1}{Vk_{\rm B}} \int_0^\infty dt\, \langle \delta J_\alpha(t)\, \delta J_\beta(0)\rangle_{\rm eq} = \lim_{V\to\infty} \frac{1}{Vk_{\rm B}} \lim_{t\to\infty} \frac{1}{2t} \langle \delta G_\alpha(t)\, \delta G_\beta(t)\rangle_{\rm eq}, \tag{3.117}$$

this latter expressing the fact that Helfand's moments perform random walks at equilibrium. This method has been implemented for the numerical calculation of transport coefficients in hard-sphere fluids by Alder et al. (1970). For molecular dynamics in the presence of periodic boundary conditions, the Helfand moments can be extended to cope with the minimum-image convention (Allen and Tildesley, 2017), so that their random walk is unbounded even if the volume V is finite. In this regard, the limit $V \to \infty$ can be taken after the limit $t \to \infty$ (Viscardy et al., 2007a,b).

3.2.10 Dissipative Hydrodynamic Equations

Finally, the dissipative current densities (3.104)–(3.106) with the heat current density (3.83) can be replaced in the balance equations (3.56) to obtain the dissipative hydrodynamic equations

$$\partial_t n_k + \nabla\cdot(n_k\mathbf{v} + \mathcal{J}_k) = 0, \tag{3.118}$$

$$\partial_t \rho + \nabla\cdot(\rho\mathbf{v}) = 0, \tag{3.119}$$

$$\partial_t(\rho\mathbf{v}) + \nabla\cdot(\rho\mathbf{v}\mathbf{v} + p\,\mathbf{1} + \boldsymbol{\Pi}) = 0, \tag{3.120}$$

$$\partial_t\left(\epsilon_0 + \frac{\rho}{2}\mathbf{v}^2\right) + \nabla\cdot\left[\left(\epsilon_0 + p + \frac{\rho}{2}\mathbf{v}^2\right)\mathbf{v} + \boldsymbol{\Pi}\cdot\mathbf{v} + \mathcal{J}_q\right] = 0. \tag{3.121}$$

They have the same form as the phenomenological balance equations (A.12)–(A.15) in the absence of reaction (i.e., for $\sigma_k = 0$), as established in Chapter 1 on the basis of nonequilibrium thermodynamics. In particular, equation (3.120) is equivalent to the Navier–Stokes equations of hydrodynamics. Therefore, the microscopic approach based on the local equilibrium distribution (3.21) and the expansion in powers of the gradients justifies the rules of nonequilibrium thermodynamics for the linear transport properties.

The dispersion relations of the five hydrodynamic modes of relaxation towards equilibrium in one-component normal fluids are given in Section 1.6.1 and shown in Figure 1.3. If the fluid contained c components that are not reacting, the spectrum of hydrodynamic modes would also include $c-1$ diffusive modes with the dispersion relations $z_\alpha = -\mathcal{D}_\alpha q^2 + o(q^2)$, describing mutual diffusion between solvent and solute species, as mentioned in Section 1.6.2. The presence of reactions can also be formulated in the framework of nonequilibrium statistical mechanics, as sketched out in Section D.2.

3.3 Phases of Matter with Broken Continuous Symmetries

3.3.1 Continuous Symmetry Breaking and Long-Range Order

Beyond uniform and isotropic fluids, there exist other phases of matter where continuous translational or rotational symmetries are broken. This is the case for crystals, which are spatially periodic structures manifesting anisotropic properties according to any one of the 230 discrete space groups of three-dimensional space (Ashcroft and Mermin, 1976). Liquid crystals are phases with symmetries intermediate between those of fluids and crystals (Kats and Lebedev, 1994; Chaikin and Lubensky, 1995). They are composed of molecules with prolate or oblate shapes. Depending on the molecular shape and the conditions of temperature and pressure, different rotational and translational symmetries may be broken in liquid crystals. Nematics are liquid crystals composed of prolate molecules aligned around a preferential direction, thus manifesting anisotropic but spatially uniform properties. Cholesteric liquid crystals are chiral nematics, which are also spatially uniform with an axis of rotational symmetry. In the smectic phases of liquid crystals, the prolate molecules form layers and the molecules may have different possible orientations with respect to the layers. In the columnar discotic phases, oblate molecules may stack into columns possibly forming a two-dimensional lattice. In each of these equilibrium phases, particular rotational or translational symmetries are spontaneously broken if the conditions of composition, temperature, and pressure are met. Continuous symmetry breaking is also known in quantum regimes, for instance, in ferromagnetic systems, superconductors, or superfluids (Anderson, 1984).

In crystals and liquid crystals, the atoms or molecules form ordered structures, which are no longer symmetric under continuous spatial translations and/or rotations. Each structure is characterized by a specific order and an associated order parameter.

In crystals, the density of atoms or molecules is periodic in space, but this periodicity disappears at the melting phase transition where the solid becomes fluid again. Therefore, the crystalline order can be characterized by the Fourier amplitudes of the periodic density, which are observed in crystallographic diffraction patterns. Remarkably, this periodicity extends to the whole crystal, which may be arbitrarily large, so crystalline phases manifest long-range order.

In nematics, the molecules are aligned, on average, in a direction emerging from thermal fluctuations. Therefore, long-range order arises in the orientation of the molecules.

A nematic order parameter, called a director, can be introduced by considering the density of the microscopic orientations of each molecule (Kats and Lebedev, 1994; Chaikin and Lubensky, 1995). The equilibrium mean value of this order parameter is equal to zero in the isotropic fluid phase because the molecules have random orientations, but nonzero in the nematic phase where a direction is selected by spontaneous rotational symmetry breaking.

Long-range order provides rigidity to the structure emerging from broken continuous symmetries (Anderson, 1984). For instance, a force exerted on the surface of a crystal can drag the whole crystal, even if the force is parallel to the surface. This rigidity is a characteristic property of crystals, which do not flow contrary to fluids. This property arises from long-range order and does not need attractive forces between the constituents, as illustrated with crystalline structures made of hard spheres near the close-packing density. In such systems, the whole structure can be displaced by exerting a force only on the row of spheres parallel to the wall of the container. Such a rigidity also manifests itself in nematic liquid crystals, where the orientation of the molecules in the whole system can also be changed by acting on the surface.

If continuous symmetries are broken, order may be locally deformed transversally to the direction of ordering, while doing arbitrarily small mechanical work on the system. In crystals, the periodic structure may be locally deformed by some spatial variations of the displacement vector, i.e., the vector giving the displacement of the atoms with respect to their mean position in the lattice. If the displacement vector was spatially uniform, the whole crystal would be translated accordingly. In nematics, the spatial direction of the mean orientation of the molecules may be locally tilted by adding a perturbation transverse to the director. Again, if this perturbation was spatially uniform, the orientational order of the nematic phase would be globally rotated into a different direction.

A fundamental aspect is that such local deformations do not significantly change the structure if the broken symmetries are continuous. Therefore, the energy cost of such deformations can be arbitrarily small if the wavelength of the deformations is large enough, leading to the existence of soft dynamical modes, called Nambu–Goldstone modes (Nambu, 1960; Goldstone, 1961; Forster, 1975; Coullet and Iooss, 1990).

3.3.2 Nambu–Goldstone Modes

As a consequence of continuous symmetry breaking, there exist Nambu–Goldstone modes that behave as the slow modes of the locally conserved quantities with vanishing dispersion relations for arbitrarily large wavelength. The equations ruling all the slow and kinetic modes $\boldsymbol{\psi} = \langle \hat{\boldsymbol{\psi}} \rangle$ may be written as

$$\partial_t \boldsymbol{\psi} + \boldsymbol{F}(\boldsymbol{\psi}, \nabla\boldsymbol{\psi}, \dots) = 0. \tag{3.122}$$

Let us suppose that there exists an equilibrium solution

$$\boldsymbol{\psi}_{\rm eq}(\mathbf{r}; \boldsymbol{x}) \qquad \text{such that} \qquad \boldsymbol{F}(\boldsymbol{\psi}_{\rm eq}, \nabla\boldsymbol{\psi}_{\rm eq}, \dots) = 0, \tag{3.123}$$

where $\boldsymbol{x}$ denotes the equilibrium values of the order parameters, e.g., the global displacement vector in crystals, or the global director in nematic liquid crystals.

Now, we may consider small perturbations of different kinds with respect to this equilibrium solution.

On the one hand, we may have an additive perturbation giving solutions of the form

$$\boldsymbol{\psi}(\mathbf{r},t) = \boldsymbol{\psi}_{\text{eq}}(\mathbf{r};\boldsymbol{x}) + \delta\boldsymbol{\psi}(\mathbf{r},t). \tag{3.124}$$

Substituting into equation (3.122) and linearizing, we obtain

$$\partial_t\,\delta\boldsymbol{\psi} + \underbrace{(\boldsymbol{F})_{\text{eq}}}_{=0} + \left(\frac{\partial \boldsymbol{F}}{\partial \boldsymbol{\psi}}\right)_{\text{eq}} \cdot \delta\boldsymbol{\psi} + \left(\frac{\partial \boldsymbol{F}}{\partial \nabla\boldsymbol{\psi}}\right)_{\text{eq}} \cdot \delta\nabla\boldsymbol{\psi} + \cdots = 0. \tag{3.125}$$

Such a mode will decay exponentially in time with a rate that does not vanish for arbitrarily large wavelength, because $(\partial \boldsymbol{F}/\partial\boldsymbol{\psi})_{\text{eq}}$ is a matrix with nonvanishing elements.

On the other hand, we may also have a perturbation where the order parameter varies locally in space and time as

$$\boldsymbol{\psi}(\mathbf{r},t) = \boldsymbol{\psi}_{\text{eq}}\,[\mathbf{r};\boldsymbol{x}(\mathbf{r},t)], \tag{3.126}$$

so that

$$\partial_t\,\boldsymbol{\psi} = \frac{\partial \boldsymbol{\psi}_{\text{eq}}}{\partial \boldsymbol{x}} \cdot \partial_t\,\boldsymbol{x} \qquad \text{and} \qquad \nabla\boldsymbol{\psi} = \nabla\boldsymbol{\psi}_{\text{eq}} + \frac{\partial \boldsymbol{\psi}_{\text{eq}}}{\partial \boldsymbol{x}} \cdot \nabla\boldsymbol{x}. \tag{3.127}$$

Substituting into equation (3.122) and linearizing, we now find

$$\frac{\partial \boldsymbol{\psi}_{\text{eq}}}{\partial \boldsymbol{x}} \cdot \partial_t\,\boldsymbol{x} + \underbrace{(\boldsymbol{F})_{\text{eq}}}_{=0} + \left(\frac{\partial \boldsymbol{F}}{\partial \nabla\boldsymbol{\psi}}\right)_{\text{eq}} \cdot \frac{\partial \boldsymbol{\psi}_{\text{eq}}}{\partial \boldsymbol{x}} \cdot \nabla\boldsymbol{x} + \cdots = 0, \tag{3.128}$$

where the term with the nonvanishing coefficient $(\partial \boldsymbol{F}/\partial\boldsymbol{\psi})_{\text{eq}}$ no longer appears. Multiplying equation (3.128) by $(\partial\boldsymbol{\psi}_{\text{eq}}/\partial\boldsymbol{x})^{\text{T}}$ and integrating over space to average out the local variations of the symmetry-breaking equilibrium solution, we get

$$\boldsymbol{\mathcal{N}} \cdot \partial_t\,\boldsymbol{x} + \boldsymbol{\mathcal{M}} \cdot \nabla\boldsymbol{x} + \cdots = 0, \tag{3.129}$$

where the tensors $\boldsymbol{\mathcal{N}} = (\mathcal{N}_{\alpha\beta})$ and $\boldsymbol{\mathcal{M}} = (\mathcal{M}_{\alpha i\beta})$ are, respectively, defined as

$$\mathcal{N}_{\alpha\beta} \equiv \frac{1}{V}\int_V \frac{\partial \boldsymbol{\psi}_{\text{eq}}^{\text{T}}}{\partial x_\alpha} \cdot \frac{\partial \boldsymbol{\psi}_{\text{eq}}}{\partial x_\beta}\, d\mathbf{r}, \tag{3.130}$$

$$\mathcal{M}_{\alpha i\beta} \equiv \frac{1}{V}\int_V \frac{\partial \boldsymbol{\psi}_{\text{eq}}^{\text{T}}}{\partial x_\alpha} \cdot \left(\frac{\partial \boldsymbol{F}}{\partial \nabla_i\boldsymbol{\psi}}\right)_{\text{eq}} \cdot \frac{\partial \boldsymbol{\psi}_{\text{eq}}}{\partial x_\beta}\, d\mathbf{r}. \tag{3.131}$$

In equation (3.129), the dots denote terms with higher spatial derivatives for $\boldsymbol{x}(\mathbf{r},t)$. Since the solution (3.126) breaks continuous symmetries, we have that $(\partial\boldsymbol{\psi}_{\text{eq}}/\partial\boldsymbol{x}) \neq 0$ and thus $\boldsymbol{\mathcal{N}} \neq 0$, so that $\boldsymbol{\mathcal{N}}^{-1}$ exists, leading to the definition of the second-rank tensor $\boldsymbol{\mathcal{V}} \equiv \boldsymbol{\mathcal{N}}^{-1} \cdot \boldsymbol{\mathcal{M}}$. If we suppose that the deformation of the local order parameter behaves as $\delta\boldsymbol{x}(\mathbf{r},t) \sim \exp(\imath\mathbf{q}\cdot\mathbf{r} - \imath\omega t)$, the frequency ω is related to the wave vector $\mathbf{q}$ according to

$$\left[\omega\,\mathbf{1} - \boldsymbol{\mathcal{V}}\cdot\mathbf{q} + O\big(\mathbf{q}^2\big)\right] \cdot \delta\boldsymbol{x} = 0, \tag{3.132}$$

showing that the dispersion relations of these other solutions vanish with the wave number as $\lim_{\mathbf{q}\to 0} \omega(\mathbf{q}) = 0$, as for the locally conserved quantities. The eigenvalues of the matrix $\boldsymbol{\mathcal{V}} \cdot \mathbf{q}$ give the propagation speeds of the modes. Therefore, the dispersion relations of the Nambu–Goldstone modes tend to zero linearly with the wave number if $\boldsymbol{\mathcal{V}} \neq 0$, and faster than linearly if $\boldsymbol{\mathcal{V}} = 0$.

These considerations show the existence of the Nambu–Goldstone modes as a consequence of continuous symmetry breaking. There are as many such modes as components for the vector $\boldsymbol{x} = (x_\alpha)$, i.e., as continuous symmetries that are broken, which is the statement of the theorem by Goldstone (1961).

3.3.3 Microscopic Order Fields

In matter phases where continuous symmetries are broken, the atoms or molecules are ordered into characteristic structures, which exist in spite of the irregular thermal motion of the atoms. Accordingly, the properties of these phases should be described using some equilibrium statistical distribution as reference state. A key point is that this equilibrium distribution should no longer be symmetric under spatial translations and rotations, in order to describe the symmetry breaking phase.

Under spatial translations shifting the particle positions as $\mathbf{r}_a \to \mathbf{r}_a + \mathbf{R}$ with $\mathbf{R} \in \mathbb{R}^3$ but leaving their momenta unchanged so that $\mathbf{p}_a \to \mathbf{p}_a$, any phase-space probability density $p(\boldsymbol{\Gamma})$ is transformed according to

$$\hat{T}^{\mathbf{R}} p = p + \mathbf{R} \cdot \{\mathbf{P}, p\} + O\big(\mathbf{R}^2\big), \tag{3.133}$$

where $\mathbf{P} = \sum_{a=1}^{N} \mathbf{p}_a$ is the total momentum and $\{\cdot, \cdot\}$ denotes the Poisson bracket if the translation $\mathbf{R}$ is infinitesimal.

Under rotations by the angle θ around the direction of the unit vector $\mathbf{n}$ including the origin of the reference frame, the positions undergo the transformations $\mathsf{O}_{\mathbf{n}}^{\theta} \cdot \mathbf{r}_a = \mathbf{r}_a + \theta\, \mathbf{n} \times \mathbf{r}_a + O(\theta^2)$, while the momenta change similarly. Therefore, the probability density is transformed under infinitesimal rotations as

$$\hat{T}_{\mathbf{n}}^{\theta} p = p + \theta \{\mathbf{n} \cdot \mathbf{L}, p\} + O\big(\theta^2\big) \tag{3.134}$$

with the total angular momentum $\mathbf{L} = \sum_{a=1}^{N} \mathbf{r}_a \times \mathbf{p}_a$.

The issue is that the grand canonical equilibrium distribution (3.34) has all the translational and rotational symmetries for the Hamiltonian functions (2.2) ruling the motion of atoms interacting by central forces in the bulk of matter (i.e., away from the walls so that the external potentials $u_a^{(\mathrm{ext})}$ are vanishing). Indeed, $\{\mathbf{P}, H\} = 0$ and $\{\mathbf{L}, H\} = 0$ for such Hamiltonian functions, so that all the statistical properties are symmetric under continuous translations and rotations if the equilibrium distribution (3.34) is used.

In order to circumvent this issue, we may consider perturbations due to some external fields having the effect of explicitly breaking the symmetries, thus leading to the emergence of the ordered phase under appropriate conditions for composition, temperature, and pressure. Since these perturbations should be arbitrarily small if the external fields have a long enough wavelength, the Hamiltonian function may be modified into

$$H_\lambda = H + \lambda\, U^{(\text{ext})} \qquad \text{with} \qquad U^{(\text{ext})} = -\int \sum_\alpha f_\alpha^{(\text{ext})}(\mathbf{r})\, \delta\hat{x}_\alpha(\mathbf{r};\boldsymbol{\Gamma})\, d\mathbf{r} \tag{3.135}$$

with external fields $f_\alpha^{(\text{ext})}(\mathbf{r})$ directly exerted on the deformations $\delta\hat{x}_\alpha(\mathbf{r};\boldsymbol{\Gamma})$ of the local order parameters, which may thus be identified in this way. Now, the corresponding grand canonical equilibrium distribution

$$p_{\text{eq},\lambda}(\boldsymbol{\Gamma}) = \frac{1}{\Xi_\lambda \Delta\boldsymbol{\Gamma}}\, \mathrm{e}^{-\beta(H_\lambda - \sum_{k=1}^{c} \mu_k N_k - \mathbf{v}\cdot\mathbf{P})} \tag{3.136}$$

is no longer symmetric under translations and rotations, since $\{\mathbf{P}, H_\lambda\} \neq 0$ and $\{\mathbf{L}, H_\lambda\} \neq 0$ in general. By explicitly breaking the continuous symmetries, the mean values of the order parameters may now be nonvanishing. In the limit $\lambda \to 0$ of arbitrarily small external perturbations, the order parameters may keep their nonvanishing values if the ordered phase is stable under the given temperature and pressure conditions. In such circumstances, we speak about spontaneous symmetry breaking.

In the presence of long-range order, the equilibrium correlation functions of the deformations in the local order parameter decay in real and Fourier spaces respectively as

$$\langle \delta\hat{x}_\alpha(\mathbf{r})\, \delta\hat{x}_\beta(\mathbf{r}')\rangle_{\text{eq},0} \sim \|\mathbf{r}-\mathbf{r}'\|^{-1} \quad \text{and} \quad \left\langle \delta\tilde{\hat{x}}_\alpha(\mathbf{q})\, \delta\tilde{\hat{x}}_\beta(-\mathbf{q})\right\rangle_{\text{eq},0} \sim \|\mathbf{q}\|^{-2}, \tag{3.137}$$

as can be calculated for the displacement vector in harmonic models of solids (Landau and Lifshitz, 1980a, 1984; Chaikin and Lubensky, 1995). In this regard, the microscopic fields $\delta\hat{x}_\alpha(\mathbf{r})$ have a singular behavior, contrary to the regular locally conserved densities (3.3)–(3.6) that have correlations of short or medium range. In the following, these fields will be more simply denoted $\hat{x}_\alpha(\mathbf{r})$. An important observation is that their gradients

$$\hat{\mathsf{u}}_{i\alpha} \equiv \nabla_i\, \hat{x}_\alpha \qquad \text{or} \qquad \hat{\mathbf{u}} \equiv \boldsymbol{\nabla}\, \hat{\boldsymbol{x}} \tag{3.138}$$

have short- or medium-ranged correlations because

$$\left\langle \delta\tilde{\hat{\mathbf{u}}}(\mathbf{q})\, \delta\tilde{\hat{\mathbf{u}}}(-\mathbf{q})\right\rangle_{\text{eq},0} \sim \|\mathbf{q}\|^{0}. \tag{3.139}$$

We note that long-range order is only possible for space dimension larger than two, as shown by Mermin and Wagner (1966). In particular, thermal fluctuations destroy crystalline long-range order in dimension two (Mermin, 1968), where the autocorrelation function of particle density has an algebraic decay of its spatial periodicity, instead of a periodicity extending to infinity in dimension three (Landau and Lifshitz, 1980a). Some phases of liquid crystals manifest a similar quasi long-range order with an algebraic decay of density autocorrelation functions (Chaikin and Lubensky, 1995), which may affect the dynamics of the corresponding soft modes (Kats and Lebedev, 1994).

Knowing the microscopic expression of the order field $\hat{x}_\alpha(\mathbf{r};\boldsymbol{\Gamma})$, its evolution equation can be deduced from the Hamilton equations of motion, giving

$$\partial_t\, \hat{x}_\alpha + \hat{\gamma}_\alpha = 0, \tag{3.140}$$

where $\hat{\gamma}_\alpha(\mathbf{r};\boldsymbol{\Gamma})$ is the local decay rate of the order field. Accordingly, the order field is not a conserved quantity, so that it does not obey a continuity equation, and should be considered *a priori* belonging to the kinetic modes. Nevertheless, the gradients (3.138) of the order fields obey the continuity equation,

$$\partial_t \hat{\mathsf{u}} + \nabla \cdot \hat{\jmath}_{\mathsf{u}} = 0 \qquad \text{with} \qquad \hat{\jmath}_{\mathsf{u}} \equiv \mathbf{1}\,\hat{\gamma}, \tag{3.141}$$

as a direct consequence of equation (3.140). Hence, the gradients of the order fields behave as the densities ruled by the fundamental local conservation laws (3.8)–(3.11).

3.3.4 Local Equilibrium Approach

From here on, we suppose for simplicity that the system is composed of a single atomic species ($c = 1$).

In order to investigate the effects of the Nambu–Goldstone modes, we should thus consider equation (3.141) on the same footing as the balance equations (3.8)–(3.11) for the locally conserved quantities. In this respect, equation (3.16) here holds with the following densities and current densities including those associated with the Nambu–Goldstone modes:

$$\hat{\mathbf{c}} = \left(\hat{\epsilon}, \hat{n}, \hat{\mathbf{g}}, \hat{\mathsf{u}}\right) \qquad \text{and} \qquad \hat{\jmath} = \left(\hat{\jmath}_\epsilon, \hat{\jmath}_n, \hat{\jmath}_{\mathbf{g}}, \hat{\jmath}_{\mathsf{u}}\right). \tag{3.142}$$

With this extension, the local equilibrium distribution (3.21) can be defined and the methods of Sections 3.2.3, 3.2.4, and 3.2.5 can be applied here as well (Mabillard and Gaspard, 2020).

Local Thermodynamics

In this way, the results of Section 3.2.6 extend to phases with broken continuous symmetries. At the leading order of the expansion in the gradients, the functionals (3.23) and (3.26) may be taken in the forms (3.57).

Here, the entropy density should obey the following Euler and Gibbs relations,

$$s = \frac{\epsilon + p}{T} - \frac{\mu}{T} n - \frac{\mathbf{v}}{T} \cdot \mathbf{g} - \frac{\boldsymbol{\phi}}{T} : \mathsf{u}, \qquad ds = \frac{1}{T} d\epsilon - \frac{\mu}{T} dn - \frac{\mathbf{v}}{T} \cdot d\mathbf{g} - \frac{\boldsymbol{\phi}}{T} : d\mathsf{u}, \tag{3.143}$$

up to terms of $O\left(\nabla^2\right)$, where ϵ and μ are defined in equation (3.60), $\mathbf{v} \cdot \mathbf{g} = v_i g_i$, and $\boldsymbol{\phi} : \mathsf{u} = \phi_{i\alpha} \mathsf{u}_{i\alpha}$ with the convention of summation over repeated indices. As a consequence, the conjugate fields are given by equations (3.62), (3.63), (3.64), and also

$$\chi_{\mathsf{u}_{i\alpha}}(\mathbf{r}, t) \equiv \frac{1}{k_{\mathrm{B}}} \frac{\delta S(\mathbf{c})}{\delta \mathsf{u}_{i\alpha}(\mathbf{r}, t)} = -\beta(\mathbf{r}, t)\, \phi_{i\alpha}(\mathbf{r}, t) + O\left(\nabla^2\right). \tag{3.144}$$

Comparing with the Hamiltonian function (3.135), we see that the applied external fields can be expressed as $f_\alpha^{(\mathrm{ext})} = -\nabla_i \phi_{i\alpha}$ for given conjugate fields $\phi_{i\alpha}$ in isothermal systems. Accordingly, the fields $\phi_{i\alpha}$ can be interpreted as contributions to the stress on the system.

Furthermore, the Massieu density $\omega = \beta p$ satisfies the following Gibbs–Duhem relation instead of (3.65):

$$d\omega = -\epsilon\, d\beta + n\, d(\beta\mu) + \mathbf{g} \cdot d(\beta \mathbf{v}) + \mathsf{u} : d(\beta \boldsymbol{\phi}). \tag{3.145}$$

Dissipativeless Time Evolution

According to equation (3.54), entropy would be conserved if the dissipative current densities (3.53) could be neglected. In this case, the local conservation equations would be given by dissipativeless equations

$$\partial_t \langle \hat{\mathbf{c}}(\mathbf{r}) \rangle_{\text{leq},\chi_t} + \nabla \cdot \langle \hat{\jmath}(\mathbf{r};\Gamma) \rangle_{\text{leq},\chi_t} = 0, \tag{3.146}$$

which is equivalent to equation (3.56) with $\mathcal{J} = 0$.

The local equilibrium mean values (3.24) of the microscopic densities define the corresponding macroscopic fields according to equations (3.37) and (3.44). As carried out in Section 3.2.7, the local equilibrium mean values (3.52) of the current densities are obtained by considering a Galilean transformation to the frame moving with the matter element, giving equations (3.66)–(3.68). Since $\langle \hat{\epsilon}_0 \rangle_{\text{leq}} = \epsilon_0$, $\langle \hat{\rho} \rangle_{\text{leq}} = \rho = mn$ with $n = \langle \hat{n} \rangle_{\text{leq}}$, and the velocity field is defined as $\mathbf{v} \equiv \langle \hat{\mathbf{g}} \rangle_{\text{leq}} / \langle \hat{\rho} \rangle_{\text{leq}}$, we obtain the local equilibrium mean values (3.70), but instead of equations (3.71)–(3.72), we have

$$\bar{\jmath}_{\mathbf{g}} = \langle \hat{\jmath}_{\mathbf{g}} \rangle_{\text{leq}} = \rho \mathbf{v}\mathbf{v} + p\,\mathbf{1} + \jmath_{\mathbf{g}}^{(\text{BS})} + O(\nabla^2), \tag{3.147}$$

$$\bar{\jmath}_{\epsilon} = \langle \hat{\jmath}_{\epsilon} \rangle_{\text{leq}} = (\epsilon + p)\,\mathbf{v} + \jmath_{\epsilon}^{(\text{BS})} + O(\nabla^2), \tag{3.148}$$

$$\bar{\jmath}_{\mathbf{u}} = \langle \hat{\jmath}_{\mathbf{u}} \rangle_{\text{leq}} = \mathbf{u}\mathbf{v} + \jmath_{\mathbf{u}}^{(\text{BS})} + O(\nabla^2), \tag{3.149}$$

where we should consider the possibility that symmetry breaking may introduce extra contributions in the local equilibrium mean current densities of energy, momentum, and the gradient $\mathbf{u}$ of the order parameter. In particular, equation (3.141) implies that

$$\jmath_{\mathbf{u}}^{(\text{BS})} = \mathbf{1}\,\gamma \qquad \text{or} \qquad \jmath_{u_{j\alpha},i}^{(\text{BS})} = \delta_{ij}\,\gamma_\alpha. \tag{3.150}$$

In general, we should expect that

$$\gamma_\alpha = -A_{i\alpha}\, v_i - B_{ij\alpha}\, \nabla_i v_j + O(\nabla^2) \tag{3.151}$$

with $B_{ij\alpha} = 0$ in crystals and $A_{i\alpha} = 0$ in nematic liquid crystals (Martin et al., 1972).

In order to determine the contributions $\jmath_{\mathbf{g}}^{(\text{BS})}$ and $\jmath_{\epsilon}^{(\text{BS})}$ to the current densities of momentum and energy, we consider the identity (3.39), which requires that the dissipativeless current densities (3.52) should conserve entropy (since they rule adiabatic, i.e., isoentropic, processes) (Mabillard and Gaspard, 2020). Therefore, we should have that

$$\nabla\beta * \bar{\jmath}_\epsilon - \nabla(\beta\mu) * \bar{\jmath}_n - \nabla(\beta\mathbf{v}) * \bar{\jmath}_{\mathbf{g}} - \nabla(\beta\boldsymbol{\phi}) * \bar{\jmath}_{\mathbf{u}} = 0, \tag{3.152}$$

up to higher-order corrections. We note that $\bar{\jmath}_n = n\mathbf{v}$ because of equation (3.70). Using the Gibbs–Duhem relation (3.145) with the differential d replaced by the gradient ∇, replacing the dissipativeless current densities by their expressions (3.147)–(3.149) with $\mathbf{g} = \rho\mathbf{v}$, their broken symmetry parts should satisfy the identity,

$$\nabla\beta * \bar{\jmath}_\epsilon^{(\text{BS})} - \nabla(\beta\mathbf{v}) * \bar{\jmath}_{\mathbf{g}}^{(\text{BS})} - \nabla(\beta\boldsymbol{\phi}) * \bar{\jmath}_{\mathbf{u}}^{(\text{BS})} = 0. \tag{3.153}$$

Expanding the gradients and using equations (3.150) and (3.151), we find

$$J_{g_j,i}^{(\mathrm{BS})} = -\phi_{i\alpha} A_{j\alpha} + B_{ij\alpha} \nabla_k \phi_{k\alpha} + O(\nabla^2), \tag{3.154}$$

$$J_{\epsilon,i}^{(\mathrm{BS})} = -\phi_{i\alpha} A_{j\alpha} v_j + B_{ij\alpha} v_j \nabla_k \phi_{k\alpha} - \phi_{i\alpha} B_{jk\alpha} \nabla_j v_k + O(\nabla^2), \tag{3.155}$$

and the condition $\nabla_i A_{j\alpha} = 0$ (Mabillard and Gaspard, 2020). The dissipativeless parts of the momentum and energy current densities are thus obtained as

$$\bar{J}_{g_j,i} = \rho v_i v_j - \sigma_{ij} + O(\nabla^2), \tag{3.156}$$

$$\bar{J}_{\epsilon,i} = \epsilon v_i - \sigma_{ij} v_j - \phi_{i\alpha} B_{jk\alpha} \nabla_j v_k + O(\nabla^2), \tag{3.157}$$

with the reversible stress tensor defined by

$$\sigma_{ij} \equiv -p\,\delta_{ij} + \phi_{i\alpha} A_{j\alpha} - B_{ij\alpha} \nabla_k \phi_{k\alpha}. \tag{3.158}$$

The results are consistent with macroscopic theory (Martin et al., 1972; Fleming and Cohen, 1976). We note that entropy is conserved by these current densities.

Accordingly, the dissipativeless macroscopic equations for the particle density, the velocity field, the internal energy density, and the gradients of the order fields can be expressed as

$$\frac{dn}{dt} = -n \nabla_i v_i, \tag{3.159}$$

$$\rho \frac{dv_i}{dt} = \nabla_j \sigma_{ji}, \tag{3.160}$$

$$\frac{d\epsilon_0}{dt} = -\epsilon_0 \nabla_i v_i + \sigma_{ij} \nabla_i v_j + \nabla_i (\phi_{i\alpha} B_{jk\alpha} \nabla_j v_k), \tag{3.161}$$

$$\frac{d\mathsf{u}_{i\alpha}}{dt} = -\mathsf{u}_{i\alpha} \nabla_j v_j + \nabla_i (A_{j\alpha} v_j + B_{jk\alpha} \nabla_j v_k), \tag{3.162}$$

in terms of the reversible stress tensor (3.158) and up to corrections of higher order in the gradients. The dissipativeless equations for the time evolution of the conjugate fields can also be deduced (Mabillard and Gaspard, 2020).

Dissipative Time Evolution and Entropy Production

In one-component matter with broken continuous symmetries, the entropy production rate at leading order in the gradients has the form

$$\frac{1}{k_{\mathrm{B}}} \frac{d_{\mathrm{i}} S}{dt} \simeq \int \left(\nabla \beta \cdot \mathcal{J}_q - \beta\, \nabla \mathbf{v} : \mathcal{J}_{\mathbf{g}} - \beta\, \nabla \boldsymbol{\phi} \,\vdots\, \mathcal{J}_{\mathsf{u}} \right) d\mathbf{r}, \tag{3.163}$$

where the heat current density is here given by

$$\mathcal{J}_q \equiv \mathcal{J}_\epsilon - \mathbf{v} \cdot \mathcal{J}_{\mathbf{g}} - \boldsymbol{\phi} : \mathcal{J}_{\mathsf{u}}, \tag{3.164}$$

$\mathcal{J}_{\mathbf{g}}$ is the dissipative part of the pressure tensor, and

$$\mathcal{J}_{\mathsf{u}} = \mathbf{1}\, \mathcal{J}_{\boldsymbol{x}} \qquad \text{or} \qquad \mathcal{J}_{\mathsf{u}_{j\alpha},i} = \delta_{ij}\, \mathcal{J}_{x_\alpha} \tag{3.165}$$

is the dissipative part of the current density associated with the order fields and introduced in equation (3.141).

The dissipative current densities can be evaluated using equations (3.85)–(3.89) (Mabillard and Gaspard, 2020). As in Section 3.2.8 for normal fluids, the dissipative current densities can be expressed at leading order in the gradients according to equation (3.94), but with an extra term involving $\beta(\mathbf{r}',\tau)\nabla'\cdot\boldsymbol{\phi}(\mathbf{r}',\tau)$. This form gives the dissipative current densities ruling viscoelasticity (DeVault and McLennan, 1965) because the time correlation functions may describe memory effects during relaxation.

Green–Kubo Formulas for the Transport Coefficients

As in normal fluids, the timescales of the memory effects may be assumed to be short with respect to the observational timescales and, moreover, the spatial range of the correlation functions can be supposed to be shorter than the wavelength of the conjugate fields. If these conditions hold, the dissipative current densities can be expressed as

$$\mathcal{J}_{q,i} = -\kappa_{ij}\nabla_j T - \chi_{ijk}\nabla_j v_k - \xi_{i\alpha}\nabla_j\phi_{j\alpha}, \tag{3.166}$$

$$\mathcal{J}_{g_j,i} = \frac{\chi_{kij}}{T}\nabla_k T - \eta_{ijkl}\nabla_k v_l - \theta_{ij\alpha}\nabla_k\phi_{k\alpha}, \tag{3.167}$$

$$\mathcal{J}_{x_\alpha} = -\frac{\xi_{i\alpha}}{T}\nabla_i T + \theta_{ij\alpha}\nabla_i v_j - \zeta_{\alpha\beta}\nabla_i\phi_{i\beta}, \tag{3.168}$$

with the following Green–Kubo formulas for the transport coefficients:

$$\kappa_{ij} \equiv \lim_{V\to\infty}\frac{1}{Vk_{\mathrm{B}}T^2}\int_0^\infty dt\,\langle\delta J_{\epsilon i}(t)\,\delta J_{\epsilon j}(0)\rangle_{\mathrm{eq}}, \tag{3.169}$$

$$\eta_{ijkl} \equiv \lim_{V\to\infty}\frac{1}{Vk_{\mathrm{B}}T}\int_0^\infty dt\,\langle\delta J_{ij}(t)\,\delta J_{kl}(0)\rangle_{\mathrm{eq}}, \tag{3.170}$$

$$\xi_{i\alpha} \equiv \lim_{V\to\infty}\frac{1}{Vk_{\mathrm{B}}T}\int_0^\infty dt\,\langle\delta J_{\epsilon i}(t)\,\delta\Gamma_\alpha(0)\rangle_{\mathrm{eq}}, \tag{3.171}$$

$$\zeta_{\alpha\beta} \equiv \lim_{V\to\infty}\frac{1}{Vk_{\mathrm{B}}T}\int_0^\infty dt\,\langle\delta\Gamma_\alpha(t)\,\delta\Gamma_\beta(0)\rangle_{\mathrm{eq}}, \tag{3.172}$$

$$\chi_{ijk} \equiv \lim_{V\to\infty}\frac{1}{Vk_{\mathrm{B}}T}\int_0^\infty dt\,\langle\delta J_{\epsilon i}(t)\,\delta J_{jk}(0)\rangle_{\mathrm{eq}}, \tag{3.173}$$

$$\theta_{ij\alpha} \equiv \lim_{V\to\infty}\frac{1}{Vk_{\mathrm{B}}T}\int_0^\infty dt\,\langle\delta J_{ij}(t)\,\delta\Gamma_\alpha(0)\rangle_{\mathrm{eq}}. \tag{3.174}$$

In these latter formulas, the microscopic total currents are defined by

$$\delta J_{\epsilon i} \equiv \int_V \delta\hat{\jmath}_{\epsilon,i}\,d\mathbf{r} - \rho^{-1}(\epsilon_0 + p)\,\delta P_i, \tag{3.175}$$

$$\delta J_{ij} \equiv \int_V \delta\hat{\jmath}_{g_j,i}\,d\mathbf{r} + \left(\frac{\partial\sigma_{ij}}{\partial\epsilon_0}\right)_{n,\mathbf{u}}\delta E + \left(\frac{\partial\sigma_{ij}}{\partial n}\right)_{\epsilon_0,\mathbf{u}}\delta N + \left(\frac{\partial\sigma_{ij}}{\partial\mathbf{u}}\right)_{n,\epsilon_0} : \int_V \delta\hat{\mathbf{u}}\,d\mathbf{r}, \tag{3.176}$$

$$\delta\Gamma_\alpha \equiv \int_V \delta\hat{\gamma}_\alpha\,d\mathbf{r} + \rho^{-1}A_{j\alpha}\,\delta P_j, \tag{3.177}$$

where the volume integrals of the microscopic current densities are given by equations (3.100) and (3.101) for energy and momentum with a similar expression for the microscopic quantity $\hat{\gamma}_\alpha$. Moreover, $\delta P_i = \int \delta\hat{g}_i \, d\mathbf{r}$, $\delta E = \int \delta\hat{\epsilon} \, d\mathbf{r}$, and $\delta N = \int \delta\hat{n} \, d\mathbf{r}$ are the fluctuations of the total momentum, energy, and particle number in the equilibrium statistical ensemble considered (Mabillard and Gaspard, 2020).

The coefficients (3.169)–(3.174) can be interpreted as follows: κ_{ij} are the heat conductivities, η_{ijkl} the viscosities, $\xi_{i\alpha}$ the coefficients coupling heat transport and the order fields, $\zeta_{\alpha\beta}$ the friction coefficients of the order fields, χ_{ijk} the coefficients coupling heat and momentum transports, and $\theta_{ij\alpha}$ the coefficients coupling momentum transport and the order fields. All these coefficients obey the spatial symmetries of the anisotropic phase according to the principle of Curie (1894). In particular, third-rank tensors such as χ_{ijk} and $\theta_{ij\alpha}$ are known to vanish in isotropic phases, but they may be nonvanishing if the phase belongs to a symmetry class compatible with piezoelectricity (Landau and Lifshitz, 1984).

Consequences of Microreversibility

As a consequence of the time-reversal symmetry of the microscopic Hamiltonian dynamics, the transport coefficients obey the Onsager–Casimir reciprocal relations (3.115). Since the energy density $\hat{\epsilon}$ and the order fields $\hat{x}_\alpha$ are even under time reversal, the current density $\hat{j}_\epsilon$ and the rates γ_α are odd. In addition, the momentum current densities $\hat{j}_{g_j,i}$ are even because $\hat{g}_j$ are odd. Therefore, microreversibility implies that $\kappa_{ij} = \kappa_{ji}$, $\eta_{ijkl} = \eta_{klij}$, $\xi_{i\alpha} = \xi_{\alpha i}$, and $\zeta_{\alpha\beta} = \zeta_{\beta\alpha}$.

Moreover, the third-rank tensors χ_{ijk} and $\theta_{ij\alpha}$ should obey the Onsager–Casimir reciprocal relations (3.115) with $\epsilon_\alpha \epsilon_\beta = -1$, which explains the changes of sign in front of $\nabla_k T$ in equation (3.167) and in front of $\nabla_i v_j$ in equation (3.168) with respect to the other terms with the same coefficients in equations (3.166) and (3.167), respectively. As a consequence of this antisymmetry, these coupling coefficients do not contribute to the entropy production and thus to dissipation, which is confirmed by calculating the entropy production rate (3.163) from equations (3.166)–(3.168), giving

$$\frac{d_{\rm i} S}{dt} \simeq \int \frac{1}{T}\Bigg(\eta_{ijkl}\, \nabla_i v_j \, \nabla_k v_l + \frac{\kappa_{ij}}{T}\, \nabla_i T \, \nabla_j T \\ + 2\,\frac{\xi_{i\alpha}}{T}\, \nabla_i T \, \nabla_j \phi_{j\alpha} + \zeta_{\alpha\beta}\, \nabla_i \phi_{i\alpha} \, \nabla_j \phi_{j\beta}\Bigg) d\mathbf{r} \geq 0. \tag{3.178}$$

Furthermore, the second law of thermodynamics implies the following inequalities on the transport coefficients: $\eta_{ijij} \geq 0$, $\kappa_{ii} \geq 0$, and $\zeta_{\alpha\alpha} \geq 0$, and $\kappa_{ii}\zeta_{\alpha\alpha} \geq (\xi_{i\alpha})^2/T$ (Haase, 1969).

Dissipative Macroscopic Equations

Finally, the macroscopic equations are given by equation (3.56) in terms of the dissipativeless and dissipative current densities and rates, which have been calculated with the expansion in powers of the gradients. The modes of relaxation towards equilibrium can be obtained by considering the linearized macroscopic equations. Keeping the terms that are linear in the gradients and the velocity, we obtain

$$\partial_t \rho \simeq -\rho\, \nabla_i v_i, \tag{3.179}$$

$$\partial_t \epsilon_0 \simeq -(\epsilon_0 + p)\nabla_i v_i + \chi_{ijk}\nabla_i \nabla_j v_k + \kappa_{ij}\nabla_i \nabla_j T + \xi_{i\alpha}\nabla_i \nabla_j \phi_{j\alpha}, \tag{3.180}$$

$$\rho\, \partial_t v_j \simeq -\nabla_j p + A_{j\alpha}\nabla_i \phi_{i\alpha} - (B_{ij\alpha} - \theta_{ij\alpha})\nabla_i \nabla_k \phi_{k\alpha} - \frac{\chi_{kij}}{T}\nabla_i \nabla_k T + \eta_{ijkl}\nabla_i \nabla_k v_l, \tag{3.181}$$

$$\partial_t x_\alpha \simeq A_{i\alpha} v_i + (B_{ij\alpha} - \theta_{ij\alpha})\nabla_i v_j + \frac{\xi_{i\alpha}}{T}\nabla_i T + \zeta_{\alpha\beta}\nabla_i \phi_{i\beta}. \tag{3.182}$$

We note that the coefficients $B_{ij\alpha}$ introduced in equation (3.151) are of the same kind as the coupling coefficients $\theta_{ij\alpha}$ and, consistently, neither contribute to dissipation. The system of linearized macroscopic equations (3.179)–(3.182) can be closed using the thermodynamic relations in order to determine the hydrodynamic modes, including the Nambu–Goldstone modes emerging from continuous symmetry breaking.

3.3.5 Liquid Crystals

Liquid crystals are composed of highly anisotropic molecules, which interact with different types of intermolecular forces. Rotational or translational symmetries may be broken in liquid crystals because of the emergence of a preferential orientation, e.g., in nematics, or two-dimensional columnar order, e.g., in some phases of discotic liquid crystals (Kats and Lebedev, 1994; Chaikin and Lubensky, 1995).

In nematics, the molecules are, on average, aligned in a direction emerging from thermal fluctuations. Thus long-range order arises in the orientation of the molecules. A nematic order parameter can be introduced by considering the density of the microscopic orientation for each molecule (Forster, 1975; Kats and Lebedev, 1994; Chaikin and Lubensky, 1995). The equilibrium mean value of this order parameter is equal to zero in the isotropic fluid phase because the molecules have random orientations, but nonvanishing in the nematic phase where a direction is selected by spontaneous ordering. Here, an external field applied to such an order parameter may induce local deformations of the orientation, which can be described by the external potentials in the Hamiltonian function (3.135).

Microscopic Order Fields

For dielectric apolar nematogens (i.e., nematic molecules), an external electric field $\mathcal{E}(\mathbf{r})$ can explicitly break the rotation symmetry. In this case, the total external potential energy in the Hamiltonian function (3.135) is given by

$$U^{(\mathrm{ext})} = -\frac{1}{2}\int \mathcal{E}_i(\mathbf{r})\, \hat{q}_{ij}(\mathbf{r};\boldsymbol{\Gamma})\, \mathcal{E}_j(\mathbf{r})\, d\mathbf{r}, \quad \text{where} \quad \hat{q}_{ij}(\mathbf{r};\boldsymbol{\Gamma}) = \sum_m \mathsf{q}_{m,ij}(\boldsymbol{\Gamma})\, \delta(\mathbf{r} - \mathbf{r}_m) \tag{3.183}$$

is the traceless second-rank tensor defining the local order parameters of some quadrupolar property associated with the nematogens, $\mathbf{r}_m$ being the position of the molecule m (Chaikin and Lubensky, 1995). This order parameter obeys the evolution equation

$$\partial_t\, \hat{q}_{ij} + \hat{\gamma}_{ij} = 0 \tag{3.184}$$

with the associated decay rate $\hat{\gamma}_{ij}$ given by

$$\hat{\gamma}_{ij} = \nabla \cdot \left[\sum_m \dot{\mathbf{r}}_m \mathrm{q}_{m,ij}\, \delta(\mathbf{r} - \mathbf{r}_m) \right] - \sum_m \dot{\mathrm{q}}_{m,ij}\, \delta(\mathbf{r} - \mathbf{r}_m), \tag{3.185}$$

where dots denote time derivatives with respect to the Hamiltonian microscopic dynamics. If the local order parameters in equation (3.183) are taken as the order fields $\hat{x}_\alpha$ with $\alpha = ij$, the corresponding microscopic rates (3.140) are given by equation (3.185).

Carrying out the change of variables $\mathbf{p}_a = \mathbf{p}_{a0} + m_a \mathbf{v}(\mathbf{r}_a)$ for the momenta of the atoms in the molecules where $\mathbf{v}(\mathbf{r})$ is the velocity field, and expanding the difference of velocities as $\mathbf{v}(\mathbf{r}_a) - \mathbf{v}(\mathbf{r}_m) = (\mathbf{r}_a - \mathbf{r}_m) \cdot \nabla \mathbf{v}(\mathbf{r}_m) + O(\nabla^2)$, where $\|\mathbf{r}_a - \mathbf{r}_m\|$ is of the order of molecular radius, the mean values of the rates (3.185) over the local equilibrium distribution give their dissipativeless parts

$$\langle \hat{\gamma}_{ij} \rangle_{\mathrm{leq}} = -A_{ijk}\, v_k - B_{ijkl}\, \nabla_k v_l + O(\nabla^2), \tag{3.186}$$

where $A_{ijk} = -\nabla_k \langle \hat{q}_{ij} \rangle_{\mathrm{leq}}$, while B_{ijkl} are the coefficients in front of the velocity gradients (Mabillard and Gaspard, 2020). For nematics, we have that $A_{ijk} = -\nabla_k \langle \hat{q}_{ij} \rangle_{\mathrm{leq}} = 0$ since the local order parameters are uniform at equilibrium. Therefore, equation (3.151) is justified for such liquid crystals on the basis of the microscopic approach.

We note that the number of independent order fields is equal to the number of continuous broken symmetries. In nematics at equilibrium, the properties are uniform in space as in normal fluids, but anisotropic because of the preferential direction given by the director. Accordingly, the rotational symmetry group is reduced from the three-dimensional rotation group O(3) to the group O(2) of rotations around the director $\mathbf{n}$. Since three-dimensional rotations are specified by three Euler's angles and rotations around an axis by a single angle, there are two continuous symmetries that are broken in nematics, so that we should expect two independent order fields, which can be defined as the perturbations of the mean molecular orientation with respect to the selected director. Since $\|\mathbf{n}\| = 1$ and $\mathbf{n} \cdot \delta\mathbf{n} = 0$, the deformations $\delta\mathbf{n}$ are perpendicular to the director and they have two components defining the two order fields $\hat{x}_\alpha$ with $\alpha = 1, 2$ (Forster, 1975).

Transport Properties

In liquid crystals, the transport properties are obtained by considering the dissipative parts of the current densities according to equations (3.166)–(3.168). The transport coefficients are thus given by equations (3.169)–(3.174). For nematics, the known Green–Kubo formulas are recovered (Masters, 1998).

In the centrosymmetric phases of liquid crystals, the coefficients χ_{ijk} and $\theta_{ij\alpha}$ describing cross effects coupling the transport of momentum to those of energy and the order fields are equal to zero. However, it is possible for these coupling coefficients to be nonvanishing in phases without the inversion symmetry $\mathbf{r} \to -\mathbf{r}$, although remaining small (Mabillard and Gaspard, 2020).

Hydrodynamic Modes

The spectrum of hydrodynamic modes is known for the different kinds of liquid crystals (Forster, 1975; Kats and Lebedev, 1994; Chaikin and Lubensky, 1995).

In nematics, two rotational continuous symmetries are broken, so that these phases have two Nambu–Goldstone modes and, consequently, seven hydrodynamic modes: the two propagative sound modes, two diffusive shear modes, two diffusive relaxation modes for the director, and the diffusive heat mode (Forster, 1975; Kats and Lebedev, 1994). The speed of the sound modes depends on the propagation direction with respect to the preferential direction.

In smectics, thermal fluctuations destroy long-range order in the layered structure and the two sound modes have anisotropic propagation speeds (Kats and Lebedev, 1994; Chaikin and Lubensky, 1995). In columnar discotic phases, the molecules form a two-dimensional lattice of columns moving with respect to each other because of thermal fluctuations, so that the three continuous translational symmetries are broken in two spatial directions and there exist seven hydrodynamic modes: six sound modes and one heat mode (Kats and Lebedev, 1994). The spectrum of hydrodynamic modes may be significantly affected by thermal fluctuations.

3.3.6 Crystals

In crystals, the continuous symmetry under the three-dimensional group of spatial translations is broken into a discrete crystallographic space group. Crystals can be classified into 14 Bravais lattices, 32 crystallographic point groups, and 230 space groups (Ashcroft and Mermin, 1976). Crystals have eight hydrodynamic modes with dispersion relations vanishing with the wave number according to the Goldstone theorem. These modes are the two longitudinal sound modes, the four transverse sound modes, the mode of heat conduction, and the mode of vacancy diffusion. This latter was identified in the 1970s when the hydrodynamics of crystals was fully established at the macroscopic level of description (Martin et al., 1972; Fleming and Cohen, 1976). These modes are damped because of energy dissipation caused by the transport properties. The statistical-mechanical methods presented in Section 3.3.4 allow us to deduce these properties from the microscopic Hamiltonian dynamics. We first need to obtain the microscopic expression of the fields describing long-range crystalline order (Szamel and Ernst, 1993; Szamel, 1997; Walz and Fuchs, 2010; Häring et al., 2015). For simplicity, monatomic crystals are here considered.

Microscopic Crystalline Order Fields

In the crystalline phases, the spatial periodicity of the atomic structure is the manifestation of long-range order leading, in particular, to crystallographic diffraction patterns. This long-range order can be characterized by an order parameter, which is known as the displacement vector. This macroscopic field gives the displacement of atoms with respect to their mean position in the lattice. If the displacement vector is uniform, the whole crystal is translated in space. Otherwise, the displacement vector describes elastic deformations in the crystal (Landau and Lifshitz, 1975). Such deformations can be performed by an external force field doing some mechanical work on the crystal and, possibly, inducing symmetry breaking. In this regard, the system is assumed to be subjected to corresponding perturbations modifying the fully symmetric Hamiltonian function H into

$$H_\lambda = H + \lambda\, U^{(\mathrm{ext})} \qquad \text{with} \qquad U^{(\mathrm{ext})} = \int u^{(\mathrm{ext})}(\mathbf{r})\, \hat{n}(\mathbf{r};\mathbf{\Gamma})\, d\mathbf{r}, \tag{3.187}$$

where $u^{(\mathrm{ext})}(\mathbf{r})$ is an external energy potential exerted on the atoms and $\hat{n}$ is their microscopic density (3.3). The translational symmetry is thus explicitly broken for the equilibrium mean values of the density since

$$\langle\{\mathbf{P}, \hat{n}(\mathbf{r};\mathbf{\Gamma})\}\rangle_{\mathrm{eq},\lambda} = \nabla\langle\hat{n}(\mathbf{r};\mathbf{\Gamma})\rangle_{\mathrm{eq},\lambda} \neq 0 \qquad \text{if} \qquad \lambda \neq 0, \tag{3.188}$$

according to equation (3.133) for the observable $A = \hat{n}(\mathbf{r};\mathbf{\Gamma})$. Under temperature and pressure conditions where the crystalline phase is thermodynamically stable, the limit $\lambda \to 0$ can be taken in order to remove the external potential and keep the equilibrium periodic density emerging by spontaneous symmetry breaking:

$$n_{\mathrm{eq}}(\mathbf{r}) \equiv \lim_{\lambda\to 0} \langle\hat{n}(\mathbf{r};\mathbf{\Gamma})\rangle_{\mathrm{eq},\lambda}. \tag{3.189}$$

Accordingly, the spatial structure formed by the atoms is periodic in space. We note that the center of mass $\mathbf{R}_{\mathrm{cm}} \equiv (1/N)\sum_{a=1}^{N} \mathbf{r}_a$ of the whole crystal can undergo spatial translations. In the symmetric probability distribution (3.34) with $c = 1$, the Hamiltonian function can always be expressed as $H = \mathbf{P}^2/(2M) + H_{\mathrm{rel}}$ in terms of the total kinetic energy of the crystal center of mass and the Hamiltonian function H_{rel} ruling the motion of its atoms relative to the center of mass. In the symmetric grand canonical ensemble (3.34), the center of mass is uniformly distributed in space with a Maxwellian velocity distribution around the mean velocity $\mathbf{v}$, so that its trajectories are free flights, $\mathbf{R}_{\mathrm{cm}}(t) = \mathbf{R}_{\mathrm{cm}}(0) + (\mathbf{v} + \mathbf{P}/M)t$. The symmetry may thus be broken in the frame moving with the center of mass, although the distribution (3.34) is symmetric.

The equilibrium periodic density (3.189) of the crystal can be expanded into lattice Fourier modes as

$$n_{\mathrm{eq}}(\mathbf{r}) = \sum_{\mathbf{G}} n_{\mathrm{eq},\mathbf{G}}\, \mathrm{e}^{\imath \mathbf{G}\cdot\mathbf{r}} \tag{3.190}$$

in terms of the vectors $\mathbf{G}$ of the reciprocal lattice. The Fourier coefficients $n_{\mathrm{eq},\mathbf{G}}$ with $\mathbf{G} \neq 0$ characterize the crystallinity of the phase, since they are vanishing in the uniform fluid phase. In this regard, the equilibrium periodic density (3.190) or its dominant Fourier coefficients can be taken as order parameters.

With the purpose of obtaining the microscopic expression for the displacement vector, which may describe the arbitrarily small deformations of the periodic crystalline order, we look for external potentials $u^{(\mathrm{ext})}(\mathbf{r})$ exerting arbitrarily small changes in the mean value

$$\left\langle U^{(\mathrm{ext})}\right\rangle_{\mathrm{eq},0} = \int u^{(\mathrm{ext})}(\mathbf{r})\, n_{\mathrm{eq}}(\mathbf{r})\, d\mathbf{r}. \tag{3.191}$$

Remarkably, this mean value is equal to zero if the external potential is taken as

$$u^{(\mathrm{ext})}(\mathbf{r}) = \mathbf{C}\cdot\nabla n_{\mathrm{eq}}(\mathbf{r}) \tag{3.192}$$

for some constant vector $\mathbf{C}$. Indeed, substituting this energy potential into equation (3.191) and integrating by parts shows that $\langle U^{(\mathrm{ext})}\rangle_{\mathrm{eq},0} = -\langle U^{(\mathrm{ext})}\rangle_{\mathrm{eq},0} = 0$. For such external potentials, the perturbation may induce crystal formation, but does not change the total energy with respect to the value of the unperturbed Hamiltonian. Since its presence costs, on average, no energy, the external potential (3.192) can lead to the construction of the crystalline order parameter.

At the macroscale, the crystal should be described in terms of macrofields that slowly vary in space on scales larger than the periodic crystalline structure. This feature can be expressed by requiring that the macrofields $f(\mathbf{r})$ have no Fourier mode outside the first Brillouin zone $\mathfrak{B}$ of their Fourier decomposition:

$$f(\mathbf{r}) = \int_{\mathbb{R}^3} \frac{d\mathbf{q}}{(2\pi)^3}\, \tilde{f}(\mathbf{q})\, e^{\imath\mathbf{q}\cdot\mathbf{r}}. \tag{3.193}$$

For a macrofield, we thus require that $\tilde{f}(\mathbf{q}) = I_{\mathfrak{B}}(\mathbf{q})\,\tilde{f}(\mathbf{q})$, where $I_{\mathfrak{B}}(\mathbf{q})$ is the indicator function of the first Brillouin zone of the crystalline lattice. In position space, this condition can be expressed as

$$f(\mathbf{r}) = \int_{\mathbb{R}^3} \Delta(\mathbf{r}-\mathbf{r}')\, f(\mathbf{r}')\, d\mathbf{r}' \qquad \text{with} \qquad \Delta(\mathbf{r}-\mathbf{r}') \equiv \int_{\mathfrak{B}} \frac{d\mathbf{q}}{(2\pi)^3}\, e^{\imath\mathbf{q}\cdot(\mathbf{r}-\mathbf{r}')}. \tag{3.194}$$

The function $\Delta(\mathbf{r}-\mathbf{r}')$ satisfies the property that $\int_{\mathbb{R}^3} \Delta(\mathbf{r}-\mathbf{r}')\, d\mathbf{r}' = 1$, as for a Dirac delta distribution. Moreover, this function is real since the first Brillouin zone is the Wigner–Seitz primitive cell of the reciprocal lattice, which is a Bravais lattice, and the point group of a Bravais lattice includes the inversion $\mathbf{r} \to -\mathbf{r}$. The convolution of some field $f(\mathbf{r})$ with the function $\Delta(\mathbf{r}-\mathbf{r}')$ is a projection onto the functional space of macrofields because $\Delta(\mathbf{r}-\mathbf{r}') = \int_{\mathbb{R}^3} \Delta(\mathbf{r}-\mathbf{r}'')\,\Delta(\mathbf{r}''-\mathbf{r}')\, d\mathbf{r}''$, as can be shown using equation (3.194).

To construct the order fields associated with continuous symmetry breaking in crystals, we consider the external potential $u^{(\mathrm{ext})}(\mathbf{r}) = \mathbf{C}(\mathbf{r}) \cdot \nabla n_{\mathrm{eq}}(\mathbf{r})$ obtained by extending the constant vector $\mathbf{C}$ in equation (3.192) into a macrofield $\mathbf{C}(\mathbf{r})$. Accordingly, the external perturbation in equation (3.187) becomes

$$U^{(\mathrm{ext})} = \int \mathbf{C}(\mathbf{r}) \cdot \nabla n_{\mathrm{eq}}(\mathbf{r}) \left[\hat{n}(\mathbf{r};\boldsymbol{\Gamma}) - n_{\mathrm{eq}}(\mathbf{r})\right] d\mathbf{r}, \tag{3.195}$$

where we have subtracted the equilibrium value for $\lambda = 0$ in equation (3.187), so that $\langle U^{(\mathrm{ext})}\rangle_{\mathrm{eq},0} = 0$ in order for this external perturbation to cost arbitrarily small energy in the limit where $\mathbf{C}(\mathbf{r})$ becomes constant in space. Since $\mathbf{C}(\mathbf{r})$ is supposed to be a macrofield, it satisfies equation (3.194), whereupon the external perturbation (3.195) can be written in the form

$$U^{(\mathrm{ext})} = \int d\mathbf{r}\, \mathbf{C}(\mathbf{r}) \cdot \int d\mathbf{r}'\, \Delta(\mathbf{r}-\mathbf{r}')\, \nabla' n_{\mathrm{eq}}(\mathbf{r}') \left[\hat{n}(\mathbf{r}';\boldsymbol{\Gamma}) - n_{\mathrm{eq}}(\mathbf{r}')\right], \tag{3.196}$$

using the property that the function $\Delta(\mathbf{r}-\mathbf{r}')$ is real.

At the macroscale, such an external perturbation of the crystal would be described as

$$U^{(\mathrm{ext})} = -\int d\mathbf{r}\, \boldsymbol{\phi}(\mathbf{r}) : \hat{\mathbf{u}}(\mathbf{r};\boldsymbol{\Gamma}) \tag{3.197}$$

with the symmetric tensor $\boldsymbol{\phi} = (\phi_{ij}) = (\phi_{ji})$ and the strain tensor $\hat{\mathsf{u}} = (\hat{\mathsf{u}}_{ij})$ with the components

$$\hat{\mathsf{u}}_{ij} \equiv \frac{1}{2}\left(\nabla_i \hat{u}_j + \nabla_j \hat{u}_i\right), \tag{3.198}$$

where $\hat{u}_i(\mathbf{r};\boldsymbol{\Gamma})$ is the displacement vector field, here supposed to be defined at the microscopic level of description. The tensor $\boldsymbol{\phi}$ describes the stress applied to the crystal by the external perturbation because the integration by parts of equation (3.197) with the definition (3.198) gives

$$U^{(\mathrm{ext})} = -\int d\mathbf{r}\, \mathbf{f}(\mathbf{r}) \cdot \hat{\mathbf{u}}(\mathbf{r};\boldsymbol{\Gamma}), \qquad \text{where} \qquad \mathbf{f}(\mathbf{r}) \equiv -\boldsymbol{\nabla} \cdot \boldsymbol{\phi}(\mathbf{r}) \tag{3.199}$$

is the force density field applied to the displacement vector $\hat{\mathbf{u}} = (\hat{u}_i)$. Comparing equations (3.196) and (3.199), we see that the vector field $\mathbf{C}(\mathbf{r})$ should correspond to the force density field $\mathbf{f}(\mathbf{r})$ and the rest to the microscopic displacement vector field $\hat{\mathbf{u}}(\mathbf{r};\boldsymbol{\Gamma})$. This latter should also satisfy the property that, under a homogeneous dilatation $\mathbf{r} \to \lambda\mathbf{r}$ of the lattice by a factor λ, the mean value of the displacement vector field should be given by $\langle \hat{\mathbf{u}}(\mathbf{r};\boldsymbol{\Gamma})\rangle = (\lambda - 1)\mathbf{r}$. Accordingly, the microscopic expression for the displacement vector field can be defined as

$$\hat{\mathbf{u}}(\mathbf{r};\boldsymbol{\Gamma}) \equiv -\boldsymbol{\mathcal{N}}^{-1} \cdot \int d\mathbf{r}'\, \Delta(\mathbf{r}-\mathbf{r}')\, \boldsymbol{\nabla}' n_{\mathrm{eq}}(\mathbf{r}') \left[\hat{n}(\mathbf{r}';\boldsymbol{\Gamma}) - n_{\mathrm{eq}}(\mathbf{r}')\right] \tag{3.200}$$

with the tensor $\boldsymbol{\mathcal{N}} = (\mathcal{N}_{ij})$ given by

$$\mathcal{N}_{ij} \equiv \frac{1}{v}\int_v \nabla_i n_{\mathrm{eq}}\, \nabla_j n_{\mathrm{eq}}\, d\mathbf{r}, \tag{3.201}$$

where v is the volume of a primitive unit cell of the lattice. This tensor here plays the same role as the tensor (3.130). Since this tensor is symmetric, the force density field can be identified with $\mathbf{f}(\mathbf{r}) = -\boldsymbol{\mathcal{N}} \cdot \mathbf{C}(\mathbf{r})$. We note that equation (3.200) can be given in terms of the lattice Fourier expansion of the equilibrium density (3.190) and the integral over the first Brillouin zone of equation (3.194). For cubic crystals where $\mathcal{N}_{ij} = \mathcal{N}\delta_{ij}$, equation (3.200) is equivalent to the known microscopic expression for the displacement vector (Szamel and Ernst, 1993; Szamel, 1997).

Now, the evolution equation of the displacement vector field (3.200) can be expressed in the form (3.140) as

$$\partial_t \hat{\mathbf{u}} + \hat{\boldsymbol{\gamma}} = 0 \tag{3.202}$$

in terms of the decay rates $\hat{\boldsymbol{\gamma}} = (\hat{\gamma}_i)$ given by

$$\hat{\boldsymbol{\gamma}}(\mathbf{r};\boldsymbol{\Gamma}) = -\frac{1}{m}\boldsymbol{\mathcal{N}}^{-1} \cdot \int d\mathbf{r}'\, \Delta(\mathbf{r}-\mathbf{r}')\, \boldsymbol{\nabla}' n_{\mathrm{eq}}(\mathbf{r}')\, \boldsymbol{\nabla}' \cdot \hat{\mathbf{g}}(\mathbf{r}';\boldsymbol{\Gamma}), \tag{3.203}$$

as deduced using equation (3.9). The analogue of equation (3.141) reads

$$\partial_t \hat{\mathsf{u}}_{ij} + \nabla_k \hat{\jmath}_{\mathsf{u}_{ij},k} = 0 \qquad \text{with} \qquad \hat{\jmath}_{\mathsf{u}_{ij},k} = \frac{1}{2}\left(\delta_{ik}\delta_{jl} + \delta_{il}\delta_{jk}\right)\gamma_l. \tag{3.204}$$

Therefore, the microscopic hydrodynamics of crystals are ruled by the local conservation equations (3.16) for the densities $\hat{\mathbf{c}} = (\hat{\epsilon}, \hat{n}, \hat{g}_i, \hat{\mathsf{u}}_{ij})$ and the corresponding current densities. Since equation (3.204) is the direct consequence of equation (3.202) for the displacement vector field combined with the definition (3.198) of the strain tensor, equations (3.9), (3.10), (3.11), and (3.202) form the minimal set of eight equations ruling the eight hydrodynamic modes, including the five modes resulting from the five fundamentally conserved quantities (i.e., mass, energy, and momentum) and the three additional Nambu–Goldstone modes generated by the spontaneous symmetry breaking of three-dimensional continuous spatial translations in crystals.

Local Equilibrium Approach in Crystals

According to the previous considerations, the local equilibrium approach can also be used for crystals. The gradients $\hat{u}_{i\alpha} = \nabla_i \hat{x}_\alpha$ of the order fields correspond to the symmetric strain tensor $\hat{\mathsf{u}}_{ij} = \hat{\mathsf{u}}_{ji}$ defined by equation (3.198), and the conjugate fields $\phi_{i\alpha}$ to the symmetric tensor ϕ_{ij}. In order to apply the local equilibrium approach to crystals, the Greek indices $\alpha, \beta, \ldots$ should thus be replaced by Latin indices $i, j, \ldots = 1,2,3 = x, y, z$, and the tensors $\hat{u}_{i\alpha}$ and $\phi_{i\alpha}$ should be symmetrized.

The entropy density introduced in equation (3.143) is here given by

$$s = \frac{1}{T}\left(\epsilon + p - \mu n - v_i\, g_i - \phi_{ij}\, \mathsf{u}_{ij}\right) \tag{3.205}$$

in terms of the mean values $(\epsilon, n, g_i, \mathsf{u}_{ij})$ of the microscopic densities with respect to the local equilibrium, the fields $\left(T^{-1}, \mu, v_i, \phi_{ij}\right)$ and the hydrostatic pressure p. Consequently, the conjugate fields are given by equations (3.62), (3.63), (3.64), and (3.144), which here reads $\chi_{\mathsf{u}_{ij}} = -\beta\phi_{ij} + O\left(\nabla^2\right)$. The thermodynamics of crystals can thus be deduced in this statistical-mechanical approach (Wallace, 1998).

The local equilibrium mean values of the rates (3.203) are directly related to the velocity field according to

$$\bar{\gamma}(\mathbf{r},t) = \langle\hat{\gamma}(\mathbf{r};\boldsymbol{\Gamma})\rangle_{\text{leq},\chi_t} = -\mathbf{v}(\mathbf{r},t), \tag{3.206}$$

so that equation (3.151) here becomes

$$\gamma_k = -A_{ik}\, v_i - B_{ijk}\, \nabla_i v_j + O\left(\nabla^2\right) \quad \text{with} \quad A_{ik} = \delta_{ik} \quad \text{and} \quad B_{ijk} = 0. \tag{3.207}$$

As a consequence, the dissipativeless parts of the momentum and energy current densities (3.156)–(3.157) here take the forms

$$\bar{\mathbf{J}}_{\mathbf{g}} = \rho\mathbf{v}\mathbf{v} - \boldsymbol{\sigma} + O\left(\nabla^2\right) \qquad \text{and} \qquad \bar{\mathbf{J}}_\epsilon = \epsilon\mathbf{v} - \boldsymbol{\sigma}\cdot\mathbf{v} + O\left(\nabla^2\right) \tag{3.208}$$

with the stress tensor $\boldsymbol{\sigma} = (\sigma_{ij})$ given by

$$\sigma_{ij} = -p\,\delta_{ij} + \phi_{ij}. \tag{3.209}$$

The dissipative current densities (3.166)–(3.168) are obtained *mutatis mutandis* with the transport coefficients given by the Green–Kubo formulas (3.169)–(3.174). In this way, we find that the linearized hydrodynamic equations for crystals have the same form as in equations (3.179)–(3.182) with the coefficients given in equation (3.207) (Mabillard and Gaspard, 2021).

Transport Properties in Crystals

In crystalline phases, the Green–Kubo formulas (3.169)–(3.174) with Greek indices replaced by Latin ones provide the transport coefficients in terms of the microscopic Hamiltonian dynamics. The microscopic total current (3.177) is thus here given by

$$\delta\Gamma_i \equiv \int_V \delta\hat{\gamma}_i \, d\mathbf{r} + \rho^{-1}\,\delta P_i, \tag{3.210}$$

where the integral of the microscopic decay rate (3.203) has the expression

$$\int_V \hat{\gamma}\, d\mathbf{r} = -\frac{1}{m}\mathcal{N}^{-1}\cdot \sum_{\mathbf{G}} n_{\mathrm{eq},\mathbf{G}}\,\mathbf{G}\mathbf{G}\cdot\sum_a \mathbf{p}_a\, e^{\imath\mathbf{G}\cdot\mathbf{r}_a}, \tag{3.211}$$

obtained using equation (3.5) and the property $\int \Delta(\mathbf{r}-\mathbf{r}')\, d\mathbf{r} = 1$ for the function introduced in equation (3.194).

Because of the Onsager–Casimir reciprocal relations (3.115), the following second- and fourth-rank tensors are symmetric: the heat conductivities $\kappa_{ij} = \kappa_{ji}$, the friction coefficients associated with the displacement vector $\zeta_{ij} = \zeta_{ji}$, and the viscosities $\eta_{ijkl} = \eta_{klij}$. The coefficients ξ_{ij} couple heat transport to the displacement vector. Moreover, possible coupling effects between the transports of momentum with heat and the displacement vector are characterized by the third-rank tensors χ_{ijk} and θ_{ijk}, respectively.

According to the principle of Curie (1894), the tensorial properties should be symmetric under the transformations of the crystallographic group. This principle applies, in particular, to the third-rank tensors χ_{ijk} and θ_{ijk}. In isotropic phases, third-rank tensors are always equal to zero, because such phases have a symmetry center. However, this is no longer the case in anisotropic phases, as illustrated by the phenomenon of piezoelectricity, which is also described by a third-rank tensor (Landau and Lifshitz, 1984). Among the 32 possible crystallographic point groups (also called classes), 20 of them are known to allow for nonvanishing third-rank tensors:

$$\mathrm{C_1,\ C_s,\ C_2,\ C_{2v},\ C_3,\ C_{3v},\ C_4,\ C_{4v},\ C_6,\ C_{6v},} \tag{3.212}$$

$$\mathrm{D_2,\ D_{2d}, C_{3h},\ D_3,\ D_{3h},\ D_4,\ S_4,\ D_6,\ T,\ T_d.} \tag{3.213}$$

The 10 classes (3.212) are compatible with pyroelectricity and the 20 classes (3.212) and (3.213) with piezoelectrictity (Landau and Lifshitz, 1984). Third-rank tensors are vanishing in the 12 other classes

$$\mathrm{C_i,\ C_{2h},\ D_{2h},\ C_{4h},\ D_{4h},\ S_6,\ D_{3d},\ C_{6h},\ D_{6h},\ T_h,\ O,\ O_h,} \tag{3.214}$$

because these point groups are either centrosymmetric (i.e., they contain the inversion $\mathbf{r} \to -\mathbf{r}$ with respect to a symmetry center), or they contain 90° rotations around axes perpendicular to the faces as for the cubic (or orthohedral) crystallographic group O. Therefore, the transport coefficients χ_{ijk} and θ_{ijk} may be nonvanishing in the 20 crystallographic classes (3.212) and (3.213).

Vacancy Concentration

In monatomic crystals at equilibrium, the atomic density macrofield is uniform and equal to the component $\mathbf{G} = 0$ of the lattice Fourier expansion (3.190) or, equivalently, to the projection (3.194) applied to the periodic equilibrium density (3.189):

$$n_{\text{eq},0} = \frac{1}{v} \int_v n_{\text{eq}}(\mathbf{r})\, d\mathbf{r} = \int \Delta(\mathbf{r} - \mathbf{r}')\, n_{\text{eq}}(\mathbf{r}')\, d\mathbf{r}'. \tag{3.215}$$

Under nonequilibrium conditions, the density macrofield may deviate with respect to its equilibrium value by two possible mechanisms: (1) the lattice dilatation or contraction corresponding to the strain $\boldsymbol{\nabla} \cdot \mathbf{u} = \nabla_i u_i$; (2) vacancies or interstitials, decreasing or increasing the occupancy of the lattice cells by atoms (Ashcroft and Mermin, 1976). To describe these latter, a macrofield giving the density of vacancies, also called vacancy concentration, is defined as

$$\hat{c}(\mathbf{r};\Gamma) \equiv -\int \Delta(\mathbf{r} - \mathbf{r}') \left[\hat{n}(\mathbf{r}';\Gamma) - n_{\text{eq},0} + n_{\text{eq},0}\, \boldsymbol{\nabla} \cdot \hat{\mathbf{u}}(\mathbf{r}';\Gamma)\right] d\mathbf{r}'. \tag{3.216}$$

The time evolution of this macrofield is driven by the local conservation equation (3.9) for the mass density $\hat{\rho} = m\hat{n}$ and by equation (3.202) for the displacement vector (3.200). The local equilibrium mean value of the vacancy concentration can be expressed as

$$c(\mathbf{r},t) \equiv \langle \hat{c}(\mathbf{r};\Gamma)\rangle_{\text{leq},\chi_t} = -\delta n(\mathbf{r},t) - n_{\text{eq},0}\, \boldsymbol{\nabla} \cdot \mathbf{u}(\mathbf{r},t), \tag{3.217}$$

where $\delta n(\mathbf{r},t)$ denotes the macrofield giving the deviation of the mean particle density with respect to its equilibrium value (3.215). Accordingly, the motion of the vacancy concentration (3.216) is driven by the evolution equations for the density $\hat{n} = \hat{\rho}/m$ and the displacement vector $\hat{\mathbf{u}}$.

We may introduce the fraction of vacancies as

$$\mathfrak{y} \equiv \frac{c}{n_{\text{eq},0}}. \tag{3.218}$$

Its evolution equation can be deduced according to $\partial_t \mathfrak{y} = \nabla_i v_i - \nabla_i \partial_t u_i$ using the continuity equation (3.179) for the mass density ρ and the macroscopic equation for the displacement vector u_i given by equation (3.182) with $x_\alpha = u_i$.

Linearized Hydrodynamic Equations in Crystals

In order to obtain the hydrodynamic modes, we may introduce, on the one hand, the vacancy fraction instead of the mass density and, on the other hand, the local entropy per unit mass $\mathfrak{s}$ instead of the internal energy density ϵ_0 according to

$$\rho\, \delta\mathfrak{y} = -\delta\rho - \rho\, \nabla_i\, \delta u_i \qquad \text{and} \qquad T\rho\, \delta\mathfrak{s} = \delta\epsilon_0 - \frac{\epsilon_0 + p}{\rho}\, \delta\rho. \tag{3.219}$$

The linear hydrodynamic equations for the eight macrofields $\boldsymbol{\psi} = (\delta\mathfrak{y}, \delta\mathfrak{s}, \delta\mathbf{v}, \delta\mathbf{u})^{\mathrm{T}}$ are thus given by

$$\partial_t \mathfrak{y} = -\frac{\xi_{ij}}{T}\, \nabla_i \nabla_j T - \zeta_{ik} \nabla_i \nabla_j \phi_{jk} + \theta_{jki} \nabla_i \nabla_j v_k, \tag{3.220}$$

$$T\rho\, \partial_t \mathfrak{s} = \kappa_{ij} \nabla_i \nabla_j T + \xi_{ik} \nabla_i \nabla_j \phi^{jk} + \chi_{ijk} \nabla_i \nabla_j v_k, \tag{3.221}$$

$$\rho\, \partial_t v_i = \underbrace{-\nabla_i p + \nabla_j \phi_{ji}} + \eta_{jikl} \nabla_j \nabla_k v_l + \theta_{jil} \nabla_j \nabla_k \phi_{kl} - \frac{\chi_{kji}}{T}\, \nabla_j \nabla_k T, \tag{3.222}$$

$$\partial_t u_i = \underbrace{v_i} + \frac{\xi_{ji}}{T} \nabla_j T + \zeta_{ik} \nabla_j \phi_{jk} - \theta_{jki} \nabla_j v_k, \tag{3.223}$$

where the underbraced terms are those ruling the elastic dynamics of the crystals, conserving entropy. According to equation (3.178), the terms with the second- and fourth-rank tensors κ_{ij}, ξ_{ij}, ζ_{ij}, and η_{ijkl} are dissipating energy and thus producing entropy while the terms with the third-rank tensors θ_{ijk} and χ_{ijk} are also conserving entropy.

In order to form a closed set of equations, the macrofields $X = (T, p, \phi_{ij})$ are linearly expanded around equilibrium in terms of the fields $(\mathfrak{y}, \mathfrak{s}, \mathsf{u}_{ij})$ as

$$\delta X = \left(\frac{\partial X}{\partial \mathfrak{y}}\right)_{\mathfrak{s},\mathsf{u}} \delta\mathfrak{y} + \left(\frac{\partial X}{\partial \mathfrak{s}}\right)_{\mathfrak{y},\mathsf{u}} \delta\mathfrak{s} + \left(\frac{\partial X}{\partial \mathsf{u}_{kl}}\right)_{\mathfrak{s},\mathfrak{y}} \delta\mathsf{u}_{kl}, \tag{3.224}$$

where $\delta\mathsf{u}_{kl} = (\nabla_k u_l + \nabla_l u_k)/2$ because the strain tensor is related to the displacement vector by equation (3.198). Therefore, the linearized hydrodynamic equations (3.220)–(3.223) can be cast into the closed matrix form

$$\partial_t \boldsymbol{\psi} = \hat{\mathbf{L}} \cdot \boldsymbol{\psi}, \tag{3.225}$$

where $\hat{\mathbf{L}}$ is a 8×8 matrix with elements involving the first, second, and third powers of the gradient operator ∇.

Hydrodynamic Modes in Crystals

The eight hydrodynamic modes of crystals can now be obtained for the solutions of equation (3.225) such as $\boldsymbol{\psi}(\mathbf{r},t) \sim \exp(\imath\mathbf{q}\cdot\mathbf{r} + zt)$. The dispersion relations $z(\mathbf{q})$ of the eight hydrodynamic modes are provided by solving the eigenvalue problem $\mathbf{L}\cdot\boldsymbol{\psi} = z\,\boldsymbol{\psi}$, where the 8×8 matrix $\mathbf{L}$ is given by replacing the gradient ∇ with $\imath\mathbf{q}$ in the matrix operator $\hat{\mathbf{L}}$ introduced in equation (3.225).

The propagation speeds of the six longitudinal and transverse sound modes in the crystal can be obtained solving the eigenvalue problem

$$\mathbf{M}\cdot\boldsymbol{\epsilon}_\sigma = \lambda_\sigma\,\boldsymbol{\epsilon}_\sigma, \qquad \sigma = 1,2,3, \tag{3.226}$$

for the 3×3 symmetric matrix

$$\mathbf{M} = (M_{ij}) \equiv C_{kilj} q_k q_l, \qquad \text{where} \qquad C_{ijkl} \equiv \frac{1}{\rho}\left(\frac{\partial \sigma_{ij}}{\partial \mathsf{u}_{kl}}\right)_{\mathfrak{s},\mathfrak{y}} \tag{3.227}$$

is the isoentropic elasticity tensor of the crystal (divided by the mass density). Since the matrix $\mathbf{M}$ is quadratic in the wave vector $\mathbf{q}$, the eigenvalues $\lambda_\sigma(\mathbf{q})$ are also quadratic, so that the sound speeds

$$v_\sigma(\mathbf{n}) \equiv \frac{1}{q}\sqrt{\lambda_\sigma(\mathbf{q})}, \qquad \sigma = 1,2,3, \tag{3.228}$$

only depend on the propagation direction of the sound wave in the crystal: $\mathbf{n} \equiv \mathbf{q}/q$ with $q \equiv \|\mathbf{q}\|$. It can be shown (Mabillard and Gaspard, 2021) that the dispersion relations of the eight hydrodynamic modes are thus given by

$$z_\alpha(\mathbf{q}) = -\imath\, v_\alpha(\mathbf{n})\, q - D_\alpha(\mathbf{n})\, q^2 + o(q^2) \qquad \text{with} \qquad \alpha = 1\text{–}8, \tag{3.229}$$

where $v_{\sigma+3} = -v_\sigma$ and $v_7 = v_8 = 0$. Therefore, the modes $\alpha = 1\text{–}6$ are the sound modes propagating in opposite directions $\pm\mathbf{q}$, while the modes $\alpha = 7, 8$ are the diffusive heat mode and the mode of vacancy diffusion.

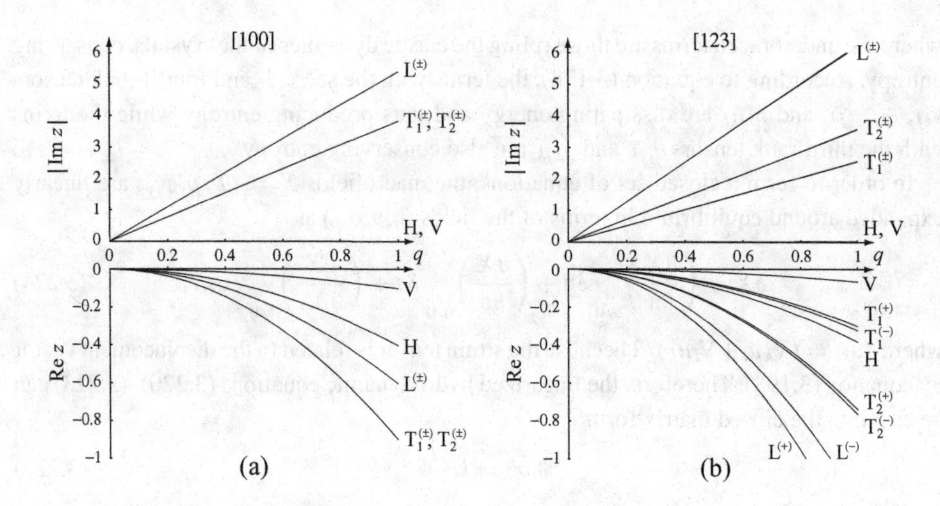

Figure 3.1 Dispersion relations of the eight hydrodynamic modes in a monatomic cubic crystal: (a) in the direction $(q_x, q_y, q_z) = (q, 0, 0)$; (b) in the direction $(q_x, q_y, q_z) = (q, 2q, 3q)/\sqrt{14}$. $L^{(\pm)}$ denote the longitudinal sound modes, $T_1^{(\pm)}$ and $T_2^{(\pm)}$ the transverse sound modes, H the heat mode, and V the vacancy diffusion mode. Reproduced with permission from Mabillard and Gaspard (2021) © IOP Publishing Ltd and SISSA Medialab.

The eight dispersion relations are shown in Figure 3.1 for cubic crystals with nonvanishing third-rank tensors χ_{ijk} and θ_{ijk}. In the opposite directions [100] and [$\bar{1}$00], we see in Figure 3.1(a) that the dispersion relations of the sound modes $L^{(\pm)}$, $T_1^{(\pm)}$, and $T_2^{(\pm)}$ are degenerate, which is known (Ashcroft and Mermin, 1976). However, in the opposite directions [123] and [$\bar{1}\bar{2}\bar{3}$], Figure 3.1(b) shows the splitting of degeneracy for the damping rates, although the absolute values of the imaginary parts remain degenerate up to second order in the wave number (Mabillard and Gaspard, 2021).

We note that amorphous solids also undergo the phenomenon of continuous translational symmetry breaking because the atoms have fixed mean positions in the solid matrix. Accordingly, amorphous solids also have eight hydrodynamic modes, including transverse sound modes.

Multicomponent Crystals

Crystals are most often composed of several atomic species, e.g., in mineralogy. In such crystals, there are as many local conservation laws as there are atomic species (in the absence of radioactivity). If the crystal contains c atomic species, there are thus $c-1$ hydrodynamic modes of mutual diffusion between these species in addition to the eight hydrodynamic modes of a monatomic crystal. The total number of hydrodynamic modes is thus equal to $7+c$ in crystals with c components. The statistical mechanics of multicomponent crystals can be developed by extending the local equilibrium approach to all the microscopic densities, as carried out for multicomponent normal fluids in Section 3.2.

Diffusion in solids is an important transport process for materials science and technology. With vacancy diffusion, all these diffusion processes generate the transport of point defects across the crystal, including impurities in addition to vacancies and interstitials.

3.4 Interfaces and Boundary Conditions

3.4.1 Interfacial Phenomena

Interfaces between two bulk phases are ubiquitous in nature and play essential roles in various phenomena. For example, fluid–fluid interfaces are formed in binary systems composed of immiscible liquids such as oil and water. At such interfaces, membranes may self-assemble in ternary systems including a surfactant. Fluid–solid interfaces exist when gases or liquids are contained in solid vessels, or if several fluid components react with each other on a solid surface in heterogeneous catalysis, or during the phenomenon of crystal growth. Solid–solid interfaces also exist in polycrystalline solids or in dry friction between two solids sliding one with respect to the other. Other examples are the interfaces between quantum phases of matter.

Interfaces have their own equilibrium and nonequilibrium properties, matching those of the adjacent phases. Surface tension is the basic equilibrium property of an interface and its microscopic expression is known in statistical mechanics (Rowlinson and Widom, 1989; Allen and Tildesley, 2017). Interfaces are also the stage of irreversible processes of transport and reaction. Atoms and molecules can accumulate at interfaces by physisorption or chemisorption and they may undergo interfacial diffusion and reaction processes. Interfacial viscosity may manifest itself in the laminar or turbulent flows of soap films. Heat transport across an interface can be characterized by the Kapitza resistance, causing a jump in the temperature profile (Pollack, 1969). Furthermore, different nonequilibrium processes may be coupled together at interfaces as in the phenomena of diffusiophoresis and thermophoresis, coupling momentum transport to those of matter and heat, respectively. These different processes taking place at interfaces can be systematically characterized within the macroscopic framework of interfacial nonequilibrium thermodynamics (Bedeaux et al., 1976; Bedeaux, 1986; Kjelstrup and Bedeaux, 2008), as summarized in Section A.9 where Table A.3 gives the list of possible interfacial irreversible processes with their affinity and associated current density contributing to entropy production.

At the atomic scale, interfaces have a structure extending over a few molecular diameters but at the macroscale, they appear discontinuous forming a two-dimensional surface, which is possibly moving with time. The interface and the two adjacent semi-infinite bulk phases thus divide the three-dimensional space into three domains. The interfacial properties are defined inside the domain of molecular width in between the bulk phases. In principle, the local equilibrium distribution introduced in Section 3.2.3 can be generalized into a statistical-mechanical approach justifying the interfacial nonequilibrium thermodynamics of Section A.9. For this purpose, the local equilibrium distribution (3.21) can be defined with conjugate fields defined in every one of the three domains and the corresponding densities are given by taking functional derivatives as in equation (3.24). The local equilibrium approach can thus be developed as for the bulk phases in the previous sections, including the effects of the interface. In this way, microscopic expressions can be obtained for the equilibrium and nonequilibrium interfacial properties.

3.4.2 Partial Slip Boundary Conditions on the Velocity Field

Interfaces also form boundaries for the bulk phases. In general, since energy or matter may accumulate at interfaces, the excess surface densities should be included among the dynamical fields describing the system, as explained in Section A.9. In such cases, the boundary conditions on the fields of the adjacent bulk phases will depend on the dynamics of the interfacial fields. However, the interfacial time evolution is often passively driven by the bulk phases. In such cases, the interfacial properties directly determine the boundary conditions on the fields of the bulk phases.

In particular, fluid–solid interfaces constitute boundaries for fluids in laminar or turbulent flows. By their geometry, the solid walls impose boundary conditions on the fluid velocity field. Often, this latter obeys stick boundary conditions at solid surfaces, requiring that the three components of the fluid and solid velocities are equal at the interface. However, this is not always the case, as known since the nineteenth century (Navier, 1827; Maxwell, 1879). Although the fluid and solid velocity components perpendicular to the interface must be equal to obey mass conservation, the fluid may slip with respect to the solid in the direction parallel to the interface. This velocity slip can be described by a linear phenomenological relation $\mathcal{J}^{\rm s}_{\rm v\|} = \mathcal{L}^{\rm s}_{\rm v,v}\mathcal{A}^{\rm s}_{\rm v\|}$ between the affinity and current density of interfacial slippage given in Table A.3 of Appendix A.

Supposing that the solid is immobile; the boundary conditions on the fluid velocity field $\mathbf{v}$ thus read

$$\mathbf{n}\cdot\mathbf{v} = 0 \qquad \text{and} \qquad (\mathbf{1}-\mathbf{nn})\cdot(\mathbf{\Pi}\cdot\mathbf{n}+\lambda\,\mathbf{v}) = 0, \tag{3.230}$$

holding at the interface $f(\mathbf{r},t) = 0$ with the unit vector perpendicular to the interface given by $\mathbf{n} = \nabla f/\|\nabla f\|$. The first boundary condition in the direction $\mathbf{n}$ perpendicular to the interface maintains mass conservation in the volume V of the fluid. Indeed, integrating the continuity equation $\partial_t\rho + \nabla\cdot(\rho\mathbf{v}) = 0$ over this volume shows that the mass $M = \int_V \rho\, d\mathbf{r}$ of the fluid should satisfy $dM/dt = -\int_{\partial V}\rho\mathbf{v}\cdot\mathbf{n}\, d\Sigma$, which is equal to zero if and only if this boundary condition holds.[8] The second boundary condition involves the sliding friction coefficient, $\lambda \equiv \mathcal{L}^{\rm s}_{\rm v,v}/T$, and the viscous part $\mathbf{\Pi}$ of the fluid pressure tensor. In the bulk of incompressible fluids such that $\nabla\cdot\mathbf{v} = 0$, this tensor is given by $\mathbf{\Pi} = -\eta\left(\nabla\mathbf{v} + \nabla\mathbf{v}^{\rm T}\right)$ in terms of the shear viscosity η of the fluid.

If the interface is planar and located at $z = z_0$, the normal unit vector is given by $\mathbf{n} = \mathbf{1}_z$ and the boundary conditions (3.230) read $v_z = 0$, $v_x = b\nabla_z v_x$, and $v_y = b\nabla_z v_y$ with the slip length $b \equiv \eta/\lambda$. As shown in Figure 3.2, the effect of velocity slip is that the parallel

[8] In this respect, the requirement that the total mass M of the fluid should be contained inside the volume V determines the location of the macroscopic interfacial discontinuity.

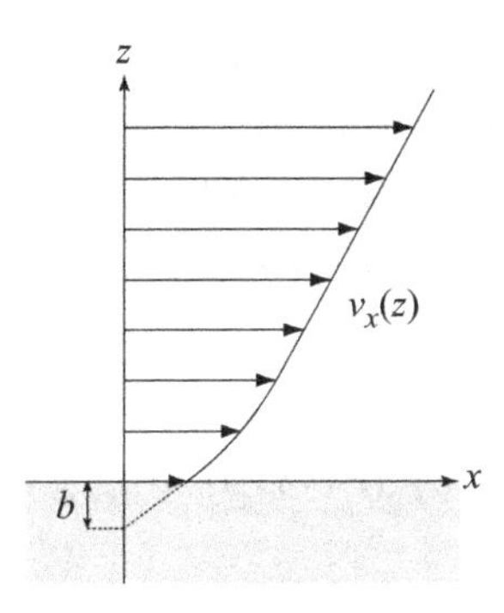

Figure 3.2 Velocity slip at a fluid–solid interface by the slip length b.

velocity component does not vanish at the interface $z = z_0$ but its extrapolation inside the solid does so at the location $z = z_0 - b$. This slip effect may be particularly large at hydrophobic surfaces (Barrat and Bocquet, 1999). Perfect slip boundary conditions hold in the limit $\lambda \to 0$ and $b \to \infty$, and stick boundary conditions in the other limit $\lambda \to \infty$ and $b \to 0$, for which $\mathbf{v} = 0$ at the interface. Partial slip boundary conditions hold for the intermediate values. When $\lambda \neq 0$, energy is dissipated and entropy produced due to sliding friction according to $d_{\rm i}S/dt|_{\rm sliding} = \int_{\partial V} \mathcal{L}^{\rm s}_{\rm v,v}(\mathcal{A}^{\rm s}_{\rm v\|})^2 d\Sigma = (\lambda/T)\int_{\partial V} \mathbf{v}^2 d\Sigma$.[9]

3.4.3 Microscopic Expression for the Sliding Friction Coefficient

Several approaches are considered regarding this issue (Ronis et al., 1977; Bocquet and Barrat, 1994; Nakano and Sasa, 2019; Duque-Zumajo et al., 2019). At the microscopic level of description, the balance equations for momentum in the fluid (3.10) should include the contribution of the interaction between the fluid particles and the solid surface according to

$$\partial_t \hat{\mathbf{g}} + \nabla \cdot \hat{\jmath}_{\mathbf{g}} = \mathbf{F}^{(\rm ext)}\, \hat{n} \tag{3.231}$$

with the positional force $\mathbf{F}^{(\rm ext)} = -\nabla u^{(\rm ext)}$, and $u^{(\rm ext)}$ representing the interaction energy potential between a fluid particle and the solid wall, where we assume that the fluid is composed of a single particle species of microscopic density $\hat{n}$. A similar modification concerns the energy balance equation of the fluid.

If the system is isothermal at the inverse temperature β and chemical potential μ with a small enough initial velocity field $\mathbf{v}(\mathbf{r})$, the initial local equilibrium distribution may be supposed of the form

$$p_0(\boldsymbol{\Gamma}) = p_{\rm eq}(\boldsymbol{\Gamma})\left[1 + \beta \int_V \mathbf{v}(\mathbf{r}) \cdot \hat{\mathbf{g}}(\mathbf{r}; \boldsymbol{\Gamma})\, d\mathbf{r}\right] \tag{3.232}$$

with the grand canonical equilibrium distribution of the Hamiltonian function (2.2), $p_{\rm eq}(\boldsymbol{\Gamma}) = (\Xi\, \Delta\boldsymbol{\Gamma})^{-1} \exp[-\beta(H - \mu N)]$. The mean value of the microscopic momentum

[9] We note that, in the limit $\lambda \to \infty$, the velocity is equal to zero at the interface, so that there is no contribution to the entropy production for stick boundary conditions.

current density tensor (3.13) can be calculated at time t for the Liouvillian time evolution (3.19). Using the momentum balance equation (3.231) and integrating by parts, the deviations of the momentum current density with respect to its equilibrium value can be expressed as

$$\begin{aligned}
\langle \delta\hat{\mathbf{J}}_{\mathbf{g}}(\mathbf{r})\rangle_t &= \int d\boldsymbol{\Gamma}\, \delta\hat{\mathbf{J}}_{\mathbf{g}}(\mathbf{r})\, \mathrm{e}^{\hat{L}t}\, p_0 \\
&\simeq -\beta \int_0^\infty dt \int_V \langle \delta\hat{\mathbf{J}}_{\mathbf{g}}(\mathbf{r},t)\, \delta\hat{\mathbf{J}}_{\mathbf{g}}(\mathbf{r}',0)\rangle_{\mathrm{eq}} : \boldsymbol{\nabla}'\mathbf{v}(\mathbf{r}')\, d\mathbf{r}' \\
&\quad +\beta \int_0^\infty dt \int_{\partial V} \langle \delta\hat{\mathbf{J}}_{\mathbf{g}}(\mathbf{r},t)\, \delta\hat{\mathbf{J}}_{\mathbf{g}}(\mathbf{r}',0)\rangle_{\mathrm{eq}} : \mathbf{v}(\mathbf{r}')\, d\boldsymbol{\Sigma}' \\
&\quad -\beta \int_0^\infty dt \int_V \langle \delta\hat{\mathbf{J}}_{\mathbf{g}}(\mathbf{r},t)\, \delta\hat{n}(\mathbf{r}',0)\rangle_{\mathrm{eq}}\, \mathbf{F}^{(\mathrm{ext})}(\mathbf{r}') \cdot \mathbf{v}(\mathbf{r}')\, d\mathbf{r}',
\end{aligned} \tag{3.233}$$

on timescales longer than the correlation time, but shorter than the hydrodynamic time, over which the velocity field varies.

In equation (3.233), the first term leads to the standard expression for the viscous part of the pressure tensor in the bulk of the fluid, $\boldsymbol{\Pi} = -\eta\left(\boldsymbol{\nabla}\mathbf{v} + \boldsymbol{\nabla}\mathbf{v}^{\mathrm{T}}\right)$, where the shear viscosity η can be computed with the Green–Kubo formula (3.107). At the interface $z = z_0$, the first term is again contributing and the second term yields a contribution that is proportional to the fluid velocity. Moreover, the correlations between the scalar and tensorial quantities, $\delta\hat{n}$ and $(\delta\hat{\mathbf{J}}_{\mathbf{g}})_{ij}$ with $i \neq j \in \{x, y\}$, are expected to vanish if the interface is isotropic on average, in which case the third term of equation (3.233) plays a negligible role. Comparing with the partial slip boundary condition in equation (3.230), the microscopic expression for the sliding friction coefficient can thus be obtained within this approach as

$$\lambda = \lim_{\Sigma\to\infty} \frac{C^{-1}}{\Sigma k_{\mathrm{B}} T} \int_0^\infty dt\, \langle \delta J_x^{\mathrm{s}}(t)\, \delta J_x^{\mathrm{s}}(0)\rangle_{\mathrm{eq}} \tag{3.234}$$

in terms of the interfacial microscopic total current $\delta J_x^{\mathrm{s}} \equiv \int_{\Sigma, z=z_0} (\delta\hat{\mathbf{J}}_{\mathbf{g}})_{xz}\, dx\, dy$, and

$$C = 1 - \frac{1}{\eta\, \Sigma k_{\mathrm{B}} T} \int_0^\infty dt\, \langle \delta J_x^{\mathrm{s}}(t)\, \delta J_{xz}(0)\rangle_{\mathrm{eq}}, \tag{3.235}$$

where $\delta J_{xz} \equiv \int_V (\delta\hat{\mathbf{J}}_{\mathbf{g}})_{xz}\, dx\, dy\, dz$. For an isotropic interface, the same value should be obtained by considering the interfacial microscopic total current in the other direction y of the solid surface.

Similar considerations apply to other interfacial transport coefficients.

3.5 Further Aspects of Microscopic Hydrodynamics

3.5.1 Hydrodynamic Long-Time Tails and Their Consequences

In the 1960s and early 1970s, the advent of powerful computers led to the discovery by Alder and Wainwright (1970) that the time autocorrelation functions entering the Green–Kubo formulas decay in time more slowly than expected from kinetic theory for dilute gases. Although Boltzmann's kinetic theory predicts their exponential decay over the intercollisional timescale, numerical simulation of hard-sphere fluids revealed algebraic decays, called long-time tails, due to fluctuating collective hydrodynamic flows on large spatiotemporal scales. These flows may arise from the thermal fluctuations of hydrodynamic modes in the system. Because of these collective slow modes, the fluid keeps the memory of its previous movements. Since their discovery, these memory effects and their consequences have been systematically investigated, in particular, using perturbative expansions in kinetic theory (Ernst and Dorfman, 1975; Dorfman et al., 1994, 2021) and mode-coupling theory (Ernst et al., 1971).

In fluids, considering autocorrelation functions like

$$C_\gamma(t) = \lim_{V\to\infty} \frac{1}{V}\, \langle \delta J_\gamma(t)\, \delta J_\gamma(0)\rangle_{\rm eq}, \tag{3.236}$$

we note that the total currents $\delta J_\gamma \equiv J_\gamma - \langle J_\gamma\rangle_{\rm eq}$ are given by microscopic expressions such as (3.101) for viscosity. On large spatiotemporal scales, fluctuating collective motions manifest themselves in the fluid, which can thus be described in terms of fluctuating hydrodynamic variables such as the small deviations $\delta\mathbf{v}(\mathbf{r},t)$ or $\delta T(\mathbf{r},t)$ of the velocity or temperature fields around equilibrium. Therefore, the particle momenta in equation (3.101) are driven on large scales by these hydrodynamic motions according to $\mathbf{p}_a = \mathbf{p}_{a0} + m_a \delta\mathbf{v}(\mathbf{r}_a,t)$. Since the microscopic total currents are, in general, nonlinear functions of the particle positions and momenta, different hydrodynamic modes can be coupled together in the perturbative evaluation of the time autocorrelation functions (3.236), giving

$$C_\gamma(t) \simeq \sum_{\alpha,\beta} \int \frac{d\mathbf{q}}{(2\pi)^d}\, A_{\alpha\beta\gamma}(\mathbf{q})\, {\rm e}^{[z_\alpha(\mathbf{q})+z_\beta(-\mathbf{q})]\,t}, \tag{3.237}$$

where $z_\alpha(\mathbf{q})$ are the dispersion relations of the hydrodynamic modes. In d-dimensional fluids, the dominant modes are the shear modes with $z(\mathbf{q}) \simeq -\nu q^2$ where $\nu = \eta/\rho$ is the kinematic shear viscosity coefficient and $q = \|\mathbf{q}\|$. Since the slowest modes are those with $q \to 0$, the time autocorrelation functions are decaying as

$$C_\gamma(t) \sim \int dq\, q^{d-1}\, {\rm e}^{-2\nu q^2 t} \sim \frac{1}{t^{d/2}} \tag{3.238}$$

with long-time tails for $t \to \infty$. Hydrodynamic fluctuations generate vortices pushing at the rear of any fluid element, thus maintaining its motion for a while in the same direction. For this reason, the algebraic decay is positive in equation (3.238) (Ernst et al., 1971).

As a consequence, the integral of the time autocorrelation function behaves as

$$\int_0^t C_\gamma(\tau)\, d\tau \sim \begin{cases} t^{1-d/2} & \text{if } \; d \neq 2, \\ \ln t & \text{if } \; d = 2, \end{cases} \tag{3.239}$$

for $t \to \infty$, so that finite transport coefficients only exist for dimensions $d > 2$.

Another consequence is that the dispersion relations of hydrodynamic modes are not given, in general, by power expansions of the wave number. For classical fluids in dimension $d = 3$, they have a nonanalytic dependence of the form

$$z(q) = a_1\, q + a_2\, q^2 + a_{5/2}\, q^{5/2} + \cdots, \tag{3.240}$$

because the time autocorrelation functions have the asymptotic time behavior $C_\gamma(t) \sim 1/t^{3/2}$ (Ernst and Dorfman, 1975). Therefore, the correction to the dispersion relation (3.2) goes as $o(q^2) \sim q^{5/2}$.

Furthermore, the transport coefficients have nonanalytic dependences on the particle density n in the dilute-gas limit $n \to 0$ according to

$$\mathcal{K}(n) = \mathcal{K}(0)\big(1 + c_1 n_* + c_2' n_*^2 \ln n_* + c_2 n_*^2 + \cdots\big), \tag{3.241}$$

where $n_* \equiv a^3 n$ is the reduced density defined in terms of the molecular radius a (Dorfman et al., 1994).

3.5.2 Hydrodynamics in Low-Dimensional Systems

Equation (3.238) also shows that the hydrodynamic long-time tails and their effects get stronger and stronger as the dimension d decreases. The dynamic fluctuations thus become more important in low-dimensional systems.

In general, a distinction should be made between the dimension d of the system and the dimension $\tilde{d}$ of the space embedding the system. For instance, a freely suspended film or membrane is a system of dimension $d = 2$ in a space of dimension $\tilde{d} = 3$. Another example is a chain, which has dimension $d = 1$ in the three-dimensional space. If $d < \tilde{d}$, the space should be assumed to be empty away from the lower-dimensional system, otherwise transport may happen outside the system itself. Moreover, if $d < \tilde{d}$, the interatomic energy potentials should be much deeper than thermal energy, so that the timescale for the molecular dissociation of the chain or film due to thermal fluctuations is significantly longer than the observational timescale. In hypothetical systems where $d = \tilde{d}$, there is no environment and the condition on the interatomic energy potentials is not required.

In one-component systems in $\tilde{d}$-dimensional space, there exist $\tilde{d}+2$ conserved quantities because of the conservation of mass, energy, and momentum. Moreover, $\tilde{d} - d$ displacement variables should be introduced to describe a d-dimensional system in a $\tilde{d}$-dimensional space. For instance, a freely suspended film in a three-dimensional space is described by a single displacement variable with respect to the planar equilibrium configuration of the film (Kats and Lebedev, 1994). Similarly, a chain is described by two displacement variables. Actually, the formation of the film or the chain breaks translational symmetries, and the displacement variables constitute the corresponding order fields. Hence, $\tilde{d} - d$ Nambu–Goldstone modes arise from the broken continuous symmetries. Consequently, there exist $2\tilde{d} - d + 2$ hydrodynamic modes in such systems.[10]

If $\tilde{d} = d$, there remain $d+2$ modes corresponding to the conservation of d-dimensional momentum, energy, and mass. Because of the divergence of equation (3.239) for $d < 2$ in the limit $t \to \infty$, the transport coefficients are infinite in low-dimensional systems, so that the local hydrodynamics ruled by partial differential equations is no longer applicable. Instead, nonlocal spatiotemporal effects manifest themselves and transport becomes anomalous.

[10] In systems with c species, there exist $2\tilde{d} - d + c + 1$ hydrodynamic modes, which is the case in a soap film containing water in addition to the surfactant species (Kats and Lebedev, 1994).

Two-Dimensional Systems

In three-dimensional space, freely suspended one-component films or membranes thus have one displacement field giving its height $z = u(x, y)$ with respect to the mean equilibrium position $z = 0$. Accordingly, such films have six hydrodynamic modes: two propagative bending modes, two propagative longitudinal sound modes, one nonpropagative shear mode, and the diffusive heat mode. The dispersion relations of such systems are reported to behave as $z(q) = a_1\, q + a_2\, q^2 + a_{5/3}\, q^{5/3} + a_3\, q^3 + \cdots$ (Kats and Lebedev, 1994).

If $\tilde{d} = d = 2$, the transport coefficients are expected to be infinite because of the logarithmic divergence of equation (3.239) in the limit $t \to \infty$ if $d = 2$. They can thus be replaced by effective transport coefficients with logarithmic dependences on either the system size or the observational timescale (Kawasaki, 1971).

One-Dimensional Systems

Recently, significant progress has been achieved in the understanding of anomalous transport and hydrodynamics in one-dimensional systems (van Beijeren, 2012; Spohn, 2014; Dhar et al., 2019; Benenti et al., 2020).

As previously mentioned, a chain freely suspended in three-dimensional space has two displacement fields, $x = u_x(z)$ and $y = u_y(z)$, giving its deviations with respect to its mean equilibrium configuration $x = y = 0$. Therefore, the chain has seven hydrodynamic modes: four bending modes, two longitudinal sound modes, and the heat mode.

For anharmonic chains in one-dimensional space ($d = \tilde{d} = 1$), the transport coefficients are infinite, so that transport becomes anomalous. Under such circumstances, the coefficients of heat conduction and sound damping diverge with the system size L according to

$$\kappa(L) \sim L^{\alpha} \tag{3.242}$$

with an exponent $0 < \alpha \leq 1$ and the corresponding time autocorrelation functions have, in general, an algebraic decay going as

$$C(t) \sim \frac{1}{t^{1-\delta}} \tag{3.243}$$

with the exponent $0 \leq \delta < 1$ (van Beijeren, 2012; Benenti et al., 2020).

In the case of generic short-ranged interaction potentials between the particles, theoretical results show that $\alpha = \delta = 1/3$ for heat conduction, while $\alpha = 1/2$ and $\delta = 1/3$ for sound damping (van Beijeren, 2012; Spohn, 2014). The hydrodynamic variables thus have a strongly nonlocal behavior. The heat mode has the dispersion relation $z(q) \sim -|q|^{5/3}$ (van Beijeren, 2012). Accordingly, heat pulses evolve in time as Lévy random walks (Dhar et al., 2019). The two counter-propagating sound modes have dispersion relations such as $|\mathrm{Im}\, z(q)| \simeq v_{\mathrm{l}}|q|$ and $\mathrm{Re}\, z(q) \sim -|q|^{3/2}$, where v_{l} is the sound velocity. The sound waves manifest scaling behavior in the universality class of the one-dimensional fluctuating equations of Burgers (1974) and Kardar et al. (1986) (Prähofer and Spohn, 2004).[11]

[11] The Burgers equation is a one-dimensional reduction of Navier–Stokes equations with pressure set equal to zero. The Kardar–Parisi–Zhang equation rules a model of stochastic surface growth.

For anharmonic chains with even interaction potentials under zero pressure, anomalous transport belongs to another universality class (van Beijeren, 2012; Spohn, 2014). These dynamical universality classes have been shown to form a whole family (Popkov et al., 2015).

Similar anomalous transport properties are observed in the numerical simulation of anharmonic chains in higher-dimensional spaces (Benenti et al., 2020).

4

Stochastic Processes

4.1 Introduction

In between the atomic and macroscopic scales, processes involving nanometric or micrometric objects are observed to be stochastic due to thermal and molecular fluctuations in their environment. Stochasticity manifests itself because of the erratic movements of the corpuscules composing matter, in particular, if the changes of energy ΔE during motion are smaller than the thermal energy $k_{\mathrm{B}}T$, i.e., if $|\Delta E| \lesssim k_{\mathrm{B}}T$. Nowadays, stochastic processes are observed and investigated in a large variety of different systems in physics, chemistry, biology, and beyond.

Early in the nineteenth century, Brown (1828) observed the random walk of micrometric pollen grains of plants in water using optical microscopy. Since the end of the nineteenth century, it has been understood that Brownian motion finds its origin in the ceaseless thermal fluctuations of the fluid surrounding the grains. At the beginning of the twentieth century, theoretical work by Einstein (1905, 1926) and von Smoluchowski (1906) showed that the Avogadro number could be measured from the Brownian motion of colloidal particles, which was validated experimentally by Perrin (1910), confirming in this way the reality of atoms. Theory was further developed to include the effects of inertia and different kinds of degrees of freedom (Langevin, 1908; Fokker, 1914; Planck, 1917). Moreover, thermal fluctuations were also observed and characterized in the rotational motion of colloidal particles, the phenomenon of critical opalescence, electric circuits, and the random oscillations of a mirror suspended in air (Perrin, 1910; Johnson, 1928; Nyquist, 1928; Mazo, 2002). At higher energies, the theory of stochastic processes has also been applied to the random emission of alpha particles from radioactive substances (Rutherford et al., 1910).

Theory was extended to chemical reactions, colloid chemistry, and nucleation (Delbrück, 1940; Kramers, 1940; Chandrasekhar, 1943; Hill and Plesner, 1965; McQuarrie, 1967; Nicolis, 1972; Schnakenberg, 1976; Nicolis and Prigogine, 1977; van Kampen, 1981; Hill, 1989), although experimental techniques were still limited to the study of large ensembles of molecules. Since the 1980s, the advent of fast electronic recording techniques and efficient laser optical methods has made possible the observation in real time of processes at the level of single atoms and single molecules. In particular, the stochastic properties of single ion channel proteins in biological membranes can be studied with the patch clamp technique developed in electrophysiology (Colquhoun and Sakmann, 1981). Moreover, the single-molecule enzymatic dynamics may be observed in real time using fluorescence microscopy (Xie, 2001). Steps and substeps in the random motion of molecular motors have been discovered thanks to the techniques of fluorescence microscopy and optical tweezers (Kitamura et al., 1999; Yasuda et al., 2001; Sowa et al., 2005; Moffitt et al., 2010). The diffusion of single atoms adsorbed on surfaces has been studied in detail using field ion and atomic force microscopies (Antczak and Ehrlich, 2004; Henss et al., 2019). Additionally, quantum jumps of a single atomic ion can be observed using laser-induced fluorescence (Cook and Kimble, 1985; Sauter et al., 1986; Bergquist et al., 1986). At a low temperature of a few kelvins, single electron transfers can be monitored in mesoscopic semiconducting circuits (Gustavsson et al., 2006; Fujisawa et al., 2006).

The experimental observations of stochastic processes raise several issues. In such experiments, trajectories are recorded by the stroboscopic observation of the system and the measurement of one or more variables with some sampling rate. These variables, which are called random variables in the theory of stochastic processes, may be the position of the colloidal particle in Brownian motion, the number of particles of some species in a compartment of the system, the number of particles transferred between reservoirs across an open system, or the number of reactive events that have occurred since the beginning of observation. These variables are recorded at regular time intervals, the statistics of their values can be established, and the issue is to know how their probabilities depend on time and control parameters. In principle, these probabilities are determined by the underlying microscopic dynamics and the question arises of the origin of the stochasticity of these processes. Another important issue is to determine the statistical correlations between successive events and how fast the memory of previous events is lost during time evolution. Statistical correlations may also exist between several variables that are jointly observed. In this regard, stochastic processes can manifest different kinds of properties. If the process ruling some variable does not depend on the time when the observation started, the process is said to be stationary. Otherwise, it is nonstationary. Furthermore, the process may evolve under equilibrium or nonequilibrium conditions. The examples of classical systems shown in Figure 4.1 illustrate these properties.

Figure 4.1(a) schematically represents the Brownian motion of a colloidal particle suspended in a fluid occupying a vessel of finite volume. The colloidal particle undergoes elastic collisions with the molecules of the fluid. The position and velocity of the colloid are supposed to be recorded over time. The position performs a random walk contained inside the volume of the vessel if the interaction between the colloid and the walls is repulsive, while the velocity of the colloid fluctuates under the effect of collisions. Over an arbitrarily long time interval, the system reaches equilibrium with a uniform probability density in the finite volume of the vessel for the colloid position and a Maxwellian probability distribution at the temperature of the fluid for the colloid velocity. Under such circumstances, the equilibrium joint probability distribution for the position and the velocity can be normalized to the unit value and is thus well defined, so that the stochastic process is stationary. In contrast, if the same Brownian motion takes place in a system of infinite spatial extension starting from a given initial position as illustrated in Figure 4.1(b), the stochastic process ruling the position of the particle is nonstationary because its probability density does not reach a normalized equilibrium distribution. Since the particle does not meet walls where it may

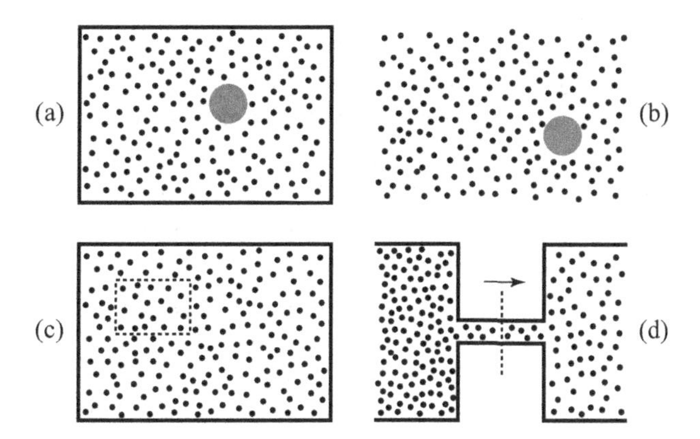

Figure 4.1 Examples of stochastic processes in systems ruled by classical dynamics: (a) Stationary Brownian motion of a colloidal particle in a fluid contained in a finite vessel; (b) Nonstationary Brownian motion of a colloidal particle in a fluid of infinite spatial extension; (c) Stationary process of the number of particles visiting a small domain (dashed line) in a fluid contained in a finite vessel; (d) Nonstationary process of the number of particles crossing a sectional area (dashed line) in a small pipe between two compartments where the fluid has different densities and/or temperatures in each compartment.

bounce back, the probability density of its position locally decreases to zero everywhere in space. However, its velocity will reach a Maxwellian probability distribution after some short time period, so that the stochastic process of its velocity is then stationary. Similarly, if a constant external force such as gravity is applied to the colloidal particle, its position will reach the stationary equilibrium Boltzmann distribution in the setup of Figure 4.1(a) because of its collisions on the wall at the bottom, but its position will continue to drift downwards in a nonequilibrium nonstationary process in the setup of Figure 4.1(b) where there is no wall for the equilibration of the position probability distribution.

In Figure 4.1(c), we consider the number of fluid particles inside a small region (shown as dashed line) in the middle of a fluid at equilibrium. As a function of time, this number fluctuates around the mean value given by the particle density multiplied by the volume of the observed region, using the equilibrium stationary probability distribution. Accordingly, this process is also stationary. This behavior also concerns the kinetic energy of the particles inside the same region. In Figure 4.1(d), a fluid of particles flows through a small pipe connecting two arbitrarily large reservoirs where the fluid has different densities in each reservoir and thus different chemical potentials keeping their values during an arbitrarily long period of time.[1] The flux of particles across the sectional area (depicted by the dashed line) goes, on average, from high to low chemical potentials. In this example, the number of

[1] We note that if the reservoirs are large but finite, their chemical potentials will slowly change in time to become equal to each other as equilibrium is reached in the whole system composed of both reservoirs after a long equilibration time, $t_{\text{equil}} \sim V_L V_R/(V_L + V_R)$, as already discussed in the last part of Section 1.7.2. A nonequilibrium steady macrostate can thus be maintained as long as $t \ll t_{\text{equil}}$. This macrostate is steady in the sense that the particle density and current density have steady profiles, but there is a flux of particles between the reservoirs, so that the particle numbers do not remain stationary inside each reservoir.

particles crossing the sectional area *per unit time* is stationary as long as the system remains in a nonequilibrium steady macrostate. However, the number of particles transferred from one reservoir to the other increases (or decreases) with time, on average, so that this variable is ruled by a nonstationary stochastic process. The stochastic process of particle transport between two reservoirs under nonequilibrium conditions is thus nonstationary, as is also the case for a colloidal particle undergoing random drift if a constant external force is driving its motion across an infinite spatially extended system. Similar considerations hold for energy exchanges between the reservoirs due to the transport of energy by the particles.

As a further example, we may envisage a particle moving in a double well potential and subjected to the thermal noise of a heat bath (Hänggi et al., 1990). The position of this particle will undergo random jumps from one well to the other, as if there would be random transitions between two discrete states. Similar stochastic processes exist when particles transiently bind and unbind from a potential well during reactions. Beyond these examples of classical systems, there are also stochastic processes in quantum systems where discrete states arise from energy quantization.

Remarkably, the theory of stochastic processes provides a unified description of the two different but complementary aspects of random time evolution. On the one hand, individual trajectories manifest randomness and thus form a statistical ensemble of irregular trajectories that differ from each other. On the other hand, this ensemble is ruled by a probability distribution that systematically and smoothly evolves in time. These two aspects can be formulated and interrelated thanks to the theory of stochastic processes. This rich and important theory, which is also referred to as the theory of random processes, is developed in the following sections, emphasizing its connections with statistical mechanics and thermodynamics.

4.2 The Joint Probability Distribution of the Stochastic Process

In a system, some variables are supposed to be measured and recorded in time t. Typically, these variables are observed to fluctuate with time, so that they are called *random variables*. They may be real variables $\mathbf{X}(t) \in \mathbb{R}^c$ such as the position and/or the velocity of a colloidal particle, and/or integer variables $\mathbf{N}(t) \in \mathbb{Z}^d$ such as the numbers of particles contained inside some regions of the system or exchanged between two regions. These variables correspond to observables $\mathbf{X}(\boldsymbol{\Gamma})$ and $\mathbf{N}(\boldsymbol{\Gamma})$ with respect to the underlying microscopic dynamics, which may be assumed to be classical and ruled by the flow, $\boldsymbol{\Phi}^t$, in phase space. In this regard, the time evolution of the random variables of the stochastic process is determined by the phase-space trajectory $\boldsymbol{\Gamma}_t = \boldsymbol{\Phi}^t \boldsymbol{\Gamma}_0$, starting from the initial condition $\boldsymbol{\Gamma}_0$ distributed according to some initial probability density (2.32), as explained in Chapter 2. Therefore, the probability distribution of the variables $\mathbf{X}(t) = \mathbf{X}(\boldsymbol{\Gamma}_t)$ and $\mathbf{N}(t) = \mathbf{N}(\boldsymbol{\Gamma}_t)$ can, in principle, be determined at every time using the underlying microscopic dynamics according to statistical mechanics. In any case, these random variables are ruled by the principles of probability theory (Kolmogorov, 1956a; Feller, 1968, 1971; van Kampen, 1981; Gardiner, 2004).

Events, or states, $\omega(t)$ observed at a given time t correspond to the observation of the variables in cells of the space of random variables, $\mathscr{C}_\omega \subset \mathbb{R}^c \times \mathbb{Z}^d$, such as

$$\mathscr{C}_{\omega(t)} = \{\mathbf{X}(t) \in [\mathbf{x}, \mathbf{x} + d\mathbf{x}]\} \cap \{\mathbf{N}(t) = \mathbf{n}\} \tag{4.1}$$

with $t \in \mathbb{R}$, $\mathbf{x} \in \mathbb{R}^c$, $d\mathbf{x} \in \mathbb{R}^c$, and $\mathbf{n} \in \mathbb{Z}^d$. These discrete events or states belong to the state space $\omega \in \mathscr{E}$, which may be finite or countably infinite. Events observed at successive times $t_1 < t_2 < \cdots < t_k$ form the *trajectory*, also called the *path* or *history*: $\{\omega_i = \omega(t_i)\}_{i=1}^k$. The stochastic process is characterized by the joint probability distribution $P(\omega_1, t_1; \omega_2, t_2; \ldots; \omega_k, t_k)$ of the events $\{\omega_i\}_{i=1}^k$ observed at the successive times $\{t_i\}_{i=1}^k$ along the trajectory (Gardiner, 2004). This joint probability distribution is also called the path or trajectory probability distribution. Such joint probabilities are obtained by calculating the statistics of the trajectories. If the joint probabilities of all the mutually exclusive events occurring at some time t_i, i.e., such that $\mathscr{C}_{\omega_i} \cap \mathscr{C}_{\omega_i'} = \emptyset$ for $\omega_i \neq \omega_i'$, and $\cup_{\omega_i} \mathscr{C}_{\omega_i} = \mathbb{R}^c \times \mathbb{Z}^d$, are summed together, we obtain the probability of the trajectory defined by skipping and discarding what happens at time t_i:

$$\sum_{\omega_i} P(\omega_1, t_1; \ldots; \omega_{i-1}, t_{i-1}; \omega_i, t_i; \omega_{i+1}, t_{i+1}; \ldots; \omega_k, t_k)$$
$$= P(\omega_1, t_1; \ldots; \omega_{i-1}, t_{i-1}; \omega_{i+1}, t_{i+1}; \ldots; \omega_k, t_k). \tag{4.2}$$

This basic property of probability theory results in the following normalization condition holding at every time t:

$$\sum_{\omega} P(\omega, t) = 1. \tag{4.3}$$

If $I_\omega(\mathbf{\Gamma})$ denotes the indicator function of the phase-space cell

$$\mathcal{C}_\omega = \{\mathbf{\Gamma} \in \mathcal{M} : \mathbf{X}(\mathbf{\Gamma}) \in [\mathbf{x}, \mathbf{x} + d\mathbf{x}] \text{ and } \mathbf{N}(\mathbf{\Gamma}) = \mathbf{n}\} \tag{4.4}$$

corresponding to the event (4.1) for the underlying microscopic dynamics, the joint probability can be expressed as

$$P(\omega_1, t_1; \ldots; \omega_k, t_k) = \int d\mathbf{\Gamma}\, p_0(\mathbf{\Gamma})\, I_{\omega_1}\left(\mathbf{\Phi}^{t_1}\mathbf{\Gamma}\right) \cdots I_{\omega_k}\left(\mathbf{\Phi}^{t_k}\mathbf{\Gamma}\right), \tag{4.5}$$

where p_0 is the probability density (2.32) of initial conditions $\mathbf{\Gamma}$ evolving in phase space under the dynamics of the flow $\mathbf{\Phi}^t$. In this regard, the description of the stochastic process in terms of joint probability distributions results from the statistical ensemble of initial conditions introduced in the framework of statistical mechanics. Another aspect is that the events ω correspond to the cells $\mathcal{C}_\omega$ in phase space, so that the stochastic process provides a coarse-grained description of the underlying microscopic dynamics. Different choices of random variables and/or coarse graining define different stochastic processes.

Within this formulation, several properties can now be defined.

The stochastic process is said to be *stationary* if

$$P(\omega_1, t_1; \ldots; \omega_k, t_k) = P(\omega_1, t_1 + \mathscr{T}; \ldots; \omega_k, t_k + \mathscr{T}) \tag{4.6}$$

for any time translation by $\mathscr{T} \in \mathbb{R}$. In particular, taking $\mathscr{T} = -t_1$ shows that the path probability does not depend on the initial time, i.e., the instant of time when the stochastic process started. If this is not the case, the stochastic process is said to be *nonstationary* and the initial time should be specified in its description.

The *conditional probabilities* for observing some future trajectory $\{\omega_i = \omega(t_i)\}_{i=j+1}^k$ provided that the past trajectory $\{\omega_i = \omega(t_i)\}_{i=1}^j$ has been observed are defined as

$$P(\omega_{j+1},t_{j+1};\ldots;\omega_k,t_k|\omega_1,t_1;\ldots;\omega_j,t_j) \equiv \frac{P(\omega_1,t_1;\ldots;\omega_j,t_j;\omega_{j+1},t_{j+1};\ldots;\omega_k,t_k)}{P(\omega_1,t_1;\ldots;\omega_j,t_j)} \quad \text{for} \quad 1 \leq j < k. \tag{4.7}$$

A stochastic process is said to be *Bernoullian* if there is memory loss of the past, i.e., if the conditional probability (4.7) of any future trajectory does not depend on the past trajectory, in which case the path probability can be factorized into the probabilities of the successive events:

$$P(\omega_1,t_1;\ldots;\omega_k,t_k) = \prod_{i=1}^{k} P(\omega_i,t_i). \tag{4.8}$$

The successive events are thus statistically independent.

A stochastic process is said to be *Markovian* of first order if the conditional probability (4.7) only depends on the previous past event. Accordingly, the past trajectory can be factorized into the conditional probabilities to observe two successive events and the probability of the first event:

$$P(\omega_1,t_1;\ldots;\omega_k,t_k) = \prod_{i=2}^{k} P(\omega_i,t_i|\omega_{i-1},t_{i-1})\, P(\omega_1,t_1). \tag{4.9}$$

Markovian stochastic processes of higher orders can be defined if there is a factorization into conditional probabilities that depends on more than one past event.

If the path probability cannot be factorized into any kind of conditional probabilities, the stochastic process is said to be *non-Markovian*.

The concept of reversibility can be introduced in the theory of stochastic processes. A stationary stochastic process is said to be *reversible* if the probability of any path is equal to the probability of the time-reversed path, i.e., if

$$P(\omega_1,t_1;\ldots;\omega_k,t_k) = P(\omega_k,\mathscr{T}-t_k;\ldots;\omega_1,\mathscr{T}-t_1) \tag{4.10}$$

for all $\mathscr{T} \in \mathbb{R}$ (Weiss, 1975; Spohn, 1991; Jiang et al., 2004). We note that, if the observation times form a monotonically increasing sequence $t_1 < \cdots < t_k$, those of the time-reversed path are also ordered, since $\mathscr{T} - t_k < \cdots < \mathscr{T} - t_1$. Equation (2.93) gives an example of reversible process. Nevertheless, we emphasize that in the theory of stochastic processes, the property of reversibility is defined for *given* events $\{\omega_j\}$ and, thus, a *given* coarse graining. For this reason, this property differs from the detailed balance conditions defined in the framework of statistical mechanics for arbitrarily small coarse graining in systems that are time-reversal symmetric in phase space and described by equilibrium probability distributions. In this respect, the property of reversibility is weaker than the detailed balance conditions and, indeed, there exist reversible nonequilibrium stochastic processes, as further discussed below.

Going back to the definition (4.1) of the random events we consider, the quantities $d\mathbf{x}$ may be supposed to be infinitesimal so that the path probability leads to the path probability density

$$P(\omega_1,t_1;\ldots;\omega_k,t_k) = p(\mathbf{x}_1,\mathbf{n}_1,t_1;\ldots;\mathbf{x}_k,\mathbf{n}_k,t_k)\, d\mathbf{x}_1\cdots d\mathbf{x}_k \tag{4.11}$$

with $d\mathbf{x}_i = d^c x_i$, which is normalized according to

$$\sum_{\mathbf{n}_1,\ldots,\,\mathbf{n}_k} \int_{\mathbb{R}^{ck}} p(\mathbf{x}_1,\mathbf{n}_1,t_1;\ldots;\mathbf{x}_k,\mathbf{n}_k,t_k)\, d\mathbf{x}_1\cdots d\mathbf{x}_k = 1, \tag{4.12}$$

for any integer k, as a consequence of equations (4.2) and (4.3).

The variables of the stochastic process may be continuous, discrete, or both. The process is called a *continuous-state stochastic process* if the variables are continuous, and a *discrete-state stochastic process* if they are discrete. Moreover, the process may evolve in continuous time $t \in \mathbb{R}$ or discrete time $t_i = i\Delta t$ with $i \in \mathbb{Z}$. In the former case, we speak about a *continuous-time stochastic process* and about a *discrete-time stochastic process* or *random chain* in the latter one.

The theory of continuous-time first-order Markov processes with either discrete or continuous variables will be developed in Sections 4.4 and 4.5 on the basis of the previous properties. Examples of stochastic processes are given in Appendix E.

4.3 Correlation and Spectral Functions

Once the path probabilities of a stochastic process are known, we may consider the functions $Y(t) \equiv Y[\mathbf{X}(t)]$ or $Y[\mathbf{N}(t)]$ of the observed quantities $\mathbf{X}$ or $\mathbf{N}$ and determine their mean values such as $\langle Y(t)\rangle$, where $\langle\cdot\rangle$ denotes the expectation value with respect to the probability distribution at some time t, or their correlation functions such as

$$C(t,t') \equiv \langle Y(t)\,Y(t')\rangle - \langle Y(t)\rangle\langle Y(t')\rangle = C(t',t), \tag{4.13}$$

for $t,t' \in \mathbb{R}$. The vanishing of such a correlation function as $|t-t'| \to \infty$ is the manifestation of the loss of memory between random events separated by a long time interval, which is reminiscent of the property of dynamical mixing considered in Chapter 2. In this regard, we expect that Bernoulli or Markov processes should arise, for instance, in systems where there is timescale separation between some fast randomizing dynamics and some resulting slow motion. We notice that correlation functions between more than two successive events may also be defined.

If the process is stationary, the correlation function (4.13) only depends on the difference between the times t and t', i.e., $C(t,t') = C(t-t')$. In this case, the frequency spectrum of the stochastic process may be characterized by the spectral density function

$$\mathcal{S}(\omega) \equiv \lim_{\mathcal{T}\to\infty} \frac{1}{\mathcal{T}} \left\langle \left| \int_0^{\mathcal{T}} \mathrm{e}^{\imath\omega t}\, \delta Y(t)\, dt \right|^2 \right\rangle, \tag{4.14}$$

where $\omega = 2\pi f$ is the angular frequency corresponding to the frequency f and $\delta Y(t) \equiv Y(t) - \langle Y(t) \rangle$. The spectral density can be expressed in terms of the correlation function as

$$S(\omega) = \int_{-\infty}^{+\infty} C(t)\, e^{\imath \omega t}\, dt = 2 \int_0^{\infty} C(t) \cos \omega t\, dt \tag{4.15}$$

according to the Wiener–Khinchin theorem (van Kampen, 1981; Gardiner, 2004). The following formula by MacDonald (1948–1949) is also useful:

$$S(\omega) = \omega \int_0^{\infty} \frac{d}{dt} \langle Z(t)^2 \rangle \sin \omega t\, dt \qquad \text{with} \qquad Z(t) \equiv \int_0^t \delta Y(t')\, dt'. \tag{4.16}$$

The integral of the spectral density over the whole frequency spectrum gives the variance of the process $Y(t)$ because

$$\frac{1}{2\pi} \int_{-\infty}^{+\infty} S(\omega)\, d\omega = C(0) = \langle Y^2 \rangle - \langle Y \rangle^2. \tag{4.17}$$

Another issue, which will be addressed in Chapter 6, is the characterization of the amount of stochasticity generated by some process.

4.4 Discrete-State Markov Processes

4.4.1 Master Equation

Let us envisage a stochastic process for the time evolution of discrete random variables $\omega(t) \in \mathscr{E}$ ruled by the joint probability distribution

$$P(\omega_1, t_1; \dots; \omega_k, t_k) = P\left[\omega(t_1) = \omega_1; \dots; \omega(t_k) = \omega_k\right]. \tag{4.18}$$

If the process is Markovian and of first order, this joint probability distribution factorizes as in equation (4.9) in terms of the conditional and one-event probability distributions, which are, respectively, normalized according to

$$\sum_{\omega_i} P(\omega_i, t_i | \omega_{i-1}, t_{i-1}) = 1 \qquad \text{and} \qquad \sum_{\omega_i} P(\omega_i, t_i) = 1. \tag{4.19}$$

If we consider the joint probability distribution (4.9) for two successive events ($k = 2$) and carry out the sum over all the possible events at the first time t_1, we obtain the following equation that the one-event probability distribution should satisfy with respect to the conditional one:

$$P(\omega_2, t_2) = \sum_{\omega_1} P(\omega_2, t_2 | \omega_1, t_1)\, P(\omega_1, t_1). \tag{4.20}$$

Using equation (4.9) for three successive events ($k = 3$) and summing over the possible events at the intermediate time t_2, we get the so-called *Chapman–Kolmogorov equation* (van Kampen, 1981; Gardiner, 2004)

$$P(\omega_3, t_3 | \omega_1, t_1) = \sum_{\omega_2} P(\omega_3, t_3 | \omega_2, t_2)\, P(\omega_2, t_2 | \omega_1, t_1), \tag{4.21}$$

which the conditional probability distribution must satisfy. These equations allow us to deduce the master equation for the first-order Markov process considered. The master equation is a differential equation ruling the time evolution of the one-event probability distribution $P(\omega,t)$. The master equation is established by considering equation (4.20) for the events $(\omega_2 = \omega, t_2 = t + \Delta t)$ and $(\omega_1 = \omega', t_1 = t)$ in the limit of an infinitesimal time interval $\Delta t \to 0$ and computing the variation of the probability distribution over this time interval:

$$\frac{1}{\Delta t}\left[P(\omega,t+\Delta t) - P(\omega,t)\right] = \sum_{\omega'} \frac{1}{\Delta t} P(\omega,t+\Delta t|\omega',t)\, P(\omega',t) - \frac{1}{\Delta t} P(\omega,t). \quad (4.22)$$

In the last term of the right-hand side we may use the normalization condition (4.19) for the conditional probability, $\sum_{\omega'} P(\omega', t+\Delta t|\omega,t) = 1$. The transition rates of the Markov process are defined as

$$W(\omega \to \omega';t) \equiv \lim_{\Delta t\to 0} \frac{1}{\Delta t} P(\omega',t+\Delta t|\omega,t), \quad (4.23)$$

as in equation (2.87) in the particular case when the system is at equilibrium. Substituting these relations into equation (4.22) and taking the limit $\Delta t \to 0$, we obtain the *master equation of the Markov process* as

$$\frac{d}{dt} P(\omega,t) = \sum_{\omega'} \left[P(\omega',t)\, W(\omega' \to \omega;t) - P(\omega,t)\, W(\omega \to \omega';t)\right]. \quad (4.24)$$

In the right-hand side, we find the gain terms increasing the probability of the state ω by transitions coming from other states $\omega' \neq \omega$ and the loss terms decreasing this probability because of transitions going out the state ω. This structure guarantees that the normalization condition $\sum_{\omega} P(\omega,t) = 1$ is preserved for all times $t \in \mathbb{R}$. Since the transitions induce jumps between the discrete states ω, such stochastic processes are also called *Markov jump processes*. The transitions are supposed to be instantaneous in time, so that any trajectory $\{\omega(t)\}_{t=-\infty}^{+\infty}$ of the process can be represented as a succession of instantaneous jumps separated by finite time intervals between the jumps.

Often the transition rates (4.23) do not depend on time, so that the master equation (4.24) is invariant under time translations and reads

$$\frac{d}{dt} P(\omega,t) = \sum_{\omega'} \left[P(\omega',t)\, W(\omega' \to \omega) - P(\omega,t)\, W(\omega \to \omega')\right]. \quad (4.25)$$

In this case, the process is stationary if the master equation admits a stationary probability distribution that can be normalized to the unit value, i.e., such that $(d/dt)P_{\rm st}(\omega) = 0$ and $\sum_{\omega} P_{\rm st}(\omega) = 1$.

In many systems of interest, the transitions between the states $\{\omega\}$ may be caused by several independent elementary mechanisms, such as different elementary reactions. In such circumstances, the transition rates are given by the sum of the rates $W\big(\omega \overset{\rho}{\to} \omega'\big)$ of the elementary mechanisms $\{\rho\}$ causing the transitions:

$$W(\omega \to \omega') = \sum_{\rho} W\big(\omega \overset{\rho}{\to} \omega'\big). \quad (4.26)$$

This is the case if the two states ω and ω' are interconnected by several distinct paths $\omega \overset{\rho}{\to} \omega'$ in the space of the underlying microscopic dynamics (as already considered in Chapter 2). In this general case, the master equation is given by

$$\frac{d}{dt} P(\omega,t) = \sum_{\rho,\,\omega'} \Big[P(\omega',t)\, W\big(\omega' \overset{\rho}{\to} \omega\big) - P(\omega,t)\, W\big(\omega \overset{\rho}{\to} \omega'\big) \Big]. \tag{4.27}$$

For such stochastic processes, the random time interval τ before the next jump from the state ω towards any other state is exponentially distributed as

$$p(\tau) = \gamma_\omega\, \mathrm{e}^{-\gamma_\omega \tau}, \qquad \text{where} \qquad \gamma_\omega \equiv \sum_{\omega'} W(\omega \to \omega') = \sum_{\rho,\,\omega'} W(\omega \overset{\rho}{\to} \omega') \tag{4.28}$$

is the rate of escape from the state ω. The mean dwell time in the state ω is thus given by $\langle \tau \rangle_\omega = 1/\gamma_\omega$.

The stochastic process can be simulated as a random sequence of jump times $t_{j+1} = t_j + \tau_j$ with $j \in \mathbb{Z}$. During the time interval $t_j < t < t_{j+1}$, the system remains in the state ω_j. At the next time t_{j+1}, a random jump through the elementary transition mechanism ρ to the state $\omega' = \omega_{j+1}$ occurs with probability

$$\mathscr{P}(\omega \overset{\rho}{\to} \omega') = \frac{W\big(\omega \overset{\rho}{\to} \omega'\big)}{\sum_{\rho,\,\omega'} W\big(\omega \overset{\rho}{\to} \omega'\big)}. \tag{4.29}$$

Such stochastic processes can be simulated exactly using the kinetic Monte Carlo algorithm of Gillespie (1976, 1977), which is based on the rules (4.28) and (4.29).

An example of a stationary Markov process is given by the random jumps of an atomic ion between several quantum states in the phenomenon of laser-induced fluorescence (Cook and Kimble, 1985).

An example of a nonstationary Markov process is the standard Poisson process ruling the random number of particles detected since the beginning of some experiment (see Section E.3). This process can be generalized to several random integer numbers $\mathbf{n} \in \mathbb{Z}^d$, which define the states ω of such processes. If every elementary mechanism ρ induces the specific transition $\mathbf{n} \to \mathbf{n}' = \mathbf{n} + \boldsymbol{\nu}_\rho$ by some integer values $\boldsymbol{\nu}_\rho \in \mathbb{Z}^d$ (such as the stoichiometric coefficients of elementary chemical reactions), we can introduce the transition rates

$$W_\rho(\mathbf{n}) \equiv W\Big(\mathbf{n} \overset{\rho}{\to} \mathbf{n} + \boldsymbol{\nu}_\rho\Big) \tag{4.30}$$

and write the master equation (4.27) in the form

$$\frac{d}{dt} P(\mathbf{n},t) = \sum_{\rho} \big[W_\rho(\mathbf{n} - \boldsymbol{\nu}_\rho)\, P(\mathbf{n} - \boldsymbol{\nu}_\rho, t) - W_\rho(\mathbf{n})\, P(\mathbf{n},t) \big]. \tag{4.31}$$

Equivalently, the master equation reads

$$\frac{d}{dt} P(\mathbf{n},t) = \sum_{\rho} \big(\mathrm{e}^{-\boldsymbol{\nu}_\rho \cdot \partial_{\mathbf{n}}} - 1 \big)\, W_\rho(\mathbf{n})\, P(\mathbf{n},t), \tag{4.32}$$

using the shift operators defined as

$$\mathrm{e}^{\pm \boldsymbol{\nu} \cdot \partial_{\mathbf{n}}} f(\mathbf{n}) = f(\mathbf{n} \pm \boldsymbol{\nu}) \tag{4.33}$$

for any function $f(\mathbf{n})$ of integer arguments $\mathbf{n} \in \mathbb{Z}^d$ (van Kampen, 1981).

The random trajectory of the process during the time interval $[0,t]$ can be formally expressed as

$$\mathbf{n}(t) = \mathbf{n}(0) + \sum_{\rho} \boldsymbol{\nu}_{\rho}\, \mathcal{P}_{\rho} \left\{ \int_0^t W_{\rho}[\mathbf{n}(t')]\, dt' \right\} \tag{4.34}$$

in terms of independent standard Poisson processes $\mathcal{P}_{\rho}$ such that $\langle \mathcal{P}_{\rho}(X) \rangle = X$ (Kurtz, 1978). Indeed, for a given trajectory $\mathbf{n}(t)$ undergoing k jumps at the jump times $t_j \in [0,t]$, the time integrals of the rates are given by $\int_0^t W_{\rho}[\mathbf{n}(t')]\, dt' = \sum_{j=1}^{k} W_{\rho}(\mathbf{n}_j)(t_{j+1} - t_j)$, where every time interval $\tau_j = t_{j+1} - t_j$ between two jumps is exponentially distributed according to equation (4.28) with the escape rate $\gamma_{\mathbf{n}_j} = \sum_{\rho} W_{\rho}(\mathbf{n}_j)$, as expected for a process simulated by Gillespie's algorithm.

Such processes are considered in the counting of particles that are, for instance, contained in some part of a system or crossing some sectional area, as illustrated in Figure 4.1(c) and (d), respectively. Related examples are processes with random transitions between discrete internal states $\{\sigma\}$ together with particle transfers inside or outside the system. In such examples, the discrete states are specified as $\omega = (\sigma, \mathbf{n})$ by the internal state σ of the system and the numbers $\mathbf{n}$ of particles of different species exchanged with the environment. The integers $\mathbf{n}$ may also count the numbers $\{n_{\rho}\}$ of transitions due to the elementary mechanisms $\{\rho\}$.

Other examples are provided by random walks for a particle on some lattice $\mathcal{L}$ of dimension d. In these cases, the discrete states ω are the lattice sites where the particle is transiently located. These sites are specified by d integers $\mathbf{n} \in \mathbb{Z}^d$ giving the particle position as $\mathbf{r} = \sum_{i=1}^{d} n_i \mathbf{a}_i$, where $\{\mathbf{a}_i\}_{i=1}^{d}$ are the primitive vectors of the lattice $\mathcal{L}$.

We note that the transition rates might be independent of time and yet the master equation (4.25) would not admit a stationary probability distribution. In particular, this is the case for the standard Poisson process (see Section E.3). For such nonstationary Markovian stochastic processes, the probability distribution continues to evolve in time without reaching an asymptotic stationary distribution.

4.4.2 Spectral Theory

The master equation (4.25) is linear in the probabilities and it can be written in the matrix form

$$\frac{d\mathbf{P}}{dt} = \mathsf{L} \cdot \mathbf{P} \tag{4.35}$$

in terms of the vector $\mathbf{P} = \{P(\omega, t)\}$ of probabilities and the matrix $\mathsf{L} = (\mathsf{L}_{\omega\omega'})$ composed of the transition rates $W(\omega' \to \omega)$ out of the diagonal for $\omega' \neq \omega$, and minus the escape rates (4.28) on the diagonal for $\omega' = \omega$:

$$\mathsf{L}_{\omega\omega'} = (1 - \delta_{\omega\omega'})\, W(\omega' \to \omega) - \gamma_{\omega}\, \delta_{\omega\omega'}. \tag{4.36}$$

Since the matrix L does not depend on time, the time evolution of the probability distribution can be expressed as

$$\mathbf{P}(t) = \sum_{\alpha} c_{\alpha}\, \mathrm{e}^{z_{\alpha} t}\, \boldsymbol{\Psi}_{\alpha} + \cdots \qquad \text{with} \qquad c_{\alpha} = \tilde{\boldsymbol{\Psi}}_{\alpha}^{\dagger} \cdot \mathbf{P}(0) \tag{4.37}$$

in terms of its eigenvalues and the associated right and left eigenvectors, such that $\mathsf{L} \cdot \boldsymbol{\Psi}_\alpha = z_\alpha \, \boldsymbol{\Psi}_\alpha$ and $\tilde{\boldsymbol{\Psi}}_\alpha^\dagger \cdot \mathsf{L} = z_\alpha \, \tilde{\boldsymbol{\Psi}}_\alpha^\dagger$.[2] In equation (4.37), the dots denote possible contributions from Jordan blocks. In general, the eigenvalues are given by complex numbers. If the state space is infinite, the spectrum may include a continuous part. As probability is conserved, the spectrum includes the null eigenvalue $z_0 = 0$.

The right and left eigenvectors are supposed to satisfy the biorthonormality conditions: $\tilde{\boldsymbol{\Psi}}_\alpha^\dagger \cdot \boldsymbol{\Psi}_\beta = \delta_{\alpha\beta}$. In this case, there exists at least one null right eigenvector defining a stationary probability distribution $\boldsymbol{\Psi}_0 = \mathbf{P}_{\rm st}$ and corresponding to the left eigenvector $\tilde{\boldsymbol{\Psi}}_0^\dagger = (1, 1, 1, \dots)$. If the null eigenvalue is simply degenerate, the process is said to be *ergodic* and there exists a unique stationary probability distribution.[3] For such processes, a general result is that any initial probability distribution will converge in time towards the stationary probability distribution. This property can be proved by considering the Kullback–Leibler divergence

$$D_{\rm KL}(P \| P_{\rm st}) \equiv \sum_\omega P(\omega,t) \ln \frac{P(\omega,t)}{P_{\rm st}(\omega)} \geq 0 \tag{4.38}$$

between the time-dependent and the stationary probability distributions. The quantity (4.38) is known to be always nonnegative and equal to zero if and only if the two distributions coincide (Cover and Thomas, 2006). Using the master equation (4.25) and the inequality $\ln x \leq x - 1$ for $x > 0$, we have that

$$\frac{dD_{\rm KL}}{dt} = \sum_{\omega \neq \omega'} P(\omega',t)\, W(\omega' \to \omega) \ln \frac{P_{\rm st}(\omega')\, P(\omega,t)}{P(\omega',t)\, P_{\rm st}(\omega)} \tag{4.39}$$

$$\leq \sum_{\omega \neq \omega'} P(\omega',t)\, W(\omega' \to \omega) \left[\frac{P_{\rm st}(\omega')\, P(\omega,t)}{P(\omega',t)\, P_{\rm st}(\omega)} - 1 \right] = \sum_\omega \frac{P(\omega,t)}{P_{\rm st}(\omega)} \frac{dP_{\rm st}(\omega)}{dt} = 0,$$

so that the time derivative of the Kullback–Leibler divergence (4.38) is always nonpositive. Since the stationary distribution is unique for ergodic processes, the quantity $D_{\rm KL}$ can only decrease to zero. Therefore, the Kullback–Leibler divergence (4.38) is a Lyapunov function, implying the convergence of any initial probability distribution towards the stationary one (Schnakenberg, 1976). As a consequence, all the eigenvalues have a nonpositive real part: $\operatorname{Re} z_\alpha \leq 0$.

4.4.3 Reversible Discrete-State Markov Processes

If the Markov jump process is stationary and reversible such that the reversibility condition (4.10) holds, the transition rates should satisfy the conditions

$$P_{\rm st}(\omega')\, W(\omega' \to \omega) = P_{\rm st}(\omega)\, W(\omega \to \omega') \tag{4.40}$$

[2] $\tilde{\boldsymbol{\Psi}}_\alpha^\dagger$ denotes the complex conjugate transpose of the column vector $\tilde{\boldsymbol{\Psi}}_\alpha$.

[3] A process with at least two sets of states without possible transition between them is not ergodic because a stationary probability distribution can exist for each one of these sets.

for every pair of states $\omega \rightleftharpoons \omega'$ and with respect to the stationary probability distribution $P_{\rm st}(\omega)$. Although similar to the detailed balance conditions (2.88), the reversibility conditions (4.40) for the stochastic process may hold even if the system is out of equilibrium because the states ω are too coarse grained to identify thermodynamic equilibrium with the conditions (4.40) alone.

For reversible stochastic processes, we may introduce the quantities

$$K(\omega \to \omega') \equiv P_{\rm st}(\omega)^{1/2}\, W(\omega \to \omega')\, P_{\rm st}(\omega')^{-1/2} = K(\omega' \to \omega), \tag{4.41}$$

which are symmetric under the permutation of ω and ω'. Using these quantities as the elements of the matrix K according to

$$\mathsf{K}_{\omega\omega'} = (1 - \delta_{\omega\omega'})\, K(\omega' \to \omega) - \gamma_\omega\, \delta_{\omega\omega'} \tag{4.42}$$

and defining the vector $\mathfrak{F} = \{\mathfrak{F}(\omega,t)\}$ with

$$\mathfrak{F}(\omega,t) \equiv \frac{P(\omega,t)}{P_{\rm st}(\omega)^{1/2}}, \tag{4.43}$$

the master equation (4.35) can be written in the following form:

$$\frac{d\mathfrak{F}}{dt} = \mathsf{K} \cdot \mathfrak{F}, \qquad \text{where} \qquad \mathsf{K} = \mathsf{K}^{\rm T}. \tag{4.44}$$

Since the matrix K is symmetric, all its eigenvalues are real numbers, i.e., $\operatorname{Im} z_\alpha = 0$. The spectrum is thus real as a consequence of the reversibility condition (4.40) and, moreover, all the eigenvalues are nonpositive: $\operatorname{Re} z_\alpha = z_\alpha \leq 0$.

4.4.4 Entropy Production

General Processes

Since the stochastic process provides a coarse-grained description of the system, the thermodynamic entropy can be evaluated with the expression (2.124), i.e.,

$$S(t) = \sum_\omega \left[S^0(\omega) - k_{\rm B} \ln P(\omega,t)\right] P(\omega,t), \tag{4.45}$$

where $S^0(\omega)$ is the thermodynamic entropy (2.125) of the system found in the coarse-grained state ω (Gaspard, 2004a). The first terms of equation (4.45) give the mean value of this entropy, $\langle S^0 \rangle_t = \sum_\omega S^0(\omega) P(\omega,t)$, and the other terms the contribution due to disorder coming from the probability distribution $P(\omega,t)$ over the coarse-grained states. This latter contribution is often referred to as the Shannon disorder,

$$D(t) \equiv -\sum_\omega P(\omega,t)\, \ln P(\omega,t) \geq 0, \tag{4.46}$$

also called Shannon's entropy (Cover and Thomas, 2006). The thermodynamic entropy is the sum of the two contributions: $S(t) = \langle S^0 \rangle_t + k_{\rm B} D(t)$. The coarser the graining, the smaller the contribution of the Shannon disorder with respect to the mean value of the entropy [i.e., $\langle S^0\rangle_t \gg k_B D(t)$]. We note that the thermodynamic entropy can equivalently be written as the mean value $S(t) = \langle S(\omega,t)\rangle_t$ of the quantity defined as

$$S(\omega,t) \equiv S^0(\omega) - k_B \ln P(\omega,t) \tag{4.47}$$

by gathering together the two contributions if the system is found in the coarse-grained state ω at the time t.

In order to determine the entropy production, every gain term of the master equation (4.27) associated with some elementary transition mechanism ρ should be regrouped with the loss term corresponding to its reversal, $-\rho$, in such a way as to identify thermodynamic equilibrium with the detailed balance conditions (2.89), which here read

$$P_{\rm eq}(\omega')\, W\big(\omega' \overset{\rho}{\to} \omega\big) = P_{\rm eq}(\omega)\, W\big(\omega \overset{-\rho}{\to} \omega'\big). \tag{4.48}$$

In this regard, the rate of the reversed process $-\rho$ should be nonvanishing, as well as the rate of the process ρ itself. Accordingly, the master equation (4.27) can be written in the form

$$\frac{d}{dt}P(\omega,t) = \sum_{\rho,\,\omega'} J\big(\omega' \overset{\rho}{\to} \omega;t\big) \tag{4.49}$$

in terms of the net rates associated with the transitions $\omega' \overset{\rho}{\to} \omega$:

$$J\big(\omega' \overset{\rho}{\to} \omega;t\big) \equiv P(\omega',t)\, W\big(\omega' \overset{\rho}{\to} \omega\big) - P(\omega,t)\, W\big(\omega \overset{-\rho}{\to} \omega'\big) = -J\big(\omega \overset{-\rho}{\to} \omega';t\big). \tag{4.50}$$

Therefore, the time derivative of the mean value $\langle X\rangle_t \equiv \sum_\omega X(\omega)\, P(\omega,t)$ of some quantity $X(\omega)$ is given by

$$\frac{d\langle X\rangle_t}{dt} = \sum_{\rho,\omega,\omega'} X(\omega)\, J\big(\omega' \overset{\rho}{\to} \omega;t\big). \tag{4.51}$$

If we consider $X(\omega) = 1$, we get $(d/dt)\sum_\omega P(\omega,t) = 0$, meaning that the normalization condition $\sum_\omega P(\omega,t) = 1$ of the probability distribution $P(\omega,t)$ is preserved for all times $t \in \mathbb{R}$, as expected.

Now, the time derivative of the thermodynamic entropy (4.45) is obtained as

$$\begin{aligned}
\frac{dS}{dt} &= \frac{d\langle S^0\rangle_t}{dt} - k_B \sum_{\rho,\omega,\omega'} J\big(\omega' \overset{\rho}{\to} \omega;t\big) \ln P(\omega,t) \\
&= \frac{d\langle S^0\rangle_t}{dt} + \frac{k_B}{2} \sum_{\rho,\omega,\omega'} J\big(\omega' \overset{\rho}{\to} \omega;t\big) \ln \frac{P(\omega',t)}{P(\omega,t)} \\
&= \frac{d\langle S^0\rangle_t}{dt} - \frac{k_B}{2} \sum_{\rho,\omega,\omega'} J\big(\omega' \overset{\rho}{\to} \omega;t\big) \ln \frac{W\big(\omega' \overset{\rho}{\to} \omega\big)}{W\big(\omega \overset{-\rho}{\to} \omega'\big)} \\
&\quad + \frac{k_B}{2} \sum_{\rho,\omega,\omega'} J\big(\omega' \overset{\rho}{\to} \omega;t\big) \ln \frac{P(\omega',t)\, W\big(\omega' \overset{\rho}{\to} \omega\big)}{P(\omega,t)\, W\big(\omega \overset{-\rho}{\to} \omega'\big)}.
\end{aligned} \tag{4.52}$$

Accordingly, the time derivative of the entropy can be expressed as

$$\frac{dS}{dt} = \frac{d_{\rm e}S}{dt} + \frac{d_{\rm i}S}{dt} \tag{4.53}$$

in terms of the rate of *entropy exchange with the environment*

$$\frac{d_{\rm e}S}{dt} \equiv \sum_{\rho,\omega,\omega'} J\left(\omega' \xrightarrow{\rho} \omega; t\right) \left[S^0(\omega) - \frac{k_{\rm B}}{2} \ln \frac{W\left(\omega' \xrightarrow{\rho} \omega\right)}{W\left(\omega \xrightarrow{-\rho} \omega'\right)} \right] \tag{4.54}$$

and the rate of *entropy production*

$$\frac{d_{\rm i}S}{dt} \equiv \frac{k_{\rm B}}{2} \sum_{\rho,\omega,\omega'} A\left(\omega' \xrightarrow{\rho} \omega; t\right) J\left(\omega' \xrightarrow{\rho} \omega; t\right), \tag{4.55}$$

where

$$A\left(\omega' \xrightarrow{\rho} \omega; t\right) \equiv \ln \frac{P(\omega',t)\, W\left(\omega' \xrightarrow{\rho} \omega\right)}{P(\omega,t)\, W\left(\omega \xrightarrow{-\rho} \omega'\right)} \tag{4.56}$$

is the (dimensionless) affinity associated with the transition $\omega' \xrightarrow{\rho} \omega$ (Luo et al., 1984; Gaspard, 2004a). The entropy production rate is always nonnegative in accordance with the second law of thermodynamics because

$$\frac{d_{\rm i}S}{dt} = \frac{k_{\rm B}}{2} \sum_{\rho,\omega,\omega'} \left[P(\omega',t)\, W\left(\omega' \xrightarrow{\rho} \omega\right) - P(\omega,t)\, W\left(\omega \xrightarrow{-\rho} \omega'\right) \right]$$
$$\times \ln \frac{P(\omega',t)\, W\left(\omega' \xrightarrow{\rho} \omega\right)}{P(\omega,t)\, W\left(\omega \xrightarrow{-\rho} \omega'\right)} \geq 0. \tag{4.57}$$

Indeed, we have the inequality $(x - y)\ln(x/y) \geq 0$ for any positive real numbers x and y, since $(x - y) > 0$ and $\ln(x/y) > 0$ if $x > y$, $(x - y) < 0$ and $\ln(x/y) < 0$ if $x < y$, and the equality holds if $x = y$. However, the entropy exchange rate (4.54) can be positive, negative, or zero. Remarkably, the time evolution of the entropy behaves similarly for Markovian stochastic processes as for macroscopic systems within the framework of thermodynamics presented in Chapter 1. The factor $\frac{1}{2}$ in the entropy production rate arises because every transition is counted twice, once in the direction $\omega' \xrightarrow{\rho} \omega$ and once in the opposite direction $\omega \xrightarrow{-\rho} \omega'$.

At equilibrium, the entropy production rate is equal to zero because of the detailed balance conditions (4.48), implying that the probabilities of opposite paths are equal. Consequently, both the current (4.50) and the affinity (4.56) are equal to zero for all the transitions $\omega' \xrightarrow{\rho} \omega$ and the entropy production rate is thus equal to zero, as expected for a system at equilibrium.

An important remark is that the Markovian stochastic process should be expressed in terms of elementary transition mechanisms (Hill and Plesner, 1965; Luo et al., 1984).

This requirement allows us to satisfy the principle of detailed balance (4.48) at equilibrium, since the latter concerns every elementary transition (Hill and Plesner, 1965). In this respect, this requirement is similar to the one at the macroscopic level of description when we consider elementary chemical reactions in the evaluation of the entropy production rate density. Otherwise, the entropy production rate would be underevaluated, as shown in Section E.4.

We note that the entropy production rate (4.57) would become infinite if one of the rates was equal to zero. In this limit, the stochastic process is said to be *fully irreversible*. This is, in particular, the case for the standard Poisson process described in Section E.3.

Isobaric-Isothermal Processes

In addition to the entropy, $S^0(\omega)$, other thermodynamic quantities can be associated with the coarse-grained state ω. The assumption is that the system should stay long enough in the state ω for the establishment of quasi equilibrium in this state before the next jump, i.e., the dwell time in the state ω is significantly longer than the time of the jump to another state.[4] If the system is under isobaric and isothermal conditions, the Gibbs free energy $G(\omega)$ and the enthalpy $H(\omega)$ can also be defined in the coarse-grained state ω, obeying the thermodynamic relation

$$G(\omega) = H(\omega) - T\, S^0(\omega). \tag{4.58}$$

The quantities $\{G(\omega)\}$ form the free-energy landscape where the stochastic process evolves in time. In general, the free-energy landscape may be bounded or unbounded from below. For open systems in contact with reservoirs, the coarse-grained states $\{\omega\}$ may depend on the internal state of the system, as well as on the state of the external reservoirs. If the reservoirs are arbitrarily large, they constitute an infinite free-energy source over the duration of the whole process, so that the free-energy landscape is unbounded from below. An example of such a system is an electric battery, which gives a constant electromotive force as long as it is charged. Another example is the Sun, which plays the role of hot heat reservoir for Earth, the cold heat reservoir being the sky at night. Such free-energy sources can drive processes away from equilibrium during long periods of time before reaching the minimum of the free-energy landscape.

Within the framework of the stochastic description, the key point is that the ratio of opposite transition rates between two states ω and ω' is related to the difference of Gibbs free energies in these states according to

$$\frac{W\left(\omega \overset{\rho}{\to} \omega'\right)}{W\left(\omega' \overset{-\rho}{\to} \omega\right)} = \mathrm{e}^{\beta[G(\omega)-G(\omega')]}, \tag{4.59}$$

[4] This assumption is in accordance with the requirement that the process is described in terms of elementary transition mechanisms. Indeed, every elementary transition goes through a transition state corresponding to a saddle in the energy landscape of the microscopic dynamics. The saddle forms a passage between two valleys corresponding to the coarse-grained states. If the transitions occur with a low rate, the saddle constitutes a bottleneck for the process, providing long enough times to stay in the valleys where quasi equilibrium can be reached in the coarse-grained states ω before the next transition.

for all the elementary mechanisms $\{\rho\}$, where $\beta = (k_B T)^{-1}$ is the inverse temperature. We note that the relations (4.59) differ from the detailed balance conditions (4.48). These latter are only satisfied if the system is at thermodynamic equilibrium globally, although the relations (4.59) may hold while the system is away from equilibrium. Indeed, the free energies $\{G(\omega)\}$ may depend on the conditions prevailing inside the reservoirs. Since these reservoirs may have different temperatures or chemical potentials, the free-energy landscape can be unbounded from below, under which circumstances equations (4.59) hold for systems driven in nonequilibrium steady states.

As a consequence of equations (4.59), the entropy exchange rate (4.54) can be expressed as

$$\frac{d_e S}{dt} = \frac{1}{T}\frac{d\langle H\rangle_t}{dt} \tag{4.60}$$

in terms of the time derivative of the mean enthalpy, which represents the heat exchanged with the environment. Moreover, the entropy production rate reads

$$\frac{d_i S}{dt} = -\frac{1}{T}\frac{d\langle G\rangle_t}{dt} + k_B \frac{dD(t)}{dt} \geq 0, \tag{4.61}$$

in terms of the Shannon disorder (4.46) for the probability distribution $P(\omega, t)$.

If the system is subjected to different conditions, other thermodynamic potentials should be considered in place of the Gibbs free energy (Gaspard, 2006).

4.4.5 Network Theory and Cycles

The network theory of Markov jump processes has been developed by Hill (1989), Schnakenberg (1976), Jiang et al. (2004), and others.

The idea is to associate a graph G with the Markov jump process. The graph is composed of *vertices* corresponding to the states ω and *edges* associated with the transitions between the states

$$e \equiv \omega \overset{\rho}{\to} \omega'. \tag{4.62}$$

In this sense, every edge has a conventional direction, but the transition may occur in both directions, either $e = \omega \overset{\rho}{\to} \omega'$ or $\bar{e} = \omega' \overset{-\rho}{\to} \omega$. In the graph G, the edge is drawn as a line joining two vertices and the line stands for both transitions $\omega \overset{\rho}{\rightleftharpoons} \omega'$ if the direction of the edge is not specified. The numbers of vertices and edges may be finite or countably infinite. Examples of graphs will be given in Section 4.4.6. Two states may be connected by several edges if several elementary mechanisms ρ allow transitions between them.

Whether the transition on some directed edge e contributes to the gain or the loss of probability for the state ω can be determined by introducing the quantity

$$\varsigma_{\omega e} \equiv \begin{cases} +1 & \text{if } \omega \text{ is the final state of } e, \\ -1 & \text{if } \omega \text{ is the starting state of } e, \\ 0 & \text{otherwise,} \end{cases} \tag{4.63}$$

which plays the same role for graphs as the stoichiometric coefficients introduced in Chapter 1 for macroscopic reaction networks. Indeed, the master equation (4.49) can be written in the form

$$\frac{d}{dt}P(\omega,t) = \sum_e \varsigma_{\omega e}\, J_e \tag{4.64}$$

in terms of the net current flowing on the edge e, which is defined as

$$J_e \equiv P(\omega,t)\, W\big(\omega \xrightarrow{\rho} \omega'\big) - P(\omega',t)\, W\big(\omega' \xrightarrow{-\rho} \omega\big) \tag{4.65}$$

according to equation (4.50). The master equation (4.64) can thus be written in matrix form as

$$\frac{d\mathbf{P}}{dt} = \varsigma \cdot \mathbf{J} \tag{4.66}$$

in terms of the vector $\mathbf{P} = \{P(\omega,t)\}$ for the probabilities, the vector $\mathbf{J} = \{J_e\}$ for the currents, and the matrix $\varsigma = (\varsigma_{\omega e})$. If the Markov jump process and thus the graph have the finite numbers Ω and E of states and edges, respectively, the matrix ς has the dimensions $\Omega \times E$. If we introduce the affinity

$$A_e \equiv \ln \frac{P(\omega,t)\, W\big(\omega \xrightarrow{\rho} \omega'\big)}{P(\omega',t)\, W\big(\omega' \xrightarrow{-\rho} \omega\big)} \tag{4.67}$$

associated with the edge e as in equation (4.56), the entropy production rate (4.55) is given by

$$\frac{d_{\rm i}S}{dt} = k_{\rm B} \sum_e A_e\, J_e \geq 0, \tag{4.68}$$

where the sum extends over all the edges of the graph, here counting every possible transition between states once.

An important purpose of network theory is to identify all the cycles of a graph and thus all the possibilities to drive the system out of equilibrium. For this purpose, a *maximal tree* $T(G)$ of the graph G is defined as satisfying the following properties (Schnakenberg, 1976):

(1) $T(G)$ is a covering subgraph of G, i.e., $T(G)$ contains all the vertices of G and all the edges of $T(G)$ are edges of G;
(2) $T(G)$ is connected;
(3) $T(G)$ contains no cycle, i.e., no cyclic sequence of edges.

In general, a given graph G has several maximal trees $T(G)$. The edges l of G that do not belong to $T(G)$ are called the *chords* of $T(G)$. If we add to $T(G)$ one of its chords l, the resulting subgraph $T(G) + l$ contains exactly one cycle, C_l, which is obtained from $T(G) + l$ by removing all the edges that are not part of the cycle. An arbitrary orientation can be assigned to each cycle C_l. The set of cycles $\{C_1, C_2, \ldots, C_l, \ldots\}$ is called a fundamental set. A maximal tree $T(G)$ together with its associated fundamental set of cycles provides a decomposition of the graph G, called *cycle decomposition*. We notice that the maximal tree $T(G)$ can be chosen arbitrarily because each cycle C_l can be redefined by linear combinations of the cycles of the fundamental set.

The orientation of an edge e inside any directed subgraph F of G can be specified by the quantity

$$\varsigma_e(F) \equiv \begin{cases} +1 & \text{if } e \text{ and } F \text{ are parallel,} \\ -1 & \text{if } e \text{ and } F \text{ are antiparallel,} \\ 0 & \text{if } e \text{ is not in } F, \end{cases} \tag{4.69}$$

where e and F are said to be parallel (respectively antiparallel) if F contains the edge e in its reference (respectively opposite) orientation. In particular, $\varsigma_e(C) = \pm 1$ if the cycle C includes the edge e in its conventional orientation or the opposite orientation. This quantity is useful to express the affinities and the currents if the system has reached some steady state (Andrieux and Gaspard, 2007b).

If there exists a stationary probability distribution such that $d\mathbf{P}_{\rm st}/dt = 0$, equation (4.66) reads

$$\boldsymbol{\varsigma} \cdot \mathbf{J} = 0, \tag{4.70}$$

meaning that the vector $\mathbf{J}$ of the currents is a null right eigenvector of the matrix $\boldsymbol{\varsigma}$, which is simply the Kirchhoff current law. If $\{\mathbf{v}_C\}$ denotes the set of these null right eigenvectors such that $\boldsymbol{\varsigma} \cdot \mathbf{v}_C = 0$, the vector of the stationary currents can be decomposed as

$$\mathbf{J} = \sum_C J_C \, \mathbf{v}_C . \tag{4.71}$$

These null right eigenvectors can be obtained as the vectors

$$\mathbf{v}_C = \{\varsigma_e(C)\} \tag{4.72}$$

associated with the cycles of some maximal tree $T(G)$ for the graph G. Indeed, the sums $\sum_e \varsigma_{\omega e} \varsigma_e(C)$ are equal to zero for any state ω of the graph and any cycle C, since the states do not gain or loose probability for the transitions of a cycle. Consequently, the current on some edge e can be decomposed as

$$J_e = \sum_C J_C \, \varsigma_e(C) \tag{4.73}$$

in terms of some coefficients J_C representing the contributions of each cycle to the current J_e. For a stationary probability distribution, equation (4.68) for the entropy production rate can thus be expressed as

$$\frac{1}{k_{\rm B}} \frac{d_{\rm i} S}{dt}\bigg|_{\rm st} = \mathbf{A} \cdot \mathbf{J} = \sum_C J_C \, \mathbf{A} \cdot \mathbf{v}_C = \sum_C A_C \, J_C \geq 0, \tag{4.74}$$

where the affinity associated with the cycle C is given in terms of the edge affinities (4.67) according to

$$A_C \equiv \mathbf{A} \cdot \mathbf{v}_C = \sum_e \varsigma_e(C) \, A_e . \tag{4.75}$$

In a steady state, the entropy production rate is thus only determined by the cycles of the graph. This property is the analogue of the result obtained in equation (1.108) at the macroscale. Using equation (4.67), the affinity of a cycle is thus given by

$$A_C = \sum_e \varsigma_e(C) \ln \frac{W(\omega \overset{\rho}{\to} \omega')}{W(\omega' \overset{-\rho}{\to} \omega)}, \tag{4.76}$$

because the stationary probabilities $P_{\rm st}(\omega)$ cancel out in the cycle. Therefore, we find that the affinity of a cycle can be obtained by considering the product of the ratios of opposite rates along the cycle according to

$$A_C = \ln \prod_{\rho \in C} \frac{W(\omega \overset{\rho}{\to} \omega')}{W(\omega' \overset{-\rho}{\to} \omega)}, \tag{4.77}$$

where $\rho \in C$ are the transitions of the cycle C. The affinity A_C of a cycle is not equal to zero if there are transitions supplying free energy and thus driving the system out of equilibrium in the cycle. Otherwise, the affinity is equal to zero, i.e., $A_C = 0$. Indeed, according to equation (4.59), the affinity of a cycle C is given by $A_C = \beta \sum_{e \in C} \Delta G_e$, where $\Delta G_e = G(\omega) - G(\omega')$ is the free energy difference in the transition. If free energy is supplied from the environment, although the internal state of the system undergoes a cyclic path, the environment undergoes transformations since free energy is consumed. Hence, the affinity of such a cycle is nonvanishing. For such processes, the states $\omega = (\sigma, \mathbf{n})$ can be labeled by the internal states σ of the system and the variables $\mathbf{n}$ describing the exchanges with the environment. The cycle C refers to a cyclic path of internal states $\sigma_1 \to \sigma_2 \to \cdots \to \sigma_c \to \sigma_1$ coming back to its initial internal state. The affinity of this cycle is given by $A_C = \beta \sum_{i=1}^{c} [G(\sigma_i, \mathbf{n}_i) - G(\sigma_{i+1}, \mathbf{n}_{i+1})]$ with $\sigma_{c+1} = \sigma_1$ but $\mathbf{n}_{c+1} \neq \mathbf{n}_1$, so that it is related to the free energy supplied to the system by the exchanges with the environment. This is, in particular, the case for a rotary molecular motor consuming fuel from the environment or subjected to an external torque, which performs mechanical work on the system (Andrieux and Gaspard, 2006b; Gaspard, 2006). In this regard, the relation (4.77) is the analogue of Kirchhoff's voltage law for Markov jump processes in steady states.

The stationary probability distribution can also be constructed with the methods of network theory (Jiang et al., 2004).

4.4.6 Examples

Random Drift on Countably Many Discrete States

A simple example is the random drift of a particle on a one-dimensional chain of sites located at the positions $x = an$ with $n \in \mathbb{Z}$. The particle may jump to the next neighboring sites with the rates $W_\pm$ for the transitions $n \to n \pm 1$, as illustrated in Figure 4.2. The integer $n \in \mathbb{Z}$ may also represent the number of particles exchanged between two reservoirs, as for the process shown in Figure 4.1(d).

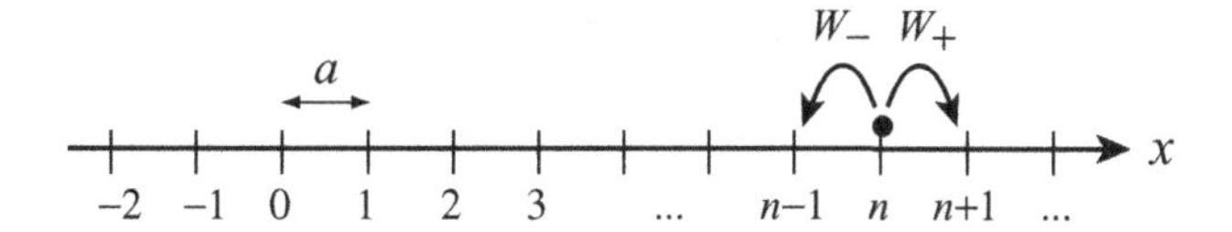

Figure 4.2 Schematic representation of a random drift on countably many discrete states, which may be lattice sites $x = an$ separated by the distance a or the numbers n of particles exchanged between two reservoirs. The rates of the jumps $n \to n \pm 1$ are denoted $W_\pm$.

The master equation ruling the time evolution of the probability $P(n,t)$ to find the system in the state $n \in \mathbb{Z}$ has the following form:

$$\frac{d}{dt} P(n,t) = W_+ P(n-1,t) + W_- P(n+1,t) - (W_+ + W_-) P(n,t). \tag{4.78}$$

The total probability $\sum_{n=-\infty}^{+\infty} P(n,t) = 1$ is preserved by this equation. Since the master equation is linear, its general solution can be obtained as

$$P(n,t) = \int_{-\pi/a}^{+\pi/a} \frac{dq}{2\pi} c(q)\, \mathrm{e}^{z(q)t}\, \mathrm{e}^{\imath qan} \tag{4.79}$$

in terms of the Fourier modes $\exp(\imath qan)$. Substituting equation (4.79) into the master equation (4.78), we obtain the dispersion relation

$$z(q) = W_+ \left(\mathrm{e}^{-\imath qa} - 1\right) + W_- \left(\mathrm{e}^{+\imath qa} - 1\right). \tag{4.80}$$

Expanding in powers of the wave number q as

$$z(q) = -\imath Vq - \mathcal{D}q^2 + O\left(q^3\right), \tag{4.81}$$

the mean drift velocity and the diffusion coefficient can be identified as

$$V = a(W_+ - W_-) \qquad \text{and} \qquad \mathcal{D} = \frac{a^2}{2}(W_+ + W_-). \tag{4.82}$$

Accordingly, the solution starting from the initial condition $P(n,0) = \delta_{n0}$ will behave as

$$P(n,t) \simeq \frac{a}{\sqrt{4\pi\mathcal{D}t}} \exp\left[-\frac{(an - Vt)^2}{4\mathcal{D}t}\right] \tag{4.83}$$

after a long enough time t. This result can also be understood as a consequence of the central limit theorem, which is discussed in Section E.1. Accordingly, the mean value of the random variable is drifting as $\langle n\rangle_t \simeq (W_+ - W_-)t$ for $t \to \infty$.

If the rates are equal, i.e., $W_+ = W_- \equiv W$, the drift velocity is equal to zero and the stochastic process is a symmetric random walk with diffusion coefficient $\mathcal{D} = a^2 W$.

In any case, the stochastic process is nonstationary because $\lim_{t\to\infty} P(n,t) = 0$, although the probability distribution remains normalized to the unit value for all times. The reason is that the probability distribution spreads over larger and larger scales since the variance is increasing with time as $\mathrm{Var}(n) = \left\langle n^2\right\rangle_t - \langle n\rangle_t^2 \simeq (W_+ + W_-)t$.

The stochastic process can be considered as the difference between two standard Poisson processes according to

$$n(t) = \mathcal{P}_+(W_+t) - \mathcal{P}_-(W_-t), \tag{4.84}$$

because the successive jumps are statistically independent. The random numbers of forward and backward jumps $n_\pm = \mathcal{P}_\pm(W_\pm t)$ are thus distributed as $P_\pm(n_\pm,t) = \mathrm{e}^{-W_\pm t}(W_\pm t)^{n_\pm}/(n_\pm!)$ and they can be combined together to determine the net number of jumps $n = n_+ - n_-$ in the forward direction. The corresponding probability distribution $P(n,t)$, which is the solution of the master equation starting from the initial condition $P(n,0) = \delta_{n0}$, can thus be obtained analytically as

$$\begin{aligned} P(n,t) &= \sum_{m=0}^{\infty} P_+(n_+ = m+n,t)\, P_-(n_- = m,t) \\ &= \mathrm{e}^{-(W_+ + W_-)t} \left(\frac{W_+}{W_-}\right)^{n/2} I_n\left(2t\sqrt{W_+W_-}\right), \end{aligned} \tag{4.85}$$

in terms of the modified Bessel functions defined as

$$I_n(z) = I_{-n}(z) = \left(\frac{z}{2}\right)^n \sum_{m=0}^{\infty} \frac{(z^2/4)^m}{m!\,(m+n)!} \tag{4.86}$$

for $z \in \mathbb{C}$ (Abramowitz and Stegun, 1972). After a long enough time, the Gaussian distribution (4.83) is reached. However, we note that for fixed values of time t, the exact distribution (4.85) decreases more slowly than the Gaussian distribution as $|n| \to \infty$.

Figure 4.3 shows trajectories for the random drift with different values for the jump rates corresponding to positive, zero, and negative values for the drift velocity, but the same diffusion coefficient.

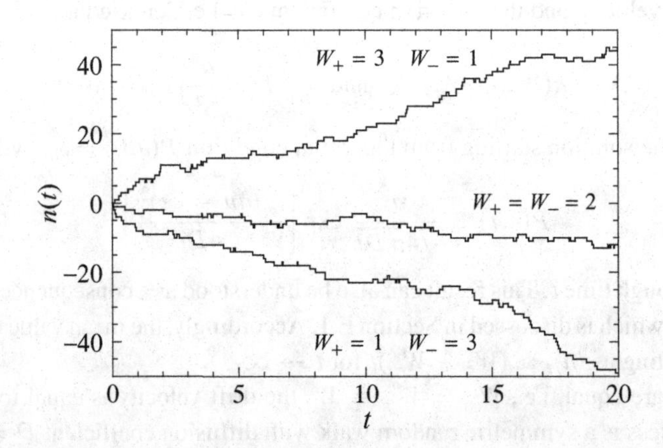

Figure 4.3 Random drifts $n(t)$ over the integer positions of Figure 4.2 generated using Gillespie's algorithm. For the given forward and backward jump rates $W_\pm$, their mean drift velocities are, respectively, $V = +2, 0, -2$ with the same diffusivity $D = 2$, and the affinity takes the values $A = \ln(W_+/W_-) = +\ln 3, 0, -\ln 3$.

Now, the thermodynamic entropy production rate is here equal to

$$\frac{d_{\rm i}S}{dt} = k_{\rm B} \sum_{n=-\infty}^{+\infty} \big[W_+ P(n,t) - W_- P(n+1,t)\big] \ln \frac{W_+ P(n,t)}{W_- P(n+1,t)} \geq 0. \tag{4.87}$$

Since the process is nonstationary, the entropy production rate continues to evolve in time, but it reaches the asymptotic value

$$\lim_{t\to\infty} \frac{d_{\rm i}S}{dt} = k_{\rm B}(W_+ - W_-) \ln \frac{W_+}{W_-} \geq 0 \tag{4.88}$$

because the probability distribution becomes broader and broader as time increases so that $\lim_{t\to\infty} P(n,t)/P(n+1,t) = 1$ according to equation (4.83).

In the case $W_+ = W_-$, where the particle undergoes a random walk over the unbounded chain, the entropy production rate (4.87) tends to zero as $d_{\rm i}S/dt \simeq k_{\rm B}/(2t)$ in the long-time limit $t \to \infty$, so that the system behaves as if it would reach equilibrium.

In contrast, if $W_+ \neq W_-$, entropy continues to be produced, so that the system remains out of equilibrium and driven by some energy supply such as an external force $F_{\rm ext}$. In this case, the entropy production rate would be given in terms the mechanical power of this force according to $d_{\rm i}S/dt = F_{\rm ext}V/T$ where $V = a(W_+ - W_-)$ is the drift velocity. As a consequence of equation (4.88), we would conclude in this case that the ratio of forward to backward transition rates should be related to the external force by $W_+/W_- = \exp(\beta a F_{\rm ext})$, where $\beta = (k_{\rm B}T)^{-1}$ is the inverse temperature and a the distance between the sites.

If the entropies of all the states are equal $S^0(n) = S^0$ for $n \in \mathbb{Z}$, the entropy (4.45) associated with the asymptotic Gaussian probability distribution (4.83) takes the value $S(t) \simeq S^0 + k_{\rm B} \ln \sqrt{4\pi {\rm e}\mathcal{D}t/a^2}$, which increases with the logarithm of time because the probability distribution spreads over the unbounded chain $n \in \mathbb{Z}$. We note that if the chain was bounded and contained $\mathcal{N}$ states, the probability distribution would reach the uniform stationary distribution $P_{\rm st}(n) = 1/\mathcal{N}$, in which case the entropy would converge towards the finite stationary value $S_{\rm st} \simeq S^0 + k_{\rm B} \ln \mathcal{N}$.

In the process where the random variable $n(t)$ represents the number of particles transported between two reservoirs, the mean current and the diffusivity are equal to $J(A) = W_-({\rm e}^A - 1)$ and $D(A) = W_-({\rm e}^A + 1)/2$, where $A = \ln(W_+/W_-) = \beta(\mu_{\rm L} - \mu_{\rm R})$ is the affinity given in terms of the difference of chemical potentials between the two reservoirs in agreement with the results of Section 1.7.2. If the chemical potentials are equal, the affinity, the current, and the entropy production are equal to zero and the system is at thermodynamic equilibrium. This process is also the basic model for electronic diodes (Shockley, 1949; Ashcroft and Mermin, 1976; Gu and Gaspard, 2018). Indeed, the affinity is related to the applied voltage V according to $A = \beta eV$ with the elementary electric charge $e = |e|$, while the electric current is given by $I = eJ$, so that the current-voltage characteristic curve of the diode is obtained as $I = eW_-({\rm e}^{\beta eV} - 1)$. The spectral density of the fluctuating current $\jmath(t) = \dot{n}(t)$ is given by $\mathcal{S}(\omega) = 2D(A) = J(A)\coth(A/2)$, which is independent of the frequency.[5] In this regard, the process is a white noise, interpolating

[5] This result can be obtained using equation (4.16) for $Z(t) = n(t) - \langle n(t)\rangle$ such that $\left\langle Z(t)^2\right\rangle = 2D(A)t$, after regularizing the integral with $\exp(-\epsilon t)$ in the limit $\epsilon \to 0$.

between thermal noise[6] with $\mathcal{S}(\omega) \simeq \lim_{A\to 0} 2J(A)/A$ near equilibrium for $A \ll 1$, and shot noise[7] with $\mathcal{S}(\omega) \simeq J(A)$ far away from equilibrium for $A \gg 1$.

According to equation (4.59), the ratio of the transition rates can be expressed as $W_+/W_- = \exp(\beta\Delta G)$ in terms of the free-energy drop $\Delta G = G(\omega) - G(\omega')$ in the transition $\omega \overset{\rho}{\to} \omega'$. For the random particle drift, the free-energy drop is the mechanical work $\Delta G = aF_{\rm ext}$ of the external force $F_{\rm ext}$ exerted during the displacement a of the particle. For the transport process of particles, it is given by the difference of chemical potentials between the particle reservoirs, $\Delta G = \mu_{\rm L} - \mu_{\rm R}$. For both processes, the free-energy landscape is unbounded from below if ΔG remains invariant in time, thus driving the system into a nonequilibrium steady state if $\Delta G \neq 0$.

In the limit where one of the rates is equal to zero, the entropy production rate (4.88) becomes infinite and the process is fully irreversible. For instance, if the rate of backward jumps is equal to zero, i.e., $W_- = 0$, the random drift reduces to the standard Poisson process, in which case the random variable $n(t) = n_+(t)$ is monotonically increasing. Indeed, such stochastic processes describe the unidirectional random motion or flow, which is a regime arbitrarily far from equilibrium.

From the viewpoint of the network theory presented in Section 4.4.5, the states of the Markov jump process are the integers $\omega = n \in \mathbb{Z}$ and the edges correspond to the transitions $e = n \to n+1$. The associated graph G is thus given in Figure 4.2, which is already a maximal tree. The graph is infinite and there is no cycle. This process has no internal state. Nevertheless, we might consider the finite chain of length $\mathcal{N}$, closing on itself to form a cycle $C_{\mathcal{N}}$ with the $\mathcal{N}$ states $\omega = 1, 2, \ldots, \mathcal{N}$. In this case, equation (4.77) shows that the affinity of this cycle is given by $A_{C_{\mathcal{N}}} = \mathcal{N} \ln(W_+/W_-) = \mathcal{N} A$ with $A = \ln(W_+/W_-)$. In this regard, this latter quantity is the affinity per lattice site, locally driving the process out of equilibrium. On the finite cyclic chain, there exists a stationary probability distribution given by $P_{\rm st}(n) = 1/\mathcal{N}$, so that the system possibly running under nonequilibrium conditions is described by the corresponding stationary stochastic process. In the case where the transition rates are equal, $W_+ = W_-$, the process is reversible, which corresponds to a vanishing asymptotic entropy production rate, as expected at equilibrium for $A = 0$.

Particle Exchange Process

An example with an infinite number of internal states consists of the Markov jump process that describes particle exchanges between a system containing the random number N of particles and two arbitrarily large reservoirs R_1 and R_2, as shown in Figure 4.4(a). Each reservoir R_j contains N_j particles, such that $N_j \gg N$ $(j = 1, 2)$. The exchanges happen at the transition rates $W_\rho(N)$ with $\rho = j$ for the transfer of one particle from the reservoir R_j to the system and $\rho = -j$ for the reverse transition $(j = 1, 2)$. Moreover, we assume

[6] Thermal noises manifest themselves in linear regimes close to equilibrium where the mean current is linear in the affinity as $J(A) = LA + O(A^2)$ with the linear response coefficient L, giving the spectral density $\mathcal{S}(\omega) = 2L$ by the fluctuation–dissipation theorem.

[7] Shot noises are observed for electronic currents in vacuum tubes. They are characterized by the proportionality between their diffusivity or their spectral density and their mean current $\mathcal{S}(\omega) = 2D = J$, as in the standard Poisson process of Section E.3. Shot noises are fully irreversible processes.

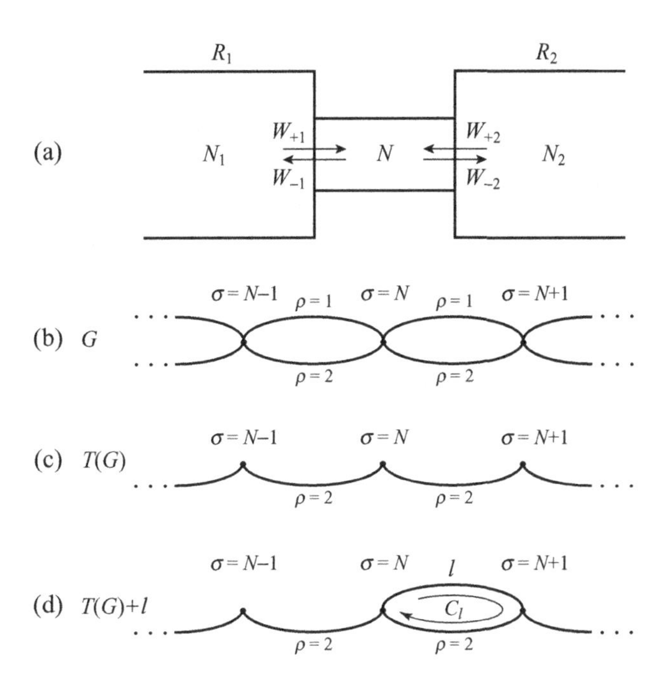

Figure 4.4 Particle exchange process between a system and two reservoirs: (a) Schematic representation of the total system composed of the system containing N particles and the reservoirs R_j with N_j particles ($j = 1,2$). The particles are exchanged between the reservoirs at the rates $W_{\pm j}$. (b) The graph G of the process. (c) Maximal tree $T(G)$. (d) Maximal tree $T(G)$ with the chord l forming the cycle C_l.

that the charging rates W_{+j} are independent of the internal state N of the system, while the discharging rates W_{-j} are proportional to the number of particles inside the system, i.e.,

$$W_{+j}(N) = k_{+j}\,\Omega \qquad \text{and} \qquad W_{-j}(N) = k_{-j}\,N, \tag{4.89}$$

for $j = 1,2$ with some rate constants $k_{\pm j}$ and the dimensionless extensivity parameter Ω proportional to the volume of the system. If the transition ρ transfers ν_ρ particles to the system from any reservoir and Δn_ρ particles from the reservoir R_1 to the system, we have

$$N_1 \to N_1 - \Delta n_\rho, \qquad N \to N + \nu_\rho, \qquad \text{and} \qquad N_2 \to N_2 - \nu_\rho + \Delta n_\rho, \tag{4.90}$$

so that the total number of particles $N_{\text{tot}} = N_1 + N + N_2$ is conserved. Since ρ and $-\rho$ are opposite transitions, the properties $\nu_{-\rho} = -\nu_\rho$ and $\Delta n_{-\rho} = -\Delta n_\rho$ are satisfied. Moreover, $\nu_1 = \nu_2 = 1$, $\Delta n_1 = 1$, and $\Delta n_2 = 0$.

The states of the Markov jump process are $\omega = (\sigma, n)$, where the internal state $\sigma = N \in \mathbb{N}$ gives the number of particles inside the system, and the integer $n \in \mathbb{Z}$ is the number of particles transferred from the reservoir R_1 to the system. The master equation ruling the time evolution of the probability $P(N,n,t)$ of finding N particles inside the system while n particles have been transferred from the reservoir R_1 at time t takes the form

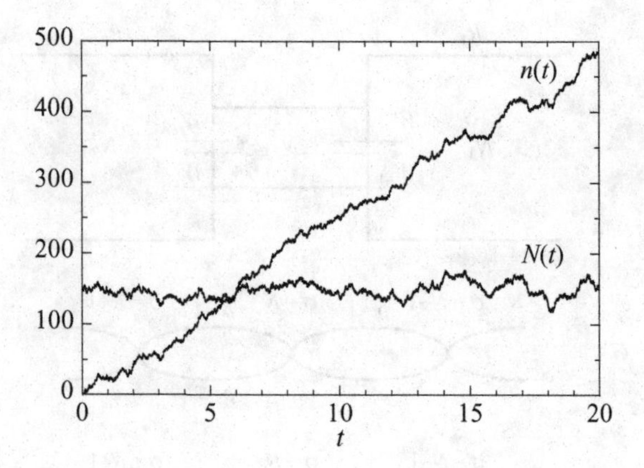

Figure 4.5 Random time evolution of the number $N(t)$ of particles inside the system and the number $n(t)$ of exchanged particles between the reservoirs versus time t in the particle exchange process of Figure 4.4 between a system and two reservoirs for the parameter values $k_{+1} = 1, k_{-1} = 0.5, k_{+2} = 0.2, k_{-2} = 0.3$, and $\Omega = 100$. The process is out of equilibrium with the values $A = \ln 3$ for the affinity (4.92), $R = 25$ for the rate (4.96), and the stationary mean value $\langle N \rangle_{\rm st} = 150$.

$$\frac{d}{dt} P(N,n,t) = \sum_{\rho=\pm 1,\pm 2} \left(e^{-\nu_\rho \partial_N} e^{-\Delta n_\rho \partial_n} - 1\right) W_\rho(N)\, P(N,n,t), \tag{4.91}$$

using the shift operators (4.33).

Since the transition rates $W_\rho(N)$ do not depend on the number n of transferred particles, a master equation of similar form rules the time evolution of the probability distribution $\mathscr{P}(N,t) \equiv \sum_{n\in\mathbb{Z}} P(N,n,t)$ for the internal states. The graph G associated with this Markov jump process is shown in Figure 4.4(b). This graph has countably many vertices $\sigma = N \in \mathbb{N}$ that are linked by edges corresponding to the two possible transitions $\sigma \overset{\rho=1}{\rightleftharpoons} \sigma'$ and $\sigma \overset{\rho=2}{\rightleftharpoons} \sigma'$ with $\sigma = N$ and $\sigma' = N+1$, for $N \in \mathbb{N}$. A maximal tree $T(G)$ is obtained by removing all the edges corresponding to the transitions $\rho = 1$. As shown in Figure 4.4(c), all the vertices are linked together by the edges corresponding to the transitions $\rho = 2$ for this maximal tree, which contains no cycle. If one edge of the transitions $\rho = 1$ is added to this maximal tree, such as the chord l in Figure 4.4(d), the new graph $T(G)+l$ contains exactly one cycle C_l. The affinity of this cycle is given according to equation (4.77) by

$$A_{C_l} = \ln \frac{W_{+1}(N)\, W_{-2}(N+1)}{W_{-1}(N+1)\, W_{+2}(N)} = \ln \frac{k_{+1}\, k_{-2}}{k_{-1}\, k_{+2}} \equiv A, \tag{4.92}$$

which is independent of the internal state N for the rates (4.89), thus defining the constant affinity A. We note that other maximal trees may also be constructed.

The time evolution of the mean number of particles inside the system, $\langle N \rangle \equiv \sum_{N\in\mathbb{N}} N \mathscr{P}(N,t)$, obeys the kinetic equation

$$\frac{d}{dt}\langle N \rangle = (k_{+1} + k_{+2})\, \Omega - (k_{-1} + k_{-2})\, \langle N \rangle. \tag{4.93}$$

The stationary probability distribution of the number N is given by the following Poisson distribution

$$\mathscr{P}_{\rm st}(N) = {\rm e}^{-\langle N\rangle_{\rm st}} \frac{\langle N\rangle_{\rm st}^N}{N!}, \qquad \text{where} \qquad \langle N\rangle_{\rm st} = \frac{k_{+1}+k_{+2}}{k_{-1}+k_{-2}}\,\Omega \tag{4.94}$$

is the stationary solution of equation (4.93).

The equilibrium state is identified as the stationary state satisfying the conditions of detailed balance, according to which $W_{+j}(N)\mathscr{P}_{\rm eq}(N) = W_{-j}(N+1)\mathscr{P}_{\rm eq}(N+1)$ for all $N \in \mathbb{N}$ and $j = 1,2$. These conditions imply that $k_{+j}\Omega = k_{-j}\langle N\rangle_{\rm eq}$ for $j = 1$ and $j = 2$, so that $k_{+1}\,k_{-2} = k_{-1}\,k_{+2}$.

In the long-time limit, the joint probability distribution factorizes as

$$P(N,n,t) \simeq_{t\to\infty} \mathscr{P}_{\rm st}(N)\,\mathcal{P}(n,t), \tag{4.95}$$

into the stationary distribution (4.94) of the number N of particles inside the system and the time-dependent distribution of the random variable n counting the particles exchanged with the reservoirs. This latter behaves as the Gaussian distribution (4.83) of the random drift with the mean current or rate given by

$$J(A) = \lim_{t\to\infty}\frac{1}{t}\langle n(t)\rangle = \frac{k_{+1}\,k_{-2}-k_{-1}\,k_{+2}}{k_{-1}+k_{-2}}\,\Omega = \frac{k_{-1}\,k_{+2}\,\Omega}{k_{-1}+k_{-2}}\left({\rm e}^A - 1\right) \tag{4.96}$$

in terms of the affinity (4.92). If J_{C_l} denotes the mean current on the chord l of the graph depicted in Figure 4.4(d) between the states $\sigma = N$ and $\sigma = N+1$, the macroscopic mean current (4.96) is obtained as the sum $J(A) = \sum_{C_l} J_{C_l}$ since all the chords are contributing to the overall current between the reservoirs.

In the steady state (4.94) reached in the long-time limit (4.95), the entropy production rate (4.74) can be expressed as

$$\frac{1}{k_{\rm B}}\frac{d_{\rm i}S}{dt}\bigg|_{\rm st} = A\,J(A) \geq 0 \tag{4.97}$$

with the macroscopic mean current (4.96). Accordingly, the entropy production rate is equal to zero at equilibrium where the conditions of detailed balance hold and the affinity is equal to zero with the mean current, but it is otherwise positive.[8]

A realization of random time evolution simulated by the algorithm of Gillespie (1976, 1977) is shown in Figure 4.5 for the system under nonequilibrium conditions. The number $N(t)$ of particles inside the system obeys a stationary process with the distribution (4.94), although the number $n(t)$ of particles exchanged between the reservoirs follows a nonstationary process with a drift at the mean rate (4.96).

[8] The stationary process for the lone random variable $N \in \mathbb{N}$ is reversible because the reversibility condition (4.40) is satisfied for the stationary probability distribution (4.94) in the form $\sum_{j=1,2} W_{+j}(N-1)\,\mathscr{P}_{\rm st}(N-1) = \sum_{j=1,2} W_{-j}(N)\,\mathscr{P}_{\rm st}(N)$, even if the conditions of detailed balance do not hold. In contrast, the stationary process for the two random variables $N \in \mathbb{N}$ and $n \in \mathbb{Z}_{\mathcal{N}}$ (where n would evolve on a finite cyclic chain of length $\mathcal{N}$ as in the aforediscussed example of random drift) is not reversible, unless the affinity (4.92) is vanishing together with the rate (4.96) and the entropy production rate (4.97), which are the equilibrium conditions. This example shows that the conditions of stochastic reversibility (4.40) are not sufficient to identify the state of thermodynamic equilibrium. This identification requires consideration of the stronger conditions of detailed balance (4.48) between every elementary transition and the corresponding reverse transition.

This example illustrates the importance of making the distinction between the random variables associated with the internal states of the system and those describing energy or particle exchanges with its environment. The process may be stationary with respect to the former, but nonstationary if the latter are included in the description, which is an essential aspect of open systems driven away from equilibrium.

4.5 Continuous-State Markov Processes

4.5.1 Generalities

Markovian processes are also defined for continuous variables $\mathbf{x}_i \in \mathbb{R}^d$ by considering the probability density $p(\mathbf{x},t)$ normalized according to $\int_{\mathbb{R}^d} p(\mathbf{x},t)\, d\mathbf{x} = 1$ and the conditional probability density defined in terms of the joint probability density of two consecutive events by

$$p(\mathbf{x}_2,t_2|\mathbf{x}_1,t_1) \equiv \frac{p(\mathbf{x}_1,t_1;\mathbf{x}_2,t_2)}{p(\mathbf{x}_1,t_1)}. \tag{4.98}$$

This latter obeys the Chapman–Kolmogorov equation (van Kampen, 1981; Gardiner, 2004),

$$p(\mathbf{x}_3,t_3|\mathbf{x}_1,t_1) = \int_{\mathbb{R}^d} p(\mathbf{x}_3,t_3|\mathbf{x}_2,t_2)\, p(\mathbf{x}_2,t_2|\mathbf{x}_1,t_1)\, d\mathbf{x}_2, \tag{4.99}$$

as a consequence of the Markovian property (4.9). Integrating equation (4.98) over $\mathbf{x}_2$, we get its normalization condition:

$$\int_{\mathbb{R}^d} p(\mathbf{x}_2,t_2|\mathbf{x}_1,t_1)\, d\mathbf{x}_2 = 1. \tag{4.100}$$

Multiplying both sides of equation (4.98) by the probability density $p(\mathbf{x}_1,t_1)$ and integrating over $\mathbf{x}_1$, we obtain the integral evolution equation

$$p(\mathbf{x}_2,t_2) = \int_{\mathbb{R}^d} p(\mathbf{x}_2,t_2|\mathbf{x}_1,t_1)\, p(\mathbf{x}_1,t_1)\, d\mathbf{x}_1. \tag{4.101}$$

These results are similar to those concerning discrete-state Markov processes. Accordingly, the probability density $p(\mathbf{x},t)$ of a continuous-state Markov process is ruled by a master equation, which can be established as for discrete-state Markov processes. However, different classes of processes can be defined depending on the properties of the conditional probability density (4.98) in the limit where $t_2 \to t_1$.

On the one hand, the transitions may induce infinitesimal jumps of the continuous variables $\mathbf{x}$ over infinitesimal time intervals, which defines the class of advection-diffusion Markov processes, including Brownian motion, random drifts, as well as deterministic processes in the limit where diffusivity disappears. On the other hand, Markov jump processes can be defined in terms of transition rates allowing the continuous variables $\mathbf{x}$ to perform jumps of finite amplitude. Furthermore, there also exist processes with finite random jumps combined with advection and/or diffusion.

4.5.2 Advection-Diffusion Processes

In many stochastic processes, the transitions happen over infinitesimal distances during infinitesimal time intervals in the space of continuous variables, in which cases we may introduce the quantities,[9]

$$\mathbf{C}(\mathbf{x},t) \equiv \lim_{\Delta t \to 0} \frac{1}{\Delta t} \int_{\mathbb{R}^d} \Delta\mathbf{x}\, p(\mathbf{x}+\Delta\mathbf{x}, t+\Delta t|\mathbf{x},t)\, d\Delta\mathbf{x}, \tag{4.102}$$

$$\mathsf{D}(\mathbf{x},t) \equiv \lim_{\Delta t \to 0} \frac{1}{2\Delta t} \int_{\mathbb{R}^d} \Delta\mathbf{x}\, \Delta\mathbf{x}^{\mathrm{T}}\, p(\mathbf{x}+\Delta\mathbf{x}, t+\Delta t|\mathbf{x},t)\, d\Delta\mathbf{x}, \tag{4.103}$$

while the higher moments are equal to zero in the same limit,

$$\lim_{\Delta t \to 0} \frac{1}{\Delta t} \int_{\mathbb{R}^d} \Delta\mathbf{x}^n\, p(\mathbf{x}+\Delta\mathbf{x}, t+\Delta t|\mathbf{x},t)\, d\Delta\mathbf{x} = 0 \qquad \text{for} \qquad n \geq 3. \tag{4.104}$$

We note that the matrix (4.103) is always symmetric, so $\mathsf{D} = \mathsf{D}^{\mathrm{T}}$, and nonnegative, i.e., $\mathsf{D} \geq 0$.

The master equation of such processes can be obtained as follows. We consider the mean value of some time-independent observable $Y(\mathbf{x})$:

$$\langle Y \rangle_t = \int_{\mathbb{R}^d} Y(\mathbf{x})\, p(\mathbf{x},t)\, d\mathbf{x}. \tag{4.105}$$

Since the probability density evolves in time according to equation (4.101), we have that

$$\langle Y \rangle_{t+\Delta t} = \int_{\mathbb{R}^d} Y(\mathbf{x}+\Delta\mathbf{x})\, p(\mathbf{x}+\Delta\mathbf{x}, t+\Delta t|\mathbf{x},t)\, p(\mathbf{x},t)\, d\mathbf{x}\, d\Delta\mathbf{x}. \tag{4.106}$$

Expanding $Y(\mathbf{x}+\Delta\mathbf{x})$ in powers of $\Delta\mathbf{x}$ up to second order and using the definitions (4.102) and (4.103), we find that

$$\frac{d}{dt}\langle Y \rangle_t = \int_{\mathbb{R}^d} \Big[\mathbf{C}(\mathbf{x},t) \cdot \partial_{\mathbf{x}} Y(\mathbf{x}) + \mathsf{D}(\mathbf{x},t) : \partial_{\mathbf{x}}^2 Y(\mathbf{x})\Big]\, p(\mathbf{x},t)\, d\mathbf{x}. \tag{4.107}$$

Integrating by parts and assuming that the observable is equal to zero at infinity, we finally obtain the master equation,

$$\partial_t\, p = -\partial_{\mathbf{x}} \cdot (\mathbf{C}\, p) + \partial_{\mathbf{x}}^2 : (\mathsf{D}\, p), \tag{4.108}$$

ruling the time evolution of the probability density $p(\mathbf{x},t)$. This is a partial differential equation of diffusive type. The $d \times d$ matrix $\mathsf{D} = \left(D^{kl}\right)$ is the tensor of diffusivities, while the vector $\mathbf{C} = \left(C^k\right)$ is associated with advection or drift in the space $\mathbf{x} = (x^k) \in \mathbb{R}^d$, hence the name advection-diffusion process. We note that the master equation is linear and can thus be expressed as

$$\partial_t\, p = \hat{L}\, p \qquad \text{with} \qquad \hat{L} p \equiv -\partial_{\mathbf{x}} \cdot (\mathbf{C}\, p) + \partial_{\mathbf{x}}^2 : (\mathsf{D}\, p). \tag{4.109}$$

Equivalently, the master equation (4.108) can be written in the form of a continuity equation expressing the local conservation of probability in the d-dimensional space of continuous variables $\mathbf{x}$ as

[9] The superscript T denotes the transpose and $\Delta\mathbf{x}^n$ the tensorial product of n vectors $\Delta\mathbf{x}$.

$$\partial_t p + \partial_{\mathbf{x}} \cdot \mathbf{J} = 0, \tag{4.110}$$

by introducing the current density

$$\mathbf{J} \equiv \mathbf{V}\, p - \mathbf{D} \cdot \partial_{\mathbf{x}}\, p \tag{4.111}$$

with the drift velocity

$$\mathbf{V} \equiv \mathbf{C} - \partial_{\mathbf{x}} \cdot \mathbf{D}. \tag{4.112}$$

The master equation (4.110) with the current density (4.111) is generally called Fokker–Planck equation (Fokker, 1914; Planck, 1917).

The process is purely diffusive if the drift velocity is equal to zero. Moreover, if the diffusivities D^{kl} do not depend on the continuous variables $\mathbf{x}$, we recover the diffusion equation in an anisotropic medium, $\partial_t p = \sum_{kl} D^{kl} \partial_k \partial_l p$. If the diffusion process is isotropic, the tensor of diffusivities is diagonal so that $D^{kl} = D\delta^{kl}$, where δ^{kl} is the Kronecker symbol, and the master equation reduces to the usual diffusion equation $\partial_t p = D\Delta p$ with the d-dimensional Laplacian operator $\Delta p \equiv \sum_k \partial_k^2 p$.

The process is purely advective if the diffusivities are all equal to zero, in which case the master equation becomes a generalized Liouvillian equation $\partial_t p + \partial_{\mathbf{x}} \cdot (\mathbf{V}\, p) = 0$. Since there is no diffusion, this process is deterministic and the master equation can be solved knowing the deterministic trajectories that are solutions of the ordinary differential equations $\dot{\mathbf{x}} = \mathbf{V}(\mathbf{x}, t)$, as discussed in Appendix B.

In general, solving the partial differential equation (4.108) requires that boundary conditions should be assumed in the stochastic problem at hand. The stochastic process may be considered in the whole space $\mathbf{x} \in \mathbb{R}^d$ extending to infinity or in some domain $\mathfrak{D} \subset \mathbb{R}^d$, in which latter case boundary conditions should be defined on its boundary $\partial\mathfrak{D}$. Reflection boundary conditions such that $\mathbf{J}|_{\partial\mathfrak{D}} = 0$ will guarantee the global conservation of probability in the domain $\mathfrak{D}$ because $dP/dt = -\int_{\partial\mathfrak{D}} \mathbf{J} \cdot d\mathbf{\Sigma} = 0$ for $P \equiv \int_{\mathfrak{D}} p\, d\mathbf{x}$. If the process takes place in a spatially periodic system such as a lattice, quasiperiodic boundary conditions may be considered in order to reduce the problem to solving the equation inside the unit cell of the lattice by Fourier analysis. If absorbing boundary conditions such that $p|_{\partial\mathfrak{D}} = 0$ are assumed, the total probability is no longer conserved and escape or leakage happens at the boundary. Different boundary conditions may be considered on different parts of the boundary. So-defined stochastic processes may be stationary or not.

4.5.3 Stochastic Differential Equations

The advection-diffusion Markov stochastic processes can be equivalently generated by the Itô stochastic differential equations

$$\frac{d\mathbf{x}}{dt} = \mathbf{A}(\mathbf{x}, t) + \sum_{\mu=1}^{n} \mathbf{B}_\mu(\mathbf{x}, t)\, \chi_\mu(t), \tag{4.113}$$

where $\mathbf{A}(\mathbf{x}, t)$ and $\mathbf{B}_\mu(\mathbf{x}, t)$ are smooth enough vector fields defined in the space of continuous variables $\mathbf{x} \in \mathbb{R}^d$, while $\chi_\mu(t)$ are statistically independent Gaussian white noises such that

$$\langle \chi_\mu(t) \rangle = 0 \qquad \text{and} \qquad \langle \chi_\mu(t)\, \chi_\nu(t') \rangle = \delta_{\mu\nu}\, \delta(t-t') \tag{4.114}$$

for $\mu, \nu = 1, 2, \ldots, n$. Moreover, it is assumed that, at time t, the noises $\chi_\mu(t)$ are statistically independent from the variables $\mathbf{x}_t$ upon integrating equation (4.113) over some time interval. Accordingly, the discretization into arbitrarily small time steps Δt leads to the iterative scheme

$$\mathbf{x}_{i+1} = \mathbf{x}_i + \mathbf{A}(\mathbf{x}_i, t_i)\, \Delta t + \sum_{\mu=1}^{n} \mathbf{B}_\mu(\mathbf{x}_i, t_i)\, G_{\mu i}\, \sqrt{\Delta t}, \tag{4.115}$$

where $t_i = i\Delta t + t_0$ and $G_{\mu i}$ are statistically independent Gaussian random variables such that $\langle G_{\mu i} \rangle = 0$ and $\langle G_{\mu i} G_{\nu j} \rangle = \delta_{\mu\nu}\delta_{ij}$ (Gardiner, 2004). Taking the mean value and the diffusivities of the displacements, $\Delta\mathbf{x} = \mathbf{x}_{i+1} - \mathbf{x}_i$, given by equation (4.115), we deduce the following expressions for (4.102) and (4.103):

$$\mathbf{C}(\mathbf{x}, t) = \mathbf{A}(\mathbf{x}, t) \qquad \text{and} \qquad \mathsf{D}(\mathbf{x}, t) = \frac{1}{2} \sum_{\mu=1}^{n} \mathbf{B}_\mu(\mathbf{x}, t)\, \mathbf{B}_\mu^{\mathrm{T}}(\mathbf{x}, t). \tag{4.116}$$

Consequently, the master equation of the process generated by the Itô stochastic differential equations is given by equation (4.108).

The number n of Gaussian white noises can always be taken equal to or lower than the number d of continuous random variables $\mathbf{x}$. Indeed, at every time step, the last term of the iterative scheme (4.115) adds a Gaussian random vector of zero mean with correlations between the d vectorial components. The corresponding correlation matrix is determined by the matrix of diffusivities in equation (4.116). Since this latter is real, symmetric, and positive, it can be diagonalized by an orthogonal transformation. The vectors $\mathbf{B}_\mu$ can be taken as the d eigenvectors of this matrix, thus reducing the number n of independent Gaussian white noises to the rank of the matrix of diffusivities. If this matrix has zero eigenvalues, the iterative scheme (4.115) is deterministic in the directions of the corresponding null eigenvectors and random only in the directions of the nonzero eigenvalues.

Stochastic differential equations are efficient methods for the numerical simulation of advection-diffusion processes. We note that there exist several possible formulations for the integration of stochastic differential equations and for stochastic integrals, as discussed in Section E.6.

4.5.4 Jump Processes

In other Markovian processes, the variables may be continuous, but the transitions generate finite jumps in the continuous state space. In order to deduce the master equation for these processes, we consider the probability density between two successive instants of time, $t_1 = t$ and $t_2 = t + \Delta t$. Taking the difference between equation (4.101) with $\mathbf{x}_2 = \mathbf{x}$ and $\mathbf{x}_1 = \mathbf{x}'$, and $p(\mathbf{x}, t)$ multiplied by equation (4.100) with $\mathbf{x}_2 = \mathbf{x}'$ and $\mathbf{x}_1 = \mathbf{x}$, we get

$$p(\mathbf{x}, t+\Delta t) - p(\mathbf{x}, t) = \int_{\mathbb{R}^d} \left[p(\mathbf{x}, t+\Delta t | \mathbf{x}', t)\, p(\mathbf{x}', t) - p(\mathbf{x}, t)\, p(\mathbf{x}', t+\Delta t | \mathbf{x}, t) \right] d\mathbf{x}'. \tag{4.117}$$

The transition rate density is defined as

$$w(\mathbf{x} \to \mathbf{x}'; t) \equiv \lim_{\Delta t \to 0} \frac{1}{\Delta t}\, p(\mathbf{x}', t + \Delta t | \mathbf{x}, t). \tag{4.118}$$

Dividing equation (4.117) by Δt and taking the limit $\Delta t \to 0$, we thus obtain the master equation:

$$\partial_t\, p(\mathbf{x}, t) = \int_{\mathbb{R}^d} \left[p(\mathbf{x}', t)\, w(\mathbf{x}' \to \mathbf{x}; t) - p(\mathbf{x}, t)\, w(\mathbf{x} \to \mathbf{x}'; t) \right] d\mathbf{x}'. \tag{4.119}$$

As for discrete-state Markov processes, the right-hand side of this master equation has a positive gain term and a negative loss term, so that the normalization condition of the probability density is preserved during the time evolution.

An example of such processes is given by the velocity of a particle undergoing random jumps at every collision on immobile scatterers randomly distributed over space. In such two-component gases composed of independent light particles moving among arbitrarily heavy particles, the kinetic energy ε is conserved and, thus, the magnitude of the particle velocity $v = \|\mathbf{v}\| = \sqrt{2\varepsilon/m}$, so that the velocity probability distribution only depends on the orientation of the particle velocity. If the gas is assumed to be homogeneous, the velocity probability distribution is the same everywhere across the whole system. These systems are called homogeneous random Lorentz gases (Lorentz, 1905; Spohn, 1991). In such d-dimensional Lorentz gases, the probability density is a function $p(\mathbf{n}, t)$ of the velocity direction $\mathbf{n} = \mathbf{v}/v$ defined on a $(d-1)$-dimensional sphere $\mathbf{n} \in \mathbb{S}^{d-1}$. The transition rate density is given by $w(\mathbf{n} \to \mathbf{n}'; t) = n_{\mathrm{s}} v \sigma_{\mathrm{diff}}(\mathbf{n}, \mathbf{n}')$ in terms of the density of scatterers n_{s}, the magnitude of velocity v, and the differential cross section $\sigma_{\mathrm{diff}}(\mathbf{n}, \mathbf{n}')$ of the collisions between the particles and the scatterers.

Other examples are provided by energy exchange processes (Nicolis and Malek Mansour, 1984; Cleuren et al., 2006). In such processes, the energy ε undergoes random jumps at some rate density $w(\varepsilon \to \varepsilon')$, ruling the probability density $p(\varepsilon, t)$.

4.5.5 Advection-Jump Processes

There also exist Markovian stochastic processes that combine random jumps, advection, and diffusion. An important class of such processes, called random flights, combine random jumps with advection. Supposing that there is no diffusion, so $\mathsf{D} = \mathsf{0}$, and that the rate densities $w(\mathbf{x} \to \mathbf{x}')$ and the drift velocity $\mathbf{V}(\mathbf{x})$ are independent of time, the master equation for such processes reads

$$\partial_t\, p(\mathbf{x}, t) + \partial_{\mathbf{x}} \cdot [\mathbf{V}(\mathbf{x})\, p(\mathbf{x}, t)] = \int_{\mathbb{R}^d} \left[p(\mathbf{x}', t)\, w(\mathbf{x}' \to \mathbf{x}) - p(\mathbf{x}, t)\, w(\mathbf{x} \to \mathbf{x}') \right] d\mathbf{x}'. \tag{4.120}$$

As examples, we may mention the inhomogeneous random Lorentz gases where a point particle performs free flights interrupted by elastic collisions on scatterers that are fixed and randomly distributed in a d-dimensional space with the density n_{s}. The random variables of these stochastic processes are the position and velocity of the particle, $\mathbf{x} = (\mathbf{r}, \mathbf{v}) \in \mathbb{R}^{2d}$. The probability density of these variables, $p(\mathbf{r}, \mathbf{v}, t)$, is ruled by the following master equation, called the linear Boltzmann equation or Boltzmann–Lorentz equation:

$$\partial_t \, p(\mathbf{r},\mathbf{v},t) + \mathbf{v}\cdot\nabla p(\mathbf{r},\mathbf{v},t) = n_s \, v \int_{\mathbb{S}^{d-1}} \sigma_{\text{diff}}(\mathbf{n},\mathbf{n}') \left[p(\mathbf{r},\mathbf{v}',t) - p(\mathbf{r},\mathbf{v},t) \right] d\Omega, \quad (4.121)$$

where $v = \|\mathbf{v}\|$, $\mathbf{n} \equiv \mathbf{v}/v$, $\mathbf{n}' \equiv \mathbf{v}'/v$, $d\Omega = d^{d-1}n'$, and $\sigma_{\text{diff}}(\mathbf{n},\mathbf{n}')$ is the differential cross section of the particle colliding on each scatterer. If the system is homogeneous, the processes of Section 4.5.4 are recovered.

4.5.6 Spectral Theory

If the vector field (4.102) and the diffusivities (4.103) do not depend on time, the linear operator $\hat{L}$ of the master equation (4.109) is time independent, so that its formal solution is given by $p_t = \exp(\hat{L}t)p_0$. In this case, the master equation may be solved in terms of the eigenvalues and eigenvectors of its operator:

$$\hat{L}\,\psi_\alpha = z_\alpha\,\psi_\alpha \qquad \text{and} \qquad \hat{L}^\dagger\,\tilde{\psi}_\alpha = z_\alpha^*\,\tilde{\psi}_\alpha, \quad (4.122)$$

and possible Jordan block structures. The right and left eigenfunctions can be taken as satisfying the biorthonormal conditions

$$\int \tilde{\psi}_\alpha^*(\mathbf{x})\,\psi_\beta(\mathbf{x})\,d\mathbf{x} = \delta_{\alpha\beta}. \quad (4.123)$$

Accordingly, the time evolution of the probability density can be decomposed as

$$p(\mathbf{x},t) = \sum_\alpha c_\alpha\,e^{z_\alpha t}\,\psi_\alpha(\mathbf{x}) + \cdots \quad \text{with} \quad c_\alpha = \int \tilde{\psi}_\alpha^*(\mathbf{x})\,p_0(\mathbf{x})\,d\mathbf{x}, \quad (4.124)$$

where p_0 is the initial probability density and the dots denote possible Jordan block contributions. Jordan blocks may arise if some eigenvalue z_α has an algebraic multiplicity m_α, but a single associated eigenfunction ψ_α, in which case the time dependence contains terms behaving as $t^{m_\alpha-1}\exp(z_\alpha t)$. If the spectrum is discrete, we have the trace formula

$$\operatorname{tr} e^{\hat{L}t} = \sum_\alpha m_\alpha\,e^{z_\alpha t} \quad (4.125)$$

and its Laplace transform

$$\int_0^\infty e^{-zt}\,\operatorname{tr} e^{\hat{L}t}\,dt = \operatorname{tr}\frac{1}{z-\hat{L}} = \sum_\alpha \frac{m_\alpha}{z-z_\alpha}. \quad (4.126)$$

In general, the eigenvalues may have a real part and an imaginary part: $z_\alpha = \operatorname{Re} z_\alpha + \imath \operatorname{Im} z_\alpha$. A general property of operators associated with master equations such as (4.108) with a positive matrix of diffusivities $\mathsf{D} > 0$ is that the real part of the eigenvalues is always nonpositive, $\operatorname{Re} z_\alpha \leq 0$, so that the probability distribution can converge towards the stationary one. The modes associated with a complex eigenvalue manifest damped oscillations.

An example of discrete spectrum is provided by the Ornstein–Uhlenbeck stochastic process presented in Section E.7. The random drift process where the drift velocity and the diffusivities are constant and uniform in equations (4.110)–(4.111) may have a continuous or discrete spectrum depending on the boundary conditions, as shown in Section E.8.

4.5.7 Reversible Continuous-State Markov Processes

For continuous-state Markov processes, the condition of stochastic reversibility (4.10) reads

$$p(\mathbf{x}',t'|\mathbf{x},t)\,p_{\rm st}(\mathbf{x}) = p(\mathbf{x},t'|\mathbf{x}',t)\,p_{\rm st}(\mathbf{x}') \tag{4.127}$$

in analogy with equation (4.40), where $p_{\rm st}(\mathbf{x})$ is the stationary probability density. In the case of advection-diffusion processes, substituting this condition with $t' = t + \Delta t$ into equation (4.106), carrying out the same expansion as the one leading to equation (4.107), and using the condition of stationarity $\hat{L}\,p_{\rm st} = 0$, we find that the time evolution of the probability density is equivalently ruled by equation (4.109) and

$$\partial_t\, p = p_{\rm st}\,\hat{L}^\dagger\left(\frac{p}{p_{\rm st}}\right), \quad \text{where} \quad \hat{L}^\dagger \equiv \mathbf{C}\cdot\partial_{\mathbf{x}} + \mathsf{D} : \partial_{\mathbf{x}}^2 \tag{4.128}$$

is the adjoint of the operator $\hat{L}$ with respect to the scalar product

$$\langle \mathfrak{u}|\mathfrak{v}\rangle \equiv \int_{\mathbb{R}^d} \mathfrak{u}^*(\mathbf{x})\,\mathfrak{v}(\mathbf{x})\,d\mathbf{x}. \tag{4.129}$$

In analogy with equations (4.43) and (4.44), we may introduce

$$\mathfrak{f}(\mathbf{x},t) \equiv \frac{p(\mathbf{x},t)}{p_{\rm st}(\mathbf{x})^{1/2}}, \tag{4.130}$$

so that the master equation (4.109) can be expressed as

$$\partial_t\,\mathfrak{f} = \hat{K}\,\mathfrak{f} \qquad \text{with} \qquad \hat{K}\,\mathfrak{f} \equiv p_{\rm st}^{-1/2}\,\hat{L}\left(p_{\rm st}^{1/2}\,\mathfrak{f}\right), \tag{4.131}$$

which is is self-adjoint with respect to the scalar product (4.129).

If the stochastic process is reversible, the operator $\hat{L}$ is thus self-adjoint with respect to the scalar product

$$(u|v) \equiv \int_{\mathbb{R}^d} \rho(\mathbf{x})\,u^*(\mathbf{x})\,v(\mathbf{x})\,d\mathbf{x}, \tag{4.132}$$

with the weighting density $\rho \equiv p_{\rm st}^{-1}$. As a consequence, the eigenvalues of a reversible stochastic process are always real, so that $z_\alpha = {\rm Re}\,z_\alpha \le 0$ and ${\rm Im}\,z_\alpha = 0$.

Expanding the reversibility condition

$$\hat{L}\,p = p_{\rm st}\,\hat{L}^\dagger\left(\frac{p}{p_{\rm st}}\right) \tag{4.133}$$

with the operator (4.109), we obtain the relation

$$\mathbf{V} = \mathbf{C} - \partial_{\mathbf{x}}\cdot\mathsf{D} = \mathsf{D}\cdot\frac{\partial_{\mathbf{x}}p_{\rm st}}{p_{\rm st}} \tag{4.134}$$

and another relation that can be deduced by taking the partial derivative of equation (4.134) with respect to $\mathbf{x}$. Therefore, the current density of the stationary distribution is equal to zero if the process is reversible, i.e., $\mathbf{J}_{\rm st} = \mathbf{V}\,p_{\rm st} - \mathsf{D}\cdot\partial_{\mathbf{x}}p_{\rm st} = 0$, which is a stronger condition than the condition of stationarity requiring that $\partial_t\,p_{\rm st} = -\partial_{\mathbf{x}}\cdot\mathbf{J}_{\rm st} = 0$.

If the stationary density can be expressed as

$$p_{\rm st}(\mathbf{x}) = \mathcal{N}^{-1}\,{\rm e}^{-\Upsilon(\mathbf{x})} \tag{4.135}$$

in terms of some potential $\Upsilon(\mathbf{x})$ and the normalization constant $\mathcal{N}$, the drift velocity (4.134) is related to the gradient of the potential according to

$$\mathbf{V} = -\mathsf{D}\cdot\partial_{\mathbf{x}}\Upsilon. \tag{4.136}$$

Consequently, the operator defined in equation (4.131) can be written as

$$\hat{K}\mathfrak{f} = \left[\mathsf{D} : (\partial_{\mathbf{x}} - \mathbf{a})^2 - \Phi\right]\mathfrak{f} \qquad \text{with} \qquad \mathbf{a} = -\frac{1}{2}\,\mathsf{D}^{-1}\cdot(\partial_{\mathbf{x}}\cdot\mathsf{D}) \tag{4.137}$$

and

$$\Phi = \frac{1}{4}\,\mathsf{D} : (\partial_{\mathbf{x}}\Upsilon)^2 - \frac{1}{2}\,\partial_{\mathbf{x}}\cdot(\mathsf{D}\cdot\partial_{\mathbf{x}}\Upsilon) - \frac{1}{4}\,\mathsf{D}^{-1} : (\partial_{\mathbf{x}}\,\mathsf{D})^2 + \frac{1}{2}\,\partial_{\mathbf{x}}^2 : \mathsf{D}. \tag{4.138}$$

If the diffusivities are uniform in space, i.e., $\partial_{\mathbf{x}}\,\mathsf{D} = 0$, we have that $\mathbf{a} = 0$ and the operator reduces to $\hat{K}\mathfrak{f} = \mathsf{D} : \partial_{\mathbf{x}}^2\,\mathfrak{f} - \Phi\,\mathfrak{f}$, which has the same mathematical form as the Hamiltonian operator of Schrödinger's equation.

The case of one-dimensional advection-diffusion processes is discussed in Section E.9.

4.6 Weak-Noise Limit in Markov Processes

4.6.1 From Discrete- to Continuous-State Processes

In the limit where the discrete variables $\mathbf{n} \in \mathbb{Z}^d$ take large values, the discrete-state process can be approximated by a continuous-state one. This approximation is carried out by considering the continuous variables $\mathbf{x} = \epsilon\mathbf{n}$ with some small parameter ϵ in the limit where the spacing between the discrete states is vanishing, i.e., $\epsilon \to 0$, so that the state space becomes a continuum. In this limit, the volume element is given by $d\mathbf{x} = \epsilon^d$ and the probability density $p(\mathbf{x},t)$ is related to the probability distribution $P(\mathbf{n},t)$ according to $P(\mathbf{n},t) = \epsilon^d\,p(\epsilon\mathbf{n},t)$, so that the normalization conditions are consistent with each other: $\sum_{\mathbf{n}\in\mathbb{Z}^d} P(\mathbf{n},t) = \int_{\mathbb{R}^d} p(\mathbf{x},t)\,d\mathbf{x} = 1$.

Using the master equation (4.32) and expanding the shift operators in powers of $\partial_{\mathbf{n}}$, we have that

$$\frac{d}{dt}P(\mathbf{n},t) = \sum_{\rho}\left(-\boldsymbol{\nu}_\rho\cdot\partial_{\mathbf{n}} + \frac{1}{2}\,\boldsymbol{\nu}_\rho\boldsymbol{\nu}_\rho : \partial_{\mathbf{n}}^2 + \cdots\right) W_\rho(\mathbf{n})\,P(\mathbf{n},t). \tag{4.139}$$

Introducing the rates

$$w_\rho(\mathbf{x}) \equiv \lim_{\epsilon\to 0}\epsilon\,W_\rho(\mathbf{x}/\epsilon), \tag{4.140}$$

we obtain the master equation (4.108) for an advection-diffusion process with the drifts and diffusivities given by

$$\mathbf{C}(\mathbf{x}) = \sum_{\rho}\boldsymbol{\nu}_\rho\,w_\rho(\mathbf{x}) \qquad \text{and} \qquad \mathsf{D}(\mathbf{x}) = \frac{\epsilon}{2}\sum_{\rho}\boldsymbol{\nu}_\rho\,\boldsymbol{\nu}_\rho\,w_\rho(\mathbf{x}), \tag{4.141}$$

if higher order terms can be neglected. For instance, if the integers $\mathbf{n}$ represent the numbers of molecules in a system of volume V, the small parameter can be taken as $\epsilon = 1/V$, so that $\mathbf{x} = \mathbf{n}/V$ are the molecular concentrations, and the master equation (4.108) with the drifts and diffusivities (4.141) is the chemical Fokker–Planck equation (Nicolis and Prigogine, 1971; Nicolis, 1972; Gillespie, 2000; Gardiner, 2004). We note that if the molecular numbers are arbitrarily large and $\epsilon \to 0$, the diffusivities are negligible with respect to the vector fields in equation (4.141), so that the process becomes deterministic in this macroscopic limit.

Now, the master equation (4.108) reads

$$\partial_t p = -\partial_{\mathbf{x}} \cdot (\mathbf{C}\, p) + \epsilon\, \partial_{\mathbf{x}}^2 : (\mathbf{Q}\, p) \qquad \text{with} \qquad \mathbf{Q} \equiv \epsilon^{-1}\, \mathbf{D}. \tag{4.142}$$

Since this is a partial differential equation of second order in the variables $\mathbf{x}$ as for the Schrödinger equation, the limit $\epsilon \to 0$ can be investigated using methods similar to the semiclassical approximation in quantum mechanics.

4.6.2 Semideterministic Approximation

The probability density is supposed to behave as

$$p(\mathbf{x}, t) = \mathrm{e}^{-\phi(\mathbf{x},t)/\epsilon} \tag{4.143}$$

for small values of the parameter ϵ. Substituting the expansion

$$\phi = \phi_0 + \epsilon\, \phi_1 + \epsilon^2\, \phi_2 + \cdots \tag{4.144}$$

into equation (4.142), we obtain the following hierarchy of equations:

$$0 = \partial_t\, \phi_0 + \mathbf{C} \cdot \partial_{\mathbf{x}}\phi_0 + \mathbf{Q} : (\partial_{\mathbf{x}}\phi_0)^2, \tag{4.145}$$

$$\begin{aligned} 0 = \partial_t\, \phi_1 + \mathbf{C} \cdot \partial_{\mathbf{x}}\phi_1 + 2\,\mathbf{Q} : (\partial_{\mathbf{x}}\phi_0)(\partial_{\mathbf{x}}\phi_1) \\ -\partial_{\mathbf{x}} \cdot \mathbf{C} - 2\,(\partial_{\mathbf{x}} \cdot \mathbf{Q}) \cdot \partial_{\mathbf{x}}\phi_0 - \mathbf{Q} : \partial_{\mathbf{x}}^2\phi_0. \end{aligned} \tag{4.146}$$

$$\vdots$$

The first equation (4.145) can be written in the form of the Hamilton–Jacobi equation

$$\partial_t \phi_0 + \mathscr{H}(\mathbf{x}, \partial_{\mathbf{x}}\phi_0, t) = 0 \tag{4.147}$$

with the Freidlin–Wentzell Hamiltonian function

$$\mathscr{H}(\mathbf{x}, \mathfrak{p}, t) = \mathbf{Q}(\mathbf{x}, t) : \mathfrak{p}^2 + \mathbf{C}(\mathbf{x}, t) \cdot \mathfrak{p} \tag{4.148}$$

and the momenta $\mathfrak{p} = \partial_{\mathbf{x}}\phi_0$. In this respect, the field $\phi_0(\mathbf{x}, t)$ is the action

$$\phi_0 = \int \mathscr{L}\, dt = \int \mathfrak{p} \cdot d\mathbf{x} - \mathscr{H}\, dt, \tag{4.149}$$

expressed in terms of the Lagrangian function (Onsager and Machlup, 1953)

$$\mathscr{L}(\mathbf{x}, \dot{\mathbf{x}}, t) = \frac{1}{4}\, \mathbf{Q}(\mathbf{x}, t)^{-1} : [\dot{\mathbf{x}} - \mathbf{C}(\mathbf{x}, t)]^2, \tag{4.150}$$

or, alternatively, with the momenta $\mathfrak{p} = \partial \mathscr{L}/\partial \dot{\mathbf{x}}$, which are canonically conjugated to the variables $\mathbf{x}$, and the Hamiltonian function (4.148). This Hamiltonian dynamical system is defined in the phase space $(\mathbf{x}, \mathfrak{p}) \in \mathbb{R}^{2d}$ with twice as many variables as the macroscopic variables $\mathbf{x}$. In this doubled phase space, time evolution is governed by Hamilton's equations

$$\begin{cases} \dot{\mathbf{x}} = \partial_{\mathfrak{p}} \mathscr{H} = \mathbf{C}(\mathbf{x}, t) + 2\, \mathbf{Q}(\mathbf{x}, t) \cdot \mathfrak{p}, \\ \dot{\mathfrak{p}} = -\partial_{\mathbf{x}} \mathscr{H} = -\partial_{\mathbf{x}} \mathbf{C}(\mathbf{x}, t) \cdot \mathfrak{p} - \partial_{\mathbf{x}} \mathbf{Q}(\mathbf{x}, t) : \mathfrak{p}^2. \end{cases} \tag{4.151}$$

We observe that the subspace of zero momenta $\mathfrak{p} = 0$ is invariant for this Hamiltonian dynamical system. This subspace is the phase space of the deterministic dynamics according to $\dot{\mathbf{x}} = \mathbf{C}(\mathbf{x}, t)$.

Additionally, equation (4.146) is the associated transport equation ruling the dynamics in the vicinity of the Hamiltonian orbits according to their linear stability properties (Gaspard, 2002b).

If the stochastic operator does not depend on time, this is also the case for the Hamiltonian function (4.148), so that there is conservation of the quantity $\mathscr{H}(\mathbf{x}, \mathfrak{p}) = \mathscr{E}$ and $\partial_t \phi_0 = -\mathscr{E}$. Since $\partial_t \phi = -\epsilon \partial_t p/p$, the Hamilton–Jacobi equation (4.145) shows that this pseudo-energy is related according to $\mathscr{E} = \epsilon z$ to the variable z such that $p \sim \exp(zt)$ in the limit $\epsilon \to 0$.

The phase portrait of a noisy dynamical system with time-independent coefficients $\mathbf{C}$ and $\mathbf{Q}$ is schematically depicted in Figure 4.6 and shows a stationary and an oscillatory attractor in the deterministic invariant subspace $\mathfrak{p} = 0$.

In the weak-noise limit, Gaussian stochastic processes can thus be analyzed in terms of the Freidlin–Wentzell Hamiltonian dynamics (4.151) deduced from the variational principle $\delta\phi_0 = 0$ for the action (4.149). As a corollary, such dynamics have a symplectic structure whether time is continuous or discrete (Fogedby and Jensen, 2005; Demaeyer and Gaspard, 2009).

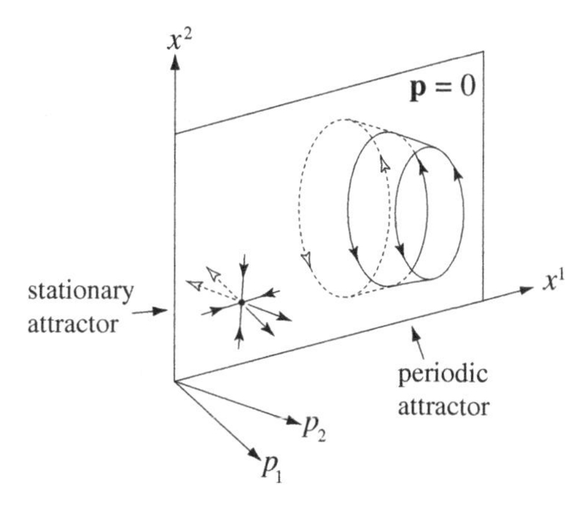

Figure 4.6 Phase portrait of stationary and oscillatory attractors in the doubled phase space of the variables $\mathbf{x} = (x^k)_{k=1}^d$ and their canonically conjugated momenta $\mathfrak{p} = (p_k)_{k=1}^d$ near the deterministic invariant subspace $\mathfrak{p} = 0$ for a noisy dynamical system.

The formalism can be extended beyond the quadratic approximation as explained in Section E.10.

4.6.3 Spectral Theory in the Weak-Noise Limit

If the stochastic operator does not depend on time, the time evolution can be decomposed in terms of the eigenvalues and associated eigenfunctions, as discussed in Section 4.5.6. Remarkably, the trace formula (4.125) or (4.126) can be evaluated in terms of the stationary or periodic orbits of the Freidlin–Wentzell Hamiltonian dynamics, providing approximations for the eigenvalues in the weak-noise limit (Gaspard, 2002b).

Noisy Stationary Attractor or Repeller

For a noisy dynamical system around a stationary attractor or repeller, the spectrum is essentially the same as for the deterministic dynamics if the stationary point is not marginally stable. If $\{\xi_i\}_{i=1}^d$ denote the eigenvalues of the Jacobian matrix $\partial_{\mathbf{x}}\mathbf{C}$ ruling the linearized dynamical system $\delta\dot{\mathbf{x}} = \partial_{\mathbf{x}}\mathbf{C}(\mathbf{x}_*)\cdot\delta\mathbf{x}$ in the vicinity of the stationary point $\mathbf{x}_*$, the eigenvalues of the stochastic operators are given by

$$z_{\mathbf{lm}} = \sum_{\mathrm{Re}\,\xi_i<0} l_i\,\xi_i - \sum_{\mathrm{Re}\,\xi_i>0} (m_i+1)\,\xi_i + O(\epsilon) \tag{4.152}$$

with the integers $l_i, m_i = 0, 1, 2, \ldots$ (Gaspard, 2002b). In the limit $\epsilon \to 0$, the Pollicott–Ruelle resonances of the deterministic dynamics are recovered (Gaspard, 1998).

Noisy Oscillatory Attractor

If the deterministic dynamical system has an attracting limit cycle,[10] this periodic orbit is continuously deformed into a family of periodic orbits away from the invariant subspace $\mathfrak{p} = 0$, as shown in Figure 4.6. Since the stochastic operator is assumed to be time independent, the Hamiltonian function does not depend on time, so that the pseudoenergy $\mathscr{E} = \mathscr{H}$ is conserved and is equal to zero in the invariant subspace $\mathfrak{p} = 0$. Therefore, the prime period T of the periodic orbit depends on this pseudoenergy and, typically, decreases with it (Gaspard, 2002b). Now, the action $\phi_0(\mathbf{x}, t)$ can be expanded around some point $(\mathbf{x}_*, t_*)$ of the deterministic limit cycle in the invariant subspace $\mathfrak{p} = 0$ according to (Gaspard, 2002a)

$$\begin{aligned}\phi_0(\mathbf{x}, t) = \phi_{0*} + \partial_t\phi_{0*}(t - t_*) + \partial_{\mathbf{x}}\phi_{0*}\cdot(\mathbf{x} - \mathbf{x}_*) + \frac{1}{2}\,\partial_t^2\phi_{0*}(t - t_*)^2 \\ + \partial_t\partial_{\mathbf{x}}\phi_{0*}\cdot(\mathbf{x} - \mathbf{x}_*)(t - t_*) + \frac{1}{2}\,\partial_{\mathbf{x}}^2\phi_{0*} : (\mathbf{x} - \mathbf{x}_*)^2 + \cdots\end{aligned} \tag{4.153}$$

In the invariant subspace $\mathfrak{p} = 0$, we have that $\phi_{0*} = 0$, $\partial_t\phi_{0*} = -\mathscr{E}_* = 0$, $\partial_{\mathbf{x}}\phi_{0*} = \mathfrak{p}_* = 0$, and

$$\partial_t^2\phi_{0*} = -\frac{1}{r\,\partial_{\mathscr{E}}T} = \frac{1}{r\,|\partial_{\mathscr{E}}T|}, \tag{4.154}$$

[10] Such a limit cycle is an oscillatory attractor.

where r is the number of repetitions of the prime period T during the time interval $[0,t]$ and $\partial_{\mathscr{E}} T$ is the derivative of the prime period with respect to the pseudoenergy in the invariant subspace where $\mathscr{E} = 0$. As the limit cycle is an attractor, the deviations $\mathbf{x} - \mathbf{x}_*$ are small, so that the term going as $(t - t_*)^2$ prevails in the expansion (4.153). Therefore, noise induces phase diffusion along the limit cycle. If the phase along the cycle is denoted $0 \leq \vartheta < T$, its conditional probability density evolves in time according to

$$\mathcal{P}(\vartheta, t|\vartheta_0, 0) \simeq \frac{1}{\sqrt{2\pi\epsilon r|\partial_{\mathscr{E}} T|}} \exp\left\{-\frac{[\vartheta - (\vartheta_0 + t - rT)]^2}{2\epsilon r|\partial_{\mathscr{E}} T|}\right\} \tag{4.155}$$

for $rT < t < (r+1)T$, as the time t and the repetition number r increase. This result shows that the diffusivity of the phase ϑ along the noisy limit cycle is given by $\mathscr{D} = \epsilon|\partial_{\mathscr{E}} T|/(2T)$. This diffusivity is enhanced if the noise amplitude increases or if the dependence of the period T on the pseudoenergy $\mathscr{E}$ becomes stronger. If the periodic orbit has the Fourier decomposition

$$\mathbf{x}(\vartheta) = \sum_{n=-\infty}^{+\infty} \mathbf{a}_n \exp\left(\imath \frac{2\pi n}{T} \vartheta\right), \tag{4.156}$$

the correlation function of the noisy limit cycle should behave for $t \to +\infty$ as

$$C(t) = \langle \mathbf{x}(t) \cdot \mathbf{x}(0) \rangle = \sum_{n=-\infty}^{+\infty} \mathbf{a}_n^2 \, \mathrm{e}^{z_n t} \tag{4.157}$$

with the eigenvalues

$$z_n = \imath \frac{2\pi n}{T} - \epsilon \frac{|\partial_{\mathscr{E}} T|}{2T} \left(\frac{2\pi n}{T}\right)^2 + O(\epsilon^2), \tag{4.158}$$

for $n \in \mathbb{Z}$. The eigenvalue $z_0 = 0$ is associated with the stationary probability density. The next ones, $z_{\pm 1} = \pm\imath\omega - 1/\tau_{\mathrm{c}}$, correspond to oscillations at the period T of the limit cycle. These oscillations are damped exponentially in time with the correlation time

$$\tau_{\mathrm{c}} = \frac{T^3}{2\pi^2|\partial_{\mathscr{E}} T|\epsilon} + O(\epsilon^0), \tag{4.159}$$

which is thus determined by the period T and its derivative with respect to the pseudoenergy $\mathscr{E}$. This correlation time diverges in the noiseless limit $\epsilon \to 0$, as expected, because the oscillations of the limit cycle are self-sustained and perfectly periodic for the deterministic dynamics. The quality factor of the noisy oscillator is thus given by

$$\mathscr{Q} \equiv 2\pi \frac{\tau_{\mathrm{c}}}{T} \simeq \frac{T^2}{\pi|\partial_{\mathscr{E}} T|\epsilon}. \tag{4.160}$$

The oscillator can act as a clock if its quality factor is large enough, i.e., if the noise amplitude ϵ is small enough, which corresponds to the reduction of phase diffusion since $\mathscr{Q} \simeq T/(2\pi\mathscr{D})$. This phenomenon of phase diffusion is important for biochemical and chemical clocks in small systems where molecular fluctuations manifest themselves (Gaspard, 2002a; Gonze et al., 2002; Barroo et al., 2015).

We note that the spectrum of the stochastic operator can be very different in the vicinity of bifurcations undergone by the attractors. Moreover, the attractors of the noiseless dynamics may become metastable in the presence of noise. This noise-induced metastability is often characterized by dwell times growing as $\tau_{\rm d} \sim \exp(\phi_0/\epsilon)$ in the noiseless limit $\epsilon \to 0$ with the action (4.149) of specific escaping orbits in the doubled phase space (Demaeyer and Gaspard, 2009, 2013).

4.7 Langevin Stochastic Processes

4.7.1 Langevin Equation for Brownian Motion

Let us consider the Brownian motion of a colloidal particle in the presence of gravity, as depicted in Figure 4.7(a).

If the particle was macroscopic, its movement would be described by Newton's equation,

$$m\frac{d^2\mathbf{r}}{dt^2} = -\gamma\,\frac{d\mathbf{r}}{dt} + \mathbf{F}_{\rm g} \qquad \text{with} \qquad \mathbf{F}_{\rm g} = -m\,g\,\mathbf{1}_x, \tag{4.161}$$

where m is the mass of the particle, which is supposed to be much heavier than the mass of the fluid displaced by the particle, $\mathbf{r}$ its position, γ its friction coefficient given by Stokes' formula $\gamma = 6\pi\eta R$ for a spherical particle of radius R in a fluid of shear viscosity η, $\mathbf{F}_{\rm g}$ the force exerted by the gravitational acceleration g, and $\mathbf{1}_x$ the unit vector in the vertical x-direction. If the velocity of the particle is small enough, the hydrodynamic flow is laminar around the particle as shown in Figure 4.7(b) and the relation between the friction force and the particle velocity is thus linear. In this regard, the friction coefficient is a linear response

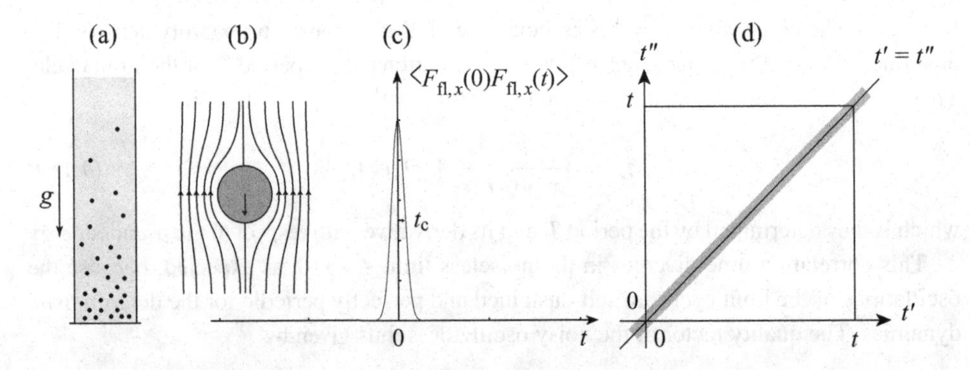

Figure 4.7 (a) Colloidal particles in a dilute suspension inside a vessel in the presence of gravity g. (b) Laminar hydrodynamic flow around a spherical colloidal particle moving downwards under the effect of gravity, leading to Stokes' formula for the friction coefficient. (c) Schematic representation of the autocorrelation function of the fluctuating Langevin force $F_{{\rm fl},x}(t)$ decreasing to zero over the timescale of the correlation time $t_{\rm c}$. (d) The domain of integration of this autocorrelation function in equation (4.171) and the zone around the diagonal line $t' = t''$, where it is nonvanishing. This zone has a width of the order of the correlation time $t_{\rm c}$.

coefficient. According to equation (4.161), the colloidal particle would thus be moving downwards with the drift velocity $\mathbf{V} = \mathbf{F}_g/\gamma$.

However, the effects of thermal fluctuations are neglected in equation (4.161), although they are essential to understand Brownian motion. Therefore, the right-hand side of equation (4.161) should include a fluctuating force in order to generate the random walk of micrometric colloidal particles, as proposed by Langevin (1908). If the diffusion resulting from thermal fluctuations is included with the drift velocity $\mathbf{V}$, the probability density $p(\mathbf{r}, t)$ for observing the particle at position $\mathbf{r}$ and time t should obey the following local conservation equation, called *Smoluchowski's master equation*,

$$\partial_t p + \nabla \cdot \jmath = 0, \qquad \text{where} \qquad \jmath = \mathbf{V} p - \mathcal{D} \nabla p \tag{4.162}$$

is the current density.[11] At equilibrium, the density should satisfy the Boltzmann distribution, $p_{\text{eq}} \sim \exp(-\beta m g x)$, where $\beta = (k_B T)^{-1}$ is the inverse temperature, as depicted in Figure 4.7(a). Since this density should be a stationary solution of equation (4.162), the diffusion coefficient $\mathcal{D}$ must satisfy Einstein's relation

$$\mathcal{D} = \frac{k_B T}{\gamma}. \tag{4.163}$$

At the microscale, the Newton equation for the colloidal particle has the form

$$m \frac{d^2 \mathbf{r}}{dt^2} = \sum_{a=1}^{N} \mathbf{f}_a + \mathbf{F}_g, \tag{4.164}$$

where $\{\mathbf{r}_a\}_{a=1}^N$ are the positions of N atoms composing the fluid and $\mathbf{f}_a$ is the force exerted on the colloidal particle by the ath atom. These interaction forces are at the origin of the friction force already appearing in equation (4.161), but also of the fluctuating force $\mathbf{F}_{\text{fl}}(t)$, which is introduced to describe the Brownian motion of the colloidal particle. Therefore, we should expect that

$$\sum_{a=1}^{N} \mathbf{f}_a = -\gamma \frac{d\mathbf{r}}{dt} + \mathbf{F}_{\text{fl}}(t). \tag{4.165}$$

According to the central limit theorem, the sum $\sum_{a=1}^N \mathbf{f}_a$ should behave at every instant of time as a Gaussian random variable since the number of atoms N is very large and the forces have short or medium range. The fact is that Gaussian random variables are uniquely determined by their mean value and their variance. Here, we have a random function of time, which is thus fully characterized by its mean value and its time autocorrelation function. Since the friction force is supposed to capture the mean value of all the microscopic forces, the fluctuating force should have a mean value equal to zero at every time: $\langle \mathbf{F}_{\text{fl}}(t) \rangle = 0$. For a spherical particle in a fluid, the fluctuations are isotropic and the three components of the fluctuating force should thus behave in the same way in the three directions of space. Moreover, the autocorrelation function of the fluctuating force should vanish asymptotically

[11] In a dilute colloidal suspension, the density $n_c(\mathbf{r}, t)$ of colloidal particles is related to the probability density $p(\mathbf{r}, t)$ describing a single particle by $n_c = N_c p$, where N_c is the total number of these particles in the suspension.

in time as $t \to \infty$. As illustrated in Figure 4.7(c), the decrease of this function is expected to happen over some correlation time t_c, which is typically shorter than the observational timescale of Brownian motion.[12] In this respect, the autocorrelation function can be represented as a Dirac delta distribution in time. Consequently, the correlation functions of the different components of the fluctuating force can be assumed to obey

$$\langle F_{\mathrm{fl},i}(t)\, F_{\mathrm{fl},j}(t')\rangle = C\, \delta_{ij}\, \delta(t-t'), \tag{4.166}$$

for $i,j = x,y,z$ and $|t-t'| \gg t_c$, with some constant C to be determined.

For this purpose, we consider the Newton equation (4.164) in the absence of gravity and with the decomposition (4.165) of the microscopic forces. This equation is given by

$$m\frac{d^2\mathbf{r}}{dt^2} = -\gamma\frac{d\mathbf{r}}{dt} + \mathbf{F}_{\mathrm{fl}}(t). \tag{4.167}$$

Integrating over the time interval $[0,t]$, we obtain

$$m\,(\dot{\mathbf{r}}_t - \dot{\mathbf{r}}_0) = -\gamma\,(\mathbf{r}_t - \mathbf{r}_0) + \int_0^t \mathbf{F}_{\mathrm{fl}}(t')\,dt'. \tag{4.168}$$

In the left-hand side, we find the difference between the particle velocity at time t and initial time $t=0$. These two velocities are randomly distributed according to a Maxwellian velocity distribution if there is equilibrium at the temperature of the fluid. Accordingly, even if the magnitude of these two velocities may take large values, these values have a stationary probability distribution because $\langle(\dot{\mathbf{r}}_t - \dot{\mathbf{r}}_0)^2\rangle \simeq \langle\dot{\mathbf{r}}_t^2\rangle + \langle\dot{\mathbf{r}}_0^2\rangle = 6k_{\mathrm{B}}T/m$ if $|t| \gg m/\gamma$. In this sense, the left-hand side remains bounded in time. In contrast, the first term of the right-hand side is not bounded since the colloidal particle undergoes a random walk and its displacement is known to behave as $\|\mathbf{r}_t - \mathbf{r}_0\| \sim \sqrt{t}$ for $t \to \infty$ and far from the walls of the vessel. As a consequence, the second term of the right-hand side should also be unbounded in order to compensate the first term and maintain the balance with the left-hand side. This latter becomes negligible over a long time interval; we should thus have that

$$\gamma\,(\mathbf{r}_t - \mathbf{r}_0) \simeq_{t\to\infty} \int_0^t \mathbf{F}_{\mathrm{fl}}(t')\,dt'. \tag{4.169}$$

Taking the square of this relation and its mean value, dividing by $6t$, we find that the diffusion coefficient of Brownian motion,

$$\mathcal{D} \equiv \lim_{t\to\infty}\frac{1}{6t}\langle(\mathbf{r}_t - \mathbf{r}_0)^2\rangle, \tag{4.170}$$

should satisfy the relation

$$\mathcal{D}\gamma^2 = \lim_{t\to\infty}\frac{1}{6t}\int_0^t dt' \int_0^t dt''\,\langle\mathbf{F}_{\mathrm{fl}}(t')\cdot\mathbf{F}_{\mathrm{fl}}(t'')\rangle = \frac{1}{6}\int_{-\infty}^{+\infty} dt'\,\langle\mathbf{F}_{\mathrm{fl}}(0)\cdot\mathbf{F}_{\mathrm{fl}}(t')\rangle. \tag{4.171}$$

Here, the double integral is calculated in the domain shown in Figure 4.7(d). The stationarity of the process implies that the autocorrelation function only depends on the time interval

[12] For micrometric colloidal particles moving in water, this correlation time is of the order of a microsecond (Kheifets et al., 2014).

$t' - t''$ between the instants of time when the fluctuating force is observed. The autocorrelation function takes nonvanishing values in the range of the correlation time t_c around the diagonal line $t' = t''$ and it tends to zero far from this diagonal line. Moreover, its transverse profile is the same all along the diagonal and the contributions near the initial and final points $t' = t'' = 0$ and $t' = t'' = t$ are negligible if the time interval $[0,t]$ is large enough. Consequently, the result given in equation (4.171) is obtained. Using the Einstein relation (4.163), we thus find that the friction coefficient is given by the time integral of the force autocorrelation function according to

$$\gamma = \frac{1}{6k_{\rm B}T} \int_{-\infty}^{+\infty} \langle \mathbf{F}_{\rm fl}(0) \cdot \mathbf{F}_{\rm fl}(t) \rangle \, dt \tag{4.172}$$

as obtained by Kirkwood (1946). Substituting assumption (4.166), we get the needed constant $C = 2\, k_{\rm B} T\, \gamma$.

If the colloidal particle moves in the energy potential $U(\mathbf{r})$ (for instance, due to the presence of a trap or the walls of the vessel) and in the external force due to gravity $\mathbf{F}_{\rm g}$, Langevin's equation takes the form of the following stochastic differential equation of second order in time,

$$m\, \frac{d^2\mathbf{r}}{dt^2} = -\nabla U + \mathbf{F}_{\rm g} - \gamma\, \frac{d\mathbf{r}}{dt} + \mathbf{F}_{\rm fl}(t), \tag{4.173}$$

where the fluctuating force is given by Gaussian white noises characterized by

$$\langle F_{{\rm fl},i}(t) \rangle = 0 \qquad \text{and} \qquad \langle F_{{\rm fl},i}(t)\, F_{{\rm fl},j}(t') \rangle = 2\,\gamma\, k_{\rm B} T\, \delta_{ij}\, \delta(t - t'), \tag{4.174}$$

for $i, j = x, y, z$. This relation between the noise amplitude and the friction coefficient governing energy dissipation is the expression of the fluctuation–dissipation theorem, which applies here because the colloidal particle moves in the regime of linear response, as further discussed below.

In the limit $\gamma = 0$, energy is conserved and the dissipationless Newton equation is recovered. If friction is weak, inertial effects are dominant and the system is said to be *underdamped*. Conversely, if friction is strong, the system is *overdamped* and inertial effects are negligible. In this overdamped regime, the Langevin equation (4.173) becomes the following stochastic differential equation of first order in time,

$$\frac{d\mathbf{r}}{dt} = \mathbf{V} + \mathbf{v}_{\rm fl}(t) \tag{4.175}$$

with the drift velocity $\mathbf{V} = \left(-\nabla U + \mathbf{F}_{\rm g}\right)/\gamma$ and the fluctuating velocity $\mathbf{v}_{\rm fl}(t) = \mathbf{F}_{\rm fl}(t)/\gamma$ given by other Gaussian white noises satisfying

$$\langle v_{{\rm fl},i}(t) \rangle = 0 \qquad \text{and} \qquad \langle v_{{\rm fl},i}(t)\, v_{{\rm fl},j}(t') \rangle = 2\,\mathcal{D}\, \delta_{ij}\, \delta(t - t'), \tag{4.176}$$

for $i, j = x, y, z$, according to the Einstein relation (4.163). In the absence of drift velocity, we thus recover a random walk of diffusion coefficient $\mathcal{D}$, as required. We note that the stochastic differential equation (4.175) with the noise (4.176) corresponds to the Smoluchowski master equation (4.162), which rules the time evolution of the probability density $p(\mathbf{r}, t)$.

4.7.2 Kramers' Master Equation

The master equation associated with the Langevin equation (4.173) is called the equation of Kramers (1940). It rules the time evolution of the joint probability density $\mathcal{P}(\mathbf{r},\mathbf{p},t)$ to observe the colloidal particle at the position $\mathbf{r}$ with the momentum $\mathbf{p} = m\dot{\mathbf{r}}$ at time t. This master equation can be established by considering the local conservation of probability in the six-dimensional phase space $\mathbf{x} = (\mathbf{r},\mathbf{p}) \in \mathbb{R}^6$ of the colloidal particle:

$$\partial_t \mathcal{P} + \text{div}_{\mathbb{R}^6} \boldsymbol{J} = 0. \tag{4.177}$$

The expression of the six-dimensional current density $\boldsymbol{J}$ can be obtained if the Langevin equation (4.173) is written in the form of a system of first-order stochastic differential equations for the position $\mathbf{r}$ and the momentum $\mathbf{p}$ of the particle:

$$\begin{cases} \dfrac{d\mathbf{r}}{dt} = \dfrac{\mathbf{p}}{m}, \\ \dfrac{d\mathbf{p}}{dt} = \mathbf{F}(\mathbf{r}) - \gamma\,\dfrac{\mathbf{p}}{m} + \mathbf{F}_{\text{fl}}(t), \end{cases} \tag{4.178}$$

where $\mathbf{F}(\mathbf{r}) = -\nabla U + \mathbf{F}_{\text{g}}$. An analogy can be made with the stochastic differential equation (4.113) associated with the master equation (4.108). In this analogy, the six-dimensional current density has three components for position and three other components for momentum: $\boldsymbol{J} = (\boldsymbol{J}_{\mathbf{r}}, \boldsymbol{J}_{\mathbf{p}})$. The components of position are given in terms of the first line in equation (4.178) according to $\boldsymbol{J}_{\mathbf{r}} = (\mathbf{p}/m)\mathcal{P}$ without diffusive contribution since there is no noise in the equation for $\dot{\mathbf{r}}$. The components of momentum are given by summing the deterministic part of the second line in equation (4.178) with the diffusive contribution due to the fluctuating force. Using the analogy with equation (4.175), the diffusivity of momentum is given by the coefficient playing the same role in the force autocorrelation function (4.174) as diffusion in the velocity autocorrelation function (4.176). Consequently, the momentum components of the current density take the form $\boldsymbol{J}_{\mathbf{p}} = (\mathbf{F} - \gamma\mathbf{p}/m)\mathcal{P} - \gamma\, k_{\text{B}}T\,\partial_{\mathbf{p}}\mathcal{P}$. In the continuity equation (4.177), the divergence is defined in phase space as $\text{div}_{\mathbb{R}^6}\boldsymbol{J} = \partial_{\mathbf{r}} \cdot \boldsymbol{J}_{\mathbf{r}} + \partial_{\mathbf{p}} \cdot \boldsymbol{J}_{\mathbf{p}}$, whereupon we obtain *Kramers' master equation,*

$$\partial_t \mathcal{P} + \frac{\mathbf{p}}{m} \cdot \partial_{\mathbf{r}} \mathcal{P} + \mathbf{F}(\mathbf{r}) \cdot \partial_{\mathbf{p}} \mathcal{P} - \partial_{\mathbf{p}} \cdot \left(\gamma\,\frac{\mathbf{p}}{m}\mathcal{P}\right) = \gamma\, k_{\text{B}}T\, \partial_{\mathbf{p}}^2 \mathcal{P}, \tag{4.179}$$

for the time evolution of the probability density $\mathcal{P}(\mathbf{r},\mathbf{p},t)$ that the colloidal particle is observed with position $\mathbf{r}$ and momentum $\mathbf{p}$ at time t.

If the colloidal particle moves in a vessel with a wall at bottom, Kramers' master equation has the Boltzmann distribution for a stationary equilibrium solution:

$$\mathcal{P}_{\text{eq}}(\mathbf{r},\mathbf{p}) = \mathcal{N}^{-1}\,\text{e}^{-\beta\mathcal{E}} \qquad \text{with} \qquad \mathcal{E} = \frac{\mathbf{p}^2}{2m} + U(\mathbf{r}) + mgx, \tag{4.180}$$

where $\beta = (k_{\text{B}}T)^{-1}$ and $\mathcal{N} = \int \exp(-\beta\mathcal{E})\, d^3r\, d^3p$ is a normalization factor. We note that equation (4.180) gives the equilibrium distribution if reflecting boundary conditions are considered at the bottom of the vessel, as depicted in Figure 4.7(a). In this case, the stochastic process is stationary and reversible.[13]

Furthermore, Kramers' master equation should reduce to the Smoluchowski master equation (4.162) on timescales longer than the thermalization time $t_{\rm r} = m/\gamma$, which is the relaxation time of the velocity distribution towards the Maxwellian equilibrium distribution.[14] In this limit, the solution of Kramers' master equation can be written in the form

$$\mathcal{P}(\mathbf{r},\mathbf{p},t) = \frac{\exp\left(-\frac{\mathbf{p}^2}{2mk_{\rm B}T}\right)}{(2\pi mk_{\rm B}T)^{3/2}} \left[p(\mathbf{r},t) + \beta\,\mathbf{p}\cdot\jmath(\mathbf{r},t)\right] \tag{4.181}$$

with $\beta = (k_{\rm B}T)^{-1}$. Substituting into equation (4.179), we obtain the evolution equation for the probability density $p(\mathbf{r},t)$:

$$\partial_t\, p + \nabla\cdot\jmath = 0. \tag{4.182}$$

Multiplying (4.179) by the momentum $\mathbf{p}$ and integrating with the Maxwellian distribution, we find the evolution equation for the current density $\jmath(\mathbf{r},t)$:

$$m\,\partial_t\,\jmath + \gamma\,\jmath + k_{\rm B}T\,\nabla p - \mathbf{F}\,p = 0, \tag{4.183}$$

which shows that the current density undergoes a fast relaxation towards Maxwell's equilibrium velocity distribution over the timescale $t_{\rm r} = m/\gamma$. In this regard, equation (4.183) can be solved to get

$$\jmath = \frac{\mathbf{F}}{\gamma}\,p - \mathcal{D}\,\nabla p + O\left({\rm e}^{-\gamma t/m}\right) \tag{4.184}$$

with the Einstein relation (4.163). Replacing in equation (4.182), we recover Smoluchowski's master equation:

$$\partial_t\, p + \nabla\cdot\left(\frac{\mathbf{F}}{\gamma}\,p\right) = \mathcal{D}\,\nabla^2 p, \tag{4.185}$$

as required, to be consistent with the overdamped Langevin equation (4.175) for a colloidal particle moving in the force field $\mathbf{F} = -\nabla U + \mathbf{F}_{\rm g}$, since the drift velocity is given by $\mathbf{V} = \mathbf{F}/\gamma$.

4.7.3 Entropy Production of Brownian Motion

If the colloidal particle evolves in a fluid of temperature T contained in a vessel of volume V, its thermodynamic equilibrium macrostate should be described by the canonical statistical ensemble corresponding to the equilibrium probability distribution (4.180). The thermodynamic potential associated with this statistical ensemble is the Helmholtz free energy, which is here given by

[13] Here, reflecting boundary conditions at the bottom wall $x = 0$ of the vessel lead to the condition $(d/dt)\int \mathcal{P}\,d^3r\,d^3p = \int_{x=0}(p_x/m)\mathcal{P}\,dy\,dz\,d^3p = 0$.

[14] For micrometric silica particles in water, this thermalization time is of the order of $t_{\rm r} \sim 10^{-6}$ s, which is thus very short with respect to the observational timescale typically ranging from milliseconds to seconds.

$$F = E - TS = k_{\rm B}T \int \mathcal{P} \ln \frac{\mathcal{P}}{\mathcal{P}_{\rm eq}} \, d^3r \, d^3p \tag{4.186}$$

in terms of the equilibrium distribution (4.180), up to a constant contribution of the fluid itself. Indeed, substituting equation (4.180) into equation (4.186), we can identify the energy as $E = \int \mathcal{E}\mathcal{P}\, d^3r\, d^3p$ and the entropy as $S = -k_{\rm B} \int \mathcal{P} \ln(\mathcal{N}\mathcal{P})\, d^3r\, d^3p$. At equilibrium, the Helmholtz free energy is supposed to reach a minimum. Now, the total free energy can be expressed in terms of the local free-energy density

$$f = k_{\rm B}T \int \mathcal{P} \ln \frac{\mathcal{P}}{\mathcal{P}_{\rm eq}} \, d^3p, \qquad \text{such that} \qquad F = \int_V f \, d^3r. \tag{4.187}$$

Its time evolution under the Kramers master equation (4.179) has the form

$$\partial_t f + \nabla \cdot \boldsymbol{J}_f = \sigma_f \tag{4.188}$$

with the free-energy current density

$$\boldsymbol{J}_f = k_{\rm B}T \int \frac{\mathbf{p}}{m} \mathcal{P} \ln \frac{\mathcal{P}}{\mathcal{P}_{\rm eq}} \, d^3p, \tag{4.189}$$

and the free-energy production rate density

$$\sigma_f = -\gamma\,(k_{\rm B}T)^2 \int \frac{1}{\mathcal{P}} \left(\partial_{\mathbf{p}}\mathcal{P} + \frac{\mathbf{p}}{mk_{\rm B}T}\mathcal{P} \right)^2 d^3p \le 0, \tag{4.190}$$

which is always nonpositive. This result implies that $dF/dt = d_{\rm e}F/dt + d_{\rm i}F/dt$ with the free-energy exchange rate $d_{\rm e}F/dt = -\int_{\partial V} \boldsymbol{J}_f \cdot d\boldsymbol{\Sigma}$ and the free-energy production rate $d_{\rm i}F/dt = \int_V \sigma_f \, d^3r \le 0$. Since the energy E is conserved according to the first law of thermodynamics (1.2) and the system is isothermal, we have that $d_{\rm i}F/dt = -T d_{\rm i}S/dt$, so that the entropy production rate is here given by

$$\frac{d_{\rm i}S}{dt} = \gamma\, k_{\rm B}^2 T \int \frac{1}{\mathcal{P}} \left(\partial_{\mathbf{p}}\mathcal{P} + \frac{\mathbf{p}}{mk_{\rm B}T}\mathcal{P} \right)^2 d^3r \, d^3p \ge 0, \tag{4.191}$$

which is nonnegative, in agreement with the second law of thermodynamics. Equilibrium is reached and the entropy production rate is equal to zero when the velocity probability distribution becomes the Maxwellian equilibrium distribution.

During relaxation towards equilibrium, the probability distribution is given by equation (4.181) with the current density (4.184), so that the entropy production rate (4.191) becomes

$$\frac{d_{\rm i}S}{dt} \simeq k_{\rm B}\,\mathcal{D} \int \frac{1}{p} \left(\nabla p - \frac{\mathbf{F}}{k_{\rm B}T}\, p \right)^2 d^3r \ge 0, \tag{4.192}$$

as expected for the diffusion of the colloidal particle in the force field $\mathbf{F}$. If this latter is uniform in space, so that the Brownian particle is drifting at the mean velocity $\mathbf{V} = \mathbf{F}/\gamma$, the entropy production rate in the long-time limit approaches the value

$$\lim_{t\to\infty} \frac{d_{\rm i}S}{dt} = \frac{1}{T}\,\mathbf{F}\cdot\mathbf{V} = \frac{\gamma}{T}\,\mathbf{V}^2 \ge 0 \tag{4.193}$$

given by the dissipated mechanical power divided by the temperature T. Therefore, Brownian motion is out of equilibrium if $\mathbf{F} = \gamma\mathbf{V} \neq 0$.

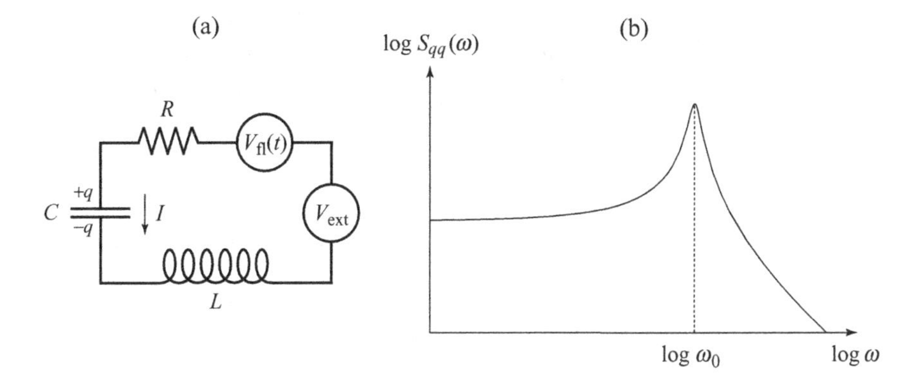

Figure 4.8 (a) Noisy LRC electric circuit with an inductor of inductance L, a capacitor of capacitance C, a resistor of resistance R with its associated Johnson–Nyquist noise $V_{\text{fl}}(t)$, and an external electromotive force V_{ext} connected in series. (b) The spectral density function of charge fluctuations in this electric circuit with $L = C = 1$ and $R = 0.1$.

4.7.4 Noisy Electric Circuits

Langevin stochastic processes also describe noisy electric circuits. This is, in particular, the case for the LRC electric circuit shown in Figure 4.8 where the external electromotive force is connected in series with an inductor, a capacitor, and a resistor with its associated fluctuating voltage due to the Johnson–Nyquist electric noise of thermal origin (Johnson, 1928; Nyquist, 1928).

Using Kirchhoff's voltage law, as presented in Section 1.7.4, we obtain the following stochastic differential equation for the electric charge q in the capacitor,

$$L\,\frac{d^2q}{dt^2} + R\,\frac{dq}{dt} + \frac{q}{C} = V_{\text{ext}} - V_{\text{fl}}(t), \tag{4.194}$$

where $V_{\text{fl}}(t)$ is the Johnson–Nyquist noise associated with dissipation in the resistor of resistance R. This noise is of thermal origin, as for the Langevin fluctuating force. In order to determine the characteristic properties of this electric noise, an analogy can thus be established with a one-dimensional damped mechanical oscillator of mass m, spring constant k, and friction coefficient γ subjected to a Langevin fluctuating force $F_{\text{fl}}(t)$ and an external force F_{ext}:

$$m\,\frac{d^2x}{dt^2} + \gamma\,\frac{dx}{dt} + kx = F_{\text{ext}} + F_{\text{fl}}(t). \tag{4.195}$$

In this case, the fluctuating force is a Gaussian white noise characterized by equation (4.174) for $i = j = x$. Because of the analogy given in Table 4.1 between the electrical and mechanical systems, the fluctuating voltage associated with the resistance R at the temperature T is therefore a Gaussian white noise characterized by

$$\langle V_{\text{fl}}(t)\rangle = 0 \qquad \text{and} \qquad \langle V_{\text{fl}}(t)\,V_{\text{fl}}(t')\rangle = 2\,R\,k_{\text{B}}T\,\delta(t-t'). \tag{4.196}$$

The fluctuation–dissipation theorem is thus satisfied by the Johnson–Nyquist electric noise.

Table 4.1. *Table of correspondence between the quantities of the noisy LRC electric circuit of charge q and current intensity I, and the Langevin mechanical process for a one-dimensional damped mechanical oscillator of position x and velocity v.*

Mechanical system	x	v	m	k	γ	$F_{\rm fl}(t)$	$F_{\rm ext}$
Electrical system	q	I	L	$1/C$	R	$-V_{\rm fl}(t)$	$V_{\rm ext}$

Since the electric current is given by $I = dq/dt$ (which is the analogue of the mechanical velocity $v = dx/dt$), the master equation ruling the time evolution of the probability density $\mathcal{P}_t(q, I)$ reads

$$\frac{\partial \mathcal{P}}{\partial t} + I\,\frac{\partial \mathcal{P}}{\partial q} - \frac{q - C V_{\rm ext}}{LC}\,\frac{\partial \mathcal{P}}{\partial I} - \frac{R}{L}\,\frac{\partial\,(I\mathcal{P})}{\partial I} = \frac{R k_{\rm B} T}{L^2}\,\frac{\partial^2 \mathcal{P}}{\partial I^2} \tag{4.197}$$

in analogy with the Kramers master equation (4.179). Accordingly, the equilibrium distribution is given by the stationary solution,

$$\mathcal{P}_{\rm eq}(q, I) \sim \exp(-\beta\mathcal{E}) \qquad \text{with} \qquad \mathcal{E} = \frac{1}{2}\,LI^2 + \frac{q^2}{2C} - V_{\rm ext} q. \tag{4.198}$$

The thermal mean values and variances of the electric charge and current are thus given by

$$\langle q\rangle_{\rm eq} = C V_{\rm ext}, \quad \langle q^2\rangle_{\rm eq} = C k_{\rm B} T, \quad \langle I\rangle_{\rm eq} = 0, \quad \text{and} \quad \langle I^2\rangle_{\rm eq} = \frac{k_{\rm B} T}{L}. \tag{4.199}$$

The mean electric current is equal to zero at equilibrium. Here, the external voltage $V_{\rm ext}$ has the sole effect of charging the capacitance to the mean value $\langle q\rangle_{\rm eq}$. The fluctuations are thermal since they are proportional to the temperature T.

The spectral analysis of the fluctuations can be performed by using the Fourier transform

$$\tilde{A}(\omega) = \int_{-\infty}^{+\infty} {\rm e}^{-\imath\omega t} A(t)\, dt = \tilde{A}^*(-\omega), \tag{4.200}$$

for arbitrary time-dependent quantity $A(t)$. In particular, taking the Fourier transform of equation (4.196), we obtain

$$\left\langle \tilde{V}_{\rm fl}(\omega)\, \tilde{V}_{\rm fl}(\omega')\right\rangle = 4\pi\, R\, k_{\rm B} T\, \delta(\omega + \omega'). \tag{4.201}$$

As a consequence, the spectral density (2.178) of the fluctuating voltage is given by

$$\mathcal{S}_{VV}(\omega) = 2\, R\, k_{\rm B} T, \tag{4.202}$$

which takes the same value for all the frequencies ω, so that this noise is said to be white. Moreover, the Fourier transform of the stochastic differential equation (4.194) in the absence of external voltage gives $\tilde{q}(\omega) = \tilde{V}_{\rm fl}(\omega)(L\omega^2 - C^{-1} - \imath R\omega)^{-1}$, so that the spectral density of the electric charge fluctuations takes the form

$$\mathcal{S}_{qq}(\omega) = \frac{2\, R\, k_{\rm B} T}{(L\omega^2 - C^{-1})^2 + R^2\omega^2}. \tag{4.203}$$

As seen in Figure 4.8(b), the spectral density of the noisy LRC circuit may depict a resonance at the frequency $\omega_0 \simeq 1/\sqrt{LC}$, in the case where the oscillator is underdamped because the damping time $\tau_{\rm d} \simeq L/R$ is longer than its period $\mathscr{T} = 2\pi/\omega_0$. In contrast, there is no resonance for the spectral density of a noisy RC circuit, which is overdamped with the relaxation time $\tau_{\rm r} \simeq 1/(RC)$. In the noisy RC circuit, as well as in the overdamped mechanical oscillator, the Langevin processes defined by equations (4.194) with $L = 0$ and (4.195) with $m = 0$ are known as Ornstein–Uhlenbeck stochastic processes (see Section E.7).

By analogy with the corresponding mechanical system, the series LRC circuit of Figure 4.8 undergoes relaxation towards equilibrium after some time, so that the entropy production rate converges to zero when equilibrium is reached. We note that, without the capacitance C, the series LR circuit with an applied external voltage would remain out of equilibrium with the nonzero stationary value of the electric current $\langle I\rangle_{\rm st} = V_{\rm ext}/R$, giving the positive entropy production rate $d_{\rm i}S/dt = V_{\rm ext}\langle I\rangle_{\rm st}/T$. In the analogous mechanical system, the particle would thus perform a drift with the mean velocity $\langle dx/dt\rangle_{\rm st} = F_{\rm ext}/\gamma$ since there is no spring to maintain its position around some mean value. If such conditions hold, both systems are thus evolving away from equilibrium with positive entropy production rate.

Such considerations extend to general noisy LRC electric circuits (Freitas et al., 2020).

4.8 Friction in Systems with Slow and Fast Degrees of Freedom

Now, the issue is to understand how the theory of stochastic processes and, in particular, the Markovian memory loss can be justified from the viewpoint of statistical mechanics. The key point is that the dynamics of systems described as Markovian stochastic processes often present slow and fast degrees of freedom evolving over different characteristic timescales separated by several orders of magnitude. Even if the time evolution of the whole system is deterministic, the motion of the slow degrees of freedom appears stochastic on timescales longer than the movements of the fast degrees of freedom.

This is, in particular, the case for the Brownian motion of a colloidal particle in a fluid since the colloidal particle is much heavier than the fluid atoms, so that their root-mean-square velocities at thermal equilibrium are in the inverse ratio of their mass, $\sqrt{\langle \mathbf{v}_a^2\rangle_{\rm eq}/\langle \mathbf{v}_b^2\rangle_{\rm eq}} = \sqrt{m_b/m_a}$, which is of the order of 10^5 to 10^6. Such large separations of timescales is also the feature of macroscopic pistons exerting the compression or dilation of gases.

In such systems, the fast degrees of freedom rapidly reach quasi equilibrium although the slow degrees of freedom continue to move subjected to the forces of interaction with the fast degrees of freedom. These forces exchange energy between the degrees of freedom, including the forces of friction dissipating energy from the slow to the fast degrees of freedom and the fluctuating forces due to the backaction of the fast degrees of freedom onto the slow ones.

Such processes have been studied in different contexts using the methods of statistical mechanics since early work on Brownian motion. These studies have led to the microscopic understanding of friction in the linear regime from small to large systems, thermalization for slow degrees of freedom in contact with faster ones, and dissipative deviations from adiabaticity (Kirkwood, 1946; Brown et al., 1987; Berry and Robbins, 1993; Jarzynski, 1992, 1993, 1995; Servantie and Gaspard, 2003).

Such problems can be formulated in general as follows.

4.8.1 General Formulation

The microscopic motion of such systems is supposed to be governed by Hamiltonian mechanics, as presented in Chapter 2. The degrees of freedom of the whole system belong to either the slow subsystem or the fast heat bath. Their positions and momenta are thus partitioned according to $\mathbf{\Gamma} = (\mathbf{\Gamma}_{\rm s}, \mathbf{\Gamma}_{\rm b})$. The Hamiltonian function has the form

$$H(\mathbf{\Gamma}_{\rm s}, \mathbf{\Gamma}_{\rm b}) = H_{\rm s}(\mathbf{\Gamma}_{\rm s}) + H_{\rm b}(\mathbf{\Gamma}_{\rm b}) + U_{\rm sb}(\mathbf{\Gamma}_{\rm s}, \mathbf{\Gamma}_{\rm b}), \tag{4.204}$$

where $H_{\rm s}$ is the Hamiltonian function of the slow subsystem, $H_{\rm b}$ the one of the fast bath, and $U_{\rm sb}$ the potential energy of interaction between the slow and fast parts.

The time evolution of the probability density $p_t(\mathbf{\Gamma}_{\rm s}, \mathbf{\Gamma}_{\rm b})$ of the whole system in phase space is ruled by the Liouville equation (2.39), which here reads

$$\partial_t p = \hat{L}_{\rm s} p + \hat{L}_{\rm b} p \tag{4.205}$$

with the Liouvillian operators of the slow subsystem and the bath defined as

$$\hat{L}_{\rm s} p \equiv \{H, p\}_{\rm s} = \{H_{\rm s}, p\}_{\rm s} + \{U_{\rm sb}, p\}_{\rm s}, \tag{4.206}$$

$$\hat{L}_{\rm b} p \equiv \{H, p\}_{\rm b} = \{H_{\rm b}, p\}_{\rm b} + \{U_{\rm sb}, p\}_{\rm b}, \tag{4.207}$$

where $\{\cdot,\cdot\}_{\rm s}$ denotes the Poisson bracket for the slow degrees of freedom and $\{\cdot,\cdot\}_{\rm b}$ the Poisson bracket for the fast ones. We also introduce the Liouvillian operators $\hat{L}_{\rm s0} p \equiv \{H_{\rm s}, p\}_{\rm s}$ for the isolated slow subsystem and $\hat{L}_{\rm sb} p \equiv \{U_{\rm sb}, p\}_{\rm s}$ for its interaction with the bath. An important remark is that the Liouvillian operator (4.207) rules the bath dynamics while keeping fixed the variables of the slow subsystem. Moreover, since the interaction potential $U_{\rm sb}$ is often assumed to depend only on the position variables, the operator (4.207) governs the fast dynamics of the bath interacting with a frozen configuration of the subsystem.

Now, the goal is to deduce the evolution equation for the marginal probability density,

$$\mathcal{P}_t(\mathbf{\Gamma}_{\rm s}) \equiv \int p_t(\mathbf{\Gamma}_{\rm s}, \mathbf{\Gamma}_{\rm b})\, d\mathbf{\Gamma}_{\rm b}, \tag{4.208}$$

of the slow subsystem from the Liouville equation (4.205). An important property of the Poisson bracket is that it is equal to zero after integration on all the corresponding phase space variables, so that

$$\int \hat{L}_{\rm b} p\, d\mathbf{\Gamma}_{\rm b} = \int \{H, p\}_{\rm b}\, d\mathbf{\Gamma}_{\rm b} = 0. \tag{4.209}$$

Therefore, integrating Liouville's equation over the bath variables gives the evolution equation

$$\partial_t \mathcal{P}_t = \int \hat{L}_{\rm s} p_t\, d\mathbf{\Gamma}_{\rm b} = \hat{L}_{\rm s0} \mathcal{P}_t + \int \hat{L}_{\rm sb} p_t\, d\mathbf{\Gamma}_{\rm b}, \tag{4.210}$$

or, equivalently,

$$\partial_t \mathcal{P}_t = \{H_s, \mathcal{P}_t\}_s + \int \{U_{sb}, p_t\}_s \, d\boldsymbol{\Gamma}_b. \tag{4.211}$$

Furthermore, equation (4.205) can be integrated in time by setting $p_t = \exp\left(\hat{L}_b t\right) X_t$ and solving for X_t to get

$$p_t = e^{\hat{L}_b t} p_0 + \int_0^t d\tau \, e^{\hat{L}_b \tau} \hat{L}_s p_{t-\tau}, \tag{4.212}$$

thus giving the probability density at time t in terms of its initial condition p_0 and its past time evolution $p_{t-\tau}$ delayed by the time $0 < \tau < t$. If the time evolution of the bath is fast enough, this latter contribution should be determined by its fast dynamics over its short characteristic timescale t_b. Substituting equation (4.212) back into equation (4.210), we find

$$\partial_t \mathcal{P}_t = \hat{L}_{s0} \mathcal{P}_t + \int \hat{L}_{sb} \, e^{\hat{L}_b t} p_0 \, d\boldsymbol{\Gamma}_b + \int_0^t d\tau \int \hat{L}_{sb} \, e^{\hat{L}_b \tau} \hat{L}_s p_{t-\tau} \, d\boldsymbol{\Gamma}_b, \tag{4.213}$$

which is reminiscent of equation (2.188) since the full description is now projected onto the slow subsystem by equation (4.208). In the absence of interaction between the slow subsystem and the bath ($\hat{L}_{sb} = 0$), the evolution equation (4.213) would reduce to the first term in its right-hand side, which is simply Liouville's equation $\partial_t \mathcal{P}_t = \hat{L}_{s0} \mathcal{P}_t = \{H_s, \mathcal{P}_t\}_s$ for the isolated subsystem of the Hamiltonian function H_s. The second term has the memory of the initial condition p_0, while the third term keeps the memory of the past over the variable time delay $0 < \tau < t$.

Since the bath is supposed to be fast, the time evolution operator $\exp(\hat{L}_b t)$ will generate the relaxation towards the quasi equilibrium probability distribution of the bath in the fixed configuration of the subsystem over a timescale longer than the bath characteristic time t_b. Accordingly, the memory of the initial condition may be expected to be lost for $t \gg t_b$ in the second term of equation (4.213), while the third term should continue to exert its effect for $t \gg t_b$.

The problem is now to obtain a closed equation for the probability density $\mathcal{P}_t$, which should be comparable with the Kramers master equation (4.179) in the case of Brownian motion. For this purpose, we note that we may define the conditional probability density of the bath with respect to the subsystem according to

$$\mathscr{P}_t(\boldsymbol{\Gamma}_b | \boldsymbol{\Gamma}_s) \equiv \frac{p_t(\boldsymbol{\Gamma}_s, \boldsymbol{\Gamma}_b)}{\mathcal{P}_t(\boldsymbol{\Gamma}_s)}, \tag{4.214}$$

using the marginal probability density (4.208). We should thus expect that, over a timescale longer than the bath characteristic time, the probability density of the whole system would evolve as

$$p_t(\boldsymbol{\Gamma}_s, \boldsymbol{\Gamma}_b) \to_{t \gg t_b} \mathcal{P}_t(\boldsymbol{\Gamma}_s) \, \mathscr{P}_{eq}(\boldsymbol{\Gamma}_b | \boldsymbol{\Gamma}_s), \tag{4.215}$$

towards the time-dependent marginal probability density of the subsystem multiplied by the quasi equilibrium conditional probability density of the bath. This latter is the stationary probability density for the Liouvillian operator (4.207), such that $\hat{L}_{\rm b}\mathscr{P}_{\rm eq} = 0$, or, equivalently, $\exp(\hat{L}_{\rm b}t)\mathscr{P}_{\rm eq} = \mathscr{P}_{\rm eq}$.[15]

4.8.2 The Case of Brownian Motion

Let us now be specific and consider the case of Brownian motion for a spherical colloidal particle in a fluid far from the walls of the vessel. The issue is to deduce the Kramers master equation (4.179) from the underlying Hamiltonian microscopic dynamics using the methods of statistical mechanics (Lebowitz and Rubin, 1963; Résibois and Lebowitz, 1965; Balescu, 1975; Résibois and De Leener, 1977). The phase-space variables of the system are the position and momentum of the colloidal particle $\mathbf{\Gamma}_{\rm s} = (\mathbf{r}, \mathbf{p}) \in \mathbb{R}^6$ and those of the N atoms composing the fluid $\mathbf{\Gamma}_{\rm b} = (\mathbf{r}_a, \mathbf{p}_a)_{a=1}^N \in \mathbb{R}^{6N}$. The Hamiltonian functions in equation (4.204) are here taken as

$$H_{\rm s} = \frac{\mathbf{p}^2}{2m} + U(\mathbf{r}), \qquad U_{\rm sb} = \sum_{a=1}^{N} u(\|\mathbf{r} - \mathbf{r}_a\|), \tag{4.216}$$

and $H_{\rm b}$ of the form (2.2). The colloidal particle is much heavier than the atoms, $m \gg m_a$ for $a = 1, 2, \ldots, N$, so that its motion is significantly slower than for the atoms.

The Liouvillian operators are thus given by

$$\hat{L}_{\rm s0}p = \{H_{\rm s}, p\}_{\rm s} = -\frac{\mathbf{p}}{m} \cdot \partial_{\mathbf{r}} p + \partial_{\mathbf{r}} U \cdot \partial_{\mathbf{p}} p, \tag{4.217}$$

$$\hat{L}_{\rm sb}p = \{U_{\rm sb}, p\}_{\rm s} = -\mathbf{F}_{\rm sb} \cdot \partial_{\mathbf{p}} p, \tag{4.218}$$

$$\hat{L}_{\rm b}p = \{H_{\rm b}, p\}_{\rm b} + \{U_{\rm sb}, p\}_{\rm b} = \{H_{\rm b}, p\}_{\rm b} + \sum_{a=1}^{N} \mathbf{f}_a \cdot \partial_{\mathbf{p}_a} p, \tag{4.219}$$

where $\mathbf{F}_{\rm sb} = -\partial_{\mathbf{r}} U_{\rm sb} = \sum_{a=1}^{N} \mathbf{f}_a$ is the sum of the forces exerted by the atoms of the fluid on the colloidal particle, $\mathbf{f}_a = -\partial_{\mathbf{r}} u(\|\mathbf{r} - \mathbf{r}_a\|)$. We note that every atom exerts on the colloidal particle the force $-\partial_{\mathbf{r}_a} u(\|\mathbf{r} - \mathbf{r}_a\|) = -\mathbf{f}_a$, which is opposite, in agreement with the principle of action-reaction. As a consequence, the evolution equation (4.211) for the marginal probability density $\mathcal{P}_t(\mathbf{r}, \mathbf{p})$ of the colloidal particle can be written as

$$\partial_t \mathcal{P} = -\frac{\mathbf{p}}{m} \cdot \partial_{\mathbf{r}} \mathcal{P} + \partial_{\mathbf{r}} U \cdot \partial_{\mathbf{p}} \mathcal{P} - \partial_{\mathbf{p}} \cdot \mathcal{J}_{\mathbf{p}} \quad \text{with} \quad \mathcal{J}_{\mathbf{p}} = \int \mathbf{F}_{\rm sb}\, p_t \, d\mathbf{\Gamma}_{\rm b}. \tag{4.220}$$

We see that the evolution equation (4.220) has the same form as the master equation (4.177) with the current density in position space given by $\boldsymbol{J}_{\mathbf{r}} = (\mathbf{p}/m)\mathcal{P}$ and the one in momentum space by $\boldsymbol{J}_{\mathbf{p}} = -\partial_{\mathbf{r}} U \, \mathcal{P} + \mathcal{J}_{\mathbf{p}}$, which is already reminiscent of the Kramers master equation (4.179).

[15] As discussed in the literature (Zwanzig, 1961; Mazo, 2002), the Zwanzig projection-operator method of Section 2.9.1 can here be applied by considering the projection operator $\hat{P} p_t(\mathbf{\Gamma}_{\rm s}, \mathbf{\Gamma}_{\rm b}) = \mathscr{P}_{\rm eq}(\mathbf{\Gamma}_{\rm b}|\mathbf{\Gamma}_{\rm s}) \int p_t(\mathbf{\Gamma}_{\rm s}, \mathbf{\Gamma}'_{\rm b})\, d\mathbf{\Gamma}'_{\rm b}$ in terms of the equilibrium conditional probability density $\mathscr{P}_{\rm eq}$ introduced in equation (4.215). This operator is a projector because it satisfies the condition $\hat{P}^2 = \hat{P}$.

Substituting equation (4.212) into the expression for $\mathcal{J}_{\mathbf{p}}$ in equation (4.220), we find

$$\mathcal{J}_{\mathbf{p}} = \mathcal{J}_{\mathbf{p}0} + \int_0^t d\tau \int \mathbf{F}_{\text{sb}}\, \mathrm{e}^{\hat{L}_{\text{b}}\tau} \hat{L}_{\text{s}} p_{t-\tau}\, d\mathbf{\Gamma}_{\text{b}}, \qquad \text{where} \qquad \mathcal{J}_{\mathbf{p}0} = \int \mathbf{F}_{\text{sb}}\, \mathrm{e}^{\hat{L}_{\text{b}}t} p_0\, d\mathbf{\Gamma}_{\text{b}} \tag{4.221}$$

is the contribution of the initial conditions. Since the bath ruled by the Liouvillian operator $\hat{L}_{\text{b}}$ is fast, we should expect the relaxation of the bath towards quasi equilibrium over a timescale longer than the short characteristic time t_{b} of the bath as in equation (4.215):

$$\mathrm{e}^{\hat{L}_{\text{b}}t} p_0 \to_{t \gg t_{\text{b}}} \mathcal{P}_0(\mathbf{r}, \mathbf{p})\, \mathscr{P}_{\text{eq}}(\mathbf{\Gamma}_{\text{b}}|\mathbf{r}) \qquad \text{with} \qquad \mathcal{P}_0(\mathbf{r}, \mathbf{p}) = \int p_0\, d\mathbf{\Gamma}_{\text{b}} \tag{4.222}$$

and the canonical equilibrium distribution of the bath at the inverse temperature $\beta = (k_{\text{B}}T)^{-1}$ with the frozen position $\mathbf{r}$ for the colloidal particle,

$$\mathscr{P}_{\text{eq}}(\mathbf{\Gamma}_{\text{b}}|\mathbf{r}) \equiv \frac{\mathrm{e}^{-\beta(H_{\text{b}}+U_{\text{sb}})}}{\int \mathrm{e}^{-\beta(H_{\text{b}}+U_{\text{sb}})}\, d\mathbf{\Gamma}'_{\text{b}}}. \tag{4.223}$$

For $t \gg t_{\text{b}}$, the contribution of initial conditions in equation (4.221) should thus evolve as $\mathcal{J}_{\mathbf{p}0} \to \langle \mathbf{F}_{\text{sb}} \rangle_{\mathbf{r},\text{eq}} \mathcal{P}_0(\mathbf{r}, \mathbf{p})$, where $\langle \mathbf{F}_{\text{sb}} \rangle_{\mathbf{r},\text{eq}} = \int \mathbf{F}_{\text{sb}} \mathscr{P}_{\text{eq}}(\mathbf{\Gamma}_{\text{b}}|\mathbf{r})\, d\mathbf{\Gamma}_{\text{b}}$ is the equilibrium mean value of the force exerted by the bath on the colloidal particle fixed in the position $\mathbf{r}$. However, this force is given by $\mathbf{F}_{\text{sb}} = -\partial_{\mathbf{r}} U_{\text{sb}}$ and the fluid is uniform at equilibrium far from the walls of the vessel, so that $\int U_{\text{sb}} \exp[-\beta(H_{\text{b}} + U_{\text{sb}})]\, d\mathbf{\Gamma}_{\text{b}}$ does not depend on the position $\mathbf{r}$ of the colloidal particle. Consequently, the mean value of the force is equal to zero, i.e., $\langle \mathbf{F}_{\text{sb}} \rangle_{\mathbf{r},\text{eq}} = 0$ and $\mathcal{J}_{\mathbf{p}0} = 0$ for $t \gg t_{\text{b}}$ in equation (4.221).

Next, the second contribution to equation (4.221) can be calculated as follows. Using the expressions (4.217) and (4.218) for the Liouvillian operator $\hat{L}_{\text{s}} = \hat{L}_{\text{s}0} + \hat{L}_{\text{sb}}$, equation (4.221) can be written as

$$\mathcal{J}_{\mathbf{p}} = \int_0^t d\tau \int d\mathbf{\Gamma}_{\text{b}}\, \mathbf{F}_{\text{sb}}\, \mathrm{e}^{\hat{L}_{\text{b}}\tau} \left(-\frac{\mathbf{p}}{m} \cdot \partial_{\mathbf{r}} + \partial_{\mathbf{r}} U \cdot \partial_{\mathbf{p}} - \mathbf{F}_{\text{sb}} \cdot \partial_{\mathbf{p}} \right) p_{t-\tau}. \tag{4.224}$$

Again, because of the fast relaxation of the bath towards the quasi equilibrium distribution (4.223) with respect to the slow motion of the heavy colloidal particle, we may assume that

$$p_{t-\tau}(\mathbf{r}, \mathbf{p}, \mathbf{\Gamma}_{\text{b}}) \simeq \mathcal{P}_{t-\tau}(\mathbf{r}, \mathbf{p})\, \mathscr{P}_{\text{eq}}(\mathbf{\Gamma}_{\text{b}}|\mathbf{r}). \tag{4.225}$$

Since $\mathscr{P}_{\text{eq}}$ does not depend on the momentum $\mathbf{p}$ of the colloidal particle $\partial_{\mathbf{p}} \mathscr{P}_{\text{eq}} = 0$, equation (4.224) becomes

$$\mathcal{J}_{\mathbf{p}} \simeq \int_0^t d\tau \int d\mathbf{\Gamma}_{\text{b}}\, \mathbf{F}_{\text{sb}}\, \mathrm{e}^{\hat{L}_{\text{b}}\tau} \times \left[\mathscr{P}_{\text{eq}} \left(-\frac{\mathbf{p}}{m} \cdot \partial_{\mathbf{r}} + \partial_{\mathbf{r}} U \cdot \partial_{\mathbf{p}} - \mathbf{F}_{\text{sb}} \cdot \partial_{\mathbf{p}} \right) - \frac{\mathbf{p}}{m} \cdot \partial_{\mathbf{r}} \mathscr{P}_{\text{eq}} \right] \mathcal{P}_{t-\tau}. \tag{4.226}$$

Again, by the translational invariance of the system far from the walls of the vessel, the normalization factor of the quasi equilibrium distribution (4.223) does not depend on the position $\mathbf{r}$, so that we have that

$$\partial_{\mathbf{r}} \mathscr{P}_{\text{eq}} = -\beta\, \partial_{\mathbf{r}} U_{\text{sb}}\, \mathscr{P}_{\text{eq}} = \beta\, \mathbf{F}_{\text{sb}}\, \mathscr{P}_{\text{eq}}. \tag{4.227}$$

Substituting this result into equation (4.226), using the time invariance $\exp\left(\hat{L}_b t\right)\mathscr{P}_{eq} = \mathscr{P}_{eq}$, and, again, the property $\langle \mathbf{F}_{sb}\rangle_{\mathbf{r},eq} = 0$, we obtain

$$\mathcal{J}_{\mathbf{p}} \simeq -\int_0^t d\tau\, \langle \mathbf{F}_{sb}(0)\,\mathbf{F}_{sb}(\tau)\rangle_{\mathbf{r},eq} \cdot \left(\partial_{\mathbf{p}} + \beta\,\frac{\mathbf{p}}{m}\right)\mathcal{P}_{t-\tau} \tag{4.228}$$

in terms of the force autocorrelation function

$$\langle \mathbf{F}_{sb}(0)\,\mathbf{F}_{sb}(\tau)\rangle_{\mathbf{r},eq} \equiv \int \mathscr{P}_{eq}\,\mathbf{F}_{sb}\,e^{\hat{L}_b\tau}\,\mathbf{F}_{sb}\,d\boldsymbol{\Gamma}_b, \tag{4.229}$$

which tends to zero for $\tau \gg t_b$. In this regard, the characteristic time t_b of the bath may thus be identified as the correlation time t_c introduced in Figure 4.7(c). This characteristic time depends on the bath dynamics. In a rarefied gas where the mean free path of the molecules is larger than the particle radius R, the characteristic time is of the order of the collision time with the colloidal particle, which is very short, so that the force autocorrelation function is essentially delta correlated, as discussed in Section E.12. However, in a fluid such as water, the mean free path of molecules is smaller than the radius R of the colloidal particle and the hydrodynamics of the fluid plays an important role. In particular, the force autocorrelation function is known to manifest long-time tails decreasing as $t^{-3/2}$ over the hydrodynamic timescale $t_{hydro} \sim R^2/\nu$, where $\nu = \eta/\rho_{fluid}$ is the kinematic viscosity of the fluid (Kheifets et al., 2014). The characteristic time of the bath is thus of the order $t_b \sim t_{hydro}$ under such circumstances. If the expression (4.228) was substituted into the evolution equation (4.220), we would obtain the master equation,

$$\partial_t \mathcal{P}_t = -\frac{\mathbf{p}}{m}\cdot\partial_{\mathbf{r}}\mathcal{P}_t + \partial_{\mathbf{p}}\cdot\left[\partial_{\mathbf{r}}U\,\mathcal{P}_t + \int_0^t d\tau\,\mathbf{K}(\tau)\cdot\left(\frac{\mathbf{p}}{m} + k_B T\,\partial_{\mathbf{p}}\right)\mathcal{P}_{t-\tau}\right] \tag{4.230}$$

with the memory kernel $\mathbf{K}(\tau) = \beta\langle \mathbf{F}_{sb}(0)\,\mathbf{F}_{sb}(\tau)\rangle_{\mathbf{r},eq}$ given by the force autocorrelation function (4.229) (Mazo, 2002). If this was the case, there would be a memory of the past over the characteristic time t_b of the bath dynamics and the stochastic process would be non-Markovian. Yet these hydrodynamic memory effects are negligible if the mass of the colloidal particle is larger than the mass of the fluid displaced by the particle, so that $m \gg 4\pi R^3\rho_{fluid}/3$ (as will be explained in Section 10.7).

If the colloidal particle is heavy enough, its characteristic timescale t_s can be significantly longer than the bath characteristic time t_b. For instance, if the colloidal particle moved in a harmonic trap of spring constant k, the energy potential would be given by $U(\mathbf{r}) = k\mathbf{r}^2/2$, so that the characteristic time of the particle would be of the order of the period of oscillation, $t_s \sim 2\pi\sqrt{m/k}$, which is typically longer than the bath characteristic time for micrometric particles, $t_s \gg t_b$. There is thus the separation of timescales required to justify that the force autocorrelation function has a fast decay as supposed in Figure 4.7(c). Since the probability density $\mathcal{P}_{t-\tau}$ of the colloidal particle has a slow evolution over the timescale $t \gg t_b$, it may be approximated by the density $\mathcal{P}_t$ at the present time in equation (4.228), which thus becomes

$$\mathcal{J}_{\mathbf{p}} \simeq -\gamma\,k_B T\left(\partial_{\mathbf{p}} + \beta\,\frac{\mathbf{p}}{m}\right)\mathcal{P}_t \tag{4.231}$$

with the friction coefficient

$$\gamma = \frac{1}{3k_{\rm B}T}\int_0^{\infty} \langle \mathbf{F}_{\rm sb}(0)\cdot\mathbf{F}_{\rm sb}(t)\rangle_{\mathbf{r},\,{\rm eq}}\, dt \qquad (4.232)$$

if the fluid is isotropic. This calculation shows that the friction coefficient is given by Kirkwood's formula in terms of the autocorrelation function of the microscopic force exerted by the fluid atoms on the colloidal particle kept in the fixed position $\mathbf{r}$, as can be evaluated using simulations by molecular dynamics based on classical Hamiltonian mechanics. The force $\mathbf{F}_{\rm sb}$ is here defined at the microscopic level of description in contrast to the Langevin fluctuating force $\mathbf{F}_{\rm fl}$ in equation (4.172), which was defined by equation (4.174) in terms of Gaussian white noises at the mesoscopic level of description in the framework of the theory of stochastic processes. Otherwise, both Kirkwood formulas (4.172) and (4.232) are equivalent because $\int_{-\infty}^{+\infty} C(t)\, dt = 2\int_0^{\infty} C(t)\, dt$, since the force autocorrelation function is even under time reversal, i.e., $C(t) = C(-t)$.

Substituting the result (4.231) into the evolution equation (4.220), we find

$$\partial_t \mathcal{P} = -\frac{\mathbf{p}}{m}\cdot\partial_{\mathbf{r}}\mathcal{P} + \partial_{\mathbf{r}}U\cdot\partial_{\mathbf{p}}\mathcal{P} + \partial_{\mathbf{p}}\cdot\left(\gamma\,\frac{\mathbf{p}}{m}\,\mathcal{P}\right) + \gamma\, k_{\rm B}T\,\partial_{\mathbf{p}}^2\mathcal{P}, \qquad (4.233)$$

which is simply the Kramers master equation (4.179). Because of the timescale separation, the description of Brownian motion as a Markovian stochastic process is thus justified if the stated assumptions are satisfied.

Nevertheless, the microscopic approach based on statistical mechanics shows that processes supposed to be Markovian on timescales longer than the correlation time should be expected to be non-Markovian on shorter timescales, as discussed for equation (4.230).

We note that the Kirkwood formula (4.232) can also be deduced in the framework of linear response theory if the colloidal particle is, for instance, assumed to be driven by a harmonic trap moving at some constant velocity.

Friction at the nanoscale can also be investigated with this approach. In particular, the friction coefficient of translational motion between two concentric carbon nanotubes has been computed using molecular dynamics and Kirkwood's formula (Servantie and Gaspard, 2003, 2006b). Friction due to their rotational motion has been similarly evaluated (Servantie and Gaspard, 2006a). In general, different systems may manifest various kinds of friction beyond the Markovian linear regime considered here.

5

Fluctuation Relations for Energy and Particle Fluxes

5.1 Microreversibility Out of Equilibrium

The contrast between microreversibility and the macroscopic time asymmetry expressed by the second law of thermodynamics is a puzzling aspect of natural phenomena. Significant progress in understanding this fundamental question has been achieved since the pioneering contributions by Onsager (1931a,b) and the development of linear response theory in the 1950s (Callen and Welton, 1951; Green, 1952b, 1954; Kubo, 1957). Furthermore, results have been obtained on the basis of microreversibility for nonlinear transport properties in systems driven out of equilibrium by mechanical forces (Bernard and Callen, 1959; Peterson, 1967; Bochkov and Kuzovlev, 1977, 1979, 1981a,b; Stratonovich, 1992, 1994). Since the 1990s, the advent of the so-called fluctuation relations has shed new lights on the relationships between microreversibility and the second law of thermodynamics in nonequilibrium systems driven arbitrarily far away from equilibrium. These relations rule the fluctuations of random quantities giving the entropy production after statistical average (Evans et al., 1993; Gallavotti and Cohen, 1995; Gallavotti, 1996; Kurchan, 1998; Lebowitz and Spohn, 1999; Maes, 1999; Maes and Netočný, 2003; Evans and Searles, 2002; Sevick et al., 2008; Seifert, 2012), the random nonequilibrium work performed on a system driven by time-dependent forces (Jarzynski, 1997; Crooks, 1998, 1999; Jarzynski, 2000, 2011), the random numbers of reactive events in chemical reaction networks (Gaspard, 2004a; Andrieux and Gaspard, 2008b,f), as well as the random exchanges of energy or particles flowing across an open system in contact with several reservoirs (Andrieux and Gaspard, 2004, 2006a, 2007b; Esposito et al., 2009; Campisi et al., 2011; Gaspard, 2013a,e).

The fluctuation relations are large-deviation properties for these random variables, which are valid from close to far from equilibrium (Gaspard, 2004a; Andrieux and Gaspard, 2008b). Nowadays, they have been established in different theoretical frameworks for thermostatted dynamical systems (Evans et al., 1993; Gallavotti and Cohen, 1995; Gallavotti, 1996; Evans and Searles, 2002), stochastic processes (Kurchan, 1998; Crooks, 1998, 1999; Lebowitz and Spohn, 1999; Maes, 1999; Maes and Netočný, 2003; Gaspard, 2004a; Andrieux and Gaspard, 2007b; Derrida, 2007; Seifert, 2012), and classical Hamiltonian systems (Jarzynski, 1997, 2000, 2011), as well as quantum systems (Andrieux and Gaspard, 2008e; Andrieux et al., 2009; Esposito et al., 2009; Campisi et al., 2011; Gaspard, 2013a).

On the one hand, the fluctuation relations find their origin in the microreversibility of the underlying dynamics governing the motion of particles composing matter. In particular, the principle of detailed balance is recovered at equilibrium as their consequence. On the other hand, they characterize the breaking of time-reversal symmetry in the statistical description of nonequilibrium systems. Although they extend the symmetry of microreversibility from equilibrium to the regimes out of equilibrium, they imply the nonnegativity of the entropy production rate, in agreement with the second law of thermodynamics. Furthermore, they provide the way to generalize the Onsager reciprocal relations and the fluctuation–dissipation theorem from the linear regime close to equilibrium towards the nonlinear regimes farther away from equilibrium. Accordingly, the fluctuation relations allow us to extend response theory from linear to nonlinear transport properties for systems driven away from equilibrium not only by mechanical forces, but also by chemical or thermal affinities, which are thermodynamic forces of statistical origin.

This chapter is organized as follows. Section 5.2 is devoted to the nonequilibrium work fluctuation relation and its consequences. In Section 5.3, the fluctuation relation is extended to energy and particle currents across classical open systems in contact with several reservoirs and ruled by Hamiltonian dynamics. The thermodynamic consequences of the fluctuation relation are discussed in Section 5.4. The implications for the linear and nonlinear response properties are presented in Section 5.5. The results are extended to systems in the presence of an external magnetizing field in Section 5.6. The fluctuation relation for currents is also given for stochastic processes in Section 5.7.

5.2 Fluctuation Relations for Time-Dependent Systems

An equality was first obtained by Jarzynski (1997) to evaluate equilibrium free energy differences using the nonequilibrium work performed on a Hamiltonian system by some time-dependent forcing. Thereafter, a fluctuation relation was deduced for this nonequilibrium work by Crooks (1998, 1999) within the theory of stochastic processes and by Jarzynski (2000) using Hamiltonian microscopic dynamics.

5.2.1 Systems Subjected to Time-Dependent Forcing

We consider a Hamiltonian system composed of a subsystem interacting with a bath, such as a biomolecule suspended in an aqueous solution forming the bath. The positions and momenta of the subsystem are denoted $\mathbf{\Gamma}_{\mathrm{s}} = (\mathbf{q}_{\mathrm{s}}, \mathbf{p}_{\mathrm{s}})$ and those of the bath as $\mathbf{\Gamma}_{\mathrm{b}} = (\mathbf{q}_{\mathrm{b}}, \mathbf{p}_{\mathrm{b}})$. The Hamiltonian function of the total system takes the form

$$H(\mathbf{\Gamma}, \lambda) = H_{\mathrm{s}}(\mathbf{\Gamma}_{\mathrm{s}}, \lambda) + H_{\mathrm{b}}(\mathbf{\Gamma}_{\mathrm{b}}) + U_{\mathrm{sb}}(\mathbf{\Gamma}_{\mathrm{s}}, \mathbf{\Gamma}_{\mathrm{b}}), \tag{5.1}$$

where $H_{\mathrm{s}}(\mathbf{\Gamma}_{\mathrm{s}}, \lambda)$ is the subsystem Hamiltonian subjected to the forcing by the time-dependent control parameter $\lambda = \lambda(t)$, $H_{\mathrm{b}}(\mathbf{\Gamma}_{\mathrm{b}})$ is the Hamiltonian of the bath, and $U_{\mathrm{sb}}(\mathbf{\Gamma}_{\mathrm{s}}, \mathbf{\Gamma}_{\mathrm{b}})$ the interaction energy between the subsystem and the bath. The Hamiltonian function (5.1) gives the total energy of the system. It is symmetric under the time-reversal transformation $\mathbf{\Theta}(\mathbf{q}, \mathbf{p}) = (\mathbf{q}, -\mathbf{p})$, thus satisfying

$$H(\mathbf{\Gamma}, \lambda) = H(\mathbf{\Theta}\mathbf{\Gamma}, \lambda) \tag{5.2}$$

for every value of the control parameter λ and for $\mathbf{\Gamma} = (\mathbf{q}, \mathbf{p})$ with $\mathbf{q} = (\mathbf{q}_s, \mathbf{q}_b)$ and $\mathbf{p} = (\mathbf{p}_s, \mathbf{p}_b)$. The total system is supposed to contain N atoms and h denotes Planck's constant.

In order to test the symmetry under time reversal, two protocols are used for the time-dependent forcing of the system and for collecting the statistics of the nonequilibrium work performed on the system. For both protocols, the forcing drives the system during the time interval $t \in [0, \tau]$. The work performed on the total system by this forcing is given by the difference of total energy between the final time $t = \tau$ and the initial time $t = 0$.

- **Forward protocol:** During this protocol, the control parameter changes in time according to $\lambda_{\rm F}(t)$ with $\lambda_{\rm F}(0) = \lambda_A$ and $\lambda_{\rm F}(\tau) = \lambda_B$. At the beginning of the protocol, the system is prepared in the equilibrium canonical ensemble at the inverse temperature $\beta = (k_{\rm B}T)^{-1}$ in the whole system for the parameter value $\lambda = \lambda_A$:

$$p_{\rm F}(\mathbf{\Gamma}, 0) = \frac{1}{h^{3N} Z_A}\, {\rm e}^{-\beta H(\mathbf{\Gamma}, \lambda_A)} \quad \text{with} \quad Z_A = {\rm e}^{-\beta F_A} \tag{5.3}$$

in terms of the Helmholtz free energy F_A. The work performed on the system along the trajectory $\mathbf{\Gamma}_{\rm F}(t)$ takes the value

$$W_{\rm F} = H\left[\mathbf{\Gamma}_{\rm F}(\tau), \lambda_B\right] - H\left[\mathbf{\Gamma}_{\rm F}(0), \lambda_A\right]. \tag{5.4}$$

Since the initial condition $\mathbf{\Gamma}_{\rm F}(0)$ is distributed according to the probability density (5.3), the work (5.4) is a random variable and its probability density is defined as

$$p_{\rm F}(W) \equiv \langle \delta(W - W_{\rm F}) \rangle_{\rm F}, \tag{5.5}$$

where the statistical average $\langle \cdot \rangle_{\rm F}$ is evaluated using the canonical distribution (5.3) for the initial conditions of the forward protocol and $\delta(x)$ denotes the density of Dirac's delta distribution. The probability density (5.5) is thus normalized according to $\int_{-\infty}^{+\infty} p_{\rm F}(W)\, dW = 1$.

- **Reverse protocol:** For this other protocol, the control parameter varies with a reverse time dependence compared to the forward protocol: $\lambda_{\rm R}(t) = \lambda_{\rm F}(\tau - t)$ so that $\lambda_{\rm R}(0) = \lambda_B$ and $\lambda_{\rm R}(\tau) = \lambda_A$. At the beginning of the reverse protocol, the system is prepared in the equilibrium canonical ensemble again at the inverse temperature β in the whole system, but for the parameter value $\lambda = \lambda_B$ and the Helmholtz free energy F_B:

$$p_{\rm R}(\mathbf{\Gamma}, 0) = \frac{1}{h^{3N} Z_B}\, {\rm e}^{-\beta H(\mathbf{\Gamma}, \lambda_B)} \quad \text{with} \quad Z_B = {\rm e}^{-\beta F_B}. \tag{5.6}$$

The work performed on the system along some trajectory $\mathbf{\Gamma}_{\rm R}(t)$ of the reverse protocol is given by

$$W_{\rm R} = H\left[\mathbf{\Gamma}_{\rm R}(\tau), \lambda_A\right] - H\left[\mathbf{\Gamma}_{\rm R}(0), \lambda_B\right]. \tag{5.7}$$

Again, this work is randomly distributed and its probability density is here defined as

$$p_{\rm R}(W) \equiv \langle \delta(W - W_{\rm R}) \rangle_{\rm R}, \tag{5.8}$$

by averaging with the canonical distribution (5.6) of its initial conditions and normalized according to $\int_{-\infty}^{+\infty} p_{\rm R}(W)\, dW = 1$.

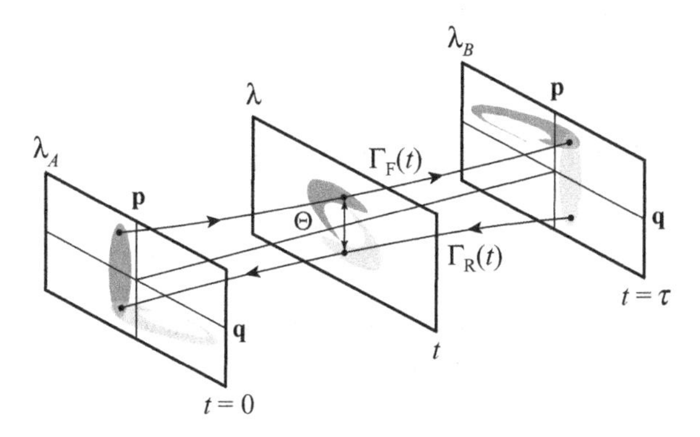

Figure 5.1 Schematic representation of probability densities evolving in phase space $\boldsymbol{\Gamma} = (\mathbf{q}, \mathbf{p})$ for the forward and reverse protocols driving a Hamiltonian system by some time-dependent forcing. The planes depict the phase space for different values of the control parameter λ and time t. $\boldsymbol{\Gamma}_{\rm F}(t)$ denotes one of the trajectories for the forward protocol and $\boldsymbol{\Gamma}_{\rm R}(t)$ one for the reverse protocol. These trajectories are mapped onto each other by time reversal $\boldsymbol{\Theta}$.

During the forward and reverse protocols, the time evolution of the phase-space probability distributions are ruled by the Liouville equation (2.39) starting from the initial distributions (5.3) and (5.6), respectively. Because of the time dependence of the forcing, the probability distributions do not stay canonical, but they evolve into nonequilibrium distributions. In particular, at the end of the forward protocol when the control parameter takes the value $\lambda = \lambda_B$, the resulting probability distribution is, in general, different from the canonical distribution (5.6) and the same feature concerns the distribution at the end of the reverse protocol. This is illustrated in Figure 5.1, which schematically shows how both probability distributions evolve in time from their initial conditions. This figure also shows a pair of trajectories, the first one belonging to the forward protocol and the second to the reverse protocol, both being related to each other by time reversal according to

$$\boldsymbol{\Gamma}_{\rm R}(t) = \boldsymbol{\Theta}\boldsymbol{\Gamma}_{\rm F}(\tau - t), \tag{5.9}$$

which is a consequence of microreversibility. We note that even if the initial distribution of each protocol is canonical and thus symmetric under time reversal, the distribution evolved by the time-dependent forcing does not, in general, preserve the symmetry by $\boldsymbol{\Theta}$. Nevertheless, the time-reversal transformation $\boldsymbol{\Theta}$ and the fact that the time dependence of the control parameter in the forward and reverse protocols are also related to each other by time reversal imply that trajectories taken in both protocols can be interconnected according to equation (5.9).

During each protocol, the work performed on the total system by the time-dependent forcing is measured, its statistics calculated, and the probability densities (5.5) and (5.8) are obtained. Setting $\boldsymbol{\Gamma}_A = \boldsymbol{\Gamma}_{\rm F}(0)$ and $\boldsymbol{\Gamma}_B = \boldsymbol{\Theta}\boldsymbol{\Gamma}_{\rm R}(0)$, equation (5.9) shows that $\boldsymbol{\Gamma}_{\rm R}(\tau) = \boldsymbol{\Theta}\boldsymbol{\Gamma}_A$ and $\boldsymbol{\Gamma}_{\rm F}(\tau) = \boldsymbol{\Gamma}_B$. Because of the time-reversal symmetry (5.2) of the total Hamiltonian function, we have that

$$W_{\rm F} = H_B - H_A \qquad \text{and} \qquad W_{\rm R} = H_A - H_B \tag{5.10}$$

with the notations

$$H_A \equiv H(\mathbf{\Gamma}_A, \lambda_A) \qquad \text{and} \qquad H_B \equiv H(\mathbf{\Gamma}_B, \lambda_B). \tag{5.11}$$

Thus, we find the relation

$$W_{\rm R} = -W_{\rm F} \tag{5.12}$$

between the values of the work performed on the system during the trajectory $\mathbf{\Gamma}_{\rm F}(t)$ of the forward protocol and the associated trajectory (5.9) of the reverse protocol.

5.2.2 Nonequilibrium Work Fluctuation Relation

By virtue of the microreversibility and Liouville's theorem, the probability densities (5.5) and (5.8) of the work performed on the total system in both protocols satisfy the *nonequilibrium work fluctuation relation*

$$\frac{p_{\rm F}(W)}{p_{\rm R}(-W)} = {\rm e}^{\beta(W-\Delta F)}, \tag{5.13}$$

where W is the work performed on the system during the time interval $t \in [0,\tau]$, and $\Delta F = F_B - F_A$ is the difference of free energy between the equilibrium canonical distributions for λ_A and λ_B.

This result is proved as follows. As a consequence of the Liouville theorem (2.15), the phase-space volume elements are preserved during the Hamiltonian time evolution of each protocol:

$$d\mathbf{\Gamma}_{\rm F}(t) = d\mathbf{\Gamma}_{\rm F}(0) \qquad \text{for } t \in [0,\tau], \tag{5.14}$$

$$d\mathbf{\Gamma}_{\rm R}(t) = d\mathbf{\Gamma}_{\rm R}(0) \qquad \text{for } t \in [0,\tau]. \tag{5.15}$$

Since $\mathbf{\Gamma}_A = \mathbf{\Gamma}_{\rm F}(0)$ and $\mathbf{\Gamma}_B = \mathbf{\Gamma}_{\rm F}(\tau)$, we can deduce that

$$d\mathbf{\Gamma}_A = d\mathbf{\Gamma}_B. \tag{5.16}$$

Now, the two probability densities (5.5) and (5.8) of the nonequilibrium work can be related to each other as follows:

$$\begin{aligned}
p_{\rm F}(W) &= \langle \delta\,[W - (H_B - H_A)] \rangle_{\rm F} \\
&= \int d\mathbf{\Gamma}_A \frac{1}{h^{3N} Z_A}\, {\rm e}^{-\beta H_A}\, \delta\,[W - (H_B - H_A)] \\
&= \int d\mathbf{\Gamma}_A \frac{1}{h^{3N} Z_A}\, {\rm e}^{-\beta(H_B - W)}\, \delta\,[W - (H_B - H_A)] \\
&= {\rm e}^{\beta W} \frac{Z_B}{Z_A} \int d\mathbf{\Gamma}_A \frac{1}{h^{3N} Z_B}\, {\rm e}^{-\beta H_B}\, \delta\,[W - (H_B - H_A)] \\
&= {\rm e}^{\beta W} \frac{Z_B}{Z_A} \int d\mathbf{\Gamma}_B \frac{1}{h^{3N} Z_B}\, {\rm e}^{-\beta H_B}\, \delta\,[-W - (H_A - H_B)] \\
&= {\rm e}^{\beta W}\, {\rm e}^{-\beta(F_B - F_A)} \langle \delta\,[-W - (H_A - H_B)] \rangle_{\rm R} \\
&= {\rm e}^{\beta W}\, {\rm e}^{-\beta \Delta F}\, p_{\rm R}(-W).
\end{aligned} \tag{5.17}$$

From the first to the second line, the statistical average is explicitly written in terms of the initial distribution (5.3) of the forward protocol. From the second to the third line, the equality $H_A = H_B - W$ resulting from Dirac's delta distribution is used. From the third to the fourth line, $e^{\beta W}$ is factorized out of the integral, as well as the partition function Z_B with the aim letting the equilibrium canonical distribution (5.6) appear. From the fourth to the fifth line, the consequence (5.16) of Liouville's theorem is used, as well as the relation (5.12) resulting from microreversibility. From the fifth to the sixth line, the ratio of partition functions is expressed in terms of the difference ΔF between the free energies introduced in equations (5.6) and (5.3) to conclude the proof at the seventh line by using the definition (5.8). The fluctuation relation (5.13) is thus established for the nonequilibrium work W.

This relation was first obtained by Crooks (1999) and was proved in the framework of Hamiltonian dynamics by Jarzynski (2000).

5.2.3 Jarzynski's Equality and Clausius' Inequality

The nonequilibrium work fluctuation relation (5.13) has for consequence an equality previously obtained by Jarzynski (1997) and giving the difference $\Delta F = F_B - F_A$ between the free energies corresponding to the equilibrium canonical distributions for λ_A and λ_B, from the statistics of nonequilibrium work in the forward protocol during the time interval τ:

$$\left\langle e^{-\beta W} \right\rangle_{\rm F} = e^{-\beta \Delta F}. \tag{5.18}$$

This equality is proved as follows,

$$\left\langle e^{-\beta W} \right\rangle_{\rm F} = \int dW \, p_{\rm F}(W) \, e^{-\beta W} = e^{-\beta \Delta F} \int dW \, p_{\rm R}(-W) = e^{-\beta \Delta F}, \tag{5.19}$$

where the last equality results from the change of variable $W \to -W$ and the normalization condition of the probability density (5.8).

The Jarzynski equality (5.18) shows that the difference of free energies between two equilibrium states at the same temperature and corresponding to the Hamiltonian functions $H(\boldsymbol{\Gamma}, \lambda_A)$ and $H(\boldsymbol{\Gamma}, \lambda_B)$ can be computed thanks to the statistics of the nonequilibrium work performed by some time-dependent forcing able to connect both Hamiltonian functions.

Remarkably, the Jensen inequality (3.46) combined with the Jarzynski equality (5.18) allows us to deduce Clausius' inequality

$$\langle W \rangle_{\rm F} \geq \Delta F, \tag{5.20}$$

according to which the mean value of the nonequilibrium work $\langle W \rangle_{\rm F}$ performed on the system by the forcing is never smaller than the free energy ΔF that the forcing has provided to the system. This well-known inequality is a consequence of the second law of thermodynamics and it is proved on the basis of Hamiltonian microscopic dynamics within the approach presented here.

Furthermore, the mean work performed during the forward protocol can be expressed as

$$\langle W \rangle_{\rm F} = \Delta F + k_{\rm B} T \int d\boldsymbol{\Gamma} \, p_{\rm F}(\boldsymbol{\Gamma}, t) \ln \frac{p_{\rm F}(\boldsymbol{\Gamma}, t)}{p_{\rm R}(\boldsymbol{\Theta}\boldsymbol{\Gamma}, \tau - t)} \tag{5.21}$$

at any intermediate time $t \in [0,\tau]$ for the phase-space probability distributions of the forward and reverse protocols taken at the same value $\lambda_{\rm F}(t) = \lambda_{\rm R}(\tau - t)$ of the control parameter (Kawai et al., 2007). This result is a consequence of the property that probability densities are preserved along Hamiltonian trajectories according to equation (2.38). Therefore, they keep their initial values, respectively given by equations (5.3) and (5.6), so that

$$p_{\rm F}[\mathbf{\Gamma}_{\rm F}(t),t] = p_{\rm F}(\mathbf{\Gamma}_A,0) = \frac{1}{h^{3N} Z_A}\, {\rm e}^{-\beta H_A}, \tag{5.22}$$

$$p_{\rm R}[\mathbf{\Gamma}_{\rm R}(t'),t'] = p_{\rm R}(\mathbf{\Theta\Gamma}_B,0) = \frac{1}{h^{3N} Z_B}\, {\rm e}^{-\beta H_B}, \tag{5.23}$$

by using the microreversibility condition (5.2). Now, taking $t' = \tau - t$ such that $\lambda_{\rm F}(t) = \lambda_{\rm R}(t')$ and equation (5.9) hold, their ratio gives $\exp(\beta W_{\rm F} - \beta \Delta F)$ with the work (5.4). Setting $\mathbf{\Gamma} = \mathbf{\Gamma}_{\rm F}(t)$ and averaging the logarithm of this ratio with the probability density $p_{\rm F}(\mathbf{\Gamma},t)$ of the forward protocol, equation (5.21) is proved.

The term with the integral in the right-hand side of equation (5.21) has the form of a Kullback–Leibler divergence $D_{\rm KL}$. Therefore, this term is always nonnegative and the Clausius inequality (5.20) is recovered. However, the Kullback–Leibler divergence could be evaluated by coarse graining the probability distributions $p_{\rm F,R}$ into $p^{\rm c}_{\rm F,R}$. After coarse graining, a smaller value would be obtained for the Kullback–Leibler divergence $D^{\rm c}_{\rm KL} \le D_{\rm KL}$, giving inequalities $\langle W\rangle_{\rm F} \ge \Delta F + k_{\rm B} T D^{\rm c}_{\rm KL}$, which are intermediate between the Clausius inequality (5.20) and the equality $\langle W\rangle_{\rm F} = \Delta F + k_{\rm B} T D_{\rm KL}$ given by equation (5.21). In this way, the effect of coarse graining on the evaluation of dissipation can be investigated (Kawai et al., 2007).

In order to obtain the continuous-time expression of the mechanical work, the Hamiltonian function (5.1) can be differentiated to give

$$dH = \frac{\partial H}{\partial \mathbf{\Gamma}} \cdot d\mathbf{\Gamma} + \frac{\partial H}{\partial \lambda}\, d\lambda = \underbrace{\frac{\partial H}{\partial \mathbf{\Gamma}} \cdot \dot{\mathbf{\Gamma}}}_{=\{H,H\}=0}\; dt + \frac{\partial H}{\partial \lambda}\, \dot{\lambda}\, dt, \tag{5.24}$$

where the first term is equal to zero because of the Hamilton equations (2.9).[1] Integrating equation (5.24) over the time interval $[0,\tau]$ gives us the following expression for the mechanical work,

$$W = H[\mathbf{\Gamma}(\tau),\lambda(\tau)] - H[\mathbf{\Gamma}(0),\lambda(0)] = \int_0^\tau dH = \int_0^\tau \frac{\partial H}{\partial \lambda}\, \dot{\lambda}\, dt \tag{5.25}$$

with $\dot{\lambda} = d\lambda/dt$, which is equivalent to equations (5.4) and (5.7) for the work in the forward and the backward protocols, respectively.

The work dissipated during some trajectory can be defined as the difference between the work performed on the system and the free energy stored in the system:

$$W_{\rm diss} \equiv W - \Delta F. \tag{5.26}$$

[1] This result can also be deduced using Hamilton's equations in their equivalent form $\dot{\mathbf{\Gamma}} = \mathbf{\Sigma} \cdot \frac{\partial H}{\partial \mathbf{\Gamma}}$ with the fundamental symplectic matrix $\mathbf{\Sigma}$. This latter being antisymmetric, the first term of equation (5.24) is thus equal to zero.

Jarzynski's equality can thus be written in the form

$$\left\langle e^{-\beta W_{\rm diss}} \right\rangle_{\rm F} = 1 \tag{5.27}$$

and Clausius' inequality as $\langle W_{\rm diss}\rangle_{\rm F} \geq 0$.

If the time-dependent forcing is such that the system reaches a nonequilibrium steady state in the long-time limit, the mean rate of the energy dissipation (5.26) over the forward protocol during the time interval τ gives the entropy production rate

$$\left.\frac{d_{\rm i}S}{dt}\right|_{\rm st} = \frac{1}{T} \lim_{\tau\to\infty} \frac{1}{\tau} \langle W_{\rm diss}\rangle_{\rm F}, \tag{5.28}$$

where T is the temperature of the system. In this way, the properties of nonequilibrium steady states can be obtained using transient processes using suitable time-dependent forcings, as will be further discussed below.

In addition, we note that Jarzynski's equality can be generalized to include the effect of feedback control (Sagawa and Ueda, 2010).

5.2.4 Free-Energy Measurements in Biomolecules

The nonequilibrium work fluctuation relation is important for the experimental measurement of free-energy differences between two conformations of polymeric biomolecules such as RNA in an aqueous solution (Collin et al., 2005). Such experiments are made possible thanks to modern techniques, which consist of linking the two ends of a biomolecule to micrometric beads and pulling on them with optical tweezers, as schematically depicted in Figure 5.2.

At the beginning of the forward protocol, the RNA molecule is folded, forming a hairpin because of pairing between complementary nucleotides, A-U and C-G. The optical tweezer exerts a force that linearly increases with time at some rate. During this protocol, the external force breaks the hydrogen bonds of the base pairs and the RNA molecule unfolds. In the reverse protocol, the molecule returns to its folded conformation. These processes manifest

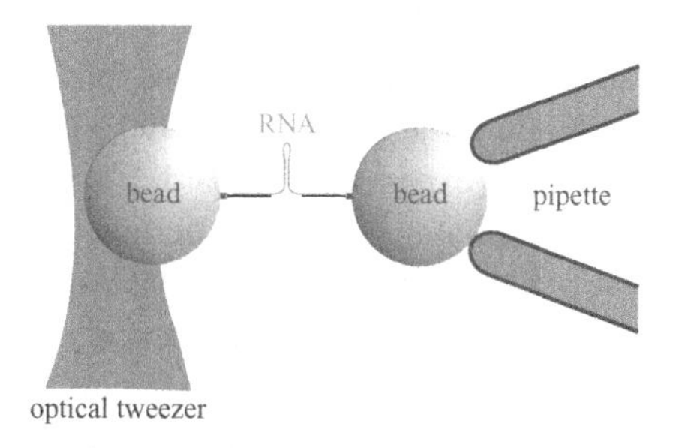

Figure 5.2 Schematic representation of an experiment for unfolding an RNA molecule in an aqueous solution with an optical tweezer (Collin et al., 2005). The RNA molecule is linked to two micrometric polystyrene beads. One of them is kept fixed with a pipette and the other one is pulled with the optical tweezer.

stochasticity because of thermal fluctuations in the aqueous solution, so that the work performed on the system by the time-dependent force is randomly distributed.

The measurement of the free-energy difference between the folded and unfolded conformations in an important issue, which can be addressed by using the fluctuation relation. The nonequilibrium work is measured several hundred times in order to obtain the probability densities (5.5) and (5.8). Since the experiment is carried out at constant pressure, the volume of the system fluctuates and the Gibbs free energy G should be used instead of the Helmholtz free energy F. The canonical distributions of equations (5.3) and (5.6) should thus be replaced by corresponding isobaric-isothermal distributions, which are defined in equation (C.17). Accordingly, the work should include the contribution from the change of volume between the final and initial conditions to get $W_{\rm F} = (H_B + pV_B) - (H_A + pV_A) = -W_{\rm R}$. Additionally, in this case the work probability densities of the forward and reserve protocols can be related to each other as in equation (5.17). Consequently, the nonequilibrium work fluctuation relation is here given by

$$\frac{p_{\rm F}(W)}{p_{\rm R}(-W)} = {\rm e}^{\beta(W-\Delta G)}, \tag{5.29}$$

where $\Delta G = G_B - G_A$ is now the difference of Gibbs free energy between the equilibrium isobaric-isothermal distributions for λ_A and λ_B.

According to equation (5.29), the equality between $p_{\rm F}(W)$ and $p_{\rm R}(-W)$ is expected precisely at the value of the nonequilibrium work giving the free-energy difference, $W = \Delta G$ because $p_{\rm F}(\Delta G) = p_{\rm R}(-\Delta G)$, independently of the pulling rate. This method is shown in Figure 5.3 (open circles) to give the value $\Delta G = 110.3\, k_{\rm B}T$ for the free-energy difference between the unfolded and folded conformations of the RNA molecule that is investigated in the experiment (Collin et al., 2005). Moreover, the evaluation of the free-energy difference is carried out for three different pulling rates. Within experimental accuracy, the same value is obtained in agreement with the prediction that the free-energy difference being an equilibrium property does not depend on the nonequilibrium driving, verifying in this way the validity of the nonequilibrium work fluctuation relation (Collin et al., 2005).

5.2.5 Electromagnetic Heating of Microplasmas

Microplasmas are small systems composed of atomic ions moving in a Penning trap over micrometric spatial distances (Bollinger and Wineland, 1990). The motion of ions is known to be chaotic with positive Lyapunov exponents (Gaspard, 2003a,b). As long as the system is not subjected to time-dependent driving, energy is conserved and these systems may be described at equilibrium using microcanonical statistical ensembles. The ions form remarkable crystalline-like configurations at low mean kinetic energy. These ordered configurations melt as their kinetic energy is increased, as seen in Figure 5.4.

In a frame rotating at the Larmor frequency associated with the magnetic field of the Penning trap, the Hamiltonian of the microplasma is given by

$$H_0 = \sum_{a=1}^{N} \left[\frac{1}{2}\, \mathbf{p}_a^2 + \left(\frac{1}{8} - \frac{\gamma^2}{4} \right) (x_a^2 + y_a^2) + \frac{\gamma^2}{2}\, z_a^2 \right] + \sum_{1\le a<b\le N} \frac{1}{r_{ab}} \tag{5.30}$$

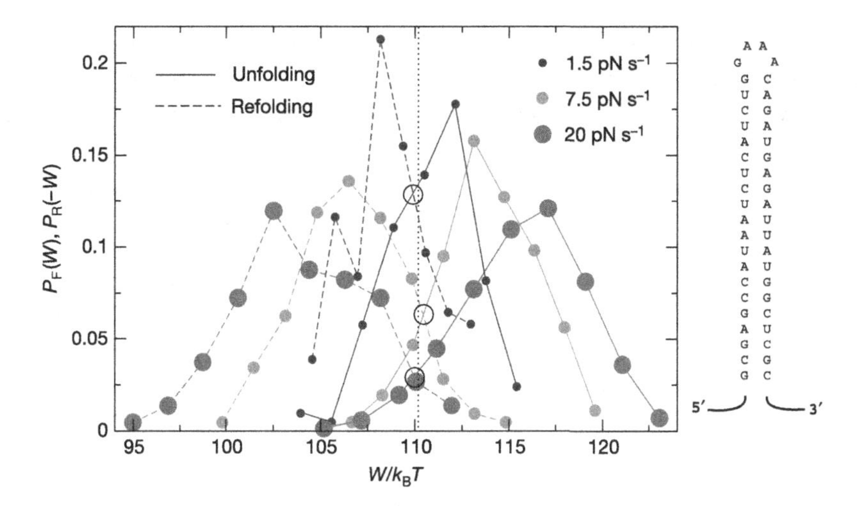

Figure 5.3 Unfolding and refolding work distributions $p_F(W)$ and $p_R(-W)$ for the RNA molecule shown on the right-hand side. The external force is varied in time at three different pulling rates r. The statistics is carried out over 130 pulls and three molecules for $r = 1.5$ pN s^{-1} (small filled circles), 380 pulls and four molecules for $r = 7.5$ pN s^{-1} (medium filled circles), and 700 pulls and three molecules $r = 20$ pN s^{-1} (large filled circles). Reprinted from Collin et al. (2005) with the permission of Springer Nature.

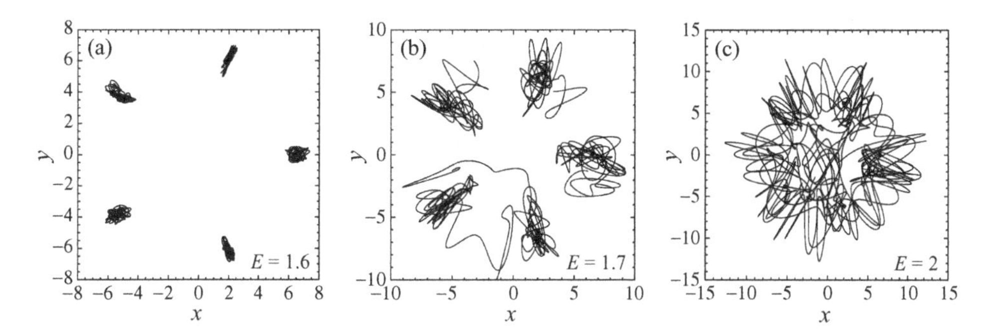

Figure 5.4 Numerical simulation of Hamiltonian trajectories for five atomic ions in an oblate Penning trap with $\gamma = 0.7$. The total angular momentum in the z-direction is equal to zero. The total energy (5.30) takes the values: (a) $E = 1.6$, (b) $E = 1.7$, (c) $E = 2$. Adapted from Gaspard (2010).

in terms of the momenta $\mathbf{p}_a = (p_{xa}, p_{ya}, p_{za})$ of the N ions, and the distances between them, $r_{ab} = \left[(x_a - x_b)^2 + (y_a - y_b)^2 + (z_a - z_b)^2\right]^{1/2}$. The parameter γ controls the geometry of the trap. The trap is elongated or prolate if $0 < |\gamma| < (1/\sqrt{6})$, spherical if $|\gamma| = (1/\sqrt{6})$, and flat or oblate if $(1/\sqrt{6}) < |\gamma| < (1/\sqrt{2})$ (Gaspard, 2003a,b, 2010).

The microplasma can be heated with an electromagnetic wave. This is the case, for instance, with the time-dependent Hamiltonian

$$H = H_0 - A \sum_{a=1}^{N} z_a \sin \omega t, \tag{5.31}$$

so that $dH/dt = \partial H/\partial t = -\omega A \sum_{a=1}^{N} z_a \cos \omega t$ and energy is no longer conserved. The nonequilibrium work fluctuation relation (5.13) can be used to understand the effects of the periodic driving. In order for the forward protocol to be identical to the reverse protocol, the driving is considered over a time interval with an odd number of half periods, e.g., $\tau = 3\pi/\omega$. In this case, the Hamiltonian (5.31) is the same at the beginning and the end of the driving, so that the forward and reverse protocols have the same probability distribution of nonequilibrium work, $p_{\rm F} = p_{\rm R} \equiv p$, and the difference of free energy is thus equal to zero, i.e., $\Delta F = 0$. In this case, the nonequilibrium work fluctuation relation (5.13) can be expressed as

$$\int_{-\infty}^{W} p(W')\, dW' = \int_{-\infty}^{W} {\rm e}^{\beta W'}\, p(-W')\, dW'. \tag{5.32}$$

The numerical verification of this result is shown in Figure 5.5 for a heated microplasma of five ions (Gaspard, 2010). The effect of heating is seen by the shift of the cumulative functions (5.32) to the right-hand side of the vertical line at $W = 0$, and the shift of $\int_{-\infty}^{W} p(-W')\, dW'$ to its left-hand side. The mean value of the work performed on the system by heating is about $\langle W \rangle = 0.16 \pm 0.05$. This mean value is relatively small with respect to the width of the distribution, which extends to negative as well as positive values of the

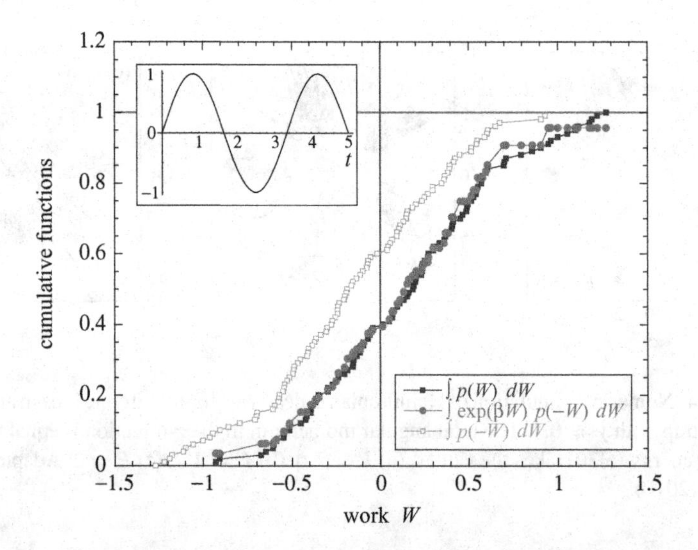

Figure 5.5 Numerical verification of the nonequilibrium work fluctuation relation (5.13) for a microplasma of five atomic ions in an oblate Penning trap with $\gamma = 0.7$ and heated by the time-dependent external field shown in inset over the time interval $\tau = 3\pi/\omega = 5$. The verification of the fluctuation relation is the coincidence of the integrals given, respectively, by the left-hand side (filled squares) and the right-hand side (filled circles) of equation (5.32). The initial distribution is canonical with temperature $T = 1$ in units where $k_{\rm B} = 1$. The final distribution is no longer canonical. The cumulative function of the negative values of the work (open squares) is shifted with respect to the other ones because of heating. Adapted from Gaspard (2010).

work W. In this regard, the system may loose energy for some trajectories. However, there are more trajectories for which the system gains energy, so that, on average, heating results due to nonequilibrium driving by the time-dependent forcing. This energy dissipation in the cyclic driving of the system is the manifestation of the phenomenon of stochastic hysteresis.

5.2.6 Energy and Angular Momentum Transfers

In the example of microplasmas ruled by the Hamiltonian function (5.30), the time-dependent forcing may transfer not only energy (i.e., work), but also angular momentum,

$$L_z = \sum_{a=1}^{N} \left(x_a \, p_{ya} - y_a \, p_{xa} \right), \tag{5.33}$$

if the time-dependent Hamiltonian has, for instance, the following form,

$$H = H_0 - A \sum_{a=1}^{N} x_a \sin \omega t. \tag{5.34}$$

Indeed, there would be a transfer of angular momentum in this case since $dL_z/dt = \{L_z, H\} = -A \sum_{a=1}^{N} y_a \sin \omega t$ because $\{L_z, x_a\} = y_a$ and $\{L_z, H_0\} = 0$ owing to the rotational symmetry of the Hamiltonian function (5.30) around the z-axis. Under time reversal, the Hamiltonian function and the angular momentum are transformed according to $H(\mathbf{\Theta\Gamma}) = H(\mathbf{\Gamma})$ and $L_z(\mathbf{\Theta\Gamma}) = -L_z(\mathbf{\Gamma})$.

In such circumstances, we may consider the joint probability distribution of energy and angular momentum that are transferred to the system by the time-dependent forcing during some time interval $t \in [0, \tau]$ (Gombert, 2017). Because of microreversibility, a fluctuation relation holds for the joint probabilities of the forward and reverse protocols defined as follows:

- **Forward protocol:** As before, the control parameter changes in time according to $\lambda_{\rm F}(t)$ from $\lambda_{\rm F}(0) = \lambda_A$ to $\lambda_{\rm F}(\tau) = \lambda_B$. Initially, the system is prepared in the equilibrium canonical ensemble at the inverse temperature β for the parameter value $\lambda = \lambda_A$ and the system rotating at the angular velocity Ω:

$$p_{{\rm F},\Omega}(\mathbf{\Gamma}, 0) = \frac{1}{h^{3N} Z_A} \, {\rm e}^{-\beta[H(\mathbf{\Gamma},\lambda_A) - \Omega\, L_z(\mathbf{\Gamma})]}. \tag{5.35}$$

The Helmholtz free energy is given, as before, by $F_A = -k_{\rm B}T \ln Z_A$. The Hamiltonian time evolution along the trajectory $\mathbf{\Gamma}_{\rm F}(t)$ leads to the transfers of energy and angular momentum, respectively, given by

$$\Delta E_{\rm F} = H_B - H_A \qquad \text{and} \qquad \Delta L_{\rm F} = L_{z,B} - L_{z,A}, \tag{5.36}$$

where the initial values of energy and angular momentum at $\mathbf{\Gamma}_A = \mathbf{\Gamma}_{\rm F}(0)$ are denoted H_A and $L_{z,A}$, and the final ones at $\mathbf{\Gamma}_B = \mathbf{\Gamma}_{\rm F}(\tau)$ by H_B and $L_{z,B}$. The joint probability density of these transfers is defined as

$$p_{{\rm F},\Omega}(\Delta E, \Delta L) \equiv \langle \delta(\Delta E - \Delta E_{\rm F})\, \delta(\Delta L - \Delta L_{\rm F}) \rangle_{{\rm F},\Omega}. \tag{5.37}$$

- **Reverse protocol:** Now, the control parameter changes in time in the reverse way, $\lambda_{\rm R}(t) = \lambda_{\rm F}(\tau - t)$, and the system is prepared with the equilibrium canonical distribution,

$$p_{{\rm R},\Omega}(\boldsymbol{\Gamma},0) = \frac{1}{h^{3N} Z_B} \, {\rm e}^{-\beta[H(\boldsymbol{\Gamma},\lambda_B) - \Omega\, L_z(\boldsymbol{\Gamma})]}. \tag{5.38}$$

Because of microreversibility, it is possible to associate the trajectories of the reverse protocol to those of the forward protocol according to equation (5.9) and the transfers of energy and angular momentum of both protocols are related to each other by

$$\Delta E_{\rm R} = -\Delta E_{\rm F} \qquad \text{and} \qquad \Delta L_{\rm R} = \Delta L_{\rm F}. \tag{5.39}$$

For the reverse protocol, the joint probability density of these transfers is defined as

$$p_{{\rm R},\Omega}(\Delta E, \Delta L) \equiv \langle \delta(\Delta E - \Delta E_{\rm R})\, \delta(\Delta L - \Delta L_{\rm R}) \rangle_{{\rm R},\Omega}. \tag{5.40}$$

By using microreversibility and Liouville's theorem, the following fluctuation relation is obtained for the transfers of energy and angular momentum,

$$\frac{p_{{\rm F},\Omega}(\Delta E, \Delta L)}{p_{{\rm R},-\Omega}(-\Delta E, \Delta L)} = {\rm e}^{\beta(\Delta E - \Omega\, \Delta L - \Delta F)}, \tag{5.41}$$

where $\Delta F = F_B - F_A = k_{\rm B}T \ln(Z_A/Z_B)$ is the difference of free energy between the equilibrium distributions for λ_A and λ_B at the angular velocity Ω. For the fluctuation relation to hold, the angular velocity should thus also be reversed. The proof can be achieved by analogy with equation (5.17) (Gombert, 2017).

We note that the transfer of energy is the work performed on the system by the external forcing, $\Delta E = W$. Therefore, the nonequilibrium work fluctuation relation (5.13) is recovered from equation (5.41) in the absence of rotation ($\Omega = 0$) for the marginal probability densities obtained by integrating the joint probability density over the variable ΔL.

5.2.7 Nonequilibrium Work Fluctuation Relation in a Magnetizing Field

The nonequilibrium work fluctuation relation can be extended to systems driven by a time-dependent forcing and subjected to an external magnetizing field $\boldsymbol{\mathcal{H}}$. Since electrodynamics is symmetric under time reversal, microreversibility holds, but reversing the motion of all the particles in the global system including the external electric circuits generating the magnetizing field has the effect of reversing the magnetizing field itself: $\boldsymbol{\mathcal{H}} \to -\boldsymbol{\mathcal{H}}$. Therefore, the Hamiltonian function describing the system of interest has the symmetry

$$H(\boldsymbol{\Gamma},\lambda;\boldsymbol{\mathcal{H}}) = H(\boldsymbol{\Theta\Gamma},\lambda;\, -\boldsymbol{\mathcal{H}}), \tag{5.42}$$

instead of the one of equation (5.2). In this respect, the reverse protocol should be performed for the time-reversed driving $\lambda_{\rm R}(t) = \lambda_{\rm F}(\tau - t)$, but also with the reversed magnetizing field $-\boldsymbol{\mathcal{H}}$. Therefore, microreversibility associates a trajectory of the reversed protocol with one of the forward protocol according to

$$\boldsymbol{\Gamma}_{\rm R}(t;\boldsymbol{\mathcal{H}}) = \boldsymbol{\Theta\Gamma}_{\rm F}(\tau - t;\, -\boldsymbol{\mathcal{H}}), \tag{5.43}$$

instead of equation (5.9). Otherwise, the forward and reverse protocols are similarly defined as in Section 5.2.1. The relation between the probability distributions of the work in the forward and reverse protocols is established as in Section 5.2.2, but here using equations (5.42) and (5.43).

Consequently, the nonequilibrium work fluctuation relation in an external magnetizing field reads

$$\frac{p_{\rm F}(W;\mathcal{H})}{p_{\rm R}(-W;-\mathcal{H})} = {\rm e}^{\beta(W-\Delta F)} \qquad \text{with} \qquad \Delta F \equiv F_B(-\mathcal{H}) - F_A(\mathcal{H}) \tag{5.44}$$

for the probability densities $p_{\rm F}$ and $p_{\rm R}$ of the work W performed on the system during the time interval $[0,\tau]$ in the forward and reverse protocols starting from equilibrium canonical distributions with λ_A and λ_B, respectively. We note that the fluctuation relation (5.13) is recovered if the magnetizing field is equal to zero, i.e., for $\mathcal{H} = \mathbf{0}$.

5.3 Fluctuation Relation for Currents in Hamiltonian Systems

5.3.1 Open System in Contact with Reservoirs

Here, we deduce the multivariate fluctuation relation for the energy and particle global currents flowing across a system driven away from equilibrium by its coupling to several reservoirs that are large enough to maintain nonequilibrium conditions over a long time interval. We note that the energy and the particle numbers are conserved quantities, so that they obey conservation laws during the whole time evolution and they can be used as resources as long as the nonequilibrium conditions are maintained.

Several species of particles $k = 1, 2, \ldots, c$ compose the total system, which extends over the domain $V \subset \mathbb{R}^3$ of physical space. The domain $V_j \subset \mathbb{R}^3$ is associated with the jth reservoir, as shown in Figure 5.6 and such that $V = \cup_{j=1}^r V_j$. In this respect, the domain V_r associated with the reference reservoir $j = r$ includes the subsystem in contact with the reservoirs. This subsystem is supposed to be small with respect to the reservoirs.

The time evolution of the total system is ruled by a Hamiltonian function such as the one given by equation (2.2). This latter presents the total energy, which can be expressed in terms of the microscopic energy density (3.6) according to

$$H(\mathbf{\Gamma}) = \int_V \hat{\epsilon}(\mathbf{r};\mathbf{\Gamma})\, d^3r = \sum_{j=1}^{r} H_j(\mathbf{\Gamma}), \tag{5.45}$$

where $V = \cup_{j=1}^r V_j$ and

$$H_j(\mathbf{\Gamma}) \equiv \int_{V_j} \hat{\epsilon}(\mathbf{r};\mathbf{\Gamma})\, d^3r \tag{5.46}$$

is the energy of the jth reservoir defined by integrating over its domain V_j. We note that the energies (5.46) depend not only on the particles inside the domain V_j, but also on some particles in the contiguous domains V_k (for $k \neq j$). However, if the interaction energy potential u_{ab} is short or medium ranged, such contributions are limited to the interfaces between the domains V_j and V_k, and they are thus small with respect to the energy in the reservoirs.

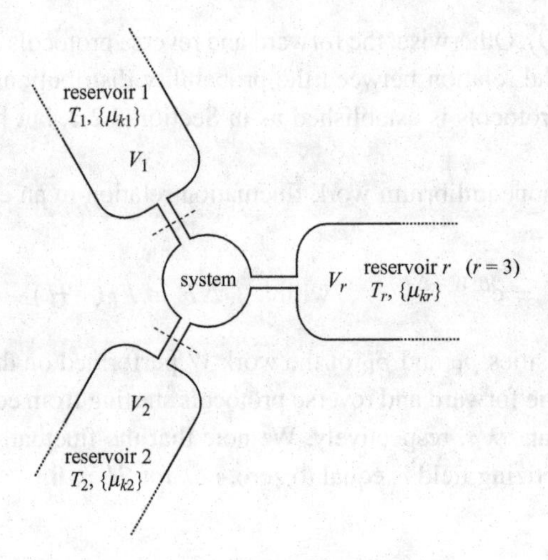

Figure 5.6 Schematic representation of a subsystem coupled to several reservoirs $j = 1, 2, \dots, r$ at the temperatures T_j and the chemical potentials μ_{kj} of species $k = 1, 2, \dots, c$. Here, V_j denotes the domain of physical space corresponding to the jth reservoir. The domain associated with the reference reservoir $j = r$ includes the subsystem. The dashed lines delimit these different domains.

Similarly, the total number of particles of species k can be written as follows in terms of the corresponding microscopic particle density defined by equation (3.3),

$$N_k(\mathbf{\Gamma}) = \int_V \hat{n}_k(\mathbf{r}; \mathbf{\Gamma})\, d^3r = \sum_{j=1}^{r} N_{kj}(\mathbf{\Gamma}), \tag{5.47}$$

where $V = \cup_{j=1}^{r} V_j$ and

$$N_{kj}(\mathbf{\Gamma}) \equiv \int_{V_j} \hat{n}_k(\mathbf{r}; \mathbf{\Gamma})\, d^3r \tag{5.48}$$

is the number of particles of species k inside the domain V_j associated with the jth reservoir. The total number of particles of any species is thus equal to $N = \sum_{k=1}^{c} N_k$.

The Hamiltonian time evolution is given by the flow $\mathbf{\Gamma}_t = \mathbf{\Phi}^t \mathbf{\Gamma}_0$ in phase space. Since the Hamiltonian function is supposed to be time independent and the particles are preserved, the total energy and the total particle numbers are conserved by the time evolution,

$$H(\mathbf{\Gamma}_0) = H(\mathbf{\Phi}^t \mathbf{\Gamma}_0) \qquad \text{and} \qquad N_k(\mathbf{\Gamma}_0) = N_k(\mathbf{\Phi}^t \mathbf{\Gamma}_0) \tag{5.49}$$

for $k = 1, 2, \dots, c$.

The Hamiltonian function, the energies of the reservoirs, as well as the particle numbers are observables that are symmetric under the time-reversal transformation (2.27), so that

$$H_j(\mathbf{\Theta\Gamma}) = H_j(\mathbf{\Gamma}) \qquad \text{and} \qquad N_{jk}(\mathbf{\Theta\Gamma}) = N_{jk}(\mathbf{\Gamma}), \tag{5.50}$$

which imply that $H(\mathbf{\Theta\Gamma}) = H(\mathbf{\Gamma})$ for the total Hamiltonian function (5.45) and $N_k(\mathbf{\Theta\Gamma}) = N_k(\mathbf{\Gamma})$ for the total particle number (5.47). Consequently, microreversibility is satisfied and

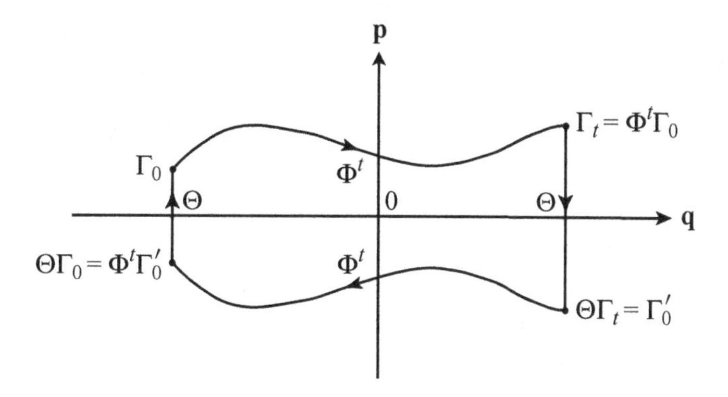

Figure 5.7 Schematic phase portrait showing a trajectory and its time reversal of the Hamiltonian flow $\mathbf{\Phi}^t$ in the phase space $\mathbf{\Gamma} = (\mathbf{q}, \mathbf{p})$ for the total system of Figure 5.6. The time-reversal transformation is denoted $\mathbf{\Theta}$, and $\mathbf{\Gamma}_0$ is the initial condition for the trajectory and $\mathbf{\Gamma}_0'$ for its time reversal.

the flow can be reversed according to equation (2.29). Figure 5.7 shows the implications of this symmetry on the trajectory evolving from the initial condition $\mathbf{\Gamma}_0$ under the action of the flow $\mathbf{\Phi}^t$, and its time reversal obtained by reversing the final point to get its initial condition $\mathbf{\Gamma}_0'$, which is then evolved by the flow $\mathbf{\Phi}^t$ up to its final point given by the image of the initial condition $\mathbf{\Gamma}_0$ upon time reversal. Therefore, the loop is closed, so that $\mathbf{\Theta\Phi}^t\mathbf{\Theta\Phi}^t = \mathbf{1}$, which is equivalent to equation (2.29) since the time-reversal transformation $\mathbf{\Theta}$ is an involution $\mathbf{\Theta}^2 = \mathbf{1}$.

5.3.2 Measuring Energy and Particle Fluxes

The initial conditions of the total system are distributed according to the grand canonical probability density (C.11) at the inverse temperatures $\beta_j = (k_B T_j)^{-1}$ and the chemical potentials μ_{kj} for the species k in the reservoirs $j = 1, 2, \ldots, r$, so that

$$p_0(\mathbf{\Gamma}) = \frac{1}{h^{3N}\,\Xi} \prod_{j=1}^{r} \mathrm{e}^{-\beta_j [H_j(\mathbf{\Gamma}) - \sum_{k=1}^{c} \mu_{kj} N_{kj}(\mathbf{\Gamma})]}. \tag{5.51}$$

For this distribution, the number N_k of particles for species k and the total number of particles $N = \sum_{k=1}^{c} N_k$ are random variables, taking different values for every initial condition $\mathbf{\Gamma}_0$. Nevertheless, every trajectory $\mathbf{\Gamma}_t$ conserves these particle numbers during their whole time evolution.[2] Therefore, for a given initial condition $\mathbf{\Gamma}_0$, the time evolution takes place in a phase space as given by equation (2.4), but for a mixture of N_k identical particles and denoted $\mathcal{M}_{\mathbf{N}}$ with $\mathbf{N} = (N_k)_{k=1}^{c}$. Its dimension is equal to $\dim \mathcal{M}_{\mathbf{N}} = 6N$, where $N = \sum_{k=1}^{c} N_k$ is the total number of particles. Accordingly, the random variable $\mathbf{\Gamma}$

[2] Here, we assume that there is no reaction that transforms the species into each other.

stands for $(\boldsymbol{\Gamma}_N, \mathbf{N})$ with $\boldsymbol{\Gamma}_N \in \mathcal{M}_{\mathbf{N}}$ and $\mathbf{N} \in \mathbb{N}^c$. The probability density (5.51) thus has the normalization

$$\int p_0(\boldsymbol{\Gamma})\, d\boldsymbol{\Gamma} = \sum_{\mathbf{N}} \int_{\mathcal{M}_{\mathbf{N}}} p_0(\boldsymbol{\Gamma}_N, \mathbf{N})\, d\boldsymbol{\Gamma}_N = \sum_{\mathbf{N}} \frac{1}{\mathbf{N}!} \int_{\mathbb{R}^{6N}} p_0(\boldsymbol{\Gamma}_N, \mathbf{N})\, d\boldsymbol{\Gamma}_N = 1, \tag{5.52}$$

where $\sum_{\mathbf{N}}(\cdot) \equiv \prod_{k=1}^{c} \sum_{N_k=0}^{\infty}(\cdot)$ and $\mathbf{N}! \equiv \prod_{k=1}^{c} N_k!$. This normalization is giving the partition function Ξ in equation (5.51).

The energies (5.46) and the particle numbers (5.48) can be measured in the reservoirs at the initial time $t = 0$ and later at the time t. Because of energy and particle fluxes between the reservoirs, these quantities change in time according to

$$\Delta E_j \equiv H_j(\boldsymbol{\Gamma}_0) - H_j(\boldsymbol{\Phi}^t \boldsymbol{\Gamma}_0), \tag{5.53}$$

$$\Delta N_{kj} \equiv N_{kj}(\boldsymbol{\Gamma}_0) - N_{kj}(\boldsymbol{\Phi}^t \boldsymbol{\Gamma}_0), \tag{5.54}$$

which represent the amounts of energy and particles that are flowing out of the jth reservoir. We use the short notation

$$\mathbf{C} = [H_j, (N_{kj})_{k=1}^{c}]_{j=1}^{r} \qquad \text{and} \qquad \Delta\mathbf{C} = [\Delta E_j, (\Delta N_{kj})_{k=1}^{c}]_{j=1}^{r}, \tag{5.55}$$

so that

$$\Delta\mathbf{C} \equiv \mathbf{C}(\boldsymbol{\Gamma}_0) - \mathbf{C}(\boldsymbol{\Phi}^t \boldsymbol{\Gamma}_0) \tag{5.56}$$

is a quantity with $r(c+1)$ components. With this notation, the initial probability density (5.51) can be written as

$$p_0(\boldsymbol{\Gamma}) = \frac{1}{h^{3N}\,\Xi}\, \mathrm{e}^{\boldsymbol{\alpha}\cdot\mathbf{C}(\boldsymbol{\Gamma})} \qquad \text{with} \qquad \boldsymbol{\alpha} \equiv [-\beta_j, (\beta_j \mu_{kj})_{k=1}^{c}]_{j=1}^{r}. \tag{5.57}$$

Moreover, the quantities defined in equation (5.55) have the time-reversal symmetry

$$\mathbf{C}(\boldsymbol{\Theta}\boldsymbol{\Gamma}) = \mathbf{C}(\boldsymbol{\Gamma}), \tag{5.58}$$

because of equation (5.50).

Now, we introduce the following probability density that the exchanges $\Delta\mathbf{C}$ have occurred during the time interval t since the initial condition,

$$p_t(\Delta\mathbf{C}) \equiv \int d\boldsymbol{\Gamma}_0\, p_0(\boldsymbol{\Gamma}_0)\, \delta[\Delta\mathbf{C} - \mathbf{C}(\boldsymbol{\Gamma}_0) + \mathbf{C}(\boldsymbol{\Phi}^t \boldsymbol{\Gamma}_0)], \tag{5.59}$$

with the Dirac delta distribution

$$\delta[\Delta\mathbf{C} - \mathbf{C}(\boldsymbol{\Gamma}_0) + \mathbf{C}(\boldsymbol{\Phi}^t \boldsymbol{\Gamma}_0)] = \prod_{j=1}^{r} \delta[\Delta E_j - H_j(\boldsymbol{\Gamma}_0) + H_j(\boldsymbol{\Phi}^t \boldsymbol{\Gamma}_0)] \times \prod_{j=1}^{r} \prod_{k=1}^{c} \delta[\Delta N_{kj} - N_{kj}(\boldsymbol{\Gamma}_0) + N_{kj}(\boldsymbol{\Phi}^t \boldsymbol{\Gamma}_0)], \tag{5.60}$$

and the normalization condition,

$$\int_{\mathbb{R}^{r(c+1)}} p_t(\Delta\mathbf{C})\, d\Delta\mathbf{C} = 1. \tag{5.61}$$

The probability density (5.59) can be successively transformed by using microreversibility and Liouville's theorem along the loop shown in Figure 5.7. First, using Liouville's theorem $d\mathbf{\Gamma}_0 = d\mathbf{\Gamma}_t$ and the inverse of the flow $\mathbf{\Gamma}_0 = \mathbf{\Phi}^{-t}\mathbf{\Gamma}_t$, equation (5.59) can be written in the form

$$p_t(\Delta\mathbf{C}) = \int d\mathbf{\Gamma}_t\, p_0(\mathbf{\Phi}^{-t}\mathbf{\Gamma}_t)\, \delta[\Delta\mathbf{C} - \mathbf{C}(\mathbf{\Phi}^{-t}\mathbf{\Gamma}_t) + \mathbf{C}(\mathbf{\Gamma}_t)]. \tag{5.62}$$

Since the time-reversal transformation maps the final condition $\mathbf{\Gamma}_t$ of the trajectory onto the initial condition $\mathbf{\Gamma}'_0$ of its reversal, we can carry out the change of integration variable $\mathbf{\Gamma}_t = \mathbf{\Theta}\mathbf{\Gamma}'_0$ to get

$$p_t(\Delta\mathbf{C}) = \int d\mathbf{\Theta}\mathbf{\Gamma}'_0\, p_0(\mathbf{\Phi}^{-t}\mathbf{\Theta}\mathbf{\Gamma}'_0)\, \delta[\Delta\mathbf{C} - \mathbf{C}(\mathbf{\Phi}^{-t}\mathbf{\Theta}\mathbf{\Gamma}'_0) + \mathbf{C}(\mathbf{\Theta}\mathbf{\Gamma}'_0)]. \tag{5.63}$$

Using the property $d\mathbf{\Theta}\mathbf{\Gamma}'_0 = d\mathbf{\Gamma}'_0$, the microreversibility $\mathbf{\Phi}^{-t}\mathbf{\Theta} = \mathbf{\Theta}\mathbf{\Phi}^t$, and the time-reversal symmetry (5.58), the probability density (5.63) becomes

$$p_t(\Delta\mathbf{C}) = \int d\mathbf{\Gamma}'_0\, p_0(\mathbf{\Phi}^t\mathbf{\Gamma}'_0)\, \delta[\Delta\mathbf{C} - \mathbf{C}(\mathbf{\Phi}^t\mathbf{\Gamma}'_0) + \mathbf{C}(\mathbf{\Gamma}'_0)]. \tag{5.64}$$

Now, using the expression (5.57) for the initial probability density with the equality $\mathbf{C}(\mathbf{\Phi}^t\mathbf{\Gamma}'_0) = \Delta\mathbf{C} + \mathbf{C}(\mathbf{\Gamma}'_0)$ resulting from the Dirac delta distribution, and the evenness of this latter, we find that

$$p_t(\Delta\mathbf{C}) = \mathrm{e}^{\boldsymbol{\alpha}\cdot\Delta\mathbf{C}} \int d\mathbf{\Gamma}'_0\, p_0(\mathbf{\Gamma}'_0)\, \delta[-\Delta\mathbf{C} - \mathbf{C}(\mathbf{\Gamma}'_0) + \mathbf{C}(\mathbf{\Phi}^t\mathbf{\Gamma}'_0)]. \tag{5.65}$$

Comparing this with definition (5.59), we obtain the following exact time-reversal symmetry relation,

$$p_t(\Delta\mathbf{C}) = \mathrm{e}^{\boldsymbol{\alpha}\cdot\Delta\mathbf{C}}\, p_t(-\Delta\mathbf{C}). \tag{5.66}$$

5.3.3 Fluctuation Relation for the Global Currents

Because of the conservation laws (5.49), the quantities (5.53) and (5.54) are not all independent. One of the reservoirs (e.g. $j = r$) may be taken as the reference reservoir. The amounts of energy and particles flowing out of the other reservoirs are thus flowing into the reference reservoir, so that

$$\Delta E_r = -\sum_{j=1}^{r-1} \Delta E_j \qquad \text{and} \qquad \Delta N_{kr} = -\sum_{j=1}^{r-1} \Delta N_{kj}. \tag{5.67}$$

Substituting these relations into the argument of the exponential function in equation (5.66) gives

$$\boldsymbol{\alpha}\cdot\Delta\mathbf{C} = \sum_{j=1}^{r-1} (\beta_r - \beta_j)\,\Delta E_j + \sum_{j=1}^{r-1}\sum_{k=1}^{c} (\beta_j\mu_{kj} - \beta_r\mu_{kr})\,\Delta N_{kj}, \tag{5.68}$$

where the following thermodynamic forces or affinities appear,

$$\text{thermal affinity:} \qquad A_{Ej} \equiv \beta_r - \beta_j, \tag{5.69}$$

$$\text{chemical affinity:} \qquad A_{kj} \equiv \beta_j \, \mu_{kj} - \beta_r \, \mu_{kr}, \tag{5.70}$$

associated with the driving reservoirs $j = 1, 2, \ldots, r-1$ and the particle species $k = 1, 2, \ldots, c$. These affinities are equal to zero at equilibrium where the temperature and the chemical potentials should be uniform across the whole system. Otherwise, they represent the control parameters driving the total system out of equilibrium. The expressions (5.69) and (5.70) are the same as the affinities (1.64) and (1.65) introduced at the macroscale.[3] Therefore, using the notation,

$$\mathbf{A} \equiv [A_{Ej}, (A_{kj})_{k=1}^{c}]_{j=1}^{r-1}, \tag{5.71}$$

$$\Delta\mathfrak{C} \equiv [\Delta E_j, (\Delta N_{kj})_{k=1}^{c}]_{j=1}^{r-1}, \tag{5.72}$$

which now have $(r-1)(c+1)$ components corresponding to the independent currents, we have that $\boldsymbol{\alpha} \cdot \Delta\mathbf{C} = \mathbf{A} \cdot \Delta\mathfrak{C}$ in the right-hand side of equation (5.66). The probability density (5.59) can be integrated over the quantities in the reference reservoir to define the probability density of the independent quantities as $p_t(\Delta\mathfrak{C}) \equiv \int p_t(\Delta\mathbf{C})\, d\Delta\mathbf{C}_r$. Integrating both sides of equation (5.66) over $\Delta\mathbf{C}_r$, we finally obtain the fluctuation relation

$$\frac{p_t(\Delta\mathfrak{C};\mathbf{A})}{p_t(-\Delta\mathfrak{C};\mathbf{A})} = \mathrm{e}^{\mathbf{A}\cdot\Delta\mathfrak{C}} \tag{5.73}$$

for the exchanges of energy and particles between the reservoirs. In equation (5.73), we have explicitly written the dependence of the probability density on the affinities $\mathbf{A}$ to emphasize that this probability distribution depends on the conditions used to drive the system with the reservoirs. If we define the random currents over the time interval t as $\mathfrak{J} \equiv \Delta\mathfrak{C}/t$,[4] equation (5.73) can be written in the following form of the *multivariate fluctuation relation for the currents*:

$$\frac{p_t(\mathfrak{J};\mathbf{A})}{p_t(-\mathfrak{J};\mathbf{A})} = \mathrm{e}^{\mathbf{A}\cdot\mathfrak{J}\,t}. \tag{5.74}$$

This fundamental result has many implications. First of all, the principle of detailed balance is recovered at equilibrium where the affinities are equal to zero. Indeed, if $\mathbf{A} = \mathbf{0}$, opposite fluctuations for the currents become equiprobable, since the fluctuation relation (5.74) reduces to $p_t(\mathfrak{J};\mathbf{0}) = p_t(-\mathfrak{J};\mathbf{0})$ in this case. Away from equilibrium, the fluctuation relation (5.74) expresses the manifestation of directionality induced by the nonequilibrium constraints on the system. Indeed, if $\mathbf{A} \neq \mathbf{0}$, the ratio of opposite fluctuations of the currents is either increasing or decreasing exponentially with time. Accordingly, the fluctuations in the direction of the affinities $\mathbf{A}$ (i.e., such that $\mathbf{A}\cdot\mathfrak{J} > 0$) become more probable than those

[3] In the present framework, there is no distinction between energy and heat currents because this would require formulating the assumption of local thermodynamic equilibrium and introducing the fluid velocity, as done in Chapter 1.

[4] The notation $\mathfrak{J}$ is used for the fluctuating currents defined at the microscopic level of description to make the distinction with respect to the mean values of the currents $\mathbf{J} = \langle\mathfrak{J}\rangle$ as considered at the macroscale, for instance, by equation (1.61) in a fluid at rest.

in the opposite direction. In this respect, a directionality manifests itself if the affinities $\mathbf{A}$ are driving the system out of equilibrium.

We note that the initial probability distribution (5.51) is not stationary. Since opposite fluctuations may have different probabilities during the time interval t according to the fluctuation relation (5.74) with $\mathbf{A} \neq \mathbf{0}$, the time-reversal symmetry is broken at the statistical level of description although microreversibility continues to be satisfied. Indeed, reversing the time $t \to -t$ gives the same relation (5.74) but for the opposite fluctuations of the currents, so that the fluctuation relation has the symmetry of time reversal. Moreover, reversing the affinities $\mathbf{A} \to -\mathbf{A}$ again gives the same relation (5.74) for the opposite fluctuations of the currents:

$$\frac{p_t(\mathfrak{J};\mathbf{A})}{p_t(-\mathfrak{J};\mathbf{A})} = \frac{p_{-t}(-\mathfrak{J};\mathbf{A})}{p_{-t}(\mathfrak{J};\mathbf{A})} = \frac{p_t(-\mathfrak{J};-\mathbf{A})}{p_t(\mathfrak{J};-\mathbf{A})} = \mathrm{e}^{\mathbf{A}\cdot\mathfrak{J}\,t}. \tag{5.75}$$

5.3.4 Cumulant Generating Function and Full Counting Statistics

Here, the reservoirs are supposed to be arbitrarily large in order to maintain nonvanishing currents in the long-time limit,[5] which requires that the equilibration time introduced in Section 1.7.2 is much longer than the timescales of exchanges between the reservoirs. If these assumptions are satisfied, the system can be maintained in a nonequilibrium stationary macrostate and we may expect that the mean values of the exchanges of energy (5.53) and particles (5.54) would behave as $\langle \Delta E_j \rangle \simeq J_{Ej} t$ and $\langle \Delta N_{kj} \rangle \simeq J_{kj} t$, where J_{Ej} and J_{kj} are the mean values of the corresponding global currents of energy and particles.

For arbitrarily large reservoirs in the conditions fixed by the affinities $\mathbf{A}$, the cumulant generating function of the fluctuating currents can be defined as

$$Q(\boldsymbol{\lambda};\mathbf{A}) \equiv \lim_{t\to\infty} -\frac{1}{t} \ln \left\langle \mathrm{e}^{-\boldsymbol{\lambda}\cdot\Delta\mathfrak{C}(t)} \right\rangle_{\mathbf{A}} \tag{5.76}$$

in terms of the so-called *counting parameters* $\boldsymbol{\lambda} = (\lambda_\alpha)_{\alpha=1}^{\chi}$ with $\chi = (r-1)(c+1)$. This function has the property

$$Q(\mathbf{0};\mathbf{A}) = 0, \tag{5.77}$$

resulting from the normalization condition $\langle 1 \rangle_{\mathbf{A}} = 1$.

In general, there are as many counting parameters as independent currents and affinities according to equations (5.71) and (5.72). The aim of introducing these auxiliary parameters is to obtain the (rescaled) statistical cumulants of the currents by taking successive derivatives of the generating function with respect to them and setting $\boldsymbol{\lambda} = \mathbf{0}$.[6] A statistical cumulant of order m can thus be defined as

$$Q_{\alpha_1\cdots\alpha_m}(\mathbf{A}) \equiv \frac{\partial^m Q}{\partial\lambda_{\alpha_1}\cdots\partial\lambda_{\alpha_m}}(\mathbf{0};\mathbf{A}) \tag{5.78}$$

with $\alpha_i \in \{1,2,\ldots,\chi\}$ for $i = 1,2,\ldots,m$.

[5] Accordingly, the limit $V_j \to \infty$ for the volumes of the reservoirs $j = 1,2,\ldots,r$ should be taken before the long-time limit $t \to \infty$.

[6] These cumulants are said to be rescaled because they are defined by dividing the standard definition (Abramowitz and Stegun, 1972) with the time and by taking the long-time limit.

In particular, the mean values of the currents are defined as[7]

$$J_\alpha(\mathbf{A}) \equiv \lim_{t\to\infty} \frac{1}{t} \langle \Delta\mathfrak{C}_\alpha(t)\rangle_{\mathbf{A}} \tag{5.79}$$

and they are given by the first cumulants:

$$J_\alpha(\mathbf{A}) = Q_\alpha(\mathbf{A}) = \frac{\partial Q}{\partial \lambda_\alpha}(\mathbf{0};\mathbf{A}). \tag{5.80}$$

Furthermore, the diffusivities of the currents are introduced as

$$D_{\alpha\beta}(\mathbf{A}) \equiv \lim_{t\to\infty} \frac{1}{2t}\left[\langle \Delta\mathfrak{C}_\alpha(t)\,\Delta\mathfrak{C}_\beta(t)\rangle_{\mathbf{A}} - \langle \Delta\mathfrak{C}_\alpha(t)\rangle_{\mathbf{A}}\langle \Delta\mathfrak{C}_\beta(t)\rangle_{\mathbf{A}}\right] \tag{5.81}$$

and they can be obtained in terms of the second cumulants according to

$$D_{\alpha\beta}(\mathbf{A}) = -\frac{1}{2}\, Q_{\alpha\beta}(\mathbf{A}) = -\frac{1}{2}\,\frac{\partial^2 Q}{\partial\lambda_\alpha\partial\lambda_\beta}(\mathbf{0};\mathbf{A}). \tag{5.82}$$

Similarly, the third cumulants are defined as

$$C_{\alpha\beta\gamma}(\mathbf{A}) \equiv Q_{\alpha\beta\gamma}(\mathbf{A}) = \frac{\partial^3 Q}{\partial\lambda_\alpha\partial\lambda_\beta\partial\lambda_\gamma}(\mathbf{0};\mathbf{A}), \tag{5.83}$$

and the fourth cumulants as

$$B_{\alpha\beta\gamma\delta}(\mathbf{A}) \equiv -\frac{1}{2}\, Q_{\alpha\beta\gamma\delta}(\mathbf{A}) = -\frac{1}{2}\,\frac{\partial^4 Q}{\partial\lambda_\alpha\partial\lambda_\beta\partial\lambda_\gamma\lambda_\delta}(\mathbf{0};\mathbf{A}). \tag{5.84}$$

Higher-order statistical cumulants can also be defined.

We note that all these cumulants are defined at equilibrium if $\mathbf{A} = \mathbf{0}$ and out of equilibrium if $\mathbf{A} \neq \mathbf{0}$. Since the partial derivatives commute in equation (5.78), the cumulants are totally symmetric under the permutations P of their indices

$$Q_{\alpha_1\cdots\alpha_m}(\mathbf{A}) = Q_{\alpha_{P(1)}\cdots\alpha_{P(m)}}(\mathbf{A}). \tag{5.85}$$

In particular, the diffusivities (5.81) are symmetric, so that $D_{\alpha\beta}(\mathbf{A}) = D_{\beta\alpha}(\mathbf{A})$, and this property extends to all the cumulants. The generating function (5.76) can thus be expanded in power series of the counting parameters as

$$Q(\boldsymbol{\lambda};\mathbf{A}) = \sum_{m=1}^{\infty}\frac{1}{m!}\sum_{\alpha_1,\ldots,\alpha_m} Q_{\alpha_1\cdots\alpha_m}(\mathbf{A})\,\lambda_{\alpha_1}\cdots\lambda_{\alpha_m}, \tag{5.86}$$

where the coefficients are given by the statistical cumulants (5.78).

The temporal decay of the probability density for given values of the currents can be expressed in terms of the rate function defined as

$$R(\mathfrak{J};\mathbf{A}) \equiv \lim_{t\to\infty} -\frac{1}{t}\ln p_t(\mathfrak{J};\mathbf{A}). \tag{5.87}$$

[7] We note that $J_\alpha(\mathbf{A}) = \lim_{t\to\infty}\Delta\mathfrak{C}_\alpha(t)/t$ for almost all trajectories if the property of ergodicity holds in the nonequilibrium steady state of affinities $\mathbf{A}$.

According to large-deviation theory, the cumulant generating function is thus given by the Legendre transform of this rate function,

$$Q(\boldsymbol{\lambda};\mathbf{A}) = \inf_{\mathfrak{J}}\,[\boldsymbol{\lambda}\cdot\mathfrak{J} + R(\mathfrak{J};\mathbf{A})], \tag{5.88}$$

as explained in Section E.2.[8] Vice versa, the rate function can be recovered with the Legendre transform of the cumulant generating function:

$$R(\mathfrak{J};\mathbf{A}) = \sup_{\boldsymbol{\lambda}}\,[Q(\boldsymbol{\lambda};\mathbf{A}) - \mathfrak{J}\cdot\boldsymbol{\lambda}]. \tag{5.89}$$

In principle, the knowledge of the cumulant generating function (5.76) is equivalent to knowing all the statistical cumulants of the fluctuating currents $\mathfrak{J}$ and, thus, their probability distribution $p_t(\mathfrak{J};\mathbf{A})$ in the long-time limit; this behavior is controlled by the large-deviation function (5.89). Therefore, this knowledge allows us to carry out what is called the *full counting statistics* of the currents (Nazarov and Blanter, 2009).

5.3.5 Symmetry Relation for the Cumulant Generating Function

According to the fluctuation relation for the currents (5.74), the rate function (5.87) should obey

$$R(-\mathfrak{J};\mathbf{A}) - R(\mathfrak{J};\mathbf{A}) = \mathbf{A}\cdot\mathfrak{J} \tag{5.90}$$

with respect to the reversal of all the currents $\mathfrak{J}$. Therefore, the rate function becomes even $R(-\mathfrak{J};\mathbf{0}) = R(\mathfrak{J};\mathbf{0})$ at equilibrium where the affinities are equal to zero, i.e., $\mathbf{A} = \mathbf{0}$, meaning that the rate function is equal for opposite fluctuations, which confirms that the principle of detailed balance is recovered at equilibrium.

A further consequence of the fluctuation relation (5.74) is the following symmetry relation for the cumulant generating function,[9]

$$Q(\boldsymbol{\lambda};\mathbf{A}) = Q(\mathbf{A}-\boldsymbol{\lambda};\mathbf{A}). \tag{5.91}$$

This result is proved by starting from definition (5.76), using either the form (5.73) or (5.74) of the fluctuation relation for the currents, and changing the integration variables $\Delta\mathfrak{C} \to -\Delta\mathfrak{C}$ to get

$$\begin{aligned} Q(\boldsymbol{\lambda};\mathbf{A}) &= \lim_{t\to\infty} -\frac{1}{t}\ln\int p_t(\Delta\mathfrak{C};\mathbf{A})\,\mathrm{e}^{-\boldsymbol{\lambda}\cdot\Delta\mathfrak{C}}\,d\Delta\mathfrak{C} \\ &= \lim_{t\to\infty} -\frac{1}{t}\ln\int p_t(-\Delta\mathfrak{C};\mathbf{A})\,\mathrm{e}^{\mathbf{A}\cdot\Delta\mathfrak{C}}\,\mathrm{e}^{-\boldsymbol{\lambda}\cdot\Delta\mathfrak{C}}\,d\Delta\mathfrak{C} \\ &= \lim_{t\to\infty} -\frac{1}{t}\ln\int p_t(\Delta\mathfrak{C};\mathbf{A})\,\mathrm{e}^{-(\mathbf{A}-\boldsymbol{\lambda})\cdot\Delta\mathfrak{C}}\,d\Delta\mathfrak{C} \\ &= Q(\mathbf{A}-\boldsymbol{\lambda};\mathbf{A}), \end{aligned} \tag{5.92}$$

[8] The correspondence with the quantities introduced in Section E.2 is made with $\boldsymbol{\beta} = -\boldsymbol{\lambda}$, $\boldsymbol{\alpha} = \mathfrak{J}$, $J(\boldsymbol{\beta}) = -Q(\boldsymbol{\lambda})$, and $I(\boldsymbol{\alpha}) = R(\mathfrak{J})$ (dropping the implicit dependence of these functions on the affinities $\mathbf{A}$).

[9] Remarkably, the same symmetry relation holds for quantum transport, as will be shown in Chapter 15.

whereupon equation (5.91) is obtained. The same result is also deduced using equation (5.88) and the form (5.90) of the fluctuation relation.

Setting $\boldsymbol{\lambda} = \mathbf{0}$ in the symmetry relation (5.91) and using the property (5.77), we find that

$$Q(\mathbf{A};\mathbf{A}) = 0. \tag{5.93}$$

Furthermore, taking the derivative of equation (5.91) with respect to λ_α, the mean current (5.80) is given by the equivalent expression

$$J_\alpha(\mathbf{A}) = -\frac{\partial Q}{\partial \lambda_\alpha}(\mathbf{A};\mathbf{A}). \tag{5.94}$$

Using the power expansion (5.86) of the cumulant generating function and setting $\boldsymbol{\lambda} = \mathbf{A}$, the mean currents can thus be expressed only in terms of the higher-order statistical cumulants as

$$J_\alpha(\mathbf{A}) = \sum_\beta D_{\alpha\beta}(\mathbf{A})\, A_\beta - \frac{1}{4}\sum_{\beta,\gamma} C_{\alpha\beta\gamma}(\mathbf{A})\, A_\beta\, A_\gamma + \frac{1}{6}\sum_{\beta,\gamma,\delta} B_{\alpha\beta\gamma\delta}(\mathbf{A})\, A_\beta\, A_\gamma\, A_\delta + O\left(\mathbf{A}^4\right). \tag{5.95}$$

In this regard, the mean currents turn out to be related to their fluctuations properties (Gaspard, 2013a). This result is valid arbitrarily far away from equilibrium. In the linear regime close to equilibrium, the mean currents are determined by the diffusivities since $J_\alpha(\mathbf{A}) = \sum_\beta D_{\alpha\beta}(\mathbf{0})A_\beta + O\left(\mathbf{A}^2\right)$, which is expected from linear response theory, e.g., Ohm's law.

Equation (5.91) is also very useful to deduce time-reversal symmetry relations for the statistical cumulants combined with the response coefficients, as shown below.

5.4 Accord with the Second Law of Thermodynamics

5.4.1 Entropy Production

Remarkably, the fluctuation relation for currents implies the nonnegativity of entropy production in accordance with the second law of thermodynamics. In order to establish this fundamental result, we note that the thermodynamic entropy of the whole system including the reservoirs is a function of the energies and particle numbers in the reservoirs $S = S\{[E_j, (N_{kj})_{k=1}^c]_{j=1}^r\}$. Since the reservoirs are large, their energies and particle numbers are much larger than in the small open system in contact with the reservoirs. Furthermore, they evolve in time because of the fluxes of energy and matter between the reservoirs, so that

$$\frac{dS}{dt} = \sum_{j=1}^{r}\left(\frac{\partial S}{\partial E_j}\dot{E}_j + \sum_{k=1}^{c}\frac{\partial S}{\partial N_{kj}}\dot{N}_{kj}\right). \tag{5.96}$$

The reservoirs are assumed to be large enough for the global currents of energy and matter to reach the stationary values $\dot{E}_j = J_{Ej}$ and $\dot{N}_{kj} = J_{kj}$ given by equations (5.79). Because of the conservation of the total energy and particle numbers, the global currents towards the reference reservoir $j = r$ are thus given by $J_{Er} = -\sum_{j=1}^{r-1} J_{Ej}$ and $J_{kr} = -\sum_{j=1}^{r-1} J_{kj}$. As the whole system is isolated, there is no exchange of entropy $d_e S/dt = 0$ elsewhere, so the time derivative (5.96) of the entropy is equal to the entropy production, $dS/dt = d_i S/dt$. Since $\partial S/\partial E_j = 1/T_j$ and $\partial S/\partial N_{kj} = -\mu_{kj}/T_j$ in terms of the temperatures and chemical potentials of the reservoirs, the entropy production thus has the rate

$$\frac{1}{k_B}\frac{d_i S}{dt} = \mathbf{A}\cdot\mathbf{J}(\mathbf{A}) = \mathbf{A}\cdot\langle\mathfrak{J}\rangle_{\mathbf{A}} = \lim_{t\to\infty}\frac{1}{t}\,\mathbf{A}\cdot\langle\Delta\mathfrak{C}(t)\rangle_{\mathbf{A}} \tag{5.97}$$

in terms of the global affinities (5.71) and where $\mathbf{A}\cdot\mathbf{J} = \sum_{\alpha=1}^{\chi} A_\alpha J_\alpha$ with $\chi = (r-1)(c+1)$. Equation (5.97) is consistent with the expression (1.63) obtained in the framework of thermodynamics, which is thus here justified on the basis of the microscopic dynamics. Since the long-time limit is taken in equation (5.97), the system will reach a stationary regime if the reservoirs are arbitrarily large. Under such circumstances, the entropy production rate (5.97) characterizes this stationary macrostate.

Since the statistical averages are carried out in equation (5.97) over the probability distribution obeying the fluctuation relation (5.73), we find that the rate of entropy production is given by the Kullback–Leibler divergence

$$\frac{1}{k_B}\frac{d_i S}{dt} = \lim_{t\to\infty}\frac{1}{t}\int d\Delta\mathfrak{C}\; p_t(\Delta\mathfrak{C};\mathbf{A})\,\ln\frac{p_t(\Delta\mathfrak{C};\mathbf{A})}{p_t(-\Delta\mathfrak{C};\mathbf{A})} \geq 0, \tag{5.98}$$

which is known to be always nonnegative (Cover and Thomas, 2006). Equivalently, equation (5.98) can be written in the form

$$\frac{1}{k_B}\frac{d_i S}{dt} = \lim_{t\to\infty}\frac{1}{2t}\int d\mathfrak{J}\,[p_t(\mathfrak{J};\mathbf{A}) - p_t(-\mathfrak{J};\mathbf{A})]\ln\frac{p_t(\mathfrak{J};\mathbf{A})}{p_t(-\mathfrak{J};\mathbf{A})} \geq 0, \tag{5.99}$$

which is obviously nonnegative because $(x-y)\ln(x/y) > 0$ for both $x > y > 0$ and $y > x > 0$, while $(x-y)\ln(x/y) = 0$ for the equality $x = y > 0$.

Another way to prove the same result is based on the *integral fluctuation relation*

$$\left\langle e^{-\mathbf{A}\cdot\Delta\mathfrak{C}(t)}\right\rangle_{\mathbf{A}} = 1, \tag{5.100}$$

which is obtained using the fluctuation relation (5.73) to get

$$\left\langle e^{-\mathbf{A}\cdot\Delta\mathfrak{C}(t)}\right\rangle_{\mathbf{A}} = \int d\Delta\mathfrak{C}\; p_t(\Delta\mathfrak{C};\mathbf{A})\,e^{-\mathbf{A}\cdot\Delta\mathfrak{C}} = \int d\Delta\mathfrak{C}\; p_t(-\Delta\mathfrak{C};\mathbf{A}) = 1, \tag{5.101}$$

after the change of variables $\Delta\mathfrak{C} \to -\Delta\mathfrak{C}$ in the last integral.[10] According to the Jensen inequality (3.46) with $x = -\mathbf{A}\cdot\Delta\mathfrak{C}(t)$, we thus find that

$$\mathbf{A}\cdot\langle\Delta\mathfrak{C}(t)\rangle_{\mathbf{A}} \geq 0, \tag{5.102}$$

which expresses the nonnegativity of the entropy production.

[10] In isothermal systems, equation (5.100) reduces to integral fluctuation relations obtained by Bochkov and Kuzovlev (1977); Jarzynski (2011). Stochastic versions for the integral fluctuation relation have also been considered (Crooks, 2000; Seifert, 2005a, 2012).

Furthermore, we note that the integral fluctuation relation (5.100) implies that the cumulant generating function (5.76) obeys the relation (5.93), which is the expression of the integral fluctuation relation in the long-time limit from the viewpoint of large-deviation theory.[11]

5.4.2 Fluctuation Relation for Nonequilibrium Directionality

We now introduce the fluctuating quantity defined as

$$\Sigma(t) \equiv \mathbf{A} \cdot \Delta\mathfrak{C}(t), \tag{5.103}$$

and its corresponding rate

$$\mathfrak{R}(t) \equiv \mathbf{A} \cdot \mathfrak{J}(t) = \frac{1}{t}\, \Sigma(t), \tag{5.104}$$

characterizing the alignment of the current fluctuations with respect to the direction of the affinities. These quantities have a clear operational status since the fluctuating currents $\mathfrak{J}(t) = \Delta\mathfrak{C}(t)/t$ are defined in terms of the variables of the microscopic Hamiltonian dynamics, as shown in Section 5.3, while the affinities can be determined by the macroscopic temperatures and chemical potentials of the reservoirs, without requiring the measurement of probability distributions prior to their own determination. The probability density of the rate (5.104) is defined as

$$p_t(\mathfrak{R}; \mathbf{A}) \equiv \int d\mathfrak{J}\; p_t(\mathfrak{J}; \mathbf{A})\, \delta(\mathfrak{R} - \mathbf{A} \cdot \mathfrak{J}), \tag{5.105}$$

which satisfies the univariate fluctuation relation

$$\frac{p_t(\mathfrak{R}; \mathbf{A})}{p_t(-\mathfrak{R}; \mathbf{A})} = \mathrm{e}^{\mathfrak{R}\, t} \tag{5.106}$$

as a consequence of the multivariate fluctuation relation (5.74) for all the currents. We note that the implication does not hold vice versa, since the multivariate fluctuation relation contains more information than the univariate one. The integral fluctuation relation (5.100) can be expressed equivalently as[12]

$$\left\langle \mathrm{e}^{-\Sigma(t)} \right\rangle_{\mathbf{A}} = 1, \tag{5.107}$$

which again implies that the entropy production rate is nonnegative,

$$\frac{1}{k_{\mathrm{B}}} \frac{d_{\mathrm{i}}S}{dt} = \langle \mathfrak{R} \rangle_{\mathbf{A}} = \lim_{t\to\infty} \frac{1}{t}\, \langle \Sigma(t) \rangle_{\mathbf{A}} \geq 0, \tag{5.108}$$

because of the Jensen inequality (3.46) with $x = -\Sigma(t)$. The nonnegativity of the entropy production is thus a direct implication of the fluctuation relation. The entropy production

[11] This long-time limit of the integral fluctuation relation was considered by Förster and Büttiker (2008), who argued that it may hold in general without the assumption of microreversibility. This statement is supported by the integral fluctuation relation (3.45), which is established without using microreversibility.

[12] This relation is reminiscent of Sasa's integral fluctuation relation (3.45) obtained in the microscopic approach to hydrodynamics of Chapter 3.

rate is equal to zero at equilibrium where the affinities are equal to zero and detailed balance holds, but it is positive away from equilibrium because of the nonequilibrium directionality due to nonzero affinities. In this regard, the fluctuation relation shows that the positivity of the entropy production out of equilibrium finds its origin in the breaking of time-reversal symmetry by the probability distribution of the currents.

5.4.3 Timescale for the Emergence of Thermodynamic Behavior

The manifestation of the nonequilibrium directionality is that the current fluctuations become aligned with the affinities. Therefore, the knowledge of the affinities in the frame defined by the reservoirs determines the direction of the most probable current fluctuations in relation to the breaking of time-reversal symmetry at the statistical level of description, as discussed in Section 2.6.

In this regard, the question arises as to whether it is possible to guess the direction of the affinities $\mathbf{A}$ imposed by the reservoirs from the sole observation of current fluctuations during the time interval $[0,t]$. Using Bayesian inference (Maragakis et al., 2008; Jarzynski, 2011), the likelihood of the hypotheses that the observed current fluctuations $\mathfrak{J}$ are going either forward (+) or backward (−) with respect to the direction of the affinities $\mathbf{A}$ can be expressed as

$$P(\pm|\mathfrak{J}) = \frac{P(\mathfrak{J}|\pm)\,P(\pm)}{P(\mathfrak{J}|+)\,P(+) + P(\mathfrak{J}|-)\,P(-)}, \tag{5.109}$$

where $P(\pm)$ are the prior probabilities of the two hypotheses and

$$P(\mathfrak{J}|\pm) \equiv P\left\{t^{-1}\Delta\mathfrak{C}(t) \in [\mathfrak{J}, \mathfrak{J} + \Delta\mathfrak{J}]\,\middle|\,\mathbf{A}\cdot\mathfrak{J} \gtrless 0\right\} \tag{5.110}$$

are the probabilities of observing the current fluctuations $\mathfrak{J}$ given that they are either aligned or not with the affinities $\mathbf{A}$. Tossing a fair coin on either hypothesis amounts to taking $P(+) = P(-) = \frac{1}{2}$. According to the multivariate fluctuation relation (5.74), the likelihood that the observed current fluctuations $\mathfrak{J}$ are indeed going forward with respect to the direction of the affinities $\mathbf{A}$ is thus given by

$$P(+|\mathfrak{J}) = \frac{1}{1 + \exp\left(-\mathbf{A}\cdot\mathfrak{J}\,t\right)} \to_{t\to+\infty} \theta(\mathbf{A}\cdot\mathfrak{J}), \tag{5.111}$$

where $\theta(x)$ is Heaviside's step function. In the limit of a long positive time, the likelihood thus reaches the unit probability for $\mathbf{A}\cdot\mathfrak{J} > 0$ if the system is out of equilibrium with $\mathbf{A} \neq 0$. In contrast, if the affinities are equal to zero, $\mathbf{A} = 0$, the likelihood remains at the value $P(+|\mathfrak{J}) \simeq \frac{1}{2}$, which confirms the absence of directionality at equilibrium. The reasoning is compatible with the overall time-reversal symmetry because a similar result holds in the limit $t \to -\infty$ for $-\mathbf{A}\cdot\mathfrak{J}$ replacing $\mathbf{A}\cdot\mathfrak{J}$. As previously mentioned, the question is pertinent only if the reservoirs and their affinities remain unknown to the observer. Since the mean value of the fluctuating quantity $\mathfrak{R} = \mathbf{A}\cdot\mathfrak{J}$ gives the entropy production rate, the result (5.111) also shows that the likelihood of antithermodynamic behavior decreases exponentially in time as $P(-|\mathfrak{J}) \simeq \exp(-\mathbf{A}\cdot\mathfrak{J}\,t)$ for $t \to +\infty$ with the mean rate given by the entropy production rate (5.108) in units of Boltzmann's constant.

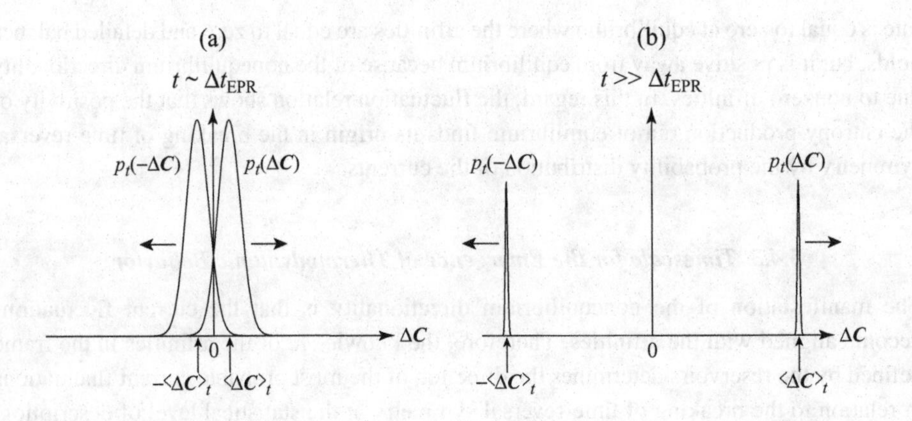

Figure 5.8 Schematic representation of time evolution for the probability densities of opposite fluctuations $\pm\Delta\mathfrak{C}$ of the exchanges (5.72) in an open system under nonequilibrium conditions characterized by the affinities (5.71): (a) on the timescale (5.112); (b) much beyond.

After beginning the statistics of current fluctuations, thermodynamic behavior is thus expected to manifest itself beyond the characteristic timescale given by the inverse of the entropy production rate,

$$\Delta t_{\rm EPR} = \langle \mathbf{A} \cdot \mathfrak{J} \rangle_{\mathbf{A}}^{-1} = \left(\frac{1}{k_{\rm B}} \frac{d_{\rm i}S}{dt} \right)^{-1}, \tag{5.112}$$

under nonequilibrium conditions such that $\mathbf{A} \neq 0$. The emergence of thermodynamic behavior is illustrated in Figure 5.8, which shows how the probability densities of opposite fluctuations, $\pm\Delta\mathfrak{C} = \pm\mathfrak{J}t$, evolve in time. On the characteristic timescale (5.112), these probability densities still have an important overlap, corresponding to the trajectories with uncertain thermodynamic behavior. However, this overlap between the probability densities quickly vanishes beyond the characteristic timescale (5.112), so that the probability of antithermodynamic behavior becomes negligible and most trajectories manifest a clear thermodynamic directionality. Since the entropy production rate is extensive, this characteristic timescale is inversely proportional to the volume of the system and it is thus tiny if the system is macroscopic and driven away from equilibrium. At every time, the fluctuation relation (5.73) is satisfied in consistency with microreversibility.

5.4.4 Entropy Production and Current Fluctuations

Using expansion (5.95) for the mean currents, the entropy production rate (5.97) is given by

$$\frac{1}{k_{\rm B}} \frac{d_{\rm i}S}{dt} = \sum_{\alpha} A_\alpha J_\alpha(\mathbf{A}) = \sum_{\alpha,\beta} D_{\alpha\beta}(\mathbf{A})\, A_\alpha\, A_\beta - \frac{1}{4} \sum_{\alpha,\beta,\gamma} C_{\alpha\beta\gamma}(\mathbf{A})\, A_\alpha\, A_\beta\, A_\gamma$$
$$+ \frac{1}{6} \sum_{\alpha,\beta,\gamma,\delta} B_{\alpha\beta\gamma\delta}(\mathbf{A})\, A_\alpha\, A_\beta\, A_\gamma\, A_\delta + O(\mathbf{A}^5). \tag{5.113}$$

Therefore, the entropy production rate can be expressed in terms of all the statistical cumulants beyond the first ones. This relation generalizes the fluctuation–dissipation theorem from the linear regime close to equilibrium, where the entropy production rate is given by $k_{\rm B}^{-1}(d_{\rm i}S/dt) \simeq \sum_{\alpha,\beta} D_{\alpha\beta}(\mathbf{0})\, A_\alpha\, A_\beta$ in terms of the diffusivities, towards the nonlinear regimes, where the contributions from higher cumulants become significant.

In this regard, we may wonder to what extent the current fluctuations may give evidence for entropy production under nonequilibrium conditions. Actually, the following lower bounds on the entropy production rate, called *thermodynamic uncertainty relations*, have been established using large-deviation theory,

$$\frac{1}{k_{\rm B}}\frac{d_{\rm i}S}{dt} \geq \frac{J(\mathbf{A})^2}{D(\mathbf{A})}, \tag{5.114}$$

which is written in terms of the mean current $J = \sum_{\alpha=1}^{\chi} c_\alpha J_\alpha$ and the diffusivity $D = \sum_{\alpha,\beta=1}^{\chi} c_\alpha c_\beta D_{\alpha\beta}$ of any linear combination $\mathfrak{J} \equiv \sum_{\alpha=1}^{\chi} c_\alpha \mathfrak{J}_\alpha$ of the fluctuating currents with some constant coefficients $\{c_\alpha\}_{\alpha=1}^{\chi}$ (Barato and Chetrite, 2015; Barato and Seifert, 2015; Pietzonka et al., 2016; Gingrich et al., 2016). Accordingly, the entropy production rate is positive as soon as some mean current $J(\mathbf{A})$ flows across the system with fluctuations characterized by the corresponding diffusivity $D(\mathbf{A})$. At equilibrium, the entropy production rate is equal to zero, so that the mean current is also equal to zero, i.e., $J(\mathbf{0}) = \langle \mathfrak{J} \rangle_{\rm eq} = 0$. This result confirms that nonvanishing mean currents generate entropy production if current diffusivities are finite. If the coefficients are taken as the affinities $c_\alpha = A_\alpha$, the fluctuating current is given by the fluctuating rate (5.104) (i.e., $\mathfrak{J} = \mathfrak{R}$) and the mean current J is identical to the entropy production rate (5.113), so that the thermodynamic uncertainty relation (5.114) implies the following upper bound on the entropy production rate,

$$\frac{1}{k_{\rm B}}\frac{d_{\rm i}S}{dt} \leq \sum_{\alpha,\beta} D_{\alpha\beta}(\mathbf{A})\, A_\alpha\, A_\beta. \tag{5.115}$$

Nevertheless, the precise value of the entropy production rate requires the knowledge of all the statistical cumulants, as shown by equation (5.113).

5.4.5 Thermodynamic Efficiencies

Because of the coupling between the different currents flowing across an open system, the possibility arises that one current can be driven in a direction that is opposite to its associated affinity, while always satisfying the second law of thermodynamics. In such circumstances, the corresponding term is negative, i.e., $A_\gamma J_\gamma(\mathbf{A}) < 0$ in the entropy production rate (5.97), although the sum of all the other terms is positive and large enough to compensate the negative term.

This phenomenon happens in particular for the energy transduction of chemical to mechanical energy in molecular motors. Such phenomena of energy transduction are made possible by the coupling between the different currents, which are the mechanical current or velocity and the chemical current or reaction rate in the example of molecular motors (Jülicher et al., 1997). Such couplings result from the dependence of the current $J_\gamma(\mathbf{A})$ on the affinities A_α with $\alpha \neq \gamma$.

The thermodynamic efficiency of energy transduction due to this coupling can be defined as

$$\eta_\gamma \equiv -\frac{A_\gamma J_\gamma(\mathbf{A})}{\sum_{\alpha(\neq\gamma)} A_\alpha J_\alpha(\mathbf{A})}, \tag{5.116}$$

which is positive in the regime of energy transduction where the current J_γ is driven in the direction opposite to its affinity A_γ by all the other currents and affinities. Since the second law is always satisfied,

$$\frac{1}{k_B}\frac{d_i S}{dt} = \sum_\alpha A_\alpha J_\alpha(\mathbf{A}) \geq 0, \tag{5.117}$$

the thermodynamic efficiency (5.116) is always lower than the unit value:

$$\eta_\gamma \leq 1. \tag{5.118}$$

Thermodynamic efficiencies may also be defined over finite time intervals in terms of the currents $\mathfrak{J}_\alpha(t) = \Delta\mathfrak{C}_\alpha(t)/t$. Since these latter are random, the so-defined finite-time efficiencies are also random variables, in contrast to the efficiency (5.116) defined in terms of the mean values of the currents. Accordingly, finite-time efficiencies are described by probability distributions. These latter have been investigated in several types of systems with coupling between two currents (Verley et al., 2014).

5.4.6 Loose versus Tight Coupling between Currents

In systems with two currents, the multivariate fluctuation relation (5.74) takes the form

$$\frac{p_t(\mathfrak{J}_1, \mathfrak{J}_2; A_1, A_2)}{p_t(-\mathfrak{J}_1, -\mathfrak{J}_2; A_1, A_2)} = e^{(A_1\mathfrak{J}_1 + A_2\mathfrak{J}_2)t}. \tag{5.119}$$

If the two currents were statistically independent, their joint probability distribution would factorize as $p_t(\mathfrak{J}_1, \mathfrak{J}_2; A_1, A_2) = p_t(\mathfrak{J}_1; A_1)\, p_t(\mathfrak{J}_2; A_2)$ into the probability distributions of each current, which would each obey a univariate fluctuation relation. Nevertheless, there may exist statistical correlations between the two currents, so that the joint probability distribution does not factorize and the multivariate fluctuation relation (5.119) is nontrivial. If the statistical correlations are moderate, the coupling between the currents is said to be *loose* (Oosawa and Hayashi, 1986). However, it is possible for the two currents to be proportional to each other as

$$\mathfrak{J}_2 = c\, \mathfrak{J}_1, \tag{5.120}$$

in which case the coupling is said to be *tight*.

In the limit of tight coupling, the joint probability density can be replaced by the marginal probability density

$$\mathcal{P}_t(\mathfrak{J}_1; A_1, A_2) \equiv \int p_t(\mathfrak{J}_1, \mathfrak{J}_2; A_1, A_2)\, d\mathfrak{J}_2 \tag{5.121}$$

and the bivariate fluctuation relation (5.119) reduces to the univariate fluctuation relation,

$$\frac{\mathcal{P}_t(\mathfrak{J}_1; A_1, A_2)}{\mathcal{P}_t(-\mathfrak{J}_1; A_1, A_2)} = e^{\tilde{A}_1 \mathfrak{J}_1 t} \tag{5.122}$$

with the effective affinity

$$\tilde{A}_1 \equiv A_1 + c\, A_2, \tag{5.123}$$

which is associated with the coupled currents. In this case, the entropy production rate (5.97) becomes

$$\frac{1}{k_{\mathrm{B}}} \frac{d_{\mathrm{i}} S}{dt} = \tilde{A}_1 \langle \mathfrak{J}_1 \rangle \geq 0, \tag{5.124}$$

so that the sole affinity $\tilde{A}_1$ drives the system out of equilibrium. If the tight-coupling condition (5.120) holds, the thermodynamic efficiency (5.116) is entirely determined by the affinities

$$\eta_1 \equiv -\frac{A_1 \langle \mathfrak{J}_1 \rangle}{A_2 \langle \mathfrak{J}_2 \rangle} = -\frac{A_1}{c\, A_2} = \frac{1}{1 - \tilde{A}_1 / A_1}. \tag{5.125}$$

The condition of tight coupling between mechanics and chemistry is encountered in the rotary motor F_1-ATPase if it is subjected to a small enough external torque, as further discussed in Chapter 12. Otherwise, the torque may decouple the mechanical and chemical currents and the motor enters in the regime of loose coupling (Gaspard and Gerritsma, 2007; Gerritsma and Gaspard, 2010).

5.4.7 Fluctuation Relation for Paths and Entropy Production

The results of Section 5.3 can be extended to paths obtained by sampling the changes of energy and particle numbers in the reservoirs at the successive times $t_i = i\Delta t$ with $i = 1, 2, \ldots, k$ and $t = t_k = k\Delta t$. As in Section 5.3.2, the initial conditions are taken in the grand canonical statistical ensemble (5.51), which can be written in the equivalent form (5.57). This initial probability distribution is symmetric under time reversal because of equation (5.58). The changes of energy and particle numbers are measured during every time interval $[t_{i-1}, t_i]$, giving

$$\Delta \mathbf{C}_i \equiv \mathbf{C}\left(\mathbf{\Phi}^{t_{i-1}} \mathbf{\Gamma}_0\right) - \mathbf{C}\left(\mathbf{\Phi}^{t_i} \mathbf{\Gamma}_0\right). \tag{5.126}$$

Their joint probability density is defined by

$$p(\Delta \mathbf{C}_1, \ldots, \Delta \mathbf{C}_k) \equiv \int d\mathbf{\Gamma}_0\, p_0(\mathbf{\Gamma}_0) \prod_{i=1}^{k} \delta\left[\Delta \mathbf{C}_i - \mathbf{C}\left(\mathbf{\Phi}^{t_{i-1}} \mathbf{\Gamma}_0\right) + \mathbf{C}\left(\mathbf{\Phi}^{t_i} \mathbf{\Gamma}_0\right)\right]. \tag{5.127}$$

The successive random variables $\{\Delta \mathbf{C}_i\}_{i=1}^{k}$ represent a path of the process and equation (5.127) the corresponding path probability density.

Carrying out the same calculations as between equations (5.59) and (5.65), we here obtain

$$p(\Delta \mathbf{C}_1, \ldots, \Delta \mathbf{C}_k) = \prod_{i=1}^{k} \mathrm{e}^{\boldsymbol{\alpha} \cdot \Delta \mathbf{C}_i} \int d\boldsymbol{\Gamma}'_0\, p_0(\boldsymbol{\Gamma}'_0) \times \prod_{i=1}^{k} \delta\left[-\Delta \mathbf{C}_i - \mathbf{C}\left(\boldsymbol{\Phi}^{t-t_i}\boldsymbol{\Gamma}'_0\right) + \mathbf{C}\left(\boldsymbol{\Phi}^{t-t_{i-1}}\boldsymbol{\Gamma}'_0\right)\right], \tag{5.128}$$

where $\boldsymbol{\Gamma}'_0 = \boldsymbol{\Theta}\boldsymbol{\Gamma}_t$ is the initial condition of the time-reversed trajectory. We note the integral in the right-hand side of this result can be interpreted as the joint probability of the time-reversed path sampled at the successive times $t'_i = t - t_{k-i}$ running backward in time. We may thus introduce a new index $j \equiv k - i + 1$, such that equation (5.128) becomes

$$p(\Delta \mathbf{C}_1, \ldots, \Delta \mathbf{C}_k) = \prod_{i=1}^{k} \mathrm{e}^{\boldsymbol{\alpha} \cdot \Delta \mathbf{C}_i} \int d\boldsymbol{\Gamma}'_0\, p_0(\boldsymbol{\Gamma}'_0) \times \prod_{j=1}^{k} \delta[-\Delta \mathbf{C}_{k-j+1} - \mathbf{C}(\boldsymbol{\Phi}^{t'_{j-1}}\boldsymbol{\Gamma}'_0) + \mathbf{C}(\boldsymbol{\Phi}^{t'_j}\boldsymbol{\Gamma}'_0)], \tag{5.129}$$

which can now be compared with definition (5.127). Accordingly, the following time-reversal symmetry relation is obtained,

$$p(\Delta \mathbf{C}_1, \ldots, \Delta \mathbf{C}_k) = \prod_{i=1}^{k} \mathrm{e}^{\boldsymbol{\alpha} \cdot \Delta \mathbf{C}_i}\, p(-\Delta \mathbf{C}_k, \ldots, -\Delta \mathbf{C}_1), \tag{5.130}$$

where the left-hand side contains the joint probability of the time-reversed path.

Using the same reasoning as in Section 5.3.3 with the energy and particle fluxes towards the reference reservoir r given by equation (5.67), we find the fluctuation relation

$$\frac{p(\Delta \mathfrak{C}_1, \ldots, \Delta \mathfrak{C}_k; \mathbf{A})}{p(-\Delta \mathfrak{C}_k, \ldots, -\Delta \mathfrak{C}_1; \mathbf{A})} = \prod_{i=1}^{k} \mathrm{e}^{\mathbf{A} \cdot \Delta \mathfrak{C}_i} = \mathrm{e}^{\Sigma(t)} \tag{5.131}$$

for the path of energy and particles exchanges $\{\Delta \mathfrak{C}_i\}_{i=1}^{k}$ between the reservoirs at the successive times $t_i = i\,\Delta t$ with $i = 1, 2, \ldots, k$ and its time reversal if the reservoirs have the thermal and chemical affinities $\mathbf{A}$ given by equations (5.69) and (5.70). The second equality in equation (5.131) results from the fact that the sum of successive exchanges gives the total exchanges $\Delta \mathfrak{C}(t) = \sum_{i=1}^{k} \Delta \mathfrak{C}_i$ over the time interval $[0, t]$ with $t = k\Delta t$ and that $\mathbf{A} \cdot \Delta \mathfrak{C}(t) = \Sigma(t)$ by the definition (5.103). Now, the mean value of this latter random variable gives the entropy production rate according to equation (5.108), whereupon

$$\frac{1}{k_{\mathrm{B}}} \frac{d_{\mathrm{i}} S}{dt} = \lim_{k \to \infty} \frac{1}{k \Delta t} \int \prod_{i=1}^{k} d\Delta \mathfrak{C}_i\, p(\Delta \mathfrak{C}_1, \ldots, \Delta \mathfrak{C}_k; \mathbf{A}) \times \ln \frac{p(\Delta \mathfrak{C}_1, \ldots, \Delta \mathfrak{C}_k; \mathbf{A})}{p(-\Delta \mathfrak{C}_k, \ldots, -\Delta \mathfrak{C}_1; \mathbf{A})} \geq 0. \tag{5.132}$$

Remarkably, the fluctuation relation (5.131) gives the same random value $\Sigma(t)$ for any number k of measurements. Therefore, the entropy production can be evaluated using the statistics of paths and their reversals independently of their sampling in the measurement of the energy and particle exchanges between the reservoirs.

Similar considerations apply to nonequilibrium work fluctuations using the protocols defined in Section 5.2. On the one hand, equation (5.21), which gives the mean value of the dissipated work (5.26), can be generalized by replacing the phase-space integral into a multiple integral over the paths $\{\boldsymbol{\Gamma}(i\,\Delta t)\}_{i=1}^{k}$ sampled with time steps Δt in the limit $k \to \infty$ (Jarzynski, 2006; Gomez-Marin et al., 2008). On the other hand, a fluctuation relation can be obtained for the paths $\{W_i\}_{i=1}^{k}$ of nonequilibrium work sampled at successive time steps, as in equation (5.131) with $\Sigma(t)$ replaced by $\beta W_{\text{diss}} = \beta\left(\sum_{i=1}^{k} W_i - \Delta F\right)$ (Gaspard, 2010).

5.5 Linear and Nonlinear Response Properties

Since the symmetry relation (5.91) is the expression of microreversibility for the current fluctuations driven by the affinities $\mathbf{A}$, we may wonder what the implications on the linear and nonlinear response properties of the currents are. As shown here in Section 5.5, such implications can be systematically deduced for cumulants and their responses up to arbitrarily high orders.

5.5.1 *Response Coefficients*

At equilibrium, the mean currents are equal to zero with the affinities, $\mathbf{J}(\mathbf{0}) = \mathbf{0}$. Therefore, the mean currents can be expanded in powers of the affinities in the vicinity of equilibrium according to

$$J_\alpha = \sum_{\beta} L_{\alpha,\beta}\, A_\beta + \frac{1}{2}\sum_{\beta,\gamma} M_{\alpha,\beta\gamma}\, A_\beta A_\gamma + \frac{1}{6}\sum_{\beta,\gamma,\delta} N_{\alpha,\beta\gamma\delta}\, A_\beta A_\gamma A_\delta + \cdots \tag{5.133}$$

in terms of the response coefficients defined as

$$L_{\alpha,\beta} \equiv \frac{\partial J_\alpha}{\partial A_\beta}(\mathbf{0}) = \frac{\partial^2 Q}{\partial\lambda_\alpha \partial A_\beta}(\mathbf{0};\mathbf{0}), \tag{5.134}$$

$$M_{\alpha,\beta\gamma} \equiv \frac{\partial^2 J_\alpha}{\partial A_\beta \partial A_\gamma}(\mathbf{0}) = \frac{\partial^3 Q}{\partial\lambda_\alpha \partial A_\beta \partial A_\gamma}(\mathbf{0};\mathbf{0}), \tag{5.135}$$

$$N_{\alpha,\beta\gamma\delta} \equiv \frac{\partial^3 J_\alpha}{\partial A_\beta \partial A_\gamma \partial A_\delta}(\mathbf{0}) = \frac{\partial^4 Q}{\partial\lambda_\alpha \partial A_\beta \partial A_\gamma \partial A_\delta}(\mathbf{0};\mathbf{0}), \tag{5.136}$$

$$\vdots$$

We note that, by definition, these coefficients are symmetric under the permutations of the indices associated with the affinities, so that $M_{\alpha,\beta\gamma} = M_{\alpha,\gamma\beta}$, $N_{\alpha,\beta\gamma\delta} = N_{\alpha,\gamma\beta\delta} = N_{\alpha,\gamma\delta\beta} = \dots$.

These response coefficients also express the coupling between different currents, leading to energy transduction, e.g., in thermoelectric and mechanochemical effects. In particular, the linear response coefficient $L_{\alpha,\beta}$ with $\alpha \neq \beta$ describes the coupling between the current

J_α and the affinity A_β. Consequently, the current J_α can be driven by the affinity A_β even though all the other affinities are equal to zero, i.e., if $A_\alpha = 0$ for $\alpha \neq \beta$. Such couplings are not possible at the level of linear response if $L_{\alpha,\beta} = 0$. In such cases, the coupling may nevertheless arise at the level of nonlinear responses, for instance, if $M_{\alpha,\beta\beta} \neq 0$ or $N_{\alpha,\beta\beta\beta} \neq 0$ for $\alpha \neq \beta$.

Since the response coefficients are given by the derivatives of the cumulant generating function with respect to the counting parameter λ_α and the affinities, the symmetry relation (5.91) can be used to find their relationships to the statistical cumulants (5.82), (5.83), (5.84), ... and their own responses with respect to the affinities. These relations can be obtained by taking successive partial derivatives of the symmetry relation (5.91) with respect to the counting parameters and the affinities.

5.5.2 Implications for the Cumulants at Equilibrium

We start taking the partial derivative of the symmetry relation (5.91) with respect to the counting parameter λ_α to get

$$\frac{\partial Q}{\partial \lambda_\alpha}(\boldsymbol{\lambda}; \mathbf{A}) = -\frac{\partial Q}{\partial \lambda_\alpha}(\mathbf{A} - \boldsymbol{\lambda}; \mathbf{A}). \tag{5.137}$$

Setting $\boldsymbol{\lambda} = \mathbf{0}$ and $\mathbf{A} = \mathbf{0}$ gives

$$J_\alpha(\mathbf{0}) = \frac{\partial Q}{\partial \lambda_\alpha}(\mathbf{0}; \mathbf{0}) = 0, \tag{5.138}$$

confirming the general result that the mean currents are equal to zero at equilibrium.

Further partial derivatives with respect to the counting parameters give

$$\frac{\partial}{\partial \lambda_{\alpha_1}} \cdots \frac{\partial}{\partial \lambda_{\alpha_m}} Q(\boldsymbol{\lambda}; \mathbf{A}) = (-1)^m \, \frac{\partial}{\partial \lambda_{\alpha_1}} \cdots \frac{\partial}{\partial \lambda_{\alpha_m}} Q(\mathbf{A} - \boldsymbol{\lambda}; \mathbf{A}), \tag{5.139}$$

so that

$$Q_{\alpha_1 \cdots \alpha_m}(\mathbf{0}) = (-1)^m \, Q_{\alpha_1 \cdots \alpha_m}(\mathbf{0}) \tag{5.140}$$

for the mth-order cumulants (5.78) at equilibrium $\mathbf{A} = \mathbf{0}$. Consequently, all the cumulants of odd order are equal to zero at equilibrium. Equation (5.138) is the particular case of this result for $m = 1$.

5.5.3 Linear Response Properties

We continue by taking the partial derivative of the relation (5.137) with respect to the affinity A_β, which gives

$$\frac{\partial^2 Q}{\partial \lambda_\alpha \partial A_\beta}(\boldsymbol{\lambda}; \mathbf{A}) = -\frac{\partial^2 Q}{\partial \lambda_\alpha \partial \lambda_\beta}(\mathbf{A} - \boldsymbol{\lambda}; \mathbf{A}) - \frac{\partial^2 Q}{\partial \lambda_\alpha \partial A_\beta}(\mathbf{A} - \boldsymbol{\lambda}; \mathbf{A}). \tag{5.141}$$

Setting $\boldsymbol{\lambda} = \mathbf{0}$ and $\mathbf{A} = \mathbf{0}$ and using the definitions (5.82) and (5.134), we find that every linear response coefficient is given by the corresponding diffusivity of the currents

$$L_{\alpha,\beta} = D_{\alpha\beta}(\mathbf{0}), \tag{5.142}$$

which is the expression of the *fluctuation–dissipation theorem* presented in Section 2.8.4 and which holds in the regime of linear response.

Now, since the diffusivities are symmetric under the permutation of their indices $D_{\alpha\beta} = D_{\beta\alpha}$, we obtain the *Onsager reciprocal relations*,

$$L_{\alpha,\beta} = L_{\beta,\alpha}, \tag{5.143}$$

as a consequence of equation (5.142) (Onsager, 1931a,b).

The other partial derivatives of the symmetry relation (5.91) of second order with respect to the counting parameters or the affinities do not give more relations.

5.5.4 Nonlinear Response Properties at Second Order

According to equation (5.140) for $m = 3$, the equilibrium value of every third cumulant (5.83) is equal to zero:

$$C_{\alpha\beta\gamma}(\mathbf{0}) = 0. \tag{5.144}$$

We proceed by taking the partial derivative of equation (5.141) with respect to the affinity A_γ, so that we obtain

$$\begin{aligned}\frac{\partial^3 Q}{\partial\lambda_\alpha \partial A_\beta \partial A_\gamma}(\boldsymbol{\lambda};\mathbf{A}) = &-\frac{\partial^3 Q}{\partial\lambda_\alpha \partial\lambda_\beta \partial\lambda_\gamma}(\mathbf{A}-\boldsymbol{\lambda};\mathbf{A}) - \frac{\partial^3 Q}{\partial\lambda_\alpha \partial\lambda_\beta \partial A_\gamma}(\mathbf{A}-\boldsymbol{\lambda};\mathbf{A}) \\ &-\frac{\partial^3 Q}{\partial\lambda_\alpha \partial A_\beta \partial\lambda_\gamma}(\mathbf{A}-\boldsymbol{\lambda};\mathbf{A}) - \frac{\partial^3 Q}{\partial\lambda_\alpha \partial A_\beta \partial A_\gamma}(\mathbf{A}-\boldsymbol{\lambda};\mathbf{A}).\end{aligned} \tag{5.145}$$

For $\boldsymbol{\lambda} = \mathbf{0}$ and $\mathbf{A} = \mathbf{0}$, we find the relation

$$M_{\alpha,\beta\gamma} = \left(\frac{\partial D_{\alpha\beta}}{\partial A_\gamma} + \frac{\partial D_{\alpha\gamma}}{\partial A_\beta}\right)_{\mathbf{A}=\mathbf{0}}, \tag{5.146}$$

which shows that the nonlinear response coefficients at second order in the affinities are given in terms of the first responses of the diffusivities (Bochkov and Kuzovlev, 1979; Stratonovich, 1992; Andrieux and Gaspard, 2004, 2006a, 2007a). These relations generalize the fluctuation–dissipation relations (5.142) to the first nonlinear response coefficients. They will be applied to effusion in Section 8.7, transistors in Section 10.6.3, and chemical reactions in Section 11.3.

5.5.5 Nonlinear Response Properties at Third Order

At the next order, we first consider the partial derivative of equation (5.139) for $m = 3$ with respect to the affinity A_δ, which gives

$$\frac{\partial^4 Q}{\partial\lambda_\alpha \partial\lambda_\beta \partial\lambda_\gamma \partial A_\delta}(\boldsymbol{\lambda};\mathbf{A}) = -\frac{\partial^4 Q}{\partial\lambda_\alpha \partial\lambda_\beta \partial\lambda_\gamma \partial\lambda_\delta}(\mathbf{A}-\boldsymbol{\lambda};\mathbf{A}) - \frac{\partial^4 Q}{\partial\lambda_\alpha \partial\lambda_\beta \partial\lambda_\gamma \partial A_\delta}(\mathbf{A}-\boldsymbol{\lambda};\mathbf{A}). \tag{5.147}$$

Setting $\boldsymbol{\lambda} = \mathbf{0}$ and $\mathbf{A} = \mathbf{0}$ leads to the following relation between the fourth cumulants (5.84) and the first responses of the third cumulants (5.83), both at equilibrium,

$$B_{\alpha\beta\gamma\delta}(\mathbf{0}) = \frac{\partial C_{\alpha\beta\gamma}}{\partial A_\delta}(\mathbf{0}). \tag{5.148}$$

Since the fourth cumulants are totally symmetric under the permutations of their four indices, the same holds for the first responses of the third cumulants. In this regard, equation (5.148) is a generalization of the Onsager reciprocal relations.

Now, if we take the partial derivative of equation (5.145) with respect to the affinity A_δ, we get

$$\begin{aligned}\frac{\partial^4 Q}{\partial\lambda_\alpha\partial A_\beta\partial A_\gamma\partial A_\delta}(\boldsymbol{\lambda};\mathbf{A}) = &-\frac{\partial^4 Q}{\partial\lambda_\alpha\partial\lambda_\beta\partial\lambda_\gamma\partial\lambda_\delta}(\mathbf{A}-\boldsymbol{\lambda};\mathbf{A}) - \frac{\partial^4 Q}{\partial\lambda_\alpha\partial\lambda_\beta\partial\lambda_\gamma\partial A_\delta}(\mathbf{A}-\boldsymbol{\lambda};\mathbf{A})\\ &-\frac{\partial^4 Q}{\partial\lambda_\alpha\partial\lambda_\beta\partial A_\gamma\partial\lambda_\delta}(\mathbf{A}-\boldsymbol{\lambda};\mathbf{A}) - \frac{\partial^4 Q}{\partial\lambda_\alpha\partial\lambda_\beta\partial A_\gamma\partial A_\delta}(\mathbf{A}-\boldsymbol{\lambda};\mathbf{A})\\ &-\frac{\partial^4 Q}{\partial\lambda_\alpha\partial A_\beta\partial\lambda_\gamma\partial\lambda_\delta}(\mathbf{A}-\boldsymbol{\lambda};\mathbf{A}) - \frac{\partial^4 Q}{\partial\lambda_\alpha\partial A_\beta\partial\lambda_\gamma\partial A_\delta}(\mathbf{A}-\boldsymbol{\lambda};\mathbf{A})\\ &-\frac{\partial^4 Q}{\partial\lambda_\alpha\partial A_\beta\partial A_\gamma\partial\lambda_\delta}(\mathbf{A}-\boldsymbol{\lambda};\mathbf{A}) - \frac{\partial^4 Q}{\partial\lambda_\alpha\partial A_\beta\partial A_\gamma\partial A_\delta}(\mathbf{A}-\boldsymbol{\lambda};\mathbf{A}).\end{aligned} \tag{5.149}$$

For $\boldsymbol{\lambda} = \mathbf{0}$ and $\mathbf{A} = \mathbf{0}$, we find

$$N_{\alpha,\beta\gamma\delta} = \left(\frac{\partial^2 D_{\alpha\beta}}{\partial A_\gamma\partial A_\delta} + \frac{\partial^2 D_{\alpha\gamma}}{\partial A_\beta\partial A_\delta} + \frac{\partial^2 D_{\alpha\delta}}{\partial A_\beta\partial A_\gamma} - \frac{1}{2}B_{\alpha\beta\gamma\delta}\right)_{\mathbf{A}=\mathbf{0}}, \tag{5.150}$$

so that the nonlinear response coefficients at third order in the affinities can be expressed in terms of the second responses of the diffusivities and the fourth cumulants (Bochkov and Kuzovlev, 1979; Stratonovich, 1992; Andrieux and Gaspard, 2004, 2007a).

These relations are the consequences of the time-reversal symmetry relation (5.91) and they generalize the fluctuation–dissipation theorem and the Onsager reciprocal relations from linear towards nonlinear response properties.

5.5.6 Nonlinear Response Properties at Higher Orders

By taking further derivatives of the symmetry relation (5.91), generalizations up to arbitrarily high orders can be obtained (Bochkov and Kuzovlev, 1979; Stratonovich, 1992; Andrieux and Gaspard, 2007a; Andrieux, 2009; Barbier and Gaspard, 2018). The cumulant generating function (5.76) can be expanded in powers series of both the counting parameters $\boldsymbol{\lambda}$ and the affinities $\mathbf{A}$ according to

$$Q\,(\boldsymbol{\lambda};\mathbf{A}) = \sum_{m,n=0}^{\infty}\frac{1}{m!\,n!}\sum_{\substack{\alpha_1\cdots\alpha_m\\ \beta_1\cdots\beta_n}} Q_{\alpha_1\cdots\alpha_m,\,\beta_1\cdots\beta_n}\lambda_{\alpha_1}\cdots\lambda_{\alpha_m}A_{\beta_1}\cdots A_{\beta_n} \tag{5.151}$$

with the coefficients

$$Q_{\alpha_1\cdots\alpha_m,\,\beta_1\cdots\beta_n} \equiv \frac{\partial^n Q_{\alpha_1\cdots\alpha_m}}{\partial A_{\beta_1}\cdots\partial A_{\beta_n}}(\mathbf{0}) = \frac{\partial^{m+n} Q}{\partial\lambda_{\alpha_1}\cdots\partial\lambda_{\alpha_m}\partial A_{\beta_1}\cdots\partial A_{\beta_n}}(\mathbf{0};\mathbf{0}), \tag{5.152}$$

which are totally symmetric with respect to the permutations of the m indices α_j and the n indices β_k since partial derivatives commute.

Making the substitution $\boldsymbol{\lambda} \to -\boldsymbol{\lambda}$ and introducing the translation operator

$$\hat{T}(\mathbf{A}) \equiv \exp(\mathbf{A}\cdot\partial_{\boldsymbol{\lambda}}), \tag{5.153}$$

the symmetry relation (5.91) can be written in the following equivalent form:

$$Q(-\boldsymbol{\lambda};\mathbf{A}) = \hat{T}(\mathbf{A})\,Q(\boldsymbol{\lambda};\mathbf{A}). \tag{5.154}$$

Replacing the cumulant generating function by its expansion (5.151) in both sides of equation (5.154) and identifying the coefficients lead to the relations

$$m \text{ odd}: \qquad Q_{\alpha_1\cdots\alpha_m,\,\beta_1\cdots\beta_n} = -\frac{1}{2}\sum_{j=1}^{n} Q^{(j)}_{\alpha_1\cdots\alpha_m,\,\beta_1\cdots\beta_n}, \tag{5.155}$$

with

$$Q^{(j)}_{\alpha_1\cdots\alpha_m,\,\beta_1\cdots\beta_n} \equiv \sum_{k_1=1}^{n}\sum_{\substack{k_2=1\\k_2>k_1}}^{n}\cdots\sum_{\substack{k_j=1\\k_j>k_{j-1}}}^{n} Q_{\alpha_1\cdots\alpha_m\beta_{k_1}\cdots\beta_{k_j},\,(\cdot)} \tag{5.156}$$

for $j \geqslant 1$, where $(\cdot)$ denotes the set of all subscripts β_k that are different from the subscripts $\beta_{k_1},\ldots,\beta_{k_j}$ present on the left of the comma, and $Q^{(0)} \equiv Q$ for $j = 0$ (Stratonovich, 1992; Andrieux and Gaspard, 2007a; Andrieux, 2009; Barbier and Gaspard, 2018). The proof is given in Section F.1. All the relations (5.155) can be ordered in terms of the integer $\mathcal{N} \equiv m+n$. In particular, equation (5.142) involving the linear response coefficients is recovered for $\mathcal{N}=2$ and $m=1$; equation (5.144) for $\mathcal{N}=3$ and $m=3$; equation (5.146) for $\mathcal{N}=3$ and $m=1$; equation (5.148) for $\mathcal{N}=4$ and $m=3$; equation (5.150) for $\mathcal{N}=4$ and $m=1$; and so on up to arbitrarily high orders.

The relations (5.155) imply that microreversibility systematically reduces the number of independent coefficients (5.152) required to determine the cumulant generating function and thus the full counting statistics of the currents. Actually, the enumeration of all the different coefficients (5.152) and all the independent relations (5.155) caused by microreversibility shows that the number of independent coefficients is divided by two for $\mathcal{N}\to\infty$ (Barbier and Gaspard, 2018). Indeed, if there are χ affinities and counting parameters, the total number of different coefficients (5.152) with $\mathcal{N}=m+n$ is given by

$$N_\chi(\mathcal{N}) = \sum_{m=1}^{\mathcal{N}}\binom{m+\chi-1}{\chi-1}\binom{\mathcal{N}-m+\chi-1}{\chi-1}, \tag{5.157}$$

since they are invariant under the $m!$ (respectively $n!$) permutations of the indices α_j (respectively β_k) with $n=\mathcal{N}-m$, and $Q_{,\beta_1\cdots\beta_n}=0$ because of the normalization condition (5.77).

Now, the number of relations (5.155) is given by a sum similar to equation (5.157), but limited to m odd. Consequently, the total number of independent coefficients with $\mathcal{N} = m + n$ is equal to the sum restricted to the terms with m even, which thus behaves as (Barbier and Gaspard, 2018)

$$\frac{N_\chi^{(\rm ind)}(\mathcal{N})}{N_\chi(\mathcal{N})} \simeq \frac{1}{2} \qquad \text{for} \qquad \mathcal{N} \gg 1. \tag{5.158}$$

These results demonstrate that microreversibility has consequences on the nonlinear transport properties up to arbitrarily high orders and, thus, arbitrarily farther away from the linear regime close to equilibrium.

5.5.7 Odd versus Even Cumulants

The relations (5.155) and (5.156) show that the odd cumulants and their responses are given by cumulants of higher orders. By recurrence, all the odd cumulants and their responses could thus be expressed in terms of even cumulants and their responses in the right-hand side of equation (5.155). This result is confirmed by introducing the parts of the cumulant generating function that are even or odd in the counting parameters as

$$Q_\pm(\boldsymbol{\lambda};\mathbf{A}) \equiv \frac{1}{2}\left[Q(\boldsymbol{\lambda};\mathbf{A}) \pm Q(-\boldsymbol{\lambda};\mathbf{A})\right]. \tag{5.159}$$

Substituting the symmetry relation (5.154) with the translation operator (5.153), equation (5.159) becomes

$$Q_\pm(\boldsymbol{\lambda};\mathbf{A}) = \frac{1}{2}\left(1 \pm e^{\mathbf{A}\cdot\partial_{\boldsymbol{\lambda}}}\right) Q(\boldsymbol{\lambda};\mathbf{A}). \tag{5.160}$$

Multiplying by $\left(1 \mp e^{\mathbf{A}\cdot\partial_{\boldsymbol{\lambda}}}\right)$ and using the identity

$$\tanh\frac{t}{2} = \frac{e^t - 1}{e^t + 1}, \tag{5.161}$$

we find the result

$$Q_-(\boldsymbol{\lambda};\mathbf{A}) = -\tanh\left(\frac{1}{2}\mathbf{A}\cdot\partial_{\boldsymbol{\lambda}}\right) Q_+(\boldsymbol{\lambda};\mathbf{A}), \tag{5.162}$$

which is an equivalent form of the symmetry relation (5.91).

Now, the expansion (5.86) of the cumulant generating function can be substituted into equation (5.162) leading to the following relations between the cumulants given by equations (5.80), (5.82), (5.83), (5.84), ...,

$$J_\alpha(\mathbf{A}) = \sum_\beta D_{\alpha\beta}(\mathbf{A})\,A_\beta - \frac{1}{12}\sum_{\beta,\gamma,\delta} B_{\alpha\beta\gamma\delta}(\mathbf{A})\,A_\beta\,A_\gamma\,A_\delta + O\left(\mathbf{A}^5\right), \tag{5.163}$$

$$C_{\alpha\beta\gamma}(\mathbf{A}) = \sum_\delta B_{\alpha\beta\gamma\delta}(\mathbf{A})\,A_\delta + O\left(\mathbf{A}^3\right), \tag{5.164}$$

$$\vdots$$

This confirms that the odd cumulants can be related to the even cumulants. With these relations, the entropy production rate (5.113) can thus be expressed as

$$\frac{1}{k_{\rm B}}\frac{d_{\rm i}S}{dt} = \sum_{\alpha,\beta} D_{\alpha\beta}(\mathbf{A})\, A_\alpha\, A_\beta - \frac{1}{12}\sum_{\alpha,\beta,\gamma,\delta} B_{\alpha\beta\gamma\delta}(\mathbf{A})\, A_\alpha\, A_\beta\, A_\gamma\, A_\delta + O\big(\mathbf{A}^6\big) \tag{5.165}$$

in terms of the even cumulants.

5.5.8 *Current Rectification*

The nonlinear response coefficients play an important role in current rectification, manifesting itself, for instance, in diodes, which allow a higher current in one direction than the other. In these devices, a single current $J(A)$ is driven by a single affinity A. The statistical cumulants defined by equations (5.82)–(5.84) for the fluctuating current can be expanded in powers of this affinity as

$$D(A) = D_0 + D_1\, A + \frac{1}{2}\, D_2\, A^2 + O\big(A^3\big), \tag{5.166}$$

$$C(A) = C_0 + C_1\, A + O\big(A^2\big), \tag{5.167}$$

$$B(A) = B_0 + O(A), \tag{5.168}$$

where $C_0 = 0$ by equation (5.144) and $B_0 = C_1$ by equation (5.148). Inserting these results into equation (5.163), we obtain

$$J(A) = D_0\, A + D_1\, A^2 + \left(\frac{1}{2}\, D_2 - \frac{1}{12}\, C_1\right) A^3 + O\big(A^4\big). \tag{5.169}$$

Rectification is due to the terms with even powers of the affinity. The first of them has the coefficient D_1, which characterizes the sensitivity of the diffusivity of current fluctuations with respect to some perturbation A away from equilibrium. This coefficient gives the main contribution to the rectification ratio

$$R \equiv \frac{|J(A) + J(-A)|}{|J(A)|} = 2\left|\frac{D_1 A}{D_0}\right| + O\big(A^2\big). \tag{5.170}$$

In this regard, rectification finds its origin in the nonequilibrium properties of current fluctuations (Gaspard, 2013a).

5.6 Fluctuation Relation for Currents in a Magnetizing Field

5.6.1 *Microreversibility for Open Systems in a Magnetizing Field*

A system of charged particles in an external magnetizing field is, in general, included in a global system also composed of magnetic coils where electric currents circulate to generate the magnetizing field. Therefore, upon time reversal, those external electric currents are also reversed, so that the magnetizing field as well as the magnetic field and the vector potential are odd under time reversal (Jackson, 1999). The external magnetizing field $\mathcal{H}$ is supposed to be static.

The Hamiltonian function has the symmetry of equation (2.30) under the time-reversal transformation $\boldsymbol{\Theta}(\mathbf{r}_a, \mathbf{p}_a) = (\mathbf{r}_a, -\mathbf{p}_a)$, and the solutions of Hamilton's equations have the symmetry given by equation (2.31). For nonrelativistic particles with masses m_a and charges q_a, the Hamiltonian function can be expressed as in equation (5.45) in terms of the microscopic energy density (3.6), but the energy of the ath particle is here given by

$$\varepsilon_a(\boldsymbol{\Gamma};\mathcal{H}) \equiv \frac{1}{2m_a}\left[\mathbf{p}_a - q_a\mathcal{A}(\mathbf{r}_a;\mathcal{H})\right]^2 + \frac{1}{2}\sum_{b(\neq a)} u_{ab} + \cdots, \tag{5.171}$$

with the static vector potential $\mathcal{A}(\mathbf{r};\mathcal{H}) = -\mathcal{A}(\mathbf{r};-\mathcal{H})$. As a consequence, the symmetry of microreversibility implies

$$H_j(\boldsymbol{\Theta\Gamma};\mathcal{H}) = H_j(\boldsymbol{\Gamma};-\mathcal{H}). \tag{5.172}$$

As in Section 5.3.1, the energy and particle numbers in the jth reservoir are defined by integrating the corresponding densities according to equations (5.46) and (5.48). The conservation laws of the quantities (5.45) and (5.47) are satisfied as a consequence of the fact that the system is the union of its parts $V = \cup_{j=1}^{r} V_j$. Accordingly, the exchanges of energy and particles during the time interval $[0,t]$ are similarly defined by equations (5.53) and (5.54), but with the time evolution $\boldsymbol{\Phi}^t$ replaced by $\boldsymbol{\Phi}^t_{\mathcal{H}}$, satisfying equation (2.31).

5.6.2 Fluctuation Relation for the Global Currents

The multivariate fluctuation relation for the currents can here also be proved as in Section 5.3.2 for the probability density (5.59) of energy and particle exchanges $\Delta\mathbf{C}$. The initial probability distribution is here also taken as the grand canonical distribution (5.51), which introduces the temperatures and the chemical potentials of the reservoirs. Because of the symmetry (5.172) and since the particle numbers do not depend on $\mathcal{H}$, the initial grand canonical probability distribution satisfies

$$p_0(\boldsymbol{\Theta\Gamma};\mathcal{H}) = p_0(\boldsymbol{\Gamma};-\mathcal{H}), \tag{5.173}$$

and, instead of equation (5.58), we here have that

$$\mathbf{C}(\boldsymbol{\Theta\Gamma};\mathcal{H}) = \mathbf{C}(\boldsymbol{\Gamma};-\mathcal{H}). \tag{5.174}$$

Accordingly, when microreversibility is considered in the steps between equations (5.63) and (5.64), the symmetry relation $\boldsymbol{\Phi}^{-t}_{\mathcal{H}}\boldsymbol{\Theta} = \boldsymbol{\Theta\Phi}^t_{-\mathcal{H}}$ should instead be used. Consequently, the probability density in the reversed magnetizing field appears in the right-hand side of equation (5.66), which here becomes

$$p_t(\Delta\mathbf{C};\mathcal{H}) = \mathrm{e}^{\boldsymbol{\alpha}\cdot\Delta\mathbf{C}}\, p_t(-\Delta\mathbf{C};-\mathcal{H}), \tag{5.175}$$

where $\boldsymbol{\alpha}$ are the parameters introduced in equation (5.57).

Finally, using the conservation laws (5.67) as in Section 5.3.3, we obtain the *multivariate fluctuation relation* in the external magnetizing field $\mathcal{H}$ given by

$$\frac{p_t(\mathfrak{J};\mathbf{A},\mathcal{H})}{p_t(-\mathfrak{J};\mathbf{A},-\mathcal{H})} = \mathrm{e}^{\mathbf{A}\cdot\mathfrak{J}\,t} \tag{5.176}$$

for the fluctuating currents $\mathfrak{J} \equiv \Delta\mathfrak{C}/t$ due to the energy and particle exchanges (5.72) and with the affinities (5.71) defined in terms of the thermal and chemical affinities (5.69) and (5.70).

The multivariate fluctuation relation leads to the following symmetry relation for the cumulant generating function (5.76) of the fluctuating currents,

$$Q(\boldsymbol{\lambda}; \mathbf{A}, \mathcal{H}) = Q(\mathbf{A} - \boldsymbol{\lambda}; \mathbf{A}, -\mathcal{H}). \tag{5.177}$$

This symmetry relation generalizes equation (5.91) to systems subjected to an external magnetizing field $\mathcal{H}$, where microreversibility maps the dynamics of the system to the one in the opposite magnetizing field $-\mathcal{H}$. The implications of this result for entropy production, and the linear and nonlinear transport properties can thus be deduced *mutatis mutandis*.

5.6.3 Transport Properties in a Magnetizing Field

As in equation (5.86), the cumulant generating function in the magnetizing field $\mathcal{H}$ can be expanded in powers of the counting parameters $\boldsymbol{\lambda}$, defining the successive cumulants (5.80), (5.82), (5.83), (5.84), ... Furthermore, the cumulants can also be expanded in powers of the affinities $\mathbf{A}$ around the equilibrium state where the affinities are equal to zero, which leads to the full expansion (5.151) in terms of the coefficients (5.152) giving the cumulants and their responses at equilibrium. In particular, the first cumulants, i.e., the mean values of the currents, can be expanded as in equation (5.133), defining the linear and nonlinear response coefficients (5.134), (5.135), (5.136), ..., here for given magnetizing field $\mathcal{H}$. By substituting these expansions into the symmetry relation (5.177), a hierarchy of relations is obtained as a consequence of microreversibility. These relations can also be obtained by taking successive derivatives of the symmetry relation (5.177) with respect to the counting parameters $\boldsymbol{\lambda}$ and the affinities $\mathbf{A}$. In this way, we find relations similar to equations (5.137), (5.141), (5.145), (5.147), (5.149), ..., but here with opposite values $\pm\mathcal{H}$ of the magnetizing field on each side. The following results can thus be deduced (Andrieux et al., 2009; Andrieux, 2009; Barbier and Gaspard, 2019). They will be applied to the case of quantum transport in Aharonov–Bohm rings in Section 15.5.4.

Linear Transport Properties

Twice differentiating the symmetry relation (5.177) with respect to $\boldsymbol{\lambda}$ or $\mathbf{A}$ and setting them equal to zero yields the Onsager–Casimir reciprocal relations (Onsager, 1931a,b; Casimir, 1945)

$$L_{\alpha,\beta}(\mathcal{H}) = L_{\beta,\alpha}(-\mathcal{H}), \tag{5.178}$$

as well as the fluctuation–dissipation theorem

$$D_{\alpha\beta}(\mathbf{0}, \mathcal{H}) = \frac{1}{2}\left[L_{\alpha,\beta}(\mathcal{H}) + L_{\alpha,\beta}(-\mathcal{H})\right] = D_{\alpha\beta}(\mathbf{0}, -\mathcal{H}) \tag{5.179}$$

in the presence of the magnetizing field $\mathcal{H}$. The latter shows that the diffusivities $D_{\alpha\beta}$ at equilibrium are fully specified by the linear response coefficients $L_{\alpha,\beta}$.

Nonlinear Transport Properties at Second Order

Remarkably, similar relations hold beyond the linear response regime. Taking third derivatives of the symmetry relation (5.177) leads to the following formulas for the third cumulants (5.83):

$$C_{\alpha\beta\gamma}(\mathbf{0},\mathcal{H}) = -C_{\alpha\beta\gamma}(\mathbf{0},-\mathcal{H}), \tag{5.180}$$

$$C_{\alpha\beta\gamma}(\mathbf{0},\mathcal{H}) = 2\,\frac{\partial D_{\alpha\beta}}{\partial A_\gamma}(\mathbf{0},\mathcal{H}) - 2\,\frac{\partial D_{\alpha\beta}}{\partial A_\gamma}(\mathbf{0},-\mathcal{H}), \tag{5.181}$$

$$C_{\alpha\beta\gamma}(\mathbf{0},\mathcal{H}) = -\left(M_{\alpha,\beta\gamma} + M_{\beta,\gamma\alpha} + M_{\gamma,\alpha\beta}\right)_{\mathcal{H}} + 2\left(\frac{\partial D_{\alpha\beta}}{\partial A_\gamma} + \frac{\partial D_{\beta\gamma}}{\partial A_\alpha} + \frac{\partial D_{\gamma\alpha}}{\partial A_\beta}\right)_{\mathbf{A}=\mathbf{0},\,\mathcal{H}}. \tag{5.182}$$

The third cumulants (5.83) characterize the magnetic field asymmetry of the fluctuations (Sánchez and Büttiker, 2004; Förster and Büttiker, 2008). At equilibrium where $\mathbf{A} = \mathbf{0}$, they are odd with respect to the magnetizing field because of equation (5.180), so that the magnetic field asymmetry disappears in the absence of a magnetizing field, i.e., for $\mathcal{H} = \mathbf{0}$, and the third cumulants then vanish at equilibrium, $C_{\alpha\beta\gamma}(\mathbf{0},\mathbf{0}) = 0$.

Moreover, equation (5.181) shows that the third cumulants at equilibrium can be expressed in terms of the first responses of the second cumulants (5.82) with respect to the affinities. This can be viewed as a generalization of the fluctuation–dissipation relation (5.179). Furthermore, the third cumulants are fully given in terms of the second response coefficients according to

$$C_{\alpha\beta\gamma}(\mathbf{0},\mathcal{H}) = \left(M_{\alpha,\beta\gamma} + M_{\beta,\gamma\alpha} + M_{\gamma,\alpha\beta}\right)_{\mathcal{H}} - \left(M_{\alpha,\beta\gamma} + M_{\beta,\gamma\alpha} + M_{\gamma,\alpha\beta}\right)_{-\mathcal{H}}, \tag{5.183}$$

which is deduced using equation (5.182) for $\pm\mathcal{H}$ combined with equations (5.180) and (5.181). The third cumulants can thus be constructed from the measurement of either the second cumulants (by means of equation (5.181)) or the first ones (by means of equation (5.183)).

Furthermore, the third derivatives of the symmetry relation (5.177) also give

$$\frac{\partial D_{\beta\gamma}}{\partial A_\alpha}(\mathbf{0},\mathcal{H}) = \frac{1}{2}\left[\left(M_{\beta,\gamma\alpha} + M_{\gamma,\alpha\beta}\right)_{\mathcal{H}} - \left(M_{\alpha,\beta\gamma}\right)_{-\mathcal{H}}\right], \tag{5.184}$$

which expresses the first responses of the diffusivities at equilibrium in terms of the second response coefficients (Stratonovich, 1992; Barbier and Gaspard, 2019).

We note that the formulas (5.180)–(5.184) for the third cumulants and the first responses of the diffusivities were obtained by Saito and Utsumi (2008) in the special cases of open systems in contact with two reservoirs where $\alpha = \beta = \gamma$ and three reservoirs where $\gamma = \beta$ or $\gamma = \alpha$.

Nonlinear Transport Properties at Third Order

From the fourth derivatives of the symmetry relation (5.177), we similarly deduce that the fourth cumulants (5.84) can be expressed as

$$B_{\alpha\beta\gamma\delta}(\mathbf{0},\boldsymbol{\mathcal{H}}) = B_{\alpha\beta\gamma\delta}(\mathbf{0},-\boldsymbol{\mathcal{H}}), \tag{5.185}$$

$$B_{\alpha\beta\gamma\delta}(\mathbf{0},\boldsymbol{\mathcal{H}}) = \frac{1}{2}\frac{\partial C_{\alpha\beta\gamma}}{\partial A_\delta}(\mathbf{0},\boldsymbol{\mathcal{H}}) + \frac{1}{2}\frac{\partial C_{\alpha\beta\gamma}}{\partial A_\delta}(\mathbf{0},-\boldsymbol{\mathcal{H}}), \tag{5.186}$$

$$\begin{aligned} B_{\alpha\beta\gamma\delta}(\mathbf{0},\boldsymbol{\mathcal{H}}) = {} & \frac{1}{2}N_{\alpha,\beta\gamma\delta}(\boldsymbol{\mathcal{H}}) + \frac{1}{2}N_{\alpha,\beta\gamma\delta}(-\boldsymbol{\mathcal{H}}) \\ & - \left(\frac{\partial^2 D_{\alpha\beta}}{\partial A_\gamma \partial A_\delta} + \frac{\partial^2 D_{\alpha\gamma}}{\partial A_\beta \partial A_\delta} + \frac{\partial^2 D_{\alpha\delta}}{\partial A_\beta \partial A_\gamma}\right)_{\mathbf{A}=\mathbf{0},\,\boldsymbol{\mathcal{H}}} \\ & + \frac{1}{2}\left(\frac{\partial C_{\alpha\beta\gamma}}{\partial A_\delta} + \frac{\partial C_{\alpha\beta\delta}}{\partial A_\gamma} + \frac{\partial C_{\alpha\gamma\delta}}{\partial A_\beta}\right)_{\mathbf{A}=\mathbf{0},\,\boldsymbol{\mathcal{H}}}, \\ B_{\alpha\beta\gamma\delta}(\mathbf{0},\boldsymbol{\mathcal{H}}) = {} & \frac{1}{2}\left(N_{\alpha,\beta\gamma\delta} + N_{\beta,\gamma\delta\alpha} + N_{\gamma,\delta\alpha\beta} + N_{\delta,\alpha\beta\gamma}\right)_{\boldsymbol{\mathcal{H}}} \\ & - \left(\frac{\partial^2 D_{\alpha\beta}}{\partial A_\gamma \partial A_\delta} + \frac{\partial^2 D_{\alpha\gamma}}{\partial A_\beta \partial A_\delta} + \frac{\partial^2 D_{\alpha\delta}}{\partial A_\beta \partial A_\gamma}\right. \\ & \left. + \frac{\partial^2 D_{\beta\gamma}}{\partial A_\alpha \partial A_\delta} + \frac{\partial^2 D_{\beta\delta}}{\partial A_\alpha \partial A_\gamma} + \frac{\partial^2 D_{\gamma\delta}}{\partial A_\alpha \partial A_\beta}\right)_{\mathbf{A}=\mathbf{0},\,\boldsymbol{\mathcal{H}}} \\ & + \frac{1}{2}\left(\frac{\partial C_{\beta\gamma\delta}}{\partial A_\alpha} + \frac{\partial C_{\gamma\delta\alpha}}{\partial A_\beta} + \frac{\partial C_{\delta\alpha\beta}}{\partial A_\gamma} + \frac{\partial C_{\alpha\beta\gamma}}{\partial A_\delta}\right)_{\mathbf{A}=\mathbf{0},\,\boldsymbol{\mathcal{H}}}. \end{aligned} \tag{5.187}$$

The fourth derivatives of equation (5.177) also give the following relations:

$$\frac{\partial^2 D_{\alpha\beta}}{\partial A_\gamma \partial A_\delta}(\mathbf{0},\boldsymbol{\mathcal{H}}) - \frac{1}{2}\frac{\partial C_{\alpha\beta\gamma}}{\partial A_\delta}(\mathbf{0},\boldsymbol{\mathcal{H}}) = \frac{\partial^2 D_{\alpha\beta}}{\partial A_\gamma \partial A_\delta}(\mathbf{0},-\boldsymbol{\mathcal{H}}) - \frac{1}{2}\frac{\partial C_{\alpha\beta\delta}}{\partial A_\gamma}(\mathbf{0},-\boldsymbol{\mathcal{H}}). \tag{5.188}$$

In the absence of a magnetizing field, $\boldsymbol{\mathcal{H}} = \mathbf{0}$, these relations reduce to those of Section 5.5.

Like equation (5.181) for the third cumulants, equation (5.186) generalizes the fluctuation–dissipation theorem (5.179) to the fourth cumulants $B_{\alpha\beta\gamma\delta}$ at equilibrium. It shows that the latter are fully specified by the first responses of the third cumulants. Furthermore, equations (5.187) express $B_{\alpha\beta\gamma\delta}$ in terms of the responses of the first, second, and third cumulants.

Similar relations can also be deduced at higher orders, as explained in Section F.2 (Barbier and Gaspard, 2019, 2020b). All these results show that the cumulants of order $n+1$ can be obtained from the measurement of the cumulants of order n or lower. These relations deduced from the multivariate fluctuation relation (5.177) thus give the statistical cumulants at equilibrium, i.e., for $\mathbf{A} = \mathbf{0}$, in terms of the statistical cumulants of lower orders and their responses with respect to the nonequilibrium constraints. The latter can be obtained through the measurement of the corresponding cumulants over some range of the affinities. Therefore, the dependence of these quantities on the magnetizing field $\boldsymbol{\mathcal{H}}$ is considerably constrained by microreversibility. These results fully rely on the time-reversal symmetry expressed in equation (5.177), independently of the actual functional form of the cumulant generating function $Q(\boldsymbol{\lambda};\mathbf{A},\boldsymbol{\mathcal{H}})$.

Relations between Even/Odd and Symmetric/Antisymmetric Parts

As in Section 5.5.7, the even and odd parts of the cumulant generating function can be defined by equation (5.159). Moreover, the symmetric and antisymmetric parts of an arbitrary function f of $\boldsymbol{\mathcal{H}}$ are defined as

$$f^{\mathrm{S,A}}(\boldsymbol{\mathcal{H}}) \equiv \frac{1}{2}\left[f(\boldsymbol{\mathcal{H}}) \pm f(-\boldsymbol{\mathcal{H}})\right], \tag{5.189}$$

so that

$$f(\pm\boldsymbol{\mathcal{H}}) = f^{\mathrm{S}}(\boldsymbol{\mathcal{H}}) \pm f^{\mathrm{A}}(\boldsymbol{\mathcal{H}}). \tag{5.190}$$

Now, substituting the fluctuation relation (5.177) into the definition (5.159) of $Q_\pm$ yields

$$Q_\pm(\boldsymbol{\lambda};\mathbf{A},\boldsymbol{\mathcal{H}}) = \frac{1}{2}\left[Q(\boldsymbol{\lambda};\mathbf{A},\boldsymbol{\mathcal{H}}) \pm \mathrm{e}^{\mathbf{A}\cdot\partial_{\boldsymbol{\lambda}}}\, Q(\boldsymbol{\lambda};\mathbf{A},-\boldsymbol{\mathcal{H}})\right]. \tag{5.191}$$

Moreover, taking the symmetric and antisymmetric parts of equations (5.191) in the magnetizing field $\boldsymbol{\mathcal{H}}$ gives the following four relations:

$$Q^{\mathrm{S}}_\pm(\boldsymbol{\lambda};\mathbf{A},\boldsymbol{\mathcal{H}}) = \frac{1}{2}\left(1 \pm \mathrm{e}^{\mathbf{A}\cdot\partial_{\boldsymbol{\lambda}}}\right) Q^{\mathrm{S}}(\boldsymbol{\lambda};\mathbf{A},\boldsymbol{\mathcal{H}}), \tag{5.192}$$

$$Q^{\mathrm{A}}_\mp(\boldsymbol{\lambda};\mathbf{A},\boldsymbol{\mathcal{H}}) = \frac{1}{2}\left(1 \pm \mathrm{e}^{\mathbf{A}\cdot\partial_{\boldsymbol{\lambda}}}\right) Q^{\mathrm{A}}(\boldsymbol{\lambda};\mathbf{A},\boldsymbol{\mathcal{H}}). \tag{5.193}$$

Multiplying (5.192) and (5.193) by $\left(1 \mp \mathrm{e}^{\mathbf{A}\cdot\partial_{\boldsymbol{\lambda}}}\right)$, eliminating $Q^{\mathrm{S,A}}$, and using the identity (5.161), these two relations give

$$Q^{\mathrm{S}}_-(\boldsymbol{\lambda};\mathbf{A},\boldsymbol{\mathcal{H}}) = -\tanh\left(\frac{1}{2}\mathbf{A}\cdot\partial_{\boldsymbol{\lambda}}\right) Q^{\mathrm{S}}_+(\boldsymbol{\lambda};\mathbf{A},\boldsymbol{\mathcal{H}}), \tag{5.194}$$

$$Q^{\mathrm{A}}_+(\boldsymbol{\lambda};\mathbf{A},\boldsymbol{\mathcal{H}}) = -\tanh\left(\frac{1}{2}\mathbf{A}\cdot\partial_{\boldsymbol{\lambda}}\right) Q^{\mathrm{A}}_-(\boldsymbol{\lambda};\mathbf{A},\boldsymbol{\mathcal{H}}), \tag{5.195}$$

which are equivalent to the symmetry relation (5.177) (Barbier and Gaspard, 2020b).

Now, expanding the hyperbolic tangent in powers of its argument leads to the following relations,

$$J^{\mathrm{S}}_\alpha(\mathbf{A},\boldsymbol{\mathcal{H}}) = \sum_\beta D^{\mathrm{S}}_{\alpha\beta}(\mathbf{A},\boldsymbol{\mathcal{H}})\, A_\beta - \frac{1}{12}\sum_{\beta,\gamma,\delta} B^{\mathrm{S}}_{\alpha\beta\gamma\delta}(\mathbf{A},\boldsymbol{\mathcal{H}})\, A_\beta\, A_\gamma\, A_\delta + O(\mathbf{A}^5), \tag{5.196}$$

$$C^{\mathrm{S}}_{\alpha\beta\gamma}(\mathbf{A},\boldsymbol{\mathcal{H}}) = \sum_\delta B^{\mathrm{S}}_{\alpha\beta\gamma\delta}(\mathbf{A},\boldsymbol{\mathcal{H}})\, A_\delta + O(\mathbf{A}^3), \tag{5.197}$$

$$\vdots$$

for the symmetric parts of the cumulants and response coefficients, and

$$\sum_\alpha J^{\rm A}_\alpha(\mathbf{A},\mathcal{H})\,A_\alpha = \frac{1}{12}\sum_{\alpha\beta\gamma} C^{\rm A}_{\alpha\beta\gamma}(\mathbf{A},\mathcal{H})\,A_\alpha\,A_\beta\,A_\gamma + O\big(\mathbf{A}^5\big), \tag{5.198}$$

$$D^{\rm A}_{\alpha\beta}(\mathbf{A},\mathcal{H}) = \frac{1}{4}\sum_\gamma C^{\rm A}_{\alpha\beta\gamma}(\mathbf{A},\mathcal{H})\,A_\gamma + O(\mathbf{A}^3), \tag{5.199}$$

$$B^{\rm A}_{\alpha\beta\gamma\delta}(\mathbf{A},\mathcal{H}) = O(\mathbf{A}), \tag{5.200}$$

$$\vdots$$

for the antisymmetric parts of the cumulants and response coefficients.

Expanding these relations in powers of the affinities $\mathbf{A}$ and setting $\mathbf{A} = \mathbf{0}$, we recover the previously obtained relations for the linear and nonlinear transport properties. At equilibrium, we thus have that

$$J^{\rm S}_\alpha(\mathbf{0},\mathcal{H}) = 0, \quad D^{\rm A}_{\alpha\beta}(\mathbf{0},\mathcal{H}) = 0, \quad C^{\rm S}_{\alpha\beta\gamma}(\mathbf{0},\mathcal{H}) = 0, \quad B^{\rm A}_{\alpha\beta\gamma\delta}(\mathbf{0},\mathcal{H}) = 0, \dots \tag{5.201}$$

In particular, the linear response coefficients can be decomposed as

$$L_{\alpha,\beta}(\mathcal{H}) = L^{\rm S}_{\alpha,\beta}(\mathcal{H}) + L^{\rm A}_{\alpha,\beta}(\mathcal{H}) \tag{5.202}$$

with

$$L^{\rm S}_{\alpha,\beta}(\mathcal{H}) = D^{\rm S}_{\alpha\beta}(\mathbf{0},\mathcal{H}) \qquad \text{and} \qquad L^{\rm A}_{\alpha,\beta}(\mathcal{H}) = -L^{\rm A}_{\alpha,\beta}(-\mathcal{H}). \tag{5.203}$$

Accordingly, the symmetric part is even in the magnetizing field, while the antisymmetric part is odd. In three-dimensional space, the antisymmetric tensor can be written in terms of an axial vector characterizing the Hall effect (Landau and Lifshitz, 1984). Moreover, the antisymmetric part behaves linearly in a weak magnetizing field, $L^{\rm A}_{\alpha,\beta}(\mathcal{H}) = O(\mathcal{H})$, and we have that $L^{\rm A}_{\alpha,\alpha}(\mathcal{H}) = 0$ as a consequence of equation (5.178). Similarly, the third cumulants behave as $C_{\alpha\beta\gamma}(\mathbf{0},\mathcal{H}) = O(\mathcal{H})$.

5.6.4 Entropy Production in a Magnetizing Field

The entropy production rate is given by equation (5.97) for the system in the magnetizing field $\mathcal{H}$. According to the multivariate fluctuation relation (5.176), the entropy production rate can thus be expressed by the following Kullback–Leibler divergence,

$$\frac{1}{k_{\rm B}}\frac{d_{\rm i}S}{dt} = \lim_{t\to\infty}\frac{1}{t}\int d\mathfrak{J}\; p_t(\mathfrak{J};\mathbf{A},\mathcal{H})\,\ln\frac{p_t(\mathfrak{J};\mathbf{A},\mathcal{H})}{p_t(-\mathfrak{J};\mathbf{A},-\mathcal{H})} \geq 0, \tag{5.204}$$

which is always nonnegative (Cover and Thomas, 2006), in agreement with the second law of thermodynamics.

In a magnetizing field $\mathcal{H}$, the entropy production rate can also be expressed as

$$\begin{aligned}\frac{1}{k_{\rm B}}\frac{d_{\rm i}S}{dt} &= \sum_\alpha A_\alpha\, J_\alpha(\mathbf{A},\mathcal{H}) = \sum_\alpha A_\alpha\, J_\alpha^{\rm S}(\mathbf{A},\mathcal{H}) + \sum_\alpha A_\alpha\, J_\alpha^{\rm A}(\mathbf{A},\mathcal{H}) \\ &= \sum_{\alpha,\beta} D_{\alpha\beta}^{\rm S}(\mathbf{A},\mathcal{H})\, A_\alpha\, A_\beta + \frac{1}{12}\sum_{\alpha,\beta,\gamma} C_{\alpha\beta\gamma}^{\rm A}(\mathbf{A},\mathcal{H})\, A_\alpha\, A_\beta\, A_\gamma \\ &\quad - \frac{1}{12}\sum_{\alpha,\beta,\gamma,\delta} B_{\alpha\beta\gamma\delta}^{\rm S}(\mathbf{A},\mathcal{H})\, A_\alpha\, A_\beta\, A_\gamma\, A_\delta + O(\mathbf{A}^5) \end{aligned} \tag{5.205}$$

in terms of the second and higher cumulants characterizing the current fluctuations. In the absence of a magnetizing field, i.e., $\mathcal{H} = \mathbf{0}$, equation (5.165) is recovered.

5.7 Fluctuation Relation for Currents in Stochastic Processes

The fluctuation relation for currents can also be proved for Markovian stochastic processes (Andrieux and Gaspard, 2007b; Andrieux, 2009; Faggionato and Di Pietro, 2011; Barato and Chetrite, 2012; Barato et al., 2012) in the long-time limit and the absence of external magnetizing field, using the cycle decomposition of the graph associated with the Markov jump process (Hill, 1989; Schnakenberg, 1976; Jiang et al., 2004).

5.7.1 Formulation

In the context of stochastic processes, the fluctuation relation for currents can be established in the form of the symmetry relation (5.91) for the cumulant generating function (5.76) obtained as the leading eigenvalue for the time evolution operator of the Markovian stochastic process modified to include the parameters $\boldsymbol{\lambda}$ counting the amounts of energy or particles exchanged between the system and its environment. The environment may be composed of several energy or particle reservoirs in contact with the system. The temperatures and chemical potentials of the reservoirs determine the affinities (5.69) and (5.70). The Markovian stochastic process ruling these exchanges is defined using the coarse-grained states $\omega = (\sigma, \mathfrak{C})$ where σ denotes the internal state of the system and $\mathfrak{C} = \Delta\mathfrak{C}(t)$ the variables counting the amounts exchanged with the environment during some time interval t. In general, these exchanges are determined by the transitions between the internal states and the transition rates depend on the state variables of the reservoirs such as their temperature and chemical potentials. If the reservoirs are large enough to keep their state variables constant in time, the amounts of energy or particles exchanged with the reservoirs are driven by the jumps between the internal states. These states are supposed to correspond to phase-space cells that are symmetric under time reversal. The transitions of this process can thus be expressed as[13]

$$\omega = (\sigma, \mathfrak{C}) \xrightarrow{\rho} \omega' = (\sigma', \mathfrak{C}') \qquad \text{with} \qquad \mathfrak{C}' = \mathfrak{C} + \Delta\mathfrak{C}_\rho, \tag{5.206}$$

[13] As explained in Chapter 3, the transitions of Markov jump processes are considered to be instantaneous, i.e., much shorter than the timescale t of measurement of the amounts $\mathfrak{C} = \Delta\mathfrak{C}(t)$.

where the amounts exchanged between the system and its environment during the transition ρ and its reversal $-\rho$ satisfy the relation

$$\Delta\mathfrak{C}_\rho = -\Delta\mathfrak{C}_{-\rho}. \tag{5.207}$$

The corresponding transition rate $W(\sigma \xrightarrow{\rho} \sigma')$ do not depend on the counting variables $\mathfrak{C}$, but on constant environmental parameters such as the temperatures and the chemical potentials of the reservoirs.

Accordingly, the time evolution of the probability $P_t(\sigma, \mathfrak{C})$ to find the system in its internal state σ having exchanged the amounts $\mathfrak{C}$ during the time interval $[0,t]$ is ruled by the master equation

$$\frac{d}{dt} P_t(\sigma, \mathfrak{C}) = \sum_{\rho,\sigma'} \left[P_t(\sigma', \mathfrak{C} - \Delta\mathfrak{C}_\rho)\, W(\sigma' \xrightarrow{\rho} \sigma) - P_t(\sigma, \mathfrak{C})\, W(\sigma \xrightarrow{-\rho} \sigma') \right]. \tag{5.208}$$

If the reservoirs are arbitrarily large, the random variables $\mathfrak{C}$ may undergo unbounded drift in time, in which case the stochastic process in the state space of the variables $\omega = (\sigma, \mathfrak{C})$ is nonstationary. However, the stochastic process in the state space of internal variables σ may be stationary. The master equation of this latter process is obtained by summing equation (5.208) over the variables $\mathfrak{C}$ to get

$$\frac{d}{dt} \mathscr{P}_t(\sigma) = \sum_{\rho,\sigma'} \left[\mathscr{P}_t(\sigma')\, W(\sigma' \xrightarrow{\rho} \sigma) - \mathscr{P}_t(\sigma)\, W(\sigma \xrightarrow{-\rho} \sigma') \right] \tag{5.209}$$

for the probability distribution defined as

$$\mathscr{P}_t(\sigma) \equiv \sum_{\mathfrak{C}} P_t(\sigma, \mathfrak{C}). \tag{5.210}$$

The master equation (5.209) may describe equilibrium or nonequilibrium processes, depending on whether the affinities characterizing the relative values of the temperatures and the chemical potentials in the reservoirs are equal to zero or not. If the affinities are equal to zero, i.e., $\mathbf{A} = \mathbf{0}$, the system is at equilibrium and the stationary probability distribution of equation (5.209) should thus satisfy the detailed balance conditions

$$\mathscr{P}_{\rm eq}(\sigma')\, W(\sigma' \xrightarrow{\rho} \sigma) = \mathscr{P}_{\rm eq}(\sigma)\, W(\sigma \xrightarrow{-\rho} \sigma') \tag{5.211}$$

for all the transitions $\{\rho\}$ and all the internal states $\{\sigma\}$. This equilibrium distribution is assumed to be the unique stationary distribution of the process if the affinities are equal to zero, i.e., $\mathbf{A} = \mathbf{0}$. According to the conditions of reversibility (4.40), the stochastic process is thus reversible at equilibrium.

Furthermore, we may introduce the generating function,

$$F_t(\sigma, \boldsymbol{\lambda}) \equiv \sum_{\mathfrak{C}} P_t(\sigma, \mathfrak{C})\, {\rm e}^{-\boldsymbol{\lambda}\cdot\mathfrak{C}}, \tag{5.212}$$

of the statistical moments of the variables $\mathfrak{C}$ by using the so-called counting parameters $\boldsymbol{\lambda}$. The time evolution of these generating functions can be deduced from the master equation (5.208), giving

$$\frac{d}{dt}\,F_t(\sigma,\boldsymbol{\lambda}) = \sum_{\rho,\sigma'} \Big[\mathrm{e}^{-\boldsymbol{\lambda}\cdot\Delta\mathfrak{C}_\rho} F_t(\sigma',\boldsymbol{\lambda})\, W(\sigma' \overset{\rho}{\to} \sigma) - F_t(\sigma,\boldsymbol{\lambda})\, W(\sigma \overset{-\rho}{\to} \sigma') \Big]. \tag{5.213}$$

We note that the probability distribution (5.210) and its master equation (5.209) are recovered by taking $\boldsymbol{\lambda} = \mathbf{0}$ since $\mathscr{P}_t(\sigma) = F_t(\sigma,\mathbf{0})$. Gathering the generating functions (5.212) into the vector $\mathbf{F}_t(\boldsymbol{\lambda}) \equiv \{F_t(\sigma,\boldsymbol{\lambda})\}$, equation (5.213) can be written into the following form,

$$\frac{d\mathbf{F}}{dt} = \mathsf{L}_{\boldsymbol{\lambda}} \cdot \mathbf{F}, \tag{5.214}$$

defining the evolution operator $\mathsf{L}_{\boldsymbol{\lambda}}$ modified by the counting parameters $\boldsymbol{\lambda}$ with the matrix elements

$$L_{\boldsymbol{\lambda}}(\sigma,\sigma') = \begin{cases} \sum_\rho \mathrm{e}^{-\boldsymbol{\lambda}\cdot\Delta\mathfrak{C}_\rho}\, W(\sigma' \overset{\rho}{\to} \sigma) & \text{for} \quad \sigma \neq \sigma', \\ -\sum_\rho W(\sigma \overset{-\rho}{\to} \sigma') & \text{for} \quad \sigma = \sigma'. \end{cases} \tag{5.215}$$

If the counting parameters are set equal to zero, i.e., $\boldsymbol{\lambda} = \mathbf{0}$, the modified operator reduces to the evolution operator of the master equation (5.209), since $d\mathscr{P}/dt = \mathsf{L}\cdot\mathscr{P}$ with $\mathsf{L} = \mathsf{L}_{\mathbf{0}}$.

Now, the cumulant generating function (5.76) can be expressed as

$$Q(\boldsymbol{\lambda}) = \lim_{t\to\infty} -\frac{1}{t} \ln \sum_\sigma F_t(\sigma,\boldsymbol{\lambda}) \tag{5.216}$$

in terms of the moment generating function (5.212), since the counting variables $\mathfrak{C}$ give the amounts $\Delta\mathfrak{C}(t)$ of energy and particles exchanged during the time interval $[0,t]$. Here, for simplicity, we have dropped the implicit dependence of the cumulant generating function on the affinities $\mathbf{A}$. According to equation (5.216), we should expect the asymptotic behavior $F_t(\sigma,\boldsymbol{\lambda}) \sim \exp[-Q(\boldsymbol{\lambda})t]$ for $t \to \infty$, suggesting that the cumulant generating function can be obtained as the leading eigenvalue of the modified operator (5.215). We can thus set up the eigenvalue problem,

$$\mathsf{L}_{\boldsymbol{\lambda}} \cdot \boldsymbol{\Psi}_{\boldsymbol{\lambda}} = -Q(\boldsymbol{\lambda})\, \boldsymbol{\Psi}_{\boldsymbol{\lambda}}, \tag{5.217}$$

$$\mathsf{L}_{\boldsymbol{\lambda}}^{\mathrm{T}} \cdot \tilde{\boldsymbol{\Psi}}_{\boldsymbol{\lambda}} = -Q(\boldsymbol{\lambda})\, \tilde{\boldsymbol{\Psi}}_{\boldsymbol{\lambda}}, \tag{5.218}$$

where $\boldsymbol{\Psi}_{\boldsymbol{\lambda}}$ and $\tilde{\boldsymbol{\Psi}}_{\boldsymbol{\lambda}}$ are, respectively, the right and left eigenvectors associated with the eigenvalue $-Q(\boldsymbol{\lambda})$. We note that the modified operator as well as the eigenvectors implicitly depend on the affinities $\mathbf{A}$.

The property behind the fluctuation relation is that there exists a positive real symmetric operator $\mathsf{M} = \mathsf{M}^{\mathrm{T}} > 0$ such that the modified operator obeys the symmetry relation,

$$\mathsf{M}^{-1} \cdot \mathsf{L}_{\boldsymbol{\lambda}} \cdot \mathsf{M} = \mathsf{L}_{\mathbf{A}-\boldsymbol{\lambda}}^{\mathrm{T}}, \tag{5.219}$$

expressed in terms of the affinities $\mathbf{A}$ (Lebowitz and Spohn, 1999; Harris and Schütz, 2007; Lacoste et al., 2008; Lacoste and Mallick, 2009; Bulnes Cuetara et al., 2011; Gaspard, 2013a,e). If this property holds, the eigenvalue and the eigenvectors satisfy the symmetry relations

$$Q(\boldsymbol{\lambda}) = Q(\mathbf{A}-\boldsymbol{\lambda}), \tag{5.220}$$

$$\tilde{\boldsymbol{\Psi}}_{\boldsymbol{\lambda}} = \mathsf{M}^{-1} \cdot \boldsymbol{\Psi}_{\mathbf{A}-\boldsymbol{\lambda}}, \tag{5.221}$$

as can be directly verified using the Perron–Frobenius theorem. Indeed, as the operator $\exp(\mathsf{L}_{\boldsymbol{\lambda}}t)$ is positive, there exists a unique leading eigenvalue $\exp[-Q(\boldsymbol{\lambda})t]$ (Gantmacher, 1959). Therefore, the symmetry relation (5.91) can be proved in this way.

We note that equation (5.219) can be written in the form

$$\mathsf{K}_{\boldsymbol{\lambda}} = \mathsf{K}^{\mathrm{T}}_{\mathbf{A}-\boldsymbol{\lambda}} \qquad \text{with} \qquad \mathsf{K}_{\boldsymbol{\lambda}} \equiv \mathsf{M}^{-1/2} \cdot \mathsf{L}_{\boldsymbol{\lambda}} \cdot \mathsf{M}^{1/2}. \tag{5.222}$$

If the system is at equilibrium with $\mathbf{A} = \mathbf{0}$, the operator (5.222) for $\boldsymbol{\lambda} = \mathbf{0}$ is symmetric, i.e., $\mathsf{K}_{\mathbf{0}} = \mathsf{K}_{\mathbf{0}}^{\mathrm{T}}$, and satisfies the condition (4.44), so that the stochastic process is reversible, as expected. If the system is out of equilibrium with $\mathbf{A} \neq \mathbf{0}$, the condition of reversibility is broken and replaced by the generalized symmetry (5.219).

In the following subsection, we shall establish the existence of the matrix M using the Hill–Schnakenberg cycle decomposition for Markov jump processes.

5.7.2 Hill–Schnakenberg Cycle Decomposition

As explained in Section 4.4.5, a graph G can be associated with the Markov jump process ruled by the master equation (5.209). The internal states $\{\sigma\}$ are the vertices of this graph and the transitions correspond to its edges, $e = \sigma \overset{\rho}{\to} \sigma'$. An example of such a graph is given in Figure 4.4.

Let us choose a maximal tree $T(G)$ of the graph G. This maximal tree connects all the states and contains no cycle. Accordingly, there exist positive real numbers $M(\sigma)$ defined for every state and satisfying

$$M(\sigma')\, W\big(\sigma' \overset{\rho}{\to} \sigma\big) = M(\sigma)\, W\big(\sigma \overset{-\rho}{\to} \sigma'\big) \qquad \forall\, \rho \in T(G), \tag{5.223}$$

i.e., for every transition ρ of the maximal tree. Starting from any state, these numbers can be calculated for the nearest neighboring states, the next-nearest ones, etc... Since the maximal tree has no cycle, this construction proceeds step by step without ambivalence.

A unique cycle C_l is defined as soon as a chord l, i.e., an edge that does not belong to the maximal tree $T(G)$, is added to form the subgraph $T(G) + l$. This cycle can be written as

$$C_l = \sigma_1 \overset{\rho_1}{\to} \sigma_2 \overset{\rho_2}{\to} \sigma_3 \to \cdots \to \sigma_c \overset{\rho_c}{\to} \sigma_1, \tag{5.224}$$

involving c states and so many edges. The affinity of this cycle is given by equation (4.77), which reads

$$A_l = \ln \prod_{i=1}^{c} \frac{W\left(\sigma_i \overset{\rho_i}{\to} \sigma_{i+1}\right)}{W\left(\sigma_{i+1} \overset{-\rho_i}{\to} \sigma_i\right)}. \tag{5.225}$$

We may suppose that the last transition ρ_c takes place on the chord l, with all the other transitions belonging to the maximal tree $T(G)$. Since the relation (5.223) holds for every transition belonging to the maximal tree, we have that

$$\frac{W\left(\sigma_i \overset{\rho_i}{\to} \sigma_{i+1}\right)}{W\left(\sigma_{i+1} \overset{-\rho_i}{\to} \sigma_i\right)} = \frac{M(\sigma_{i+1})}{M(\sigma_i)} \qquad \text{for } i = 1,2,\ldots,c-1. \tag{5.226}$$

Substituting into equation (5.225) and setting $\sigma_1 = \sigma$, $\sigma_c = \sigma'$, and $\rho_c = l$, the affinity of the cycle C_l formed by the chord $\sigma' \overset{l}{\to} \sigma$ is thus given by

$$A_l = \ln \frac{M(\sigma')}{M(\sigma)} \frac{W\left(\sigma' \overset{l}{\to} \sigma\right)}{W\left(\sigma \overset{-l}{\to} \sigma'\right)}. \tag{5.227}$$

This result holds for every chord l. Moreover, we have that $A_l = -A_{-l}$ between the affinities of opposite chords, l and $\bar{l} = -l$.

During a random path of the stochastic process, the instantaneous current on the chord l can be introduced according to

$$\mathscr{J}_l(t) \equiv \sum_{i=-\infty}^{+\infty} \varsigma_l(e_i)\, \delta(t - t_i), \tag{5.228}$$

where t_i is the time of the random transition e_i, and $\varsigma_l(e_i)$ is defined by equation (4.69). The integral of this instantaneous current gives the signed number of transitions that have occurred on the chord l during the time interval $[0,t]$,

$$n_l(t) \equiv \int_0^t \mathscr{J}_l(t')\, dt'. \tag{5.229}$$

The set $\boldsymbol{n} = \{n_l\}$ of all these numbers associated with all the chords can here be taken as the counting variables $\mathfrak{C}$ for the currents on the chords. The jumps $\Delta\mathfrak{C}_\rho$ of the counting variables are thus replaced by $\Delta\boldsymbol{n}_\rho = \{\Delta n_{\rho l}\}$ with $\Delta n_{\rho l} = \pm 1$ if the transition ρ corresponds to the chord l or its opposite $-l$, otherwise $\Delta n_{\rho l} = 0$.

If the operator $\mathbf{M}$ had the matrix elements

$$M(\sigma,\sigma') = M(\sigma)\, \delta_{\sigma\sigma'}, \tag{5.230}$$

the symmetry relation (5.219) would read

$$\frac{1}{M(\sigma)} L_{\boldsymbol{\lambda}}(\sigma,\sigma')\, M(\sigma') = L_{\boldsymbol{A}-\boldsymbol{\lambda}}(\sigma',\sigma) \tag{5.231}$$

with affinities $\boldsymbol{A} = \{A_l\}$ and counting parameters $\boldsymbol{\lambda} = \{\lambda_l\}$ defined for all the chords. Given the definition (5.215) of the modified operator with $\Delta\mathfrak{C}_\rho$ replaced by $\Delta\boldsymbol{n}_\rho$, we should thus have that

$$M(\sigma') \sum_\rho \mathrm{e}^{-\boldsymbol{\lambda}\cdot\Delta\boldsymbol{n}_\rho}\, W\left(\sigma' \overset{\rho}{\to} \sigma\right) = M(\sigma) \sum_\rho \mathrm{e}^{(\boldsymbol{A}-\boldsymbol{\lambda})\cdot\Delta\boldsymbol{n}_\rho}\, W\left(\sigma \overset{-\rho}{\to} \sigma'\right) \tag{5.232}$$

for any pair of internal states, σ' and σ. Among all the transitions ρ connecting these two states, at least one transition belongs to the maximal tree $T(G)$ since this latter connects all the states. The sum over the transitions ρ can thus be split into the sum over the transitions belonging to the maximal tree and the sum over those that do not belong to $T(G)$. For all the transitions of the maximal tree $\rho \in T(G)$, we have that $\Delta \boldsymbol{n}_\rho = 0$ because none of these transitions is a chord. Consequently, equation (5.232) becomes

$$M(\sigma') \sum_{\rho \in T(G)} W\big(\sigma' \xrightarrow{\rho} \sigma\big) + M(\sigma') \sum_{\rho \notin T(G)} \mathrm{e}^{-\boldsymbol{\lambda}\cdot\Delta \boldsymbol{n}_\rho}\, W\big(\sigma' \xrightarrow{\rho} \sigma\big)$$
$$= M(\sigma) \sum_{\rho \in T(G)} W\big(\sigma \xrightarrow{-\rho} \sigma'\big) + M(\sigma) \sum_{\rho \notin T(G)} \mathrm{e}^{(\boldsymbol{A}-\boldsymbol{\lambda})\cdot\Delta \boldsymbol{n}_\rho}\, W\big(\sigma \xrightarrow{-\rho} \sigma'\big). \quad (5.233)$$

Since the numbers $M(\sigma)$ satisfy the relations (5.223) for every transition $\rho \in T(G)$ of the maximal tree, the corresponding terms cancel each other in equation (5.233), which becomes

$$M(\sigma') \sum_{\rho \notin T(G)} \mathrm{e}^{-\boldsymbol{\lambda}\cdot\Delta \boldsymbol{n}_\rho}\, W\big(\sigma' \xrightarrow{\rho} \sigma\big) = M(\sigma) \sum_{\rho \notin T(G)} \mathrm{e}^{(\boldsymbol{A}-\boldsymbol{\lambda})\cdot\Delta \boldsymbol{n}_\rho}\, W\big(\sigma \xrightarrow{-\rho} \sigma'\big). \quad (5.234)$$

By construction, every transition $\rho \notin T(G)$ that is not contained in the maximal tree defines a chord l, for which $\boldsymbol{\lambda} \cdot \Delta \boldsymbol{n}_\rho = \lambda_l$ and $\boldsymbol{A} \cdot \Delta \boldsymbol{n}_\rho = A_l$, or the opposite $\bar{l} = -l$ of a chord, for which $\boldsymbol{\lambda} \cdot \Delta \boldsymbol{n}_\rho = -\lambda_l$ and $\boldsymbol{A} \cdot \Delta \boldsymbol{n}_\rho = -A_l$. Therefore, equation (5.234) takes the form

$$\sum_l \mathrm{e}^{-\lambda_l} \Big[M(\sigma') W\big(\sigma' \xrightarrow{l} \sigma\big) - \mathrm{e}^{A_l}\, M(\sigma)\, W\big(\sigma \xrightarrow{-l} \sigma'\big)\Big]$$
$$+ \sum_{\bar{l}} \mathrm{e}^{\lambda_l} \Big[M(\sigma') W\big(\sigma' \xrightarrow{-l} \sigma\big) - \mathrm{e}^{-A_l}\, M(\sigma)\, W\big(\sigma \xrightarrow{l} \sigma'\big)\Big] = 0. \quad (5.235)$$

Now, the terms of the first line are all equal to zero because of equation (5.227). Similarly, all the terms of the second line are also equal to zero if we use equation (5.227) for the opposite chords $\bar{l} = -l$. Hence, equation (5.235) is identically equal to zero, whereupon the symmetry relation (5.219) is proved.

Thus, we have demonstrated the symmetry relation for the basic currents defined on all the chords associated with the chosen maximal tree. Since these chords correspond to the transitions of the Markovian process, these basic currents are defined for the transitions between the internal states $\{\sigma\}$ of the system. These basic currents are not necessarily associated with energy or particle exchanges with the reservoirs because many transitions of the stochastic process occur inside the system, although the exchanges usually involve transitions at the boundaries between the system and the reservoirs. Since these latter transitions are those that determine the global currents flowing across the system due to its contacts with the reservoirs, we now have to relate the basic to the global currents in order to complete the result.

5.7.3 Fluctuation Relation for the Global Currents

The global exchanges of energy and particles from the reservoirs $j = 1, 2, \ldots, r-1$ to the reference reservoir $j = r$ across the system can be monitored with the counting variables (5.72) for c species of particles.

The maximal tree $T(G)$ of the graph G might be chosen as corresponding to partial equilibrium with the reference reservoir $j = r$, so that equation (5.223) defining the matrix M

would represent the conditions of local detailed balance for the transitions at the boundary between the system and the reference reservoir. These transitions do not contribute to the jumps of the counting variables because $\Delta\mathfrak{C}_\rho = \mathbf{0}$ for $\rho \in T(G)$.

In contrast, the other transitions $\rho \notin T(G)$ may contribute to the jumps. Since every transition $\rho \notin T(G)$ corresponds to some chords l of the graph G, we may introduce their associated jumps $\Delta\mathfrak{C}_l$. If a transition occurs at time t_i on the edge e_i, the instantaneous energy and particle currents can thus be expressed as

$$\mathscr{J}(t) = \sum_{i=-\infty}^{+\infty} \Delta\mathfrak{C}_{e_i}\, \delta(t - t_i) \qquad \text{with} \qquad \Delta\mathfrak{C}_{e_i} = \sum_l \Delta\mathfrak{C}_l\, \varsigma_l(e_i) \tag{5.236}$$

along a random path of the stochastic process. Therefore, the counting variables are given at time t by

$$\Delta\mathfrak{C}(t) = \int_0^t \mathscr{J}(t')\, dt'. \tag{5.237}$$

Their cumulant generating function is given by equation (5.76). The instantaneous currents (5.236) are thus related to the currents (5.228) on the chords according to

$$\mathscr{J}(t) = \sum_l \Delta\mathfrak{C}_l\, \mathscr{J}_l(t). \tag{5.238}$$

Now, the dimensionless affinity given by equation (5.225) for the cycle C_l formed by the chord l is determined by the thermal and chemical affinities $\mathbf{A}$ introduced in equations (5.69), (5.70), and (5.71), according to the expression

$$A_l = \mathbf{A} \cdot \Delta\mathfrak{C}_l. \tag{5.239}$$

The counting parameters $\boldsymbol{\lambda} = \{\lambda_l\}$ for the transitions on the chords are similarly related to the counting parameters of energy and particle exchanges by

$$\lambda_l = \boldsymbol{\lambda} \cdot \Delta\mathfrak{C}_l \qquad \text{with} \qquad \boldsymbol{\lambda} = \left[\lambda_{Ej}, (\lambda_{kj})_{k=1}^{c}\right]_{j=1}^{r-1}. \tag{5.240}$$

Substituting these expressions into equation (5.231), we thus prove the operatorial symmetry relation (5.219) and, as a consequence, the symmetry (5.220) of the cumulant generating function $Q(\boldsymbol{\lambda})$ for the global currents. This proof is carried out using Hill–Schnakenberg network theory in the framework of the theory of Markovian stochastic processes (Andrieux and Gaspard, 2007b; Andrieux, 2009).

We note that the conditions (5.223) defining the diagonal elements $M(\sigma)$ of the matrix M on the maximal tree $T(G)$ are similar to the detailed balance conditions (5.211) holding at global equilibrium with $\mathbf{A} = \mathbf{0}$ on the whole graph G. Accordingly, the diagonal elements could be taken as the stationary probabilities of the system at equilibrium with the reference reservoir $j = r$: $M(\sigma) = \mathscr{P}_{\text{eq}}(\sigma)$. Of course, the detailed balance conditions do not hold on all the chords of the maximal tree if the affinities are not equal to zero, $\mathbf{A} \neq \mathbf{0}$.

Finally, the fluctuation relation for currents (5.73) is here obtained in the long-time limit using large-deviation theory. The Legendre transform (5.89) of the cumulant generating function given by the leading eigenvalue (5.217) provides the rate function $R(\mathfrak{J})$. This large-deviation rate function controls the exponential decay of the probability that the energy and particle exchanges $\Delta\mathfrak{C}(t)$ take their values in the range $[\mathfrak{J}t, (\mathfrak{J} + \Delta\mathfrak{J})t]$ according to

$$P\left[\Delta\mathfrak{C}(t) \simeq \mathfrak{J}t\right] \sim \mathrm{e}^{-R(\mathfrak{J})t} \tag{5.241}$$

for $t \to \infty$, up to a prefactor with a subexponential dependence on time. Now, the symmetry (5.220) of the cumulant generating function $Q(\boldsymbol{\lambda})$ can be substituted into equation (5.89) to find the same symmetry relation for the rate function as equation (5.90). Consequently, the ratio of the probabilities of opposite fluctuations of the currents behaves as

$$\frac{p_t(\mathfrak{J})}{p_t(-\mathfrak{J})} \simeq \frac{P\left[\Delta\mathfrak{C}(t) \simeq \mathfrak{J}t\right]}{P\left[\Delta\mathfrak{C}(t) \simeq -\mathfrak{J}t\right]} \sim \frac{\mathrm{e}^{-R(\mathfrak{J})t}}{\mathrm{e}^{-R(-\mathfrak{J})t}} = \mathrm{e}^{\mathbf{A}\cdot\mathfrak{J}\,t} \tag{5.242}$$

for $t \to \infty$, which proves the fluctuation relation for the global currents, holding asymptotically in the long-time limit for systems driven by the global affinities $\mathbf{A}$ (Andrieux and Gaspard, 2006a, 2007b).

The condition for this result to hold is the existence of the rate function (5.89). In spatially extended systems, dynamical phase transitions may take place, restricting the domain of existence of the rate function, as discussed, in particular, by Harris and Schütz (2007).

For Markovian stochastic processes, the multivariate fluctuation relation for the currents (5.242) is established in the long-time limit. Indeed, the cumulant generating function of the currents is given by equations (5.217)–(5.218) as the leading eigenvalue of the modified evolution operator, and its Legendre transform gives the rate function (5.89), which is a large-deviation property holding asymptotically in time for $t \to \infty$. In this regard, the result is weaker than the fluctuation relation for currents (5.74) obtained on the basis of Hamiltonian dynamics. The advantage of the stochastic approach is that the mesoscopic level of description is often used for small systems, where the random variables ruled by a stationary stochastic process can thus be identified.

The fluctuation relation for currents is also demonstrated in semi-Markov processes satisfying the condition of direction-time independence mentioned in Section E.11 and used to describe the transport of ions through membranes (Andrieux and Gaspard, 2008c).

5.7.4 Implications

As a consequence of the multivariate fluctuation relation (5.242), here the results of Sections 5.4 and 5.5 also hold for the fluctuating currents in Markovian stochastic processes. In particular, the entropy production rate (5.97) is always nonnegative according to equation (5.99), in agreement with the second law of thermodynamics.

Furthermore, the statistical cumulants and their responses can be deduced from the cumulant generating function given by the leading eigenvalue (5.217). Because of its symmetry relation (5.220), the fluctuation–dissipation relations (5.142) and the Onsager reciprocal relations (5.143) are satisfied among the linear transport properties, and their generalizations given by equations (5.144), (5.146), (5.148), (5.150), ... towards nonlinear transport properties of arbitrarily high orders also hold for the fluxes of energy and particles between reservoirs at different temperatures and chemical potentials.

5.7.5 Examples

Random Drift on Countably Many Discrete States

For the random drift ruled by the master equation (4.78), the probability distribution (4.85) for observing n steps during the time interval t obeys the fluctuation relation

$$\frac{P(n,t)}{P(-n,t)} = \mathrm{e}^{An} \qquad \text{with} \qquad A = \ln \frac{W_+}{W_-}, \tag{5.243}$$

exactly holding at every time t as a consequence of the symmetry (4.86) for the modified Bessel functions. For this process, the generating function (5.212) only depends on the counting parameter λ and the time t, so that its evolution equation can be directly deduced, giving $F(\lambda,t) = \exp[-Q(\lambda)t]$ with

$$Q(\lambda) = W_+ \left(1 - \mathrm{e}^{-\lambda}\right) + W_- \left(1 - \mathrm{e}^{+\lambda}\right). \tag{5.244}$$

This cumulant generating function obeys the symmetry relation (5.220) with the affinity introduced in equation (5.243). Therefore, the mean current (5.80), the diffusivity (5.82), and the higher cumulants (5.83) and (5.84) are given by

$$J(A) = C(A) = W_-\left(\mathrm{e}^A - 1\right) \quad \text{and} \quad D(A) = B(A) = \frac{W_-}{2}\left(\mathrm{e}^A + 1\right). \tag{5.245}$$

In this example, all the linear and nonlinear response coefficients (5.134), (5.135), (5.136), ... take the same value: $L = M = N = \cdots = W_-$.

As a consequence of the fluctuation relation (5.243), expression (5.142) for the fluctuation–dissipation theorem is satisfied, as well as the higher-order relations (5.144), (5.146), (5.148), (5.150), ... Furthermore, the entropy production rate obtained by equation (5.98) from the fluctuation relation is consistently identical to the asymptotic value (4.88).

We note that the exact value of the affinity given in equation (5.243) differs from its approximation $A \simeq aV/\mathcal{D}$ that would be obtained using the asymptotic Gaussian distribution (4.83). This result illustrates the fact that the fluctuation relation is a large-deviation property, essentially depending on the tails of the probability distribution (4.85) for $|n| \to \infty$.

Particle Exchange Process

Similar results hold for the particle exchange process governed by the master equation (4.91) with the transition rates (4.89). Here, the generating function (5.212) can be written in the form,

$$F(N,\lambda,t) = \mathrm{e}^{-Q(\lambda)t}\,\frac{1}{N!}\,C(\lambda)^N. \tag{5.246}$$

Substituting this assumption into equation (5.213), we find that $C(\lambda) = \Omega(k_{+1}\mathrm{e}^{-\lambda} + k_{+2})/(k_{-1}+k_{-2})$, whereupon the cumulant generating function is here given by

$$Q(\lambda) = \frac{\Omega}{k_{-1}+k_{-2}}\left[k_{+1}k_{-2}\left(1-\mathrm{e}^{-\lambda}\right) + k_{-1}k_{+2}\left(1-\mathrm{e}^{+\lambda}\right)\right]. \tag{5.247}$$

Consequently, the symmetry relation (5.220) is satisfied, but here with the affinity (4.92). This result can also be established from the symmetry (5.219) of the modified operator

(5.215), using the diagonal matrix $\mathsf{M} = [M(N)\,\delta_{NN'}]_{N,N'=0}^{\infty}$ composed of the elements given by the Poisson distribution $M(N) = \mathrm{e}^{-\alpha}\alpha^N/N!$ with $\alpha = k_{+2}\Omega/k_{-2}$, which is associated to the equilibrium stationary distribution with the reservoir $j = 2$ in Figure 4.4.

Since the cumulant generating functions (5.244) and (5.247) have the same dependence on the counting parameter λ, the random drift and the particle exchange process share the same linear and nonlinear transport properties, although with their corresponding rates $W_{\pm}$. In particular, we recover the overall rate (4.96) from the mean current $J = \partial_\lambda Q(0)$ and, thus, the entropy production rate (4.97).

Stochastic Model of a Transistor

As a further example, we consider the following Markov jump process for the transfer of electric charges between three ports, representing the collector, the base, and the emitter of an electronic transistor,

$$\text{collector} \underset{W_{EC}}{\overset{W_{CE}}{\rightleftharpoons}} \text{emitter}, \qquad \text{base} \underset{W_{EB}}{\overset{W_{BE}}{\rightleftharpoons}} \text{emitter}, \qquad \text{collector} \underset{W_{BC}}{\overset{W_{CB}}{\rightleftharpoons}} \text{base}, \tag{5.248}$$

with the transition rates $\{W_{ij}\}_{i,j=C,B,E}$ (Gu and Gaspard, 2019). These latter satisfy the conditions (4.59) in terms of the voltages $V_{ij} = V_i - V_j$ applied between the ports of the transistor, so that $W_{ij}/W_{ji} = \exp(\beta e V_{ij})$, where $\beta = (k_{\mathrm{B}}T)^{-1}$ is the inverse temperature, e the elementary electric charge, and $i, j = C, B, E$. As the ports are in contact with the corresponding reservoirs of charge carriers, the applied voltages drive electric currents across the transistor under such nonequilibrium conditions.

This process is related to the Ebers–Moll transport model of bipolar junction transistors (Ebers and Moll, 1954; Sedra and Smith, 2004). In this case, the rates are given by $W_{EB} = J_s/\beta_F$, $W_{CB} = J_s/\beta_R$, and $W_{EC} = J_s \exp(\beta e V_{BC})$, where β_F and β_R are, respectively, the forward and reverse common-emitter current gains, and J_s is the reverse saturation current.

The number of charges transferred from the collector (respectively the base) to the rest of the system is denoted Z_C (respectively Z_B). For the model (5.248), the joint probability distribution $P(Z_C, Z_B, t)$ of observing the charge transfers Z_C and Z_B during the time interval [0, t] evolves according to the master equation

$$\begin{aligned}\frac{d}{dt}P(Z_C, Z_B, t) = {} & W_{CE}\,P(Z_C - 1, Z_B, t) + W_{EC}\,P(Z_C + 1, Z_B, t) \\ & + W_{BE}\,P(Z_C, Z_B - 1, t) + W_{EB}\,P(Z_C, Z_B + 1, t) \\ & + W_{CB}\,P(Z_C - 1, Z_B + 1, t) + W_{BC}\,P(Z_C + 1, Z_B - 1, t) \\ & - (W_{CE} + W_{EC} + W_{BE} + W_{EB} + W_{CB} + W_{BC})\,P(Z_C, Z_B, t).\end{aligned} \tag{5.249}$$

Therefore, the generating function (5.212) is here given by $F(\boldsymbol{\lambda}, t) = \exp[-Q(\boldsymbol{\lambda})\,t]$ with the counting parameters $\boldsymbol{\lambda} = (\lambda_C, \lambda_B)$ of the two electric currents flowing across the transistor. Here, the cumulant generating function has the expression

$$\begin{aligned}Q(\lambda_C, \lambda_B) = {} & W_{CE}\left(1 - \mathrm{e}^{-\lambda_C}\right) + W_{EC}\left(1 - \mathrm{e}^{+\lambda_C}\right) \\ & + W_{BE}\left(1 - \mathrm{e}^{-\lambda_B}\right) + W_{EB}\left(1 - \mathrm{e}^{+\lambda_B}\right) \\ & + W_{CB}\left(1 - \mathrm{e}^{-\lambda_C + \lambda_B}\right) + W_{BC}\left(1 - \mathrm{e}^{+\lambda_C - \lambda_B}\right).\end{aligned} \tag{5.250}$$

The symmetry relation (5.220) is also satisfied, here with the affinities $\mathbf{A} = (A_C, A_B)$ associated with the two applied voltages driving the transistor away from equilibrium: $A_C = \beta e V_{CE}$ and $A_B = \beta e V_{BE}$.

Furthermore, the stochastic process (5.248) is generated by independent standard Poisson processes for the random numbers $\{N_{ij}\}$ of charges transferred from the ith to the jth port:

$$P(N_{ij},t) = \mathrm{e}^{-\langle N_{ij}\rangle_t} \frac{\langle N_{ij}\rangle_t^{N_{ij}}}{N_{ij}!} \tag{5.251}$$

with $\langle N_{ij}\rangle_t = W_{ij}t$. The number of charges transferred from the collector and the base are, respectively, given by

$$Z_C(\mathbf{N}) = N_{CE} - N_{EC} + N_{CB} - N_{BC} \quad \text{and} \quad Z_B(\mathbf{N}) = N_{BE} - N_{EB} - N_{CB} + N_{BC}, \tag{5.252}$$

so that the solution of the master equation (5.249) starting from $Z_C = Z_B = 0$ at time $t = 0$ can be expressed as

$$P(Z_C, Z_B, t) = \sum_{\mathbf{N}} \delta_{Z_C, Z_C(\mathbf{N})}\, \delta_{Z_B, Z_B(\mathbf{N})} \prod_{i \neq j} P(N_{ij}, t) \tag{5.253}$$

in terms of the Poisson distributions (5.251). Since the summation variables N_{ij} can be replaced by N_{ji} and the affinities are related to the rates according to $W_{ij} = W_{ji}\exp(A_i - A_j)$ with $A_E = 0$, we find that the probability distribution (5.253) is in fact exactly obeying the fluctuation relation

$$\frac{P(Z_C, Z_B, t)}{P(-Z_C, -Z_B, t)} = \mathrm{e}^{A_C Z_C + A_B Z_B} \tag{5.254}$$

at every instant of time t (Gu and Gaspard, 2020).

Therefore, the linear and nonlinear transport properties of the transistor obey the fluctuation–dissipation theorem (5.142), the Onsager reciprocal relation (5.143), as well as their generalizations given by equations (5.144), (5.146), (5.148), (5.150), ... Moreover, the entropy production rate takes the value (5.97), which corresponds to the electric power dissipated in the transistor.

In conclusion, the fluctuation relations find their origin in the property of microreversibility and they represent a fundamental advance in nonequilibrium statistical mechanics because they are valid under arbitrarily strong nonequilibrium constraints from the linear regimes close to equilibrium towards the nonlinear regimes farther away from equilibrium. Furthermore, they imply the nonnegativity of the entropy production rate, in accordance with the second law of thermodynamics.

6

Path Probabilities, Temporal Disorder, and Irreversibility

6.1 Introduction

The time evolution of observable quantities is a key aspect of natural phenomena. Different dynamical behaviors may manifest themselves and these can be observed, recorded, and characterized in order to determine the properties of the system. In general, observable quantities can be measured at two or more successive times. If measurements are repeated many times, paths or trajectories can be recorded, providing data on the history of the system, as schematically represented in Figure 6.1. By increasing the number of successive observations, more and more information is recorded on the dynamics of the system. This information allows us to determine how these observable quantities may change in time and to know if the time evolution of the observable quantities is regular or irregular. The occurrences of their changes or jumps can be characterized by considering their statistics and evaluating their rates from the recorded trajectories. These rates may concern the dynamical activity inside the system, the fluxes of energy or particles between the system and its environment, the energy that is dissipated, or the thermodynamic entropy that is produced.

In order to determine whether a signal is periodic, quasiperiodic, or random, repeated observations over many successive instants of time are required. In this way, patterns can be detected by the statistical analysis of time evolution, in particular, by evaluating the probabilities of their possible occurrences. *Dynamical randomness*, also known as *temporal disorder*, corresponds to the absence of regularities persisting over longer and longer periods of time. Accordingly, the number of different possible paths is expected to increase exponentially with time. Since these paths form a tree growing exponentially in time, their probability is quickly distributed over more and more possibilities. As a consequence, the probability of observing a given path decreases exponentially with the length of its duration, as in the coin tossing random process where the probability of observing some sequence $\boldsymbol{\omega} = \omega_1 \omega_2 \cdots \omega_k$ of k successive heads or tails $\omega_i = \pm 1$ decreases as $P(\boldsymbol{\omega}) = 1/2^k$ at the rate $h = \ln 2$. Comparing different random processes, this rate is larger for higher dynamical randomness. Hence, temporal disorder in the process can be characterized by the mean decrease rate of the path probabilities. This rate is the temporal analogue of the thermodynamic entropy per unit volume in equilibrium statistical mechanics, which characterizes disorder in the equilibrium configurations of a system in the space of its

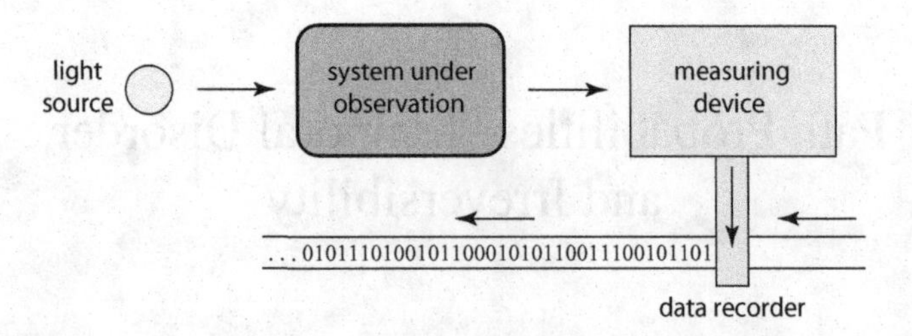

Figure 6.1 Schematic representation of the observation of some system by a measuring device and the recording of the data produced by measurement.

microstates. In this regard, the rate characterizing temporal disorder is often called *entropy per unit time*. Similarly, spatiotemporal disorder may be characterized by an entropy per unit time and unit volume. Such characteristic properties measure the degree of randomness in the dynamical activity of the system (Gaspard and Wang, 1993). The entropy per unit time is positive if the number of random events grows linearly with time, as for the coin tossing process, and equal to zero if this number has a sublinear growth in time, as for periodic or quasiperiodic signals.

Moreover, symmetries may also be detected in signals by comparing the statistics of the paths with that of the symmetry transformed paths. If some asymmetry exists in the dynamics, the former and the latter are expected to have different probabilities and mean rates for their decrease. The mean decrease rate for the probabilities of the symmetry transformed paths can be defined as a *coentropy per unit time*, generalizing the coentropy (2.121) to path probabilities. Asymmetry can thus be detected by the difference between the coentropy and the entropy, i.e., by a positive Kullback–Leibler divergence, as explained in Section 2.7.3.

In this perspective, the time-reversal transformation is of particular importance, since time asymmetry is expected to be the signature of irreversibility due to some nonequilibrium driving of the system. If elementary transitions are observed during time evolution, we may wonder whether time asymmetry in the path probabilities is related to thermodynamic entropy production in the system. As shown in Chapter 5 on the basis of statistical mechanics, the probability distributions of opposite fluctuations in energy and particle exchanges obey fluctuation relations and the thermodynamic entropy production is given by considering the ratio of these probability distributions. Remarkably, these results extend to path probabilities, allowing us to evaluate the thermodynamic entropy production by comparing the statistical properties of the typical paths with those of their time reversal.

These fundamental issues will be addressed in this chapter.

6.2 Path Probabilities

6.2.1 General Processes

In order to define path probabilities, the system of interest should undergo a stroboscopic observation at successive times $t_i = t_0 + i\tau$ with $i \in \mathbb{Z}$, where $\tau = \Delta t$ is the sampling time. The measuring device detects the observables with some finite resolution and records data in the form of symbolic sequences composed of bits or digits, as schematically shown in Figure 6.1. The symbols ω_i denote data recorded every time t_i and corresponding to some small domain in the state space of the system, such as the phase space of microdynamics. From this viewpoint, the observation of the system with a finite resolution amounts to a partition $\mathcal{P} = \{\mathcal{C}_i\}$ of its phase space $\mathcal{M}$ into cells $\mathcal{C}_i$ that are assumed to be disjoint and to cover the whole phase space: $\mathcal{C}_i \cap \mathcal{C}_j = \emptyset$ for $i \neq j$ and $\mathcal{M} = \cup_i \mathcal{C}_i$. A phase-space trajectory $\mathbf{\Gamma}_t = \mathbf{\Phi}^t \mathbf{\Gamma}_0$ of the system is thus recorded as a sequence of symbols ω_i corresponding to the successive cells that are visited at the times $t_i = t_0 + i\tau$: $\boldsymbol{\omega} = \omega_1\omega_2 \cdots \omega_k$ if $\mathbf{\Phi}^{t_i}\mathbf{\Gamma}_0 \in \mathcal{C}_{\omega_i}$ for $i = 1, 2, \dots, k$. This sequence represents the path or history of the system during the lapse of time $k\tau$.

The probability of a given path or history can be expressed in terms of the initial probability density $p_0(\mathbf{\Gamma})$ at time $t = 0$ according to

$$P(\omega_1\omega_2 \cdots \omega_k) = \int_{\mathcal{C}_{\boldsymbol{\omega}}} p_0(\mathbf{\Gamma})\, d\mathbf{\Gamma} \qquad \text{with} \quad \mathcal{C}_{\boldsymbol{\omega}} = \cap_{i=1}^{k} \mathbf{\Phi}^{-t_i} \mathcal{C}_{\omega_i}, \tag{6.1}$$

where $t_i = t_0 + i\tau$. The set of all these path probabilities fully determines the statistics of time evolution carried out on the system by the measuring device with the sampling time τ. The smaller the resolution of the measuring device the higher the quantity of information recorded on the history of the system, possibly down to the observation of the phase-space trajectories for the particles composing the system. However, devices usually measure the observable quantities and the phase-space trajectories with some limited accuracy, so that information known on the time evolution of the system is, in general, coarse grained, as previously discussed in Section 2.3.1.

The observation of the system by the measuring device is thus defining a stochastic process with the path probabilities (6.1). By construction, this process is discrete in time, but the sampling time τ may take arbitrarily small values if the underlying motion is continuous in time. Similarly, the states $\{\omega_i\}$ of the process are also discrete corresponding to some finite resolution in the measurement of the observable quantities, but here also the discretization can be performed into arbitrarily small cells if these quantities are continuous. Instead, if the quantities take discrete values such as particle numbers or occupancies as in the examples of Section 4.4.6, there is no need to improve the resolution in the observation of the states, which thus remain discrete.

Often, the path probabilities of a process can be expressed as

$$P(\boldsymbol{\omega}) \sim \mathrm{e}^{-\mathcal{A}(\boldsymbol{\omega})} \tag{6.2}$$

in terms of the so-called *action functional* $\mathcal{A}(\boldsymbol{\omega})$. This grows linearly in time if the process manifests dynamical randomness and the path probabilities decrease exponentially in time. For the processes of relaxation towards equilibrium, such action functionals are given by the time integral of some Lagrangian function $\mathcal{A} = \int \mathcal{L}\, dt$, as considered by Onsager and Machlup (1953) and Machlup and Onsager (1953).

The path probabilities (6.1) obey the property (4.2), so that summing them over all the states $\omega_2 \cdots \omega_k$ gives the probability $P(\omega_1)$ of the first observed state ω_1 at time t_1. Instead, summing the probabilities over all the states $\omega_1 \cdots \omega_{k-1}$ should provide the stationary probability $P_{\rm st}(\omega_k)$ of the state ω_k in the limit $k \gg 1$, if it exists. Moreover, the conditional probability of observing the state ω_k at time $t_k = t_1 + (k-1)\tau$ if the state ω_1 was observed at time t_1 is given by summing over all the paths joining the initial and final states according to

$$P(\omega_k, t_k|\omega_1, t_1) = \sum_{\omega_2 \cdots \omega_{k-1}} P(\omega_1 \omega_2 \cdots \omega_k)/P(\omega_1, t_1). \tag{6.3}$$

Accordingly, the mean value of some observable quantity $X(\omega)$ at time $t_k = t_1 + (k-1)\tau$ can thus be evaluated as

$$\langle X \rangle_k \equiv \sum_{\omega_k} X(\omega_k) P(\omega_k) = \sum_{\omega_1, \omega_k} X(\omega_k)\, P(\omega_k, t_k|\omega_1, t_1)\, P(\omega_1, t_1) \tag{6.4}$$

and its stationary mean value as $\langle X \rangle_{\rm st} = \lim_{k\to\infty} \langle X \rangle_k$.

The path probabilities may contain information on statistical correlations possibly existing between successive times during the process. Instead, if the process is Bernoullian or Markovian, the path probabilities should factorize as in equation (4.8) or (4.9), respectively.

Let us consider here below several specific classes of stochastic processes and their path probabilities.

6.2.2 Discrete-State Markov Processes

We suppose that the process is Markovian with continuous time and discrete states and ruled by the master equation (4.25) with transition rates that are constant in time. Accordingly, the probability of some path observed with the sampling time τ can be expressed as

$$P(\omega_1 \omega_2 \cdots \omega_k) = \prod_{i=2}^{k} P_\tau(\omega_i|\omega_{i-1})\, P(\omega_1) \tag{6.5}$$

in terms of the conditional probability $P_\tau(\omega_i|\omega_{i-1}) \equiv P(\omega_i, t_i|\omega_{i-1}, t_{i-1})$ with $\tau = t_i - t_{i-1}$. This conditional probability and the associated stationary probability distribution $P_{\rm st}(\omega) = P(\omega)$ should satisfy the conditions (4.19) and obey equation (4.20).

If the master equation is written into the matrix form (4.35), the conditional probability can be equivalently written as

$$P_\tau(\omega_i|\omega_{i-1}) = \left(e^{\mathsf{L}\tau}\right)_{\omega_i\omega_{i-1}} = \delta_{\omega_i\omega_{i-1}} + (\mathsf{L})_{\omega_i\omega_{i-1}}\, \tau + O\left(\tau^2\right) \tag{6.6}$$

with the matrix elements given by equation (4.36) in terms of the transition rates and the escape rates introduced in equation (4.28).

Since the Chapman–Kolmogorov equation (4.21) is satisfied, the sum (6.3) of the path probabilities is given by

$$P(\omega_k, t_k|\omega_1, t_1) = \sum_{\omega_2 \cdots \omega_{k-1}} \prod_{i=2}^{k} P_\tau(\omega_i|\omega_{i-1}) = \left[e^{\mathsf{L}(t_k - t_1)}\right]_{\omega_k\omega_1}, \tag{6.7}$$

as required.

These considerations can be extended to Markov jump processes ruled by the master equation (4.27) with the transition rates $W(\omega \overset{\rho}{\to} \omega')$ associated with the elementary transition mechanisms $\{\rho\}$. This master equation can be equivalently expressed in the form (5.208) in terms of the coarse-grained states (5.206). For such processes, the random paths should be denoted

$$\boldsymbol{\omega} = \omega_1 \overset{\rho_1}{\longrightarrow} \omega_2 \overset{\rho_2}{\longrightarrow} \cdots \overset{\rho_{k-1}}{\longrightarrow} \omega_k, \tag{6.8}$$

where $\{\rho_1, \ldots, \rho_{k-1}\}$ is the sequence of elementary transition mechanisms that may cause jumps between the states ω_i observed at successive times $t_i = t_1 + (i-1)\tau$. Since the observation is stroboscopic with the sampling time τ, the system may remain in the same state during some time interval τ, which is denoted as $\rho_i = \emptyset$ if $\omega_{i+1} = \omega_i$. The probability of such a path can be written as

$$P(\boldsymbol{\omega}) = \prod_{i=1}^{k-1} P_\tau\left(\omega_i \overset{\rho_i}{\to} \omega_{i+1}\right) P(\omega_1) \tag{6.9}$$

with the transition probabilities given in the limit $\tau \to 0$ by

$$P_\tau\left(\omega_i \overset{\rho_i}{\to} \omega_{i+1}\right) = \begin{cases} W(\omega_i \overset{\rho_i}{\to} \omega_{i+1})\tau + O\left(\tau^2\right) & \text{if} \quad \omega_i \neq \omega_{i+1}, \\ 1 - \gamma_{\omega_i}\tau + O\left(\tau^2\right) & \text{if} \quad \omega_i = \omega_{i+1}, \end{cases} \tag{6.10}$$

where γ_{ω_i} denotes the escape rate introduced in equation (4.28).

If l transitions happen at the successive times $\mathscr{T}_1 < \mathscr{T}_2 < \cdots < \mathscr{T}_l$ during the time interval $[t_1, t_k]$ giving the trajectory $\omega_1 \overset{\rho_1}{\longrightarrow} \omega_2 \overset{\rho_2}{\longrightarrow} \cdots \overset{\rho_l}{\longrightarrow} \omega_{l+1} = \omega_k$, the path probability (6.9) can be approximated by

$$P(\boldsymbol{\omega}) \simeq \mathrm{e}^{-\gamma_{\omega_{l+1}}(\mathscr{T}_{l+1}-\mathscr{T}_l)} \prod_{j=1}^{l} \left[W\left(\omega_j \overset{\rho_j}{\to} \omega_{j+1}\right) \tau\, \mathrm{e}^{-\gamma_{\omega_j}(\mathscr{T}_j - \mathscr{T}_{j-1})}\right] P(\omega_1, t_1) \tag{6.11}$$

with $\mathscr{T}_0 = t_1$, $\mathscr{T}_{l+1} = t_k$, and small enough time steps τ. Indeed, the probability that one of the l transitions occurs in the time interval $[\mathscr{T}_j, \mathscr{T}_j + \tau]$ is given by the first line of equation (6.10), while the second line leads to the exponentials involving the time intervals $[\mathscr{T}_{j-1}, \mathscr{T}_j]$ between the transitions, if the trajectory stays in the same state during successive small enough time steps τ inside these time intervals.

6.2.3 Continuous-State Markov Processes

Similar considerations apply to the continuous-state Markov processes of Section 4.5. In the case of advection–diffusion processes generated by the Itô stochastic differential equations (4.113) and ruled by the master equation (4.108) with the vector field and diffusivities (4.116), the conditional probability density for evolving from $\mathbf{x}_i$ to $\mathbf{x}_{i+1}$ in the space $\mathbb{R}^d$ during arbitrarily short time steps τ is given by the Gaussian distribution

$$p_\tau(\mathbf{x}_{i+1}|\mathbf{x}_i) \simeq \frac{1}{\sqrt{(4\pi\tau)^d \det \mathbf{D}_i}} \exp\left[-\frac{1}{4\tau}\mathbf{D}_i^{-1} : (\mathbf{x}_{i+1} - \mathbf{x}_i - \mathbf{C}_i\,\tau)^2\right] \tag{6.12}$$

with $\mathbf{C}_i = \mathbf{C}(\mathbf{x}_i, t_i)$ and $\mathbf{D}_i = \mathbf{D}(\mathbf{x}_i, t_i)$. This result is obtained using the discretization (4.115) and the change of random variables from $\{G_{\mu i}\}$ to $\mathbf{x}$. As a consequence, the solution of the master equation (4.108) starting from some initial distribution p_0 at time t_0 can be expressed as

$$p(\mathbf{x},t) = \int p(\mathbf{x},t|\mathbf{x}_0,t_0)\, p_0(\mathbf{x}_0,t_0)\, d\mathbf{x}_0, \tag{6.13}$$

where

$$p(\mathbf{x},t|\mathbf{x}_0,t_0) = \lim_{k\to\infty} \int \prod_{i=0}^{k-1} p_\tau(\mathbf{x}_{i+1}|\mathbf{x}_i) \prod_{i=1}^{k-1} d\mathbf{x}_i \tag{6.14}$$

with $\tau = (t-t_0)/k$ and $\mathbf{x}_k = \mathbf{x}$. In the limit $k \to \infty$, this Green function or propagator can be symbolically written as the *path integral* (Feynman and Hibbs, 1965)

$$p(\mathbf{x},t|\mathbf{x}_0,t_0) = \int \exp\left[-\int_{t_0}^{t} \mathcal{L}(\mathbf{x},\dot{\mathbf{x}},t')\, dt'\right] \mathcal{D}\mathbf{x}(t) \tag{6.15}$$

with the Lagrangian function

$$\mathcal{L}(\mathbf{x},\dot{\mathbf{x}},t) = \frac{1}{4}\,\mathbf{D}(\mathbf{x},t)^{-1} : [\dot{\mathbf{x}} - \mathbf{C}(\mathbf{x},t)]^2 \tag{6.16}$$

and the path element of integration $\mathcal{D}\mathbf{x}(t) \simeq (1/\upsilon_0)\prod_{i=1}^{k-1}(d\mathbf{x}_i/\upsilon_i)$, where $\upsilon_i = \sqrt{(4\pi\tau)^d \det \mathbf{D}_i}$ and $\tau = (t-t_0)/k$.

The action functional of the process is given by $\mathcal{A}[\mathbf{x}(t)] = \int \mathcal{L}\, dt$. The conditional probability that the path $\mathbf{x}(t)$ starts from the initial condition $\mathbf{x}_0$ at time t_0 and passes by the cells $[\mathbf{x}_i, \mathbf{x}_i + d\mathbf{x}_i]$ at the successive times $t_i = t_0 + i\tau$ with $1 \le i \le k$ can be written in the form

$$P[\mathbf{x}(t)|\mathbf{x}_0,t_0] \simeq \mathrm{e}^{-\mathcal{A}[\mathbf{x}(t)]}\, d\mathbf{x}\, \mathcal{D}\mathbf{x}(t), \tag{6.17}$$

including the volume element $d\mathbf{x} = d\mathbf{x}_k$ of the final point $\mathbf{x} = \mathbf{x}(t_k)$ of the path. If the initial condition $\mathbf{x}_0$ is distributed according to the probability density $p_0(\mathbf{x}_0,t_0)$, the path probability itself is thus given by

$$P[\mathbf{x}(t)] \simeq \mathrm{e}^{-\mathcal{A}[\mathbf{x}(t)]}\, p_0(\mathbf{x}_0,t_0)\, d\mathbf{x}\, \mathcal{D}\mathbf{x}(t)\, d\mathbf{x}_0. \tag{6.18}$$

These path probabilities typically decrease with time because the Lagrangian (6.16) is non-negative. We note the consistency with the Lagrangian function (4.150). In the weak-noise limit, the paths that dominate the path integral become concentrated around the deterministic trajectory given by the solution of $\dot{\mathbf{x}} = \mathbf{C}(\mathbf{x},t)$, as explained in Section 4.6.2.

Similar results can be obtained for other Markovian processes.

6.3 Path Probabilities, Time Reversal, and Fluctuation Relations

6.3.1 Fluctuation Relations and Entropy Production

The results of Chapter 5 show that remarkable relations arise from microreversibility if we compare the probability distributions of energy and particle exchanges during some time evolution with those of the corresponding time-reversed history. These fluctuation relations hold not only for exchanges occurring between two time points, but also for paths measured at many successive times, as established in Section 5.4.7. In particular, the fluctuation relation (5.131) can be written in the form

$$\frac{P(\boldsymbol{\omega})}{P(\boldsymbol{\omega}^{\rm R})} = {\rm e}^{\Sigma(\boldsymbol{\omega})}, \tag{6.19}$$

where $P(\boldsymbol{\omega})$ denotes the probability of the path

$$\boldsymbol{\omega} = \omega_1 \omega_2 \cdots \omega_k \tag{6.20}$$

sampled at the successive times $t_i = i\tau$ for $1 \leq i \leq k$ with the sampling time τ, and $P\left(\boldsymbol{\omega}^{\rm R}\right)$ is the probability of its time reversal

$$\boldsymbol{\omega}^{\rm R} = \omega_k \cdots \omega_2 \omega_1 \tag{6.21}$$

such that $\omega_i^{\rm R} = \omega_{k-i+1}$, and $\Sigma(\boldsymbol{\omega}) = \Sigma(t) = \mathbf{A} \cdot \Delta\boldsymbol{\mathfrak{C}}(t)$ by equation (5.103). Each coarse-grained state $\omega_i = (\sigma_i, \boldsymbol{\mathfrak{C}}_i)$ includes the internal state σ_i of the open system and the state variables $\boldsymbol{\mathfrak{C}}_i$ of the environment, which monitor energy and particle fluxes across the open system, which is possibly composed of several reservoirs. In this respect, the path (6.20) and its time reversal (6.21) could be written as

$$\boldsymbol{\omega} = (\sigma_1, \boldsymbol{\mathfrak{C}}_1) \to (\sigma_2, \boldsymbol{\mathfrak{C}}_2) \to \cdots \to (\sigma_k, \boldsymbol{\mathfrak{C}}_k), \tag{6.22}$$

$$\boldsymbol{\omega}^{\rm R} = (\sigma_k, \boldsymbol{\mathfrak{C}}_k) \to \cdots \to (\sigma_2, \boldsymbol{\mathfrak{C}}_2) \to (\sigma_1, \boldsymbol{\mathfrak{C}}_1). \tag{6.23}$$

The path probability distribution P in equation (6.19) is normalized according to $\sum_{\boldsymbol{\omega}} P(\boldsymbol{\omega}) = \sum_{\boldsymbol{\omega}} P\left(\boldsymbol{\omega}^{\rm R}\right) = 1$. This probability distribution depends on the affinities $\mathbf{A}$ defined by equations (5.69) and (5.70).

At equilibrium where the affinities are equal to zero, i.e., $\mathbf{A} = \mathbf{0}$, the identity $\Sigma = 0$ is exactly satisfied. Therefore, $P_{\rm eq}(\boldsymbol{\omega}) = P_{\rm eq}(\boldsymbol{\omega}^{\rm R})$ in agreement with the principle of detailed balance presented in Section 2.5.2.

Here, we suppose for simplicity that the partition defining the coarse-grained states is composed of self-reverse cells and that the control parameters take the same values for the statistics of the paths and their time reversal. More general assumptions may be considered, in particular, performing the statistics of the time-reversed paths with control parameters taking time-reversed values with respect to those used for the paths, as in the forward and backward protocols for the nonequilibrium work fluctuation relation or in the presence of an external magnetizing field, as discussed in Section 5.2.

The fluctuation relation (6.19) has for corollary the following identity,

$$\frac{p_t(\Sigma)}{p_t(-\Sigma)} = {\rm e}^{\Sigma}, \tag{6.24}$$

expressed in terms of the probability density of the random variable $\Sigma = \mathbf{A} \cdot \Delta\boldsymbol{\mathfrak{C}}$,

$$p_t(\Sigma) \equiv \sum_{\boldsymbol{\omega}} P(\boldsymbol{\omega})\, \delta\left[\Sigma - \Sigma(\boldsymbol{\omega})\right], \tag{6.25}$$

which is normalized as $\int_{-\infty}^{+\infty} p_t(\Sigma)\, d\Sigma = 1$ (Jiang et al., 2004; Seifert, 2005a, 2012; Van den Broeck and Esposito, 2015). The fluctuation relation (6.24) is deduced from equation (6.19) as follows,

$$p_t(\Sigma) = \sum_{\omega} \mathrm{e}^{\Sigma(\omega)} P\left(\omega^{\mathrm{R}}\right) \delta\left[\Sigma - \Sigma(\omega)\right] = \mathrm{e}^{\Sigma} \sum_{\omega} P\left(\omega^{\mathrm{R}}\right) \delta\left[\Sigma - \Sigma(\omega)\right]$$
$$= \mathrm{e}^{\Sigma} \sum_{\omega} P(\omega)\, \delta\left[\Sigma - \Sigma\left(\omega^{\mathrm{R}}\right)\right] = \mathrm{e}^{\Sigma}\, p_t(-\Sigma), \tag{6.26}$$

where we have used the property $\Sigma\left(\omega^{\mathrm{R}}\right) = -\Sigma(\omega)$, which is the direct consequence of equation (6.19).

Therefore, the fluctuating quantity Σ satisfies the integral fluctuation relation

$$\left\langle \mathrm{e}^{-\Sigma(\omega)} \right\rangle = 1 \tag{6.27}$$

and the inequality $\langle \Sigma(\omega) \rangle \geq 0$, obtained using the Jensen inequality (3.46). The proof is similar to equation (5.101).

Another remarkable result is that if the stationarity condition holds, the random variable $\mathrm{e}^{-\Sigma(\omega)}$ is a so-called *martingale*, i.e., its mean value at time t' conditioned on observations until some previous time $t < t'$ is equal to its value at the time t, as shown in Section E.5 (Chetrite and Gupta, 2011; Chetrite and Touchette, 2015; Neri et al., 2017). Among other generic properties, the random variable $\Sigma(t)$ can thus be decomposed under nonequilibrium conditions as $\Sigma(t) = \tau(t) + M(t)$ into a monotonically increasing random process $\tau(t)$ and a martingale $M(t)$ with zero mean (Pigolotti et al., 2017; Neri et al., 2017; Neri, 2020).[1]

Now, equation (5.132) for the entropy production rate can be written in terms of the path probability distribution as

$$\frac{1}{k_{\mathrm{B}}} \frac{d_{\mathrm{i}}S}{dt} = \lim_{k\to\infty} \frac{1}{k\tau} \langle \Sigma(\omega) \rangle = \lim_{k\to\infty} \frac{1}{k\tau} \sum_{\omega} P(\omega) \ln \frac{P(\omega)}{P(\omega^{\mathrm{R}})} \geq 0. \tag{6.28}$$

At equilibrium, we note that $\Sigma = 0$ always holds, so that $p_t(\Sigma) = \delta(\Sigma) = \delta(-\Sigma)$ and the entropy production rate is equal to zero.

6.3.2 Large-Deviation Properties

The large-deviation properties of the probabilities of the paths and their time reversal can be characterized by several generating functions defined in the long-time limit. They obey symmetry relations as a consequence of the fluctuation relation (Kurchan, 1998; Lebowitz and Spohn, 1999; Maes, 1999; Jiang et al., 2004).

First of all, we may introduce the cumulant generating function of the random variable $\Sigma(t)$ as

$$\mathcal{Q}(\eta) \equiv \lim_{t\to\infty} -\frac{1}{t} \ln \left\langle \mathrm{e}^{-\eta\Sigma(t)} \right\rangle, \tag{6.29}$$

[1] Since the random variable $\tau(t)$ as well as its mean value $\langle \tau(t) \rangle = \langle \Sigma(t) \rangle$ are monotonically increasing with time t in nonequilibrium processes, their time dependence is compatible with what is expected for entropy production according to the second law of thermodynamics.

where the statistical average is carried out over the probability distribution $P(\boldsymbol{\omega})$. The entropy production rate (6.28) is thus given by

$$\frac{1}{k_{\mathrm{B}}} \frac{d_{\mathrm{i}} S}{dt} = \frac{d\mathcal{Q}}{d\eta}(0). \tag{6.30}$$

The fluctuation relation (6.19) or (6.24) implies that the generating function (6.29) obeys the symmetry relation

$$\mathcal{Q}(\eta) = \mathcal{Q}(1 - \eta) \tag{6.31}$$

in the stationary state of affinities $\mathbf{A}$. We note that the link with the cumulant generating function (5.76) for the currents is established by setting $\boldsymbol{\lambda} = \eta \mathbf{A}$, giving $\mathcal{Q}(\eta) = Q(\eta\mathbf{A}; \mathbf{A})$. Therefore, the symmetry relation (5.91) directly implies equation (6.31).

We may also introduce the corresponding rate function

$$\mathcal{R}(\varsigma) \equiv \lim_{t\to\infty} -\frac{1}{t} \ln P\left\{\Sigma(t)/t \in [\varsigma, \varsigma + d\varsigma]\right\}, \tag{6.32}$$

which is the Legendre–Fenchel transform of the cumulant generating function

$$\mathcal{R}(\varsigma) = \sup_{\eta} \left[\mathcal{Q}(\eta) - \varsigma\, \eta\right]. \tag{6.33}$$

Because of the fluctuation relation, the rate function has the symmetry

$$\mathcal{R}(-\varsigma) - \mathcal{R}(\varsigma) = \varsigma. \tag{6.34}$$

The entropy production rate is given by $\varsigma_0 = d\mathcal{Q}/d\eta(0)$, satisfying $\mathcal{R}(\varsigma_0) = 0$ and $(d\mathcal{R}/d\varsigma)(\varsigma_0) = 0$.

Next, the joint cumulant generating function of the probability distributions $P(\boldsymbol{\omega})$ and $P(\boldsymbol{\omega}^{\mathrm{R}})$ can be defined as

$$\mathcal{J}(\alpha, \beta) \equiv \lim_{t\to\infty} -\frac{1}{t} \ln\left\langle P(\boldsymbol{\omega})^{\alpha} P\left(\boldsymbol{\omega}^{\mathrm{R}}\right)^{\beta}\right\rangle, \tag{6.35}$$

where again $\langle X \rangle = \sum_{\boldsymbol{\omega}} P(\boldsymbol{\omega}) X(\boldsymbol{\omega})$. The symmetry

$$\mathcal{J}(\alpha, \beta) = \mathcal{J}(\beta - 1, \alpha + 1) \tag{6.36}$$

is satisfied by the generating function (6.35) in consistency with the fluctuation relation (6.19) (Gaspard, 2007a). The cumulant generating function (6.29) is the special case $\mathcal{Q}(\eta) = \mathcal{J}(-\eta, \eta)$ and the symmetry (6.31) is implied by equation (6.36).

6.3.3 Markov Jump Processes with Constant Transition Rates

The previous considerations can be applied to the paths of the Markov jump processes presented in Section 6.2.2. We compare the probability (6.11) for the following path having l transitions during the time interval $[t_1, t_k]$,

$$\boldsymbol{\omega} = \omega_1 \xrightarrow{\rho_1} \omega_2 \xrightarrow{\rho_2} \cdots \xrightarrow{\rho_l} \omega_{l+1}, \tag{6.37}$$

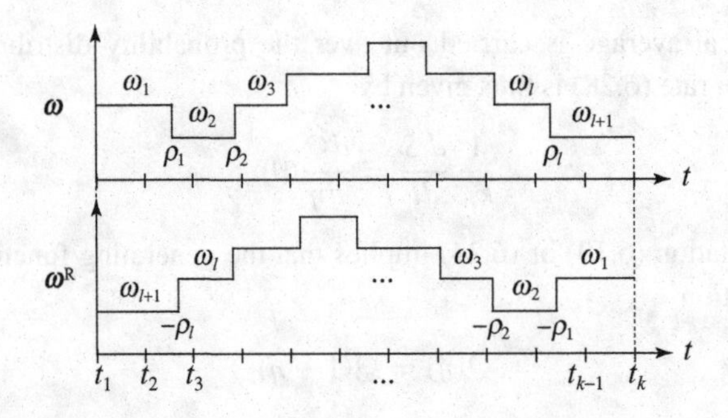

Figure 6.2 Typical path $\boldsymbol{\omega}$ and its time reversal $\boldsymbol{\omega}^{\rm R}$ sampled at the successive times $t_i = t_1 + (i-1)\tau$ with $1 \leq i \leq k$.

where $\omega_{l+1} = \omega_k$, and the probability of its time reversal

$$\boldsymbol{\omega}^{\rm R} = \omega_{l+1} \overset{-\rho_l}{\longrightarrow} \cdots \overset{-\rho_2}{\longrightarrow} \omega_2 \overset{\rho_1}{\longrightarrow} \omega_1. \tag{6.38}$$

These paths are depicted in Figure 6.2. We note that, for $1 \leq j \leq l$, the transition $\omega_j \overset{\rho_j}{\to} \omega_{j+1}$ occurs at the time $\mathscr{T}_j$ and the opposite transition $\omega_{j+1} \overset{-\rho_j}{\to} \omega_j$ at the time $(t_1 + t_k) - \mathscr{T}_j$.

Using equation (6.11), the ratio of these probabilities is given by

$$\Sigma(\boldsymbol{\omega}) \equiv \ln \frac{P(\boldsymbol{\omega})}{P(\boldsymbol{\omega}^{\rm R})} = Z(\boldsymbol{\omega}) + \ln \frac{P(\omega_1, t_1)}{P(\omega_k, t_k)} \tag{6.39}$$

with

$$Z(\boldsymbol{\omega}) \equiv \sum_{j=1}^{l} \ln \frac{W\left(\omega_j \overset{\rho_j}{\to} \omega_{j+1}\right)}{W\left(\omega_{j+1} \overset{-\rho_j}{\to} \omega_j\right)}, \tag{6.40}$$

because all the exponential factors cancel each other since they correspond to the same time intervals between the jumps in the path and its time reversal. The quantity (6.40) was introduced by Lebowitz and Spohn (1999), who proved that it obeys the symmetry relations (6.31) and (6.34) in the long-time limit, as is the case for the quantity (6.39). Furthermore, the fluctuation relation (6.24) is exactly satisfied at every time by the quantity (6.39) because it has the form of equation (6.19) (Jiang et al., 2004; Seifert, 2005a, 2012).

In order to interpret the identity (6.39), we consider its statistical average over the probability distribution $P(\boldsymbol{\omega})$ of the process. Because of equation (6.28), the entropy production during the time interval $[t_1, t_k]$ of length $t = t_k - t_1 = (k-1)\tau$ is given by

$$\Delta_{\rm i} S(t) = k_{\rm B} \langle \Sigma(\boldsymbol{\omega}) \rangle. \tag{6.41}$$

Accordingly, the entropy production rate is obtained as $d_{\rm i}S/dt = \lim_{t\to\infty} \Delta_{\rm i}S(t)/t$ if the stationarity condition holds. Next, we note that the difference of entropy undergone by the open system during the time interval t can be obtained as

$$\Delta S(t) = \langle S(\omega_k, t_k) - S(\omega_1, t_1) \rangle, \tag{6.42}$$

where $S(\omega,t)$ is the entropy (4.47), including the two contributions, $S^0(\omega)$ given by equation (2.125) and $-k_B \ln P(\omega,t)$ due to the disorder of the probability distribution $P(\omega,t)$. In general, the entropy difference (6.42) can be decomposed as $\Delta S(t) = \Delta_e S(t) + \Delta_i S(t)$ into the entropy production (6.41) and the entropy exchange $\Delta_e S(t)$, which is the entropy transported from the environment to the system. According to equations (4.47) and (6.39), this latter can thus be expressed as

$$\Delta_e S(t) = -k_B \langle Z(\boldsymbol{\omega})\rangle + \left\langle S^0(\omega_k) - S^0(\omega_1)\right\rangle. \tag{6.43}$$

At equilibrium, the identities $\Delta_i S = 0$, $\Delta S = 0$, and $\Delta_e S = 0$ are satisfied.

If the condition of stationarity is satisfied, we have that $\left\langle S^0(\omega_1)\right\rangle_{st} = \left\langle S^0(\omega_k)\right\rangle_{st}$ and $\langle \ln[P(\omega_1,t_1)/P(\omega_k,t_k)]\rangle_{st} = 0$, so that the entropy exchange rate is equal to

$$\frac{d_e S}{dt} = \lim_{t\to\infty} -\frac{k_B}{t}\langle Z(\boldsymbol{\omega})\rangle = -\frac{d_i S}{dt} \leq 0, \tag{6.44}$$

which is in agreement with the expectation (1.15) from thermodynamics. Moreover, equation (4.54) is recovered for the entropy exchange rate and equation (4.55) for the entropy production rate. In particular, if the quantity (6.40) is considered as a function of time, its time derivative can be written as

$$\frac{dZ}{dt} = \sum_{j=-\infty}^{+\infty} \delta(t - \mathscr{T}_j)\ln \frac{W\left(\omega_j \overset{\rho_j}{\to} \omega_{j+1}\right)}{W\left(\omega_{j+1} \overset{-\rho_j}{\to} \omega_j\right)}, \tag{6.45}$$

where $\mathscr{T}_j$ is the time of the transition $\omega_j \overset{\rho_j}{\to} \omega_{j+1}$. This latter occurs with the rate $W\left(\omega_j \overset{\rho_j}{\to} \omega_{j+1}\right)$ and, thus, with the net rate (4.50) of the master equation (4.49). Consequently, the statistical average of equation (6.45) corresponds to the second terms in equation (4.54), while the first terms are obtained by similar considerations from the jumps of $S^0[\omega(t)]$ along the trajectory. Furthermore, the time derivative of the quantity (6.39) can be expressed as

$$\frac{d\Sigma}{dt} = \frac{dZ}{dt} + \sum_{j=-\infty}^{+\infty} \delta(t - \mathscr{T}_j)\ln \frac{P(\omega_j, \mathscr{T}_j)}{P(\omega_{j+1}, \mathscr{T}_j)}, \tag{6.46}$$

where $P(\omega,t) = \exp\{k_B^{-1}[S^0(\omega) - S(\omega,t)]\}$ according to equation (4.47). Since the statistical average of equation (6.46) gives the entropy production rate (4.57), we may say that the results of Section 4.4.4 for infinitesimal time intervals dt are here extended to finite time intervals t.

We note that, in the work by Seifert (2005a, 2012), the entropy production obeying the second law of thermodynamics $\Delta_i S \geq 0$ is denoted $\Delta S_{tot} \geq 0$, $k_B\langle Z(\boldsymbol{\omega})\rangle$ corresponds to ΔS_m, and $k_B\langle \ln[P(\omega_1,t_1)/P(\omega_k,t_k)]\rangle$ to $\Delta S(t)$, since the contribution from $S^0(\omega)$ is not considered. Under the latter assumption, $\Delta S_m = -\Delta_e S$ would be the entropy transported from the system to its environment.

For isobaric-isothermal processes, the ratios of opposite transition rates are related by equation (4.59) to jumps in the Gibbs free energy of the system including its environment, so that

$$Z(\boldsymbol{\omega}) = \frac{1}{k_{\rm B}T}\sum_{j=1}^{l}\left[G(\omega_j) - G(\omega_{j+1})\right] = \frac{1}{k_{\rm B}T}\left[G(\omega_1) - G(\omega_k)\right]. \tag{6.47}$$

If the environment supplies free energy to drive the process away from equilibrium, we have that $\Delta G(t) = \langle G(\omega_k) - G(\omega_1)\rangle < 0$, implying that the fluctuating quantity (6.47) increases on average as $\langle Z(\boldsymbol{\omega})\rangle = -\Delta G(t)/(k_{\rm B}T)$. Since the Gibbs free energy of each coarse-grained state is related to its enthalpy, its entropy, and the temperature T by equation (4.58), the entropy exchange (6.43) with the environment becomes

$$\Delta_{\rm e}S(t) = \frac{1}{T}\langle H(\omega_k) - H(\omega_1)\rangle = \frac{1}{T}\,\Delta H(t), \tag{6.48}$$

which is given in terms of the heat ΔH transferred from the environment to the system, as expected from equation (4.60).

The considerations are thus consistent with thermodynamics for the stochastic processes ruled by the master equation (4.49), for which the paths (6.8) allow us to determine the energy and particle fluxes with the environment that are driving the system out of equilibrium.

6.3.4 Markov Jump Processes with Time-Varying Transition Rates

The previous results can be generalized to Markov jump processes driven by external time-dependent forcings (Crooks, 1998, 1999, 2000; Seifert, 2005a, 2012; Harris and Schütz, 2007; Van den Broeck and Esposito, 2010a,b,c). Such processes are ruled by master equations such as (4.24) with transition rates varying in time. Denoting λ_t to be some time-dependent control parameter, the transition rates can be written as $W\big(\omega \overset{\rho}{\to} \omega';\lambda_t\big)$ with $0 \le t \le \mathcal{T}$.

If some path is supposed to have l transitions $\omega_j \overset{\rho_j}{\to} \omega_{j+1}$ at every time $\mathscr{T}_j$ when $\lambda_{\mathscr{T}_j} \equiv \lambda_j$, its probability is given by

$$P(\boldsymbol{\omega}) \simeq {\rm e}^{-\int_{\mathscr{T}_l}^{\mathscr{T}_{l+1}}\gamma_{\omega_{l+1}}(\lambda_t)\,dt}\prod_{j=1}^{l}\left[W\left(\omega_j \overset{\rho_j}{\to} \omega_{j+1};\lambda_j\right)\tau\,{\rm e}^{-\int_{\mathscr{T}_{j-1}}^{\mathscr{T}_j}\gamma_{\omega_j}(\lambda_t)\,dt}\right]P(\omega_1,t_1) \tag{6.49}$$

with $\mathscr{T}_0 = t_1 = 0$ and $\mathscr{T}_{l+1} = t_k = \mathcal{T}$. We note that the path probability (6.11) is recovered if the transition rates are invariant in time.

As already mentioned, the time-reversed process should be defined with the reversed time dependence $\lambda_t^{\rm R} \equiv \lambda_{\mathcal{T}-t}$, defining the path probability $P^{\rm R}(\boldsymbol{\omega})$. For the time-reversed path $\boldsymbol{\omega}^{\rm R}$, the reversed transition $\omega_{j+1} \overset{-\rho_j}{\to} \omega_j$ happens at the time $\mathcal{T} - \mathscr{T}_j$ when the control parameter takes the value $\lambda^{\rm R}_{\mathcal{T}-\mathscr{T}_j} = \lambda_{\mathscr{T}_j} = \lambda_j$. The initial state of the time-reversed path is the final state $\omega_{l+1} = \omega_k$ of the path $\boldsymbol{\omega}$ and it is assumed to have the probability $P(\omega_k, t_k)$.

Here, we consider the ratio between the path probability (6.49) and the probability of the time-reversed path $\boldsymbol{\omega}^{\rm R}$ for the time-reversed process, defining the quantity

$$\Sigma(\boldsymbol{\omega}) \equiv \ln \frac{P(\boldsymbol{\omega})}{P^{\rm R}(\boldsymbol{\omega}^{\rm R})} = Z(\boldsymbol{\omega}) + \ln \frac{P(\omega_1, t_1)}{P(\omega_k, t_k)}. \tag{6.50}$$

Because of the identity

$$\int_{\mathscr{T}_{j-1}}^{\mathscr{T}_j} \gamma_{\omega_j}(\lambda_t)\, dt = \int_{T-\mathscr{T}_j}^{T-\mathscr{T}_{j-1}} \gamma_{\omega_j}\left(\lambda_t^{\rm R}\right) dt, \tag{6.51}$$

the exponential factors again cancel each other and we find

$$Z(\boldsymbol{\omega}) = \sum_{j=1}^{l} \ln \frac{W\left(\omega_j \overset{\rho_j}{\to} \omega_{j+1}; \lambda_j\right)}{W\left(\omega_{j+1} \overset{-\rho_j}{\to} \omega_j; \lambda_j\right)}. \tag{6.52}$$

If the same forward and backward protocols as in Section 5.2 are applied to an isothermal system subjected to time-dependent forcing, the quantity (6.50) corresponds to $\Sigma = \beta(W - \Delta F)$, where W is the nonequilibrium work performed on the system and ΔF the free energy stored in the system during the forward protocol. For this case, the fluctuation relation satisfied by the quantity (6.50) is equivalent to the nonequilibrium work fluctuation relation (5.13) (Crooks, 1998, 1999, 2000).

In this context, an important issue is to understand the effects of deviations with respect to the instantaneous stationary distribution $P_{\rm st}(\omega; \lambda_t)$, which satisfies the following condition for the master equation (4.49) with the transition rates of time t:

$$0 = \sum_{\rho,\, \omega'} \left[P_{\rm st}(\omega'; \lambda_t)\, W\left(\omega' \overset{\rho}{\to} \omega; \lambda_t\right) - P_{\rm st}(\omega; \lambda_t)\, W\left(\omega \overset{-\rho}{\to} \omega'; \lambda_t\right)\right]. \tag{6.53}$$

This distribution would be the stationary one if the rates were constant in time, which is not the case in general. To investigate this issue, the quantity (6.50) can be decomposed as

$$\Sigma(\boldsymbol{\omega}) = \Sigma_{\rm a}(\boldsymbol{\omega}) + \Sigma_{\rm na}(\boldsymbol{\omega}) \tag{6.54}$$

into

$$\Sigma_{\rm a}(\boldsymbol{\omega}) \equiv \sum_{j=1}^{l} \ln \frac{P_{\rm st}(\omega_j; \lambda_j)\, W\left(\omega_j \overset{\rho_j}{\to} \omega_{j+1}; \lambda_j\right)}{P_{\rm st}(\omega_{j+1}; \lambda_j)\, W\left(\omega_{j+1} \overset{-\rho_j}{\to} \omega_j; \lambda_j\right)}, \tag{6.55}$$

$$\Sigma_{\rm na}(\boldsymbol{\omega}) \equiv \sum_{j=1}^{l} \ln \frac{P_{\rm st}(\omega_{j+1}; \lambda_j)}{P_{\rm st}(\omega_j; \lambda_j)} + \ln \frac{P(\omega_1, t_1)}{P(\omega_k, t_k)}, \tag{6.56}$$

which are referred to as the adiabatic and nonadiabatic contributions, respectively, to the quantity (6.50) (Van den Broeck and Esposito, 2010a,b,c; Seifert, 2012). This is justified by the fact that $\Sigma_{\rm na}(\boldsymbol{\omega}) = 0$ if the transition rates are constant in time, so that the process evolves in the steady state of probability distribution $P_{\rm st}(\omega) = P(\omega, t_1) = P(\omega, t_k)$. However, we should expect that the quantity $\Sigma_{\rm na}(\boldsymbol{\omega})$ becomes different from zero if the control parameter λ_t depends on time, especially if the process is switching from one steady state in the remote past towards another steady state in the future with $\lim_{t\to-\infty}\lambda_t = \lambda_- \neq \lim_{t\to+\infty}\lambda_t = \lambda_+$ (Oono and Paniconi, 1998; Hatano and Sasa, 2001).

Remarkably, we may consider the time evolution of the dual process defined with the transition rates (Crooks, 2000),

$$\tilde{W}(\omega \xrightarrow{\rho} \omega';\lambda) \equiv \frac{P_{\rm st}(\omega';\lambda)}{P_{\rm st}(\omega;\lambda)}\, W(\omega' \xrightarrow{-\rho} \omega;\lambda). \tag{6.57}$$

If the path probability is denoted $\tilde{P}(\boldsymbol{\omega})$ for the dual process and $\tilde{P}^{\rm R}(\boldsymbol{\omega})$ for the time-reversed dual process, the quantities (6.55) and (6.56) can be expressed as

$$\Sigma_{\rm a}(\boldsymbol{\omega}) = \ln \frac{P(\boldsymbol{\omega})}{\tilde{P}(\boldsymbol{\omega})} \qquad \text{and} \qquad \Sigma_{\rm na}(\boldsymbol{\omega}) = \ln \frac{P(\boldsymbol{\omega})}{\tilde{P}^{\rm R}(\boldsymbol{\omega}^{\rm R})}, \tag{6.58}$$

if $\tilde{P}^{\rm R}(\boldsymbol{\omega}^{\rm R})$ is starting from $P(\omega_k, t_k)$, as well as $P^{\rm R}(\boldsymbol{\omega}^{\rm R})$. Since the quantities (6.50) and (6.58) have the same form as the quantity Σ considered in equation (6.19), the following fluctuation relations are exactly satisfied,

$$\frac{p_t(\Sigma)}{p_t^{\rm R}(-\Sigma)} = {\rm e}^{\Sigma}, \qquad \frac{p_t(\Sigma_{\rm a})}{\tilde{p}_t(-\Sigma_{\rm a})} = {\rm e}^{\Sigma_{\rm a}}, \qquad \frac{p_t(\Sigma_{\rm na})}{\tilde{p}_t^{\rm R}(-\Sigma_{\rm na})} = {\rm e}^{\Sigma_{\rm na}}, \tag{6.59}$$

where the probability densities p_t, $p_t^{\rm R}$, $\tilde{p}_t$, and $\tilde{p}_t^{\rm R}$ are, respectively, defined as in equation (6.25) with the path probability distributions $P(\boldsymbol{\omega})$, $P^{\rm R}(\boldsymbol{\omega})$, $\tilde{P}(\boldsymbol{\omega})$, and $\tilde{P}^{\rm R}(\boldsymbol{\omega})$ (Van den Broeck and Esposito, 2010a). As a consequence, they obey the integral fluctuation relations

$$\langle {\rm e}^{-\Sigma} \rangle = 1, \qquad \langle {\rm e}^{-\Sigma_{\rm a}} \rangle = 1, \qquad \langle {\rm e}^{-\Sigma_{\rm na}} \rangle = 1, \tag{6.60}$$

where the statistical average is carried out with respect to the path probability distribution $P(\boldsymbol{\omega})$. Their corresponding mean values are thus always nonnegative. Taking the ensemble average of the decomposition (6.54), we thus find that

$$\Delta_{\rm i} S = \Delta_{\rm i} S_{\rm a} + \Delta_{\rm i} S_{\rm na} \geq 0 \qquad \text{with} \qquad \Delta_{\rm i} S_{\rm a} \geq 0 \qquad \text{and} \qquad \Delta_{\rm i} S_{\rm na} \geq 0, \tag{6.61}$$

where $\Delta_{\rm i} S = k_{\rm B}\langle \Sigma(\boldsymbol{\omega})\rangle$ is the entropy production during the time interval $[0, \mathcal{T}]$, while $\Delta_{\rm i} S_{\rm a} = k_{\rm B}\langle \Sigma_{\rm a}(\boldsymbol{\omega})\rangle$ and $\Delta_{\rm i} S_{\rm na} = k_{\rm B}\langle \Sigma_{\rm na}(\boldsymbol{\omega})\rangle$ are the adiabatic and nonadiabatic contributions to the entropy production (Van den Broeck and Esposito, 2010b,c).

If the transition rates of the process remain constant in time, the system undergoes relaxation towards a steady state. In this case, the nonadiabatic contribution is equal to zero, i.e., $\Delta_{\rm i} S_{\rm na} = k_{\rm B}\langle \Sigma_{\rm na}(\boldsymbol{\omega})\rangle = 0$, and the entropy production is given by the adiabatic contribution $\Delta_{\rm i} S = \Delta_{\rm i} S_{\rm a}$. Under isothermal conditions, this adiabatic contribution corresponds to the so-called *housekeeping heat*, $Q_{\rm hk} \equiv k_{\rm B} T \Sigma_{\rm a}(\boldsymbol{\omega})$, which gives the mean heat $\langle Q_{\rm hk}\rangle = T\Delta_{\rm i} S_{\rm a}$ dissipated inside the system and transported to the environment in order to maintain the system in the steady state during the time interval $[0, \mathcal{T}]$ (Oono and Paniconi, 1998; Hatano and Sasa, 2001). The associated fluctuating quantity obeys the integral fluctuation relation $\langle \exp(-\beta Q_{\rm hk})\rangle = 1$ (Speck and Seifert, 2005).

However, if the control parameter switches from the value λ_- in the remote past to the value λ_+ in the future, the process is evolving from one steady state to another one (Oono and Paniconi, 1998). During this process, the nonadiabatic contribution takes the positive mean value $\Delta_{\rm i} S_{\rm na} = k_{\rm B}\langle \Sigma_{\rm na}(\boldsymbol{\omega})\rangle > 0$, which represents the extra entropy production required for switching from one steady state to the other. Moreover, the entropy of the system may change by the amount $\Delta S = S(\lambda_+) - S(\lambda_-)$. Therefore, the entropy exchange transported from the environment to the system is given by $\Delta_{\rm e} S = \Delta S - \Delta_{\rm i} S$. Because of the switching, the total dissipated heat is expected to have two contributions, on the one hand, the housekeeping heat $Q_{\rm hk}$ as long as the system is maintained in one steady state and the *excess heat* $Q_{\rm ex}$ due to the switching: $Q = Q_{\rm hk} + Q_{\rm ex}$. Since $\langle Q\rangle = -T\Delta_{\rm e} S$ and the mean housekeeping heat is given by $\langle Q_{\rm hk}\rangle = T\Delta_{\rm i} S_{\rm a}$, the mean excess heat is thus equal to $\langle Q_{\rm ex}\rangle = T(\Delta_{\rm i} S_{\rm na} - \Delta S)$. As a consequence of the integral fluctuation relation (6.60) for $\Sigma_{\rm na}$, the mean excess heat satisfies the inequality $\langle Q_{\rm ex}\rangle \geq -T\Delta S$ (Hatano and Sasa, 2001).

Stochastic processes with variables that are odd under time reversal can also be considered (Spinney and Ford, 2012; Lee et al., 2013).

6.3.5 Langevin Processes

The previous results can be extended to Langevin processes ruled by the master equation (4.108) (Seifert, 2005a, 2012; Chernyak et al., 2006). Here, we again suppose that the random variables $\mathbf{x}$ include the observables of energy and particle fluxes that are driving the system out of equilibrium. Furthermore, the system is assumed to be subjected to some time-dependent forcing by the control parameter λ_t, which is another possible source of nonequilibrium drive.

In order to perform the time-reversal transformation in path integrals, the convenient choice of time discretization is the Stratonovich midpoint rule with $\alpha = 1/2$, as explained in Section E.6.[2] In this case, the path probabilities are expressed using the action functional

$$\mathcal{A}[\mathbf{x}(t)] = \int_0^{\mathcal{T}} \mathcal{L}^{(1/2)}(\mathbf{x}, \dot{\mathbf{x}}, t)\, dt, \tag{6.62}$$

where the Lagrangian function is given by

$$\mathcal{L}^{(1/2)}(\mathbf{x}, \dot{\mathbf{x}}, t) = \frac{1}{4}\,\mathsf{D}^{-1} : (\dot{\mathbf{x}} - \mathbf{V})^2 + \Lambda^{(1/2)}(\mathbf{x}, t) \tag{6.63}$$

in terms of the velocity drift (4.112), the diffusivity matrix (4.103), and the function (E.39). The drift velocity and the diffusivities may change with time because of the time-dependent control parameter λ_t. With the action (6.62), the path probability of this process can thus be written in the form (6.18) with the probability density $p_0 = p(\mathbf{x}_0, 0)$ of the initial conditions $\mathbf{x}_0$.

The time-reversed process is defined with the control parameter $\lambda_t^{\rm R} = \lambda_{\mathcal{T}-t}$. The time reversal of the path $\mathbf{x}(t)$ is given by $\mathbf{x}^{\rm R}(t) = \mathbf{x}(\mathcal{T} - t)$, so that $\mathbf{x}^{\rm R}(0) = \mathbf{x}(\mathcal{T}) = \mathbf{x}_{\mathcal{T}}$ and $\mathbf{x}^{\rm R}(\mathcal{T}) = \mathbf{x}(0) = \mathbf{x}_0$. The initial conditions of the time-reversed process are thus sampled with the probability density $p(\mathbf{x}_{\mathcal{T}}, \mathcal{T})$. For the Stratonovich midpoint rule, the discretized

[2] Other choices are possible by considering the discretization rule $\alpha^{\rm R} = 1 - \alpha$ for the time-reversed process (Aron et al., 2016).

integrals of the paths and their time reversal have similar forms with $\dot{\mathbf{x}}^{\rm R} = -\dot{\mathbf{x}}$. Taking the ratio of the probability P of the path $\mathbf{x}(t)$ for the driving λ_t to the probability $P^{\rm R}$ of its time reversal $\mathbf{x}^{\rm R}(t)$ for the time-reversed driving $\lambda_t^{\rm R}$, we obtain the quantity

$$\Sigma[\mathbf{x}(t)] = \ln \frac{P[\mathbf{x}(t)]}{P^{\rm R}[\mathbf{x}^{\rm R}(t)]} = \int_0^{\mathcal{T}} \left(\mathbf{D}^{-1}\cdot\mathbf{V}\right)\circ\dot{\mathbf{x}}\,dt + \ln\frac{p(\mathbf{x}_0,0)}{p(\mathbf{x}_{\mathcal{T}},\mathcal{T})}, \tag{6.64}$$

which is the analogue of equation (6.50). This quantity obeys the fluctuation relation (6.24) if λ_t is independent of time (Seifert, 2005a) and, in general, the first of the three fluctuation relations (6.59).

Let us denote by $p_{\rm st}(\mathbf{x};\lambda)$ the stationary probability density (4.135) with some potential $\Upsilon(\mathbf{x};\lambda)$, such that $\partial_t p_{\rm st} = \hat{L} p_{\rm st} = 0$, where $\hat{L}$ is the linear operator of the master equation (4.108) at the value λ of the control parameter. Let us suppose that the initial conditions of the process and its time reversal are sampled according to $p(\mathbf{x}_0,0) = p_{\rm st}(\mathbf{x}_0;\lambda_0)$ and $p(\mathbf{x}_{\mathcal{T}},\mathcal{T}) = p_{\rm st}(\mathbf{x}_{\mathcal{T}};\lambda_{\mathcal{T}})$, respectively. With these assumptions, the quantity (6.64) becomes

$$\Sigma[\mathbf{x}(t)] = \int_0^{\mathcal{T}} \left(\mathbf{D}^{-1}\cdot\mathbf{V}\right)\circ\dot{\mathbf{x}}\,dt + \Delta\Upsilon, \tag{6.65}$$

with

$$\Delta\Upsilon = \Upsilon(\mathbf{x}_{\mathcal{T}};\lambda_{\mathcal{T}}) - \Upsilon(\mathbf{x}_0;\lambda_0) = \int_0^{\mathcal{T}} d\Upsilon = \int_0^{\mathcal{T}} \left(\partial_{\mathbf{x}}\Upsilon\circ\dot{\mathbf{x}} + \partial_\lambda\Upsilon\,\dot{\lambda}\right)dt. \tag{6.66}$$

Introducing the new vector field

$$\mathbf{A} \equiv \mathbf{D}^{-1}\cdot\mathbf{V} + \partial_{\mathbf{x}}\Upsilon, \tag{6.67}$$

the quantity (6.65) reads

$$\Sigma[\mathbf{x}(t)] = \Sigma_{\rm a}[\mathbf{x}(t)] + \Sigma_{\rm na}[\mathbf{x}(t)] \tag{6.68}$$

with

$$\Sigma_{\rm a}[\mathbf{x}(t)] \equiv \int_0^{\mathcal{T}} \mathbf{A}\circ d\mathbf{x} \qquad \text{and} \qquad \Sigma_{\rm na}[\mathbf{x}(t)] \equiv \int_0^{\mathcal{T}} \partial_\lambda\Upsilon\,\dot{\lambda}\,dt. \tag{6.69}$$

As for Markov jump processes, $\Sigma_{\rm na} = 0$ for autonomous processes with $\dot{\lambda} = 0$, in which case $\Sigma = \Sigma_{\rm a}$. Accordingly, equation (6.69) defines the adiabatic and nonadiabatic contributions to the quantity (6.64).

Here, the *dual process* is defined with the conditional probability

$$\tilde{p}_\tau(\mathbf{x}'|\mathbf{x}) \equiv \frac{p_{\rm st}(\mathbf{x}')}{p_{\rm st}(\mathbf{x})}\, p_\tau(\mathbf{x}|\mathbf{x}') \tag{6.70}$$

for small enough time steps τ (Chernyak et al., 2006). According to the methods of Section 4.5.7, the operator ruling the time evolution of this dual process is given by

$$\hat{\tilde{L}}\,\tilde{p} \equiv p_{\rm st}\,\hat{L}^\dagger\left(\frac{\tilde{p}}{p_{\rm st}}\right) \tag{6.71}$$

in terms of the adjoint operator defined in equation (4.128). The master equation of the dual process can thus be expressed as

$$\partial_t \tilde{p} = \hat{\tilde{L}}\, \tilde{p} = -\partial_{\mathbf{x}} \cdot \left(\tilde{\mathbf{V}}\, \tilde{p} - \mathbf{D} \cdot \partial_{\mathbf{x}} \tilde{p} \right) \tag{6.72}$$

with the same diffusivities $\mathbf{D}$ as for the process itself, but a different velocity drift defined as

$$\tilde{\mathbf{V}} \equiv -\mathbf{V} - 2\,\mathbf{D} \cdot \partial_{\mathbf{x}} \Upsilon . \tag{6.73}$$

Consequently, the path probabilities can be defined for the dual process and for the time-reversal dual process as in equation (6.58), so that the three fluctuation relations (6.59) hold here, as well as the integral fluctuation relations (6.60) (Hatano and Sasa, 2001; Chernyak et al., 2006; Van den Broeck and Esposito, 2010a,c). The quantity (6.64) can thus be decomposed as in equation (6.54) into its adiabatic and nonadiabatic contributions.

The integral fluctuation relation $\langle \exp(-\Sigma_{\rm na}) \rangle = 1$ has been experimentally verified (Trepagnier et al., 2004), confirming the prediction of Hatano and Sasa (2001) that the mean excess heat $\langle Q_{\rm ex} \rangle$ and the entropy change ΔS satisfy the inequality $\langle Q_{\rm ex} \rangle \geq -T\Delta S$ for an isothermal process at temperature T.

In the case of the overdamped Langevin process for a Brownian particle moving in the energy potential $U(\mathbf{r};\lambda_t)$ and subjected to an external force $\mathbf{F}_{\rm ext}$, the stochastic Langevin equation reads

$$\gamma \frac{d\mathbf{r}}{dt} = -\boldsymbol{\nabla} U + \mathbf{F}_{\rm ext} + \mathbf{F}_{\rm fl}(t), \tag{6.74}$$

where $\mathbf{x} = \mathbf{r}$ is the position of the particle, γ its friction coefficient, and $\mathbf{F}_{\rm fl}(t)$ the Langevin fluctuating force given by Gaussian white noises satisfying equation (4.174). The diffusivity matrix is thus given by $D_{ij} = \mathcal{D}\,\delta_{ij}$ with the diffusion coefficient $\mathcal{D} = k_{\rm B}T/\gamma$. Accordingly, the master equation of this Langevin process has the form

$$\partial_t\, p + \boldsymbol{\nabla} \cdot (\mathbf{V}\, p) = \mathcal{D}\, \boldsymbol{\nabla}^2\, p \tag{6.75}$$

with the drift velocity $\mathbf{V} = (-\boldsymbol{\nabla} U + \mathbf{F}_{\rm ext})/\gamma$. The stationary probability density of this process can be expressed as $p_{\rm st} = \mathcal{N} \exp(-\Upsilon)$ with $\Upsilon = \beta[U(\mathbf{r};\lambda) - F(\lambda)]$, where $\beta = (k_{\rm B}T)^{-1}$ is the inverse temperature and $F(\lambda)$ is the free energy associated with Boltzmann's probability distribution for the particle position in the potential $U(\mathbf{r};\lambda)$. Hence, the vector field (6.67) is here given by the external force according to $\mathbf{A} = \beta \mathbf{F}_{\rm ext}$, thus representing the mechanical affinity. For overdamped Langevin processes, the quantities (6.69) can thus be expressed as

$$\Sigma_{\rm a}[\mathbf{r}(t)] = \beta \int_0^{\mathcal{T}} \mathbf{F}_{\rm ext} \circ d\mathbf{r} \qquad \text{and} \qquad \Sigma_{\rm na}[\mathbf{r}(t)] = \beta \int_0^{\mathcal{T}} \partial_\lambda (U - F)\, \dot{\lambda}\, dt, \tag{6.76}$$

which obey the fluctuation relations (6.59) together with $\Sigma = \Sigma_{\rm a} + \Sigma_{\rm na}$.

On the one hand, the system may undergo relaxation towards equilibrium if the energy potential does not depend on time and the external force does not generate unbounded motion for the Brownian particle. On the other hand, the system can be driven out of equilibrium either by a time-dependent forcing with the control parameter λ_t if the Brownian particle is trapped in the potential $U(\mathbf{r};\lambda_t)$, or by the constant external force $\mathbf{F}_{\rm ext}$ if the potential extends in position space with bounded variations $U_{\rm min} \leq U(\mathbf{r}) \leq U_{\rm max}$, or by a combination of both nonequilibrium drivings.

Accordingly, there exist four types of processes whether dynamics are autonomous (i.e., time-independent) or not, and if the affinity (6.67) is equal to zero or not. If the affinity is equal to zero, the condition of reversibility (4.136) is satisfied, so that the process is reversible and detailed balance is expected to hold. The drift velocity can thus be expressed in terms of the gradient of a potential, so that the external force is conservative. Therefore, the following classification can be established (Chernyak et al., 2006):

- *Autonomous dynamics* $\dot{\lambda} = 0$, *conservative force* $\mathbf{A} = 0$. The system undergoes relaxation towards equilibrium, where $\Sigma = 0$, $\Sigma_{\rm a} = 0$, and $\Sigma_{\rm na} = 0$.
- *Nonautonomous dynamics* $\dot{\lambda} \neq 0$, *conservative force* $\mathbf{A} = 0$. The system is driven out of equilibrium by the time-dependent forcing, so that $\Sigma = \Sigma_{\rm na}$ and $\Sigma_{\rm a} = 0$. If the system is in contact with a single heat reservoir at the inverse temperature β, the conditions are similar to those used for establishing the nonequilibrium work fluctuation relation (5.13) and $\Sigma = \Sigma_{\rm na} = \beta(W - \Delta F) = \beta W_{\rm diss}$ in terms of the dissipated work (5.26). Under these conditions, the experimental verification of the fluctuation relation (6.59) for Σ has been demonstrated (Blickle et al., 2006; Joubaud et al., 2008).
- *Autonomous dynamics* $\dot{\lambda} = 0$, *nonconservative force* $\mathbf{A} \neq 0$. The system is driven out of equilibrium by the nonconservative external force, so that $\Sigma = \Sigma_{\rm a}$ and $\Sigma_{\rm na} = 0$. In the case of Brownian motion in the time-independent periodic potential $U(x) = U_0 \sin(kx)$ and the uniform external force $\mathbf{F}_{\rm ext}$ according to the overdamped Langevin equation (6.74), we have that $\Sigma = \Sigma_{\rm a} = \beta \mathbf{F}_{\rm ext} \cdot \Delta\mathbf{r}$ is the mechanical work performed by the external force during the displacement $\Delta\mathbf{r}$ of the particle in the unit of thermal energy $\beta^{-1} = k_{\rm B}T$. The corresponding fluctuation relation (6.24) has also been experimentally verified (Speck et al., 2007).
- *Nonautonomous dynamics* $\dot{\lambda} \neq 0$, *nonconservative force* $\mathbf{A} \neq 0$. The system is driven out of equilibrium by both the time-dependent forcing and the nonconservative force, so that $\Sigma = \Sigma_{\rm a} + \Sigma_{\rm na}$.

We may also consider the time derivative of the quantity (6.64) with respect to time $t = \mathcal{T}$

$$\frac{d\Sigma}{dt} = \left(\mathsf{D}^{-1} \cdot \mathbf{V}\right) \circ \dot{\mathbf{x}} - \frac{d}{dt} \ln p(\mathbf{x}_t, t), \tag{6.77}$$

where the Stratonovich midpoint rule is again used for time discretization, so that the stochastic differential equation reads

$$\dot{\mathbf{x}} = \mathbf{V} + \sqrt{2\mathsf{D}} \circ \boldsymbol{\chi}(t) \tag{6.78}$$

in terms of the drift velocity (4.112), the diffusivities in equation (4.116), and d Gaussian white noises satisfying equation (4.114). Moreover, we have that

$$\frac{d}{dt} \ln p(\mathbf{x}_t, t) = p^{-1} \left(\partial_t p + \partial_{\mathbf{x}} p \circ \dot{\mathbf{x}}\right) \tag{6.79}$$

according to equation (E.34) with $\alpha = 1/2$. Using the current density (4.111), the time derivative can be written in the form (Seifert, 2005a)

$$\frac{d\Sigma}{dt} = p^{-1}\left(\mathbf{D}^{-1}\cdot\mathbf{J}\right)\circ\dot{\mathbf{x}} - p^{-1}\,\partial_t p. \tag{6.80}$$

The mean value of this time derivative with respect to the probability density $p(\mathbf{x},t)$ gives the entropy production rate according to

$$\frac{1}{k_{\mathrm{B}}}\frac{d_{\mathrm{i}}S}{dt} = \left\langle\frac{d\Sigma}{dt}\right\rangle = \int p^{-1}\,\mathbf{J}^{\mathrm{T}}\cdot\mathbf{D}^{-1}\cdot\mathbf{J}\,d\mathbf{x}, \tag{6.81}$$

because $\dot{\mathbf{x}} \simeq \mathbf{J}/p$ in the statistical average $\langle\cdot\rangle = \int(\cdot)\,p\,d\mathbf{x}$ (Seifert, 2005a). We recover the entropy production rate (4.192) for the overdamped Langevin process of Brownian motion in the velocity field $\mathbf{V} = \mathbf{F}/\gamma$ with the diffusivities $\mathbf{D} = \mathcal{D}\mathbf{1}$, where $\mathcal{D} = k_{\mathrm{B}}T/\gamma$ is the diffusion coefficient.

Additionally, for Langevin processes, the random variable $\mathrm{e}^{-\Sigma(t)}$ is a martingale, so that $\Sigma(t)$ obeys several important generic properties (Chetrite and Gupta, 2011; Chetrite and Touchette, 2015; Pigolotti et al., 2017; Neri et al., 2017; Neri, 2020). In particular, the infimum law holds, according to which the mean value of its infimum $\Sigma_{\mathrm{inf}}(\mathcal{T}) \equiv \inf_{t\in[0,\mathcal{T}]}\Sigma(t)$ is always equal to $\langle\Sigma_{\mathrm{inf}}(\mathcal{T})\rangle = -1$. Moreover, the first-passage times $\mathscr{T}_{\pm}$ when the quantity $\Sigma(t)$ reaches for the first time the values $\Sigma(\mathscr{T}_{\pm}) = \pm\varsigma$ have probability densities also obeying a fluctuation relation. These remarkable results provide important clues on the extreme-value statistics of the random variable $\Sigma(t)$ underlying entropy production (Neri et al., 2017; Neri, 2020).

All these considerations constitute the basis of *stochastic thermodynamics* (Seifert, 2012; Van den Broeck, 2013).

6.4 Temporal Disorder

6.4.1 Entropy per Unit Time

The previous examples of path probabilities show that they are often exponentially decaying with time. This behavior is reminiscent of random processes such as coin tossing, for which path probabilities decay exponentially as 2^{-k} with their length k.[3] This exponential decay is characteristic of dynamical randomness, i.e., temporal disorder.

In this regard, we may introduce the mean decay rate of the path probabilities as

$$h(\mathcal{P},\tau) \equiv \lim_{k\to\infty} -\frac{1}{k\tau}\sum_{\omega_1\omega_2\cdots\omega_k} P(\omega_1\omega_2\cdots\omega_k)\,\ln P(\omega_1\omega_2\cdots\omega_k), \tag{6.82}$$

where $\mathcal{P}$ denotes the partition of the state space into cells $\{\mathcal{C}_i\}$ and τ is the sampling time of the paths. The rate of temporal disorder (6.82) is called *entropy per unit time* or *dynamical entropy*. This concept was introduced by Shannon and Weaver (1949), applied to the theory of stochastic processes (Kolmogorov, 1956b; Gaspard and Wang, 1993), developed in

[3] Such an exponential decay is also the feature of Gibbsian equilibrium probability distributions at positive temperature as the volume of the system increases, which is in relation with the positivity of their entropy density.

dynamical systems theory (Kolmogorov, 1959; Sinai, 1959; Cornfeld et al., 1982; Eckmann and Ruelle, 1985), and extended to nonequilibrium statistical mechanics, as well as quantum systems (Gaspard, 1998; Ernst et al., 1995; Lecomte et al., 2005, 2007; Pressé et al., 2013; Connes et al., 1987; Narnhofer and Thirring, 1987; Benatti et al., 1998).

If the process is ergodic, the Shannon–McMillan–Breiman theorem holds, according to which

$$P(\omega_1\omega_2\cdots\omega_k) \sim \mathrm{e}^{-ht} \qquad \text{with} \qquad t = k\tau \to \infty \tag{6.83}$$

for almost all paths $\boldsymbol{\omega} = \omega_1\omega_2\cdots\omega_k$, where $h = h(\mathcal{P},\tau)$ (Billingsley, 1978; Cornfeld et al., 1982). Such paths, having the mean statistical properties of the process, are said to be *typical*.

The entropy per unit time represents the data accumulation rate required to reconstruct the typical paths or trajectories of the process. In this sense, this quantity represents the rate of production of information by the process.

The entropy per unit time (6.82) is the analogue of the entropy per unit volume defined in statistical mechanics to characterize disorder in the phase space of a microscopic Hamiltonian system. Similarly, the spatiotemporal disorder in the time evolution of a spatially extended system can be characterized by an entropy per unit time and unit volume.

We note that the entropy per unit time (6.82) is also the mean growth rate of the action functional introduced in equation (6.2)

$$h(\mathcal{P},\tau) = \lim_{k\to\infty} \frac{1}{k\tau}\langle \mathcal{A}(\boldsymbol{\omega})\rangle, \tag{6.84}$$

where $\langle\cdot\rangle$ denotes the statistical average over the path probability distribution $P(\boldsymbol{\omega})$. In this regard, the entropy per unit time is equal to zero if the number of random events grows more slowly than time itself, as for periodic and quasiperiodic signals, or processes where random events have intermittent or sporadic occurrences in time (Gaspard and Wang, 1988).[4]

6.4.2 Kolmogorov–Sinai Entropy per Unit Time

In general, the entropy per unit time depends on the conditions of observation given by the partition $\mathcal{P}$ together with the sampling time τ. The Kolmogorov–Sinai (KS) entropy per unit time is defined as the supremum of the dynamical entropy (6.82) over all the possible partitions:

$$h_{\mathrm{KS}} = \mathrm{Sup}_{\mathcal{P}}\, h(\mathcal{P},\tau), \tag{6.85}$$

which is thus independent of the partition and the sampling time (Eckmann and Ruelle, 1985; Cornfeld et al., 1982).

For chaotic dynamical systems manifesting sensitivity to initial conditions in bounded phase-space domains, the KS entropy is given by the sum of positive Lyapunov exponents according to equation (B.14). Therefore, the entropy per unit time is positive and finite,

[4] For such processes with long memory, the path probabilities do not have the Gibbsian property of exponential decay in time.

meaning that dynamical instability induces dynamical randomness with the same degree as in the coin tossing process, i.e., as if finitely many discrete states would be visited every finite time intervals. Examples for such chaotic systems are the hard-sphere gases at thermodynamic equilibrium, where the KS entropy per unit time can be estimated using Krylov's formula (2.25) as

$$h_{\rm KS} \sim N \, \frac{\bar{v}}{\ell} \, \ln \frac{\ell}{d}, \tag{6.86}$$

for a dilute gas of N hard spheres of diameter d moving in a finite volume with mean velocity $\bar{v}$ and mean free path ℓ (Dellago et al., 1996; van Beijeren et al., 1997). Another example is given by the Hamiltonian motion of atomic ions in a Penning trap shown in Figure 5.4(c), which is chaotic because its maximum Lyapunov exponent is equal to $\lambda_{\rm max} = 0.037 \pm 0.001$ at the total energy $E = 2$, so that its KS entropy per unit time (6.85) is positive (Gaspard, 2003a).

The rate of temporal disorder given by the KS entropy per unit time (6.86) is due to thermal agitation. It is positive even if the system is at equilibrium with zero thermodynamic entropy production rate. Therefore, the entropy per unit time and the thermodynamic entropy production rate are fundamentally different physical concepts, the former characterizing the temporal disorder in random processes and the latter the increase of disorder in the phase space of the microscopic Hamiltonian system.

The same degree of randomness as in chaotic dynamics manifests itself in Markov chains, which are defined by the transition probabilities $P(\omega \to \omega')$ such that $\sum_{\omega'} P(\omega \to \omega') = 1$ and the associated stationary probabilities $P_{\rm st}(\omega)$. If the random transitions between the states ω are supposed to occur with some finite positive rate r, the KS entropy per unit time of the Markov chain is given by

$$h_{\rm KS} = -r \sum_{\omega, \omega'} P_{\rm st}(\omega) \, P(\omega \to \omega') \, \ln P(\omega \to \omega'), \tag{6.87}$$

which is typically finite and positive as for chaotic dynamical systems. Bernoulli chains are special cases, for which the transition probabilities are equal to the stationary probabilities, so that

$$h_{\rm KS} = -r \sum_{\omega} P_{\rm st}(\omega) \, \ln P_{\rm st}(\omega). \tag{6.88}$$

In the coin tossing random process, the stationary probabilities are equal to $P_{\rm st}(\omega) = 1/2$, giving $h_{\rm KS} = \ln 2$ per step. For these processes, states and times are discrete.

Kac's ring model also has a positive KS entropy per unit time in the very same infinite size limit where it manifests exponential relaxation towards stationarity, as shown in Section E.13.

Remarkably, the KS entropy per unit time (6.85) or, more generally, the entropy per unit time (6.82) for a given partition, can be introduced to characterize dynamical randomness, even if Lyapunov exponents cannot be defined for the system of interest, or if dynamical instability is not directly related to dynamical randomness as in leaky dynamical systems, where equation (B.41) holds instead of equation (B.14). In this regard, the concept of entropy per unit time is appropriate to extend the characterization of temporal disorder beyond the class of chaotic nonlinear dynamical systems, in particular, towards mesoscopic processes and microscopic dynamics, possibly ruled by quantum mechanics (Gaspard, 1994, 1998; Castiglione et al., 2008).

6.4.3 (ε, τ)-Entropy per Unit Time

Many random processes have an infinite KS entropy per unit time. This is already the case for ideal gases in infinite volumes with noninteracting particles moving in free flights with a Maxwellian velocity distribution, as well as in infinite harmonic crystals, which are ergodic and mixing (Cornfeld et al., 1982) as mentioned in Section 2.4. This is also the case for Poisson processes or noises of Wiener type. For such random processes, the partitions used to define the entropy per unit time (6.82) should be restricted to those $\mathcal{P}_\varepsilon$ into cells $\{\mathcal{C}_i\}$ with a small but finite radius ε corresponding to the resolution in the measurement of the signals. With this condition, equation (6.82) defines the so-called (ε, τ)-*entropy per unit time*:

$$h(\varepsilon, \tau) \equiv h(\mathcal{P}_\varepsilon, \tau). \tag{6.89}$$

The KS entropy per unit time is obtained in the limit $\varepsilon \to 0$. The advantage of this definition is that stochastic processes can be characterized according to their degree of randomness in every cell of size ε and/or time step τ by the way the (ε, τ)-entropy per unit time increases as $\varepsilon \to 0$ and/or $\tau \to 0$ (Gaspard and Wang, 1993).

The stochastic processes where randomness is generated continuously in time should have an (ε, τ)-entropy per unit time increasing as the sampling time τ tends to zero, and those with randomness continuously distributed in state space as the resolution ε goes to zero. In this way, the stochastic processes can be classified according to their degree of randomness, as shown in Table 6.1. Let us consider here below a few important cases.

Table 6.1. *Temporal disorder of random processes characterized by the scaling behavior of their entropy per unit time (6.82) versus the sampling time τ or the resolution ε (Gaspard and Wang, 1993). Here, $h_{\rm KS}$ denotes the KS entropy per unit time (6.85). For random flights, n is the number of continuous state variables. For fractional noises, H is the Hölder exponent of the random functions of time defined in equation (6.92).*

Random Process	Entropy per Unit Time h
Periodic or quasiperiodic	0
Sporadic randomness	0
Discrete states, discrete times	$0 < h_{\rm KS} < \infty$
Discrete states, continuous times	$\sim \ln(1/\tau)$
Continuous states, discrete times	$\sim \ln(1/\varepsilon)$
Random flights	$\sim \ln(1/\tau\varepsilon^n)$
Diffusive noises	$\sim (1/\varepsilon)^2$
Fractional noises	$\sim (1/\varepsilon)^{1/H}$

6.4.4 Temporal Disorder of Markov Jump Processes

For Markovian processes ruled by the master equation (4.27), the states are discrete but randomness is generated continuously in time. Indeed, such processes can be simulated by the algorithm of Gillespie (1976, 1977), in which, after every jump, the random time lapse before the next jump is exponentially distributed and the next jump is randomly chosen between the discrete states allowed by the transitions.

If the process is sampled at every time step τ, the path probabilities are given by equation (6.9). The process is thus similar to a Markov chain at every time step τ, which is assumed to be arbitrarily small. Accordingly, the entropy per unit time is given by equation (6.87) with the transition probabilities (6.10) and the rate $r = 1/\tau$ of the random events. Consequently, the τ-entropy per unit time of the Markov jump process is given by

$$h(\tau) = \ln\left(\frac{\mathrm{e}}{\tau}\right) \sum_{\rho,\omega\neq\omega'} P_{\rm st}(\omega)\, W\big(\omega \xrightarrow{\rho} \omega'\big) - \sum_{\rho,\omega\neq\omega'} P_{\rm st}(\omega)\, W\big(\omega \xrightarrow{\rho} \omega'\big) \ln W\big(\omega \xrightarrow{\rho} \omega'\big) + O(\tau) \tag{6.90}$$

in terms of the transition rates and the stationary probability distribution (Gaspard and Wang, 1993; Gaspard, 1998, 2004b). Such processes are characterized by an entropy per unit time increasing as $\ln(1/\tau)$ for decreasing values of τ because dynamical randomness is generated continuously in time. In particular, the formula (6.90) can be used to evaluate temporal disorder in Markov jump processes for particles transported by diffusion in a medium between two reservoirs at different chemical potentials (Gaspard, 2005) and to processes of chemical reactions (Gaspard, 2004b; Andrieux and Gaspard, 2008f).

6.4.5 Temporal Disorder of Continuous-State Processes

For continuous-state processes, the resolution ε should be introduced in the state space of the random variables.

Continuous-State Discrete-Time Processes

Let us consider a process where d random variables $\mathbf{x} \in \mathbb{R}^d$ distributed with some probability density $p(\mathbf{x})$ occur with the finite rate r. If the state space is partitioned into cells $\omega = [\mathbf{x}, \mathbf{x} + \Delta\mathbf{x}]$ with the finite resolution $\varepsilon^d = \Delta^d x$, this process behaves as a Bernoulli process, so that its entropy per unit time is given by equation (6.88) with the stationary probabilities $P_{\rm st}(\omega) = p(\mathbf{x})\Delta^d x$, leading to

$$h = r \ln \frac{1}{\Delta^d x} - r \int p(\mathbf{x}) \ln p(\mathbf{x})\, d\mathbf{x} + O\big(\Delta^d x\big), \tag{6.91}$$

which increases as $\ln(1/\varepsilon)$ for decreasing ε. In the limit $\varepsilon \to 0$, the KS entropy per unit time is thus infinite. Infinite ideal gases are examples of such processes if the observation of the particles is performed in some sectional area Σ perpendicular to the x-direction. If the gas is at equilibrium with the particle density n, the particles cross this sectional area from both sides with mean rate $r = (n\sqrt{2k_{\rm B}T/\pi m})\Sigma$. The crossing events occur at random times, positions inside the sectional area, and velocities measured with the respective resolutions $\tau \propto \Delta x$, Δy, Δz, and $\Delta^3 v$. Therefore, the entropy per unit time of this process is evaluated as $h \simeq r \ln\left(1/\Delta^3 r \Delta^3 v\right)$ (Gaspard, 1994). Since the resolutions in positions and velocities can be arbitrarily small for classical systems, the KS entropy per unit time is infinite for such gases (Cornfeld et al., 1982).

Further processes are random flights, which are Markov jump processes over a continuous state space, as considered in Sections 4.5.4 and 4.5.5. A physical example is the random flight of a particle undergoing elastic collisions and ruled by the Boltzmann–Lorentz equation (4.121). The mean free path of the particle is given by $\ell = 1/(n_s \sigma_{tot})$, where n_s is the density of scatterers and $\sigma_{tot} = \int \sigma_{diff}\, d\Omega$ the total cross section. Since the particle moves at the constant velocity $v = \|\mathbf{v}\|$, the mean collision frequency is equal to $\nu = v/\ell = v\, n_s \sigma_{tot}$. At every elastic collision, the particle scatters in a new random direction according to the differential cross section. Since the new direction is specified by $d-1$ angles, the entropy per unit time goes as $h \simeq \nu \ln\left[1/(\tau \Delta^{d-1}\Omega)\right]$, where $\varepsilon^{d-1} = \Delta^{d-1}\Omega$ is the element of solid angle for the new direction after collision, and τ is the sampling time. The number of continuous state variables is here equal to $n = d - 1$.

Continuous-State Continuous-Time Processes

These stochastic processes include the Wiener process (E.30) and the Ornstein–Uhlenbeck process (E.54), modeling Brownian motion and diffusive noises, as well as their generalizations for fractional noises (Mandelbrot, 1982). Almost all the trajectories $X(t)$ of such processes behave as

$$X(t+\tau) - X(t) \sim \tau^H \qquad \text{for} \qquad \tau \ll 1, \tag{6.92}$$

where the Hölder exponent satisfies $0 < H < 1$. Diffusive noises have the exponent $H = 1/2$.

The temporal disorder of these processes has a still higher degree than for processes with discrete times because randomness is generated not only in the continuous state space, but also on arbitrarily small time intervals. In order to evaluate their temporal disorder, such processes should, moreover, be discretized in time with the time step τ. The trajectories $X(t)$ are thus sampled at the discrete times $t_i = t_0 + i\tau$ with $i = 1, 2, \dots, k$ and the path probabilities can be defined by considering the tubes of trajectories $|X(t_i) - x_i| < \varepsilon$ around some reference trajectory sampled at the values $\{x_i\}_{i=1}^{k}$.

For the processes (6.92), the (ε, τ)-entropy per unit time is scaling as

$$h(\varepsilon, \tau) \simeq \frac{1}{\tau}\, \psi\left(\frac{c\,\tau}{\varepsilon^{1/H}}\right) \tag{6.93}$$

with some function $\psi(\xi)$ and some constant c, if τ is small enough. This (ε, τ)-entropy per unit time forms a family of functions $F(\varepsilon, h; \tau) \equiv \psi - \tau h$ depending on the sampling time τ as a parameter. The envelope of this family is found by eliminating τ between the equations $F(\varepsilon, h; \tau) = 0$ and $\partial_\tau F(\varepsilon, h; \tau) = 0$, giving

$$h_{\text{envelope}}(\varepsilon) \simeq \frac{A}{\varepsilon^{1/H}} \tag{6.94}$$

with the constant $A = c\,\psi'(\xi_0)$, where ξ_0 is the root of $\psi(\xi_0) = \xi_0\,\psi'(\xi_0)$, for $\varepsilon \to 0$. This envelope scales as $h_{\text{envelope}}(\varepsilon) \sim 1/\varepsilon^2$ for diffusive noises, where $H = 1/2$. These results are reported in Table 6.1.

6.4.6 Temporal Disorder of the Ornstein–Uhlenbeck Process

The Ornstein–Uhlenbeck process is generated by the overdamped Langevin stochastic equation $dx/dt + x/\tau_{\text{r}} = v_{\text{fl}}(t)$, where τ_{r} is the relaxation time and $v_{\text{fl}}(t)$ a Gaussian white noise such that $\langle v_{\text{fl}}(t)\rangle = 0$ and $\langle v_{\text{fl}}(t)v_{\text{fl}}(t')\rangle = 2\mathcal{D}\delta(t-t')$ with the diffusion coefficient $\mathcal{D}$. This process describes the one-dimensional motion of a particle immersed in a viscous fluid and trapped in the harmonic potential $U(x) = kx^2/2$ with the spring constant k. For the friction coefficient γ, the relaxation time is thus given by $\tau_{\text{r}} = \gamma/k$ and the diffusion coefficient by $\mathcal{D} = k_{\text{B}}T/\gamma$. If the spring constant is equal to zero ($k = 0$), the particle is no longer trapped and it undergoes Brownian motion because the relaxation time is infinite.

As calculated in Section E.7, the (ε,τ)-entropy per unit time of this process is given by

$$h(\varepsilon,\tau) = \frac{1}{\tau}\ln\sqrt{\frac{\mathrm{e}\pi\mathcal{D}\tau_{\text{r}}}{2\,\varepsilon^2}\left(1-\mathrm{e}^{-2\tau/\tau_{\text{r}}}\right)} + O\left(\varepsilon^2/\tau\right), \tag{6.95}$$

if the condition $\varepsilon \ll \sqrt{\mathcal{D}\tau_{\text{r}}\left(1-\mathrm{e}^{-2\tau/\tau_{\text{r}}}\right)}$ is satisfied (Andrieux et al., 2008).

In the limit where the sampling time is much longer than the relaxation time, $\tau \gg \tau_{\text{r}}$, the (ε,τ)-entropy per unit time goes as $h(\varepsilon,\tau) \sim (1/\tau)\ln(1/\varepsilon)$, as for a process with independent identical Gaussian random variables sampled according to the stationary distribution (E.56) at every time step τ.

In the other limit, $\tau \ll \tau_{\text{r}}$, equation (6.95) becomes

$$h(\varepsilon,\tau) = \frac{1}{\tau}\ln\sqrt{\frac{\mathrm{e}\pi\mathcal{D}\tau}{\varepsilon^2}} + O\left(\varepsilon^2/\tau\right), \tag{6.96}$$

if $\varepsilon \ll \sqrt{2\mathcal{D}\tau}$. In this limit, the (ε,τ)-entropy per unit time has the form (6.93) with the Hölder exponent $H = 1/2$, as expected for the Wiener process modeling the Brownian motion of the particle in the absence of trap for a vanishing spring constant in the limit $\tau_{\text{r}} \to \infty$ (Gaspard and Wang, 1993; Gaspard, 1998).

6.4.7 Temporal Disorder of Brownian Motion

The temporal disorder of Brownian motion has been experimentally measured by observing a single micrometric polystyrene particle moving in suspension in water (Gaspard et al., 1998; Briggs et al., 2001). The motion of the particle was tracked and recorded with a video camera over a total time interval of about 2430 s with the sampling time $\Delta t = 1/60$ s and the smallest resolution $\varepsilon_{\min} = 25$ nm. The particle is maintained horizontally between a polished silicon wafer and a glass coverslip separated by 50 μm, so that its Brownian motion is essentially two-dimensional with diffusion coefficient $\mathcal{D} = 0.124$ μm²/s.

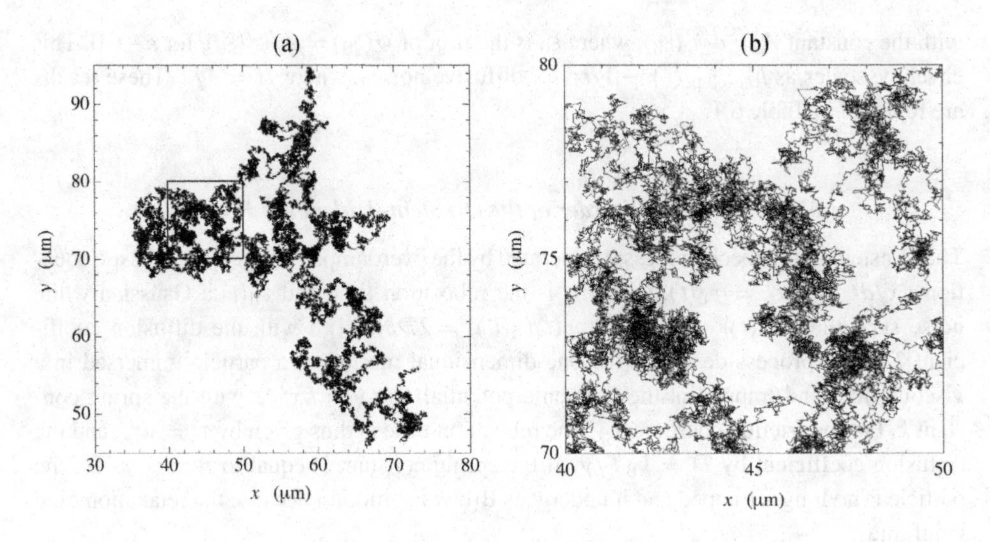

Figure 6.3 (a) Brownian trajectory of a polystyrene particle with 2.5 μm diameter in water at 22°C, observed by a CCD camera with sampling time 1/60 s. (b) Zoom of panel (a). Data from Gaspard et al. (1998) and Briggs et al. (2001).

Figure 6.3 shows a portion of the two-dimensional trajectory and a zoom, demonstrating that dynamical randomness continues to manifest itself on smaller spatial scales in Brownian motion.

The degree of dynamical randomness or temporal disorder is characterized in Figure 6.4 by its (ε, τ)-entropy per unit time (6.89) as a function of the resolution ε, as measured from experimental data. The behavior expected from equation (6.96) is observed as the value of the sampling time τ is changed. Indeed, the shorter the sampling time, the higher the (ε, τ)-entropy per unit time. Since the (ε, τ)-entropy has the form

$$h(\varepsilon, \tau) = \frac{1}{\tau}\, \psi\left(\frac{\mathcal{D}\tau}{\varepsilon^2}\right), \tag{6.97}$$

the envelope of this family of functions scales as

$$h_{\text{envelope}}(\varepsilon) \sim \frac{\mathcal{D}}{\varepsilon^2}, \tag{6.98}$$

which is observed in Figure 6.4. This scaling is characteristic of pure diffusion without trap potential (Gaspard and Wang, 1993; Gaspard, 1998; Gaspard et al., 1998; Briggs et al., 2001).

The smaller the resolution ε, the higher the temporal disorder. Therefore, the degree of dynamical randomness is significantly higher for Brownian motion than for chaotic dynamics, as can already be visually comprehended by comparing the trajectory of Figure 6.3 with the chaotic motion of a few atomic ions in Figure 5.4(c). Indeed, in finite chaotic systems, the trajectories are ruled by deterministic ordinary differential equations, so that dynamical randomness cannot be generated at a rate faster than the dynamical instability characterized

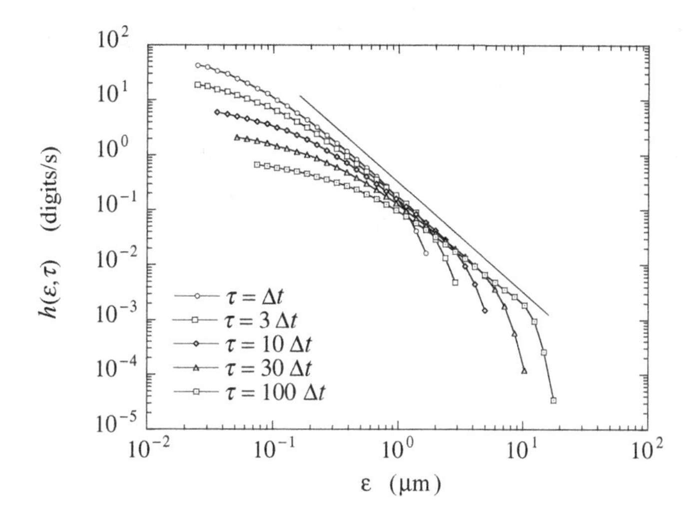

Figure 6.4 Temporal disorder of the Brownian motion of Figure 6.3, as characterized by its (ε, τ)-entropy per unit time versus the resolution ε for different values of the sampling time $\tau = \Delta t, 10\Delta t, 30\Delta t, 100\Delta t$ with $\Delta t = 1/60$ s. The statistics were carried out with a time series of 145 612 positions over a total time interval of about 2430 s. The straight line gives the slope -2, in accordance with the scaling (6.98). Adapted from Gaspard et al. (1998).

by the positive Lyapunov exponents. Besides, Brownian motion can be described as resulting from the Hamiltonian dynamics of the micrometric particle interacting with surrounding water molecules. Since the spatial and temporal scales of this Hamiltonian motion are very small, dynamical randomness is generated at a huge rate because of the many collisions of the small water molecules with the large colloidal particle. This rate can be evaluated as the KS entropy per unit time given by the positive Lyapunov exponents, which are computed in Hamiltonian models of Brownian motion (Gaspard and van Beijeren, 2002).

In the framework of the mathematical theory of stochastic processes, Brownian motion is described as a Wiener process, assuming that dynamical randomness is generated on arbitrarily small spatial and temporal scales $\varepsilon \sim \sqrt{\mathcal{D}\tau}$ for $\tau \to 0$, leading to an infinite KS entropy per unit time.

Since the (ε, τ)-entropy per unit time (6.98) depends on the diffusion coefficient $\mathcal{D}$ determined by Stokes' formula, the dynamical randomness of Brownian motion is influenced by the hydrodynamic flow around the colloidal particle, which is generating recollisions with water molecules. We note that such recollisions do not happen for the Rayleigh flight of Section E.12. In this regard, the huge temporal disorder observed for Brownian motion arises from the interaction of the micrometric particle with the surrounding fluid undergoing ceaseless thermal fluctuations at positive temperature.

Temporal disorder can be quantitatively characterized for stochastic processes, allowing us to know how their dynamical randomness scales with the sampling time τ, the resolution ε, the total time interval of observation $\mathscr{T} = k\tau$ in equation (6.82), or the volume of the system.

6.5 Time Asymmetry in Temporal Disorder and Entropy Production

As previously discussed, differences are expected under nonequilibrium conditions between the probabilities of the paths $\boldsymbol{\omega} = \omega_1\omega_2\cdots\omega_k$ and their time reversal $\boldsymbol{\omega}^{\rm R} = \omega_k\cdots\omega_2\omega_1$. In particular, the decay rate of the probability $P(\boldsymbol{\omega}^{\rm R})$ may differ from the one of $P(\boldsymbol{\omega})$ for the typical paths of the process. In such circumstances, the time-reversal symmetry is broken in temporal disorder and the process manifests irreversibility.

6.5.1 Time-Reversed Coentropy per Unit Time

In order to characterize this broken symmetry, we may introduce the *time-reversed coentropy per unit time*

$$h^{\rm R} \equiv \lim_{k\to\infty} -\frac{1}{k\tau}\sum_{\boldsymbol{\omega}} P(\boldsymbol{\omega}) \ln P\left(\boldsymbol{\omega}^{\rm R}\right), \tag{6.99}$$

considering the mean decay of the probability of the time-reversed path $\boldsymbol{\omega}^{\rm R}$ averaged over the probability distribution $P(\boldsymbol{\omega})$ of the paths themselves (Gaspard, 2004b). This definition is introduced by analogy with the coentropy (2.121) (Gaspard, 2014). The quantity (6.99) is equivalently given by $h^{\rm R} = \lim_{k\to\infty}\left\langle \mathcal{A}\left(\boldsymbol{\omega}^{\rm R}\right)\right\rangle/(k\tau)$ in terms of the action functional introduced in equation (6.2).

We note that if the statistical average was taken with respect to the probability distribution $P\left(\boldsymbol{\omega}^{\rm R}\right)$ of the time-reversed paths instead of the one for the paths themselves, the entropy per unit time (6.82) would be recovered because $-\sum_{\boldsymbol{\omega}} P\left(\boldsymbol{\omega}^{\rm R}\right)\ln P\left(\boldsymbol{\omega}^{\rm R}\right) = -\sum_{\boldsymbol{\omega}} P(\boldsymbol{\omega})\ln P(\boldsymbol{\omega})$. Indeed, the set of all the paths $\boldsymbol{\omega}$ coincides with the set of all the time-reversed paths $\boldsymbol{\omega}^{\rm R}$, so that changing the summation variable $\boldsymbol{\omega}$ into $\boldsymbol{\omega}' = \boldsymbol{\omega}^{\rm R}$ gives the aforementioned identity.

The entropy and coentropy per unit time can be expressed in terms of the cumulant generating function (6.35) according to $h = \partial_\alpha \mathcal{J}(0,0)$ and $h^{\rm R} = \partial_\beta \mathcal{J}(0,0)$.

6.5.2 Temporal Disorder of Typical and Time-Reversed Paths

Now, the difference between the dynamical coentropy (6.99) and the entropy (6.82) is given by

$$h^{\rm R} - h = \lim_{k\to\infty}\frac{1}{k\tau}\, D_{\rm KL}\left[P(\boldsymbol{\omega})\| P\left(\boldsymbol{\omega}^{\rm R}\right)\right] \geq 0 \tag{6.100}$$

in terms of the Kullback–Leibler divergence between the probability distributions of the paths and their time-reversal

$$D_{\rm KL}\left[P(\boldsymbol{\omega})\| P\left(\boldsymbol{\omega}^{\rm R}\right)\right] \equiv \sum_{\boldsymbol{\omega}} P(\boldsymbol{\omega})\ln\frac{P(\boldsymbol{\omega})}{P\left(\boldsymbol{\omega}^{\rm R}\right)}, \tag{6.101}$$

which is known to be always nonnegative (Cover and Thomas, 2006). The difference (6.100) is equal to zero if the process is reversible, satisfying the condition (4.10). Otherwise, it is positive. Therefore, this difference, i.e., the Kullback–Leibler divergence between the

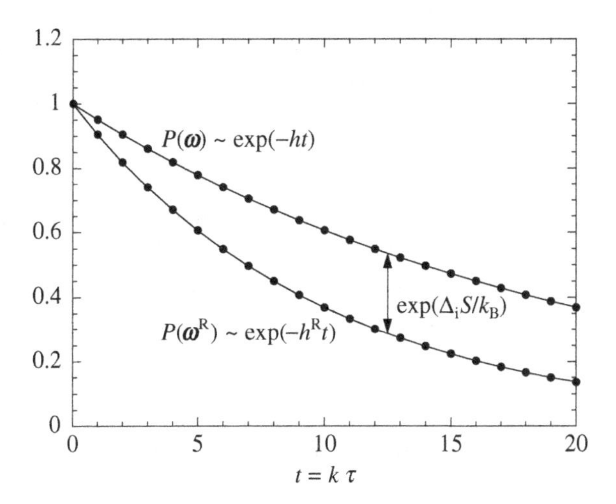

Figure 6.5 Schematic representation of the probabilities of a typical path and its time reversal versus time $t = k\tau$, where τ is the sampling time. These probabilities, respectively, decay with the rates of temporal disorder h and time-reversed one $h^{\rm R}$. Their ratio is determined by the entropy production $\Delta_{\rm i}S$ during time t.

probability distributions of the paths and time-reversed paths, is that of characterizing irreversibility in the process. This quantity has been used to detect the signature of irreversibility for different kinds of random process in relation or not with energy dissipation (Porporato et al., 2007; Roldán and Parrondo, 2010; Roldán, 2014; Provata et al., 2014).

If the process is ergodic, the probabilities of typical paths and their time reversal should thus decay as

$$P(\omega) \sim {\rm e}^{-ht} \qquad \text{and} \qquad P\left(\omega^{\rm R}\right) \sim {\rm e}^{-h^{\rm R}t} \qquad \text{with} \qquad t = k\tau \to \infty, \tag{6.102}$$

for almost all the paths ω. These probabilities decay at the same rate $h = h^{\rm R}$ if the process is reversible. However, in the presence of time asymmetry in temporal disorder, the inequality $h^{\rm R} > h$ holds, so that the probabilities of the time-reversed typical paths decay faster than the probabilities of the typical paths themselves. Therefore, a typical path is more probable than its time reversal if the process is not reversible, as shown in Figure 6.5 (Gaspard, 2007a,b).

6.5.3 Entropy Production, Temporal Disorder, and Time Reversal

Next, the results of Chapter 5 and, in particular, equation (5.132) show that the entropy production rate can be obtained in terms of the Kullback–Leibler divergence between the probabilities of the paths and their time reversal if the paths measure the energy and particle fluxes with the reservoirs, which drive the system away from equilibrium. Since this Kullback–Leibler divergence is related by equation (6.100) to the difference between the temporal disorder rates h and h^{R} in the paths and their time reversal, the thermodynamic entropy production rate can be expressed as

$$\frac{1}{k_{\mathrm{B}}} \frac{d_{\mathrm{i}}S}{dt}\bigg|_{\mathrm{st}} = h^{\mathrm{R}} - h \geq 0 \tag{6.103}$$

for a fine enough partition $\mathcal{P}$ and arbitrarily small sampling time τ if the stationarity condition holds (Gaspard, 2004b). Accordingly, the Kullback–Leibler divergence (6.101) multiplied by Boltzmann's constant k_{B} represents the thermodynamic entropy production $\Delta_{\mathrm{i}}S$ during the total time interval t. Therefore, the ratio of the probabilities of a typical path $\boldsymbol{\omega}$ and its time reversal $\boldsymbol{\omega}^{\mathrm{R}}$ should increase exponentially with time t according to

$$\frac{P(\boldsymbol{\omega})}{P(\boldsymbol{\omega}^{\mathrm{R}})} \sim \frac{\exp(-ht)}{\exp\left(-h^{\mathrm{R}}t\right)} \sim \exp\left(\frac{1}{k_{\mathrm{B}}} \frac{d_{\mathrm{i}}S}{dt}\bigg|_{\mathrm{st}} t\right), \tag{6.104}$$

which is illustrated in Figure 6.5 where $\Delta_{\mathrm{i}}S = (d_{\mathrm{i}}S/dt)_{\mathrm{st}}t > 0$. If the system is at equilibrium, the entropy production rate is equal to zero, the temporal disorder rates are equal, i.e., $h = h^{\mathrm{R}}$, and detailed balance holds, according to which the paths and their time reversal have equal probabilities.

In order to obtain the thermodynamic entropy production using equation (6.103), it is essential that the path $\boldsymbol{\omega}$ contains the relevant information on the elementary mechanisms that fuel the nonequilibrium process with free energy and produce entropy. In general, equation (6.103) calculated using partial information on the process only provides a lower bound on the entropy production rate (Roldán and Parrondo, 2010), as shown with equation (E.23) in Section E.4.

6.5.4 The Case of Markov Jump Processes

The relationship (6.103) between time asymmetry in temporal disorder and thermodynamic entropy production can be established, in particular, for the Markov jump processes ruled by the master equation (4.49) and having their thermodynamic entropy production rate given by equation (4.57). If the stationarity condition is satisfied, the temporal disorder of these processes is characterized by the τ-entropy per unit time (6.90). The corresponding time-reversed dynamical coentropy (6.99) has the following expression obtained by taking the transition rate of the opposite transition inside the logarithm (Gaspard, 2004b),

$$\begin{aligned} h^{\mathrm{R}}(\tau) = \ln\left(\frac{\mathrm{e}}{\tau}\right) &\sum_{\rho,\,\omega\neq\omega'} P_{\mathrm{st}}(\omega)\, W\left(\omega \xrightarrow{\rho} \omega'\right) \\ &- \sum_{\rho,\,\omega\neq\omega'} P_{\mathrm{st}}(\omega)\, W\left(\omega \xrightarrow{\rho} \omega'\right) \ln W\left(\omega' \xrightarrow{-\rho} \omega\right) + O(\tau). \end{aligned} \tag{6.105}$$

Taking the difference between this quantity and (6.90), the terms increasing as $\ln(1/\tau)$ in the limit $\tau \to 0$ cancel each other and we find

$$h^{\rm R}(\tau) - h(\tau) = \frac{1}{2} \sum_{\rho,\omega\neq\omega'} \left[P_{\rm st}(\omega)\, W\big(\omega \xrightarrow{\rho} \omega'\big) - P_{\rm st}(\omega')\, W\big(\omega' \xrightarrow{-\rho} \omega\big)\right]$$
$$\times \ln \frac{P_{\rm st}(\omega)\, W\big(\omega \xrightarrow{\rho} \omega'\big)}{P_{\rm st}(\omega')\, W\big(\omega' \xrightarrow{-\rho} \omega\big)} + O(\tau), \tag{6.106}$$

after using the condition of stationarity in order to insert the ratio of stationary probabilities inside the logarithm. In the limit $\tau \to 0$, this difference gives the expression (4.57) of the thermodynamic entropy production rate for these processes:

$$\frac{1}{k_{\rm B}} \frac{d_{\rm i}S}{dt}\bigg|_{\rm st} = \lim_{\tau\to 0} \left[h^{\rm R}(\tau) - h(\tau)\right] \geq 0, \tag{6.107}$$

which is always nonnegative, in agreement with the second law of thermodynamics. If the process is at equilibrium, the conditions (4.48) of detailed balance are satisfied, so that temporal disorder is symmetric under time reversal since $h^{\rm R}(\tau) = h(\tau)$ and the thermodynamic entropy production is equal to zero, as expected.

6.5.5 Nonequilibrium Temporal Ordering

A corollary of the relation (6.103) is that the probabilities of the typical paths decay more slowly in time than those of their time reversal:

$$h^{\rm R} \geq h \geq 0. \tag{6.108}$$

Consequently, it is less probable that the time reversal of a typical path will be observed than the path itself if the process evolves away from equilibrium, as shown in Figure 6.5. Therefore, temporal disorder is smaller for typical paths than for their time reversal, meaning that temporal ordering manifests itself under nonequilibrium conditions. Accordingly, we have the following (Gaspard, 2007a,b):

Theorem of nonequilibrium temporal ordering: *In nonequilibrium steady states, the typical paths are more ordered in time than their corresponding time reversal in the sense that their temporal disorder rate h is smaller than the rate $h^{\rm R}$ characterizing the temporal disorder of the time-reversed paths.*

There is no temporal ordering at equilibrium where $h = h^{\rm R}$, so that the paths and their time reversal appear equally erratic in both time directions by virtue of the principle of detailed balance. However, *dynamical order* manifests itself away from equilibrium. This nonequilibrium temporal ordering is a corollary of the second law of thermodynamics because of the relation (6.103) to the thermodynamic entropy production rate. This result is compatible with Boltzmann's interpretation of the second law, according to which phase-space disorder increases in time, since temporal ordering is generated by the growth of phase-space disorder.

This theorem can be illustrated with the simple process of Figure 6.6, which considers a Markov chain with three states $\omega \in \{1,2,3\}$ (Gaspard, 2007b; Andrieux, 2009). In this example, the transition probabilities are taken as $p = P(\omega \to \omega + 1, \text{modulo } 3)$ and $1 - p = P(\omega \to \omega - 1, \text{modulo } 3)$. The stationary probability distribution is thus given by $P_{\text{st}}(\omega) = 1/3$ for the three states ω. For this Markov chain, temporal disorder is characterized by the KS dynamical entropy

$$h = h_{\text{KS}} = -\sum_{\omega,\omega'} P_{\text{st}}(\omega)\, P(\omega \to \omega') \ln P(\omega \to \omega') = -p \ln p - (1-p) \ln(1-p), \tag{6.109}$$

and the time-reversed coentropy

$$h^{\text{R}} = -\sum_{\omega,\omega'} P_{\text{st}}(\omega)\, P(\omega \to \omega') \ln P(\omega' \to \omega) = -p \ln(1-p) - (1-p) \ln p. \tag{6.110}$$

The thermodynamic entropy production given by equation (6.103) thus takes the value

$$\frac{1}{k_{\text{B}}} \Delta_{\text{i}} S = h^{\text{R}} - h = (1 - 2p) \ln \frac{1-p}{p} \geq 0. \tag{6.111}$$

Each of these quantities is evaluated per time step. They are plotted in Figure 6.6 as a function of the probability $0 \leq p \leq 1$. The equilibrium macrostate is identified as the state where the thermodynamic entropy production (6.111) is equal to zero, which happens for $p = 1/2$. In this macrostate, the transitions have equal probabilities to happen in opposite directions $\omega \to \omega \pm 1$, so that there is no directionality and no time asymmetry in temporal disorder since $h = h^{\text{R}} = \ln 2$. In contrast, dynamical order manifests itself as soon as $p \neq 1/2$, in which case the thermodynamic entropy production becomes positive and the rate h smaller than h^{R}, as seen in Figure 6.6. In the extreme cases where $p = 0$ and

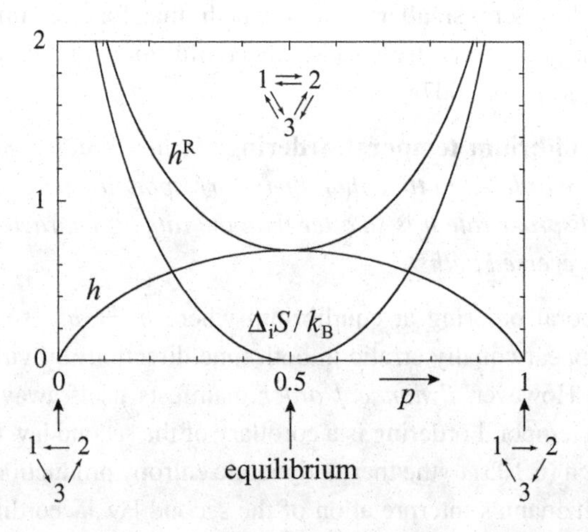

Figure 6.6 Time asymmetry in temporal disorder characterized by the entropy h and coentropy h^{R} per time step Δt and the entropy production $\Delta_{\text{i}} S / k_{\text{B}} = h^{\text{R}} - h$ in the three-state Markov chain model defined with the probabilities p for the transitions $\{1 \to 2, 2 \to 3, 3 \to 1\}$ and $1 - p$ for the reversed transitions (Gaspard, 2007b; Andrieux, 2009).

$p = 1$, nonequilibrium temporal ordering is complete and the process fully irreversible because time evolution is then cyclic running either in the direction 123123... or in the opposite direction 321321.... For such cyclic processes, the temporal disorder rate is equal to zero, i.e., $h = 0$, but $h^{\rm R} = \infty$, since the time-reversed cyclic path has zero probability of occurring. This example shows that temporal ordering as well as directionality are key features in nonequilibrium processes (Gaspard, 2007a,b).

6.5.6 Landauer's Principle

The previous results shed a new light on the principle of Landauer (1961), according to which the erasure of information in the memory of a computer dissipates energy and generates entropy production. Indeed, such an operation is logically irreversible on data, hence producing entropy if implemented as a physical process. Quantitatively, Landauer's principle says that erasing one bit of information generates at least $k_{\rm B} \ln 2$ of entropy production and, thus, dissipates a minimal energy equal to $k_{\rm B}T \ln 2$ for information processing devices functioning at temperature T. These considerations have resolved the paradox of Maxwell's demon. This paradox, formulated by Maxwell (1871), assumes that a demon could sort gas molecules between two compartments according to their kinetic energy, thus, hypothetically acting in contradiction to the second law of thermodynamics. However, the information registered by the demon about the molecules should be erased for its action to continue and the irreversibility of this erasure produces entropy such that the second law is always satisfied (Landauer, 1961; Bennett, 1973, 1982).

Remarkably, the minimal thermodynamic entropy production of the information erasure can be directly evaluated with the relation (6.103). Let us suppose that information is physically recorded in the memory of some device under the form of the sequence of bits $\sigma_1\sigma_2\cdots\sigma_i\cdots$ with $\sigma_i = 0$ or 1. The recorded data are described by some probability distribution $P(\sigma_1\sigma_2\cdots\sigma_k)$ giving the occurrence of the sequences in the memory of the device. Often, there may exist statistical correlations among the bits σ_i. The information contained in the sequence can be characterized by the Shannon entropy per bit of information defined as

$$I \equiv \lim_{k\to\infty} -\frac{1}{k} \sum_{\sigma_1\sigma_2\cdots\sigma_k} P(\sigma_1\sigma_2\cdots\sigma_k) \ln P(\sigma_1\sigma_2\cdots\sigma_k). \tag{6.112}$$

In general, the inequality $I \leq \ln 2$ holds with the equality satisfied if the bits are randomly distributed with equal probability and no statistical correlations. We note that the quantity (6.112) is the minimum information required to store the sequence per symbol (Cover and Thomas, 2006).

The erasure of the aforementioned sequence can be thought of as a spacetime process, in which the sequence is moving step by step to the right-hand side of an eraser located at a fixed position and successively setting to $\sigma = 0$ all the bits of the sequence. At every instant of time, the state of the system is given by $\omega_i = \cdots\sigma_{i+2}\sigma_{i+1}\sigma_i \cdot 00000\cdots$. When looking forward in time, the process of erasure is deterministic, transforming the state ω_i into ω_{i+1}, which does not generate dynamical randomness since the outcome is unique every time a bit is erased. Consequently, the entropy per time step (6.82) is equal to zero, $h = 0$. However, the backward process restores data and, thus, generates dynamical randomness at the rate $h^{\rm R} = I$ given by the information (6.112). Therefore, the relation (6.103) shows that the minimal thermodynamic entropy production of the erasure process is given by

$$\Delta_{\rm i} S = k_{\rm B}\big(h^{\rm R} - h\big) = k_{\rm B}\, I \quad \text{per bit.} \tag{6.113}$$

Landauer's result is recovered in the particular case of statistically independent random bits of equal probability, for which $\Delta_{\rm i} S = k_{\rm B} I = k_{\rm B} \ln 2$. Therefore, the thermodynamic entropy production of information processing devices can be understood from the viewpoint of time asymmetry in temporal disorder (Gaspard, 2004b; Andrieux and Gaspard, 2008a).

Landauer's principle has been experimentally verified for a one-bit memory constructed with a single micrometric colloidal particle subjected to the thermal fluctuations of some surrounding fluid and trapped in a time-dependent double-well potential generated by an optical tweezer (Bérut et al., 2012, 2015). The states $\sigma = 0$ and $\sigma = 1$ of the memory correspond to finding the particle in the left- and right-hand potential wells, respectively. The height of the barrier between the wells is controlled by the laser intensity of the optical tweezer. Initially, the particle is set in either one or the other potential well with equal probability. The erasure protocol starts by lowering the barrier to a value of the order of the thermal energy. Afterwards, the double well is tilted by exerting a viscous drag force on the colloidal particle during some time interval $[0, t]$ in order to drive the particle to the left-hand potential well and, thus, set the memory to $\sigma = 0$. Finally, the barrier between the wells is increased back to its initial value.[5] Before and after the erasure, the system is at equilibrium, so that the aforementioned mean housekeeping heat is equal to zero, $\langle Q_{\rm hk}\rangle = 0$. The entropy of the memory is equal to $S_- = k_{\rm B} \ln 2$ before erasure, and $S_+ = 0$ thereafter, so that the entropy is changed by $\Delta S = S_+ - S_- = -k_{\rm B} \ln 2$. The excess heat of erasure $Q_{\rm ex}$ can be measured by observing the random position $x(t)$ of the colloidal particle and its statistics can be carried out by repeating the protocol many times. Because of the inequality by Hatano and Sasa (2001), the mean excess heat should satisfy $\langle Q_{\rm ex}\rangle \geq -T\Delta S = k_{\rm B} T \ln 2$, which is Landauer's lower bound on dissipated heat due to one-bit erasure. The experiment shows that $\langle Q_{\rm ex}\rangle \simeq k_{\rm B} T \ln 2 + B/t$ with the protocol duration t and $B \simeq 8{-}10\, k_{\rm B} T$ s for $0 < t < 40$ s, confirming the validity of Landauer's principle (Bérut et al., 2012, 2015). At the temperature $T = 300$ K of the experiment, Landauer's lower bound has the value $k_{\rm B} T \ln 2 \simeq 3 \times 10^{-21}$ J, which is much smaller than the typical energies of about $1000\, k_{\rm B} T$ that are dissipated per elementary operation in semiconductor computers. Landauer's bound is the ultimate physical limit on logically irreversible operations in computation, linking information and thermodynamics as anticipated by Szilard (1929) and Brillouin (1951).

6.5.7 The Time-Symmetric Part of Path Probability Decay Rates

Under nonequilibrium conditions, the probabilities of the paths and their time reversal do not decay with the same rate, as shown by equation (6.102), and the thermodynamic entropy

[5] Since the barrier height $E^{\ddagger}$ is finite, the memory has a finite lifetime equal to the mean dwell time of the particle in one potential well, which is the Kramers time given by $\tau_{\rm K} \sim \exp\big(\beta E^{\ddagger}\big)$ with the inverse temperature $\beta = (k_{\rm B} T)^{-1}$ (Kramers, 1940; Chandrasekhar, 1943). In the experiment of Bérut et al. (2015), $E^{\ddagger} \simeq 8\, k_{\rm B} T$ and $\tau_{\rm K} \simeq 3000$ s.

production rate is given by the time-antisymmetric part of the decay rate according to equation (6.103). The time-symmetric part is also of interest (Maes and van Wieren, 2006; Maes, 2020). In this regard, the action functional (6.2) can be decomposed into its symmetric and antisymmetric parts as

$$P(\boldsymbol{\omega}) \sim \mathrm{e}^{-\Phi(\boldsymbol{\omega})+\frac{1}{2}\Sigma(\boldsymbol{\omega})} \qquad \text{and} \qquad P\left(\boldsymbol{\omega}^{\mathrm{R}}\right) \sim \mathrm{e}^{-\Phi(\boldsymbol{\omega})-\frac{1}{2}\Sigma(\boldsymbol{\omega})}, \tag{6.114}$$

where $\Sigma(\boldsymbol{\omega}) = \mathcal{A}\left(\boldsymbol{\omega}^{\mathrm{R}}\right) - \mathcal{A}(\boldsymbol{\omega})$ is the time-antisymmetric part of the action functional introduced in equation (6.19) and

$$\Phi(\boldsymbol{\omega}) \equiv \frac{1}{2}\left[\mathcal{A}(\boldsymbol{\omega}) + \mathcal{A}\left(\boldsymbol{\omega}^{\mathrm{R}}\right)\right] \tag{6.115}$$

is its time-symmetric part, called *frenesy* by Maes (2020). Since the mean growth rates of the action functionals $\mathcal{A}(\boldsymbol{\omega})$ and $\mathcal{A}\left(\boldsymbol{\omega}^{\mathrm{R}}\right)$ are given by the dynamical entropy h and coentropy per unit time h^{R}, respectively, the mean rate of frenesy can thus be expressed as

$$\mathscr{F} \equiv \lim_{k\to\infty} \frac{1}{k\tau}\langle \Phi(\boldsymbol{\omega})\rangle = \frac{1}{2}\left(h + h^{\mathrm{R}}\right), \tag{6.116}$$

while $\mathscr{R} \equiv (1/k_{\mathrm{B}})d_{\mathrm{i}}S/dt = h^{\mathrm{R}} - h$ gives the thermodynamic entropy production rate. Accordingly, the probabilities of typical paths and their time reversal decay as

$$P(\boldsymbol{\omega}) \sim \mathrm{e}^{-ht} = \mathrm{e}^{-\left(\mathscr{F}-\frac{1}{2}\mathscr{R}\right)t}, \qquad P\left(\boldsymbol{\omega}^{\mathrm{R}}\right) \sim \mathrm{e}^{-h^{\mathrm{R}}t} = \mathrm{e}^{-\left(\mathscr{F}+\frac{1}{2}\mathscr{R}\right)t}. \tag{6.117}$$

At equilibrium where $h_{\mathrm{eq}} = h^{\mathrm{R}}_{\mathrm{eq}}$ and $\mathscr{R} = 0$, the mean rate of frenesy is thus equal to the entropy per unit time (6.82), $\mathscr{F}_{\mathrm{eq}} = h_{\mathrm{eq}}$. For dilute hard-sphere gases at equilibrium, the mean rate of frenesy can thus be estimated with the KS entropy per unit time (6.86). Therefore, the mean rate (6.116) can be very large even if the system is at equilibrium, where the thermodynamic entropy production rate is equal to zero.

Often, the path probabilities are compared with those of some reference process. If this latter is the equilibrium process, we have that

$$\frac{P(\boldsymbol{\omega})}{P_{\mathrm{eq}}(\boldsymbol{\omega})} \simeq \mathrm{e}^{-\Delta h\, t} \simeq \mathrm{e}^{-\left(\Delta\mathscr{F}-\frac{1}{2}\mathscr{R}\right)t}, \qquad \frac{P\left(\boldsymbol{\omega}^{\mathrm{R}}\right)}{P_{\mathrm{eq}}(\boldsymbol{\omega})} \simeq \mathrm{e}^{-\Delta h^{\mathrm{R}}\, t} \simeq \mathrm{e}^{-\left(\Delta\mathscr{F}+\frac{1}{2}\mathscr{R}\right)t}, \tag{6.118}$$

where

$$\Delta h \equiv h - h_{\mathrm{eq}}, \qquad \Delta h^{\mathrm{R}} \equiv h^{\mathrm{R}} - h_{\mathrm{eq}}, \qquad \Delta\mathscr{F} \equiv \mathscr{F} - h_{\mathrm{eq}}, \tag{6.119}$$

because $\Delta\mathscr{R} \equiv \mathscr{R} - \mathscr{R}_{\mathrm{eq}} = \mathscr{R}$ since $\mathscr{R}_{\mathrm{eq}} = 0$. The three quantities (6.119), including the mean rate of excess in frenesy $\Delta\mathscr{F}$, are equal to zero at equilibrium.

6.6 Analogy with Other Symmetry-Breaking Phenomena

A remarkable analogy exists between time-reversal symmetry breaking in the statistical description of nonequilibrium systems and other symmetry breaking phenomena at equilibrium in condensed matter physics (Kurchan, 2010; Gaspard, 2012a,b; Lacoste and Gaspard, 2014, 2015). In order to set up this analogy, we consider a system of N interacting spins $\boldsymbol{\sigma} = \{\sigma_a\}_{a=1}^{N}$ in an external magnetizing field $\boldsymbol{\mathcal{H}}$. The Hamiltonian function is given by

$$H_N(\boldsymbol{\sigma};\boldsymbol{\mathcal{H}}) = H_N(\boldsymbol{\sigma};\mathbf{0}) - \boldsymbol{\mathcal{H}}\cdot\mathbf{M}_N(\boldsymbol{\sigma}), \qquad \text{where} \qquad \mathbf{M}_N(\boldsymbol{\sigma}) = \sum_{a=1}^{N}\boldsymbol{\sigma}_a \tag{6.120}$$

is the total magnetization, which is the order parameter of the system. In the absence of the magnetizing field, the system is assumed to be symmetric under orthogonal transformations R, which can be rotations with $\det\mathsf{R} = +1$ or mirror reflections with $\det\mathsf{R} = -1$. These transformations act on the spins according to $\boldsymbol{\sigma}\to\boldsymbol{\sigma}^{\mathsf{R}} = \{\mathsf{R}\cdot\boldsymbol{\sigma}_a\}_{a=1}^{N}$, so that the symmetry of the Hamiltonian function is expressed as $H_N\left(\boldsymbol{\sigma}^{\mathsf{R}};\mathbf{0}\right) = H_N(\boldsymbol{\sigma};\mathbf{0})$. However, the presence of an external magnetizing field explicitly breaks the symmetry, giving $H_N\left(\boldsymbol{\sigma}^{\mathsf{R}};\boldsymbol{\mathcal{H}}\right) = H_N\left(\boldsymbol{\sigma};\mathsf{R}^{-1}\cdot\boldsymbol{\mathcal{H}}\right)$ because $\mathsf{R}^{\mathrm{T}} = \mathsf{R}^{-1}$ for orthogonal transformations.

The system is assumed to be in the following canonical equilibrium distribution at the inverse temperature β,

$$P_{\rm eq}(\boldsymbol{\sigma}) = \frac{1}{Z_N(\beta;\boldsymbol{\mathcal{H}})}\,{\rm e}^{-\beta H_N(\boldsymbol{\sigma};\boldsymbol{\mathcal{H}})}. \tag{6.121}$$

This distribution can be compared with the symmetry transformed distribution $P_{\rm eq}^{\mathsf{R}}(\boldsymbol{\sigma}) = P_{\rm eq}\left(\boldsymbol{\sigma}^{\mathsf{R}}\right)$. Evaluating its thermodynamic entropy (2.112) and coentropy (2.121) and taking their difference, we obtain the following expression for the Kullback–Leibler divergence (2.122),

$$D_{\rm KL}\left(P_{\rm eq}\|P_{\rm eq}^{\mathsf{R}}\right) = \frac{1}{k_{\rm B}}\left(S^{\mathsf{R}} - S\right) = \beta\,\boldsymbol{\mathcal{H}}\cdot(\mathbf{1}-\mathsf{R})\cdot\langle\mathbf{M}_N(\boldsymbol{\sigma})\rangle_{\boldsymbol{\mathcal{H}}} \geq 0, \tag{6.122}$$

which is always nonnegative. This relation is the equilibrium analogue of equation (6.103) for the entropy production rate in nonequilibrium systems.

The analogy goes farther because the fluctuations of the order parameter can be shown to obey fluctuation relations in equilibrium finite systems subjected to an external field. In particular, we may consider the magnetization density $\mathfrak{m} \equiv \mathbf{M}_N(\boldsymbol{\sigma})/V$, where the volume is given by $V = Nv$ in terms of the mean volume v occupied by one spin. The probability density of this random variable is defined by

$$p_V(\mathfrak{m};\boldsymbol{\mathcal{H}}) \equiv \langle\delta\left[\mathfrak{m} - \mathbf{M}_N(\boldsymbol{\sigma})/V\right]\rangle_{\boldsymbol{\mathcal{H}}}, \tag{6.123}$$

where $\langle\cdot\rangle_{\boldsymbol{\mathcal{H}}}$ denotes the average with respect to the equilibrium canonical distribution (6.121).[6] Comparing with the probability density of the symmetry transformed random variable $\mathsf{R}\cdot\mathfrak{m}$, we find the following equilibrium fluctuation relation (Lacoste and Gaspard, 2014, 2015),

$$\frac{p_V(\mathfrak{m};\boldsymbol{\mathcal{H}})}{p_V(\mathsf{R}\cdot\mathfrak{m};\boldsymbol{\mathcal{H}})} = {\rm e}^{\beta\boldsymbol{\mathcal{H}}\cdot(\mathbf{1}-\mathsf{R})\cdot\mathfrak{m}\,V}, \tag{6.124}$$

which is the equilibrium analogue of the multivariate fluctuation relation (5.74) for currents. We note that, here, the orthogonal transformation R can be discrete such a mirror reflection or continuous such as a rotation.

[6] An equivalent formulation can be carried out using the magnetization per spin defined as $\mathbf{m} \equiv \mathbf{M}_N(\boldsymbol{\sigma})/N = (V/N)\mathfrak{m}$.

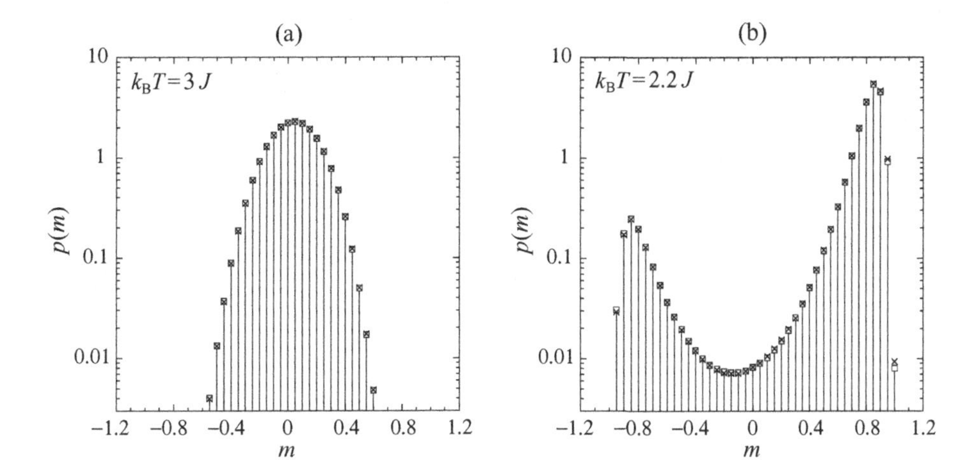

Figure 6.7 Probability distribution $p(m)$ (squares) of the magnetization per spin $m = M/N$ for the two-dimensional Ising model (6.125) on a 20×20 square lattice with the magnetizing field $\mathcal{H} = 0.01$: (a) in the paramagnetic phase and (b) in the ferromagnetic phase. The crosses show the predictions of the fluctuation relation (6.126). Reproduced with permission from Gaspard (2012b) © IOP Publishing Ltd and SISSA Medialab.

The equilibrium fluctuation relation (6.124) is shown to hold in Figure 6.7 for the two-dimensional Ising model on a square lattice, for which the Hamiltonian function is given by

$$H_N(\boldsymbol{\sigma}; \mathcal{H}) = \sum_{i=1}^{L} \sum_{j=1}^{L} \left[-J\, \sigma_{i,j} \left(\sigma_{i+1,j} + \sigma_{i,j+1} \right) - \mathcal{H}\, \sigma_{i,j} \right], \tag{6.125}$$

where J is the interaction parameter, $N = L^2$ the number of spins, and $\sigma_{i,L+1} = \sigma_{i,1}$ and $\sigma_{L+1,j} = \sigma_{1,j}$ for $i, j = 1, 2, \ldots, L$. In the absence of magnetizing field $\mathcal{H}$, this model is symmetric under spin reversal $\mathsf{R} = -\mathbf{1}$. Since the magnetization per spin $m = M/N$ is aligned in the z-direction, the fluctuation relation (6.124) is equivalent to (Goldenfeld, 1992)

$$p_N(m; \mathcal{H}) = p_N(-m; \mathcal{H}) \exp(2\beta \mathcal{H} m N), \tag{6.126}$$

which is satisfied above and below the critical temperature $k_B T_c = 2.269\, J$, as seen in Figure 6.7.

Figure 6.8 shows that the equilibrium fluctuation relation (6.124) is also satisfied in the case of the Curie–Weiss model of magnetism defined using the Hamiltonian function,

$$H_N(\boldsymbol{\sigma}; \boldsymbol{\mathcal{H}}) = -\frac{J}{2N} \mathbf{M}_N(\boldsymbol{\sigma})^2 - \boldsymbol{\mathcal{H}} \cdot \mathbf{M}_N(\boldsymbol{\sigma}). \tag{6.127}$$

Since this model is symmetric under continuous rotations R, the symmetry relation (6.124) holds for all the magnetization vectors having the same norm, $\|\mathbf{m}\| = \|\mathsf{R} \cdot \mathbf{m}\|$. In this respect, the relation (6.124) is called *isometric fluctuation relation* (Hurtado et al., 2011).

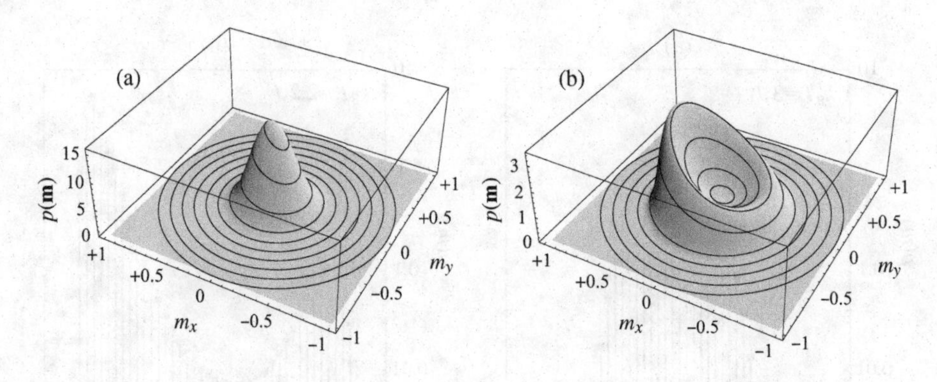

Figure 6.8 Probability density $p(\mathbf{m})$ of the magnetization per spin $\mathbf{m} = (m_x, m_y, m_z = 0)$ for the three-dimensional Curie–Weiss model (6.127) in the magnetizing field $\boldsymbol{\mathcal{H}} = (\mathcal{H}, 0, 0)$ with $\mathcal{H} = 0.005$ and $N = 100$ at the rescaled inverse temperatures (a) $\beta J = 2.7$ in the paramagnetic phase and (b) $\beta J = 3.3$ in the ferromagnetic phase. The lines depict the predictions of the isometric fluctuation relation (6.124) at $\|\mathbf{m}\| = 0.1, 0.2, \ldots, 1.0$. Reproduced with permission from Lacoste and Gaspard (2015) © IOP Publishing Ltd and SISSA Medialab.

In Figure 6.8, the validity of the isometric fluctuation relation is checked by the coincidence of the lines at constant values of $\|\mathbf{m}\|$ with the surface of the distribution above and below the critical temperature $k_{\rm B} T_{\rm c} = J/3$.

Furthermore, we may introduce the cumulant generating function

$$\Gamma(\boldsymbol{\lambda}; \boldsymbol{\mathcal{H}}) \equiv \lim_{N\to\infty} -\frac{1}{N} \ln \left\langle {\rm e}^{-\boldsymbol{\lambda}\cdot\mathbf{M}_N(\sigma)} \right\rangle_{\boldsymbol{\mathcal{H}}} . \tag{6.128}$$

This latter obeys the following symmetry relation as a consequence of the equilibrium fluctuation relation,

$$\Gamma(\boldsymbol{\lambda}; \boldsymbol{\mathcal{H}}) = \Gamma(2\beta\boldsymbol{\mathcal{H}} - \boldsymbol{\lambda}; \boldsymbol{\mathcal{H}}), \tag{6.129}$$

which is the analogue of the symmetry relation (5.91) for the cumulant generating function (5.76) of the currents (Lacoste and Gaspard, 2014, 2015).

The analogy between the equilibrium and nonequilibrium relations is thus very complete, which emphasizes their common origins in symmetry breaking at the statistical level of description due to external constraints on the system. The similarities between the properties of symmetry breaking in equilibrium and nonequilibrium systems are shown in Table 6.2. In nonequilibrium systems, the discrete time reversal symmetry is broken by the external constraints imposed by the mechanical, thermal, or chemical affinities in a manner that is similar to the breaking of the discrete symmetry under spin reversal or the continuous symmetry under spin rotations by the external magnetizing field in equilibrium systems. These results extend the considerations of Section 2.6.1 on discrete symmetry breaking.

Table 6.2. *Fluctuation relations and nonnegative quantities characterizing symmetry breaking in equilibrium and nonequilibrium systems. Here,* $\beta = (k_{\rm B}T)^{-1}$ *denotes the inverse temperature and* μ_k *the chemical potentials, and* V *is the volume of the equilibrium system. The fluctuating magnetization density* $\mathfrak{m} = \mathbf{M}/V$, *the thermodynamic entropy density* $s = S/V$, *and the corresponding coentropy* $s^{\mathsf{R}} = S^{\mathsf{R}}/V$ *are equilibrium quantities, which are intensive in space. The fluctuating currents* $\mathfrak{J}$, *the rate* h *of temporal disorder, and the rate* $h^{\rm R}$ *of time-reversed disorder are intensive in time, as well as the thermodynamic entropy production rate* $d_{\rm i}S/dt$. *Lastly,* t *is the time interval, during which the fluctuating currents are measured. The statistical averages with respect to the equilibrium and nonequilibrium probability distributions are denoted* $\langle\cdot\rangle$.

	Equilibrium Systems	Nonequilibrium Systems
Broken symmetry	spin reversal or rotations R	time reversal R
External driving	magnetizing field $\boldsymbol{\mathcal{H}}$	affinities $\mathbf{A} = \beta - \beta',\ \beta'\mu'_k - \beta\mu_k$
Fluctuating order parameter	fluctuating magnetization density $\mathfrak{m}$	fluctuating currents $\mathfrak{J}$
Fluctuation relation	$\dfrac{p_V(\mathfrak{m};\boldsymbol{\mathcal{H}})}{p_V(\mathsf{R}\cdot\mathfrak{m};\boldsymbol{\mathcal{H}})} = \exp\left[\beta\boldsymbol{\mathcal{H}}\cdot(\mathsf{1}-\mathsf{R})\cdot\mathfrak{m}\,V\right]$	$\dfrac{p_t(\mathfrak{J};\mathbf{A})}{p_t(-\mathfrak{J};\mathbf{A})} = \exp(\mathbf{A}\cdot\mathfrak{J}\,t)$
Nonnegative quantity	$\dfrac{1}{k_{\rm B}}\left(s^{\mathsf{R}} - s\right) = \beta\boldsymbol{\mathcal{H}}\cdot(\mathsf{1}-\mathsf{R})\cdot\langle\mathfrak{m}\rangle_{\boldsymbol{\mathcal{H}}} \geq 0$	$\dfrac{1}{k_{\rm B}}\dfrac{d_{\rm i}S}{dt} = h^{\rm R} - h = \mathbf{A}\cdot\langle\mathfrak{J}\rangle_{\mathbf{A}} \geq 0$

7

Driven Brownian Particles and Related Systems

7.1 Stochastic Energetics

The breaking of time reversal symmetry in nonequilibrium fluctuations can be experimentally investigated for several types of systems and, in particular, driven Brownian particles (Wang et al., 2002; Trepagnier et al., 2004; Blickle et al., 2006; Speck et al., 2007; Andrieux et al., 2007, 2008; Joubaud et al., 2008). In these systems, a micrometric colloidal particle is suspended in a fluid such as water and dragged by an optical trap moving with some speed, as schematically shown in Figure 7.1. This latter exerts work onto the system, which, on average, undergoes heating. The motion of the micrometric particle can be tracked with a microscope and its trajectory can thus be recorded.

From the microscopic viewpoint, the time evolution of this system is ruled by the Hamiltonian function

$$H = \frac{\mathbf{p}^2}{2m} + U(\mathbf{r} - \mathbf{u}\,t) + \sum_{a=1}^{N} \left(\frac{\mathbf{p}_a^2}{2m_a} + u_a \right) + \sum_{1 \le a < b \le N} u_{ab}^{(2)}, \tag{7.1}$$

where $\mathbf{r}$, $\mathbf{p}$, and m denote the position, momentum, and mass of the micrometric particle, U is the energy potential of the optical trap moving with the constant speed $\mathbf{u}$ in the reference frame; $\mathbf{r}_a$, $\mathbf{p}_a$, and m_a are the positions, momenta, and masses of N atoms composing the fluid; $u_a = u_a(\|\mathbf{r} - \mathbf{r}_a\|)$ is the interaction energy potential between the colloidal particle and the ath atom; and $u_{ab}^{(2)} = u_{ab}^{(2)}(\|\mathbf{r}_a - \mathbf{r}_b\|)$ is the binary interaction energy potential between the ath and bth atoms.

For this Hamiltonian system, the Newton equations (2.10) can be deduced and, especially, the microscopic equation of motion for the colloidal particle

$$m\,\frac{d^2\mathbf{r}}{dt^2} = -\nabla U(\mathbf{r} - \mathbf{u}\,t) + \sum_{a=1}^{N} \mathbf{f}_a \qquad \text{with} \qquad \mathbf{f}_a = -\nabla u_a, \tag{7.2}$$

which is the force exerted by the ath atom on the colloidal particle, $\nabla = \partial_{\mathbf{r}}$ denoting the gradient. The total energy of the system given by the Hamiltonian function (7.1) can be decomposed as

$$E_{\text{tot}} \equiv H = E + E_{\text{fluid}} \tag{7.3}$$

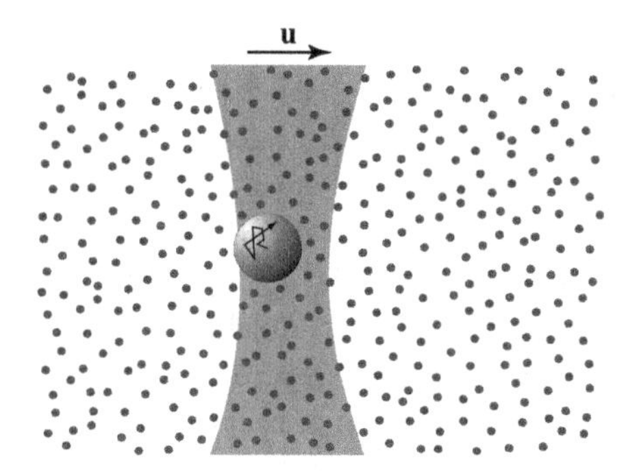

Figure 7.1 Brownian particle driven by an optical trap moving at the speed **u** with respect to the fluid.

into the energy of the subsystem formed by the colloidal particle moving in the optical trap

$$E \equiv \frac{1}{2} m \mathbf{v}^2 + U(\mathbf{r} - \mathbf{u}\, t), \tag{7.4}$$

where $\mathbf{v} = \dot{\mathbf{r}} = \mathbf{p}/m$ is the particle velocity, and the energy of the fluid E_{fluid} given by the other terms of the Hamiltonian function. Since the whole system is isolated, the power of the mechanical work performed on the system by moving the optical trap can be defined as

$$\frac{đW}{dt} = \frac{dE_{\text{tot}}}{dt} = \frac{\partial U}{\partial t} = -\mathbf{u} \cdot \nabla U. \tag{7.5}$$

Now, the time derivative of the energy (7.4) of the colloidal particle is given by

$$\frac{dE}{dt} = -\mathbf{u} \cdot \nabla U + \mathbf{v} \cdot \left(\sum_{a=1}^{N} \mathbf{f}_a \right). \tag{7.6}$$

Using the standard expression of the first law of thermodynamics

$$dE = đW - đQ, \tag{7.7}$$

we can thus identify the heat released per unit time by the colloidal particle to the surrounding fluid as

$$\frac{đQ}{dt} = -\mathbf{v} \cdot \left(\sum_{a=1}^{N} \mathbf{f}_a \right). \tag{7.8}$$

This mechanical quantity is defined at the microscopic level of description and is thus fluctuating on the same timescale as the Brownian motion of the colloidal particle.

If the particle is dragged by the optical trap over some time interval $[0,t]$, we may consider the time integral of the microscopic mechanical observables. In particular, equation (7.7) becomes $\Delta E = W - Q$ with $\Delta E = E(t) - E(0)$, $W \equiv \int_0^t đW$, and $Q \equiv \int_0^t đQ$. If the fluid is large enough, it plays the role of a heat bath and the process is isothermal at the temperature, T, of the surrounding fluid. Under such conditions, the energy (7.4) of the colloidal particle may reach a stationary mean value $\langle E \rangle$ such that $\langle \Delta E \rangle = 0$.

As shown in Section 4.7.1, the sum of the molecular forces in equation (7.2) can be decomposed according to equation (4.165) into the viscous friction force $-\gamma\mathbf{v}$ and the Langevin fluctuating force $\mathbf{F}_{\rm fl}(t)$ given by the Gaussian white noises (4.174). Since this latter has a zero mean value, the statistical average of the molecular forces exerted on the colloidal particle is given by $-\gamma\langle\mathbf{v}\rangle = -\gamma\mathbf{u}$ in the stationary regime. The statistical average of the Newton equation (7.2) shows that $\langle\boldsymbol{\nabla}U\rangle = -\gamma\mathbf{u}$. Therefore, the mean power (7.5) provided by the moving optical trap to the colloidal particle is equal to $\langle đW/dt\rangle = \gamma\mathbf{u}^2$. According to equation (7.7), the work performed on the colloidal particle by the moving trap is thus dissipated on average in the form of heat in the fluid at the mean rate

$$\left\langle\frac{đW}{dt}\right\rangle = \left\langle\frac{đQ}{dt}\right\rangle = T\,\frac{d_{\rm i}S}{dt} = \gamma\,\mathbf{u}^2 \geq 0, \tag{7.9}$$

which gives the entropy production rate of the driven Brownian particle. If the optical trap is harmonic and isotropic, its energy potential is given by $U = (k/2)(\mathbf{r}-\mathbf{u}\,t)^2$ where k is the spring constant of the harmonic trap. In this case, the mean position of the colloidal particle is shifted with respect to the center of the optical trap by the drag effect of the fluid viscosity according to $\langle\mathbf{r}\rangle = \mathbf{u}\,t - (\gamma/k)\,\mathbf{u}$.

In the framework of the theory of stochastic processes, the expression (7.8) for the rate of heat transferred to the colloidal particle can be transposed into

$$\frac{đQ}{dt} = \mathbf{v}\circ[\gamma\,\mathbf{v} - \mathbf{F}_{\rm fl}(t)] \tag{7.10}$$

with Stratonovich's time discretization because the sum of molecular forces should be evaluated at the midpoint of the integration time step in the Newton equation (7.2).[1] Consequently, the work and heat undergone by the colloidal particle during the time interval $[0,t]$ are given by

$$W = -\int_0^t \mathbf{u}\cdot\boldsymbol{\nabla}U(\mathbf{r}-\mathbf{u}\,t')\,dt', \tag{7.11}$$

$$Q = \int_0^t \mathbf{v}\circ\left[\gamma\,\mathbf{v} - \mathbf{F}_{\rm fl}(t')\right]dt', \tag{7.12}$$

this latter being defined with the Stratonovich midpoint rule.

For the driven colloidal particle, the energy balance and the thermodynamic relation (7.7) are thus well founded at the micro- and mesoscopic levels of description, as shown in the *stochastic energetics* of Sekimoto (1997, 1998, 2010).

We note that the Hamiltonian function (7.1) has the form (5.1) where $H_{\rm s}$ is given by equation (7.4) with the time-dependent control parameters $\boldsymbol{\lambda}_t \equiv \mathbf{u}\,t$, $U_{\rm sb} = \sum_{a=1}^N u_a$, and $H_{\rm b}$ is the Hamiltonian function of the fluid without the colloidal particle, so that the nonequilibrium work fluctuation relation (5.13) is satisfied for this system if the reverse protocol is defined with $\boldsymbol{\lambda}_t^{\rm R} \equiv \mathbf{u}\,(\tau - t)$. This fluctuation relation holds for any finite time

[1] Indeed, in order to preserve the symplectic character and the microreversibility of Hamiltonian dynamics, the time discretization of Newton's equation $m d\mathbf{v}/dt = \mathbf{F}(t)$ should be done with the so-called leapfrog algorithm as $\mathbf{v}(t+\tau) = \mathbf{v}(t) + \mathbf{F}(t+\tau/2)\tau/m$, if τ is the time step (Frenkel and Smit, 2002; Allen and Tildesley, 2017).

interval $[0, \tau]$ and initial probability distributions taken as equilibrium canonical ensembles. In this regard, this fluctuation relation applies to a transient process, although the stationary process can be reached for long enough time. Furthermore, the quantities (7.11) and (7.12) obey fluctuation relations in the long-time limit if the potential $U(\mathbf{r}')$ increases fast enough for $\|\mathbf{r}'\| \to \infty$ (van Zon and Cohen, 2003a,b, 2004). For the work (7.11), the fluctuation relation has the asymptotic form,

$$\varsigma = \lim_{t\to\infty} \frac{1}{t} \ln \frac{p_t(\varsigma)}{p_t(-\varsigma)} \tag{7.13}$$

with $\varsigma = \beta W_t/t$, which can be inferred from the generic form $p_t(W)/p_t(-W) = C(W;t)\exp(\beta W)$ holding in the limit $t \to \infty$ with some subexponential function of time such that $\lim_{t\to\infty}(1/t)\ln C(W;t) = 0$. A similar asymptotic form holds for the heat (7.12) with $\varsigma = \beta Q_t/t$, if the extra condition $-\langle\varsigma\rangle < \varsigma < +\langle\varsigma\rangle$ is satisfied (van Zon and Cohen, 2003a, 2004).

Similar considerations extend to other kinds of fluctuating systems such as driven electric circuits or torsion pendula in viscous fluids at room temperature (van Zon et al., 2004; Garnier and Ciliberto, 2005; Joubaud et al., 2007; Ciliberto et al., 2013; Ciliberto, 2017).

7.2 Driven Brownian Particle

The time asymmetry of nonequilibrium fluctuations can be demonstrated experimentally for a Brownian particle trapped by an optical tweezer by measuring the rates of temporal disorder in the paths of the Brownian particle and their time reversal (Andrieux et al., 2007, 2008).

In this experiment, a polystyrene particle with a diameter of 2 μm is suspended in a 20% glycerol-water solution at the temperature $T = 298$ K. The trap potential is harmonic with a spring constant equal to $k = 9.62 \times 10^{-6}$ kg s^{-2}. The optical trap is fixed and the fluid moves with respect to the trap at the speed u in the x-direction, so that $\mathbf{u} = -u\mathbf{1}_x$. The particle position is measured in the x-direction using interferometry with a He-Ne laser providing a resolution of 10^{-11} m.

The motion of the Brownian particle is overdamped and ruled in the x-direction by the Langevin equation,

$$\gamma \frac{dx}{dt} = F(x + ut) + F_{\rm fl}(t), \tag{7.14}$$

where γ is the viscous friction coefficient, $F = -\partial_x U = -k(x + ut)$, and $F_{\rm fl}(t)$ is a Gaussian white noise such that

$$\langle F_{\rm fl}(t)\rangle = 0 \qquad \text{and} \qquad \langle F_{\rm fl}(t)\, F_{\rm fl}(t')\rangle = 2\,\gamma\, k_{\rm B}T\, \delta(t - t'). \tag{7.15}$$

Setting $z \equiv x + ut$, equation (7.14) becomes

$$\gamma \frac{dz}{dt} = F(z) + \gamma u + F_{\rm fl}(t). \tag{7.16}$$

For this process, the work done on the particle by the external drive during the time interval $[0, t]$ and the heat released into the surrounding fluid are, respectively, given by

$$W_t = -\int_0^t u\, F(z_{t'})\, dt', \tag{7.17}$$

$$Q_t = \int_0^t (\dot{z}_{t'} - u) \circ F(z_{t'})\, dt', \tag{7.18}$$

which are related to the change in potential energy $\Delta U_t \equiv U(z_t) - U(z_0)$ with $U(z) = kz^2/2$ according to

$$\Delta U_t = W_t - Q_t. \tag{7.19}$$

The thermodynamic entropy production rate is here evaluated as

$$\frac{d_{\rm i} S}{dt} = \lim_{t\to\infty} \frac{1}{t} \frac{\langle Q_t \rangle}{T} = \lim_{t\to\infty} \frac{1}{t} \frac{\langle W_t \rangle}{T} = \frac{1}{T} \gamma\, u^2 \geq 0, \tag{7.20}$$

where the statistical average $\langle \cdot \rangle$ is carried out over the forward process. As a consequence, equilibrium corresponds to the absence of driving if $u = 0$.

We note that this overdamped Langevin process can be mapped onto the Ornstein–Uhlenbeck stochastic process of Section E.7. In the experiment, the relaxation time has the value $\tau_{\rm r} = \gamma/k = 3.05 \times 10^{-3}$ s. In the stationary regime, the probability density is Gaussian with the mean position $\langle z \rangle = u\tau_{\rm r}$, which is shifted with respect to the minimum of the harmonic potential because of viscous drag by the fluid moving at the speed u, and the standard deviation $\sigma = \langle (z - \langle z \rangle)^2 \rangle^{1/2} = 20.7$ nm. Therefore, the diffusion coefficient is $\mathcal{D} = \sigma^2/\tau_{\rm r} = 0.14\ \mu\text{m}^2\ \text{s}^{-1}$.

The variable z_t is acquired with the sampling frequency $f = 8192$ Hz. Figure 7.2(a) shows two examples of paths for the trapped Brownian particle driven by the fluid moving at the speeds $\pm u$ and Figure 7.2(b) verifies that the stationary probability density of the

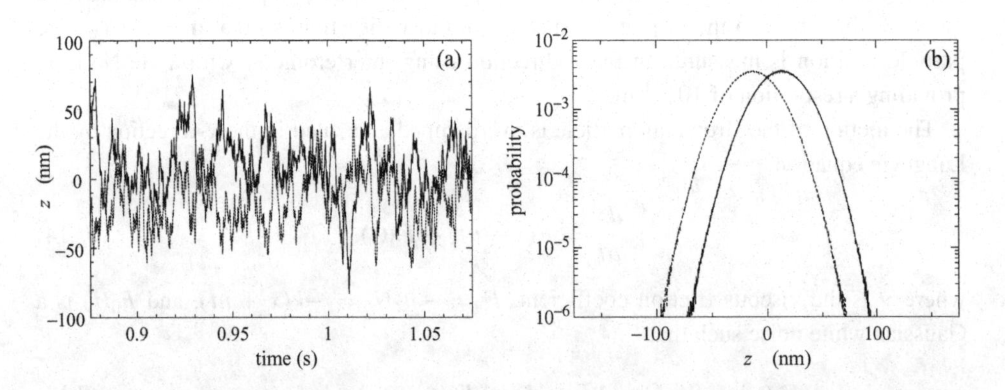

Figure 7.2 (a) Typical paths z_t for the trapped Brownian particle in the fluid moving at speed u for the forward process (solid line) and $-u$ for the reversed process (dashed line) with $u = 4.24 \times 10^{-6}$ m s^{-1}. (b) Gaussian probability densities for the forward and reversed experiments with the mean positions $\langle z \rangle = \pm u\tau_{\rm r} = \pm 12.9$ nm. Reproduced with permission from Andrieux et al. (2008) © IOP Publishing Ltd and SISSA Medialab.

position is Gaussian as expected. For each value of the speed, time series containing up to 2×10^7 points are recorded to measure their statistical properties.

For the overdamped Langevin process (7.16), the path probability conditioned to the value z_0 of the initial condition is given by

$$P[z_t|z_0] \propto \exp\left\{-\frac{1}{4\gamma k_B T}\int_0^t [\gamma\,\dot{z} - F(z) - \gamma\,u\,]^2\,dt'\right\}, \tag{7.21}$$

and the corresponding path probability by $P[z_t] \propto P[z_t|z_0]\,p_{\rm st}(z_0)$, where $p_{\rm st}(z)$ is the aforementioned stationary probability density. In order to extract the heat dissipated along the path, we consider the probability $P_+[z_t|z_0]$ of a given path over the probability $P_-\left[z_t^{\rm R}|z_0^{\rm R}\right]$ to observe its time reversal $z_t^{\rm R}$ having also reversed the sign of the driving speed u. The reversed path is thus defined by $z_{t'}^{\rm R} \equiv z_{t-t'}$, so that $\dot{z}_{t'}^{\rm R} = -\dot{z}_{t-t'}$, whereupon we have that

$$\ln\frac{P_+[z_t|z_0]}{P_-\left[z_t^{\rm R}|z_0^{\rm R}\right]} = \frac{Q_t}{k_B T}, \tag{7.22}$$

after using the path probability (7.21) and the expression (7.18) for the heat transferred from the colloidal particle to the fluid. Therefore, the thermodynamic entropy production rate (7.20) can be obtained as

$$\begin{aligned}\frac{d_{\rm i}S}{dt} &= \lim_{t\to\infty}\frac{k_B}{t}\left\langle\ln\frac{P_+[z_t|z_0]}{P_-\left[z_t^{\rm R}|z_0^{\rm R}\right]}\right\rangle = \lim_{t\to\infty}\frac{k_B}{t}\left\langle\ln\frac{P_+[z_t]}{P_-\left[z_t^{\rm R}\right]}\right\rangle \\ &= \lim_{t\to\infty}\frac{k_B}{t}\int P_+[z_t]\,\ln\frac{P_+[z_t]}{P_-\left[z_t^{\rm R}\right]}\,\mathcal{D}z_t \geq 0,\end{aligned} \tag{7.23}$$

where $\mathcal{D}z_t$ denotes the integral element of the whole path z_t. This expression can be equivalently written in the form of equation (6.103) by measuring the probabilities of the paths and their time reversal with the resolution ε and the sampling time τ. Accordingly, the path probabilities $P_+[z_t]$ are evaluated as the probabilities

$$P_+(z;\varepsilon,\tau,k) = {\rm Prob}\,\{|Z(t_i) - z_i| \leq \varepsilon;\ \text{for } t_i = i\tau \text{ with } i = 1,2,\ldots,k\}, \tag{7.24}$$

where $z = \{z_i\}_{i=1}^k$ denotes one among M reference paths $\{z_m\}_{m=1}^M$ sampled in the full time series. The mean decay rate of this path probability defines the (ε,τ)-entropy per unit time for the process:

$$h(\varepsilon,\tau) = \lim_{k,M\to\infty}\frac{1}{\tau}\frac{1}{M}\sum_{m=1}^{M}\ln\frac{P_+(z_m;\varepsilon,\tau,k)}{P_+(z_m;\varepsilon,\tau,k+1)}. \tag{7.25}$$

Similarly, the time-reversed (ε,τ)-coentropy per unit time is defined as

$$h^{\rm R}(\varepsilon,\tau) = \lim_{k,M\to\infty}\frac{1}{\tau}\frac{1}{M}\sum_{m=1}^{M}\ln\frac{P_-\left(z_m^{\rm R};\varepsilon,\tau,k\right)}{P_-\left(z_m^{\rm R};\varepsilon,\tau,k+1\right)}, \tag{7.26}$$

where P_- is the probability distribution of the reversed process with the opposite driving speed $-u$ and $z_m^{\rm R}$ denotes the time reversal of the reference path z_m of the forward process. The thermodynamic entropy production rate (7.23) can thus be expressed as

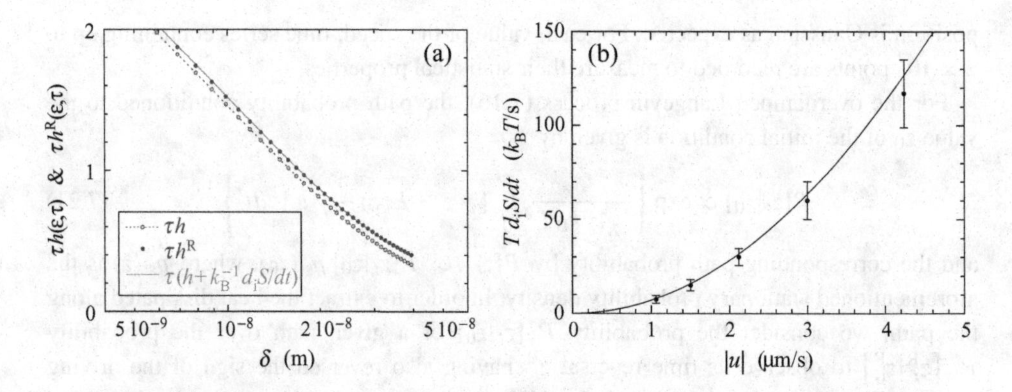

Figure 7.3 (a) The rates of temporal disorder characterized by the (ε, τ)-entropy per unit time (7.25) (dashed line with open circles) and the time-reversed (ε, τ)-coentropy per unit time (7.26) (solid circles) for the Brownian particle driven by the speed $u = 4.24 \times 10^{-6}$ m s^{-1} versus $\delta = \varepsilon/\sqrt{1 - \exp(-2\tau/\tau_r)}$ for the sampling time $\tau = 4/f = 2048^{-1}$ s. The rates are scaled by the sampling time τ. The prediction of equation (7.27) is shown as the solid line. (b) The thermodynamic entropy production rate of the Brownian particle versus the driving speed u. The dots are the results of equation (7.27) calculated with the differences between the (ε, τ)-entropies per unit time and the solid line is the expectation (7.20). Reproduced with permission from Andrieux et al. (2008) © IOP Publishing Ltd and SISSA Medialab.

$$\frac{d_i S}{dt} = \lim_{\varepsilon, \tau \to 0} k_B \left[h^R(\varepsilon, \tau) - h(\varepsilon, \tau) \right] \geq 0 \tag{7.27}$$

in terms of the temporal disorder rates (7.25) and (7.26). Since the process is equivalent to the Ornstein–Uhlenbeck process, the (ε, τ)-entropy per unit time (7.25) is expected to behave as (6.95) for small enough values of the spatial resolution ε.

These expectations are confirmed by the experiment, as shown in Figure 7.3. We see in Figure 7.3(a) that the temporal disorder measured with the (ε, τ)-entropy per unit time (7.25) increases as the spatial resolution ε decreases, as expected from equation (6.95), predicting $\tau h(\varepsilon, \tau) \sim \ln(1/\delta)$ with $\delta = \varepsilon/\sqrt{1 - \exp(-2\tau/\tau_r)}$ for $\delta \to 0$. Remarkably, the time-reversed (ε, τ)-coentropy per unit time (7.26) is larger by the amount predicted by equation (7.27). Therefore, the thermodynamic entropy production rate can be measured with the difference between the (ε, τ)-entropies per unit time characterizing temporal disorder in the paths and their time reversal, as seen in Figure 7.3(b) for different values of the driving speed u. This result demonstrates that thermodynamic entropy production is related to time asymmetry in temporal disorder. Furthermore, Figure 7.3(a) shows that irreversibility can be observed down to nanometric resolution.

Figure 7.3(a) also shows that the positive entropy production is a small effect on top of a substantial temporal disorder. The mean rate of frenesy defined by equation (6.116) would be the curve $\mathscr{F} = (h + h^R)/2$ in between the curves of the quantities $h = h(\varepsilon, \tau)$ and $h^R = h^R(\varepsilon, \tau)$, characterizing temporal disorder in the paths and their time reversal.

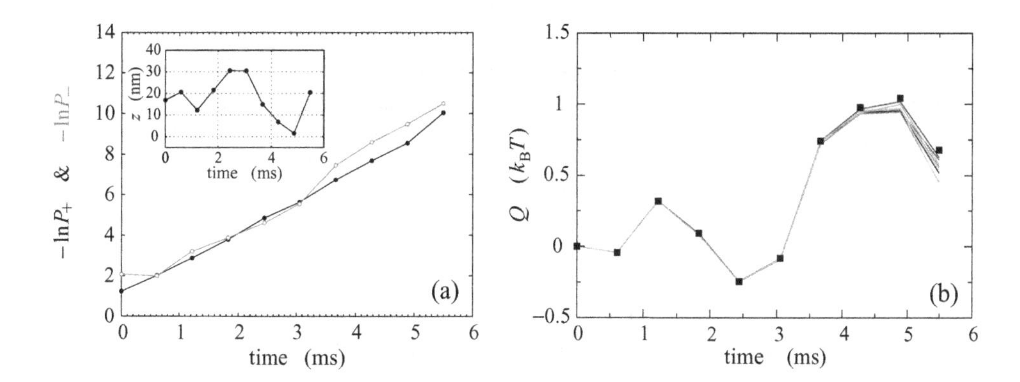

Figure 7.4 Measure of the heat released by the Brownian particle to the surrounding fluid along the trajectory depicted in the inset of panel (a), which is randomly selected from the time series of Figure 7.2. The fluid speeds are $\pm u = \pm 4.24 \times 10^{-6}$ m s^{-1}. (a) The conditional probabilities (7.28) and (7.29) of the corresponding forward (filled circles) and reversed (open circles) paths for $\varepsilon = 8.4$ nm. At time zero, the conditional probabilities are not defined and, instead, the stationary probabilities are plotted for indication. (b) The heat given by equation (7.22) for different values of $\varepsilon = n \times 0.558$ nm with $12 \leq n \leq 19$ in the range 6.7–10.6 nm (solid lines), and compared with the values directly calculated with equation (7.18) (squares). Reproduced with permission from Andrieux et al. (2008) © IOP Publishing Ltd and SISSA Medialab.

Maxwell's demon vividly illustrates the paradox that the dissipated heat is always non-negative at the macroscopic level, although this may no longer be the case if considered for individual trajectories observed on microscopic scales. The resolution of Maxwell's paradox can be remarkably demonstrated with the path probabilities provided by experimental data. Indeed, the heat transferred from the Brownian particle to the fluid along an individual trajectory is given by equation (7.22), where the conditional path probabilities can be evaluated in terms of the path probabilities (7.24) according to

$$P_+(z;\varepsilon,\tau,k|z_1) = P_+(z;\varepsilon,\tau,k)/P_+(z_1;\varepsilon,\tau,1), \tag{7.28}$$

$$P_-\left(z^{\rm R};\varepsilon,\tau,k|z_k\right) = P_-\left(z^{\rm R};\varepsilon,\tau,k\right)/P_-(z_k;\varepsilon,\tau,1), \tag{7.29}$$

where $P_\pm(z;\varepsilon,\tau,1) \simeq p_{\rm st}(z)\,dz$ with the stationary probability density $p_{\rm st}(z)$ and $dz = 2\varepsilon$. As shown in Figure 7.4, the heat (7.22) can be measured along individual paths with this method. In Figure 7.4(a), the conditional probabilities (7.28) and (7.29) are shown as a function of time for the selected trajectory z depicted in the inset. These conditional probabilities present a mean exponential decrease modulated by the fluctuations. In Figure 7.4(b), the heat transferred from the particle to the fluid is computed for this selected trajectory by using equation (7.22) and plotted as solid lines. We see the very good agreement with the expected values directly calculated with equation (7.18) for the selected trajectory. As seen in Figure 7.4(b), this fluctuating heat may take negative values for an individual trajectory, although the mean value of the heat obtained by averaging over the forward process and shown in Figure 7.3(b) is always nonnegative, in agreement with the second law of thermodynamics. The nonnegativity of the thermodynamic entropy production results from this averaging, which solves Maxwell's paradox (Andrieux et al., 2008).

7.3 Analogous Electric Circuits

Overdamped Langevin processes also describe the stochastic time evolution of noisy electric circuits with resistors and capacitors. In particular, the Langevin stochastic equation (7.14) for the driven Brownian particle also rules charge transport in the RC electric circuit of Figure 7.5 with a resistor R and a capacitor C in parallel. This circuit is driven out of equilibrium by a constant current source I, which splits into the current intensities in the resistor, I_R, and in the capacitor, I_C, so that $I = I_R + I_C$. Denoting q_R to be the electric charge passing in the resistor, q_C the charge in the capacitor, and V the voltage on the parallel edges of the circuit, we have that $V = RI_R + V_{\rm fl}(t)$ and $q_C = CV$ with $I_R = \dot{q}_R = I - \dot{q}_C$. Consequently, the evolution equation for the charge $q \equiv q_R$ is given by

$$R\,\frac{dq}{dt} = -\frac{1}{C}\,(q - I\,t) - V_{\rm fl}(t), \tag{7.30}$$

where $V_{\rm fl}(t)$ is the Johnson–Nyquist fluctuating voltage characterized by equation (4.196). This stochastic differential equation is analogous to equation (7.14) with the fluctuating force (7.15). The correspondence between the mechanical and electrical quantities is given in Table 7.1. Again, the stochastic process can be mapped onto the Ornstein–Uhlenbeck process of Section E.7 with the relaxation time $\tau_{\rm r} = RC$, the diffusivity $D = k_{\rm B}T/R$, and $x = q_C - \tau_{\rm r}I = I(t - \tau_{\rm r}) - q_R$. The stationary probability density of the electric charge in the capacitor is given by the Gaussian distribution

$$p_{\rm st}(q_C) = \frac{1}{\sqrt{2\pi\sigma^2}}\,\exp\left[-\frac{(q_C - \tau_{\rm r}I)^2}{2\sigma^2}\right] \qquad \text{with} \qquad \sigma^2 = k_{\rm B}TC, \tag{7.31}$$

and the mean charge passing in the resistor by $\langle q\rangle = \langle q_R\rangle = I(t - \tau_{\rm r})$. Here, the thermodynamic entropy production rate is given by Joule's law:

$$\frac{d_{\rm i}S}{dt} = \lim_{t\to\infty}\frac{1}{t}\frac{\langle Q_t\rangle}{T} = \lim_{t\to\infty}\frac{1}{t}\frac{\langle W_t\rangle}{T} = \frac{1}{T}\,R\,I^2 \geq 0. \tag{7.32}$$

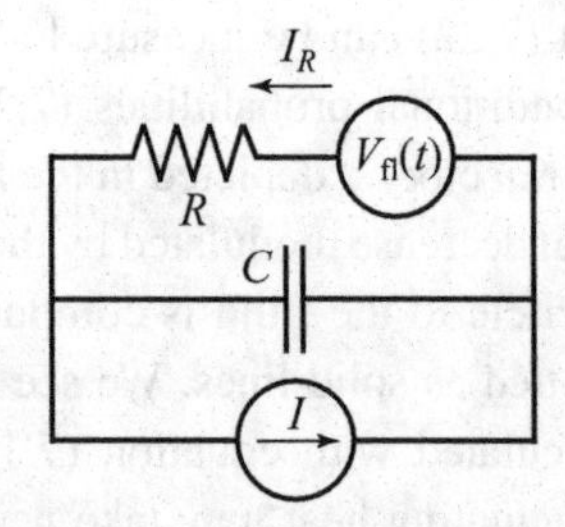

Figure 7.5 Noisy RC electric circuit with a capacitor of capacitance C, a resistor of resistance R with its associated Johnson–Nyquist noise $V_{\rm fl}(t)$, and subjected to a constant current source I. The current intensity through the resistor is denoted I_R.

Table 7.1. *Analogy between a driven Brownian particle and an RC electric circuit. For the Brownian particle, z is its position, ż its velocity, k the harmonic strength of the optical trap, γ the viscous friction coefficient,* $F_{\rm fl}(t)$ *the Langevin fluctuating force, and u the fluid speed. For the electric circuit,* $q_R = q$ *is the electric charge passing through the resistor during time t,* $I_R = \dot{q}$ *the corresponding current intensity, C the capacitance, R the resistance,* $V_{\rm fl}(t)$ *the fluctuating electric potential of the Johnson–Nyquist noise, and I the mean current source.*

Driven Brownian particle	z	$\dot{z}$	k	γ	$F_{\rm fl}(t)$	u
Driven RC electric circuit	$q - It$	$\dot{q} - I$	C^{-1}	R	$-V_{\rm fl}(t)$	$-I$

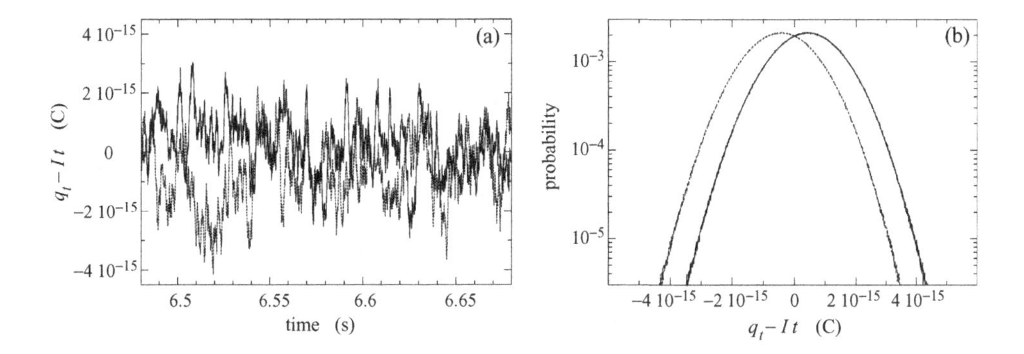

Figure 7.6 (a) Typical paths $q_t - It$ for the Johnson–Nyquist noise in the RC electric circuit driven by the current I (solid line) and opposite current $-I$ (dashed line) with $I = 1.67 \times 10^{-13}$ A. (b) Gaussian probability densities for the forward and reversed experiments. The unit of electric charge is the coulomb (C). Reproduced with permission from Andrieux et al. (2008) © IOP Publishing Ltd and SISSA Medialab.

The connection between time asymmetry in the temporal disorder of electric fluctuations and the thermodynamic entropy production can also be experimentally demonstrated (Andrieux et al., 2007, 2008). In the experiment, the resistance is $R = 9.22$ MΩ and the capacitance $C = 278$ pF, so that the relaxation time of the circuit is $\tau_{\rm r} = RC = 2.56 \times 10^{-3}$ s. The temperature is $T = 298$ K, which determines the magnitude of the thermal electric fluctuations (4.196). The standard deviation of electric charge in the capacitor is $\sigma = \sqrt{k_{\rm B}TC} = 6.7 \times 10^3\, e$, where $e = 1.602 \times 10^{-19}$ C is the elementary electric charge. The diffusivity is $D = \sigma^2/\tau_{\rm r} = 1.75 \times 10^{10}\, e^2\, {\rm s}^{-1}$.

For the driven electric circuit, pairs of time series for opposite drivings $\pm I$ are recorded with 2×10^7 points each. The sampling frequency is again $f = 8192$ Hz. Figure 7.6 shows an example of a pair of such paths with the corresponding probability distribution of charge fluctuations in agreement with equation (7.31).

The (ε, τ)-entropy per unit time and time-reversed coentropy per unit time are computed from the recorded time series with equations (7.25) and (7.26). The scaled entropies per unit time are plotted in Figure 7.7(a) as a function of $\delta = \varepsilon/\sqrt{1 - \exp(-2\tau/\tau_{\rm r})}$. Here again, the scaled entropy per unit time $\tau h(\varepsilon, \tau)$ is verified to depend only on δ for $\delta \to 0$, as expected from equation (6.95). Moreover, the scaled time-reversed entropy per

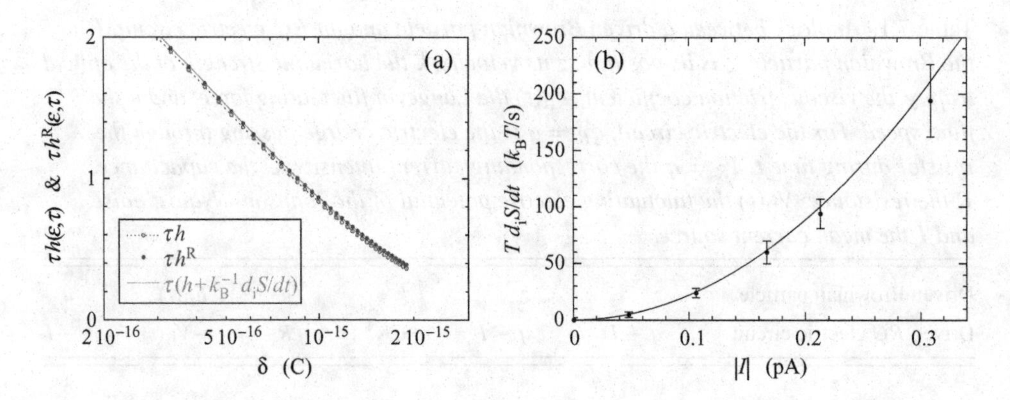

Figure 7.7 (a) The rates of temporal disorder characterized by the (ε, τ)-entropy per unit time (dashed line with open circles) and time-reversed coentropy per unit time (solid circles) for the RC electric circuit driven by the current $I = 1.67 \times 10^{-13}$ A versus $\delta = \varepsilon/\sqrt{1-\exp(-2\tau/\tau_r)}$ for the sampling time $\tau = 4/f = 2048^{-1}$ s. The rates are scaled by the sampling time τ. The prediction of equation (7.27) is shown as the solid line. (b) Thermodynamic entropy production rate of the RC electric circuit versus the injected current I. The dots are the results of equation (7.27) calculated with the differences between the (ε, τ)-entropies per unit time and the solid line is the Joule law, $T d_i S/dt = RI^2$. Reproduced with permission from Andrieux et al. (2008) © IOP Publishing Ltd and SISSA Medialab.

unit time is shifted upward by the thermodynamic entropy production rate according to $\tau h^R \simeq \tau(h + k_B^{-1} d_i S/dt)$, as predicted by the main result (7.27). This shift continues to exist as δ decreases, so that irreversibility is here observed down to fluctuations as small as a few thousand electrons. Finally, Figure 7.7(b) shows the difference between the rates of temporal disorder h^R and h for several values of the driving current $|I|$ and the result is in agreement with the Joule law (7.32) as predicted by equation (7.27). Therefore, the thermodynamic entropy production rate is here also demonstrated to find its origin in the time asymmetry of temporal disorder.

A comparison can be made with the stationary fluctuation relations (7.13) for the work (7.11) and the heat (7.12). Here, the work is given by $W_t = I \int_0^t V(t')\,dt'$, where I is the constant current source and $V(t)$ is the fluctuating voltage across the capacitor. The heat has the value $Q_t = W_t - \Delta U_t$ with the energy potential of the capacitor $U = q_C^2/(2C)$. The corresponding stationary fluctuation relations (7.13) are experimentally verified using $\varsigma = \beta W_t/t$ for the work and $\varsigma = \beta Q_t/t$ for the heat if $|\varsigma| < \langle \varsigma \rangle$ (Garnier and Ciliberto, 2005). Introducing the same function $\mathcal{R}(\varsigma)$ as defined by equation (6.32), the asymptotic fluctuation relations are equivalent to symmetry relations identical to equation (6.34). Since the function $\mathcal{R}(\varsigma)$ is equal to zero with its first derivative for the thermodynamic entropy production rate $\langle \varsigma \rangle = k_B^{-1} d_i S/dt$ in the units of Boltzmann's constant, we have that $\langle \varsigma \rangle = \mathcal{R}(-\langle \varsigma \rangle)$, which can be compared with the thermodynamic entropy production rate expressed by equation (7.27), giving $\langle \varsigma \rangle = \mathcal{R}(-\langle \varsigma \rangle) = \lim_{\varepsilon,\tau\to 0} \left[h^R(\varepsilon,\tau) - h(\varepsilon,\tau)\right]$. Since the temporal disorder rate $h(\varepsilon,\tau)$ is always nonnegative, the inequality $h^R(\varepsilon,\tau) \geq \mathcal{R}(-\langle \varsigma \rangle)$ is, in general, satisfied. Moreover, $h(\varepsilon,\tau)$ is typically a large positive quantity.

The greater the temporal disorder, the larger its rate $h(\varepsilon, \tau)$, as seen in Figures 7.3(a) and 7.7(a). In contrast, the function $\mathcal{R}(\varsigma)$ is of the same order of magnitude as $k_B^{-1} d_i S/dt$ and, thus, significantly smaller than $h(\varepsilon, \tau)$. This shows that the rates of temporal disorder for the paths and their time reversal can measure not only the irreversibility of the process but also its dynamical activity on much finer temporal and spatial scales in the mesoscopic description than the function $\mathcal{R}(\varsigma)$ expressing the work and heat fluctuation relations in the form (6.34).

7.4 Driven Langevin Processes

In this section, fluctuation relations are considered for the Langevin processes of Section 4.7 that rule the random motion of a particle in some time-independent bounded energy potential $U_{\min} \leq U(\mathbf{r}) \leq U_{\max}$ and driven away from equilibrium by some external force $\mathbf{F}_{\text{ext}}$, which is supposed to be uniform. The total positional force exerted on the particle is thus given by

$$\mathbf{F} = -\nabla U + \mathbf{F}_{\text{ext}} = -\nabla U_{\text{tot}} \qquad \text{with} \qquad U_{\text{tot}}(\mathbf{r}) = U(\mathbf{r}) - \mathbf{F}_{\text{ext}} \cdot \mathbf{r}. \tag{7.33}$$

Moreover, the particle is subjected to a dissipative friction force and the associated fluctuating Langevin force at the temperature T of the system. For instance, the particle may be a colloidal particle moving in water under the effect of gravitation (in which case $\mathbf{F}_{\text{ext}} = \mathbf{F}_{\text{g}} = -mg\mathbf{1}_x$) or an ion of charge q moving in the periodic lattice of a crystal subjected to an electric field $\boldsymbol{\mathcal{E}}$ (in which case $\mathbf{F}_{\text{ext}} = q\,\boldsymbol{\mathcal{E}}$). The control parameter of nonequilibrium driving is the mechanical affinity

$$\mathbf{A} \equiv \frac{\mathbf{F}_{\text{ext}}}{k_B T}, \tag{7.34}$$

as already identified in Section 6.3.5. Accordingly, the external force has the effect of generating the random drift of the particle at some mean velocity that should be determined by solving the problem. The distinction is made between overdamped and underdamped processes, whether the inertial effects can be neglected or not.

7.4.1 Underdamped Processes

In this regime, the stochastic differential equation of the Langevin process has the form

$$m\,\ddot{\mathbf{r}} + \gamma\,\dot{\mathbf{r}} = \mathbf{F}(\mathbf{r}) + \mathbf{F}_{\text{fl}}(t), \tag{7.35}$$

where m is the particle mass, γ the friction coefficient, $\mathbf{F}(\mathbf{r})$ the force (7.33), and $\mathbf{F}_{\text{fl}}(t)$ the Langevin fluctuating force given by the Gaussian white noises (4.174). The master equation ruling the probability density $\mathcal{P}(\mathbf{r}, \mathbf{p}, t)$ that the particle has the position $\mathbf{r}$ and the momentum $\mathbf{p} = m\dot{\mathbf{r}}$ at time t is given by the Kramers equation (4.179), which here reads

$$\partial_t \mathcal{P} = \hat{L}\,\mathcal{P} \equiv -\frac{\mathbf{p}}{m} \cdot \partial_{\mathbf{r}} \mathcal{P} - \mathbf{F} \cdot \partial_{\mathbf{p}} \mathcal{P} + \partial_{\mathbf{p}} \cdot \left(\gamma\,\frac{\mathbf{p}}{m}\,\mathcal{P}\right) + \gamma\, k_B T\, \partial_{\mathbf{p}}^2 \mathcal{P}. \tag{7.36}$$

The solution of this master equation can be expressed as a path integral, as explained in Chapter 6 with $\mathbf{x} = (\mathbf{r}, \mathbf{p})$. Since the stochastic differential equation (7.35) is of second order in time, the corresponding tensor of diffusivity does not have full rank, but is positive only for momentum. Accordingly, the conditional probability to observe the trajectory $\mathbf{r}(t)$ starting from the initial conditions $(\mathbf{r}_0, \mathbf{p}_0)$ at time zero is here given by

$$P[\mathbf{r}(t)|\mathbf{r}_0, \mathbf{p}_0, 0] \propto \exp\left[-\frac{1}{4\gamma k_\mathrm{B} T}\int (m\,\ddot{\mathbf{r}} + \gamma\,\dot{\mathbf{r}} - \mathbf{F})^2\,dt\right], \tag{7.37}$$

using the Lagrangian function (6.63) for Stratonovich's midpoint rule.[2]

Time reversal in the phase space of the particle is defined by the transformation

$$\boldsymbol{\Theta}(\mathbf{r}, \mathbf{p}) = (\mathbf{r}, -\mathbf{p}), \tag{7.38}$$

reversing momentum, but leaving position unchanged. Now, we may consider the ratio between the conditional path probability (7.37) and the probability to observe the time-reversed trajectory $\mathbf{r}(-t)$ provided that its initial conditions are the final conditions $(\mathbf{r}, \mathbf{p})$ of the forward trajectory. Remarkably, this ratio only depends on the initial and final conditions, so this ratio also gives the ratio of the corresponding conditional probabilities obtained by integrating over all the paths joining these initial and final conditions. In this way, we find the fluctuation relation

$$\frac{\mathcal{P}(\mathbf{r}, \mathbf{p}, t|\mathbf{r}_0, \mathbf{p}_0, 0)}{\mathcal{P}(\mathbf{r}_0, -\mathbf{p}_0, t|\mathbf{r}, -\mathbf{p}, 0)} = \mathrm{e}^{\beta(W-\Delta K-\Delta U)}, \tag{7.39}$$

where $\beta = (k_\mathrm{B} T)^{-1}$ is the inverse temperature,

$$W = \mathbf{F}_\mathrm{ext} \cdot \Delta\mathbf{r} \qquad \text{with} \qquad \Delta\mathbf{r} = \mathbf{r} - \mathbf{r}_0 \tag{7.40}$$

is the work performed by the external force $\mathbf{F}_\mathrm{ext}$ on the particle, and

$$\Delta K = \frac{\mathbf{p}^2}{2m} - \frac{\mathbf{p}_0^2}{2m} \qquad \text{and} \qquad \Delta U = U(\mathbf{r}) - U(\mathbf{r}_0) \tag{7.41}$$

are the changes in kinetic and potential energies between the initial and final conditions. We note that these latter can be gathered into the change $\Delta E = \Delta K + \Delta U$ of the total energy of the particle defined as $E = \mathbf{p}^2/(2m) + U(\mathbf{r})$ without the contribution of the external force.

If we introduce the function

$$\eta(\mathbf{r}, \mathbf{p}) \equiv \exp(-\beta E) = \exp\left\{-\beta\left[\frac{\mathbf{p}^2}{2m} + U(\mathbf{r})\right]\right\}, \tag{7.42}$$

the fluctuation relation (7.39) can equivalently be written as

$$\mathcal{P}(\mathbf{r}, \mathbf{p}, t|\mathbf{r}_0, \mathbf{p}_0, 0)\,\eta(\mathbf{r}_0, \mathbf{p}_0) = \mathcal{P}(\mathbf{r}_0, -\mathbf{p}_0, t|\mathbf{r}, -\mathbf{p}, 0)\,\eta(\mathbf{r}, -\mathbf{p})\,\mathrm{e}^{\mathbf{A}\cdot(\mathbf{r}-\mathbf{r}_0)}, \tag{7.43}$$

in terms of the mechanical affinity (7.34). If the external force is equal to zero and the energy potential $U(\mathbf{r})$ is confining, the function η is proportional to the canonical equilibrium

[2] For simplicity, the contribution of $\int \Lambda^{(1/2)} dt$ with $\Lambda^{(1/2)} = -3\gamma/(2m)$ given by equation (E.39) is dropped in the proportionality relation since the associated factor is independent of the trajectory.

probability density for the particle in the thermal bath of the surrounding medium, so that the fluctuation relation (7.43) becomes the expression (2.85) for the principle of detailed balance, holding at thermodynamic equilibrium.

The conditional probability density that the particle undergoes the displacement from $\mathbf{r}_0$ to $\mathbf{r}$ during the time interval t can be defined as

$$p(\mathbf{r},t|\mathbf{r}_0,0) \equiv \int \mathcal{P}(\mathbf{r},\mathbf{p},t|\mathbf{r}_0,\mathbf{p}_0,0)\, p_{\rm st}(\mathbf{p}_0)\, d^3p\, d^3p_0, \tag{7.44}$$

where $p_{\rm st}(\mathbf{p}) = \exp\left[-\mathbf{p}^2/(2mk_{\rm B}T)\right]/(2\pi mk_{\rm B}T)^{3/2}$ is the Maxwellian stationary probability density for momentum. As a corollary of equation (7.39), the conditional probability distribution (7.44) obeys the fluctuation relation

$$\frac{p(\mathbf{r},t|\mathbf{r}_0,0)}{p(\mathbf{r}_0,t|\mathbf{r},0)} = {\rm e}^{\beta(W-\Delta U)}, \tag{7.45}$$

with the work (7.40) and $\Delta U = U(\mathbf{r}) - U(\mathbf{r}_0)$.

If we define the current associated with the displacement of the particle as $\mathfrak{J} \equiv \Delta\mathbf{r}/t$, the multivariate fluctuation relation (5.242) for the vectorial current can be proved by using a method similar to that in Section 5.7. Indeed, the cumulant generating function for the current

$$Q(\boldsymbol{\lambda};\mathbf{A}) \equiv \lim_{t\to\infty} -\frac{1}{t}\ln\left\langle {\rm e}^{-\boldsymbol{\lambda}\cdot\Delta\mathbf{r}(t)}\right\rangle_{\mathbf{A}} \tag{7.46}$$

can be obtained as the leading eigenvalue

$$\hat{L}_{\boldsymbol{\lambda}}\,\Psi_{\boldsymbol{\lambda}} = -Q(\boldsymbol{\lambda};\mathbf{A})\,\Psi_{\boldsymbol{\lambda}} \tag{7.47}$$

of the evolution operator modified into

$$\hat{L}_{\boldsymbol{\lambda}} \equiv {\rm e}^{-\boldsymbol{\lambda}\cdot\mathbf{r}}\,\hat{L}\,{\rm e}^{\boldsymbol{\lambda}\cdot\mathbf{r}} \tag{7.48}$$

in order to include the counting parameters $\boldsymbol{\lambda}$. This modified operator is supposed to obey the symmetry relation

$$\eta^{-1}\,\hat{\Theta}\,\hat{L}_{\boldsymbol{\lambda}}\big(\hat{\Theta}\,\eta\,\Psi\big) = \hat{L}^{\dagger}_{\mathbf{A}-\boldsymbol{\lambda}}\,\Psi, \tag{7.49}$$

where $\hat{\Theta}$ is the time-reversal operator such that $\hat{\Theta}\Psi(\mathbf{r},\mathbf{p}) = \Psi(\mathbf{r},-\mathbf{p})$ induced by the transformation (7.38), η is some function of position and momentum, and Ψ is an arbitrary function. Consequently, its leading eigenvalue satisfies the symmetry relation

$$Q(\boldsymbol{\lambda};\mathbf{A}) = Q(\mathbf{A}-\boldsymbol{\lambda};\mathbf{A}) \tag{7.50}$$

as proved in Section 5.7 (Kurchan, 1998; Lebowitz and Spohn, 1999).

Now, for the operator (7.36) of Kramers' master equation, the modified operator (7.48) is given by

$$\hat{L}_{\boldsymbol{\lambda}} = \hat{L} - \boldsymbol{\lambda}\cdot\frac{\mathbf{p}}{m} \tag{7.51}$$

and it obeys the symmetry relation (7.49) with the mechanical affinity (7.34) and the function (7.42). Therefore, the multivariate fluctuation relation (5.242) is established for the current $\mathfrak{J} = \Delta\mathbf{r}(t)/t$. As a consequence, all the remarkable linear and nonlinear response properties of Section 5.5 are satisfied for the Langevin process. In particular, the fluctuation–dissipation theorem and the Onsager reciprocal relations are valid in the regime of linear response and, furthermore, their generalizations to the nonlinear response properties are also holding.

7.4.2 Overdamped Processes

For overdamped processes, the inertial term $m\ddot{\mathbf{r}}$ becomes negligible in front of the friction term $\gamma\dot{\mathbf{r}}$ in the Langevin equation (7.35), which becomes a first-order differential equation. The corresponding master equation is the Smoluchowski equation (4.162) or (4.185),

$$\partial_t\, p = \hat{L}\, p \equiv \mathcal{D}\,\nabla^2 p - \beta\mathcal{D}\,\nabla\cdot(\mathbf{F}\, p), \tag{7.52}$$

which rules the time evolution of the probability density $p(\mathbf{r},t)$ of finding the particle at the position $\mathbf{r}$ at current time t. Again, the force is given by equation (7.33).

Here, the fluctuation relation (7.45) is also satisfied for the work (7.40) and $\Delta U = U(\mathbf{r}) - U(\mathbf{r}_0)$.

Furthermore, the multivariate fluctuation relation for the current can also be proved in the overdamped regime using the Smoluchowski equation (7.52). Indeed, the modified operator (7.48) is here given by

$$\hat{L}_{\boldsymbol{\lambda}} = \hat{L} + 2\mathcal{D}\,\boldsymbol{\lambda}\cdot\nabla + \mathcal{D}\,\boldsymbol{\lambda}^2 - \beta\mathcal{D}\,\boldsymbol{\lambda}\cdot\mathbf{F}, \tag{7.53}$$

where $\hat{L} = \hat{L}_{\mathbf{0}}$ is the unmodified operator (7.52), and it obeys the symmetry relation

$$\eta^{-1}\,\hat{L}_{\boldsymbol{\lambda}}\,(\eta\,\psi) = \hat{L}^{\dagger}_{\mathbf{A}-\boldsymbol{\lambda}}\,\psi \tag{7.54}$$

for any function $\psi(\mathbf{r})$ with respect to the mechanical affinity (7.34) and the Boltzmann factor $\eta = \exp[-\beta U(\mathbf{r})]$ (Lacoste and Mallick, 2009). As a consequence of this symmetry, equation (7.50) is satisfied and, thus, the multivariate fluctuation relation (5.242) as well as all the linear and nonlinear response properties of Section 5.5.

7.4.3 Examples

Figure 7.8 shows three examples of driven Langevin processes (Risken, 1989).

The first one is a particle of mass m and friction coefficient γ moving in the one-dimensional periodic potential $U(x) = U_0 \cos kx$ of spatial periodicity $\lambda = 2\pi/k$, in the uniform external force field $F_{\rm ext}$, and subjected to thermal fluctuations. For this example, the Langevin stochastic equation is given by

$$m\,\ddot{x} + \gamma\,\dot{x} - F_0\,\sin kx = F_{\rm ext} + F_{\rm fl}(t) \quad \text{and} \quad \langle F_{\rm fl}(t)\,F_{\rm fl}(t')\rangle = 2\gamma k_{\rm B}T\,\delta(t-t'), \tag{7.55}$$

where $F_0 = U_0 k$.

The second example is the noisy pendulum of Figure 7.8(b) ruled by the Langevin equation,

$$m\,\ell^2\,\ddot{\theta} + \zeta\,\dot{\theta} + mg\ell\,\sin\theta = \tau_{\rm ext} + \tau_{\rm fl}(t) \quad \text{and} \quad \langle\tau_{\rm fl}(t)\,\tau_{\rm fl}(t')\rangle = 2\zeta k_{\rm B}T\,\delta(t-t'), \tag{7.56}$$

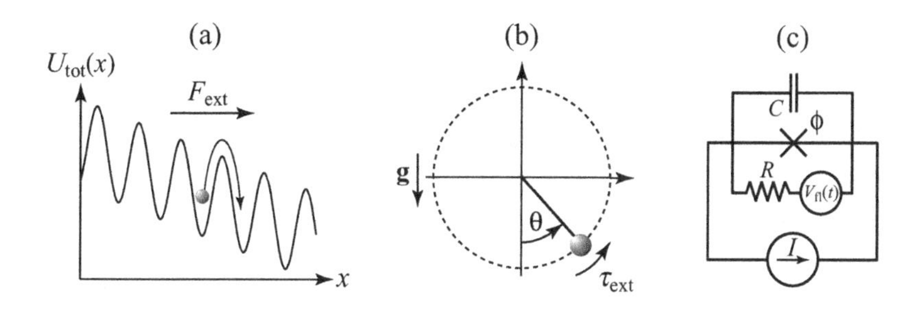

Figure 7.8 One-dimensional driven Langevin processes: (a) Brownian particle moving in a periodic potential tilted by the uniform external force F_{ext}; (b) noisy pendulum in the uniform gravitation field of acceleration $\mathbf{g}$ and driven by the constant external torque τ_{ext}; (c) noisy Josephson tunneling junction driven by the constant current source I.

where ℓ is the length of the pendulum, ζ the rotational friction coefficient, g the gravitational acceleration, τ_{ext} an applied external torque, and $\tau_{fl}(t)$ the fluctuating torque associated with friction.

The third example is the superconducting Josephson tunneling junction of Figure 7.8(c) governed by a similar stochastic differential equation,

$$\frac{\hbar C}{2e}\ddot{\phi} + \frac{\hbar}{2eR}\dot{\phi} + I_0 \sin\phi = I + \frac{V_{fl}(t)}{R} \quad \text{and} \quad \langle V_{fl}(t)\, V_{fl}(t')\rangle = 2Rk_B T\, \delta(t-t'), \tag{7.57}$$

where ϕ is the quantum phase of the junction characterized by its capacitance C, its resistance R, its maximum Josephson current I_0, and driven by the constant current intensity I, $\hbar$ denoting Planck's constant and e the elementary electric charge. This equation is obtained by applying Kirchhoff's laws to the equivalent electric circuit of Figure 7.8(c). Assuming that the junction is subjected to the voltage V, the current intensity of the source splits as $I = I_C + I_R + I_S$ into the current intensities $I_C = C\dot{V}$ for the capacitor, $I_R = [V - V_{fl}(t)]/R$ for the resistor, and $I_S = I_0 \sin\phi$ for the supercurrent, where the quantum phase ϕ determines the voltage according to $V = \hbar\dot{\phi}/(2e)$, whereupon equation (7.57) is inferred (Ambegaokar and Halperin, 1969; Ivanchenko and Zil'berman, 1969).

The three equations (7.55), (7.56), and (7.57) can be mapped onto each other by setting $kx = \theta = \phi$ and rescaling time with the relaxation time $\tau_r = m/\gamma = m\ell^2/\zeta = RC$, which leads to

$$\theta'' + \theta' = a_0 \sin\theta + a_{ext} + b\,\chi(t) \quad \text{and} \quad \langle\chi(t)\,\chi(t')\rangle = \delta(t-t'). \tag{7.58}$$

For this system, the fluctuation relation (7.45) reads

$$p(\theta,t|\theta_0,0) = p(\theta_0,t|\theta,0)\exp\left[\alpha_{ext}(\theta-\theta_0) - \alpha_0(\cos\theta - \cos\theta_0)\right], \tag{7.59}$$

with $\alpha_{ext} \equiv 2a_{ext}/b^2$ and $\alpha_0 \equiv 2a_0/b^2$. In the three examples, the representative particle undergoes stochastic motion in the tilted periodic potential shown in Figure 7.8(a). As long as there are barriers hindering the motion of the particle if $|a_{ext}| < |a_0|$, the mean drift velocity remains very small, being proportional to the probability of jumping over the barriers by thermal activation, $\langle\dot{\theta}\rangle \propto a_{ext}\exp(-4|a_0|/b^2)$. However, as soon as the

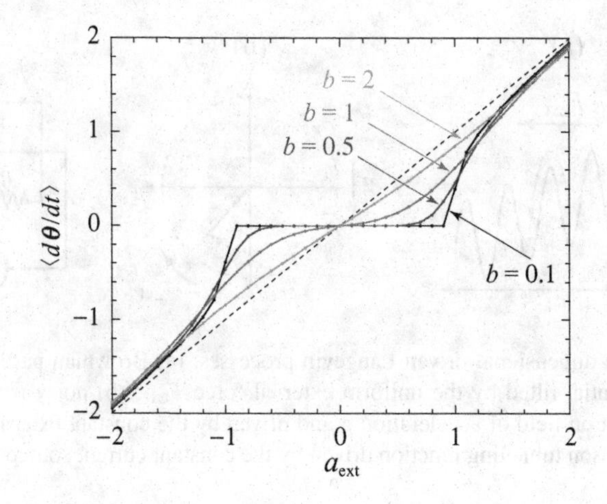

Figure 7.9 Mean drift $\langle\theta'\rangle$ versus the external drive a_{ext} in the one-dimensional Langevin process (7.58) numerically simulated for $a_0 = 1$ and the shown values of the parameter b. The mean drift is approaching the dashed line $\langle\theta'\rangle = a_{\text{ext}}$ in the limit $b \to \infty$.

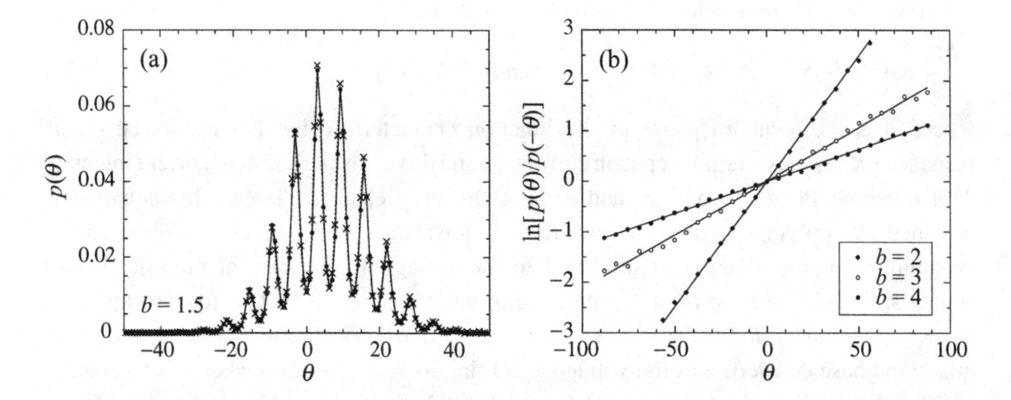

Figure 7.10 Fluctuation relation for the displacement from $\theta_0 = 0$ to θ over the rescaled time $t = 100$ in the one-dimensional Langevin process of Figure 7.9 with $a_{\text{ext}} = 0.1$: (a) the probability density $p(\theta)$ (circles) and the expectation $p(-\theta)\exp(\alpha_{\text{ext}}\theta)$ of the fluctuation relation (crosses) for $b = 1.5$; (b) the logarithm of the ratio $p(\theta)/p(-\theta)$ compared to $\alpha_{\text{ext}}\theta$ (solid lines) for $b = 2, 3, 4$.

external force exceeds the thresholds where the barriers disappear in the tilted potential, the mean drift velocity becomes significantly larger and $\langle\theta'\rangle \simeq a_{\text{ext}}$ for $|a_{\text{ext}}| \gg |a_0|$. This behavior is seen in Figure 7.9. As temperature increases (i.e., for $b^2 \gtrsim 4|a_0|$), thermal activation becomes easier and the hindering effect of the barriers is decreasing in the interval $|a_{\text{ext}}| < |a_0|$.

The fluctuation relation for the current is satisfied, as shown in Figure 7.10 in the case where the system is driven out of equilibrium by the work performed on the system with the external source. This work is given by $W = F_{\rm ext}\Delta x$, $W = \tau_{\rm ext}\Delta\theta$, and $W = I\hbar\Delta\phi/(2e)$, respectively, for the three examples.

7.5 Stochastic Motion of a Charged Particle in Electric and Magnetic Fields

7.5.1 Langevin Stochastic Equation and Kramers' Master Equation

In this section, we consider the stochastic motion of a particle with electric charge q and mass m evolving in a spatially periodic energy potential $U(\mathbf{r})$, with a uniform external magnetic field $\boldsymbol{\mathcal{B}}$ inducing the Lorentz velocity-dependent force,[3] and driven away from equilibrium by a uniform external electric field $\boldsymbol{\mathcal{E}}$. This particle is, for instance, a charged Brownian particle moving in an optical lattice or an ion in a crystal. The potential thus has the crystallographic spatial periodicity of the lattice $\mathcal{L}$: $U(\mathbf{r}) = U(\mathbf{r}+\mathbf{a})$ for all the lattice vectors $\mathbf{a} \in \mathcal{L}$.

The Langevin stochastic differential equation of motion is given by

$$m\,\ddot{\mathbf{r}} = -\boldsymbol{\nabla} U + q\,\boldsymbol{\mathcal{E}} + q\,\dot{\mathbf{r}}\times\boldsymbol{\mathcal{B}} - \gamma\,\dot{\mathbf{r}} + \mathbf{F}_{\rm fl}(t), \tag{7.60}$$

where $\mathbf{r} = (x, y, z)$ is the position of the particle, γ its friction coefficient, and $\mathbf{F}_{\rm fl}(t)$ the Langevin force (4.174) due to thermal fluctuations at the temperature T of the medium.

In the absence of external electric field $\boldsymbol{\mathcal{E}} = 0$, the particle is at equilibrium. If the electric field is switched on so that $\boldsymbol{\mathcal{E}} \neq 0$, the particle undergoes a random drift. The mechanical affinity controlling the nonequilibrium drive of this system is again given by equation (7.34) with $\mathbf{F}_{\rm ext} = q\,\boldsymbol{\mathcal{E}}$:

$$\mathbf{A} = \frac{q\,\boldsymbol{\mathcal{E}}}{k_{\rm B}T}. \tag{7.61}$$

This formulation was considered in order to study orbital diamagnetism (Jayannavar and Kumar, 1981) and the nonequilibrium work fluctuation relation for a charged Brownian particle in a time-dependent potential or magnetic field (Jayannavar and Sahoo, 2007; Saha and Jayannavar, 2008). Here, the fluctuation relation is considered for the current in steady states, so that the energy potential U as well as the external electric and magnetic fields are assumed to be time independent.

In the absence of dissipation and fluctuations, the motion of the particle would be ruled by the Hamiltonian function

$$H_{\boldsymbol{\mathcal{E}},\boldsymbol{\mathcal{B}}} = \frac{1}{2m}\left(\mathbf{p} - \frac{q}{2}\,\boldsymbol{\mathcal{B}}\times\mathbf{r}\right)^2 - q\,\boldsymbol{\mathcal{E}}\cdot\mathbf{r} + U(\mathbf{r}), \tag{7.62}$$

where

$$\mathbf{p} = m\,\dot{\mathbf{r}} + \frac{q}{2}\,\boldsymbol{\mathcal{B}}\times\mathbf{r} \tag{7.63}$$

is the momentum of the particle. Under the time-reversal transformation (7.38), the external magnetic field should also be reversed to get the following symmetry of the Hamiltonian function,

[3] In a medium with magnetic permeability μ, the magnetic and magnetizing fields are linearly related to each other as $\boldsymbol{\mathcal{B}} = \mu\boldsymbol{\mathcal{H}}$.

$$\hat{\Theta}\, H_{\mathcal{E},\mathcal{B}}\, \hat{\Theta} = H_{\mathcal{E},-\mathcal{B}}. \tag{7.64}$$

Now, Kramers' master equation, which governs the time evolution of the probability density $\mathcal{P}(\mathbf{r}, \mathbf{p}, t)$, can here be written as

$$\frac{\partial \mathcal{P}}{\partial t} = \hat{L}_{\mathbf{A},\mathcal{B}}\, \mathcal{P} \equiv \{H_{\mathcal{E},\mathcal{B}}, \mathcal{P}\} + \gamma\, \frac{\partial}{\partial \mathbf{p}} \cdot \left(\frac{\partial H_{\mathcal{E},\mathcal{B}}}{\partial \mathbf{p}}\, \mathcal{P}\right) + \gamma\, k_{\mathrm{B}} T\, \frac{\partial^2 \mathcal{P}}{\partial \mathbf{p}^2}. \tag{7.65}$$

7.5.2 Microreversibility and Multivariate Fluctuation Relation

In order to obtain the cumulant generating function (7.46) for the particle displacement $\Delta\mathbf{r}(t)$ as the eigenvalue problem (7.47), the linear operator of Kramers' equation (7.65) is modified as in equation (7.48) in order to include the counting parameters $\boldsymbol{\lambda} = (\lambda_x, \lambda_y, \lambda_z)$, which here gives

$$\hat{L}_{\mathbf{A},\mathcal{B},\boldsymbol{\lambda}}\, \Psi = \hat{L}_{\mathbf{A},\mathcal{B}}\, \Psi - \boldsymbol{\lambda} \cdot \frac{\partial H_{\mathcal{E},\mathcal{B}}}{\partial \mathbf{p}}\, \Psi \tag{7.66}$$

for an arbitrary differentiable function Ψ.

In the presence of an external magnetic field $\mathcal{B} \neq \mathbf{0}$, but zero electric field, so $\mathcal{E} = \mathbf{0}$, the linear operator (7.65) obeys the time-reversal symmetry relation

$$\mathrm{e}^{\beta H_{0,\mathcal{B}}}\, \hat{\Theta}\, \hat{L}_{\mathbf{0},-\mathcal{B}} \left(\hat{\Theta}\, \mathrm{e}^{-\beta H_{0,\mathcal{B}}}\, \Psi\right) = \hat{L}^{\dagger}_{\mathbf{0},\mathcal{B}}\, \Psi \tag{7.67}$$

with the inverse temperature $\beta = (k_{\mathrm{B}} T)^{-1}$. This identity can be verified by direct calculation. We note that the magnetic field is reversed in the left-hand side of this identity.

Now, if the affinity (7.61) is switched on, so $\mathbf{A} \neq \mathbf{0}$, the modified operator satisfies

$$\mathrm{e}^{\beta H_{0,\mathcal{B}}}\, \hat{\Theta}\, \hat{L}_{\mathbf{A},-\mathcal{B},\boldsymbol{\lambda}} \left(\hat{\Theta}\, \mathrm{e}^{-\beta H_{0,\mathcal{B}}}\, \Psi\right) = \hat{L}^{\dagger}_{\mathbf{A},\mathcal{B},\mathbf{A}-\boldsymbol{\lambda}}\, \Psi. \tag{7.68}$$

We note that the Hamiltonian function in the Boltzmann factor continues to be taken with zero electric field. The reason for this is that the Boltzmann factor defines the reference equilibrium state, so the electric field should be equal to zero.

Since equation (7.67) holds, the symmetry (7.68) can be verified by focusing on the operator

$$\hat{K}_{\mathbf{A},\mathcal{B},\boldsymbol{\lambda}}\, \Psi \equiv \hat{L}_{\mathbf{A},\mathcal{B},\boldsymbol{\lambda}}\, \Psi - \hat{L}_{\mathbf{0},\mathcal{B}}\, \Psi = -q\, \mathcal{E} \cdot \frac{\partial \Psi}{\partial \mathbf{p}} - \boldsymbol{\lambda} \cdot \frac{\partial H_{\mathcal{E},\mathcal{B}}}{\partial \mathbf{p}}\, \Psi, \tag{7.69}$$

for which we have that

$$\mathrm{e}^{\beta H_{0,\mathcal{B}}}\, \hat{\Theta}\, \hat{K}_{\mathbf{A},-\mathcal{B},\boldsymbol{\lambda}} \left(\hat{\Theta}\, \mathrm{e}^{-\beta H_{0,\mathcal{B}}}\, \Psi\right) = q\, \mathcal{E} \cdot \frac{\partial \Psi}{\partial \mathbf{p}} - (\mathbf{A} - \boldsymbol{\lambda}) \cdot \frac{\partial H_{\mathcal{E},\mathcal{B}}}{\partial \mathbf{p}}\, \Psi = \hat{K}^{\dagger}_{\mathbf{A},\mathcal{B},\mathbf{A}-\boldsymbol{\lambda}}\, \Psi, \tag{7.70}$$

because of expression (7.61) for the mechanical affinity and the identity $\partial_{\mathbf{p}}^{\dagger} = -\partial_{\mathbf{p}}$. In this way, the symmetry (7.68) of the modified operator is proved.

As a consequence, the cumulant generating function obeys the symmetry relation

$$Q(\boldsymbol{\lambda}; \mathbf{A}, \mathcal{B}) = Q(\mathbf{A} - \boldsymbol{\lambda}; \mathbf{A}, -\mathcal{B}), \tag{7.71}$$

which implies the asymptotic multivariate fluctuation relation for the particle displacement,

$$\frac{p_t(\Delta\mathbf{r};\mathbf{A},\boldsymbol{\mathcal{B}})}{p_t(-\Delta\mathbf{r};\mathbf{A},-\boldsymbol{\mathcal{B}})} \sim_{t\to\infty} \mathrm{e}^{\mathbf{A}\cdot\Delta\mathbf{r}}. \tag{7.72}$$

Furthermore, the linear and nonlinear galvanomagnetic transport properties satisfy the time-reversal symmetry relations of Section 5.5 with $\boldsymbol{\mathcal{H}} = \boldsymbol{\mathcal{B}}$, including the Onsager–Casimir reciprocal relations (5.178) as well as their generalizations (5.180)–(5.188) beyond the linear response regime (Andrieux et al., 2009; Gaspard, 2013a).

7.6 Heat Transport Driven by Thermal Reservoirs

7.6.1 Langevin Stochastic Equations and Master Equation

Heat transport can be considered in harmonic or anharmonic systems in contact with several thermal baths at different temperatures (Eckmann et al., 1999; Maes et al., 2003). In such systems, fluctuation relations can be deduced for heat exchange between two or more reservoirs (Jarzynski et al., 2004; Baiesi et al., 2006; Saito and Dhar, 2011; Agarwalla et al., 2012; Liu et al., 2013; Noh and Park, 2012; Fogedby and Imparato, 2012; Gaspard, 2013a). The control parameters of heat transport are given by the thermal affinities (1.64) defined in terms of the reservoir temperatures. The dynamics inside the system can be assumed to be Hamiltonian and the contact with the reservoirs ruled by Langevin stochastic processes.

Let the Hamiltonian system contain N particles of masses m_a, positions $\mathbf{r}_a \in \mathbb{R}^3$, and momenta $\mathbf{p}_a \in \mathbb{R}^3$ with $a = 1, 2, \ldots, N$. The total energy of the isolated system is given by the Hamiltonian function

$$H = \sum_{a=1}^{N} \frac{\mathbf{p}_a^2}{2m_a} + U(\mathbf{r}_1, \mathbf{r}_2, \ldots, \mathbf{r}_N). \tag{7.73}$$

Moreover, the particles $a = j = 1, 2, \ldots, r$ are supposed to be coupled to heat reservoirs at the temperatures $\{T_j\}_{j=1}^r$ as Brownian particles with friction coefficients $\{\gamma_j\}_{j=1}^r$. Accordingly, the Newton equations ruling their motion should include a friction force proportional to their velocity, $-\gamma_j \dot{\mathbf{r}}_j$, and a Langevin fluctuating force $\mathbf{F}_{\mathrm{fl},j}(t)$ taken as the following Gaussian white noises,

$$\langle \mathbf{F}_{\mathrm{fl},j}(t)\rangle = 0 \qquad \text{and} \qquad \langle \mathbf{F}_{\mathrm{fl},j}(t)\,\mathbf{F}_{\mathrm{fl},j'}(t')\rangle = 2\,\gamma_j\,k_{\mathrm{B}}T_j\,\delta_{jj'}\,\delta(t-t')\,\mathbf{1} \tag{7.74}$$

for $j, j' = 1, 2, \ldots, r$, where $\mathbf{1}$ is the 3×3 unit matrix. Newton's equations are thus given by

$$m_a\,\ddot{\mathbf{r}}_a = -\frac{\partial U}{\partial \mathbf{r}_a} \qquad \text{for} \quad a \neq j = 1, 2, \ldots, r \tag{7.75}$$

and

$$m_j\,\ddot{\mathbf{r}}_j = -\frac{\partial U}{\partial \mathbf{r}_j} - \gamma_j\,\dot{\mathbf{r}}_j + \mathbf{F}_{\mathrm{fl},j}(t) \qquad \text{for} \quad j = 1, 2, \ldots, r. \tag{7.76}$$

If $p = p(\mathbf{\Gamma}, t)$ denotes the probability density in the phase space $\mathbf{\Gamma} = (\mathbf{r}_a, \mathbf{p}_a)_{a=1}^N \in \mathbb{R}^{6N}$ of the system, the Fokker–Planck equation ruling the Langevin stochastic process (7.75)–(7.76) is given by

$$\partial_t p = \{H, p\} + \sum_{j=1}^{r} \hat{L}_j p \equiv \hat{L}_0 p, \tag{7.77}$$

where $\{\cdot,\cdot\}$ is the Poisson bracket (2.14) and

$$\hat{L}_j p = \gamma_j \frac{\partial}{\partial \mathbf{p}_j} \cdot \left(\frac{\mathbf{p}_j}{m_j} p \right) + \gamma_j k_{\mathrm{B}} T_j \frac{\partial^2 p}{\partial \mathbf{p}_j^2} \tag{7.78}$$

are the operators ruling the Langevin stochastic processes of heat exchange with the reservoirs.

7.6.2 Microreversibility and Multivariate Fluctuation Relation

The Fokker–Planck operator (7.77) can be modified to include the counting parameters $\boldsymbol{\lambda} = \{\lambda_j\}_{j=1}^{r-1}$ of the heat transfers from the heat reservoirs $1 \le j \le r-1$ to the Hamiltonian system and the reference reservoir $j = r$. For this purpose, we use the path-integral formulation. Extending equation (7.37), the probability of the path $\mathbf{R}(t) = \{\mathbf{r}_a(t)\}_{a=1}^N$ can be expressed as

$$P\left[\mathbf{R}(t)\right] \propto \exp\left[-\sum_{j=1}^{r} \frac{1}{4\gamma_j k_{\mathrm{B}} T_j} \int_0^t \left(m_j \ddot{\mathbf{r}}_j + \frac{\partial U}{\partial \mathbf{r}_j} + \gamma_j \dot{\mathbf{r}}_j \right)^2 dt' \right]. \tag{7.79}$$

In principle, this path probability includes the Liouvillian contribution from the Hamiltonian system, which is deterministic and preserves phase-space volumes. The probability weight given to the paths by equation (7.79) corresponds to the operators (7.78) in the Fokker–Planck equation (7.77).

Now, we consider the moment generating function of the heat currents, which is defined as

$$\left\langle \mathrm{e}^{-\boldsymbol{\lambda}\cdot\mathbf{Q}(t)} \right\rangle = \int \mathrm{e}^{-\boldsymbol{\lambda}\cdot\mathbf{Q}(t)}\, P\left[\mathbf{R}(t)\right]\, \mathcal{D}\mathbf{R}(t), \tag{7.80}$$

where $\mathbf{Q} = \{Q_j\}_{j=1}^{r-1}$ denote the heat transfers from the reservoirs to the Hamiltonian system during the time interval $[0, t]$. Using Stratonovich's stochastic integration, the heat transfers are given by the following integrals of the instantaneous heat currents (Maes et al., 2003),

$$Q_j(t) = \int_0^t \dot{\mathbf{r}}_j \circ \left[-\gamma_j \dot{\mathbf{r}}_j + \mathbf{F}_{\mathrm{fl},j}(t') \right] dt', \tag{7.81}$$

which are equivalent to

$$Q_j(t) = \int_0^t \left[\dot{\mathbf{r}}_j \cdot \left(m_j \ddot{\mathbf{r}}_j + \frac{\partial U}{\partial \mathbf{r}_j} \right) + 3\, k_{\mathrm{B}} T_j \frac{\gamma_j}{m_j} \right] dt'. \tag{7.82}$$

Combining with the path probability (7.79) and setting $\lambda_r = 0$, we get

$$
\begin{aligned}
&P\,[\mathbf{R}(t)]\,\mathrm{e}^{-\boldsymbol{\lambda}\cdot\mathbf{Q}(t)} \\
&\quad \propto \exp\Bigg\{ -\sum_{j=1}^{r} \frac{1}{4\gamma_j k_{\mathrm{B}} T_j} \int_0^t dt' \left[m_j \ddot{\mathbf{r}}_j + \frac{\partial U}{\partial \mathbf{r}_j} + \gamma_j \dot{\mathbf{r}}_j \,(1 + 2\,\lambda_j k_{\mathrm{B}} T_j) \right]^2 \\
&\qquad + \sum_{j=1}^{r-1} \int_0^t dt' \left[\lambda_j (1 + \lambda_j k_{\mathrm{B}} T_j)\, \gamma_j \dot{\mathbf{r}}_j^2 - 3\,\lambda_j k_{\mathrm{B}} T_j \frac{\gamma_j}{m_j} \right] \Bigg\}.
\end{aligned}
\tag{7.83}
$$

Provided that the path probability (7.79) corresponds to the Fokker–Planck operator (7.77), the modified path probability (7.83) should thus correspond to the following operator:

$$
\hat{L}_{\boldsymbol{\lambda}} = \hat{L}_0 + \sum_{j=1}^{r-1} \hat{K}_{j,\lambda_j}, \tag{7.84}
$$

where $\hat{L}_0$ is the Fokker–Planck operator (7.77) and

$$
\hat{K}_{j,\lambda_j} \equiv 2\,\lambda_j\, k_{\mathrm{B}} T_j\, \gamma_j \frac{\mathbf{p}_j}{m_j} \cdot \frac{\partial}{\partial \mathbf{p}_j} + \lambda_j (1 + \lambda_j k_{\mathrm{B}} T_j)\, \gamma_j \frac{\mathbf{p}_j^2}{m_j^2} + 3\,\lambda_j k_{\mathrm{B}} T_j \frac{\gamma_j}{m_j}. \tag{7.85}
$$

In equation (7.84), the sum is restricted to $r-1$ terms because the reference reservoir $j = r$ does not need a counting parameter. Moreover, the identity $\frac{\partial}{\partial \mathbf{p}_j} \cdot (\mathbf{p}_j\, p) = \mathbf{p}_j \cdot \frac{\partial p}{\partial \mathbf{p}_j} + 3\,p$ has been used, which explains that the last term of equation (7.85) comes with a plus sign in place of the minus sign in the last term of equation (7.83). Using equation (7.77), the modified operator (7.84) can thus be written in the form

$$
\hat{L}_{\boldsymbol{\lambda}} = \{H, \cdot\} + \sum_{j=1}^{r} \hat{L}_j + \sum_{j=1}^{r-1} \hat{K}_{j,\lambda_j} \tag{7.86}
$$

with the operators (7.85). This modified operator obeys the symmetry relation

$$
\mathrm{e}^{\beta_r H} \hat{\Theta}\, \hat{L}_{\boldsymbol{\lambda}} \left(\hat{\Theta}\, \mathrm{e}^{-\beta_r H}\, \Psi \right) = \hat{L}^{\dagger}_{\mathbf{A}-\boldsymbol{\lambda}} \Psi, \tag{7.87}
$$

where $\hat{\Theta}\Psi(\mathbf{R}, \mathbf{P}) = \Psi(\mathbf{R}, -\mathbf{P})$ is the time-reversal operator and $\beta_r = (k_{\mathrm{B}} T_r)^{-1}$ the inverse temperature of the reference reservoir. This symmetry relation is proved as follows. First, the Hamiltonian part of the modified operator (7.86) satisfies the symmetry

$$
\mathrm{e}^{\beta_r H} \hat{\Theta} \left\{ H, \hat{\Theta}\, \mathrm{e}^{-\beta_r H}\, \Psi \right\} = -\{H, \Psi\} \tag{7.88}
$$

since the Liouvillian operator $\{H, \cdot\}$ is anti-Hermitian, the Hamiltonian function is even under time reversal while the Poisson bracket is odd and, furthermore, the Boltzmann factor $\mathrm{e}^{-\beta_r H}$ has a vanishing Poisson bracket with the Hamiltonian function since it is a constant of Hamiltonian motion. Secondly, the operator (7.78) with $j = r$ satisfies the symmetry

$$
\mathrm{e}^{\beta_r H} \hat{\Theta}\, \hat{L}_r \left(\hat{\Theta}\, \mathrm{e}^{-\beta_r H}\, \Psi \right) = \hat{L}_r^{\dagger} \Psi \tag{7.89}
$$

because the Hamiltonian function depends on the momentum $\mathbf{p}_r$ as $H = \mathbf{p}_r^2/(2m_r) + \cdots$. Thirdly, the following identity is satisfied,

$$e^{\beta_r H}\hat{\Theta}\left(\hat{L}_j + \hat{K}_{j,\lambda_j}\right)\left(\hat{\Theta}\, e^{-\beta_r H}\Psi\right) = \hat{L}_j^\dagger \Psi + \hat{K}_{j,A_j-\lambda_j}^\dagger \Psi \tag{7.90}$$

with the thermal affinity

$$A_j = \beta_r - \beta_j = \frac{1}{k_{\rm B} T_r} - \frac{1}{k_{\rm B} T_j} \tag{7.91}$$

for every $j = 1, 2, \ldots, r-1$. Hence, the symmetry relation (7.87) holds for the modified operator. As a consequence, its leading eigenvalue giving the cumulant generating function for the heat currents $\mathfrak{J} = \mathbf{Q}/t$ obeys the same symmetry relation as (7.50), whereupon the asymptotic multivariate fluctuation relation (5.242) is proved for heat transport. Therefore, the linear and nonlinear heat transport properties satisfy all the symmetry relations of Section 5.5.

Nonlinear transport properties are important for thermal rectifiers (Chang et al., 2006; Li et al., 2012). Such thermal diodes are connected to two thermal reservoirs. There is thus a single heat current $J(A)$, which is a nonlinear function of the single thermal affinity $A = (k_{\rm B}T_2)^{-1} - (k_{\rm B}T_1)^{-1}$. Thermal rectification is characterized by the ratio (5.170).

8

Effusion Processes

8.1 The Kinetic Process of Effusion

Effusion is the collisionless flow of particles through a small hole in a thin wall between two particle reservoirs at different chemical potentials or temperatures (Knudsen, 1909; Present, 1958; Pathria, 1972; Balian, 1991). Kinetic energy is carried by the particles, so that the effusive flow also generates energy exchange between the reservoirs. The conditions for effusion are met if the mean free path is much larger than the diameter of the hole and if this latter is much larger than the thickness of the wall. Accordingly, the particles pass through the hole in free flight. In the effusive flow, the gas can thus be considered as ideal, in which case the mean free path of the particles is infinite.

Effusion is probably the most elementary of irreversible phenomena. Its nonequilibrium properties can be exactly deduced from statistical mechanics. The stationary probability distribution of the effusion process can be rigorously constructed as a so-called Poisson suspension, illustrating the time-reversal symmetry breaking at the statistical level of description under nonequilibrium conditions, as discussed in Section 2.6. The master equation and the multivariate fluctuation relation for energy and particle currents can be proved on the basis of microscopic dynamics, implying the time-reversal symmetry relations of Section 5.5 for the linear and nonlinear response properties of effusion. The expression of the thermodynamic entropy production rate can thus be consistently established from statistical mechanics. Furthermore, the properties of mass separation by effusion can be obtained using the same methods.

8.2 Stationary Distribution Function

Figure 8.1 schematically represents the effusion of an ideal gas through a small hole of area σ in an arbitrarily thin wall separating the gas into two reservoirs at different densities and/or temperatures. The ideal gas is composed of independent monatomic particles of mass m moving in free flight and eventually interrupted by an elastic collision on the thin wall W. The free flights are the solutions of Hamilton's equations for the particle position $\mathbf{r} \in \mathbb{R}^3$ and momentum $\mathbf{p} \in \mathbb{R}^3$ with the Hamiltonian function given by $H = \mathbf{p}^2/(2m)$. Moreover, the elastic collision on the thin wall W reflects the x-component of the momentum according to the collision rule of specular reflection on the wall W. The motion of

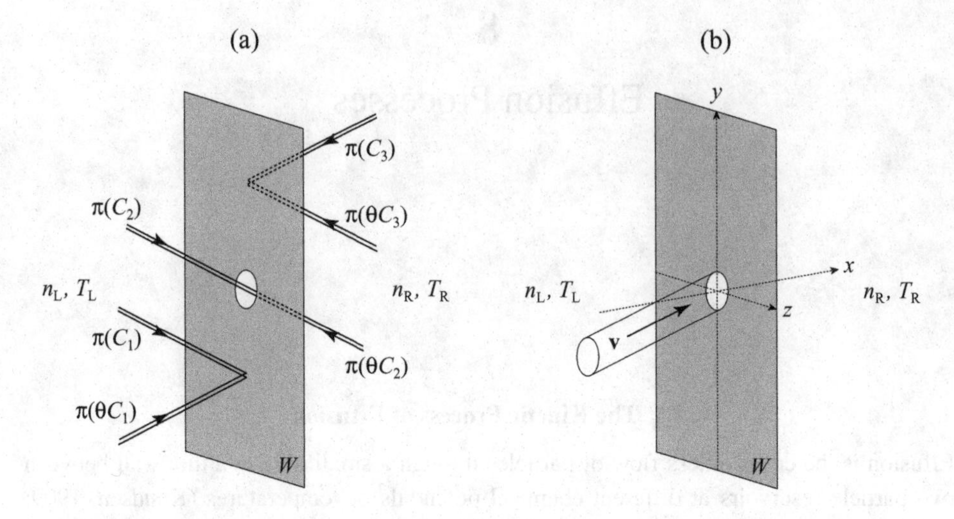

Figure 8.1 (a) Schematic representation of an effusion process in the physical position space $\mathbf{r} = (x, y, z) \in (\mathbb{R}^3 \setminus W)$ where the wall W separates the position space between the two reservoirs where the gas of noninteracting particles has different densities and temperatures. The wall W has a hole of area σ. The figure depicts the projection $\pi(\mathbf{r}, \mathbf{p}) = \mathbf{r}$ for the three types of phase-space trajectories C_i and their time reversal ΘC_i (with $i = 1, 2, 3$). (b) Representation of the cylindrical volume containing the particles with the velocity $\mathbf{v} = \mathbf{p}/m$ crossing the boundary between both reservoirs through the hole during some time interval $[0, t]$.

every particle is thus ruled by the corresponding Hamiltonian flow $\boldsymbol{\Phi}^t$ in the one-particle phase space $\mathfrak{M} = \{(\mathbf{r}, \mathbf{p}) \in (\mathbb{R}^3 \setminus W) \otimes \mathbb{R}^3\}$. This one-particle Hamiltonian flow is symplectic, obeys Liouville's theorem, and is symmetric under the time-reversal transformation $\Theta(\mathbf{r}, \mathbf{p}) = (\mathbf{r}, -\mathbf{p})$, thus satisfying microreversibility: $\Theta\, \boldsymbol{\Phi}^t\, \Theta = \boldsymbol{\Phi}^{-t}$.

In the left-hand reservoir, particles are coming from $x = -\infty$ with velocities distributed according to a Maxwellian distribution at the temperature $T_{\rm L}$ and the density $n_{\rm L}$. In the right-hand reservoir, other particles arrive from $x = +\infty$ with velocities distributed according to a Maxwellian distribution at the temperature $T_{\rm R}$ and the density $n_{\rm R}$. The one-particle distribution function thus has the Maxwellian form:

$$f(\mathbf{r}, \mathbf{p}) = \frac{n_j}{(2\pi m k_{\rm B} T_j)^{3/2}} \exp\left(-\frac{\mathbf{p}^2}{2 m k_{\rm B} T_j}\right), \tag{8.1}$$

where the particle density n_j and the temperature T_j take the values associated with the incoming reservoir $j = {\rm L}$ or $j = {\rm R}$ of the trajectory $C = \{\boldsymbol{\Phi}^t(\mathbf{r}, \mathbf{p}), t \in \mathbb{R}\}$ issued from the phase-space point $(\mathbf{r}, \mathbf{p})$. As represented in Figure 8.1(a), there are three types of possible trajectories: (1) The trajectories $\{C_1\}$ and their reversals $\{\Theta C_1\}$ undergoing a specular reflection on the left-hand side of the wall. (2) The trajectories $\{C_2\}$ and their reversals $\{\Theta C_2\}$ crossing the wall through its small hole. (3) The trajectories $\{C_3\}$ and their reversals $\{\Theta C_3\}$ undergoing a specular reflection on the right-hand side of the wall. The key point is that the trajectories $\{C_2\}$ crossing the hole move towards a reservoir at a different density and/or temperature than where they come from if the system is out of equilibrium. Therefore, the one-particle distribution function (8.1) may take different values for the trajectories $\{C_2\}$ than for their time reversals $\{\Theta C_2\}$. Accordingly, we have that

$$n_j = n_{\rm L}, \quad T_j = T_{\rm L} \qquad \text{for} \qquad (\mathbf{r},\mathbf{p}) \in C_1,\ \Theta C_1,\ C_2; \tag{8.2}$$

$$n_j = n_{\rm R}, \quad T_j = T_{\rm R} \qquad \text{for} \qquad (\mathbf{r},\mathbf{p}) \in C_3,\ \Theta C_3,\ \Theta C_2. \tag{8.3}$$

The distribution function is stationary with respect to the Hamiltonian flow, so that $f\left[\boldsymbol{\phi}^t(\mathbf{r},\mathbf{p})\right] = f(\mathbf{r},\mathbf{p})$. Nevertheless, it is not symmetric under time reversal, i.e., $f[\Theta(\mathbf{r},\mathbf{p})] \neq f(\mathbf{r},\mathbf{p})$, if the reservoirs have different densities or temperatures: $n_{\rm L} \neq n_{\rm R}$ or $T_{\rm L} \neq T_{\rm R}$. For monatomic particles, the chemical potentials are related to the densities by $\mu_j = k_{\rm B} T_j \ln\left(a n_j / T_j^{3/2}\right)$ with $j = {\rm L}$ or $j = {\rm R}$, and some constant a.

8.3 Poisson Suspension

Because of the infinite spatial extension of the reservoirs, the ideal gas contains an infinite number of particles and its microstate is given by the point

$$\boldsymbol{\Gamma} = (\mathbf{r}_1, \mathbf{p}_1, \mathbf{r}_2, \mathbf{p}_2, \ldots, \mathbf{r}_N, \mathbf{p}_N, \ldots) \in \mathcal{M} = \mathfrak{M}^{\infty}. \tag{8.4}$$

In this infinite-particle system, the Hamiltonian flow defined by $\boldsymbol{\Phi}^t \boldsymbol{\Gamma} = \left(\boldsymbol{\phi}^t \mathbf{x}_1, \boldsymbol{\phi}^t \mathbf{x}_2, \ldots, \boldsymbol{\phi}^t \mathbf{x}_N, \ldots\right)$ also satisfies microreversibility:

$$\boldsymbol{\Theta}\, \boldsymbol{\Phi}^t\, \boldsymbol{\Theta} = \boldsymbol{\Phi}^{-t}, \tag{8.5}$$

where $\boldsymbol{\Theta}$ is the time-reversal transformation acting on the microstate (8.4) by reversing all the momenta: $\mathbf{p}_a \to -\mathbf{p}_a$ for $a = 1, 2, 3, \ldots, N, \ldots$.

For the infinite-particle system, a stationary probability distribution P can be constructed as a Poisson suspension on the basis of the stationary distribution function (8.1) (Cornfeld et al., 1982). For any six-dimensional domain $\mathfrak{D} \subset \mathfrak{M}$ of the one-particle phase space, we consider the random events

$$\mathcal{A}_{\mathfrak{D},N} = \{\boldsymbol{\Gamma} \in \mathcal{M} : \text{number}(\boldsymbol{\Gamma} \cap \mathfrak{D}) = N\}, \tag{8.6}$$

for which there are N particles in the domain $\mathfrak{D}$. The mean number of particles in the domain $\mathfrak{D}$ is given by

$$\nu(\mathfrak{D}) = \int_{\mathfrak{D}} f(\mathbf{r},\mathbf{p})\, d\mathbf{r}\, d\mathbf{p} \tag{8.7}$$

in terms of the one-particle distribution function (8.1). The number N of particles in the domain $\mathfrak{D}$ is supposed to be a random variable of Poisson distribution:[1]

$$P\left(\mathcal{A}_{\mathfrak{D},N}\right) = \frac{\nu(\mathfrak{D})^N}{N!}\, {\rm e}^{-\nu(\mathfrak{D})}. \tag{8.8}$$

[1] We note that the Poisson distribution (C.16) for the particle number of an ideal gas in a volume V and described by the grand canonical ensemble is recovered if the domain $\mathfrak{D}$ is corresponding to the volume V and any value for the momentum $\mathbf{p} \in \mathbb{R}^3$.

Furthermore, random events with given particle numbers in disjoint domains are assumed to be statistically independent, so that

$$P\left(\mathcal{A}_{\mathfrak{D}_1,N_1} \cap \mathcal{A}_{\mathfrak{D}_2,N_2}\right) = P\left(\mathcal{A}_{\mathfrak{D}_1,N_1}\right) P\left(\mathcal{A}_{\mathfrak{D}_2,N_2}\right) \quad \text{if} \quad \mathfrak{D}_1 \cap \mathfrak{D}_2 = \emptyset. \tag{8.9}$$

Both equations (8.8) and (8.9) define the probability distribution of the so-called Poisson suspension (Cornfeld et al., 1982). We have the

Lemma: *The probability to find N particles in a domain $\mathfrak{D} = \mathfrak{D}_1 \cup \mathfrak{D}_2$ composed of two disjoint domains $\mathfrak{D}_1 \cap \mathfrak{D}_2 = \emptyset$ is again a Poisson distribution with an average equal to $\nu(\mathfrak{D}) = \nu(\mathfrak{D}_1) + \nu(\mathfrak{D}_2)$.*

Indeed, since the domains $\mathfrak{D}_1$ and $\mathfrak{D}_2$ are disjoint, the number of particles in the domain $\mathfrak{D} = \mathfrak{D}_1 \cup \mathfrak{D}_2$ is equal to $N = N_1 + N_2$, where N_1 is the number of particles in the domain $\mathfrak{D}_1$ and N_2 the number in $\mathfrak{D}_2$. Both N_1 and N_2 are distributed according to Poisson distributions of averages $\nu(\mathfrak{D}_1)$ and $\nu(\mathfrak{D}_2)$ and the statistical independence (8.9) implies that

$$P\left(\mathcal{A}_{\mathfrak{D},N}\right) = \sum_{N_1=0}^{N} P\left(\mathcal{A}_{\mathfrak{D}_1,N_1} \cap \mathcal{A}_{\mathfrak{D}_2,N-N_1}\right) = \frac{\nu(\mathfrak{D})^N}{N!}\, \mathrm{e}^{-\nu(\mathfrak{D})}$$

with $\nu(\mathfrak{D}) = \nu(\mathfrak{D}_1) + \nu(\mathfrak{D}_2)$, thus proving the lemma.

As a consequence, the distribution P is normalized to unity. Indeed, considering a countable partition of the one-particle phase space into disjoint domains, $\mathfrak{M} = \cup_i \mathfrak{D}_i$ with $\mathfrak{D}_i \cap \mathfrak{D}_j = \emptyset$ for $i \neq j$, the total probability of all the possible corresponding events is thus given by

$$\sum_{\{N_i\}} P(\cap_i \mathcal{A}_{\mathfrak{D}_i,N_i}) = \prod_i \sum_{N_i=0}^{\infty} P(\mathcal{A}_{\mathfrak{D}_i,N_i}) = \prod_i \sum_{N_i=0}^{\infty} \frac{\nu(\mathfrak{D}_i)^{N_i}}{N_i!}\, \mathrm{e}^{-\nu(\mathfrak{D}_i)} = 1. \tag{8.10}$$

Therefore, the Poisson suspension defines a probability distribution.

Moreover, this probability distribution is stationary for the time evolution of the Hamiltonian flow $\mathbf{\Phi}^t$ of the infinite system:

$$P\left(\mathbf{\Phi}^t \mathcal{A}\right) = P\left(\mathcal{A}\right) \tag{8.11}$$

for any random event $\mathcal{A}$. This is the consequence of the stationarity of the one-particle distribution function (8.1), $f\left[\boldsymbol{\phi}^t(\mathbf{r},\mathbf{p})\right] = f(\mathbf{r},\mathbf{p})$, which implies the invariance $\nu\left(\boldsymbol{\phi}^t\mathfrak{D}\right) = \nu\left(\mathfrak{D}\right)$ of the distribution (8.7).

The flow $\mathbf{\Phi}^t$ in the infinite phase space $\mathcal{M}$ and the stationary probability distribution of the Poisson suspension defines the infinite-particle dynamical system $\left(\mathbf{\Phi}^t, \mathcal{M}, P\right)$. Although the flow $\mathbf{\Phi}^t$ has the microreversibility (8.5), the dynamical system is not symmetric under time reversal because

$$P\left(\mathbf{\Theta}\mathcal{A}\right) \neq P\left(\mathcal{A}\right) \quad \text{if} \quad T_{\mathrm{L}} \neq T_{\mathrm{R}} \quad \text{or} \quad n_{\mathrm{L}} \neq n_{\mathrm{R}}. \tag{8.12}$$

The time-reversal symmetry is thus broken at the statistical level of description in terms of the stationary probability distribution P under nonequilibrium conditions, as described in Table 2.2.

8.4 Energy and Particle Fluxes

Remarkably, the statistics of the energy and particle currents between the two reservoirs can be directly established in order to obtain their mean values. The random exchanges of energy and particles from the left-hand to the right-hand reservoirs during the time interval $[0,t]$ are defined as

$$\Delta E = \int \varepsilon \, dN_{\varepsilon+} - \int \varepsilon \, dN_{\varepsilon-}, \tag{8.13}$$

$$\Delta N = \int dN_{\varepsilon+} - \int dN_{\varepsilon-}, \tag{8.14}$$

where $dN_{\varepsilon+}$ (respectively, $dN_{\varepsilon-}$) denotes the random number of particles with a positive (respectively, negative) x-component v_x of their velocity $\mathbf{v} = \mathbf{p}/m$, a kinetic energy such that $\frac{1}{2}m\mathbf{v}^2 \in [\varepsilon, \varepsilon + d\varepsilon]$, and going through the hole in the wall during the time interval $[0,t]$. In the one-particle phase space, all the particles contributing to the number $dN_{\varepsilon+}$ belong to the domain $\mathfrak{D}_{\varepsilon+}$, in which the particles with a given momentum $\mathbf{p} = m\dot{\mathbf{r}}$ are contained in the cylinder shown in Figure 8.1(b). Since the basis of this cylinder is the hole of area σ and its height is $v_x t$, its volume is equal to $\sigma v_x t$. Similar considerations hold for the particles in the domain $\mathfrak{D}_{\varepsilon-}$ and contributing to the number $dN_{\varepsilon-}$. According to the distribution function (8.1), the mean numbers of these particles are thus given by

$$\nu_{\varepsilon+} = \nu(\mathfrak{D}_{\varepsilon+}) = \int_{\mathfrak{D}_{\varepsilon+}} f(\mathbf{r},\mathbf{p}) \, d\mathbf{r} \, d\mathbf{p} = t \, w_{\mathrm{L}}(\varepsilon) \, d\varepsilon, \tag{8.15}$$

$$\nu_{\varepsilon-} = \nu(\mathfrak{D}_{\varepsilon-}) = \int_{\mathfrak{D}_{\varepsilon-}} f(\mathbf{r},\mathbf{p}) \, d\mathbf{r} \, d\mathbf{p} = t \, w_{\mathrm{R}}(\varepsilon) \, d\varepsilon, \tag{8.16}$$

with the rate densities

$$w_j(\varepsilon) = \frac{\sigma \, n_j}{\sqrt{2\pi m k_{\mathrm{B}} T_j}} \, \frac{\varepsilon}{k_{\mathrm{B}} T_j} \exp\left(-\frac{\varepsilon}{k_{\mathrm{B}} T_j}\right) \qquad \text{for} \qquad j = \mathrm{L}, \mathrm{R}. \tag{8.17}$$

The ratio of these mean numbers satisfies the relation

$$\frac{\nu_{\varepsilon+}}{\nu_{\varepsilon-}} = \frac{w_{\mathrm{L}}(\varepsilon)}{w_{\mathrm{R}}(\varepsilon)} = \mathrm{e}^{\varepsilon A_E + A_N} \tag{8.18}$$

in terms of the thermal and chemical affinities of the effusion process:

$$A_E = \frac{1}{k_{\mathrm{B}} T_{\mathrm{R}}} - \frac{1}{k_{\mathrm{B}} T_{\mathrm{L}}}, \tag{8.19}$$

$$A_N = \frac{\mu_{\mathrm{L}}}{k_{\mathrm{B}} T_{\mathrm{L}}} - \frac{\mu_{\mathrm{R}}}{k_{\mathrm{B}} T_{\mathrm{R}}} = \ln\left[\frac{n_{\mathrm{L}}}{n_{\mathrm{R}}} \left(\frac{T_{\mathrm{R}}}{T_{\mathrm{L}}}\right)^{3/2}\right]. \tag{8.20}$$

Now, the mean values of the net energy and particle currents flowing from the left- to the right-hand reservoir can be expressed as

$$J_E = \lim_{t\to\infty} \frac{1}{t} \langle \Delta E \rangle_t = \int_0^\infty d\varepsilon \, \varepsilon \, [w_{\mathrm{L}}(\varepsilon) - w_{\mathrm{R}}(\varepsilon)], \tag{8.21}$$

$$J_N = \lim_{t\to\infty} \frac{1}{t} \langle \Delta N \rangle_t = \int_0^\infty d\varepsilon \, [w_{\mathrm{L}}(\varepsilon) - w_{\mathrm{R}}(\varepsilon)]. \tag{8.22}$$

Integrating over the energy ε, we find

$$J_E = 2\,\sigma\, n_{\rm L}\sqrt{\frac{(k_{\rm B}T_{\rm L})^3}{2\pi m}} - 2\,\sigma\, n_{\rm R}\sqrt{\frac{(k_{\rm B}T_{\rm R})^3}{2\pi m}}, \tag{8.23}$$

$$J_N = \sigma\, n_{\rm L}\sqrt{\frac{k_{\rm B}T_{\rm L}}{2\pi m}} - \sigma\, n_{\rm R}\sqrt{\frac{k_{\rm B}T_{\rm R}}{2\pi m}}. \tag{8.24}$$

We thus recover the known expression for the net effusion rate

$$J_N = \frac{1}{4}\,\sigma\, n_{\rm L}\,\langle v\rangle_{\rm L} - \frac{1}{4}\,\sigma\, n_{\rm R}\,\langle v\rangle_{\rm R}, \tag{8.25}$$

where

$$\langle v\rangle_j = \langle\|\mathbf{v}\|\rangle_j = \sqrt{\frac{8k_{\rm B}T_j}{\pi m}} \tag{8.26}$$

is the mean value of the particle speed (Knudsen, 1909; Present, 1958; Pathria, 1972; Balian, 1991).

Similar considerations also apply if there is an energy barrier $\varepsilon_{\rm a}$ in the hole such that the particles can only cross the hole if their kinetic energy is higher than the energy barrier. In this case, the net effusion rate is given by

$$J_N = \frac{\sigma\, n_{\rm L}(\varepsilon_{\rm a}+k_{\rm B}T_{\rm L})}{\sqrt{2\pi m k_{\rm B}T_{\rm L}}}\exp\left(-\frac{\varepsilon_{\rm a}}{k_{\rm B}T_{\rm L}}\right) - \frac{\sigma\, n_{\rm R}(\varepsilon_{\rm a}+k_{\rm B}T_{\rm R})}{\sqrt{2\pi m k_{\rm B}T_{\rm R}}}\exp\left(-\frac{\varepsilon_{\rm a}}{k_{\rm B}T_{\rm R}}\right), \tag{8.27}$$

which involves the Boltzmann factors $\exp(-\beta_j\varepsilon_{\rm a})$ with the inverse temperature $\beta_j = (k_{\rm B}T_j)^{-1}$ in each reservoir. We thus obtain *Arrhenius' law*, according to which thermally activated fluxes above some energy barrier are proportional to the probability for the particle energy to reach the *activation energy* $\varepsilon_{\rm a}$.

8.5 Entropy Production of Effusion

For the system described by the Poisson suspension, the coarse-grained entropy of some domain partitioned into small cells $\mathfrak{D} = \cup_i \mathfrak{C}_i$ can be evaluated as follows using equation (2.112). Here, the cells of the infinite-particle phase space $\mathcal{M} = \mathfrak{M}^\infty$ are given by the events $\mathcal{A}_{\mathfrak{C}_i,N_i} \subset \mathcal{M}$ corresponding to the cells $\mathfrak{C}_i \subset \mathfrak{M}$ of the one-particle phase space $\mathfrak{M}$. The cells $\mathfrak{C}_i$ are supposed to be small enough that they contain on average less than one particle, $\nu(\mathfrak{C}_i) \ll 1$. In general, the entropy of the gas inside the domain $\mathfrak{D}$ coarse grained with the partition $\{\mathfrak{C}_i\}$ is given by

$$S(\mathfrak{D}|\{\mathfrak{C}_i\}) = -k_{\rm B}\sum_{\{N_i\}} P(\mathcal{A}_{\mathfrak{C}_i,N_i})\ln P(\mathcal{A}_{\mathfrak{C}_i,N_i}) \tag{8.28}$$

according to equation (2.129) (Gaspard, 1997, 1998; Dorfman et al., 2002). Since the cells are small enough to contain less than one particle on average, the sum can be calculated with

$$P(\mathcal{A}_{\mathfrak{C}_i,0}) = {\rm e}^{-\nu(\mathfrak{C}_i)} \qquad \text{and} \qquad P(\mathcal{A}_{\mathfrak{C}_i,1}) = \nu(\mathfrak{C}_i)\,{\rm e}^{-\nu(\mathfrak{C}_i)}, \tag{8.29}$$

giving the expression

$$S(\mathfrak{D}|\{\mathfrak{C}_i\}) \simeq k_{\rm B} \sum_{\mathfrak{C}_i \subset \mathfrak{D}} \nu(\mathfrak{C}_i) \ln \frac{\rm e}{\nu(\mathfrak{C}_i)}, \tag{8.30}$$

up to a negligible term vanishing as $\sum_i O\left[\nu(\mathfrak{C}_i)^2 \ln \nu(\mathfrak{C}_i)\right]$ for $\nu(\mathfrak{C}_i) \ll 1$. The expression (8.30) can now be expressed in terms of the one-particle distribution function. If all the cells have the same volume $\Delta\mathbf{x} = \Delta^3 r \Delta^3 p$, the mean particle numbers in the cells of the partition take the values

$$\nu(\mathfrak{C}_i) = \int_{\mathfrak{C}_i} f(\mathbf{x})\, d\mathbf{x} \simeq f(\mathbf{x}_i)\, \Delta\mathbf{x} \tag{8.31}$$

for some point $\mathbf{x}_i = (\mathbf{r}_i, \mathbf{p}_i) \in \mathfrak{C}_i \subset \mathfrak{M}$. Therefore, the entropy is obtained as

$$S(\mathfrak{D}|\{\mathfrak{C}_i\}) \simeq k_{\rm B} \int_{\mathfrak{D}} d\mathbf{x}\, f(\mathbf{x}) \ln \frac{\rm e}{f(\mathbf{x})\Delta\mathbf{x}} + O(\Delta\mathbf{x}) \tag{8.32}$$

in the limit $\Delta\mathbf{x} \to 0$.

The domain $\mathfrak{D}$ is assumed to extend over the left- and right-hand reservoirs $\mathfrak{D}_{\rm L}$ and $\mathfrak{D}_{\rm R}$ of fixed volumes $V_{\rm L}$ and $V_{\rm R}$. These volumes are supposed to be large enough that the effusion process through the small hole of area σ does not significantly change the thermodynamic conditions in the reservoirs over the timescale of observation. Under such circumstances, the thermodynamic entropy of the system takes the value

$$S = S_{\rm L}(E_{\rm L}, N_{\rm L}) + S_{\rm R}(E_{\rm R}, N_{\rm R}), \tag{8.33}$$

where the entropies of the left- and right-hand reservoirs are given by the Sackur–Tetrode formula (2.115) with the thermal energies in the reservoirs replaced by $k_{\rm B}T_j = 2E_j/(3N_j)$ for $j = {\rm L}$ and $j = {\rm R}$. Since the domain $\mathfrak{D} = \mathfrak{D}_{\rm L} \cup \mathfrak{D}_{\rm R}$ extends over the whole system, this latter is isolated, so that there is no entropy exchange $d_{\rm e}S/dt = 0$. Therefore, the time derivative of the entropy (8.33)

$$\frac{dS}{dt} = \frac{\partial S}{\partial E_{\rm L}}\,\dot{E}_{\rm L} + \frac{\partial S}{\partial N_{\rm L}}\,\dot{N}_{\rm L} + \frac{\partial S}{\partial E_{\rm R}}\,\dot{E}_{\rm R} + \frac{\partial S}{\partial N_{\rm R}}\,\dot{N}_{\rm R} \tag{8.34}$$

gives the thermodynamic entropy production rate $d_{\rm i}S/dt = \langle dS/dt\rangle$ after statistical averaging in the stationary state. According to the conservation of energy and particles, the transfer rates are related to the net energy and particle currents (8.23) and (8.24) by

$$J_E = \langle \dot{E}_{\rm R}\rangle = -\langle \dot{E}_{\rm L}\rangle \qquad \text{and} \qquad J_N = \langle \dot{N}_{\rm R}\rangle = -\langle \dot{N}_{\rm L}\rangle. \tag{8.35}$$

Using the Gibbs thermodynamic relation (1.12), the entropy production rate (8.34) thus becomes

$$\frac{1}{k_{\rm B}}\frac{d_{\rm i}S}{dt} = A_E\, J_E + A_N\, J_N \geq 0 \tag{8.36}$$

in terms of the affinities (8.19)–(8.20) and the mean currents (8.23)–(8.24). In accordance with the second law of thermodynamics, this expression can be confirmed to always be nonnegative and to vanish at equilibrium if the temperature and the particle density are uniform. In this way, the expression (1.63) is deduced for the thermodynamic entropy production rate of the effusion process.

If the wall between both reservoirs had some thickness and the hole formed a pipe with some corrugation, transport in the hole would become diffusive for a long enough pipe, as discussed in Chapter 13. In the further case where the gas is not ideal but dilute or rarefied, the effusion process should be described in terms of Boltzmann's equation presented in Chapter 9.

8.6 Master Equation

Since the random variables (8.13) and (8.14) are defined as Poisson processes as in equation (4.34), the master equation ruling the time evolution of their probability density $p_t(\Delta E, \Delta N)$ can be directly deduced as

$$\frac{d}{dt} p_t(\Delta E, \Delta N) = \int_0^\infty d\varepsilon \, w_{\rm L}(\varepsilon) \left[p_t(\Delta E - \varepsilon, \Delta N - 1) - p_t(\Delta E, \Delta N) \right] + \int_0^\infty d\varepsilon \, w_{\rm R}(\varepsilon) \left[p_t(\Delta E + \varepsilon, \Delta N + 1) - p_t(\Delta E, \Delta N) \right] \tag{8.37}$$

in terms of the rate densities (8.17) of the effusion process (Cleuren et al., 2006; Gaspard and Andrieux, 2011b; Gaspard, 2014).

We note that this stochastic process is not stationary because the joint probability density $p_t(\Delta E, \Delta N)$ does not converge towards a normalized distribution for $t \to \infty$ under nonequilibrium or equilibrium conditions, the diffusivities of the random variables ΔE and ΔN being positive. In contrast, the Poisson suspension of Section 8.3 describes effusion as a stationary process in the infinite-particle phase space including all the observables of the system.

8.7 Fluctuation Relation for Energy and Particle Currents

In the effusion process, the multivariate fluctuation relation (5.73) is exactly satisfied for energy and particle exchanges $\Delta\mathfrak{C} = (\Delta E, \Delta N)$ with the affinities (8.19) and (8.20).

The corresponding cumulant generating function, which is defined as

$$Q(\boldsymbol{\lambda}; \mathbf{A}) \equiv \lim_{t\to\infty} -\frac{1}{t} \ln \left\langle e^{-\lambda_E \Delta E - \lambda_N \Delta N} \right\rangle_t \tag{8.38}$$

in terms of the counting parameters $\boldsymbol{\lambda} = (\lambda_E, \lambda_N)$, can be directly obtained from the Poisson suspension. The axis of energy $0 < \varepsilon < \infty$ is divided into small intervals $[\varepsilon, \varepsilon + d\varepsilon]$ of size $d\varepsilon$. In every one of these intervals, the numbers of particles exchanged from the left- to the right-hand reservoir or vice versa constitute Poisson random variables, $n_{\varepsilon+}$ and $n_{\varepsilon-}$, of mean values given by equations (8.15) and (8.16), respectively. Therefore, the moment generating function can be calculated as

$$
\begin{aligned}
\left\langle e^{-\lambda_E \Delta E - \lambda_N \Delta N} \right\rangle_t &= \sum_{\Delta N} \int d\Delta E \; p_t(\Delta E, \Delta N) \, e^{-\lambda_E \Delta E - \lambda_N \Delta N} \\
&= \sum_{\{n_{\varepsilon+}, n_{\varepsilon-}\}} \mathscr{P}(\{n_{\varepsilon+}, n_{\varepsilon-}\}) \, e^{-\sum_\varepsilon (\varepsilon \lambda_E + \lambda_N)(n_{\varepsilon+} - n_{\varepsilon-})}
\end{aligned}
\tag{8.39}
$$

in terms of the multiple Poisson distribution

$$
\mathscr{P}(\{n_{\varepsilon+}, n_{\varepsilon-}\}) = \prod_\varepsilon \frac{(\nu_{\varepsilon+})^{n_{\varepsilon+}}}{(n_{\varepsilon+})!} \, e^{-\nu_{\varepsilon+}} \, \frac{(\nu_{\varepsilon-})^{n_{\varepsilon-}}}{(n_{\varepsilon-})!} \, e^{-\nu_{\varepsilon-}}, \tag{8.40}
$$

where $\prod_\varepsilon$ denotes the product over all the intervals $[\varepsilon, \varepsilon + d\varepsilon]$ partitioning the energy axis. Summing over the random numbers $\{n_{\varepsilon+}, n_{\varepsilon-}\}$, we get

$$
\left\langle e^{-\lambda_E \Delta E - \lambda_N \Delta N} \right\rangle_t = \prod_\varepsilon \exp\left[\nu_{\varepsilon+} \left(e^{-\varepsilon \lambda_E - \lambda_N} - 1 \right) + \nu_{\varepsilon-} \left(e^{\varepsilon \lambda_E + \lambda_N} - 1 \right) \right]. \tag{8.41}
$$

Using equations (8.15)–(8.16) and taking the limit $d\varepsilon \to 0$, the generating function (8.38) becomes

$$
Q(\lambda_E, \lambda_N; A_E, A_N) = \int_0^\infty d\varepsilon \left[w_{\rm L}(\varepsilon) \left(1 - e^{-\varepsilon \lambda_E - \lambda_N} \right) + w_{\rm R}(\varepsilon) \left(1 - e^{\varepsilon \lambda_E + \lambda_N} \right) \right]. \tag{8.42}
$$

After integrating over the energy ε, the cumulant generating function for energy and particle currents in effusion is finally obtained as

$$
\begin{aligned}
Q(\lambda_E, \lambda_N; A_E, A_N) = \sigma \, n_{\rm L} \sqrt{\frac{k_{\rm B} T_{\rm L}}{2\pi m}} \left[1 - \frac{e^{-\lambda_N}}{(1 + k_{\rm B} T_{\rm L} \lambda_E)^2} \right] \\
+ \sigma \, n_{\rm R} \sqrt{\frac{k_{\rm B} T_{\rm R}}{2\pi m}} \left[1 - \frac{e^{+\lambda_N}}{(1 - k_{\rm B} T_{\rm R} \lambda_E)^2} \right].
\end{aligned}
\tag{8.43}
$$

As a consequence of equation (8.18), the cumulant generating function (8.43) obeys the symmetry relation

$$
Q(\lambda_E, \lambda_N; A_E, A_N) = Q(A_E - \lambda_E, A_N - \lambda_N; A_E, A_N), \tag{8.44}
$$

implying the multivariate fluctuation relation for energy and particle currents (Cleuren et al., 2006; Gaspard and Andrieux, 2011b; Gaspard, 2013a).

This demonstration can be directly carried out from the microscopic Hamiltonian dynamics and the stationary probability distribution of the infinite-particle dynamical system without using the master equation of the corresponding stochastic process. The result clearly shows that the multivariate fluctuation relation characterizes time-reversal symmetry breaking by the invariant probability distribution under nonequilibrium conditions, while microreversibility is always satisfied by the underlying Hamiltonian dynamics.

We note that the fluctuation relation for energy and particle currents can also be obtained from the master equation (Cleuren et al., 2006). Effusion can also be considered with momentum transfer and in relativistic ideal gases (Wood et al., 2007; Cleuren et al., 2008).

As a consequence of the symmetry relation (8.44), the linear and nonlinear properties demonstrated in Section 5.5 are satisfied for effusion. In particular, the linear response coefficients coupling the energy and particle fluxes obey the fluctuation–dissipation formulas (5.142) and the Onsager reciprocal relations (5.143) because $L_{E,N} = L_{N,E} = D_{EN}(0,0) = 2rk_BT$ with $r = \sigma n\sqrt{k_BT/(2\pi m)}$, $n = n_L = n_R$, and $T = T_L = T_R$. Next, the nonlinear response coefficients $M_{E,EN} = 6r(k_BT)^2$ and $M_{E,NN} = 2rk_BT$ coupling the energy current to the chemical affinity satisfy the relations (5.146) since the relevant first responses of the diffusivities $R_{\alpha\beta,\gamma} = \partial_{A_\gamma} D_{\alpha\beta}(0,0)$ are given by $R_{EE,N} = R_{EN,E} = 3r(k_BT)^2$ and $R_{EN,N} = rk_BT$ (Gaspard and Andrieux, 2011b). These properties are essential because effusion is a nonlinear transport process, the mean currents (8.23) and (8.24) being nonlinear functions of the affinities (8.19) and (8.20).

8.8 Fluctuation Relation for the Isothermal Particle Current

In isothermal systems, effusion can generate a particle current between both reservoirs if their densities are different. In stationary regimes, this particle current obeys the fluctuation relation

$$\frac{p_t(\Delta N; A)}{p_t(-\Delta N; A)} = \exp(A\,\Delta N) \qquad \text{for} \qquad A = A_N = \ln\frac{n_L}{n_R}, \tag{8.45}$$

which is the chemical affinity (8.20) with $T_L = T_R$.

The proof of this fluctuation relation is based on the results obtained for the random drift in Section 4.4.6. Indeed, the probability distribution $p_t(\Delta N; A)$ that the net number $n = \Delta N$ of particles has been exchanged from the left- to the right-hand reservoir during the time interval t is ruled by the same master equation as (4.78) for $P(n,t)$ with the transition rates

$$W_+ = \int_0^\infty d\varepsilon\, w_L(\varepsilon) = \sigma\, n_L \sqrt{\frac{k_BT}{2\pi m}}, \tag{8.46}$$

$$W_- = \int_0^\infty d\varepsilon\, w_R(\varepsilon) = \sigma\, n_R \sqrt{\frac{k_BT}{2\pi m}}. \tag{8.47}$$

Accordingly, effusion is described as the difference (4.84) between the two Poisson processes of corresponding rates and the probability distribution is given by equation (4.85) in terms of the modified Bessel function (4.86). Taking the ratio between the probabilities to observe the opposite exchanges $\pm n = \pm\Delta N$, the Bessel functions cancel each other by the property given in equation (4.86), so that $P(n,t)/P(-n,t) = \exp(An)$ with the affinity $A = \ln(W_+/W_-) = A_N$ for the rates (8.46)–(8.47), hence the result (8.45). The fluctuation relation (8.45) corresponds to the symmetry relation $Q(\lambda; A) = Q(A-\lambda; A)$ of the cumulant generating function (8.43) with $\lambda_E = 0$, $\lambda_N = \lambda$, and $T_L = T_R = T$.

For this isothermal effusion process, the thermal affinity (8.19) is equal to zero, so that the entropy production rate (8.36) is equal to $d_iS/dt = k_B A_N J_N$ with the net mean current $J_N = W_+ - W_-$ given by equation (8.24) in stationary regimes. These results are consistent with the asymptotic value of the entropy production rate (4.88).

8.9 Temporal Disorder and Entropy Production

The temporal disorder of the effusion process can be characterized in terms of the entropy per unit time (6.82) of the Poisson suspension (Gaspard, 1994, 2014). The time evolution of the system can be monitored by observing the position and momentum of every particle incident on both sides of the wall at $x = 0$ on a large but finite area Σ including the small hole σ. The surface area Σ can be partitioned into small cells $\Delta^2\Sigma = \Delta y \Delta z$, which play the role of detectors measuring the position and momentum of every particle with a given resolution $\Delta^3 r\, \Delta^3 p$. The knowledge of all these events allows us to reconstruct the trajectories of the particles since they move in free flight possibly interrupted by an elastic collision on the wall, as shown in Figure 8.1. The number of particles incident on the cell $[\mathbf{r}_i, \mathbf{r}_i + \Delta\mathbf{r}]$ with the momentum $[\mathbf{p}_i, \mathbf{p}_i + \Delta\mathbf{p}]$ during the time interval $[0, t]$ is equal to

$$N_i = t \left|\frac{p_{xi}}{m}\right| \Delta^2\Sigma\, f(\mathbf{r}_i, \mathbf{p}_i)\, \Delta^3 p, \tag{8.48}$$

where $f(\mathbf{r}_i, \mathbf{p}_i)$ is the distribution function (8.1) for the trajectory C followed by the corresponding particles. The initial conditions of these N_i particles are uniformly distributed in space inside the volume $t\,|p_{xi}/m|\,\Delta^2\Sigma$. If space is discretized into cells of volume $\Delta^3 r = \Delta x \Delta y \Delta z$, the number of different positions is given by

$$M_i(t) = \left|\frac{p_{xi}}{m}\right| \frac{t\,\Delta^2\Sigma}{\Delta x \Delta y \Delta z} = \left|\frac{p_{xi}}{m}\right| \frac{t}{\Delta x}. \tag{8.49}$$

The entropy per unit time is calculated as the rate of exponential growth with the time interval $[0, t]$ for the number of possible configurations of the N_i particles in the M_i positions (Gaspard, 1994):

$$h = \lim_{t\to\infty} \frac{1}{t} \ln \prod_i \frac{1}{N_i!}\, M_i(t)^{N_i}. \tag{8.50}$$

Since the particles incoming on each side of the surface area Σ are distributed according to the temperature and density of the corresponding reservoir, the entropy per unit time is obtained as

$$\begin{aligned} h = \Sigma \int_{p_x>0} d^3p \left|\frac{p_x}{m}\right| f_{\rm L}(\mathbf{p}) \ln \frac{\rm e}{f_{\rm L}(\mathbf{p})\, \Delta^3 r\, \Delta^3 p} \\ + \Sigma \int_{p_x<0} d^3p \left|\frac{p_x}{m}\right| f_{\rm R}(\mathbf{p}) \ln \frac{\rm e}{f_{\rm R}(\mathbf{p})\, \Delta^3 r\, \Delta^3 p}, \end{aligned} \tag{8.51}$$

where $f_{\rm L}(\mathbf{p})$ and $f_{\rm R}(\mathbf{p})$ are the single-particle distribution functions (8.1) for the left- and right-hand reservoirs, respectively. This entropy per unit time gives the accumulation rate of information by the detectors monitoring the process. Since these detectors have the finite resolution $\Delta^3 r\, \Delta^3 p$, the accumulation rate of information depends on this resolution. The fact that the entropy per unit time increases logarithmically as the resolution is refined means that the random process is continuous in the corresponding variables $(\mathbf{r}, \mathbf{p})$ according to the distribution functions (8.1).

Now, the coentropy per unit time (6.99) can be similarly calculated taking into account the fact that the trajectories belong to the subsets (8.2) and (8.3). The time-reversal symmetry permutes the density and temperature values only for the trajectories going through the small hole of area σ in the wall W. For these trajectories, the single-particle distribution function appearing in the logarithm should have the density and temperature of the time-reversed trajectory. However, the distribution function outside the logarithm should remain the same because the statistics continues to be carried out over the typical trajectories of the process (and not their time reversal). With these considerations, the time-reversed coentropy per unit time is given by

$$h^{\mathrm{R}} = \int_{p_x>0} d^3p \left|\frac{p_x}{m}\right| f_{\mathrm{L}}(\mathbf{p}) \left[(\Sigma - \sigma) \ln \frac{\mathrm{e}}{f_{\mathrm{L}}(\mathbf{p})\, \Delta^3 r\, \Delta^3 p} + \sigma \ln \frac{\mathrm{e}}{f_{\mathrm{R}}(\mathbf{p})\, \Delta^3 r\, \Delta^3 p} \right]$$
$$+ \int_{p_x<0} d^3p \left|\frac{p_x}{m}\right| f_{\mathrm{R}}(\mathbf{p}) \left[(\Sigma - \sigma) \ln \frac{\mathrm{e}}{f_{\mathrm{R}}(\mathbf{p})\, \Delta^3 r\, \Delta^3 p} + \sigma \ln \frac{\mathrm{e}}{f_{\mathrm{L}}(\mathbf{p})\, \Delta^3 r\, \Delta^3 p} \right]. \tag{8.52}$$

Taking the difference between the coentropy (8.52) and the entropy (8.51) per unit time, we find the contribution of the trajectories with probabilities that are not time-reversal symmetric:

$$h^{\mathrm{R}} - h = \frac{\sigma}{2} \int_{\mathbb{R}^3} d^3p \left|\frac{p_x}{m}\right| [f_{\mathrm{L}}(\mathbf{p}) - f_{\mathrm{R}}(\mathbf{p})] \ln \frac{f_{\mathrm{L}}(\mathbf{p})}{f_{\mathrm{R}}(\mathbf{p})} \geq 0, \tag{8.53}$$

where the symmetry $p_x \to -p_x$ of the Maxwellian distributions has been used. The ratio of the single-particle distribution functions of both reservoirs can be written in terms of the affinities (8.19)–(8.20) as

$$\frac{f_{\mathrm{L}}(\mathbf{p})}{f_{\mathrm{R}}(\mathbf{p})} = \exp\left(\varepsilon A_E + A_N\right), \tag{8.54}$$

where $\varepsilon = \mathbf{p}^2/(2m)$ is the kinetic energy of the particle. In addition, the single-particle distribution functions multiplied by the velocity $|p_x/m|$ and integrated over some surface represent the fluxes of particles of given momentum through this surface. Integrating these fluxes over the corresponding values of the momentum gives the transition rates (8.17). Accordingly, we find

$$h^{\mathrm{R}} - h = \int_0^{\infty} d\varepsilon\, [w_{\mathrm{L}}(\varepsilon) - w_{\mathrm{R}}(\varepsilon)]\, (\varepsilon A_E + A_N) = A_E\, J_E + A_N\, J_N = \frac{1}{k_{\mathrm{B}}} \frac{d_{\mathrm{i}} S}{dt} \tag{8.55}$$

in terms of the mean energy and particle currents (8.23) and (8.24), whereupon the thermodynamic entropy production (8.36) is obtained (Gaspard, 2014). This result confirms that the thermodynamic entropy production is given by equation (6.103) in terms of the time asymmetry of temporal disorder under nonequilibrium conditions.

8.10 Mass Separation by Effusion

The previous results can be extended to the effusion of a binary mixture of particles with different masses $m_1 \neq m_2$. In this case, effusion may generate mass separation if the flow between the two reservoirs is larger for one species than the other. Since the mixing entropy is reduced during this process, mass separation requires energy supply. The effusive flow provides this energy by the coupling between the particle and energy exchanges. For the effusion of a binary mixture, there exist three mean currents for the energy and the two particle species. On the one hand, the mean energy current is given by $J_E = J_{E1} + J_{E2}$ with

$$J_{Ek} = \frac{2\sigma}{\sqrt{2\pi m_k}} \left(y_{k\mathrm{L}}\, p_{\mathrm{L}} \sqrt{k_{\mathrm{B}} T_{\mathrm{L}}} - y_{k\mathrm{R}}\, p_{\mathrm{R}} \sqrt{k_{\mathrm{B}} T_{\mathrm{R}}} \right) \tag{8.56}$$

for $k = 1, 2$ in terms of the mole fractions $y_{kj} \equiv [n_k/(n_1 + n_2)]_j$, the pressures p_j, and the temperatures T_j in the two reservoirs $j = \mathrm{L}, \mathrm{R}$. On the other hand, the mean particle currents have the form

$$J_k = \frac{\sigma}{\sqrt{2\pi m_k}} \left(y_{k\mathrm{L}} \frac{p_{\mathrm{L}}}{\sqrt{k_{\mathrm{B}} T_{\mathrm{L}}}} - y_{k\mathrm{R}} \frac{p_{\mathrm{R}}}{\sqrt{k_{\mathrm{B}} T_{\mathrm{R}}}} \right) \tag{8.57}$$

for the species $k = 1, 2$. The nonequilibrium control parameters are the thermal affinity (8.19) and the chemical affinities A_1 and A_2, defined as in equation (8.20) for the two species.

For the process of mass separation, it is convenient to introduce the total and differential particle currents and the corresponding chemical affinities:

$$J_N \equiv J_1 + J_2, \qquad A_N \equiv \frac{1}{2}(A_1 + A_2), \tag{8.58}$$

$$J_D \equiv J_1 - J_2, \qquad A_D \equiv \frac{1}{2}(A_1 - A_2). \tag{8.59}$$

The entropy production rate of this process can thus be expressed as

$$\frac{1}{k_{\mathrm{B}}} \frac{d_{\mathrm{i}} S}{dt} = A_E\, J_E + A_N\, J_N + A_D\, J_D \geq 0. \tag{8.60}$$

The multivariate fluctuation relation for all the currents in the effusive flow reads

$$\frac{p_t(\Delta E, \Delta N, \Delta D; A_E, A_N, A_D)}{p_t(-\Delta E, -\Delta N, -\Delta D; A_E, A_N, A_D)} = \exp\left(A_E\, \Delta E + A_N\, \Delta N + A_D\, \Delta D\right), \tag{8.61}$$

where the random variable ΔE is defined as in equation (8.13), $\Delta N \equiv \Delta N_1 + \Delta N_2$, and $\Delta D \equiv \Delta N_1 - \Delta N_2$ (Gaspard and Andrieux, 2011b).[2]

[2] In the case where there is one species of particles, the fluctuation relation (8.61) reduces to the one mentioned in Section 8.7 because $\Delta N_2 = 0$ implies $\Delta N = \Delta D$ and $A_N + A_D = A_1$.

The thermodynamic efficiency of mass separation can be defined as

$$\eta \equiv -\frac{A_D J_D}{A_E J_E + A_N J_N}, \tag{8.62}$$

which always satisfies the inequality $\eta \leq 1$ because of the second law of thermodynamics (8.60) and also $\eta > 0$ in the regime of mass separation where $A_D J_D < 0$ since the differential current J_D is opposed to the affinity A_D and driven by the exchanges of energy and total particle number. Since effusion is a nonlinear transport process, the thermodynamic efficiency can be significantly larger than if the mean currents were supposed to be linearly related to the affinities. Furthermore, mass separation by effusion can be shown to be more efficient for a large chemical affinity A_N than a large thermal affinity A_E (Gaspard and Andrieux, 2011b).

9

Processes in Dilute and Rarefied Gases

9.1 Length Scales in Dilute and Rarefied Gases

In this chapter, we consider the kinetics of gases at low density. In rare gases, the particles are monatomic and their motion is ruled by the Hamiltonian function (2.2) for identical particles of mass m. The interaction between the particles is assumed to have short or medium range d. At low density n, the mean distance between nearest neighboring particles[1] $\mathcal{R} \simeq 0.554\, n^{-1/3}$ is much larger than the interaction range d, so that the condition $nd^3 \ll 1$ is fulfilled. The equilibrium properties of such a gas can thus be obtained by neglecting the interaction in a similar way to an ideal gas where the equations of state for pressure and energy are given by $p = nk_{\rm B}T$ and $e = 3nk_{\rm B}T/2$, respectively.

However, ideal gases have infinite transport coefficients since the particles have ballistic motion in free flights between the walls of the container. In real gases, even though the density is so low that the condition $nd^3 \ll 1$ is satisfied, the free flights of the particles may be shorter than the size of the container because of the total cross section $\sigma_{\rm tot} \propto d^2$ for binary collisions between the particles. Accordingly, the free flight of a particle is limited to its next binary collision with another particle in the gas. The mean free path is evaluated to be

$$\ell \simeq \frac{1}{\sqrt{2}\,\sigma_{\rm tot} n} \tag{9.1}$$

in gases (Present, 1958; Reichl, 1998). If the density n is low enough, the mean free path can thus be much larger than the interaction range d, so that the mean intercollision time $\tau_{\rm intercoll} = \ell/\bar{v}$ is much longer than the collision time $\tau_{\rm coll} \sim d/\bar{v}$, where $\bar{v} = \langle \|\mathbf{v}\| \rangle_{\rm eq} = \sqrt{8k_{\rm B}T/(\pi m)}$ is the mean velocity of the particles:

$$\frac{\tau_{\rm intercoll}}{\tau_{\rm coll}} \sim \frac{\ell}{d} \gg 1. \tag{9.2}$$

This condition is consistent with the previous one, $nd^3 \ll 1$. In the low-density limit, the mean free path is also much greater than the mean distance between nearest neighboring

[1] This mean distance is given by $\mathcal{R} = \Gamma(4/3)\,(4\pi n/3)^{-1/3}$ in a random distribution of particles with density $n = N/V$, where $\Gamma(z)$ is the Gamma function (Chandrasekhar, 1943).

particles, $\ell \gg \mathcal{R} \simeq 0.554\, n^{-1/3}$. In air at standard pressure and temperature, the mean free path of nitrogen molecules of diameter $d \simeq 0.35$ nm is about $\ell \simeq 75$ nm, the mean distance between nearest neighbors about $\mathcal{R} \simeq 1.9$ nm, their mean velocity $\bar{v} \simeq 475$ m/s, and the mean intercollision time $\tau_{\text{intercoll}} \simeq 1.6 \times 10^{-10}$ s, so that the conditions $\ell \gg d$ and $\ell \gg \mathcal{R}$ are indeed satisfied.

The dimensionless *Knudsen number* is defined as the ratio $\text{Kn} \equiv \ell/L$ between the mean free path (9.1) and the size L of the container, the ducts, or the obstacles in the gaseous flow. *Dilute gases* are characterized by a mean free path smaller than the size L and $\text{Kn} \ll 1$. Conversely, the mean free path can be larger than the size L in *rarefied gases*, where $\text{Kn} \gg 1$ (Cercignani, 2000). Effusion is an example of transport phenomenon in a rarefied gas (Present, 1958; Steckelmacher, 1986).

The kinetics of dilute and rarefied gases are ruled by the famous Boltzmann equation for the time evolution of the velocity distribution function of the particles. Since the free flights of the particles are interrupted by binary collisions, these latter change the particle velocities. The rates of these changes determine Boltzmann's kinetic equation, as for chemical reactions $\text{A} + \text{B} \rightleftharpoons \text{C} + \text{D}$ where the species here correspond to particle populations with different velocities. Since its inception by Boltzmann (1872, 1896), several methods of derivation have been proposed in the literature (Hirschfelder et al., 1954; Grad, 1958; Chapman and Cowling, 1960; Balescu, 1975; Résibois and De Leener, 1977; Landau and Lifshitz, 1981; Akhiezer and Peletminskii, 1981; Cercignani, 1988; Liboff, 1990; Spohn, 1991; Reichl, 1998; Pottier, 2009). In particular, its derivation often starts from the BBGKY hierarchy presented in Section 2.3.5 and supposes that the density is low enough to neglect the three-particle distribution function in equation (2.63) for the two-particle one because $f_3/f_2 = O(n)$. Accordingly, equation (2.63) reduces to Liouville's equation for the motion of two interacting particles, which is described by the differential cross section everywhere binary collisions happen. Once the two-body problem is solved, the collision term of equation (2.62) can be expressed in terms of this cross section using Boltzmann's *Stosszahlansatz*, according to which there are no statistical correlations between the positions and velocities of any pair of particles in the gas: $f_2(\mathbf{x}_1, \mathbf{x}_2, t) \simeq f_1(\mathbf{x}_1, t) f_1(\mathbf{x}_2, t)$. This assumption has been much debated, pointing out the fact that correlations between two particles are generated during their interaction, in particular, upon binary collisions. Further considerations show that such correlations are dispersed over large distances and among many particles in the phase space of low-density gases, leading to a loss of memory of previous collisions.

A derivation, which is partly inspired by the method of Grad (1958), is here proposed based on a phase-space picture for low-density gases. This approach allows us to describe the gas kinetics as successive scattering events, during which particles are incoming and outgoing binary collisions.

In addition, the fluctuating Boltzmann equation is used to deduce a fluctuation relation for the energy and particle currents in nonequilibrium dilute or rarefied gases (Gaspard, 2013c,e).

The applications of Boltzmann's equation are briefly presented.

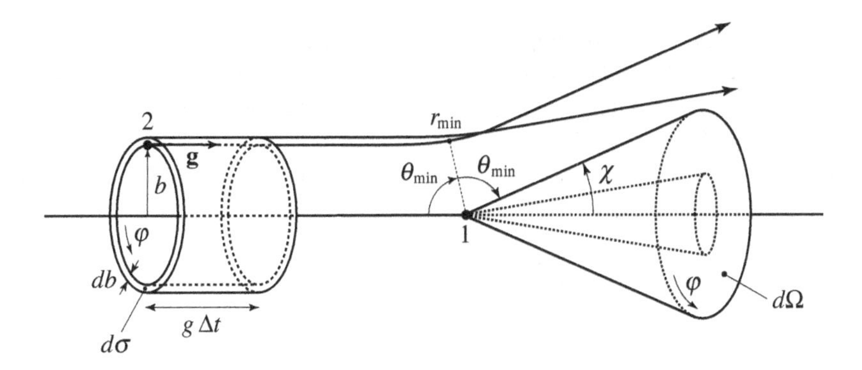

Figure 9.1 Binary collision between two particles in the frame of the relative position $\mathbf{r} = \mathbf{r}_2 - \mathbf{r}_1$. The relative velocity is $\mathbf{g} = \dot{\mathbf{r}}_2 - \dot{\mathbf{r}}_1$, b is the impact parameter, φ the azimuthal angle of the orbital plane, $d\sigma = 2\pi b\,db$ the cross-sectional area of collisions with scattering angle χ and $d\Omega$ the corresponding spherical angle, and $r_{\min}$ is the minimum interparticle distance during collision and $\theta_{\min}$ the corresponding angle. Finally, $g\,\Delta t$ with $g = \|\mathbf{g}\|$ is the height of the cylinder of basis $d\sigma$ containing particles incoming binary collisions into the spherical angle $d\Omega$ during the time lapse Δt.

9.2 Boltzmann's Kinetic Equation

9.2.1 Scattering Cross Sections

As previously mentioned, the time evolution of the velocity distribution is determined by distant binary collisions that interrupt long free flights. In the bulk of the gas, i.e., far from the walls of the container, these events are ruled by the following two-particle Hamiltonian function written for the particles $a = 1$ and $b = 2$:

$$h_{12} = \frac{\mathbf{p}_1^2}{2m} + \frac{\mathbf{p}_2^2}{2m} + u^{(2)}(r_{12}) = \frac{\mathbf{P}^2}{2M} + \frac{\mathbf{p}^2}{2\mu} + u^{(2)}(r), \tag{9.3}$$

where $\mathbf{P} \equiv \mathbf{p}_1 + \mathbf{p}_2$ is the momentum of the center $\mathbf{R} \equiv (\mathbf{r}_1 + \mathbf{r}_2)/2$ of mass $M = 2m$, and $\mathbf{p} \equiv (\mathbf{p}_2 - \mathbf{p}_1)/2$ the momentum of the relative position $\mathbf{r} \equiv \mathbf{r}_2 - \mathbf{r}_1$ of reduced mass $\mu = m/2$.

We note that the corresponding Liouville equation $\partial_t f_2 = \{h_{12}, f_2\}$ is equivalent to equation (2.63) for the two-particle distribution function f_2 with $f_3 = 0$ and $\mathbf{F}_1^{(\text{ext})} = \mathbf{F}_2^{(\text{ext})} = 0$. The solution of this Liouville equation is thus given by $f_2(\mathbf{x}_1, \mathbf{x}_2, t) = f_2\left[\boldsymbol{\phi}_{12}^{-t}(\mathbf{x}_1, \mathbf{x}_2); 0\right]$, where $\boldsymbol{\phi}_{12}^t$ is the flow of the Hamiltonian dynamics (9.3) in the twelve-dimensional phase space $(\mathbf{x}_1, \mathbf{x}_2) \in \mathbb{R}^{12}$ of the two-body mechanical problem.

This problem can be separated into the free flight of the center of mass $\mathbf{R} = \mathbf{R}_0 + \mathbf{P}_0 t/M$ and the dynamics of the relative motion between the two particles. This latter is schematically represented in Figure 9.1, which shows a scattering event between the two particles in the frame of the relative position $\mathbf{r} \equiv \mathbf{r}_2 - \mathbf{r}_1$, where the origin is the center $r = 0$ of the interaction potential $u^{(2)}(r)$ corresponding to the location of particle no. 1. Since angular momentum is conserved in this relative motion, the trajectory $\mathbf{r}(t)$ belongs to a plane that includes the line passing through the interaction center in the direction of the vector $\mathbf{g} \equiv \dot{\mathbf{r}}_2 - \dot{\mathbf{r}}_1$ of the relative velocity. The impact parameter b is the distance between this line through $r = 0$ and the incoming trajectory. The trajectory of impact parameter b is scattered into the direction making the angle $\chi = \pi - 2\theta_{\text{min}}$, where θ_{min} corresponds to the point of minimum distance r_{min} between the trajectory and the interaction center $r = 0$. In spherical coordinates, the energy of the relative motion can be written as

$$E = \frac{\mathbf{p}^2}{2\mu} + u^{(2)}(r) = \frac{1}{2}\,\mu\,\dot{r}^2 + \frac{L^2}{2\mu r^2} + u^{(2)}(r), \tag{9.4}$$

where $L = b\mu g$ is the angular momentum fixed by the incoming trajectory together with the energy $E = \mu g^2/2$. Moreover, the time derivative of the polar angle θ is related to the angular momentum L according to $\dot{\theta} = L/(\mu r^2)$ as a consequence of angular momentum conservation (Goldstein, 1950). Therefore, the polar angle θ_{min} at the point of closest approach and, thus, the scattering angle χ can be obtained for the interaction potential $u^{(2)}$ in terms of the impact parameter b and the magnitude of the relative velocity g: $\chi = \chi(b, g)$.

As shown in Figure 9.1, the trajectories incoming with impact parameters $[b, b + db]$ in the ring of area $d\sigma = 2\pi b db$ are scattered into the solid angle equal to $d\Omega = |2\pi \sin\chi d\chi|$, where $2\pi = \int d\varphi$. The differential cross section is defined as their ratio as (Goldstein, 1950)

$$\sigma_{\text{diff}} \equiv \frac{d\sigma}{d\Omega} = \left| \frac{b}{\sin\chi} \frac{db}{d\chi} \right|. \tag{9.5}$$

The total cross section is given by integration over the solid angle according to

$$\sigma_{\text{tot}} \equiv \int \sigma_{\text{diff}}\, d\Omega = \int \frac{d\sigma}{d\Omega}\, d\Omega. \tag{9.6}$$

The angular dependence of the differential cross section provides the signature of the interaction potential. In particular, for the collision between two hard spheres of diameter d, the impact parameter is related to the scattering angle by $b = d \sin\theta_{\text{min}} = d \cos(\chi/2)$, so that the differential cross section is given by $\sigma_{\text{diff}} = d^2/4$, which has no angular dependence. The corresponding total cross section is equal to the area of a disk of diameter d, $\sigma_{\text{tot}} = \pi d^2$, as expected for such hard spheres undergoing elastic collisions.

9.2.2 Derivation of Boltzmann's Equation

In low-density gases, the particles are more frequently in free flight than colliding. The collisions thus happen in the small phase-space regions where the particles interact. The interaction range is here assumed to be finite and limited to the distance d, so that the interaction forces are equal to zero as

$$\mathbf{F}_{ab} = -\frac{\partial u^{(2)}(r_{ab})}{\partial \mathbf{r}_a} = 0 \qquad \text{for} \qquad r_{ab} = \|\mathbf{r}_a - \mathbf{r}_b\| > d. \tag{9.7}$$

The particles are thus in free flight in the large phase-space domain where the conditions (9.7) are satisfied. In the position space $(\mathbf{r}_a)_{a=1}^N \in \mathbb{R}^{3N}$ of N identical particles, the $3N$-dimensional volume where trajectories are free flights is evaluated as

$\left(1 - 2\pi n d^3/3\right)\left(V^N/N!\right)$, if V is the three-dimensional volume of the container. In this regard, the phase space (2.4) can be partitioned as

$$\mathcal{M} = \mathcal{M}_{\rm f} + \mathcal{M}_{\rm c} \tag{9.8}$$

into the following part, where motion is ballistic,

$$\mathcal{M}_{\rm f} = \{\boldsymbol{\Gamma} \in \mathcal{M} \text{ such that } \|\mathbf{r}_a - \mathbf{r}_b\| > d \text{ for all } 1 \leq a, b \leq N\}, \tag{9.9}$$

and the complementary subset $\mathcal{M}_{\rm c}$, where collisions happen.

We may introduce the microscopic distribution function for the position $\mathbf{r}$ and momentum $\mathbf{p}$ of the particles as

$$\hat{f}(\mathbf{r}, \mathbf{p}; \boldsymbol{\Gamma}) \equiv \sum_{a=1}^{N} \delta^3(\mathbf{r} - \mathbf{r}_a)\, \delta^3(\mathbf{p} - \mathbf{p}_a), \tag{9.10}$$

which is normalized to the total number of particles, $N = \int \hat{f}(\mathbf{r}, \mathbf{p}; \boldsymbol{\Gamma})\, d^3r\, d^3p$. The microscopic particle density defined by equation (3.3) is recovered by integrating over momentum: $\hat{n}(\mathbf{r}; \boldsymbol{\Gamma}) = \int \hat{f}(\mathbf{r}, \mathbf{p}; \boldsymbol{\Gamma})\, d^3p$. The one-particle distribution function (2.52) is thus given by averaging over the phase-space probability distribution:

$$f(\mathbf{r}, \mathbf{p}) = \langle \hat{f}(\mathbf{r}, \mathbf{p}; \boldsymbol{\Gamma}) \rangle = \int_{\mathcal{M}} \hat{f}(\mathbf{r}, \mathbf{p}; \boldsymbol{\Gamma})\, p(\boldsymbol{\Gamma})\, d\boldsymbol{\Gamma}. \tag{9.11}$$

Now, the one-particle distribution function can be decomposed as the phase space (9.8) into the distribution functions for the particles in free flight and those in collisions:

$$f(\mathbf{r}, \mathbf{p}) = f_{\rm f}(\mathbf{r}, \mathbf{p}) + f_{\rm c}(\mathbf{r}, \mathbf{p}) \tag{9.12}$$

with

$$f_{\rm x}(\mathbf{r}, \mathbf{p}) \equiv \int_{\mathcal{M}_{\rm x}} \hat{f}(\mathbf{r}, \mathbf{p}; \boldsymbol{\Gamma})\, p(\boldsymbol{\Gamma})\, d\boldsymbol{\Gamma} \qquad \text{for} \quad \rm x = f, c. \tag{9.13}$$

In low-density gases, we expect that $f_{\rm c}/f_{\rm f} = O(nd^3) \ll 1$, so that the full one-particle distribution function is well approximated by its part in the free-flight zones:

$$f(\mathbf{r}, \mathbf{p}) = f_{\rm f}(\mathbf{r}, \mathbf{p})\, [1 + O(nd^3)]. \tag{9.14}$$

Now, the time evolution of this part of the one-particle distribution function can be deduced from the Liouville equation (2.37) as follows,

$$\begin{aligned}
\partial_t f_{\rm f} = -\int_{\mathcal{M}_{\rm f}} \hat{f}\, \mathrm{div}(\dot{\boldsymbol{\Gamma}} p)\, d\boldsymbol{\Gamma} &= \int_{\mathcal{M}_{\rm f}} [p\, \dot{\boldsymbol{\Gamma}} \cdot \mathrm{grad}\, \hat{f} - \mathrm{div}(\hat{f}\, \dot{\boldsymbol{\Gamma}} p)]\, d\boldsymbol{\Gamma} \\
&= \int_{\mathcal{M}_{\rm f}} p\, \dot{\boldsymbol{\Gamma}} \cdot \mathrm{grad}\, \hat{f}\, d\boldsymbol{\Gamma} - \int_{\partial\mathcal{M}_{\rm f}} \hat{f}\, p\, \dot{\boldsymbol{\Gamma}} \cdot d\boldsymbol{\Sigma}
\end{aligned} \tag{9.15}$$

with the element $d\boldsymbol{\Sigma}$ of the $(6N - 1)$-dimensional hypersurface $\partial\mathcal{M}_{\rm f}$ where $\|\mathbf{r}_a - \mathbf{r}_b\| = d$ and oriented towards the exterior of the domain $\mathcal{M}_{\rm f}$. Since there is no interaction in the phase-space domain $\mathcal{M}_{\rm f}$, the first term gives the time evolution without the binary interactions. Furthermore, the integral over the hypersurface $\partial\mathcal{M}_{\rm f}$ can be decomposed into the contribution of the particles incoming collisions on the side $\partial\mathcal{M}_{\rm f}^{(-)}$ where $\dot{\boldsymbol{\Gamma}}\cdot d\boldsymbol{\Sigma}>0$ and that of the particles outgoing collisions on the other side $\partial\mathcal{M}_{\rm f}^{(+)}$ where $\dot{\boldsymbol{\Gamma}}\cdot d\boldsymbol{\Sigma}<0$. They, respectively, define the loss and gain terms,

$$C^{(-)} \equiv \int_{\partial\mathcal{M}_{\rm f}^{(-)}} \hat{f}\, p\, \underbrace{\dot{\boldsymbol{\Gamma}}\cdot d\boldsymbol{\Sigma}}_{>0} \quad \text{and} \quad C^{(+)} \equiv -\int_{\partial\mathcal{M}_{\rm f}^{(+)}} \hat{f}\, p\, \underbrace{\dot{\boldsymbol{\Gamma}}\cdot d\boldsymbol{\Sigma}}_{<0}, \tag{9.16}$$

which are both nonnegative. Consequently, equation (9.15) becomes

$$\frac{\partial f_{\rm f}}{\partial t} + \frac{\mathbf{p}}{m}\cdot\frac{\partial f_{\rm f}}{\partial \mathbf{r}} + \mathbf{F}^{(\rm ext)}\cdot\frac{\partial f_{\rm f}}{\partial \mathbf{p}} = C^{(+)} - C^{(-)}. \tag{9.17}$$

The hypersurface $\partial\mathcal{M}_{\rm f}$ is composed of the pieces of the $(6N-1)$-dimensional hypersurfaces $\|\mathbf{r}_a - \mathbf{r}_b\| = d$ for $1 \le a,b \le N$. We note that the $6N$-dimensional domains $\|\mathbf{r}_a - \mathbf{r}_b\| < d$ may overlap in the phase-space regions where collisions between more than two particles are possible. In the position space $\mathbb{R}^{3N}$, these regions are even smaller than those where binary collisions happen. However, these regions are enclosed inside the whole hypersurface $\partial\mathcal{M}_{\rm f}$, which includes the zones of collisions between two and more particles. Now, on the hypersurface $\|\mathbf{r}_a - \mathbf{r}_b\| = d$, the element of hypersurface integration can be obtained by considering the relative position $\mathbf{r}_{ab} \equiv \mathbf{r}_a - \mathbf{r}_b$ and the center of mass $\mathbf{R}_{ab} \equiv (\mathbf{r}_a + \mathbf{r}_b)/2$ of the particles a and b instead of their positions in the phase-space variables $\boldsymbol{\Gamma}$. In these coordinates, the hypersurface corresponds to the locus where $r_{ab} = \|\mathbf{r}_{ab}\| = d$. Therefore, the vectorial hypersurface element can be written as $d\boldsymbol{\Sigma} = \mathbf{n}\, d\Sigma$ in terms of the $6N$-dimensional unit vector $\mathbf{n}$ normal to the hypersurface and $d\Sigma = dS_{ab}\, d\mathbf{R}_{ab}\, d\mathbf{p}_a\, d\mathbf{p}_b \prod_{c\neq a,b} d\mathbf{r}_c\, d\mathbf{p}_c$, where $dS_{ab} = r_{ab}^2 \sin\theta_{ab}\, d\theta_{ab}\, d\varphi_{ab}$ is the two-dimensional surface element of the sphere $r_{ab} = \|\mathbf{r}_{ab}\| = d$. The unit vector is given by $\mathbf{n} = (-\mathbf{n}_{ab}, 0, 0, \dots)$ with the three-dimensional unit vector $\mathbf{n}_{ab} \equiv \mathbf{r}_{ab}/\|\mathbf{r}_{ab}\|$. As a consequence, we have that

$$\dot{\boldsymbol{\Gamma}}\cdot d\boldsymbol{\Sigma} = -\mathbf{n}_{ab}\cdot\dot{\mathbf{r}}_{ab}\, dS_{ab}\, d\mathbf{R}_{ab}\, d\mathbf{p}_a\, d\mathbf{p}_b \prod_{c\neq a,b} d\mathbf{r}_c\, d\mathbf{p}_c. \tag{9.18}$$

Let us suppose that the variables $(\mathbf{r},\mathbf{p})$ of the distribution function (9.10) correspond to the position and momentum $(\mathbf{r}_1,\mathbf{p}_1)$ of particle no. 1. This particle may have binary collisions with the $N-1$ other particles on the hypersurfaces $\|\mathbf{r}_1 - \mathbf{r}_a\| = d$ with $a = 2,3,\dots,N$. All these collisions have identical dynamics, so that the loss term can be written in terms of the contribution of the binary collision of particle no. 1 with a generic particle no. 2 with variables $(\mathbf{r}_2,\mathbf{p}_2)$ as

$$C^{(-)} = -N(N-1)\int_{\Sigma_{12}^{(-)}} p\, \mathbf{n}_{21}\cdot\dot{\mathbf{r}}_{21}\, dS_{21}\, d\mathbf{p}_2 \prod_{a\neq 1,2} d\mathbf{r}_a\, d\mathbf{p}_a, \tag{9.19}$$

where the factor N is given by integrating the microscopic distribution function $\hat{f}$ over $d\mathbf{R}_{12}$ and $d\mathbf{p}_1$, and $\Sigma_{12}^{(-)}$ is the hypersurface $\|\mathbf{r}_1 - \mathbf{r}_2\| = d$ where the two particles are incoming the binary collision with $\mathbf{n}_{21}\cdot\dot{\mathbf{r}}_{21} < 0$.

As previously mentioned, the trajectories are dispersed in all directions at every collision, so that the two colliding particles, no. 1 and no. 2, will typically have their next collisions with two new particles, no. 3 and no. 4, whereupon the possible correlations between the particles no. 1 and no. 2 are later shared with more particles. These correlations tend to disappear due to the dispersion of the particles at large distances from each other because the free flights are longer than the mean distance between nearest neighboring particles in low-density gases, $\ell \simeq (\sqrt{2}\,\sigma_{\rm tot}n)^{-1} \gg \mathcal{R} \simeq 0.554\,n^{-1/3}$. The lower the density, the larger the number of collisions before a possible recollision between the particles no. 1 and no. 2. During these collisions between uncorrelated particles, the correlations arising from the first collision between these two particles have had time to become evanescent before the recollision. In this regard, the particles incoming collisions may be assumed to be uncorrelated in low-density gases.

Accordingly, the loss term (9.19) can be evaluated by counting the collisions responsible for the loss of particles with momentum $\mathbf{p} = \mathbf{p}_1$ near the position $\mathbf{r} = \mathbf{r}_1$. As shown in Figure 9.1, the particles no. 2 incoming the collision with impact parameter between $[b, b + db]$ during the time interval Δt are contained inside the cylinder of volume

$$d\mathbf{r}_2 = 2\pi\, b\, db\, g\, \Delta t = d\sigma\, g\, \Delta t = \sigma_{\rm diff}\, d\Omega\, g\, \Delta t. \tag{9.20}$$

The number of these particles with momentum inside $[\mathbf{p}_2, \mathbf{p}_2 + d\mathbf{p}_2]$ is equal to $f(\mathbf{r}_2, \mathbf{p}_2, t)\, d\mathbf{r}_2\, d\mathbf{p}_2 = f(\mathbf{r}_2, \mathbf{p}_2, t)\, d\sigma\, g\, \Delta t\, d\mathbf{p}_2$. These particles no. 2 collide with the number $f(\mathbf{r}_1, \mathbf{p}_1, t)\, d\mathbf{r}_1\, d\mathbf{p}_1$ of particles no. 1 in the volume $d\mathbf{r}_1$ with their momentum inside $[\mathbf{p}_1, \mathbf{p}_1 + d\mathbf{p}_1]$. The number of particles no. 1 of momentum $\mathbf{p} = \mathbf{p}_1$ in the volume $d\mathbf{r} = d\mathbf{r}_1$ that are lost during the time interval Δt is thus given by

$$C^{(-)}\, d\mathbf{r}_1\, d\mathbf{p}_1\, \Delta t \simeq \int f(\mathbf{r}_1, \mathbf{p}_1, t)\, f(\mathbf{r}_2, \mathbf{p}_2, t)\, d\mathbf{r}_1\, d\mathbf{p}_1\, d\mathbf{r}_2\, d\mathbf{p}_2, \tag{9.21}$$

where $d\mathbf{r}_2$ is the volume (9.20) and the integral is taken over all the momenta $\mathbf{p}_2$ since all these particles no. 2 contribute to the loss of particles no. 1 with momentum $\mathbf{p} = \mathbf{p}_1$. Dividing by the arbitrary volume element $d\mathbf{r}_1 d\mathbf{p}_1$, the loss term is therefore evaluated as

$$C^{(-)} \simeq \int d\mathbf{p}_2\, d\Omega\, \sigma_{\rm diff}\, g\, f(\mathbf{r}_1, \mathbf{p}_1, t)\, f(\mathbf{r}_2, \mathbf{p}_2, t). \tag{9.22}$$

The gain term in equation (9.16) can be obtained by exchanging incoming and outgoing particles $(\mathbf{r}_1, \mathbf{p}_1, \mathbf{r}_2, \mathbf{p}_2) \to (\mathbf{r}'_1, \mathbf{p}'_1, \mathbf{r}'_2, \mathbf{p}'_2)$, to get

$$C^{(+)}\, d\mathbf{r}_1\, d\mathbf{p}_1\, \Delta t \simeq \int f(\mathbf{r}'_1, \mathbf{p}'_1, t)\, f(\mathbf{r}'_2, \mathbf{p}'_2, t)\, d\mathbf{r}'_1\, d\mathbf{p}'_1\, d\mathbf{r}'_2\, d\mathbf{p}'_2. \tag{9.23}$$

Since the Hamiltonian flow $(\mathbf{r}'_1, \mathbf{p}'_1, \mathbf{r}'_2, \mathbf{p}'_2) = \boldsymbol{\Phi}^t_{12}(\mathbf{r}_1, \mathbf{p}_1, \mathbf{r}_2, \mathbf{p}_2)$ of the two-body system obeys Liouville's theorem, we have that

$$d\mathbf{r}'_1\, d\mathbf{p}'_1\, d\mathbf{r}'_2\, d\mathbf{p}'_2 = d\mathbf{r}_1\, d\mathbf{p}_1\, d\mathbf{r}_2\, d\mathbf{p}_2. \tag{9.24}$$

As a consequence, the gain term is given by

$$C^{(+)} \simeq \int d\mathbf{p}_2\, d\Omega\, \sigma_{\rm diff}\, g\, f(\mathbf{r}'_1, \mathbf{p}'_1, t)\, f(\mathbf{r}'_2, \mathbf{p}'_2, t), \tag{9.25}$$

which is similar to the loss term (9.22) but with the prime variables of the outgoing particles instead of those of the incoming ones. Moreover, we may suppose that $\mathbf{r}_1 \simeq \mathbf{r}_2 \simeq \mathbf{r}'_1 \simeq \mathbf{r}'_2 \simeq \mathbf{r}$, since the collision happens in a small volume around the position $\mathbf{r} = \mathbf{r}_1$ where the distribution function is considered. Replacing the loss and gain terms back into equation (9.17) and again using the approximation (9.14) valid for low enough density $nd^3 \ll 1$, we finally obtain Boltzmann's equation:

$$\begin{aligned}\frac{\partial f}{\partial t} + \frac{\mathbf{p}}{m}\cdot\frac{\partial f}{\partial \mathbf{r}} + \mathbf{F}^{(\text{ext})}\cdot\frac{\partial f}{\partial \mathbf{p}} \\ = \int d\mathbf{p}_2\, d\Omega\, \sigma_{\text{diff}}\, g\left[f(\mathbf{r},\mathbf{p}',t)\, f(\mathbf{r},\mathbf{p}'_2,t) - f(\mathbf{r},\mathbf{p},t)\, f(\mathbf{r},\mathbf{p}_2,t)\right],\end{aligned} \tag{9.26}$$

where σ_{diff} is the differential cross section (9.5) and $g = \|\dot{\mathbf{r}}_2 - \dot{\mathbf{r}}_1\| = \|\mathbf{p}_2 - \mathbf{p}_1\|/m$ is the relative velocity between the colliding particles.

By solving the two-body problem in Section 9.2.1, the particle momenta after the collision are uniquely determined by their momenta before the collision, the impact parameter b, and the azimuthal angle φ, as shown in Figure 9.1:

$$\begin{cases}\mathbf{p}'_1 = \mathbf{p}'_1(\mathbf{p}_1,\mathbf{p}_2;b,\varphi), \\ \mathbf{p}'_2 = \mathbf{p}'_2(\mathbf{p}_1,\mathbf{p}_2;b,\varphi),\end{cases} \tag{9.27}$$

or, equivalently, using the scattering angle χ instead of the impact parameter b. As the total momentum and energy are conserved in the elastic binary collision, these momenta satisfy the conservation laws

$$\mathbf{p}_1 + \mathbf{p}_2 = \mathbf{p}'_1 + \mathbf{p}'_2 \qquad \text{and} \qquad \mathbf{p}_1^2 + \mathbf{p}_2^2 = \mathbf{p}_1'^2 + \mathbf{p}_2'^2. \tag{9.28}$$

The outgoing momenta can also be expressed according to

$$\begin{cases}\mathbf{p}'_1 = \mathbf{p}_1 + (\boldsymbol{\epsilon}_{21}\cdot\mathbf{p}_{21})\,\boldsymbol{\epsilon}_{21}, \\ \mathbf{p}'_2 = \mathbf{p}_2 - (\boldsymbol{\epsilon}_{21}\cdot\mathbf{p}_{21})\,\boldsymbol{\epsilon}_{21},\end{cases} \tag{9.29}$$

where $\boldsymbol{\epsilon}_{21} \equiv \mathbf{r}_{21}/\|\mathbf{r}_{21}\|$ is the unit vector in the direction of the relative position at the point of closest approach and $\mathbf{p}_{21} \equiv \mathbf{p}_2 - \mathbf{p}_1$.

Because of momentum and energy conservation (9.28) in binary collisions, the Boltzmann equation (9.26) can be equivalently written as

$$\begin{aligned}\frac{\partial f}{\partial t} + \frac{\mathbf{p}}{m}\cdot\frac{\partial f}{\partial \mathbf{r}} + \mathbf{F}^{(\text{ext})}\cdot\frac{\partial f}{\partial \mathbf{p}} \\ = \int d\mathbf{p}_2\, d\mathbf{p}'\, d\mathbf{p}'_2\, w(\mathbf{p},\mathbf{p}_2|\mathbf{p}',\mathbf{p}'_2)\left[f(\mathbf{r},\mathbf{p}',t)\, f(\mathbf{r},\mathbf{p}'_2,t) - f(\mathbf{r},\mathbf{p},t)\, f(\mathbf{r},\mathbf{p}_2,t)\right]\end{aligned} \tag{9.30}$$

in terms of the transition rate density

$$w(\mathbf{p}_1,\mathbf{p}_2|\mathbf{p}'_1,\mathbf{p}'_2) \equiv \frac{8}{m}\,\sigma_{\text{diff}}\,\delta^3(\mathbf{p}_1+\mathbf{p}_2-\mathbf{p}'_1-\mathbf{p}'_2)\,\delta\!\left(\mathbf{p}_1^2+\mathbf{p}_2^2-\mathbf{p}_1'^2-\mathbf{p}_2'^2\right). \tag{9.31}$$

This latter has the following symmetries:

$$\text{time reversal:} \quad w\left(\mathbf{p}_1,\mathbf{p}_2|\mathbf{p}_1',\mathbf{p}_2'\right) = w\left(-\mathbf{p}_1',-\mathbf{p}_2'|-\mathbf{p}_1,-\mathbf{p}_2\right), \tag{9.32}$$

$$\text{orthogonal:} \quad w\left(\mathbf{p}_1,\mathbf{p}_2|\mathbf{p}_1',\mathbf{p}_2'\right) = w\left(\mathbf{R}\cdot\mathbf{p}_1,\mathbf{R}\cdot\mathbf{p}_2\middle|\mathbf{R}\cdot\mathbf{p}_1',\mathbf{R}\cdot\mathbf{p}_2'\right), \tag{9.33}$$

$$\text{inversion:} \quad w\left(\mathbf{p}_1,\mathbf{p}_2|\mathbf{p}_1',\mathbf{p}_2'\right) = w\left(\mathbf{p}_1',\mathbf{p}_2'|\mathbf{p}_1,\mathbf{p}_2\right), \tag{9.34}$$

where $\mathbf{R}$ is a matrix of the orthogonal group O(3) including spatial rotations and reflections (Huang, 1987). We note that the inversion symmetry results from the time-reversal symmetry combined with the spatial inversion of orthogonal matrix $\mathbf{R} = -\mathbf{1}$, where $\mathbf{1}$ is the 3×3 identity matrix.

Boltzmann's equation provides a coarse-grained description, since the phase-space time evolution is projected onto the six-dimensional space of one-particle position and momentum.

9.2.3 H-Theorem

A famous property of Boltzmann's equation is that the entropy production rate density can be shown to be always nonnegative, in agreement with the second law of thermodynamics. In low-density gases described in terms of the one-particle distribution function, the entropy is given by equation (2.117). The entropy density at position $\mathbf{r}$ and time t can thus be defined as

$$s(\mathbf{r},t) \equiv k_{\mathrm{B}} \int_{\mathbb{R}^3} d\mathbf{p}\, f(\mathbf{r},\mathbf{p},t) \ln \frac{\mathrm{e}}{h^3 f(\mathbf{r},\mathbf{p},t)}, \tag{9.35}$$

so that $S(t) = \int_V s(\mathbf{r},t)\, d\mathbf{r}$. Boltzmann's H quantity was defined as $H(t) \equiv -S(t)/k_{\mathrm{B}}$. According to the time evolution of Boltzmann's equation, the entropy density (9.35) obeys the balance equation

$$\partial_t s + \nabla \cdot \boldsymbol{J}_s = \sigma_s \tag{9.36}$$

with the entropy current density

$$\boldsymbol{J}_s = k_{\mathrm{B}} \int_{\mathbb{R}^3} d\mathbf{p}\, \frac{\mathbf{p}}{m}\, f(\mathbf{r},\mathbf{p},t) \ln \frac{\mathrm{e}}{h^3 f(\mathbf{r},\mathbf{p},t)}, \tag{9.37}$$

and the nonnegative entropy production rate density

$$\sigma_s = \frac{k_{\mathrm{B}}}{4} \int d\mathbf{p}\, d\mathbf{p}_2\, d\Omega\, \sigma_{\mathrm{diff}}\, g \left(f' f_2' - f f_2\right) \ln \frac{f' f_2'}{f f_2} \geq 0, \tag{9.38}$$

where f_a (respectively f_a') denotes the distribution function at the position $\mathbf{r}$ and momentum $\mathbf{p}_a$ (respectively $\mathbf{p}_a'$) for $a=1,2$.[2] This expression is reminiscent of the entropy production rate density (1.38) for reactions of the type $\mathrm{A} + \mathrm{B} \rightleftharpoons \mathrm{C} + \mathrm{D}$, here for particles with different momenta.

[2] The expression (9.38) is nonnegative because $(x - y)\ln(x/y) \geq 0$ for positive real numbers x and y (Balescu, 1975).

As a consequence of the nonnegativity of the entropy production rate density (9.38), the time evolution of the one-particle distribution function undergoes local relaxation towards the Maxwell–Boltzmann equilibrium distribution (2.119) with uniform particle density n and temperature T if the system is isolated. At equilibrium, the following condition of detailed balance, also called reciprocity, is satisfied,

$$w\left(\mathbf{p}_1,\mathbf{p}_2|\mathbf{p}_1',\mathbf{p}_2'\right) f_{\mathrm{eq}}(\mathbf{p}_1') f_{\mathrm{eq}}(\mathbf{p}_2') = w\left(-\mathbf{p}_1',-\mathbf{p}_2'|-\mathbf{p}_1,-\mathbf{p}_2\right) f_{\mathrm{eq}}(-\mathbf{p}_1) f_{\mathrm{eq}}(-\mathbf{p}_2), \tag{9.39}$$

because of the time-reversal symmetry (9.32) and the energy conservation in equation (9.28).

However, this is no longer the case if the system is open and in contact with particle reservoirs at different densities and temperatures because of entropy exchange between the system and its environment through the current density (9.37). Indeed, integrating the entropy balance equation (9.36) over the volume V of the system, we obtain

$$\frac{dS}{dt} = \frac{d_{\mathrm{e}}S}{dt} + \frac{d_{\mathrm{i}}S}{dt} \tag{9.40}$$

with the entropy exchange and production rates

$$\frac{d_{\mathrm{e}}S}{dt} = -\int_{\partial V} \mathbf{J}_s \cdot d\mathbf{\Sigma} \qquad \text{and} \qquad \frac{d_{\mathrm{i}}S}{dt} = \int_V \sigma_s \, d^3r \geq 0, \tag{9.41}$$

where $d\mathbf{\Sigma}$ is the three-dimensional vector surface element in the direction of the system exterior. Hence, the entropy produced inside the system $d_{\mathrm{i}}S/dt > 0$ can be released into the environment if the system is open. For nonequilibrium stationary macrostates, the entropy exchange rate is thus given by $d_{\mathrm{e}}S/dt = -d_{\mathrm{i}}S/dt < 0$, which is the case for gaseous flows between particle reservoirs at different densities and temperatures.

9.2.4 Transport Properties

In dilute gaseous flows, local equilibrium can be reached on the intercollisional timescale, after which the one-particle distribution function takes the form

$$f_{\mathrm{leq}}(\mathbf{r},\mathbf{p},t) = \frac{n(\mathbf{r},t)}{[2\pi m k_{\mathrm{B}} T(\mathbf{r},t)]^{3/2}} \exp\left\{-\frac{[\mathbf{p}-m\mathbf{v}(\mathbf{r},t)]^2}{2mk_{\mathrm{B}}T(\mathbf{r},t)}\right\}, \tag{9.42}$$

where n, T, and $\mathbf{v}$ are, respectively, the density, temperature, and velocity fields in the gas. Substituting this expression into Boltzmann's equation, allows the macroscopic equations of hydrodynamics to be deduced with explicit expressions for the transport coefficients in terms of the microscopic properties of the particles and their mutual interactions (Hirschfelder et al., 1954; Chapman and Cowling, 1960; Balescu, 1975; Résibois and De Leener, 1977; Landau and Lifshitz, 1981; Cercignani, 1988; Liboff, 1990; Reichl, 1998). In particular, the viscosities and heat conductivity for a dilute gas of hard spheres of diameter d and mass m have the following approximate values according to Boltzmann's equation,

$$\eta \simeq \frac{5\,m}{16\,d^2}\sqrt{\frac{k_{\mathrm{B}}T}{\pi m}}, \qquad \zeta = 0, \qquad \kappa \simeq \frac{75\,k_{\mathrm{B}}}{64\,d^2}\sqrt{\frac{k_{\mathrm{B}}T}{\pi m}}, \tag{9.43}$$

which are known with an accuracy of a few percent (Résibois and De Leener, 1977). The kinematic viscosity $\nu = \eta/\rho$ (where $\rho = mn$ is the mass density) and the heat diffusivity defined as $\chi = \kappa/c_{\mathfrak{v}}$ (with the heat capacity per unit volume $c_{\mathfrak{v}} = 3nk_{\rm B}/2$) are thus proportional to the product $\ell\bar{v}$ of the mean free path ℓ with the mean particle velocity $\bar{v}$ up to a dimensionless numerical value, which is a general result in dilute gases (Present, 1958).

9.2.5 Gas Mixtures and Random Lorentz Gases

The Boltzmann equation can be generalized to low-density gas mixtures composed of several atomic species. Each species is described by its own one-particle distribution function $f_k(\mathbf{r}_k, \mathbf{p}_k, t)$ and the elastic binary collisions between the species k and l are characterized by the differential cross sections $d\sigma_{kl}/d\Omega$ with $k, l = 1, 2, \ldots, c$. Accordingly, the kinetics of the gas mixture is ruled by c coupled Boltzmann equations,

$$\frac{\partial f_k}{\partial t} + \frac{\mathbf{p}_k}{m_k} \cdot \frac{\partial f_k}{\partial \mathbf{r}_k} + \mathbf{F}_k^{(\rm ext)} \cdot \frac{\partial f_k}{\partial \mathbf{p}_k} = \sum_{l=1}^{c} \int d\mathbf{p}_l \, d\sigma_{kl} \, g_{kl} \left(f_k' \, f_l' - f_k \, f_l \right), \tag{9.44}$$

where $g_{kl} \equiv \|\dot{\mathbf{r}}_l - \dot{\mathbf{r}}_k\| = \|\mathbf{p}_l/m_l - \mathbf{p}_k/m_k\|$ and f_k (respectively f_k') are the distribution functions with unprime (respectively prime) momenta (Hirschfelder et al., 1954; Chapman and Cowling, 1960).

Random Lorentz gases arise in two-component mixtures where one species is much lighter than the other. The lighter particles are swifter than the heavier ones. These latter may thus be supposed to be quasi immobile for the transport properties of the light particles. In the limit where the mass ratio $m_{\rm heavy}/m_{\rm light}$ is arbitrarily large, the light particles undergo elastic collisions on fixed scatterers. Moreover, the light particles may be assumed to be so small that they do not interact between them. Therefore, they form a gas of independent particles conserving their energy upon their elastic collisions on the fixed scatterers. Their free-flight velocity $g = v$ is thus constant along their whole trajectory. In such circumstances, the distribution function of the heavy particles is invariant and given by $f_{\rm heavy}(\mathbf{r}, \mathbf{p}) = n_{\rm s} \, \delta^3(\mathbf{p})$, where $n_{\rm s}$ is the uniform density of their random distribution in space. The differential cross section of the binary collisions between light and heavy particles is denoted $\sigma_{\rm diff}$, and the Boltzmann equation for the light particles reduces to

$$\left(\frac{\partial}{\partial t} + \frac{\mathbf{p}}{m} \cdot \frac{\partial}{\partial \mathbf{r}} + \mathbf{F}^{(\rm ext)} \cdot \frac{\partial}{\partial \mathbf{p}} \right) f(\mathbf{r}, \mathbf{p}, t) = n_{\rm s} \, v \int \sigma_{\rm diff} \left[f(\mathbf{r}, \mathbf{p}', t) - f(\mathbf{r}, \mathbf{p}, t) \right] d\Omega, \tag{9.45}$$

where $m = m_{\rm light}$ is the mass of the light particles and $\mathbf{p}'$ is the momentum that is incoming the collision in the gain term. This equation is linear, since the particles are independent of each other, so that we may introduce the corresponding probability density $p(\mathbf{r}, \mathbf{p}, t) = f(\mathbf{r}, \mathbf{p}, t)/N$, which is normalized to one, instead of the particle number N. Therefore, we recover the Boltzmann–Lorentz equation (4.121), which rules the time evolution of the position and velocity probability distribution (Lorentz, 1905; Spohn, 1991). Reactions occurring in random Lorentz gases can also be taken into account (Mátyás and Gaspard, 2005).

9.2.6 Applications of Boltzmann's Equation

Boltzmann's kinetic equation has many applications beside the calculation of transport properties in low-density monatomic gases. It can be extended not only to polyatomic gases as previously mentioned, but also to molecular gases by including the rotational and vibrational degrees of freedom of the molecules in the one-particle distribution function (Hirschfelder et al., 1954; Chapman and Cowling, 1960). Reactions can also be described by including the terms due to the inelastic collisions in addition to those of the elastic collisions (Brun, 2009).

In low-density gases, Boltzmann's equation shows that the macroscopic equations of hydrodynamics, including the Navier–Stokes, heat, and diffusion–reaction equations, have a domain of validity limited to gaseous flows where local equilibrium is maintained over spatial scales smaller than the size of the container, ducts, or obstacles (Prigogine, 1949; Hirschfelder et al., 1954; Chapman and Cowling, 1960; Cercignani, 1988).

The predictions of hydrodynamics may become invalid if the velocity distribution manifests significant deviations with respect to the local equilibrium Maxwellian distribution because the macrofields strongly vary over spatial scales shorter than the mean free path as in rarefied gases or shock waves (Cercignani, 2000; Gaspard and Lutsko, 2004). This is especially the case for hypersonic gaseous flows with velocity much larger than the sound of speed.

Boltzmann's equation is also considered in the transport theory of dense gases, plasmas, neutrons, relativistic particles, and electromagnetic radiation, as well as in astrophysics and cosmology. Furthermore, Boltzmann's equation can be generalized to quantum systems (Uehling and Uhlenbeck, 1933; Kadanoff and Baym, 1962; Balescu, 1975; Akhiezer and Peletminskii, 1981; Calzetta and Hu, 2008).

9.2.7 Gas-Surface Interactions

The interactions of the gas with the surface of the container, ducts, or obstacles often play important roles, especially in rarefied gases, where heat transfer between the gas and the surface may occur (Cercignani, 1988, 2000). In general, the interactions of the gas with the surface of a solid depend on many different aspects. The scattering of particles with surfaces has been much studied and many processes are known beyond elastic collisions: adsorption, desorption, diffusion on the surface, as well as reactions (Kreuzer and Gortel, 1986).[3]

The temperature may be different in the gas than the solid surface, inducing heat transfer between them. Here, the solid is supposed to have a high enough heat conductivity that its temperature is uniform. Let us assume that surface adsorption and desorption occur while there is local chemical equilibrium between the surface and the gas.[4] Accordingly, only

[3] Depending on the pressure and temperature conditions, gas particles may also diffuse into the solid to form thermodynamic phases that are more stable. The solid could thus become a sink or a source of gas particles, in which case the difference of chemical potentials of the particles between the gas and the solid would generated an extra particle current. However, diffusive transport is slow in solids, so that these processes are here neglected.

[4] In the absence of local chemical equilibrium, particle surface coverages should also be included in the description as mentioned in Section A.9.

thermal exchanges are envisaged between the gas and the surface. Such surface processes may be taken into account as boundary conditions or by modifying the left-hand side of the Boltzmann equation (9.26) to include the contribution of gas-surface interactions. This latter may be supposed to have the form

$$\left.\frac{\partial f}{\partial t}\right|_{\rm S} = \frac{1}{m}\int_{\rm S} d^2\Sigma\, \delta^3(\mathbf{r}-\mathbf{r}_{\rm S})\Big[\int_{\mathbf{n}\cdot\mathbf{p}'<0} d\mathbf{p}'\, |\mathbf{n}\cdot\mathbf{p}'|\, p_{\mathbf{r}}(\mathbf{p}|\mathbf{p}')\, f(\mathbf{r},\mathbf{p}',t) \\ -(\mathbf{n}\cdot\mathbf{p})\,\theta(\mathbf{n}\cdot\mathbf{p})\, f(\mathbf{r},\mathbf{p},t)\Big], \tag{9.46}$$

where $\mathbf{r}_{\rm S}$ is the position of a point on the surface S, $\mathbf{n}$ is a unit vector normal to the surface at this position, $p_{\mathbf{r}}(\mathbf{p}|\mathbf{p}')$ is the scattering probability density that a particle impinging the surface at the position $\mathbf{r}$ with the momentum $\mathbf{p}'$ will be scattered to the momentum $\mathbf{p}$ such that $\mathbf{n}\cdot\mathbf{p} > 0$, and $\theta(x)$ is Heaviside's function (Cercignani, 2000). This probability density does not depend on the position $\mathbf{r}$ if the surface is homogeneous and the scattering process is the same everywhere. In general, this function is normalized according to

$$\int_{\mathbf{n}\cdot\mathbf{p}>0} p_{\mathbf{r}}(\mathbf{p}|\mathbf{p}')\, d\mathbf{p} = 1 \qquad \text{if} \qquad \mathbf{n}\cdot\mathbf{p}' < 0 \tag{9.47}$$

and it satisfies the following two properties. First, the preservation of equilibrium at the temperature $T_{\rm S}$ of the surface:

$$|\mathbf{n}\cdot\mathbf{p}|\, p_{\rm S}(\mathbf{p}) = \int_{\mathbf{n}\cdot\mathbf{p}'<0} d\mathbf{p}'\, |\mathbf{n}\cdot\mathbf{p}'|\, p_{\mathbf{r}}(\mathbf{p}|\mathbf{p}')\, p_{\rm S}(\mathbf{p}'), \tag{9.48}$$

where

$$p_{\rm S}(\mathbf{p}) = \frac{1}{(2\pi m k_{\rm B} T_{\rm S})^{3/2}} \exp\left(-\frac{\mathbf{p}^2}{2mk_{\rm B}T_{\rm S}}\right) \tag{9.49}$$

is the corresponding Maxwellian equilibrium distribution. The second is the property of reciprocity:

$$p_{\mathbf{r}}(\mathbf{p}|\mathbf{p}')\, |\mathbf{n}\cdot\mathbf{p}'|\, p_{\rm S}(\mathbf{p}') = p_{\mathbf{r}}(-\mathbf{p}'| - \mathbf{p})\, |\mathbf{n}\cdot\mathbf{p}|\, p_{\rm S}(-\mathbf{p}), \tag{9.50}$$

which is implied by the time-reversal symmetry of the underlying microscopic dynamics and the condition of local thermodynamic equilibrium of the surface at the temperature $T_{\rm S}$ of the Maxwellian equilibrium distribution $p_{\rm S}(\mathbf{p})$ (Cercignani, 2000). The total number of particles is always conserved by the contribution (9.46).

In the special case where gas particles undergo elastic collisions on the surface, so that momentum obeys the law of specular reflection $\mathbf{p}' = \mathbf{p} - 2(\mathbf{n}\cdot\mathbf{p})\,\mathbf{n}$, the scattering probability density is given by

$$p_{\mathbf{r}}(\mathbf{p}|\mathbf{p}') = \delta\left[\mathbf{p}' - \mathbf{p} + 2(\mathbf{n}\cdot\mathbf{p})\,\mathbf{n}\right] \tag{9.51}$$

uniformly on the surface, which satisfies all the aforementioned properties. In this case, kinetic energy is conserved because $\mathbf{p}'^2 = \mathbf{p}^2$ at every collision, so that there is no heat transfer between the gas and the surface.

However, the scattering of particles on the surface may be inelastic with energy transfer occurring between the adsorption and the desorption of the particle. If local thermodynamic equilibrium at the surface temperature is assumed, the particles are expected to be desorbed with the corresponding Maxwellian distribution. A simple model achieving this condition is provided by the kernel

$$p_{\mathbf{r}}(\mathbf{p}|\mathbf{p}') = \frac{|\mathbf{n}\cdot\mathbf{p}|}{2\pi(mk_{\mathrm{B}}T_{\mathrm{S}})^2}\exp\left(-\frac{\mathbf{p}^2}{2mk_{\mathrm{B}}T_{\mathrm{S}}}\right), \tag{9.52}$$

which is independent of the incoming momentum $\mathbf{p}'$ and also satisfies the required properties. The kernel (9.52) does not conserve the total energy, contrary to (9.51).

More realistic gas-surface kernels are considered in the literature (Cercignani, 2000). In principle, the form of the kernel $p_{\mathbf{r}}(\mathbf{p}|\mathbf{p}')$ should be determined by studying the quantum scattering of the particles with the surface on the basis of their atomic structure (Kreuzer and Gortel, 1986).

Furthermore, the surfaces may also move with respect to each other, for instance, in Couette–Taylor flows (Cercignani, 2000). In such cases, the relative velocities of the surfaces are extra parameters that control the nonequilibrium drive of the gaseous flow.

9.3 Fluctuating Boltzmann Equation and Fluctuation Relation

9.3.1 Gaseous Flows in Open Systems

Boltzmann's kinetic equation can be established by the mass action law of binary collisions evolving the populations of particles with different velocities. In this regard, Boltzmann's equation is deterministic like the other equations of the macroscopic level of description. Because of its nonlinearity, Boltzmann's equation is not the master equation of a stochastic process.

Since the 1940s, a fluctuating Boltzmann equation has been proposed in order to rule the one-particle distribution function $f(\mathbf{r},\mathbf{p},t)$ as the random variable of a Langevin stochastic process (Siegert, 1949; Ludwig, 1962; Bixon and Zwanzig, 1969; Fox and Uhlenbeck, 1970a,b; van Kampen, 1974; Logan and Kac, 1976; Onuki, 1978; Ernst and Cohen, 1981; Brenig and Van den Broeck, 1980; van Kampen, 1981; Gardiner, 2004). This formulation provides a description between the microscopic level of description where the one-particle distribution function is defined by equation (9.10) and governed by the phase-space dynamics, and the more macroscopic one in terms of the mean distribution $\langle\hat{f}(\mathbf{r},\mathbf{p},t)\rangle$ ruled by Boltzmann's equation. This formulation is appropriate to investigate the fluctuation properties of low-density gaseous flows, in particular, for the energy and particle currents flowing between particle reservoirs.

Figure 9.2 shows examples of such open systems where the gas is flowing between several reservoirs, possibly at different densities or temperatures. The gas can also be in contact with surfaces at their own given temperatures. The gaseous flow of one species in an open system connected to r reservoirs and containing s surfaces at so many different temperatures is characterized by $2r+s$ control parameters, namely, the $r+s$ temperatures of the reservoirs and the surfaces, and the r particle densities of the reservoirs. If the reservoir

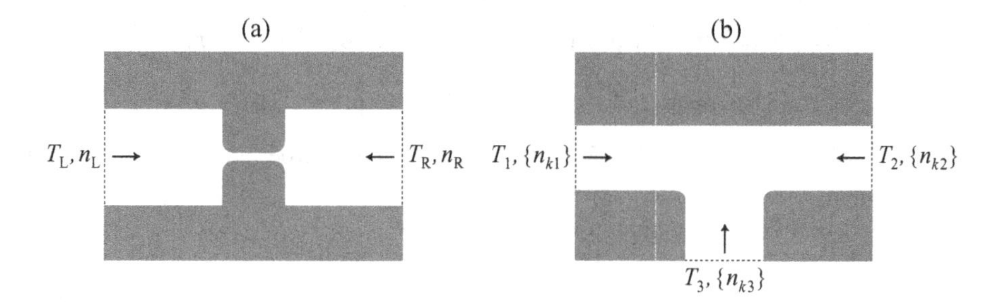

Figure 9.2 Schematic representation of dilute or rarefied gases flowing through pores or pipes between (a) two and (b) three reservoirs at given temperatures and particle densities. In gas mixtures, several particle species k and so many particle densities may exist.

$j = r$ plays the role of reference, the nonequilibrium conditions are thus determined by the $2r + s - 2$ following thermal and chemical affinities,

$$A_{Ej} \equiv \beta_r - \beta_j \qquad \text{for} \quad 1 \leq j \leq r + s - 1, \tag{9.53}$$

$$A_{Nj} \equiv \beta_j \mu_j - \beta_r \mu_r = \ln \frac{\beta_j^{3/2} n_j}{\beta_r^{3/2} n_r} \qquad \text{for} \quad 1 \leq j \leq r - 1, \tag{9.54}$$

where $\beta_j = (k_B T_j)^{-1}$ are the inverse temperatures and μ_j the chemical potentials.

9.3.2 Fluctuating Boltzmann Equation

The fluctuating Boltzmann equation is the following stochastic differential equation of Langevin type for the one-particle distribution function $f(\mathbf{r}, \mathbf{p}, t)$,

$$\frac{\partial f}{\partial t} = \mathscr{A}^{\mathrm{C}}(f) + \mathscr{A}^{\mathrm{F}}(f) + \mathscr{A}^{\mathrm{R}}(f) + \mathscr{A}^{\mathrm{S}}(f) + g(\mathbf{r}, \mathbf{p}, t), \tag{9.55}$$

where $\mathscr{A}^{\mathrm{C,F,R,S}}(f)$ are the deterministic rates due to binary collisions, one-particle Hamiltonian dynamics, particle exchanges with the reservoirs, and gas-surface interactions, while $g(\mathbf{r}, \mathbf{p}, t)$ is a Gaussian white noise field defined in $(\mathbf{r}, \mathbf{p}) \in \mathbb{R}^6$ and satisfying

$$\langle g(\mathbf{r}, \mathbf{p}, t) \rangle = 0, \tag{9.56}$$

$$\langle g(\mathbf{r}, \mathbf{p}, t)\, g(\mathbf{r}', \mathbf{p}', t') \rangle = 2 \left[\mathscr{D}^{\mathrm{C}}(f) + \mathscr{D}^{\mathrm{F}}(f) + \mathscr{D}^{\mathrm{R}}(f) + \mathscr{D}^{\mathrm{S}}(f) \right] \delta(t - t'), \tag{9.57}$$

where the diffusivities $\mathscr{D}^{\mathrm{C,F,R,S}}(f)$ depend on the distribution function $f(\mathbf{r}, \mathbf{p}, t)$ itself.

The binary collisions contribute by Boltzmann's collision term in equation (9.30):

$$\mathscr{A}^{\mathrm{C}}(f) = \int d\mathbf{p}_2\, d\mathbf{p}_1'\, d\mathbf{p}_2'\, w(\mathbf{p}_1, \mathbf{p}_2 | \mathbf{p}_1', \mathbf{p}_2') \left(f_1' f_2' - f_1 f_2 \right), \tag{9.58}$$

where f_1 denotes the distribution function at the position $\mathbf{r} = \mathbf{r}_1$ and momentum $\mathbf{p} = \mathbf{p}_1$. The corresponding diffusivity is given by

$$\mathscr{D}^{\rm C}(f) = \frac{1}{2}\,\delta^3(\mathbf{r}-\mathbf{r}') \int d\mathbf{p}_2\, d\mathbf{p}_3\, d\mathbf{p}_4\, w\left(\mathbf{p}_1,\mathbf{p}_2|\mathbf{p}_3,\mathbf{p}_4\right)(f_1 f_2 + f_3 f_4)$$
$$\times[\delta^3(\mathbf{p}_1-\mathbf{p}')+\delta^3(\mathbf{p}_2-\mathbf{p}')-\delta^3(\mathbf{p}_3-\mathbf{p}')-\delta^3(\mathbf{p}_4-\mathbf{p}')], \tag{9.59}$$

where $f_a = f(\mathbf{r},\mathbf{p}_a,t)$ for $a = 1,2,3,4$ (Brenig and Van den Broeck, 1980; van Kampen, 1981; Gardiner, 2004).

The one-particle Hamiltonian motion is governed by

$$\mathscr{A}^{\rm F}(f) = -\frac{\mathbf{p}}{m}\cdot\frac{\partial f}{\partial \mathbf{r}} - \mathbf{F}^{(\rm ext)}\cdot\frac{\partial f}{\partial \mathbf{p}} \tag{9.60}$$

and $\mathscr{D}^{\rm F}(f) = 0$ because it is noiseless.

Particle exchanges between the system and r reservoirs at given densities and temperatures can be described by

$$\mathscr{A}^{\rm R}(f) = \sum_{j=1}^{r}\frac{1}{m}\int_{{\rm S}_j} d^2\Sigma\,\delta^3(\mathbf{r}-\mathbf{r}_{{\rm S}_j})(\mathbf{n}\cdot\mathbf{p})\theta(\mathbf{n}\cdot\mathbf{p})\left[f_j(\mathbf{p}) - f(\mathbf{r},\mathbf{p},t)\right], \tag{9.61}$$

$$\mathscr{D}^{\rm R}(f) = \frac{1}{2}\,\delta^3(\mathbf{r}-\mathbf{r}')\,\delta^3(\mathbf{p}-\mathbf{p}')$$
$$\times\sum_{j=1}^{r}\frac{1}{m}\int_{{\rm S}_j} d^2\Sigma\,\delta^3(\mathbf{r}-\mathbf{r}_{{\rm S}_j})(\mathbf{n}\cdot\mathbf{p})\theta(\mathbf{n}\cdot\mathbf{p})\left[f_j(\mathbf{p}) + f(\mathbf{r},\mathbf{p},t)\right], \tag{9.62}$$

where ${\rm S}_j$ is the boundary between the system and the jth reservoir, $\mathbf{n}$ is a unit vector normal to the boundary and pointing towards the interior of the system, $\theta(\mathbf{n}\cdot\mathbf{p})$ is the Heaviside function selecting the incoming momenta, and

$$f_j(\mathbf{p}) = \frac{n_j}{(2\pi m k_{\rm B} T_j)^{3/2}}\exp\left(-\frac{\mathbf{p}^2}{2mk_{\rm B}T_j}\right) \tag{9.63}$$

is the Maxwell–Boltzmann distribution function of the gas coming from the jth reservoir at density n_j and temperature T_j (Gaspard, 2013c,e).

The gas-surface interactions are similarly described with the deterministic rate $\mathscr{A}^{\rm S}(f) = \sum_j \partial f/\partial t\big|_{{\rm S}_j}$ obtained by summing equation (9.46) over the surfaces ${\rm S}_j$ surrounding the gaseous flow, and the corresponding diffusivity $\mathscr{D}^{\rm S}(f)$.

The master equation of Fokker–Planck type associated with the fluctuating Boltzmann equation (9.55)–(9.57) is the following functional master equation that rules the time evolution of the probability density functional, $\mathscr{P}[f(\mathbf{r},\mathbf{p})]$, that the random one-particle distribution function would be given by the function $f(\mathbf{r},\mathbf{p})$ at time t (Brenig and Van den Broeck, 1980):

$$\frac{\partial}{\partial t}\mathscr{P}(f) = -\int d\mathbf{r}\,d\mathbf{p}\,\frac{\delta}{\delta f(\mathbf{r},\mathbf{p})}\,\mathscr{A}(f)\,\mathscr{P}(f)$$
$$+\int d\mathbf{r}\,d\mathbf{p}\,d\mathbf{r}'\,d\mathbf{p}'\,\frac{\delta^2}{\delta f(\mathbf{r},\mathbf{p})\,\delta f(\mathbf{r}',\mathbf{p}')}\,\mathscr{D}(f)\,\mathscr{P}(f), \tag{9.64}$$

where $\mathscr{A}(f)$ and $\mathscr{D}(f)$ are the sums of the deterministic rates $\mathscr{A}^{\rm C,F,R,S}(f)$ and the corresponding diffusivities, respectively.

9.3.3 The Coarse-Grained Master Equation

A method to deal with the formidable master equation (9.64) is to coarse grain the system into fictitious cells of volume $\Delta r^3 \Delta p^3$, which are cubic in the position and momentum three-dimensional spaces and centered around the points $(\mathbf{r}_\alpha, \mathbf{p}_\alpha)$. The random number of particles in the cell α is given by

$$N_\alpha \equiv \int_\alpha f(\mathbf{r},\mathbf{p})\, d\mathbf{r}\, d\mathbf{p} \simeq f(\mathbf{r}_\alpha, \mathbf{p}_\alpha)\, \Delta r^3 \Delta p^3. \tag{9.65}$$

If $P(\mathbf{N})$ denotes their probability distribution, the mean particle numbers $\langle \mathbf{N} \rangle = \sum_{\mathbf{N}} \mathbf{N}\, P(\mathbf{N})$ are related to the one-particle distribution function $\langle f \rangle$ ruled by Boltzmann's equation according to $\langle N_\alpha \rangle \simeq \langle f(\mathbf{r}_\alpha, \mathbf{p}_\alpha) \rangle\, \Delta r^3 \Delta p^3$.

At equilibrium in low-density gases, this probability distribution is expected to be given by the multiple Poisson distribution

$$P_{\text{eq}}(\mathbf{N}) = \prod_\alpha \mathrm{e}^{-\langle N_\alpha \rangle_{\text{eq}}} \frac{\langle N_\alpha \rangle_{\text{eq}}^{N_\alpha}}{N_\alpha !} \qquad \text{with} \qquad \langle N_\alpha \rangle_{\text{eq}} = f_{\text{eq}}(\mathbf{p}_\alpha)\, \Delta r^3\, \Delta p^3, \tag{9.66}$$

where f_{eq} is the Maxwell–Boltzmann equilibrium distribution (2.119).

Under time reversal, the cell α is mapped onto the time-reversed cell α^Θ centered around the point $\boldsymbol{\Theta}(\mathbf{r}_\alpha, \mathbf{p}_\alpha) = (\mathbf{r}_\alpha, -\mathbf{p}_\alpha)$. The equilibrium distribution (9.66) is symmetric under time reversal because $\langle N_{\alpha^\Theta} \rangle_{\text{eq}} = \langle N_\alpha \rangle_{\text{eq}}$ by the invariance of the Maxwellian distribution under momentum reversal $\mathbf{p} \to -\mathbf{p}$.

For nonequilibrium conditions, the time evolution of the probability distribution $P(\mathbf{N}, t)$ is ruled by the master equation

$$\frac{dP}{dt} = \hat{L}\, P \qquad \text{with} \qquad \hat{L} = \hat{L}^{\text{C}} + \hat{L}^{\text{F}} + \hat{L}^{\text{R}} + \hat{L}^{\text{S}}, \tag{9.67}$$

where the different contributions are given by

$$\text{binary collisions:}\quad \hat{L}^{\text{C}} P = \sum_{\lambda\mu\rho\sigma} W^{\text{C}}_{\lambda\mu\rho\sigma} \left(\hat{E}_\lambda^{-1} \hat{E}_\mu^{-1} \hat{E}_\rho^{+1} \hat{E}_\sigma^{+1} - 1 \right) N_\rho N_\sigma P, \tag{9.68}$$

$$\text{free flights:}\quad \hat{L}^{\text{F}} P = \sum_{\lambda\rho} W^{\text{F}}_{\lambda\rho} \left(\hat{E}_\lambda^{-1} \hat{E}_\rho^{+1} - 1 \right) N_\rho P, \tag{9.69}$$

$$\text{reservoir exchanges:}\quad \hat{L}^{\text{R}} P = \sum_{\lambda} W^{\text{R, in}}_{\lambda} \left(\hat{E}_\lambda^{-1} - 1 \right) P + \sum_{\lambda} W^{\text{R, out}}_{\lambda} \left(\hat{E}_\lambda^{+1} - 1 \right) N_\lambda P, \tag{9.70}$$

$$\text{surface collisions:}\quad \hat{L}^{\text{S}} P = \sum_{\lambda\rho} W^{\text{S}}_{\lambda\rho} \left(\hat{E}_\lambda^{-1} \hat{E}_\rho^{+1} - 1 \right) N_\rho P. \tag{9.71}$$

These operators are expressed in terms of the rising and lowering operators

$$\hat{E}_\alpha^{\pm 1} P(\dots, N_\alpha, \dots) = \exp(\pm \partial_{N_\alpha})\, P(\dots, N_\alpha, \dots) = P(\dots, N_\alpha \pm 1, \dots), \tag{9.72}$$

adding or removing one particle in the cell α (van Kampen, 1974, 1981).

The rates of the transitions $\rho\sigma \to \lambda\mu$ due to binary collisions are given by

$$W^{\rm C}_{\lambda\mu\rho\sigma} = \frac{1}{2\Delta r^3 \Delta p^6} \int_\lambda d\mathbf{p}_1 \int_\mu d\mathbf{p}_2 \int_\rho d\mathbf{p}'_1 \int_\sigma d\mathbf{p}'_2 \, w\left(\mathbf{p}_1, \mathbf{p}_2 | \mathbf{p}'_1, \mathbf{p}'_2\right) \delta_{\mathbf{r}_\mu, \mathbf{r}_\lambda} \delta_{\mathbf{r}_\rho, \mathbf{r}_\lambda} \delta_{\mathbf{r}_\sigma, \mathbf{r}_\lambda}, \tag{9.73}$$

which do not change the positions but modify the momenta according to the collision rule (Brenig and Van den Broeck, 1980; van Kampen, 1981). These rates satisfy the

$$\text{time-reversal symmetry:} \qquad W^{\rm C}_{\lambda\mu\rho\sigma} = W^{\rm C}_{\rho^\Theta\sigma^\Theta\lambda^\Theta\mu^\Theta}, \tag{9.74}$$

$$\text{inversion symmetry:} \qquad W^{\rm C}_{\lambda\mu\rho\sigma} = W^{\rm C}_{\rho\sigma\lambda\mu}, \tag{9.75}$$

as a consequence of equations (9.32) and (9.34). Moreover, since the binary collision terms (9.68) of the master equation are summed over four indices, the following symmetries also hold (Logan and Kac, 1976),

$$W^{\rm C}_{\lambda\mu\rho\sigma} = W^{\rm C}_{\mu\lambda\rho\sigma} = W^{\rm C}_{\lambda\mu\sigma\rho} = W^{\rm C}_{\mu\lambda\sigma\rho}. \tag{9.76}$$

The transition rates (9.73) also satisfy the property of reciprocity:

$$W^{\rm C}_{\lambda\mu\rho\sigma} \langle N_\rho \rangle_{\rm eq} \langle N_\sigma \rangle_{\rm eq} = W^{\rm C}_{\rho^\Theta\sigma^\Theta\lambda^\Theta\mu^\Theta} \langle N_{\lambda^\Theta} \rangle_{\rm eq} \langle N_{\mu^\Theta} \rangle_{\rm eq}, \tag{9.77}$$

as a consequence of equation (9.39). Therefore, the transitions due to binary collisions obey the principle of detailed balance at equilibrium (van Kampen, 1981; Gardiner, 2004).

The rates of free-flight transitions $\rho \to \lambda$ are given by

$$W^{\rm F}_{\lambda\rho} = \frac{1}{m\,\Delta r} (\mathbf{n}_{\lambda\rho} \cdot \mathbf{p})\,\theta(\mathbf{n}_{\lambda\rho} \cdot \mathbf{p})\, \delta_{\mathbf{r}_\lambda, \mathbf{r}_\rho + \Delta r\, \mathbf{n}_{\lambda\rho}} \delta_{\mathbf{p}_\lambda, \mathbf{p}_\rho}, \tag{9.78}$$

where $\mathbf{n}_{\lambda\rho}$ is a unit vector directed from the center $\mathbf{r}_\rho$ of the cell ρ to the center $\mathbf{r}_\lambda$ of λ (Brenig and Van den Broeck, 1980; Gardiner, 2004). The Kronecker delta in position expresses the fact that the transition happens between nearest-neighboring cells. The other Kronecker delta means that momentum remains constant during free flights. In the case where the particle undergoes an acceleration, this other Kronecker delta should be replaced by the appropriate expression taking into account the force field $\mathbf{F}^{(\rm ext)}(\mathbf{r})$ exerted on the particle. Since one-particle motion is time-reversal symmetric, the corresponding rates have the

$$\text{time-reversal symmetry:} \qquad W^{\rm F}_{\lambda\rho} = W^{\rm F}_{\rho^\Theta\lambda^\Theta}, \tag{9.79}$$

holding because $\mathbf{n}_{\lambda\rho} = -\mathbf{n}_{\rho\lambda}$. Moreover, the transition rates (9.78) satisfy the following property of reciprocity at equilibrium:

$$W^{\rm F}_{\lambda\rho} \langle N_\rho \rangle_{\rm eq} = W^{\rm F}_{\rho^\Theta\lambda^\Theta} \langle N_{\lambda^\Theta} \rangle_{\rm eq}. \tag{9.80}$$

Accordingly, the transitions due to free flights also obey the principle of detailed balance at equilibrium.

Similar expressions can be deduced for the transition rates $W^{\rm R,in}_\lambda$, $W^{\rm R,out}_\lambda$, and $W^{\rm S}_{\lambda\rho}$ from the rates (9.61) for particle exchanges with the reservoirs and (9.46) for gas-surface interactions. We note that the master equation (9.67) describes the process as a combination

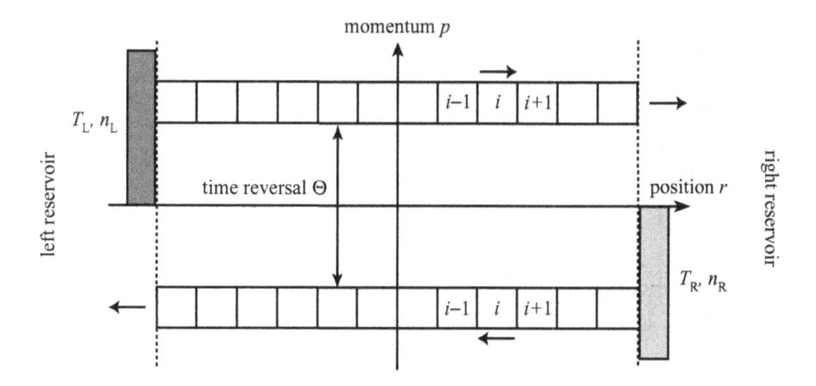

Figure 9.3 Schematic phase portrait of free motion between two reservoirs from which the particles are incoming with different temperatures and densities.

of Poisson processes as in equation (4.34). In the limit where the particle numbers $\{N_\alpha\}$ are large enough, the process may become Gaussian and ruled by the fluctuating Boltzmann equation of Langevin type (9.55).

The exchange of particles between two reservoirs is schematically represented in Figure 9.3 for particles of momenta $\pm p$ jumping between the cells $\alpha = (i, \pm)$ with $i = 1, 2, \ldots, l$, which are located in between the left- and right-hand reservoirs. The time-reversal transformation maps every cell with a given momentum $\pm p$ onto the cell with the opposite momentum $\mp p$: $\boldsymbol{\Theta}(i, \pm) = (i, \mp)$. Restricting the momenta to $\pm p$ for simplicity, the flow operator $\hat{L}^{\text{FR}} = \hat{L}^{\text{F}} + \hat{L}^{\text{R}}$ including the contributions of the free flights (9.69) and the exchanges with the reservoirs (9.70) reads

$$\hat{L}^{\text{FR}} P = \frac{p}{m\Delta r}\Bigg[\big(\hat{E}_{1,+}^{-1} - 1\big)\langle N_+\rangle_{\text{L}}\, P + \sum_{i=1}^{l-1}\big(\hat{E}_{i,+}^{+1}\hat{E}_{i+1,+}^{-1} - 1\big)N_{i,+}P$$
$$+\big(\hat{E}_{l,+}^{+1} - 1\big)N_{l,+}\, P + \big(\hat{E}_{1,-}^{+1} - 1\big)N_{1,-}\, P$$
$$+\sum_{i=1}^{l-1}\big(\hat{E}_{i+1,-}^{+1}\hat{E}_{i,-}^{-1} - 1\big)N_{i+1,-}\, P + \big(\hat{E}_{l,-}^{-1} - 1\big)\langle N_-\rangle_{\text{R}}\, P\Bigg]. \tag{9.81}$$

In the first term of this expression, $\langle N_+\rangle_{\text{L}} = f_{\text{L}}\Delta r^3\Delta p^3$ denotes the mean number of particles coming into the cell $(1, +)$ with the momentum $+p$ from the left-hand reservoir at the temperature T_{L} and density n_{L}. In the last term, $\langle N_-\rangle_{\text{R}} = f_{\text{R}}\Delta r^3\Delta p^3$ is the mean number of particles coming into the cell $(l, -)$ with the momentum $-p$ from the right-hand reservoir. These mean numbers are expressed in terms of the equilibrium Maxwell–Boltzmann distribution function (9.63) in the corresponding reservoir. Taking the right-hand reservoir as reference, the corresponding thermal and chemical affinities are defined as

$$A_E \equiv \beta_{\text{R}} - \beta_{\text{L}} \qquad \text{and} \qquad A_N \equiv \beta_{\text{L}}\mu_{\text{L}} - \beta_{\text{R}}\mu_{\text{R}} = \ln\frac{\beta_{\text{L}}^{3/2} n_{\text{L}}}{\beta_{\text{R}}^{3/2} n_{\text{R}}}. \tag{9.82}$$

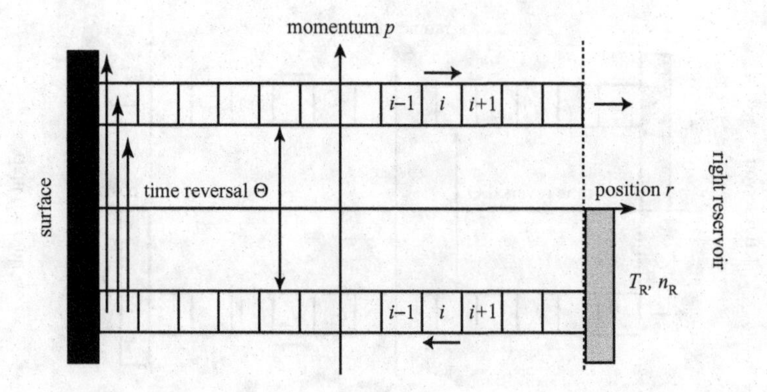

Figure 9.4 Schematic phase portrait of free motion from the right reservoir, elastic or inelastic collisions with a surface, and the motion back to the right reservoir. At the surface, the different vertical arrows show the different possible transitions due to elastic or inelastic collisions. Elastic collisions induce transitions at constant kinetic energy $\varepsilon = \mathbf{p}^2/(2m)$, which is not the case for inelastic collisions.

Accordingly, the logarithm of the ratio of the aforementioned mean particle numbers is given by

$$\ln \frac{\langle N_+ \rangle_{\rm L}}{\langle N_- \rangle_{\rm R}} = \ln \frac{f_{\rm L}}{f_{\rm R}} = A_N + \varepsilon\, A_E \equiv A, \tag{9.83}$$

where $\varepsilon = p^2/(2m)$ is the kinetic energy of the particles.

The phase portrait of gas-surface interactions is schematically represented in Figure 9.4 with a diagram similar to that of Figure 9.3. To fix the ideas, the surface is located at $z = 0$, so that the normal unit vector is $\mathbf{n} = (0, 0, 1)$. If we discretize position and momentum, the term (9.46) describing gas-surface interactions in the coarse-grained master equation is given by

$$\hat{L}^{\rm S} P = \sum_{\mathbf{r}} \delta_{z,0} \sum_{\substack{\mathbf{p} \\ p_z>0}} \sum_{\substack{\mathbf{p}' \\ p'_z<0}} \Delta p^3 \, \frac{|p'_z|}{m\Delta r}\, p_{\mathbf{r}}(\mathbf{p}|\mathbf{p}') \left(\hat{E}^{+1}_{\mathbf{r}\mathbf{p}'}\, \hat{E}^{-1}_{\mathbf{r}\mathbf{p}} - 1 \right) N_{\mathbf{r}\mathbf{p}}\, P, \tag{9.84}$$

expressing the adsorption of a particle of momentum $\mathbf{p}'$ with $p'_z < 0$ and its desorption for the momentum $\mathbf{p}$ with $p_z > 0$ distributed according to some function $p_{\mathbf{r}}(\mathbf{p}|\mathbf{p}')$, which should satisfy the three properties (9.47), (9.48), and (9.50) inducing thermalization at the temperature $T_{\rm S}$ of the surface. Accordingly, the total number of particles is conserved by this operator, although the total energy is not if collisions are inelastic on the surface. The thermal affinity of energy exchanges between the right-hand reservoir and the surface is defined as

$$A_E^{\rm S} \equiv \beta_{\rm R} - \beta_{\rm S}, \tag{9.85}$$

where $\beta_{\rm S} = (k_{\rm B} T_{\rm S})^{-1}$ is the inverse temperature of the surface.

The stationary solution of the master equation (9.67) provides the probability distribution $P_{\mathbf{A}}(\mathbf{N})$ of the nonequilibrium steady state corresponding to nonvanishing values $\mathbf{A} = \left(A_N, A_E, A_E^{\mathrm{S}}\right)$ for the affinities.

9.3.4 The Modified Coarse-Grained Operator

Now, the linear operator of the master equation (9.67) can be modified, as explained in Section 5.7, in order to include the parameters counting particle and energy exchanges between the reservoirs and the surfaces.

For a transition $\mathbf{N} \to \mathbf{N}'$ such that a particle of kinetic energy $\varepsilon_{\mathbf{p}} = \mathbf{p}^2/(2m)$ enters or exits the system through the cell $\alpha = \mathbf{rp}$ in contact with the left-hand reservoir, the off-diagonal elements of the modified operator are given in terms of the operator $\hat{L}$ of the master equation (9.67) by

$$\left(\hat{L}_{\lambda}\right)_{\mathbf{N}'\mathbf{N}} = \left(\hat{L}\right)_{\mathbf{N}'\mathbf{N}}\, \mathrm{e}^{-(\lambda_N+\varepsilon_{\mathbf{p}}\lambda_E)\,\Delta N_{\mathbf{rp}}} \qquad \text{with} \qquad \Delta N_{\mathbf{rp}} = N'_{\mathbf{rp}} - N_{\mathbf{rp}} = \pm 1 \tag{9.86}$$

and the counting parameters λ_N and λ_E. The diagonal elements are not modified:

$$\left(\hat{L}_{\lambda}\right)_{\mathbf{NN}} = - \sum_{\mathbf{N}'(\neq\mathbf{N})} \left(\hat{L}\right)_{\mathbf{N}'\mathbf{N}}. \tag{9.87}$$

In particular, the operator (9.81), which rules the inflow, outflow, and free flights of particles between the two reservoirs, is modified into

$$\begin{aligned}
\hat{L}^{\mathrm{FR}}_{\lambda_N,\lambda_E}\Psi = \frac{p}{m\Delta r}\Bigg[&\left(\mathrm{e}^{-\lambda}\,\hat{E}^{-1}_{1,+} - 1\right)\langle N_+\rangle_{\mathrm{L}}\,\Psi + \sum_{i=1}^{l-1}\left(\hat{E}^{+1}_{i,+}\,\hat{E}^{-1}_{i+1,+} - 1\right)N_{i,+}\,\Psi \\
&+ \left(\hat{E}^{+1}_{l,+} - 1\right)N_{l,+}\,\Psi + \left(\mathrm{e}^{+\lambda}\,\hat{E}^{+1}_{1,-} - 1\right)N_{1,-}\,\Psi \\
&+ \sum_{i=1}^{l-1}\left(\hat{E}^{+1}_{i+1,-}\,\hat{E}^{-1}_{i,-} - 1\right)N_{i+1,-}\,\Psi + \left(\hat{E}^{-1}_{l,-} - 1\right)\langle N_-\rangle_{\mathrm{R}}\,\Psi\Bigg]
\end{aligned} \tag{9.88}$$

with $\lambda = \lambda_N + \varepsilon\lambda_E$ and where $\Psi(\mathbf{N})$ is an arbitrary function of the particle numbers $\mathbf{N}$. The flow operator (9.81) is thus modified by inserting a factor $\exp(\mp\lambda)$ in the terms ruling incoming or outgoing particles from or to the left-hand reservoir, respectively.

Similarly, the operator (9.84) governing gas-surface interactions is modified by introducing an extra counting parameter λ_E^{S} for the energy exchanged during inelastic collisions with the surface, leading to

$$\hat{L}^{\mathrm{S}}_{\lambda_E^{\mathrm{S}}}\Psi = \sum_{\mathbf{r}}\delta_{z,0}\sum_{\substack{\mathbf{p}\\ p_z>0}}\sum_{\substack{\mathbf{p}'\\ p'_z<0}}\Delta p^3\,\frac{|p'_z|}{m\Delta r}\,p_{\mathbf{r}}(\mathbf{p}|\mathbf{p}')\left[\mathrm{e}^{\lambda_E^{\mathrm{S}}(\varepsilon'-\varepsilon)}\hat{E}^{+1}_{\mathbf{rp}'}\,\hat{E}^{-1}_{\mathbf{rp}} - 1\right]N_{\mathbf{rp}}\,\Psi, \tag{9.89}$$

where $\varepsilon = \mathbf{p}^2/(2m)$ and $\varepsilon' = \mathbf{p}'^2/(2m)$.

In contrast, the collision operator (9.68) is not modified because the binary collisions locally exchange particles among all the cells corresponding to the *same* position in the bulk of the gas so that $\Delta N_{\mathbf{rp}} = 0$ in equation (9.86) for these transitions.

Therefore, the operator is modified into

$$\hat{L}_{\lambda_N,\lambda_E,\lambda_E^{\rm S}} = \hat{L}^{\rm C} + \hat{L}^{\rm FR}_{\lambda_N,\lambda_E} + \hat{L}^{\rm S}_{\lambda_E^{\rm S}}. \tag{9.90}$$

We notice the additive structure due to the statistical independence between the different types of transitions. Accordingly, every operator can be treated separately.

9.3.5 The Symmetry of the Modified Operator

The remarkable result is that the modified coarse-grained operator obeys the symmetry relation

$$\eta^{-1}\,\hat{\Theta}\,\hat{L}_{\boldsymbol{\lambda}}(\hat{\Theta}\,\eta\,\Psi) = \hat{L}^{\dagger}_{\mathbf{A}-\boldsymbol{\lambda}}\,\Psi, \tag{9.91}$$

where $\hat{\Theta}\,\Psi(\{N_\alpha\}) = \Psi(\{N_{\alpha^\Theta}\})$ is the time-reversal operator, $\eta(\mathbf{N})$ is the multiple Poisson distribution (9.66) at the temperature $T_{\rm R}$ and density $n_{\rm R}$ of the reference right-hand reservoir, $\boldsymbol{\lambda} = (\lambda_N, \lambda_E, \lambda_E^{\rm S})$ are the counting parameters, $\mathbf{A} = (A_N, A_E, A_E^{\rm S})$ the affinities, and Ψ an arbitrary function. The different parts of the modified operator (9.90) obey the symmetry relations

$$\eta^{-1}\hat{\Theta}\,\hat{L}^{\rm C}(\hat{\Theta}\,\eta\,\Psi) = \hat{L}^{\rm C\dagger}\,\Psi, \tag{9.92}$$

$$\eta^{-1}\hat{\Theta}\,\hat{L}^{\rm FR}_{\lambda_N,\lambda_E}(\hat{\Theta}\,\eta\,\Psi) = \hat{L}^{\rm FR\dagger}_{A_N-\lambda_N,A_E-\lambda_E}\,\Psi, \tag{9.93}$$

$$\eta^{-1}\hat{\Theta}\,\hat{L}^{\rm S}_{\lambda_E^{\rm S}}(\hat{\Theta}\,\eta\,\Psi) = \hat{L}^{\rm S\dagger}_{A_E^{\rm S}-\lambda_E^{\rm S}}\,\Psi, \tag{9.94}$$

in terms of the affinities (9.82) and (9.85). The symmetry (9.92) results from the time-reversal symmetry (9.32) for the rate density of Boltzmann's equation itself, which thus also concerns the transition rates (9.73), as proved in Section F.3.

The symmetry (9.93) of the modified operator (9.88) is obtained by using the properties of the lowering and rising operators (9.72),

$$\left(\hat{E}_\alpha^{\pm 1}\right)^\dagger = \hat{E}_\alpha^{\mp 1}, \qquad \hat{E}_\alpha^{-1}\,\eta\,\Psi = \frac{N_\alpha}{\langle N_\alpha\rangle_{\rm R}}\,\eta\,\hat{E}_\alpha^{-1}\Psi, \qquad \hat{E}_\alpha^{+1}\,\eta\,\Psi = \frac{\langle N_\alpha\rangle_{\rm R}}{N_\alpha+1}\,\eta\,\hat{E}_\alpha^{+1}\,\Psi, \tag{9.95}$$

where $\langle N_\alpha\rangle_{\rm R}$ denotes the mean value of the particle number N_α with respect to the multiple Poisson distribution $\eta(\mathbf{N})$ given by equation (9.66) at the density $n_{\rm R}$ and the temperature $T_{\rm R}$ of the reference right-hand reservoir. This distribution has the symmetry $\hat{\Theta}\eta = \eta$ under the time-reversal transformation $\hat{\Theta}N_{i,\pm} = N_{i,\mp}$ reversing the momenta $\pm p$ at every position $i = 1, 2, \ldots, l$. Therefore, the transformation in the left-hand side of equation (9.93) has the effect of permuting the terms of positive momentum with those of negative momentum. Furthermore, the boundary term with ${\rm e}^{-\lambda}\langle N_+\rangle_{\rm L}$ has this coefficient transformed into ${\rm e}^{-\lambda}\langle N_+\rangle_{\rm L}/\langle N_{1,-}\rangle_{\rm R}$ and the boundary term with ${\rm e}^{+\lambda}$ into ${\rm e}^{+\lambda}\langle N_{1,+}\rangle_{\rm R}$. Now, the left-hand reservoir is at the density $n_{\rm L}$ and temperature $T_{\rm L}$ while the right-hand reservoir is used as reference, so that $\langle N_{i,\pm}\rangle_{\rm R} = f_{\rm R}\Delta r^3\Delta p^3$ for all $i = 1, 2, \ldots, l$. Hence, we find the required consistency between the two following equalities:

$$\mathrm{e}^{-\lambda}\,\frac{\langle N_+\rangle_{\rm L}}{\langle N_{1,-}\rangle_{\rm R}} = \mathrm{e}^{A-\lambda}, \qquad \mathrm{e}^{+\lambda}\,\langle N_{1,+}\rangle_{\rm R} = \mathrm{e}^{-(A-\lambda)}\,\langle N_+\rangle_{\rm L}, \tag{9.96}$$

where A is the affinity defined by equation (9.83). Consequently, the two boundary terms with the counting parameter λ may exchange their role if we carry out the transformation $\lambda \to A - \lambda$. We thus find the symmetry relation (9.93).

The symmetry (9.94) of the modified operator (9.89) holds because the function $p_{\mathbf{r}}(\mathbf{p}|\mathbf{p}')$ satisfies the property of reciprocity (9.50), finding its origin in the microreversibility of gas-surface interactions.

For an open system in contact with two reservoirs and a thermalizing surface, the symmetry relation (9.91) can thus be proved with the counting parameters $\boldsymbol{\lambda} = \left(\lambda_N, \lambda_E, \lambda_E^{\rm S}\right)$ and the affinities $\mathbf{A} = \left(A_N, A_E, A_E^{\rm S}\right)$ (Gaspard, 2013c,e).

9.3.6 Fluctuation Relation for Energy and Particle Currents

As in Chapter 5, the cumulant generating function for the particle and energy currents can be defined by equation (5.76) with $\Delta\mathfrak{C} = \left(\Delta N, \Delta E, \Delta E^{\rm S}\right)$ and the counting parameters $\boldsymbol{\lambda} = \left(\lambda_N, \lambda_E, \lambda_E^{\rm S}\right)$. As explained in Section 5.7, the cumulant generating function is given by the leading eigenvalue of the modified operator according to

$$\hat{L}_{\boldsymbol{\lambda}}\,\Psi_{\boldsymbol{\lambda}} = -Q(\boldsymbol{\lambda};\mathbf{A})\,\Psi_{\boldsymbol{\lambda}}. \tag{9.97}$$

Since the modified operator obeys the symmetry relation (9.91), the cumulant generating function satisfies

$$Q(\boldsymbol{\lambda};\mathbf{A}) = Q(\mathbf{A}-\boldsymbol{\lambda};\mathbf{A}), \tag{9.98}$$

from which we can deduce the fluctuation relation for the currents

$$\frac{p_t\left(\Delta N, \Delta E, \Delta E^{\rm S}; A_N, A_E, A_E^{\rm S}\right)}{p_t\left(-\Delta N, -\Delta E, -\Delta E^{\rm S}; A_N, A_E, A_E^{\rm S}\right)} \sim_{t\to\infty} \mathrm{e}^{A_N\,\Delta N + A_E\,\Delta E + A_E^{\rm S}\,\Delta E^{\rm S}} \tag{9.99}$$

with large-deviation theory. As a consequence, the properties of Chapter 5 hold for the linear and nonlinear transport properties and the entropy production rate is given by

$$\frac{1}{k_{\rm B}}\,\frac{d_{\rm i}S}{dt} = A_N\,J_N + A_E\,J_E + A_E^S\,J_E^S \geq 0 \tag{9.100}$$

in terms of the mean currents $\mathbf{J} = \left(J_N, J_E, J_E^S\right) = \lim_{t\to\infty}(1/t)\langle\Delta\mathfrak{C}(t)\rangle_{\mathbf{A}}$ because of equations (5.97) and (5.98). The results extend to low-density gaseous flows between more than two reservoirs and in contact with several thermalizing surfaces (Gaspard, 2013c,e).

These considerations can be further developed to find the large-deviation action of path probabilities for the stochastic time evolution of the one-particle distribution function $f(\mathbf{r},\mathbf{p},t)$ generated by the fluctuating Boltzmann equation (Bouchet, 2020).

9.4 Integral Fluctuation Relation

The deviations of the exact phase-space probability distribution (2.45) with respect to its factorization (2.116) into a product of one-particle distribution functions can be characterized as follows in isolated systems. If the particles are statistically uncorrelated as assumed by Boltzmann, the phase-space probability distribution may be written in the factorized form

$$p_{\rm B}(\boldsymbol{\Gamma};f) \equiv C_N \prod_{a=1}^{N} f(\mathbf{r}_a,\mathbf{p}_a) \tag{9.101}$$

with $C_N = N!/N^N$. The entropy $S(f)$ of this probability distribution is given by equation (2.117).

If such a factorized distribution is supposed to hold at the initial time $t = 0$ with the one-particle distribution function $f_0(\mathbf{r},\mathbf{p})$, its Hamiltonian time evolution is given by

$$p_t(\boldsymbol{\Gamma}) = {\rm e}^{\hat{L}t} p_{\rm B}(\boldsymbol{\Gamma};f_0) = p_{\rm B}(\boldsymbol{\Gamma}_{-t};f_0), \tag{9.102}$$

where $\boldsymbol{\Gamma}_{-t} = \boldsymbol{\Phi}^{-t}\boldsymbol{\Gamma}$. At time t, the one-particle distribution function can be defined as

$$f_t(\mathbf{r},\mathbf{p}) \equiv \int \hat{f}(\mathbf{r},\mathbf{p};\boldsymbol{\Gamma})\, p_t(\boldsymbol{\Gamma})\, d\boldsymbol{\Gamma} = N \int p_t(\mathbf{x},\mathbf{x}_2,\dots,\mathbf{x}_N)\, d\mathbf{x}_2\cdots d\mathbf{x}_N, \tag{9.103}$$

where $\mathbf{x} = (\mathbf{r},\mathbf{p})$ and $\mathbf{x}_a = (\mathbf{r}_a,\mathbf{p}_a)$ for $2 \le a \le N$, because of the total symmetry (2.51) of the probability distribution for the N identical particles.

The exact time evolution of the phase-space distribution does not preserve the factorized form (9.101) into a product of time-dependent one-particle distribution functions $f_t(\mathbf{r},\mathbf{p})$. Nevertheless, the deviations can be expressed as

$$p_t(\boldsymbol{\Gamma}) = p_{\rm B}(\boldsymbol{\Gamma};f_t)\, {\rm e}^{\Sigma_t(\boldsymbol{\Gamma})} \tag{9.104}$$

with

$$\Sigma_t(\boldsymbol{\Gamma}) \equiv -\int_0^t d\tau\, \partial_\tau \ln p_{\rm B}(\boldsymbol{\Gamma}_{\tau-t};f_\tau) = \ln \frac{p_{\rm B}(\boldsymbol{\Gamma}_{-t};f_0)}{p_{\rm B}(\boldsymbol{\Gamma};f_t)}. \tag{9.105}$$

Accordingly, the mean value (2.36) of any observable $A(\boldsymbol{\Gamma})$ over the exact time-evolved phase-space distribution can be written as

$$\langle A(\boldsymbol{\Gamma})\rangle_t = \left\langle A(\boldsymbol{\Gamma})\, {\rm e}^{\Sigma_t(\boldsymbol{\Gamma})}\right\rangle_{{\rm B},f_t}, \tag{9.106}$$

where $\langle\cdot\rangle_{{\rm B},f}$ denotes the mean value with respect to the factorized distribution (9.101). Setting $A(\boldsymbol{\Gamma}) = {\rm e}^{-\Sigma_t(\boldsymbol{\Gamma})}$, we obtain the result that the exact Hamiltonian time evolution obeys the following integral fluctuation relation,

$$\left\langle {\rm e}^{-\Sigma_t(\boldsymbol{\Gamma})}\right\rangle_t = 1. \tag{9.107}$$

As a consequence of the Jensen inequality (3.46), the mean value of the quantity (9.105) can be expressed as Kullback–Leibler divergences:

$$\langle \Sigma_t(\boldsymbol{\Gamma})\rangle_t = D_{\rm KL}\big[p_t(\boldsymbol{\Gamma})\| p_{\rm B}(\boldsymbol{\Gamma};f_t)\big] = D_{\rm KL}\left[p_{\rm B}(\boldsymbol{\Gamma};f_0)\| p_{\rm B}\left(\boldsymbol{\Phi}^t\boldsymbol{\Gamma};f_t\right)\right] \ge 0, \tag{9.108}$$

which are always nonnegative. Because of the definition (9.105), this quantity is related to the difference between the values of the entropy (2.117) at the time t and $t = 0$ according to

$$k_{\rm B}\,\langle \Sigma_t(\boldsymbol{\Gamma})\rangle_t = S(f_t) - S(f_0) = \Delta_{\rm i} S \ge 0, \tag{9.109}$$

which is the entropy production during the time interval $[0,t]$ since the system is isolated, so that $\Delta_{\rm e}S = 0$ and $\Delta S = \Delta_{\rm i}S$. For low-density gases, we thus find results similar to equations (3.45) and (3.47) for hydrodynamics.

In the domain of validity of Boltzmann's equation, the entropy production (9.109) can be evaluated according to $\Delta_{\rm i}S = \int_0^t d_{\rm i}S = \int_0^t \int_V \sigma_s \, d^3r \, dt \geq 0$ in terms of the entropy production rate density (9.38), as shown in Section 9.2.3. If the particle velocity distribution is close to local equilibrium in the gaseous flow, the value of the entropy production (9.109) is the one given by nonequilibrium thermodynamics in Chapter 1 or by the microscopic approach to hydrodynamics in Chapter 3. Away from local equilibrium, the velocity distribution might be undergoing thermalization by the kinetic modes of relaxation towards local Maxwellian equilibrium, in which case the entropy production rate density (9.38) would be larger than its local-equilibrium value (1.24).

10
Fluctuating Chemohydrodynamics

10.1 The Principles of Fluctuating Chemohydrodynamics

Fluctuating hydrodynamics finds its origin in the work of Landau and Lifshitz (1957). Since then, this stochastic theory has proved to be powerful for the study of fluids and other continuous media at intermediate scales between the microscopic and macroscopic levels of description (Landau and Lifshitz, 1980b; Fox and Uhlenbeck, 1970a; Spohn, 1991; Ortiz de Zárate and Sengers, 2006; Bertini et al., 2015; Jülicher et al., 2018). The theory is referred to as fluctuating chemohydrodynamics if fluid mechanics is combined with diffusion–reaction processes. The general idea is similar to the one used for the fluctuating Boltzmann equation in Section 9.3.

The approach consists in the stochastic extension of the partial differential equations of hydrodynamics in the same way as Langevin's stochastic equations extend Newton's deterministic equations with Stokes' friction force to Brownian motion. This extension is important to describe the properties of continuous media on spatial scales where thermal fluctuations are important. These properties can be observed, in particular, using light scattering, in which case the intensity of scattered light is related to the spatiotemporal Fourier transform of the density autocorrelation function $\langle \hat{n}(\mathbf{r},t)\, \hat{n}(\mathbf{r}',t')\rangle$ (Berne and Pecora, 1976; Boon and Yip, 1980). At equal times $t = t'$, the density fluctuations depend on the isothermal compressibility κ_T according to equation (C.30), and $(k_{\rm B} T \kappa_T)^{1/3}$ is the characteristic spatial scale of the equilibrium fluctuations.[1] At different times $t \neq t'$, the density autocorrelation function characterizes the dynamic properties of fluctuations.

In fluctuating hydrodynamics, the fluid is assumed to be in local equilibrium and described in terms of the same fields as in hydrodynamics, i.e., the fluid velocity and the densities of energy and particles, but considered as random fields. This extension can be introduced by comparing the macroscopic partial differential equation (1.16) for some macrofield $x(\mathbf{r},t)$ of atomic mechanical origin to its microscopic version

$$\partial_t\, \hat{x}(\mathbf{r};\boldsymbol{\Gamma}_t) + \nabla \cdot \hat{\jmath}_x(\mathbf{r};\boldsymbol{\Gamma}_t) = \hat{\sigma}_x(\mathbf{r};\boldsymbol{\Gamma}_t) \qquad \text{with} \qquad \boldsymbol{\Gamma}_t = \boldsymbol{\Phi}^t \boldsymbol{\Gamma}_0, \tag{10.1}$$

where $\hat{x}$, $\hat{\jmath}_x$, and $\hat{\sigma}_x$ are, respectively, the corresponding microscopic density, current density, and rate density, as considered in Chapter 3.

[1] Light scattering is especially striking in critical opalescence, where compressibility becomes infinite.

On spatiotemporal scales larger than the interaction range and the mean intercollisional time, the differences

$$\delta\mathcal{J}_x(\mathbf{r},t) \equiv \hat{\jmath}_x(\mathbf{r};\boldsymbol{\Gamma}_t) - \langle \hat{\jmath}_x(\mathbf{r};\boldsymbol{\Gamma})\rangle_t, \qquad \delta\sigma_x(\mathbf{r},t) \equiv \hat{\sigma}_x(\mathbf{r};\boldsymbol{\Gamma}_t) - \langle \hat{\sigma}_x(\mathbf{r};\boldsymbol{\Gamma})\rangle_t, \tag{10.2}$$

between the macroscopic and microscopic current and rate densities are noise fields, obeying specific probability laws that should be deduced from the underlying microscopic dynamics (Spohn, 1991; Bertini et al., 2015). These probability laws are assumed to depend on the fields themselves (such as the local energy and particle densities) because the continuous medium is assumed to be in local thermodynamic equilibrium. Accordingly, the microscopic field ruled by equation (10.1) is approximated by the random field $x(\mathbf{r},t)$ governed by the stochastic partial differential equation

$$\partial_t x + \nabla\cdot\left(\jmath_x + \delta\mathcal{J}_x\right) = \sigma_x + \delta\sigma_x, \tag{10.3}$$

where $\jmath_x$ and σ_x are functions that depend on the different random fields $\{x\}$ and their gradients $\{\nabla x\}$, while $\delta\mathcal{J}_x$ and $\delta\sigma_x$ are the noise fields (10.2). In the case where x is a locally conserved quantity, we have that $\jmath_x = \bar{\jmath}_x + \mathcal{J}_x$ in terms of the dissipativeless and dissipative current densities defined by equations (3.52) and (3.53), respectively.

Generalized Langevin stochastic equations ruling the fields can be deduced from the microscopic dynamics using the theory of Mori (1965), as explained in Section 2.9.2. Accordingly, stochastic forcing by the fast degrees of freedom obeys the generalized fluctuation–dissipation theorem (2.194) with the memory kernel (2.193). For the hydrodynamic fields of normal fluids, this memory kernel has short-ranged spatiotemporal correlations, so that their correlation functions may be supposed to be given by Dirac delta distributions as in equation (4.174) for Langevin's fluctuating force. Similarly, the fluctuations (10.2) can be assumed to be Gaussian if these quantities are considered on spatial scales such that the corresponding volumes contain enough particles for the central limit theorem to apply. In such circumstances, the system may be considered to evolve in the linear regime close to local equilibrium where the fluctuation–dissipation theorem holds (Callen and Welton, 1951). As explained in Section 2.8.4, the amplitude of the fluctuations should thus be determined by the linear response coefficients ruling the relaxation and dissipation of nonequilibrium deviations with respect to local equilibrium.

In the linear regime close to local equilibrium, equation (1.28) for the current and rate densities $\{\mathcal{J}_\alpha\}$ of Table 1.2 is thus replaced by its stochastic version,

$$\mathcal{J}_\alpha = \sum_\beta \mathcal{L}_{\alpha,\beta}\,\mathcal{A}_\beta + \delta\mathcal{J}_\alpha, \tag{10.4}$$

where the coefficients $\mathcal{L}_{\alpha,\beta}$ are the linear response coefficients and the fluctuations $\{\delta\mathcal{J}_\alpha\}$ are assumed to be Gaussian white noise fields obeying the fluctuation–dissipation theorem, so that

$$\langle\delta\mathcal{J}_\alpha(\mathbf{r},t)\rangle = 0, \tag{10.5}$$

$$\langle\delta\mathcal{J}_\alpha(\mathbf{r},t)\,\delta\mathcal{J}_\beta(\mathbf{r}',t')\rangle = k_{\rm B}\,\underbrace{\left(\mathcal{L}_{\alpha,\beta}+\mathcal{L}_{\beta,\alpha}\right)}_{2\mathcal{L}^{\rm S}_{\alpha,\beta}}\,\delta^3(\mathbf{r}-\mathbf{r}')\,\delta(t-t'). \tag{10.6}$$

We note that the coefficients $\mathcal{L}_{\alpha,\beta}$ may have possible spatiotemporal dependences, such as a diffusion coefficient depending on the particle density, or a heat conductivity on the temperature.

Accordingly, the path probability for the stochastic process generated by these Gaussian white noise fields is given by

$$P[\delta\mathcal{J}(\mathbf{r},t)] \propto \exp\left[-\frac{1}{4k_{\mathrm{B}}}\int \left(\mathcal{L}^{\mathrm{S}}\right)^{-1} : \delta\mathcal{J}^2\, d\mathbf{r}\, dt\right] \tag{10.7}$$

with the vector $\delta\mathcal{J} = (\delta\mathcal{J}_\alpha)$ and the symmetric matrix $\mathcal{L}^{\mathrm{S}} = \left(\mathcal{L}^{\mathrm{S}}_{\alpha,\beta}\right)$, which determines the entropy production rate density (1.32) at the macroscopic level of description. The approach is thus reminiscent of the theory by Onsager and Machlup (1953). In general, the path probability (10.7) should be supplemented by the constraints due to the local conservation laws (Bertini et al., 2015). Since the fields are random variables, the entropy should be defined by considering their probability distribution, allowing us to establish the balance of entropy during time evolution and the entropy production of the stochastic process in fluctuating chemohydrodynamics (Mazur, 1999; Rubí and Mazur, 2000). In the noiseless limit, which would formally correspond to the limit $k_{\mathrm{B}} \to 0$, equation (1.28) is recovered and the entropy production rate density is thus given by equation (1.32).

These principles can be extended to interfacial processes by considering their surface current densities $\mathcal{J}^{\mathrm{s}}_\alpha$ and affinities $\mathcal{A}^{\mathrm{s}}_\alpha$, as discussed in Section A.9 (Bedeaux et al., 1977; Bedeaux, 1986; Gaspard and Kapral, 2018b, 2019b).

Furthermore, because of microreversibility, the linear response coefficients $\mathcal{L}_{\alpha,\beta}$ with $\alpha \neq \beta$ obey the Onsager–Casimir reciprocal relations (2.173). If both affinities $\mathcal{A}_\alpha$ and $\mathcal{A}_\beta$ have the same parity under time reversal, i.e., if $\epsilon_\alpha\epsilon_\beta = +1$, the Onsager reciprocal relations $\mathcal{L}_{\alpha,\beta} = \mathcal{L}_{\beta,\alpha}$ are satisfied. Otherwise, if $\epsilon_\alpha\epsilon_\beta = -1$, the relations $\mathcal{L}_{\alpha,\beta} = -\mathcal{L}_{\beta,\alpha}$ hold, which may be the case for specific coupling phenomena in some matter phases with broken continuous symmetries as reported in Section 3.3 (Mabillard and Gaspard, 2020) and in the interfacial phoretic phenomena mentioned in Section A.9 (Gaspard and Kapral, 2018b,c, 2019b). In this latter case, the antisymmetric linear response coefficients $\mathcal{L}_{\alpha,\beta} = -\mathcal{L}_{\beta,\alpha}$ do not contribute to the thermodynamic entropy production given by equation (1.24) and there are no correlations between the corresponding noises $\delta\mathcal{J}_\alpha$ and $\delta\mathcal{J}_\beta$ because of equation (10.6).

In the case of normal fluids in global equilibrium, the results of Section C.2 are recovered, according to which the fluctuations of energy and particle densities are Gaussian and short ranged.[2]

Remarkably, the fluctuations of chemohydrodynamic fields may have long-ranged spatial correlations in nonequilibrium steady macrostates, although they are short-ranged at equilibrium (Onuki, 1978; Nicolis and Malek Mansour, 1984; Spohn, 1991; Dorfman et al., 1994; Ortiz de Zárate and Sengers, 2006). Another important issue is that, although the fluctuation–dissipation theorem is locally obeyed in the continuous medium, the global

[2] In continuous media with broken continuous symmetries, the gradients (3.138) of the local order parameters can be included in the framework of fluctuating hydrodynamics since they have short- or medium-ranged fluctuations according to equation (3.139) at global equilibrium.

transport properties between the boundaries of the system may be nonlinear in nonequilibrium steady macrostates, as shown in Section 1.7 for diffusion and electric conduction. As a consequence of microreversibility, these nonlinear transport properties satisfy time-reversal symmetry relations, which can be deduced from the fluctuation relation for currents (5.74).

In this chapter, fluctuating chemohydrodynamics is considered for diffusive particle transport, diffusion-influenced surface reactions, ion transport, the transport of charge carriers in diodes and transistors, and driven Brownian motion. The fluctuation relation for currents and its implications are considered for each of these processes.

10.2 Transport by Diffusion

10.2.1 Stochastic Diffusion Equation

Transport by diffusion without or with an external field was considered in Sections 1.7.2 and 1.7.3 at the macroscale in the absence of fluctuations. The same problem is here extended to include the effect of fluctuations at the mesoscopic level of description.

The medium where diffusion takes place is assumed to be isotropic, isothermal, and at rest with zero velocity, so that there is no transport by advection. According to equations (10.4)–(10.6), Gaussian white noises should be added to the diffusive current density. Since this latter is vectorial as well as the associated affinity, we have that

$$\mathcal{J} = \mathcal{L} \cdot \mathcal{A} + \delta \mathcal{J} \qquad \text{with} \qquad \mathcal{A} = \nabla(-\mu/T). \tag{10.8}$$

If the density of particles is low enough, the chemical potential is given by $\mu = \mu^0 + k_{\mathrm{B}} T \ln\left(n/n^0\right) + u$ with the external energy potential $u(\mathbf{r})$ for the diffusing particles. Therefore, the affinity can be expressed as $\mathcal{A} = -k_{\mathrm{B}}\, n^{-1}(\nabla n + \beta n \nabla u)$ with $\beta = (k_{\mathrm{B}} T)^{-1}$. Since Fick's law, $\mathcal{J} = -\mathcal{D} \nabla n$, should hold at the macroscale in the absence of external energy potential, we find that $k_{\mathrm{B}} \mathcal{L} = \mathcal{D} n \mathbf{1}$, where $\mathbf{1}$ is the 3×3 unit matrix. In general, the diffusion coefficient may depend on the particle density, $\mathcal{D}(n)$.

If the velocity of the medium is equal to zero, i.e., $\mathbf{v} = 0$, the particle current density is given by its diffusive part, $\jmath \equiv n\mathbf{v} + \mathcal{J} = \mathcal{J}$. Accordingly, we shall use the notation $\delta \jmath \equiv \delta \mathcal{J}$ for the fluctuations of the particle current density. As a consequence of equations (10.5)–(10.6), these fluctuations are given by the vectorial Gaussian white noise field,

$$\langle \delta \jmath(\mathbf{r},t) \rangle = 0 \qquad \text{and} \qquad \langle \delta \jmath(\mathbf{r},t)\, \delta \jmath(\mathbf{r}',t') \rangle = 2\, \mathcal{D}\, n\, \delta^3(\mathbf{r}-\mathbf{r}')\, \delta(t-t')\, \mathbf{1} \tag{10.9}$$

in terms of the local particle density $n = n(\mathbf{r},t)$. The stochastic diffusion equation has thus the form

$$\partial_t\, n + \nabla \cdot \jmath = 0 \tag{10.10}$$

with the current density

$$\jmath = -\mathcal{D}\, \nabla n - \beta \mathcal{D}\, n\, \nabla u + \delta \jmath = -\mathcal{D}\, \mathrm{e}^{-\beta u}\, \nabla\left(\mathrm{e}^{\beta u} n\right) + \delta \jmath. \tag{10.11}$$

As in Figure 1.5, diffusion is considered between two reservoirs with particle densities n_{L} and n_{R} and separated by the distance L in the x-direction. In the transverse directions, the conductive medium has the cross-sectional area Σ. After some time, the system reaches

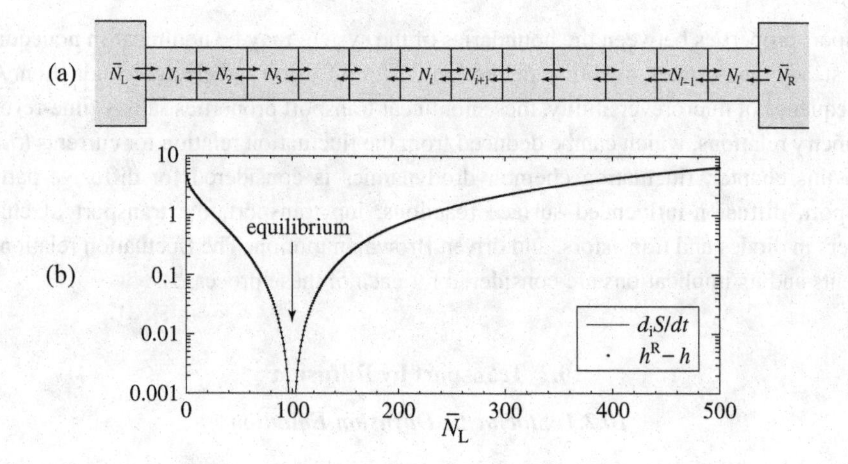

Figure 10.1 (a) Schematic representation of stochastic transport by diffusion in a channel extending between two reservoirs with densities n_L and n_R. The channel is discretized into l cells with volume ΔV and containing the numbers $(N_i)_{i=1}^l$ of particles. Here, $\bar{N}_L = n_L \Delta V$ and $\bar{N}_R = n_R \Delta V$ are the corresponding mean particle numbers in the left- and right-hand reservoirs. (b) Entropy production rate in units where $k_B = 1$ versus the mean particle number $\bar{N}_L$ for diffusion with the rate constant $k = 1$ in a channel of length $l = 99$ for the fixed value $\bar{N}_R = 100$. The thermodynamic equilibrium happens at $\bar{N}_L = \bar{N}_R$ where the entropy production rate vanishes. Reproduced with permission from Gaspard (2005) © IOP Publishing Ltd and Deutsche Physikalische Gesellschaft.

a stationary macrostate, where the mean current is given by equation (1.68). There is a flux of particles and the system is out of equilibrium if the densities take different values $n_L \neq n_R$. Otherwise, the flux is vanishing and the stationary macrostate corresponds to thermodynamic equilibrium.

If the diffusion coefficient is independent of the density, the stochastic diffusion equation is linear in the density $n(\mathbf{r}, t)$ and its solution can be expressed by the principle of linear superposition. Otherwise, the problem is nonlinear, which requires other methods.

10.2.2 Space Discretization and Master Equation

A useful method is to discretize the interval $0 \leq x \leq L$ into l cells with size $\Delta x = L/l$ and volume $\Delta V = \Sigma \Delta x$, as shown in Figure 10.1(a). The random state of the system is thus described by the numbers $\mathbf{N} = (N_i)_{i=1}^l \in \mathbb{N}^l$ of particles in the l cells. Moreover, the particle exchanges with the environment are counted with the number $\mathcal{N} \in \mathbb{Z}$ of particles incoming the channel at the boundary $x = 0$ from the left-hand reservoir at the density n_L.

The particles undergo random jumps between the cells, which induces a Markov jump process for diffusion between the reservoirs, as schematically represented by

$$\bar{N}_L \underset{-0}{\overset{+0}{\rightleftharpoons}} N_1 \underset{-1}{\overset{+1}{\rightleftharpoons}} N_2 \underset{-2}{\overset{+2}{\rightleftharpoons}} \cdots \underset{-(i-1)}{\overset{+(i-1)}{\rightleftharpoons}} N_i \underset{-i}{\overset{+i}{\rightleftharpoons}} N_{i+1} \underset{-(i+1)}{\overset{+(i+1)}{\rightleftharpoons}} \cdots \underset{-(l-1)}{\overset{+(l-1)}{\rightleftharpoons}} N_l \underset{-l}{\overset{+l}{\rightleftharpoons}} \bar{N}_R, \tag{10.12}$$

where $+i$ denotes the elementary step of a particle jumping from the ith to the $(i+1)$th cell and $-i$ the reversed step.

The probability $P(\mathbf{N}, \mathcal{N}, t)$ that the cells contain the particle numbers $\mathbf{N}$ and that $\mathcal{N}$ particles have arrived from the left-hand reservoir at time t is ruled by the following master equation,

$$\frac{d}{dt} P(\mathbf{N}, \mathcal{N}, t) = \sum_{i=0}^{l} \Big[\left(\mathrm{e}^{-\delta_{i0}\partial_{\mathcal{N}}} \hat{E}_i^{+1} - 1 \right) W_{+i}(\mathbf{N})\, P(\mathbf{N}, \mathcal{N}, t) + \left(\mathrm{e}^{+\delta_{i0}\partial_{\mathcal{N}}} \hat{E}_i^{-1} - 1 \right) W_{-i}(\mathbf{N})\, P(\mathbf{N}, \mathcal{N}, t) \Big], \tag{10.13}$$

expressed in terms of the rising and lowering operators

$$\hat{E}_i^{+1} \equiv \exp\left(\partial_{N_i} - \partial_{N_{i+1}}\right) \qquad \text{and} \qquad \hat{E}_i^{-1} = \left(\hat{E}_i^{+1}\right)^{\dagger} \tag{10.14}$$

and the transition rates

$$W_{\pm i}(\mathbf{N}) \equiv W\left(\mathbf{N} \overset{\pm i}{\to} \mathbf{N} + \boldsymbol{\nu}_i\right) \qquad \text{with} \qquad (\boldsymbol{\nu}_i)_j = -\delta_{i,j} + \delta_{i+1,j}. \tag{10.15}$$

In order to satisfy the mass action law and the thermodynamic condition (4.59), the transition rates are assumed to have the form,

$$W_{+i}(\mathbf{N}) = \frac{\mathcal{D}}{\Delta x^2}\, \psi(\Delta U_{i,i+1})\, N_i, \tag{10.16}$$

$$W_{-i}(\mathbf{N}) = \frac{\mathcal{D}}{\Delta x^2}\, \psi(\Delta U_{i+1,i})\, N_{i+1}, \tag{10.17}$$

with the function

$$\psi(\Delta U) = \frac{\beta\, \Delta U}{\mathrm{e}^{\beta \Delta U} - 1}, \tag{10.18}$$

satisfying the identity

$$\psi(\Delta U) = \mathrm{e}^{-\beta \Delta U}\, \psi(-\Delta U), \tag{10.19}$$

which guarantees that detailed balance is fulfilled at equilibrium for the inverse temperature $\beta = (k_{\mathrm{B}}T)^{-1}$.[3] The argument of this function is the change of energy U in the transition

$$\Delta U_{i,i+1} = -\Delta U_{i+1,i} = U(\mathbf{N} + \boldsymbol{\nu}_i) - U(\mathbf{N}). \tag{10.20}$$

If the diffusion coefficient $\mathcal{D}$ depends on the density, this latter is taken as the value $n_{i,i+1} = (N_i + N_{i+1})/(2\Delta V)$ in between the cells i and $i+1$: $\mathcal{D} = \mathcal{D}(n_{i,i+1})$. The rates $W_{\pm i}$ are proportional to the particle numbers N_i or N_{i+1} in order to satisfy the mass action law. Accordingly, the transition rate is equal to zero if there is no particle in the cell, from which the particle should undergo the transition.

[3] Several other functions $\psi(\Delta U)$ satisfy the identity (10.19) and $\lim_{\Delta U \to 0} \psi(\Delta U) = 1$, such as the Metropolis function, $\psi(\Delta U) = \min\{1, \exp(-\beta \Delta U)\}$, and the Kawasaki function, $\psi(\Delta U) = 2/[1 + \exp(\beta \Delta U)]$ (Spohn, 1991).

Here, the energy of the particles can be taken as $U(\mathbf{N}) = \sum_{i=0}^{l+1} u_i N_i$ with the external energy potential $u_i = u(x_i)$ of the particles in the ith cell, whereupon the change of energy in the transition $\mathbf{N} \to \mathbf{N} + \boldsymbol{\nu}_i$ has the value $\Delta U_{i,i+1} = u_{i+1} - u_i$.

In this scheme, the boundary conditions are implemented by noting that the numbers N_0 and N_{l+1} are not random variables and they should thus have fixed values related to the invariant densities of the reservoirs. Moreover, the external energy potential is fixed to the values $u_0 = u_{\mathrm{L}}$ and $u_{l+1} = u_{\mathrm{R}}$ in the reservoirs. The particle number required by the mass action law is thus taken to be $N_0 = n_{\mathrm{L}} \Delta V \equiv \bar{N}_{\mathrm{L}}$ in the rate (10.16) with $i = 0$ and $N_{l+1} = n_{\mathrm{R}} \Delta V \equiv \bar{N}_{\mathrm{R}}$ in the rate (10.17) with $i = l$. If the diffusion coefficient depends on the density, this latter should nevertheless be evaluated by counting the particle exchanged with the reservoir in the expression of the diffusion coefficient at the boundaries, thus taking $n_{0,1} = \left(\bar{N}_{\mathrm{L}} + 1 + N_1\right)/(2\Delta V)$ in the rate (10.16) with $i = 0$ and $n_{l,l+1} = \left(N_l + \bar{N}_{\mathrm{R}} + 1\right)/(2\Delta V)$ in rate (10.17) with $i = l$.

As discussed in Chapter 4, the stochastic process for the coarse-grained states $\omega = (\mathbf{N}, \mathcal{N})$ is nonstationary. Since the transition rates do not depend on the counting variable $\mathcal{N}$, we may introduce the probability distribution of the internal states of the system as $\mathscr{P}(\mathbf{N}, t) \equiv \sum_{\mathcal{N}=-\infty}^{+\infty} P(\mathbf{N}, \mathcal{N}, t)$, which obeys the reduced master equation

$$\frac{d}{dt}\mathscr{P}(\mathbf{N},t) = \sum_{i=0}^{l} \left[\left(\hat{E}_i^{+1} - 1\right) W_{+i}(\mathbf{N})\, \mathscr{P}(\mathbf{N},t) + \left(\hat{E}_i^{-1} - 1\right) W_{-i}(\mathbf{N})\, \mathscr{P}(\mathbf{N},t) \right], \tag{10.21}$$

which is deduced from equation (10.13). The solution of this reduced master equation typically converges towards a stationary probability distribution $\mathscr{P}_{\mathrm{st}}(\mathbf{N})$, in which case the stochastic process of the internal random variables $\mathbf{N}$ is stationary.

With these assumptions, the network theory of Section 4.4.5 applies and the affinity empowering particle diffusion between the reservoirs can be obtained with the cycle corresponding to the successive transitions of one particle from the left- to the right-hand reservoir:

$$\begin{aligned} C: &\quad \mathbf{N}_0 \overset{+0}{\to} \mathbf{N}_1 \overset{+1}{\to} \mathbf{N}_2 \overset{+2}{\to} \cdots \overset{+(l-1)}{\to} \mathbf{N}_l \overset{+l}{\to} \mathbf{N}_0, \\ \text{where} &\quad \mathbf{N}_i = \mathbf{N}_0 + \mathbf{1}_i \qquad \text{for} \qquad 1 \le i \le l, \end{aligned} \tag{10.22}$$

$\mathbf{1}_i$ denoting the unit vector with one particle in the ith cell, such that $(\mathbf{1}_i)_j = \delta_{i,j}$. During this cycle, one particle has been transferred from the left- to the right-hand reservoir, so that $\mathcal{N} \overset{C}{\to} \mathcal{N} + 1$. According to equation (4.77), the affinity of the cycle (10.22) is given by

$$A_C = \ln \prod_{i=0}^{l} \frac{W_{+i}(\mathbf{N}_i)}{W_{-i}(\mathbf{N}_i)} = \ln \frac{n_{\mathrm{L}}}{n_{\mathrm{R}}} + \beta(u_{\mathrm{L}} - u_{\mathrm{R}}) = \beta(\mu_{\mathrm{L}} - \mu_{\mathrm{R}}), \tag{10.23}$$

as expected for the chemical potential $\mu = \mu^0 + k_{\mathrm{B}} T \ln\left(n/n^0\right) + u$ in an isothermal system.

In the limit of large particle numbers in the cells, i.e., $N_i \gg 1$, the Markov jump process can be approximated by a Langevin stochastic process (Gaspard, 2005). In this approximation, the operators $\exp(\pm\partial_X)$ are truncated into $\left(1 \pm \partial_X + \frac{1}{2}\partial_X^2\right)$ in the master equation

(10.21), so that we get the following Fokker–Planck equation for the time evolution of the probability density $\mathcal{P}(\mathbf{N}, t)$,

$$\frac{\partial \mathcal{P}}{\partial t} = -\sum_i \frac{\partial}{\partial N_i}(C_i\, \mathcal{P}) + \sum_{i,j} \frac{\partial^2}{\partial N_i \partial N_j}(D_{ij}\, \mathcal{P}) \tag{10.24}$$

with

$$C_i = W_{-i} - W_{+i} + W_{+(i-1)} - W_{-(i-1)}, \tag{10.25}$$

$$D_{ii} = \frac{1}{2}\left(W_{-i} + W_{+i} + W_{+(i-1)} + W_{-(i-1)}\right), \tag{10.26}$$

$$D_{i,i+1} = -\frac{1}{2}\,(W_{-i} + W_{+i}). \tag{10.27}$$

This Fokker–Planck equation corresponds to the Langevin process defined by the Itô stochastic differential equations

$$\frac{dN_i}{dt} = \mathscr{J}_{i-1} - \mathscr{J}_i \tag{10.28}$$

expressed in terms of the random particle fluxes

$$\mathscr{J}_i = W_{+i} - W_{-i} + \sqrt{W_{+i} + W_{-i}}\; \chi_i(t) \tag{10.29}$$

with the independent Gaussian white noises characterized by

$$\langle \chi_i(t) \rangle = 0 \qquad \text{and} \qquad \langle \chi_i(t)\, \chi_j(t') \rangle = \delta_{ij}\, \delta(t - t'), \tag{10.30}$$

as explained in Section 4.5.3. Accordingly, the total particle number $N_{\rm tot} \equiv \sum_{i=1}^{l} N_i$ evolves because of particle exchanges with the reservoirs: $dN_{\rm tot}/dt = \mathscr{J}_0 - \mathscr{J}_l$.

In the continuum limit $\Delta x \to 0$, the particle density is obtained according to $n(x_i, t) = N_i(t)/\Delta V$. Using the approximation $\psi(\Delta U) \simeq \exp(-\beta \Delta U/2)$ for the function (10.18), the random flux (10.29) becomes

$$\mathscr{J}_i \simeq -\frac{\Sigma \mathcal{D}}{\Delta x}\, \mathrm{e}^{-\beta \frac{u_i + u_{i+1}}{2}} \left(\mathrm{e}^{\beta u_{i+1}} n_{i+1} - \mathrm{e}^{\beta u_i} n_i\right) + \sqrt{\frac{\Sigma \mathcal{D}}{\Delta x}\,(n_i + n_{i+1})}\; \chi_i(t). \tag{10.31}$$

Therefore, the current density $j_x = \mathscr{J}_i/\Sigma$ is indeed given by equation (10.11) with the Gaussian white noise fields (10.9) in the limit $\Delta x \to 0$. Consequently, the stochastic diffusion equation (10.10), which expresses the local conservation of the particle number, is recovered from the stochastic differential equations (10.28) in the continuum limit.

The mean value of the particle number in the ith cell at time t can be obtained using

$$\langle N_i \rangle = \sum_{\mathbf{N}} N_i\, \mathscr{P}(\mathbf{N}, t) = \sum_{\mathbf{N}, \mathcal{N}} N_i\, P(\mathbf{N}, \mathcal{N}, t). \tag{10.32}$$

In particular, its time evolution can be deduced from the master equation (10.21) giving

$$\frac{d}{dt}\langle N_i \rangle = J_{i-1} - J_i \qquad \text{with} \qquad J_i \equiv \langle W_{+i} \rangle - \langle W_{-i} \rangle \tag{10.33}$$

for $1 \le i \le l$, so that we recover the local conservation law of particle number. The stochastic description of the transport process is thus consistent with the macroscopic diffusion equation.

10.2.3 Fluctuation Relation for the Current

In order to establish the fluctuation relation for the particle current from the left- to the right-hand reservoir, we proceed as in Section 5.7, introducing the generating function

$$F(\mathbf{N},\lambda,t) \equiv \sum_{\mathcal{N}=-\infty}^{+\infty} \mathrm{e}^{-\lambda\mathcal{N}} P(\mathbf{N},\mathcal{N},t), \tag{10.34}$$

which is ruled by

$$\partial_t F(\mathbf{N},\lambda,t) = \hat{L}_\lambda F(\mathbf{N},\lambda,t), \qquad \text{where} \qquad \hat{L}_\lambda \equiv \mathrm{e}^{-\lambda\mathcal{N}}\hat{L}\,\mathrm{e}^{+\lambda\mathcal{N}} \tag{10.35}$$

is the modified operator given by

$$\hat{L}_\lambda F = \sum_{i=0}^{l}\left[\left(\mathrm{e}^{-\delta_{i0}\lambda}\hat{E}_i^{+1} - 1\right) W_{+i}\,F + \left(\mathrm{e}^{+\delta_{i0}\lambda}\hat{E}_i^{-1} - 1\right) W_{-i}\,F\right]. \tag{10.36}$$

For long enough time, the function (10.34) is expected to decrease exponentially as $F(\mathbf{N},\lambda,t) \sim \exp(-Q\,t)$ with a rate Q giving the cumulant generating function of the current fluctuations,

$$Q(\lambda;A) \equiv \lim_{t\to\infty} -\frac{1}{t}\ln\left\langle \mathrm{e}^{-\lambda\mathcal{N}(t)}\right\rangle_A, \tag{10.37}$$

where $A = A_C$ is the affinity (10.23). The mean current of particles flowing from the left- to the right-hand reservoir and the corresponding diffusivity can thus be obtained as

$$J(A) = \frac{\partial Q}{\partial\lambda}(0;A) = \lim_{t\to\infty}\frac{1}{t}\,\langle\mathcal{N}(t)\rangle_A, \tag{10.38}$$

$$D(A) = -\frac{1}{2}\frac{\partial^2 Q}{\partial\lambda^2}(0;A) = \lim_{t\to\infty}\frac{1}{2t}\left\langle[\mathcal{N}(t) - \langle\mathcal{N}(t)\rangle_A]^2\right\rangle_A. \tag{10.39}$$

The remarkable property is that the modified operator (10.36) obeys the symmetry relation

$$\eta^{-1}\hat{L}_\lambda(\eta\,\Psi) = \hat{L}^{\dagger}_{A-\lambda}\Psi, \tag{10.40}$$

where $A = A_C$ is the affinity (10.23), Ψ an arbitrary function of the particle numbers $\mathbf{N}$, and $\eta = \eta(\mathbf{N})$ the equilibrium distribution at the chemical potential of the right-hand reservoir with $u_i = u_{\mathrm{R}}$ and $\langle N_i\rangle = \bar{N}_{\mathrm{N}}$. This equilibrium distribution satisfies the conditions of detailed balance

$$W_{+i}(\mathbf{N})\,\eta(\mathbf{N}) = W_{-i}(\mathbf{N}+\boldsymbol{\nu}_i)\,\eta(\mathbf{N}+\boldsymbol{\nu}_i) \qquad \text{for} \quad 1 \le i \le l, \tag{10.41}$$

but not for $i = 0$, where instead

$$W_{+0}^{(\mathrm{R})}(\mathbf{N})\,\eta(\mathbf{N}) = W_{-0}^{(\mathrm{R})}(\mathbf{N}+\boldsymbol{\nu}_0)\,\eta(\mathbf{N}+\boldsymbol{\nu}_0) \tag{10.42}$$

with the transition rates $W_{\pm 0}^{(\mathrm{R})}$ corresponding to the left-hand reservoir at the same density and potential as the right-hand reservoir. Accordingly, we find that

$$\eta^{-1}\hat{L}_\lambda(\eta\,\Psi) - \hat{L}^\dagger_{A-\lambda}\Psi \tag{10.43}$$
$$= \mathrm{e}^{+\lambda}\left[\frac{W^{(\mathrm{R})}_{+0}(\mathbf{N})}{W^{(\mathrm{R})}_{-0}(\mathbf{N}+\boldsymbol{\nu}_0)} - \frac{W_{+0}(\mathbf{N})}{W_{-0}(\mathbf{N}+\boldsymbol{\nu}_0)}\,\mathrm{e}^{-A}\right] W_{-0}(\mathbf{N}+\boldsymbol{\nu}_0)\,\Psi(\mathbf{N}+\boldsymbol{\nu}_0)$$
$$+\,\mathrm{e}^{-\lambda}\left[\frac{W^{(\mathrm{R})}_{-0}(\mathbf{N})}{W^{(\mathrm{R})}_{+0}(\mathbf{N}-\boldsymbol{\nu}_0)} - \frac{W_{-0}(\mathbf{N})}{W_{+0}(\mathbf{N}-\boldsymbol{\nu}_0)}\,\mathrm{e}^{+A}\right] W_{+0}(\mathbf{N}-\boldsymbol{\nu}_0)\,\Psi(\mathbf{N}-\boldsymbol{\nu}_0),$$

which is identically vanishing if

$$A = \ln \frac{W_{+0}(\mathbf{N})\, W^{(\mathrm{R})}_{-0}(\mathbf{N}+\boldsymbol{\nu}_0)}{W_{-0}(\mathbf{N}+\boldsymbol{\nu}_0)\, W^{(\mathrm{R})}_{+0}(\mathbf{N})} \qquad \text{for all } \mathbf{N} \in \mathbb{N}^l. \tag{10.44}$$

This is the case for the previously defined transition rates $W_{\pm 0}$ at the left-hand boundary given by equations (10.16) and (10.17) with $i = 0$. Furthermore, the values of the affinity given by equations (10.23) and (10.44) coincide, so that the symmetry (10.40) is indeed satisfied for the modified operator (10.36). As shown in Section 5.7, the leading eigenvalue of the modified operator (10.36) giving the cumulant generating function (10.37) thus has the symmetry

$$Q(\lambda; A) = Q(A - \lambda; A), \tag{10.45}$$

so that the fluctuation relation for the current holds in the long-time limit,

$$\frac{\mathcal{P}_t(\mathcal{N}; A)}{\mathcal{P}_t(-\mathcal{N}; A)} \sim_{t\to\infty} \exp(A\,\mathcal{N}), \tag{10.46}$$

where $\mathcal{P}_t(\mathcal{N}; A) = \sum_{\mathbf{N}} P(\mathbf{N}, \mathcal{N}, t)$ is the probability distribution that $\mathcal{N}$ particles are transferred from the left- to the right-hand reservoir during the time interval t and A is the affinity (10.23) driving the process away from equilibrium (Andrieux and Gaspard, 2006a).

As a consequence of the fluctuation relation (10.46), the entropy production rate (5.98) is here given by

$$\frac{1}{k_\mathrm{B}}\frac{d_\mathrm{i}S}{dt} = A\,J(A) \geq 0 \tag{10.47}$$

in terms of the mean current (10.38) in the stationary regime.

10.2.4 The Case of Homogeneous Diffusion

If the diffusion coefficient is independent of the density $\partial_n \mathcal{D} = 0$ and there is no external energy potential, i.e., $u = 0$, the transition rates are given by $W_{+i} = kN_i$ and $W_{-i} = kN_{i+1}$ with the rate constant $k = \mathcal{D}/\Delta x^2$. In this simple case, the cumulant generating function can be obtained by solving equation (10.35) under the assumption that the generating function can be expressed as

$$F(\mathbf{N},\lambda,t) = \mathrm{e}^{-Qt} \prod_{i=1}^{l} \frac{C_i^{N_i}}{N_i!} \tag{10.48}$$

with some coefficients C_i to be determined (Andrieux and Gaspard, 2006a). Substituting this expression into equation (10.35), the relations $C_{i-1} - 2C_i + C_{i+1} = 0$ are obtained for $1 \leq i \leq l$ with $C_0 = \mathrm{e}^{-\lambda}\bar{N}_\mathrm{L}$ and $C_{l+1} = \bar{N}_\mathrm{R}$. These relations are solved taking $C_i = C_0 + i(C_{l+1} - C_0)/(l+1)$, so that the cumulant generating function is found to be equal to

$$Q(\lambda; A) = \Sigma \frac{\mathcal{D}}{L} \left[n_\mathrm{L} \left(1 - \mathrm{e}^{-\lambda}\right) + n_\mathrm{R} \left(1 - \mathrm{e}^{+\lambda}\right)\right], \tag{10.49}$$

satisfying the symmetry relation (10.45) with the affinity $A = \ln(n_\mathrm{L}/n_\mathrm{R})$ given by equation (10.23) in the absence of external energy potential. Accordingly, the mean current (10.38) and the diffusivity (10.39) here have the values

$$J(A) = \Sigma \frac{\mathcal{D}}{L} (n_\mathrm{L} - n_\mathrm{R}) = \Sigma \frac{\mathcal{D}}{L} n_\mathrm{R} (\mathrm{e}^A - 1), \tag{10.50}$$

$$D(A) = \Sigma \frac{\mathcal{D}}{2L} (n_\mathrm{L} + n_\mathrm{R}) = \Sigma \frac{\mathcal{D}}{2L} n_\mathrm{R} (\mathrm{e}^A + 1). \tag{10.51}$$

Both are proportional to the cross-sectional area Σ and the diffusion coefficient $\mathcal{D}$. As expected, the mean current (10.50) satisfies Fick's law in the form (1.70). We note that the affinity of such processes can be expressed in terms of the ratio between the mean current and the diffusivity as

$$A = \ln \frac{2D + J}{2D - J} = \frac{J}{D} + \frac{1}{12} \left(\frac{J}{D}\right)^3 + O\left[\left(\frac{J}{D}\right)^5\right] \tag{10.52}$$

for $-2D \leq J \leq +2D$. The affinity can be approximated as $A \simeq J/D$ if the mean current is significantly smaller than the diffusivity, $|J| \ll D$. This is the case, in particular, if the probability distribution $\mathcal{P}_t(\mathcal{N}; A)$ is Gaussian, so that the cumulant generating function (10.49) is quadratic, $Q = \lambda(J - D\lambda)$, and the affinity is equal to $A = J/D$. However, the probability distribution $\mathcal{P}_t(\mathcal{N}; A)$ becomes non-Gaussian if the mean current is close to the limit values $J = \pm 2D$, in which cases the affinity is a strongly nonlinear function of the mean current.

In the case of homogeneous diffusion, the current fluctuation relation (10.46) has been proved by Andrieux and Gaspard (2006a) and experimentally confirmed by Seitaridou et al. (2007) under the name of flux fluctuation theorem. In this experiment, the distribution of fluxes is measured with fluorescent polystyrene particles of diameter 0.29 μm diffusing in a microfluidic device with diffusion coefficient $\mathcal{D} \simeq 1.3\ \mu\mathrm{m}^2/\mathrm{s}$. The measurement is performed every time interval $\Delta t = 10$ s within bins of size $\Delta x \simeq 5$ μm. The probability distribution of the fluxes is observed to be Gaussian. In this case, the affinity is given by the ratio of the mean current (10.50) over the diffusivity (10.51), $A = J/D = 2(n_\mathrm{L} - n_\mathrm{R})/(n_\mathrm{L} + n_\mathrm{R})$, as confirmed by the experiment of Seitaridou et al. (2007).

The entropy production rate of transport by diffusion is given by equation (10.47) in agreement with the macroscopic value (1.73). The entropy production rate can also be obtained from the time asymmetry of temporal disorder arising from fluctuations in particle transport according to equation (6.103), as shown in Figure 10.1(b), where the solid line gives the expectation from thermodynamics,

$$\frac{1}{k_{\rm B}}\frac{d_{\rm i}S}{dt} = \int_V \mathcal{D}\,\frac{(\nabla n)^2}{n}\,d^3r = \Sigma\,\frac{\mathcal{D}}{L}\,(n_{\rm L}-n_{\rm R})\ln\frac{n_{\rm L}}{n_{\rm R}} \geq 0, \tag{10.53}$$

and the dots depict the difference $h^{\rm R} - h$ between the τ-entropies per unit time (6.90) and (6.105) in the limit $\tau \to 0$ (Gaspard, 2005).

10.3 Finite-Time Fluctuation Relation

For Markov jump processes where the transition rates are linearly dependent on the variables $\mathbf{N}$ (or if they are constant), the counting statistics can be exactly obtained by solving equation (10.35) with the modified operator (10.36). Several examples of such processes are known, including diffusive transport and reaction (Andrieux and Gaspard, 2008f; Gaspard and Kapral, 2018a; Gaspard et al., 2018; Gu and Gaspard, 2020). In these processes, the random number $\mathcal{N}$ of particles exchanged with the environment during the time interval t, when starting with $\mathcal{N} = 0$ and the stationary probability distribution $\mathscr{P}_{\rm st}(\mathbf{N})$ for the random variables $\mathbf{N}$, can be expressed as the difference between two standard Poisson processes

$$\mathcal{N}(t) = \mathcal{P}_+\left(W_t^{(+)}t\right) - \mathcal{P}_-\left(W_t^{(-)}t\right) \tag{10.54}$$

with finite-time rates $W_t^{(\pm)}$. Accordingly, the probability distribution of the counting statistics is given by

$$\mathcal{P}_t(\mathcal{N}) = {\rm e}^{-t\left(W_t^{(+)}+W_t^{(-)}\right)}\left(\frac{W_t^{(+)}}{W_t^{(-)}}\right)^{\mathcal{N}/2} I_{\mathcal{N}}\left(2t\sqrt{W_t^{(+)}W_t^{(-)}}\right), \tag{10.55}$$

in terms of the modified regular Bessel function $I_{\mathcal{N}}(z)$ (Abramowitz and Stegun, 1972). Since $I_{\mathcal{N}}(z) = I_{-\mathcal{N}}(z)$, this probability distribution remarkably obeys the *finite-time fluctuation relation*

$$\frac{\mathcal{P}_t(\mathcal{N})}{\mathcal{P}_t(-\mathcal{N})} = \exp(A_t\,\mathcal{N}), \tag{10.56}$$

exactly holding at every time t with the time-dependent affinity

$$A_t = \ln\frac{W_t^{(+)}}{W_t^{(-)}}. \tag{10.57}$$

This result is proved in Section F.4.

The asymptotic fluctuation relation (10.46) is recovered for long enough time because the time-dependent affinity converges towards its asymptotic value given by equation (10.23): $\lim_{t\to\infty} A_t = A = A_C$.[4]

The finite-time rates have the explicit forms

$$W_t^{(\pm)} = W_\infty^{(\pm)} + \frac{1}{t}\,\Upsilon(t), \tag{10.58}$$

where $W_\infty^{(\pm)}$ are the asymptotic rates and $\Upsilon(t)$ is some time-dependent function characterizing the process.

The results (10.54)–(10.55) and, thus, the finite-time fluctuation relation (10.56) hold in particular for the random drift of Section 4.4.6 since the probability distribution of that process is given by equation (4.85) with the invariant rates $W_t^{(\pm)} = W_\pm$ and $\Upsilon(t) = 0$ in this special case. For other processes, the finite-time rates (10.58) depend on time because of a nontrivial function $\Upsilon(t)$ (Andrieux and Gaspard, 2008f; Gaspard and Kapral, 2018a; Gaspard et al., 2018; Gu and Gaspard, 2020).

Introducing the time-dependent cumulant generating function

$$Q_t(\lambda) \equiv -\frac{1}{t}\ln \sum_{\mathcal{N}=-\infty}^{+\infty} e^{-\lambda\mathcal{N}}\,\mathcal{P}_t(\mathcal{N}) \tag{10.59}$$

with the counting parameter λ, the distribution (10.55) implies that

$$Q_t(\lambda) = W_t^{(+)}\left(1 - e^{-\lambda}\right) + W_t^{(-)}\left(1 - e^{+\lambda}\right). \tag{10.60}$$

As a consequence of the finite-time fluctuation relation (10.56), the following symmetry relation is satisfied at every time,

$$Q_t(\lambda) = Q_t(A_t - \lambda), \tag{10.61}$$

in terms of the finite-time affinity (10.57). The mean current and the diffusivity at time t are thus given by

$$J_t = \frac{\partial Q_t}{\partial\lambda}(0) = W_\infty^{(+)} - W_\infty^{(-)}, \tag{10.62}$$

$$D_t = -\frac{1}{2}\frac{\partial^2 Q_t}{\partial\lambda^2}(0) = \frac{1}{2}\left(W_\infty^{(+)} + W_\infty^{(-)}\right) + \frac{1}{t}\,\Upsilon(t), \tag{10.63}$$

so that the mean current does not depend on time in contrast to the diffusivity.

In the long-time limit, the cumulant generating function (10.37) is recovered from equation (10.59), so that

$$Q(\lambda; A) = \lim_{t\to\infty} Q_t(\lambda) = W_\infty^{(+)}\left(1 - e^{-\lambda}\right) + W_\infty^{(-)}\left(1 - e^{+\lambda}\right). \tag{10.64}$$

[4] Here, the counting statistics starts when the system is in the stationary regime with fixed particle densities in the reservoirs. These assumptions are different from those used to obtain the fluctuation relation (5.73) in the framework of statistical mechanics, where the counting statistics is supposed to start from the initial phase-space distribution (5.51). These differences explain that the fluctuation relation (10.56) exactly holds with a time-dependent affinity, although the fluctuation relation (5.73) is also exactly satisfied but with the invariant affinity (5.70). The results are consistent with each other because the same value is obtained for the affinity in the long-time limit: $A_\infty = \beta(\mu_L - \mu_R)$.

In the case of homogeneous diffusion in the bounded domain of length L and cross-sectional area Σ between the two reservoirs at the particle densities $n_{\rm L}$ and $n_{\rm R}$ with the diffusion coefficient $\mathcal{D}$, the asymptotic values of the finite-time rates are given by

$$W_\infty^{(+)} = \Sigma\,\frac{\mathcal{D}}{L}\,n_{\rm L} \qquad \text{and} \qquad W_\infty^{(-)} = \Sigma\,\frac{\mathcal{D}}{L}\,n_{\rm R} \tag{10.65}$$

because of equation (10.49). For this diffusion process with counting performed at the boundary with the left-hand reservoir, the time dependence arises from the function

$$\Upsilon(t) = \Sigma\, n_{\rm L} \int_0^L u(x,0)\,[u(x,0) - u(x,t)]\,dx, \tag{10.66}$$

where the function $u(x,t)$ is the solution of the deterministic diffusion equation with the following boundary and initial conditions,

$$\partial_t u(x,t) = \mathcal{D}\,\partial_x^2 u(x,t), \qquad u(0,t) = u(L,t) = 0, \qquad u(x,0) = 1 - \frac{x}{L}, \tag{10.67}$$

as shown in Section F.4. Consequently, the function (10.66) is given by

$$\Upsilon(t) = \Sigma\, n_{\rm L}\, L\, \tau\, g(\tau), \tag{10.68}$$

where

$$g(\tau) = \frac{2}{\tau}\sum_{n=1}^{\infty} \frac{1}{(n\pi)^2}\left[1 - {\rm e}^{-(n\pi)^2\tau}\right] \qquad \text{with} \quad \tau \equiv \frac{\mathcal{D}}{L^2}\,t, \tag{10.69}$$

such that

$$g(\tau) \simeq_{\tau\to\infty} \frac{1}{3\tau}\left(1 - \frac{6}{\pi^2}\,{\rm e}^{-\pi^2\tau}\right) \qquad \text{and} \qquad g(\tau) \simeq_{\tau\to 0} \frac{2}{\sqrt{\pi\tau}}. \tag{10.70}$$

Therefore, the time-dependent affinity of the homogeneous diffusion process has the form

$$A_\tau = \ln \frac{1 + g(\tau)}{(n_{\rm R}/n_{\rm L}) + g(\tau)}, \tag{10.71}$$

which is shown in Figure 10.2(a) for two values of the density ratio $n_{\rm L}/n_{\rm R}$. This affinity behaves as

$$A_\tau \simeq_{\tau\to\infty} \ln\frac{n_{\rm L}}{n_{\rm R}} - \frac{n_{\rm L} - n_{\rm R}}{3\,n_{\rm R}\,\tau} + O\left(\frac{1}{\tau^2}\right) \qquad \text{and} \qquad A_\tau \simeq_{\tau\to 0} \frac{n_{\rm L} - n_{\rm R}}{2\,n_{\rm L}}\sqrt{\pi\tau}. \tag{10.72}$$

In the long-time limit, the asymptotic value of the affinity is recovered, $A_\infty = \ln(n_{\rm L}/n_{\rm R})$. The convergence to this asymptotic value goes as $A_\tau = A_\infty + O(1/\tau)$ over a timescale longer than the diffusion time, $t \gg t_{\rm diff} = L^2/\mathcal{D}$. Accordingly, the asymptotic value reached after stationarity is settled for the counting statistics at the left-hand boundary by diffusion over the whole length L of the system. At early times, the time-dependent affinity increases from zero as $A_\tau = O(\sqrt{\tau})$ in the case of homogeneous diffusion, as seen in Figure 10.2(a).

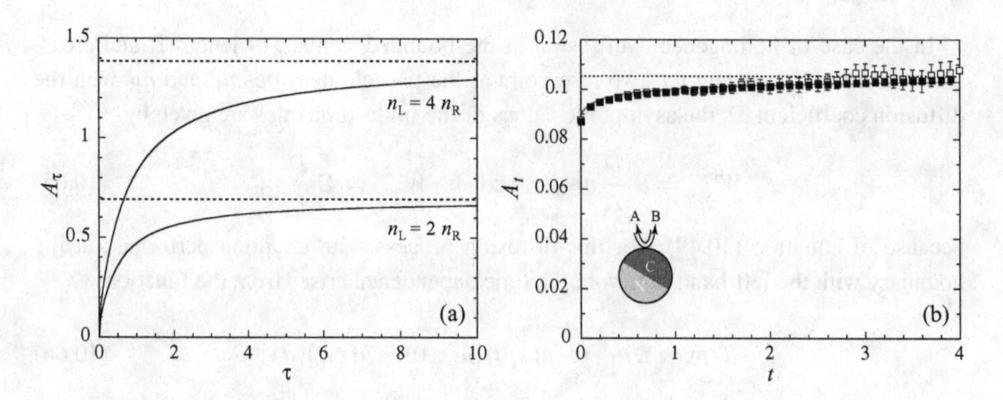

Figure 10.2 Time-dependent affinity $A_t = \ln[\mathcal{P}_t(\mathcal{N})/\mathcal{P}_t(-\mathcal{N})]$ given by: (a) equation (10.71) for one-dimensional diffusion between two reservoirs at the densities $n_{\rm L}$ and $n_{\rm R}$ with the rescaled time $\tau = \mathcal{D}t/L^2$ and the dashed lines representing the asymptotes at the values $A_\infty = \ln(n_{\rm L}/n_{\rm R})$; (b) equation (10.57) for the diffusion-influenced surface reaction of Section 10.4 on a Janus catalytic particle with the parameter values $\mathrm{Da} = 2\kappa^{\rm S} R/\mathcal{D} = 0.4$, $L/R = 10$, $\bar{n}_{\rm A} = 527.2$, $\bar{n}_{\rm B} = 474.5$, and $A_\infty = 0.105$, the open squares showing the affinity directly measured with $\ln[\mathcal{P}_t(\mathcal{N})/\mathcal{P}_t(-\mathcal{N})]$, and the filled squares the affinity obtained using Gaussian fits to the probability distribution $\mathcal{P}_t(\mathcal{N})$. The solid line gives the theoretical result and the dashed line the asymptotic value of the affinity (10.80). Reproduced with permission from Gaspard et al. (2018) © IOP Publishing Ltd and SISSA Medialab.

10.4 Diffusion-Influenced Surface Reactions

10.4.1 Stochastic Description

The stochastic formulation can also be considered for diffusion–reaction processes such as the diffusion-influenced surface reaction $\mathrm{A} \rightleftharpoons \mathrm{B}$ between the molecular species A and B diffusing in a three-dimensional domain V (Gaspard and Kapral, 2018a; Gaspard et al., 2018). The boundary of this domain may be composed of three types of surfaces: $\partial V = \Sigma_{\rm cat} \cup \Sigma_{\rm inert} \cup \Sigma_{\rm res}$. The reaction takes place at the catalytic surface $\Sigma_{\rm cat}$. The molecules A and B are reflected at the inert surface $\Sigma_{\rm inert}$. Moreover, they enter and exit the domain V at the surface $\Sigma_{\rm res}$ in contact with a reservoir, where the molecular species have the fixed densities (or concentrations) $\bar{n}_{\rm A}$ and $\bar{n}_{\rm B}$.

The density n_k of the species k is ruled by the fluctuating diffusion equation

$$\partial_t n_k + \nabla \cdot \mathbf{\mathcal{J}}_k = 0, \qquad \text{where} \qquad \mathbf{\mathcal{J}}_k = -\mathcal{D}_k \nabla n_k + \delta\mathbf{\mathcal{J}}_k \tag{10.73}$$

with the Gaussian white noise fields characterized by

$$\langle \delta\mathbf{\mathcal{J}}_k(\mathbf{r},t)\rangle = 0, \qquad \langle \delta\mathbf{\mathcal{J}}_k(\mathbf{r},t)\,\delta\mathbf{\mathcal{J}}_l(\mathbf{r}',t')\rangle = 2\,\mathcal{D}_k\, n_k(\mathbf{r},t)\,\delta_{kl}\,\delta(\mathbf{r}-\mathbf{r}')\,\delta(t-t')\,\mathbf{1}, \tag{10.74}$$

$\mathcal{D}_k$ being the diffusion coefficient of the molecular species k, $\mathbf{1}$ the 3×3 unit matrix, and $k,l \in \{\mathrm{A},\mathrm{B}\}$. The boundary conditions are given by

$$\text{if } \mathbf{r} \in \Sigma_{\rm cat}: \quad \mathcal{D}_{\rm A}\, \partial_\perp n_{\rm A}(\mathbf{r},t) = -\mathcal{D}_{\rm B}\, \partial_\perp n_{\rm B}(\mathbf{r},t)$$
$$= \kappa_+^{\rm s} n_{\rm A}(\mathbf{r},t) - \kappa_-^{\rm s} n_{\rm B}(\mathbf{r},t) + \xi^{\rm s}(\mathbf{r},t), \tag{10.75}$$
$$\text{if } \mathbf{r} \in \Sigma_{\rm inert}: \quad \partial_\perp n_k(\mathbf{r},t) = 0, \tag{10.76}$$
$$\text{if } \mathbf{r} \in \Sigma_{\rm res}: \quad n_k(\mathbf{r},t) = \bar{n}_k, \tag{10.77}$$

where $\partial_\perp = \mathbf{n}\cdot\nabla$ is the gradient in the direction normal to the surface and oriented towards the interior of the domain V and $\kappa_\pm^{\rm s}$ are the surface reaction rate constants. Since the surface reaction is a dissipative process contributing to the entropy production, there is an associated surface Gaussian white noise field characterized by

$$\langle \xi^{\rm s}(\mathbf{r},t)\rangle = 0, \tag{10.78}$$
$$\delta^{\rm s}(\mathbf{r})\, \langle \xi^{\rm s}(\mathbf{r},t)\, \xi^{\rm s}(\mathbf{r}',t')\rangle\, \delta^{\rm s}(\mathbf{r}') = (\kappa_+^{\rm s}\, n_{\rm A} + \kappa_-^{\rm s}\, n_{\rm B})\, \delta^{\rm s}(\mathbf{r})\, \delta(\mathbf{r}-\mathbf{r}')\, \delta(t-t'),$$

$\delta^{\rm s}(\mathbf{r})$ being the surface delta distribution (A.37), which is nonvanishing if $\mathbf{r} \in \Sigma_{\rm cat}$ (Bedeaux et al., 1976).

10.4.2 Fluctuation Relation for the Reactive Events

The counting statistics for the number $\mathcal{N}$ of reactive events A $\rightarrow$ B occurring during the time interval t can be expressed as the difference (10.54) between two standard Poisson processes with finite-time rates if counting starts from $\mathcal{N} = 0$ with the stationary probability distribution for the molecular densities $n_{\rm A}$ and $n_{\rm B}$. Accordingly, the probability distribution $\mathcal{P}_t(\mathcal{N})$ is given by equation (10.55) and the finite-time fluctuation relation (10.56) holds for the counting statistics of the reactive events with the time-dependent affinity (10.57). Here, the finite-time rates also have the form (10.58), where

$$W_\infty^{(+)} = \Sigma_{\rm cat}\, a_0\, \kappa_+^{\rm s}\, \bar{n}_{\rm A} \qquad \text{and} \qquad W_\infty^{(-)} = \Sigma_{\rm cat}\, a_0\, \kappa_-^{\rm s}\, \bar{n}_{\rm B} \tag{10.79}$$

are the asymptotic rates expressed in terms of the area $\Sigma_{\rm cat}$ of the catalytic surface and some dimensionless number a_0 calculated by solving the deterministic diffusion–reaction equations, and combined with a related function $\Upsilon(t)$ (Gaspard and Kapral, 2018a; Gaspard et al., 2018). Accordingly, the mean reaction rate is here given by $J = \Sigma_{\rm cat} a_0 (\kappa_+^{\rm s}\, \bar{n}_{\rm A} - \kappa_-^{\rm s}\, \bar{n}_{\rm B})$ according to equation (10.62) with the rates (10.79). The time-dependent affinity (10.57) converges towards its asymptotic value

$$A_\infty = \ln \frac{W_\infty^{(+)}}{W_\infty^{(-)}} = \ln \frac{\kappa_+^{\rm s}\, \bar{n}_{\rm A}}{\kappa_-^{\rm s}\, \bar{n}_{\rm B}}, \tag{10.80}$$

which depends on the surface reaction constants $\kappa_\pm^{\rm s}$ and the molecular densities of the reservoirs. The process is out of equilibrium if the affinity (10.80) is not equal to zero and, thus, at equilibrium if the molecular densities are in the Guldberg–Waage ratio (1.39), $\bar{n}_{\rm A}/\bar{n}_{\rm B} = \kappa_-^{\rm s}/\kappa_+^{\rm s}$, such that detailed balance holds and the mean reaction rate is equal to zero.

The time-dependent affinity (10.57) is shown in Figure 10.2(b) for the process of diffusion-influenced surface reaction on a Janus particle of radius R composed of catalytic and inert hemispheres with equal reaction rate constants $\kappa_{\pm}^{\rm s} \equiv \kappa^{\rm s}$ and molecular diffusion coefficients $\mathcal{D}_{\rm A} = \mathcal{D}_{\rm B} \equiv \mathcal{D}$. The molecular species diffuse between the surface of the Janus particle and a concentric spherical reservoir at a distance L from the center of the Janus particle. The surface reaction process is characterized by the dimensionless Damköhler number

$$\mathrm{Da} \equiv R\left(\frac{\kappa_+^{\rm s}}{\mathcal{D}_{\rm A}} + \frac{\kappa_-^{\rm s}}{\mathcal{D}_{\rm B}}\right) = 2\,R\,\frac{\kappa^{\rm s}}{\mathcal{D}} = 0.4. \tag{10.81}$$

The catalytic surface area is given by $\Sigma_{\rm cat} = 2\pi R^2$ and the numerical constant by $a_0 \simeq 0.792$. The mean reaction rate has the value $J \simeq 26.2$ and the asymptotic diffusivity $D_\infty \simeq 249.3$. As seen in Figure 10.2(b), the time-dependent affinity converges towards its asymptotic value (10.80). Contrary to the process of pure diffusion, the affinity at zero time is not equal to zero, but here takes a positive value (Gaspard et al., 2018).

For the diffusion-influenced surface reaction, the entropy production rate is obtained as

$$\begin{aligned}\frac{1}{k_{\rm B}}\frac{d_{\rm i}S}{dt} &= \int_V \left[\mathcal{D}_{\rm A}\,\frac{(\nabla n_{\rm A})^2}{n_{\rm A}} + \mathcal{D}_{\rm B}\,\frac{(\nabla n_{\rm B})^2}{n_{\rm B}}\right] d^3r \\ &\quad + \int_{\rm cat} \left(\kappa_+^{\rm s} n_{\rm A} - \kappa_-^{\rm s} n_{\rm B}\right) \ln \frac{\kappa_+^{\rm s} n_{\rm A}}{\kappa_-^{\rm s} n_{\rm B}}\, d\Sigma = A_\infty\, J \geq 0\end{aligned} \tag{10.82}$$

in terms of the asymptotic affinity (10.80) and the mean reaction rate J. The entropy production rate can also be evaluated using the finite-time fluctuation relation (10.56) as

$$\frac{1}{k_{\rm B}}\frac{d_{\rm i}S}{dt} = \lim_{t\to\infty} \frac{1}{t} \sum_{\mathcal{N}=-\infty}^{+\infty} \mathcal{P}_t(\mathcal{N}) \ln \frac{\mathcal{P}_t(\mathcal{N})}{\mathcal{P}_t(-\mathcal{N})} = \lim_{t\to\infty} A_t\, J = A_\infty\, J \geq 0, \tag{10.83}$$

which is in agreement with the thermodynamic value (10.82) (Gaspard and Kapral, 2018a).

10.5 Ion Transport

10.5.1 The Stochastic Nernst–Planck–Poisson Problem

Ion transport in electrolytes is described as a phenomenon of electrodiffusion, arising from a diffusive part due to concentration gradients and a drift part due to the electric field generated not only by external means, but also by the electric charges themselves. In the quasistatic limit of the Maxwell equations of electromagnetism, the electric field is self-consistently determined by the charge distribution as described by the Poisson equation. Accordingly, the local deviations from electroneutrality have long-ranged effects because of the Coulomb interaction between the electric charges. At the noiseless macroscale, this Nernst–Planck–Poisson problem is ruled by equations (1.76)–(1.78) of Section 1.7.3.

Moreover, the electric current, which is measured at the boundaries of the electrolytic solution, should include the additional contribution due to the temporal variations of the electric field, which is known as the displacement current (Jackson, 1999; Blanter and Büttiker, 2000). This contribution is especially important for time-dependent currents in the study of current fluctuations.

Ion transport is described by the Nernst–Planck–Poisson problem in many natural or artificial systems, such as biological transmembrane ion channels (Nonner and Eisenberg, 1998), solid-state nanopores (Smeets et al., 2008), and nanofluidic diodes (Constantin and Siwy, 2007).

At the nanoscale, the motion of ions is subjected to molecular fluctuations and presents a stochastic behavior. These random fluctuations can be successfully described at the mesoscopic level by adding noise to the ionic current density according to equations (10.9)–(10.11), such that the fluctuation–dissipation theorem is satisfied. The fluctuating Nernst–Planck equations are coupled to the Poisson equation ruling the instantaneous long-ranged Coulomb interaction between the electric charges in random motion.

Considering positively charged ions moving in a fixed negatively charged background, the electric charge density is given by $\rho_e = e\,(n - n_0)$, where $e = |e|$ is the elementary electric charge, n the density of mobile ions, and n_0 the density of immobile negatively charged ions. In the quasistatic limit, the electric field $\boldsymbol{\mathcal{E}}$ is related to the electric potential Φ according to $\boldsymbol{\mathcal{E}} = -\nabla\Phi$. The coupled Nernst–Planck–Poisson equations thus read as

$$\partial_t n + \nabla \cdot \boldsymbol{\jmath} = 0 \qquad \text{with} \qquad \boldsymbol{\jmath} = -\mathcal{D}\,\nabla n - \beta e \mathcal{D}\, n\, \nabla\Phi + \delta\boldsymbol{\jmath} \qquad \text{and} \tag{10.84}$$

$$\nabla^2\Phi = -\frac{e}{\epsilon}\,(n - n_0), \tag{10.85}$$

where $\mathcal{D}$ is the diffusion coefficient of the ions, $\delta\boldsymbol{\jmath}(\mathbf{r},t)$ are Gaussian white noise fields characterized by equation (10.9), and ϵ the electric permittivity of the medium. The local conservation of electric charge (1.77) is thus satisfied by the stochastic description. In this scheme, deviations with respect to electroneutrality manifest themselves over spatial scales smaller than the Debye screening length (1.81). We note that the electric field is constant and invariant in the limit of a large enough electric permittivity, such that the Debye length is much larger than the length L of the conductive channel.

If the conductive channel is in contact with two reservoirs at the positions $x = 0$ and $x = L$, the ion density and the electric potential have the following boundary conditions,

$$n(0,t) = n_{\rm L}, \qquad \Phi(0,t) = \Phi_{\rm L}, \qquad n(L,t) = n_{\rm R}, \qquad \Phi(L,t) = \Phi_{\rm R}, \tag{10.86}$$

so that the affinity driving the system out of equilibrium is related to the applied voltage V according to

$$A = \beta e V = \beta\,(\mu_{\rm L} - \mu_{\rm R}) = \ln\frac{n_{\rm L}}{n_{\rm R}} + \beta e\,(\Phi_{\rm L} - \Phi_{\rm R}). \tag{10.87}$$

On the one hand, the current of mobile ionic particles flowing through some section Σ across the conductive channel and its mean value are defined as

$$\mathscr{J} \equiv \int_\Sigma \boldsymbol{\jmath} \cdot d\boldsymbol{\Sigma} \qquad \text{and} \qquad J \equiv \langle \mathscr{J} \rangle. \tag{10.88}$$

We note that this instantaneous particle current can also be expressed as

$$\mathscr{J}(t) = \sum_{k=-\infty}^{+\infty} \varsigma_k \, \delta(t - t_k), \tag{10.89}$$

where $\varsigma_k = \pm 1$ depending on whether the particle crossing the section Σ at time t_k is moving in the direction of the vector $d\mathbf{\Sigma}$ or the opposite direction.

On the other hand, the measured electric current density includes the displacement current and is given by

$$\imath \equiv e\, \jmath + \epsilon\, \partial_t \boldsymbol{\mathcal{E}}, \qquad \text{obeying} \qquad \nabla \cdot \imath = 0 \tag{10.90}$$

at all times, as a consequence of Maxwell's equations (Jackson, 1999). The measured electric current intensity and its mean value are thus defined as

$$\mathcal{I} \equiv \int_{\Sigma} \imath \cdot d\mathbf{\Sigma} \qquad \text{and} \qquad I \equiv \langle \mathcal{I} \rangle. \tag{10.91}$$

In a stationary macrostate, the mean values of the electric and particle currents are related to each other by

$$I = e\, J \tag{10.92}$$

because the time average of the displacement current is equal to zero in this case. However, the fluctuations in the charge distribution generate a fluctuating displacement current, so that $\mathcal{I}(t)$ and $e\,\mathscr{J}(t)$ are not equal in general. In particular, the random variables defined by integrating the currents over the time interval $[0,t]$ as

$$Z(t) \equiv \int_0^t \mathscr{J}(t')\, dt' \qquad \text{and} \qquad \tilde{Z}(t) \equiv \frac{1}{e} \int_0^t \mathcal{I}(t')\, dt' \tag{10.93}$$

have different statistical properties and probability distributions $P(Z,t)$ and $P(\tilde{Z},t)$.

10.5.2 Space Discretization and Master Equation

The stochastic Nernst–Planck–Poisson problem can be numerically simulated by the same method of space discretization as presented in Section 10.2.2. Accordingly, ion transport can be described as the Markov jump process of equation (10.12), which is ruled by the master equation (10.13) with the transition rates (10.16) and (10.17). The difference is that the changes of energy (10.20) in the transitions are determined by the total electrostatic energy of the system. This latter is given by

$$U = \frac{\epsilon}{2} \int \boldsymbol{\mathcal{E}}^2 \, d^3 r = \frac{\epsilon \Sigma}{2 \Delta x} \sum_{i=0}^{l} (\Phi_{i+1} - \Phi_i)^2, \tag{10.94}$$

after discretization of the channel into l cells of size Δx in the longitudinal direction, cross-sectional area Σ, and volume $\Delta V = \Sigma \Delta x$, where $\Phi_0 = \Phi_{\rm L}$ and $\Phi_{l+1} = \Phi_{\rm R}$. In addition, the discretized Poisson equation of equation (10.85) reads

$$\frac{\Phi_{i+1} - 2\Phi_i + \Phi_{i-1}}{\Delta x^2} = -\frac{e}{\epsilon} \left(\frac{N_i}{\Delta V} - n_0 \right) \qquad \text{for} \qquad 1 \le i \le l. \tag{10.95}$$

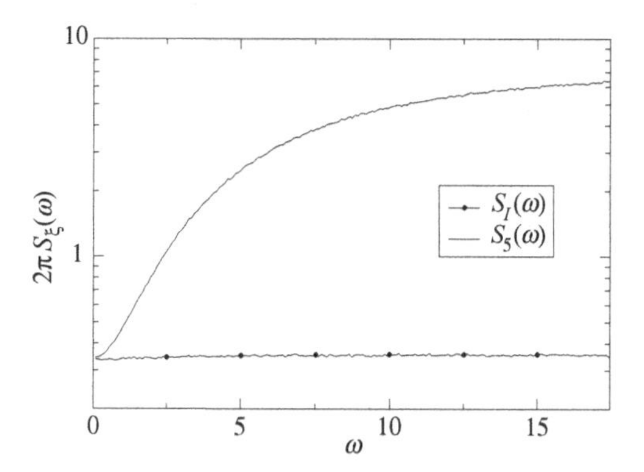

Figure 10.3 Ion transport: Power spectra of the particle current (10.89) evaluated at $i = 5$ (solid line) and of the electric current (10.96) in units of e^2 (line with dots). The channel is composed of $l = 20$ cells and is constrained to the boundary conditions $N_{\rm L} = 5$, $N_{\rm R} = 45$, and $\Phi_{\rm L} = \Phi_{\rm R}$. Reproduced with permission from Andrieux and Gaspard (2009) © IOP Publishing Ltd and SISSA Medialab.

In this discretized model, the instantaneous particle current $\mathcal{J}_i(t)$ at the boundary between the ith and $(i + 1)$th cells is given by equation (10.89) evaluated at the position $x_i = i\Delta x$ for $0 \leq i \leq l$. The instantaneous electric current intensity (10.91) is thus obtained as

$$\mathcal{I}(t) = \frac{e}{l+1} \sum_{i=0}^{l} \mathcal{J}_i(t), \tag{10.96}$$

which is the average of the instantaneous particle currents in all the cells (Andrieux and Gaspard, 2009).

The stochastic process can be simulated by the algorithm of Gillespie (1976, 1977). The numerical results of Figures 10.3 and 10.4 are obtained using $k^{-1} = \Delta x^2/\mathcal{D}$ as the time unit, the value $L/\ell_{\rm D} = 50$ for the ratio of the channel length L to the Debye screening length (1.81), and $n_0 \Delta V = 25$ fixed negatively charged ions in each cell.

The spectral density functions (4.14) of the particle and electric currents are shown in Figure 10.3 (Andrieux and Gaspard, 2009). On the one hand, the power spectrum of the particle current increases with the frequency ω because of the long-ranged Coulomb interaction between the ions. On the other hand, the power spectrum of the electric current is essentially uniform in frequency. At zero frequency, both power spectra coincide in relation to the identity (10.92) between their mean values. This figure illustrates the fact that the electric current defined with equation (10.90) has significantly different statistical properties than the particle current (10.88). Moreover, the inclusion of the displacement current in the electric current has the effect of averaging the fluctuations of the local particle currents.

10.5.3 Fluctuation Relations for the Currents

For the discretized model of ion transport, the affinity (10.23) of the cycle (10.22) is given by equation (10.87) since the function (10.18) used in the definition of the transition rates

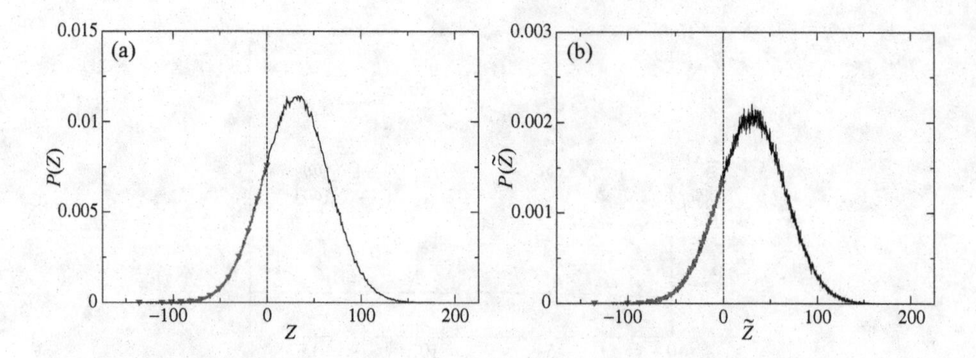

Figure 10.4 Ion transport: (a) Probability distribution of the integrated particle current, $Z = \int_0^t \mathscr{J}_i(t')\,dt'$, evaluated at $i = 2$. The comparison is performed with the prediction of the fluctuation relation for the negative part of the distribution (thick line with triangles). (b) Probability distribution of the integrated electric current, $\tilde{Z} = (1/e)\int_0^t \mathcal{I}(t')\,dt'$, which is also compared with the prediction of the fluctuation relation for the negative part of the distribution (thick line with triangles). The probability distributions are evaluated at time $t = 273$. The channel is composed of $l = 10$ cells and subjected to the boundary conditions $N_{\rm L} = N_{\rm R} = 25$ and $\beta e(\Phi_{\rm L} - \Phi_{\rm R}) = 0.05$, so that the affinity (10.87) takes the value $A = 1/20$. Reproduced with permission from Andrieux and Gaspard (2009) © IOP Publishing Ltd and SISSA Medialab.

(10.16) and (10.17) satisfies the identity (10.19), guaranteeing detailed balance at thermodynamic equilibrium. Accordingly, the fluctuation relation (10.46) holds for the probability distribution $P(Z,t)$ of the particle current with the affinity (10.87). Furthermore, the demonstration can be extended to the electric current (10.96). Accordingly, both the particle and electric integrated currents (10.93) obey the fluctuation relations

$$\frac{P(Z,t)}{P(-Z,t)} \sim_{t\to\infty} \exp(A\,Z) \qquad \text{and} \qquad \frac{P(\tilde{Z},t)}{P\left(-\tilde{Z},t\right)} \sim_{t\to\infty} \exp\left(A\,\tilde{Z}\right) \tag{10.97}$$

with the same affinity (10.87) for long enough time t (Andrieux and Gaspard, 2009).[5] The predictions of these fluctuation relations are numerically verified in Figure 10.4.

These results show that the fluctuation relation is also valid in systems with long-ranged interaction between the particles.

10.6 Diodes and Transistors

10.6.1 Stochastic Approach to Charge Transport

In semiconductors, the charge carriers are electrons and holes (Ashcroft and Mermin, 1976). They are generated by the thermally activated reaction

$$\emptyset \underset{k_-}{\overset{k_+}{\rightleftharpoons}} \mathrm{e}^- + \mathrm{h}^+, \tag{10.98}$$

[5] We note that the voltage here is fixed and the current is fluctuating, although the current was fixed and the voltage fluctuating in the RC electric circuit of Section 7.3.

where k_+ and k_- are, respectively, the electron-hole generation and recombination rate constants and $\emptyset$ denotes the ground state. If the semiconductor was at equilibrium, the electron and hole densities would obey the condition

$$n_{\rm eq} p_{\rm eq} = \nu^2, \qquad \text{where} \qquad \nu = \sqrt{\frac{k_+}{k_-}} \tag{10.99}$$

is the intrinsic carrier density. This equilibrium condition is reminiscent of the Guldberg–Waage relation (1.39) with the equilibrium constant $K_{\rm eq} = \nu^2$. Since electron-hole generation requires thermal activation over the gap energy $E_{\rm g}$ of the semiconductor, the equilibrium constant goes as $K_{\rm eq} \propto \exp(-\beta E_{\rm g})$ with the inverse temperature $\beta = (k_{\rm B}T)^{-1}$. The gap energy is known to take the value $E_{\rm g} = 0.67$ eV for germanium and $E_{\rm g} = 1.12$ eV for silicium at room temperature (Ashcroft and Mermin, 1976).

Semiconductors can be doped with impurities that are electron donors or acceptors. Accordingly, n-type semiconductors are obtained by doping with donors and p-type with acceptors. Although electrons and holes are mobile, impurities are essentially immobile inside the semiconductor crystal. In p-type semiconductors, the majority of charge carriers are holes and the minority carriers are electrons. The situation is opposite in n-type semiconductors. The charge density is given by

$$\rho_{\rm e} = e(p - n + d - a) \tag{10.100}$$

in terms of the elementary electric charge $e = |e|$, the hole density p, the electron density n, the donor density d, and the acceptor density a. The charge density determines the electric potential Φ according to the Poisson equation

$$\nabla^2 \Phi = -\frac{\rho_{\rm e}}{\epsilon} \tag{10.101}$$

and, thus, the electric field $\boldsymbol{\mathcal{E}} = -\boldsymbol{\nabla}\Phi$.

The stochastic time evolution of the electron and hole densities can be described in terms of the following fluctuating diffusion–reaction equations,

$$\partial_t n + \boldsymbol{\nabla} \cdot \boldsymbol{j}_n = \sigma_n \qquad \text{and} \qquad \partial_t p + \boldsymbol{\nabla} \cdot \boldsymbol{j}_p = \sigma_p \tag{10.102}$$

with the current densities

$$\boldsymbol{j}_n = -\mathcal{D}_n \boldsymbol{\nabla} n - \beta e \mathcal{D}_n n \boldsymbol{\mathcal{E}} + \delta\boldsymbol{j}_n \quad \text{and} \quad \boldsymbol{j}_p = -\mathcal{D}_p \boldsymbol{\nabla} p + \beta e \mathcal{D}_p p \boldsymbol{\mathcal{E}} + \delta\boldsymbol{j}_p, \tag{10.103}$$

the reaction rate densities

$$\sigma_n = \sigma_p = k_+ - k_- n\, p + \delta\sigma, \tag{10.104}$$

and uncorrelated Gaussian white noise fields $\delta\boldsymbol{j}_n$, $\delta\boldsymbol{j}_p$, and $\delta\sigma$ characterized by the nonvanishing correlation functions

$$\langle \delta\boldsymbol{j}_n(\mathbf{r},t)\, \delta\boldsymbol{j}_n(\mathbf{r}',t')\rangle = 2\,\mathcal{D}_n\, n\, \delta^3(\mathbf{r}-\mathbf{r}')\,\delta(t-t')\,\mathbf{1}, \tag{10.105}$$

$$\langle \delta\boldsymbol{j}_p(\mathbf{r},t)\, \delta\boldsymbol{j}_p(\mathbf{r}',t')\rangle = 2\,\mathcal{D}_p\, p\, \delta^3(\mathbf{r}-\mathbf{r}')\,\delta(t-t')\,\mathbf{1}, \tag{10.106}$$

$$\langle \delta\sigma(\mathbf{r},t)\, \delta\sigma(\mathbf{r}',t')\rangle = (k_+ + k_- n\, p)\; \delta^3(\mathbf{r}-\mathbf{r}')\,\delta(t-t'). \tag{10.107}$$

The local conservation of electric charge is satisfied because the continuity equation (1.77) holds with the charge density (10.100) and the charge current density $\mathbf{j}_e = e(\mathbf{j}_p - \mathbf{j}_n)$. As previously mentioned, the measured electric current density is obtained by adding the contribution from the displacement current, $\boldsymbol{\imath} = \mathbf{j}_e + \epsilon\, \partial_t \boldsymbol{\mathcal{E}}$. Therefore, the electric current intensity and its mean value are also defined by equation (10.91). The instantaneous charge current in the same section Σ is given by $\mathscr{I}_e \equiv \int_\Sigma \mathbf{j}_e \cdot d\boldsymbol{\Sigma}$ and its mean value is equal to the mean electric current intensity $I = \langle \mathscr{I}_e \rangle = e(J_p - J_n)$, where $J_p = \langle \mathscr{I}_p \rangle$ and $J_n = \langle \mathscr{I}_n \rangle$.

In the mean-field approximation, the standard equations are recovered for the electron and hole densities, the current densities, and the electric potential in the semiconductor (Ashcroft and Mermin, 1976; Gu and Gaspard, 2018, 2019).

If space is discretized into cells with volume ΔV as in Section 10.2.2, the stochastic process can be approximated by a Markov jump process for the numbers $\mathbf{N} = (N_i)_{i=1}^{l}$ and $\mathbf{P} = (P_i)_{i=1}^{l}$ of electrons and holes in the cells $1 \le i \le l$ (Gu and Gaspard, 2018, 2019). The probability distribution to find the system in the coarse-grained state $(\mathbf{N}, \mathbf{P})$ at time t is ruled by a master equation similar to equation (10.21) with transition rates like those of equations (10.16) and (10.17) for the jumps due to diffusion, and given by

$$W_i^{(+)} = k_+ \Delta V \qquad \text{and} \qquad W_i^{(-)} = k_- N_i P_i / \Delta V \tag{10.108}$$

for electron-hole generation and recombination in the ith cell. In order to take into account the Coulomb interaction between the electric charges, the electrostatic energy (10.94) should be used in equations (10.16) and (10.17). In the continuum limit, the fluctuating diffusion–reaction equations (10.102)–(10.107) are recovered. If the identity (10.19) holds for the function $\psi(\Delta U)$, the conditions of detailed balance are obeyed at equilibrium. The asymptotic fluctuation relation can thus be satisfied for the transport of electrons and holes in semiconductor devices (Gu and Gaspard, 2018, 2019). We note that the transition rates depend nonlinearly on the random variables N_i and P_i, so that the stationary distribution of the Markov jump process is not Poissonian.

10.6.2 Diodes

The diode consists of a p-n junction, which is composed of a p-type semiconductor doped with acceptors in contact with a n-type semiconductor with donors, as shown in Figure 10.5 (Ashcroft and Mermin, 1976). The p-n junction is considered as a three-dimensional channel of length L with its coordinate x extending from $-L/2$ to $+L/2$, and of cross-sectional area Σ in the transverse directions. The position is denoted $\mathbf{r} = (x, y, z)$. The acceptor density $a(\mathbf{r})$ and the donor density $d(\mathbf{r})$ are uniform in the p- and n-sides, respectively, and given by

$$a(\mathbf{r}) = a\,\theta(-x), \qquad d(\mathbf{r}) = d\,\theta(x), \tag{10.109}$$

where $\theta(x)$ is Heaviside's step function. The electric charge density (10.100) thus gives the electric potential by the Poisson equation. According to electroneutrality, the inhomogeneity

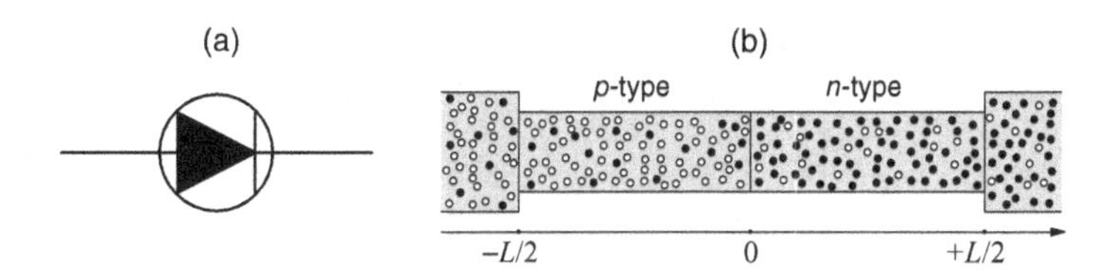

Figure 10.5 Schematic representation of (a) the diode and (b) the p-n junction. In the p-n junction, the filled (respectively, open) circles represent electrons (respectively, holes). Reprinted with permission from Gu and Gaspard (2018) © American Physical Society.

(10.109) of the acceptor and donor densities induces the global asymmetric distribution of mobile electrons and holes across the junction, leading to current rectification by the diode.

The time evolution of the electron and hole densities is ruled by the stochastic partial differential equations (10.102)–(10.107) with the boundary conditions

$$\Phi(x = -L/2) = \Phi_{\rm L}, \quad n(x = -L/2) = n_{\rm L}, \quad p(x = -L/2) = p_{\rm L}, \tag{10.110}$$

$$\Phi(x = +L/2) = \Phi_{\rm R}, \quad n(x = +L/2) = n_{\rm R}, \quad p(x = +L/2) = p_{\rm R}. \tag{10.111}$$

The diode is driven out of equilibrium if the applied voltage is nonvanishing,

$$V = \Phi_{\rm L} - \Phi_{\rm R} - \frac{1}{\beta e} \ln \frac{n_{\rm L}}{n_{\rm R}} = \Phi_{\rm L} - \Phi_{\rm R} + \frac{1}{\beta e} \ln \frac{p_{\rm L}}{p_{\rm R}}, \tag{10.112}$$

in which case an electric current is flowing in the diode. The equilibrium macrostate is recovered if the applied voltage is zero, i.e., $V = 0$.

The integrated charge and electric currents in the cross-sectional area Σ at $x = 0$ during the time interval $[0, t]$ can be defined as in equation (10.93) with $\mathscr{J} = \mathscr{J}_p - \mathscr{J}_n$. According to the results of Section 10.5, the probability distributions of these random currents obey the fluctuation relations of equation (10.97).

The diode model can be simulated numerically by discretizing space into cells with size Δx and volume ΔV, as explained above. A simulation method that is faster than the Gillespie algorithm is provided by the corresponding Langevin stochastic process obtained by adapting equations (10.28)–(10.30) to electron-hole transport and reaction (Gu and Gaspard, 2018). Dimensionless quantities can be defined by rescaling time with the intrinsic carrier lifetime $\tau = (k_+ k_-)^{-1/2}$, space with the intrinsic carrier diffusion length $\ell_{\rm diff} = (\mathcal{D}\tau)^{1/2}$ (supposing equal diffusion coefficients $\mathcal{D}_n = \mathcal{D}_p \equiv \mathcal{D}$), the densities with the intrinsic carried density introduced in equation (10.99), and the electric potential or voltage according to $V_* \equiv \beta e V$. The properties of the p-n junction are thus determined by the dimensionless parameter defined as

$$\alpha \equiv \left(\frac{\ell_{\rm diff}}{\ell_{\rm D}}\right)^2 = \frac{\beta e^2 \mathcal{D}}{\epsilon\, k_-} \tag{10.113}$$

in terms of the Debye screening length (1.81) with $n_0 = \nu$, as well as by the boundary conditions. In order for the electric field to remain uniform in contact with the reservoirs,

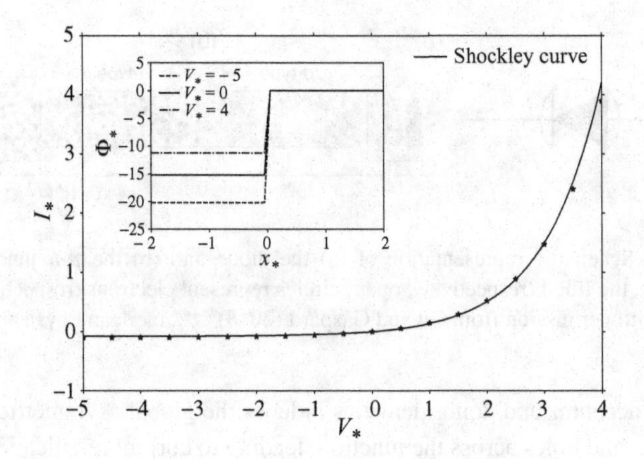

Figure 10.6 The current-voltage characteristic curve of the *p*-*n* junction for $\alpha = 100$ and $c_{\rm major}$:$c_{\rm minor}$ $= 4 \times 10^6$. Here, I_* denotes the dimensionless mean current and V_* the dimensionless applied voltage. The inset shows the electric potential profiles for several values of the applied voltage. The dots show the results of numerical simulations. The line is the Shockley curve (10.115) fitted to simulation data. The simulations are performed using Langevin's algorithm for $l = 40$ cells with volume $\Delta V = 8 \times 10^5$. Reprinted with permission from Gu and Gaspard (2018) © American Physical Society.

the boundary conditions $p_{\rm L} = n_{\rm L} + a(-L/2)$ and $n_{\rm R} = p_{\rm R} + d(+L/2)$ are considered. Moreover, we may suppose that $a = d$, $n_{\rm L} = p_{\rm R}$, and $n_{\rm R} = p_{\rm L}$, so that the diode is symmetric under inversion and permutation between electrons and holes. With these assumptions, the diode is characterized by the ratio of majority to minority carrier concentrations:

$$\frac{c_{\rm major}}{c_{\rm minor}} \equiv \frac{p_{\rm L}}{n_{\rm L}} = \frac{n_{\rm R}}{p_{\rm R}}. \tag{10.114}$$

The current-voltage characteristic curve of the diode is shown in Figure 10.6 in the case of a large concentration ratio (10.114). As seen in the inset of Figure 10.6, the rescaled electric potential $\Phi_* = \beta e \Phi$ has a sharp step at the junction if such a condition holds. In this case, the diode can be described with the simple Markov jump process (4.78) in terms of the transition rates $W_\pm$ for the electric charges to jump in one direction or the opposite at $x = 0$, so that the mean electric current $I = eJ$ is given by equation (5.245) with the affinity $A = \beta e V$. Therefore, the I–V curve is described by

$$I(V) = I_{\rm s}\big(e^{\beta e V} - 1\big), \tag{10.115}$$

where $I_{\rm s} \propto \exp(-\beta E_{\rm g})$ is the saturation current at negative voltage (Shockley, 1949; Ashcroft and Mermin, 1976).

Figure 10.7 shows that the predictions of the fluctuation relation for the electric current $\tilde{Z}_*$ defined in equation (10.93) are confirmed in the long-time limit. The ratio of the probabilities for the opposite fluctuations of electric current has a logarithm that is linear in the random variable $\tilde{Z}_*$, as seen in Figure 10.7(b). The slope given at finite time t by

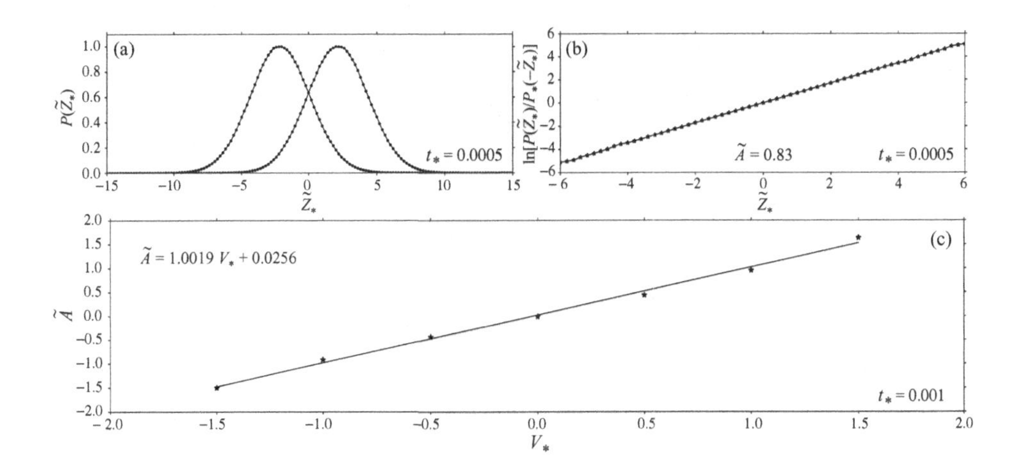

Figure 10.7 Full-counting statistics of the total electric current including the contribution of the displacement current in the diode for $A = V_* = \beta eV = 1$, $\alpha = 100$, $c_{\rm major}{:}c_{\rm minor} = 4$, and $l_* = 4$: (a) The probability distributions $P(\pm\tilde{Z}_*)$ at time $t_* = 0.0005$ versus the rescaled charge number $\tilde{Z}_* = \tilde{Z}/(\ell^3\nu)$ with peak value normalized to one. The statistics is carried out over 10^6 random trajectories. (b) $\ln[P(\tilde{Z}_*)/P(-\tilde{Z}_*)]$ versus $\tilde{Z}_*$ showing the linearity with the slope $\tilde{A} \simeq 0.83$ at time $t_* = 0.0005$. (c) The affinities computed with fitted Gaussian distributions versus the dimensionless applied voltage V_*, checking the linear dependence $\tilde{A} \simeq V_*$ with a unit slope (solid line) at time $t_* = 0.001$, as predicted by the fluctuation relation. The simulations are performed using Langevin's algorithm for $l = 40$ cells with volume $\Delta V = 800$. Reprinted with permission from Gu and Gaspard (2018) © American Physical Society.

$\tilde{A}_t = \left(1/\tilde{Z}_*\right) \ln\left[P\left(\tilde{Z}_*, t\right)/P\left(-\tilde{Z}_*, t\right)\right]$ is lower than its asymptotic value $\lim_{t\to\infty} \tilde{A}_t = \beta eV$. Since the probability distribution is nearly Gaussian, the time-dependent affinity can be evaluated according to $\tilde{A}_t = \ln\left[\left(\sigma_t^2 + \mu_t\right)/\left(\sigma_t^2 - \mu_t\right)\right]$ in terms of its mean value μ_t and its variance σ_t^2. For a long enough time, the asymptotic value is obtained, as observed in Figure 10.7(c) confirming the prediction of the fluctuation relation for the electric current.

10.6.3 Transistors

The bipolar n-p-n junction transistor is one of the most common transistors (Shockley et al., 1951; Ebers and Moll, 1954; Sedra and Smith, 2004). It consists of three small doped semiconducting regions, respectively typed as n, p, and n, thus forming two junctions. Each doped region has a port and the three ports are in contact with charge reservoirs. They are, respectively, called the *collector*, *base*, and *emitter*, as shown in Figure 10.8.

The transistor can be modeled by assuming that the semiconducting material extends from $x = -L/2$ to $x = +L/2$ and is divided into three parts. The part from $x = -L/2$ to $x = -L_p/2$ is of n-type, the one from $x = -L_p/2$ to $x = +L_p/2$ of p-type, and the one from $x = +L_p/2$ to $x = +L/2$ of n-type. The collector is in contact at $x = -L/2$, the emitter at $x = +L/2$, and the base along a length L_B symmetrically located around the

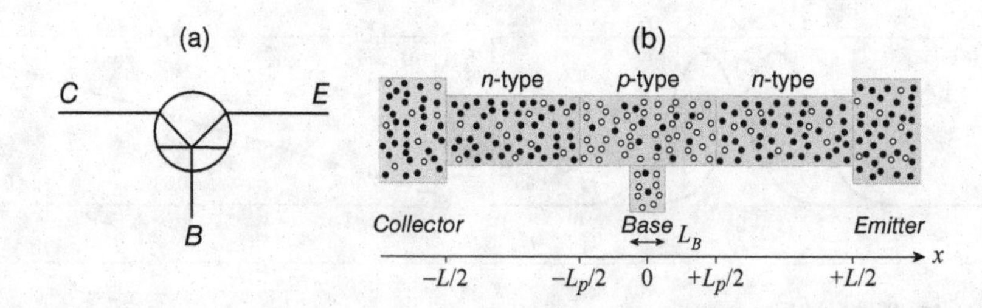

Figure 10.8 Schematic representation of (a) the transistor and (b) the bipolar n-p-n double junction. In panel (b), the filled (respectively, open) circles represent electrons (respectively, holes). The three reservoirs, *collector*, *base*, and *emitter*, fix the values of the electron density, the hole density, and the electric potentials at their contact with the transistor. Reprinted with permission from Gu and Gaspard (2019) © American Physical Society.

origin $x = 0$. In addition, the bipolar n-p-n double junction has the cross-sectional area Σ in the transverse directions. The areas of the contacts with the collector and emitter are assumed to be equal, so $\Sigma_C = \Sigma_E = \Sigma$, while Σ_B denotes the area of the contact with the base. The donor and acceptor densities are assumed to be uniform in the different types of semiconductor according to

$$d(\mathbf{r}) = d\,\theta\left(-x - L_p/2\right) + d\,\theta\left(x - L_p/2\right), \quad a(\mathbf{r}) = a\,\theta\left(x + L_p/2\right)\,\theta\left(-x + L_p/2\right) \tag{10.116}$$

in terms of two constant values a and d, combined with Heaviside's step function $\theta(x)$. The charge density is again given by equation (10.100), which determines the electric field by the Poisson equation.

The stochastic motion of electrons and holes in the transistor can also be described using equations (10.102)–(10.107). The electron and hole densities and the electric potential have fixed boundary values at the contacts with the three reservoirs, which are, respectively, n_C, p_C, Φ_C at the collector; n_B, p_B, Φ_B at the base; and n_E, p_E, Φ_E at the emitter. The two independent voltages driving the transistor out of equilibrium can be taken as those applied to the collector and the base with respect to the emitter:

$$V_C \equiv \Phi_C - \Phi_E + \frac{1}{\beta e}\ln\frac{p_C}{p_E} \quad \text{and} \quad V_B \equiv \Phi_B - \Phi_E + \frac{1}{\beta e}\ln\frac{p_B}{p_E}, \tag{10.117}$$

which induce two currents across the transistor. The corresponding affinities are given by

$$A_C \equiv \beta e V_C \quad \text{and} \quad A_B \equiv \beta e V_B, \tag{10.118}$$

which are dimensionless. The equilibrium macrostate is reached if both affinities, i.e., both applied voltages, are equal to zero.

In transistors, there are thus twice as many integrated charge and electric currents as in the diode, which are respectively defined at the collector and the base in the direction of entrance into the transistor according to equation (10.93) with $\mathscr{I} = \mathscr{I}_p - \mathscr{I}_n$ at the sections Σ_C and Σ_B: Z_C, Z_B, $\tilde{Z}_C$, and $\tilde{Z}_B$.

The results of Section 10.5 extend to these random currents, so that their probability distributions obey the following fluctuation relations in the long-time limit,

$$\frac{P(Z_C, Z_B, t)}{P(-Z_C, -Z_B, t)} \sim_{t\to\infty} \exp\left(A_C\, Z_C + A_B\, Z_B\right), \tag{10.119}$$

$$\frac{P\left(\tilde{Z}_C, \tilde{Z}_B, t\right)}{P\left(-\tilde{Z}_C, -\tilde{Z}_B, t\right)} \sim_{t\to\infty} \exp\left(A_C\, \tilde{Z}_C + A_B\, \tilde{Z}_B\right), \tag{10.120}$$

with the same affinities (10.118). Consequently, the linear and nonlinear transport properties of Section 5.5 are satisfied for the transistor (Gu and Gaspard, 2019).

Furthermore, the mean values of the charge and electric currents coincide $I_C = eJ_C = \lim_{t\to\infty} e\langle Z_C(t)\rangle/t$ and $I_B = eJ_B = \lim_{t\to\infty} e\langle Z_B(t)\rangle/t$ as previously mentioned, so that the entropy production rate is given by

$$\frac{1}{k_{\rm B}}\frac{d_{\rm i}S}{dt} = A_C\, J_C + A_B\, J_B = \beta\,(V_C\, I_C + V_B\, I_B) \geq 0, \tag{10.121}$$

according to equation (5.97) or, equivalently, equation (5.98), giving the electric power dissipated in the transistor, as expected.

We note that if the ratios of majority to minority carrier concentrations are large enough, the bipolar junction transistor can be described by the Ebers–Moll model (5.248), in which case the finite-time fluctuation relation (5.254) holds (Gu and Gaspard, 2020).

As for the diode, the numerical simulation of the transistor can be performed by spatial discretization into cells with size Δx in the longitudinal direction, cross-sectional area Σ, and volume $\Delta V = \Sigma\Delta x$ (Gu and Gaspard, 2019). In the transverse direction in contact with the base, the size is taken as Δy. There are l_n cells in each n-type region, l_p cells in the p-type region, and l_B cells in contact with the base. The coarse-grained state of the system is specified by the numbers of electrons and holes in the cells: $(\mathbf{N}, \mathbf{P})$ with $\mathbf{N} = (N_i)_{i=1}^{l}$ and $\mathbf{P} = (P_i)_{i=1}^{l}$ with $l = 2l_n + l_p$. The random transitions between these states are generated by the diffusion of electrons and holes between the cells and their reaction (10.98). The associated stochastic process is a Markov jump process ruled by a master equation similar to equation (10.21) with the transition rates (10.16) and (10.17) for diffusion, and (10.108) for reaction (Gu and Gaspard, 2019). The transition rates depend on the changes in the total electrostatic energy (10.94) of the system. If the numbers of electrons and holes in the cells are large enough (i.e., $N_i \gg 1$ and $P_i \gg 1$), the Markov jump process can be approximated by a Langevin stochastic process ruled by equations such as (10.28)–(10.30). At the boundaries in contact with the reservoirs, the electron and hole densities satisfy the conditions $n_C p_C = n_B p_B = n_E p_E = \nu^2$. Moreover, the system is assumed to be symmetric with respect to $x = 0$ as in Figure 10.8 by setting $n_C = n_E$ and $p_C = p_E$. In the aforementioned units, the parameters take the following values:

$$\begin{aligned}
&e = 1,\ \beta = 1,\ \epsilon = 0.01,\ \mathcal{D}_p = \mathcal{D}_n \equiv \mathcal{D} = 0.01,\ k_+ = k_- = 0.01,\ \nu = 1,\\
&n_C = n_E = 10,\ n_B = 0.1,\ p_C = p_E = 0.1,\ p_B = 10,\\
&\Delta x = 0.1,\ \Delta y = 0.2,\ l_n = 10,\ l_p = 3,\ l_B = 1,\ \Sigma = 10^4,\ \Sigma_B = 5\times 10^3.
\end{aligned} \tag{10.122}$$

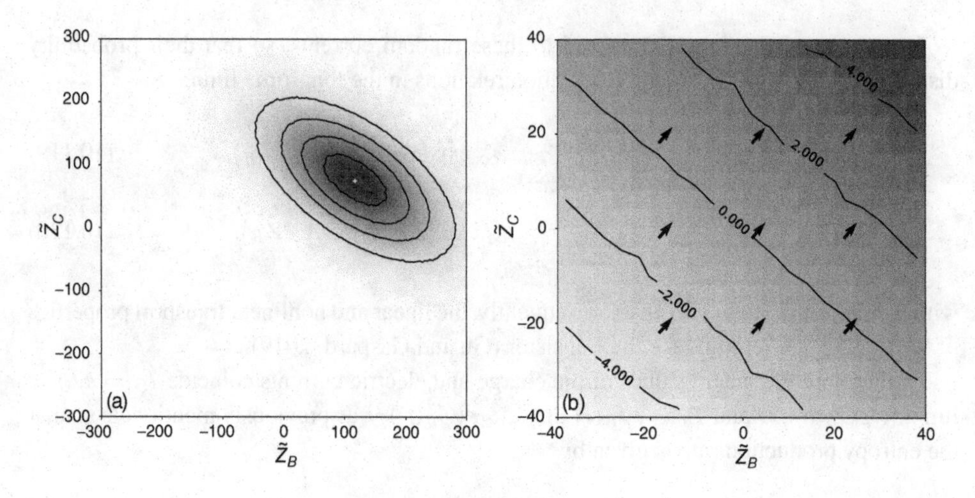

Figure 10.9 Full counting statistics of the total electric charges $\tilde{Z}_C$ and $\tilde{Z}_B$ including the displacement currents in the transistor: (a) Joint probability distribution $P(\tilde{Z}_C, \tilde{Z}_B, t)$ at time $t = 20$. This distribution is centered on the mean values $\langle \tilde{Z}_B(t) \rangle = 117.4$ and $\langle \tilde{Z}_C(t) \rangle = 75.2$. (b) The corresponding function $\ln\left[P(\tilde{Z}_C, \tilde{Z}_B, t)/P(-\tilde{Z}_C, -\tilde{Z}_B, t)\right]$ versus $\tilde{Z}_C$ and $\tilde{Z}_B$ at the same time $t = 20$, giving the finite-time affinities $\tilde{A}_B(t) = 0.066$ and $\tilde{A}_C(t) = 0.075$. The affinities are set to the value $A_C = A_B = 0.1$. The simulation is carried out for 3×10^7 trajectories with the time step $dt = 0.1$. Reprinted with permission from Gu and Gaspard (2019) © American Physical Society.

Accordingly, the transistor has the total length $L = (2l_n + l_p)\Delta x = 2.3$, each cell the volume $\Delta V = \Sigma \Delta x = 10^3$, the parameter (10.113) the value $\alpha = 100$, the intrinsic carrier diffusion length $\ell_{\rm diff} = 1$, the Debye screening length $\ell_{\rm D} = 0.1$, and the intrinsic carrier lifetime $\tau = (k_+ k_-)^{-1/2} = 10^2$.

In order to test the fluctuation relation (10.120) for the electric current including the displacement current, the probability distribution $P(\tilde{Z}_C, \tilde{Z}_B, t)$ is simulated numerically as shown in Figure 10.9(a) at the time $t = 20$ under nonequilibrium conditions where the applied voltages correspond to the values $A_C = A_B = 0.1$ for the affinities (10.118). As long as this distribution has a statistically significant overlap with the opposite distribution $P(-\tilde{Z}_C, -\tilde{Z}_B, t)$, the logarithm of their ratio can be evaluated, giving the plot in Figure 10.9(b), which is numerically consistent with the linear function $\tilde{A}_C(t)\tilde{Z}_C + \tilde{A}_B(t)\tilde{Z}_B$. The coefficients $\tilde{A}_C(t)$ and $\tilde{A}_B(t)$ define effective time-dependent affinities, which should converge towards their asymptotic values (10.118), as predicted by the fluctuation relation (10.120). As shown in Figure 10.10, the time-dependent affinities increase towards the expected values $A_C = A_B = 0.1$. The behavior is similar to the one shown in Figure 10.2 for diffusion and diffusion-influenced surface reaction.

Furthermore, we may also consider the cumulant generating function

$$Q(\boldsymbol{\lambda}; \mathbf{A}) \equiv \lim_{t\to\infty} -\frac{1}{t} \ln \int P_{\mathbf{A}}(Z_C, Z_B, t)\, {\rm e}^{-\lambda_C Z_C - \lambda_B Z_B}\, dZ_C\, dZ_B, \tag{10.123}$$

Table 10.1. *Transistor model with the parameter values (10.122). The linear response coefficients $L_{\alpha,\beta}$ defined by equation (5.134) and the equilibrium current diffusivities $D_{\alpha\beta}(0,0)$ versus $\alpha,\beta \in \{B,C\}$. The last column tests the fluctuation–dissipation relation $L_{\alpha,\beta} = D_{\alpha\beta}(0,0)$ and, thus, the Onsager reciprocal relation $L_{B,C} = L_{C,B}$ in the transistor model (Gu and Gaspard, 2019).*

α	β	$L_{\alpha,\beta}$	$D_{\alpha\beta}(0,0)$	$L_{\alpha,\beta} - D_{\alpha\beta}(0,0)$
B	B	112.60 ± 0.02	113.16 ± 0.49	-0.56
B	C	-56.30 ± 0.02	-56.34 ± 0.49	0.04
C	B	-56.29 ± 0.02	-56.34 ± 0.49	0.06
C	C	93.11 ± 0.02	92.99 ± 1.04	0.12

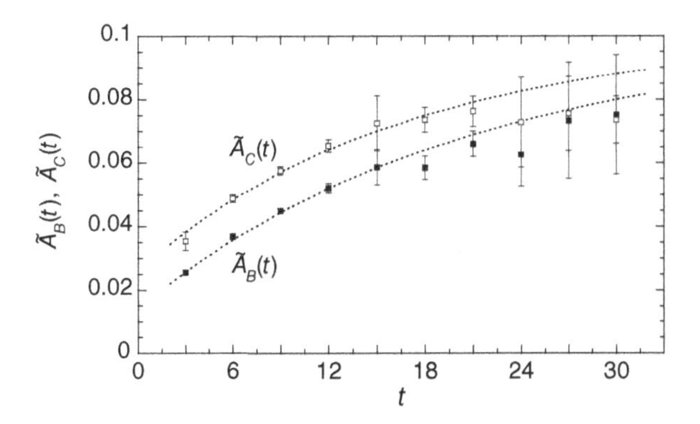

Figure 10.10 The finite-time affinities $\tilde{A}_C(t)$ and $\tilde{A}_B(t)$ versus time t for the transistor in the same conditions as in Figure 10.9. These affinities are obtained by fitting $\ln[P(\tilde{Z}_C,\tilde{Z}_B,t)/P(-\tilde{Z}_C,-\tilde{Z}_B,t)]$ to the linear function $\tilde{A}_C(t)\,\tilde{Z}_C+\tilde{A}_B(t)\,\tilde{Z}_B$. The dashed lines show the convergence to the asymptotic value of the affinities $\tilde{A}_C(\infty) = \tilde{A}_B(\infty) = 0.1$. Reprinted with permission from Gu and Gaspard (2019) © American Physical Society.

where $\mathbf{A} = (A_C, A_B)$ are the affinities (10.118) and $\boldsymbol{\lambda} = (\lambda_C, \lambda_B)$ the counting parameters of the transferred electric charges (Z_C, Z_B). As a consequence of the fluctuation relation (10.119), the cumulant generating function satisfies the symmetry relation (5.91), as proved with equation (5.92). Therefore, the linear and nonlinear transport properties of the transistor should obey the fluctuation–dissipation relations (5.142), the Onsager reciprocal relations (5.143), and their generalizations (5.146). The validity of these predictions is confirmed in Table 10.1 for the linear response properties and in Table 10.2 for the nonlinear ones at second order in the affinities.

Since the scheme is consistent with the laws of electricity, thermodynamics, and microreversibility, the fluctuation relation holds for the two electric currents that are coupled together in the double junction of the transistor. As a corollary of microreversibility, the

Table 10.2. *Transistor model with the parameter values (10.122). The nonlinear response coefficients $M_{\alpha,\beta\gamma}$ defined by equation (5.135) and the nonequilibrium responses of the diffusivities $R_{\alpha\beta,\gamma} \equiv \partial_{A_\gamma} D_{\alpha\beta}(0,0)$ versus $\alpha, \beta, \gamma \in \{B, C\}$. The last column tests the prediction $M_{\alpha,\beta\gamma} = R_{\alpha\beta,\gamma} + R_{\alpha\gamma,\beta}$ of the fluctuation relation in the transistor model (Gu and Gaspard, 2019).*

α	β	γ	$M_{\alpha,\beta\gamma}$	$R_{\alpha\beta,\gamma}$	$R_{\alpha\gamma,\beta}$	$M_{\alpha,\beta\gamma} - R_{\alpha\beta,\gamma} - R_{\alpha\gamma,\beta}$
B	B	B	90.07 ± 0.67	45.07 ± 4.64	45.07 ± 4.64	-0.07
B	C	B	-44.78 ± 0.11	-22.47 ± 4.64	-22.33 ± 4.63	0.03
B	C	C	42.06 ± 0.67	20.99 ± 4.64	20.99 ± 4.64	0.08
C	B	B	-45.33 ± 0.62	-22.47 ± 4.64	-22.47 ± 4.64	-0.38
C	C	B	68.75 ± 0.10	47.41 ± 9.90	20.99 ± 4.64	0.35
C	C	C	-67.39 ± 0.62	-33.64 ± 9.90	-33.64 ± 9.90	-0.10

Onsager reciprocal relations and their generalizations towards the nonlinear regimes away from equilibrium are satisfied in the transistor (Gu and Gaspard, 2019).

10.7 Fluctuating Hydrodynamics and Brownian Motion

Remarkably, the theory of Brownian motion can be formulated on the basis of fluctuating hydrodynamics (Vladimirsky and Terletsky, 1945; Fox and Uhlenbeck, 1970a; Zwanzig and Bixon, 1970; Hauge and Martin-Löf, 1973; Dufty, 1974; Bedeaux and Mazur, 1974; Hills, 1975; Bedeaux et al., 1977; Clercx and Schram, 1992; Berg-Sørensen and Flyvbjerg, 2005). In this framework, the Langevin fluctuating force exerted on the Brownian particle arises from the thermal fluctuations of the fluid velocity, which determines the viscous pressure tensor applied to the surface of the particle. Accordingly, the Langevin fluctuating force should be generalized to include the memory effects induced by the hydrodynamic flow around the Brownian particle. Its velocity autocorrelation function thus decays as $t^{-3/2}$ with time t, which is referred to as a hydrodynamic long-time tail (Alder and Wainwright, 1970; Dorfman et al., 1994). These effects have been observed in experiments using photon correlation dynamic laser light scattering (Paul and Pusey, 1981) and direct particle tracking (Franosch et al., 2011; Kheifets et al., 2014). The generalized Langevin process is therefore non-Markovian, but still Gaussian as in the standard Langevin process.

In the following, this theory is presented and the fluctuation relation (7.39) is shown to hold despite the non-Markovianity of the generalized Langevin process (Gaspard, 2020a).

10.7.1 *Brownian Particle in a Fluctuating Fluid*

We consider a spherical Brownian particle of radius R in an incompressible viscous fluid and subjected to the gravitational field of acceleration $\mathbf{g} = -g\mathbf{1}_x$. In the framework of fluctuating hydrodynamics (Landau and Lifshitz, 1980b; Ortiz de Zárate and Sengers, 2006), the velocity field of an incompressible fluid obeys the condition $\nabla \cdot \mathbf{v} = 0$ as a consequence of the local conservation of mass and $d\rho_{\rm f}/dt = 0$. The fluctuating Navier–Stokes equations can be expressed as

$$\rho_{\rm f}\,(\partial_t \mathbf{v} + \mathbf{v}\cdot\nabla\mathbf{v}) = -\nabla\cdot\mathsf{P} + \boldsymbol{f}_{\rm g}, \tag{10.124}$$

in terms of the fluid density $\rho_{\rm f}$, the gravitational force density $\boldsymbol{f}_{\rm g} = \rho_{\rm f}\,\mathbf{g}$, and the pressure tensor

$$\mathsf{P} = p\,\mathbf{1} - \eta\big(\nabla\mathbf{v} + \nabla\mathbf{v}^{\rm T}\big) + \boldsymbol{\pi}, \tag{10.125}$$

where p is the hydrostatic pressure, $\mathbf{1}$ is the 3×3 unit matrix, η is the shear viscosity of the fluid, the superscript T denotes the transpose, and $\boldsymbol{\pi} = (\pi_{ij})$ is a fluctuating pressure tensor with components given by Gaussian white noise fields satisfying

$$\begin{aligned} &\langle \pi_{ij}(\mathbf{r},t)\rangle = 0 \qquad \text{and} \\ &\langle \pi_{ij}(\mathbf{r},t)\,\pi_{kl}(\mathbf{r}',t')\rangle = 2\,\eta\,k_{\rm B}T\,\left(\delta_{ik}\delta_{jl} + \delta_{il}\delta_{jk}\right)\delta(\mathbf{r}-\mathbf{r}')\,\delta(t-t'), \end{aligned} \tag{10.126}$$

where $k_{\rm B}$ is Boltzmann's constant and T is the temperature.

The mass center of the Brownian particle is denoted $\mathbf{R}(t)$, its velocity $\mathbf{V}(t) = d\mathbf{R}(t)/dt$, and its angular velocity $\boldsymbol{\Omega}(t)$. Inside the solid particle, the velocity field is given by

$$\mathbf{v}(\mathbf{r},t) = \mathbf{V}(t) + \boldsymbol{\Omega}(t)\times[\mathbf{r}-\mathbf{R}(t)] \tag{10.127}$$

for $\|\mathbf{r}-\mathbf{R}(t)\| < R$ (Mazur and Bedeaux, 1974). At the interface $\|\mathbf{r}-\mathbf{R}(t)\| = R$ between the solid particle and the fluid, the following boundary conditions are considered (Bedeaux and Mazur, 1974; Albano et al., 1975; Bedeaux et al., 1977). On the one hand, the fluid velocity field in the direction $\mathbf{n}$ normal to the interface should satisfy

$$\mathbf{n}\cdot\mathbf{v}(\mathbf{r},t) = \mathbf{n}\cdot\mathbf{V}(t). \tag{10.128}$$

On the other hand, in the tangential directions $\mathbf{1}_{\|} = \mathbf{1} - \mathbf{nn}$, the boundary conditions on the velocity field are expressed for the slip velocity as (Bedeaux et al., 1977)

$$\begin{aligned} \lambda\,\mathbf{1}_{\|}\cdot\mathbf{v}_{\rm slip} &\equiv \lambda\,\mathbf{1}_{\|}\cdot\{\mathbf{v}(\mathbf{r},t) - \mathbf{V}(t) - \boldsymbol{\Omega}(t)\times[\mathbf{r}-\mathbf{R}(t)]\} \\ &= \mathbf{1}_{\|}\cdot\Big\{\eta\left[\nabla\mathbf{v}(\mathbf{r},t) + \nabla\mathbf{v}(\mathbf{r},t)^{\rm T}\right]\cdot\mathbf{n} + \boldsymbol{f}^{\rm s}_{\rm fl}(\mathbf{r},t)\Big\}, \end{aligned} \tag{10.129}$$

in terms of the sliding friction coefficient λ and the interfacial Gaussian white noise field satisfying

$$\begin{aligned} &\langle \boldsymbol{f}^{\rm s}_{\rm fl}(\mathbf{r},t)\rangle = 0 \qquad \text{and} \\ &\delta^{\rm s}(\mathbf{r},t)\,\langle \boldsymbol{f}^{\rm s}_{\rm fl}(\mathbf{r},t)\,\boldsymbol{f}^{\rm s}_{\rm fl}(\mathbf{r}',t')\rangle\,\delta^{\rm s}(\mathbf{r}',t') = 2\,k_{\rm B}T\,\lambda\,\delta^{\rm s}(\mathbf{r},t)\,\delta(\mathbf{r}-\mathbf{r}')\,\delta(t-t')\,\mathbf{1}_{\|}, \end{aligned} \tag{10.130}$$

where $\delta^{\rm s}(\mathbf{r},t)$ is the interfacial Dirac distribution (Bedeaux et al., 1976). The sliding friction causes the partial slippage of the velocity field at the interface with the *slip length* defined as

$$b \equiv \frac{\eta}{\lambda}. \tag{10.131}$$

As explained in Section A.9, sliding friction is the irreversible process due to interfacial slippage between the fluid and solid phases at the slip velocity $\mathbf{v}_{\text{slip}} \equiv \mathbf{v}^+ - \mathbf{v}^-$, where $\mathbf{v}^+$ is the fluid velocity and $\mathbf{v}^-$ the solid velocity. In the linear regime, the corresponding affinity and diffusive current density given in Table A.3 are related to each other according to $\mathcal{J}^{\text{s}}_{\text{v}\|} = \mathcal{L}^{\text{s}}_{\text{v,v}} \mathcal{A}^{\text{s}}_{\text{v}\|} + \delta \mathcal{J}^{\text{s}}_{\text{v}\|}$ with the linear response coefficient $\mathcal{L}^{\text{s}}_{\text{v,v}} = T\lambda$, so that equations (10.129) and (10.130) are obtained.

In the framework of fluctuating hydrodynamics, the stochastic differential equation for Brownian motion is obtained from Newton's equation,

$$m_{\text{s}} \frac{d\mathbf{V}(t)}{dt} = -\int_{\Sigma(t)} \mathbf{P}(\mathbf{r},t) \cdot \mathbf{n}\, d\Sigma + \mathbf{F}_{\text{g}}, \tag{10.132}$$

by integrating the fluid pressure tensor (10.125) over the moving surface $\Sigma(t)$ of the particle (Bedeaux and Mazur, 1974; Mazur and Bedeaux, 1974), where $m_{\text{s}} = \left(4\pi R^3/3\right) \rho_{\text{s}}$ is its mass and $\mathbf{F}_{\text{g}} = m_{\text{s}}\mathbf{g}$ is the gravitational force exerted on the particle.

At large distances from the particle, the fluid is at equilibrium, so that the mean value of the fluid velocity is equal to zero, i.e., $\langle \mathbf{v} \rangle = 0$, and the mean hydrostatic pressure is given by $\langle p \rangle = p_0 - \rho_{\text{f}} g x$. Consequently, the surface integral of the pressure tensor in equation (10.132) gives Archimedes' buoyant force $\mathbf{F}_{\text{A}} = -m_{\text{f}}\mathbf{g}$, where $m_{\text{f}} = \left(4\pi R^3/3\right) \rho_{\text{f}}$ is the mass of the fluid displaced by the particle. The particle is thus subjected to the total external force $\mathbf{F}_{\text{ext}} = \mathbf{F}_{\text{g}} + \mathbf{F}_{\text{A}} = (m_{\text{s}} - m_{\text{f}})\mathbf{g} = (m_{\text{f}} - m_{\text{s}}) g \mathbf{1}_x$.

For micrometric colloidal particles of radius $R = 10^{-6}$ m moving with velocities $V \sim 10^{-5}$ m/s in water solutions where the kinematic viscosity is $\nu = \eta/\rho_{\text{f}} \simeq 10^{-6}$ m^2/s, the Reynolds number is much smaller than one, $\text{Re} \equiv RV/\nu \simeq 10^{-5}$, so that the Navier–Stokes equations (10.124) can be linearized by neglecting the advective term $\mathbf{v} \cdot \nabla$ in comparison with the time derivative ∂_t. The nonlinear term $\boldsymbol{\Omega} \times \mathbf{R}$ can be similarly neglected in equations (10.127) and (10.129).

The problem is reformulated by introducing an induced force density field $\boldsymbol{f}_{\text{ind}}(\mathbf{r},t)$ to represent the effects of the boundary conditions on the velocity field, so that the linearized equations of motion of the fluid become

$$\rho_{\text{f}}\, \partial_t \mathbf{v}(\mathbf{r},t) = -\nabla \cdot \mathbf{P}(\mathbf{r},t) + \boldsymbol{f}_{\text{ind}}(\mathbf{r},t) + \boldsymbol{f}_{\text{g}} \tag{10.133}$$

with the fluctuating pressure tensor (10.125) (Mazur and Bedeaux, 1974). The induced force density field is chosen to vanish in the fluid and to comply with the constraints that the velocity field is given by equation (10.127) and the hydrostatic pressure is fixed inside the solid particle. In order to satisfy the boundary conditions (10.128) and (10.129), the induced force density field is singular at the interface between the solid and the fluid.

Since the equations are linear at low values for the Reynolds number, they can be solved by taking the Fourier transform

$$\mathbf{v}(\mathbf{r},\omega) = \int_{-\infty}^{+\infty} dt\, e^{\imath\omega t}\, \mathbf{v}(\mathbf{r},t) \tag{10.134}$$

for the velocity field. Introducing the Green function,

$$G(\mathbf{r},\omega) = \frac{1}{4\pi\eta r}\exp(-\alpha r) \qquad \text{with} \qquad \alpha = \sqrt{-\imath\omega/\nu}, \qquad \operatorname{Re}\alpha > 0, \tag{10.135}$$

the solution of equation (10.133) is given by

$$\mathbf{v}(\mathbf{r},\omega) = \mathbf{v}_0(\mathbf{r},\omega) + \int d\mathbf{r}' \Big\{ G(\mathbf{r}-\mathbf{r}',\omega)\,\mathbf{1} \tag{10.136}$$

$$+ \frac{1}{\alpha^2}\nabla'\nabla'\left[G(\mathbf{r}-\mathbf{r}',0) - G(\mathbf{r}-\mathbf{r}',\omega)\right]\Big\} \cdot f_{\mathrm{ind}}(\mathbf{r}',\omega)$$

in terms of the fluctuating fluid velocity $\mathbf{v}_0$ in the absence of the particle (Mazur and Bedeaux, 1974). We note that the unperturbed fluid velocity $\mathbf{v}_0$ is a Gaussian white noise field characterized by the correlation functions (C.32).

The relation between the induced force density and the unperturbed velocity field $\mathbf{v}_0$ can be obtained by averaging the Fourier transforms of the boundary conditions (10.128) and (10.129) over the surface of the spherical particle. Inserting the expression of the induced force density into equation (10.136), allows the velocity field and the pressure tensor to be deduced. The surface integral of the pressure tensor can thus be evaluated to find the force exerted by the fluid on the particle, including the effects of viscous friction, sliding friction, and thermal fluctuations (Mazur and Bedeaux, 1974; Bedeaux and Mazur, 1974; Albano et al., 1975; Bedeaux et al., 1977).

In this way, the Fourier transform of equation (10.132) can be expressed as

$$-\imath\omega m_{\mathrm{s}}\mathbf{V}(\omega) = -\gamma(\omega)\,\mathbf{V}(\omega) + \mathbf{F}_{\mathrm{fl}}(\omega) + 2\pi\,\delta(\omega)\,\mathbf{F}_{\mathrm{ext}} \tag{10.137}$$

in terms of the frequency-dependent friction coefficient

$$\gamma(\omega) = 6\pi\eta R\left[\frac{(1+2b/R)(1+R\alpha)}{(1+3b/R)+b\alpha} + \frac{R^2\alpha^2}{9}\right] \qquad \text{with} \quad \alpha = \sqrt{-\imath\omega/\nu} \tag{10.138}$$

and the fluctuating force $\mathbf{F}_{\mathrm{fl}}(\omega)$ obeying the generalized fluctuation–dissipation theorem (Bedeaux and Mazur, 1974; Bedeaux et al., 1977)

$$\langle\mathbf{F}_{\mathrm{fl}}(\omega)\rangle = 0 \qquad \text{and} \qquad \langle\mathbf{F}_{\mathrm{fl}}(\omega)\,\mathbf{F}^*_{\mathrm{fl}}(\omega')\rangle = 4\pi\operatorname{Re}\gamma(\omega)\,k_{\mathrm{B}}T\,\delta(\omega-\omega')\,\mathbf{1}. \tag{10.139}$$

Exploiting that the last term goes as α^2 in equation (10.138), the frequency-dependent friction coefficient can be rewritten as

$$\gamma(\omega) = \gamma_{\mathrm{d}}(\omega) - \imath\omega\frac{m_{\mathrm{f}}}{2} \tag{10.140}$$

in terms of the mass m_{f} of the fluid displaced. Moreover, since $\operatorname{Re}\gamma(\omega) = \operatorname{Re}\gamma_{\mathrm{d}}(\omega)$, the last term of equation (10.140) does not contribute to damping. Accordingly, the effects of dissipation are only described by the contribution

$$\gamma_{\mathrm{d}}(\omega) = \gamma\,\frac{1+R\sqrt{-\imath\omega/\nu}}{1+B\sqrt{-\imath\omega/\nu}}, \tag{10.141}$$

where

$$\gamma \equiv \gamma(0) = \gamma_{\rm d}(0) = 6\pi\eta R \, \frac{1+2b/R}{1+3b/R} \tag{10.142}$$

is the friction coefficient at zero frequency (Sutherland, 1905) and

$$B \equiv \frac{b}{1+3b/R}. \tag{10.143}$$

The Stokes friction coefficient is recovered for zero slip length $b = 0$.

Furthermore, the last term in equation (10.140) gives a contribution similar to the inertial term in the left-hand side of equation (10.137). Therefore, these terms can be gathered to get

$$-\,\imath\omega m\, \mathbf{V}(\omega) = -\gamma_{\rm d}(\omega)\, \mathbf{V}(\omega) + \mathbf{F}_{\rm fl}(\omega) + 2\pi\, \delta(\omega)\, \mathbf{F}_{\rm ext} \tag{10.144}$$

with the total mass

$$m \equiv m_{\rm s} + \frac{m_{\rm f}}{2}, \tag{10.145}$$

including the mass of the solid particle itself together with half the mass of the fluid displaced (Corrsin and Lumley, 1956; Dufty, 1974; Maxey and Riley, 1983).

10.7.2 Generalized Langevin Equation

According to the previous results, the stochastic motion of the Brownian particle should be described by the following generalized Langevin equation (Hauge and Martin-Löf, 1973; Dufty, 1974),

$$m\, \frac{d\mathbf{V}(t)}{dt} = -\int_0^t \Gamma(t-t')\, \mathbf{V}(t')\, dt' + \mathbf{F}_{\rm fl}(t) + \mathbf{F}_{\rm ext} \tag{10.146}$$

with a memory kernel $\Gamma(t-t')$ determined by the frequency-dependent friction coefficient (10.141), and a fluctuating force given by Gaussian colored noises characterized by

$$\langle \mathbf{F}_{\rm fl}(t)\rangle = 0 \qquad \text{and} \qquad \langle \mathbf{F}_{\rm fl}(t)\, \mathbf{F}_{\rm fl}(t')\rangle = k_{\rm B}T\, \Gamma(|t-t'|)\, \mathbf{1}. \tag{10.147}$$

Moreover, the fluctuating force is not correlated with the initial condition of the velocity, $\langle \mathbf{V}(0)\, \mathbf{F}_{\rm fl}(t)\rangle = 0$ for $t > 0$ (Dufty, 1974; Balakrishnan, 1979). Accordingly, the stochastic process is Gaussian, but not Markovian.

Since equation (10.146) is linear, it can be solved by direct integration (Dufty, 1974). Afterwards, the position is found by integrating $d\mathbf{R}(t)/dt = \mathbf{V}(t)$. Therefore, the particle velocity and position are given by

$$\mathbf{V}(t) = K(t)\, \mathbf{V}_0 + L(t)\, \frac{\mathbf{F}_{\rm ext}}{m} + \frac{1}{m} \int_0^t dt'\, K(t-t')\, \mathbf{F}_{\rm fl}(t'), \tag{10.148}$$

$$\mathbf{R}(t) = \mathbf{R}_0 + L(t)\, \mathbf{V}_0 + M(t)\, \frac{\mathbf{F}_{\rm ext}}{m} + \frac{1}{m} \int_0^t dt' \int_0^{t'} dt''\, K(t'-t'')\, \mathbf{F}_{\rm fl}(t''), \tag{10.149}$$

in terms of the initial velocity and position, $\mathbf{V}_0$ and $\mathbf{R}_0$, the function $K(t)$ satisfying

$$m\,\frac{dK(t)}{dt} = -\int_0^t \Gamma(t-t')\,K(t')\,dt' \qquad \text{with} \qquad K(0)=1, \tag{10.150}$$

and its integrals

$$L(t) \equiv \int_0^t K(t')\,dt' \qquad \text{and} \qquad M(t) \equiv \int_0^t L(t')\,dt'. \tag{10.151}$$

Defining the Laplace transform of an arbitrary function $\psi(t)$ of time as

$$\tilde{\psi}(z) \equiv \int_0^\infty \mathrm{e}^{-zt}\,\psi(t)\,dt, \tag{10.152}$$

equation (10.150) leads to

$$\tilde{K}(z) = \left[z + \frac{1}{m}\,\tilde{\Gamma}(z)\right]^{-1}, \qquad \text{where} \qquad \tilde{\Gamma}(z) = \gamma\,\frac{1+R\sqrt{z/\nu}}{1+B\sqrt{z/\nu}} \tag{10.153}$$

is the Laplace transform of the memory kernel $\Gamma(t)$ given by the analytic continuation $z=-\imath\omega$ for the noninertial part (10.141) of the frequency-dependent friction coefficient. Introducing the parameters[6]

$$\tau \equiv \frac{m}{\gamma}, \qquad \lambda \equiv \frac{\nu\tau}{R^2}, \qquad \text{and} \qquad \mu \equiv \frac{\nu\tau}{B^2}, \tag{10.154}$$

the function $K(t)$ is obtained from the inverse Laplace transform of $\tilde{K}(z)$ in equation (10.153),

$$K(t) = \int_{c-\imath\infty}^{c+\imath\infty} \frac{dz}{2\pi\imath}\,\mathrm{e}^{zt}\,\tilde{K}(z), \tag{10.155}$$

where the constant c should exceed the real part of all the singularities of $\tilde{K}(z)$. This latter presents a branch cut on the negative half of the real axis in the complex plane of the variable z. Deforming the integration contour around this branch cut and setting $z=-r/\tau$, we find the following integral representation for $t\geq 0$:

$$K(t) = \frac{1}{\pi}\left(\frac{1}{\sqrt{\lambda}} - \frac{1}{\sqrt{\mu}}\right)\int_0^\infty \frac{\sqrt{r}\,\mathrm{e}^{-rt/\tau}}{(r-1)^2 + r\left(\frac{1}{\sqrt{\lambda}} - \frac{r}{\sqrt{\mu}}\right)^2}\,dr. \tag{10.156}$$

Defining $a \equiv \pi^{-1/2}\left(\lambda^{-1/2} - \mu^{-1/2}\right)$, the function (10.156) and those given by equation (10.151) behave as follows for time $t\to\infty$ (Gaspard, 2020a),

$$K(t) \simeq \frac{a}{2}\left(\frac{\tau}{t}\right)^{3/2}, \quad L(t) \simeq \tau\left[1 - a\left(\frac{\tau}{t}\right)^{1/2}\right], \quad M(t) \simeq \tau t\left[1 - 2a\left(\frac{\tau}{t}\right)^{1/2}\right]. \tag{10.157}$$

[6] As mentioned in Section 4.7.2, the parameter $\tau = \tau_{\mathrm{r}} = m/\gamma$ is the thermalization time of the velocity distribution towards Maxwellian equilibrium.

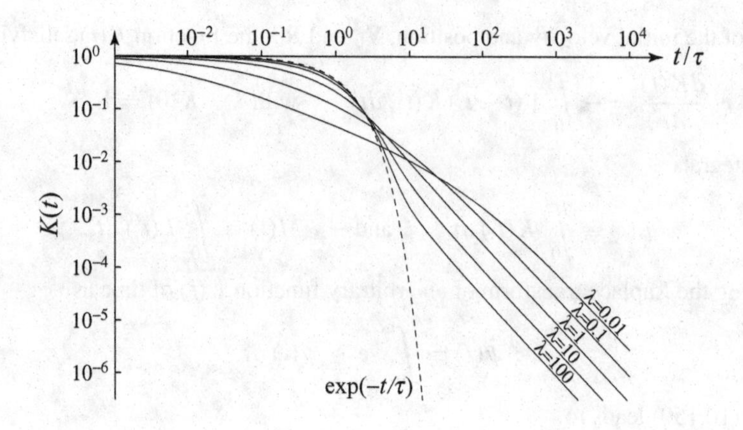

Figure 10.11 The function $K(t)$ of the generalized Langevin process versus the rescaled time t/τ for stick boundary conditions ($b = 0$, $\mu = \infty$) and the parameter values $\lambda = 0.01$–100 (solid lines). The dashed line shows the exponential decay expected for the standard Langevin process. Reprinted from Gaspard (2020a) with the permission of Elsevier.

Figure 10.11 shows the function $K(t)$ as a function of time for different values of the parameter λ. We observe the long-time tail $K(t) \sim t^{-3/2}$ due to hydrodynamics, which generates the persistence of motion and the memory of the initial velocity.

If the slip length is equal to zero, i.e., $b = 0$, the memory kernel of equation (10.146) can be obtained with the inverse Laplace transform of $\tilde{\Gamma}(z)$ in equation (10.153) with $B = 0$, giving (Dufty, 1974)

$$\Gamma(t) = 6\pi\eta R \left[2\,\delta(t) - \frac{R}{2\sqrt{\pi\nu}\,t^{3/2}} \right] \theta(t), \tag{10.158}$$

where $\theta(t)$ is the Heaviside unit step function, so that the factor 2 is required with the Dirac distribution $\delta(t)$ in order to recover Stokes' friction coefficient (10.142) with $b = 0$. The second term in equation (10.158) is the hydrodynamic long-time tail, leading to the equation of motion for a particle in a flow by Corrsin and Lumley (1956) and Maxey and Riley (1983).

10.7.3 Standard Langevin Process

As understood by Lorentz (1921), the standard Langevin process is valid in the limit where the fluid has a much lower mass density than the solid particle:

$$\frac{m_{\rm f}}{m_{\rm s}} = \frac{\rho_{\rm f}}{\rho_{\rm s}} \ll 1. \tag{10.159}$$

If this condition holds, the functions in equation (10.153) reduce to $\tilde{\Gamma}(z) = \gamma$ and $\tilde{K}(z) = (z + 1/\tau)^{-1}$. Therefore, there is a simple pole in the complex plane of the variable z upon integrating (10.155) and we find

$$K(t) = {\rm e}^{-t/\tau}, \quad L(t) = \tau \left(1 - {\rm e}^{-t/\tau}\right), \quad M(t) = \tau t - \tau^2 \left(1 - {\rm e}^{-t/\tau}\right), \tag{10.160}$$

so that the classic results are recovered (Chandrasekhar, 1943). As seen in Figure 10.11, the function $K(t)$ indeed converges towards the exponential function in the limit $\lambda \to \infty$ corresponding to the Lorentz condition (10.159). In this case, the effects of the long-time tail $t^{-3/2}$ are thus negligible in the memory kernel $\Gamma(t)$, which is given by equation (10.158) for $b = 0$, and the generalized Langevin equation (10.146) with the Gaussian colored noises (10.147) reduces to the standard Langevin equation (4.173) with the Gaussian white noises (4.174).

10.7.4 Conditional and Joint Probability Densities

The stochastic process described by the generalized Langevin equation has Gaussian conditional and joint probability densities to move from the initial condition $(\mathbf{R}_0, \mathbf{V}_0)$ at time $t = 0$ to the phase-space point $(\mathbf{R}, \mathbf{V})$ at time t (Dufty, 1974). In particular, the conditional probability density has the Gaussian form,

$$\mathcal{P}(\mathbf{R}, \mathbf{V}, t|\mathbf{R}_0, \mathbf{V}_0, 0) = \frac{1}{8\pi^3(FG - H^2)^{3/2}} \exp\left[-\frac{G\mathbf{X}^2 - 2H\mathbf{X}\cdot\mathbf{Y} + F\mathbf{Y}^2}{2(FG - H^2)}\right] \tag{10.161}$$

with the variables

$$\mathbf{X} \equiv \mathbf{R} - \mathbf{R}_0 - L(t)\,\mathbf{V}_0 - M(t)\,\frac{\mathbf{F}_{\text{ext}}}{m}, \qquad \mathbf{Y} \equiv \mathbf{V} - K(t)\,\mathbf{V}_0 - L(t)\,\frac{\mathbf{F}_{\text{ext}}}{m}, \tag{10.162}$$

and the time-dependent coefficients

$$F = \frac{k_{\text{B}}T}{m}\left[2\,M(t) - L(t)^2\right], \quad G = \frac{k_{\text{B}}T}{m}\left[1 - K(t)^2\right], \quad H = \frac{k_{\text{B}}T}{m}L(t)\,[1 - K(t)], \tag{10.163}$$

expressed in terms of the functions defined by equations (10.150) and (10.151). If the Lorentz condition (10.159) is satisfied, the conditional probability density of the standard Langevin process is recovered (Chandrasekhar, 1943).

The conditional probability distribution of the velocity is obtained by integrating the density (10.161) over the position $\mathbf{R}$ to find

$$\mathscr{P}(\mathbf{V}, t|\mathbf{V}_0, 0) = \int \mathcal{P}(\mathbf{R}, \mathbf{V}, t|\mathbf{R}_0, \mathbf{V}_0, 0)\, d\mathbf{R} = \frac{1}{(2\pi G)^{3/2}} \exp\left(-\frac{\mathbf{Y}^2}{2G}\right), \tag{10.164}$$

which converges in the long-time limit towards the Maxwell equilibrium probability density,

$$p_{\text{eq}}(\mathbf{V}) \equiv \lim_{t\to\infty} \mathscr{P}(\mathbf{V}, t|\mathbf{V}_0, 0) = \left(\frac{m}{2\pi k_{\text{B}}T}\right)^{3/2} \exp\left(-\frac{m\mathbf{V}^2}{2k_{\text{B}}T}\right), \tag{10.165}$$

corresponding to the kinetic energy of the total mass (10.145). Thermalization thus takes place for the colloidal particle of mass m_{s} accompanied with the mass $m_{\text{f}}/2$ of fluid because of the inertial effect induced by hydrodynamics (Dufty, 1974).

The joint probability density to move from the position $\mathbf{R}_0$ and the velocity $\mathbf{V}_0$ distributed according to the equilibrium Maxwell density (10.165) at time $t = 0$ to the phase-space point $(\mathbf{R}, \mathbf{V})$ at time t can be defined as

$$\mathcal{P}(\mathbf{R},\mathbf{V},t;\mathbf{R}_0,\mathbf{V}_0,0) \equiv \mathcal{P}(\mathbf{R},\mathbf{V},t|\mathbf{R}_0,\mathbf{V}_0,0)\, p_{\text{eq}}(\mathbf{V}_0). \tag{10.166}$$

Therefore, the conditional probability density to move from the initial position $\mathbf{R}_0$ towards the position $\mathbf{R}$ at time t is given by

$$\begin{aligned} p(\mathbf{R},t|\mathbf{R}_0,0) &\equiv \int \mathcal{P}(\mathbf{R},\mathbf{V},t;\mathbf{R}_0,\mathbf{V}_0,0)\, d\mathbf{V}\, d\mathbf{V}_0 \\ &= \left[\frac{\beta m}{4\pi M(t)}\right]^{3/2} \exp\left\{-\frac{\beta m}{4\, M(t)}\left[\mathbf{R}-\mathbf{R}_0 - M(t)\frac{\mathbf{F}_{\text{ext}}}{m}\right]^2\right\}, \end{aligned} \tag{10.167}$$

where $\beta = (k_{\text{B}}T)^{-1}$.

Furthermore, the velocity autocorrelation function has the form

$$\langle \Delta\mathbf{V}(0)\cdot\Delta\mathbf{V}(t)\rangle = 3\,\frac{k_{\text{B}}T}{m}\,K(t) \tag{10.168}$$

with $\Delta\mathbf{V}(t) = \mathbf{V}(t) - \langle \mathbf{V}(t)\rangle$, which tends to zero as $t^{-3/2}$ in the limit $t \to \infty$. This hydrodynamic long-time memory has the effect of pushing the particle in the direction of its velocity because the fluid forms a vortex flow pattern around the particle (Alder and Wainwright, 1970; Zwanzig and Bixon, 1970).

10.7.5 Fluctuation Relation for the Generalized Langevin Process

In order to investigate the consequences of microreversibility on the generalized Langevin process, we compare two different paths that are mapped onto each other by time reversal, which consists of exchanging the initial and final positions while reversing the velocities.

The forward and time-reversed paths have the following joint probability densities:

$$(\mathbf{R}_0,\mathbf{V}_0,0) \xrightarrow{\text{F}} (\mathbf{R},\mathbf{V},t): \qquad \mathcal{P}_{\text{F}} \equiv \mathcal{P}(\mathbf{R},\mathbf{V},t;\mathbf{R}_0,\mathbf{V}_0,0), \tag{10.169}$$

$$(\mathbf{R},-\mathbf{V},0) \xrightarrow{\text{R}} (\mathbf{R}_0,-\mathbf{V}_0,t): \qquad \mathcal{P}_{\text{R}} \equiv \mathcal{P}(\mathbf{R}_0,-\mathbf{V}_0,t;\mathbf{R},-\mathbf{V},0), \tag{10.170}$$

as defined by equation (10.166). Under the effect of the external force, these probability densities move in opposite directions, so that their overlap rapidly decreases as time increases.

The remarkable result is that the following fluctuation relation holds for the generalized Langevin process defined by equations (10.146) and (10.147) (Gaspard, 2020a):

$$\frac{\mathcal{P}(\mathbf{R},\mathbf{V},t;\mathbf{R}_0,\mathbf{V}_0,0)}{\mathcal{P}(\mathbf{R}_0,-\mathbf{V}_0,t;\mathbf{R},-\mathbf{V},0)} = e^{\beta\mathbf{F}_{\text{ext}}\cdot(\mathbf{R}-\mathbf{R}_0)} \tag{10.171}$$

for $t \geq 0$. The ratio of the joint probability densities of the forward and reversed paths is thus independent of hydrodynamic effects. If the external force vanishes, i.e., $\mathbf{F}_{\text{ext}} = 0$, we recover the principle of detailed balance, as required by microreversibility under equilibrium conditions.

The fluctuation relation (10.171) has the same form as (7.43) with the mechanical affinity $\mathbf{A} = \beta\mathbf{F}_{\text{ext}}$ and $U = 0$. Using the definition (10.166) of the joint probability density and the Maxwell distribution (10.165), the fluctuation relation can be expressed in the equivalent form (7.39) in terms of the work $W \equiv \mathbf{F}_{\text{ext}} \cdot (\mathbf{R} - \mathbf{R}_0)$ performed by the external force during the displacement of the particle from $\mathbf{R}_0$ to $\mathbf{R}$, the change in the kinetic energy of the particle $\Delta K \equiv m\left(\mathbf{V}^2 - \mathbf{V}_0^2\right)/2$, and $\Delta U = 0$, thus establishing the energy balance of stochastic energetics (Sekimoto, 2010). Furthermore, the conditional probability density in position space (10.167) satisfies the fluctuation relation (7.45) independently of the specificities of the generalized Langevin process.

Since $\lim_{t\to\infty} M(t)/t = \tau = m/\gamma$ as a consequence of equation (10.157), the probability distribution (10.167) shows that the particle undergoes a random drift in the long-time limit with the asymptotic mean drift velocity

$$\mathbf{V}_{\text{drift}} = \lim_{t\to\infty} \frac{1}{t} \langle \Delta\mathbf{R}(t) \rangle = \frac{1}{\gamma} \mathbf{F}_{\text{ext}}, \tag{10.172}$$

and the diffusion coefficient

$$\mathcal{D} = \lim_{t\to\infty} \frac{1}{6t} \left\langle [\Delta\mathbf{R}(t) - \langle \Delta\mathbf{R}(t) \rangle]^2 \right\rangle = \frac{k_{\text{B}}T}{\gamma}, \tag{10.173}$$

where $\Delta\mathbf{R}(t) \equiv \mathbf{R}(t) - \mathbf{R}(0)$ and γ is the friction coefficient (10.142) including the effect of the slip length (10.131). In the absence of external force, the particle thus performs a random walk with diffusion coefficient (10.173), so that the hydrodynamic long-time tail does not modify the formula of Einstein (1905) and Sutherland (1905) for the diffusion coefficient.

The fluctuation relation for the conditional probability distribution (10.167) implies that the entropy production rate is given by

$$\frac{1}{k_{\text{B}}} \frac{d_{\text{i}}S}{dt} = \lim_{t\to\infty} \frac{1}{t} \int p(\mathbf{R}, t|\mathbf{R}_0, 0) \ln \frac{p(\mathbf{R}, t|\mathbf{R}_0, 0)}{p(\mathbf{R}_0, t|\mathbf{R}, 0)}\, d\mathbf{R} = \beta\, \mathbf{F}_{\text{ext}} \cdot \mathbf{V}_{\text{drift}} \geq 0, \tag{10.174}$$

in accordance with the second law of thermodynamics.

11

Reactions

11.1 Stochastic Approach to Reactive Systems

A reaction is the transformation of some particle species into other ones. Depending on the energy scale, the system may undergo molecular, atomic, or nuclear reactions. An example is the reaction between an atom A and a diatomic molecule BC, leading to their transformation into the molecule AB and the atom C. Such a reaction and its reversal are governed by the Hamiltonian dynamics of the three nuclei A, B, and C on a Born–Oppenheimer potential energy surface determined by the fast electrons. If the motion of the nuclei is collinear, the surface can be represented as a function of the distances between the particles, as shown in Figure 11.1. In classical mechanics, the elementary reactive events correspond to trajectories passing from one valley to the other through the energy barrier. This barrier between the two valleys is located at the saddle point of the surface, which is the bottleneck of the reaction. The forward or reversed reaction is characterized by an activation energy E_{a} given by the difference between the energies of the saddle point and the bottom of the incoming valley. According to Arrhenius' law, reactive events may occur if the thermal energy is high enough to have a significant Boltzmann factor $\exp(-E_{\mathrm{a}}/k_{\mathrm{B}}T)$ for the trajectory to pass over the energy barrier. In gases, such thermally activated reactive events occur upon random inelastic collisions, as described by Boltzmann's kinetic equation.

At the macroscale, the reactive events are so numerous that the process can be described by deterministic kinetic equations, as explained in Chapter 1. However, the reactive events follow stochastic processes in systems of finite size. The smaller the size, the larger the fluctuations relative to the average behavior. With the advent of field emission and atomic force microscopies, fluorescence microscopy, optical tweezers, and the patch clamp technique in electrophysiology, chemical and biochemical reactions can nowadays be observed in real time at the nanoscale (McEwen et al., 2009; Barroo et al., 2015; Grosfils et al., 2015; Yasuda et al., 2001; Moffitt et al., 2010), and even at the level of a single ion channel protein (Colquhoun and Sakmann, 1981) or a single enzyme (Xie, 2001). With such techniques, stochastic signals can be recorded to measure the transition rates between the observed coarse-grained states and their probabilities, as well as to carry out the counting statistics of the reactive events.

At equilibrium, the transition rates and the probabilities obey the principle of detailed balance. Out of equilibrium, the counting statistics of reactive events satisfy fluctuation relations, which are valid arbitrarily far from equilibrium. These general rules find their

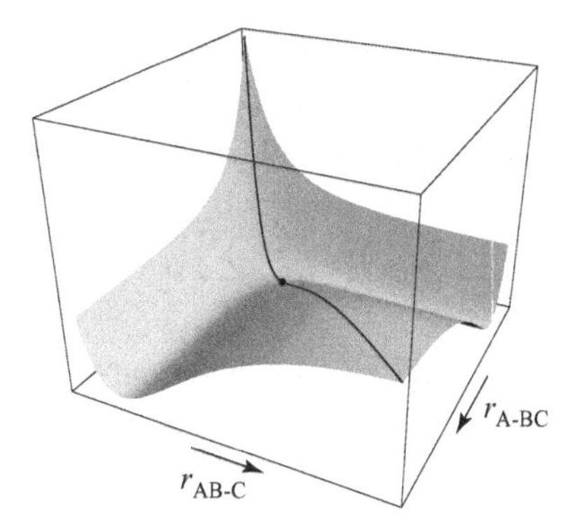

Figure 11.1 Schematic representation of the potential energy surface for three atoms in the collinear elementary reaction A+BC $\rightleftharpoons$ AB+C. Here, $r_{\text{AB}-\text{C}}$ is the distance between the diatomic molecule AB and the atom C, and $r_{\text{A}-\text{BC}}$ is the distance between the atom A and the diatomic molecule BC. The solid line depicts the barrier separating the two valleys and the dot is the saddle point or bottleneck of the reaction.

origin in the microreversibility of the underlying Hamiltonian dynamics, as explained in Chapters 2 and 5.

Because of the fluctuations manifesting themselves in finite systems, nonequilibrium phenomena such as bistability or regular oscillations turn out to be noisy and characterized by finite lifetimes, which may become arbitrarily long if the system is large enough to be described by the deterministic macroscopic dynamics.

An example from heterogeneous catalysis is the reaction of water production from dihydrogen and dioxygen on nanosized rhodium crystals, as observed by field ion microscopy (McEwen et al., 2009; Barroo et al., 2015; Grosfils et al., 2015). In this far-from-equilibrium system, bistability and oscillations take place on a surface with a radius of curvature equal to 10–20 nm, where a few thousand atoms can undergo adsorption, transport, and reactions, before desorption. As a consequence of molecular and thermal fluctuations, bistability turns into metastability with finite dwell times in the coarse-grained states that would be stable at the macroscale and the oscillations are noisy with a correlation time proportional to the number of atoms in the reaction.

At the nanoscale, the stochastic approach is thus mandatory for the description of reactive systems. In this context, a fundamental issue is to formulate the kinetics and the thermodynamics of reactions within the theory of stochastic processes (McQuarrie, 1967; Nicolis and Prigogine, 1971; Nicolis, 1972; Nicolis and Prigogine, 1977; Gardiner, 1979, 2004; van Kampen, 1981; Luo et al., 1984; Gaspard, 2004a; De Decker, 2015; Rao and Esposito, 2018a).

11.2 Reaction Networks

We consider a dilute solution of the solute species $\{\text{X}_k\}_{k=1}^{d}$ and $\{\text{Y}_l\}_{l=1}^{e}$ in a solvent under isobaric and isothermal conditions. The solution is supposed to be well mixed by mechanical stirring as in Figure 1.7 or by a large enough diffusivity for the solute species, implying that their concentrations are uniform in the system. The solute species evolve in time according to the following network of elementary forward and backward reactions:

$$\sum_{k=1}^{d} \nu_{kr}^{(+)} \mathrm{X}_k + \sum_{l=1}^{e} \tilde{\nu}_{lr}^{(+)} \mathrm{Y}_l \underset{W_{-r}}{\overset{W_{+r}}{\rightleftharpoons}} \sum_{k=1}^{d} \nu_{kr}^{(-)} \mathrm{X}_k + \sum_{l=1}^{e} \tilde{\nu}_{lr}^{(-)} \mathrm{Y}_l \tag{11.1}$$

with $r = 1, 2, \ldots, m$, where $\nu_{kr}^{(\pm)}$ and $\tilde{\nu}_{lr}^{(\pm)}$ are nonnegative integers giving the respective numbers of particles that are incoming or outgoing the reaction. The stoichiometric coefficients of the different species in the reaction r are thus given by

$$\nu_{kr} \equiv \nu_{kr}^{(-)} - \nu_{kr}^{(+)} \qquad \text{and} \qquad \tilde{\nu}_{lr} \equiv \tilde{\nu}_{lr}^{(-)} - \tilde{\nu}_{lr}^{(+)}. \tag{11.2}$$

The number of the particles of species X_k (respectively, Y_l) is denoted X_k (respectively, Y_l). The time evolution of the particle numbers $\mathbf{X}(t) \equiv \{X_k(t)\}_{k=1}^{d} \in \mathbb{N}^d$ and $\mathbf{Y}(t) \equiv \{Y_l(t)\}_{l=1}^{e} \in \mathbb{N}^e$ can be expressed in terms of the numbers $\{n_{\pm r}(t)\}_{r=1}^{m} \in \mathbb{Z}^{2m}$ of forward and backward reactive events that have occurred since the beginning of the process at time $t = 0$. These counting variables allow us to determine the fluxes of matter in the reaction network (11.1) as well as the composition of the system at every instant of time according to

$$X_k(t) = X_k(0) + \sum_{r=1}^{m} \nu_{kr} \left[n_{+r}(t) - n_{-r}(t) \right] \qquad (k = 1, 2, \ldots, d), \tag{11.3}$$

$$Y_l(t) = Y_l(0) + \sum_{r=1}^{m} \tilde{\nu}_{lr} \left[n_{+r}(t) - n_{-r}(t) \right] \qquad (l = 1, 2, \ldots, e), \tag{11.4}$$

in terms of the stoichiometric coefficients (11.2). Since every reaction is assumed to be elementary, the corresponding forward and backward reactive events pass through the same bottleneck, so that the time evolution of the particle numbers only depends on the net numbers of reactive events defined as $n_r(t) \equiv n_{+r}(t) - n_{-r}(t)$, using the forward direction as convention.

The reason for making the distinction between the species $\{\mathrm{X}_k\}_{k=1}^{d}$ and $\{\mathrm{Y}_l\}_{l=1}^{e}$ is that the former are ruled by the stochastic process, while the latter are assumed to be chemostatted at uniform and invariant concentrations $\{y_l = [\mathrm{Y}_l]\}_{l=1}^{e}$ during the whole observation of the stochastic process. This chemostatting may drive the process away from equilibrium if the fixed concentrations do not satisfy the Guldberg–Waage equilibrium conditions (1.39). However, the species $\{\mathrm{X}_k\}_{k=1}^{d}$, which may be called the intermediate species, are supposed to evolve in time according to the reaction network (11.1). This distinction between evolving and chemostatted species is often assumed in the literature on nonequilibrium reaction networks (Schlögl, 1971, 1972; Nicolis and Prigogine, 1971; Nicolis, 1972; Nicolis and Prigogine, 1977; Luo et al., 1984; Epstein and Pojman, 1998; Kondepudi and Prigogine, 1998; Polettini and Esposito, 2014; Rao and Esposito, 2016).

The species $\{\mathrm{Y}_l\}_{l=1}^{e}$ can be chemostatted, for instance, using semipermeable membranes that are permeable to these species, but impermeable to the intermediate species $\{\mathrm{X}_k\}_{k=1}^{d}$ (Fermi, 1937). Reactions thus run in the volume V that is delimited by the semipermeable membranes and contains the intermediate species $\{\mathrm{X}_k\}_{k=1}^{d}$. This volume may be called the reactor since the reaction network (11.1) is evolving therein. The chemostatted species

$\{Y_l\}_{l=1}^e$ can be supplied from outside the reactor and evacuated to the exterior of the system in such a way that their concentrations $\{y_l\}_{l=1}^e$ are held fixed inside the system. Consequently, the total particle numbers $\mathbf{Y}(t) = \{Y_l(t)\}_{l=1}^e$ of the chemostatted species in the interior and the exterior of the system change in time according to equation (11.4). Actually, these numbers drift in time as long as the reaction network is maintained in a nonequilibrium regime. In general, the chemostatted species should be supplied from separate reservoirs as in Figure 1.4. Such reservoirs may be called chemostats. If the chemostatted species do not react with each other, they may be supplied from a single exterior compartment, in which case the intermediate species $\{X_k\}_{k=1}^d$ play the role of catalysts for the reactions.[1]

In practice, all the species can also evolve in a single container, called a batch reactor, where the species $\{Y_l\}_{l=1}^e$ can be assumed to be chemostatted if they do not react with each other, if they are in excess with respect to the other species $Y_l \gg X_k$, and if their reactions with the intermediate species are slow enough for their concentrations to remain essentially invariant during the reactive process. If these conditions are satisfied, the depletion of the pool of chemostatted species can be negligible over a time interval that is long with respect to the timescale of the reactions undergone by the intermediate species $\{X_k\}_{k=1}^d$. Ultimately, after a long enough time, the system will reach the thermodynamic equilibrium macrostate, where the concentrations of all the species are in the Guldberg–Waage equilibrium ratios (1.39), possibly constrained by the conservation laws that might exist in the reaction network (11.1) (Polettini and Esposito, 2014; Rao and Esposito, 2016). Such assumptions are considered in the study of many classes of reactions. For instance, the substrate and product species of enzymatic reactions in biochemical kinetics are often supposed to have fixed concentrations (Segel, 1975; Wachtel et al., 2018).

In such reactive systems, a stochastic process thus rules the time evolution of the particle numbers $\mathbf{X}(t) \equiv \{X_k(t)\}_{k=1}^d \in \mathbb{N}^d$ for the intermediate species. These numbers evolve in time because reactive events randomly occur with the transition rates $W_{\pm r}$ of the forward and backward reactions (11.1), respectively. These rates are denoted as

$$W_r(\mathbf{X}'|\mathbf{X}) \qquad \text{for the transition} \qquad \mathbf{X} \xrightarrow{r} \mathbf{X}' = \mathbf{X} + \boldsymbol{\nu}_r, \tag{11.5}$$

where $\boldsymbol{\nu}_r = \{\nu_{kr}\}_{k=1}^d$ are the stoichiometric coefficients (11.2), such that $\boldsymbol{\nu}_r = -\boldsymbol{\nu}_{-r}$ with $r = \pm 1, \pm 2, \ldots, \pm m$.

According to the *mass action law* of kinetics, the transition rates of the forward and backward elementary reactions are given in dilute solutions by

$$W_{\pm r}(\mathbf{X} \pm \boldsymbol{\nu}_r|\mathbf{X}) = V\, \tilde{k}_{\pm r} \prod_{k=1}^{d} \prod_{j=1}^{\nu_{kr}^{(\pm)}} \frac{X_k - j + 1}{n^0\, V} \qquad \text{for} \qquad 1 \leq r \leq m, \tag{11.6}$$

in terms of the particle numbers $\{X_k\}_{k=1}^d$ of the intermediate species involved in the reaction (11.1), the integer numbers $\nu_{kr}^{(\pm)}$ of the particles incoming the reaction, the reference concentration n^0 equal to one mole per liter, and the following effective rate constants,

$$\tilde{k}_{\pm r} \equiv k_{\pm r} \prod_{l=1}^{e} \left(\frac{y_l}{n^0}\right)^{\tilde{\nu}_{lr}^{(\pm)}}, \tag{11.7}$$

[1] In heterogeneous catalysis, these intermediate species should include the empty sites of the catalytic surface.

depending on the invariant concentrations $\{y_l\}_{l=1}^{e}$ of reactant or product species (Nicolis and Prigogine, 1971; Nicolis, 1972; Nicolis and Prigogine, 1977; Luo et al., 1984; Gaspard, 2004a).[2] We note that the transition rates (11.6) are extensive with respect to the dimensionless *extensivity parameter* $\Omega \equiv n^0 V$. Moreover, they do not depend on the counting variables $\{n_r\}_{r=1}^{m}$ if the species $\{\mathrm{Y}_l\}_{l=1}^{e}$ are chemostatted.

11.2.1 Chemical Master Equations

With the transition rates (11.6), we can define a Markovian stochastic process for the time evolution of the joint random variables $\mathbf{X} = \{X_k\}_{k=1}^{d} \in \mathbb{N}^d$ and $\mathbf{n} = \{n_r\}_{r=1}^{m} \in \mathbb{Z}^m$ with $n_r \equiv n_{+r} - n_{-r}$. The algorithm of Gillespie (1976, 1977) provides an exact Monte Carlo method for the numerical simulation of such Markovian stochastic processes.

If $\Delta\mathbf{n}_r = -\Delta\mathbf{n}_{-r}$ denotes the jump of the counters $\mathbf{n}(t)$ upon the reactive event r, the master equation ruling the time evolution of the joint probability distribution $P(\mathbf{X}, \mathbf{n}, t)$ is given by

$$\frac{d}{dt} P(\mathbf{X}, \mathbf{n}, t) = \sum_{r=\pm 1}^{\pm m} \left[W_r(\mathbf{X}|\mathbf{X} - \boldsymbol{\nu}_r)\, P(\mathbf{X} - \boldsymbol{\nu}_r, \mathbf{n} - \Delta\mathbf{n}_r, t) - W_{-r}(\mathbf{X} - \boldsymbol{\nu}_r|\mathbf{X})\, P(\mathbf{X}, \mathbf{n}, t) \right] \tag{11.8}$$

with the stoichiometric coefficients $\boldsymbol{\nu}_r = -\boldsymbol{\nu}_{-r}$ (Andrieux and Gaspard, 2008f).

Since the transition rates (11.6) are independent of the counters $\mathbf{n}$ under the stated assumptions, summing equation (11.8) over $\mathbf{n}$ leads to the standard chemical master equation

$$\frac{d}{dt} \mathscr{P}(\mathbf{X}, t) = \sum_{r=\pm 1}^{\pm m} \left[W_r(\mathbf{X}|\mathbf{X} - \boldsymbol{\nu}_r)\, \mathscr{P}(\mathbf{X} - \boldsymbol{\nu}_r, t) - W_{-r}(\mathbf{X} - \boldsymbol{\nu}_r|\mathbf{X})\, \mathscr{P}(\mathbf{X}, t) \right] \tag{11.9}$$

for the marginal probability distribution of the particle numbers of intermediate species

$$\mathscr{P}(\mathbf{X}, t) = \sum_{\mathbf{n}} P(\mathbf{X}, \mathbf{n}, t), \tag{11.10}$$

which is normalized according to $\sum_{\mathbf{X}} \mathscr{P}(\mathbf{X}, t) = 1$ (Nicolis and Prigogine, 1971; Nicolis, 1972; Nicolis and Prigogine, 1977; Schnakenberg, 1976; Gardiner, 1979; van Kampen, 1981; Gardiner, 2004). We note that the probability distribution (11.10) of the intermediate species may reach a stationary distribution $\mathscr{P}_{\mathrm{st}}(\mathbf{X})$, corresponding to the equilibrium or nonequilibrium stationary macrostate.

The reaction network is at thermodynamic equilibrium if the following conditions of detailed balance are satisfied,

$$W_r(\mathbf{X}|\mathbf{X} - \boldsymbol{\nu}_r)\, \mathscr{P}_{\mathrm{eq}}(\mathbf{X} - \boldsymbol{\nu}_r) = W_{-r}(\mathbf{X} - \boldsymbol{\nu}_r|\mathbf{X})\, \mathscr{P}_{\mathrm{eq}}(\mathbf{X}), \tag{11.11}$$

[2] Transition rates have other dependences on the particle numbers in dense solutions and heterogeneous catalysis (Kondepudi and Prigogine, 1998; Grosfils et al., 2015).

for all the elementary reactions $1 \leq r \leq m$ and all the states $\mathbf{X} \in \mathbb{N}^d$. In this case, the stationary probability distribution is Poissonian because the solution is dilute:

$$\mathscr{P}_{\rm eq}(\mathbf{X}) = \prod_{k=1}^{d} {\rm e}^{-\langle X_k\rangle_{\rm eq}} \frac{\langle X_k\rangle_{\rm eq}^{X_k}}{X_k!}, \tag{11.12}$$

which can be verified by substitution into the master equation (11.9), using the transition rates (11.6) of the mass action law, and the Guldberg–Waage conditions (1.39) for the equilibrium concentrations $x_{k,{\rm eq}} = \langle X_k\rangle_{\rm eq}/V$ (Nicolis and Prigogine, 1977).

In the long-time limit, the joint probability distribution, which is the solution of the master equation (11.8), is expected to factorize as

$$P(\mathbf{X}, \mathbf{n}, t) \simeq_{t\to\infty} \mathscr{P}_{\rm st}(\mathbf{X})\, \mathcal{P}_t(\mathbf{n}) \tag{11.13}$$

into the stationary probability distribution $\mathscr{P}_{\rm st}(\mathbf{X})$ of the particle numbers and the nonstationary distribution $\mathcal{P}_t(\mathbf{n})$ of the counters $\mathbf{n}$. These latter are drifting at the mean reaction rates if the process is out of equilibrium, or their mean drift rates are equal to zero and they follow a random walk if the process is at equilibrium. According to the central limit theorem, the distribution $\mathcal{P}_t(\mathbf{n})$ becomes a broader and broader Gaussian distribution for $t \to \infty$ in finite networks. In this regard, the particle numbers $\mathbf{X}$ and the counters $\mathbf{n}$, respectively, behave as the random variables N and n in the particle exchange process ruled by the master equation (4.91).

We note that the Markov jump process of the chemical master equation (11.9) can be approximated by the corresponding Langevin process ruled by the Fokker–Planck equation (4.142) with the drifts and diffusivities (4.141) in the limit $\epsilon = \Omega^{-1} \to 0$ where the particle numbers are large enough, i.e., $X_k \gg 1$ (Gillespie, 2000; Gaspard, 2002a).

11.2.2 Kinetic Equations

At the macroscale, the particle fluctuations become negligible in front of the mean particle numbers of the different species. In this limit, the macroscopic kinetic equations are recovered for the concentrations of the intermediate species defined as

$$x_k = [{\rm X}_k] \equiv \frac{1}{V}\langle X_k\rangle, \qquad \text{where} \qquad \langle X_k\rangle \equiv \sum_{\mathbf{X}} X_k\, \mathscr{P}(\mathbf{X}, t)\,. \tag{11.14}$$

The macroscopic reaction rates can be similarly defined as

$$w_{\pm r} \equiv \frac{1}{V}\sum_{\mathbf{X}} W_{\pm r}(\mathbf{X} \pm \boldsymbol{\nu}_r|\mathbf{X})\, \mathscr{P}(\mathbf{X}, t) \tag{11.15}$$

for $r = 1, 2, \ldots, m$. They can be equivalently expressed as $w_{\pm r} = V^{-1} d\langle n_{\pm r}(t)\rangle/dt$ in terms of the variables $n_{\pm r}(t)$ counting the forward and backward reactive events and using the statistical average with respect to the joint probability distribution $P(\mathbf{X}, \mathbf{n}, t)$.

If the volume V is large enough, the marginal distribution $\mathscr{P}(\mathbf{X}, t)$ may be assumed to peak around the most probable values for the particle numbers. In this limit, the reaction rates corresponding to the transition rates (11.6) become the known macroscopic reaction rates established with the mass action law,

$$w_{\pm r}(\mathbf{x}) = \tilde{k}_{\pm r} \prod_{k=1}^{d} \left(\frac{x_k}{n^0}\right)^{\nu_{kr}^{(\pm)}} \qquad \text{for} \qquad 1 \le r \le m, \tag{11.16}$$

with the rate constants (11.7) and the most probable values for the concentrations $\mathbf{x} = \{x_k\}_{k=1}^{d}$ (de Groot and Mazur, 1984; Nicolis, 1995).

As a consequence of equation (11.3), the macroscopic kinetic equation for the concentration x_k is given by

$$\frac{dx_k}{dt} = \sum_{r=1}^{m} \nu_{kr} \left(w_{+r} - w_{-r}\right) \tag{11.17}$$

in terms of the stoichiometric coefficients (11.2). Similarly, equation (11.4) leads to

$$\frac{d}{dt}\langle Y_l \rangle = V \sum_{r=1}^{m} \tilde{\nu}_{lr} \left(w_{+r} - w_{-r}\right) \tag{11.18}$$

for the mean production rate of species Y_l caused by the reaction network (11.1). Since the species Y_l is supposed to be chemostatted, equation (11.18) also gives the rate of transport towards the exterior in order to keep its concentration at the constant value y_l inside the reactor. In the case of a batch reactor, the rates (11.18) should be small enough so that $\langle Y_l \rangle_t \simeq \langle Y_l \rangle_0$ during the time interval $[0, t]$ of the process.

The coupled kinetic equations (11.17) can be written in the form

$$\frac{d\mathbf{x}}{dt} = \boldsymbol{\nu} \cdot \mathbf{w} \tag{11.19}$$

using the vector $\mathbf{w} = \{w_r\}_{r=1}^{m}$ of the net reaction rates $w_r \equiv w_{+r} - w_{-r}$ and the matrix $\boldsymbol{\nu} \equiv \{\boldsymbol{\nu}_r\}_{r=1}^{m} = (\nu_{kr})_{k=1,\ldots,d;\, r=1,\ldots,m}$ with the stoichiometric coefficients (11.2). This set of first-order differential equations defines the dynamical system ruling the macroscopic time evolution of the particle concentrations in the phase space $\mathbf{x} \in \mathbb{R}^d$ (Nicolis, 1995). Since this dynamical system is typically dissipative in the sense that the phase-space volumes are contracted on average, the trajectories converge in time towards an attractor of zero Lebesgue measure in phase space. This attractor may be a fixed point corresponding to a macroscopic stationary state, a limit cycle describing periodic oscillations, a torus if the oscillations are quasiperiodic, or a fractal attractor if the system is chaotic and manifesting sensitivity to initial conditions characterized by positive Lyapunov exponents (Nicolis, 1995). These dissipative dynamical systems can be analyzed using the methods of Appendix B.

Furthermore, we may also carry out the stoichiometric analysis of the reaction network with the methods of Section 1.8.2. Even if the system is time dependent, we can always consider the time average (1.109). A basic property is that the time derivative df/dt of any function $f = f(\mathbf{x})$ has a time average equal to zero, so that

$$\overline{\frac{d\mathbf{x}}{dt}} = \boldsymbol{\nu} \cdot \overline{\mathbf{w}} = 0. \tag{11.20}$$

Consequently, the time average of the reaction rates belongs to the right null space of the stoichiometric matrix $\boldsymbol{\nu}$. This subspace is spanned by the right null eigenvectors $\mathbf{e}_\gamma = \{e_{r\gamma}\}_{r=1}^m$ such that

$$\boldsymbol{\nu} \cdot \mathbf{e}_\gamma = 0 \qquad \text{for} \qquad \gamma = 1, 2, \ldots, o = \dim \ker \boldsymbol{\nu}. \tag{11.21}$$

Each right null eigenvector defines a so-called stoichiometric cycle γ, which is a sequence of elementary reactions (weighted by the components $e_{r\gamma}$ of the eigenvector) such that the particle numbers $\mathbf{X}$ of intermediate species come back to their initial values once the cycle is closed, since $\Delta\mathbf{X} = \boldsymbol{\nu} \cdot \mathbf{e}_\gamma = 0$ by equation (11.3). Instead, the cycle may change the pool of chemostatted species according to $\Delta\mathbf{Y} = \tilde{\boldsymbol{\nu}} \cdot \mathbf{e}_\gamma \neq 0$ because of equation (11.4). We note that different sets can be obtained for the right null eigenvectors by using linear combinations. Moreover, the left null space of the stoichiometric matrix contains all the constants of motion $L = V\, \boldsymbol{\mathfrak{l}}^{\mathrm{T}} \cdot \mathbf{x}$ such that $dL/dt = 0$. Accordingly, this space is spanned by the left null eigenvectors, $\boldsymbol{\mathfrak{l}}^{\mathrm{T}} \cdot \boldsymbol{\nu} = 0$. If $l = \dim \operatorname{coker} \boldsymbol{\nu}$ is the number of constants of motion, the rank of the stoichiometric matrix is given by

$$\operatorname{rank} \boldsymbol{\nu} = d - l = m - o \tag{11.22}$$

with the numbers d of intermediate species, m of reactions, and o of stoichiometric cycles (Polettini and Esposito, 2014).

Because of equation (11.20), the time average of the reaction rates can thus be decomposed according to

$$\overline{\mathbf{w}} = \sum_{\gamma=1}^{o} \mathbf{e}_\gamma \, J_\gamma \qquad \text{or} \qquad \overline{w}_r = \sum_{\gamma=1}^{o} e_{r\gamma} \, J_\gamma \tag{11.23}$$

onto the basis of the o right null eigenvectors (11.21). The coefficients J_γ of this decomposition define the rates associated with the o cycles of the reaction network.

The affinity associated with the cycle γ is defined as

$$A_\gamma \equiv \ln \prod_{r=1}^{m} \left(\frac{w_{+r}}{w_{-r}} \right)^{e_{r\gamma}}. \tag{11.24}$$

We note that all these affinities are equal to zero if the reaction network is at thermodynamic equilibrium where $w_{+r} = w_{-r}$ for all the reactions due to the conditions of detailed balance. An important property is that these affinities do not depend on the concentrations $\mathbf{x}$ of the intermediate species if the reaction rates obey the mass action law (11.16). Indeed, the ratios of these opposite reaction rates only depend on $x_k^{\nu_{kr}}$. Since $\sum_{r=1}^m \nu_{kr}\, e_{r\gamma} = 0$ because of equation (11.21), we find

$$A_\gamma = \sum_{r=1}^{m} e_{r\gamma} \, \ln \frac{\tilde{k}_{+r}}{\tilde{k}_{-r}}, \tag{11.25}$$

thus establishing the property (Gaspard, 2020a).

11.2.3 Entropy Production

The random paths of the process ruled by the master equation (11.8) can be denoted

$$\mathcal{X}(t) = (\mathbf{X}_0, \mathbf{n}_0) \to (\mathbf{X}_1, \mathbf{n}_1) \to \cdots \to (\mathbf{X}_j, \mathbf{n}_j), \tag{11.26}$$

or, more shortly,

$$\mathcal{X}(t) = \mathbf{X}_0 \xrightarrow{r_1} \mathbf{X}_1 \xrightarrow{r_2} \cdots \xrightarrow{r_j} \mathbf{X}_j, \tag{11.27}$$

where $\{r_1, r_2, \ldots, r_j\}$ is the sequence of elementary reactive events occurring along the path. If the reactive event r happens, the corresponding counter is incremented by the unit value according to $n_r \xrightarrow{r} n_r + 1$ (for $r = \pm 1, \pm 2, \ldots, \pm m$), so that equations (11.26) and (11.27) provide equivalent descriptions of the path.

According to equation (6.11), the probability of the path (11.27) can be expressed in the limit $\tau \to 0$ as

$$P[\mathcal{X}(t)] = \mathrm{e}^{-\gamma(\mathbf{X}_j)(t-t_j)} \prod_{i=1}^{j} \left[W_{r_i}(\mathbf{X}_i|\mathbf{X}_{i-1})\, \tau\, \mathrm{e}^{-\gamma(\mathbf{X}_{i-1})(t_i - t_{i-1})} \right] P(\mathbf{X}_0, \mathbf{n}_0, t_0) \tag{11.28}$$

in terms of the escape rates from the states $\mathbf{X}$ given by

$$\gamma(\mathbf{X}) \equiv \sum_{r=\pm 1}^{\pm m} W_r(\mathbf{X} + \boldsymbol{\nu}_r|\mathbf{X}) \tag{11.29}$$

if the transitions $\mathbf{X}_i = \mathbf{X}_{i-1} + \boldsymbol{\nu}_{r_i}$ occur at the successive times $\{t_i\}_{i=1}^{j}$ during the time interval $[t_0, t]$.

The time reversal of the path (11.27) can be written as

$$\mathcal{X}^{\mathrm{R}}(t) = \mathbf{X}_j \xrightarrow{-r_j} \cdots \xrightarrow{-r_3} \mathbf{X}_2 \xrightarrow{-r_2} \mathbf{X}_1 \xrightarrow{-r_1} \mathbf{X}_0 \tag{11.30}$$

in terms of the reversed sequence defined by also reversing the elementary reactions. The probability of this reversed path is given by an expression similar to equation (11.28).

As shown in Chapter 6, the entropy production rate can be obtained by considering the ratio (6.19) between the probabilities of the path and its time reversal. Because of equation (11.28), we here have that

$$\Sigma(t) \equiv \ln \frac{P[\mathcal{X}(t)]}{P[\mathcal{X}^{\mathrm{R}}(t)]} = Z(t) + \ln \frac{P(\mathbf{X}_0, \mathbf{n}_0, t_0)}{P(\mathbf{X}_j, \mathbf{n}_j, t)} \tag{11.31}$$

in terms of the quantity

$$Z(t) \equiv \ln \prod_{i=1}^{j} \frac{W_{r_i}(\mathbf{X}_i|\mathbf{X}_{i-1})}{W_{-r_i}(\mathbf{X}_{i-1}|\mathbf{X}_i)}, \tag{11.32}$$

involving the transition rates of the forward and backward elementary reactions that have occurred during the path (11.27). The time derivatives of the quantities (11.31) and (11.32) can be expressed in terms of the successive jumps of the Markov reactive process, as in equations (6.46) and (6.45), respectively. In particular, we get

$$\frac{dZ}{dt} = \sum_{i=-\infty}^{+\infty} \delta(t - t_i) \ln \frac{W_{r_i}(\mathbf{X}_i|\mathbf{X}_{i-1})}{W_{-r_i}(\mathbf{X}_{i-1}|\mathbf{X}_i)}. \tag{11.33}$$

Now, the jumps caused by the rth reaction are occurring with the net rates given by

$$J_r = W_r(\mathbf{X}|\mathbf{X} - \boldsymbol{\nu}_r)\, P(\mathbf{X} - \boldsymbol{\nu}_r, \mathbf{n} - \Delta\mathbf{n}_r, t) - W_{-r}(\mathbf{X} - \boldsymbol{\nu}_r|\mathbf{X})\, P(\mathbf{X}, \mathbf{n}, t) \tag{11.34}$$

according to the master equation (11.8). The rates (11.34) correspond to the net rates (4.50) in the framework of Section 4.4.4 for the coarse-grained states $\omega = (\mathbf{X}, \mathbf{n})$. Consequently, the mean value of the time derivative (11.33) over the path probability (11.28) is given by

$$\left\langle \frac{dZ}{dt} \right\rangle = \frac{1}{k_{\mathrm{B}}} \frac{d_{\mathrm{i}}S}{dt} - \frac{dD}{dt} \tag{11.35}$$

in terms of the entropy production rate (4.57) and the time derivative of the Shannon disorder (4.46). We note that the time derivative of the last term in equation (11.31) has the mean value equal to dD/dt, so that

$$\left\langle \frac{d\Sigma}{dt} \right\rangle = \left\langle \frac{dZ}{dt} \right\rangle + \frac{dD}{dt} = \frac{1}{k_{\mathrm{B}}} \frac{d_{\mathrm{i}}S}{dt} \tag{11.36}$$

with the following expression for the entropy production rate,

$$\begin{aligned} \frac{1}{k_{\mathrm{B}}} \frac{d_{\mathrm{i}}S}{dt} = \sum_{r=1}^{m} \sum_{\mathbf{X},\mathbf{n}} \big[& W_r(\mathbf{X}|\mathbf{X} - \boldsymbol{\nu}_r)\, P(\mathbf{X} - \boldsymbol{\nu}_r, \mathbf{n} - \Delta\mathbf{n}_r, t) \\ & - W_{-r}(\mathbf{X} - \boldsymbol{\nu}_r|\mathbf{X})\, P(\mathbf{X}, \mathbf{n}, t) \big] \\ & \times \ln \frac{W_r(\mathbf{X}|\mathbf{X} - \boldsymbol{\nu}_r)\, P(\mathbf{X} - \boldsymbol{\nu}_r, \mathbf{n} - \Delta\mathbf{n}_r, t)}{W_{-r}(\mathbf{X} - \boldsymbol{\nu}_r|\mathbf{X})\, P(\mathbf{X}, \mathbf{n}, t)} \geq 0. \end{aligned} \tag{11.37}$$

In the macroscopic limit $V \to \infty$, the probability distribution $P(\mathbf{X}, \mathbf{n}, t)$ typically peaks around the most probable values $\mathbf{X} = V\mathbf{x}$ and $\mathbf{n}$, which are, respectively, ruled by the kinetic equations (11.17) and $dn_r/dt = w_r$, whereupon the entropy production rate (11.37) becomes

$$\frac{1}{k_{\mathrm{B}}} \frac{d_{\mathrm{i}}S}{dt} = V \sum_{r=1}^{m} (w_{+r} - w_{-r}) \ln \frac{w_{+r}}{w_{-r}} \geq 0, \tag{11.38}$$

which corresponds to the expression (1.96) known for the entropy production rate density of reaction networks (De Donder and Van Rysselberghe, 1936; Prigogine, 1967; Nicolis, 1979; de Groot and Mazur, 1984; Kondepudi and Prigogine, 1998; Rao and Esposito, 2016). The sum is taken over all the elementary reactions in the network. The entropy production rate (11.38) is, in general, nonnegative, vanishing at the equilibrium concentrations $\mathbf{x}_{\mathrm{eq}}$ such that the macroscopic conditions of detailed balance are satisfied: $w_{+r}(\mathbf{x}_{\mathrm{eq}}) = w_{-r}(\mathbf{x}_{\mathrm{eq}})$.

Furthermore, according to the asymptotic behavior (11.13), equation (11.37) converges to the stationary value

$$\frac{1}{k_{\rm B}} \frac{d_{\rm i}S}{dt}\bigg|_{\rm st} = \sum_{r=1}^{m} \sum_{\mathbf{X}} \left[W_r(\mathbf{X}|\mathbf{X}-\boldsymbol{\nu}_r)\, \mathscr{P}_{\rm st}(\mathbf{X}-\boldsymbol{\nu}_r) - W_{-r}(\mathbf{X}-\boldsymbol{\nu}_r|\mathbf{X})\, \mathscr{P}_{\rm st}(\mathbf{X}) \right]$$

$$\times \ln \frac{W_r(\mathbf{X}|\mathbf{X}-\boldsymbol{\nu}_r)\, \mathscr{P}_{\rm st}(\mathbf{X}-\boldsymbol{\nu}_r)}{W_{-r}(\mathbf{X}-\boldsymbol{\nu}_r|\mathbf{X})\, \mathscr{P}_{\rm st}(\mathbf{X})} \geq 0, \tag{11.39}$$

which is the known expression for the entropy production rate of stochastic reactive processes (Luo et al., 1984; Gaspard, 2004a), again with the sum over the elementary reactions. According to the central limit theorem for the distribution $\mathcal{P}_t(\mathbf{n})$, we have that $D(t) \simeq D_{\rm st} + k_{\rm B} \ln \sqrt{(4\pi e t)^m \det \mathsf{D}}$, where $D_{\rm st}$ is the Shannon disorder of the stationary distribution $\mathscr{P}_{\rm st}(\mathbf{X})$ and D the matrix of diffusivities for the m counters $\mathbf{n}$. The counters $\mathbf{n} \in \mathbb{Z}^m$ thus behave as the random variable $n \in \mathbb{Z}$ of the random drift presented in Section 4.4.6, so that their asymptotic probability distribution $\mathcal{P}_t(\mathbf{n})$ becomes broader and broader and its Shannon disorder increases accordingly. Since this increase is logarithmic in time, its time derivative is asymptotically vanishing, as in the examples of Section 4.4.6. Hence, $dD/dt = k_{\rm B} m/(2t)$ and $\lim_{t\to\infty} dD/dt = 0$, so that the mean values (11.35) and (11.36) become equal in the long-time limit. We also note that the quantities $\Sigma(t)$ and $Z(t)$ are extensive in time in the case of the present Markov jump processes under nonequilibrium conditions. Accordingly, the mean value of their time derivative is identical to their mean value divided by time in the long-time limit and the stationary value of the entropy production rate is given by

$$\frac{1}{k_{\rm B}} \frac{d_{\rm i}S}{dt}\bigg|_{\rm st} = \lim_{t\to\infty} \frac{1}{t} \langle \Sigma(t) \rangle = \lim_{t\to\infty} \frac{1}{t} \langle Z(t) \rangle \geq 0, \tag{11.40}$$

confirming the result (6.44) for the reactive processes.

The issue of evaluating the entropy production in the Langevin approximation of the chemical master equation has also been considered (Horowitz, 2015).

In order to recover the macroscopic entropy production rate (1.96), the successive elementary reactions followed during the stochastic evolution should be specified, as in the paths (11.26) and (11.27). If the sequence of elementary reactions was omitted as in the path $\mathbf{X}_0 \to \mathbf{X}_1 \to \cdots \to \mathbf{X}_j$ of the internal states of the system, the entropy production rate might be underevaluated, as shown in Section E.4.

The entropy production rate is finite if all the rate constants $\{k_{\pm r}\}_{r=1}^m$ and all the concentrations $\{y_l\}_{l=1}^e$ of the chemostatted species are positive, so that all the effective rate constants (11.7) are positive. If any one of them is equal to zero, the entropy production rate is infinite, i.e., $d_{\rm i}S/dt = \infty$, and the reaction network is said to be *fully irreversible*. In such cases, the affinities depending on the vanishing effective rate constants are infinite. Nevertheless, the stochastic process continues to make sense from the viewpoint of kinetics because the transition rates and the mean currents remain finite in fully irreversible regimes.

11.2.4 Cycle Decomposition of the Entropy Production

At the mesoscopic level of description, the stationary value (11.39) of the entropy production rate can be decomposed using the Hill–Schnakenberg network theory presented in Section 4.4.5 for the Markov jump process ruled by the master equation (11.9).

A graph G is associated with the process, which is composed of vertices corresponding to the states $\mathbf{X} \in \mathbb{N}^d$ and edges associated with the transitions introduced in equation (11.5) (Andrieux and Gaspard, 2007b). A fundamental set of cycles $\{C\}$ can be constructed on the graph G, as explained in Section 4.4.5. For the cycle $C = \{\mathbf{X}_i, r_i\}_{i=1}^{c}$ of length c, the affinity (4.77) is here given by

$$A_C = \ln \prod_{i=1}^{c} \frac{W_{r_i}(\mathbf{X}_i|\mathbf{X}_{i-1})}{W_{-r_i}(\mathbf{X}_{i-1}|\mathbf{X}_i)}, \tag{11.41}$$

with $\mathbf{X}_0 = \mathbf{X}_c$.

The remarkable result obtained by Schnakenberg (1976) is that these affinities do not depend on the states $\{\mathbf{X}_i\}_{i=1}^{c}$ followed along each cycle for the transition rates (11.6) given by the mass action law:

$$A_C = \ln \prod_{i=1}^{c} \frac{\tilde{k}_{+r_i}}{\tilde{k}_{-r_i}} = \ln \prod_{i=1}^{c} \left[\frac{k_{+r_i}}{k_{-r_i}} \prod_{l=1}^{e} \left(\frac{n^0}{y_l} \right)^{\nu_{lr_i}} \right]. \tag{11.42}$$

These affinities vanish at equilibrium where the Guldberg–Waage conditions (1.39) are satisfied for the chemostatted species:

$$\prod_{l=1}^{e} \left(\frac{y_l}{n^0} \right)_{\text{eq}}^{\nu_{lr}} = \frac{k_{+r}}{k_{-r}} \qquad \text{with} \qquad 1 \leq r \leq m, \tag{11.43}$$

which correspond to the equalities $\tilde{k}_{+r} = \tilde{k}_{-r}$. Accordingly, the affinities (11.25) are equal to zero for the same equilibrium conditions (11.43). Otherwise, the system is driven away from equilibrium. Depending on the number e of chemostatted species and their stoichiometric coefficients $\tilde{\nu}_{lr}$, the equations (11.43) may determine conditions on the equilibrium concentrations $y_{l,\text{eq}}$, or compatibility conditions on the constants $K_r = k_{+r}/k_{-r}$ for the existence of an equilibrium macrostate in the reaction network.

Since the affinity (11.42) is independent of the states $\{\mathbf{X}\}$ of the process, it only depends on the sequence of reactions $\{r_i\}_{i=1}^{c}$ followed along the cycle C. Moreover, the state $\mathbf{X}$ comes back to its initial condition after the cycle $C = \{\mathbf{X}_i, r_i\}_{i=1}^{c}$ is closed, so that the numbers $\mathbf{n}_C = \{n_{Cr}\}_{r=1}^{m}$ of elementary reactions followed in the cycle should be a right null eigenvector for the stoichiometric matrix according to equation (11.3): $\boldsymbol{\nu} \cdot \mathbf{n}_C = 0$. This vector should thus be a linear combination of the set of right null eigenvectors (11.21): $\mathbf{n}_C = \sum_{\gamma=1}^{o} \varsigma_{C\gamma} \mathbf{e}_\gamma$. Consequently, the affinity (11.42) of the cycle C can be written as the following linear combination of the macroscopic affinities (11.25),

$$A_C = \sum_{r=1}^{m} n_{Cr} \ln \frac{\tilde{k}_{+r}}{\tilde{k}_{-r}} = \sum_{\gamma=1}^{o} \varsigma_{C\gamma} A_\gamma. \tag{11.44}$$

In the stationary regime, a mean current J_C is associated with the cycle C according to equations (4.71) and (4.73) and the entropy production rate is thus given by equation (4.74). Since the graph contains countably many vertices, the sum extends over countably many cycles $\{C\}$ in the expression (4.74). Because of the decomposition (11.44), the entropy production rate (4.74) can thus be written as

$$\frac{1}{k_B}\frac{d_i S}{dt}\bigg|_{st} = \sum_C A_C J_C = \sum_{\gamma=1}^{o} A_\gamma J_\gamma \tag{11.45}$$

in terms of the finite set of macroscopic currents defined as

$$J_\gamma \equiv \sum_C \varsigma_{C\gamma} J_C \tag{11.46}$$

by summing over all the cycles $\{C\}$. In the expression (11.45), the currents are averaged over the stationary probability distribution $\mathscr{P}_{st}(\mathbf{X})$.

At the macroscopic level of description, the solutions of the kinetic equations (11.17) may depend on time over arbitrarily long timescales for instance in oscillatory regimes. In such regimes, the macroscopic entropy production rate (11.38) is also depending on time. In order to compare the mesoscopic and macroscopic descriptions, we may consider the time average of the macroscopic entropy production rate (11.38). Remarkably, this time average can also be decomposed into the o stoichiometric cycles (11.21) of the stoichiometric analysis, if the reaction rates obey the mass action law (11.16). Indeed, the macroscopic entropy production rate (11.38) can be written in the following form using the reaction rates (11.16),

$$\frac{1}{k_B}\frac{d_i S}{dt} = V \sum_{r=1}^{m} (w_{+r} - w_{-r}) \ln \frac{\tilde{k}_{+r}}{\tilde{k}_{-r}} + \frac{df}{dt}, \tag{11.47}$$

where $f \equiv V \sum_{k=1}^{d} x_k \ln(e n^0 / x_k)$. Since the time average is equal to zero for the total derivative of the function f with respect to time, using the decomposition (11.23) and the expression (11.25) for the affinities, we find

$$\frac{1}{k_B}\overline{\frac{d_i S}{dt}} = \sum_{\gamma=1}^{o} A_\gamma J_\gamma. \tag{11.48}$$

Hence, for reaction kinetics obeying the mass action law, the time average of the macroscopic entropy production rate can always be decomposed as the sum of the affinities and rates associated with the o cycles defined by the right null eigenvectors (11.21) of the stoichiometric matrix (Polettini and Esposito, 2014; Rao and Esposito, 2018b; Blokhuis et al., 2018; Gaspard, 2020a). In this way, the mesoscopic decomposition (11.45) holds at the macroscopic level of description.

We note that the macroscopic affinities (11.25) are equal to zero at equilibrium if the Guldberg–Waage conditions (11.43) are satisfied. As mentioned in Chapter 1, some moieties $L = \sum_{k=1}^{d} l_k X_k + \sum_{l=1}^{e} l_l Y_l$ may be conserved by the evolution equations (11.3) and (11.4) of the reaction network in the sense that $L(t) = L(0)$ for all times t. These conservation laws impose constraints such that some of the macroscopic affinities $\{A_\gamma\}$ may be equal to zero once the conditions of compatibility with the existence of equilibrium are satisfied by the rate constants $\{k_{\pm r}\}$. Indeed, the macroscopic affinities (11.25) can be expressed as

$$A_\gamma = \sum_{r=1}^{m} e_{r\gamma} \ln \frac{k_{+r}}{k_{-r}} + \sum_{l=1}^{e} \sum_{r=1}^{m} \tilde{\nu}_{lr}\, e_{r\gamma} \ln \frac{n^0}{y_l}, \tag{11.49}$$

because of equation (11.7). Therefore, the affinity A_γ does not depend on the concentrations y_l if $\sum_{r=1}^{m} \tilde{\nu}_{lr}\, e_{r\gamma} = 0$ holds for $1 \le l \le e$. If this is the case, the existence of equilibrium requires that the rate constants satisfy the compatibility condition $\prod_{r=1}^{m} (k_{+r}/k_{-r})^{e_{r\gamma}} = 1$, so that this affinity is vanishing, i.e., $A_\gamma = 0$.[3]

11.2.5 Fluctuation Relations

The reaction network is driven out of equilibrium by the nonvanishing macroscopic affinities (11.49) supplying free energy to the open system. At the mesoscopic level of description, the currents associated with these affinities are fluctuating according to the random numbers of reactive events driving the stochastic time evolution of the different species by equations (11.3) and (11.4).

Let us denote the nonvanishing macroscopic affinities as $\{A_\alpha\}_{\alpha=1}^{\chi}$ and their total number as $\chi \le o$. These affinities correspond to stoichiometric cycles defined by right null eigenvectors $\{\mathbf{e}_\alpha\}_{\alpha=1}^{\chi}$ satisfying equation (11.21) with respect to the stoichiometric matrix. Among all the counters $\{n_r\}_{r=1}^{m}$ of the reaction network (11.1), we may, in general, select the counters $\mathbf{n} = \{n_\alpha\}_{\alpha=1}^{\chi}$ that are associated with the χ stoichiometric cycles and count the reactive events feeding the reaction network with chemostatted species.

The generating function of the statistical moments of these counters can be defined as

$$F(\mathbf{X}, \boldsymbol{\lambda}, t) \equiv \sum_{\mathbf{n}} P(\mathbf{X}, \mathbf{n}, t)\, \mathrm{e}^{-\boldsymbol{\lambda}\cdot\mathbf{n}} \tag{11.50}$$

by introducing the counting parameters $\boldsymbol{\lambda} = \{\lambda_\alpha\}_{\alpha=1}^{\chi}$. We note that the marginal probability distribution (11.10) is recovered as $\mathscr{P}(\mathbf{X}, t) = F(\mathbf{X}, \mathbf{0}, t)$ from the generating function (11.50) for $\boldsymbol{\lambda} = \mathbf{0}$.

The time evolution of this generating function can be directly deduced from the full master equation (11.8) since the rates do not depend on the counters, giving

$$\partial_t F(\mathbf{X}, \boldsymbol{\lambda}, t) = \sum_{r=\pm 1}^{\pm m} \left[\mathrm{e}^{-\boldsymbol{\lambda}\cdot\Delta\mathbf{n}_r} W_r(\mathbf{X}|\mathbf{X} - \boldsymbol{\nu}_r)\, F(\mathbf{X} - \boldsymbol{\nu}_r, \boldsymbol{\lambda}, t) - W_{-r}(\mathbf{X} - \boldsymbol{\nu}_r|\mathbf{X})\, F(\mathbf{X}, \boldsymbol{\lambda}, t)\right]. \tag{11.51}$$

This equation can also be inferred from the reduced master equation (11.9), multiplying the gain terms by exponential factors involving the counting parameters $\boldsymbol{\lambda}$ according to equation (5.215).

[3] This case will be illustrated in Section 11.6.

The cumulant generating function of the reactive currents can be defined as

$$Q(\boldsymbol{\lambda};\mathbf{A}) \equiv \lim_{t\to\infty} -\frac{1}{t} \ln \sum_{\mathbf{X},\mathbf{n}} P(\mathbf{X},\mathbf{n},t)\, \mathrm{e}^{-\boldsymbol{\lambda}\cdot\mathbf{n}} = \lim_{t\to\infty} -\frac{1}{t} \ln \sum_{\mathbf{X}} F(\mathbf{X},\boldsymbol{\lambda},t), \tag{11.52}$$

for the affinities $\mathbf{A}$. The methods of Section 5.7 can thus be used to establish the symmetry relation

$$Q(\boldsymbol{\lambda};\mathbf{A}) = Q(\mathbf{A}-\boldsymbol{\lambda};\mathbf{A}) \tag{11.53}$$

and the multivariate fluctuation relation for the reactive currents

$$\frac{\mathcal{P}_t(\mathbf{n};\mathbf{A})}{\mathcal{P}_t(-\mathbf{n};\mathbf{A})} \sim_{t\to\infty} \exp(\mathbf{A}\cdot\mathbf{n}) \tag{11.54}$$

in its asymptotic form (Andrieux and Gaspard, 2007b). As a consequence, the mean reaction rates $\mathbf{J}(\mathbf{A})$ and the higher cumulants satisfy the expressions (5.142) of the fluctuation–dissipation theorem and the Onsager reciprocal relations (5.143) in the linear regime close to equilibrium, as well as their generalizations for the nonlinear response coefficients and the responses of the cumulants, in particular, given by equations (5.146), (5.148), and (5.150) (Andrieux and Gaspard, 2004).

Another consequence is that the quantity $\tilde{\Sigma}(t) \equiv \sum_{\alpha=1}^{\chi} A_\alpha n_\alpha$ obeys the asymptotic univariate fluctuation relation $p_t(\tilde{\Sigma}) \sim p_t(-\tilde{\Sigma}) \exp\tilde{\Sigma}$ for $t\to\infty$, so that its cumulant generating function defined as in equation (6.29) has the symmetry of equation (6.31). The univariate fluctuation relation $p_t(Z) \sim p_t(-Z)\exp Z$ is also satisfied by the quantity (11.32) (Gaspard, 2004a). Furthermore, the univariate fluctuation relation (6.24) holds exactly for the quantity (11.31) according to the results of Section 6.3.1 (Seifert, 2005a).

Fluctuation relations have been investigated for several types of reactive systems (Gaspard, 2004a; Andrieux and Gaspard, 2004; Seifert, 2005b, 2011; Min et al., 2005; Qian and Xie, 2006; Andrieux and Gaspard, 2008b,f; Rao and Esposito, 2018a; Xiao and Zhou, 2018).

11.3 Linear Reaction Networks

Reaction networks where the transition rates are some linear functions of the random variables $\{X_k\}$ have remarkable properties, as suggested by the results of Section 10.3 and Section F.4. This is, in particular, the case for the reaction network,

$$\mathrm{Y}_r \underset{W_{-r}}{\overset{W_{+r}}{\rightleftharpoons}} \mathrm{X} \qquad \text{with} \qquad r = 1,2,\ldots,m, \tag{11.55}$$

where X is the single intermediate species and Y_r are chemostatted. The number m of reactions coincides with the number e of chemostatted species in this network: $e = m$. The transition rates are defined as

$$W_{+r} = k_{+r}\, y_r\, \Omega \qquad \text{and} \qquad W_{-r}(X) = k_{-r}\, X \qquad \text{for} \qquad 1 \le r \le m, \tag{11.56}$$

in terms of the extensivity parameter $\Omega = n^0 V$ and with concentration units where $n^0 = 1$. Moreover, we may introduce the random variables $\mathbf{n} = \{n_i\}_{i=1}^{m-1}$ counting the particle numbers for the chemostatted species $\{Y_i\}_{i=1}^{m-1}$. The last chemostatted species Y_m is here considered as the product of the reactions used as reference in the definition of the independent affinities and currents. The master equation (11.8) ruling the time evolution of the probability distribution $P(X,\mathbf{n},t)$ is thus given by

$$\frac{d}{dt}P(X,\mathbf{n},t) = \sum_{i=1}^{m-1}\left[\left(e^{-\partial_X-\partial_{n_i}}-1\right)W_{+i}+\left(e^{+\partial_X+\partial_{n_i}}-1\right)W_{-i}(X)\right]P(X,\mathbf{n},t)$$
$$+\left[\left(e^{-\partial_X}-1\right)W_{+m}+\left(e^{+\partial_X}-1\right)W_{-m}(X)\right]P(X,\mathbf{n},t). \quad (11.57)$$

Since the transition rates (11.56) are constant or linear in the random variable X, its mean value obeys the linear kinetic equation

$$\frac{d}{dt}\langle X\rangle = \sum_{r=1}^{m}\left(k_{+r}\, y_r\, \Omega - k_{-r}\,\langle X\rangle\right). \quad (11.58)$$

This property would not hold if the transition rates were some nonlinear functions of the variables, in which case the ordinary differential equations for the mean values would be coupled to the evolution equations of higher statistical moments.

The linear dependence of the transition rates on the variable X has the other consequence that the stationary solution of the reduced master equation (11.9) for the distribution $\mathscr{P}(X,t)=\sum_{\mathbf{n}} P(X,\mathbf{n},t)$ is given by the Poisson distribution

$$\mathscr{P}_{\rm st}(X) = e^{-\langle X\rangle_{\rm st}}\frac{\langle X\rangle_{\rm st}}{X!}, \qquad \text{where} \qquad \langle X\rangle_{\rm st} = \frac{\sum_{r=1}^{m} k_{+r}y_r}{\sum_{r=1}^{m} k_{-r}}\,\Omega \quad (11.59)$$

is the stationary value of equation (11.58). The key point is that this result here holds even if the process is out of equilibrium.

The graph associated with the Markov jump process ruled by the reduced master equation (11.9) deduced from equation (11.57) is shown in Figure 11.2. The affinities of the reactive

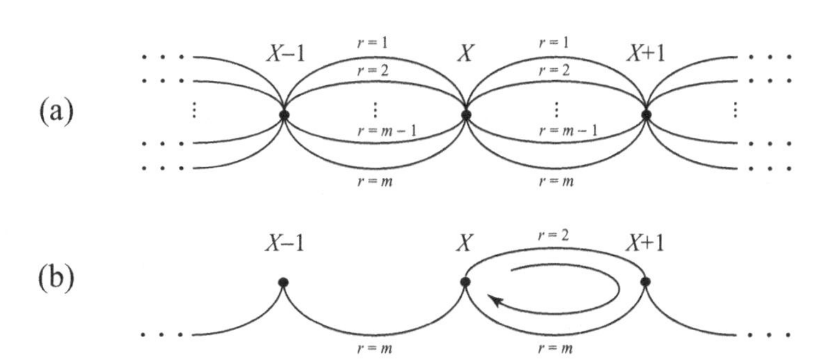

Figure 11.2 (a) Graph G associated with the linear reaction network (11.55). (b) Subgraph $T(G)+l$ composed of the maximal tree $T(G)$ formed by all the edges of the reaction $r=m$ and the chord $l = X \overset{r=2}{\rightarrow} X+1$, forming the stoichiometric cycle used to define the macroscopic affinity (11.60) for $r=2$. Reprinted with permission from Andrieux and Gaspard (2008f) © American Physical Society.

process can be obtained by considering equation (4.77) for cycles such as the one depicted in Figure 11.2(b):

$$A_i \equiv \ln \frac{k_{+i}\, y_i\, k_{-m}}{k_{-i}\, k_{+m}\, y_m} \qquad \text{for} \qquad 1 \le i \le m-1, \tag{11.60}$$

which are vanishing at equilibrium when all the Guldberg–Waage conditions hold, so that $(y_i/y_m)_{\rm eq} = K_m/K_i$ with the equilibrium constants $K_r \equiv k_{+r}/k_{-r}$.

The multivariate fluctuation relation (11.54) can here be established in the long-time limit with respect to these affinities by taking the following form for the moment generating function (11.50),

$$F(X, \boldsymbol{\lambda}, t) = {\rm e}^{-Qt}\, \frac{C^X}{X!}, \tag{11.61}$$

where the cumulant generating function Q and the coefficient C are determined by substituting this form into equation (11.51) with the counting parameters $\boldsymbol{\lambda} = (\lambda_1, \lambda_2, \ldots, \lambda_{m-1})$. Setting $\lambda_m = 0$, the cumulant generating function (11.52) can thus be expressed as

$$Q(\boldsymbol{\lambda}; \mathbf{A}) = \sum_{i,j=1}^{m} W_{ij}\left(1 - {\rm e}^{-\lambda_i + \lambda_j}\right) \tag{11.62}$$

with the rates

$$W_{ij} \equiv \frac{k_{+i}\, y_i\, k_{-j}}{\sum_{r=1}^{m} k_{-r}}\, \Omega = \frac{k_{+m}\, y_m}{k_{-m}}\, \frac{k_{-i}\, k_{-j}\, \Omega}{\sum_{r=1}^{m} k_{-r}}\, {\rm e}^{A_i}. \tag{11.63}$$

Accordingly, the mean currents and the diffusivities are given by

$$J_i = \sum_{j=1}^{m} (W_{ij} - W_{ji}), \qquad D_{ii} = \frac{1}{2} \sum_{j=1, j\neq i}^{m} (W_{ij} + W_{ji}), \qquad \text{and}$$

$$D_{ij} = -\frac{1}{2}(W_{ij} + W_{ji}) = D_{ji} \qquad \text{for} \qquad i, j = 1, 2, \ldots, m-1. \tag{11.64}$$

In terms of the affinities (11.60), the mean currents take the following forms:

$$J_i(\mathbf{A}) = \frac{k_{+m}\, y_m}{k_{-m}}\, \frac{k_{-i}\, \Omega}{\sum_{r=1}^{m} k_{-r}} \left[\sum_{j=1}^{m-1} k_{-j}\left({\rm e}^{A_i} - {\rm e}^{A_j}\right) + k_{-m}\left({\rm e}^{A_i} - 1\right) \right]. \tag{11.65}$$

Although the transition rates and the kinetic equation (11.58) are linear functions of the particle number X, the mean reaction rates have a strongly nonlinear dependence on the affinities, which is in accordance with the phenomenological relation (1.37) for the reaction rate density.

The symmetry relation (11.53) is satisfied with the affinities (11.60), so that the multivariate fluctuation relation (11.54) holds (Andrieux and Gaspard, 2004, 2008f). Among the implications of the multivariate fluctuation relation presented in Chapter 5, let us mention the Onsager reciprocal relations (5.143) in the linear regime close to equilibrium,

$$L_{i,j} = L_{j,i} = -\frac{k_{+m}\, y_m}{k_{-m}}\, \frac{k_{-i}\, k_{-j}\, \Omega}{\sum_{r=1}^{m} k_{-r}}, \tag{11.66}$$

as well their nonlinear generalizations (5.146) reading as

$$M_{i,jj} = 2R_{ij,j} = L_{i,j} \qquad \text{and} \qquad M_{i,ij} = R_{ii,j} + R_{ij,i} = 0 \tag{11.67}$$

for $i \neq j$, where $R_{ij,k} \equiv \partial_{A_k} D_{ij}(\mathbf{0})$ are the first responses of the diffusivities (Andrieux and Gaspard, 2004, 2008f).

Furthermore, as a consequence of the Poissonian character of the process, the finite-time multivariate fluctuation relation

$$\frac{\mathcal{P}_t(\mathbf{n})}{\mathcal{P}_t(-\mathbf{n})} = \exp(\mathbf{A}_t \cdot \mathbf{n}) \tag{11.68}$$

holds exactly at every time t for the probability distribution $\mathcal{P}_t(\mathbf{n}) = \sum_{X=0}^{\infty} P(X, \mathbf{n}, t)$ of the variables $\mathbf{n} = \{n_i\}_{i=1}^{m-1}$ counting the reactive events during the time lapse t, starting from the stationary distribution for the random variable X. This result is proved in Section F.5 (Andrieux and Gaspard, 2008f; Gu and Gaspard, 2020). The time-dependent affinities $\mathbf{A}_t$ converge in the limit $t \to \infty$ towards the values (11.60), which is consistent with the asymptotic form (11.54) of the fluctuation relation.

In the stationary state, the entropy production rate of the linear reaction network (11.55) is given by

$$\frac{1}{k_{\mathrm{B}}} \frac{d_{\mathrm{i}}S}{dt}\bigg|_{\mathrm{st}} = \sum_{i=1}^{m-1} A_i \, J_i(\mathbf{A}) \qquad \text{with} \qquad J_i(\mathbf{A}) = \Omega(k_{+i}\, y_i - k_{-i}\, x_{\mathrm{st}}) \geq 0, \tag{11.69}$$

where $x_{\mathrm{st}} = \langle X \rangle_{\mathrm{st}}/\Omega$ is the concentration corresponding to the mean value of the stationary distribution (11.59) in units where $n^0 = 1$. We note that the expression (11.69) coincides with the macroscopic value

$$\frac{1}{k_{\mathrm{B}}} \frac{d_{\mathrm{i}}S}{dt}\bigg|_{\mathrm{st}} = \Omega \sum_{r=1}^{m} (k_{+r}\, y_r - k_{-r}\, x_{\mathrm{st}}) \ln \frac{k_{+r}\, y_r}{k_{-r}\, x_{\mathrm{st}}}, \tag{11.70}$$

after using the definition (11.60) of the affinities.

In the linear regime close to equilibrium, the Markov jump process can be approximated by the corresponding chemical Langevin process ruled by the following stochastic differential equations,

$$\frac{dX}{dt} = \sum_{r=1}^{m} \mathscr{J}_r, \qquad \frac{dn_r}{dt} = \mathscr{J}_r, \qquad \mathscr{J}_r = k_{+r}\, y_r\, \Omega - k_{-r}\, X + \delta \mathscr{J}_r(t), \tag{11.71}$$

$\delta \mathscr{J}_r(t)$ being Gaussian white noises characterized by $\langle \delta \mathscr{J}_r(t) \rangle = 0$ and

$$\langle \delta \mathscr{J}_r(t)\, \delta \mathscr{J}_{r'}(t') \rangle = 2\, D_r\, \delta_{rr'}\, \delta(t - t'), \quad \text{where} \quad D_r = \frac{1}{2}\,(k_{+r}\, y_r\, \Omega + k_{-r}\, X) \tag{11.72}$$

is the diffusivity of the rth reaction. Since $dX/dt = \sum_{r=1}^{m} dn_r/dt$ in the reaction network (11.55), the path of the reactive process has the probability

$$P[\mathbf{n}(t)] \sim \exp\left[-\frac{1}{4} \int \mathsf{D}^{-1} : (\dot{\mathbf{n}} - \mathfrak{J})^2\, dt\right], \tag{11.73}$$

where $\mathbf{D} = (D_r \delta_{rr'})$ is the matrix of diffusivities and $\mathfrak{J} \simeq (k_{+r} y_r \Omega - k_{-r} X)_{r=1}^m$ the vector of extensive reaction rates up to negligible corrections for $\Omega \to \infty$. According to equation (11.31), the ratio of the probabilities of the path and its time reversal gives

$$\Sigma(t) = \ln \frac{P[\mathbf{n}(t)]}{P\left[\mathbf{n}^{\mathrm{R}}(t)\right]} \simeq \int \mathbf{D}^{-1} : \mathfrak{J}\, \dot{\mathbf{n}}\, dt, \tag{11.74}$$

since $\dot{\mathbf{n}}^{\mathrm{R}} = -\dot{\mathbf{n}}$. Consequently, equation (11.40) gives the entropy production rate

$$\frac{1}{k_{\mathrm{B}}} \frac{d_{\mathrm{i}} S}{dt}\bigg|_{\mathrm{st}} \simeq \mathbf{D}^{-1} : \mathfrak{J}^2\bigg|_{\mathrm{st}} = 2\Omega \sum_{r=1}^{m} \frac{(k_{+r}\, y_r - k_{-r}\, x_{\mathrm{st}})^2}{k_{+r}\, y_r + k_{-r}\, x_{\mathrm{st}}} \geq 0, \tag{11.75}$$

which is the quadratic approximation of the expression (11.70), as expected in the linear regime and the macroscopic limit $\Omega \to \infty$.

In the case with $m = 2$ reactions in the network (11.55), there is a single independent affinity driving the system away from equilibrium. The corresponding Markov jump process ruled by the master equation (11.57) with $m = 2$ is the same stochastic process as the particle exchange model of Section 4.4.6, which is also discussed in Section E.4. According to equation (11.63), the mean current (11.65) has the dependence $J(A) = W_{21}\left(\mathrm{e}^A - 1\right)$ on the affinity A, which is the same as in equation (4.96) for the particle exchange model, equation (5.245) for the random drift model of Section 4.4.6, equation (1.72) or (10.50) for homogeneous diffusion, and the current-voltage characteristic (10.115) for the electronic diode. The common feature of these systems is that there is a free energy difference larger than the thermal energy $k_{\mathrm{B}}T$ in the flux of particles between the reservoirs. In diffusive transport, this difference develops globally between the boundaries over spatial scales larger than the mean free path. In electronic diodes, this difference arises across the micrometric or submicrometric semiconducting junction. In chemical reactions, this difference happens over the atomic size of the molecular transformation. These systems can function as current rectifiers with a quality measured by the rectification ratio (5.170).

In the case with $m = 3$ reactions in the network (11.55), there are two independent affinities and mean currents, so that the reactive process is similar to the transistor model (5.248) (Andrieux and Gaspard, 2004; Gu and Gaspard, 2020).

Similar results hold for linear reaction networks with more than a single intermediate species.

11.4 Bistable Reaction Networks

Because of autocatalysis or cross-catalysis, reaction networks have transition rates that depends nonlinearly on the variables $\{X_k\}$. Far enough from equilibrium, such reaction networks may undergo bifurcations at the macroscale, as mentioned in Section 1.9. At such bifurcations, new stationary macrostates can emerge, leading to multistability in the deterministic kinetic equations of the reactive process. In the presence of fluctuations, the stable states of the deterministic description become metastable and characterized by mean dwell times before jumping towards other states (Nicolis and Prigogine, 1977; Demaeyer and Gaspard, 2009, 2013).

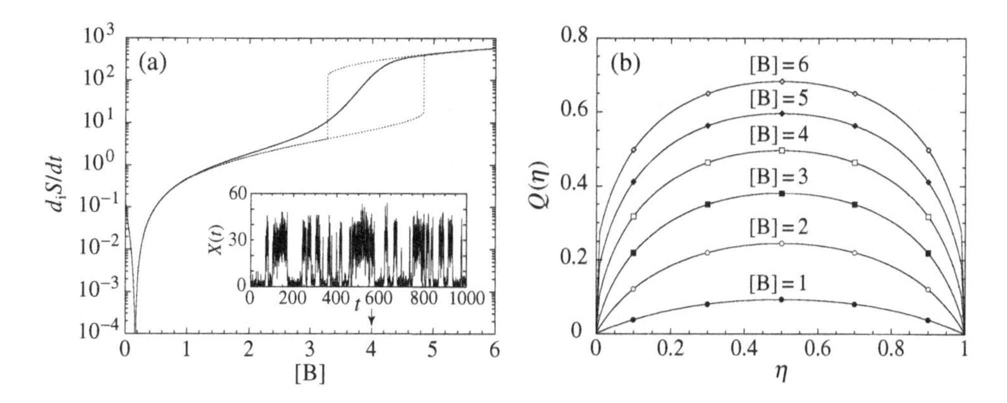

Figure 11.3 Schlögl model (11.76) with the parameter values $k_{+1}[\mathrm{A}] = 0.5$, $k_{-1} = 3$, and $k_{+2} = k_{-2} = 1$: (a) Entropy production rate versus [B], obtained by the macroscopic formula (11.38) (dashed lines) and by the stochastic description with equation (11.39) for $\Omega = 10$ (solid line). The thermodynamic equilibrium is located at $[\mathrm{B}]_{\mathrm{eq}} = 1/6$. Inset: Stochastic time evolution of the number $X(t)$ of molecules X for $[\mathrm{B}] = 4$ and $\Omega = 10$. (b) Cumulant generating function (11.81) of the fluctuating quantity (11.32) versus η for $[\mathrm{B}] = 1, 2, \ldots, 6$ and $\Omega = 10$. Reprinted from Gaspard (2004a) with the permission of AIP Publishing.

As an example, the phenomenon of bistability manifests itself far from equilibrium in the following trimolecular reaction network (Schlögl, 1971, 1972):

$$\mathrm{A} \underset{W_{-1}}{\overset{W_{+1}}{\rightleftharpoons}} \mathrm{X}, \qquad 3\,\mathrm{X} \underset{W_{-2}}{\overset{W_{+2}}{\rightleftharpoons}} 2\,\mathrm{X} + \mathrm{B}, \tag{11.76}$$

involving the chemostatted species A and B, and the intermediate species X. The transition rates are defined according to the mass action law (11.6) as

$$W_{+1}(X+1|X) = k_{+1}\,[\mathrm{A}]\,\Omega, \qquad W_{-1}(X-1|X) = k_{-1}\,X, \tag{11.77}$$
$$W_{+2}(X-1|X) = k_{+2}\,X\,\frac{X-1}{\Omega}\,\frac{X-2}{\Omega}, \qquad W_{-2}(X+1|X) = k_{-2}\,[\mathrm{B}]\,X\,\frac{X-1}{\Omega},$$

where [A] and [B] denote the concentrations of the chemostatted species in units where $n^0 = 1$, and $\Omega = n^0 V$ is the extensivity parameter. This network is autocatalytic because the intermediate species X accelerates its own synthesis in the reaction $r = -2$. We may also introduce a random variable n counting the number of chemostatted molecules A that are consumed upon the reactive events, such that $\Delta n_{\pm 1} = \pm 1$ and $\Delta n_{\pm 2} = 0$. For this stochastic process, the probability distribution $P(X,n,t)$ is ruled by the master equation (11.8) with $m = 2$. The marginal probability distribution $\mathscr{P}(X,t) = \sum_{n=-\infty}^{+\infty} P(X,n,t)$ obeys the reduced master equation (11.9).

Since the transition rates have nonlinear dependences on the variable X, the stationary distribution $\mathscr{P}_{\mathrm{st}}(X)$ is not Poissonian under nonequilibrium conditions (Nicolis, 1972; Nicolis and Prigogine, 1977; Gaspard, 2004a; Andrieux and Gaspard, 2007b). At equilibrium, the Poissonian distribution (11.12) is still the stationary solution of the reduced master equation.

According to the Hill–Schnakenberg network theory of Section 4.4.5, the graph associated with the Markov jump process of the model (11.76) over the internal states $X \in \mathbb{N}$ is the same as in Figure 11.2 with $m = 2$. Therefore, the reaction network is driven away from equilibrium by the single affinity defined as

$$A \equiv \ln \frac{W_{+1}(X+1|X)\, W_{+2}(X|X+1)}{W_{-1}(X|X+1)\, W_{-2}(X+1|X)} = \ln \frac{k_{+1}\, k_{+2}\, [\mathrm{A}]}{k_{-1}\, k_{-2}\, [\mathrm{B}]}, \tag{11.78}$$

which is independent of the number $X \in \mathbb{N}$.

The macroscopic kinetic equation for the concentration $x = [\mathrm{X}] = \langle X \rangle / \Omega$ of the intermediate species is given by

$$\frac{d}{dt}[\mathrm{X}] = \underbrace{k_{+1}[\mathrm{A}]}_{w_{+1}} - \underbrace{k_{-1}[\mathrm{X}]}_{w_{-1}} - \underbrace{k_{+2}[\mathrm{X}]^3}_{w_{+2}} + \underbrace{k_{-2}[\mathrm{B}][\mathrm{X}]^2}_{w_{-2}}, \tag{11.79}$$

which is obtained from the reduced master equation by neglecting the effects of fluctuations in the limit $\Omega \to \infty$. The equilibrium macrostate is determined using the detailed balance conditions $w_{+r} = w_{-r}$, so that the equilibrium concentrations are mutually related as

$$[\mathrm{X}]_{\mathrm{eq}} = \frac{k_{+1}}{k_{-1}}\, [\mathrm{A}]_{\mathrm{eq}} = \frac{k_{-2}}{k_{+2}}\, [\mathrm{B}]_{\mathrm{eq}}. \tag{11.80}$$

Away from equilibrium, the stationarity condition $d[\mathrm{X}]_{\mathrm{st}}/dt = 0$ for the kinetic equation (11.79) shows that the stationary concentration $[\mathrm{X}]_{\mathrm{st}}$ is given by the roots of a cubic polynomial. With generic values for the control parameters, there are thus one or three stationary concentrations. The system is monostable in the former case, and bistable in the latter, the third root being unstable. For the parameter values of Figure 11.3, the regime of bistability extends over the interval $3.28 < [\mathrm{B}] < 4.84$. An example of random trajectory simulated by the algorithm of Gillespie (1976, 1977) for $[\mathrm{B}] = 4$ in the bistable regime is shown in the inset of Figure 11.3(a), where we observe that the variable $X(t)$ randomly switches between lower and upper values distributed around the two metastable states of the reactive process. The weaker the noise, the longer the dwell times in the two metastable states. As mentioned in Section 4.6, the dwell times are growing exponentially with the extensivity parameter as $\tau_{\mathrm{d}} \sim \exp(\Omega \phi_0)$ (Demaeyer and Gaspard, 2009, 2013).

The thermodynamic entropy production rate of the model is plotted in Figure 11.3 as a function of the control parameter [B]. The dashed lines depict its macroscopic values given by equation (11.38), which is equal to zero at the equilibrium macrostate (11.80) and forms a hysteretic loop in the bistable regime. The mean stochastic value computed using the formula (11.39) is shown as the solid line, which essentially interpolates between the two macroscopic values in the bistable regime. Indeed, the macroscopic values are determined by the kinetic equation, which may admit multiple solutions, although the master equation (11.9) has a unique solution giving the stationary distribution $\mathscr{P}_{\mathrm{st}}(X)$. In the bistable regime, this distribution is bimodal with two peaks located around the two stable stationary solutions of the macroscopic kinetic equations. The respective weights of these peaks in the bimodal distribution give the probabilities of the two macrostates in the regime of bistability (Ge and Qian, 2009). In this way, the two levels of description are shown to be compatible with each other.

The entropy production (11.39) can also be computed as the mean growth rate of the quantity (11.32) according to equation (11.40). Along the paths of the reactive process, the quantity (11.32) fluctuates and obeys the fluctuation relation $p_t(Z) \sim p_t(-Z) \exp Z$, for $t \to \infty$, as previously mentioned. As a consequence, its cumulant generating function should satisfy the following symmetry relation,

$$\mathcal{Q}(\eta) \equiv \lim_{t\to\infty} -\frac{1}{t} \ln \langle e^{-\eta Z(t)} \rangle = \mathcal{Q}(1-\eta). \tag{11.81}$$

This symmetry is observed in Figure 11.3 not only close to equilibrium, but also in the regime of bistability far away from equilibrium. For the present Markov jump process, equation (11.81) is also the cumulant generating function (6.29) of the quantity (11.31) satisfying the fluctuation relation (6.24). Therefore, the verification of the symmetry (11.81) in the bistable reaction network confirms the validity of the fluctuation relation arbitrarily far from equilibrium (Gaspard, 2004a).

11.5 Noisy Chemical Clocks

Far from equilibrium, oscillatory dynamics may exist in nonlinear reaction networks because of autocatalysis or cross-catalysis. Such time-dependent regimes have been experimentally observed, in particular, in the Belousov–Zhabotinsky and related reactions, heterogeneous catalysis, and biochemical reactions (Bergé et al., 1984; Scott, 1991; Epstein and Pojman, 1998). If the oscillations are periodic, these reactions are called *chemical clocks*. The Brusselator is a reaction network, which has been proposed as a model of such nonequilibrium oscillations (Prigogine and Lefever, 1968). The reversible Brusselator reaction network is defined as

$$\mathrm{A} \underset{W_{-1}}{\overset{W_{+1}}{\rightleftharpoons}} \mathrm{X}, \qquad \mathrm{B}+\mathrm{X} \underset{W_{-2}}{\overset{W_{+2}}{\rightleftharpoons}} \mathrm{Y}+\mathrm{C}, \qquad 2\,\mathrm{X}+\mathrm{Y} \underset{W_{-3}}{\overset{W_{+3}}{\rightleftharpoons}} 3\,\mathrm{X}, \tag{11.82}$$

where A, B, and C are the chemostatted species, while X and Y are the two evolving intermediate species. According to the mass action law, the transition rates of this model are given by

$$W_{+1}(X+1,Y|X,Y) = k_{+1}\,[\mathrm{A}]\,\Omega, \tag{11.83}$$

$$W_{-1}(X-1,Y|X,Y) = k_{-1}X, \tag{11.84}$$

$$W_{+2}(X-1,Y+1|X,Y) = k_{+2}\,[\mathrm{B}]\,X, \tag{11.85}$$

$$W_{-2}(X+1,Y-1|X,Y) = k_{-2}\,[\mathrm{C}]\,Y, \tag{11.86}$$

$$W_{+3}(X+1,Y-1|X,Y) = k_{+3}\,\frac{X(X-1)Y}{\Omega^2}, \tag{11.87}$$

$$W_{-3}(X-1,Y+1|X,Y) = k_{-3}\,\frac{X(X-1)(X-2)}{\Omega^2}, \tag{11.88}$$

where X and Y are the particle numbers of the intermediate species; [A], [B], and [C] the concentrations of the chemostatted species; $k_{\pm r}$ the rate constants; and Ω is the extensivity parameter. The corresponding Markov jump process is ruled by the master equation (11.9).

The macroscopic kinetic equations for the concentrations $x = [\mathrm{X}]$ and $y = [\mathrm{Y}]$ have the forms:

$$\frac{dx}{dt} = k_{+1}\,[\mathrm{A}] - k_{-1}x - k_{+2}\,[\mathrm{B}]\,x + k_{-2}\,[\mathrm{C}]\,y + k_{+3}\,x^2\,y - k_{-3}\,x^3, \tag{11.89}$$

$$\frac{dy}{dt} = k_{+2}\,[\mathrm{B}]\,x - k_{-2}\,[\mathrm{C}]\,y - k_{+3}\,x^2\,y + k_{-3}\,x^3. \tag{11.90}$$

These equations admit a unique steady state for the following concentrations,

$$x_{\mathrm{st}} = \frac{k_{+1}}{k_{-1}}\,[\mathrm{A}] \quad \text{and} \quad y_{\mathrm{st}} = x_{\mathrm{st}}\,\frac{k_{+2}\,[\mathrm{B}] + k_{-3}\,x_{\mathrm{st}}^2}{k_{-2}\,[\mathrm{C}] + k_{+3}\,x_{\mathrm{st}}^2}. \tag{11.91}$$

The equilibrium macrostate is obtained using the conditions of detailed balance, giving

$$x_{\mathrm{eq}} = \frac{k_{+1}}{k_{-1}}\,[\mathrm{A}], \qquad y_{\mathrm{eq}} = \frac{k_{-3}\,k_{+1}}{k_{+3}\,k_{-1}}\,[\mathrm{A}], \qquad \text{and} \qquad k_{+3}k_{+2}\,[\mathrm{B}]_{\mathrm{eq}} = k_{-3}k_{-2}\,[\mathrm{C}]_{\mathrm{eq}}. \tag{11.92}$$

Since there is a single condition on the concentrations of the three chemostatted species, we should expect that there exists a single affinity driving the reaction network (11.82) away from equilibrium and a single corresponding current or rate. In order to identify this affinity, we may use the Hill–Schnakenberg network theory of Section 4.4.5. Here, the graph G of the Markov process is composed of the vertices $(X,Y) \in \mathbb{N}^2$ connected by the edges due to the transitions (11.83)–(11.88). In this graph, the cycle from (X,Y) to $(X-1,Y+1)$ by the reaction $+2$ and back to (X,Y) by the reaction $+3$ has the affinity

$$A \equiv \ln \frac{W_{+2}(X-1,Y+1|X,Y)\,W_{+3}(X,Y|X-1,Y+1)}{W_{-2}(X,Y|X-1,Y+1)\,W_{-3}(X-1,Y+1|X,Y)} = \ln \frac{k_{+3}\,k_{+2}\,[\mathrm{B}]}{k_{-3}\,k_{-2}\,[\mathrm{C}]}, \tag{11.93}$$

which is independent of the concentration [A] and equal to zero under the equilibrium condition (11.92). This affinity is associated with the overall reaction $\mathrm{B} \to \mathrm{C}$, so that we may introduce a corresponding counter n. The probability distribution $P(X,Y,n,t)$ is governed by the master equation (11.8) and the marginal distribution $\mathscr{P}(X,Y,t) = \sum_{n=-\infty}^{+\infty} P(X,Y,n,t)$ by the reduced master equation (11.9), both with the transition rates (11.83)–(11.88).

The linear stability analysis of the kinetic equations (11.89)–(11.90) shows that the steady state (11.91) is a stable node in the vicinity of equilibrium to become a stable focus further away for equilibrium. Next, this latter undergoes a Hopf bifurcation leading to a stable periodic attractor representing the oscillations of the macroscopic concentrations (Nicolis, 1995).

In this oscillatory regime, the Markov jump process can be simulated by the algorithm of Gillespie (1976, 1977) for different values of the extensivity parameter $\Omega = n^0 V$. As shown in Figure 11.4, the random trajectories become more and more regular and periodic as the extensivity parameter increases, which is expected since the deterministic time evolution governed by the macroscopic kinetic equations is reached in this limit. The issue is that, for finite values of the extensivity parameter Ω, the trajectories remain random, so that

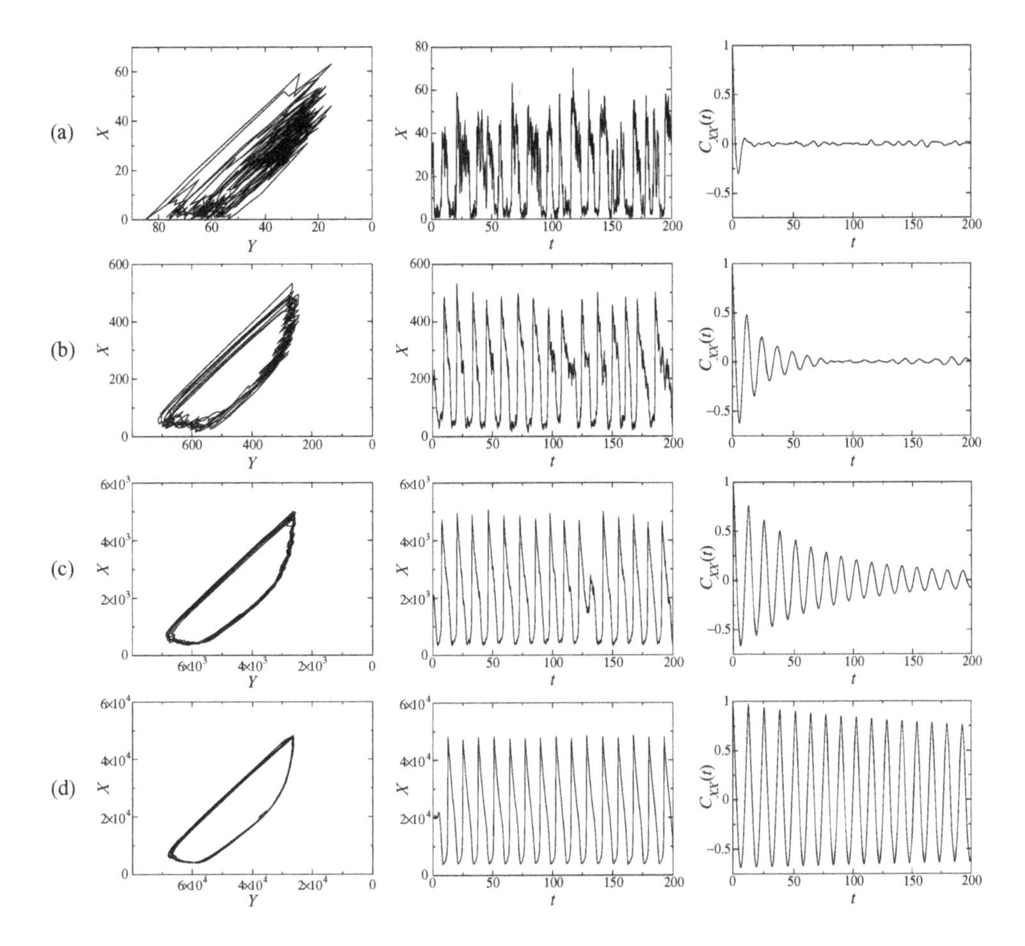

Figure 11.4 Reversible Brusselator model (11.82): Simulation of the oscillatory regime by the Gillespie algorithm. The parameter values are [B] = 7, [A] = [C] = 1, $k_{+1} = 0.5$, $k_{+2} = k_{+3} = 1$, and $k_{-1} = k_{-2} = k_{-3} = 0.25$. From top to bottom, the extensivity parameter takes the values: (a) $\Omega = 10$, (b) $\Omega = 100$, (c) $\Omega = 1000$, and (d) $\Omega = 10000$. For each value of Ω, the first column shows the phase portrait in the plane of the molecule numbers X and Y, the second one the number X as a function of time, and the third one the autocorrelation function (11.94) of the number X, which is normalized to unity. Reprinted from Andrieux and Gaspard (2008b) with the permission of AIP Publishing.

phase diffusion manifests itself in the oscillations and the autocorrelation functions of the concentrations typically present damped oscillations as

$$C_{XX}(t) = \langle X\rangle_{\rm st}^{-2}\,\langle X(t)\,X(0)\rangle_{\rm st} - 1 \sim \exp\left(-\frac{t}{\tau_{\rm c}}\right)\cos\left(\frac{2\pi}{T}\,t+\phi\right), \tag{11.94}$$

where $\tau_{\rm c}$ is the correlation time, T the period, and ϕ a phase. As explained in Section 4.6, the correlation time can be calculated using the Hamilton–Jacobi method in the weak-noise limit, leading to the formula (4.159) with $\epsilon = 1/\Omega$ (Gaspard, 2002a). Hence, the correlation time becomes infinite in the macroscopic limit $\Omega \to \infty$, in which case we recover a strictly periodic behavior. Otherwise the oscillations are affected by the molecular fluctuations and are irregular to some degree. The stochastic process can be considered as a chemical clock if the correlation time is longer than the period of oscillations. This condition is essentially equivalent to requiring that the quality factor (4.160) is larger than the unit value. According to the simulations of Figure 11.4, the minimum number of reactive molecules required for having a chemical clock may be evaluated to a few hundred (Gaspard, 2002a). Therefore, chemical clocks can exist at the nanoscale, as in the reaction of water production from dihydrogen and dioxygen on rhodium field emitter tips with a radius of curvature equal to 10–20 nm (McEwen et al., 2009). This reaction is observed to be oscillatory with a regular period of about 40 s, although with a few thousand adsorbed atoms on the rhodium surface tip. Such nanometric chemical clocks are also observed during the reaction of NO_2 with H_2 on platinum (Barroo et al., 2015). Similar considerations hold for the reaction networks of circadian rhythms in a single cell, where the number of regulatory molecules could be low enough to possibly affect the robustness of biochemical clocks (Gonze et al., 2002). We note that chemical clocks may exist even though the reaction network is functioning in a fully irreversible regime, where the reversed reactions have negligible rates and the entropy production rate is infinite (Gaspard, 2002a; Gonze et al., 2002; McEwen et al., 2009).

As previously mentioned, the nonequilibrium fluxes in the Brusselator reaction network is the feature of a single current corresponding to the overall reaction B $\to$ C and monitored with the counter n. If the forward and reversed reactions have positive rates, this counter obeys the fluctuation relation

$$\mathcal{P}_t(-n) \sim \mathcal{P}_t(n) \exp(-A\, n) \tag{11.95}$$

in the long-time limit $t \to \infty$, as proved in Section 11.2.5. This prediction is verified in Figure 11.5 for the Brusselator. In panel (a), the probability distribution $\mathcal{P}_t(n)$ is observed

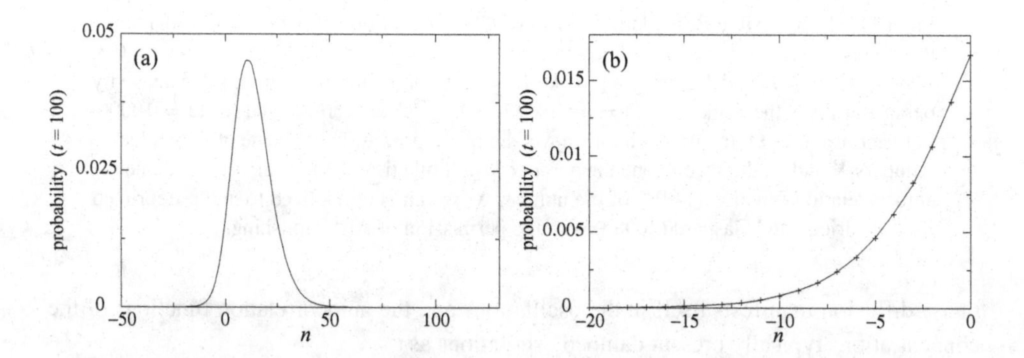

Figure 11.5 Reversible Brusselator model (11.82): (a) Probability distribution function for the cumulated current n over the time interval $t = 100$. The concentration [B] is fixed so that the affinity (11.93) takes the value $A = \ln 1.5$. The other parameters take the values $k_{+1} = 0.5$, $k_{+2} = k_{+3} = 1$, $k_{-1} = k_{-2} = k_{-3} = 0.25$, [A] = [C] = 1, and $\Omega = 1$. (b) Probabilities of the negative events (solid line) compared with the predictions (crosses) of the fluctuation relation (11.95). Reprinted from Andrieux and Gaspard (2008b) with the permission of AIP Publishing.

to have an asymmetry caused by the nonlinearity of the transition rates. In spite of this nonGaussian character of the distribution, the fluctuation relation (11.95) is nevertheless satisfied, as seen in Figure 11.5(b). Moreover, the symmetry (11.81) of the fluctuation relation for the quantity (11.32) is also satisfied from close to far from equilibrium in the oscillatory regime (Andrieux and Gaspard, 2008b). Since the entropy production is caused by the single current B $\to$ C, the growth of the fluctuating quantity (11.32) is related to the counter $n(t)$ according to $Z(t) \simeq A\, n(t)$ with the affinity (11.93), so that the fluctuation relation of the former is essentially equivalent to the one of the latter.

Similar results hold for reaction networks manifesting chemical chaos (Gaspard, 2020b).

11.6 Enzymatic Kinetics

Enzymes are proteins with the function of catalyzing specific biochemical reactions. For this purpose, enzymes have catalytic sites where substrate molecules S are transformed into product molecules P. The specificity of the enzymatic activity provides the control of the locus and timing of its functionality in biological organisms. Enzymes may have multiple binding sites, allowing cooperativity between the catalytic reaction of substrate S into product P and the presence of some other species with the role of activator or inhibitor for the reaction. Cooperativity is generated by conformal changes in the enzyme upon the binding of the activator or inhibitor to some dedicated site. These conformal changes may enhance or reduce the catalytic action of the other binding site achieving the catalysis of S into P, which is the feature of the so-called allosteric enzymes (Segel, 1975; Nicolis and Prigogine, 1977; Goldbeter, 1996; Chowdhury, 2013). Enzymes have nanometric sizes and they typically function in aqueous solutions where isothermal conditions are well maintained at the nanoscale.[4]

Figure 11.6(a) shows a simplified kinetic model for such an allosteric enzyme with two distinct sites, the one for the catalytic reaction S $\to$ P on its left-hand side and the other one for the binding of activator A or inhibitor I on its right-hand side. These two sites may be empty (0) or occupied (1) so that the enzyme has four internal states E_{00}, E_{10}, E_{01}, and E_{11}, which are, respectively, denoted $\sigma = 1, 2, 3, 4$. The enzymatic kinetics proceeds by the six following reactions,

$$\mathrm{E}_{00} + \mathrm{S} \underset{k_{-1}}{\overset{k_{+1}}{\rightleftharpoons}} \mathrm{E}_{10} \underset{k_{-2}}{\overset{k_{+2}}{\rightleftharpoons}} \mathrm{E}_{00} + \mathrm{P}, \tag{11.96}$$

$$\mathrm{E}_{01} + \mathrm{S} \underset{k_{-3}}{\overset{k_{+3}}{\rightleftharpoons}} \mathrm{E}_{11} \underset{k_{-4}}{\overset{k_{+4}}{\rightleftharpoons}} \mathrm{E}_{01} + \mathrm{P}, \tag{11.97}$$

$$\mathrm{E}_{00} + \mathrm{U} \underset{k_{-5}}{\overset{k_{+5}}{\rightleftharpoons}} \mathrm{E}_{01}, \qquad \mathrm{E}_{10} + \mathrm{U} \underset{k_{-6}}{\overset{k_{+6}}{\rightleftharpoons}} \mathrm{E}_{11}, \tag{11.98}$$

[4] Since water has the thermal diffusivity $D_{\rm th} \simeq 1.5 \times 10^{-7}$ m^2/s at 300 K (Probstein, 2003), the timescale of heat conduction over a distance of 10 nm is about $t_{\rm th} \sim 7 \times 10^{-10}$ s, while the mean lapse of time between the reactive events is typically in the range 10^{-6} s to 10^{-3} s, which justifies the isothermal assumption.

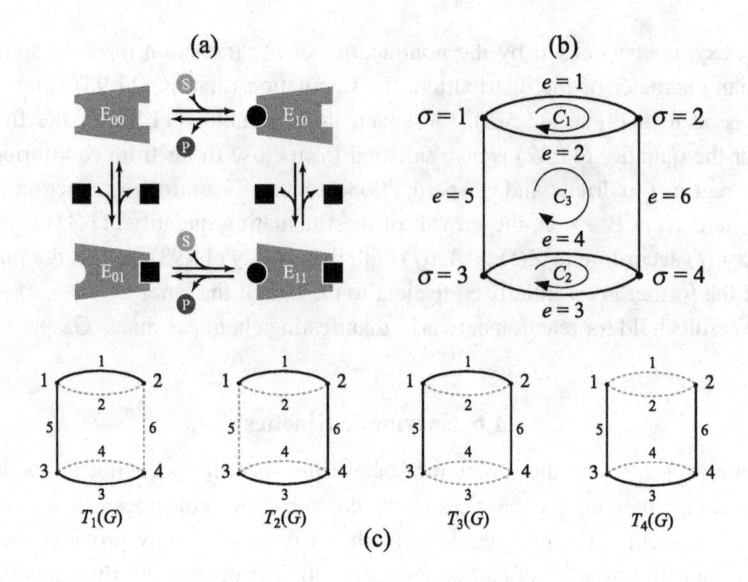

Figure 11.6 Model of enzymatic kinetics: (a) The enzyme has two distinct sites, one for the catalysis of substrate S into product P (circles), and the other one for the occupancy of an activator or inhibitor inducing allostery (squares). (b) Graph G associated with the Markov jump process describing the kinetics of this enzyme with its four states σ and its six edges e. The three cycles C_l of the graph are also shown. (c) Four examples of maximal trees $T_i(G)$ associated with the graph G.

where U = A or I, and $k_{\pm r}$ are the reaction rate constants. The environment of the enzyme is supposed to be large enough that the concentrations of substrate [S], product [P], and activator or inhibitor [U] remain constant in time during the whole stochastic process. Accordingly, the species S, P, and U are chemostatted, while the four internal states of the enzyme play the role of the intermediate species in the framework of Section 11.2. However, the reactive process is considered here for a single enzyme molecule instead of an ensemble of enzyme molecules. If the solution is dilute enough, the enzyme molecules do not interact with each other, so that the concentrations of the four states of the enzyme are proportional to the probabilities of finding a single enzyme molecule in its different states during the process. Moreover, the diffusivity of the chemostatted species is assumed to be fast enough that the process is limited by the reactions and not by diffusion.

The kinetics of this reaction network can thus be described as a Markov jump process for the internal states $\{\sigma\}$ of the enzyme together with an extra random variable representing the state of the environment supplying substrate molecules to drive the enzymatic activity. This random variable may be taken as the number n of substrate molecules that enter the catalytic site of the enzyme. Accordingly, the coarse-grained states of the stochastic process are given by $\omega = (\sigma, n)$ with $\sigma \in \{1,2,3,4\}$ and $n \in \mathbb{Z}$. The time evolution of the probabilities $P(\sigma,n,t)$ is ruled by the following coupled master equations,

$$\frac{d}{dt}P(1,n) = -J_1(n) + J_2(n) - J_5(n), \tag{11.99}$$

$$\frac{d}{dt}P(2,n) = J_1(n-1) - J_2(n) - J_6(n), \tag{11.100}$$

$$\frac{d}{dt}P(3,n) = -J_3(n) + J_4(n) + J_5(n), \tag{11.101}$$

$$\frac{d}{dt}P(4,n) = J_3(n-1) - J_4(n) + J_6(n), \tag{11.102}$$

expressed in terms of the probability currents of the six reactions:

$$J_1(n) = k_{+1}\,[\mathrm{S}]\,P(1,n) - k_{-1}\,P(2,n+1), \tag{11.103}$$

$$J_2(n) = k_{+2}\,P(2,n) - k_{-2}\,[\mathrm{P}]\,P(1,n), \tag{11.104}$$

$$J_3(n) = k_{+3}\,[\mathrm{S}]\,P(3,n) - k_{-3}\,P(4,n+1), \tag{11.105}$$

$$J_4(n) = k_{+4}\,P(4,n) - k_{-4}\,[\mathrm{P}]\,P(3,n), \tag{11.106}$$

$$J_5(n) = k_{+5}\,[\mathrm{U}]\,P(1,n) - k_{-5}\,P(3,n), \tag{11.107}$$

$$J_6(n) = k_{+6}\,[\mathrm{U}]\,P(2,n) - k_{-6}\,P(4,n), \tag{11.108}$$

where the dependence on time t is omitted for simplicity. As a consequence of the master equations, the total probability $\sum_{\sigma,n} P(\sigma,n,t) = 1$ is conserved.

An important property is that the stochastic process for the coarse-grained states $\omega = (\sigma,n)$ is nonstationary. However, summing over the number n, we obtain a stationary stochastic process for the probabilities $\mathscr{P}(\sigma,t) = \sum_{n=-\infty}^{+\infty} P(\sigma,n,t)$ of the internal states $\{\sigma\}$. The master equations for these probabilities are similar to equations (11.99)–(11.108), but without the dependence on the number n. The Markov jump process on the internal states can be analyzed using the network theory of Section 4.4.5 (Hill, 1989; Schnakenberg, 1976). The graph G associated with the process is shown in Figure 11.6(b). This graph has four vertices corresponding to the four internal states, and six edges corresponding to the six reactions. The states $\sigma = 1$ and $\sigma = 2$ are linked by two edges because of the reactions (11.96), and similarly for the states $\sigma = 3$ and $\sigma = 4$ because of the reactions (11.97). However, the states $\sigma = 1$ and $\sigma = 3$ are linked by only one edge, as well as the states $\sigma = 2$ and $\sigma = 4$, because of the reactions (11.98). Maximal trees are constructed from the graph G by removing as many edges as possible in order to still connect all the vertices, but without forming loops. Figure 11.6(c) shows four examples of maximal trees $T_i(G)$. The graph admits twelve maximal trees. Four of them are similar to $T_1(G)$, four to $T_2(G)$, two to $T_3(G)$, and two to $T_4(G)$.

The stationary probability distribution corresponding to thermodynamic equilibrium is identified with the detailed balance conditions, according to which the currents (11.103)–(11.108) summed over the integer variable n should be equal to zero, i.e., $\sum_{n=-\infty}^{+\infty} J_e(n)|_{\mathrm{eq}} = 0$. These conditions give

$$\frac{\mathscr{P}_{\mathrm{eq}}(2)}{\mathscr{P}_{\mathrm{eq}}(1)} = K_1\,[\mathrm{S}] = \frac{[\mathrm{P}]}{K_2}, \qquad \frac{\mathscr{P}_{\mathrm{eq}}(4)}{\mathscr{P}_{\mathrm{eq}}(3)} = K_3\,[\mathrm{S}] = \frac{[\mathrm{P}]}{K_4}, \tag{11.109}$$

$$\frac{\mathscr{P}_{\mathrm{eq}}(3)}{\mathscr{P}_{\mathrm{eq}}(1)} = K_5\,[\mathrm{U}], \qquad \frac{\mathscr{P}_{\mathrm{eq}}(4)}{\mathscr{P}_{\mathrm{eq}}(2)} = K_6\,[\mathrm{U}], \tag{11.110}$$

with the equilibrium constants $K_r \equiv k_{+r}/k_{-r}$. The compatibility between these four relations at equilibrium implies that the equilibrium concentrations of the substrate and product species should satisfy the following Guldberg–Waage condition,

$$\left.\frac{[\mathrm{P}]}{[\mathrm{S}]}\right|_{\mathrm{eq}} = K_1 K_2 = K_3 K_4, \tag{11.111}$$

and, moreover, that the further relation $K_5/K_6 = K_1/K_3$ should hold between these equilibrium constants. Accordingly, the following compatibility conditions,

$$\frac{K_1}{K_3} = \frac{K_4}{K_2} = \frac{K_5}{K_6}, \tag{11.112}$$

should be satisfied by the equilibrium constants of the six reactions. If these conditions are satisfied, an equilibrium state exists for any given concentrations [S] and [U] (or equivalently, for any given concentrations [P] and [U]).

However, the system is driven out of equilibrium if the substrate and production concentrations do not satisfy the Guldberg–Waage condition (11.111). In order to obtain the possible affinities driving the system into a nonequilibrium steady state, we consider the cycle decomposition using the maximal tree $T_1(G)$ in Figure 11.6(c). The cycle C_1 shown in Figure 11.6(b) is formed by closing with the edge $e = 2$, the cycle C_2 with the edge $e = 4$, and the cycle C_3 with $e = 6$.[5] The affinities of these cycles are respectively given by

$$A_{C_1} = \ln \frac{K_1 K_2 [\mathrm{S}]}{[\mathrm{P}]}, \tag{11.113}$$

$$A_{C_2} = \ln \frac{K_3 K_4 [\mathrm{S}]}{[\mathrm{P}]}, \tag{11.114}$$

$$A_{C_3} = \ln \frac{K_1 K_6}{K_3 K_5}, \tag{11.115}$$

according to equation (4.75) or (4.77). Now, the equilibrium compatibility conditions (11.112) imply that the affinity of the second cycle is identical to the one of the first cycle, $A_{C_1} = A_{C_2}$, and, moreover, that the affinity of the third cycle is equal to zero, whereupon there is only one affinity driving the system away from equilibrium, i.e., the one given by equation (11.113).[6]

In order to complete the study of the enzymatic kinetics, we need to come back to the master equations for the probabilities $P(\sigma,n,t)$. They can be decomposed in Fourier modes as in equation (4.79). After a long enough time t, the probabilities undergo the factorization as

$$P(\sigma,n,t) \simeq_{t\to\infty} \mathscr{P}_{\mathrm{st}}(\sigma)\, \mathcal{P}_t(n) \tag{11.116}$$

in terms of the stationary probability distribution $\mathscr{P}_{\mathrm{st}}(\sigma)$ of the Markov jump process over the internal states $\{\sigma\}$, and the probability $\mathcal{P}_t(n)$ that n substrate molecules have entered

[5] We note that these three cycles can also be obtained by considering the null right eigenvectors (4.70) of the matrix composed of the elements (4.63).

[6] In this regard, we note that the reaction network (11.96)–(11.98) is conserving the quantities $E_{00} + E_{01} + E_{10} + E_{11}$, $S + P + E_{10} + E_{11}$, and $U + E_{01} + E_{11}$.

the catalytic site. The master equation for this latter can be obtained by summing the coupled master equations (11.99)–(11.102) now over the internal states $\{\sigma\}$, which gives the same master equation (4.78) as for the random drift of Section 4.4.6, but here with the jump rates

$$W_+ \equiv k_{+1}\,[\mathrm{S}]\,\mathscr{P}_{\mathrm{st}}(1) + k_{+3}\,[\mathrm{S}]\,\mathscr{P}_{\mathrm{st}}(3), \tag{11.117}$$

$$W_- \equiv k_{-1}\,\mathscr{P}_{\mathrm{st}}(2) + k_{-3}\,\mathscr{P}_{\mathrm{st}}(4). \tag{11.118}$$

Consequently, the probability distribution $\mathcal{P}_t(n)$ behaves as given in equation (4.83) after a long enough time (with $a = 1$ since particles instead of displacement steps are counted). The overall reaction rate, i.e., the substrate consumption rate by the reaction network, is thus given by $\langle \dot{n}\rangle_{\mathrm{st}} = W_+ - W_-$. As a consequence of expressions (11.117) and (11.118) for the jump rates, the overall reaction rate is equivalently given by $\langle \dot{n}\rangle_{\mathrm{st}} = W_+ - W_- = J_{C_1} + J_{C_2}$, where

$$J_{C_1} = k_{+1}\,[\mathrm{S}]\,\mathscr{P}_{\mathrm{st}}(1) - k_{-1}\,\mathscr{P}_{\mathrm{st}}(2), \tag{11.119}$$

$$J_{C_2} = k_{+3}\,[\mathrm{S}]\,\mathscr{P}_{\mathrm{st}}(3) - k_{-3}\,\mathscr{P}_{\mathrm{st}}(4), \tag{11.120}$$

are the mean values of the probability currents on the edges, respectively, closing the cycles C_1 and C_2 in the graph of Figure 11.6(b) for the stationary Markov jump process over the internal states.

Figure 11.7 shows a random time evolution for the enzymatic process simulated by Gillespie's algorithm. The number $n(t)$ of consumed substrate molecules is depicted jointly with the dichotomic random occupancies of the two enzymatic sites. These latter random

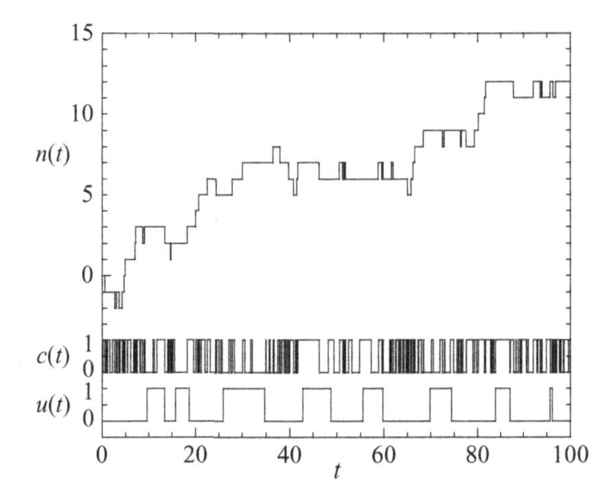

Figure 11.7 Random time evolution for the enzymatic model (11.96)–(11.98) with the parameter values $[\mathrm{S}] = 1$, $[\mathrm{P}] = 2$, $[\mathrm{U}] = 0.2$, $k_{+1} = 1$, $k_{+2} = 2$, $k_{+3} = 0.4$, $k_{+4} = 0.8$, $k_{+5} = 0.5$, $k_{+6} = 0.7$, and $k_{-r} = k_{+r}/2$ for $r = 1$–6. The system is out of equilibrium with the affinities $A_{C_1} = A_{C_2} = \ln 2$ and $A_{C_3} = 0$. The mean overall reaction rate has the value $\langle \dot{n}\rangle_{\mathrm{st}} = J_{C_1} + J_{C_2} \simeq 0.15$. Here, $n(t)$ is the number of consumed substrate molecules versus time t, $c(t)$ the occupancy of the catalytic site, and $u(t)$ the occupancy of the U binding site.

variables have the mean stationary values $\langle c\rangle_{\rm st} = \mathscr{P}_{\rm st}(2) + \mathscr{P}_{\rm st}(4)$ and $\langle u\rangle_{\rm st} = \mathscr{P}_{\rm st}(3) + \mathscr{P}_{\rm st}(4)$. The parameter values are such that the reaction proceeds away from equilibrium, as evidenced by the mean positive slope of the random function $n(t)$ of time, giving the mean overall reaction rate $\langle \dot{n}\rangle_{\rm st} = J_{C_1} + J_{C_2}$. We note that fuel consumption driving the system out of equilibrium is monitored by the nonstationary variable $n(t)$, which behaves as the random drifts of Figure 4.3 depending on the value of the affinity $A = A_{C_1} = A_{C_2}$.

At equilibrium, the stationary probability distribution satisfies the detailed balance conditions (11.109)–(11.110), whereupon the rates are equal $W_+ = W_-$ and the overall reaction rate is equal to zero, i.e., $\langle \dot{n}\rangle_{\rm st} = 0$, as required.

The entropy production rate of the enzymatic kinetics is given by equation (4.57) where the sum extends over the six reactions $r = 1$–6 and the coarse-grained states $\omega = (\sigma, n)$. After a long enough time, the factorization (11.116) and the property $\lim_{t\to\infty} \mathcal{P}_t(n)/\mathcal{P}_t(n+1) = 1$ resulting from equation (4.83) have for consequence that the entropy production rate converges towards its asymptotic value given by considering the stationary Markov jump process over the internal states $\{\sigma\}$. Using equation (4.74) and the results $A_{C_1} = A_{C_2}$ and $A_{C_3} = 0$ for the affinities of the process, we thus obtain the asymptotic entropy production rate

$$\lim_{t\to\infty} \frac{1}{k_{\rm B}} \frac{d_{\rm i}S}{dt} = \sum_i A_{C_i}\, J_{C_i} = A_{C_1}\left(J_{C_1} + J_{C_2}\right) = A_{C_1}\, \langle \dot{n}\rangle_{\rm st} \geq 0, \tag{11.121}$$

with the mean overall reaction rate $\langle \dot{n}\rangle_{\rm st} = J_{C_1} + J_{C_2}$. This entropy production rate is equal to zero at equilibrium and positive in nonequilibrium steady states for the enzymatic kinetics.

If the enzyme is considered as an open reactor supplied with substrate and product molecules by the surrounding solution, the mean rate of free energy consumed to drive the enzymatic activity can be evaluated as follows. If the solution is dilute, the Gibbs free energy of the species participating in the overall reaction $\rm S \to P$ can be expressed as $\mathcal{G} = \mu_{\rm S} N_{\rm S} + \mu_{\rm P} N_{\rm P}$ in terms of the numbers N_k of the molecules of species k, and their chemical potentials $\mu_k = \mu_k^0 + k_{\rm B}T \ln\left([k]/n^0\right)$, where $[k]$ denotes their density or concentration (k = S or P). If the solution is large enough to keep constant the chemical potentials $\mu_{\rm S}$ and $\mu_{\rm P}$, the time derivative of this free energy is given by $\dot{\mathcal{G}} = \mu_{\rm S}\dot{N}_{\rm S} + \mu_{\rm P}\dot{N}_{\rm P}$. In the long-time limit, the overall reaction rate reaches the stationary value $\langle \dot{n}\rangle_{\rm st} = J_{C_1} + J_{C_2} = -\left\langle \dot{N}_{\rm S}\right\rangle_{\rm st} = \left\langle \dot{N}_{\rm P}\right\rangle_{\rm st}$ given by the currents (11.119) and (11.120). Under such nonequilibrium conditions, the free energy of the solution decreases in time as $\left\langle \dot{\mathcal{G}}\right\rangle_{\rm st} = \mu_{\rm S}\left\langle \dot{N}_{\rm S}\right\rangle_{\rm st} + \mu_{\rm P}\left\langle \dot{N}_{\rm P}\right\rangle_{\rm st} = -\Delta\mu\, \langle \dot{n}\rangle_{\rm st}$, where $\Delta\mu \equiv \mu_{\rm S} - \mu_{\rm P}$ is the chemical potential difference or free energy of the reaction $\rm S \to P$. This latter is related to the affinity (11.113) according to $\Delta\mu = k_{\rm B}T A_{C_1}$, which vanishes at chemical equilibrium when the Guldberg–Waage condition (11.111) is satisfied, as required. In nonequilibrium steady states, the net free-energy power supplied to the enzyme by the surrounding solution is thus obtained as

$$\left.\frac{d_{\rm e}G}{dt}\right|_{\rm st} = -\left\langle \dot{\mathcal{G}}\right\rangle_{\rm st} = k_{\rm B}T\, A_{C_1}\, \langle \dot{n}\rangle_{\rm st} = T\left.\frac{d_{\rm i}S}{dt}\right|_{\rm st} \geq 0 \tag{11.122}$$

in terms of the entropy production rate (11.121).

For the enzymatic kinetics (11.96)–(11.98), the probability distribution $\mathcal{P}_t(n) \equiv \sum_\sigma P(\sigma,n,t)$ that n reactive events S $\to$ P have occurred and its cumulant generating function $Q(\lambda;A) \equiv \lim_{t\to\infty}(-1/t)\ln\langle\exp(-\lambda n)\rangle_t$ obey the symmetry of the fluctuation relation

$$\frac{\mathcal{P}_t(n)}{\mathcal{P}_t(-n)} \sim_{t\to\infty} \exp(An) \qquad \text{or, equivalently,} \qquad Q(\lambda;A) = Q(A-\lambda;A) \tag{11.123}$$

with the affinity $A = A_{C_1} = A_{C_2}$ driving the kinetics out of equilibrium. This result can be established using the method of Section 5.7. Indeed, the master equations (11.99)–(11.108) can be expressed in matrix form as

$$\frac{d}{dt}\mathbf{P}(n,t) = \left(\mathsf{L}^{(+)}\,\mathrm{e}^{-\partial_n} + \mathsf{L}^{(0)} + \mathsf{L}^{(-)}\,\mathrm{e}^{+\partial_n}\right)\cdot\mathbf{P}(n,t) \tag{11.124}$$

for the probability vector $\mathbf{P}(n,t) \equiv \{P(\sigma,n,t)\}_{\sigma=1}^4$. The time evolution of the generating vector $\mathbf{F}(\lambda,t) \equiv \sum_{n=-\infty}^{+\infty}\mathbf{P}(n,t)\,\mathrm{e}^{-\lambda n}$ is ruled by equation (5.214) with the modified operator here given by the 4×4 matrix

$$\mathsf{L}_\lambda = \mathsf{L}^{(+)}\,\mathrm{e}^{-\lambda} + \mathsf{L}^{(0)} + \mathsf{L}^{(-)}\,\mathrm{e}^{+\lambda}, \tag{11.125}$$

where the only nonvanishing elements of the matrices $\mathsf{L}^{(\pm)}$ are $\mathsf{L}^{(+)}(2,1) = k_{+1}[\mathrm{S}]$, $\mathsf{L}^{(+)}(4,3) = k_{+3}[\mathrm{S}]$, $\mathsf{L}^{(-)}(1,2) = k_{-1}$, and $\mathsf{L}^{(-)}(3,4) = k_{-3}$. As a consequence of the compatibility conditions (11.112), this modified operator satisfies the symmetry

$$\mathsf{M}^{-1}\cdot\mathsf{L}_\lambda\cdot\mathsf{M} = \mathsf{L}_{A-\lambda}^{\mathrm{T}} \tag{11.126}$$

with the diagonal matrix M composed of the elements $\{M(\sigma)\}_{\sigma=1}^4$ with $M(1)=1$, $M(2) = [\mathrm{P}]/K_2$, $M(3) = K_5[\mathrm{U}]$, and $M(4) = K_5[\mathrm{U}][\mathrm{P}]/K_4$.[7] Therefore, the cumulant generating function obeys the symmetry relation of equation (11.123) and the fluctuation relation holds in the limit $t\to\infty$.

This cumulant generating function is plotted in Figure 11.8 for different values of the control parameters, confirming the symmetry (11.123) of the fluctuation relation for these parameter values. The slope of the cumulant generating function at $\lambda = 0$ gives the mean rate of the overall reaction S $\to$ P, $\langle\dot{n}\rangle_{\mathrm{st}} = \partial_\lambda Q(0;A) = J_{C_1} + J_{C_2}$. In Figure 11.8, the values of the rate constants are taken by assuming that the species U is an activator of the enzymatic reaction S $\to$ P, so that the slope at $\lambda = 0$ increases with the concentration [U] in panel (a). As seen in Figure 11.8(b), the mean reaction rate also increases with the affinity, while the symmetry of the fluctuation relation is always satisfied.

In general, the fluctuation relation (11.123) allows us to determine the thermodynamic driving force of enzymatic kinetics at the nanoscale of a single enzyme (Min et al., 2005; Qian and Xie, 2006).

In the absence of activator or inhibitor, i.e., for [U] = 0, the kinetic scheme (11.96)–(11.98) reduces to the sole equation (11.96) holding for unireactant enzymes (Segel, 1975). Indeed, the stationary probabilities of the internal states E_{01} and E_{11} and thus the probability

[7] We note that, at equilibrium, the matrix (11.125) with $\lambda = 0$ satisfies the reversibility condition $\mathsf{L}_0^{\mathrm{T}} = \mathsf{M}^{-1}\cdot\mathsf{L}_0\cdot\mathsf{M}$. Out of equilibrium, the relation $\mathsf{L}^{(+)\mathrm{T}} = \mathrm{e}^A\,\mathsf{M}^{-1}\cdot\mathsf{L}^{(-)}\cdot\mathsf{M}$ is also satisfied, so that equation (11.126) can be deduced.

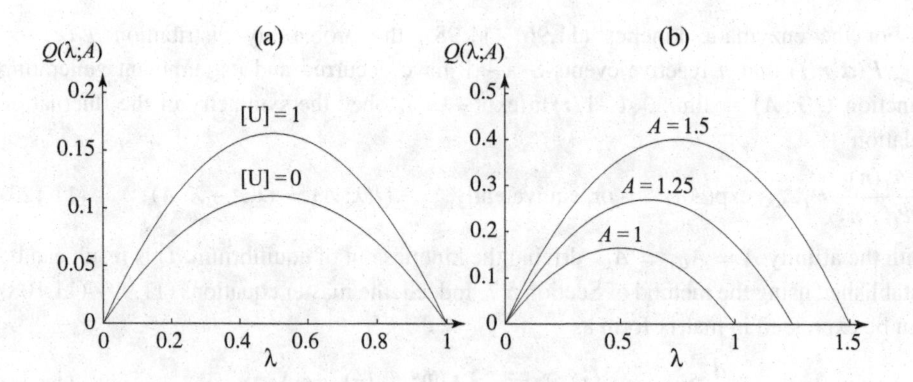

Figure 11.8 The cumulant generating function $Q(\lambda; A)$ of the counter n in the enzymatic model (11.96)–(11.98) with the parameter values $[\mathrm{P}] = 1$, $k_{\pm 1} = k_{\pm 5} = 1$, $k_{\pm 2} = k_{\pm 3} = k_{\pm 6} = 2$, and $k_{\pm 4} = 4$: (a) for the affinity $A = 1$ and two values of [U]; (b) for $[\mathrm{U}] = 1$ and three values of the affinity A.

current (11.120) are equal to zero in this case: $\mathscr{P}_{\rm st}(3) = \mathscr{P}_{\rm st}(4) = 0$ and $J_{C_2} = 0$. Accordingly, the asymptotic value of the overall reaction rate reduces to $\langle \dot{n} \rangle_{\rm st} = J_{C_1}$ and only the cycle C_1 in Figure 11.6(b) is active. In such nonequilibrium steady regimes, the probabilities of the states E_{00} and E_{10} are given by

$$\mathscr{P}_{\rm st}(1) = \frac{k_{-1} + k_{+2}}{k_{+1}[\mathrm{S}] + k_{-1} + k_{+2} + k_{-2}[\mathrm{P}]}, \tag{11.127}$$

$$\mathscr{P}_{\rm st}(2) = \frac{k_{+1}[\mathrm{S}] + k_{-2}[\mathrm{P}]}{k_{+1}[\mathrm{S}] + k_{-1} + k_{+2} + k_{-2}[\mathrm{P}]}, \tag{11.128}$$

and the overall reaction rate by the probability current (11.119) here equal to

$$\langle \dot{n} \rangle_{\rm st} = J_{C_1} = \frac{k_{+1}k_{+2}[\mathrm{S}] - k_{-1}k_{-2}[\mathrm{P}]}{k_{+1}[\mathrm{S}] + k_{-1} + k_{+2} + k_{-2}[\mathrm{P}]}, \tag{11.129}$$

which is the expression of the stationary reaction rate for the kinetics of unireactant enzymes including forward and reverse reactions (Segel, 1975). The famous reaction rate of Michaelis and Menten (1913) is recovered if the product concentration is equal to zero, i.e., $[\mathrm{P}] = 0$. In this limit, the kinetics is fully irreversible since the entropy production rate (11.121) becomes infinite. At equilibrium, the rate (11.129) is equal to zero together with the affinity (11.113) and the entropy production rate (11.121), because of the Guldberg–Waage condition (11.111).

Other types of enzymatic kinetics can be similarly described (Segel, 1975; Wachtel et al., 2018). Similar considerations apply to the kinetics of transmembrane ion channels (Schnakenberg, 1976; Andrieux and Gaspard, 2006a).

11.7 Copolymerization Processes

Copolymers are macromolecules composed of several species of monomeric units forming a chain, which may, in principle, be arbitrarily long. Examples are styrene-butadiene rubber, polyolefins such as ethylene-octene copolymers, polypeptide chains of amino-acids forming proteins, or DNA and RNA molecules composed of nucleotides (Pauling, 1970; Moore, 1972). If all the monomeric units are of the same species, the macromolecule is a pure polymer. The primary structure of a copolymer is characterized not only by the number, but also by the sequence of monomeric units in the chain, which may or may not contain information.

Copolymers are grown in reactions called copolymerization by the formation of covalent bonds between monomers or small polymers. An important mechanism consists of the successive attachments of monomers to one end of the growing chain, either by the presence of a catalyst, or because this end is chemically active. The monomers arrive from the surrounding solution with rates depending on their concentrations. The growth process may continue as long as the solution is not depleted of monomers and if there is no reaction terminating the copolymerization, which is said to be living under such circumstances. These copolymerization processes evolve out of equilibrium since the chain lengths are, on average, growing. At equilibrium, the chain lengths fluctuate and their distribution is stationary because of detailed balance between monomeric attachments and detachments, and the entropy production rate is equal to zero. As the copolymerization process is driven away from equilibrium, for instance, by increasing the concentrations of monomers, their attachments are more frequent and the chains grow because of less frequent monomeric detachments, so that the chain length has a positive mean velocity or elongation rate. In the fully irreversible limit, the detachment rates become negligible, the growth velocity is thus determined only by the attachment rates, and the entropy production rate is infinite. We note that if the concentrations of the monomers are too low, copolymers cannot grow and, instead, they may undergo depolymerization and dissolve into the surrounding solution if they would be initially present in such solutions.[8]

Once the primary structure is synthesized, polymers or copolymers may form secondary structures characterized by the three-dimensional configuration of their chain, depending on weak bonds between monomeric units on different parts of the chain and the properties of the surrounding solution. Here, the focus is on the formation of the primary structure of copolymers during copolymerization.

Copolymerization can proceed freely as for the formation of styrene-butadiene rubber, or be directed by a template as for the biological copolymers that are DNA, RNA, and proteins, as shown in Figure 11.9. In the former case, the sequence of monomeric units is more or less random depending on the attachment and detachment rates of the monomers randomly arriving from the surrounding solution. In the latter case, the copolymer grows along a template, so that the monomeric units can form preliminary pairs with the units of the template, before being attached to the growing chain. By this mechanism, the growing chain may have a sequence that is correlated with the sequence of the template, possibly achieving the copy of the template with some degree of fidelity depending on the frequency

[8] In such depolymerization processes, the mean growth velocity is negative and information erasure is ruled by Landauer's principle (Andrieux and Gaspard, 2013).

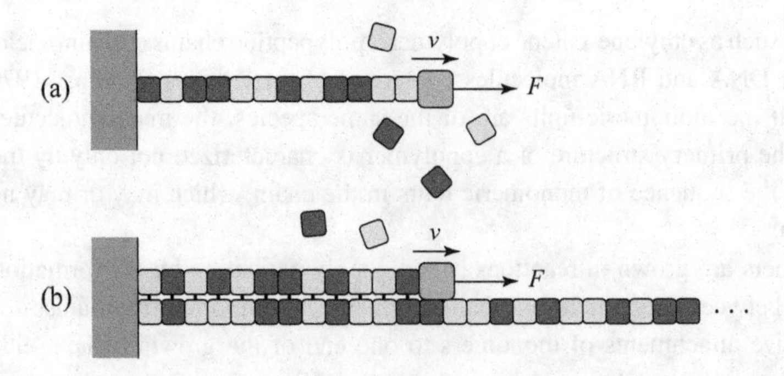

Figure 11.9 Schematic representations of copolymerization processes by a catalyst for the incorporation of monomers into the growing chain: (a) free copolymerization, (b) template-directed copolymerization. Here, v denotes the mean growth velocity and F an external force exerted on the catalyst.

of copying errors. In this regard, template-directed copolymerization is a basic mechanism for information transmission at the molecular scale.

Remarkably, the entropy production rate of copolymerization processes is fundamentally related to the statistical properties of the copolymer sequences (Andrieux and Gaspard, 2008d).

11.7.1 Free Copolymerization

We first consider free copolymerization, as depicted in Figure 11.9(a). The copolymer is assumed to be composed of M different species of monomeric units $m \in \{1, 2, \ldots, M\}$. The sequence of monomeric units is denoted ω and $l = |\omega|$ is its length counted by the number of units in the sequence. In living copolymerization processes, the copolymers are subjected to attachment and detachment reactions forming the following reaction network,

$$\omega' = m_1 m_2 \cdots m_{l-1} + m_l \underset{W_{-m_l|\omega}}{\overset{W_{+m_l|\omega'}}{\rightleftharpoons}} \omega = m_1 m_2 \cdots m_{l-1} m_l, \tag{11.130}$$

where m_j denotes the monomeric unit at the jth location in the chain, and m_l is the monomer that is attached to the chain ω' with the attachment rate $W_{+m_l|\omega'}$ or detached from the chain ω with the detachment rate $W_{-m_l|\omega}$. In general, these rates depend on the monomeric species involved in the reaction, as well as on the other monomeric units at the end of the copolymer. The rates also depend on the concentrations of monomers in the solution. Since the copolymers grow by the attachment of monomers, their concentration will decrease if the reaction proceeds in a closed reactor, where the system will ultimately reached equilibrium. In order to maintain the nonequilibrium conditions of the growth, the reactor should be fed from outside, or the rates should be low enough and the monomers in excess for their concentrations to remain quasi invariant during the period of observation. Under such low conversion conditions, the concentrations of the monomers may be assumed to have fixed values controlling the nonequilibrium constraints with respect to chemical equilibrium.

If the solution is dilute enough, the copolymer chains are isolated from each other, so that the concentrations $[\omega]$ of the sequences ω are proportional to the corresponding probabilities $P(\omega)$ according to $[\omega] = N P(\omega)/V$, where V is the volume of the solution and N the total number of growing copolymer chains. With these assumptions, the kinetic equations of the copolymerization process are equivalent to the master equations

$$\frac{d}{dt} P_t(\omega) = \sum_{\omega'} \left[P_t(\omega')\, W(\omega' \to \omega) - P_t(\omega)\, W(\omega \to \omega') \right], \tag{11.131}$$

ruling the time evolution of the probability distribution $P_t(\omega)$ to find a single copolymer with the sequence ω at time t and preserving the normalization condition $\sum_\omega P_t(\omega) = 1$. In this way, the copolymerization process is considered as a stochastic Markov jump process.

In equation (11.131), $W(\omega' \to \omega)$ and $W(\omega \to \omega')$ denote the transition rates of the forward and reversed reactions (11.130). Under isobaric and isothermal conditions, these transition rates should obey the conditions (4.59) in terms of the Gibbs free energy $G(\omega)$ of the copolymer with the sequence ω and the length $l = |\omega|$ having a quasi equilibrium configuration in the solution during the time lapse between two successive reactive events changing its sequence. The enthalpy $H(\omega)$ as well as the entropy $S^0(\omega)$ can be similarly defined for the copolymer ω, and the three thermodynamic quantities are related to each other by equation (4.58). The full thermodynamic entropy of the system is given by equation (4.45), including the contribution of Shannon disorder (4.46) in the statistical ensemble of probability distribution $P_t(\omega)$ for the copolymers with different sequences and lengths. In this regard, we note that the monomeric and thermodynamic exchanges between the copolymer and the surrounding solution composing its environment can be monitored with the coarse-grained states specified by the copolymer sequence ω because the transitions (11.130) between them uniquely determine the fluxes of monomers from the solution to the copolymer and back.

Since the monomeric concentrations are assumed to be invariant, the attachment and detachment rates remain constant in time, so that the process may reach a regime of steady growth characterized by a constant mean elongation rate

$$\mathscr{R} = \frac{d\langle l \rangle_t}{dt} = \lim_{t\to\infty} \frac{1}{t} \langle l \rangle_t \qquad \text{with} \qquad \langle l \rangle_t = \sum_{\omega} P_t(\omega)\, |\omega|. \tag{11.132}$$

In this regime, the probability distribution can be factorized as

$$P_t(\omega) \simeq p_t(l)\, \mu_l(\omega) \qquad \text{for} \qquad t \to \infty \tag{11.133}$$

into the probability distribution of the lengths $p_t(l) \equiv \sum_{\omega:\, |\omega|=l} P_t(\omega)$ and the one of the sequences ω provided that the length is equal to $l = |\omega|$, which is denoted $\mu_l(\omega)$ and normalized according to $\sum_\omega \mu_l(\omega) = 1$ (Coleman and Fox, 1963a,b). The key point of this factorization is that the former depends on time, but not the latter. On the one hand, by the central limit theorem, the time-dependent length distribution is Gaussian corresponding to a mean drift at the elongation rate (11.132) with some diffusivity of the length l. On the other hand, the invariant sequence distribution $\mu_l(\omega)$ characterizes the statistical properties of the copolymer sequences, such as the occurrence frequencies of the different monomeric units and subsequences like doublets, triplets, etc. Moreover, the copolymers have thermodynamic properties such as the mean Gibbs free energy, enthalpy, and entropy per monometric unit, respectively defined as

$$\begin{pmatrix} g \\ h \\ s^0 \end{pmatrix} = \lim_{l\to\infty} \frac{1}{l} \sum_{\omega} \mu_l(\omega) \begin{pmatrix} G(\omega) \\ H(\omega) \\ S^0(\omega) \end{pmatrix} \tag{11.134}$$

and related to each other according to $g = h - T\, s^0$ as a consequence of equation (4.58).

In the regime of steady growth where the factorization (11.133) holds, the entropy (4.45) is given in the long-time limit by

$$S(t) \simeq \langle l \rangle_t \big[s^0 + \mathscr{D}(\omega) \big], \tag{11.135}$$

where

$$\mathscr{D}(\omega) \equiv \lim_{l\to\infty} -\frac{1}{l} \sum_{\omega} \mu_l(\omega) \ln \mu_l(\omega) \geq 0 \tag{11.136}$$

is the Shannon disorder per monomeric unit in the copolymer sequence, which is always bounded according to $0 \leq \mathscr{D}(\omega) \leq \ln M$. The contribution of the length distribution $p_t(l)$ to the Shannon disorder increases more slowly than linearly with time, so that its time derivative becomes negligible after a long enough time: $\lim_{t\to\infty}(d/dt)[-\sum_l p_t(l) \ln p_t(l)] = 0$. Therefore, the time derivative of the entropy is equal to

$$\frac{dS}{dt} = \mathscr{R}\big[s^0 + \mathscr{D}(\omega) \big] \tag{11.137}$$

with the elongation rate (11.132). Furthermore, the entropy exchange can be expressed by equation (4.60) in terms of the enthalpy $H(\omega)$, so that

$$\frac{d_e S}{dt} = \frac{1}{T} \frac{d\langle H \rangle_t}{dt} = \mathscr{R} \frac{h}{T}, \tag{11.138}$$

since $\langle H \rangle_t \simeq \langle l \rangle_t h$ in terms of the mean enthalpy per monomeric unit h. Now, using $d_i S/dt = dS/dt - d_e S/dt$, the entropy production rate can be obtained as

$$\frac{1}{k_B} \frac{d_i S}{dt} = \mathscr{R} \left[-\frac{g}{k_B T} + \mathscr{D}(\omega) \right] \geq 0, \tag{11.139}$$

where $g = h - T\, s^0$ is the mean Gibbs free energy per monomeric unit (Andrieux and Gaspard, 2008d). The same expression can also be obtained from equation (4.57).

We note that the entropy production rate (11.139) has the standard form $d_i S/dt = k_B \mathscr{A} \mathscr{R}$, where a rate or current here given by the elongation rate (11.132) is multiplied by an affinity, which can here be identified as $\mathscr{A} \equiv \varepsilon + \mathscr{D}(\omega)$ with the mean driving force $\varepsilon \equiv -\beta\, g$ introduced by Bennett (1979) and the inverse temperature $\beta = (k_B T)^{-1}$. This affinity per monomeric unit is the sum of two contributions. The first one ε is the contribution from the free energy supplied by chemical reactions, which is positive if the free energy landscape is favorable to the copolymer growth. The second one $\mathscr{D}(\omega)$ is the entropic contribution due to disorder in the copolymer sequence. Since the entropy production rate is always nonnegative according to the second law of thermodynamics, the affinity must be positive, i.e., $\mathscr{A} > 0$, if the copolymer is growing with $\mathscr{R} > 0$. Consequently, there are two scenarios for the copolymer to grow. In the first scenario, the free energy landscape is favorable with a positive mean driving force, so that the affinity is positive because the Shannon disorder is always nonnegative, and the copolymer grows by the free energy supply from reactions. In the second scenario, the free energy landscape is adverse, but the Shannon disorder (11.136) is large enough for the affinity to be positive if the mean driving force is negative and bounded as $-\mathscr{D}(\omega) < \varepsilon < 0$. In this second scenario, the copolymer grows by the entropic effect of sequence disorder. At equilibrium, the affinity tends to zero with the elongation rate and the mean driving force is related to the Shannon disorder according to $\varepsilon_{\text{eq}} = -\mathscr{D}_{\text{eq}}(\omega)$ (Andrieux and Gaspard, 2008d).

An example of such copolymerization processes is when the attachment and detachment rates only depend on the monomer involved in the reaction, so that the transition rates are given by $W_{\pm m_l}$ in the reaction network (11.130) without dependence on either ω or ω'. For this particularly simple kinetics, the growing copolymer is a Bernoulli chain and the probability distribution of its sequence factorizes as

$$\mu_l(\omega) = \prod_{j=1}^{l} \mu_1(m_j), \tag{11.140}$$

so that the Shannon disorder (11.136) is given by $\mathscr{D}(\omega) = -\sum_{m=1}^{M} \mu_1(m) \ln \mu_1(m)$ in terms of the probability distribution $\{\mu_1(m)\}_{m=1}^{M}$ of the monomeric units in the chain. Otherwise, the sequence may be a Markov chain, or even non-Markovian (Coleman and Fox, 1963b; Gaspard, 2016a).

In addition, the copolymer may be subjected to an external force, as shown in Figure 11.9(a). This force exerts a mechanical work $W = F x(\omega)$, where $x(\omega)$ is the physical length of the copolymer measured from its anchor point to its growing end. The mean growth velocity is thus given by $v = (d/dt)\langle x \rangle_t$ with $\langle x \rangle_t = \sum_{\omega} P_t(\omega)\, x(\omega)$. The mean size of a monomeric unit in the chain can thus be evaluated as $\delta = v/\mathscr{R}$. In such circumstances, the Gibbs free energy per monomeric unit can be decomposed as $g = -F\delta + g_{\text{c}}$ into its mechanical part $-F\delta$ and its chemical part g_c. Consequently, the affinity per monomeric unit reads $\mathscr{A} = \beta F \delta - \beta g_{\text{c}} + \mathscr{D}(\omega)$. According to the second law, the growth is possible if this affinity is positive, which implies that the external force should be larger than the stall force where elongation is stopped, i.e., $F \geq F_{\text{st}} \equiv \delta^{-1}[g_{\text{c}} - k_{\text{B}} T\, \mathscr{D}(\omega)]$. If the stall force is negative, there exists a range of values for the external force $F_{\text{st}} \leq F < 0$, where the growth of the copolymer can perform mechanical work by pulling a load. As a corollary of the previous results, this mechanical work is possible either because the chemical free energy is favorable if $g_{\text{c}} < 0$, or by the entropic effect of disorder in the copolymer sequence if $0 \leq g_{\text{c}} < k_{\text{B}} T\, \mathscr{D}(\omega)$ (Gaspard, 2015a).

Furthermore, fluctuation relations can be established for copolymerization processes. For kinetics yielding the Bernoulli chains (11.140), we may consider the probability distribution $\mathcal{P}_t(\Delta\mathbf{N})$ that the copolymer has incorporated $\Delta\mathbf{N} = \{\Delta N_m\}_{m=1}^{M} \in \mathbb{Z}^M$ monomeric units of the different species during the time interval $[0,t]$. This distribution obeys the following multivariate fluctuation relation,

$$\frac{\mathcal{P}_t(\Delta\mathbf{N})}{\mathcal{P}_t(-\Delta\mathbf{N})} \sim_{t\to\infty} \exp\left[\beta(W - \Delta G_c) + \Delta D\right], \tag{11.141}$$

where $\beta = (k_B T)^{-1}$ is the inverse temperature, W the mechanical work performed by the external force F, ΔG_c the chemical free energy change, and ΔD the change in Shannon sequence disorder upon the length variation by $\Delta l = \sum_{m=1}^{M} \Delta N_m$. The random variables $\Delta\mathbf{N}$ should count the subsequences of length $k+1$ for kinetics generating Markov chains of kth order (Gaspard, 2015a).

11.7.2 Template-Directed Copolymerization

The previous considerations extend to template-directed copolymerization processes such as DNA replication, transcription of DNA into RNA, and translation of messenger RNA into protein, which are the fundamental mechanisms of information processing in molecular biology (Alberts et al., 1998). The catalysts of these processes are respectively called DNA polymerase, RNA polymerase, and ribosome.

The reaction network of template-directed copolymerization can be schematically written in the following form,

$$\begin{array}{lcl} \omega' = m_1 m_2 \cdots m_{l-1} \quad + m_l & \underset{W_{-m_l|\alpha\omega}}{\overset{W_{+m_l|\alpha\omega'}}{\rightleftharpoons}} & \omega = m_1 m_2 \cdots m_{l-1} m_l \\ \alpha = n_1\, n_2 \cdots n_{l-1}\, n_l\, n_{l+1} \cdots & & \alpha = n_1\, n_2 \cdots n_{l-1}\, n_l\, n_{l+1} \cdots, \end{array} \tag{11.142}$$

where ω is the sequence of the growing copolymer and α the given template sequence, as shown in Figure 11.9(b). The template sequence is composed of the units $n_j \in \{1,2,\ldots,M\}$ from as many different species as the monomeric species composing the copolymer ω. Here, the transition rates $W_{\pm m_l|\alpha\omega}$ depend not only on the monomer m_l undergoing attachment or detachment and the previously incorporated monomeric units of ω, but also on the template sequence α and, in particular, on the template unit n_l forming a pair with the reacting monomer m_l. The template sequence is characterized by the probability distribution $\nu_l(\alpha) = \nu_l(n_1 n_2 \cdots n_l)$, which is normalized according to $\sum_\alpha \nu_l(\alpha) = 1$ for all the chain lengths l. Otherwise, the process can be described in terms of the master equation (11.131) under assumptions similar to those used for free copolymerization.

If the monomeric concentrations are invariant during copolymerization, the process will evolve into a regime of steady growth with a constant elongation rate (11.132) and a factorization similar to equation (11.133) for the time-dependent probability distribution, so that

$$P_t(\omega) \simeq p_t(l)\, \mu_l(\omega|\alpha) \qquad \text{for} \qquad t \to \infty \tag{11.143}$$

in terms of the probability distribution $p_t(l)$ of the length $l = |\omega|$ of the growing copolymer ω and the stationary conditional probability $\mu_l(\omega|\alpha)$ to find the sequence ω of length $l = |\omega|$

in the copolymer, provided that the template has the sequence α. In the same way as for free copolymerization, the entropy production rate is found to be given by

$$\frac{1}{k_\mathrm{B}}\frac{d_\mathrm{i}S}{dt} = \mathscr{R}\left[-\frac{g}{k_\mathrm{B}T} + \mathscr{D}(\omega|\alpha)\right] = \mathscr{R}\left[-\frac{g}{k_\mathrm{B}T} + \mathscr{D}(\omega) - \mathscr{I}(\omega,\alpha)\right] \geq 0 \qquad (11.144)$$

in terms of the conditional Shannon disorder per monomeric unit of the copy ω with respect to the template α defined as

$$\mathscr{D}(\omega|\alpha) = \lim_{l\to\infty} -\frac{1}{l}\sum_{\alpha,\omega} \nu_l(\alpha)\,\mu_l(\omega|\alpha)\,\ln\mu_l(\omega|\alpha) \geq 0. \qquad (11.145)$$

In addition, we can introduce the overall Shannon disorder of the copy ω as

$$\mathscr{D}(\omega) = \lim_{l\to\infty} -\frac{1}{l}\sum_{\omega} \mu_l(\omega)\,\ln\mu_l(\omega) \geq 0, \qquad \text{where} \qquad \mu_l(\omega) = \sum_{\alpha} \nu_l(\alpha)\,\mu_l(\omega|\alpha) \qquad (11.146)$$

is the probability distribution of the copy whatever the template sequence can be, and the mutual information between the copy and the template as

$$\mathscr{I}(\omega,\alpha) = \lim_{l\to\infty} \frac{1}{l}\sum_{\alpha,\omega} \mu_l(\omega,\alpha)\,\ln\frac{\mu_l(\omega,\alpha)}{\nu_l(\alpha)\,\mu_l(\omega)} \geq 0 \qquad (11.147)$$

in terms of the joint probability $\mu_l(\omega,\alpha) = \nu_l(\alpha)\,\mu_l(\omega|\alpha)$. By a standard formula of information theory (Cover and Thomas, 2006), the mutual information can be written as

$$\mathscr{I}(\omega,\alpha) = \mathscr{D}(\omega) - \mathscr{D}(\omega|\alpha) \geq 0, \qquad (11.148)$$

hence the second equality in equation (11.144) (Andrieux and Gaspard, 2008d).

The larger the percentage of errors in the copy sequence ω with respect to the template sequence α, the larger the conditional Shannon disorder (11.145) and the lower the mutual information. This latter thus characterizes the fidelity in the copying process $\alpha \to \omega$.

Here, the multivariate fluctuation relation (11.141) extends into

$$\frac{\mathcal{P}_t(\Delta\mathbf{N})}{\mathcal{P}_t(-\Delta\mathbf{N})} \sim_{t\to\infty} \exp\left[\beta(W - \Delta G_\mathrm{c}) + \Delta D - \Delta I\right], \qquad (11.149)$$

where ΔD is the change in overall Shannon disorder and ΔI the change in mutual information upon length variation Δl.

These considerations apply in particular to DNA replication (Andrieux and Gaspard, 2008d; Gaspard, 2016a,b).

12
Active Processes

12.1 Active versus Passive Nonequilibrium Processes

In nonequilibrium processes, an essential distinction should be made between passive and active transports. In the former, fluxes go downhill in the same direction as the corresponding thermodynamic forces or affinities. In the latter, the fluxes are driven uphill in the opposite direction by the coupling to different thermodynamic forces, which are supplying free energy.

Examples can be found in cellular biology, where different kinds of membrane proteins are known for the transport of ions such as H^+, Na^+, K^+, or Ca^{2+} between the interior and the exterior of cells, or between two intracellular compartments (Alberts et al., 1998).

On the one hand, there exist ion channels for passive transport downhill electrochemical gradients across membranes, as schematically illustrated in Figure 12.1(a). The thermodynamic force or affinity driving the current $J_{\rm e}$ of ions X^+ in the channel is given by

$$A_{\rm e} = \beta\,(\mu_1 - \mu_2) = \beta e \Delta\Phi + \ln \frac{[X^+]_1}{[X^+]_2}, \tag{12.1}$$

where $\mu_j = \mu_j^0 + k_{\rm B}T \ln\left([X^+]_j/n^0\right) + e\Phi_j$ are the electrochemical potentials of the ionic species at the concentrations $[X^+]_j$ and electric potentials Φ_j in the two compartments $j = 1, 2$ separated by the membrane, $\Delta\Phi = \Phi_1 - \Phi_2$ is the potential difference, and $\beta = (k_{\rm B}T)^{-1}$ is the inverse temperature. For this passive transport, the entropy production rate can thus be expressed as

$$\frac{1}{k_{\rm B}} \frac{d_{\rm i}S}{dt} = A_{\rm e}\, J_{\rm e} \geq 0, \tag{12.2}$$

so that the electrochemical current $J_{\rm e}$ must go in the same direction as the associated affinity $A_{\rm e}$ as a consequence of the second law of thermodynamics.

On the other hand, there also exist carrier proteins, called ATPases, that are able to pump ions across the membrane using the free energy supplied by the hydrolysis of adenosine triphosphate (ATP), which is the main fuel of the cellular machinery. The energy barrier of hydrolysis is significantly lowered in the catalytic site of the carrier protein, so that the free energy is locally delivered inside the protein. In this way, catalysis by specific proteins can achieve local control of the cellular functions. ATP hydrolysis is the following reaction,

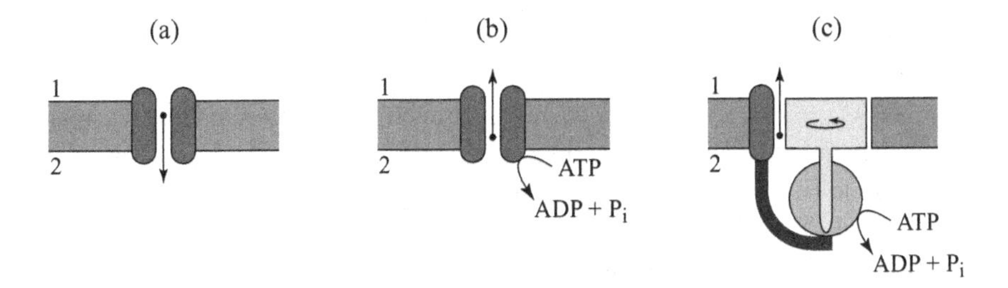

Figure 12.1 Schematic representation of membrane transport proteins: (a) ion channel for passive transport; (b) ion pump for active transport; (c) rotary ion pump for active transport. The aqueous solutions on the two sides of the membrane are, respectively, denoted $j = 1, 2$.

$$\mathrm{ATP}^{4-} + \mathrm{H_2O} \longrightarrow \mathrm{ADP}^{3-} + \mathrm{P}_{\mathrm{i}}^{2-} + \mathrm{H}^+, \tag{12.3}$$

releasing adenosine diphosphate (ADP) and inorganic phosphate $\left(\mathrm{P}_{\mathrm{i}}^{2-} = \mathrm{HPO}_4^{2-}\right)$, and providing the Gibbs free energy $\Delta\mu^0 \simeq 30.5\,\mathrm{kJ/mole} = 50\,\mathrm{pN\,nm} = 12.2\,k_{\mathrm{B}}T$ for an aqueous solution with standard solute concentrations $n^0 = 1$ mole per liter (1 M), $T = 298$ K, and $\mathrm{pH} = -\log_{10}\left([\mathrm{H}^+]/n^0\right) = 7$ (Moore, 1972; Alberts et al., 1998). With other values for the concentrations, ATP hydrolysis provides the Gibbs free energy given by $\Delta\mu = \mu_{\mathrm{ATP}} - \mu_{\mathrm{ADP}} - \mu_{\mathrm{P_i}}$ in terms of the chemical potentials, so that active transport can be powered by the chemical affinity

$$A_{\mathrm{c}} = \beta\,\Delta\mu = \ln \frac{[\mathrm{ATP}]\,n^0}{K_{\mathrm{eq}}[\mathrm{ADP}]\,[\mathrm{P_i}]}, \tag{12.4}$$

where $K_{\mathrm{eq}} = \exp(-\beta\Delta\mu^0) \simeq 4.9 \times 10^{-6}$ is the equilibrium constant, including water concentration. Since the standard free energy of ATP hydrolysis is about one order of magnitude larger than the thermal energy $k_{\mathrm{B}}T \simeq 4.1$ pN nm, the carrier protein can undergo unidirectional internal movements with an amplitude larger than the thermal fluctuations. The conformational changes of the carrier protein couple together the chemical reaction (12.3) at the rate J_{c} with the ionic current J_{e}. Ions can thus be pumped uphill across the membrane using the chemical energy of ATP hydrolysis, as illustrated in Figure 12.1(b). Since two currents are involved, the entropy production rate is here given by

$$\frac{1}{k_{\mathrm{B}}}\frac{d_{\mathrm{i}}S}{dt} = A_{\mathrm{e}}\,J_{\mathrm{e}} + A_{\mathrm{c}}\,J_{\mathrm{c}} \geq 0. \tag{12.5}$$

The dependences of these currents on the associated affinities (12.1) and (12.4) can be expressed as

$$\begin{cases} J_{\mathrm{e}}(A_{\mathrm{e}}, A_{\mathrm{c}}) = L_{\mathrm{e,e}}\,A_{\mathrm{e}} + L_{\mathrm{e,c}}\,A_{\mathrm{c}} + \frac{1}{2}\,M_{\mathrm{e,ee}}\,A_{\mathrm{e}}^2 + M_{\mathrm{e,ec}}\,A_{\mathrm{e}}\,A_{\mathrm{c}} + \frac{1}{2}\,M_{\mathrm{e,cc}}\,A_{\mathrm{c}}^2 + \cdots \\ J_{\mathrm{c}}(A_{\mathrm{e}}, A_{\mathrm{c}}) = L_{\mathrm{c,e}}\,A_{\mathrm{e}} + L_{\mathrm{c,c}}\,A_{\mathrm{c}} + \frac{1}{2}\,M_{\mathrm{c,ee}}\,A_{\mathrm{e}}^2 + M_{\mathrm{c,ec}}\,A_{\mathrm{e}}\,A_{\mathrm{c}} + \frac{1}{2}\,M_{\mathrm{c,cc}}\,A_{\mathrm{c}}^2 + \cdots \end{cases} \tag{12.6}$$

in terms of the linear and nonlinear response coefficients introduced in equation (5.133). Therefore, the electrochemical current $J_{\mathrm{e}} \neq 0$ can be driven by the free energy of the chemical reaction ($A_{\mathrm{c}} \neq 0$) in the absence of electrochemical potential ($A_{\mathrm{e}} = 0$) according to

$$J_e(0, A_c) = L_{e,c}\, A_c + \frac{1}{2}\, M_{e,cc}\, A_c^2 + \cdots \tag{12.7}$$

if $L_{e,c} \neq 0$, $M_{e,cc} \neq 0, \ldots$, which are the response coefficients characterizing the coupling. This phenomenon is similar to the thermoelectric coupling in Seebeck and Peltier effects (Haase, 1969; de Groot and Mazur, 1984). With this coupling, the active transport of ions may be possible even in the direction opposite to the electrochemical gradient such that $A_e J_e < 0$. Indeed, if the contribution of the chemical reaction is positive and large enough, i.e., $A_c J_c > 0$, the entropy production rate (12.5) remains nonnegative, so that the second law of thermodynamics is always satisfied. This coupling is the basic mechanism of active processes (Nicolis, 1979; Hill, 1989; Chowdhury, 2013).

In addition, the active transport of ions between intracellular compartments is also carried out by larger proteins like the V_oV_1-ATPase, which is made of a rotary proton or sodium pump V_o located inside the membrane and the motor V_1-ATPase driving the rotation of the pump with the chemical energy of ATP hydrolysis from the membrane side, as shown in Figure 12.1(c). These ion pumps also function with the coupling (12.7) between the electrochemical current J_e and the chemical affinity (12.4). We note that the functioning of such rotary molecular machines can, in principle, be reversed, so that ATP synthesis can be powered by the electrochemical affinity (12.1), giving $A_c J_c < 0$ from a large enough positive value for $A_e J_e > 0$. This is the case for the F_oF_1-ATPase, also called ATP synthase, which plays the key role of supplying cells with their main fuel (Alberts et al., 1998).

Active processes also manifest themselves in molecular motors coupling chemical reactions to mechanical motion for the motility of biological cells, which is the topic of Section 12.2. Polymerization and copolymerization can also be considered as active processes since mechanochemical coupling may arise in the elongation of macromolecules, as discussed in Section 11.7. Furthermore, nonequilibrium processes coupling motion to reaction also exist in artificial micrometric or submicrometric active particles, which are described in Section 12.3.

12.2 Molecular Motors

12.2.1 Mechanochemical Coupling and Energy Transduction

In biological cells, several kinds of molecular motors made of proteins have been discovered at the nanoscale (Alberts et al., 1998; Kolomeisky and Fisher, 2007; Chowdhury, 2013). They are powered by ATP hydrolysis or other free energy sources. Some are rotary like the bacterial flagellar motor (Sowa et al., 2005), which is bound to a membrane and powered by a transmembrane electrochemical affinity (12.1). Eukaryotic cells have linear motors powered by ATP hydrolysis, such as myosins moving on actin filaments, and kinesins and dyneins moving on microtubules.

In particular, myosin-actin motors perform muscular contraction (Alberts et al., 1998). They work with the cross-bridge mechanism, which is schematically represented in Figure 12.2. Myosin heads have a catalytic site for ATP hydrolysis. In the absence of ATP, myosin is strongly bound to actin in a *rigor* configuration. Upon ATP arrival in the catalytic

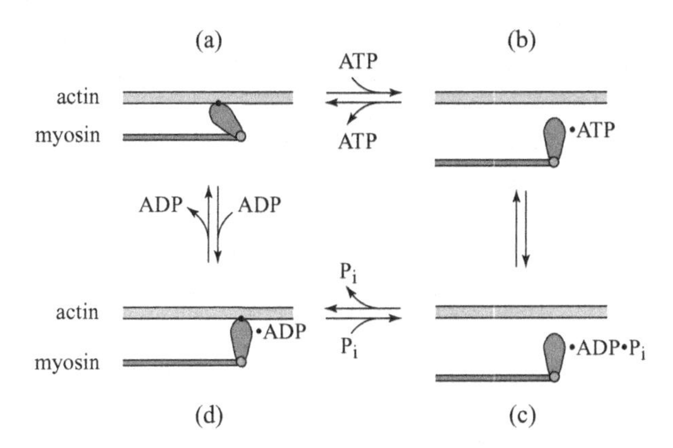

Figure 12.2 Schematic representation of the cross-bridge mechanism for the myosin-actin molecular motor.

site, myosin unbinds from actin, catalyzes ATP hydrolysis, and changes its configuration. Thereafter, inorganic phosphate leaves the catalytic site and myosin can weakly bind again to actin. Finally, upon the release of ADP, the power strokes that generate force take place, moving the myosin head by about 5 nm with respect to the actin filament (Kitamura et al., 1999).

Similarly, kinesin-microtubule and dynein-microtubule motors work with two catalytic heads for ATP hydrolysis (Alberts et al., 1998; Svoboda et al., 1993; Schnitzer et al., 2000). In particular, they drive the intracellular active transport of cargoes such as vesicles. Dynein-microtubule motors also generate the beating of cilia and flagella in eukaryotic cells.

The functioning of molecular motors powered by ATP hydrolysis is based on the mechanochemical coupling between the chemical reaction (12.3) and the mechanical motion by an external force in linear motors or an external torque in rotary motors (Chowdhury, 2013). The mechanical affinity and the associated current are, respectively, given by the following expressions for

$$\text{linear motors:} \qquad A_{\rm m} = \beta\, F_{\rm ext}, \qquad J_{\rm m} = V = \langle \dot{x} \rangle, \tag{12.8}$$

where $F_{\rm ext}$ is the external force exerted in the x-direction of the filament and V the mean velocity of the motor in that direction. Similarly, we have for

$$\text{rotary motors:} \qquad A_{\rm m} = \beta\, \tau_{\rm ext}, \qquad J_{\rm m} = \Omega = \langle \dot{\theta} \rangle, \tag{12.9}$$

where $\tau_{\rm ext}$ is the external torque exerted on the shaft of the motor and Ω its mean angular velocity. Accordingly, the entropy production rate of molecular motors is given by

$$\frac{1}{k_{\rm B}} \frac{d_{\rm i}S}{dt} = A_{\rm m}\, J_{\rm m} + A_{\rm c}\, J_{\rm c} \geq 0 \tag{12.10}$$

with the chemical affinity (12.4) and the mean rate $J_{\rm c} = R$ for consuming ATP molecules in the reaction (12.3). The chemical and mechanical mean currents are thus given by equation (12.6) with $(A_{\rm e}, J_{\rm e})$ replaced by $(A_{\rm m}, J_{\rm m})$. If the response coefficients $L_{\rm m,c}$, $M_{\rm m,cc}, \ldots$ are not equal to zero, the mechanochemical coupling is effective for energy transduction and the motion of the motor can be powered by the chemical reaction in agreement with the second law of thermodynamics.

At thermodynamic equilibrium, all the affinities are equal to zero as well as the associated currents and the entropy production rate (12.10), so that the concentrations of ATP, ADP, and inorganic phosphate are in their Guldberg–Waage ratio,

$$\left.\frac{[\mathrm{ATP}]\, n^0}{[\mathrm{ADP}]\,[\mathrm{P_i}]}\right|_{\mathrm{eq}} = K_{\mathrm{eq}} \simeq 4.9 \times 10^{-6}. \tag{12.11}$$

Under physiological conditions, these concentrations are maintained away from this equilibrium ratio by the metabolism, so that free energy is available for powering the molecular motor and the chemical affinity (12.4) may take a positive value of about $A_{\mathrm{c}} \simeq 21$.[1] In the limit where the products ADP and $\mathrm{P_i}$ of ATP hydrolysis have vanishingly small concentrations, the chemical affinity (12.4) as well as the entropy production rate (12.10) become infinite. In this fully irreversible regime, the reaction rate $R = J_{\mathrm{c}}$ and the mean velocity of the motor $V = J_{\mathrm{m}}$ are nevertheless finite. In particular, the velocity typically depends on ATP concentration as in the kinetics of Michaelis and Menten (1913) according to

$$V = \frac{V_{\max}(F_{\mathrm{ext}})\,[\mathrm{ATP}]}{[\mathrm{ATP}] + K_{\mathrm{M}}(F_{\mathrm{ext}})}, \tag{12.12}$$

where the maximal velocity $V_{\max}$ and the dissociation constant K_{M} here depend on the external force F_{ext} (Schnitzer et al., 2000).

Since molecular motors are nanometric, they are subjected to molecular and thermal fluctuations, so that they undergo irregular movements and reactive events occurring at random times. Therefore, the displacement x of the motor along the filament and the number n of ATP molecules that are consumed during some time interval $[0,t]$ are random variables ruled by some stochastic process. This latter can be based on the model of the flashing ratchet, coupling the transitions due to the chemical reaction and the motion of the motor (Astumian, 1997; Jülicher et al., 1997). Because of microreversibility, their joint probability distribution obeys the following bivariate fluctuation relation,

$$\frac{\mathcal{P}_t(x,n;A_{\mathrm{m}},A_{\mathrm{c}})}{\mathcal{P}_t(-x,-n;A_{\mathrm{m}},A_{\mathrm{c}})} \sim_{t\to\infty} \exp(A_{\mathrm{m}}\,x + A_{\mathrm{c}}\,n), \tag{12.13}$$

expressed in terms of the mechanical affinity introduced in equation (12.8) and the chemical affinity (12.4) (Lacoste and Mallick, 2009; Gerritsma and Gaspard, 2010). As a consequence, the entropy production rate is given by equation (12.10) and is nonnegative. Furthermore, the fluctuation relation (12.13) has fundamental implications for the properties of mechanochemical coupling in molecular motors.

In Section 12.2.2, these results will be presented in detail for the $\mathrm{F_1}$-ATPase rotary motor.

[1] The physiological concentrations are reported to take values around $[\mathrm{ATP}] \simeq 10^{-3}$ M, $[\mathrm{ADP}] \simeq 10^{-4}$ M, and $[\mathrm{P_i}] \simeq 10^{-3}$ M (Kinosita et al., 2004).

12.2.2 The F_1-ATPase Rotary Molecular Motor

Stochastic Process

F_1-ATPase is the hydrophilic part of the ATP synthase named F_oF_1-ATPase. *In vivo*, the hydrophobic part F_o is embedded in a membrane and driven into rotation by a transmembrane ionic current, as shown in Figure 12.1(c). The F_o rotor is attached to a γ-shaft rotating inside the F_1 part, which forms a barrel composed of six alternated subunits $(\alpha\beta)_3$. Each β-subunit has a catalytic site for ATP synthesis, i.e., reverse ATP hydrolysis. During its rotation, the γ-shaft induces mechanical deformations in the F_1 part, which provides energy to perform the rotary catalysis of ATP synthesis from ADP and P_i. ATP synthase is a major fuel supplier in prokaryotic and eukaryotic cells. Since the mechanochemical coupling is reversible in F_oF_1-ATPase, this molecular machine can also function as an ion pump using the chemical energy of ATP hydrolysis if there is plenty of fuel.

In vitro, a colloidal bead can be attached to the γ-shaft in the place of the F_o rotor, while the F_1 part is fixed to a overglass, as illustrated in Figure 12.3(a). The bead attached to the γ-shaft is large enough to observe its rotation with optical microscopy (Yasuda et al., 2001; Kinosita et al., 2004). The surrounding aqueous solution contains ATP to drive rotation. The F_1-ATPase nanomotor has a mechanical power of about 10^{-18} W for high ATP concentration. The experiments show that rotation proceeds by random steps and substeps upon ATP arrival and binding to the catalytic sites of the β-subunits, its hydrolysis, and

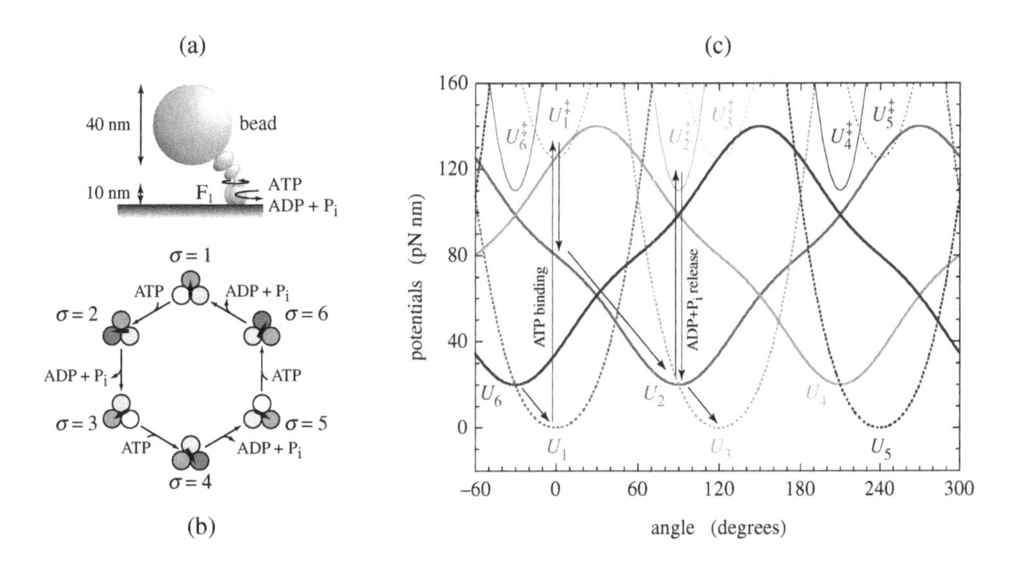

Figure 12.3 (a) Schematic representation of the F_1-ATPase molecular motor fixed to a coated coverslip and with a bead of 40 nm diameter attached to its γ-shaft. (b) Diagram of the cycle of the motor. Each $\alpha\beta$-subunit is schematically depicted by a circle and the γ-subunit by an arrow. The γ-subunit rotates by about 90° after ATP binding to an empty catalytic β-subunit (first substep) and by about 30° during the release of ADP and P_i (second substep). (c) The potentials of the chemical states $U_\sigma(\theta)$ and the transition states $U_\sigma^\ddagger(\theta)$. Reprinted from Gaspard and Gerritsma (2007) with the permission of Elsevier.

the release of the products ADP and P_i. The transition rates of these events are accurately measured under specific conditions and the experimental data can be used for modeling the stochastic kinetics and thermodynamics of the F_1-ATPase rotary motor (Wang and Oster, 1998; Gaspard and Gerritsma, 2007; Gerritsma and Gaspard, 2010).

The different occupancies of the three catalytic sites correspond to the chemical states $\{\sigma\}$ of the motor. In each chemical state, the γ-shaft can make an angle $\theta \in [0, 2\pi[$. Since the rotation of the γ-shaft induces mechanical deformations in the hexameric barrel $(\alpha\beta)_3$, there is a free energy potential $U_\sigma(\theta)$ depending on the angle θ for each chemical state σ. The internal state of F_1-ATPase is thus specified by the chemical state σ and the angle θ of the γ-shaft. *A priori*, each catalytic site can be empty or occupied by ATP, ADP·P_i, or ADP without counting further possibilities, so that there are 4^3 possible states or more. However, the experimental observations of Yasuda et al. (2001) provide evidence for six successive transitions associated with the following steps and substeps: ATP binding generates a rotation by 80°–90° and the release of ADP or P_i a further rotation by 40°–30°, completing a third of revolution. In addition, the statistics of dwell times reveals that there exists a supplementary transition in between ATP binding and product release, which is caused by ATP hydrolysis. Accordingly, there are nine experimentally observed states (Yasuda et al., 2001). For the purpose of modeling, these nine states can be lumped into the six states corresponding to the observed steps and substeps in the rotation of the motor shaft, giving the cycle depicted in Figure 12.3(b). The free energy potentials $U_\sigma(\theta)$ of the chemical states $\sigma = 1$–6 can be fitted to experimental data and they are shown in Figure 12.3(c) together with the potentials $U_\sigma^\ddagger(\theta)$ of the transition states, which form the bottlenecks of the reactions between the chemical states. The set of these potentials is symmetric under the rotations $\theta \to \theta + 2\pi/3$ because of the threefold symmetry of the hexamer $(\alpha\beta)_3$, but there is no symmetry under reflection or inversion in relation to the chirality of protein molecules.

The resulting stochastic process is thus a combination of overdamped Langevin processes for the random rotation of the angle θ in each potential $U_\sigma(\theta)$ and Markov jump processes for the transitions between the six chemical states (Jülicher et al., 1997). The whole stochastic process can be described in terms of the probability densities $p_\sigma(\theta, n, t)$ of finding the F_1-ATPase in the chemical state σ with the angle θ for its shaft, and the number n of ATP molecules consumed at time t. The master equations ruling the time evolution of these probability densities are given by

$$\partial_t \, p_\sigma(\theta, n, t) + \partial_\theta \, j_\sigma(\theta, n, t) = \sum_{\sigma'} \left[p_{\sigma'}(\theta, n - \nu_{\sigma' \to \sigma}, t) \, w_{\sigma' \to \sigma}(\theta) - p_\sigma(\theta, n, t) \, w_{\sigma \to \sigma'}(\theta) \right] \tag{12.14}$$

with the probability current densities

$$j_\sigma = -D \, \partial_\theta \, p_\sigma + \beta D \, (-\partial_\theta \, U_\sigma + \tau_{\text{ext}}) \, p_\sigma, \tag{12.15}$$

expressed in terms of the rotational diffusivity D, the inverse temperature $\beta = (k_B T)^{-1}$, the external torque τ_{ext} exerted on the shaft, the transition rate densities $w_{\sigma \to \sigma'}(\theta)$, and the stoichiometric coefficients $\nu_{\sigma \to \sigma'}$ counting the consumption of ATP molecules in the different transitions (Gaspard and Gerritsma, 2007). The diffusivity is related by Einstein's formula $D = k_{\rm B}T/\gamma$ to the rotational viscous friction of the bead attached to the shaft in the aqueous solution. The transition rate densities are determined using Arrhenius' law to obtain their dependence on the temperature, the potentials $U_\sigma(\theta)$ and $U_\sigma^\ddagger(\theta)$, and the chemical potentials of ATP, ADP, and P_i in the surrounding solution, where their concentrations are assumed to be uniform. As a consequence of the formula (1.36) for chemical potentials in dilute solutions, the transition rate densities depend on ATP, ADP, and P_i concentrations as required by the mass action law. For the kinetic scheme

$$\text{ATP} + \text{M}_\sigma(\theta) \rightleftharpoons \text{M}_{\sigma+1}(\theta) \rightleftharpoons \text{M}_{\sigma+2}(\theta) + \text{ADP} + \text{P}_\text{i} \tag{12.16}$$

with $\sigma = 1, 3, 5$ (modulo 6), the transition rate densities and the corresponding stoichiometric coefficients are given as follows,

$$w_{1\to2}(\theta) = k_0\,\mathrm{e}^{-\beta\left[U_1^\ddagger(\theta)-U_1(\theta)-\mu_{\rm ATP}^0\right]}\,[\text{ATP}], \qquad \nu_{1\to2} = +1, \tag{12.17}$$

$$w_{2\to1}(\theta) = k_0\,\mathrm{e}^{-\beta\left[U_1^\ddagger(\theta)-U_2(\theta)\right]}, \qquad \nu_{2\to1} = -1, \tag{12.18}$$

$$w_{2\to3}(\theta) = \tilde{k}_0\,\mathrm{e}^{-\beta\left[U_2^\ddagger(\theta)-U_2(\theta)\right]}, \qquad \nu_{2\to3} = 0, \tag{12.19}$$

$$w_{3\to2}(\theta) = \tilde{k}_0\,\mathrm{e}^{-\beta\left[U_2^\ddagger(\theta)-U_3(\theta)-\mu_{\rm ADP}^0-\mu_{\rm P_i}^0\right]}\,[\text{ADP}]\,[\text{P}_\text{i}], \qquad \nu_{3\to2} = 0, \tag{12.20}$$

$$\vdots$$

where the first and second are the rates of ATP binding and unbinding, the third the rate of ATP hydrolysis and product release, and the fourth the rate of product binding and ATP synthesis (Gaspard and Gerritsma, 2007). The successive potentials are related to each other according to $U_{\sigma+2}(\theta) = U_\sigma(\theta - 2\pi/3)$ and $U_{\sigma+2}^\ddagger(\theta) = U_\sigma^\ddagger(\theta - 2\pi/3)$ (modulo 6) to fulfill the threefold symmetry.

We note that the reversed transitions should be included in the kinetic scheme in order to evaluate the thermodynamic quantities. Otherwise, if their rates were assumed to be zero, the detailed balance conditions could not be satisfied at equilibrium and the entropy production rate would be infinite.

The motor is driven away from equilibrium if the external torque is not equal to zero, i.e., $\tau_{\rm ext} \neq 0$, or if the uniform concentrations [ATP], [ADP], and [P_i] do not satisfy the Guldberg–Waage equilibrium condition (12.11). In this regard, the external torque and these concentrations are the control parameters of the nonequilibrium driving.

This stochastic process can be numerically simulated by the algorithm of Gillespie (1976, 1977) after discretizing the angle into a hundred bins. Figure 12.4 shows stochastic trajectories for different values of ATP concentration and the mean rotation rate compared with experimental data (Yasuda et al., 2001). The dependence on ATP concentration is similar to equation (12.12), which is a characteristic feature of the Michaelis–Menten kinetics. For low ATP concentrations, the rotation is slow because the motor spends time waiting for the arrival of ATP in its catalytic sites. In this regime, the rotation rate is proportional to ATP concentration and, between the jumps, the angle fluctuates around the minima of the potentials U_1, U_3, and U_5, as seen in Figure 12.4(a). Instead, for large values of ATP concentration, the rotation is limited by the reaction of hydrolysis and product release, which

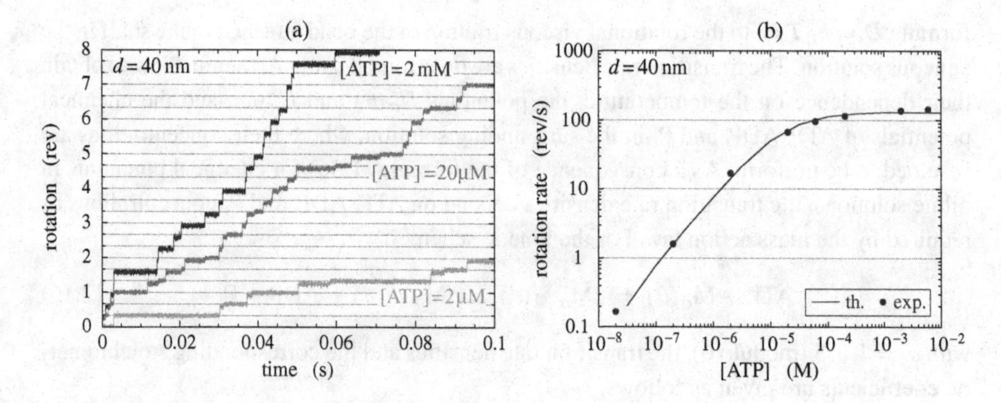

Figure 12.4 (a) Stochastic trajectories of rotation for the γ-shaft of the F_1-ATPase motor in the model (12.14). The number of revolutions $\theta(t)/2\pi$ is plotted versus time t in seconds for [ATP] $= 2$ μM, 20 μM, 2 mM, and $[\mathrm{ADP}][\mathrm{P_i}] = 0$. (b) Mean rotation rate for the γ-shaft of the F_1-ATPase motor in revolutions per second, versus the ATP concentration [ATP] in mole per liter for $[\mathrm{ADP}][\mathrm{P_i}] = 0$. The solid line is the model and the circles are the experimental data of Yasuda et al. (2001). In panels (a) and (b), the diameter of the bead is $d = 40$ nm, the temperature is of 23°C, and the external torque is zero. Reprinted from Gaspard and Gerritsma (2007) with the permission of Elsevier.

does not depend on the concentrations. Therefore, the mean rotation rate reaches saturation at the value $\Omega_{\max} \simeq 130$ rev/s for ATP concentration larger than the dissociation constant $K_{\mathrm{M}} \simeq 16$ μM. At high ATP concentration, the motor thus spends most of its time waiting for hydrolysis and product release around the minima of the potentials U_2, U_4, and U_6, as also seen in Figure 12.4(a).

Different regimes appear if the external torque τ_{ext} is varied. If the torque, is opposed to the rotation generated by ATP hydrolysis, it may reach a critical value called the stall torque $\tau_{\text{ext, stall}}$, for which the rotation is stopped. For more strongly opposed values of the external torque, the rotation is reversed and ATP synthesis becomes possible. *In vitro*, ATP synthesis can be mechanically driven with a magnetic bead in a rotating external magnetic field (Itoh et al., 2004). *In vivo*, it is the F_o part of the F_oF_1-ATPase that exerts the torque, yielding ATP synthesis.

Characterization of the Mechanochemical Coupling

The mechanochemical coupling can be characterized by the number of ATP molecules consumed or synthesized per revolution of the motor, as shown in Figure 12.5(a). This number is equal to three and the coupling is tight for torque values in the range $\tau_{\text{ext, stall}} \simeq -26\,\text{pN nm} \lesssim \tau_{\text{ext}} \lesssim +35\,\text{pN nm}$, but the coupling is loose outside this range, i.e., where the ratio ATP/rev falls off (Oosawa and Hayashi, 1986). In the range of tight coupling, the rotation and reaction rates are related according to $J_{\mathrm{m}} = (2\pi/3)J_{\mathrm{c}}$ because the potentials $U_\sigma(\theta) - \tau_{\text{ext}}\theta$ form barriers opposed to the free rotation of the shaft without transitions between the chemical states. Instead, outside this range, the external torque $|\tau_{\text{ext}}|$ is so large

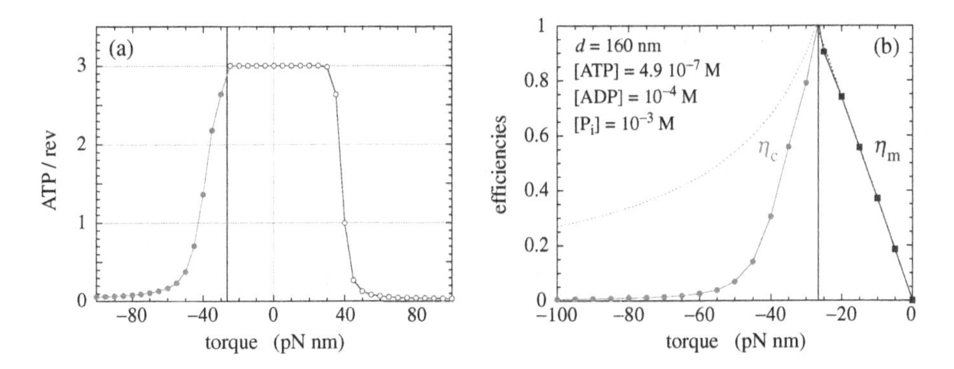

Figure 12.5 (a) The mean number of ATP molecules consumed or synthesized per revolution versus the external torque $\tau_{\rm ext}$ for a bead of diameter $d = 160$ nm, the concentrations $[{\rm ATP}] = 4.9\ 10^{-7}$ M, $[{\rm ADP}] = 10^{-4}$ M, $[{\rm P_i}] = 10^{-3}$ M, and the temperature of 23°C. The vertical solid line is located at the stall torque $\tau_{\rm ext,\,stall} \simeq -26$ pN nm. Reprinted from Gaspard and Gerritsma (2007) with the permission of Elsevier. (b) Chemical efficiency (12.21) and mechanical efficiency (12.22) versus the external torque $\tau_{\rm ext}$ in the continuous-angle model (12.14) (circles and squares joined by a solid line) and compared with the prediction (12.23) of tight coupling (dashed lines). The vertical solid line indicates the stall torque. The conditions are the same as in panel (a). Republished from Gerritsma and Gaspard (2010) with the permission of World Scientific.

that the barriers no longer exist for the potentials $U_\sigma(\theta) - \tau_{\rm ext}\theta$, so that the rotation becomes decoupled from the chemical reactions (Gaspard and Gerritsma, 2007).

Energy transduction by the F_1-ATPase can be characterized by thermodynamic efficiencies such as (1.123) or (1.125), which can in general be defined according to equation (5.116). In the regime of ATP synthesis where $A_{\rm c} > 0$, $A_{\rm m} < 0$, $J_{\rm c} = \langle \dot{n} \rangle_{\rm st} < 0$, and $J_{\rm m} = \langle \dot{\theta} \rangle_{\rm st} < 0$, the chemical efficiency can be characterized by the ratio

$$\eta_{\rm c} \equiv -\frac{A_{\rm c}\, J_{\rm c}}{A_{\rm m}\, J_{\rm m}} = -\frac{\Delta\mu\, \langle \dot{n} \rangle_{\rm st}}{\tau_{\rm ext} \langle \dot{\theta} \rangle_{\rm st}}, \tag{12.21}$$

such that $0 \leq \eta_{\rm c} \leq 1$ because of the second law (12.10) (Jülicher et al., 1997). Similarly, in the regime of rotation powered by ATP where $A_{\rm c} > 0$, $A_{\rm m} < 0$, $J_{\rm c} = \langle \dot{n} \rangle_{\rm st} > 0$, and $J_{\rm m} = \langle \dot{\theta} \rangle_{\rm st} > 0$, the mechanical efficiency can be defined as the inverse of the chemical efficiency,

$$\eta_{\rm m} \equiv -\frac{A_{\rm m}\, J_{\rm m}}{A_{\rm c}\, J_{\rm c}} = -\frac{\tau_{\rm ext} \langle \dot{\theta} \rangle_{\rm st}}{\Delta\mu\, \langle \dot{n} \rangle_{\rm st}}, \tag{12.22}$$

such that $0 \leq \eta_{\rm m} \leq 1$. These efficiencies are shown in Figure 12.5(b) as a function of the external torque. They can reach the maximal unit value near the stall torque, where the rotation is slowed down and the losses are thus minimal. Instead, the efficiencies take lower values if the speed of the motor increases.

If the condition of tight coupling is satisfied, the chemical and mechanical efficiencies should only depend on the free energy of the reaction and on the external torque according to

$$\text{tight coupling:} \quad \langle\dot{\theta}\rangle_{\rm st} = \frac{2\pi}{3}\langle\dot{n}\rangle_{\rm st} \quad \Longrightarrow \quad \eta_{\rm c} = \frac{1}{\eta_{\rm m}} = -\frac{3\,\Delta\mu}{2\pi\,\tau_{\rm ext}}. \tag{12.23}$$

As seen in Figure 12.5(b), this is the case for the mechanical efficiency (12.22) in the regime of tight coupling for $\tau_{\rm ext,\,stall} < \tau_{\rm ext} < 0$. However, the chemical efficiency (12.21) is significantly smaller than predicted by the condition (12.23) in the regime of loose coupling for $\tau_{\rm ext} < \tau_{\rm ext,\,stall}$.

The tight mechanochemical coupling of the F_1-ATPase and the resulting maximal free-energy transduction efficiency have been experimentally observed (Toyabe et al., 2011).

If the tight-coupling condition (12.23) is satisfied, the entropy production rate (12.10) becomes

$$\frac{1}{k_{\rm B}}\frac{d_{\rm i}S}{dt} = A\,\langle\dot{n}\rangle_{\rm st} \geq 0 \qquad \text{with} \qquad \langle\dot{n}\rangle_{\rm st} = J_{\rm c} \tag{12.24}$$

and the *mechanochemical affinity* defined as

$$A \equiv A_{\rm c} + \frac{2\pi}{3}A_{\rm m} = \beta\left(\Delta\mu + \frac{2\pi}{3}\tau_{\rm ext}\right) = \ln\frac{[{\rm ATP}]\,n^0}{\tilde{K}_{\rm eq}(\tau_{\rm ext})\,[{\rm ADP}]\,[{\rm P_i}]}, \tag{12.25}$$

where $\tilde{K}_{\rm eq}(\tau_{\rm ext}) = K_{\rm eq}\exp(-2\pi\beta\tau_{\rm ext}/3)$ is the equilibrium constant shifted by the external torque. Indeed, since the reaction is controlled by catalysis inside the F_1-ATPase stressed by the external torque, this latter performs a mechanical power, which has the effect of displacing the thermodynamic equilibrium. This effective equilibrium corresponds to the following Guldberg–Waage ratio obtained if the mechanochemical affinity (12.25) is equal to zero, i.e., $A = 0$,

$$\left.\frac{[{\rm ATP}]\,n^0}{[{\rm ADP}]\,[{\rm P_i}]}\right|_{\rm eq} = \tilde{K}_{\rm eq}(\tau_{\rm ext}) = K_{\rm eq}{\rm e}^{-\frac{2\pi}{3}\beta\tau_{\rm ext}}. \tag{12.26}$$

Moreover, the entropy production rate (12.24) shows that there is a single flux instead of two driving the F_1-ATPase away from equilibrium if the coupling is tight rather than loose.

In order to identify the regimes of tight coupling for the F_1-ATPase molecular motor, the lines where the chemical and mechanical currents are equal to zero are plotted in Figure 12.6 in the plane of the associated affinities. In this plane, the actual thermodynamic equilibrium corresponds to the origin where the torque and $\Delta\mu$ are equal to zero. The line $\Omega = \langle\dot{\theta}\rangle_{\rm st} = 0$ is giving the values of the stall torque as a function of the chemical affinity $A_{\rm c} = \beta\Delta\mu$. The line $R = \langle\dot{n}\rangle_{\rm st} = 0$ separates the domain of ATP consumption where $\langle\dot{n}\rangle_{\rm st} > 0$ from the domain of ATP synthesis where $\langle\dot{n}\rangle_{\rm st} < 0$. In domain I where $\langle\dot{\theta}\rangle_{\rm st} > 0$ and $\langle\dot{n}\rangle_{\rm st} > 0$, the motor is powered by the chemical reaction and the mechanical efficiency is given by equation (12.22) with $0 \leq \eta_{\rm m} \leq 1$. In domain II where $\langle\dot{\theta}\rangle_{\rm st} < 0$ and $\langle\dot{n}\rangle_{\rm st} < 0$, the F_1-ATPase is synthesizing ATP using the mechanical power of the external torque and the chemical efficiency is defined by equation (12.21) with $0 \leq \eta_{\rm c} \leq 1$. In the third domain where $\langle\dot{\theta}\rangle_{\rm st} < 0$ and $\langle\dot{n}\rangle_{\rm st} > 0$, the quantities (12.21) and (12.22) are negative and ATP is consumed but the torque is too strongly opposed for the spent chemical energy to perform useful mechanical work.

Furthermore, the plot also shows the line $\Delta\mu = -2\pi\tau_{\rm ext}/3$ where the mechanochemical affinity (12.25) is equal to zero. In the region around the actual thermodynamic equilibrium,

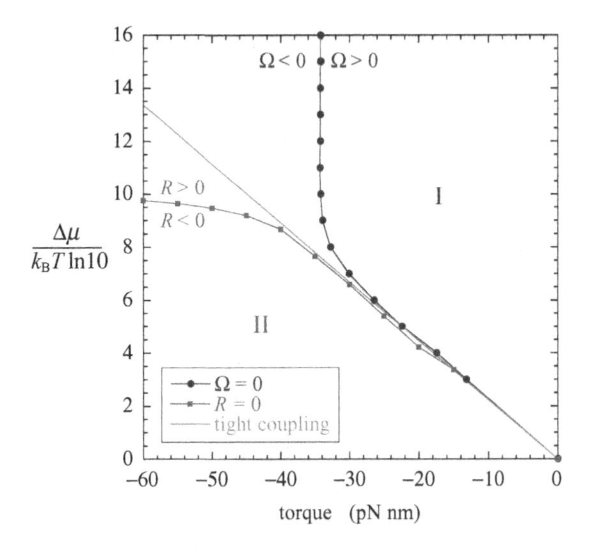

Figure 12.6 Chemical potential difference $\Delta\mu$ in units of $k_{\rm B}T \ln 10$ versus the external torque $\tau_{\rm ext}$ for the situations where the rotation rate $\Omega = \langle\dot{\theta}\rangle_{\rm st}$ (circles) and the ATP consumption rate $R = \langle\dot{n}\rangle_{\rm st}$ (squares) are equal to zero in the continuous model (12.14) and compared with the straight line $\Delta\mu = -2\pi\,\tau_{\rm ext}/3$ where the mechanochemical affinity (12.25) is equal to zero, i.e., $A = 0$. The concentrations are taken as $[{\rm ATP}] = 4.9 \times 10^{-11} \exp(0.8\,\beta\,\Delta\mu)\,{\rm M}$ and $[{\rm ADP}][{\rm P_i}] = 10^{-5} \exp(-0.2\,\beta\,\Delta\mu)\,{\rm M}^2$. The bead attached to the γ-shaft has the diameter $d = 80$ nm and the temperature is of 23°C. Republished from Gerritsma and Gaspard (2010) with the permission of World Scientific.

the two lines $\Omega = \langle\dot{\theta}\rangle_{\rm st} = 0$ and $R = \langle\dot{n}\rangle_{\rm st} = 0$ coincide approximately with the line $A = 0$, predicted by the condition (12.23). Therefore, the motor is functioning in the tight-coupling regime in this region, but in the loose-coupling regime in the region where the lines $\Omega = \langle\dot{\theta}\rangle_{\rm st} = 0$ and $R = \langle\dot{n}\rangle_{\rm st} = 0$ no longer coincide.

Bivariate Fluctuation Relation

In general, the random chemical and mechanical currents of the stochastic process ruled by the master equations (12.14) obey a bivariate fluctuation relation, which can be proved as follows. Since the stochastic process is a combination of Markovian jumps between the discrete chemical states $\{\sigma\}$ and overdamped Langevin motion for the continuous angle θ, we need to combine the methods of Section 5.7 with those of Section 7.4.2. For this purpose, generating functions can be introduced as follows,

$$f_\sigma(\theta, \lambda_{\rm m}, \lambda_{\rm c}, t) \equiv \sum_{l,n=-\infty}^{+\infty} {\rm e}^{-\lambda_{\rm m}(\theta+2\pi l)}\,{\rm e}^{-\lambda_{\rm c} n}\; p_\sigma(\theta + 2\pi l, n, t) \tag{12.27}$$

with the counting parameters $\boldsymbol{\lambda} = (\lambda_{\rm m}, \lambda_{\rm c})$, respectively, for mechanical rotation and ATP consumption, and the angle $0 \le \theta < 2\pi$ (Lacoste and Mallick, 2009). These functions are ruled by $\partial_t\, \boldsymbol{f} = \hat{\mathbf{L}}_{\boldsymbol{\lambda}} \cdot \boldsymbol{f}$ with $\boldsymbol{f} = \{f_\sigma\}_{\sigma=1}^{6}$ and the modified linear operator defined as

$$\hat{\mathbf{L}}_{\lambda_m,\lambda_c} \equiv e^{-\lambda_m\theta-\lambda_c n}\,\hat{\mathbf{L}}\,e^{+\lambda_m\theta+\lambda_c n}, \tag{12.28}$$

where $\hat{\mathbf{L}}$ is the 6×6 matrix of linear operators such that the set of master equations (12.14) has the form $\partial_t\, \mathbf{p} = \hat{\mathbf{L}} \cdot \mathbf{p}$ with $\mathbf{p} = \{p_\sigma\}_{\sigma=1}^6$. Defining $\tau_\sigma \equiv -\partial_\theta U_\sigma + \tau_{\text{ext}}$, the modified operator (12.28) is explicitly given by

$$\begin{aligned}\left(\hat{\mathbf{L}}_{\boldsymbol{\lambda}} \cdot f\right)_\sigma &= D\,\partial_\theta^2 f_\sigma - \beta D\,\partial_\theta\,(\tau_\sigma\, f_\sigma) + 2\,D\,\lambda_m\,\partial_\theta f_\sigma + D\,\lambda_m^2\, f_\sigma - \beta D\,\lambda_m\,\tau_\sigma\, f_\sigma \\ &\quad + \sum_{\sigma'} \left(e^{-\lambda_c \nu_{\sigma'\to\sigma}}\, f_{\sigma'}\, w_{\sigma'\to\sigma} - f_\sigma\, w_{\sigma\to\sigma'}\right),\end{aligned} \tag{12.29}$$

which is reminiscent of both equations (5.215) and (7.53). This operator obeys the symmetry relation

$$\mathbf{M}^{-1} \cdot \hat{\mathbf{L}}_{\boldsymbol{\lambda}} \cdot \mathbf{M} = \hat{\mathbf{L}}_{\mathbf{A}-\boldsymbol{\lambda}}^{\dagger} \tag{12.30}$$

with the mechanical and chemical affinities $\mathbf{A} = (A_m, A_c)$ and the 6×6 diagonal matrix $\mathbf{M}$ composed of the elements $M_{\sigma\sigma'} = \delta_{\sigma\sigma'}\, p_{\sigma,\text{eq}}(\theta)$, expressed in terms of the equilibrium probability densities $p_{\sigma,\text{eq}}(\theta) \sim \exp[-\beta U_\sigma(\theta)]$ satisfying the detailed balance conditions

$$p_{\sigma',\text{eq}}(\theta)\, w_{\sigma'\to\sigma}(\theta) = p_{\sigma,\text{eq}}(\theta)\, w_{\sigma\to\sigma'}(\theta) \qquad \text{for every} \qquad \theta \in [0, 2\pi[, \tag{12.31}$$

and corresponding to the chemical equilibrium with respect to the concentrations [ADP] and [P_i] of the product species and the absence of external torque, i.e., $\tau_{\text{ext}} = 0$. As in equation (5.217), the leading eigenvalue of the modified operator (12.28) gives the cumulant generating function

$$Q(\boldsymbol{\lambda}; \mathbf{A}) \equiv \lim_{t\to\infty} -\frac{1}{t} \ln \left\langle e^{-\lambda_m\theta(t)-\lambda_c n(t)} \right\rangle_{\mathbf{A}} = \lim_{t\to\infty} -\frac{1}{t} \ln \sum_{\sigma=1}^{6} \int_0^{2\pi} d\theta\, f_\sigma(\theta, \boldsymbol{\lambda}, t). \tag{12.32}$$

As a consequence of equation (12.30), this cumulant generating function has the fundamental symmetry (5.220), so that the following bivariate fluctuation relation is satisfied,

$$\frac{\mathcal{P}_t(\theta, n; A_m, A_c)}{\mathcal{P}_t(-\theta, -n; A_m, A_c)} \sim_{t\to\infty} \exp(A_m\,\theta + A_c\, n) \tag{12.33}$$

for the probability density $\mathcal{P}_t(\theta, n) \equiv \sum_{\sigma=1}^{6} p_\sigma(\theta, n, t)$ of observing the cumulated rotation angle θ and the number n of ATP binding during the time interval $[0, t]$.

We note that the affinities can be obtained by considering the cycles of the Markovian stochastic process. On the one hand, the affinity (4.77) for the cycle $1 \to 2 \to 3 \to 4 \to 5 \to 6 \to 1$ of chemical states at fixed angle θ is given by

$$A_C = \ln \prod_{\sigma=1}^{6} \frac{w_{\sigma\to\sigma+1}(\theta)}{w_{\sigma+1\to\sigma}(\theta)} = 3\, A_c \tag{12.34}$$

with the chemical affinity (12.4), since there are three reactions of ATP hydrolysis along this cycle. On the other hand, the mechanical affinity can be obtained with the cycle of the angle θ from $\theta = 0$ to $\theta = 2\pi$ on any single potential $U_\sigma(\theta)$, leading to the following affinity,

$$A_{C'} = \beta \oint \tau_\sigma(\theta)\, d\theta = 2\pi\, \beta\, \tau_{\rm ext} = 2\pi\, A_{\rm m} \tag{12.35}$$

in terms of the mechanical affinity introduced in equation (12.9) (Gaspard, 2006). Therefore, the framework is consistent with the network theory of Section 4.4.5 and the underlying microreversibility, giving the response properties of Chapter 5.

Univariate Fluctuation Relation in the Tight-Coupling Regime

In the tight-coupling regime, because of the barriers formed by the potentials, the number of revolutions remains on average proportional to the number of ATP molecules that are consumed, $\theta(t) \simeq (2\pi/3)\, n(t)$, and the condition (12.23) holds. As explained in Section 5.4.6, the consequence of tight coupling is that, over long time intervals, the joint probability distribution for the angle θ and the counter n of ATP hydrolysis can be reduced to the probability distribution of one of them, e.g., the counter according to $\mathcal{P}_t(n) \equiv \int \mathcal{P}_t(\theta, n)\, d\theta$ and the bivariate fluctuation relation (12.33) becomes the univariate fluctuation relation:

$$\frac{\mathcal{P}_t(n; A)}{\mathcal{P}_t(-n; A)} \sim_{t\to\infty} \exp(A\, n), \tag{12.36}$$

where A is the mechanochemical affinity defined by equation (12.25) (Andrieux and Gaspard, 2006b; Gerritsma and Gaspard, 2010).

The implications are that the multiple linear and nonlinear response coefficients $L_{\alpha,\beta}$, $M_{\alpha,\beta\gamma}, \dots$ can be expressed at every order in terms of the single one of the expansion $J = LA + (1/2)MA^2 + \cdots$,

$$L_{\alpha,\beta} = (2\pi/3)^{\mathcal{N}_{\rm m}(\alpha,\beta)}\, L, \qquad M_{\alpha,\beta\gamma} = (2\pi/3)^{\mathcal{N}_{\rm m}(\alpha,\beta,\gamma)}\, M, \qquad \dots, \tag{12.37}$$

where the exponent $\mathcal{N}_{\rm m}$ is the number of mechanical labels in the subscripts $\alpha, \beta, \gamma, \dots$. Therefore, the Onsager reciprocal relations (5.143) are directly verified and, moreover, the linear response coefficients satisfy the relation $L_{\rm c,c}L_{\rm m,m} = L_{\rm c,m}L_{\rm m,c}$, which expresses the condition of tight coupling among the linear response coefficients. Similar results hold for the nonlinear response coefficients and the responses of higher cumulants.

Discrete-State Model in the Tight-Coupling Regime

In the regime of tight coupling, the continuous-angle model (12.14) can be reduced to a Markov jump process with the discrete chemical states by introducing their probabilities $P_\sigma(n,t) \equiv \int p_\sigma(\theta, n, t)\, d\theta$. Moreover, in the kinetic scheme (12.16), there are essentially two different states, the state $\sigma = 1$ before ATP binding and the state $\sigma = 2$ after ATP hydrolysis and the release of the products. Under such circumstances, the stochastic process is ruled by the master equations

$$\frac{dP_1}{dt} = (W_{-1} + W_{+2})\, P_2 - (W_{+1} + W_{-2})\, P_1, \tag{12.38}$$

$$\frac{dP_2}{dt} = (W_{+1} + W_{-2})\, P_1 - (W_{-1} + W_{+2})\, P_2, \tag{12.39}$$

which preserve the normalization condition $P_1 + P_2 = 1$. The transitions have the following rates and corresponding stoichiometric coefficients and angular jumps,

$$W_{+1} = k_{+1}\,[\mathrm{ATP}], \qquad \nu_{+1} = +1, \quad \Delta\theta_{+1} = +\frac{\pi}{2}, \tag{12.40}$$

$$W_{-1} = k_{-1}, \qquad \nu_{-1} = -1, \quad \Delta\theta_{-1} = -\frac{\pi}{2}, \tag{12.41}$$

$$W_{+2} = k_{+2}, \qquad \nu_{+2} = 0, \quad \Delta\theta_{+2} = +\frac{\pi}{6}, \tag{12.42}$$

$$W_{-2} = k_{-2}\,[\mathrm{ADP}]\,[\mathrm{P_i}], \qquad \nu_{-2} = 0, \quad \Delta\theta_{-2} = -\frac{\pi}{6}, \tag{12.43}$$

where the transition $\rho = +1$ corresponds to ATP binding, $\rho = -1$ to ATP unbinding, $\rho = +2$ to ATP hydrolysis and product release, and $\rho = -2$ to product binding and ATP synthesis (Andrieux and Gaspard, 2006b). The rate constants $k_{\pm\rho}$ have a dependence on temperature and external torque, which can be obtained on the basis of the continuous-angle model (Gerritsma and Gaspard, 2010).

In the stationary regime, the probabilities of the two states can be expressed as

$$P_1^{(\mathrm{st})} = \frac{k_{-1} + k_{+2}}{k_{+1}[\mathrm{ATP}] + k_{-1} + k_{+2} + k_{-2}[\mathrm{ADP}][\mathrm{P_i}]}, \tag{12.44}$$

$$P_2^{(\mathrm{st})} = \frac{k_{+1}[\mathrm{ATP}] + k_{-2}[\mathrm{ADP}][\mathrm{P_i}]}{k_{+1}[\mathrm{ATP}] + k_{-1} + k_{+2} + k_{-2}[\mathrm{ADP}][\mathrm{P_i}]}, \tag{12.45}$$

so that the mean rotation and reaction rates are given by

$$\Omega = \frac{2\pi}{3}\,\langle \dot{n} \rangle_{\mathrm{st}} = \frac{2\pi}{3}\,\frac{k_{+1}k_{+2}[\mathrm{ATP}] - k_{-1}k_{-2}[\mathrm{ADP}][\mathrm{P_i}]}{k_{+1}[\mathrm{ATP}] + k_{-1} + k_{+2} + k_{-2}[\mathrm{ADP}][\mathrm{P_i}]}. \tag{12.46}$$

As expected, the tight-coupling condition (12.23) is satisfied and the rates are equal to zero at the effective equilibrium where the concentrations are in the ratio (12.26) with the effective equilibrium constant $\tilde{K}_{\mathrm{eq}} = n^0 k_{-1}k_{-2}/(k_{+1}k_{+2})$. In the absence of ADP and $\mathrm{P_i}$, the Michaelis–Menten expression (12.12) holds exactly for the angular velocity in the discrete-state model with the maximal angular velocity $\Omega_{\max} = 2\pi k_{+2}/3$, which is determined by the intrinsic rate (12.42) of ATP hydrolysis and product release, and the dissociation constant is equal to $K_{\mathrm{M}} = (k_{-1} + k_{+2})/k_{+1}$.

In order to calculate the cumulant generating function defined as follows for the counter of reactive events,

$$Q(\lambda; A) \equiv \lim_{t\to\infty} -\frac{1}{t}\ln\left\langle \mathrm{e}^{-\lambda n(t)} \right\rangle_A, \tag{12.47}$$

the master equations (12.38)–(12.39) can be expressed in the matrix form (11.124) and the linear operator can be modified as in equation (11.125) to obtain the following 2×2 matrix,

$$\mathsf{L}_\lambda = \begin{pmatrix} -W_{+1} - W_{-2} & \mathrm{e}^{+\lambda}\,W_{-1} + W_{+2} \\ \mathrm{e}^{-\lambda}\,W_{+1} + W_{-2} & -W_{-1} - W_{+2} \end{pmatrix}. \tag{12.48}$$

According to equation (5.217), the cumulant generating function is thus given by

$$\mathcal{Q}(\lambda; A) = \frac{1}{2}\left(W_{+1} + W_{-1} + W_{+2} + W_{-2}\right) - \left[\frac{1}{4}\left(W_{+1} + W_{-1} + W_{+2} + W_{-2}\right)^2 + W_{+1}W_{+2}\left(e^{-\lambda} - 1\right) + W_{-1}W_{-2}\left(e^{+\lambda} - 1\right)\right]^{\frac{1}{2}}, \tag{12.49}$$

which satisfies the symmetry relation $\mathcal{Q}(\lambda; A) = \mathcal{Q}(A - \lambda; A)$ with the mechanochemical affinity (12.25), so that the univariate fluctuation relation (12.36) holds for the discrete-state model (Andrieux and Gaspard, 2006b). The mean reaction rate in equation (12.46) is recovered as the first cumulant $J_c = \partial_\lambda \mathcal{Q}(0; A)$ and its second cumulant is giving the associated diffusivity $D_c = (-1/2)\partial_\lambda^2 \mathcal{Q}(0; A)$ (Kolomeisky and Fisher, 2007).

Remarkably, the univariate fluctuation relation (12.36) can be verified on a relatively short timescale if an external torque close to the value of the stall torque is exerted, as shown in Figure 12.7. Here, the rotation is counted by the number s of substeps, which is twice the number n of consumed ATP molecules, so that the univariate fluctuation relation (12.36) reads $P(s,t) \sim P(-s,t)\exp(sA/2)$ with the mechanochemical affinity (12.25). This result is confirmed in Figure 12.7 slightly below and above the value of the stall torque $\tau_{\text{ext, stall}}$ for the given concentrations. Away from the stall torque, the timescale for the verification of the fluctuation relation can significantly increase (Andrieux and Gaspard, 2006b). The reason is that in the absence of opposed external torque, the motion is quasi unidirectional by forward steps and substeps. However, the backward steps and substeps should have a

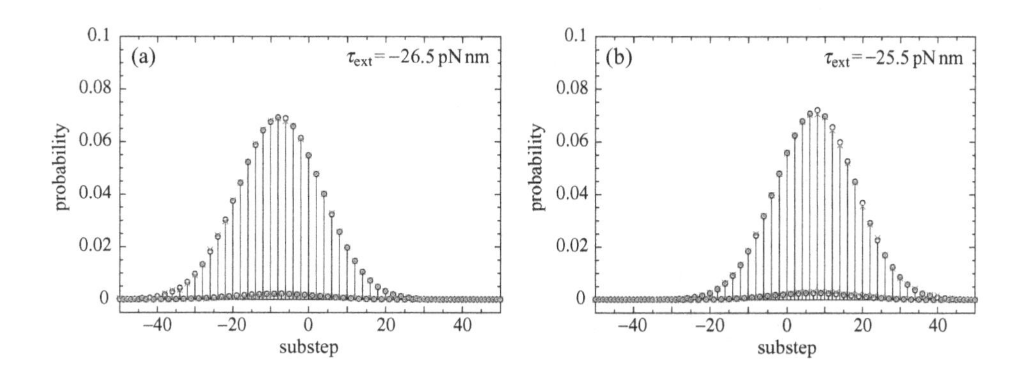

Figure 12.7 Probability distribution $P(s,t)$ (open circles) versus the number s of substeps in the rotation of the F_1 motor during the time interval $t = 10$ s compared with the prediction $P(-s)\,e^{sA/2}$ (crosses) of the univariate fluctuation relation for [ATP] $= 3 \times 10^{-6}$ M, [ADP][P_i] $= 10^{-6}$ M^2, and the torques: (a) $\tau_{\text{ext}} = -26.5$ pN nm; (b) $\tau_{\text{ext}} = -25.5$ pN nm. The diameter of the bead is $d = 80$ nm and the temperature of 23°C. For the given concentrations, the mechanochemical affinity (12.25) is equal to zero for the value $\tau_{\text{ext, stall}} = -26.02$ pN nm of the external torque. The counting statistics is obtained using direct simulation by the Gillespie algorithm over 10^6 random trajectories of the discrete-state model (12.38)–(12.39). Republished from Gerritsma and Gaspard (2010) with the permission of World Scientific.

significant probability in order to have some overlap between the probability distributions $P(\pm s,t)$ to test the fluctuation relation, which is the case in the vicinity of the stall torque (Gerritsma and Gaspard, 2010).

12.3 Active Particles

12.3.1 Self-Propulsion by Catalytic Reaction and Diffusiophoresis

Artificial active particles with sizes ranging from a few hundred nanometers to micrometers can be made of silica spheres with some portion of their surface coated by platinum for the following catalytic reaction with an aqueous solution of hydrogen peroxide,

$$2\,H_2O_2 \longrightarrow 2\,H_2O + O_2. \tag{12.50}$$

This reaction provides free energy to drive the propulsion up to velocities of the order of 10 μm/s when these synthetic motors are immersed in H_2O_2 solutions (Valadares et al., 2010). If the orientation of these silica-platinum particles is not maintained (e.g., by some external torque), they undergo rotational diffusion and the manifestation of their propulsion is the enhancement of their translational diffusion.

These solid colloidal particles have no moving parts and the mechanism of their propulsion is self-diffusiophoresis, which is isothermal. This interfacial phenomenon results from the different interaction forces of the molecular species composing the solution with the atoms of the particle surface. The mechanism can be explained for a spherical Janus particle with catalytic and noncatalytic hemispheres, as shown in Figure 12.8. We may consider the generic surface reaction

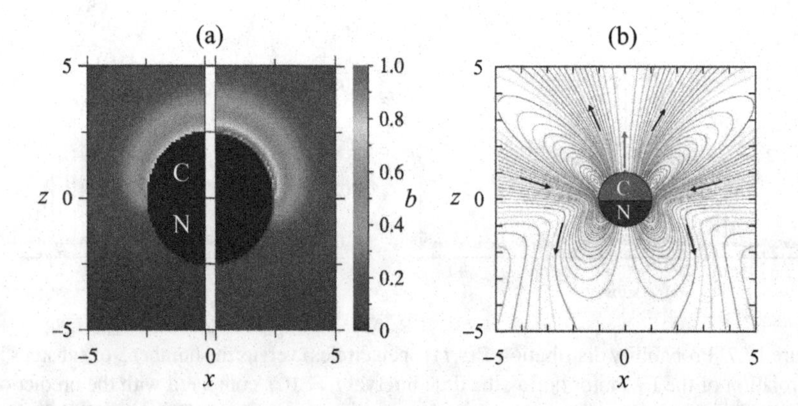

Figure 12.8 (a) Rescaled concentration field $b = [B]/([A] + [B])$ of the product species B around the Janus particle by theory (left) and simulation (right). The upper hemisphere is catalytic. The vertical axis z corresponds to the axis of the Janus particle. (b) Streamlines of the velocity field (12.66) around a Janus particle in the laboratory frame where the particle is propelled by self-diffusiophoresis. The motor here behaves as a pusher with $\Upsilon a_2 < 0$. The vertical arrow denotes the direction of particle motion and the other arrows the fluid flow. Republished with permission of the Royal Society from Reigh et al. (2016).

$$\mathrm{A} + \mathrm{C} \underset{\kappa_-^{\mathrm{s}}}{\overset{\kappa_+^{\mathrm{s}}}{\rightleftharpoons}} \mathrm{B} + \mathrm{C}, \tag{12.51}$$

where the rate constants $\kappa_{\pm}^{\mathrm{s}}$ are positive on the catalytic portion C of the Janus particle, and zero otherwise. It is assumed that there is an excess of species A in the solution, in which case A is the fuel or reactant and B the product. At large distances from the Janus particle, the concentrations of these species are supposed to take the uniform and invariant values $\overline{[\mathrm{A}]}$ and $\overline{[\mathrm{B}]}$. Accordingly, the chemical affinity powering the propulsion is given by

$$A_{\mathrm{c}} = \beta\,\Delta\mu = \ln \frac{\kappa_+^{\mathrm{s}}\,\overline{[\mathrm{A}]}}{\kappa_-^{\mathrm{s}}\,\overline{[\mathrm{B}]}}. \tag{12.52}$$

The molecular species $k \in \{\mathrm{A},\mathrm{B}\}$ are transported in the solution by molecular diffusion characterized by the diffusion coefficients, which take the typical values $\mathcal{D}_k \simeq 10^{-9}$ m^2/s in aqueous solutions.

In the vicinity of the catalytic hemisphere, the interfacial reaction (12.51) causes a depletion of reactant A and an excess of product B, as illustrated in Figure 12.8(a). As a consequence of the concentration gradients along the surface of the particle, the molecular interaction forces of the solution with the catalytic hemisphere are different than with the noncatalytic one. Therefore, a slippage is induced between the fluid and solid velocities from one hemisphere to the other and a fluid flow is generated as shown in Figure 12.8(b), which gives rise to the propulsion of the particle in the fluid.

In order to evaluate the propulsion velocity of the particle by self-diffusiophoresis, the velocity and concentration fields should be determined around the particle. The problem is thus to solve the fluctuating Navier–Stokes equations (10.124) for the velocity field coupled to the fluctuating advection–diffusion equations for the concentration fields,

$$\partial_t\,[k] + \nabla \cdot \mathcal{J}_k = 0, \qquad \text{with} \qquad \mathcal{J}_k = [k]\,\mathbf{v} - \mathcal{D}_k \nabla [k] + \delta \mathcal{J}_k, \tag{12.53}$$

where $[k] \equiv n_k$ denotes the concentration or density of species k and $\delta\mathcal{J}_k$ the fluctuating diffusive current density characterized by equation (10.74). These stochastic partial differential equations are coupled together by boundary conditions at the surface of the moving Janus particle. This particle is assumed to be spherical with radius R, so that its surface $\Sigma(t)$ is given by $\|\mathbf{r} - \mathbf{R}(t)\| = R$, where $\mathbf{R}(t)$ is the center of the particle, assumed to coincide with the center of mass. Moreover, the orientation of the particle axis is specified by the unit vector $\mathbf{u}(t)$ pointing in the direction of the catalytic hemisphere.

The boundary conditions are obtained by considering all the interfacial processes, namely, the catalytic surface reaction, the sliding friction due to the partial slip of the velocity field between the fluid and the solid, and the diffusive transport of species A and B along the interface. The surface reaction (12.51) locally runs with the rate surface density

$$w^{\mathrm{s}} = \kappa_+^{\mathrm{s}}\,[\mathrm{A}] - \kappa_-^{\mathrm{s}}\,[\mathrm{B}] + \xi^{\mathrm{s}}(\mathbf{r},t), \tag{12.54}$$

where $\xi^{\mathrm{s}}(\mathbf{r},t)$ is the interfacial Gaussian white noise field associated with the surface reaction and characterized by equation (10.78). The associated affinity is given by equation (12.52) far from the particle where the concentrations are uniform. The stoichiometric coefficients of both species in the reaction are given by $\nu_A = -1$ and $\nu_B = +1$. In the regime of linear response, the other processes are coupled together according to

$$\mathcal{J}^s_{v\|} = \mathcal{L}^s_{v,v} \mathcal{A}^s_{v\|} + \sum_k \mathcal{L}^s_{v,k} \mathcal{A}^s_{k\|} + \delta\mathcal{J}^s_{v\|}, \tag{12.55}$$

$$\mathcal{J}^s_{k\|} = \mathcal{L}^s_{k,v} \mathcal{A}^s_{v\|} + \sum_l \mathcal{L}^s_{k,l} \mathcal{A}^s_{l\|} + \delta\mathcal{J}^s_{k\|}, \tag{12.56}$$

where $\mathcal{A}^s_{\alpha\|}$ and $\mathcal{J}^s_{\alpha\|}$ are the affinity and current density for interfacial slippage if $\alpha = \mathrm{v}$ and for the interfacial diffusive transport of species k if $\alpha = k$, as they are defined in Table A.3 of Section A.9. The corresponding fluctuations $\delta\mathcal{J}^s_{\alpha\|}$ are Gaussian white noise fields characterized by equations (10.5) and (10.6) according to the fluctuation–dissipation theorem. The interface is assumed to be in local equilibrium with the solution, so that they share the same values for the chemical potentials and the temperature locally along the solid surface: $\mu^s_k = \mu_k$ and $T^s = T$. In equation (12.55), the current density of interfacial slippage can be expressed as $\mathcal{J}^s_{v\|} = \mathbf{n}\cdot\mathbf{\Pi}\cdot\mathbf{1}_\|$ in terms of the viscous pressure tensor $\mathbf{\Pi} = -\eta\left(\nabla\mathbf{v} + \nabla\mathbf{v}^{\mathrm{T}}\right)$, the unit vector $\mathbf{n}$ normal to the surface, and $\mathbf{1}_\| = \mathbf{1} - \mathbf{nn}$ (Bedeaux et al., 1976; Gaspard and Kapral, 2018b,c). The associated affinity is given by $\mathcal{A}^s_{v\|} = -\mathbf{v}_{\mathrm{slip}}/T$, where $\mathbf{v}_{\mathrm{slip}}$ is the slip velocity between the fluid and the solid. As in Section 10.7.1, the sliding friction coefficient is identified as $\lambda \equiv \mathcal{L}^s_{v,v}/T$, giving the slip length (10.131). Diffusiophoresis is characterized by the coefficients $\mathcal{L}^s_{v,k}$ coupling the current density $\mathcal{J}^s_{v\|}$ to the affinity of species k, which is given by $\mathcal{A}^s_{k\|} = -k_B(\nabla_\|[k])/[k]$, using local equilibrium and the expression for chemical potentials in dilute solutions. In equation (12.56), the coefficients $\mathcal{L}^s_{k,l}$ determine the interfacial diffusion coefficients as mentioned in Section A.9.

Because of microreversibility, the linear response coefficients obey the following Onsager–Casimir reciprocal relations,

$$\mathcal{L}^s_{k,l} = \mathcal{L}^s_{l,k} \quad \text{and} \quad \mathcal{L}^s_{v,k} = -\mathcal{L}^s_{k,v}, \tag{12.57}$$

because the affinities $\mathcal{A}^s_{k\|}$ and $\mathcal{A}^s_{l\|}$ have the same parity under time reversal, although $\mathcal{A}^s_{v\|}$ and $\mathcal{A}^s_{k\|}$ have opposite parities according to Table A.3. In this respect, the coefficients of diffusiophoresis do not contribute to entropy production and there are no statistical correlations between the fluctuations $\delta\mathcal{J}^s_{v\|}$ and $\delta\mathcal{J}^s_{k\|}$ according to the expression (10.6) of the fluctuation–dissipation theorem.

The strength of diffusiophoresis can be evaluated in terms of the molecular interaction potentials u_k between the molecules of species k in the solution and the solid surface. These potentials have a range δ of a few nanometers, localized in the thin layer $R < r < R + \delta$ of the solution in the close vicinity of the solid surface. In this thin layer, the Navier–Stokes equations (10.124) are modified by the following force density of molecular origin, $\boldsymbol{f}_{\mathrm{fs}} = -\sum_k [k]\,\nabla u_k$, in addition to the force density $\boldsymbol{f}_{\mathrm{g}}$, which instead extends over the whole solution. Solving the Navier–Stokes equations with the additional force density (Anderson, 1989; Gaspard and Kapral, 2018c), the linear response coefficient of diffusiophoresis can be expressed in terms of the following diffusiophoretic constants,

$$b_k \equiv k_{\rm B} \frac{\mathcal{L}_{{\rm v}k}}{\lambda[k]} = \frac{k_{\rm B}T}{\eta} \int_R^{R+\delta} (r - R + b) \left[{\rm e}^{-\beta u_k(r)} - 1 \right] dr. \tag{12.58}$$

For a solid particle moving in solution with the translational velocity $\mathbf{V}(t) = d\mathbf{R}(t)/dt$ and the rotational velocity $\boldsymbol{\Omega}(t)$, the boundary condition on the fluid velocity in the tangential direction $\mathbf{1}_{\parallel}$ is thus given according to equation (12.55) by the following expression,

$$\begin{aligned} \lambda\, \mathbf{1}_{\parallel} \cdot \mathbf{v}_{\rm slip} &\equiv \lambda\, \mathbf{1}_{\parallel} \cdot \{\mathbf{v}(\mathbf{r},t) - \mathbf{V}(t) - \boldsymbol{\Omega}(t) \times [\mathbf{r} - \mathbf{R}(t)]\} \\ &= \mathbf{1}_{\parallel} \cdot \left\{ \eta \left(\nabla\mathbf{v} + \nabla\mathbf{v}^{\rm T} \right) \cdot \mathbf{n} - \lambda \sum_k b_k \nabla[k] + f_{\rm fl}^{\rm s} \right\}_{\mathbf{r},t} , \end{aligned} \tag{12.59}$$

where $f_{\rm fl}^{\rm s} = \delta \mathcal{J}_{{\rm v}\parallel}^{\rm s}$ is the interfacial fluctuating force characterized by equation (10.130). In the absence of diffusiophoresis, i.e., for $b_k = 0$, the boundary condition (10.129) is recovered. In the direction $\mathbf{n}$ perpendicular to the surface, the boundary condition (10.128) still holds.

Furthermore, the concentration fields of reactant A and product B should satisfy the boundary conditions

$$-\mathcal{D}_k\, \mathbf{n} \cdot \nabla[k] = \nu_k\, w^{\rm s} - \Sigma_k^{\rm s} \qquad \text{for} \qquad \|\mathbf{r} - \mathbf{R}(t)\| = R, \tag{12.60}$$

where $w^{\rm s}$ is the rate surface density (12.54) and $\Sigma_k^{\rm s} = \partial_t \Gamma_k + \nabla_{\parallel} \cdot (\Gamma_k \mathbf{v}^{\rm s} + \mathcal{J}_{k\parallel}^{\rm s})$, as a consequence of equation (12.56) and equation (A.40) (Gaspard and Kapral, 2018b). The counting statistics of the reactive events A $\rightarrow$ B can be carried out with the counter n of reactant molecules A that are consumed such that $dn/dt = -dN_{\rm A}/dt$, where $N_{\rm A} = \int [{\rm A}]\, d^3r$ is the total number of A molecules in the solution.

In principle, the translational and rotational velocities thus evolve in time together with the counter n of reactive events according to

$$m \frac{d\mathbf{V}}{dt} = -\int_{\Sigma(t)} \mathsf{P}(\mathbf{r},t) \cdot \mathbf{n}\, d\Sigma + \mathbf{F}_{\rm ext}, \tag{12.61}$$

$$\mathsf{I} \cdot \frac{d\boldsymbol{\Omega}}{dt} = -\int_{\Sigma(t)} \Delta\mathbf{r} \times \left[\mathsf{P}(\mathbf{r},t) \cdot \mathbf{n} \right] d\Sigma + \mathbf{T}_{\rm ext}, \tag{12.62}$$

$$\frac{dn}{dt} = -\int_{\Sigma(t)} \mathcal{J}_{\rm A} \cdot \mathbf{n}\, d\Sigma, \tag{12.63}$$

where $m = (4\pi R^3/3)\rho_{\rm s}$ is the mass of the solid particle, $\Sigma(t)$ its moving surface of surface element $d\Sigma$, P the pressure tensor (10.125) of the fluid, $\mathbf{F}_{\rm ext}$ an external force exerted on the particle, $(\mathsf{I})_{ij} = \int_{V(t)} \rho_{\rm s} \left(\Delta\mathbf{r}^2 \delta_{ij} - \Delta r_i \Delta r_j \right) d^3r$ the inertia moments of the solid particle, $\Delta\mathbf{r} \equiv \mathbf{r} - \mathbf{R}(t)$, $\mathbf{T}_{\rm ext}$ an external torque exerted on the particle, and $\mathcal{J}_{\rm A}$ the diffusive current density of molecular species A (Gaspard and Kapral, 2018b, 2019b). The time evolution of the orientation of the Janus particle is thus ruled by

$$\frac{d\mathbf{u}}{dt} = \boldsymbol{\Omega} \times \mathbf{u}. \tag{12.64}$$

The timescales of the different processes taking place around a micrometric active particle moving in a dilute aqueous solution are given in Table 12.1. The translational and

Table 12.1. *Timescales of a spherical colloidal particle of radius $R = 1\,\mu\mathrm{m}$ mostly composed of silica ($\rho_\mathrm{s} = 2200\,\mathrm{kg/m^3}$) moving in an aqueous solution at temperature 20°C, with the self-diffusiophoretic velocity $V_\mathrm{sd} = 10\,\mu\mathrm{m/s}$ and stick boundary conditions ($b = 0$), together with the timescales of the velocity and concentration fields around it. The density of water is about $\rho_\mathrm{f} \simeq 10^3\,\mathrm{kg/m^3}$, its shear viscosity $\eta \simeq 10^{-3}\,\mathrm{N\,s/m^2}$, its kinematic viscosity $\nu = \eta/\rho_\mathrm{f} \simeq 10^{-6}\,\mathrm{m^2/s}$, the molecular diffusion coefficients about $\mathcal{D}_k \simeq 10^{-9}\,\mathrm{m^2/s}$, and the sound velocity $v_\mathrm{sound} \simeq 1440\,\mathrm{m/s}$. The translational and rotational diffusivities of the micrometric particle are denoted D_t and D_r and evaluated by the Stokes–Einstein formulas.*

Timescale	Order of Magnitude
Sound	$t_\mathrm{sound} \sim \frac{R}{v_\mathrm{sound}} \simeq 7\times10^{-10}$ s
Rotational thermalization	$t_\mathrm{rot\ therm} = \frac{I}{\gamma_\mathrm{r}} = \frac{R^2}{15\eta}\rho_\mathrm{s} \simeq 1.5\times10^{-7}$ s
Translational thermalization	$t_\mathrm{transl\ therm} = \frac{m}{\gamma_\mathrm{t}} = \frac{2R^2}{9\eta}\rho_\mathrm{s} \simeq 5\times10^{-7}$ s
Hydrodynamics	$t_\mathrm{hydro} \sim \frac{R^2}{\eta}\rho_\mathrm{f} = \frac{R^2}{\nu} \simeq 10^{-6}$ s
Molecular diffusion	$t_\mathrm{mol\ diff} \sim \frac{R^2}{\mathcal{D}_k} \simeq 10^{-3}$ s
Propulsion	$t_\mathrm{prop} = \frac{R}{V_\mathrm{sd}} \simeq 10^{-1}$ s
Rotational diffusion	$t_\mathrm{rot\ diff} = \frac{1}{2D_\mathrm{r}} \simeq 3$ s
Translational diffusion	$t_\mathrm{transl\ diff} \sim \frac{R^2}{D_\mathrm{t}} \simeq 5$ s

rotational diffusivities of the micrometric particle are thus significantly slower than molecular diffusion and hydrodynamics. If this case, the Lorentz condition (10.159) is valid and the hydrodynamic long-time tail effects are negligible. Accordingly, the velocity and concentration fields can be obtained in the static approximation, assuming that these fields are quasi stationary in the frame of the particle. With this approximation, the concentration fields are given by

$$[k] = \overline{[k]} + \frac{\nu_k R}{\mathcal{D}_k}\left(\kappa_+^\mathrm{s}\overline{[\mathrm{A}]} - \kappa_-^\mathrm{s}\overline{[\mathrm{B}]}\right) f(r,\theta) \quad \text{with} \quad f(r,\theta) = \sum_{l=0}^{\infty} a_l\, P_l(\cos\theta)\left(\frac{R}{r}\right)^{l+1}, \tag{12.65}$$

where θ is the angle between the axis $\mathbf{u}$ and the position vector $\mathbf{r} - \mathbf{R}(t)$, $P_l(\xi)$ are the Legendre polynomials (Abramowitz and Stegun, 1972), and a_l are dimensionless coefficients calculated by solving the diffusion equation $\nabla^2 f = 0$ with the boundary condition $R\,\partial_r f|_R = (\mathrm{Da}\, f - 1)_R$ on the catalytic hemisphere, $\partial_r f|_R = 0$ on the noncatalytic one, and $f|_\infty = 0$, expressed in terms of the Damköhler number (10.81) (Gaspard and Kapral, 2018b). With the boundary condition (12.59) in the static approximation, the velocity field can thus be obtained giving

$$\mathbf{v} = -\frac{3}{2}\,\Upsilon\, a_2\left(\frac{R}{r}\right)^2\left[3(\mathbf{n}\cdot\mathbf{u})^2 - 1\right]\mathbf{n} + O(r^{-3}), \tag{12.66}$$

where $\mathbf{n} = [\mathbf{r} - \mathbf{R}(t)]/\|\mathbf{r} - \mathbf{R}(t)\|$,

$$\Upsilon \equiv \left(\frac{b_{\rm B}}{\mathcal{D}_{\rm B}} - \frac{b_{\rm A}}{\mathcal{D}_{\rm A}}\right)\left(\kappa_+^{\rm s}\overline{[{\rm A}]} - \kappa_-^{\rm s}\overline{[{\rm B}]}\right) \tag{12.67}$$

is the parameter characterizing the effect of diffusiophoresis on the velocity field, and a_2 is the third coefficient of the expansion of the function $f(r,\theta)$ in equation (12.65) (Reigh et al., 2016; Campbell et al., 2019). The active particle behaves as a puller if $\Upsilon a_2 > 0$ and a pusher if $\Upsilon a_2 < 0$. The velocity and concentration fields are shown in the example of Figure 12.8 (Reigh et al., 2016).

Moreover, in the same approximation, the translation friction coefficient $\gamma_{\rm t} = \gamma$ is given by equation (10.142) and the rotational one by

$$\gamma_{\rm r} = \frac{8\pi\eta R^3}{1 + 3b/R} \tag{12.68}$$

for a spherical particle of radius R and hydrophobicity characterized by the slip length b (Felderhof, 1976; Gaspard and Kapral, 2018b). The corresponding diffusivities are given by the Stokes–Einstein formulas $D_{\rm t} = k_{\rm B}T/\gamma_{\rm t}$ and $D_{\rm r} = k_{\rm B}T/\gamma_{\rm r}$.

Accordingly, equations (12.61)–(12.63) become

$$m\,\frac{d\mathbf{V}}{dt} = -\gamma_{\rm t}\,\mathbf{V} + \mathbf{F}_{\rm sd} + \mathbf{F}_{\rm ext} + \mathbf{F}_{\rm fl}(t), \tag{12.69}$$

$$\mathsf{I}\cdot\frac{d\boldsymbol{\Omega}}{dt} = -\gamma_{\rm r}\,\boldsymbol{\Omega} + \mathbf{T}_{\rm sd} + \mathbf{T}_{\rm ext} + \mathbf{T}_{\rm fl}(t), \tag{12.70}$$

$$\frac{dn}{dt} = J_{\rm c} + J_{\rm sd} + J_{\rm fl}(t), \tag{12.71}$$

in terms of the friction coefficients, the self-diffusiophoretic force $\mathbf{F}_{\rm sd}$, the fluctuating force $\mathbf{F}_{\rm fl}(t)$, the self-diffusiophoretic torque $\mathbf{T}_{\rm sd}$, the mean reaction rate $J_{\rm c}$ on the whole catalytic surface of the particle, the contribution $J_{\rm sd}$ of self-diffusiophoresis to the reaction rate, and the noise $J_{\rm fl}(t)$ on the reaction rate. The self-diffusiophoretic force $\mathbf{F}_{\rm sd}$ can be obtained using the surface average $\overline{(\cdot)}^{\rm s} = \left(4\pi R^2\right)^{-1}\int_{r=R}(\cdot)\,d\Sigma$ according to

$$\mathbf{F}_{\rm sd} = \frac{6\pi\eta R}{1 + 3b/R}\sum_k \overline{b_k\,\mathbf{1}_\parallel\cdot\nabla[k]}^{\rm s} = \frac{4\pi\eta R}{1 + 3b/R}\,\Upsilon\,a_1\mathbf{u}, \tag{12.72}$$

where b_k are the diffusiophoretic constants (12.58), Υ is given by equation (12.67), a_1 is the second coefficient for $f(r,\theta)$ in equation (12.65), and $\mathbf{u}$ is the unit vector giving the orientation of the Janus particle. Because of the axisymmetry of the Janus particle, the self-diffusiophoretic torque is equal to zero, i.e., $\mathbf{T}_{\rm sd} = 0$. The mean reaction rate can also be evaluated, leading to

$$J_{\rm c} = \Gamma\left(\kappa_+^{\rm s}\overline{[{\rm A}]} - \kappa_-^{\rm s}\overline{[{\rm B}]}\right) \tag{12.73}$$

with $\Gamma = 4\pi R^2 a_0$, where a_0 is the first coefficient of the function introduced in equation (12.65) (Gaspard and Kapral, 2018b). We note that the self-diffusiophoretic contribution $J_{\rm sd}$ to the reaction rate is given, in principle, by the surface integral of the last term $\Sigma_{\rm A}^{\rm s}$ in equation (12.60). We shall see below that microreversibility directly fixes the form of $J_{\rm sd}$.

We note that in the linear regime close to chemical equilibrium, the affinity (12.52) can be expressed as $A_c = J_c/D_c$ in terms of the mean reaction rate (12.73) and the corresponding diffusivity

$$D_c = \frac{1}{2}\Gamma\left(\kappa_+^s\overline{[A]} + \kappa_-^s\overline{[B]}\right). \tag{12.74}$$

12.3.2 Overdamped Langevin Process for Motion and Reaction

As seen in Table 12.1, the timescales for the thermalization of the translational and rotational velocities of the micrometric particle are much shorter than the corresponding diffusion timescales. Therefore, the translational and rotational Langevin processes proceed in the overdamped regime for a micrometric particle, which justifies that the inertial terms with the mass m and the inertia tensor I can be neglected in the Langevin equations (12.69) and (12.70). Because of equation (12.64), the stochastic equation for the orientation of the Janus particle is thus given by

$$\frac{d\mathbf{u}}{dt} = [\boldsymbol{\Omega}_{\rm ext} + \boldsymbol{\Omega}_{\rm fl}(t)] \times \mathbf{u} \quad \text{with} \quad \boldsymbol{\Omega}_{\rm ext} \equiv \frac{\mathbf{T}_{\rm ext}}{\gamma_{\rm r}} \quad \text{and} \quad \boldsymbol{\Omega}_{\rm fl}(t) \equiv \frac{\mathbf{T}_{\rm fl}(t)}{\gamma_{\rm r}}, \tag{12.75}$$

which is decoupled from the equations for the position $\mathbf{R}(t)$ of the Janus particle and the counter $n(t)$ of the reaction. However, the overdamped equations (12.69) and (12.71) are coupled together and they can be expressed as

$$\frac{d\mathbf{R}}{dt} = \mathbf{V}_{\rm sd} + \beta D_{\rm t}\mathbf{F}_{\rm ext} + \mathbf{V}_{\rm fl}(t), \tag{12.76}$$

$$\frac{dn}{dt} = J_{\rm c} + J_{\rm sd} + J_{\rm fl}(t), \tag{12.77}$$

where the self-diffusiophoretic velocity is given by

$$\mathbf{V}_{\rm sd} \equiv \frac{\mathbf{F}_{\rm sd}}{\gamma_{\rm t}} = V_{\rm sd}\,\mathbf{u} \qquad \text{with} \qquad V_{\rm sd} = \frac{2\,\Upsilon\, a_1}{3(1+2b/R)} = \chi\, J_{\rm c} \tag{12.78}$$

and the self-diffusiophoretic parameter

$$\chi \equiv \frac{V_{\rm sd}}{J_{\rm c}} = \frac{a_1}{6\pi a_0 R^2(1+2b/R)}\left(\frac{b_{\rm B}}{\mathcal{D}_{\rm B}} - \frac{b_{\rm A}}{\mathcal{D}_{\rm A}}\right). \tag{12.79}$$

The coupled overdamped Langevin equations (12.76) and (12.77) rule the time evolution in the linear regime close to equilibrium, so that they should be written in terms of the vector $\mathbf{X} = (\mathbf{R}, n)$, containing the position $\mathbf{R}$ of the particle and the counter n of the reaction, the corresponding mechanical and chemical affinities $\mathbf{A} = (\mathbf{A}_{\rm m} = \beta\,\mathbf{F}_{\rm ext}, A_{\rm c} = J_{\rm c}/D_{\rm c})$, and the matrix L of the linear response coefficients relating the velocity and the reaction rate to the affinities. In particular, equation (12.76) can be written as $d\mathbf{R}/dt = D_{\rm t}\mathbf{A}_{\rm m} + \chi\, D_{\rm c}\mathbf{u}\, A_{\rm c} + \mathbf{V}_{\rm fl}(t)$, which is thus completely known, contrary to equation (12.77). Since the mechanical and chemical affinities are even under time reversal, the matrix of linear response should be symmetric $\mathsf{L} = \mathsf{L}^{\rm T}$ in order to satisfy the Onsager reciprocal relations and, thus, microreversibility. Therefore, the two coupled overdamped Langevin equations can be expressed as

$$\frac{d\mathbf{X}}{dt} = \mathsf{L} \cdot \mathbf{A} + \delta\mathbf{J}(t) = \begin{pmatrix} D_{\rm t}\,\mathbf{1} & \chi\, D_{\rm c}\,\mathbf{u} \\ \chi\, D_{\rm c}\,\mathbf{u} & D_{\rm c} \end{pmatrix} \cdot \mathbf{A} + \delta\mathbf{J}(t). \tag{12.80}$$

Consequently, the contribution of self-diffusiophoresis to the reaction rate (12.77) is identified as $J_{\rm sd} = \beta\chi D_{\rm c}\mathbf{u} \cdot \mathbf{F}_{\rm ext}$, which depends on the self-diffusiophoretic parameter (12.79) and on the external force. The off-diagonal elements of the matrix L characterize the mechanochemical coupling generated by self-diffusiophoresis.

In equation (12.80), $\delta\mathbf{J}(t) = [\mathbf{V}_{\rm fl}(t), J_{\rm fl}(t)]$ are correlated Gaussian white noises satisfying

$$\langle \mathbf{V}_{\rm fl}(t)\rangle = 0, \qquad \langle \mathbf{V}_{\rm fl}(t)\,\mathbf{V}_{\rm fl}(t')\rangle = 2D_{\rm t}\,\delta(t-t')\,\mathbf{1}, \tag{12.81}$$

$$\langle J_{\rm fl}(t)\rangle = 0, \qquad \langle J_{\rm fl}(t)\,J_{\rm fl}(t')\rangle = 2D_{\rm c}\,\delta(t-t'), \tag{12.82}$$

$$\text{and} \qquad \langle \mathbf{V}_{\rm fl}(t)\,J_{\rm fl}(t')\rangle = 2\chi\, D_{\rm c}\,\mathbf{u}\,\delta(t-t'), \tag{12.83}$$

as a consequence of the fluctuation–dissipation theorem. We note that the fluctuating angular velocity $\boldsymbol{\Omega}_{\rm fl}(t)$ is given by other Gaussian white noises characterized by $\langle\boldsymbol{\Omega}_{\rm fl}(t)\rangle = 0$ and $\langle\boldsymbol{\Omega}_{\rm fl}(t)\,\boldsymbol{\Omega}_{\rm fl}(t')\rangle = 2D_{\rm r}\,\delta(t-t')\,\mathbf{1}$ in terms of the rotational diffusivity $D_{\rm r}$ and having no correlation with $\mathbf{V}_{\rm fl}(t)$ and $J_{\rm fl}(t)$. The second law is satisfied if the diffusivities are such that $D_{\rm t} \geq 0$, $D_{\rm c} \geq 0$, and $D_{\rm t} \geq \chi^2 D_{\rm c}$.

12.3.3 Enhancement of Diffusion

In the absence of external force $\mathbf{F}_{\rm ext} = 0$ and external torque $\mathbf{T}_{\rm ext} = 0$, the propulsion of the Janus particle by the reaction with a nonvanishing mean rate $J_{\rm c} \neq 0$ has the effect of enhancing the random walk of the Janus particle. Indeed, its orientation undergoes isotropic rotational diffusion according to equation (12.75) with $\boldsymbol{\Omega}_{\rm ext} = 0$, so that the unit vector $\mathbf{u}(t)$ has equilibrium fluctuations with

$$\langle \mathbf{u}(t)\rangle_{\rm eq} = 0 \qquad \text{and} \qquad \langle \mathbf{u}(t)\cdot\mathbf{u}(t')\rangle_{\rm eq} = \exp\left(-2D_{\rm r}|t-t'|\right), \tag{12.84}$$

where $D_{\rm r}$ is the rotational diffusivity. Under such circumstances, the variance of the particle displacement can be calculated by integrating equation (12.76) over the time interval $[0,t]$ and using the time-dependent correlation functions (12.81) and (12.84), giving the effective translational diffusivity by

$$D_{\rm t}^{\rm (eff)} = D_{\rm t} + \frac{V_{\rm sd}^2}{6\,D_{\rm r}} \tag{12.85}$$

in terms of the self-diffusiophoretic velocity (12.78) for the Janus particle (Kapral, 2013). At chemical equilibrium where the mean reaction rate (12.73) is equal to zero, i.e., $J_{\rm c} = 0$, the effective diffusivity (12.85) reduces to the diffusion coefficient (4.163) of Brownian motion with the friction coefficient $\gamma_{\rm t} = \gamma$ given by equation (10.142): $D_{\rm t}^{\rm (eff)} = D_{\rm t} = \mathcal{D} = k_{\rm B}T/\gamma_{\rm t}$. If the chemical reaction is active, the effective diffusivity of the Janus particle is enhanced quadratically with the mean reaction rate since $V_{\rm sd} = \chi J_{\rm c}$. This enhancement is inversely proportional to the rotational diffusivity $D_{\rm r}$ because the spatial extension of the random walk is reduced by frequent changes in the orientation.

12.3.4 Mechanochemical Coupling and Efficiencies

In order for the Janus particle to move with a nonzero mean velocity, the propulsion by self-diffusiophoresis should be active with $V_{\rm sd} \neq 0$ and its orientation be maintained on average in the same direction so that $\langle \mathbf{u} \rangle \neq 0$, or an external force $\mathbf{F}_{\rm ext} \neq 0$ should be exerted on its center of mass. Indeed, taking the statistical average of equation (12.80) gives $\langle \dot{\mathbf{X}} \rangle = \mathsf{L} \cdot \mathbf{A}$ because $\langle \delta \mathbf{J}(t) \rangle = 0$. Therefore, equations (12.76)–(12.77) become

$$\langle \dot{\mathbf{R}} \rangle = V_{\rm sd} \, \langle \mathbf{u} \rangle + \beta D_{\rm t} \, \mathbf{F}_{\rm ext}, \tag{12.86}$$

$$\langle \dot{n} \rangle = J_{\rm c} + \beta \chi D_{\rm c} \, \langle \mathbf{u} \rangle \cdot \mathbf{F}_{\rm ext}, \tag{12.87}$$

which are nonzero if the stated assumptions hold. The entropy production rate of the self-diffusiophoretic motor is thus given by

$$\frac{1}{k_{\rm B}} \frac{d_{\rm i} S}{dt} = \mathbf{A}_{\rm m} \cdot \langle \dot{\mathbf{R}} \rangle + A_{\rm c} \, \langle \dot{n} \rangle \geq 0. \tag{12.88}$$

The orientation can be controlled if the Janus particle has a magnetic dipole $\mu \mathbf{u}$ and is subjected to an external magnetic field $\boldsymbol{\mathcal{B}}_{\rm ext}$, whereupon the external torque $\mathbf{T}_{\rm ext} = \mu \mathbf{u} \times \boldsymbol{\mathcal{B}}_{\rm ext}$ is exerted on the orientation $\mathbf{u}$ of the particle. This external torque corresponds to the rotational energy potential $U_{\rm r} = -\mu \boldsymbol{\mathcal{B}}_{\rm ext} \cdot \mathbf{u}$. To be specific, the external magnetic field may be supposed to be aligned in the z-direction $\boldsymbol{\mathcal{B}}_{\rm ext} = (0, 0, \mathcal{B}_{\rm ext})$, as well as the external force $\mathbf{F}_{\rm ext} = (0, 0, F_{\rm ext})$.

Since equation (12.75) describes the equilibrium thermal fluctuations of the orientation, the unit vector $\mathbf{u}$ has the stationary probability distribution $\mathscr{P}_{\rm eq}(\mathbf{u}) = \mathcal{N}^{-1} \exp(-\beta U_{\rm r}) = \mathcal{N}^{-1} \exp(\beta \mu \mathcal{B}_{\rm ext} \cos \theta)$, where θ is the angle between the particle orientation and the external magnetic field (de Groot and Mazur, 1984). As a consequence, the mean values of the three components of the unit vector $\mathbf{u} = (u_x, u_y, u_z)$ are given by

$$\langle u_x \rangle = \langle u_y \rangle = 0 \qquad \text{and} \qquad \langle u_z \rangle = \coth(\beta \mu \mathcal{B}_{\rm ext}) - (\beta \mu \mathcal{B}_{\rm ext})^{-1}. \tag{12.89}$$

In this case, the particle is moving in the z-direction with the mean velocity $\langle \dot{z} \rangle = V_{\rm sd} \langle u_z \rangle + \beta D_{\rm t} F_{\rm ext}$. Therefore, the propulsion by the chemical reaction becomes possible in the z-direction if the particle is oriented by the external magnetic field $\mathcal{B}_{\rm ext}$, even if there is no external force $F_{\rm ext} = 0$, as illustrated in Figure 12.9 showing random trajectories for the rescaled displacement, the reaction counter, and the orientation u_z.

If the external force is opposed to the displacement, the chemical reaction performs mechanical work with the power $-F_{\rm ext} \langle \dot{z} \rangle > 0$ by the self-diffusiophoretic energy transduction of the chemical free energy supplied with the power $k_{\rm B} T A_{\rm c} J_{\rm c} > 0$. The mechanical efficiency of this energy transduction can here also be defined by equation (12.22) in terms of the mechanical affinity $A_{\rm m} = \beta F_{\rm ext}$ and the associated mean current $J_{\rm m} = \langle \dot{z} \rangle$. This efficiency satisfies the condition $0 \leq \eta_{\rm m} \leq 1$ if the external force takes its value in the range $F_{\rm ext, stall} < F_{\rm ext} < 0$, extending between the stall force $F_{\rm ext, stall} \equiv -F_{\rm sd} \langle u_z \rangle = -\gamma_{\rm t} V_{\rm sd} \langle u_z \rangle$ and zero. This regime where the motor can perform mechanical work with the chemical free energy corresponds to domain I in Figure 12.10.

Furthermore, the mean reaction rate (12.87) reads $\langle \dot{n} \rangle = J_{\rm c} + \beta \chi D_{\rm c} \, \langle u_z \rangle \, F_{\rm ext}$. If the Janus particle is oriented by an external magnetic field and subjected to an external force, the last term is nonzero and the mean reaction rate can thus be modified by external mechanical

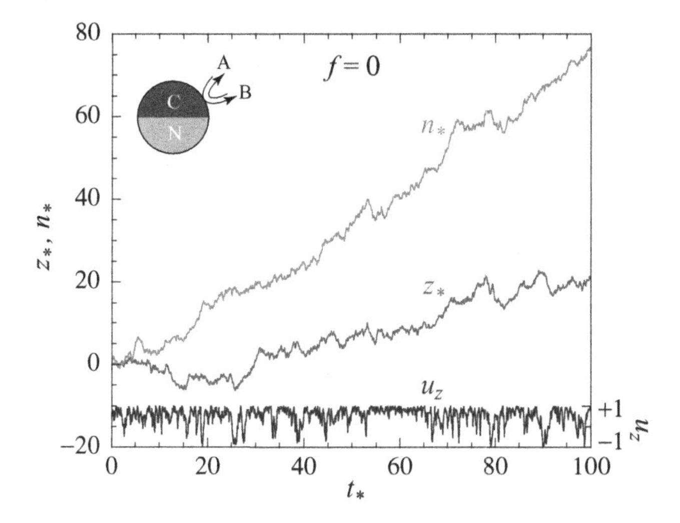

Figure 12.9 Active Janus particle subjected to an external force and magnetic field oriented in the z-direction: The random rescaled displacement $z_* = z\sqrt{D_r/D_t}$, the rescaled number of reactive events $n_* = n\sqrt{D_r/D_c}$, and the particle orientation u_z, versus the rescaled time $t_* = D_r t$ for the parameter values $\beta\mu\mathcal{B}_{ext} = 2$, $J_c/\sqrt{D_c D_r} = 0.8$, and $\chi\sqrt{D_c/D_t} = 0.8$ and a zero external force, i.e., $f = \beta F_{ext}\sqrt{D_t/D_r} = 0$. Adapted from Gaspard and Kapral (2019b).

driving. This effect is reciprocal to propulsion by self-diffusiophoresis according to equation (12.80) implied by microreversibility. The presence of this effect is confirmed by particle-based simulations with a microscopically reversible kinetics, as shown in Figure 12.11 (Huang et al., 2018; Gaspard and Kapral, 2019b).

The mean reaction rate $\langle \dot{n} \rangle$ can even become negative for a large enough external force opposed to the particle displacement in the range $F_{ext} < F_{ext,0} = -A_c/(\beta\chi\langle u_z \rangle) = -\Delta\mu/(\chi\langle u_z \rangle)$. In this regime corresponding to domain II in Figure 12.10, the external mechanical driving of the Janus particle performs the synthesis of the fuel A from the product B. This process of fuel synthesis can be characterized by the chemical efficiency (12.21). As seen in Figure 12.11(b), the simulations confirm that the rescaled mean reaction rate $\langle \dot{n}_* \rangle$ may indeed become negative at chemical equilibrium for $A_c = 0$ and negative mechanical affinity $A_m = \beta F_{ext} = f\sqrt{D_r/D_t}$ (Huang et al., 2018; Gaspard and Kapral, 2019b). Fuel synthesis is possible because of the mechanochemical coupling generated by self-diffusiophoresis.

Domains I and II are schematically plotted in Figure 12.10(b) in the plane of the mechanical and chemical affinities, where the origin corresponds to thermodynamic equilibrium. At given values for the affinities, the efficiencies reach their maximal value $\eta^{(max)}$ in between the boundaries of these domains (Gaspard and Kapral, 2017, 2018b). The comparison with Figure 12.6 shows that active Janus particles and molecular motors share similar properties. In both active processes, domains I and II can be identified in the plane of the chemical and mechanical affinities. A difference is that the mechanochemical coupling is tight for the rotary molecular motor, so that domains I and II are next to each other in Figure 12.6. This is

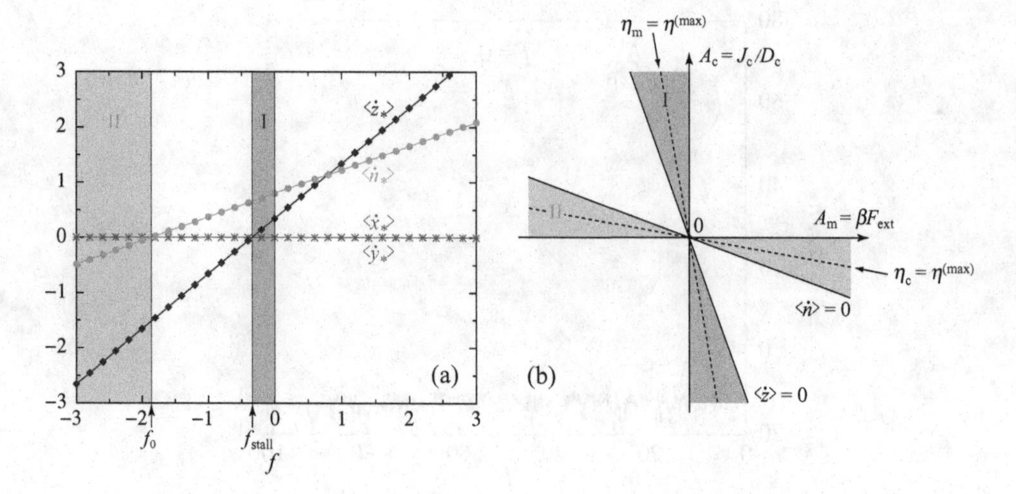

Figure 12.10 (a) Janus particle subjected to an external force and magnetic field oriented in the z-direction (Gaspard and Kapral, 2017): The mean values of the fluctuating rescaled velocities $\dot{\mathbf{r}}_* = \dot{\mathbf{r}}/\sqrt{D_t D_r}$ and rate $\dot{n}_* = \dot{n}/\sqrt{D_c D_r}$ versus the rescaled external force $f = \beta F_{\text{ext}}\sqrt{D_t/D_r}$ for the parameter values $\beta\mu\mathcal{B}_{\text{ext}} = 2$, $J_c/\sqrt{D_c D_r} = 0.8$, and $\chi\sqrt{D_c/D_t} = 0.8$. The dots show the results of numerical simulation as in Figure 12.9 over the time interval $t_* = 10$. Here, f_{stall} denotes the rescaled stall force and f_0 the threshold between fuel synthesis and consumption. The Janus particle is propelled against the external force in domain I. Fuel synthesis happens in domain II. (b) Schematic representation of the different regimes of the active particle in the plane of the mechanical and chemical affinities for $\chi > 0$. In domain I, self-diffusiophoretic mechanical work is powered by the reaction. In domain II, the synthesis of fuel from product is induced by the external force. For $\chi < 0$, the slopes of the lines $\langle \dot{z} \rangle = 0$ and $\langle \dot{n} \rangle = 0$ should instead be positive. Reprinted from Gaspard and Kapral (2017, 2018b) with the permission of AIP Publishing.

not the case for active particles propelled by self-diffusiophoresis, where the mechanochemical coupling is loose, so that domains I and II separate from each other, as shown in Figure 12.10(b).

12.3.5 Mechanochemical Bivariate Fluctuation Relation

For synthetic motors propelled by self-diffusiophoresis, the active stochastic process is also ruled by a bivariate fluctuation relation for the joint probability distribution of the random displacement and supplied chemical free energy, describing the energy transduction generated by mechanochemical coupling (Gaspard and Kapral, 2017, 2018b, 2019b). The joint probability density $\mathcal{P}_t(\mathbf{R}, n; \mathbf{A}_m, A_c)$ of finding the active particle at position $\mathbf{R}$ after n reactive events have occurred during the time interval t obeys the mechanochemical fluctuation relation

$$\frac{\mathcal{P}_t(\mathbf{R}, n; \mathbf{A}_m, A_c)}{\mathcal{P}_t(-\mathbf{R}, -n; \mathbf{A}_m, A_c)} \sim_{t\to\infty} \exp(\mathbf{A}_m \cdot \mathbf{R} + A_c\, n) \tag{12.90}$$

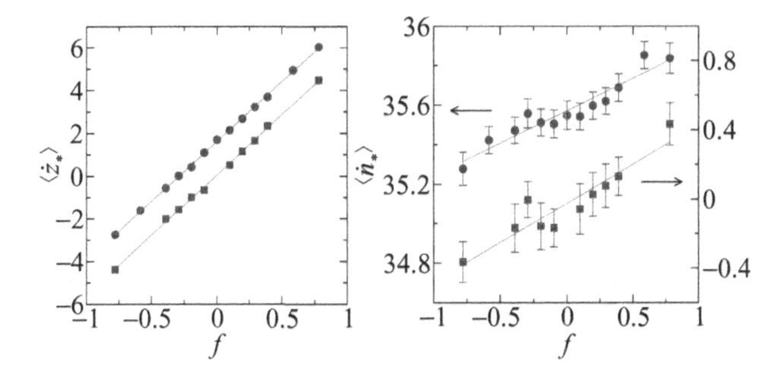

Figure 12.11 Particle-based simulations with microscopically reversible kinetics of the Janus motor subjected to an external force $F_{\rm ext}$ and a magnetic field both oriented in the z-direction: Plots of the dependence on the rescaled force f of the rescaled average motor velocity in the z-direction, $\langle \dot{z}_* \rangle$ (left), and of the rescaled reaction rate, $\langle \dot{n}_* \rangle$ (right) for systems (a) out of chemical equilibrium with $\overline{[{\rm A}]} = 10$ and $\overline{[{\rm B}]} = 9$ (circles), and (b) at chemical equilibrium with $\overline{[{\rm A}]} = 10$ and $\overline{[{\rm B}]} = 10$ (squares) for equal rate constants $\kappa_+^{\rm S} = \kappa_-^{\rm S}$. The results for (a) and (b) systems were obtained from averages over 200 and 100 realizations of the dynamics, respectively. The fits to the data are indicated by (a) upper and (b) lower lines. Reprinted from Gaspard and Kapral (2019b).

with respect to the mechanical affinity $\mathbf{A}_{\rm m} = \beta \mathbf{F}_{\rm ext}$ and the chemical affinity $A_{\rm c} = J_{\rm c}/D_{\rm c}$.

The fluctuation relation (12.90) can be deduced from the following Fokker–Planck equation for the time evolution of the joint probability density $p(\mathbf{X}, \mathbf{u}; t)$ associated with the Langevin equations (12.75) and (12.80),

$$\partial_t p = -\partial_{\mathbf{X}} \cdot (\mathsf{L} \cdot \mathbf{A}\, p - \mathsf{L} \cdot \partial_{\mathbf{X}} p) + \hat{L}_{\rm r}\, p \equiv \hat{L}\, p, \tag{12.91}$$

where the matrix L is given by equation (12.80), $\mathbf{A} = (\mathbf{A}_{\rm m}, A_{\rm c})$, and the operator

$$\hat{L}_{\rm r}\, p = \frac{D_{\rm r}}{\sin\theta} \left\{ \partial_\theta \big[\sin\theta \, (\partial_\theta\, p + \beta\mu\mathcal{B}_{\rm ext} \sin\theta\, p) \big] + \frac{1}{\sin\theta}\, \partial_\phi^2\, p \right\} \tag{12.92}$$

governs the rotational diffusion of the orientation $\mathbf{u} = (\sin\theta\ \cos\phi,\ \sin\theta\ \sin\phi,\ \cos\theta)$ in the external magnetic field $\boldsymbol{\mathcal{B}}_{\rm ext} = (0, 0, \mathcal{B}_{\rm ext})$. The fluctuation relation (12.90) can be proved by introducing the cumulant generating function

$$Q(\boldsymbol{\lambda}; \mathbf{A}) \equiv \lim_{t\to\infty} -\frac{1}{t} \ln \big\langle {\rm e}^{-\boldsymbol{\lambda}\cdot\mathbf{X}(t)} \big\rangle_{\mathbf{A}} \tag{12.93}$$

with the counting parameters $\boldsymbol{\lambda} = (\boldsymbol{\lambda}_{\rm m}, \lambda_{\rm c})$ used to generate the cumulants by successive derivations (Gaspard and Kapral, 2017). The statistical average $\langle \cdot \rangle_{\mathbf{A}}$ is taken with respect to the probability distribution $p(\mathbf{X}, \mathbf{u}; t)$, which evolves in time according to $p = {\rm e}^{\hat{L} t}\, p_0$ from the initial distribution p_0. Accordingly, the average in equation (12.93) can be transformed as follows,

$$\big\langle {\rm e}^{-\boldsymbol{\lambda}\cdot\mathbf{X}(t)} \big\rangle_{\mathbf{A}} = \int {\rm e}^{-\boldsymbol{\lambda}\cdot\mathbf{X}}\, {\rm e}^{\hat{L} t}\, p_0\, d\mathbf{X}\, d\mathbf{u} = \int {\rm e}^{\hat{L}_{\boldsymbol{\lambda}} t}\, {\rm e}^{-\boldsymbol{\lambda}\cdot\mathbf{X}}\, p_0\, d\mathbf{X}\, d\mathbf{u}, \tag{12.94}$$

by introducing the modified operator

$$\hat{L}_{\boldsymbol{\lambda}} \equiv {\rm e}^{-\boldsymbol{\lambda}\cdot\mathbf{X}} \hat{L}\, {\rm e}^{+\boldsymbol{\lambda}\cdot\mathbf{X}} = -\mathbf{A} \cdot \mathsf{L} \cdot (\partial_{\mathbf{X}} + \boldsymbol{\lambda}) + (\partial_{\mathbf{X}} + \boldsymbol{\lambda}) \cdot \mathsf{L} \cdot (\partial_{\mathbf{X}} + \boldsymbol{\lambda}) + \hat{L}_{\rm r}. \tag{12.95}$$

In the long-time limit, the expression (12.94) will decay at a rate given by the leading eigenvalue of the modified operator that is the solution of the eigenvalue equation (5.217). Now, the modified operator (12.95) has the remarkable symmetry

$$\eta^{-1}\,\hat{L}_{\boldsymbol{\lambda}}\,\eta = \hat{L}^{\dagger}_{\mathbf{A}-\boldsymbol{\lambda}} \qquad \text{with} \qquad \eta = \exp(\beta\mu\mathcal{B}_{\text{ext}}\cos\theta), \tag{12.96}$$

where the adjoint of the operator is defined with respect to the scalar product $\langle g|f\rangle \equiv \int g^* f\, d\mathbf{X}\, d\mathbf{u}$. The symmetry (12.96) is established by noting that $\eta^{-1}\,\hat{L}_{\text{r}}\,\eta = \hat{L}^{\dagger}_{\text{r}}$ and by using $\partial^{\dagger}_{\mathbf{X}} = -\partial_{\mathbf{X}}$ in the first two terms of equation (12.95). As a consequence, the cumulant generating function obeys the symmetry relation (5.220), which implies the bivariate fluctuation relation (12.90) for the marginal probability distribution $\mathcal{P}_t(\mathbf{X}) \equiv \int p(\mathbf{X},\mathbf{u};t)\, d\mathbf{u}$ by using large-deviation theory, as explained in Section 5.7 (Lebowitz and Spohn, 1999; Lacoste and Mallick, 2009). Figure 12.12 shows that the mechanochemical fluctuation relation is indeed satisfied for the coupled overdamped Langevin processes governed by equations (12.75) and (12.80). According to equation (5.98), the mechanochemical fluctuation relation (12.90) implies the nonnegativity of the entropy production rate (12.88).

The mechanochemical fluctuation relation (12.90) has further consequences. Often, only the position is observed while the rate is very large. In the presence of an external force and an external magnetic field in the z-direction such that equation (12.89) holds, an effective fluctuation relation can be obtained for the displacement along the z-direction (Falasco et al., 2016). Since the probability distribution becomes Gaussian after a long enough time by the central limit theorem, the marginal distribution $\mathscr{P}_t(z) \equiv \int \mathcal{P}_t(\mathbf{R},n)\, dx\, dy\, dn$ satisfies the effective fluctuation relation

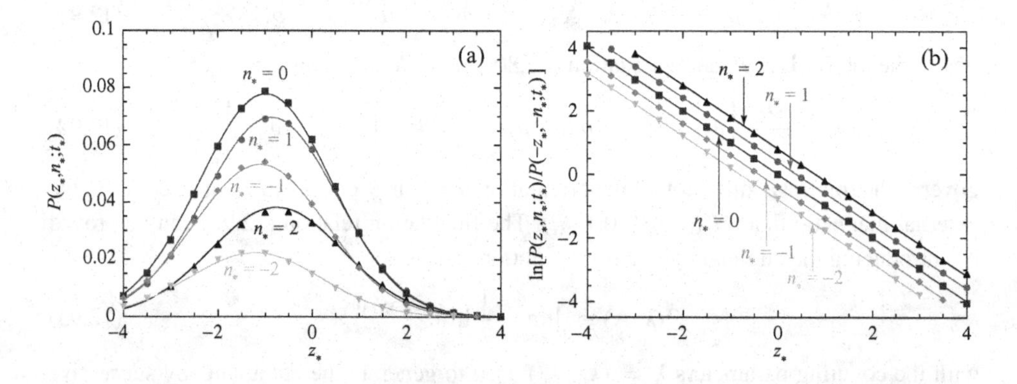

Figure 12.12 Janus particle subjected to an external force and magnetic field oriented in the z-direction: (a) Probability density $P(z_*, n_*; t_*)$ with $n_* = n\sqrt{D_{\text{r}}/D_{\text{c}}} = 2, 1, 0, -1, -2$ versus the rescaled displacement $z_* = z\sqrt{D_{\text{r}}/D_{\text{t}}}$ at the rescaled time $t_* = D_{\text{r}}t = 1$ for the parameter values $\beta\mu\mathcal{B}_{\text{ext}} = 1$, $\beta F_{\text{ext}}\sqrt{D_{\text{t}}/D_{\text{r}}} = -1$, $J_{\text{c}}/\sqrt{D_{\text{c}}D_{\text{r}}} = 0.4$, and $\chi\sqrt{D_{\text{c}}/D_{\text{t}}} = 0.4$. (b) Verification of the mechanochemical fluctuation relation (12.90) with the same conditions. The probability ratio is calculated if $P(z_*, n_*; t_*)$ and $P(-z_*, -n_*; t_*)$ are larger than 10^{-4}. The dots are the results of a numerical simulation with an ensemble of 10^7 trajectories and an integration with the time step $dt_* = 10^{-3}$. The lines depict the theoretical expectations. Reprinted from Gaspard and Kapral (2017) with the permission of AIP Publishing.

$$\frac{\mathscr{P}_t(z)}{\mathscr{P}_t(-z)} \sim_{t\to\infty} \exp\left(\frac{F_{\rm eff}\, z}{k_{\rm B} T_{\rm eff}}\right), \tag{12.97}$$

which is expressed in terms of the effective force $F_{\rm eff} = F_{\rm ext} + F_{\rm sd}\langle u_z\rangle$ resulting from the external and diffusiophoretic forces, and the effective temperature $T_{\rm eff} = T\left[1 + (V_{\rm sd}^2/D_{\rm t}) \int_0^\infty C_{zz}(t)\, dt\right]$ where $C_{zz}(t) \equiv \langle [u_z(0) - \langle u_z\rangle][u_z(t) - \langle u_z\rangle]\rangle$ is the time-dependent autocorrelation function of the orientation along the z-direction. This result justifies the introduction of the concept of effective temperature $T_{\rm eff}$ to describe the enhancement of fluctuations due to nonequilibrium activity in the system. For instance, in the absence of an external magnetic field, i.e., $\mathcal{B}_{\rm ext} = 0$, the mean value of the orientation is equal to zero, i.e, $\langle u_z\rangle = 0$, and the time integral of the autocorrelation function is equal to $\int_0^\infty C_{zz}(t)\, dt = (6D_{\rm r})^{-1}$ because of equation (12.84). As a consequence, the effective temperature is proportional to the effective diffusivity (12.85) according to $T_{\rm eff} = T\left(D_{\rm t}^{\rm (eff)}/D_{\rm t}\right)$, which characterizes the enhancement of diffusivity due to the activity of the self-propulsion by the nonequilibrium reaction.

Similar considerations hold for Janus particles propelled by self-thermophoresis (Gaspard and Kapral, 2019a).

12.3.6 Collective Dynamics

Natural and synthetic motors such as myosin-actin proteins and self-propelled colloidal particles have remarkable properties associated with their collective behavior, including active self-assembly processes into dynamical clusters (Marchetti et al., 2013; Saha et al., 2014; Cates and Tailleur, 2015; Liebchen et al., 2015; Bechinger et al., 2016; Huang et al., 2019), which are reminiscent of the dissipative structures and dynamics existing on larger scales as reported in Section 1.9.

As for the anisotropic molecules composing liquid crystals, the active particles have an orientation given in general by the three Euler angles (θ, ϕ, ψ) in addition to the position $\mathbf{r} = (x, y, z)$ of their center. Accordingly, a system with many such active particles should be described in terms of their distribution function $f(\mathbf{r}, \theta, \phi, \psi; t)$, giving the mean density of particles with the orientation (θ, ϕ, ψ) at position $\mathbf{r}$ and the time t. Since the active particles are usually conserved in the systems of interest, their distribution function is ruled by a local conservation equation, $\partial_t f + {\rm div}\, \boldsymbol{j}_f = 0$, expressed in terms of the divergence and the corresponding current density $\boldsymbol{j}_f$ in the six-dimensional space of coordinates $(x, y, z, \theta, \phi, \psi)$. If the active particles are involved in reactions with molecular species, diffusion–reaction equations such as equation (1.42) should be included in the description.

As long as there is no global fluid flow in this colloidal suspension, the coupling to the Navier–Stokes equations is considered locally around every one of the active particles to determine their propulsion with respect to the suspension, which is the case in many experimental systems (Bechinger et al., 2016). However, the colloidal suspension may also undergo global fluid flows, in which cases the Navier–Stokes equations should be extended to include the effects of activity. In the presence of active particles having local velocity fields decreasing as r^{-2} as in equation (12.66), the pressure tensor is modified by extra terms due to the activity generated by the chemical reactions. As a consequence, new properties manifest themselves in the rheology of active colloidal suspensions (Marchetti et al., 2013; Jülicher et al., 2018).

In conclusion, the concepts of mechanochemical coupling and the corresponding bivariate fluctuation relation bring a unified perspective on the properties of molecular motors and active particles powered by transduction between the mechanical and chemical forms of energy. These concepts play a key role in the formulation of chemohydrodynamics and thermodynamics for active matter in consistency with microreversibility (Gaspard and Kapral, 2020).

13

Transport in Hamiltonian Dynamical Models

13.1 Mathematical Foundations of Transport Properties

The macroscopic description of matter in the frameworks of hydrodynamics and thermodynamics is based on the existence of positive and finite transport coefficients and on related concepts such as the hydrodynamic modes of relaxation towards equilibrium and the associated local entropy production. An essential issue is to understand the foundations of these concepts in the microscopic dynamics of atoms and molecules that compose matter.

In the linear regime, the relationship between the microscopic and macroscopic descriptions is established for the response and transport properties by the Green–Kubo formulas, but the conditions for the mathematical existence and unicity of the response and transport properties and the related concepts still need to be rigorously determined. In particular, a transport coefficient may be infinite if the integral of the time-dependent autocorrelation function does not converge in the corresponding Green–Kubo formula. Since the variety of possible dynamical systems is beyond imagination, the analysis of specific models where the existence of transport properties can be rigorously proved is particularly useful.

Experimental observations show that the microscopic time evolution of atoms and molecules is ruled by quantum mechanics and electrodynamics. At room temperature, a good approximation is provided by classical mechanics, which satisfies Liouville's theorem, as a consequence of quantum unitarity, and the time-reversal symmetry of electrodynamics. The question thus arises as to whether the existence of positive and finite transport coefficients is compatible with the underlying Hamiltonian dynamics satisfying these assumptions.

The transport property of diffusion can be studied in detail in the so-called Lorentz gases, where independent particles undergo elastic collisions on fixed scatterers. In these models, the conditions for the existence of positive and finite diffusion coefficients can be determined and the hydrodynamic modes of exponential relaxation as well as the nonequilibrium stationary distributions can be rigorously obtained. In this respect, the property of dynamical mixing and the concept of Pollicott–Ruelle resonance introduced in Chapter 2 play fundamental roles. The hydrodynamic modes can be constructed as the generalized eigenmodes associated with Pollicott–Ruelle resonances. These eigenmodes are mathematical distributions with fractal cumulative functions and the known expression for the entropy production of diffusion can be deduced from the fractal character of these eigenmodes.

For heat conduction, there exist many-particle billiard models where Fourier's law and the existence of heat conductivity can be established with great rigor. Furthermore, the coupling between two transport processes can also be studied in exactly solvable models, as reported in Section 13.5.

13.2 Diffusion of Noninteracting Particles

13.2.1 Spatially Periodic Lorentz Gases

The transport process of diffusion can be understood in terms of Hamiltonian dynamics in the so-called Lorentz gases. These systems were introduced by Lorentz (1905) in order to model the transport of electrons in solids. Since the electrons are much lighter than the ions forming the lattice of the solid, the ions may be supposed to be fixed upon electron collisions, as if the ions had an arbitrarily large mass. The mobile particles are assumed to be independent of each other and to move in the potential energy landscape of the ions with the dynamics governed by the Hamiltonian function

$$H = \frac{\mathbf{p}^2}{2m} + \sum_{l \in \mathcal{L}} u(\|\mathbf{r} - l\|). \tag{13.1}$$

For each particle, energy is conserved, but not momentum. Several models can be considered. The hard-disk or hard-sphere Lorentz gases have the hard-core potential $u(r) = 0$ if $r > \rho$ and $u(r) = \infty$ if $r < \rho$. Consequently, the particles move in free flight interrupted by specular reflections at the elastic collisions on the boundary of the scatterers. Lorentz gases with a Yukawa potential $u(r) = -r^{-1}\exp(-\alpha r)$ can also be considered. In two-dimensional Lorentz gases, the scattering centers $\{l\}$ may form square or triangular lattices $\mathcal{L}$, among other possibilities. We note that the Hamiltonian function (13.1) has the time-reversal symmetry (2.28), in agreement with the principle of microreversibility.

Two-dimensional Lorentz gases composed of independent point-like particles undergoing elastic collisions on hard disks fixed in the plane have been proved to be ergodic and mixing with respect to the equilibrium distribution (2.69) (Bunimovich and Sinai, 1980a). Moreover, the motion of the point particles has been proved to be diffusive if there is no possible straight flight across the whole lattice, i.e., if the horizon is finite for the point particles (Bunimovich and Sinai, 1980b). This is the case in the triangular lattice if the hard disks have a large enough radius with respect to the interdisk distance. In these models, diffusion is deterministic and yet the trajectories are irregular on long timescales as in random walks. The mechanism underlying dynamical mixing and deterministic diffusion is the sensitivity to initial conditions due to the defocusing character of elastic collisions on hard disks. These two-dimensional systems have one positive Lyapunov exponent, as well as a negative one, because volumes are preserved by the dynamics. Indeed, their dynamics have Hamiltonian character since the free flights and the specular reflections (B.33) result from the hard-core potential. Deterministic diffusion and dynamical mixing have also been proved for the Hamiltonian motion of a point particle in a square lattice $\mathcal{L}$ of Yukawa potentials at high enough energy (Knauf, 1987). These dynamical systems are chaotic and manifest stretching and folding for volumes in phase space, inducing dynamical mixing and diffusion on long timescales.

In Lorentz gases, the particles are independent, so the gas dynamics can be expressed in terms of the flow $\boldsymbol{\Phi}^t$ in the one-particle phase space $\mathbf{x} = (\mathbf{r}, \mathbf{p}) \in \mathfrak{M} = \mathbb{R}^{2f}$, where f is the number of degrees of freedom of the Lorentz gas, i.e., the dimension of the physical space where motion takes place. For spatially periodic Lorentz gases, the position $\mathbf{r} \in \mathbb{R}^f$ can be decomposed as $\mathbf{r} = \boldsymbol{r} + \boldsymbol{l}$ into the position $\boldsymbol{r} \in \mathbb{T}^f$ defined in the torus of the primitive cell of the lattice $\mathcal{L}$, and the lattice vector $\boldsymbol{l} \in \mathcal{L}$. Consequently, the one-particle phase space can be similarly decomposed as $\mathbf{x} = (\boldsymbol{x}, \boldsymbol{l}) \in \mathfrak{M} = \mathfrak{X} \times \mathcal{L}$ with $\boldsymbol{x} = (\boldsymbol{r}, \mathbf{p}) \in \mathfrak{X} = \mathbb{T}^f \times \mathbb{R}^f$.[1] Accordingly, the flow admits the decomposition

$$\boldsymbol{\Phi}^t \mathbf{x} = \boldsymbol{\Phi}^t(\boldsymbol{x}, \boldsymbol{l}) = \left\{ \boldsymbol{\phi}^t \boldsymbol{x}, \boldsymbol{L}\left[\boldsymbol{\Phi}^t(\boldsymbol{x}, \boldsymbol{l})\right] \right\}, \tag{13.2}$$

where $\boldsymbol{L}(\mathbf{x}) = \boldsymbol{L}(\boldsymbol{x}, \boldsymbol{l}) = \boldsymbol{l}$ denotes the lattice vector of the cell where the particle is initially located. In the transport process across the lattice, an important role is played by the lattice displacement of the particle during some time interval $[0, t]$:

$$\boldsymbol{d}(\boldsymbol{x}, t) \equiv \boldsymbol{L}\left[\boldsymbol{\Phi}^t(\boldsymbol{x}, \boldsymbol{l})\right] - \boldsymbol{l}, \tag{13.3}$$

which is independent of the lattice vector $\boldsymbol{l}$ of initial conditions.

For Lorentz gases, the Liouvillian dynamics can be expressed in terms of the time evolution for the one-particle probability density given by

$$p_t(\mathbf{x}) = p_0\left(\boldsymbol{\Phi}^{-t} \mathbf{x}\right) \tag{13.4}$$

because the particles are noninteracting. Since the energy of each particle is conserved, ergodicity holds for the equilibrium microcanonical probability density, $p_{\rm eq}(\mathbf{x}) = \delta[E - H(\mathbf{x})]/\kappa(E)$, where E is the initial energy. For hard-disk or hard-sphere Lorentz gases, the energy is directly related to the particle speed by $v = \|\mathbf{v}\| = \sqrt{2E/m}$, so that the magnitude of velocity is conserved. Therefore, relaxation is only possible towards the microcanonical probability distribution instead of the canonical one. This equilibration process may manifest itself in gases with noninteracting particles. In contrast, thermalization requires infinitely many interacting particles to be achieved, which is not the case for Lorentz gases.

Diffusive transport is characterized by diffusion coefficients defined in terms of the position $\mathbf{r}_t \in \mathbb{R}^f$ of the particle in the Lorentz gas with Einstein's formula (Einstein, 1926), which can be written as

$$\mathcal{D}_{ij} = \lim_{t\to\infty} \frac{1}{2t} \langle (r_{it} - r_{i0})(r_{jt} - r_{j0}) \rangle = \lim_{t\to\infty} \frac{1}{2t} \int_{-t}^{+t} d\tau \, (t - |\tau|) \, \langle v_i(\tau) v_j(0) \rangle \tag{13.5}$$

with $i, j = 1, \ldots, f$, where $\langle \cdot \rangle$ denotes an average over some ensemble of initial conditions, for instance, located in some lattice cell such as $\boldsymbol{l} = 0$ and the last equality is obtained using $r_{it} = r_{i0} + \int_0^t v_i(t') \, dt'$ with the particle velocity $v_i = \dot{r}_i$. The statistical ensemble can be

[1] For hard-disk or hard-sphere Lorentz gases, the primitive cell of the one-particle position space is limited to the billiard table $\mathcal{Q} \subset \mathbb{T}^f$, on the boundary of which the elastic collisions happen, so that the one-particle phase space is given by $\boldsymbol{x} = (\boldsymbol{r}, \mathbf{p}) \in \mathfrak{X} = \mathcal{Q} \times \mathbb{R}^f$ in these cases.

taken as the equilibrium distribution inside this lattice cell. We thus obtain the Green–Kubo formula for the diffusivities

$$\mathcal{D}_{ij} = \frac{1}{2}\int_{-\infty}^{+\infty} \langle v_i(t)v_j(0)\rangle_{\text{eq}}\, dt \tag{13.6}$$

if the integral converges in the long-time limit $|t| \to \infty$. The diffusion coefficients $\mathcal{D}_{ii}$ are positive and finite if the particle motion behaves as a random walk. The coefficients $\{\mathcal{D}_{ij}\}_{i,j=1}^{f}$ form a tensor, which is diagonal if diffusion is isotropic, i.e., $\mathcal{D}_{ij} = \mathcal{D}\,\delta_{ij}$. This is the case for square and triangular two-dimensional lattices and for cubic three-dimensional lattices.

If N noninteracting particles are moving in the Lorentz gas, the one-particle distribution function is given by $f(\mathbf{x}) = N p(\mathbf{x})$, which is normalized according to $\int_{\mathbb{R}^{2f}} f(\mathbf{x})\, d\mathbf{x} = N$. This distribution function evolves in time according to $f_t(\mathbf{x}) = f_0\left(\boldsymbol{\Phi}^{-t}\mathbf{x}\right)$, as a consequence of equation (13.4). We note that this latter result also holds for unnormalized one-particle distribution functions $f(\mathbf{x})$ describing a gas with some particle density, $n(\mathbf{r}) = \int_{\mathbb{R}^f} f(\mathbf{r},\mathbf{p})\, d\mathbf{p}$, which is everywhere positive in the lattice.

In hard-disk Lorentz gases, the trajectory of the point particle can be represented in the so-called Birkhoff variables $(\theta, \sin\chi)$, where θ is the angle of impact and χ the angle between the incident velocity and the direction normal to the disk upon collisions. The free flights and the successive collisions induce a mapping in the variables $(\theta, \sin\chi, \boldsymbol{l})$ where $\boldsymbol{l} \in \mathcal{L}$ is the lattice vector giving the disk of the collision:

$$(\theta_{k+1}, \sin\chi_{k+1}, \boldsymbol{l}_{k+1}) = \boldsymbol{\varphi}(\theta_k, \sin\chi_k, \boldsymbol{l}_k), \tag{13.7}$$

which is area preserving in the rectangles with $0 \le \theta < 2\pi$ and $-1 < \sin\chi < +1$. Because of the defocusing character of collisions on hard disks, this mapping manifests stretching in the unstable direction and shrinking in the stable one. A similar mapping can be introduced for Lorentz gases with Yukawa potentials.

13.2.2 Multibaker Model of Deterministic Diffusion

Important dynamical properties in such Lorentz gases can be understood with the *multibaker map* (Gaspard, 1998):

$$(x_{t+1}, y_{t+1}, l_{t+1}) = \boldsymbol{\varphi}(x_t, y_t, l_t), \tag{13.8}$$

where

$$\boldsymbol{\varphi}(x, y, l) = \begin{cases} \left(2x, \dfrac{y}{2}, l-1\right) & \text{if } \; 0 \le x \le \dfrac{1}{2}, \\ \left(2x-1, \dfrac{y+1}{2}, l+1\right) & \text{if } \; \dfrac{1}{2} < x \le 1, \end{cases} \tag{13.9}$$

acting at discrete values of time $t \in \mathbb{Z}$ on a lattice of unit squares with $0 \le x \le 1, 0 \le y \le 1$, and $l \in \mathbb{Z}$. In the analogy with the Lorentz gases, the coordinates (x_t, y_t, l_t) of the particle correspond to the variables $(\theta_k, \sin\chi_k, \boldsymbol{l}_k)$ and the discrete time t to the collision number k. The map (13.9) has the effect of stretching every square by a factor two in the x-direction

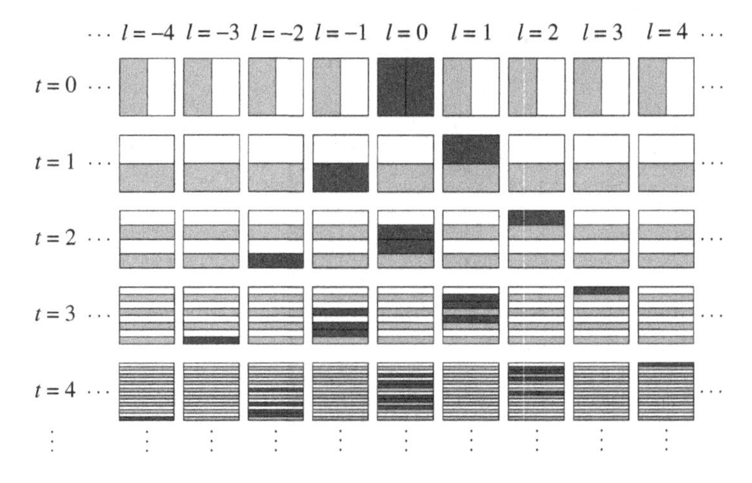

Figure 13.1 The effect of the multibaker map (13.9) on probability initially distributed in the square $l = 0$.

while shrinking it by the same factor in the y-direction, cutting it in two halves, and mapping these halves to the next neighboring squares, either on the left-hand side if $x < 1/2$ or the right-hand side if $x > 1/2$, as shown in Figure 13.1. Therefore, the point particle jumps from square to square under the effect of this map. The multibaker map preserves the phase-space areas, which is consistent with Liouville's theorem. It is symmetric under the time-reversal transformation

$$\boldsymbol{\vartheta}(x, y, l) = (1 - y, 1 - x, l), \tag{13.10}$$

so that microreversibility holds: $\boldsymbol{\varphi}\, \boldsymbol{\vartheta} = \boldsymbol{\vartheta}\, \boldsymbol{\varphi}^{-1}$. Moreover, the Lyapunov exponents of this map are $\lambda_{\pm} = \pm \ln 2$, so that this system manifests sensitivity to initial conditions. The motion induced by the multibaker map is thus chaotic and unbounded, as for the previously mentioned Lorentz gases. As seen in Figure 13.1, if all the probability is initially distributed in the square $l = 0$, this probability will spread on more and more squares as time increases, generating the transport process of diffusion.

The probability density to find the particle at the phase-space point (x, y, l) is normalized according to $\sum_{l=-\infty}^{+\infty} \int_0^1 dx \int_0^1 dy\, p_t(x, y, l) = 1$. The multibaker map induces its dynamical evolution according to

$$p_{t+1}(x, y, l) = p_t\big[\boldsymbol{\varphi}^{-1}(x, y, l)\big] \equiv \hat{M}\, p_t(x, y, l) \tag{13.11}$$

with the discrete time $t \in \mathbb{Z}$, which defines the so-called Perron–Frobenius operator $\hat{M}$ acting on the probability density. We note that, for continuous-time autonomous systems, Liouville's equation $\partial_t p = \hat{L} p$ should be integrated over time to get the Perron–Frobenius operator $\hat{M}^t = \exp\big(\hat{L} t\big)$.

For the multibaker map, it is possible to explicitly construct the diffusive modes governing the relaxation towards thermodynamic equilibrium. The probability $P_t(l)$ of finding the particle in the lth square at time t is given by integrating the probability density on the square:

$$P_t(l) = \int_0^1 dx \int_0^1 dy\; p_t(x, y, l). \tag{13.12}$$

Over long and positive times, the density (13.11) is stretched in every square into a density composed of many horizontal filaments, which leads to a quasi uniform density in the x-direction, as shown in Figure 13.1. At every iteration of the map, the probability to occupy the lth square is divided by two under the multibaker map (13.9) and it is mapped half-half on the next neighboring squares $l \pm 1$:

$$P_{t+1}(l) = \frac{1}{2}\, P_t(l+1) + \frac{1}{2}\, P_t(l-1), \tag{13.13}$$

as for a random walk. This linear equation can be solved using Fourier modes $P_t(l) \sim \exp\left(z_q^{(0)} t + \imath q l\right)$ to obtain the dispersion relation

$$z_q^{(0)} = \ln \cos q = -\frac{q^2}{2} - \frac{q^4}{12} + \cdots \quad \text{for} \quad -\pi < q \leq +\pi, \tag{13.14}$$

giving the relaxation rate $-z_q^{(0)}$ of the modes of wave number q. These modes have a diffusive behavior because their relaxation rate tends to zero as the square of the wave number q according to $z_q^{(0)} = -\mathcal{D} q^2 + O\left(q^4\right)$ with the diffusion coefficient $\mathcal{D} = \lim_{t\to\infty} \langle (l_t - l_0)^2 \rangle / (2t) = 1/2$. This result is in agreement with the fact that the particle has equal probability of jumping to the left- or the right-hand side, so that its motion is a random walk of diffusion coefficient $\mathcal{D} = 1/2$. The general solution of equation (13.13) can thus be decomposed into Fourier modes and expressed in terms of the initial probability distribution $\{P_0(l)\}_{l=-\infty}^{+\infty}$ as

$$P_t(l) = \int_{-\pi}^{+\pi} \frac{dq}{2\pi}\, c_q\, \mathrm{e}^{z_q^{(0)} t}\, \mathrm{e}^{\imath q l} \qquad \text{with} \qquad c_q = \sum_{l=-\infty}^{+\infty} P_0(l)\, \mathrm{e}^{-\imath q l}. \tag{13.15}$$

Remarkably, all the relaxation modes of the multibaker map, including its diffusive and kinetic modes, can be constructed as the generalized eigenmodes of the Perron–Frobenius operator (13.11),

$$\hat{M}\, \psi_q^{(m)} = \mathrm{e}^{z_q^{(m)}}\, \psi_q^{(m)}, \tag{13.16}$$

in terms of the Pollicott–Ruelle resonances

$$z_q^{(m)} = \ln \cos q - m \ln 2, \quad \text{for} \quad -\pi < q \leq +\pi \quad \text{and} \quad m = 0, 1, 2, \ldots, \tag{13.17}$$

corresponding to the eigenmodes $\psi_q^{(m)}(x, y, l) = \mathrm{e}^{\imath q l}\, \psi_q^{(m)}(x, y)$. These latter are not functions, but distributions of Gelfand–Schwartz type, which should be integrated once or several times in phase space in order to represent them as functions (Gaspard, 1998). Therefore, the general solution of Liouvillian dynamics can be decomposed as

$$p_t(x,y,l) = \sum_{m=0}^{\infty} \int_{-\pi}^{+\pi} \frac{dq}{2\pi}\, c_q^{(m)}\, \mathrm{e}^{z_q^{(m)} t}\, \mathrm{e}^{\imath q l}\, \psi_q^{(m)}(x,y) \tag{13.18}$$

for positive times $t > 0$, which defines the forward semigroup introduced in Section 2.4.4.

The modes with $m \geq 1$ are the kinetic modes since they decay even if their wave number is equal to zero according to equation (13.17), so that they are not associated with the conservation law of the particle number, in contrast to the diffusive modes, which should have a decay rate that vanishes with the wave number.

In this respect, the modes with $m = 0$ are the diffusive modes of dispersion relation (13.14) because $\lim_{q\to 0} z_q^{(0)} = 0$. Therefore, the dispersion relation of the diffusion turns out to be given by the leading Pollicott–Ruelle resonance of the Liouvillian dynamics. Since the multibaker map stretches distributions in the unstable x-direction, as seen in Figure 13.1, we expect that the associated eigendistributions $\psi_q^{(0)}(x,y)$ are uniform in this direction, but singular in the stable y-direction. To represent such a singular distribution, it should be integrated in the y-direction in order to define a function, called a cumulative function:

$$F_q(y) = \int_0^y dy'\, \psi_q^{(0)}(x,y'), \tag{13.19}$$

which is thus independent of the variable x. These functions correspond to Fourier modes, so that they have complex values and quasiperiodic extensions on the lattice, which are generated by multiplying them with the factor $\exp(\imath q l)$. These functions are known as de Rham functions and they are shown in Figure 13.2(a) in the complex plane of their real and imaginary parts versus the wave number q (Gaspard, 1998). These cumulative functions form fractals in the complex plane because their Hausdorff dimension is given by

$$d_{\mathrm{H}}(q) = 1 + \frac{\mathcal{D}}{\lambda}\, q^2 + O(q^4), \tag{13.20}$$

where $\mathcal{D}$ is the diffusion coefficient and λ the Lyapunov exponent (Gaspard et al., 2001). The fractal character of the diffusive modes finds its origin in the underlying chaotic dynamics. These functions describe relaxation towards the uniform distribution corresponding to thermodynamic equilibrium for long positive times. They are thus associated with a particular temporal direction, so that time-reversal symmetry breaking here manifests itself at the statistical level of description in terms of the Liouvillian eigenmodes and their cumulative functions. The functions associated with the backward semigroup holding for long negative times are obtained by applying the time-reversal transformation (13.10), which exchanges the unstable and stable directions. The rates of the backward semigroup are the Pollicott–Ruelle antiresonances $-z_q^{(m)}$. The associated eigendistributions are now uniform in the y-direction and singular in the x-direction. They are thus physically distinct from the eigendistributions of the forward semigroup. Although the forward and backward semigroups are distinct, they are compatible with each other because they hold in different time domains: the semiaxis of positive times for the forward semigroup and the one of negative times for the backward semigroup. Moreover, the semigroup decompositions do not converge if time is extended to the other semiaxis. Since the Liouvillian dynamics can

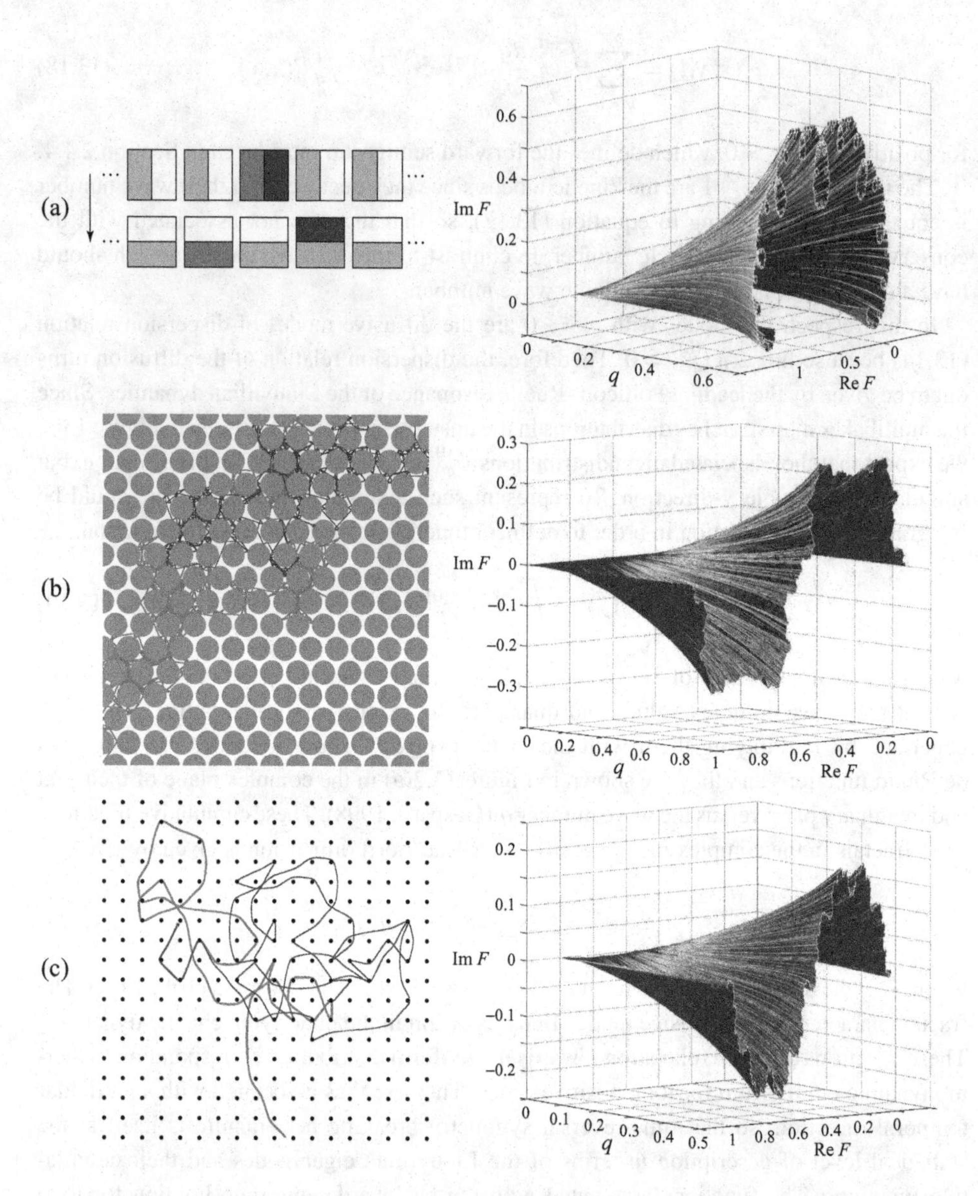

Figure 13.2 Models of deterministic diffusion and their diffusive modes represented by their cumulative function ($\text{Re}\, F_q, \text{Im}\, F_q$) versus the wave number q: (a) The multibaker map (13.9) with diffusion coefficient $\mathcal{D} = 1/2$; (b) The two-dimensional Lorentz gas where the point particle of unit mass and velocity moves in a triangular lattice of hard disks of radius a with interdisk distance $d = 2.3a$ corresponding to a finite horizon and diffusion coefficient $\mathcal{D} \simeq 0.25$. (c) The two-dimensional Lorentz gas where the particle of unit mass moves in a square lattice of Yukawa potentials with $\alpha = 2$, energy $E = 3$, and diffusion coefficient $\mathcal{D} \simeq 2.5$. The sensitivity to initial conditions manifests itself in the divergence between two trajectories, as shown in the left panels. The fractal geometry of the diffusive modes is seen in the right panels, showing the curves formed in the complex plane by the cumulative functions for $\boldsymbol{q} = (q, 0)$ with $q \neq 0$.

be represented by either the forward or the backward semigroup, there is thus time-reversal symmetry breaking in this semigroup representation.

Furthermore, it is also possible to construct the distributions describing nonequilibrium steady macrostates corresponding to a linear gradient of density in the lattice. Such a distribution can be obtained by taking the derivative with respect to the wave number according to

$$\psi_{\rm neq}(x,y) = -\imath\, \partial_q \psi_q^{(0)}\Big|_{q=0}. \tag{13.21}$$

Indeed, at the macroscopic level of description, the same derivative of some Fourier mode $\psi_q = \exp(\imath q x)$ would give a steady macrostate that is linear in the concentration, $\psi_{\rm neq} = -\imath\,\partial_q \psi_q\big|_{q=0} = x$, as expected for a diffusion process obeying Fick's law and with a diffusion coefficient that is independent of the density. Here, the corresponding Liouvillian mode is similarly obtained. Consequently, the nonequilibrium steady macrostate is described by a distribution that is uniform in the unstable x-direction, but singular in the stable y-direction. Again, such a singular distribution can be presented by its cumulative function

$$T(y) = \int_0^y dy'\, \psi_{\rm neq}(x,y'), \tag{13.22}$$

which is known as the Takagi function (Tasaki and Gaspard, 1995). Integrating over both variables gives $\int_0^{x'} dx' \int_0^y dy'\, \psi_{\rm neq}(x',y') = x T(y)$. This function is shown in Figure 13.3(a) for $x = 0.2, 0.4, 0.6, 0.8, 1$. The time-reversed nonequilibrium steady macrostate can be described by a distribution that is uniform in the stable y-direction, but singular in the unstable x-direction. Therefore, there is also time-reversal symmetry breaking for the nonequilibrium stationary distribution $\psi_{\rm neq}(x,y)$, which is an example of the mechanism presented in Table 2.2.

We note that, if there arc N independent particles moving under the dynamics of the multibaker map, the same considerations apply to their distribution function $f_t(x,y,l) = N p_t(x,y,l)$, which is normalized to the number N of particles instead of the unit value. This distribution function can be used to define the infinite dynamical system $(\boldsymbol{\varphi}^t, \mathcal{M}, P)$ as a Poisson suspension of stationary probability distribution P corresponding to the particle distribution $\nu(\mathfrak{D}) = \int_{\mathfrak{D}} dx dy$. As shown in Section 13.2.5, the expected entropy production of diffusive processes can be deduced from this distribution.

Moreover, a fluctuation relation holds for the current in the multibaker map (Gaspard and Andrieux, 2011a).

13.2.3 Diffusive Modes in Spatially Periodic Lorentz Gases

The previous results obtained for the multibaker map can be extended to the previously mentioned Lorentz gases (Gaspard et al., 2001). These dynamical systems are considered on a lattice, so that they are invariant under the discrete space group of spatial translations $\{\boldsymbol{l}\}$ of the lattice $\mathcal{L}$. Accordingly, any function $G(\mathbf{x}) = G(\boldsymbol{x}, \boldsymbol{l})$ defined in the global phase space can be decomposed into Fourier modes with the *lattice Fourier transform* and its inverse as

$$G(\boldsymbol{x},\boldsymbol{l}) = \int_{\mathfrak{B}} \frac{d\boldsymbol{q}}{|\mathfrak{B}|}\, e^{\imath \boldsymbol{q}\cdot\boldsymbol{l}}\, \tilde{G}(\boldsymbol{x},\boldsymbol{q}) \qquad \text{and} \qquad \tilde{G}(\boldsymbol{x},\boldsymbol{q}) = \sum_{\boldsymbol{l}\in\mathcal{L}} e^{-\imath \boldsymbol{q}\cdot\boldsymbol{l}}\, G(\boldsymbol{x},\boldsymbol{l}), \tag{13.23}$$

where the wave vector $\boldsymbol{q}$ belongs the first Brillouin zone $\mathfrak{B}$ of the reciprocal lattice and $|\mathfrak{B}|$ denotes its volume. Therefore, the mean value of some observable $A(\mathbf{x}) = A(\boldsymbol{x},\boldsymbol{l})$ over some probability distribution of density $p(\mathbf{x}) = p(\boldsymbol{x},\boldsymbol{l})$ can be decomposed as

$$\langle A\rangle = \sum_{\boldsymbol{l}\in\mathcal{L}} \int_{\mathfrak{X}} d\boldsymbol{x}\, A(\boldsymbol{x},\boldsymbol{l})\, p(\boldsymbol{x},\boldsymbol{l}) = \int_{\mathfrak{B}} \frac{d\boldsymbol{q}}{|\mathfrak{B}|} \int_{\mathfrak{X}} d\boldsymbol{x}\, \tilde{A}^*(\boldsymbol{x},\boldsymbol{q})\, \tilde{p}(\boldsymbol{x},\boldsymbol{q}). \tag{13.24}$$

The invariance of the Liouvillian dynamics under the lattice space group means that the Liouvillian operator $\hat{L}$, as well as the Perron–Frobenius operator $\hat{M}^t = \exp(\hat{L}t)$, commute with the translation operators $\hat{T}^{\boldsymbol{l}}$, i.e.,

$$\left[\hat{L},\hat{T}^{\boldsymbol{l}}\right] = 0 \qquad \text{and} \qquad \left[\hat{M}^t,\hat{T}^{\boldsymbol{l}}\right] = 0 \tag{13.25}$$

for $\boldsymbol{l} \in \mathcal{L}$. Consequently, the Perron–Frobenius operator can be decomposed into evolution operators acting on the subspaces of given wave vector as

$$\langle A\rangle_t = \sum_{\boldsymbol{l}\in\mathcal{L}} \int_{\mathfrak{X}} d\boldsymbol{x}\, A(\boldsymbol{x},\boldsymbol{l})\, \hat{M}^t\, p_0(\boldsymbol{x},\boldsymbol{l}) = \int_{\mathfrak{B}} \frac{d\boldsymbol{q}}{|\mathfrak{B}|} \int_{\mathfrak{X}} d\boldsymbol{x}\, \tilde{A}^*(\boldsymbol{x},\boldsymbol{q})\, \hat{Q}_{\boldsymbol{q}}^t\, \tilde{p}_0(\boldsymbol{x},\boldsymbol{q}) \tag{13.26}$$

with

$$\tilde{p}_t(\boldsymbol{x},\boldsymbol{q}) = \hat{Q}_{\boldsymbol{q}}^t\, \tilde{p}_0(\boldsymbol{x},\boldsymbol{q}) = e^{-\imath \boldsymbol{q}\cdot\boldsymbol{d}(\boldsymbol{\phi}^{-t}\boldsymbol{x},t)}\, \tilde{p}_0\left(\boldsymbol{\phi}^{-t}\boldsymbol{x},\boldsymbol{q}\right), \tag{13.27}$$

in terms of the displacement lattice vector (13.3).

Another consequence of equation (13.25) is that the Perron–Frobenius and translation operators admit common eigenmodes, so that

$$\hat{M}^t\, \psi_{\boldsymbol{q}} = e^{z_{\boldsymbol{q}} t}\, \psi_{\boldsymbol{q}} \qquad \text{and} \qquad \hat{T}^{\boldsymbol{l}}\, \psi_{\boldsymbol{q}} = e^{\imath \boldsymbol{q}\cdot\boldsymbol{l}}\, \psi_{\boldsymbol{q}}. \tag{13.28}$$

The Pollicott–Ruelle resonances $z_{\boldsymbol{q}}$ are here some functions of the wave vector. The leading resonance vanishes with the wave vector, thus giving the dispersion relation of diffusion as

$$z_{\boldsymbol{q}} = -\mathcal{D}\boldsymbol{q}^2 + O\left(\boldsymbol{q}^4\right), \tag{13.29}$$

where $\mathcal{D}$ is the diffusion coefficient, if diffusion is isotropic. Accordingly, the asymptotic time evolution of the mean value (13.26) can be decomposed as

$$\langle A\rangle_t = \int_{\mathfrak{B}} \frac{d\boldsymbol{q}}{|\mathfrak{B}|} \langle A|\psi_{\boldsymbol{q}}\rangle\, e^{z_{\boldsymbol{q}} t}\, \langle\tilde{\psi}_{\boldsymbol{q}}|p_0\rangle + \cdots, \tag{13.30}$$

where

$$\langle A|\psi_{\boldsymbol{q}}\rangle = \int_{\mathfrak{X}} d\boldsymbol{x}\, \tilde{A}^*(\boldsymbol{x},\boldsymbol{q})\, \psi_{\boldsymbol{q}}(\boldsymbol{x}), \qquad \langle\tilde{\psi}_{\boldsymbol{q}}|p_0\rangle = \int_{\mathfrak{X}} d\boldsymbol{x}\, \tilde{\psi}_{\boldsymbol{q}}^*(\boldsymbol{x})\, \tilde{p}_0(\boldsymbol{x},\boldsymbol{q}), \tag{13.31}$$

and the dots denote contributions beyond the leading Pollicott–Ruelle resonance, which are expected to decay faster in time.

Since the distributions $\psi_{\boldsymbol{q}}$ are also the eigenmodes of the reduced Perron–Frobenius operator (13.27), they satisfy

$$\hat{Q}^t_{\boldsymbol{q}} \psi_{\boldsymbol{q}}(\boldsymbol{x}) = \mathrm{e}^{z_{\boldsymbol{q}} t} \psi_{\boldsymbol{q}}(\boldsymbol{x}), \tag{13.32}$$

which implies that

$$\psi_{\boldsymbol{q}}(\boldsymbol{x}) = \hat{R}^t_{\boldsymbol{q}} \, \psi_{\boldsymbol{q}}(\boldsymbol{x}) \qquad \text{with} \qquad \hat{R}^t_{\boldsymbol{q}} \, \psi_{\boldsymbol{q}}(\boldsymbol{x}) \equiv \mathrm{e}^{-z_{\boldsymbol{q}} t} \, \mathrm{e}^{-\imath \boldsymbol{q}\cdot \boldsymbol{d}(\boldsymbol{\phi}^{-t}\boldsymbol{x},t)} \psi_{\boldsymbol{q}}\left(\boldsymbol{\phi}^{-t}\boldsymbol{x}\right) \tag{13.33}$$

for $\boldsymbol{x} \in \mathfrak{X}$, $\boldsymbol{q} \in \mathfrak{B}$, and an arbitrary positive value of time t. This important result can be interpreted saying that the eigenmode $\psi_{\boldsymbol{q}}$ is the fixed point of the renormalization semigroup operator $\hat{R}^t_{\boldsymbol{q}}$ defined by equation (13.33). This operation corresponds to the forward semigroup of time evolution, so that the eigenmode $\psi_{\boldsymbol{q}}$ is expected to be smooth in the unstable directions, but singular in the stable directions of the phase space $\mathfrak{X}$.

We note that the dispersion relation of diffusion can equivalently be expressed as

$$z_{\boldsymbol{q}} = \lim_{t\to\infty} \frac{1}{t} \ln\langle \exp[\imath \boldsymbol{q} \cdot (\mathbf{r}_t - \mathbf{r}_0)]\rangle \tag{13.34}$$

in terms of the Van Hove intermediate incoherent scattering function (Van Hove, 1954), where $\langle\cdot\rangle$ denotes the average over an ensemble of initial conditions and $\mathbf{r} = \boldsymbol{r} + \boldsymbol{l}$ is the position of the particle in the lattice. If diffusion is isotropic, the diffusion coefficient is thus given by the Einstein or Green–Kubo formula:

$$\mathcal{D} = -\frac{1}{2}\, \partial^2_{q_i} z_{\boldsymbol{q}}|_{\boldsymbol{q}=0} = \lim_{t\to\infty} \frac{1}{2t} \left\langle (r_{it} - r_{i0})^2 \right\rangle = \int_0^\infty \langle v_i(0) v_i(t)\rangle_{\mathrm{eq}}\, dt, \tag{13.35}$$

where $v_i = \dot{r}_i$ is the velocity of the particle in the direction $i = x, y, z$ of the physical space $\mathbf{r} \in \mathbb{R}^3$ and the statistical average $\langle\cdot\rangle_{\mathrm{eq}}$ is carried out over the equilibrium distribution in the primitive cell of the lattice.

In chaotic systems such as the Lorentz gases, the associated diffusive eigenmode $\psi_{\boldsymbol{q}}$ is a Gelfand–Schwartz distribution, which can only be depicted using its cumulative function

$$F_{\boldsymbol{q}}(\theta) = \frac{\int_0^\theta \psi_{\boldsymbol{q}}(\boldsymbol{x}_{\theta'})\, d\theta'}{\int_0^{2\pi} \psi_{\boldsymbol{q}}(\boldsymbol{x}_{\theta'})\, d\theta'}, \tag{13.36}$$

defined by integrating over some curve $\mathcal{C}_\theta$ in the phase space at given energy (with $0 \leq \theta < 2\pi$) and normalized to take the unit value at $\theta = 2\pi$. Since the eigenmode $\psi_{\boldsymbol{q}}$ of the forward semigroup can be obtained by applying the evolution operator over an arbitrary long time from an initial function of wave vector $\boldsymbol{q}$, the cumulative function (13.36) can be equivalently defined as (Gaspard et al., 2001)

$$F_{\boldsymbol{q}}(\theta) = \lim_{t\to\infty} \frac{\int_0^\theta \exp\left[\imath \boldsymbol{q} \cdot (\mathbf{r}_t - \mathbf{r}_0)_{\theta'}\right] d\theta'}{\int_0^{2\pi} \exp\left[\imath \boldsymbol{q} \cdot (\mathbf{r}_t - \mathbf{r}_0)_{\theta'}\right] d\theta'}, \tag{13.37}$$

where $\mathbf{r}_t$ is the trajectory starting from the initial conditions $(\mathbf{r}_0, \mathbf{p}_0)_\theta$. The cumulative functions of the diffusive modes are shown in Figure 13.2(b) and (c) for the hard-disk and Yukawa-potential Lorentz gases. If the wave vector is equal to zero, the cumulative function (13.37) is equal to $F_{\boldsymbol{q}=0}(\theta) = \theta/(2\pi)$, corresponding to the microcanonical distribution in the energy shell. For nonzero wave vectors, the cumulative function forms a fractal curve in the complex plane $(\mathrm{Re}\, F_{\boldsymbol{q}}, \mathrm{Im}\, F_{\boldsymbol{q}})$. Its Hausdorff dimension d_{H} can be calculated as follows. Since the denominator behaves as $\exp(z_{\boldsymbol{q}} t)$ according to equation (13.34), the cumulative function (13.37) can be written in the long-time limit as a sum of contributions from trajectories $\{j\}$:

$$F_{\boldsymbol{q}}(\theta) = \lim_{t\to\infty} \sum_j \Delta F_{\boldsymbol{q},t}^{(j)} \tag{13.38}$$

with

$$\Delta F_{\boldsymbol{q},t}^{(j)} = \left|\Lambda_t^{(j)}\right|^{-1} \exp\left(-z_{\boldsymbol{q}} t\right) \exp\left[\imath \boldsymbol{q}\cdot\left(\mathbf{r}_t^{(j)} - \mathbf{r}_0^{(j)}\right)\right], \tag{13.39}$$

where $\Lambda_t^{(j)}$ is the stretching factor of the jth trajectory over the time interval t, which gives the finite-time Lyapunov exponent as $\lambda^{(j)} = \left(\ln\left|\Lambda_t^{(j)}\right|\right)/t$. Accordingly, the cumulative function can be approximated over a long but finite time interval by a polygonal curve formed by a sequence of complex vectors (13.39), as in the construction of the fractal curves of von Koch's type (Mandelbrot, 1982). In the limit $t \to \infty$, the polygon converges to a fractal curve characterized by the Hausdorff dimension d_{H} given by the condition

$$\sum_j \left|\Delta F_{\boldsymbol{q},t}^{(j)}\right|^{d_{\mathrm{H}}} \sim 1. \tag{13.40}$$

This condition can be rewritten in terms of the so-called topological pressure

$$\mathscr{P}(\beta) \equiv \lim_{t\to\infty} \frac{1}{t} \ln\left\langle \left|\Lambda_t^{(j)}\right|^{1-\beta}\right\rangle, \tag{13.41}$$

where the statistical average is defined in terms of the stretching factors as $\langle A\rangle \simeq \sum_j \left|\Lambda_t^{(j)}\right|^{-1} A^{(j)}$, $A^{(j)}$ being the quantity of interest evaluated for the jth trajectory. Therefore, the Hausdorff dimension is given by the root of

$$\mathscr{P}(d_{\mathrm{H}}) = d_{\mathrm{H}}\, \mathrm{Re}\, z_{\boldsymbol{q}}. \tag{13.42}$$

Since there is no escape of particles from the system, the escape rate is equal to zero and $\gamma = -\mathscr{P}(1) = 0$. In addition, the mean Lyapunov exponent is given by $\lambda = -\mathscr{P}'(1)$. The Hausdorff dimension can thus be expanded in powers of the wave vector as $d_{\mathrm{H}}(\boldsymbol{q}) = 1 + (\mathcal{D}/\lambda)\boldsymbol{q}^2 + O\left(\boldsymbol{q}^4\right)$, whereupon the diffusion coefficient can be obtained from the Hausdorff dimension and the Lyapunov exponent by the formula

$$\mathcal{D} = \lambda \lim_{\boldsymbol{q}\to 0} \frac{1}{\boldsymbol{q}^2} \left[d_{\mathrm{H}}(\boldsymbol{q}) - 1\right]. \tag{13.43}$$

This formula has been analytically deduced and numerically verified for the Lorentz gases (Gaspard et al., 2001). We note that the formula (13.43) has a form very similar to the chaos-transport relationship of the escape-rate formalism, expressing the diffusion coefficient in terms of the Lyapunov exponent and the dimension of fractal repellers containing all the nonescaping trajectories in open Lorentz gases with absorbing boundary conditions (Gaspard, 1998).

The previous considerations for the probability density $p(\mathbf{x})$ to find the particle at some phase-space point $\mathbf{x}$ in the lattice extend to any one-particle distribution function $f(\mathbf{x})$, since both have the same time evolution, as previously mentioned. As a consequence of the asymptotic spectral decomposition (13.30), the particle density in any lattice cell, $n_t(\boldsymbol{l}) = \int_{\mathcal{X}} f_t(\boldsymbol{x},\boldsymbol{l})\, d\boldsymbol{x}/\upsilon$ where $\upsilon = \int_{\mathbb{T}^f} d\boldsymbol{r}$ is the volume of the primitive cell, obeys the diffusion equation

$$\partial_t n_t(\boldsymbol{l}) \simeq \mathcal{D}\, \partial_{\boldsymbol{l}}^2 n_t(\boldsymbol{l}) \tag{13.44}$$

in the long-time limit and over large spatial scales, showing that the Lorentz gases indeed manifest deterministic diffusion.

13.2.4 Nonequilibrium Stationary Distribution

Alike in the multibaker map, the nonequilibrium stationary distribution corresponding to a gradient of concentration maintained in finite Lorentz gases extending between two reservoirs of particles at different chemical potentials can also be obtained from the diffusive modes at small values for the wave vector according to

$$\Psi_{\rm neq} = -\imath\, \mathbf{g} \cdot \partial_{\boldsymbol{q}} \Psi_{\boldsymbol{q}} \big|_{\boldsymbol{q}=0} \tag{13.45}$$

in terms of the concentration gradient $\mathbf{g} = \partial_{\boldsymbol{l}} n$ (Tasaki and Gaspard, 1995). In hard-disk or hard-sphere Lorentz gases, this leads to the nonequilibrium steady state described by the phase-space density

$$\Psi_{\rm neq}(\mathbf{x}) = \mathbf{g} \cdot \left[\mathbf{r}(\mathbf{x}) + \int_0^{-\infty} \mathbf{v}\left(\boldsymbol{\Phi}^t\mathbf{x}\right) dt\right] \upsilon\, p(\mathbf{x}|E), \tag{13.46}$$

where $\mathbf{v} = \dot{\mathbf{r}} = \mathbf{p}/m$ is the velocity of the particle, $\boldsymbol{\Phi}^t$ is the one-particle flow acting on the phase-space point $\mathbf{x} = (\mathbf{r},\mathbf{p})$, and $p(\mathbf{x}|E) = \delta[E - H(\mathbf{x})]/\kappa(E)$ is the microcanonical probability density at the energy E of the particle. This latter density is normalized according to $\int_{\mathcal{X}} p(\mathbf{x}|E)\, d\mathbf{x} = 1$. The phase-space densities (13.46) are singular and they admit cumulative functions known as generalized Takagi functions (Gaspard, 1998). These functions are typically self-similar. They are depicted in Figure 13.3 for the hard-disk and Yukawa-potential Lorentz gas (Gaspard, 2001). Again, these cumulative functions are smooth in the unstable phase-space directions but singular in the stable directions, although the time-reversed functions are singular in the unstable directions and smooth in the stable ones. The nonequilibrium steady distribution (13.46), as well as the eigenmodes (13.32) associated with the Pollicott–Ruelle resonances, thus break the time-reversal symmetry, which is another example of the mechanism presented in Table 2.2 for time-reversal symmetry breaking at the statistical level of description.

The nonequilibrium stationary distribution (13.46) implies Fick's law. Indeed, for a steady macrostate (13.46) with a concentration gradient $\mathbf{g}$, the current density j_i being proportional to the particle velocity v_i in the direction $i = x, y, z$, its nonequilibrium statistical average is given by

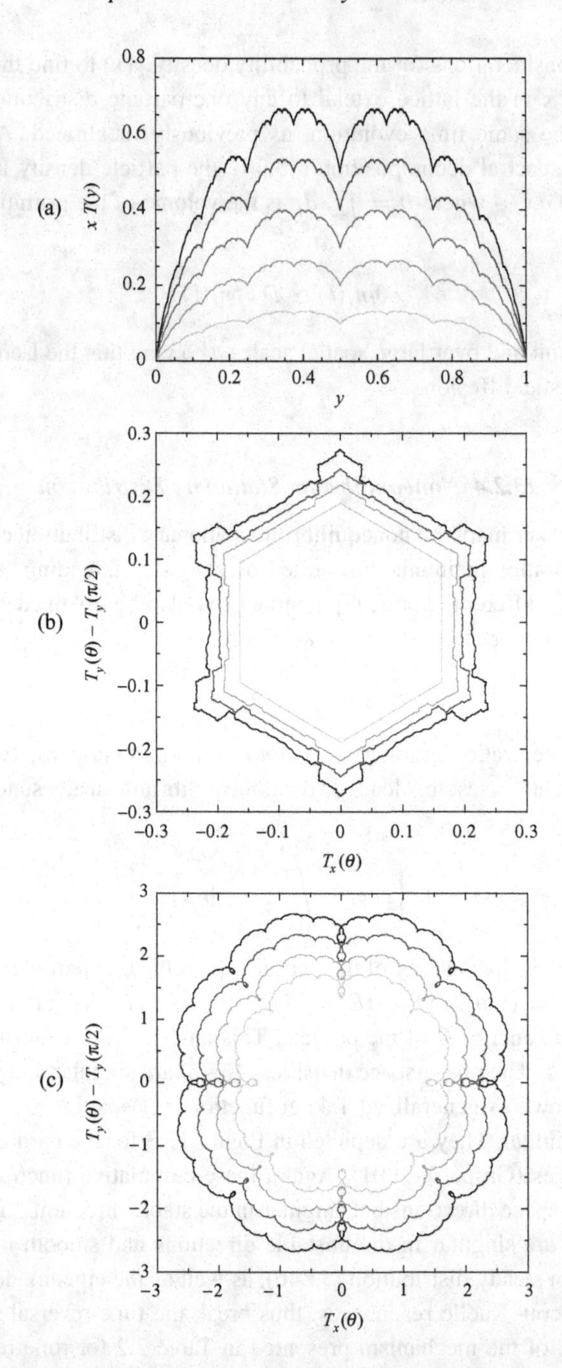

Figure 13.3 Nonequilibrium stationary distributions represented by their cumulative function for the models of deterministic diffusion: (a) The multibaker map (13.9); (b) The two-dimensional Lorentz gas where the point particle of unit mass and velocity moves in a triangular lattice of hard disks of radius a with interdisk distances $d = 2.001a$, $d = 2.1a$, $d = 2.2a$, and $d = 2.3a$. (c) The two-dimensional Lorentz gas where the particle of unit mass moves in a square lattice of Yukawa potentials with $\alpha = 2$ and energies $E = 2$, $E = 3$, $E = 4$, and $E = 5$.

$$\langle J_i \rangle_{\text{neq}} = \frac{1}{v} \int_{\mathfrak{X}} v_i(\mathbf{x})\, \psi_{\text{neq}}(\mathbf{x})\, d\mathbf{x} = - \sum_{j=x,y,z} g_j \int_0^{\infty} \langle v_i(0) v_j(t) \rangle_{\text{eq}}\, dt = -\mathcal{D}\, g_i, \quad (13.47)$$

where we have used the result that $\langle v_i r_j \rangle_{\text{eq}} = 0$. Accordingly, we recover Fick's law and the Green–Kubo formula for the diffusion coefficient $\mathcal{D}$ (Gaspard, 1998).

In the case of hard-disk or hard-sphere Lorentz gases, the expression (13.46) for the nonequilibrium stationary distribution can be obtained by direct reasoning. Let us consider diffusion in an open Lorentz gas between two chemostats at the particle densities n_{L} and n_{R}, and separated by the distance L, as schematically shown in Figure 13.4. The density inside the system can only take either the value n_{L} corresponding to the reservoir on the left-hand side or n_{R} from the reservoir on the right-hand side. In order to determine which value, we have to integrate the trajectory backward in time until the time of entrance in the system, $\mathscr{T}(\mathbf{x}) < 0$. The value is n_{L} (respectively n_{R}) if the particle trajectory enters from the left-hand (respectively right-hand) side. With the concentration gradient $\mathbf{g} = (n_{\text{R}} - n_{\text{L}})\mathbf{1}_x / L$, where $\mathbf{1}_x$ is the unit vector in the x-direction, we can write the stationary phase-space density in the form

$$f_{\text{neq}}(\mathbf{x}) = \left\{ \frac{n_{\text{R}} + n_{\text{L}}}{2} + \mathbf{g} \cdot \left[\mathbf{r}(\mathbf{x}) + \int_0^{\mathscr{T}(\mathbf{x})} \mathbf{v}\left(\boldsymbol{\phi}^t \mathbf{x}\right) dt \right] \right\} v\, p(\mathbf{x}|E), \quad (13.48)$$

again using the microcanonical distribution at the given energy E of the particles. Indeed, the integral of the particle velocity $\mathbf{v}$ backward in time until the time of entrance gives the position of entrance $\mathbf{r}\left(\boldsymbol{\phi}^{\mathscr{T}(\mathbf{x})}\mathbf{x}\right) = \pm(L/2)\mathbf{1}_x$ minus the current position $\mathbf{r}(\mathbf{x})$, which cancels the first term in the bracket. We end up with the result that $f_{\text{neq}}(\mathbf{x}) = n_j v p(\mathbf{x}|E)$ with $j = \text{L}$ or $j = \text{R}$, i.e., whether the trajectory enters from the left- or right-hand side (Gaspard, 1998). For a Lorentz gas of finite size L, the density of this nonequilibrium stationary distribution is thus a piecewise constant function over phase-space domains, depending on which side the particle enters from. As the size L increases, these domains become finer and finer in the bulk of the Lorentz gas. In the limit $L \to \infty$ where the reservoirs are separated by an arbitrarily large distance, the time of entrance goes to infinity: $\mathscr{T}(\mathbf{x}) \to \infty$. Consequently, the distribution (13.48) becomes the singular distribution $f_{\text{neq}}(\mathbf{x}) = f_{\text{eq}}(\mathbf{x}) + \psi_{\text{neq}}(\mathbf{x})$ expressed in terms of the equilibrium part $f_{\text{eq}}(\mathbf{x}) = n_{\text{eq}} v p(\mathbf{x}|E)$ at the uniform density $n_{\text{eq}} = (n_{\text{R}} + n_{\text{L}})/2$ corresponding to equilibrium, and the nonequilibrium part given by equation (13.46). These results confirm the singular character of the nonequilibrium stationary distributions.

We note that the stationary distribution (13.48) defines the Poisson suspension of an infinite gas of independent particles, describing a steady macrostate of concentration gradient $\mathbf{g}$ obeying Fick's law (13.47). In this case, the number of noninteracting particles in the one-particle phase-space domain $\mathfrak{D}$ is given by $\nu_{\text{neq}}(\mathfrak{D}) = \int_{\mathfrak{D}} f_{\text{neq}}(\mathbf{x})\, d\mathbf{x} = \nu_{\text{eq}}(\mathfrak{D}) + \delta\nu_{\text{neq}}(\mathfrak{D})$ with $\delta\nu_{\text{neq}}(\mathfrak{D}) = \int_{\mathfrak{D}} \psi_{\text{neq}}(\mathbf{x})\, d\mathbf{x}$.

13.2.5 Entropy Production of Diffusion

The study of diffusive modes as the leading eigenmodes of the Liouvillian operator shows that these modes are typically given in terms of singular distributions without a density

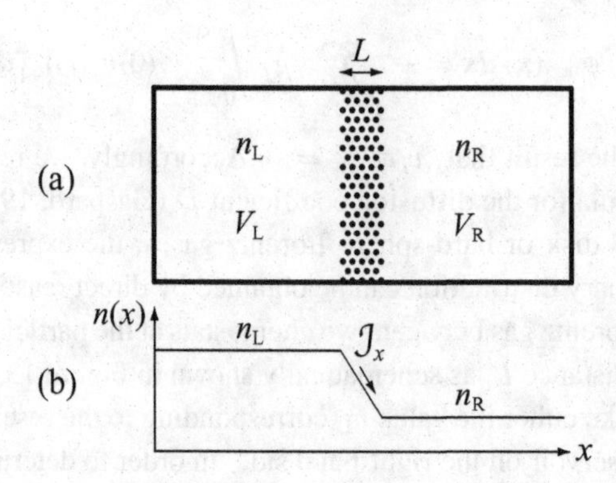

Figure 13.4 (a) Schematic representation of diffusion in a finite Lorentz gas of size L and cross-sectional area Σ between two large reservoirs of given volumes V_L and V_R such that $V_L, V_R \gg \Sigma L$. These reservoirs contain gas at the densities n_L and n_R, respectively. (b) Macroscopic density profile $n(x)$ with the current density $\mathcal{J}_x$ in the x-direction.

function. Since the works by Gelfand, Schwartz, and others in the 1950s, it is known that such distributions acquire a mathematical meaning if they are evaluated for some test functions belonging to certain classes of functions. The test function used to evaluate a distribution is arbitrary although the distribution is not. Examples of test functions are the indicator functions of the cells of some partition in phase space. In this regard, the singular character of the diffusive modes justifies the introduction of a coarse-graining procedure. This reasoning goes along with the need to carry out a coarse graining in order to determine the thermodynamic entropy production.

In particular, the entropy production rate can be evaluated for diffusion in Lorentz gases. We consider the geometry shown in Figure 13.4 with a finite Lorentz gas extending over a length L between two arbitrarily large particle reservoirs. The particles have a given energy E and they have a diffusive motion of finite diffusion coefficient $\mathcal{D}(E)$ inside the finite Lorentz gas. Because of the spatial periodicity of the lattice composing the finite Lorentz gas, the motion inside this Lorentz gas evolves in the one-particle phase space composed of the images $\mathfrak{X}_l = \hat{T}^l \mathfrak{X}$ of the basic phase space $\mathfrak{X}$ corresponding to a primitive cell of the lattice, where the lattice vector $\boldsymbol{l} = (l_x, l_y, l_z)$ is restricted to $|l_x| < L/2$ with a large enough size L and $l_y, l_z \in \mathbb{Z}$. Since the reservoirs are supposed to be much larger than the finite Lorentz gas, a nonequilibrium steady macrostate maintains itself over a long intermediate timescale before the system reaches global equilibrium, according to equations (1.74) and (1.75).

The phase-space domains $\mathfrak{X}_l$ of the lattice are partitioned into small enough disjoint cells $\{\mathfrak{C}_i\}$, such that $\mathfrak{X}_l = \cup_i \mathfrak{C}_i$ and of equal volume with respect to the microcanonical distribution. The number of gas particles found in the cell $\mathfrak{C}_i$ at time t is given by

$$\nu_t(\mathfrak{C}_i) = \int_{\mathfrak{C}_i} f_t(\mathbf{x})\, d\mathbf{x} \tag{13.49}$$

in terms of the one-particle distribution function $f_t(\mathbf{x})$. This number evolves in time as

$$\nu_{t+\tau}(\mathfrak{C}_i) = \nu_t(\boldsymbol{\Phi}^{-\tau}\mathfrak{C}_i). \tag{13.50}$$

We consider the Poisson suspension associated with this distribution function.

According to equation (8.30), the entropy of the gas in the domain $\mathfrak{X}_l$ coarse grained with the partition $\{\mathfrak{C}_i\}$ is given by

$$S_t(\mathfrak{X}_l|\{\mathfrak{C}_i\}) = k_{\mathrm{B}} \sum_{\mathfrak{C}_i \subset \mathfrak{X}_l} \nu_t(\mathfrak{C}_i) \ln \frac{\mathrm{e}}{\nu_t(\mathfrak{C}_i)}. \tag{13.51}$$

Because of equation (13.50), this entropy obeys the relation $S_{t-\tau}(\boldsymbol{\Phi}^{-\tau}\mathfrak{X}_l|\{\mathfrak{C}_i\}) = S_t(\mathfrak{X}_l|\{\boldsymbol{\Phi}^{\tau}\mathfrak{C}_i\})$. Therefore, the time variation of the coarse-grained entropy over a time interval τ

$$\Delta^\tau S = S_t(\mathfrak{X}_l|\{\mathfrak{C}_i\}) - S_{t-\tau}(\mathfrak{X}_l|\{\mathfrak{C}_i\}) \tag{13.52}$$

can be separated into the *entropy exchange* with the environment of the domain $\mathfrak{X}_l$,

$$\Delta_{\mathrm{e}}^\tau S \equiv S_{t-\tau}(\boldsymbol{\Phi}^{-\tau}\mathfrak{X}_l|\{\mathfrak{C}_i\}) - S_{t-\tau}(\mathfrak{X}_l|\{\mathfrak{C}_i\}), \tag{13.53}$$

and the *entropy production*,

$$\Delta_{\mathrm{i}}^\tau S \equiv \Delta^\tau S - \Delta_{\mathrm{e}}^\tau S = S_t(\mathfrak{X}_l|\{\mathfrak{C}_i\}) - S_t(\mathfrak{X}_l|\{\boldsymbol{\Phi}^{\tau}\mathfrak{C}_i\}), \tag{13.54}$$

which depends on the partition $\{\mathfrak{C}_i\}$, as in equations (2.135) and (2.136).

Now, the entropy production can be calculated using the decomposition of the time evolution in terms of the diffusive modes (Gaspard, 1998; Gilbert et al., 2000; Dorfman et al., 2002). The idea of this *ab initio* derivation is the following. If the system is close to equilibrium, the particle number (13.49) can be expanded according to

$$\nu_t(\mathfrak{C}_i) = \nu_{\mathrm{eq}}(\mathfrak{C}_i) + \delta\nu_t(\mathfrak{C}_i) \qquad \text{with} \qquad |\delta\nu_t(\mathfrak{C}_i)| \ll \nu_{\mathrm{eq}}(\mathfrak{C}_i). \tag{13.55}$$

Over the intermediate timescale $t_{\mathrm{relax}} \ll t \ll t_{\mathrm{equil}}$ allowed by equation (1.75), the time-dependent distribution reaches the previously mentioned nonequilibrium distribution (13.48), so that

$$\delta\nu_t(\mathfrak{C}_i) \simeq \delta\nu_{\mathrm{neq}}(\mathfrak{C}_i) = \boldsymbol{g} \cdot \int_{\mathfrak{C}_i} \boldsymbol{\psi}_{\mathrm{neq}}(\boldsymbol{x})\, d\boldsymbol{x}, \quad \text{where} \quad \boldsymbol{\psi}_{\mathrm{neq}}(\mathbf{x}) = -\imath\, \partial_{\boldsymbol{q}} \psi_{\boldsymbol{q}}(\boldsymbol{x})\big|_{\boldsymbol{q}=0} \tag{13.56}$$

and $\boldsymbol{g} = \partial_l n / n_{\mathrm{eq}}$ is the relative concentration gradient defined with respect to the local equilibrium density n_{eq}. Expanding the entropy production (13.54) up to terms of second order in $\delta\nu_{\mathrm{neq}}$, we obtain

$$\Delta_{\mathrm{i}}^\tau S = k_{\mathrm{B}} \sum_{\boldsymbol{\Phi}^\tau \mathfrak{C}_i \subset \mathfrak{X}_l} \frac{[\delta\nu_{\mathrm{neq}}(\boldsymbol{\Phi}^\tau \mathfrak{C}_i)]^2}{2\nu_{\mathrm{eq}}(\mathfrak{C}_i)} - k_{\mathrm{B}} \sum_{\mathfrak{C}_i \subset \mathfrak{X}_l} \frac{[\delta\nu_{\mathrm{neq}}(\mathfrak{C}_i)]^2}{2\nu_{\mathrm{eq}}(\mathfrak{C}_i)}, \tag{13.57}$$

up to corrections of the order of $\delta\nu_{\mathrm{neq}}^3$. Using the expression (13.56), the entropy production over the time interval $[0,\tau]$ and the volume $\upsilon = \int_{\mathbb{T}^f} d\mathbf{r}$ takes the value expected from nonequilibrium thermodynamics,

$$\Delta_{\rm i}^{\tau} S \simeq k_{\rm B}\, \mathcal{D}\, \tau\, \frac{(\partial_l n)^2}{n}\, \upsilon, \tag{13.58}$$

where $\mathcal{D}$ is the coefficient coefficient (13.35) and n is the local particle density, up to corrections that are cubic in the gradient (Dorfman et al., 2002).

This result is a consequence of the singular character of the nonequilibrium steady state coming from the diffusive modes. Indeed, it can be shown that the entropy production would vanish if the steady state had a density given by a function instead of a nontrivial distribution (Gaspard, 1997, 1998). The nonequilibrium distribution (13.48) between two particle reservoirs separated by a finite distance L is still absolutely continuous with respect to the microcanonical distribution, although very different and much more complicated than it. For finite L, there is thus a characteristic scale below which the continuity of the distribution manifests itself. For a partition into cells larger than this scale, the entropy production takes the thermodynamic value (13.58). For smaller cells, it instead vanishes. The key point is that this characteristic scale is exponentially decreasing with the distance to the physical boundaries. Therefore, in the limit $L \to \infty$, the stationary measure looses its absolute continuity and becomes singular, so that the thermodynamic value is obtained for cells of arbitrarily small size (Gilbert et al., 2000; Dorfman et al., 2002). This shows that the thermodynamic entropy production does not depend on the chosen partition in the large-system limit.

We note that the fluctuation relation also holds in the Lorentz gases under the nonequilibrium conditions of Figure 13.4 according to the results of Section 5.3.

13.3 Diffusion of a Tracer Particle in a Many-Particle System

The diffusion of a dilute tracer species in a fluid can be simulated using molecular dynamics with periodic boundary conditions. This dynamical system is composed of particles $a = 0, 1, 2, \ldots, N$ moving according to Hamilton's or Newton's equations with their positions limited, for instance, to a cubic torus such that $0 \le r_{ia} < L$ and changed by $\pm L$ if they cross the boundaries. The phase space of the system is defined by $\boldsymbol{\Gamma} = (\boldsymbol{r}_0, \mathbf{p}_0, \boldsymbol{r}_1, \mathbf{p}_1, \boldsymbol{r}_2, \mathbf{p}_2, \ldots, \boldsymbol{r}_N, \mathbf{p}_N) \in \mathcal{M}$ with $\boldsymbol{r}_a \in \mathbb{T}^d$ and $\mathbf{p}_a \in \mathbb{R}^d$, where $(\boldsymbol{r}_0, \mathbf{p}_0) = (\boldsymbol{r}, \mathbf{p})$ are the position and momentum of the tracer particle, so that the phase-space dimension is equal to $\dim \mathcal{M} = 2(N+1)d$. The fluid particles are identical with the same mass $m_a = m$ for $a = 1, 2, \ldots, N$, while the tracer particle has the mass m_0. Therefore, some phase-space integral can be written as

$$\int_{\mathcal{M}} d\boldsymbol{\Gamma}\,(\cdot) = \int_{\mathbb{R}^{2d}} d\boldsymbol{r}\, d\mathbf{p}\, \frac{1}{N!} \int_{\mathbb{R}^{2Nd}} d\boldsymbol{r}_1\, d\mathbf{p}_1\, d\boldsymbol{r}_2\, d\mathbf{p}_2 \cdots d\boldsymbol{r}_N\, d\mathbf{p}_N\,(\cdot). \tag{13.59}$$

In the global system composed of the original system and all its copies forming a lattice $\mathcal{L}$, the position of the tagged image of the tracer particle is given by $\mathbf{r} = \boldsymbol{r} + \boldsymbol{l}$ with the lattice vector $\boldsymbol{l} \in \mathcal{L}$. Since dynamics with periodic boundary conditions are invariant under continuous spatial translations, the total momentum $\mathbf{P} = \sum_{a=0}^{N} \mathbf{p}_a$ is conserved, as well as the total energy given by the Hamiltonian function. If the system is ergodic, an equilibrium probability distribution is given by the stationary probability distribution (2.70). The system is also supposed to have the property of dynamical mixing.

The transport process of diffusion can be studied by the motion of the position $\mathbf{r}_t = \boldsymbol{r}_t + \boldsymbol{l}_t$ of the tagged tracer particle in the lattice. The problem is analogous to the deterministic diffusion of a particle in spatially periodic Lorentz gases, so that the considerations of previous section also apply here by replacing $\mathbf{x} = (\boldsymbol{x}, \boldsymbol{l})$ with $(\boldsymbol{\Gamma}, \boldsymbol{l})$. The time evolution is ruled by the phase-space flow

$$\Phi^t(\boldsymbol{\Gamma}, \boldsymbol{l}) = \left[\boldsymbol{\Phi}^t \boldsymbol{\Gamma}, \boldsymbol{l} + \boldsymbol{d}(\boldsymbol{\Gamma}, t)\right], \tag{13.60}$$

where the lattice vector $\boldsymbol{d}(\boldsymbol{\Gamma}, t) \in \mathcal{L}$ of displacement is defined as in equation (13.3) for the Lorentz gases.

The Liouvillian dynamics of the system can be described in terms of the probability density $p(\boldsymbol{\Gamma}, \boldsymbol{l})$, which is normalized as $\sum_{\boldsymbol{l} \in \mathcal{L}} \int_{\mathcal{M}} d\boldsymbol{\Gamma}\, p(\boldsymbol{\Gamma}, \boldsymbol{l}) = 1$. Its time evolution obeys

$$p_t(\boldsymbol{\Gamma}, \boldsymbol{l}) = p_0\left[\Phi^{-t}(\boldsymbol{\Gamma}, \boldsymbol{l})\right] = p_0\left[\boldsymbol{\Phi}^{-t} \boldsymbol{\Gamma}, \boldsymbol{l} + \boldsymbol{d}(\boldsymbol{\Gamma}, -t)\right]. \tag{13.61}$$

Here, the Liouvillian dynamics can also be decomposed into diffusive modes constructed as eigendistributions $\psi_{\boldsymbol{q}}$ of the Perron–Frobenius operator as in equation (13.28), so that the dispersion relation of diffusion is given by the leading Pollicott–Ruelle resonance of the dynamics (13.60). The probability that the tracer particle has moved to the position $\boldsymbol{l} = \mathbf{r} - \boldsymbol{r}$ during the time t is given by $P_t(\boldsymbol{l}) = \int_{\mathcal{M}} d\boldsymbol{\Gamma}\, p_t(\boldsymbol{\Gamma}, \boldsymbol{l})$. The time evolution of this probability distribution can thus be investigated with the same methods as for spatially periodic Lorentz gases. In particular, the dispersion relation of diffusion is here also given by equation (13.34) in terms of the Van Hove intermediate incoherent scattering function (Van Hove, 1954). For cubic lattices, diffusion is isotropic and the diffusion coefficient can be obtained from equation (13.35) if the integral of the velocity autocorrelation function converges. If the diffusion coefficient $\mathcal{D}$ exists and is positive, the density of tracer particles is thus ruled by the diffusion equation (13.44). Performing the same calculation as for the spatially periodic Lorentz gases, we also find the entropy production (13.58) for the diffusion of tracer particles (Dorfman et al., 2002).

13.4 Many-Particle Billiard Models of Heat Conduction

13.4.1 Conducting and Insulating Phases in Lattice Billiards

The phenomenological relation for heat conduction is the Fourier law (1.30), which expresses the heat current density in terms of the temperature gradient and the heat conductivity κ. In principle, this latter is given by the Green–Kubo formula (3.109) involving the time-dependent autocorrelation function of the microscopic total current of energy in the system. The integral of this autocorrelation function should converge for the heat conductivity to be finite. As discussed in Section 3.5.2, the simplest models, which are one-dimensional, do not fulfill this condition and have an infinite heat conductivity. Therefore, the question arises as to whether there exist models where Fourier's law can be established and the heat conductivity can be shown to be positive and finite. The answer

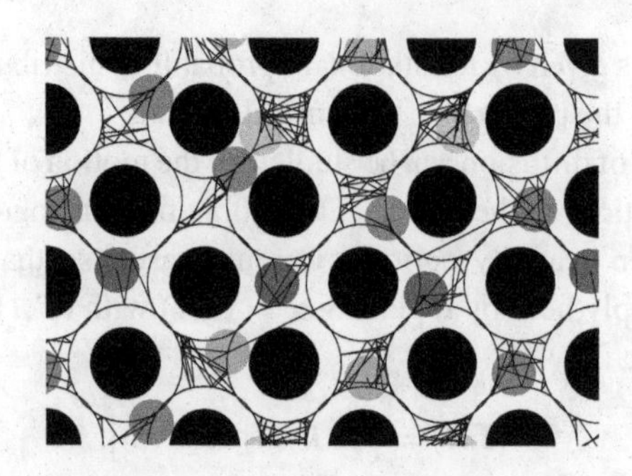

Figure 13.5 Example of a many-disk billiard model of heat conduction. Reprinted with permission from Gaspard and Gilbert (2008a) © American Physical Society.

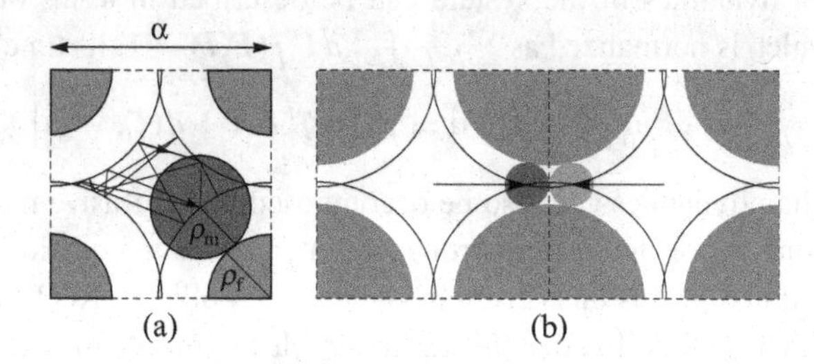

Figure 13.6 Geometry of the many-disk billiard model of heat conduction: (a) motion of a mobile disk in its cage; (b) binary collision between two next-neighboring mobile disks. Reproduced with permission from Gaspard and Gilbert (2008b) © IOP Publishing Ltd and Deutsche Physikalische Gesellschaft.

to this question is the affirmative in a class of many-particle billiard models, where hard balls are locally confined in cages but exchanging kinetic energy through binary collisions between next-neighboring balls (Gaspard and Gilbert, 2008a,b, 2009, 2017).

Two- and three-dimensional models of this kind have been investigated. In two dimension, the cages form a spatially extended lattice composed of fixed hard disks of radius ρ_f, as illustrated in Figure 13.5 in the case of a square lattice. Inside every cage, there is a mobile disk of radius ρ_m undergoing elastic collisions on the fixed disks. The geometry of the lattice is such that the mobile disks cannot leave their cage, but can have binary collisions on next-neighboring mobile disks. Figure 13.6(a) shows an example of geometry where every mobile disk is confined in a lattice cell of size α, which is made of four quarters of fixed disks. Since the hard disks may not overlap, the sum of radii, $\rho \equiv \rho_m + \rho_f$, should satisfy the condition $\rho < \alpha/\sqrt{2}$. Moreover, the mobile disk remains confined in its cage if $\rho > \alpha/2$. Therefore, there is no mass transport by diffusion in the system and the diffusion coefficient of mobile disks is equal to zero.

In addition, next-neighboring mobile disks are allowed to undergo binary collisions. As shown in Figure 13.6(b), there is a critical value for the radius of mobile disks $\rho_c = \sqrt{\rho^2 - (\alpha/2)^2}$, where such binary collisions become possible. Accordingly, the billiard conducts for $\rho_m > \rho_c$ and is insulating for $\rho_m < \rho_c$. The heat conductivity is equal to zero in the insulating phase. Therefore, the heat conductivity should vanish as the critical geometry of the billiard is reached from the conducting phase. Near criticality, the frequency $\nu_{\rm b}$ of binary collisions between next-neighboring disks goes to zero with the heat conductivity. In contrast, the frequency $\nu_{\rm w}$ of the collisions of a mobile disk with the fixed disks forming its cage remains positive across the transition between the conducting and insulating phases. Accordingly, there is the separation of timescales, $\nu_{\rm b} \ll \nu_{\rm w}$, near criticality.

Since all the collisions between the mobile and fixed disks are defocusing, the many-particle billiard is chaotic with positive Lyapunov exponents. In particular, the motion of a mobile disk inside its cage has the properties of dynamical mixing and, thus, ergodicity. Since the binary collisions are rare with respect to the collisions on the walls of each cage, there is a loss of memory of previous binary collisions near criticality. Accordingly, the exchanges of kinetic energy upon binary collisions between the mobile disks can be considered as a Markovian process in the limit of the critical geometry where $\nu_{\rm b} \ll \nu_{\rm w}$.

13.4.2 From Liouville's Equation to the Master Equation

If the many-particle billiard contains N mobile disks of mass m, the Liouville equation (2.39) ruling the time evolution of the probability density $p(\mathbf{\Gamma},t)$ in the phase space of positions and momenta $\mathbf{\Gamma} = (\mathbf{r}_a,\mathbf{p}_a)_{a=1}^N$ is given by

$$\partial_t p = \sum_{a=1}^{N} \left(-\frac{\mathbf{p}_a}{m} \cdot \frac{\partial}{\partial \mathbf{r}_a} + \sum_{k=1}^{n} \hat{K}^{(a,k)} \right) p + \sum_{1 \leq a < b \leq N} \hat{B}^{(a,b)} p, \tag{13.62}$$

where $\hat{K}^{(a,k)}$ denotes the pseudo-Liouvillian operator for the collisions of the ath mobile disk on the wall formed by the kth fixed disk in the corresponding lattice cell composed of n such walls and $\hat{B}^{(a,b)}$ is the pseudo-Liouvillian operator for the binary collisions between the ath and bth mobile disks. According to the collision rules obtained in Section B.3.3, these operators are defined, respectively, as

$$\begin{aligned} \hat{K}^{(a,k)} p(\dots,\mathbf{r}_a,\mathbf{p}_a,\dots) = \frac{\rho}{m} \int_{\boldsymbol{\epsilon}_k\cdot\mathbf{p}_a>0} & d\boldsymbol{\epsilon}_k \, (\boldsymbol{\epsilon}_k \cdot \mathbf{p}_a)\, \delta(\mathbf{r}_{ak} - \rho\, \boldsymbol{\epsilon}_k) \\ & \times \left[p(\dots,\mathbf{r}_a,\mathbf{p}_a - 2\,\boldsymbol{\epsilon}_k\, \boldsymbol{\epsilon}_k \cdot \mathbf{p}_a,\dots) - p(\dots,\mathbf{r}_a,\mathbf{p}_a,\dots)\right], \end{aligned} \tag{13.63}$$

where $\boldsymbol{\epsilon}_k$ is the unit vector normal to the kth fixed disk at the positions $\mathbf{R}_k$ and $\mathbf{r}_{ak} = \mathbf{r}_a - \mathbf{R}_k$, and

$$\begin{aligned} \hat{B}^{(a,b)} p(\dots,\mathbf{r}_a,\mathbf{p}_a,\dots,\mathbf{r}_b,\mathbf{p}_b,\dots) &= \frac{2\rho_{\rm m}}{m} \int_{\boldsymbol{\epsilon}_{ab}\cdot\mathbf{p}_{ab}>0} d\boldsymbol{\epsilon}_{ab} \, (\boldsymbol{\epsilon}_{ab} \cdot \mathbf{p}_{ab}) \\ \times \Big[\delta(\mathbf{r}_{ab} - 2\rho_{\rm m}\, \boldsymbol{\epsilon}_{ab})\, p(\dots,\mathbf{r}_a,\mathbf{p}_a - \boldsymbol{\epsilon}_{ab}\, \boldsymbol{\epsilon}_{ab}\cdot\mathbf{p}_{ab},&\dots,\mathbf{r}_b,\mathbf{p}_b + \boldsymbol{\epsilon}_{ab}\,\boldsymbol{\epsilon}_{ab}\cdot\mathbf{p}_{ab},\dots) \\ - \delta(\mathbf{r}_{ab} + 2\rho_{\rm m}\, \boldsymbol{\epsilon}_{ab})\, p(\dots,\mathbf{r}_a,\mathbf{p}_a,\dots,\mathbf{r}_b,\mathbf{p}_b,\dots)\Big], \end{aligned} \tag{13.64}$$

where $\boldsymbol{\epsilon}_{ab}$ is the unit vector in equation (B.38), $\mathbf{r}_{ab} = \mathbf{r}_a - \mathbf{r}_b$, and $\mathbf{p}_{ab} = \mathbf{p}_a - \mathbf{p}_b$.

Because of dynamical mixing due to the wall collisions, the phase-space probability density $p(\boldsymbol{\Gamma},t)$ undergoes relaxation to local equilibrium over the timescale $\nu_{\rm w}^{-1}$, i.e., well before the next binary collision occurring on the longer timescale $\nu_{\rm b}^{-1}$. This local equilibrium is described by the following probability density of finding the mobile disks with given values for their kinetic energy $\varepsilon_a = \mathbf{p}_a^2/(2m)$,

$$p(\varepsilon_1, \varepsilon_2, \ldots, \varepsilon_N, t) \equiv \int \prod_{a=1}^{N} \delta\left(\varepsilon_a - \frac{\mathbf{p}_a^2}{2m}\right) p(\boldsymbol{\Gamma},t)\, d\boldsymbol{\Gamma}, \tag{13.65}$$

where $d\boldsymbol{\Gamma} = \prod_{a=1}^{N} d^d r_a\, d^d p_a$ for a d-dimensional billiard. As justified by the separation of timescales $\nu_{\rm b} \ll \nu_{\rm w}$, the local equilibrium distribution is ruled by the Markovian master equation,

$$\begin{aligned} \partial_t p(\ldots, \varepsilon_a, \ldots, \varepsilon_b, \ldots, t) = &\sum_{1\leq a<b\leq N} \int d\eta \\ &\times \Big[w(\varepsilon_a, \varepsilon_b | \varepsilon_a + \eta, \varepsilon_b - \eta)\, p(\ldots, \varepsilon_a + \eta, \ldots, \varepsilon_b - \eta, \ldots, t) \\ &\quad - w(\varepsilon_a - \eta, \varepsilon_b + \eta | \varepsilon_a, \varepsilon_b)\, p(\ldots, \varepsilon_a, \ldots, \varepsilon_b, \ldots, t)\Big], \end{aligned} \tag{13.66}$$

where $w(\varepsilon_a - \eta, \varepsilon_b + \eta | \varepsilon_a, \varepsilon_b)$ is the transition rate density that the energy η is exchanged between the ath and bth particles as the result of their binary collision (Gaspard and Gilbert, 2008a,b). Near criticality, this transition rate density can be evaluated using the corresponding probability flux in the phase space of the two particles, which gives the expression

$$\begin{aligned} w(\varepsilon_a - \eta, \varepsilon_b + \eta | \varepsilon_a, \varepsilon_b) \simeq &\frac{2\,\rho_{\rm m}}{(2\pi)^2 |\mathcal{B}_\rho|^2 m} \int d\phi_{ab}\, d\mathbf{R}_{ab} \int_{\boldsymbol{\epsilon}_{ab}\cdot\mathbf{p}_{ab}>0} d\mathbf{p}_a\, d\mathbf{p}_b \\ &\times \boldsymbol{\epsilon}_{ab}\cdot\mathbf{p}_{ab}\, \delta\left(\varepsilon_a - \frac{\mathbf{p}_a^2}{2m}\right) \delta\left(\varepsilon_b - \frac{\mathbf{p}_b^2}{2m}\right) \delta\left\{\eta - \frac{1}{2m}\left[(\boldsymbol{\epsilon}_{ab}\cdot\mathbf{p}_a)^2 - (\boldsymbol{\epsilon}_{ab}\cdot\mathbf{p}_b)^2\right]\right\}, \end{aligned} \tag{13.67}$$

where the particles a and b are assumed to be in contact and located in their respective cells, $\mathbf{R}_{ab} \equiv (\mathbf{r}_a + \mathbf{r}_b)/2$ is their center of mass, ϕ_{ab} is the angle of the unit vector $\boldsymbol{\epsilon}_{ab} = (\cos\phi_{ab}, \sin\phi_{ab})$, and $|\mathcal{B}_\rho|$ is the area of the billiard table for the motion of every mobile disk in its cage (Gaspard and Gilbert, 2008a,b).

For the square lattice of Figure 13.6, the integrals over the center of mass and the orientation of the unit vector take the following value near criticality,

$$\int d\phi_{ab}\, d\mathbf{R}_{ab} = \frac{128\rho_{\rm c}}{3\alpha^2}(\rho_{\rm m} - \rho_{\rm c})^3 + \frac{256\rho_{\rm c}^2}{3\alpha^4}(\rho_{\rm m} - \rho_{\rm c})^4 + \cdots, \tag{13.68}$$

and the binary collision frequency is evaluated as

$$\nu_{\rm b} \simeq \sqrt{\frac{k_{\rm B}T}{\pi m}}\, \frac{2\,\rho_{\rm m}}{|\mathcal{B}_\rho|^2} \int d\phi_{ab}\, d\mathbf{R}_{ab}. \tag{13.69}$$

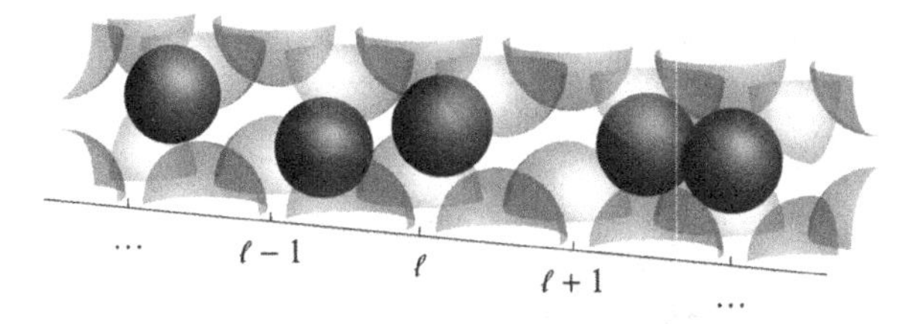

Figure 13.7 One-dimensional many-sphere billiard model of heat conduction. Reproduced with permission from Gaspard and Gilbert (2017) © IOP Publishing Ltd and SISSA Medialab.

13.4.3 Fourier's Law and Heat Conductivity

The evolution equation for the mean kinetic energies of the particles, which define the local values of the temperature as $\langle \varepsilon_a \rangle = (d/2)k_{\rm B}T_a$, can be deduced from the master equation (13.66), giving

$$\partial_t \langle \varepsilon_a \rangle = -\sum_b \langle J(\varepsilon_a, \varepsilon_b) \rangle \tag{13.70}$$

with the energy current defined as

$$J(\varepsilon_a, \varepsilon_b) = \int d\eta\, \eta\, w(\varepsilon_a - \eta, \varepsilon_b + \eta | \varepsilon_a, \varepsilon_b) = -J(\varepsilon_b, \varepsilon_a), \tag{13.71}$$

which expresses the local conservation of energy. The energy density is defined as $e = \langle \varepsilon_a \rangle / \upsilon$, where υ is the volume of a primitive cell of the lattice and the heat capacity per unit volume is equal to $c_\upsilon = (\partial e/\partial T)_\upsilon = k_{\rm B} d/(2\upsilon)$. Expanding in the gradient of the local temperature, Fourier's law can be deduced, which is reading as $\langle J \rangle = -\kappa\, \partial_x T$ with the heat conductivity κ.

In the critical limit $\rho_{\rm m} \to \rho_{\rm c}$ where $\nu_{\rm b} \ll \nu_{\rm w}$, the heat conductivity is given by $\kappa \simeq A\alpha^2 \nu_{\rm b}$ with the binary collision frequency (13.69) and a dimensionless constant A, which is very close to the unit value (Gaspard and Gilbert, 2008a,b). This constant has been accurately evaluated using the variational formula by Spohn (1991), yielding $A \simeq 1 - 0.000\,893\,56 \pm 10^{-7}$ for the one-dimensional chain of hard disks of Figure 13.6, and $A \simeq 1 - 0.000\,372\,72 \pm 10^{-7}$ for the one-dimensional chain of hard spheres of Figure 13.7 (Gaspard and Gilbert, 2017).

For these many-particle billiard models, Fourier's law can thus be established and the heat conductivity obtained with very high accuracy in the critical limit.

13.5 Models for Mechanothermal Coupling

Exactly solvable models have also been proposed for motors propelled by a temperature difference between two thermal reservoirs (Van den Broeck et al., 2005). In these models, there are gases at different temperatures in two separated reservoirs and an object is constrained to move horizontally without rotation or vertical displacement upon elastic collisions with the particles of the reservoirs, as shown in Figure 13.8. The moving object is composed

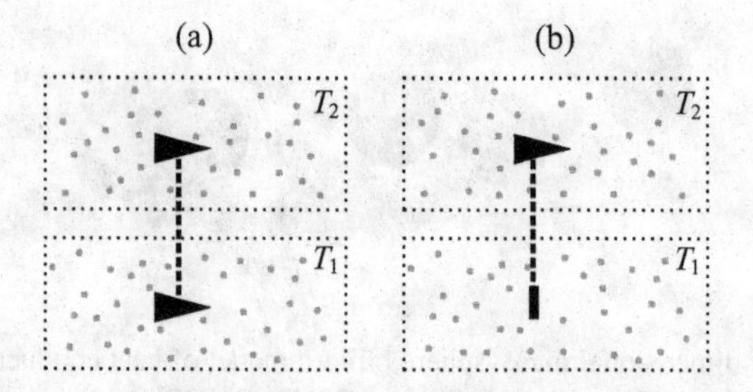

Figure 13.8 Schematic representation of (a) Triangula and (b) Triangulita models for mechanothermal coupling. Reproduced with permission from Van den Broeck et al. (2005) © IOP Publishing Ltd and Deutsche Physikalische Gesellschaft. CC BY-NC-SA.

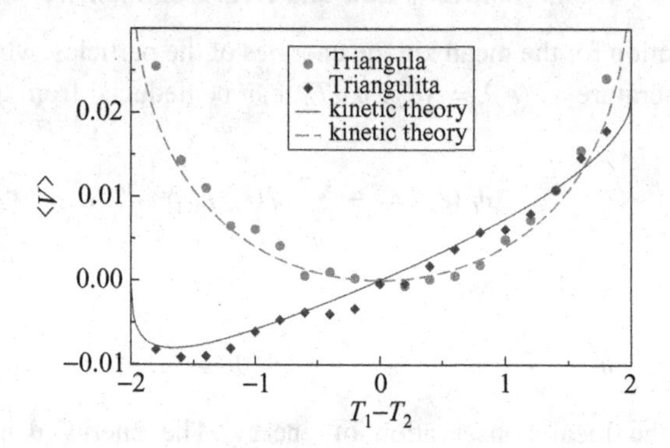

Figure 13.9 Mean velocity of Triangula and Triangulita motors versus the temperature difference for $T_1 + T_2 = 2$, system size 1200 × 300, particle numbers $N_1 = N_2 = 800$, mass ratio $M/m = 20$, apex angle $\theta_0 = \pi/18$, and vertical cross section $L = 1$. The lines are obtained using kinetic theory. Reproduced with permission from Van den Broeck et al. (2005) © IOP Publishing Ltd and Deutsche Physikalische Gesellschaft. CC BY-NC-SA.

of two rigidly linked identical triangles in the Triangula model and one triangle linked to a vertical bar in the Triangulita model. The former is thus symmetric with respect to the exchange of both reservoirs, while the latter is not.

In the limit of ideal gases of particles in the two reservoirs, the probability distribution of the motor velocity V is ruled by the Markovian master equation,

$$\partial_t \, p(V,t) = \int dV' \left[w(V|V')\, p(V',t) - w(V'|V)\, p(V,t) \right], \tag{13.72}$$

where the transition rate densities $w(V|V')$ can be determined by kinetic theory in terms of the fluxes of gas particles colliding on the borders of the triangles and/or the bar of the moving object (Van den Broeck et al., 2005). There is an excellent agreement with molecular dynamics simulations, as seen in Figure 13.9.

For such a motor, there are two currents given by the heat current J_q between the two reservoirs and the mechanical current $J_{\rm m} = V$ defined as the motor velocity. In this respect, there exist two corresponding affinities. The first is the thermal affinity $A_q = \beta_2 - \beta_1$ with

$\beta_j = (k_B T_j)^{-1}$ and the second is the mechanical affinity $A_{\rm m} = (\beta_1 + \beta_2)F_{\rm ext}/2$ in the presence of an external force $F_{\rm ext}$ exerted horizontally on the motor. In general, the currents can be expressed in terms of the affinities according to the following expansions with the linear and nonlinear response coefficients,

$$\begin{cases} J_{\rm m} = L_{\rm m,m}\, A_{\rm m} + L_{{\rm m},q}\, A_q + \frac{1}{2}\, M_{\rm m,mm}\, A_{\rm m}^2 + M_{{\rm m,m}q}\, A_{\rm m}\, A_q + \frac{1}{2}\, M_{{\rm m},qq}\, A_q^2 + \cdots \\ J_q = L_{q,{\rm m}}\, A_{\rm m} + L_{q,q}\, A_q + \frac{1}{2}\, M_{q,{\rm mm}}\, A_{\rm m}^2 + M_{q,{\rm m}q}\, A_{\rm m}\, A_q + \frac{1}{2}\, M_{q,qq}\, A_q^2 + \cdots . \end{cases} \tag{13.73}$$

If there is no external force, the motor velocity becomes

$$J_{\rm m}\big|_{A_{\rm m}=0} = L_{{\rm m},q}\, A_q + \frac{1}{2}\, M_{{\rm m},qq}\, A_q^2 + \cdots . \tag{13.74}$$

As seen in Figure 13.9, the symmetry of the Triangula motor implies that the linear response coefficient expressing the mechanothermal coupling is equal to zero, i.e., $L_{{\rm m},q} = 0$, so that the response of the Triangula motor to the thermal affinity is nonlinear with a nonvanishing coefficient $M_{{\rm m},qq} \neq 0$. In contrast, the Triangulita motor has a nonvanishing linear response coefficient $L_{{\rm m},q} \neq 0$ because of its asymmetry. This example shows the importance of considering not only the linear but also the nonlinear response coefficients in equation (5.133).

14
Quantum Statistical Mechanics

14.1 Introduction

At the microscopic scale, quantum mechanics is established as the basic theory ruling the motion of particles composing matter. In quantum mechanics, microstates are wave functions belonging to some Hilbert space, instead of points in the classical phase space. Classical mechanics is no longer a good approximation if the thermal de Broglie wavelength $\lambda_{\rm dB}$ becomes equal to or larger than the mean distance $\mathcal{R}$ between nearest neighboring particles,

$$\lambda_{\rm dB} \sim \frac{h}{\sqrt{mk_{\rm B}T}} \gtrsim \mathcal{R} \sim \left(\frac{V}{N}\right)^{1/3}, \tag{14.1}$$

which happens for either high particle density $n = N/V$ or low temperature T. In these regimes, quantum-mechanical phenomena manifest themselves in the form of wave interferences, quantum coherences, and quantization effects, e.g., in quantum Hall effects. For low temperatures, remarkable quantum phases exist such as Bose–Einstein condensates, superfluids, and superconductors (Anderson, 1984; Leggett, 2006).

Under these conditions, the question arises to understand how the general results of previous chapters are modified by quantum mechanics. This issue concerns in particular the Green–Kubo formulas for the transport coefficients, as well as the master equations ruling the time evolution of probability distributions and the corresponding stochastic differential equations of Langevin type (Kadanoff and Baym, 1962; Caldeira and Leggett, 1983; Gardiner and Zoller, 2000).

In quantum systems, the observables are no longer given by functions defined in the classical phase space, but by noncommuting Hermitian operators defined in the Hilbert space. Accordingly, probability distributions should be generalized into density operators, also known as statistical operators, describing the quantum coherences by their off-diagonal elements. Therefore, the mathematical structure of mechanics is fundamentally different, although deep analogies prevail for the formulation of statistical-mechanical properties between the classical and quantum frameworks. The aim of this chapter is to give an overview of these issues.

14.2 Quantum Mechanics

14.2.1 Quantum Microstates and Observables

In quantum mechanics, the microstates should satisfy the *principle of superposition*, so that they are defined as vectors in some complex vector space of finite or infinite dimension, which is equipped with a scalar product (Dirac, 1958). If $|\Psi\rangle$ and $|\Phi\rangle$ denote two microstates, their scalar product is the complex number $\langle\Phi|\Psi\rangle = \langle\Psi|\Phi\rangle^*$. The norm of the vector $|\Psi\rangle$ is defined as the nonnegative real number $\|\Psi\| \equiv \sqrt{\langle\Psi|\Psi\rangle}$. Observable quantities are represented by Hermitian operators acting on the microstates. The observables generate a noncommutative algebra, including the functions of operators. The value of an observable $\hat{A}$ provided that the system is in the microstate $|\Psi\rangle$ is the real number defined by the normalized quadratic form, $\langle\hat{A}\rangle_\Psi \equiv \langle\Psi|\hat{A}|\Psi\rangle/\langle\Psi|\Psi\rangle$. Here, the vectors are supposed to be normalized to the unit value $\|\Psi\| = 1$, so that the normalization factor $\langle\Psi|\Psi\rangle = 1$ is not required. In the particular case where the observable is the projector $\hat{\mathsf{P}} = |\Phi\rangle\langle\Phi|$ onto some vector $|\Phi\rangle$, the value of this observable is given by the nonnegative real number $P = \langle\Psi|\hat{\mathsf{P}}|\Psi\rangle = |\langle\Phi|\Psi\rangle|^2$, which may be interpreted as the probability $0 \leq P \leq 1$ of finding the system in the microstate $|\Phi\rangle$. Similarly, the probability density of finding the real number α as eigenvalue for the observable $\hat{A}$ is given by $p(\alpha) = \langle\Psi|\delta(\alpha - \hat{A})|\Psi\rangle$.

For a system with N particles of masses $\{m_a\}_{a=1}^N$, an observable is defined as some function of the operators $\{\hat{\mathbf{r}}_a, \hat{\mathbf{p}}_a, \hat{\mathbf{S}}_a\}_{a=1}^N$, respectively representing their position, momentum, and spin components, and obeying the following commutation relations,

$$[\hat{r}_{ai}, \hat{p}_{bj}] = \imath\,\hbar\,\delta_{ij}\,\delta_{ab} \qquad \text{and} \qquad [\hat{S}_{ai}, \hat{S}_{bj}] = \imath\,\hbar\,\epsilon_{ijk}\,\hat{S}_{ak}\,\delta_{ab}, \tag{14.2}$$

where $[\cdot,\cdot]$ is the commutator between two operators, $\imath = \sqrt{-1}$, δ_{ij} the Kronecker unit tensor, ϵ_{ijk} the Levi-Civita totally antisymmetric tensor with $i, j, k = x, y, z$, and $1 \leq a, b \leq N$. For N interacting nonrelativistic spinless particles, the Hamiltonian operator, which represents the total energy of the system, is given by the function (2.2) with the positions and momenta replaced by their respective operators. The Hamiltonian operator may depend on time if the system is subjected to a time-dependent electromagnetic field, for instance.

The microstates can be represented by wave functions after taking their scalar product with the simultaneous eigenstates of some complete set of commuting observables chosen among the basic observables (14.2). If the position operators are chosen together with the z-component of the spin operators, the wave functions are given by $\Psi(\boldsymbol{\xi}) = \langle\boldsymbol{\xi}|\Psi\rangle$ where $\boldsymbol{\xi} \equiv (\mathbf{r}_a, \sigma_a)_{a=1}^N$ denote the eigenvalues of the corresponding operators.

14.2.2 Time Evolution of the Quantum Microstates

The time evolution of the microstates is ruled by the Schrödinger equation

$$\imath\hbar\,\partial_t|\Psi\rangle = \hat{H}\,|\Psi\rangle, \tag{14.3}$$

where $\hbar = h/(2\pi) = 1.054571817 \times 10^{-34}$ J s is the reduced Planck constant, and $\hat{H}$ is the Hamiltonian operator.

Since the Schrödinger equation (14.3) is linear, the principle of superposition is preserved during the time evolution of the microstates. Moreover, the time evolution is unitary in the quantum state space, so that $\langle \Phi_t | \Psi_t \rangle = \langle \Phi_0 | \Psi_0 \rangle$. If the system is autonomous, i.e., its Hamiltonian operator is time independent, the solution of Schrödinger's equation can be expressed as

$$|\Psi_t\rangle = \hat{\mathrm{U}}^t \, |\Psi_0\rangle, \qquad \text{where} \qquad \hat{\mathrm{U}}^t = \mathrm{e}^{-\frac{\mathrm{i}}{\hbar}\hat{H}t} \tag{14.4}$$

is the unitary operator of time evolution satisfying

$$\hat{\mathrm{U}}^t \, \hat{\mathrm{U}}^{t'} = \hat{\mathrm{U}}^{t+t'}, \qquad (\hat{\mathrm{U}}^t)^{-1} = \hat{\mathrm{U}}^{-t}, \qquad \text{and} \qquad \hat{\mathrm{U}}^0 = \hat{1} \tag{14.5}$$

with the identity operator $\hat{1}$ defined in the quantum state space. Equations (14.4) and (14.5) are the quantum analogues of equations (2.20) and (2.21). Therefore, the time-evolved microstate $|\Psi_t\rangle$ is uniquely determined by its initial condition $|\Psi_0\rangle$, as in classical mechanics (Wigner, 1963).

Another consequence of Schrödinger's equation is that the distance between two microstates is preserved by the unitary time evolution, $\|\tilde{\Psi}_t - \Psi_t\| = \|\tilde{\Psi}_0 - \Psi_0\|$, so that there is no sensitivity to initial conditions in quantum systems with finitely many degrees of freedom.

14.2.3 Time Reversal

In quantum mechanics, time reversal is generated by the antiunitary operator $\hat{\Theta}$ such that

$$\hat{\Theta}\, \hat{\mathbf{r}}_a \, \hat{\Theta}^{-1} = +\hat{\mathbf{r}}_a, \qquad \hat{\Theta}\, \hat{\mathbf{p}}_a \, \hat{\Theta}^{-1} = -\hat{\mathbf{p}}_a, \qquad \text{and} \qquad \hat{\Theta}\, \hat{\mathbf{S}}_a \, \hat{\Theta}^{-1} = -\hat{\mathbf{S}}_a \tag{14.6}$$

for $1 \leq a \leq N$. For spinless particles, this operator takes the complex conjugate of the wave function $\hat{\Theta}\Psi = \Psi^*$, in which case the time-reversal transformation is an involution, i.e., $\hat{\Theta}^2 = \hat{1}$.

The symmetry of the dynamics under time reversal is expressed by

$$\hat{\Theta}\, \hat{H} \, \hat{\Theta}^{-1} = \hat{H}, \tag{14.7}$$

which is the quantum analogue of equation (2.28). As a consequence, if $|\Psi_t\rangle$ is a solution of the Schrödinger equation (14.3), $\hat{\Theta}|\Psi_{-t}\rangle$ is also a solution, which is the expression of microreversibility. These two solutions do not, in general, coincide. Accordingly, the selection of the initial condition may break the time-reversal symmetry among the solutions of Schrödinger's equation, although this equation is always symmetric, as explained for classical mechanics in Section 2.6.1.

In the presence of an external magnetizing field $\mathcal{H}$, microreversibility corresponds to the symmetry

$$\hat{\Theta}\, \hat{H}(\mathcal{H}) \hat{\Theta}^{-1} = \hat{H}(-\mathcal{H}) \tag{14.8}$$

in analogy with equation (2.30).

14.2.4 Quantum Fields

In many-body quantum systems, the observables associated with all the particles of the same species can be expressed in terms of annihilation-creation field operators $\hat{\psi}(\mathbf{r},\sigma)$ and $\hat{\psi}^{\dagger}(\mathbf{r},\sigma)$, which are defined for every space point $\mathbf{r} \in \mathbb{R}^3$ and spin component $\sigma = -s, -s+1, \dots, s-1, s$, where s is the spin of the particles. These field operators obey canonical commutation or anticommutation rules

$$\hat{\psi}(\mathbf{r},\sigma)\,\hat{\psi}^{\dagger}(\mathbf{r}',\sigma') - \theta\,\hat{\psi}^{\dagger}(\mathbf{r}',\sigma')\,\hat{\psi}(\mathbf{r},\sigma) = \delta^3(\mathbf{r}-\mathbf{r}')\,\delta_{\sigma\sigma'}, \tag{14.9}$$

whether the particles are

$$\text{bosons if } \theta = +1 \qquad \text{or} \qquad \text{fermions if } \theta = -1. \tag{14.10}$$

These rules deeply influence the statistical properties of systems with identical particles, especially for high density and low temperature. Fermions obey the Fermi–Dirac statistics and Pauli's exclusion principle, so that the same one-body quantum state cannot be occupied by more than one fermion. In contrast, the same one-body state can be occupied by an arbitrarily large number of bosons, which is called the Bose–Einstein statistics. The annihilation-creation operators act on Fock's space, which is the space of microstates with a variable number of particles, including the vacuum microstate with no particle at all. Fock's space also contains the coherent microstates, which are the linear superpositions of microstates with different numbers of particles, for instance, describing quantum condensates in superfluids or laser electromagnetic fields.

In this framework, the total number of particles is given by

$$\hat{N} = \sum_{\sigma=-s}^{+s} \int d\mathbf{r}\,\hat{\psi}^{\dagger}(\mathbf{r},\sigma)\,\hat{\psi}(\mathbf{r},\sigma) \tag{14.11}$$

and the Hamiltonian operator by

$$\begin{aligned}\hat{H} &= \sum_{\sigma=-s}^{+s} \int d\mathbf{r}\,\hat{\psi}^{\dagger}(\mathbf{r},\sigma)\,\hat{h}\,\hat{\psi}(\mathbf{r},\sigma) \\ &+ \frac{1}{2}\sum_{\sigma,\sigma'=-s}^{+s} \int d\mathbf{r}\,d\mathbf{r}'\,\hat{\psi}^{\dagger}(\mathbf{r},\sigma)\,\hat{\psi}^{\dagger}(\mathbf{r}',\sigma')\,u^{(2)}(\mathbf{r},\sigma;\mathbf{r}',\sigma')\,\hat{\psi}(\mathbf{r}',\sigma')\,\hat{\psi}(\mathbf{r},\sigma) + \cdots,\end{aligned} \tag{14.12}$$

where

$$\hat{h} = -\frac{\hbar^2}{2m}\nabla^2 + u^{(\text{ext})}(\mathbf{r},\sigma) \tag{14.13}$$

is the one-body Hamiltonian operator for particles with mass m in the external energy potential $u^{(\text{ext})}$ (Kadanoff and Baym, 1962; Thirring, 1983). In the subspace containing the fixed number N of spinless particles, the Hamiltonian operator (14.12) reduces to the Hamiltonian given by the function (2.2).

According to the famous *spin-statistics theorem*, particles with integer spin are bosons and those with half-odd integer spin are fermions (Weinberg, 1995).

14.3 Statistical Ensembles and Their Time Evolution

14.3.1 Statistical Ensemble and Statistical Operator

As in classical mechanics, the preparation of initial conditions is, in general, performed by a different and larger system that the one ruled by the Schrödinger equation (14.3). The state of this preparing device cannot be exactly the same every time a new initial condition is prepared. Therefore, the initial microstates of the system form a statistical ensemble if the preparation is repeated many times. The statistical ensemble of microstates is thus given by $\{|\Psi^{(j)}\rangle\}_{j=1}^{\infty}$, which are evolving in time according to Schrödinger's equation. Assuming that $\|\Psi^{(j)}\| = 1$, the mean value of some observable $\hat{A}$ over this ensemble is given by

$$\langle \hat{A} \rangle = \mathrm{tr}\,\hat{\rho}\,\hat{A}, \qquad \text{where} \qquad \hat{\rho} = \lim_{\mathcal{N}\to\infty} \frac{1}{\mathcal{N}} \sum_{j=1}^{\mathcal{N}} |\Psi^{(j)}\rangle\langle\Psi^{(j)}| \tag{14.14}$$

is the so-called *statistical operator* and "tr" denotes the trace over the quantum state space. The statistical operator is Hermitian and satisfies the normalization condition $\mathrm{tr}\,\hat{\rho} = 1$.

14.3.2 Time Evolution of the Statistical Operator

Because of the Schrödinger equation (14.3), the time evolution of the statistical operator is ruled by the so-called Landau–von Neumann equation,

$$\imath\hbar\,\partial_t\,\hat{\rho} = \left[\hat{H}, \hat{\rho}\,\right], \tag{14.15}$$

which can also be expressed in a form similar to the classical Liouville equation (2.39) according to

$$\partial_t\,\hat{\rho} = \widehat{\mathcal{L}}\,\hat{\rho}, \qquad \text{where} \qquad \widehat{\mathcal{L}}(\cdot) \equiv \frac{1}{\imath\hbar}\,[\hat{H}, \,\cdot\,] \tag{14.16}$$

is the quantum Liouvillian superoperator.

As a consequence of equation (14.4), the time evolution of the statistical operator over the time interval $[0,t]$ in an autonomous system is given by

$$\hat{\rho}_t = \mathrm{e}^{-\frac{\imath}{\hbar}\hat{H}t}\,\hat{\rho}_0\,\mathrm{e}^{+\frac{\imath}{\hbar}\hat{H}t}. \tag{14.17}$$

The mean value of some observable thus evolves in time as the quantum analogue of equation (2.46), i.e.,

$$\langle \hat{A} \rangle_t = \mathrm{tr}\,\hat{\rho}_t\,\hat{A} = \mathrm{tr}\,\hat{\rho}_0\,\hat{A}_t, \qquad \text{where} \qquad \hat{A}_t \equiv \mathrm{e}^{+\frac{\imath}{\hbar}\hat{H}t}\,\hat{A}\,\mathrm{e}^{-\frac{\imath}{\hbar}\hat{H}t} \tag{14.18}$$

is the operator expressed in the so-called Heisenberg picture (Dirac, 1958).

14.3.3 Wigner Function and the Classical Limit

For quantum systems composed of spinless particles, the correspondence with classical mechanics can be established using the Wigner transform of observables into functions defined in the classical phase space $\mathbf{\Gamma} = (\mathbf{q}, \mathbf{p})$ with $\mathbf{q} = (\mathbf{r}_a)_{a=1}^{N}$ and $\mathbf{p} = (\mathbf{p}_a)_{a=1}^{N}$ as

$$A_{\rm W}(\mathbf{q},\mathbf{p}) \equiv \int e^{\frac{i}{\hbar}\mathbf{p}\cdot\mathbf{y}} \left\langle \mathbf{q} - \frac{\mathbf{y}}{2} \right| \hat{A} \left| \mathbf{q} + \frac{\mathbf{y}}{2} \right\rangle d\mathbf{y}, \tag{14.19}$$

where $|\mathbf{q}\rangle$ denotes a simultaneous eigenstate of the position operators such that $\langle \mathbf{q}|\mathbf{q}'\rangle = \delta^{3N}(\mathbf{q} - \mathbf{q}')$ (Wigner, 1932). Accordingly, the mean value (14.14) can be expressed as

$$\langle \hat{A} \rangle = \int A_{\rm W}(\mathbf{\Gamma})\, W(\mathbf{\Gamma})\, d\mathbf{\Gamma}, \qquad \text{where} \qquad W(\mathbf{\Gamma}) \equiv \frac{1}{h^{3N}}\, \rho_{\rm W}(\mathbf{\Gamma}) \tag{14.20}$$

is the so-called Wigner function, which is real but may be negative. In the classical limit, the Wigner function can be identified with the classical probability density $\lim_{\hbar\to 0} W(\mathbf{\Gamma}) = p(\mathbf{\Gamma})$, so that the classical mean value (2.36) is recovered. Now, the Wigner transform of the commutator between two observables is related by

$$\left[\hat{A}, \hat{B}\right]_{\rm W} = \imath\, \hbar\, \{A_{\rm W}, B_{\rm W}\} + O\left(\hbar^3\right) \tag{14.21}$$

to the Poisson bracket (2.14) between the Wigner transforms of these observables. Consequently, the Landau–von Neumann equation (14.15) leads to the classical Liouville equation (2.39) in the classical limit $\hbar \to 0$.

14.3.4 Equilibrium Statistical Ensembles

As in classical systems, thermodynamic equilibrium manifests invariant statistical properties. Accordingly, equilibrium systems are described by stationary statistical operators that commute with the Hamiltonian operator as well as with other conserved quantities. This is in particular the case for the *canonical statistical operator*

$$\hat{\rho}_{\rm eq} = \frac{1}{Z}\, e^{-\beta \hat{H}} \qquad \text{with} \qquad Z = \operatorname{tr} e^{-\beta \hat{H}} \qquad \text{and} \qquad F = -\beta^{-1} \ln Z, \tag{14.22}$$

describing a closed equilibrium system at the inverse temperature $\beta = (k_{\rm B}T)^{-1}$ and the Helmholtz free energy F. Another example is the *grand canonical statistical operator*

$$\hat{\rho}_{\rm eq} = \frac{1}{\Xi}\, e^{-\beta\left(\hat{H} - \mu \hat{N}\right)} \quad \text{with} \quad \Xi = \operatorname{tr} e^{-\beta\left(\hat{H} - \mu \hat{N}\right)} \quad \text{and} \quad J = -\beta^{-1} \ln \Xi, \tag{14.23}$$

describing an open equilibrium system at the inverse temperature $\beta = (k_{\rm B}T)^{-1}$, the chemical potential μ, and the grand potential J, in terms of the Hamiltonian operator $\hat{H}$ and the particle number operator $\hat{N}$. In the classical limit, the corresponding Wigner functions (14.20) lead to the canonical and grand canonical probability distributions (C.8) and (C.11) presented in Appendix C. There exist further equilibrium statistical ensembles, depending on the conditions constraining the system.

We note that these equilibrium statistical operators are symmetric under time reversal as a consequence of the symmetry (14.7) for the Hamiltonian operator and a similar symmetry for the particle number operator.

Even though the system is found at equilibrium, time-dependent correlation functions can be considered, for instance, between two observables $\hat{X}$ and $\hat{Y}$ separated by some time interval t. Such correlation functions satisfy the so-called *Kubo–Martin–Schwinger (KMS) relation*,

$$\left\langle \hat{X}(0)\,\hat{Y}(t) \right\rangle_{\text{eq},\beta} = \left\langle \hat{Y}(t - \imath\hbar\beta)\,\hat{X}(0) \right\rangle_{\text{eq},\beta}, \tag{14.24}$$

where $\langle\cdot\rangle_{\text{eq},\beta} = \text{tr}\,\hat{\rho}_{\text{eq}}(\cdot)$ with the equilibrium canonical statistical operator (14.22).

14.3.5 Entropy

In quantum systems, the entropy of a statistical ensemble represented by the statistical operator $\hat{\rho}$ is given by the von Neumann formula

$$S \equiv -k_{\text{B}}\,\text{tr}\,\hat{\rho}\,\ln\hat{\rho}, \tag{14.25}$$

which provides the precise evaluation of the thermodynamic entropy in the equilibrium statistical ensembles. Furthermore, the third law of thermodynamics can be proved with this definition (Pathria, 1972; Diu et al., 1989; Balian, 1991). For an ideal monatomic gas in the canonical ensemble (14.22), the entropy (14.25) is well approximated in the classical limit by the Sackur–Tetrode formula (2.115).

In isolated nonequilibrium systems, the entropy (14.25) remains invariant during time evolution. Coarse graining is thus required in order to expect some time dependence for the entropy. This is in particular the case for a subset among the set of all the degrees of freedom inside an isolated system, or for a system in contact with some environment.

In a composite system with parts A and B, the whole system may be in the pure state $|\Psi\rangle$, so that its entropy (14.25) is equal to zero. However, each part is described by a reduced statistical operator obtained by tracing out the degrees of freedom of the other part, as either $\hat{\rho}_A = \text{tr}_B\,|\Psi\rangle\,\langle\Psi|$ or $\hat{\rho}_B = \text{tr}_A\,|\Psi\rangle\,\langle\Psi|$. For these reduced statistical operators, the von Neumann entropy (14.25) takes the same value in both parts. This results from the Schmidt decomposition of the pure state as $|\Psi\rangle = \sum_n \Psi_n |a_n\rangle\,|b_n\rangle$ into orthonormal states $\{|a_n\rangle\}$ and $\{|b_n\rangle\}$ defined in the quantum state spaces of the parts A and B, respectively. Therefore, the reduced statistical operators are given by $\hat{\rho}_A = \sum_n |a_n\rangle P_n \langle a_n|$ and $\hat{\rho}_B = \sum_n |b_n\rangle P_n \langle b_n|$ with $P_n \equiv |\Psi_n|^2$, so that $S_A = S_B = -k_{\text{B}} \sum_n P_n \ln P_n$. Typically, this entropy is thus positive, meaning that there is entanglement between the two parts A and B. The point is that the entropy of a reduced statistical operator can change with time. This happens because both parts get entangled by their mutual interaction, so that their respective entropies become positive. As a consequence of the entanglement, the time evolution of some part of a system may thus be nonunitary, even if the whole system has a unitary time evolution.

14.3.6 Ergodic Properties

The properties of ergodicity and dynamical mixing can also be defined for quantum systems, but the situation is significantly different than in classical systems (Thirring, 1983). Indeed, quantum systems with finitely many degrees of freedom in binding potentials have a discrete energy spectrum $\{E_n\}$, so that the time evolution of the mean values (14.18) is essentially quasiperiodic in time at the Bohr frequencies $\{\omega_{mn} = (E_m - E_n)/\hbar\}$.[1] Therefore, finite quantum systems do not have the property of dynamical mixing.

In order to manifest relaxation towards equilibrium and dynamical mixing, the operator generating time evolution should have a continuous spectrum. Because of the linearity of Schrödinger's equation, the quantum system should thus be infinite in order for this condition to be fulfilled. Such infinite quantum systems are spatially extended with a positive particle density and a zero or positive temperature in their reference state (Thirring, 1983). The property of dynamical mixing may thus hold in infinite quantum systems, where the memory of the initial state can be lost, as anticipated by Gibbs (1902). This property is crucial for the transport properties to be well defined in quantum systems, as much as in classical systems.

The quantum analogues of the Pollicott–Ruelle resonances are also considered in infinite quantum systems (Jakšić and Pillet, 1996a,b).

The infinite-system limit can be taken starting from a large but finite system. In such finite systems, the level density (2.8) may be so high that the energy spectrum is quasi continuous. Indeed, the level density exponentially increases with the number of degrees of freedom. Since the mean spacing between the energy levels goes as $\langle \Delta E \rangle = \langle E_{n+1} - E_n \rangle \simeq \mathscr{D}(E)^{-1}$ with $E = (E_n + E_{n+1})/2$, the discrete character of the energy spectrum cannot manifest itself before the very long timescale given by $t \gg \hbar/\langle \Delta E \rangle \simeq \hbar \mathscr{D}(E)$. In the infinite-system limit, time-dependent correlation functions such as (14.24) characterizing relaxation or transport properties may thus converge towards functions that decay with time.

14.3.7 Local Equilibrium Approach

The local equilibrium approach developed in Chapter 3 for classical systems can be extended to quantum systems (Mori, 1958). Indeed, the densities (3.3)–(3.6) can be defined in quantum mechanics with the corresponding Hermitian operators obtained by Hermitian symmetrization:

$$\hat{n}_k(\mathbf{r}) = \sum_{a \in \mathcal{S}_k} \delta(\mathbf{r} - \hat{\mathbf{r}}_a), \tag{14.26}$$

$$\hat{\mathbf{g}}(\mathbf{r}) = \frac{1}{2} \sum_{a=1}^{N} \left[\hat{\mathbf{p}}_a \, \delta(\mathbf{r} - \hat{\mathbf{r}}_a) + \delta(\mathbf{r} - \hat{\mathbf{r}}_a) \, \hat{\mathbf{p}}_a \right], \tag{14.27}$$

$$\hat{\epsilon}(\mathbf{r}) = \frac{1}{2} \sum_{a=1}^{N} \left[\hat{\varepsilon}_a \, \delta(\mathbf{r} - \hat{\mathbf{r}}_a) + \delta(\mathbf{r} - \hat{\mathbf{r}}_a) \, \hat{\varepsilon}_a \right], \tag{14.28}$$

where $\hat{\varepsilon}_a$ is the Hermitian operator corresponding to the particle energy (3.7), while the mass density operator is defined as $\hat{\rho}(\mathbf{r}) = \sum_{k=1}^{c} m_k \, \hat{n}_k(\mathbf{r})$.

In analogy with equation (3.21), the local equilibrium statistical operator can thus be introduced according to

[1] Indeed, the Bohr frequencies determine the eigenvalues of the quantum Liouvillian superoperator (14.16) according to $\widehat{\mathcal{L}} |n\rangle\langle m| = \imath \omega_{mn} |n\rangle\langle m|$.

$$\hat{\rho}_{\rm leq}(\chi) = \exp\left[-\Omega(\chi) - \int \chi(\mathbf{r}) \cdot \hat{\mathbf{c}}(\mathbf{r})\, d\mathbf{r}\right] \tag{14.29}$$

in terms of the densities, which are given by $\hat{\mathbf{c}} = (\hat{\epsilon}, \hat{n}_k, \hat{\mathbf{g}})$ for normal fluids, the conjugate fields $\chi(\mathbf{r})$, and $\Omega = \ln \operatorname{tr} \exp(-\int \chi \cdot \hat{\mathbf{c}}\, d\mathbf{r})$.

As in equation (3.40) for classical systems, the time evolution of the statistical operator starting from the local equilibrium described by the conjugate fields χ_0 can be expressed as

$$\hat{\rho}_t = \hat{\mathsf{U}}^t\, \hat{\rho}_{\rm leq}(\chi_0)\, \hat{\mathsf{U}}^{t\dagger} = \hat{\mathcal{T}}^t\, \hat{\rho}_{\rm leq}(\chi_t) \tag{14.30}$$

with the unitary time evolution operator introduced in equation (14.4) and some nonunitary operator $\hat{\mathcal{T}}^t$ transforming the local equilibrium statistical operator $\hat{\rho}_{\rm leq}(\chi_t)$ associated with the time-evolved conjugate fields χ_t into the exact statistical operator $\hat{\rho}_t$. As a consequence, the mean value (14.18) can be written as

$$\langle \hat{A} \rangle_t = \left\langle \hat{\mathcal{T}}^{t\dagger}\, \hat{A} \right\rangle_{{\rm leq},\, \chi_t} \tag{14.31}$$

in terms of the mean value with respect to the statistical operator $\hat{\rho}_{\rm leq}(\chi_t)$ and the adjoint of the operator $\hat{\mathcal{T}}^t$. Equation (14.31) is the quantum analogue of the classical relation (3.43).

The local equilibrium approach can be developed in order to determine the transport properties of quantum systems (Mori, 1958; Zubarev, 1966; Akhiezer and Peletminskii, 1981; Dufty et al., 2020).

14.3.8 Path Probabilities in Quantum Systems

Path probabilities can be introduced in quantum mechanics by considering projection operators onto some subspace of the quantum state space. The subspace can be associated with a few among all the degrees of freedom. For instance, in the case of Brownian motion, the projection operator associated with the observation of the position $\hat{\mathbf{r}}$ of the Brownian particle in the cell $\mathscr{C}_\omega \in \mathbb{R}^3$ can be defined as

$$\hat{\mathsf{P}}_\omega \equiv \int_{\mathscr{C}_\omega} d\mathbf{r}\, \delta^3(\mathbf{r} - \hat{\mathbf{r}}) \otimes \hat{\mathsf{I}}_1 \otimes \hat{\mathsf{I}}_2 \otimes \cdots \otimes \hat{\mathsf{I}}_N = \hat{\mathsf{P}}_\omega^\dagger, \tag{14.32}$$

where $\hat{\mathsf{I}}_a$ with $1 \leq a \leq N$ are the identity operators of the particles of the fluid surrounding the Brownian particle. The so-defined operator is a projection operator because $\hat{\mathsf{P}}_\omega^2 = \hat{\mathsf{P}}_\omega$.

We assume that the initial ensemble of the system is described by the statistical operator $\hat{\rho}_0 = \sum_j |\Psi_0^{(j)}\rangle P_0^{(j)} \langle \Psi_0^{(j)}|$. The position of the Brownian particle is observed at the successive times $t_1 < t_2 < \cdots < t_k$ in the cells $\{\omega_i\}_{i=1}^k$. Starting from the initial microstate $|\Psi_0^{(j)}\rangle$ for the whole system, the final microstate after the successive observations is given by

$$|\Psi_\omega^{(j)}\rangle = \hat{\mathsf{P}}_{\omega_k}\, \hat{\mathsf{U}}^{t_k - t_{k-1}} \cdots \hat{\mathsf{P}}_{\omega_2}\, \hat{\mathsf{U}}^{t_2 - t_1}\, \hat{\mathsf{P}}_{\omega_1}\, \hat{\mathsf{U}}^{t_1}\, |\Psi_0^{(j)}\rangle. \tag{14.33}$$

The probability of the path $\boldsymbol{\omega} = \{\omega(t_i)\}_{i=1}^{k}$ can thus be defined as

$$P(\omega_1,t_1;\ldots;\omega_k,t_k) \equiv \sum_j P_0^{(j)} \langle \Psi_{\boldsymbol{\omega}}^{(j)} | \Psi_{\boldsymbol{\omega}}^{(j)} \rangle \qquad (14.34)$$
$$= \mathrm{tr}\, \hat{\mathsf{P}}_{\omega_k}\, \hat{\mathsf{U}}^{t_k-t_{k-1}} \cdots \hat{\mathsf{P}}_{\omega_1}\, \hat{\mathsf{U}}^{t_1}\, \hat{\rho}_0\, \hat{\mathsf{U}}^{-t_1}\, \hat{\mathsf{P}}_{\omega_1} \cdots \hat{\mathsf{U}}^{-t_k+t_{k-1}}\, \hat{\mathsf{P}}_{\omega_k},$$

which is the quantum analogue of equation (4.5).

Typically, such path probabilities exponentially decrease with the number k of successive observations (Gaspard, 1992). The coherence between two such quantum histories $\boldsymbol{\omega}$ and $\boldsymbol{\omega}'$ can be characterized by the matrix composed of the elements $D(\boldsymbol{\omega}',\boldsymbol{\omega}) \equiv \sum_j P_0^{(j)} \langle \Psi_{\boldsymbol{\omega}'}^{(j)} | \Psi_{\boldsymbol{\omega}}^{(j)} \rangle$. The diagonal elements of this matrix are the path probabilities (14.34), while the off-diagonal elements are expected to be evanescent in classical regimes as the result of decoherence (Gaspard, 2000; Callens et al., 2004).

14.4 Functional Time-Reversal Symmetry Relation and Response Theory

Functional time-reversal symmetry relations can be established providing a unified framework to derive the quantum versions of nonequilibrium work relations and response theory, including the Green–Kubo formulas and the Onsager–Casimir reciprocal relations in the linear regime close to thermodynamic equilibrium (Bochkov and Kuzovlev, 1977; Stratonovich, 1994; Andrieux and Gaspard, 2008e).

14.4.1 Functional Time-Reversal Symmetry Relation

We consider a quantum system ruled by some Hamiltonian operator $\hat{H}(t;\mathcal{H})$, which depends on the time t and the external magnetizing field $\mathcal{H}$. This Hamiltonian operator is assumed to obey the symmetry of microreversibility (14.8) with respect to the time-reversal operator $\hat{\Theta}$ at every instant of time.

Since the system is driven by time-dependent forces, a comparison should be carried out between a forward and a reverse protocol, as in Chapter 5 (Andrieux and Gaspard, 2008e).

- **Forward protocol:** In this protocol, the Hamiltonian operator $\hat{H}(t;\mathcal{H})$ acts on the system during the time interval $0 \le t \le \tau$. The system is initially in thermal equilibrium at the inverse temperature $\beta = (k_{\mathrm{B}}T)^{-1}$ described by the canonical statistical operator

$$\hat{\rho}(0;\mathcal{H}) = \frac{1}{Z(0)}\, \mathrm{e}^{-\beta \hat{H}(0;\mathcal{H})} \qquad (14.35)$$

  with the partition function $Z(0) = \mathrm{tr}\, \mathrm{e}^{-\beta \hat{H}(0;\mathcal{H})}$ and the free energy $F(0) = -k_{\mathrm{B}}T \ln Z(0)$. Because of the symmetry (14.8), the partition function is even in the magnetizing field, $Z(0;\mathcal{H}) = Z(0;-\mathcal{H})$. The system evolves from the initial time $t = 0$ until the final time $t = \tau$ under the unitary operator that is the solution of Schrödinger's equation:

$$\imath\hbar\, \partial_t\, \hat{\mathsf{U}}_{\mathrm{F}}(t;\mathcal{H}) = \hat{H}(t;\mathcal{H})\, \hat{\mathsf{U}}_{\mathrm{F}}(t;\mathcal{H}) \qquad (14.36)$$

with the initial condition $\hat{\mathsf{U}}_{\mathrm{F}}(0;\boldsymbol{\mathcal{H}}) = \hat{\mathsf{I}}$. The mean value of any observable $\hat{X}$ at time t can be expressed as

$$\langle \hat{X}_{\mathrm{F}}(t)\rangle = \mathrm{tr}\,\hat{\rho}(0)\,\hat{X}_{\mathrm{F}}(t) \qquad \text{with} \qquad \hat{X}_{\mathrm{F}}(t) \equiv \hat{\mathsf{U}}_{\mathrm{F}}^{\dagger}(t;\boldsymbol{\mathcal{H}})\,\hat{X}\,\hat{\mathsf{U}}_{\mathrm{F}}(t;\boldsymbol{\mathcal{H}}) \tag{14.37}$$

in the Heisenberg picture.

- **Reverse protocol:** Here, the time dependence as well as the external magnetizing field are reversed in the Hamiltonian operator, which is thus given by $\hat{H}(\tau - t; -\boldsymbol{\mathcal{H}})$. Now, the system starts from the following canonical statistical operator,

$$\hat{\rho}(\tau; -\boldsymbol{\mathcal{H}}) = \frac{1}{Z(\tau)}\,\mathrm{e}^{-\beta\hat{H}(\tau;-\boldsymbol{\mathcal{H}})} \tag{14.38}$$

with the partition function $Z(\tau) = \mathrm{tr}\,\mathrm{e}^{-\beta\hat{H}(\tau;-\boldsymbol{\mathcal{H}})}$ and the free energy $F(\tau) = -k_{\mathrm{B}}T\ln Z(\tau)$. In the reverse protocol, the time evolution operator is determined by

$$\imath\hbar\,\partial_t\,\hat{\mathsf{U}}_{\mathrm{R}}(t; -\boldsymbol{\mathcal{H}}) = \hat{H}(\tau - t; -\boldsymbol{\mathcal{H}})\,\hat{\mathsf{U}}_{\mathrm{R}}(t; -\boldsymbol{\mathcal{H}}) \tag{14.39}$$

with the initial condition $\hat{\mathsf{U}}_{\mathrm{R}}(0; -\boldsymbol{\mathcal{H}}) = \hat{\mathsf{I}}$, so that the mean value of an observable can be expressed as

$$\left\langle \hat{X}_{\mathrm{R}}(t)\right\rangle = \mathrm{tr}\,\hat{\rho}(\tau)\,\hat{X}_{\mathrm{R}}(t) \qquad \text{with} \qquad \hat{X}_{\mathrm{R}}(t) \equiv \hat{\mathsf{U}}_{\mathrm{R}}^{\dagger}(t; -\boldsymbol{\mathcal{H}})\,\hat{X}\,\hat{\mathsf{U}}_{\mathrm{R}}(t; -\boldsymbol{\mathcal{H}}). \tag{14.40}$$

First, the symmetry (14.8) under time reversal has the following consequence:

Lemma: *The forward and reversed time evolution unitary operators are related to each other according to*

$$\hat{\Theta}\,\hat{\mathsf{U}}_{\mathrm{F}}(\tau - t;\boldsymbol{\mathcal{H}})\,\hat{\mathsf{U}}_{\mathrm{F}}^{\dagger}(\tau;\boldsymbol{\mathcal{H}})\,\hat{\Theta}^{-1} = \hat{\mathsf{U}}_{\mathrm{R}}(t; -\boldsymbol{\mathcal{H}}) \qquad \textit{with} \qquad 0 \le t \le \tau\,. \tag{14.41}$$

This lemma is proved by first substituting $\tau - t$ for t in equation (14.36) and multiplying the result by $\hat{\mathsf{U}}_{\mathrm{F}}^{\dagger}(t;\boldsymbol{\mathcal{H}})\hat{\Theta}^{-1}$ from the right and by $\hat{\Theta}$ from the left to get

$$\imath\hbar\,\partial_t\,\hat{\Theta}\hat{\mathsf{U}}_{\mathrm{F}}(\tau - t;\boldsymbol{\mathcal{H}})\hat{\mathsf{U}}_{\mathrm{F}}^{\dagger}(\tau;\boldsymbol{\mathcal{H}})\hat{\Theta}^{-1} = \hat{H}(\tau - t; -\boldsymbol{\mathcal{H}})\,\hat{\Theta}\hat{\mathsf{U}}_{\mathrm{F}}(\tau - t;\boldsymbol{\mathcal{H}})\hat{\mathsf{U}}_{\mathrm{F}}^{\dagger}(\tau;\boldsymbol{\mathcal{H}})\hat{\Theta}^{-1}, \tag{14.42}$$

after using the antiunitarity of the time-reversal operator $\hat{\Theta}$ and the symmetry (14.8) of the Hamiltonian operator. Thus, the operator $\hat{\Theta}\hat{\mathsf{U}}_{\mathrm{F}}(\tau - t;\boldsymbol{\mathcal{H}})\hat{\mathsf{U}}_{\mathrm{F}}^{\dagger}(\tau;\boldsymbol{\mathcal{H}})\hat{\Theta}^{-1}$ obeys the same evolution equation (14.39) as $\hat{\mathsf{U}}_{\mathrm{R}}(t; -\boldsymbol{\mathcal{H}})$. Since they also satisfy the same initial condition $\hat{\Theta}\hat{\mathsf{U}}_{\mathrm{F}}(\tau;\boldsymbol{\mathcal{H}})\hat{\mathsf{U}}_{\mathrm{F}}^{\dagger}(\tau;\boldsymbol{\mathcal{H}})\hat{\Theta}^{-1} = \hat{\mathsf{U}}_{\mathrm{R}}(0; -\boldsymbol{\mathcal{H}}) = \hat{\mathsf{I}}$, the identity (14.41) is thus proven.

With this lemma, we have the next result (Andrieux and Gaspard, 2008e):

Theorem: *If the time-independent observable $\hat{X}$ has the parity $\epsilon_X = \pm 1$ under time reversal, i.e., $\hat{\Theta}\hat{X}\hat{\Theta}^{-1} = \epsilon_X\hat{X}$, and $\xi(t)$ is an arbitrary function of time, the following functional time-reversal symmetry relation holds:*

$$\left\langle \mathrm{e}^{\int_0^\tau dt\,\xi(t)\,\hat{X}_{\mathrm{F}}(t)}\,\mathrm{e}^{-\beta\hat{H}_{\mathrm{F}}(\tau)}\,\mathrm{e}^{\beta\hat{H}(0)}\right\rangle_{\mathrm{F},\boldsymbol{\mathcal{H}}} = \mathrm{e}^{-\beta\Delta F}\left\langle \mathrm{e}^{\epsilon_X\int_0^\tau dt\,\xi(\tau - t)\,\hat{X}_{\mathrm{R}}(t)}\right\rangle_{\mathrm{R},-\boldsymbol{\mathcal{H}}} \tag{14.43}$$

with $\hat{H}_{\rm F}(\tau) = \hat{\mathsf{U}}_{\rm F}^{\dagger}(\tau;\mathcal{H})\,\hat{H}(\tau;\mathcal{H})\,\hat{\mathsf{U}}_{\rm F}(\tau;\mathcal{H})$ *and the difference* $\Delta F = F(\tau) - F(0)$ *between the free energies of the initial equilibrium ensembles* (14.38) *and* (14.35) *in the reverse and forward protocols.*

This functional relation is obtained from the lemma, implying that the observable $\hat{X}$ in the Heisenberg picture of the forward protocol can be written as

$$\hat{X}_{\rm F}(t) = \epsilon_X\, \hat{\mathsf{U}}_{\rm F}^{\dagger}(\tau;\mathcal{H})\,\hat{\Theta}^{-1}\hat{X}_{\rm R}(\tau - t)\,\hat{\Theta}\,\hat{\mathsf{U}}_{\rm F}(\tau;\mathcal{H}), \tag{14.44}$$

which is deduced multiplying the definition of $\hat{X}_{\rm F}(t)$ introduced in equation (14.37) by $\hat{\mathsf{U}}_{\rm F}^{\dagger}(\tau;\mathcal{H})\hat{\mathsf{U}}_{\rm F}(\tau;\mathcal{H}) = \hat{\mathsf{I}}$ from the left and the right, inserting $\hat{\Theta}\hat{\Theta}^{-1} = \hat{\mathsf{I}}$ between the evolution operators, and using $\hat{\Theta}\hat{X}\hat{\Theta}^{-1} = \epsilon_X \hat{X}$ together with equation (14.41). Integrating over the time interval $0 \leq t \leq \tau$ with an arbitrary function $\xi(t)$ and taking the exponential of both sides, equation (14.44) becomes

$$\mathrm{e}^{\int_0^\tau dt\,\xi(t)\,\hat{X}_{\rm F}(t)} = \hat{\mathsf{U}}_{\rm F}^{\dagger}(\tau;\mathcal{H})\hat{\Theta}^{-1}\,\mathrm{e}^{\epsilon_X \int_0^\tau dt\,\xi(\tau - t)\,\hat{X}_{\rm R}(t)}\,\hat{\Theta}\,\hat{\mathsf{U}}_{\rm F}(\tau;\mathcal{H}), \tag{14.45}$$

after the change of integration variables $t \to \tau - t$ in the right-hand side. Starting from the left-hand side of equation (14.43) and using the definition of $\hat{H}_{\rm F}(\tau)$, we find

$$\mathrm{tr}\,\hat{\rho}(0;\mathcal{H})\,\mathrm{e}^{\int_0^\tau dt\,\xi(t)\,\hat{X}_{\rm F}(t)}\,\mathrm{e}^{-\beta\hat{H}_{\rm F}(\tau)}\,\mathrm{e}^{\beta\hat{H}(0)} = \frac{Z(\tau)}{Z(0)}\,\mathrm{tr}\,\hat{\rho}(\tau;\,-\mathcal{H})\,\mathrm{e}^{\epsilon_X \int_0^\tau dt\,\xi(\tau-t)\,\hat{X}_{\rm R}(t)}, \tag{14.46}$$

hence the functional relation (14.43), which is thus proved.

We note that related results have previously been considered in the case where there is no change in free energy $\Delta F = 0$ (Bochkov and Kuzovlev, 1977; Stratonovich, 1994).

14.4.2 Quantum Nonequilibrium Work Relation

The functional relation (14.43) allows us to recover, in particular, the quantum version of the Jarzynski equality as a special case of equation (14.43) if $\xi(t) = 0$:

$$\left\langle \mathrm{e}^{-\beta\hat{H}_{\rm F}(\tau)}\,\mathrm{e}^{\beta\hat{H}(0)} \right\rangle_{\rm F,\mathcal{H}} = \mathrm{e}^{-\beta\Delta F}. \tag{14.47}$$

In the left-hand side, the quantity inside the average can be interpreted in terms of the work performed on the system in the scheme where energy is measured at the initial and final times (Kurchan, 2000; Tasaki, 2000; Esposito et al., 2009; Campisi et al., 2011). In the classical limit where the operators commute, the two exponential functions combine into the exponential of the classical work $W_{\rm cl} = [H_{\rm F}(\tau) - H(0)]_{\rm cl}$ and the classical Jarzynski equality (5.18) is recovered for the nonequilibrium work performed on the system during the forward protocol (Jarzynski, 1997).

14.4.3 Response Theory

Furthermore, the functional relation (14.43) can be used to derive the response coefficients and their symmetry under time reversal (Bochkov and Kuzovlev, 1977; Stratonovich, 1994;

Andrieux and Gaspard, 2008e). For this purpose, the functional derivatives of the relation (14.43) are taken with respect to the arbitrary function $\xi(t)$. In this way, we can obtain the Kubo formulas for the linear response coefficients in quantum systems (Kubo, 1957).

In analogy with Section 2.8 for classical systems, we here consider a perturbation of the form

$$\hat{H}(t) = \hat{H}_0 - \hat{A}\,\mathcal{F}(t), \tag{14.48}$$

where $\hat{H}_0$ is the Hamiltonian operator of the unperturbed system and $\hat{A}$ is the observable coupling the system to the time-dependent external forcing $\mathcal{F}(t)$ such that $\mathcal{F}(t) = 0$ for $t \leq 0$ and $t \geq \tau$. The functional derivative of equation (14.43) is taken with respect to $\xi(\tau)$ around $\xi = 0$, which yields

$$\left\langle \hat{X}_{\rm F}(\tau)\, {\rm e}^{-\beta \hat{H}_{\rm F}(\tau)}\, {\rm e}^{\beta \hat{H}_0} \right\rangle_{{\rm F},\mathcal{H}} = \epsilon_X \left\langle \hat{X}_{\rm R}(0)\right\rangle_{{\rm R},-\mathcal{H}} = \left\langle \hat{X}\right\rangle_{{\rm eq},\mathcal{H}}, \tag{14.49}$$

because $\Delta F = 0$ since $\mathcal{F}(\tau) = \mathcal{F}(0) = 0$ and, moreover, the time-reversal symmetry holds at equilibrium for the operator introduced in equation (14.40), $\epsilon_X \left\langle \hat{X}\right\rangle_{{\rm eq},-\mathcal{H}} = \left\langle \hat{X}\right\rangle_{{\rm eq},\mathcal{H}}$. Since $d\hat{H}_{\rm F}/dt = \left(\partial \hat{H}/\partial t\right)_{\rm F} = -\hat{A}_{\rm F}\dot{\mathcal{F}}$, the time-dependent Hamiltonian operator defined in the Heisenberg picture can be expressed as

$$\hat{H}_{\rm F}(\tau) = \hat{H}_{\rm F}(0) + \int_0^\tau \frac{d\hat{H}_{\rm F}}{dt}\, dt = \hat{H}_0 + \hat{U} \qquad \text{with} \qquad \hat{U} = \int_0^\tau \dot{\hat{A}}_{\rm F}(t)\, \mathcal{F}(t)\, dt, \tag{14.50}$$

after an integration by parts with the boundary values $\mathcal{F}(\tau) = \mathcal{F}(0) = 0$. Now, we have the identity

$$ {\rm e}^{-\beta\left(\hat{H}_0 + \hat{U}\right)}\, {\rm e}^{\beta \hat{H}_0} = 1 - \int_0^\beta d\eta\, {\rm e}^{-\eta\left(\hat{H}_0 + \hat{U}\right)}\, \hat{U}\, {\rm e}^{\eta \hat{H}_0}, \tag{14.51}$$

which can be proved by differentiating with respect to β and checking the integration constant at $\beta = 0$. Replacing the identity (14.51) into equation (14.49) and keeping the term of first order in the perturbation $\hat{U}$, we obtain

$$\left\langle \hat{X}_{\rm F}(\tau)\right\rangle_{{\rm F},\mathcal{H}} = \left\langle \hat{X}\right\rangle_{{\rm eq},\mathcal{H}} + \int_0^\beta d\eta \left\langle \hat{X}_{\rm F}(\tau)\, {\rm e}^{-\eta \hat{H}_0}\, \hat{U}\, {\rm e}^{\eta \hat{H}_0}\right\rangle_{{\rm eq},\mathcal{H}} + O\left(\hat{U}^2\right). \tag{14.52}$$

Setting $\hat{X} = \hat{B}$, using the expression for $\hat{U}$ given in equation (14.50), replacing t by t', and τ by t, the first response of the observable $\hat{B}$ is finally given by

$$\left\langle \hat{B}_{\rm F}(t)\right\rangle_{{\rm F},\mathcal{H}} = \left\langle \hat{B}\right\rangle_{{\rm eq},\mathcal{H}} + \int_0^t \phi_{BA}(t-t';\mathcal{H})\, \mathcal{F}(t')\, dt' + O\left(\mathcal{F}^2\right) \tag{14.53}$$

in terms of the following *response function*,

$$\phi_{BA}(t;\mathcal{H}) \equiv \int_0^\beta d\eta \left\langle \dot{\hat{A}}(-\imath\hbar\eta)\, \hat{B}(t)\right\rangle_{{\rm eq},\mathcal{H}}, \tag{14.54}$$

where $\langle\cdot\rangle_{{\rm eq},\mathcal{H}}$ is the mean value with respect to the canonical equilibrium statistical operator (14.22) for the unperturbed Hamiltonian operator $\hat{H}_0$. In the classical limit $\hbar \to 0$, the dependence on η disappears, so that the integral gives β and equation (2.147) is recovered for the linear response with the classical response function (2.151). We note that the response function (14.54) may have several equivalent forms according to the KMS relation (14.24).

Furthermore, the generalized susceptibility (2.155) can also be introduced for quantum systems subjected to a periodic forcing at the frequency ω. As a consequence of microreversibility, the response function and the generalized susceptibility satisfy the same symmetry relations (2.170) as for classical systems. Therefore, the linear response coefficients obey the Onsager–Casimir reciprocal relations (2.173) in quantum as well as classical systems.

An important example of linear response property is the electric conductivity discussed in Section 2.8.2 for classical systems. In this example, the Hamiltonian operator (14.48) is expressed in terms of the electric field $\mathcal{E}_j$ and the operator $\hat{A}$ defined by equation (2.160), while the linear response is measured for the electric current density (2.161). Ohm's law is thus obtained as $\langle \hat{j}_{ei} \rangle = \sigma_{ij} \mathcal{E}_j$ with the electric conductivity given by the following formula of Kubo (1957),

$$\sigma_{ij} = \frac{1}{V} \int_0^{\infty} dt \int_0^{\beta} d\eta \, \left\langle \hat{J}_i(t) \, \hat{J}_j(\imath\hbar\eta) \right\rangle_{\rm eq}, \tag{14.55}$$

where V is the volume of the system containing N charged particles and

$$\hat{J}_i \equiv \sum_{a=1}^{N} e_a \dot{\hat{r}}_{ai} \tag{14.56}$$

is the total electric current in the system.[2]

In the classical limit $\hbar \to 0$, the Green–Kubo formula (2.165) is recovered. In this regard, the Kubo formula (14.55) for the electric conductivity can be used as a template in order to extend the classical Green–Kubo formulas into their quantum-mechanical version. The procedure consists of replacing the total current at time zero in the classical formula by the integral over η, as from equation (2.165) to equation (14.55). In this way, the quantum-mechanical version of the Green–Kubo formulas of Chapter 3 can be inferred. This procedure is justified using the quantum extension of the local equilibrium distribution (3.21), as shown by Mori (1958).

The example of electric conductivity also illustrates the fact that the property of dynamical mixing is important in order to obtain finite transport properties with quantum Green–Kubo formulas. Indeed, these formulas give the transport coefficients by integrating the time-dependent correlation function from $t = 0$ to $t = \infty$, so that the integral over time should converge. For convergence to hold, the correlation function should thus decrease fast enough, as in classical systems.

We note that the frequency-dependent electric conductivity can also be defined for quantum systems. In the presence of an external magnetizing field, the Onsager–Casimir reciprocal relation (2.171) is satisfied.

[2] Equation (14.55) is deduced from (14.54) using the cyclic property of the trace and the change of integration variable $\eta \to \beta - \eta$.

14.4.4 Fluctuation–Dissipation Theorem

Quantum systems also satisfy the fluctuation–dissipation theorem in the linear regime close to equilibrium (Callen and Welton, 1951). The deduction of this theorem is similar to the one in Section 2.8.4 for classical systems, but the expressions differ due to the quantum effects.

The power dissipated by a periodic external forcing is always given by equation (2.177) in terms of the imaginary part of the generalized susceptibility (2.176) of the coupling operator $\hat{A}$ onto itself, but its quantum-mechanical expression has the form

$$\begin{aligned} \operatorname{Im} \chi_{AA}(\omega) &= \frac{1}{2\hbar} \int_{-\infty}^{+\infty} dt\, \mathrm{e}^{\imath\omega t} \left\langle \left[\hat{A}(t), \hat{A}(0)\right] \right\rangle_{\mathrm{eq},\beta} \\ &= \frac{1}{2\hbar}\left(1 - \mathrm{e}^{-\beta\hbar\omega}\right) \int_{-\infty}^{+\infty} dt\, \mathrm{e}^{\imath\omega t} \left\langle \hat{A}(t)\, \hat{A}(0) \right\rangle_{\mathrm{eq},\beta}. \end{aligned} \tag{14.57}$$

The quantum expression for the spectral density function (2.178) is given by

$$\begin{aligned} \mathcal{S}_{AA}(\omega) &= \frac{1}{2} \int_{-\infty}^{+\infty} dt\, \mathrm{e}^{\imath\omega t} \left\langle \hat{A}(0)\, \hat{A}(t) + \hat{A}(t)\, \hat{A}(0) \right\rangle_{\mathrm{eq},\beta} \\ &= \frac{1}{2}\left(1 + \mathrm{e}^{-\beta\hbar\omega}\right) \int_{-\infty}^{+\infty} dt\, \mathrm{e}^{\imath\omega t} \left\langle \hat{A}(t)\, \hat{A}(0) \right\rangle_{\mathrm{eq},\beta}. \end{aligned} \tag{14.58}$$

As a consequence, this function is related to the function (14.57) characterizing energy dissipation according to

$$\mathcal{S}_{AA}(\omega) = \hbar \coth \frac{\beta\hbar\omega}{2} \operatorname{Im} \chi_{AA}(\omega), \tag{14.59}$$

which is the quantum analogue of the classical relation (2.181). Accordingly, the dissipated power can be given in terms of the spectral density function characterizing the fluctuations of the observable $\hat{A}$ as

$$P = \frac{\omega}{2\hbar} \tanh \frac{\beta\hbar\omega}{2}\, \mathcal{S}_{AA}(\omega)\, \mathcal{F}_0^2 + O\left(\mathcal{F}_0^3\right), \tag{14.60}$$

which is the quantum expression of the fluctuation–dissipation theorem (Callen and Welton, 1951). In the classical limit $\hbar \to 0$, the classical expression (2.182) is recovered. Because of the quantum effects, dissipation is lower than classically expected for high frequencies such that $\hbar\omega \gg k_{\mathrm{B}}T$.

14.5 Quantum Master Equations

14.5.1 Overview

Since the pioneering work of Pauli (1928), master equations can be deduced from quantum mechanics for subsystems evolving in time inside a large system. Such master equations rule some probability distribution or reduced statistical operator for the subsystem of interest. In general, the diagonal elements of the statistical operator describe the population of quantum levels, i.e., their probability distribution, and the off-diagonal elements characterize the quantum coherences between the levels.

The basic idea for deducing a master equation is to use second-order time-dependent perturbation theory, giving the transition rates according to the Fermi golden rule (Cohen-Tannoudji et al., 1992). Since the master equation is meant to describe a few among all the degrees of freedom, a systematic method consists of projecting the statistical operator of the whole system onto the subsystem and, thus, reducing the unitary time evolution governed by the Landau–von Neumann equation (14.15) into a nonunitary evolution ruled by the master equation. For this purpose, the Zwanzig projection-operator method of Section 2.9.1 can be extended to quantum systems, leading in general to non-Markovian master equations for the reduced statistical operator. If there is a separation of timescales in the system, a Markovian approximation may be considered for long enough time. This program is usually carried out in the weak-coupling approximation, which is well justified for light-matter interaction (Cohen-Tannoudji et al., 1992). In this way, master equations can be derived for quantum optics, laser science, photosynthesis, photochemistry, and nuclear magnetic resonance (Gardiner and Zoller, 2000; de Vega and Alonso, 2017; Rice and Zhao, 2000; Ishizaki and Fleming, 2009; Redfield, 1965). Such master equations describe, in particular, the laser-induced quantum jumps of a single ion (Cook and Kimble, 1985; Bergquist et al., 1986; Sauter et al., 1986). Related methods can be used to obtain master equations in condensed matter for quantum Brownian motion (Caldeira and Leggett, 1983), as well as quantum Boltzmann kinetic equations (Uehling and Uhlenbeck, 1933; Balescu, 1975). For these latter equations, second-order perturbation theory only leads to the kinetic equation with the differential cross section evaluated at second order, which is called the Born approximation in quantum scattering theory (Joachain, 1975). In order to obtain the kinetic equation with the full expression of the differential cross section, the complete perturbation series for binary scattering should be taken into account, which can be justified in the low-density limit. In this way, the Boltzmann or Boltzmann–Lorentz kinetic equations can be deduced from quantum mechanics.

In Sections 14.5.2 and 14.5.3, the general lines of the deduction will be given for a subsystem interacting with a bath in order to understand the subtleties already existing in the weak-coupling problem (Gaspard and Nagaoka, 1999b; de Vega and Alonso, 2017).

14.5.2 Weak-Coupling Master Equation

We consider a system similar to the one in Section 4.8.1 for a subsystem interacting with a bath by the Hamiltonian operator

$$\hat{H} = \hat{H}_0 + \lambda\,\hat{U} = \hat{H}_{\rm s} + \hat{H}_{\rm b} + \lambda\,\hat{U} \qquad \text{with} \qquad \hat{U} = \hat{U}^\dagger = \sum_\alpha \hat{S}_\alpha\,\hat{B}_\alpha, \tag{14.61}$$

where $\hat{H}_{\rm s}$ and $\hat{H}_{\rm b}$ are the Hamiltonian operators respectively describing the subsystem and the bath, λ is the coupling parameter, and $\hat{U}$ is the interaction energy potential given in terms of the subsystem and bath operators $\hat{S}_\alpha$ and $\hat{B}_\alpha$, which may be assumed to be Hermitian (Gaspard and Nagaoka, 1999b). The time evolution is ruled by the Landau–von Neumann equation (14.15), so that the statistical operator for the whole system can be expressed as

$$\hat{\rho}(t) = {\rm e}^{-\imath\hat{H}t}\,\hat{\rho}(0)\,{\rm e}^{\imath\hat{H}t} \tag{14.62}$$

in units where $\hbar = 1$. The initial condition is assumed to be factorized into the product of statistical operators for the subsystem and the bath as follows,

$$\hat{\rho}(0) = \hat{\rho}_{\rm s}(0)\,\hat{\rho}_{\rm b,eq} \quad \text{with} \quad \hat{\rho}_{\rm b,eq} = \frac{1}{Z_{\rm b}}\,{\rm e}^{-\beta \hat{H}_{\rm b}} \quad \text{and} \quad Z_{\rm b} = {\rm tr}_{\rm b}\,{\rm e}^{-\beta \hat{H}_{\rm b}}, \tag{14.63}$$

where ${\rm tr}_{\rm b}$ denotes a partial trace over the bath quantum states. The purpose is to study the reduced statistical operator of the subsystem, which is defined by tracing out the bath degrees of freedom as

$$\hat{\rho}_{\rm s}(t) \equiv {\rm tr}_{\rm b}\,\hat{\rho}(t) = {\rm tr}_{\rm b}\,{\rm e}^{-\imath \hat{H}t}\,\hat{\rho}_{\rm s}(0)\,\hat{\rho}_{\rm b,eq}\,{\rm e}^{\imath \hat{H}t} \equiv \widehat{\mathcal{M}}^t\,\hat{\rho}_{\rm s}(0), \tag{14.64}$$

defining the evolution superoperator $\widehat{\mathcal{M}}^t$. This latter is linear and preserves the normalization of the reduced statistical operator ${\rm tr}_{\rm s}\,\hat{\rho}_{\rm s}(t) = {\rm tr}\,\hat{\rho}(t) = 1$, as well as its Hermitian character. In general, this superoperator is nonunitary because of the reduction in the description of the system.

In order to determine this evolution superoperator in the weak-coupling approximation, we introduce the interaction representation where

$$\hat{\rho}_{\rm I}(t) \equiv {\rm e}^{\imath \hat{H}_0 t}\,\hat{\rho}(t)\,{\rm e}^{-\imath \hat{H}_0 t} \quad \text{and} \quad \hat{U}_{\rm I}(t) \equiv {\rm e}^{\imath \hat{H}_0 t}\,\hat{U}\,{\rm e}^{-\imath \hat{H}_0 t}, \tag{14.65}$$

so that the Landau–von Neumann equation becomes

$$\partial_t\,\hat{\rho}_{\rm I}(t) = -\imath\left[\lambda\,\hat{U}_{\rm I}(t), \hat{\rho}_{\rm I}(t)\right] \equiv \widehat{\mathcal{L}}_{\rm I}(t)\,\hat{\rho}_{\rm I}(t), \tag{14.66}$$

which defines the Liouvillian in the interaction representation $\widehat{\mathcal{L}}_{\rm I}(t)$. Integrating equation (14.66) over the time interval $[0,t]$ and expanding in a power series of the coupling parameter λ, we obtain the following perturbative expansion for the density operator,

$$\hat{\rho}_{\rm I}(t) = \hat{\rho}(0) + \int_0^t dt_1\,\widehat{\mathcal{L}}_{\rm I}(t_1)\,\hat{\rho}(0) + \int_0^t dt_1 \int_0^{t_1} dt_2\,\widehat{\mathcal{L}}_{\rm I}(t_1)\,\widehat{\mathcal{L}}_{\rm I}(t_2)\,\hat{\rho}(0) + O\big(\lambda^3\big). \tag{14.67}$$

In the interaction representation, the reduced statistical operator (14.64) can be expressed as

$$\hat{\rho}_{\rm s}(t) = {\rm e}^{-\imath \hat{H}_{\rm s}t}\,\hat{\rho}_{\rm sI}(t)\,{\rm e}^{\imath \hat{H}_{\rm s}t} \equiv {\rm e}^{\widehat{\mathcal{L}}_{\rm s}t}\,\hat{\rho}_{\rm sI}(t) \quad \text{with} \quad \hat{\rho}_{\rm sI}(t) \equiv {\rm tr}_{\rm b}\,\hat{\rho}_{\rm I}(t). \tag{14.68}$$

The mean value of the interaction potential over the bath equilibrium ensemble is assumed to be equal to zero, i.e., ${\rm tr}_{\rm b}\big(\hat{\rho}_{\rm b,eq}\hat{U}\big) = 0$, otherwise the interaction potential and subsystem Hamiltonian should be redefined according to $\hat{U} \to \hat{U} - {\rm tr}_{\rm b}\big(\hat{\rho}_{\rm b,eq}\hat{U}\big)$ and $\hat{H}_{\rm s} \to \hat{H}_{\rm s} + \lambda\,{\rm tr}_{\rm b}\big(\hat{\rho}_{\rm b,eq}\hat{U}\big)$, while the bath Hamiltonian is unchanged, $\hat{H}_{\rm b} \to \hat{H}_{\rm b}$. The purpose of this redefinition is to include the first-order perturbation inside the zeroth-order Hamiltonian of the subsystem and, hence, to deal directly with the second-order perturbation.

Now, the time evolution of the reduced statistical operator can be obtained by replacing the perturbative series (14.67) into the definition (14.64), using the assumption (14.63) for the initial condition, and the changes of time variables $\tau = t_1 - t_2$ and $t' = t_1$. Since the first-order term disappears because of the preceding redefinition, we get the following second-order expression,

$$\hat{\rho}_s(t) = \widehat{\mathcal{M}}^t\,\hat{\rho}_s(0) = e^{\widehat{\mathcal{L}}_s t}\,\hat{\rho}_s(0) + \int_0^t dt' \int_0^{t'} d\tau\, e^{\widehat{\mathcal{L}}_s (t-t')}\,\widehat{\mathcal{C}}(\tau)\,e^{\widehat{\mathcal{L}}_s t'}\,\hat{\rho}_s(0) + O(\lambda^3) \quad (14.69)$$

with the correlation superoperator defined as

$$\widehat{\mathcal{C}}(\tau) \equiv \mathrm{tr_b}\left[\widehat{\mathcal{L}}_{\mathrm{I}}(0)\,\widehat{\mathcal{L}}_{\mathrm{I}}(-\tau)\,\hat{\rho}_{\mathrm{b,eq}}\right] = \mathrm{tr_b}\left(\widehat{\mathcal{L}}_{\mathrm{I}}\,e^{\widehat{\mathcal{L}}_0\tau}\,\widehat{\mathcal{L}}_{\mathrm{I}}\,e^{-\widehat{\mathcal{L}}_0\tau}\,\hat{\rho}_{\mathrm{b,eq}}\right), \quad (14.70)$$

where $\widehat{\mathcal{L}}_0 = \widehat{\mathcal{L}}_s + \widehat{\mathcal{L}}_b$ is the zeroth-order Liouvillian. The superoperator (14.70) still acts on the subsystem degrees of freedom because the trace is only carried out on the bath degrees of freedom. Taking the time derivative of equation (14.69), we find the following differential equation for the reduced statistical operator of the subsystem

$$\partial_t\,\hat{\rho}_s(t) = \left[\widehat{\mathcal{L}}_s + \int_0^t d\tau\,\widehat{\mathcal{C}}(\tau) + O(\lambda^3)\right]\hat{\rho}_s(t). \quad (14.71)$$

We note that this master equation is non-Markovian because its right-hand side explicitly depends on time.

If the bath is large enough, its dynamics may be assumed to have the property of dynamical mixing, so that its time correlation functions decay to zero if the time τ is longer than the bath correlation time t_b. On timescales $t \gg t_b$, the integral over time in equation (14.71) can thus converge, leading to the following master equation,

$$\partial_t\,\hat{\rho}_s^{\mathrm{R}}(t) = \left[\widehat{\mathcal{L}}_s + \int_0^\infty d\tau\,\widehat{\mathcal{C}}(\tau) + O(\lambda^3)\right]\hat{\rho}_s^{\mathrm{R}}(t) \equiv \widehat{\mathcal{L}}_s^{\mathrm{R}}\,\hat{\rho}_s^{\mathrm{R}}(t), \quad (14.72)$$

which is known as the master equation of Redfield (1965). Therefore, the evolution generator becomes time independent after a long enough time,

$$\widehat{\mathcal{L}}_s^{\mathrm{R}} = \lim_{t\to\infty} \partial_t \widehat{\mathcal{M}}^t\,\widehat{\mathcal{M}}^{-t}, \quad (14.73)$$

so that the Redfield master equation is Markovian. Its explicit form is given by

$$\partial_t\,\hat{\rho}_s^{\mathrm{R}} = -\imath\left[\hat{H}_s, \hat{\rho}_s^{\mathrm{R}}\right] + \lambda^2 \sum_\alpha \left(\hat{T}_\alpha\,\hat{\rho}_s^{\mathrm{R}}\,\hat{S}_\alpha + \hat{S}_\alpha^\dagger\,\hat{\rho}_s^{\mathrm{R}}\,\hat{T}_\alpha^\dagger - \hat{S}_\alpha\,\hat{T}_\alpha\,\hat{\rho}_s^{\mathrm{R}} - \hat{\rho}_s^{\mathrm{R}}\,\hat{T}_\alpha^\dagger\,\hat{S}_\alpha^\dagger\right) + O(\lambda^3) \quad (14.74)$$

in terms of the operators

$$\hat{T}_\alpha \equiv \sum_\beta \int_0^\infty d\tau\, C_{\alpha\beta}(\tau)\,e^{-\imath\hat{H}_s\tau}\,\hat{S}_\beta\,e^{\imath\hat{H}_s\tau} \quad (14.75)$$

and the bath correlation functions

$$C_{\alpha\beta}(\tau) \equiv \mathrm{tr_b}\,\hat{\rho}_{\mathrm{b,eq}}\,e^{\imath\hat{H}_b\tau}\,\hat{B}_\alpha\,e^{-\imath\hat{H}_b\tau}\,\hat{B}_\beta. \quad (14.76)$$

The Redfield master equation preserves the Hermiticity of the reduced statistical operator, $\hat{\rho}_s^{\mathrm{R}}(t) = \hat{\rho}_s^{\mathrm{R}}(t)^\dagger$, as well as its normalization, $\mathrm{tr_s}\,\hat{\rho}_s^{\mathrm{R}}(t) = 1$.

A subtlety is that the initial condition $\hat{\rho}_s(0)$ in equation (14.69) has disappeared by taking the time derivative to get equation (14.71). Therefore, the initial condition of the Markovian master equation (14.72) is not necessarily the same as for the non-Markovian master equation (14.71), as further discussed in Section 14.5.3.

14.5.3 Slippage of Initial Conditions

Since the reduced statistical operator defined by equation (14.64) is Hermitian and normalized to the unit value, its can be diagonalized as

$$\hat{\rho}_s(t) = \sum_i |\psi_i(t)\rangle\, P_i(t)\, \langle\psi_i(t)| \qquad \text{with} \qquad \det \hat{\rho}_s(t) = \prod_i P_i(t) \geq 0, \tag{14.77}$$

where its eigenvalues $P_i(t)$ represent the probabilities of finding the subsystem in the associated eigenstates $|\psi_i(t)\rangle$. These eigenvalues satisfy the conditions $\sum_i P_i(t) = 1$ and $0 \leq P_i(t) \leq 1$. Therefore, the determinant of the reduced statistical operator is always nonnegative. This property, which is called the positivity of the statistical operator, is preserved by the exact time evolution (14.64) and, thus, also by the non-Markovian master equation (14.71), which is supposed to be exact up to second order in the coupling parameter λ.

The issue is that the bath correlation functions (14.76) may decay after some correlation time t_b that is different from the relaxation timescales of the Redfield master equation (14.74). Indeed, these relaxation times are determined by the time-independent operators $\lambda\hat{S}_\alpha$ and $\lambda\hat{T}_\alpha$, which are defined according to equation (14.75) by integrating the bath correlation functions over time. This is, in particular, the case if the bath is fast and the subsystem slow, so that the bath correlation time t_b is shorter than the subsystem relaxation times. In this case, the exact evolution superoperator defined in equation (14.64) and, thus, the eigenvalues $P_i(t)$ may have a fast time evolution on the bath correlation time t_b followed by a slow relaxation towards their stationary equilibrium values, which is controlled by the time-independent Redfield generator (14.73) valid for long enough time. On the timescales $t \gg t_b$, we have in general that

$$\hat{\rho}_s(t) = \widehat{\mathcal{M}}^t\, \hat{\rho}_s(0) \simeq e^{\widehat{\mathcal{L}}_s^{R} t}\, \widehat{\mathcal{S}}\, \hat{\rho}_s(0) \qquad \text{with} \qquad \widehat{\mathcal{S}} \simeq e^{-\widehat{\mathcal{L}}_s^{R} t}\, \widehat{\mathcal{M}}^t, \tag{14.78}$$

because the initial conditions $\hat{\rho}_s^{R}(0) = \widehat{\mathcal{S}}\, \hat{\rho}_s(0)$ of the Redfield master equation may differ from those $\hat{\rho}_s(0)$ of the original problem. We should thus expect a slippage of initial conditions, which can be expressed with some superoperator $\widehat{\mathcal{S}}$. This slippage superoperator preserves the Hermitian character and the normalization of the reduced statistical operator. At the second order of perturbation theory, it is given by

$$\hat{\rho}_s^{R}(0) = \widehat{\mathcal{S}}\, \hat{\rho}_s(0) \equiv \hat{\rho}_s(0) - \int_0^\infty d\tau \int_0^\tau dt\, e^{-\widehat{\mathcal{L}}_{st}}\, \widehat{\mathcal{C}}(\tau)\, e^{\widehat{\mathcal{L}}_{st}}\, \hat{\rho}_s(0) + O(\lambda^3), \tag{14.79}$$

which can be deduced from equation (14.69) (Gaspard and Nagaoka, 1999b).

As illustrated in Figure 14.1, the slippage of initial conditions is required in order to preserve the positivity of the reduced statistical operator by the Redfield Markovian evolution.

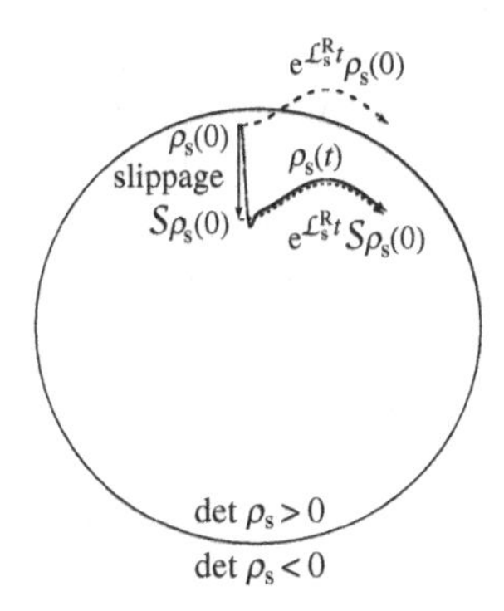

Figure 14.1 Schematic representation of the set of reduced statistical operators (14.77), showing the exact non-Markovian time evolution (14.64) (solid line) and the slippage of initial conditions (14.78) followed by the Redfield Markovian evolution (short-dashed line). Without the slippage, the Redfield Markovian evolution would exit the set of admissible statistical operators (14.77) (long-dashed line). Reprinted from Gaspard and Nagaoka (1999b) with the permission of AIP Publishing.

If the initial condition is too close to the border of admissible reduced statistical operators, the positivity cannot be maintained if the slippage of initial conditions is not considered before using the Redfield Markovian evolution (Suárez et al., 1992). The slippage of initial conditions is the temporal analogue of the spatial slippage for the fluid velocity with respect to a solid surface, which is known in hydrodynamics as discussed in Section 3.4.2 (Navier, 1827; Maxwell, 1879).

In the special case where the baths are delta-correlated, i.e., $C_{\alpha\beta}(t) = 2D_{\alpha\beta}\delta(t)$, the slippage superoperator (14.79) becomes the identity $\widehat{\mathcal{S}} = \widehat{\mathcal{I}}$, so that the slippage of initial conditions is no longer required to preserve positivity. In this case, the Redfield master equation satisfies the conditions for a time-independent Markovian master equation preserving positivity (Lindblad, 1976). These conditions are often met in light-matter interaction processes (Cohen-Tannoudji et al., 1992), although there exist photonic systems where non-Markovian effects can manifest themselves (de Vega and Alonso, 2017).

14.6 Stochastic Schrödinger Equations

Quantum mechanics and classical mechanics are ruled by differential equations, whose solutions are uniquely determined by their initial condition, as discussed for classical systems in Section 2.3.1. Initial conditions are usually prepared by a different system than the one ruled by the equation of motion, so that any physically admissible microstate can, in principle, be prepared as initial condition (Wigner, 1963). In such deterministic theories constructed on real or complex numbers, randomness may arise in the selection of initial conditions because the preparing device has its own initial conditions, which, in general, cannot be perfectly controlled. In the same way as different initial conditions determine different trajectories in classical Brownian motion, we should thus expect that a quantum subsystem interacting with a large bath may also manifest randomness for the same reason. Therefore, a parallel can be established with the complementary description of classical Brownian motion in terms of the Langevin stochastic differential equations, on the one hand, and the corresponding Kramers master equation on the other hand. For the same phenomenon, the former rules individual trajectories and the latter the statistical ensembles of such trajectories.

This scheme can be extended from classical to quantum systems and stochastic Schrödinger equations can be associated with Markovian or non-Markovian quantum master equations for quantum optics, light-matter interaction processes, and condensed matter (Gardiner and Zoller, 2000; de Vega et al., 2005). Such stochastic Schrödinger equations describe individual random trajectories for the quantum microstates of a subsystem interacting with a large bath and they can be deduced starting from the deterministic Schrödinger equation for the total system including the subsystem and the bath in the same way as the Langevin stochastic differential equation (4.173) can be deduced from the Newton equation (4.164) for Brownian motion.

In the following, a non-Markovian stochastic Schrödinger equation corresponding to the master equation (14.71) is deduced in the weak-coupling approximation. This deduction starts from the Schrödinger equation (14.3) with the Hamiltonian operator (14.61) for the interaction between the subsystem and the bath (Gaspard and Nagaoka, 1999a). The position representation is considered for the degrees of freedom of the subsystem $\mathbf{q}_{\rm s}$ and the bath $\mathbf{q}_{\rm b}$. We introduce the eigenfunctions $\chi_l(\mathbf{q}_{\rm b})$ of energy eigenvalues $\mathscr{E}_l$ for the bath Hamiltonian operator:

$$\hat{H}_{\rm b}\, \chi_l(\mathbf{q}_{\rm b}) = \mathscr{E}_l\, \chi_l(\mathbf{q}_{\rm b}) \qquad \text{with} \qquad \|\chi_l\| = 1, \tag{14.80}$$

forming a complete orthonormal basis in the Hilbert space of the bath. The wave function of the total system can thus be expanded in this basis according to

$$\Psi(\mathbf{q}_{\rm s}, \mathbf{q}_{\rm b}, t) = \sum_l \phi_l(\mathbf{q}_{\rm s}, t)\, \chi_l(\mathbf{q}_{\rm b}), \tag{14.81}$$

where the coefficients $\phi_l(\mathbf{q}_{\rm s}, t)$ depend on the subsystem positions and

$$P_l(t) = \int d\mathbf{q}_{\rm s}\, |\phi_l(\mathbf{q}_{\rm s}, t)|^2 = \|\phi_l(t)\|^2 \tag{14.82}$$

is the probability of finding the total system in the eigenfunction χ_l for the bath. The mean value of a subsystem observable $\hat{A}_{\rm s}$ is thus given as follows in terms of the reduced statistical operator,

$$\langle \Psi | \hat{A}_{\rm s} | \Psi \rangle = {\rm tr}_{\rm s}\, \hat{\rho}_{\rm s}\, \hat{A}_{\rm s} \qquad \text{with} \qquad \hat{\rho}_{\rm s} \equiv \sum_l |\phi_l\rangle\langle\phi_l|. \tag{14.83}$$

The stochastic Schrödinger equation is assumed to rule the time evolution of the functions $\phi_l(\mathbf{q}_{\rm s}, t)$, which can be considered as the unnormalized subsystem wave functions. Now, we may use the projection-operator method of Feshbach (1962) in order to project the total wave function onto one among the many subsystem wave functions with the following projection operator or its complement,

$$(\hat{\mathsf{P}}\Psi)(\mathbf{q}_s, \mathbf{q}_b) \equiv \phi_l(\mathbf{q}_s)\, \chi_l(\mathbf{q}_b), \tag{14.84}$$

$$(\hat{\mathsf{Q}}\Psi)(\mathbf{q}_s, \mathbf{q}_b) \equiv \sum_{m(\neq l)} \phi_m(\mathbf{q}_s)\, \chi_m(\mathbf{q}_b), \tag{14.85}$$

where $\hat{\mathsf{P}}^2 = \hat{\mathsf{P}} = \hat{\mathsf{P}}^\dagger$, $\hat{\mathsf{Q}}^2 = \hat{\mathsf{Q}} = \hat{\mathsf{Q}}^\dagger$, $\hat{\mathsf{Q}}\hat{\mathsf{P}} = \hat{\mathsf{P}}\hat{\mathsf{Q}} = 0$, and $\hat{\mathsf{P}} + \hat{\mathsf{Q}} = \hat{\mathsf{I}}$. Therefore, the Schrödinger equation (14.3) of the total system is transformed into the coupled equations,

$$\begin{cases} \imath\, \partial_t\, \hat{\mathsf{P}}\Psi = \hat{\mathsf{P}}\hat{H}\hat{\mathsf{P}}\Psi + \hat{\mathsf{P}}\hat{H}\hat{\mathsf{Q}}\Psi, \\ \imath\, \partial_t\, \hat{\mathsf{Q}}\Psi = \hat{\mathsf{Q}}\hat{H}\hat{\mathsf{P}}\Psi + \hat{\mathsf{Q}}\hat{H}\hat{\mathsf{Q}}\Psi. \end{cases} \tag{14.86}$$

The former leads to the equation for the selected subsystem wave function $\phi_l(\mathbf{q}_s, t)$ multiplied by the stationary eigenfunction $\chi_l(\mathbf{q}_b)$, and the latter is the equation for the rest of the total wave function. The latter can be solved and substituted into the former to obtain the following closed equation for the selected subsystem wave function,

$$\begin{aligned} \imath\, \partial_t\, \hat{\mathsf{P}}\Psi(t) &= \hat{\mathsf{P}}\hat{H}\hat{\mathsf{P}}\, \hat{\mathsf{P}}\Psi(t) + \hat{\mathsf{P}}\hat{H}\hat{\mathsf{Q}}\, \mathrm{e}^{-\imath\hat{\mathsf{Q}}\hat{H}\hat{\mathsf{Q}}t}\, \hat{\mathsf{Q}}\Psi(0) \\ &\quad - \imath \int_0^t d\tau\, \hat{\mathsf{P}}\hat{H}\hat{\mathsf{Q}}\, \mathrm{e}^{\imath\hat{\mathsf{Q}}\hat{H}\hat{\mathsf{Q}}(\tau - t)}\, \hat{\mathsf{Q}}\hat{H}\hat{\mathsf{P}}\, \hat{\mathsf{P}}\Psi(\tau), \end{aligned} \tag{14.87}$$

where the first term in the right-hand side gives the unitary time evolution due to the Hamiltonian operator $\hat{\mathsf{P}}\hat{H}\hat{\mathsf{P}}$, the second term gives an inhomogeneous time-dependent noisy forcing determined by the initial condition $\Psi(0)$ of the total wave function, and the third term is a non-Markovian contribution generating the effect of damping. The structure of this Schrödinger equation is similar to the generalized Langevin equation (2.197) obtained with Mori's projection-operator method, except that the equation here concerns the subsystem wave function instead of observables.

The bath is assumed to be large with respect to the subsystem. Thus, the energy eigenvalues of the bath form a dense spectrum with a high density of states $\mathscr{D}_b(\mathscr{E})$. Almost all the eigenfunctions can be considered having thermal mean values for quasilocal bath operators $\hat{A}_b$, i.e., such that

$$\langle \chi_l | \hat{A}_b | \chi_l \rangle \simeq \mathrm{tr}_b\, \hat{\rho}_{b,\mathrm{eq}}\, \hat{A}_b \qquad \text{with} \qquad \hat{\rho}_{b,\mathrm{eq}} = \frac{1}{Z_b}\, \mathrm{e}^{-\beta\hat{H}_b} \tag{14.88}$$

and $Z_b = \mathrm{tr}_b \exp\left(-\beta\hat{H}_b\right)$. This thermalization property for individual eigenfunctions holds for classically chaotic systems (Srednicki, 1994), where almost all the eigenfunctions behave as equilibrium microcanonical distributions. Since many-body systems in equilibrium microcanonical ensembles lead to Boltzmann's thermal distributions according to equation (C.7), the result (14.88) follows. Similarly, the bath correlation functions (14.76) are given by

$$\langle \chi_l | \hat{B}_\alpha(t)\, \hat{B}_\beta(t') | \chi_l \rangle \simeq \mathrm{tr}_b\, \hat{\rho}_{b,\mathrm{eq}} \hat{B}_\alpha(t)\, \hat{B}_\beta(t') \equiv C_{\alpha\beta}(t - t'). \tag{14.89}$$

For simplicity, the mean values of the coupling operators $\hat{B}_\alpha$ are assumed to be equal to zero for the eigenfunction χ_l: $\langle \chi_l | \hat{B}_\alpha | \chi_l \rangle = 0$. Otherwise, the subsystem Hamiltonian can be redefined as $\hat{H}_s \to \hat{H}_s + \lambda \sum_\alpha \langle \chi_l | \hat{B}_\alpha | \chi_l \rangle \hat{S}_\alpha$, with a corresponding redefinition of the interaction potential into $\hat{U} \to \hat{U} - \sum_\alpha \langle \chi_l | \hat{B}_\alpha | \chi_l \rangle \hat{S}_\alpha$.

Accordingly, the selected wave function $\phi_l(\mathbf{q}_s, t)$ can be assumed to be the typical representative of a statistical ensemble describing the system at the inverse temperature $\beta = (k_B T)^{-1}$ (Gaspard and Nagaoka, 1999a). As a consequence of this statistical typicality, all the subsystem wave functions in the ensemble are expected to behave similarly. The statistical ensemble is introduced by considering the following initial condition for the total wave function,

$$\Psi(\mathbf{q}_s, \mathbf{q}_b, 0) = \sum_l \phi_l(\mathbf{q}_s, 0)\, \chi_l(\mathbf{q}_b) = \psi(\mathbf{q}_s, 0) \sum_l c_l\, \chi_l(\mathbf{q}_b) \tag{14.90}$$

with coefficients such that

$$|c_l|^2 = \frac{\mathrm{e}^{-\beta \mathscr{E}_l}}{Z_b}, \quad \text{for instance given by} \quad c_l = \sqrt{\frac{\mathrm{e}^{-\beta \mathscr{E}_l}}{Z_b}}\, \mathrm{e}^{\imath \theta_l} \tag{14.91}$$

in terms of statistically independent uniformly distributed random phases $\theta_l \in [0, 2\pi[$. As a consequence, the noises in the second term of the right-hand side of equation (14.87) can be expressed as

$$\eta_\alpha(t) \equiv \sum_{m(\neq l)} \frac{c_m}{c_l} \langle \chi_l | \hat{B}_\alpha(t) | \chi_m \rangle \quad \text{with} \quad \frac{c_m}{c_l} = \mathrm{e}^{-\beta(\mathscr{E}_m - \mathscr{E}_l)/2}\, \mathrm{e}^{\imath(\theta_m - \theta_l)}. \tag{14.92}$$

Since the bath energy spectrum is very dense, there are many terms in the sums defining these noises, which are thus Gaussian and characterized as follows,

$$\overline{\eta_\alpha(t)} = 0, \qquad \overline{\eta_\alpha(t)\eta_\beta(t')} = 0, \qquad \text{and} \qquad \overline{\eta^*_\alpha(t)\eta_\beta(t')} = C_{\alpha\beta}(t - t'), \tag{14.93}$$

the bars denoting mean values with respect to the distribution of random phases $\{\theta_l\}$. The deduction from equation (14.87) therefore leads to the following non-Markovian stochastic Schrödinger equation (Gaspard and Nagaoka, 1999a),

$$\begin{aligned} \imath\, \partial_t\, \psi(t) &= \hat{H}_s \psi(t) + \lambda \sum_\alpha \eta_\alpha(t)\, \hat{S}_\alpha\, \psi(t) \\ &\quad - \imath \lambda^2 \int_0^t d\tau \sum_{\alpha\beta} C_{\alpha\beta}(\tau)\, \hat{S}_\alpha\, \mathrm{e}^{-\imath \hat{H}_s \tau}\, \hat{S}_\beta\, \psi(t - \tau) + O(\lambda^3). \end{aligned} \tag{14.94}$$

The non-Markovian character of this equation finds its origin in equation (14.87) given by the Feshbach projection-operator method.

If we consider a statistical ensemble of solutions $\{|\psi_k(t)\rangle\}$ for equation (14.94), the reduced statistical operator of the subsystem can be obtained as

$$\hat{\rho}_s(t) = \sum_k P_k(0)\, \overline{\frac{|\psi_k(t)\rangle\langle\psi_k(t)|}{\langle\psi_k(t)|\psi_k(t)\rangle}} \tag{14.95}$$

with the initial probabilities $P_k(0)$. Using the time evolution generated by the stochastic Schrödinger equation (14.94), the non-Markovian master equation (14.71) is recovered at the second order of perturbation theory, showing the consistency of the stochastic differential equation with this master equation (Gaspard and Nagaoka, 1999a).

We note that the individual subsystem wave functions can be interpreted as follows. The value of the subsystem observable $\hat{A}_s$ provided that the subsystem and the bath are, respectively, in the microstates ϕ_l and χ_l is given by

$$\langle \hat{A}_s \rangle_l \equiv \frac{\left\langle \hat{A}_s \otimes \hat{\mathsf{P}}_l \right\rangle}{\left\langle \hat{\mathsf{I}}_s \otimes \hat{\mathsf{P}}_l \right\rangle} = \frac{\int \phi_l^*(\mathbf{q}_s)\, \hat{A}_s\, \phi_l(\mathbf{q}_s)\, d\mathbf{q}_s}{\int |\phi_l(\mathbf{q}_s)|^2\, d\mathbf{q}_s} \qquad \text{with} \qquad \hat{\mathsf{P}}_l \equiv |\chi_l\rangle \langle \chi_l|. \tag{14.96}$$

For the whole system in the microstate (14.81), the value of $\hat{A}_s$ is thus equal to

$$\left\langle \hat{A}_s \right\rangle = \sum_l P_l \left\langle \hat{A}_s \right\rangle_l \qquad \text{with} \qquad P_l \equiv \left\langle \hat{\mathsf{I}}_s \otimes \hat{\mathsf{P}}_l \right\rangle. \tag{14.97}$$

The corresponding quantum uncertainties are defined as $\sigma_l \equiv \left[\langle \hat{A}_s^2 \rangle_l - \langle \hat{A}_s \rangle_l^2 \right]^{1/2}$ and $\sigma \equiv \left[\langle \hat{A}_s^2 \rangle - \langle \hat{A}_s \rangle^2 \right]^{1/2}$. Because of Schwarz's inequality, these uncertainties satisfy the following inequality, $\sum_l P_l\, \sigma_l^2 \leq \sigma^2$. Therefore, the subsystem wave functions $\phi_l(\mathbf{q}_s)$ can be subjected to dynamical localization caused by the interaction with the bath. In the case where the probability density $p(\mathbf{q}_s) = \sum_l |\phi_l(\mathbf{q}_s)|^2$ is broadly distributed, we should expect that the densities $P_l^{-1} |\phi_l(\mathbf{q}_s)|^2$ of the individual microstates are randomly localized as the result of this interaction.

Similar non-Markovian stochastic Schrödinger equations have been obtained using related methods (Strunz et al., 1999; de Vega et al., 2005; de Vega and Alonso, 2017).

14.7 The Case of the Spin-Boson Model

In order to illustrate the previous results, we consider the spin-boson model defined by the following total Hamiltonian,

$$\hat{H} = -\frac{\Delta}{2}\, \hat{\sigma}_z + \hat{H}_b + \lambda\, \hat{\sigma}_x\, \hat{B} \tag{14.98}$$

in terms of the Pauli matrices $(\hat{\sigma}_x, \hat{\sigma}_y, \hat{\sigma}_z)$ and the energy splitting Δ between the two levels of the subsystem (Suárez et al., 1992). The bath is a collection of harmonic oscillators, which are coupled to the two-level subsystem. The bath Hamiltonian and the coupling operator are given, respectively, by

$$\hat{H}_b = \frac{1}{2} \sum_\alpha \left(\hat{p}_\alpha^2 + \omega_\alpha^2\, \hat{q}_\alpha^2 \right) \qquad \text{and} \qquad \hat{B} = \sum_\alpha c_\alpha\, \hat{q}_\alpha. \tag{14.99}$$

The bath is characterized by its spectral strength, which is taken as

$$J(\omega) = \sum_\alpha \frac{c_\alpha^2}{2\omega_\alpha}\, \delta(\omega - \omega_\alpha) = \frac{\omega^3}{\omega_c^2}\, \exp(-\omega/\omega_c), \tag{14.100}$$

where ω_c is a cutoff frequency (Suárez et al., 1992).

The two-level subsystem can be considered as a spin one half. Therefore, the reduced statistical operator of the subsystem can be decomposed in the basis formed by the Pauli matrices according to $\hat{\rho}_s = (1 + x\,\hat{\sigma}_x + y\,\hat{\sigma}_y + z\,\hat{\sigma}_z)/2$ with three real coefficients (x, y, z), which are called the Bloch variables. The set of physically admissible reduced statistical operators thus corresponds to the ball of unit radius such that $x^2 + y^2 + z^2 \leq 1$ in the space (x, y, z). We notice that the normalization $\mathrm{tr}_s\,\hat{\rho}_s = 1$ is always satisfied.

In the Bloch representation, the non-Markovian master equation (14.71) obtained at the second order of perturbation theory is given for the spin-boson model by the following set of time-dependent, linear, ordinary differential equations,

$$\begin{cases} \dot{x} = \Delta\, y, \\ \dot{y} = -\,[\Delta + h(t)]\, x - g(t)\, y, \\ \dot{z} = -f(t) - g(t)\, z, \end{cases} \tag{14.101}$$

where the dot denotes the time derivative and

$$\begin{cases} f(t) = 4\,\lambda^2 \int_0^t d\tau\ \sin\Delta\tau\ \mathrm{Im}\, C(\tau), \\ g(t) = 4\,\lambda^2 \int_0^t d\tau\ \cos\Delta\tau\ \mathrm{Re}\, C(\tau), \\ h(t) = 4\,\lambda^2 \int_0^t d\tau\ \sin\Delta\tau\ \mathrm{Re}\, C(\tau), \end{cases} \tag{14.102}$$

with the bath correlation function (14.76) in the following form,

$$C(\tau) \equiv \mathrm{tr}_b\, \hat{\rho}_b^{\mathrm{eq}}\, \mathrm{e}^{\imath \hat{H}_b \tau}\, \hat{B}\, \mathrm{e}^{-\imath \hat{H}_b \tau}\, \hat{B} = \int_0^\infty d\omega\, J(\omega) \left(\coth \frac{\beta\omega}{2} \cos\omega\tau - \imath \sin\omega\tau \right). \tag{14.103}$$

With the spectral strength (14.100), the three functions (14.102) converge in the long-time limit towards the finite values $f(\infty)$, $g(\infty)$, and $h(\infty)$. Actually, this convergence is effective after the bath correlation time, which is here evaluated as the inverse of the cutoff frequency, $t_b \sim \omega_c^{-1}$. Thereafter, the coefficients of equations (14.101) become time independent, which corresponds to the Markovian regime of the Redfield master equation (14.74). In this regime, the reduced statistical operator converges towards equilibrium according to $\hat{\rho}_s(t) - \hat{\rho}_{s,\mathrm{eq}} \sim \exp(z_\alpha t)$ with the Liouvillian eigenvalues $z_0 = 0$, $z_{1,2} = \pm\imath[\Delta + h(\infty)/2] - g(\infty)/2 + O\left(\lambda^4\right)$, and $z_3 = -g(\infty) + O\left(\lambda^4\right)$, in agreement with rigorous evaluations (Jakšić and Pillet, 1996a,b). Since the interaction of the spin with the bath is weak, the timescale of relaxation towards equilibrium $t_{\mathrm{relax}} \sim g(\infty)^{-1} = O\left(\lambda^{-2}\right)$ is much longer than the bath correlation time $t_b \sim \omega_c^{-1}$. Consequently, the slippage of initial conditions is much shorter than the relaxation towards equilibrium, which is schematically represented in Figure 14.1.

For the spin-boson model, the non-Markovian stochastic Schrödinger equation (14.94) can be numerically integrated generating random trajectories as shown in Figure 14.2(a). Taking the statistical average over a large enough ensemble of random trajectories gives equivalent results as the solution of equations (14.101), as seen in Figure 14.2(b).

The noise correlation function is depicted in Figure 14.3(a), showing that it has the duration of the bath correlation time $t_b \sim \omega_c^{-1}$. It is over this timescale that the non-Markovian behavior manifests itself, as seen in Figure 14.3(b) for the variable z.

In conclusion, the stochastic Schrödinger equation provides a description that is consistent with the non-Markovian master equation preserving positivity. The asymptotic Markovian equation can only be used after the slippage of initial conditions at early time.

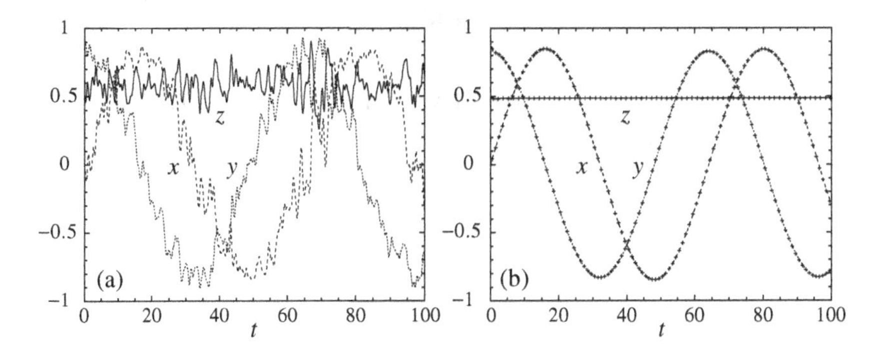

Figure 14.2 (a) Individual trajectory of the non-Markovian stochastic Schrödinger equation for the spin-boson model with the splitting energy $\Delta = 0.1$, the inverse temperature $\beta = 10$, the cutoff frequency $\omega_c = 1$, the coupling parameter $\lambda = 0.1$, and from the initial conditions $\left[x(0) = 0, y(0) = \sqrt{3}/2, z(0) = 1/2\right]$. The trajectory is represented in terms of the three Bloch variables (x, y, z) versus the time t. (b) The statistical average (pluses) is taken over a statistical ensemble of 10^4 individual trajectories and the lines show the corresponding solution of the non-Markovian equations (14.101). Reprinted from Gaspard and Nagaoka (1999a) with the permission of AIP Publishing.

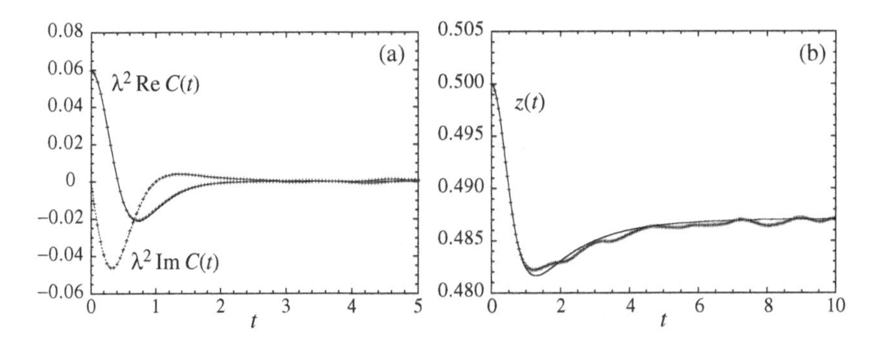

Figure 14.3 Dynamical properties of the noise of the thermal bath of harmonic oscillators in the spin-boson model with the inverse temperature $\beta = 10$, the cutoff frequency $\omega_c = 1$, and the coupling parameter $\lambda = 0.1$: (a) Correlation function versus the time t. The solid (respectively, dashed) line depicts the real (respectively, imaginary) part of $\lambda^2 C(t)$ calculated by equation (14.103). The pluses are the results of a statistical computation of the correlation function using a long time sequence of the noise. (b) The statistical average (pluses) is here taken over a statistical ensemble of 2×10^5 individual trajectories and only the variable z is depicted to show the slippage of initial conditions over the bath relaxation time $t_b \sim \omega_c^{-1} = 1$. Reprinted from Gaspard and Nagaoka (1999a) with the permission of AIP Publishing.

We note that nonlinear stochastic Schrödinger equations are also considered, especially in the high temperature regime (de Vega et al., 2005). Similar methods apply to quantum Brownian motion and other quantum processes (Strunz et al., 1999; de Vega and Alonso, 2017).

15

Transport in Open Quantum Systems

15.1 Energy and Particle Fluxes in Open Quantum Systems

In open quantum systems, microreversibility implies that energy and particle fluxes obey symmetry relations similar to those obtained in Chapter 5 for classical systems (Tobiska and Nazarov, 2005; Saito and Utsumi, 2008; Andrieux et al., 2009; Esposito et al., 2009; Campisi et al., 2011). These results are fundamental for quantum transport in semiconducting mesoscopic devices (Datta, 1995; Ferry et al., 2009; Nazarov and Blanter, 2009), where the full counting statistics of electron transfers can be experimentally achieved, giving evidence for irreversibility on the level of single-electron tunneling (Küng et al., 2012, 2013). In Aharonov–Bohm rings, symmetry relations between response coefficients beyond the linear regime can also be tested (Nakamura et al., 2010, 2011). Moreover, similar considerations apply to the quantum transport of ultracold atoms guided by optical traps (Brantut et al., 2013; Krinner et al., 2015).

If the mean free path is larger than the size of the circuit, the quantum transport of bosons and fermions can be formulated within the scattering approach (Landauer, 1957; Blanter and Büttiker, 2000), which can describe the effects of quantum coherence and interference. In this approach, the cumulant generating function for the full counting statistics of energy and particle fluxes can be obtained in terms of the scattering matrix for the independent particles moving in the mesoscopic circuit (Levitov and Lesovik, 1993; Klich, 2003; Lesovik and Sadovskyy, 2011). In addition, the concept of temporal disorder introduced in Chapter 6 can also be extended to these quantum systems and the time asymmetry in temporal disorder may again be related to the entropy production rate (Gaspard, 2014, 2015b,c). Furthermore, the stochastic approach can be developed for transport by tunneling in quantum dots and quantum point contacts, where the multivariate current fluctuation relation of Section 5.7 is satisfied.

15.2 Fluctuation Relation for Energy and Particle Fluxes

The multivariate fluctuation relation for energy and particle fluxes can be established for quantum systems with methods similar to those used in Chapter 5 for classical systems (Andrieux et al., 2009).

15.2.1 Time-Dependent Driving of Open Quantum Systems

We consider an open quantum system in contact with several reservoirs $1 \leq j \leq r$ of energy and particles, as illustrated in Figure 5.6. The reservoirs are supposed to be large but finite, possibly containing several species $1 \leq k \leq c$ of particles. The total system may be subjected to a time-dependent driving and an external magnetizing field $\mathcal{H}$. Its Hamiltonian operator has the following symmetry under time reversal

$$\hat{\Theta}\, \hat{H}(t;\mathcal{H})\, \hat{\Theta}^{-1} = \hat{H}(t;\, -\mathcal{H}), \tag{15.1}$$

holding for every time t.

In order to obtain the full counting statistics for energy and particle fluxes between the reservoirs, we may use an approach reminiscent of scattering theory, in which the incoming and outgoing quantum microstates are prepared before some initial time and measured after some final time in the absence of interactions between the parts composing the system. The initial and final times are possibly separated by a long time interval, during which the quantum coherence of the system is fully preserved. Accordingly, the total Hamiltonian operator is assumed to be of the following form,

$$\hat{H}(t;\mathcal{H}) = \begin{cases} \sum_{j=1}^{r} \hat{H}_j & \text{if} \quad t \leq 0, \\ \sum_{j=1}^{r} \hat{H}_j + \hat{U}(t) & \text{if} \quad 0 < t < \tau, \\ \sum_{j=1}^{r} \hat{\hat{H}}_j & \text{if} \quad \tau \leq t, \end{cases} \tag{15.2}$$

where $\hat{H}_j$ (respectively, $\hat{\hat{H}}_j$) are the Hamiltonian operators of the decoupled reservoirs before (respectively, after) the interaction with the operator $\hat{U}(t)$. As in Figure 5.6, the open system itself is included in the reference reservoir $j = r$, so that the Hamiltonian operators $\hat{H}_r$ and $\hat{\hat{H}}_r$ rule the open system interacting with the reservoir $j = r$ itself, before and after the interaction $\hat{U}(t)$. All the uncoupled Hamiltonian operators commute with each other, as well as with the particle number operators $\hat{N}_j$ in the reservoirs $1 \leq j \leq r$. Accordingly, the simultaneous measurement of the energies and particles numbers in the reservoirs is possible during the periods of time $t \leq 0$ and $\tau \leq t$.[1]

The multivariate fluctuation relation for energy and particle fluxes is established on the basis of microreversibility by comparing some forward and reverse protocols, as in Section 14.4.1.

- **Forward protocol:** During this protocol, the time evolution of the system is ruled by the Hamiltonian operator (15.2). At time $t = 0$, the initial ensemble is given by the following product of grand canonical statistical operators at the inverse temperatures β_j and chemical potentials μ_{kj} in the reservoirs $1 \leq j \leq r$:

$$\hat{\rho}(0;\mathcal{H}) = \left[\prod_{j=1}^{r} \mathrm{e}^{-\beta_j\left(\hat{H}_j - \sum_{k=1}^{c} \mu_{kj}\, \hat{N}_{kj} - \Phi_j\right)} \right]_{\mathcal{H}}, \tag{15.3}$$

[1] Since these periods of time are semi-infinite, such energies and particle numbers can be measured with arbitrarily high accuracy.

where $\Phi_j = -\beta_j^{-1} \ln \Xi_j = J_j$ is the grand thermodynamic potential of the jth reservoir as in equation (14.23).[2] The time evolution is ruled by the same unitary operator $\hat{\mathrm{U}}_{\mathrm{F}}(t;\boldsymbol{\mathcal{H}})$ as in equation (14.36). In order to carry out the full counting statistics of energy and particle exchanges between the reservoirs, an initial measurement is simultaneously performed for the energies and particle numbers contained in all the reservoirs during the period of time $t \leq 0$, giving the common eigenvector $|\Psi_i\rangle$ associated with the eigenvalues E_{ji} and N_{kji}:

$$\text{for } t \leq 0: \qquad \hat{H}_j\,|\Psi_i\rangle = E_{ji}\,|\Psi_i\rangle, \qquad \hat{N}_{kj}\,|\Psi_i\rangle = N_{kji}\,|\Psi_i\rangle, \tag{15.4}$$

for $1 \leq j \leq r$ and $1 \leq k \leq c$. A final similar measurement is performed after interaction during the period of time $\tau \leq t$, giving

$$\text{for } \tau \leq t: \qquad \hat{\tilde{H}}_j\,|\tilde{\Psi}_f\rangle = \tilde{E}_{jf}\,|\tilde{\Psi}_f\rangle, \qquad \hat{N}_{kj}\,|\tilde{\Psi}_f\rangle = \tilde{N}_{kjf}\,|\tilde{\Psi}_f\rangle. \tag{15.5}$$

As a consequence, the amounts of energy and particles that are flowing out of the jth reservoir during the time interval $[0,\tau]$ can be defined as

$$\Delta E_j \equiv E_{ji} - \tilde{E}_{jf} \qquad \text{and} \qquad \Delta N_{kj} \equiv N_{kji} - \tilde{N}_{kjf}, \tag{15.6}$$

which are the quantum analogues of equations (5.53)–(5.54).

Introducing the same set of observables as in equation (5.55)[3]

$$\hat{\mathbf{C}} \equiv \left[\hat{H}_j, \left(\hat{N}_{kj}\right)_{k=1}^{c}\right]_{j=1}^{r}, \tag{15.7}$$

the initial and final measurements (15.4)–(15.5) can be equivalently written as

$$\hat{\mathbf{C}}\,|\Psi_i\rangle = \mathbf{C}_i\,|\Psi_i\rangle \qquad \text{and} \qquad \hat{\tilde{\mathbf{C}}}\,|\tilde{\Psi}_f\rangle = \tilde{\mathbf{C}}_f\,|\tilde{\Psi}_f\rangle, \tag{15.8}$$

where $\mathbf{C}_i$ and $\tilde{\mathbf{C}}_f$ denote the initial and final sets of eigenvalues. The energy and particle cumulated fluxes out of the reservoirs can thus be expressed as

$$\Delta\mathbf{C} \equiv \mathbf{C}_i - \tilde{\mathbf{C}}_f = \left[\Delta E_j, (\Delta N_{kj})_{k=1}^{c}\right]_{j=1}^{r}, \tag{15.9}$$

in accordance with equation (5.56).

The protocol is repeated many times to calculate the statistics of energy and particle fluxes and obtain their probability density according to

$$p_{\mathrm{F}}(\Delta\mathbf{C};\boldsymbol{\mathcal{H}}) \equiv \sum_{fi} \delta(\Delta\mathbf{C} - \mathbf{C}_i + \tilde{\mathbf{C}}_f)$$
$$\times \underbrace{\left|\langle\tilde{\Psi}_f(\boldsymbol{\mathcal{H}})|\hat{\mathrm{U}}_{\mathrm{F}}(\tau;\boldsymbol{\mathcal{H}})|\Psi_i(\boldsymbol{\mathcal{H}})\rangle\right|^2}_{P_\tau(f|i)} \underbrace{\langle\Psi_i(\boldsymbol{\mathcal{H}})|\hat{\rho}(0;\boldsymbol{\mathcal{H}})|\Psi_i(\boldsymbol{\mathcal{H}})\rangle}_{P(i)}, \tag{15.10}$$

where $P(i)$ is the probability of observing the microstate $|\Psi_i\rangle$ upon the initial measurement when the initial statistical operator is given by equation (15.3), $P_\tau(f|i)$ is the

[2] Because of the symmetry (15.1), these grand potentials are even functions of the magnetizing field, $\Phi_j(\boldsymbol{\mathcal{H}}) = \Phi_j(-\boldsymbol{\mathcal{H}})$.

[3] Here, the bracket $[\cdot]$ simply denotes a list of elements including lists.

probability of the transition from the initial microstate $|\Psi_i\rangle$ to the final microstate $|\tilde{\Psi}_f\rangle$ during the time interval τ, and

$$\delta(\Delta\mathbf{C} - \mathbf{C}_i + \tilde{\mathbf{C}}_f) \equiv \prod_{j=1}^{r} \delta\left(\Delta E_j - E_{ji} + \tilde{E}_{jf}\right) \prod_{j=1}^{r}\prod_{k=1}^{c} \delta\left(\Delta N_{kj} - N_{kji} + \tilde{N}_{kjf}\right). \tag{15.11}$$

- **Reverse protocol:** In this protocol, the time evolution runs backward in time according to the Hamiltonian operator $\hat{H}(\tau - t;\, -\mathcal{H})$ with the reversed time dependence and external magnetizing field. Here, the system starts from the following grand canonical statistical operator,

$$\hat{\rho}(\tau;\, -\mathcal{H}) = \left[\prod\nolimits_{j=1}^{r} \mathrm{e}^{-\beta_j\left(\hat{\tilde{H}}_j - \sum_{k=1}^{c} \mu_{kj}\, \hat{N}_{kj} - \tilde{\Phi}_j\right)}\right]_{-\mathcal{H}}, \tag{15.12}$$

at the same inverse temperatures and chemical potentials as in the forward protocol, but with the Hamiltonian operators $\hat{\tilde{H}}_j$ of the reservoirs for $\tau \leq t$ in equation (15.2) and the corresponding grand thermodynamic potentials $\tilde{\Phi}_j = -\beta_j^{-1} \ln \tilde{\Xi}_j = \tilde{J}_j$. The time evolution of the reverse protocol is ruled by the unitary operator $\hat{\mathrm{U}}_{\mathrm{R}}(t;\, -\mathcal{H})$ introduced in equation (14.39).

During the reverse protocol, the energy and particle fluxes are also obtained by two measurements as in the forward protocol, but with the results (15.5) for the initial measurement and (15.4) for the final measurement. Accordingly, the probability density of observing the energy and particle amounts $\Delta\mathbf{C}$ flowing out of the reservoirs in the reverse protocol is given by

$$p_{\mathrm{R}}(\Delta\mathbf{C};\, -\mathcal{H}) \equiv \sum_{fi} \delta(\Delta\mathbf{C} - \tilde{\mathbf{C}}_f + \mathbf{C}_i) \left|\left\langle\Psi_i(-\mathcal{H})\middle|\hat{\mathrm{U}}_{\mathrm{R}}(\tau;\, -\mathcal{H})\middle|\tilde{\Psi}_f(-\mathcal{H})\right\rangle\right|^2 \times \left\langle\tilde{\Psi}_f(-\mathcal{H})\middle|\hat{\rho}(\tau;\, -\mathcal{H})\middle|\tilde{\Psi}_f(-\mathcal{H})\right\rangle, \tag{15.13}$$

since the initial microstates of the reverse protocol are now $\left|\tilde{\Psi}_f(-\mathcal{H})\right\rangle$ associated with the eigenvalues $\tilde{\mathbf{C}}_f$ and the final ones are $\left|\Psi_i(-\mathcal{H})\right\rangle$ associated with $\mathbf{C}_i$.

15.2.2 Consequences of Microreversibility

The probability densities of the fluxes for the forward and reverse protocols can be related to each other as a consequence of microreversibility. Indeed, the lemma (14.41) for $t = \tau$ gives

$$\hat{\Theta}\, \hat{\mathrm{U}}_{\mathrm{F}}^{\dagger}(\tau;\mathcal{H})\, \hat{\Theta}^{-1} = \hat{\mathrm{U}}_{\mathrm{R}}(\tau;\, -\mathcal{H}). \tag{15.14}$$

Starting from the probability density (15.10) for the forward protocol, inserting the initial statistical operator (15.3), and using the Dirac delta distributions (15.11) to replace the initial eigenvalues into the final ones, we obtain

$$p_{\rm F}(\Delta\mathbf{C};\mathcal{H}) = \prod_{j=1}^{r} {\rm e}^{-\beta_j(\Delta E_j - \sum_{k=1}^{c}\mu_{kj}\,\Delta N_{kj} + \Delta\Phi_j)} \sum_{fi} \delta(\Delta\mathbf{C} - \mathbf{C}_i + \tilde{\mathbf{C}}_f)$$
$$\times \left|\langle\tilde{\Psi}_f(\mathcal{H})|\hat{\rm U}_{\rm F}(\tau;\mathcal{H})|\Psi_i(\mathcal{H})\rangle\right|^2 \left\langle\tilde{\Psi}_f(-\mathcal{H})|\hat{\rho}(\tau;\,-\mathcal{H})|\tilde{\Psi}_f(-\mathcal{H})\right\rangle \tag{15.15}$$

with the difference of grand thermodynamic potentials

$$\Delta\Phi_j \equiv \tilde{\Phi}_j(-\mathcal{H}) - \Phi_j(\mathcal{H}) \tag{15.16}$$

and the final statistical operator (15.12). According to equation (15.14), the transition probabilities of the forward and reverse protocols are related to each other by

$$\left|\langle\tilde{\Psi}_f(\mathcal{H})|\hat{\rm U}_{\rm F}(\tau;\mathcal{H})|\Psi_i(\mathcal{H})\rangle\right|^2 = \left|\langle\Psi_i(-\mathcal{H})|\hat{\rm U}_{\rm R}(\tau;\,-\mathcal{H})|\tilde{\Psi}_f(-\mathcal{H})\rangle\right|^2 \tag{15.17}$$

because the time-reversal transformation $\hat{\Theta}$ maps the eigenstates onto those in the reversed magnetizing field by equation (15.1). Since Dirac delta distributions are even in their argument, the sum in the right-hand side of equation (15.15) is equal to the probability density (15.13) with the reversed variables $-\Delta\mathbf{C}$, so that we find the following time-reversal symmetry relation,

$$p_{\rm F}(\Delta\mathbf{C};\mathcal{H}) = {\rm e}^{-\sum_{j=1}^{r}\beta_j(\Delta E_j - \sum_{k=1}^{c}\mu_{kj}\,\Delta N_{kj} + \Delta\Phi_j)}\, p_{\rm R}(-\Delta\mathbf{C};\,-\mathcal{H}). \tag{15.18}$$

15.2.3 Nonequilibrium Work Quantum Fluctuation Relation

In the case where the system is closed and subjected to a time-dependent forcing, the Hamiltonian operator is the same as in equation (15.2) with a single part, i.e., $r = 1$, and no particle fluxes. The time-dependent forcing by the interaction operator $\hat{U}(t)$ performs mechanical work on this system. Here again, the two-time measurement scheme can be used to obtain the full counting statistics of this work. Upon the initial and final measurements, we may suppose that the energy eigenvalues E_i and $\tilde{E}_f$ are observed so that the mechanical work performed on the system between the two measurements can be defined as

$$W \equiv \tilde{E}_f - E_i = -\Delta E. \tag{15.19}$$

Otherwise, the same considerations apply as in the previous Sections 15.2.1 and 15.2.2. In this particular case, the time-reversal symmetry relation (15.18) gives the quantum version of the nonequilibrium work fluctuation relation,

$$p_{\rm F}(W;\mathcal{H}) = {\rm e}^{\beta(W-\Delta F)}\, p_{\rm R}(-W;\,-\mathcal{H}), \tag{15.20}$$

where $\Delta F = \tilde{F}(-\mathcal{H}) - F(\mathcal{H})$, which is the quantum analogue of the classical formula (5.13) or (5.44) in the absence or the presence of an external magnetizing field (Kurchan, 2000; Tasaki, 2000).

As a consequence, the Jarzynski equality (5.18) also holds for quantum systems, as proved by the same reasoning as in equation (5.19). This quantum version of Jarzynski's equality is equivalent to the formula (14.47). Indeed, this latter can be expressed as

$$\left\langle e^{-\beta \hat{H}_{\mathrm{F}}(\tau)}\, e^{\beta \hat{H}(0)} \right\rangle_{\mathrm{F},\mathcal{H}} = \int dW\, e^{-\beta W}\, p_{\mathrm{F}}(W;\mathcal{H}) \tag{15.21}$$

in terms of the probability density (15.10) of the forward protocol in the particular case where $r = 1$:

$$p_{\mathrm{F}}(W;\mathcal{H}) = \sum_{fi} \delta\big(W - \tilde{E}_f + E_i\big)\, \left| \langle \tilde{\Psi}_f(\mathcal{H}) | \hat{\mathsf{U}}_{\mathrm{F}}(\tau;\mathcal{H}) | \Psi_i(\mathcal{H}) \rangle \right|^2 \frac{e^{-\beta E_i}}{Z(0;\mathcal{H})}, \tag{15.22}$$

since the mean value in the left-hand side of equation (15.21) can be decomposed using the completeness relations on the bases of eigenstates $|\Psi_i(\mathcal{H})\rangle$ and $|\tilde{\Psi}_f(\mathcal{H})\rangle$ along with the definition $\hat{H}_{\mathrm{F}}(\tau) = \hat{\mathsf{U}}_{\mathrm{F}}^{\dagger}(\tau;\mathcal{H})\, \hat{H}(\tau;\mathcal{H})\, \hat{\mathsf{U}}_{\mathrm{F}}(\tau;\mathcal{H})$. Therefore, the quantum Jarzynski relation (14.47) is deduced from equation (15.21) combined with the nonequilibrium work fluctuation relation (15.20).

15.2.4 Full Counting Statistics of Energy and Particle Fluxes

In order to establish the fluctuation relation for energy and particle fluxes between the reservoirs, we may consider protocols that are symmetric under the reversal $t \to \tau - t$ with the interaction operator $\hat{U}(t) = \hat{U}(\tau - t)$ and $\hat{\tilde{H}}_j = \hat{H}_j$. In this case, there is no difference between the forward and reverse protocols, so that the differences (15.16) are equal to zero, i.e., $\Delta\Phi_j = 0$.

Furthermore, the interaction operator $\hat{U}$ may be assumed to remain invariant during the time interval $[0, \tau]$. If the reservoirs are large enough, the equilibration time t_{equil} needed to reach global equilibrium across the whole system can be much longer than the relaxation time t_{relax} towards a regime of nonequilibrium stationarity according to equation (1.75). Under such circumstances, the full counting statistics of the energy and particle fluxes in this nonequilibrium regime can be obtained by considering a time interval in the range $t_{\mathrm{relax}} \ll \tau \ll t_{\mathrm{equil}}$. The total Hamiltonian and the unitary time evolution operators are thus given by

$$\hat{H}(\mathcal{H}) = \sum_{j=1}^{r} \hat{H}_j + \hat{U} \qquad \text{and} \qquad \hat{\mathsf{U}}(t;\mathcal{H}) = e^{-\imath \hat{H}(\mathcal{H}) t} \qquad \text{for} \qquad 0 \le t \le \tau, \tag{15.23}$$

in units where $\hbar = 1$. Therefore, the Hamiltonian and particle numbers of the reservoirs are observables evolving in time according to

$$\hat{H}_j(t) = \hat{\mathsf{U}}^{\dagger}(t;\mathcal{H})\, \hat{H}_j\, \hat{\mathsf{U}}(t;\mathcal{H}) \qquad \text{and} \qquad \hat{N}_{kj}(t) = \hat{\mathsf{U}}^{\dagger}(t;\mathcal{H})\, \hat{N}_{kj}\, \hat{\mathsf{U}}(t;\mathcal{H}). \tag{15.24}$$

Consequently, the probability densities (15.10) and (15.13) are identical and they reduce to $p_{\mathrm{F}} = p_{\mathrm{R}} \equiv p_{\tau}$ with

$$p_{\tau}(\Delta\mathbf{C};\mathcal{H}) \equiv \sum_{fi} \delta(\Delta\mathbf{C} - \mathbf{C}_i + \mathbf{C}_f) \left| \langle \Psi_f(\mathcal{H}) | \hat{\mathsf{U}}(\tau;\mathcal{H}) | \Psi_i(\mathcal{H}) \rangle \right|^2 \langle \Psi_i(\mathcal{H}) | \hat{\rho}(0;\mathcal{H}) | \Psi_i(\mathcal{H}) \rangle. \tag{15.25}$$

Introducing the counting parameters $\boldsymbol{\lambda} = [\lambda_{Ej}, (\lambda_{kj})_{k=1}^{c}]_{j=1}^{r}$, the moment generating function can be defined as

$$\mathcal{G}_\tau(\boldsymbol{\lambda}; \mathcal{H}) \equiv \int d\Delta\mathbf{C}\, e^{-\boldsymbol{\lambda}\cdot\Delta\mathbf{C}}\, p_\tau(\Delta\mathbf{C}; \mathcal{H}). \tag{15.26}$$

According to the expression (15.25) for the probability density and using the completeness relation of the eigenstates $|\Psi_n(\mathcal{H})\rangle$ with $n = i$ or $n = f$, the moment generating function can be written equivalently in terms of the operators (15.7) as

$$\mathcal{G}_\tau(\boldsymbol{\lambda}; \mathcal{H}) = \operatorname{tr} \hat{\rho}(0; \mathcal{H})\, \hat{\mathrm{U}}^\dagger(\tau; \mathcal{H})\, e^{\boldsymbol{\lambda}\cdot\hat{\mathbf{C}}}\, \hat{\mathrm{U}}(\tau; \mathcal{H})\, e^{-\boldsymbol{\lambda}\cdot\hat{\mathbf{C}}} = \operatorname{tr} \hat{\rho}(0; \mathcal{H})\, e^{\boldsymbol{\lambda}\cdot\hat{\mathbf{C}}(\tau)}\, e^{-\boldsymbol{\lambda}\cdot\hat{\mathbf{C}}(0)} \tag{15.27}$$

with the operators $\hat{\mathbf{C}}(t)$ defined by equation (15.24). By construction, the moment generating function is real if the counting parameters are real numbers.

In addition, we may also define the cumulant generating function as

$$\mathcal{Q}(\boldsymbol{\lambda}; \mathcal{H}) \equiv \lim_{\tau\to\infty} -\frac{1}{\tau} \ln \mathcal{G}_\tau(\boldsymbol{\lambda}; \mathcal{H}) \qquad \text{with} \qquad \boldsymbol{\lambda} = \left[\lambda_{Ej}, (\lambda_{kj})_{k=1}^{c}\right]_{j=1}^{r}. \tag{15.28}$$

A remarkable property is that this cumulant generating function is invariant under the translations $[\lambda_{Ej}, (\lambda_{kj})_{k=1}^{c}]_{j=1}^{r} \to [\lambda_{Ej} + \xi_E, (\lambda_{kj} + \xi_k)_{k=1}^{c}]_{j=1}^{r}$ of the counting parameters within each set of quantities that are conserved during $0 \le t \le \tau$, namely the energy and the particle numbers of the different species:

$$\mathcal{Q}(\boldsymbol{\lambda}; \mathcal{H}) = \mathcal{Q}(\boldsymbol{\lambda} + \boldsymbol{\xi}; \mathcal{H}) \qquad \text{with} \qquad \boldsymbol{\xi} = \left[\xi_E, (\xi_k)_{k=1}^{c}\right]_{j=1}^{r}, \tag{15.29}$$

where ξ_E and ξ_k ($1 \le k \le c$) are real numbers (Andrieux et al., 2009; Benoist et al., 2020). This invariance property holds under the assumptions that the interaction $\hat{U}$ between the reservoirs is bounded in the thermodynamic limit and satisfies ultraviolet regularity conditions such that transitions are exponentially suppressed towards high energy (Benoist et al., 2020). With these assumptions, the moment generating function (15.27) and the function with the translated counting parameters can be shown to be related to each other by the following inequalities,

$$L\, \mathcal{G}_\tau(\boldsymbol{\lambda} + \boldsymbol{\xi}; \mathcal{H}) \le \mathcal{G}_\tau(\boldsymbol{\lambda}; \mathcal{H}) \le K\, \mathcal{G}_\tau(\boldsymbol{\lambda} + \boldsymbol{\xi}; \mathcal{H}), \tag{15.30}$$

where the constants K and L are independent of the time τ and remain finite in the thermodynamic limit, where the reservoirs are arbitrarily large for given temperatures and chemical potentials. As a consequence, the translational invariance (15.29) holds for the cumulant generating function (15.28).

Now, taking $\xi_E = -\lambda_{Er}$ and $\xi_k = -\lambda_{kr}$, we see that the cumulant generating function only depends on the counting parameters for the energy and particle fluxes from the reservoirs $1 \le j \le r - 1$ towards the reference reservoir $j = r$ because $\mathcal{Q}(\boldsymbol{\lambda}; \mathcal{H}) = \mathcal{Q}(\boldsymbol{\lambda} - \boldsymbol{\lambda}_r; \mathcal{H})$ with $\boldsymbol{\lambda}_r \equiv [\lambda_{Er}, (\lambda_{kr})_{k=1}^{c}]_{j=1}^{r}$. Hence, the counting parameters may be redefined as $\lambda_{Ej} - \lambda_{Er} \to \lambda_{Ej}$ and $\lambda_{kj} - \lambda_{kr} \to \lambda_{kj}$ for $1 \le j \le r - 1$ and their number can thus be reduced from $(c + 1)r$ to $(c + 1)(r - 1)$. Therefore, the cumulant generating function (15.28) may be expressed with the $(c + 1)(r - 1)$ counting parameters

$$\boldsymbol{\lambda} \equiv \left[\lambda_{Ej}, (\lambda_{kj})_{k=1}^{c}\right]_{j=1}^{r-1} \qquad \text{as} \qquad Q(\boldsymbol{\lambda}; \mathbf{A}, \mathcal{H}) \equiv \mathcal{Q}(\boldsymbol{\lambda} - \boldsymbol{\lambda}_r; \mathcal{H}), \tag{15.31}$$

where the dependence on the affinities (5.69)–(5.70) is now explicitly written down. We note that the cumulant generating function also depends on the inverse temperature β_r and the chemical potentials μ_{kr} of the reference reservoir $j = r$. We point out that the redefinition of the counting parameters is equivalent to setting the counting parameters associated with the reference reservoir equal to zero: $\lambda_{Er} = 0$ and $\lambda_{kr} = 0$. The cumulant generating function (15.31) is thus given by

$$Q(\boldsymbol{\lambda}; \mathbf{A}, \boldsymbol{\mathcal{H}}) = \lim_{\tau\to\infty} -\frac{1}{\tau} \ln \mathrm{tr}\, \hat{\rho}(0; \boldsymbol{\mathcal{H}})\, \mathrm{e}^{\imath \hat{H}(\boldsymbol{\mathcal{H}})\tau}\, \mathrm{e}^{\boldsymbol{\lambda}\cdot\hat{\boldsymbol{\mathfrak{C}}}}\, \mathrm{e}^{-\imath \hat{H}(\boldsymbol{\mathcal{H}})\tau}\, \mathrm{e}^{-\boldsymbol{\lambda}\cdot\hat{\boldsymbol{\mathfrak{C}}}} \tag{15.32}$$

in terms of the $(c+1)(r-1)$ observables

$$\hat{\boldsymbol{\mathfrak{C}}} \equiv \left[\hat{H}_j, \left(\hat{N}_{kj}\right)_{k=1}^{c}\right]_{j=1}^{r-1} \tag{15.33}$$

and the corresponding counting parameters $\boldsymbol{\lambda}$ defined in equation (15.31).

Accordingly, the statistical cumulants of the full counting statistics can be deduced from their generating function (15.32) as explained in Section 5.3.4. If the reservoirs are arbitrarily large, the fluxes across the open quantum system can reach a steady regime. The mean values of the currents are thus obtained as

$$\mathbf{J}(\mathbf{A}, \boldsymbol{\mathcal{H}}) = \frac{\partial Q}{\partial \boldsymbol{\lambda}}(\mathbf{0}; \mathbf{A}, \boldsymbol{\mathcal{H}}) = \lim_{\tau\to\infty} \frac{1}{\tau} \left\langle \hat{\boldsymbol{\mathfrak{C}}}(0) - \hat{\boldsymbol{\mathfrak{C}}}(\tau) \right\rangle_{\mathbf{A}, \boldsymbol{\mathcal{H}}} \tag{15.34}$$

with

$$\left\langle \hat{\boldsymbol{\mathfrak{C}}}(\tau) \right\rangle_{\mathbf{A}, \boldsymbol{\mathcal{H}}} = \mathrm{tr}\, \hat{\rho}(0; \boldsymbol{\mathcal{H}})\, \mathrm{e}^{\imath \hat{H}(\boldsymbol{\mathcal{H}})\tau}\, \hat{\boldsymbol{\mathfrak{C}}}\, \mathrm{e}^{-\imath \hat{H}(\boldsymbol{\mathcal{H}})\tau}. \tag{15.35}$$

The diffusivities of the currents, as well as the higher cumulants, can be similarly deduced using equations (5.82), (5.83), (5.84), ... These cumulants are defined in steady regimes specified by the affinities $\mathbf{A}$. They can be expanded in powers of the affinities around the thermodynamic equilibrium $\mathbf{A} = \mathbf{0}$ in order to define the response coefficients, as in equation (5.133) for the mean currents.

Now, the mean currents out of the reference reservoir $j = r$ can also be evaluated by differentiating the full cumulant generating function (15.28) with the full moment generating function (15.27) with respect to the associated counting parameters λ_{Er} and λ_{kr} with $1 \le k \le c$. Because of the translational invariance (15.29), these mean currents are given by

$$J_{Er}(\mathbf{A}, \boldsymbol{\mathcal{H}}) = -\sum_{j=1}^{r-1} J_{Ej}(\mathbf{A}, \boldsymbol{\mathcal{H}}) \qquad \text{and} \qquad J_{kr}(\mathbf{A}, \boldsymbol{\mathcal{H}}) = -\sum_{j=1}^{r-1} J_{kj}(\mathbf{A}, \boldsymbol{\mathcal{H}}) \tag{15.36}$$

in terms of the mean currents (15.34) out of the other reservoirs $1 \le j \le r-1$. These relations express the conservation of the total energy and the total particle numbers of the different species in steady regimes and they are reminiscent of the classical relations (5.67), although holding on average in the long-time limit.

15.2.5 Microreversibility and Full Counting Statistics

If the conditions stated in Section 15.2.4 hold (i.e., such that $\Delta\Phi_j = 0$), the time-reversal symmetry relation (15.18) becomes

$$p_\tau(\Delta\mathbf{C}; \mathcal{H}) = \mathrm{e}^{\boldsymbol{\alpha}\cdot\Delta\mathbf{C}}\, p_\tau(-\Delta\mathbf{C}; -\mathcal{H}), \tag{15.37}$$

which is the quantum version of equation (5.66) with the same parameters $\boldsymbol{\alpha} = \left[-\beta_j, (\beta_j\mu_{kj})_{k=1}^{c}\right]_{j=1}^{r}$ as introduced in equation (5.57). Therefore, the moment generating function (15.26) satisfies the relation

$$\mathcal{G}_\tau(\boldsymbol{\lambda}; \mathcal{H}) = \mathcal{G}_\tau(\boldsymbol{\alpha} - \boldsymbol{\lambda}; -\mathcal{H}) \tag{15.38}$$

by reason of microreversibility. Consequently, the cumulant generating function (15.28) obeys the relation

$$\mathcal{Q}(\boldsymbol{\lambda}; \mathcal{H}) = \mathcal{Q}(\boldsymbol{\alpha} - \boldsymbol{\lambda}; -\mathcal{H}). \tag{15.39}$$

As discussed in Section 5.3.3 for classical systems, the issue is that the parameters $\boldsymbol{\alpha} = \left[-\beta_j, (\beta_j\mu_{kj})_{k=1}^{c}\right]_{j=1}^{r}$ are not the independent thermodynamic forces or affinities (5.69)–(5.70), so that the relation (15.39) does not yet have the form of the symmetry relation (5.177) for the currents. The reason is that the quantities $\Delta\mathbf{C}$ in equation (15.37) are the fluxes going out of all the r reservoirs, although the symmetry relation should concern the fluxes between the reservoirs $1 \le j \le r-1$ and the reference reservoir $j = r$. With the fluxes defined for classical systems in Section 5.3, the conservation relations (5.67) are satisfied, so that the fluctuation relation (5.73) is directly implied by equation (5.66). However, for quantum systems, the interaction is switched on and off in the two-time measurement scheme and this time dependence has for consequence that the conservation relations (5.67) do not strictly hold. Nevertheless, the cumulant generating function (15.28) has the remarkable property of the translational invariance (15.29), so that it can be equivalently expressed by equation (15.31) in terms of the $(c+1)(r-1)$ parameters counting the energy and particle fluxes out of the reservoirs $1 \le j \le r-1$ towards the reference reservoir. Now, the transformation $\boldsymbol{\lambda} \to \boldsymbol{\alpha} - \boldsymbol{\lambda}$ for the $(c+1)r$ counting parameters in equation (15.39) induces the substitution $\boldsymbol{\lambda} - \boldsymbol{\lambda}_r \to \boldsymbol{\alpha} - \boldsymbol{\alpha}_r - \boldsymbol{\lambda} + \boldsymbol{\lambda}_r$ with $\boldsymbol{\alpha}_r = \left[-\beta_r, (\beta_j\mu_{kr})_{k=1}^{c}\right]_{j=1}^{r}$ and, thus, $\boldsymbol{\lambda} \to \mathbf{A} - \boldsymbol{\lambda}$ for the $(c+1)(r-1)$ counting parameters of the cumulant generating function (15.31), where $\mathbf{A} = \left[\beta_r - \beta_j, (\beta_j\mu_{kj} - \beta_r\mu_{kr})_{k=1}^{c}\right]_{j=1}^{r-1}$ are the $(c+1)(r-1)$ affinities (5.69)–(5.70).

Therefore, the symmetry relation (15.39) implied by microreversibility becomes

$$\mathcal{Q}(\boldsymbol{\lambda}; \mathbf{A}, \mathcal{H}) = \mathcal{Q}(\mathbf{A} - \boldsymbol{\lambda}; \mathbf{A}, -\mathcal{H}) \tag{15.40}$$

for the cumulant generating function (15.32) of the independent currents corresponding to the thermal and chemical affinities (5.69)–(5.70) driving the system out of equilibrium (Andrieux et al., 2009; Benoist et al., 2020). This fundamental result is the quantum version of the symmetry relation (5.177), which reduces to the symmetry relation (5.91) in the absence of external magnetizing field $\mathcal{H} = 0$.

Most remarkably, the linear and nonlinear response properties of Section 5.5 for $\mathcal{H} = 0$ and Section 5.6.3 for $\mathcal{H} \neq 0$ thus hold for quantum as well as classical open systems, as a consequence of microreversibility.

15.3 Scattering Approach to Quantum Transport

15.3.1 Open Quantum Systems with Independent Particles

In quantum systems with independent particles, there is no interaction between the particles, which evolve in some external energy potential. This is, in particular, the case for the quantum transport of electrons in mesoscopic semiconducting circuits (Datta, 1995; Blanter and Büttiker, 2000; Nazarov and Blanter, 2009; Lesovik and Sadovskyy, 2011), or ultracold atoms in optical traps (Brantut et al., 2013; Krinner et al., 2015, 2017), where the mean free path is larger than the distance between the walls of the device, so that the motion is essentially ballistic. Under such circumstances, the time evolution is ruled by the Hamiltonian operator (14.12) without the interaction energy potentials $u^{(n)} = 0$ for $n \geq 2$, but with the external energy potential $u^{(\text{ext})}$ in the one-body Hamiltonian (14.13) to describe the interaction of the particles with the walls. This external potential may form a pipe, a barrier, or a crossroad between two or more channels, where the motion is free in one direction or more, as shown in Figure 15.1. The example of Figure 15.1(a) is the motion of particles through a small hole in a wall between two reservoirs, as already considered for classical effusion in Chapter 8. Another example is an energy barrier forming a constriction between two particle reservoirs, as shown in Figure 15.1(b). The circuit may also connect more than two reservoirs, as in Figure 15.1(c). Since the particles are independent, they undergo scattering because of the external energy potential $u^{(\text{ext})}$ and the transport properties of such systems can be obtained using the scattering approach pioneered by Landauer (1957) and developed since then. The great virtue of the scattering approach is that the transport properties are obtained including the quantum coherence described by the scattering operator.

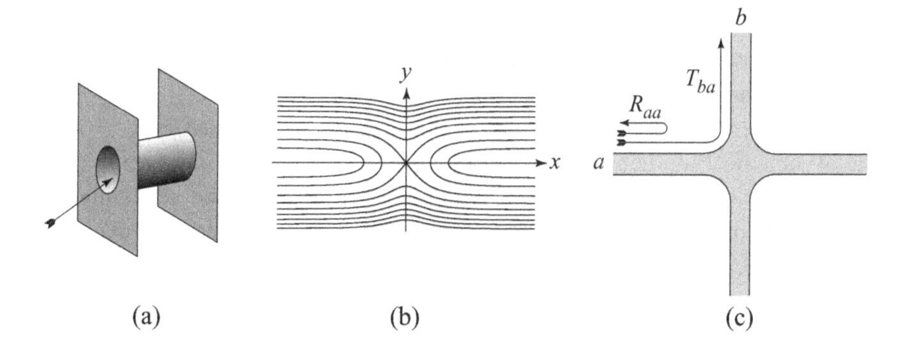

Figure 15.1 Examples of conducting devices: (a) Effusion of particles through a small hole in a wall separating two reservoirs. (b) Potential energy surface with a constriction between two channels. (c) Circuit with four channels connected to so many terminals or reservoirs. Adapted from Gaspard (2015c).

15.3.2 The Scattering Operator

As shown in Figure 15.1, the circuit can have several terminals of infinite spatial extension, which may be considered as the reservoirs $j = 1, 2, \ldots, r$. Each terminal constitutes a wave guide for the independent particles incoming or outgoing the scattering region formed by the constriction or the connector between the terminals. In every terminal, there exist a spatial coordinate $r_\parallel$ parallel to the axis of the wave guide and other coordinates $\mathbf{r}_\perp$ transverse to it. Asymptotically for $r_\parallel \to \infty$, the energy potential may be assumed to be independent of the longitudinal coordinate $r_\parallel$:

$$u^{(\mathrm{ext})}(\mathbf{r}) \simeq_{r_\parallel \to \infty} u_j(\mathbf{r}_\perp), \tag{15.41}$$

where $u_j(\mathbf{r}_\perp)$ is the potential transverse to the wave guide. This is the case, in particular, for the potential energy surface of Figure 15.1(b) given by $u^{(\mathrm{ext})}(x, y) = u_0(\cosh \alpha x)^{-2} + m\omega^2 y^2/2$ with the harmonic potential $m\omega^2 y^2/2$ in the transverse direction $y = r_\perp$ and the energy barrier $u_0(\cosh \alpha x)^{-2}$ in the longitudinal direction $x = r_\parallel$.

Asymptotically in the wave guide, the one-body Hamiltonian operator (14.13) has the following stationary microstates,

$$\left[-\frac{\hbar^2}{2m}\nabla^2 + u_j(\mathbf{r}_\perp)\right]\phi = \varepsilon\,\phi \qquad \text{with} \qquad \phi(\mathbf{r}, \sigma) = \exp\left(\frac{\imath}{\hbar}\, p\, r_\parallel\right)\, \varphi_{j\mathbf{n}}(\mathbf{r}_\perp)\, \chi_\varsigma(\sigma), \tag{15.42}$$

where p is the momentum in the longitudinal direction of the wave guide j, $\mathbf{n}$ are quantum numbers labeling the transverse modes, and $\chi_\varsigma(\sigma)$ is a spinor with $\varsigma, \sigma = -s, -s+1, \ldots, s-1, s$. The transverse modes correspond to the different possible channels for transport (Nazarov and Blanter, 2009). The associated energy eigenvalues are given by

$$\varepsilon = \varepsilon_{0j\mathbf{n}} + \frac{p^2}{2m}. \tag{15.43}$$

The channel $a = j\mathbf{n}$ is open at energies larger than the threshold $\varepsilon_{0j\mathbf{n}}$. For the potential energy surface of Figure 15.1(b), the thresholds are given by $\varepsilon_{0n} = \hbar\omega(n + 1/2)$ with $n = 0, 1, 2, \ldots$. For a wave guide with a square cross-section, Figure 15.2 shows the dispersion relations (15.43) and the number of open channels as energy increases.

As a consequence, the one-body Hamiltonian can be decomposed as $\hat{h} = \hat{h}_0 + \hat{u}$ into the asymptotic Hamiltonian $\hat{h}_0$ given by equation (15.42) and the energy potential $\hat{u}$ describing the interaction between the different terminals in the scattering region. In the example of Figure 15.1(b), the asymptotic Hamiltonian can be expressed as $\hat{h}_0 = -\hbar^2(\partial_x^2 + \partial_y^2)/(2m) + m\omega^2 y^2/2$ and the interaction as $u(x) = u_0(\cosh \alpha x)^{-2}$. The propagating modes in the wave guides are thus scattered by the barrier $u(x)$. If the incoming wave function is a plane wave in the positive x-direction, the scattering produces the outgoing reflected and transmitted waves according to

$$\psi_{pn\varsigma}(x, \sigma) = \begin{cases} \left(\mathrm{e}^{\imath px/\hbar} + r_{pn}\, \mathrm{e}^{-\imath px/\hbar}\right) \varphi_n(y)\, \chi_\varsigma(\sigma) & \text{for} \quad x < 0, \\ t_{pn}\, \mathrm{e}^{\imath px/\hbar}\, \varphi_n(y)\, \chi_\varsigma(\sigma) & \text{for} \quad x > 0, \end{cases} \tag{15.44}$$

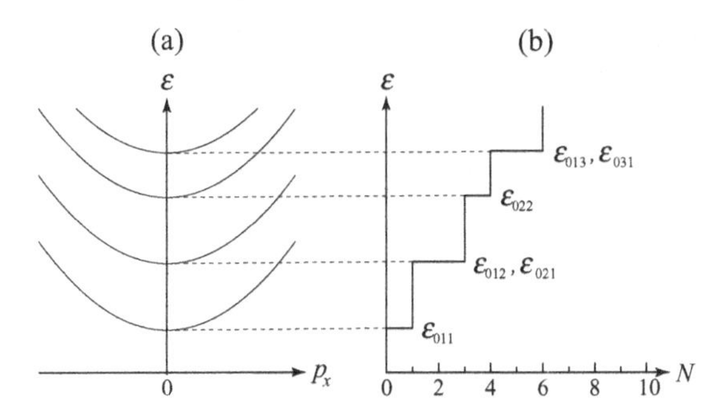

Figure 15.2 Wave guide with hard walls and a square cross-section of sides ℓ: (a) Energy spectrum of the modes (15.42) versus the momentum $p = p_x$ in the propagating x-direction given by $\varepsilon = \varepsilon_{0n_yn_z} + p_x^2/(2m)$ with $\varepsilon_{0n_yn_z} = (\pi\hbar/\ell)^2\,(n_y^2 + n_z^2)/(2m)$ for $n_y, n_z = 1, 2, 3, \ldots$. (b) Diagram showing the number N of open channels as the one-body energy ε increases (vertical axis). Adapted from Gaspard (2015c).

where r_{pn} and t_{pn} denote the reflection and transmission amplitudes depending on the momentum $p = p_x$ and the quantum number $n = 0, 1, 2, \ldots$. In every channel n, the amplitudes of the outgoing wave functions are thus determined by those of the incoming wave functions as

$$\phi_{\text{out}} = \hat{S}(\varepsilon)\,\phi_{\text{in}}, \qquad \text{where} \qquad \hat{S}(\varepsilon) = \begin{pmatrix} r_{pn} & t_{pn} \\ t_{pn} & r_{pn} \end{pmatrix} \tag{15.45}$$

is the scattering matrix, which depends on the one-body energy $\varepsilon = \varepsilon_{0n} + p^2/(2m)$ (Nazarov and Blanter, 2009). In this example, the transmission probability through the barrier is thus given by $T_n(\varepsilon) = |t_{pn}|^2$ and the reflection probability by $R_n(\varepsilon) = |r_{pn}|^2 = 1 - T_n(\varepsilon)$. The scattering matrix is 2×2 because the energy potential is separable in the longitudinal and transverse directions. In this case, the transmission coefficient can be written as $T_n(\varepsilon) = T(\varepsilon - \varepsilon_{0n})$ in terms of the transmission coefficient $T(\varepsilon)$ of the one-dimensional barrier $u(x)$. Examples of such transmission probabilities are shown in Figure 15.3. The transmission probability converges to the unit value at high energy well above the height of the barrier. For energies lower than the height of the barrier, transmission proceeds by tunneling. For the inverted harmonic potential $u(x) = u_0 - m\gamma^2x^2/2$, the transmission probability is given by

$$T(\varepsilon) = \left[1 + \exp\left(-2\pi\,\frac{\varepsilon - u_0}{\hbar\gamma}\right)\right]^{-1}, \tag{15.46}$$

which is shown in Figure 15.3(d) (Nazarov and Blanter, 2009; Lesovik and Sadovskyy, 2011). For nonseparable potentials, the scattering matrix is infinite and couples together the transverse modes.

In the space of all the asymptotic one-body microstates, the scattering operator can be defined in the long-time limit $t \to \infty$ by considering the backward evolution of some incoming wave packet with the asymptotic Hamiltonian $\hat{h}_0$ over the time interval $t/2$, followed by the forward evolution with the full Hamiltonian $\hat{h}$ during the time interval t, and,

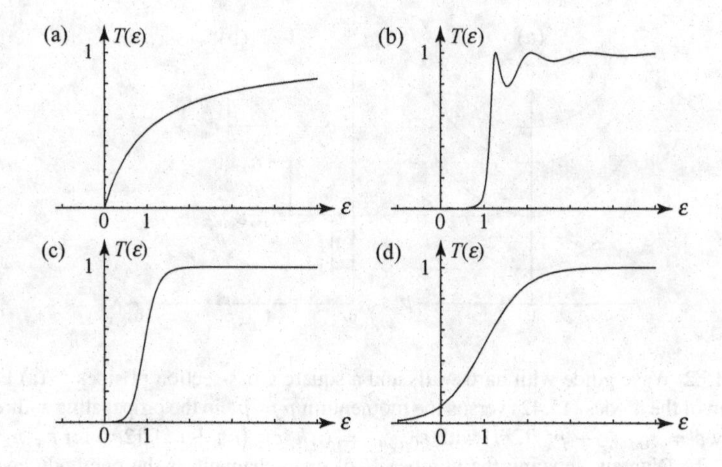

Figure 15.3 The transmission probability $T(\varepsilon)$ versus the energy ε for different types of barriers: (a) the delta barrier $u(x) = g\,\delta(x)$; (b) the square barrier $u(x) = u_0\theta(x)\theta(\ell-x)$; (c) the barrier $u(x) = u_0(\cosh\alpha x)^{-2}$; (d) the inverted harmonic barrier $u(x) = u_0 - m\gamma^2 x^2/2$. Adapted from Gaspard (2015c).

again, the backward evolution with the asymptotic Hamiltonian $\hat{h}_0$ over the time interval $t/2$, bringing back the wave packet into the space of asymptotic one-body microstates:

$$\hat{S} \equiv \lim_{t\to\infty} \mathrm{e}^{\frac{i}{2\hbar}\hat{h}_0 t}\,\mathrm{e}^{-\frac{i}{\hbar}\hat{h}t}\,\mathrm{e}^{\frac{i}{2\hbar}\hat{h}_0 t} = \int d\varepsilon\,\hat{S}(\varepsilon)\,\delta\big(\varepsilon - \hat{h}_0\big). \tag{15.47}$$

In this equation, the last identity gives the decomposition of the scattering operator into the scattering matrices $\hat{S}(\varepsilon)$ acting on the subspaces of given energy ε, such as the scattering matrix in equation (15.45). The scattering operator and the scattering matrices are known to be unitary (Joachain, 1975). Unitarity implies that $\sum_b |S_{ab}(\varepsilon)|^2 = \sum_b |S_{ba}(\varepsilon)|^2 = 1$ for scattering events $a \to b$ between the channels a and b. Accordingly, the transmission and reflection probabilities are defined, respectively, by

$$T_{ba}(\varepsilon) \equiv |S_{ba}(\varepsilon)|^2 \quad (\text{for } a \neq b) \quad \text{and} \quad R_{aa}(\varepsilon) \equiv |S_{aa}(\varepsilon)|^2 = 1 - \sum_{b(\neq a)} T_{ba}(\varepsilon). \tag{15.48}$$

15.3.3 Full Counting Statistics and Cumulant Generating Function

For quantum systems composed of independent particles, the cumulant generating function (15.32) for the full counting statistics characterizing the transport of energy and particles can be directly expressed in terms of the scattering matrix and the Bose–Einstein or Fermi–Dirac distribution functions, depending on the quantum statistics of the particles. Here, we assume that the system contains a single species of particles (i.e., $c = 1$).

In general, the field operators introduced in Section 14.2.4 can be decomposed as $\hat{\psi}(\mathbf{r},\sigma) = \sum_\nu \phi_\nu(\mathbf{r},\sigma)\,\hat{a}_\nu$ into the one-body wave functions $\phi_\nu(\mathbf{r},\sigma)$ and the annihilation-creation operators $\hat{a}_\nu$ and $\hat{a}_\nu^\dagger$, obeying the following commutation or anticommutation relations

$$\hat{a}_\nu \, \hat{a}^\dagger_{\nu'} - \theta \, \hat{a}^\dagger_{\nu'} \, \hat{a}_\nu = \delta_{\nu\nu'} \tag{15.49}$$

with $\theta = +1$ for bosons, or $\theta = -1$ for fermions.

For independent particles, the Hamiltonian operator and the particle numbers are quadratic functions of the annihilation-creation operators. Interestingly, quadratic many-body operators $\hat{X}$ can be written as follows in terms of the corresponding one-body operator $\hat{x} = (x_{\nu\nu'})$:

$$\hat{X} = \sum_{\nu,\nu'} x_{\nu\nu'} \, \hat{a}^\dagger_\nu \, \hat{a}_{\nu'} = \Gamma(\hat{x}). \tag{15.50}$$

According to the formula of Klich (2003), the trace of the product of two exponential functions of quadratic many-body operators such as (15.50) can be expressed as the following determinant of the corresponding one-body operators:

$$\mathrm{tr}\, \mathrm{e}^{\hat{X}_1} \, \mathrm{e}^{\hat{X}_2} = \mathrm{tr}\, \mathrm{e}^{\Gamma(\hat{x}_1)} \, \mathrm{e}^{\Gamma(\hat{x}_2)} = \det\left(1 - \theta \, \mathrm{e}^{\hat{x}_1} \, \mathrm{e}^{\hat{x}_2}\right)^{-\theta}. \tag{15.51}$$

Now, the moment generating function in equation (15.32) with $t = \tau$ is equivalently given by

$$\mathcal{G}_t(\boldsymbol{\lambda}) = \mathrm{tr}\, \hat{\rho}(0) \, \mathrm{e}^{\imath \hat{H} t} \, \mathrm{e}^{\boldsymbol{\lambda}\cdot\hat{\mathfrak{C}}} \, \mathrm{e}^{-\imath \hat{H} t} \, \mathrm{e}^{-\boldsymbol{\lambda}\cdot\hat{\mathfrak{C}}} = \frac{\mathrm{tr}\, \mathrm{e}^{\hat{Y}_t} \, \mathrm{e}^{-\hat{X}-\hat{Y}}}{\mathrm{tr}\, \mathrm{e}^{-\hat{X}}} \tag{15.52}$$

in terms of the initial statistical operator (15.3), the total Hamiltonian (15.23), the observables $\hat{\mathfrak{C}} \equiv \left(\hat{H}_j, \hat{N}_j\right)_{j=1}^{r-1}$, and the corresponding counting parameters $\boldsymbol{\lambda} \equiv \left(\lambda_{Ej}, \lambda_{Nj}\right)_{j=1}^{r-1}$; or with the following operators,

$$\hat{X} = \sum_{j=1}^{r} \beta_j \left(\hat{H}_j - \mu_j \hat{N}_j\right) \qquad \text{and} \qquad \hat{Y} = \sum_{j=1}^{r-1} \left(\lambda_{Ej} \hat{H}_j + \lambda_{Nj} \hat{N}_j\right) = \boldsymbol{\lambda} \cdot \hat{\mathfrak{C}}, \tag{15.53}$$

which commute $\left[\hat{X}, \hat{Y}\right] = 0$, and $\hat{Y}_t = \mathrm{e}^{\imath \hat{H} t/\hbar} \, \hat{Y} \, \mathrm{e}^{-\imath \hat{H} t/\hbar}$. Introducing the Bose–Einstein and Fermi–Dirac distribution functions of all the reservoirs as

$$\hat{f} = \frac{1}{\mathrm{e}^{\hat{x}} - \theta} \qquad \text{with} \qquad \hat{x} = \sum_{j=1}^{r} \beta_j \left(\hat{h}_j - \mu_j \hat{n}_j\right) \tag{15.54}$$

and using the formula (15.51) and the property that $\det\left(\hat{a}\hat{b}\right) = \det \hat{a} \det \hat{b}$, the moment generating function (15.52) becomes

$$\mathcal{G}_t(\boldsymbol{\lambda}) = \det\left[\left(1 - \theta \, \mathrm{e}^{-\hat{x}}\right)^{-1} \left(1 - \theta \, \mathrm{e}^{\hat{y}_t} \, \mathrm{e}^{-\hat{x}-\hat{y}}\right)\right]^{-\theta} = \det\left[1 - \theta \hat{f} \left(\mathrm{e}^{\hat{y}_t} \, \mathrm{e}^{-\hat{y}} - 1\right)\right]^{-\theta}. \tag{15.55}$$

Considering a long but finite time interval $[0, t]$, the time evolution can be expressed for $t \to \infty$ in terms of the scattering operator (15.47) as

$$\mathrm{e}^{\hat{y}_t} = \mathrm{e}^{\imath \hat{h} t} \, \mathrm{e}^{\hat{y}} \, \mathrm{e}^{-\imath \hat{h} t} \simeq \mathrm{e}^{\imath \hat{h}_0 t/2} \, \hat{S}^\dagger \, \mathrm{e}^{\imath \hat{h}_0 t/2} \, \mathrm{e}^{\hat{y}} \, \mathrm{e}^{-\imath \hat{h}_0 t/2} \, \hat{S} \, \mathrm{e}^{-\imath \hat{h}_0 t/2} \tag{15.56}$$

with the asymptotic one-body Hamiltonian $\hat{h}_0 = \sum_{j=1}^{r} \hat{h}_j$ in units where $\hbar = 1$. Since $\hat{h}_0$ commutes with $\hat{y} = \sum_{j=1}^{r-1} \left(\lambda_{Ej}\hat{h}_j + \lambda_{Nj}\hat{n}_j\right)$ and the distributions (15.54), the moment generating function takes the form

$$\mathcal{G}_t(\boldsymbol{\lambda}) \simeq \det\left[1 - \theta \hat{f}\left(\hat{S}^\dagger \, \mathrm{e}^{\hat{y}} \, \hat{S} \, \mathrm{e}^{-\hat{y}} - 1\right)\right]^{-\theta} \tag{15.57}$$

in the long-time limit $t \to \infty$. In principle, the determinant is defined in the Hilbert space of the one-body dynamics, but the finite time interval constrains the one-body energy spectrum to be discrete with a spacing equal to $\Delta\varepsilon = 2\pi\hbar/t$, so that the sum over the discrete energy levels and the spin states $\sigma = -s, -s+1, \ldots, s-1, s$ leads to

$$\lim_{t\to\infty} \frac{1}{t} \sum_{\varepsilon,\sigma} (\cdot) = g_s \int \frac{d\varepsilon}{2\pi\hbar} \, (\cdot), \tag{15.58}$$

where $g_s = 2s + 1$ is the spin multiplicity. Decomposing the one-body operators over energy as in equation (15.47) for the scattering operator, the moment generating function can thus be written as

$$\mathcal{G}_t(\boldsymbol{\lambda}) \simeq \prod_{\varepsilon,\sigma} \det\left\{1 - \theta \hat{f}(\varepsilon)\left[\hat{S}^\dagger(\varepsilon) \, \mathrm{e}^{\varepsilon\hat{\lambda}_E + \hat{\lambda}_N} \, \hat{S}(\varepsilon) \, \mathrm{e}^{-\varepsilon\hat{\lambda}_E - \hat{\lambda}_N} - 1\right]\right\}^{-\theta} \tag{15.59}$$

in terms of the $r \times r$ scattering matrix $\hat{S}(\varepsilon)$, the diagonal matrix $\hat{f}(\varepsilon) = [f_j(\varepsilon)\,\delta_{jj'}]$ with its elements given by the Bose–Einstein or Fermi–Dirac distribution functions of the reservoirs,

$$f_j(\varepsilon) = \frac{1}{\mathrm{e}^{\beta_j(\varepsilon-\mu_j)} - \theta}, \tag{15.60}$$

and the diagonal matrices composed with the counting parameters according to $\hat{\lambda}_E = (\lambda_{Ej}\,\delta_{jj'})$ and $\hat{\lambda}_N = (\lambda_{Nj}\,\delta_{jj'})$ with $\lambda_{Er} = 0$ and $\lambda_{Nr} = 0$ for the reference reservoir $j = r$.

Inserting the result (15.59) into the expression (15.32) and using equation (15.58), the cumulant generating function is given for independent particles by

$$Q(\boldsymbol{\lambda}; \mathbf{A}) = \theta \, g_s \int \frac{d\varepsilon}{2\pi\hbar} \ln\det\left\{1 - \theta \hat{f}(\varepsilon)\left[\hat{S}^\dagger(\varepsilon) \, \mathrm{e}^{\varepsilon\hat{\lambda}_E + \hat{\lambda}_N} \, \hat{S}(\varepsilon) \, \mathrm{e}^{-\varepsilon\hat{\lambda}_E - \hat{\lambda}_N} - 1\right]\right\}. \tag{15.61}$$

Here, the translational invariance (15.29) can be directly verified with the transformations $\exp\left(\varepsilon\hat{\lambda}_E + \hat{\lambda}_N\right) \to \exp\left(\varepsilon\hat{\lambda}_E + \hat{\lambda}_N + \xi\,\hat{1}\right)$ for any function $\xi(\varepsilon)$ of the energy ε and the identity matrix $\hat{1}$. Since $\exp\left(\xi\hat{1}\right)$ commutes with both $\hat{f}(\varepsilon)$ and $\hat{S}(\varepsilon)$, there is cancellation between the factors $\exp\left(\xi\hat{1}\right)$ and $\exp\left(-\xi\hat{1}\right)$, whereupon the translational invariance is proved. Now, taking $\xi = -\varepsilon\lambda_{Er} - \lambda_{Nr}$, we can directly conclude that the cumulant generating function has the form (15.31).

Using equation (15.34) with the property that $\ln\det(\cdot) = \mathrm{tr}\ln(\cdot)$, the mean energy and particle currents are obtained as

$$\begin{pmatrix} J_{Ej} \\ J_{Nj} \end{pmatrix} = g_s \int \frac{d\varepsilon}{2\pi\hbar} \begin{pmatrix} \varepsilon \\ 1 \end{pmatrix} \sum_{i,\mathbf{m},\mathbf{n}} \left[\delta_{ji}\,\delta_{\mathbf{nm}} - |S_{j\mathbf{n},i\mathbf{m}}(\varepsilon)|^2\right] f_i(\varepsilon) \tag{15.62}$$

$$= g_s \int \frac{d\varepsilon}{2\pi\hbar} \begin{pmatrix} \varepsilon \\ 1 \end{pmatrix} \sum_{i(\neq j),\mathbf{m}\neq\mathbf{n}} \left[T_{i\mathbf{m},j\mathbf{n}}(\varepsilon)\, f_j(\varepsilon) - T_{j\mathbf{n},i\mathbf{m}}(\varepsilon)\, f_i(\varepsilon)\right],$$

holding for bosons and fermions and known as the Landauer–Büttiker formulas (Landauer, 1957; Datta, 1995; Blanter and Büttiker, 2000; Nazarov and Blanter, 2009). The diffusivities characterizing noise, as well as the higher cumulants, can also be obtained.

In the case of two terminals ($j = \mathrm{L}, \mathrm{R}$) separated by a barrier or a constriction characterized by the scattering matrix (15.45) and the transmission probability $T_n(\varepsilon) = |t_{pn}|^2$, the cumulant generating function (15.61) is given for bosons if $\theta = +1$ and fermions if $\theta = -1$ by

$$Q(\boldsymbol{\lambda};\mathbf{A}) = \theta\, g_s \sum_{n=0}^{\infty} \int_{\varepsilon_{0n}}^{\infty} \frac{d\varepsilon}{2\pi\hbar} \ln\Big\{1 - \theta\, T_n \Big[f_{\mathrm{L}}\,(1+\theta\, f_{\mathrm{R}})\left(\mathrm{e}^{-\varepsilon\lambda_E - \lambda_N} - 1\right)$$

$$+ f_{\mathrm{R}}\,(1+\theta\, f_{\mathrm{L}})\left(\mathrm{e}^{\varepsilon\lambda_E + \lambda_N} - 1\right)\Big]\Big\}, \tag{15.63}$$

where $f_j = f_j(\varepsilon)$ are the distribution functions (15.60) of the reservoirs. This formula was first obtained for fermions by Levitov and Lesovik (1993). Since there are two terminals, the thermodynamic forces driving the system out of equilibrium are given by the following thermal and chemical affinities,

$$A_E = \beta_{\mathrm{R}} - \beta_{\mathrm{L}} \qquad \text{and} \qquad A_N = \beta_{\mathrm{L}}\,\mu_{\mathrm{L}} - \beta_{\mathrm{R}}\,\mu_{\mathrm{R}}, \tag{15.64}$$

which correspond to the counting parameters λ_E and λ_N of the two independent currents of energy and particles. In this circuit, the mean particle current and its diffusivity have the following expressions,

$$J_N = \frac{\partial Q}{\partial \lambda_N}(\mathbf{0};\mathbf{A}) = g_s \sum_{n=0}^{\infty} \int_{\varepsilon_{0n}}^{\infty} \frac{d\varepsilon}{2\pi\hbar}\, T_n\,(f_{\mathrm{L}} - f_{\mathrm{R}}), \tag{15.65}$$

$$D_{NN} = -\frac{1}{2}\,\frac{\partial^2 Q}{\partial \lambda_N^2}(\mathbf{0};\mathbf{A}) \tag{15.66}$$

$$= \frac{g_s}{2} \sum_{n=0}^{\infty} \int_{\varepsilon_{0n}}^{\infty} \frac{d\varepsilon}{2\pi\hbar}\, T_n\Big[f_{\mathrm{L}}(1+\theta\, f_{\mathrm{R}}) + f_{\mathrm{R}}(1+\theta\, f_{\mathrm{L}}) + \theta\, T_n (f_{\mathrm{L}} - f_{\mathrm{R}})^2\Big].$$

Similar results hold for the energy current.

15.3.4 Cumulant Generating Function and Microreversibility

The time-reversal symmetry relation (15.40) for the cumulant generating function can be directly established in the scattering approach (Gaspard, 2013a). Indeed, for spinless particles or for each spin component in systems without spin-orbit interaction, the time-reversal symmetry implies that the scattering matrix in the presence of an external magnetizing field $\mathcal{H}$ obeys

$$\hat{S}^{\mathrm{T}}_{\mathcal{H}} = \hat{S}_{-\mathcal{H}}, \tag{15.67}$$

where the superscript T denotes the transpose (Datta, 1995). Therefore, the determinant in the cumulant generating function (15.61) can be transformed as

$$\begin{aligned}
&\det\left\{1-\theta\hat{f}(\varepsilon)\left[\hat{S}^{\dagger}_{\mathcal{H}}(\varepsilon)\,\mathrm{e}^{\varepsilon\hat{\lambda}_E+\hat{\lambda}_N}\,\hat{S}_{\mathcal{H}}(\varepsilon)\,\mathrm{e}^{-\varepsilon\hat{\lambda}_E-\hat{\lambda}_N}-1\right]\right\} \qquad (15.68)\\
&=\det\left[1+\theta\hat{f}(\varepsilon)\right]\det\left[1-\frac{\theta\hat{f}(\varepsilon)}{1+\theta\hat{f}(\varepsilon)}\,\hat{S}^{\dagger}_{\mathcal{H}}(\varepsilon)\,\mathrm{e}^{\varepsilon\hat{\lambda}_E+\hat{\lambda}_N}\,\hat{S}_{\mathcal{H}}(\varepsilon)\,\mathrm{e}^{-\varepsilon\hat{\lambda}_E-\hat{\lambda}_N}\right]\\
&=\det\left[1+\theta\hat{f}(\varepsilon)\right]\det\left[1-\theta\,\hat{S}^{\dagger}_{\mathcal{H}}(\varepsilon)\,\mathrm{e}^{\varepsilon\hat{\lambda}_E+\hat{\lambda}_N}\,\hat{S}_{\mathcal{H}}(\varepsilon)\,\mathrm{e}^{-\varepsilon\hat{\lambda}_E-\hat{\lambda}_N-\hat{\beta}(\varepsilon-\hat{\mu})}\right]\\
&=\det\left[1+\theta\hat{f}(\varepsilon)\right]\det\left[1-\theta\,\hat{S}^{\mathrm{T}}_{\mathcal{H}}(\varepsilon)\,\mathrm{e}^{\varepsilon\hat{\lambda}_E+\hat{\lambda}_N}\,\hat{S}^{*}_{\mathcal{H}}(\varepsilon)\,\mathrm{e}^{-\varepsilon\hat{\lambda}_E-\hat{\lambda}_N-\hat{\beta}(\varepsilon-\hat{\mu})}\right]\\
&=\det\left[1+\theta\hat{f}(\varepsilon)\right]\det\left[1-\theta\,\hat{S}_{-\mathcal{H}}(\varepsilon)\,\mathrm{e}^{\varepsilon\hat{\lambda}_E+\hat{\lambda}_N}\,\hat{S}^{\dagger}_{-\mathcal{H}}(\varepsilon)\,\mathrm{e}^{-\varepsilon\hat{\lambda}_E-\hat{\lambda}_N-\hat{\beta}(\varepsilon-\hat{\mu})}\right]\\
&=\det\left\{1-\theta\hat{f}(\varepsilon)\left[\hat{S}^{\dagger}_{-\mathcal{H}}(\varepsilon)\,\mathrm{e}^{-\varepsilon\hat{\lambda}_E-\hat{\lambda}_N-\hat{\beta}(\varepsilon-\hat{\mu})}\,\hat{S}_{-\mathcal{H}}(\varepsilon)\,\mathrm{e}^{\varepsilon\hat{\lambda}_E+\hat{\lambda}_N+\hat{\beta}(\varepsilon-\hat{\mu})}-1\right]\right\}\\
&=\det\left\{1-\theta\hat{f}(\varepsilon)\left[\hat{S}^{\dagger}_{-\mathcal{H}}(\varepsilon)\,\mathrm{e}^{\varepsilon(\hat{A}_E-\hat{\lambda}_E)+(\hat{A}_N-\hat{\lambda}_N)}\,\hat{S}_{-\mathcal{H}}(\varepsilon)\,\mathrm{e}^{-\varepsilon(\hat{A}_E-\hat{\lambda}_E)-(\hat{A}_N-\hat{\lambda}_N)}-1\right]\right\}.
\end{aligned}$$

From the first to the second line, the property $\det(\hat{a}\hat{b}) = \det\hat{a}\,\det\hat{b}$ is used. From the second to the third line, the transformation is based on the following identity,

$$\frac{f_j(\varepsilon)}{1+\theta f_j(\varepsilon)} = \mathrm{e}^{-\beta_j(\varepsilon-\mu_j)}, \qquad \text{i.e.,} \qquad \frac{\hat{f}(\varepsilon)}{1+\theta\hat{f}(\varepsilon)} = \mathrm{e}^{-\hat{\beta}(\varepsilon-\hat{\mu})}, \tag{15.69}$$

where $\hat{\beta} = (\beta_j\,\delta_{jj'})$ and $\hat{\mu} = (\mu_j\,\delta_{jj'})$ are the $r\times r$ diagonal matrices containing the inverse temperatures and the chemical potentials on their diagonal, and also the cyclic property $\det(\hat{a}\hat{b}) = \det(\hat{b}\hat{a})$. From the third to the fourth line, the changes use the property that the determinants are real numbers, so that $\det(\hat{a}\hat{b}) = \det(\hat{a}^*\hat{b}^*)$, as well as the facts that the matrices $\hat{f}(\varepsilon)$, $\hat{\lambda}_E$, $\hat{\lambda}_N$, $\hat{\beta}$, and $\hat{\mu}$ are real, and that $(\hat{a}^{\dagger})^* = \hat{a}^{\mathrm{T}}$. From the fourth to the fifth line, the time-reversal symmetry (15.67) is applied. From the fifth to the sixth line, the identity (15.69) is again used and the determinant is reordered. The sixth line expresses the symmetry of the determinant under the transformation $(\hat{\lambda}_E, \hat{\lambda}_N, \mathcal{H}) \to (-\hat{\beta}-\hat{\lambda}_E, \hat{\beta}\hat{\mu}-\hat{\lambda}_N, -\mathcal{H})$. Since the determinant and, thus, the cumulant generating function satisfy the invariance property (15.29), the translation $(\hat{\lambda}_E, \hat{\lambda}_N) \to (\hat{\lambda}_E - \beta_r\hat{1}, \hat{\lambda}_N + \beta_r\mu_r\hat{1})$ can be considered to finally obtain the seventh line, where $\hat{A}_E = (A_{Ej}\,\delta_{jj'})$ and $\hat{A}_N = (A_{Nj}\,\delta_{jj'})$ are the diagonal matrices composed with the thermal and chemical affinities respectively given by

$$A_{Ej} = \beta_r - \beta_j \qquad \text{and} \qquad A_{Nj} = \beta_j\,\mu_j - \beta_r\,\mu_r. \tag{15.70}$$

Hence, the cumulant generating function is symmetric under the transformation $(\boldsymbol{\lambda}, \mathcal{H}) \to (\mathbf{A}-\boldsymbol{\lambda}, -\mathcal{H})$, which proves the symmetry relation (15.40) directly from the time-reversal symmetry (15.67) of the scattering matrix in the presence of an external magnetizing field.

As a consequence of microreversibility, the linear and nonlinear transport properties of Section 5.6.3 are thus satisfied, including the Onsager–Casimir reciprocal relations as well as their generalizations.

15.4 Temporal Disorder and Entropy Production

The concept of temporal disorder introduced in Section 6.4 and the relationship of its time asymmetry to the entropy production rate explained in Section 6.5 can be extended to quantum systems with independent particles (Gaspard, 2014, 2015b,c).

15.4.1 Characterization of Temporal Disorder in Quantum Systems

The entropy per unit time of Kolmogorov (1959) and Sinai (1959) characterizing temporal disorder in classical dynamics and stochastic processes has been generalized to quantum systems by Connes et al. (1987) and Narnhofer and Thirring (1987). This CNT entropy per unit time is equal to zero in quantum systems with a discrete energy spectrum because their time evolution is quasiperiodic (with a finite number of incommensurate frequencies) or almost periodic (with countably many incommensurate frequencies), so that they do not manifest temporal disorder. Yet the CNT entropy per unit may be positive for infinite quantum systems at positive particle density and temperature, which have a continuous spectrum.

The expression of the CNT entropy per unit can be obtained by considerations similar to those of Sections 6.4 and 8.9 for quantum systems with quasi-free independent particles ruled by the Hamiltonian operator (14.12) with $u^{(n)} = 0$ for $n \geq 2$ and the one-body Hamiltonian (14.13) with a uniform or spatially periodic external potential $u^{(\mathrm{ext})}$. This is the case in wave guides where the one-body microstates are given by equation (15.42) with the continuous energy spectrum (15.43).

The entropy per unit time characterizing temporal disorder can be defined as

$$h \equiv \lim_{t\to\infty} -\frac{1}{t} \sum_{\boldsymbol{\omega}} P(\boldsymbol{\omega}) \ln P(\boldsymbol{\omega}), \tag{15.71}$$

where $\boldsymbol{\omega}$ are the paths or histories of the system during the time interval $[0,t]$. If the entropy per unit time is positive, i.e., $h > 0$, the probabilities of these histories decay exponentially in time as $P(\boldsymbol{\omega}) \sim \exp(-ht)$ and the time evolution thus manifests dynamical randomness, as in a coin-tossing process. For quantum gases with independent particles in quasi free motion, the history $\boldsymbol{\omega}$ followed by the many-body system during the finite but long time interval $[0,t]$ can be determined in terms of the occupation numbers of the one-body microstates ϕ_ν with $\nu = pj\mathbf{n}\varsigma$ by bosonic or fermionic particles. These microstates have the value p for the longitudinal momentum in the channel $a = j\mathbf{n}$ and the spin orientation ς. Their energy eigenvalue $\varepsilon_\nu = \varepsilon$ is given by equation (15.43). The particles with their momentum between $p > 0$ and $p + \Delta p > 0$ have the velocity $|d\varepsilon/dp| = |p/m|$, so that they travel the distance $\Delta r_\parallel = t|d\varepsilon/dp|$. The number of possible one-body microstates ϕ_ν that can be occupied by a particle with momentum in the interval $[p, p+\Delta p]$ is given by

$$g_s \frac{\Delta r_\parallel \Delta p}{2\pi\hbar} = t\, g_s \left|\frac{d\varepsilon}{dp}\right| \frac{\Delta p}{2\pi\hbar} \tag{15.72}$$

with the spin multiplicity $g_s = 2s + 1$.

We consider the scattering $a \to b$ between the channels a and b, as illustrated in Figure 15.1(c). Among the microstates (15.72) in the incoming channel a, the fraction that is scattered into the outgoing channel b is given by $T_{ba}(\varepsilon) = |S_{ba}(\varepsilon)|^2$ according to equation (15.48). Therefore, the number of such possible microstates is evaluated as

$$M_{ba}(\varepsilon) = t\, g_s \left|\frac{d\varepsilon}{dp}\right| \frac{\Delta p}{2\pi\hbar} |S_{ba}(\varepsilon)|^2. \tag{15.73}$$

These microstates can be occupied by random numbers $\{N_\alpha\}$ of particles, which have the probability distribution

$$P(N) = \frac{f^N}{(1+\theta f)^{N+\theta}}, \tag{15.74}$$

where $\theta = +1$ and $N \in \{0, 1, 2, 3, \ldots\}$ for bosons, $\theta = -1$ and $N \in \{0, 1\}$ for fermions, and $f = \langle N \rangle = \sum_N N\, P(N)$ is the mean number of particles. This latter is given by the Bose–Einstein or Fermi–Dirac function (15.60) of the incoming reservoir: $f_\alpha = f_j(\varepsilon)$. The entropy associated with the distribution (15.74) has the expression

$$-\sum_N P(N) \ln P(N) = -f \ln f + (f+\theta) \ln(1+\theta f). \tag{15.75}$$

Since the different one-body microstates are occupied independently of each other, the probability of the history $\boldsymbol{\omega}$ factorizes as

$$P(\boldsymbol{\omega}) = \prod_\alpha P_\alpha(N_\alpha) \tag{15.76}$$

into the probabilities P_α of the particle numbers $\{N_\alpha\}$ occupying the corresponding one-body microstates (15.73). Substituting this probability into the definition (15.71), the entropy per unit time can be expressed as

$$h = \sum_\alpha \frac{dM_\alpha}{dt} \left[-\sum_N P_\alpha(N) \ln P_\alpha(N) \right], \tag{15.77}$$

because $\lim_{t\to\infty}(M_\alpha/t) = dM_\alpha/dt$ for $M_\alpha = M_{ba}(\varepsilon)$ given by equation (15.73). This entropy per unit time is always nonnegative, i.e., $h \geq 0$ and it can be interpreted as the rate of information production in recording typical histories. Using the expressions (15.73) and (15.75), summing over all the possible transitions $a \to b$, and taking the limit $\Delta p \to 0$, we find that the CNT entropy per unit time is given by

$$h = g_s \sum_{j\mathbf{n}} \int_{\varepsilon_{0j\mathbf{n}}}^{\infty} \frac{d\varepsilon}{2\pi\hbar} \left\{ -f_j(\varepsilon) \ln f_j(\varepsilon) + \left[f_j(\varepsilon)+\theta\right] \ln\left[1+\theta\, f_j(\varepsilon)\right] \right\}, \tag{15.78}$$

since $\sum_b |S_{ba}(\varepsilon)|^2 = 1$ and $d\varepsilon = |d\varepsilon/dp|dp$ (Narnhofer and Thirring, 1987; Benatti et al., 1998).

The CNT entropy per unit time is equal to zero at a temperature of absolute zero, where there is no longer any temporal disorder.

15.4.2 Temporal Disorder and Time Reversal

A time-reversed coentropy per unit time can be defined by considering the time-reversed one-body paths $\alpha^{\rm R} = b \to a$ corresponding to the transitions $\alpha = a \to b$ between the channels $a = j\mathbf{n}$ and $b = i\mathbf{m}$. The time-reversed paths are weighted by the mean occupation number $f_b = f_i$ of the reservoir i corresponding to the outgoing channel $b = i\mathbf{m}$ of the forward path $\alpha = a \to b$. Accordingly, the random particle numbers in the time-reversed paths $\alpha^{\rm R}$ have the probability distributions $P_\alpha^{\rm R}(N)$ corresponding to the Bose–Einstein or Fermi–Dirac distribution function f_i of the channel $b = i\mathbf{m}$. In analogy with Section 6.5, the time-reversed coentropy per unit time is here given by

$$h^{\rm R} = \sum_\alpha \frac{dM_\alpha}{dt} \left[-\sum_N P_\alpha(N) \ln P_\alpha^{\rm R}(N) \right], \tag{15.79}$$

by averaging again over the forward process.

The crucial point is that some particles are reflected back into the same reservoir, while others are transmitted to a different reservoir, as shown in Figure 15.1(c). For the former particles, the path is self-reverse, i.e., $\alpha^{\rm R} = \alpha = a \to a$, so that the distribution function is the same for the path α as for the time-reversed path $\alpha^{\rm R}$. For the latter particles, the time-reversed path $\alpha^{\rm R} = b \to a$ differs from the forward path $\alpha = a \to b$ because $a \neq b$, so that the particles are exchanged between two reservoirs at different temperatures or chemical potentials, for which $f_j \neq f_i$.[4] Therefore, the time-reversed coentropy per unit time has the following expression,

$$\begin{aligned} h^{\rm R} = g_s \sum_{j\mathbf{n}} \int_{\varepsilon_{0j\mathbf{n}}}^{\infty} \frac{d\varepsilon}{2\pi\hbar}\, R_{j\mathbf{n},j\mathbf{n}}(\varepsilon) \left\{ -f_j(\varepsilon) \ln f_j(\varepsilon) + \left[f_j(\varepsilon) + \theta \right] \ln \left[1 + \theta f_j(\varepsilon) \right] \right\} \\ + g_s \sum_{i\mathbf{m} \neq j\mathbf{n}} \int_{\varepsilon_{0j\mathbf{n}}}^{\infty} \frac{d\varepsilon}{2\pi\hbar}\, T_{i\mathbf{m},j\mathbf{n}}(\varepsilon) \left\{ -f_j(\varepsilon) \ln f_i(\varepsilon) + \left[f_j(\varepsilon) + \theta \right] \ln \left[1 + \theta f_i(\varepsilon) \right] \right\}, \end{aligned} \tag{15.80}$$

in terms of the transmission and reflection probabilities (15.48) (Gaspard, 2015b).

15.4.3 Time Asymmetry of Temporal Disorder and Entropy Production

Here also, the entropy production rate is given by equation (6.103) as the difference between the time-reversed coentropy per unit time (15.80) and the CNT entropy per unit time (15.78). Indeed, the transmission and reflection probabilities satisfy the relations (15.48) as a consequence of the unitarity of the scattering matrix. Again using unitarity and the identity (15.69), their difference becomes

$$h^{\rm R} - h = g_s \sum_{i\mathbf{m} \neq j\mathbf{n}} \int_{\varepsilon_{0j\mathbf{n}}}^{\infty} \frac{d\varepsilon}{2\pi\hbar}\, \beta_j(\mu_j - \varepsilon) \left[T_{i\mathbf{m},j\mathbf{n}}(\varepsilon)\, f_j(\varepsilon) - T_{j\mathbf{n},i\mathbf{m}}(\varepsilon)\, f_i(\varepsilon) \right]. \tag{15.81}$$

[4] This discussion is reminiscent of the one in Section 2.6.1 for classical systems.

Now, the expressions (15.62) for the mean energy and particle currents and the conservation laws (15.36) imply that the thermodynamic entropy production is indeed recovered as

$$\frac{1}{k_{\mathrm{B}}}\frac{d_{\mathrm{i}}S}{dt} = h^{\mathrm{R}} - h = \sum_{j=1}^{r-1}\left(A_{Ej}\,J_{Ej} + A_{Nj}\,J_{Nj}\right) \geq 0, \tag{15.82}$$

in terms of the thermal and chemical affinities (15.70) (Gaspard, 2014, 2015b,c). This expression is always nonnegative in accordance with the second law of thermodynamics, because the difference $h^{\mathrm{R}} - h$ involves the Kullback–Leibler divergence $D_{\mathrm{KL}}\left(P_\alpha \| P_\alpha^{\mathrm{R}}\right)$, which is always nonnegative. The result (15.82) confirms that thermodynamic entropy is produced due to time asymmetry in the temporal disorder of nonequilibrium processes.

15.5 Transport of Fermions

15.5.1 Generalities

Fermions obey the Pauli exclusion principle and the Fermi–Dirac statistics, so that their distribution function is given by equation (15.60) with $\theta = -1$ in each reservoir. At low temperature, the Fermi–Dirac distribution function is step-like, as shown in Figure 15.4(a), nearly equal to the unit value for all the occupied one-body states below the chemical potential μ_j and to zero for the unoccupied one-body states with energies above μ_j. This distribution decreases over an energy interval of the order of the thermal energy $\beta^{-1} = k_{\mathrm{B}}T$.

Let us consider a two-terminal circuit as in Figure 15.1(b) with a separable potential where the transport of independent particles is described by the scattering matrix (15.45) in

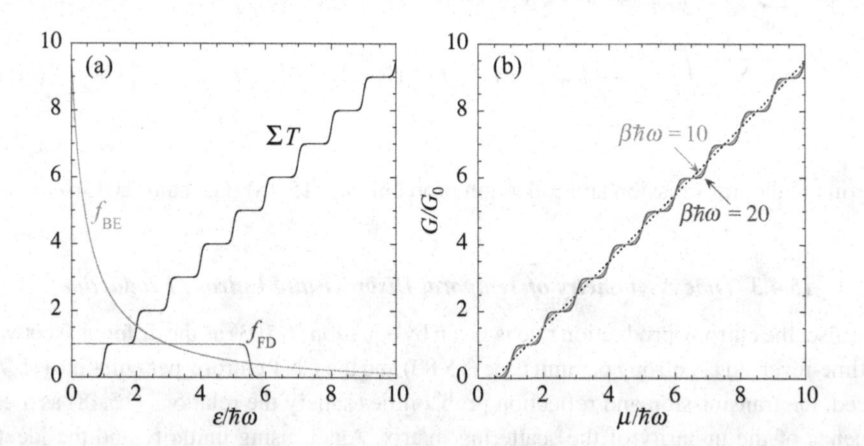

Figure 15.4 Quantization of conductance in the two-terminal circuit formed by the potential $u^{(\mathrm{ext})}(x,y) = m\omega^2 y^2/2 + u(x)$ with the inverted harmonic barrier $u(x) = u_0 - m\gamma^2 x^2/2$ of transmission probability (15.46): (a) The sum of transmission probabilities $\sum T = \sum_n T_n(\varepsilon)$ with the Bose–Einstein f_{BE} and Fermi–Dirac f_{FD} distributions versus the rescaled energy $\varepsilon/(\hbar\omega)$. (b) The fermionic conductance (15.83) in units of the conductance quantum $G_0 = g_s/(2\pi\hbar)$ versus the rescaled chemical potential $\mu/(\hbar\omega)$ for two different rescaled inverse temperatures $\beta\hbar\omega$ (solid lines). The dashed line depicts the average conductance $G_{\mathrm{av}}/G_0 \simeq (\mu - u_0)/(\hbar\omega)$. Adapted from Gaspard (2015c).

every open channel. If the potential is transversally harmonic as $u_j(y) = m\omega^2 y^2/2$ in the two terminals $j = \mathrm{L,R}$, the channels are open at the energy thresholds $\varepsilon_{0n} = \hbar\omega(n + 1/2)$ with $n = 0, 1, 2, \ldots$. If the energy barrier is the inverted harmonic potential $u(x) = u_0 - m\gamma^2 x^2/2$, the transmission probability can be expressed as $T_n(\varepsilon) = T(\varepsilon - \varepsilon_{0n})$ in terms of the function (15.46).

In this circuit, the mean particle current is given by equation (15.65) with Fermi–Dirac distributions $f_j = \left[e^{\beta_j(\varepsilon-\mu_j)} + 1\right]^{-1}$. If the system is isothermal, the chemical affinity can be written as $A_N = \beta(\mu_\mathrm{L} - \mu_\mathrm{R}) = -\beta e V$ for electrons and the applied voltage V. The corresponding electric current is thus equal to $I = -eJ_N$ and $g_s = 2$. In general, the current-voltage characteristic function $I(V)$ predicted by the Landauer–Büttiker formula (15.65) is nonlinear. Close enough to equilibrium, the linear regime can be observed where Ohm's law is recovered $I \simeq GV$, $G = 1/R$ being the electric conductance, which is the inverse of the resistance. According to equation (15.65), the electric conductance can thus be expressed as

$$G = 2\,e^2 \sum_{n=0}^{\infty} \int_{\varepsilon_{0n}}^{\infty} \frac{d\varepsilon}{2\pi\hbar}\, T_n(\varepsilon) \left(-\frac{\partial f}{\partial \varepsilon}\right), \tag{15.83}$$

where $f = f_\mathrm{L} = f_\mathrm{R} = \left[e^{\beta(\varepsilon-\mu)} + 1\right]^{-1}$ is the common Fermi–Dirac distribution of the reservoirs at the chemical potential μ and the inverse temperature β (Landauer, 1957; Datta, 1995; Nazarov and Blanter, 2009). At a temperature of absolute zero, the Fermi–Dirac distribution becomes $f(\varepsilon) = \theta(\varepsilon_\mathrm{F} - \varepsilon)$, where $\theta(x)$ is the Heaviside function and ε_F is the Fermi energy, so that the conductance becomes

$$G = G_0 \sum_{n=0}^{\infty} T_n(\varepsilon_\mathrm{F}) \simeq \mathscr{N}(\varepsilon_\mathrm{F})\, G_0, \qquad \text{where} \qquad G_0 \equiv \frac{e^2}{\pi\hbar} \tag{15.84}$$

is the conductance quantum and $\mathscr{N}(\varepsilon_\mathrm{F})$ is the number of open channels at the Fermi energy.

The quantization of conductance results from the convolution of the steps in the total transmission coefficient $\sum_n T_n(\varepsilon)$ with the step-like Fermi–Dirac distribution function f_FD at low enough temperature, as shown in Figure 15.4(a). The conductance is plotted in Figure 15.4(b) as a function of the rescaled chemical potential $\mu/(\hbar\omega)$, illustrating its quantization. This phenomenon can be observed for electrons in semiconducting devices (Nazarov and Blanter, 2009), as well as for ultracold fermionic ^{6}Li atoms (Krinner et al., 2015). As the temperature increases, the quantization steps become smoother and this phenomenon tends to disappear.

We note that the heat conductance and the thermoelectric linear response coefficient can be similarly obtained (Gaspard, 2013d).

15.5.2 Full Counting Statistics and Microreversibility

Beyond the mean current (15.65), the flux of particles manifests fluctuations and their full counting statistics can be characterized by their statistical cumulants and, more generally, by the cumulant generating function (15.63) with $\theta = -1$, which is known as the formula of Levitov and Lesovik (1993).

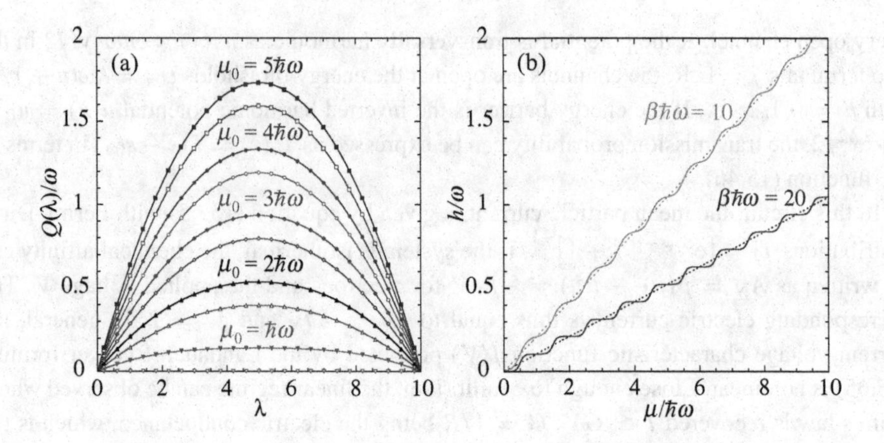

Figure 15.5 Transport of fermions in the two-terminal circuit of Figure 15.4: (a) Cumulant generating function (15.63) with $A_E = 0$ and $\lambda_E = 0$ versus the counting parameter $\lambda_N = \lambda$ for fermions crossing the inverted harmonic barrier of parameters $u_0 = \hbar\omega/2$ and $\gamma/\omega = 0.25$. The rescaled inverse temperature is $\beta\hbar\omega = 20$. The chemical potentials of the left- and right-hand reservoirs are taken as $\mu_{L,R} = \mu_0 \pm \hbar\omega/4$ with $\mu_0/(\hbar\omega) = 0.5 - 5$. (b) The rescaled CNT entropy per unit time h/ω given by equation (15.87) versus the rescaled chemical potential $\mu/(\hbar\omega)$ for a fermionic ideal gas with $g_s = 2$ at equilibrium in the same circuit for two different rescaled inverse temperatures $\beta\hbar\omega = 10, 20$ (solid lines). The dashed lines depict the average value (15.88) of the CNT entropy per unit time. Adapted from Gaspard (2015c).

As a consequence of microreversibility, the cumulant generating function has the fundamental symmetry

$$Q(\lambda_E, \lambda_N; A_E, A_N) = Q(A_E - \lambda_E, A_N - \lambda_N; A_E, A_N) \tag{15.85}$$

with respect to the thermal and chemical affinities (15.64).

The cumulant generating function (15.63) is shown in Figure 15.5(a) for the particle current in the same two-terminal circuit as in Figure 15.4(b) under the isothermal condition $A_E = 0$. The circuit is driven out of equilibrium by the chemical affinity $A_N = \beta(\mu_L - \mu_R) = \beta\hbar\omega/2 = 10$. As observed in Figure 15.5(a), the cumulant generating function is symmetric under the reflection $\lambda_N \to A_N - \lambda_N$, which is the prediction of the time-reversal symmetry relation (15.85) (Tobiska and Nazarov, 2005).

Since the number of open channels increases with energy, the mean current (15.65) becomes larger and larger as the median value $\mu_0 = (\mu_L + \mu_R)/2$ of the chemical potentials increases. This explains why the slope of the cumulant generating function at $\lambda_N = \lambda = 0$ increases with μ_0 in Figure 15.5(a).

15.5.3 Temporal Disorder

According to the Fermi–Dirac statistics (Pathria, 1972; Huang, 1987), the number of possible histories followed by N_α fermions occupying M_α possible microstates grows exponentially with the time interval t as

$$\prod_{\alpha} \frac{M_\alpha!}{N_\alpha!\,(M_\alpha - N_\alpha)!} \sim \exp(h\,t), \tag{15.86}$$

so that the CNT entropy per unit time h characterizing temporal disorder in the quasi-free transport of fermions is given by equation (15.78) with $\theta = -1$. In the aforementioned two-terminal circuit, the sums extend over the reservoirs $j = \mathrm{L}, \mathrm{R}$ and the channels $n = 0, 1, 2, \dots$, and the energy ε is integrated upwards from the channel thresholds $\varepsilon_{0n} = \hbar\omega(n + 1/2)$. If the circuit is at equilibrium with uniform temperature and chemical potential, the CNT entropy per unit time (15.78) reduces to

$$h = 2\,g_s \sum_{n=0}^{\infty} \int_{\varepsilon_{0n}}^{\infty} \frac{d\varepsilon}{2\pi\hbar} [-f\,\ln f - (1-f)\,\ln(1-f)], \tag{15.87}$$

with the Fermi–Dirac distribution $f = f_{\mathrm{L}} = f_{\mathrm{R}} = \left[\mathrm{e}^{\beta(\varepsilon-\mu)} + 1\right]^{-1}$. This quantity is plotted in Figure 15.5(b) as a function of the rescaled chemical potential for the same two-terminal circuit as in Figure 15.4. As the chemical potential increases, the temporal disorder also increases because more and more channels are open. Furthermore, the step-like increase of temporal disorder in Figure 15.5(b) has an origin similar to the quantization of conductance in Figure 15.4(b) (Gaspard, 2015c).

Since the mean number of open channels at the chemical potential μ is given by $\mathscr{N}_{\mathrm{av}}(\mu) \simeq \mu/(\hbar\omega)$ in the two-terminal circuit, the mean value of the CNT entropy can be evaluated as

$$h_{\mathrm{av}} \simeq g_s \frac{\pi k_{\mathrm{B}} T}{3\hbar} \frac{\mu}{\hbar\omega} \qquad \text{with} \qquad k_{\mathrm{B}} T = \beta^{-1}, \tag{15.88}$$

which confirms that temporal disorder increases with temperature. Moreover, it vanishes with temperature since there is no disorder in a quantum gas at absolute zero temperature.

At equilibrium, the CNT entropy per unit time is equal to the associated coentropy, $h = h^{\mathrm{R}}$. They differ out of equilibrium $h^{\mathrm{R}} > h$ and their difference gives the thermodynamic entropy production by equation (15.82).

15.5.4 Quantum Transport in Aharonov–Bohm Rings

The manifestation of quantum coherence for electron transport in Aharonov–Bohm rings is amazing (Datta, 1995; Nazarov and Blanter, 2009). These circuits are made with multiple terminals connected to a conducting ring in an external magnetic field $\boldsymbol{\mathcal{B}}$. Assuming the independence of electrons, the Hamiltonian operator of this system is given by equation (14.12) with anticommuting quantum fields $\hat{\psi}(\mathbf{r}, \sigma)$, vanishing interaction potentials, i.e., $u^{(n)} = 0$ for $n \geq 2$, and the one-body Hamiltonian operator

$$\hat{h} = \frac{1}{2m} \left(-\imath\hbar\nabla + \frac{e}{2} \boldsymbol{\mathcal{B}} \times \mathbf{r}\right)^2 + u(\mathbf{r}), \tag{15.89}$$

where m and $-e$ are the mass and the electric charge of the electrons, $\boldsymbol{\mathcal{B}}$ is a uniform magnetic field, and $u(\mathbf{r})$ is an external potential confining the electrons inside the circuit.

For simplicity, we may suppose that the electrons are moving in a single open channel, so that space is one-dimensional along the direction of the wires composing the circuit.

If the ring is circular of radius R, the position of the electron varies in the interval $x \in [0, 2\pi R[$ and the one-body Hamiltonian operator can be written as

$$\hat{h}_{\rm ring} = -\frac{\hbar^2}{2m}\left(\frac{d}{dx} + \imath\,\frac{\phi}{2\pi R}\right)^2, \qquad \text{where} \qquad \phi \equiv \frac{e}{\hbar}\oint_{\rm ring} \boldsymbol{\mathcal{B}}\cdot d\boldsymbol{\Sigma} \tag{15.90}$$

is the dimensionless magnetic flux across the ring. In the terminals $1 \le j \le r$, the position can be defined as $x \in [0, \infty[$ and the one-body Hamiltonian operator can be expressed as for one-dimensional free motion,

$$\hat{h}_j = -\frac{\hbar^2}{2m}\frac{d^2}{dx^2}. \tag{15.91}$$

The wires of the terminals are connected to the ring at vertices where the wave function is required to be continuous and to obey Neumann boundary conditions (Barbier and Gaspard, 2020a). Accordingly, the time-independent Schrödinger equation $\hat{h}\psi = \varepsilon\psi$ can be solved by considering the wave functions $\psi(x) = a\exp(\imath kx) + b\exp(-\imath kx)$ on each wire of the circuit for the energy $\varepsilon = \hbar^2 k^2/(2m)$. Using the boundary conditions, the coefficients can be calculated to obtain the $r \times r$ scattering matrix $\hat{S}_k(\phi)$. Because of microreversibility, this scattering matrix has the symmetry $\hat{S}_k^{\rm T}(\phi) = \hat{S}_k(-\phi)$ of equation (15.67).

An example of Aharonov–Bohm ring with two opposite terminals is considered in Figure 15.6(a). Under isothermal conditions, there is a single particle current in this circuit,

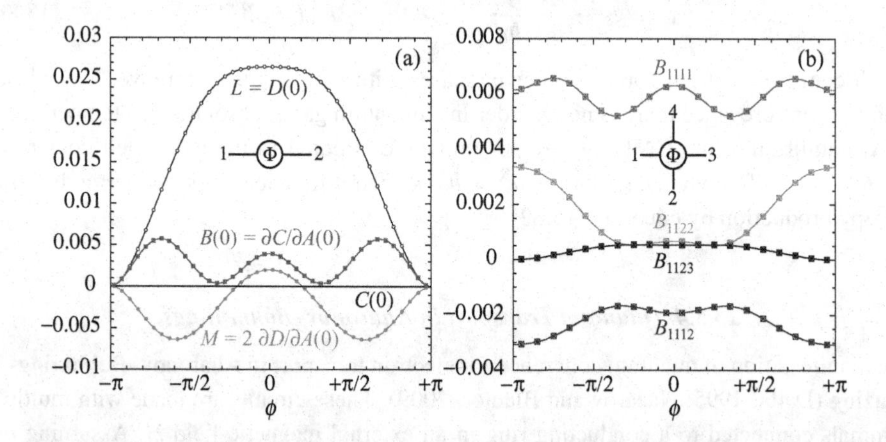

Figure 15.6 (a) Aharonov–Bohm ring with two opposite terminals: tests of the relations (5.179), (5.182), and (5.186), together with the vanishing of the third cumulant (5.83) at equilibrium for $\beta\mu = 10$, versus the dimensionless magnetic flux ϕ introduced in equation (15.90). The lines depict the left-hand side of the shown relations and the symbols their right-hand side. (b) Aharonov–Bohm ring with four terminals: tests of the relations (5.186)–(5.187) giving the fourth cumulants for $\beta\mu = 10$, versus the dimensionless magnetic flux ϕ. The lines depict the fourth cumulants (5.84). The right-hand sides of the aforementioned relations are depicted as open squares, crosses, and pluses, respectively. Reprinted with permission from Barbier and Gaspard (2020a) © American Physical Society.

which is driven away from equilibrium by the single affinity $A_N = \beta(\mu_L - \mu_R)$. The full counting statistics of this particle current is determined by the cumulant generating function (15.63) with $\theta = -1$, $A_E = 0$, and $\lambda_E = 0$. The time-reversal symmetry relation $Q(\lambda; A, \phi) = Q(A - \lambda; A, -\phi)$ with $A = A_N$, $\lambda = \lambda_N$, and the dimensionless magnetic flux ϕ introduced in equation (15.90) has several implications on the linear and nonlinear transport properties of the circuit, as discussed in Section 5.6.[5] Since there is a single current, the subscripts are all equal, i.e., $\alpha = \beta = \cdots = 1$, and can thus be dropped. In particular, the relation (5.179) expressing the fluctuation–dissipation theorem reduces to $L(\phi) = L(-\phi) = D(0, \phi)$. Since the (unique) third cumulant (5.83) is equal to zero in this circuit, i.e., $C(0, \phi) = 0$, equation (5.182) gives the relation $M(\phi) = 2\partial_A D(0, \phi)$. In addition, equation (5.186) becomes $B(0, \phi) = \partial_A C(0, \phi)$. The validity of these relations, which are implied by microreversibility, is numerically confirmed in Figure 15.6(a) (Barbier and Gaspard, 2020a).

Another example is shown in Figure 15.6(b), where several fourth cumulants (5.84) are plotted as a function of the magnetic flux for electronic transport in the Aharonov–Bohm ring with four terminals depicted as inset. In this circuit, there are three independent currents and affinities, so that the subscripts may take the values $\alpha, \beta, \cdots \in \{1, 2, 3\}$. Figure 15.6(b) confirms the validity of the time-reversal symmetry relations (5.186)–(5.187) giving the fourth cumulants $B_{\alpha\beta\gamma\delta}(0, \phi)$. Similar verifications have been performed for the Onsager–Casimir reciprocal relations (5.178), as well as for the relations (5.181)–(5.183) giving the third cumulants (5.83) (Barbier and Gaspard, 2020a).

The linear and nonlinear transport properties thus satisfy the time-reversal symmetry relations of Section 5.6.3 in the presence of an external magnetic field (Saito and Utsumi, 2008; Utsumi and Saito, 2009; Barbier and Gaspard, 2020a).

Nonlinear transport properties have been experimentally tested for an Aharonov–Bohm ring in an external magnetic field (Nakamura et al., 2010, 2011). In these experiments, the ring has a diameter of about 500 nm and is fabricated by local oxidation on a GaAs/AlGaAs heterostructure two-dimensional electron gas. The electric current is observed as a function of the applied voltage V and the external magnetic field $\boldsymbol{\mathcal{B}}$ for temperature T below one kelvin. The noise power is measured with a cross-correlation technique between the output signals from two sets of resonant circuit and amplifier in series. In general, the noise power is characterized by the spectral density function of the fluctuating electric current:

$$\mathcal{S}_{II}(\omega) \equiv \frac{1}{2} \int_{-\infty}^{+\infty} dt\, e^{\imath\omega t} \left\langle \Delta\hat{I}(0)\, \Delta\hat{I}(t) + \Delta\hat{I}(t)\, \Delta\hat{I}(0) \right\rangle_{\mathbf{A}} \tag{15.92}$$

with $\Delta\hat{I}(t) \equiv \hat{I}(t) - \langle\hat{I}\rangle_{\mathbf{A}}$ in the nonequilibrium stationary regime of affinities $\mathbf{A}$. At zero frequency, this spectral density is related to the current diffusivity $D_{NN}(\mathbf{A}, \boldsymbol{\mathcal{B}}) = -(1/2)\partial_{\lambda_N}^2 Q(\boldsymbol{\lambda} = \mathbf{0}; \mathbf{A}, \boldsymbol{\mathcal{B}})$ according to

$$\mathcal{S}_{II}(\omega = 0) = 2e^2\, D_{NN}. \tag{15.93}$$

[5] We note that the magnetic field is proportional to the magnetizing field according to $\boldsymbol{\mathcal{B}} = \mu\boldsymbol{\mathcal{H}}$ in a medium with magnetic permeability μ, so that they are simultaneously reversed.

The electric current $I = -eJ_N$ as well as the zero-frequency spectral density $\mathcal{S} \equiv \mathcal{S}_{II}(\omega = 0)$ can be expanded in powers of the applied voltage V as

$$I = G_1 V + \frac{1}{2!} G_2 V^2 + \frac{1}{3!} G_3 V^3 + \cdots, \tag{15.94}$$

$$\mathcal{S} = \mathcal{S}_0 + \mathcal{S}_1 V + \frac{1}{2!} \mathcal{S}_2 V^2 + \cdots, \tag{15.95}$$

in terms of linear and nonlinear response coefficients, which are related to those defined in equations (5.133) and (5.152) according to $G_n = e(\beta e)^n \partial_A^n J_N(A = 0)$ and $\mathcal{S}_n = 2e^2(\beta e)^n \partial_A^n D_{NN}(A = 0)$ with $A = A_N$ for $n = 0, 1, 2, \ldots$. In particular, the electric conductance is given by $G = G_1 = \beta e^2 L_{NN}$ in terms of the linear response coefficient $L_{NN} = \partial_A J_N(A = 0)$. In the linear regime close to equilibrium, we recover the Onsager–Casimir reciprocal relation and the Johnson–Nyquist relation,

$$G_1(\mathcal{B}) = G_1(-\mathcal{B}) \qquad \text{and} \qquad \mathcal{S}_0(\mathcal{B}) = 2\, k_{\mathrm{B}} T\, G_1(\mathcal{B}), \tag{15.96}$$

as a consequence of equations (5.178) and (5.179), respectively. At next order in the applied voltage, equation (5.184) gives

$$\mathcal{S}_1(\mathcal{B}) = k_{\mathrm{B}} T \left[2\, G_2(\mathcal{B}) - G_2(-\mathcal{B}) \right], \tag{15.97}$$

which is semiquantitatively validated in the experiments of Nakamura et al. (2010, 2011), bringing support to the nonequilibrium fluctuation relation for the current.

15.6 Transport of Bosons

15.6.1 Generalities

Bosons obey the Bose–Einstein statistics. Their distribution function is given by equation (15.60) with $\theta = +1$ in each reservoir, which diverges if the chemical potential μ_j belongs to the spectrum of particle energy ε. Therefore, the chemical potential of every reservoir should always be lower than the minimum energy of the spectrum: $\mu_j < \min\{\varepsilon\}$.

In the two-terminal circuit of Figure 15.4, this condition requires that $\mu_j < \varepsilon_{00} = \hbar\omega/2$ for $j = \mathrm{L}$ and $j = \mathrm{R}$. An example of Bose–Einstein distribution is shown in Figure 15.4(a). Contrary to the Fermi–Dirac distribution, the Bose–Einstein distribution is not step-like at low temperature so that, for bosons, the conductance does not manifest a phenomenon of quantization as for fermions. The mean particle current is again given by equation (15.65) but with Bose-Einstein distribution functions $f_j = \left[e^{\beta_j(\varepsilon - \mu_j)} - 1 \right]^{-1}$.

15.6.2 Full Counting Statistics and Microreversibility

For bosons, the full counting statistics of the particle and energy currents is characterized by the cumulant generating function (15.63) with $\theta = +1$, which has the symmetry (15.85) as a consequence of microreversibility. This cumulant generating function is plotted in Figure 15.7(a) as a function of the counting parameter $\lambda = \lambda_N$ for conditions where the thermal and chemical affinities take the values $A_E = 0$ and $A_N = 0.5$ and for different

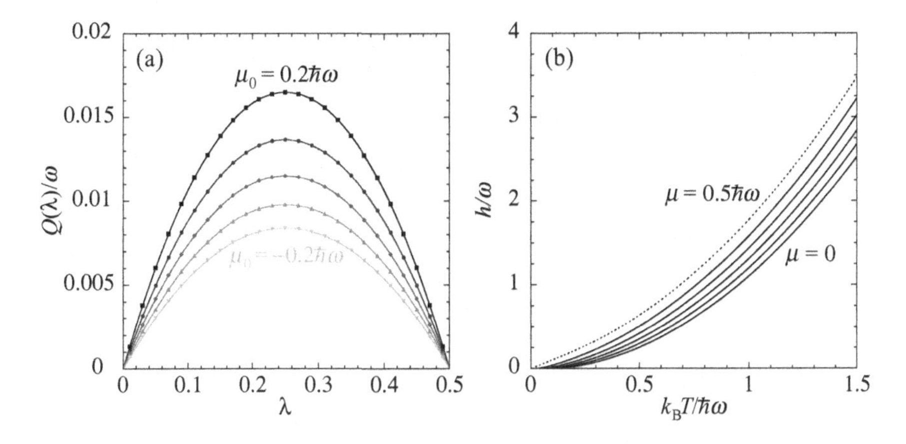

Figure 15.7 Transport of bosons in the two-terminal circuit of Figure 15.4: (a) Cumulant generating function (15.63) with $A_E = 0$ and $\lambda_E = 0$ versus the counting parameter $\lambda_N = \lambda$ for bosons crossing the inverted harmonic barrier of parameters $u_0 = \hbar\omega/2$ and $\gamma/\omega = 0.25$. The rescaled inverse temperature is $\beta\hbar\omega = 1$. The chemical potentials of the left- and right-hand reservoirs are taken as $\mu_{\mathrm{L,R}} = \mu_0 \pm \hbar\omega/4$ with $\mu_0/(\hbar\omega) = -0.2, -0.1, 0, 0.1, 0.2$. (b) The rescaled CNT entropy per unit time h/ω given by equation (15.99) versus the rescaled temperature $k_{\mathrm{B}}T/(\hbar\omega)$ for a bosonic ideal gas with $g_s = 1$ at equilibrium in the same circuit for the following values of the rescaled chemical potential: $\mu/(\hbar\omega) = 0, 0.1, 0.2, 0.3, 0.4$ (solid lines) and $\mu/(\hbar\omega) = 0.5$ (dashed line). Adapted from Gaspard (2015c).

values of the mean chemical potential $\mu_0 = (\mu_{\mathrm{L}} + \mu_{\mathrm{R}})/2$ between the two reservoirs. First of all, the plot confirms the symmetry $\lambda_N \to A_N - \lambda_N$ implied by microreversibility. Moreover, the generating function increases with μ_0 because the energy levels are more fully occupied as the mean chemical potential μ_0 becomes larger, so that the slope at $\lambda_N = \lambda = 0$ also increases and, thus, the mean current. For $\mu_0 = 0.2\hbar\omega$, the chemical potential of the left-hand reservoir $\mu_{\mathrm{L}} = \mu_0 + 0.25\hbar\omega$ is close to the minimum of the energy spectrum $\varepsilon_{00} = 0.5\hbar\omega$, but without reaching it as required by the aforementioned condition.

15.6.3 Temporal Disorder

For the Bose–Einstein statistics (Pathria, 1972; Huang, 1987), the number of possible histories followed by N_α bosons occupying M_α possible microstates increases with time t as

$$\prod_\alpha \frac{(N_\alpha + M_\alpha - 1)!}{N_\alpha!\,(M_\alpha - 1)!} \sim \exp(h\,t), \tag{15.98}$$

so that equation (15.78) with $\theta = +1$ gives the CNT entropy per unit time h characterizing temporal disorder in the quasi-free transport of bosons.

In the two-terminal circuit at equilibrium with vanishing thermal and chemical affinities (15.64), the CNT entropy per unit time (15.78) is equal to

$$h = 2\, g_s \sum_{n=0}^{\infty} \int_{\varepsilon_{0n}}^{\infty} \frac{d\varepsilon}{2\pi\hbar} [-f \ln f + (1+f)\ln(1+f)], \tag{15.99}$$

with the Bose–Einstein distribution function $f = f_{\rm L} = f_{\rm R}$ and $\varepsilon_{0n} = \hbar\omega(n+1/2)$. This quantity characterizing temporal disorder is plotted in Figure 15.7(b) as a function of temperature. Here also, the temporal disorder is equal to zero at a temperature of absolute zero and increases with the temperature, as well as with the chemical potential.

Out of equilibrium, the thermodynamic entropy production (15.82) is positive, so that the time-reversed coentropy per unit time (15.80) becomes larger than the CNT entropy: $h^{\rm R} > h$.

15.7 Transport in the Classical Limit

The transport of electrons or atoms through a constriction at low temperature is the quantum analogue of the effusion of a classical ideal gas through a small hole in a wall separating two reservoirs, which has been studied by Knudsen (1909) and others (Present, 1958; Pathria, 1972). As explained in Chapter 8, the full counting statistics of the energy and particle fluxes in this classical kinetic process is characterized by a cumulant generating function satisfying the same symmetry relation (8.44) as their quantum analogues. Actually, the relationship can be fully established in the classical limit where the de Broglie wavelength of the particles is much shorter than the characteristic sizes of the circuit, which is formally expressed as $\hbar \to 0$.

15.7.1 Full Counting Statistics and Microreversibility

We consider the ballistic motion of particles through a cylindrical pipe between two reservoirs, as illustrated in Figure 15.1(a). The perpendicular section can have any shape. The transverse classical motion can even be chaotic. In the classical limit, the mean number of transverse modes is given by

$$\mathcal{N}_{\rm av}(\varepsilon) \simeq \int \frac{d\mathbf{r}_\perp d\mathbf{p}_\perp}{(2\pi\hbar)^2}\, \theta\left(\varepsilon - h^0_{\perp{\rm cl}}\right), \tag{15.100}$$

where $h^0_{\perp{\rm cl}}$ is the classical one-body Hamiltonian ruling the transverse motion and $\theta(\varepsilon-\varepsilon')$ is the Heaviside step function, while the channels all open at the minimum of the Hamiltonian function $h^0_{\perp{\rm cl}}$. For particles with mass m undergoing elastic collisions on the surface of the pipe, the Hamiltonian function would be $h^0_{\perp{\rm cl}} = \mathbf{p}_\perp^2/(2m)$, giving

$$\mathcal{N}_{\rm av}(\varepsilon) \simeq \frac{m}{2\pi\hbar^2}\, \varepsilon\, \sigma, \tag{15.101}$$

where σ is the cross-sectional area of the pipe. Moreover, the transmission probability through an open channel reaches the unit value in the classical limit, $T_n(\varepsilon) \simeq 1$.

In addition, for classical ideal gases at the temperature T_j and chemical potential μ_j, the fugacity is given by

$$e^{\beta_j \mu_j} = \frac{1}{g_s} \left(\frac{2\pi \hbar^2}{m k_B T_j} \right)^{3/2} n_j \tag{15.102}$$

in terms of the particle density n_j (Huang, 1987). In the classical limit, the fugacities thus become much smaller than one, i.e., $e^{\beta_j \mu_j} \ll 1$, so that the Bose–Einstein and Fermi–Dirac distribution functions can both be approximated by Boltzmann's distribution

$$f_j \simeq e^{\beta_j \mu_j} \, e^{-\beta_j \varepsilon}, \tag{15.103}$$

and $1+\theta f_j \simeq 1$ in the same approximation. Thus, the cumulant generating function (15.63) precisely takes the form (8.42) with the rate densities (8.17) for the effusion of a classical gas through a small hole of area σ (Gaspard, 2013a,e, 2015c).

Since the ratio of the rate densities (8.17) is given by equation (8.18), the cumulant generating function satisfies the symmetry relation (8.44), which is exactly the same as its quantum version (15.85). The correspondence is thus established in the classical limit.

15.7.2 Temporal Disorder

A similar result holds for temporal disorder in the transport of bosons and fermions. In the two-terminal circuit, the CNT entropy per unit time (15.78) for particle transport between the two reservoirs is expressed as

$$h = g_s \sum_{j=\mathrm{L,R}} \sum_{n=0}^{\infty} \int_{\varepsilon_{0n}}^{\infty} \frac{d\varepsilon}{2\pi\hbar} \big[-f_j \ln f_j + (f_j + \theta) \ln(1 + \theta f_j) \big]. \tag{15.104}$$

In the classical limit, the Bose–Einstein and Fermi–Dirac distribution functions are approximated by the Boltzmann distribution (15.103) with the fugacity (15.102) or, equivalently, as

$$f_j(\varepsilon) \simeq \frac{(2\pi\hbar)^3}{g_s} \mathcal{F}_j(\mathbf{p}), \qquad \text{where} \qquad \mathcal{F}_j(\mathbf{p}) = \frac{n_j}{(2\pi m k_B T_j)^{3/2}} \exp\left(-\frac{\mathbf{p}^2}{2 m k_B T_j} \right) \tag{15.105}$$

is the Maxwell distribution function, because $\varepsilon = \mathbf{p}^2/(2m)$. Since $e^{\beta_j \mu_j} \ll 1$, we have that $(f_j + \theta) \ln(1 + \theta f_j) = f_j + O(f_j^2)$. Using the mean number of open channels given by equation (15.101), we obtain

$$h = \sigma \sum_{j=\mathrm{L,R}} \int_{p_x > 0} d^3 p \left| \frac{p_x}{m} \right| \mathcal{F}_j(\mathbf{p}) \ln \frac{g_s \, e}{(2\pi\hbar)^3 \mathcal{F}_j(\mathbf{p})}, \tag{15.106}$$

in terms of the Maxwell distribution functions (15.105) for the left- and right-hand reservoirs whether $p_x > 0$ or $p_x < 0$, respectively.

For spinless particles with $g_s = 1$ and the cross-sectional area $\Sigma = \sigma$, this expression is the same as the entropy per unit time (8.51) of the classical effusion process but with $\Delta^3 r \Delta^3 p = (2\pi\hbar)^3$. Therefore, the constant of the entropy per unit time is determined by quantum mechanics in the same way as for the thermodynamic entropy per unit volume with the Sackur–Tetrode formula (2.115). The entropy per unit time thus finds an absolute value in quantum mechanics for the characterization of temporal disorder (Gaspard, 2015c).[6]

The classical correspondence can be similarly established for the time-reversed coentropy per unit time and its difference with respect to the entropy per unit time gives the thermodynamic entropy production as in equation (8.55). The relationship is thus complete between the classical effusion process and their quantum analogues in the quasi-free transport of bosons and fermions.

15.8 Stochastic Approach to Electron Transport in Mesoscopic Devices

15.8.1 General Formulation

The quantization of electron energy can be achieved in so-called quantum dots, which are artificial atoms fabricated with GaAs/AlGaAs semiconductor heterostructures (Ferry et al., 2009; Nazarov and Blanter, 2009). They are a few hundred nanometers in size and they are weakly coupled by tunneling to nanowires, allowing electron transport from and to reservoirs. In these mesoscopic devices, the mean free path due to impurities and crystal vibrations can be significantly larger than the size of the quantum dots, so that electron transport manifests quantum-mechanical coherence at low temperature. In particular, electrons may occupy quantized energy levels in quantum dots according to Pauli's exclusion principle. Such energy levels are coupled to the dense energy spectra of the reservoirs by tunneling, so that they form resonant levels characterized with an energy width and a lifetime. If tunneling is weak enough, the rates of transitions from and to the reservoirs can be evaluated using Fermi's golden rule at the second order of perturbation theory. If $\hat{H}_0$ denotes the unperturbed Hamiltonian and $\hat{U}$ the perturbation, the total Hamiltonian $\hat{H} = \hat{H}_0 + \hat{U}$ generates transitions $i \to f$ between the eigenstates $\hat{H}_0|i\rangle = E_i|i\rangle$ and $\hat{H}_0|f\rangle = E_f|f\rangle$ with the following rate,

$$w_{i\to f} \simeq \frac{2\pi}{\hbar} \left|\langle f|\hat{U}|i\rangle\right|^2 \delta(E_f - E_i), \tag{15.107}$$

neglecting higher-order corrections in the perturbation $\hat{U}$. For transitions to many final states, this rate should be integrated with their level density.

In this way, electron transport in mesoscopic devices can be described as Markov stochastic processes ruled by a master equation such as

$$\frac{d}{dt}\mathbf{P}(\mathbf{Z},t) = \hat{\mathsf{L}} \cdot \mathbf{P}(\mathbf{Z},t) \qquad \text{with} \qquad \hat{\mathsf{L}} = \sum_{\nu} \mathsf{L}^{(\nu)}\, \mathrm{e}^{-\boldsymbol{\nu}\cdot\partial_{\mathbf{Z}}} \tag{15.108}$$

for the probability distribution $\mathbf{P}(\mathbf{Z},t) = \{P_\sigma(\mathbf{Z},t)\}$ to find the quantum dots in the different possible coarse-grained states $\{\sigma\}$ after the transfers of $\mathbf{Z}$ electrons between the reservoirs during the time interval $[0,t]$.[7] The rates of the transitions yielding the transfers

[6] We note that replacing the Fermi–Dirac or Bose–Einstein statistics by the Boltzmann statistics in equation (15.86) or (15.98) yields $\prod_\alpha \frac{M_\alpha^{N_\alpha}}{N_\alpha!} \sim \exp(ht)$, which gives the entropy per unit time (8.50) of a classical ideal gas, as considered in Section 8.9.

[7] In the framework of quantum mechanics, these probabilities can be defined as $P_\sigma(\mathbf{Z},t) = \mathrm{tr}\,\hat{\rho}(t)\,\hat{\mathsf{P}}_{\sigma\mathbf{N}}$, where $\hat{\mathsf{P}}_{\sigma\mathbf{N}}$ is the projection operator into the microstate σ for the quantum dots and the electron numbers $\mathbf{N}$ for the reservoirs other than the

$\mathbf{Z} \to \mathbf{Z} + \boldsymbol{\nu}$ are composing the matrices $\mathsf{L}^{(\boldsymbol{\nu})}$. The normalization condition $\sum_{\sigma,\mathbf{Z}} P_\sigma(\mathbf{Z},t) = 1$ is preserved during time evolution.

If the system is isothermal, the transitions of the process are those corresponding to electron transfers between the reservoirs, so that the system is driven out of equilibrium by so many chemical affinities $\mathbf{A} = \{A_\alpha\}_{\alpha=1}^{\chi}$ as possible independent electron transfers $\mathbf{Z} = \{Z_\alpha\}_{\alpha=1}^{\chi}$. If the mesoscopic device is composed of one circuit allowing electron transport between $1 \leq j \leq r$ reservoirs, the reservoir $j = r$ can be chosen as the reference reservoir, so that the independent affinities are given by $A_j = \beta(\mu_j - \mu_r)$ and their number is equal to $\chi = r - 1$. However, the mesoscopic device may include several circuits without the possibility of electron transfers between them, as is the case if the occupancy of some quantum dots is monitored by the capacitive coupling to parallel circuits. Capacitive coupling between two circuits is induced by Coulomb interaction between the electric charges inside these circuits, although there is no electron transport between them. In such devices, the circuits are disconnected for electron transport, so that the electron number is conserved in every circuit. If each circuit $1 \leq i \leq c$ is composed of r_i reservoirs, the number of independent chemical affinities is equal to $\chi = \sum_{i=1}^{c}(r_i - 1)$.

The full counting statistics of electron transport can be characterized by the cumulant generating function

$$Q(\boldsymbol{\lambda};\mathbf{A}) \equiv \lim_{t\to\infty} -\frac{1}{t} \ln \left\langle e^{-\boldsymbol{\lambda}\cdot\mathbf{Z}(t)} \right\rangle_{\mathbf{A}}, \tag{15.109}$$

where $\langle\cdot\rangle_{\mathbf{A}}$ denotes the statistical average with respect to the probability distribution $\mathbf{P}(\mathbf{Z},t)$ in the nonequilibrium regime of chemical affinities $\mathbf{A}$. The cumulant generating function (15.109) can be obtained in terms of the leading eigenvalue of the matrix

$$\mathsf{L}_{\boldsymbol{\lambda}} \equiv e^{-\boldsymbol{\lambda}\cdot\mathbf{Z}}\,\hat{\mathsf{L}}\,e^{+\boldsymbol{\lambda}\cdot\mathbf{Z}} = \sum_{\boldsymbol{\nu}} \mathsf{L}^{(\boldsymbol{\nu})}\,e^{-\boldsymbol{\nu}\cdot\boldsymbol{\lambda}}, \tag{15.110}$$

including the counting parameters $\boldsymbol{\lambda} = \{\lambda_\alpha\}_{\alpha=1}^{\chi}$.

As explained in Chapter 5, microreversibility implies that the matrix (15.110) satisfies the following symmetry,

$$\mathsf{M}^{-1} \cdot \mathsf{L}_{\boldsymbol{\lambda}} \cdot \mathsf{M} = \mathsf{L}_{\mathbf{A}-\boldsymbol{\lambda}}^{\mathrm{T}} \tag{15.111}$$

with respect to the diagonal matrix M composed of the Boltzmann factors associated with the reference reservoirs. As a consequence, the cumulant generating function obeys the symmetry relation

$$Q(\boldsymbol{\lambda};\mathbf{A}) = Q(\mathbf{A}-\boldsymbol{\lambda};\mathbf{A}), \tag{15.112}$$

so that the linear and nonlinear response properties of Section 5.5 can be deduced for electron transport in mesoscopic devices. The mean electron and electric currents can thus be expressed as

reference reservoirs, and $\hat{\rho}(t) = e^{-\imath\hat{H}t}\hat{\rho}(0)\,e^{\imath\hat{H}t}$ is the statistical operator at time t evolved from the initial canonical operator $\hat{\rho}(0)$ when those reservoirs had the electron numbers $\mathbf{N}_0$, such that $\mathbf{Z} = \mathbf{N}_0 - \mathbf{N}$.

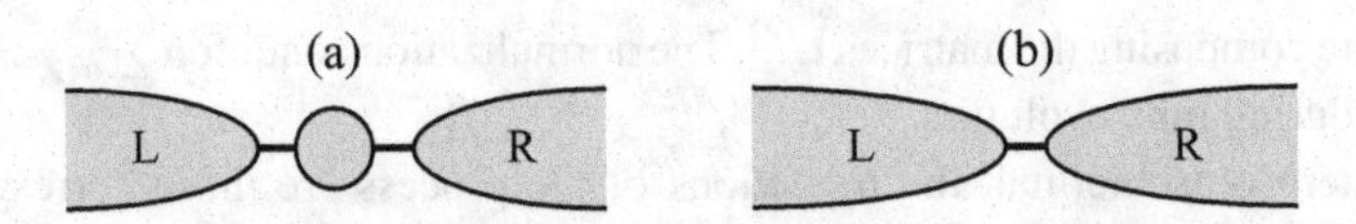

Figure 15.8 (a) Quantum dot and (b) quantum point contact driven by two reservoirs.

$$\mathbf{J}(\mathbf{A}) = \frac{\partial Q}{\partial \boldsymbol{\lambda}}(\mathbf{0};\mathbf{A}) \qquad \text{and} \qquad \mathbf{I}(\mathbf{V}) = -e\,\mathbf{J}(\mathbf{A}) \qquad \text{with} \qquad \mathbf{A} = -\beta e \mathbf{V} \tag{15.113}$$

in terms of the applied voltages $\mathbf{V} = \{V_\alpha\}_{\alpha=1}^{\chi}$. Accordingly, the entropy production rate is given by

$$\frac{1}{k_{\mathrm{B}}}\frac{d_{\mathrm{i}}S}{dt} = \mathbf{A}\cdot\mathbf{J}(\mathbf{A}) = \beta\,\mathbf{V}\cdot\mathbf{I}(\mathbf{V}) = \beta\,P \geq 0, \tag{15.114}$$

where P is the power dissipated by all the electric currents and $\beta = (k_{\mathrm{B}}T)^{-1}$.

15.8.2 Quantum Dot

First, we consider a quantum dot in contact with two reservoirs $j = \mathrm{L}, \mathrm{R}$ at the chemical potentials μ_j, as schematically represented in Figure 15.8(a). At the microscopic level of description, electron transport in this device is ruled by the Hamiltonian operator $\hat{H} = \hat{H}_{\mathrm{L}} + \hat{T}_{\mathrm{LD}} + \hat{H}_{\mathrm{D}} + \hat{T}_{\mathrm{RD}} + \hat{H}_{\mathrm{R}}$, where $\hat{H}_j$ are the reservoir Hamiltonian operators, $\hat{H}_{\mathrm{D}} = \sum_n \varepsilon_n \hat{a}_n^\dagger \hat{a}_n$ is the Hamiltonian of the quantum dot, and the operators $\hat{T}_{\mathrm{LD}}$ and $\hat{T}_{\mathrm{RD}}$ are governing electron tunneling between the reservoirs and the quantum dot. This tunneling is assumed to be small enough to have a perturbative effect with respect to the Hamiltonian operator $\hat{H}_0 = \hat{H}_{\mathrm{L}} + \hat{H}_{\mathrm{D}} + \hat{H}_{\mathrm{R}}$, describing uncoupled reservoirs and the quantum dot.[8] The circuit is driven away from equilibrium by the unique chemical affinity $A = \beta(\mu_{\mathrm{L}} - \mu_{\mathrm{R}})$.

The charging and discharging rates of the energy levels ε_n can be evaluated using Fermi's golden rule at the second order of perturbation theory as

$$a_{jn} = \Gamma_{jn}\, f_j(\varepsilon_n) \qquad \text{and} \qquad b_{jn} = \Gamma_{jn}\,[1 - f_j(\varepsilon_n)], \tag{15.115}$$

where $f_j(\varepsilon) = \left[e^{\beta_j(\varepsilon-\mu_j)} + 1\right]^{-1}$ is the Fermi–Dirac distribution function of the jth reservoir at the level energy $\varepsilon = \varepsilon_n$ and Γ_{jn} is a coefficient, which is quadratic in the tunneling amplitudes $\hat{T}_{j\mathrm{D}}$ according to Fermi's golden rule (15.107). The charging rate is proportional to the Fermi–Dirac distribution because the transition is possible if the energy levels of the reservoir are occupied at the energy ε_n of the empty level in the quantum dot. Instead, the discharging rate goes as $(1 - f_j)$ since the transition requires that the reservoir energy levels are unoccupied for tunneling out of the quantum dot to be possible.

The quantum dot is supposed to have a single energy level ε_n, which may be occupied or not by a single electron. In this circuit, electrons are thus transported one by one between the reservoirs and the quantum dot. The transition matrix (15.110) including the parameter

[8] The capacitive coupling between the quantum dot and the reservoirs is here neglected for simplicity. These effects are considered in Section 15.8.5.

λ counting the electron transfers from the reservoir L to the quantum dot has the following form,

$$\mathbf{L}_\lambda = \begin{pmatrix} -a_{\rm L} - a_{\rm R} & b_{\rm L}{\rm e}^{+\lambda} + b_{\rm R} \\ a_{\rm L}{\rm e}^{-\lambda} + a_{\rm R} & -b_{\rm L} - b_{\rm R} \end{pmatrix}, \tag{15.116}$$

where the first and second lines and columns are associated with the empty or occupied resonant level, respectively. Therefore, the cumulant generating function is given by

$$Q(\lambda; A) = \frac{1}{2}\,(a_{\rm L} + a_{\rm R} + b_{\rm L} + b_{\rm R}) - \left[\frac{1}{4}\,(a_{\rm L} + a_{\rm R} + b_{\rm L} + b_{\rm R})^2 + a_{\rm L}\, b_{\rm R}\left({\rm e}^{-\lambda} - 1\right) + a_{\rm R}\, b_{\rm L}\left({\rm e}^{+\lambda} - 1\right)\right]^{1/2}, \tag{15.117}$$

which satisfies the symmetry relation $Q(\lambda; A) = Q(A - \lambda; A)$ with the affinity

$$A \equiv \ln \frac{a_{\rm L} b_{\rm R}}{a_{\rm R} b_{\rm L}} = \beta\,(\mu_{\rm L} - \mu_{\rm R}), \tag{15.118}$$

as can be verified using the expressions (15.115) for the charging and discharging rates and the identity (15.69) with $\theta = -1$ for the Fermi–Dirac distributions of the reservoirs.

15.8.3 Quantum Point Contact

A quantum point contact is a tunnel junction between two reservoirs, as schematically represented in Figure 15.8(b). The Hamiltonian of this circuit can be expressed as $\hat{H} = \hat{H}_{\rm L} + \hat{T}_{\rm LR} + \hat{H}_{\rm R}$ in terms of the reservoir Hamiltonian operators $\hat{H}_{\rm L}$ and $\hat{H}_{\rm R}$ and the tunneling operator $\hat{T}_{\rm LR}$, which couples the reservoirs and generates transport.

Here also, the transition rate for electron transfers between the reservoirs can be evaluated using Fermi's golden rule (15.107). If the electron energy undergoes the transition $\varepsilon \to \varepsilon - \Delta\varepsilon$ and the tunnel junction has the electric conductance G, the rate is given by

$$W(\Delta\varepsilon) = \frac{G}{e^2} \int d\varepsilon\, f(\varepsilon)\,[1 - f(\varepsilon - \Delta\varepsilon)] = \frac{G}{e^2}\frac{\Delta\varepsilon}{{\rm e}^{\beta\Delta\varepsilon} - 1}, \tag{15.119}$$

where $f(\varepsilon) = \left({\rm e}^{\beta\varepsilon} + 1\right)^{-1}$, since the initial level should be occupied and the final one unoccupied. This transition rate satisfies the condition $W(\Delta\varepsilon) = {\rm e}^{-\beta\Delta\varepsilon}\, W(-\Delta\varepsilon)$, as for the similar function (10.18). We note that such transition rates have a bosonic dependence on the energy change $\Delta\varepsilon$.

Accordingly, the probability distribution $P(Z,t)$ for the transfer of Z electrons in the direction L $\to$ R during the time interval $[0,t]$ is ruled by the following master equation,

$$\frac{d}{dt} P(Z,t) = \left[W^{(+)}\left({\rm e}^{-\partial_Z} - 1\right) + W^{(-)}\left({\rm e}^{+\partial_Z} - 1\right)\right] P(Z,t) \tag{15.120}$$

as in equation (4.78) with the transition rates

$$W^{(\pm)} = \frac{G}{e^2}\frac{\mp\Delta\mu}{{\rm e}^{\mp\beta\Delta\mu} - 1} \qquad \text{and} \qquad \Delta\mu = \mu_{\rm L} - \mu_{\rm R}, \tag{15.121}$$

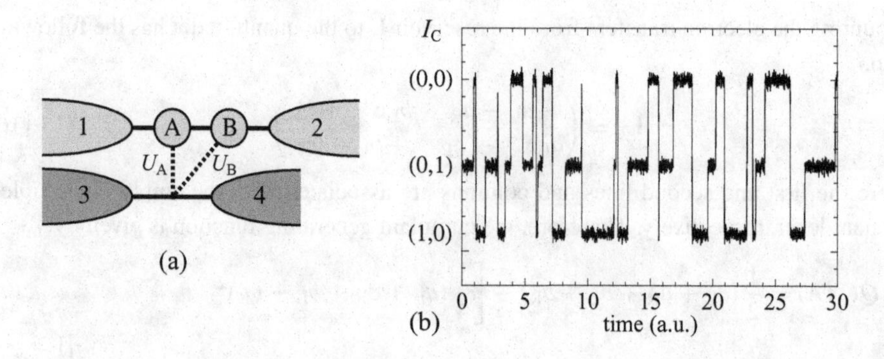

Figure 15.9 (a) Schematic representation of a double quantum dot capacitively coupled to a quantum point contact. (b) The current in the quantum point contact versus time.

since the energy change is $\Delta\varepsilon = \mp\Delta\mu$ in the respective transitions. Here, the fluctuation relation (5.243) with $n = Z$ is satisfied for the affinity $A = \ln(W^{(+)}/W^{(-)}) = \beta\Delta\mu$.

15.8.4 Double Quantum Dot with Quantum Point Contact

The occupancy of a quantum dot can be observed in real time by its capacitive coupling to a quantum point contact. In this case, the two circuits of Figure 15.8 are coupled in parallel by the Coulomb interaction between them. If the quantum dot is occupied by an electron, the Coulomb repulsion exerted on nearby electrons raises the energy barrier in the quantum point contact and, thus, lowers the corresponding electron current. In such mesoscopic devices, the statistics can be carried out for the quantum dot occupancy (Gustavsson et al., 2006).

However, the occupancy of the quantum dot does not contain information on the direction taken by the electrons moving between the reservoirs. In order to perform the bidirectional counting of electron transfers, a mesoscopic device with a double quantum dot (DQD) and an asymmetric capacitive coupling to a quantum point contact (QPC) can be considered (Fujisawa et al., 2006), as illustrated in Figure 15.9(a). In such a device, the current in the QPC is modulated by the different possible occupancies of the DQD. If the quantum dot closer to the QPC is occupied, the Coulomb repulsion is higher and the current lower, than if the other quantum dot is occupied. Accordingly, the observation of the current in the QPC can determine the direction of the jumps between the two quantum dots and, thus, the direction of electron transfers between the reservoirs connected to the DQD. The numerical simulation of a random time evolution of the QPC current is shown in Figure 15.9(b) for conditions where the DQD may be in the three following coarse-grained states: $(0,0)$ if both A and B are empty, $(1,0)$ if A is occupied, and $(0,1)$ if B is occupied. If there is a mean current $1 \to 2$ in the DQD, the most probable transitions should thus be $(0,0) \to (1,0) \to (0,1) \to \dots$, as seen in Figure 15.9(b).[9]

[9] Depending on the applied voltages, the transitions may also take place between the states $(1,0)$, $(0,1)$, and $(1,1)$.

In the experiment by Fujisawa et al. (2006), the occupancy of the DQD modulates the QPC current by about 15%. Moreover, the applied voltages, the currents, the powers, and the affinities in the DQD and the QPC are, respectively, given by

$$\text{double quantum dot:} \tag{15.122}$$
$$V_{\rm D} \simeq 0.3\ \text{mV}, \quad I_{\rm D} \simeq 6.5 \times 10^{-17}\ \text{A}, \quad P_{\rm D} \simeq 2.0 \times 10^{-20}\ \text{W}, \quad A_{\rm D} \simeq 27,$$
$$\text{quantum point contact:} \tag{15.123}$$
$$V_{\rm C} \simeq 0.8\ \text{mV}, \ I_{\rm C} \simeq 1.2 \times 10^{-8}\ \text{A}, \ P_{\rm C} \simeq 9.6 \times 10^{-12}\ \text{W}, \ A_{\rm C} \simeq 71,$$

with the reported electronic temperature $T = 130$ mK. Therefore, the current in the QPC is much higher than in the DQD. Indeed, their current ratio takes the value $I_{\rm C}/I_{\rm D} \simeq 1.8 \times 10^8$ and the entropy production rate (15.114) of the process is largely dominated by the contribution from the QPC. The huge thermodynamic cost of measurement can be evaluated by the ratio of the respective entropy production rates or dissipated powers: $(d_{\rm i}S_{\rm C}/dt)/(d_{\rm i}S_{\rm D}/dt) = P_{\rm C}/P_{\rm D} \simeq 4.8 \times 10^8$. The backaction of the QPC current onto the DQD is thus expected to be significant.

In order to model electron transport in the mesoscopic circuit, the following Hamiltonian may be considered,

$$\hat{H} = \hat{H}_{\rm AB} + \sum_{j=1}^{4} \hat{H}_j + \hat{T}_{\rm 1A} + \hat{T}_{\rm 2B} + \hat{T}_{34} + \hat{U}_{\rm ABC} + \hat{H}_{\rm vibr} + \hat{H}_{\rm em}, \tag{15.124}$$

where $\hat{H}_{\rm AB}$ is the Hamiltonian of the DQD, $\hat{H}_j$ are the reservoir Hamiltonians, $\hat{T}_{\rm 1A}$ and $\hat{T}_{\rm 2B}$ are the tunneling operators between the DQD and the reservoirs $j = 1$ and $j = 2$, $\hat{T}_{34}$ is the tunneling operator of the QPC between the reservoirs $j = 3$ and $j = 4$, $\hat{U}_{\rm ABC}$ is the Coulomb interaction between the DQD and the QPC, $\hat{H}_{\rm vibr}$ describes the effects of crystal vibration, and $\hat{H}_{\rm em}$ those of dielectric polarization. If there are two energy levels $\varepsilon_{\rm A}$ and $\varepsilon_{\rm B}$ in the DQD, its Hamiltonian operator can be expressed as $\hat{H}_{\rm AB} = \varepsilon_{\rm A}\,\hat{a}_{\rm A}^{\dagger}\hat{a}_{\rm A} + \varepsilon_{\rm B}\,\hat{a}_{\rm B}^{\dagger}\hat{a}_{\rm B} + t_{\rm AB}\big(\hat{a}_{\rm A}^{\dagger}\hat{a}_{\rm B} + \hat{a}_{\rm B}^{\dagger}\hat{a}_{\rm A}\big)$ in terms of the anticommuting annihilation-creation operators $\hat{a}_{\upsilon}$ and $\hat{a}_{\upsilon}^{\dagger}$, and the tunneling amplitude $t_{\rm AB}$ between the two dots. After diagonalization, the energy levels of the DQD are thus given by

$$\varepsilon_{\pm} = \frac{1}{2}\left[\varepsilon_{\rm A} + \varepsilon_{\rm B} \pm \sqrt{(\varepsilon_{\rm A} - \varepsilon_{\rm B})^2 + 4\,t_{\rm AB}^2}\right]. \tag{15.125}$$

The operators $\hat{H}_j$, $\hat{T}_{\rm 1A}$, $\hat{T}_{\rm 2B}$, and $\hat{T}_{34}$ can also be expressed as quadratic forms of the annihilation-creation operators. However, the Coulomb interaction is described by the quartic operator $\hat{U}_{\rm ABC} = \big(U_{\rm A}\,\hat{a}_{\rm A}^{\dagger}\hat{a}_{\rm A} + U_{\rm B}\,\hat{a}_{\rm B}^{\dagger}\hat{a}_{\rm B}\big)\,\hat{V}_{34}$, where $U_{\rm A}$ and $U_{\rm B}$ are the coupling strengths of the Coulomb interaction and $\hat{V}_{34}$ is a quadratic form of the annihilation-creation operators for the reservoirs $j = 3$ and $j = 4$ (Bulnes Cuetara et al., 2011, 2013). The operator $\hat{U}_{\rm ABC}$ rules the capacitive coupling between the DQD and QPC circuits, so that there is no transport of electrons between them. The effects of the tunneling and Coulomb interaction operators are taken into account with Fermi's golden rule at the second order of perturbation theory.

Neglecting the fluctuations due to crystal vibration and dielectric polarization, the charging and discharging rates of the energy levels $\varepsilon_\pm$ of the DQD are given by equation (15.115) with $j = 1,2$ and $\varepsilon_n = \varepsilon_\pm$. The double occupancy is assumed to be negligible because of Coulomb repulsion.

Tunneling in the QPC between the reservoirs $j = 3,4$ causes electron transfers $Z_C \to Z_C \pm 1$ and their rates are given according to equation (15.121) by

$$d_\sigma^{(\pm)} = \frac{G_\sigma}{e^2} \frac{\mp\Delta\mu_C}{e^{\mp\beta\Delta\mu_C} - 1} \qquad \text{with} \qquad \Delta\mu_C \equiv \mu_3 - \mu_4, \tag{15.126}$$

where G_σ is the electric conductance of the QPC if the DQD is found in the microstate σ.

Furthermore, the electron transfers $Z_C \to Z_C \pm 1$ between the reservoirs $j = 3,4$ of the QPC circuit may also induce transitions between the microstates $|+\rangle$ and $|-\rangle$ of the DQD with the following rates,

$$c_{\sigma\sigma'}^{(\pm)} = \Lambda \frac{\omega_{\sigma\sigma'} \mp \Delta\mu_C}{e^{\beta(\omega_{\sigma\sigma'} \mp \Delta\mu_C)} - 1} \qquad \text{with} \qquad \omega_{\sigma\sigma'} \equiv \varepsilon_\sigma - \varepsilon_{\sigma'} \tag{15.127}$$

and a coefficient $\Lambda \propto (U_A - U_B)^2$ due to the asymmetric Coulomb interaction between the QPC and the quantum dots A and B (Bulnes Cuetara et al., 2013). These transitions contribute to the backaction of the QPC onto the DQD.

Accordingly, the probability distribution $\mathbf{P}(\mathbf{Z},t) = \{P_\sigma(Z_D, Z_C, t)\}$ of finding the device in the microstate $\sigma \in \{0, +, -\}$ after the transfers of Z_D and Z_C electrons respectively in the DQD and the QPC is ruled by the master equation (15.108) with the linear operator

$$\hat{\mathbf{L}} = \hat{\mathbf{L}}_D + \hat{\mathbf{L}}_C, \tag{15.128}$$

where the transitions in the DQD are governed by

$$\hat{\mathbf{L}}_D = \begin{pmatrix} -a_{1+} - a_{2+} - a_{1-} - a_{2-} & b_{1+}\,\hat{E}_D^{+1} + b_{2+} & b_{1-}\,\hat{E}_D^{+1} + b_{2-} \\ a_{1+}\,\hat{E}_D^{-1} + a_{2+} & -b_{1+} - b_{2+} & 0 \\ a_{1-}\,\hat{E}_D^{-1} + a_{2-} & 0 & -b_{1-} - b_{2-} \end{pmatrix}, \tag{15.129}$$

and those due to the QPC by

$$\hat{\mathbf{L}}_C = \begin{pmatrix} \hat{L}_{0C} & 0 & 0 \\ 0 & \hat{L}_{+C} - c_{-+}^{(+)} - c_{-+}^{(-)} & c_{+-}^{(+)}\,\hat{E}_C^{-1} + c_{+-}^{(-)}\,\hat{E}_C^{+1} \\ 0 & c_{-+}^{(+)}\,\hat{E}_C^{-1} + c_{-+}^{(-)}\,\hat{E}_C^{+1} & \hat{L}_{-C} - c_{+-}^{(+)} - c_{+-}^{(-)} \end{pmatrix} \tag{15.130}$$

expressed in terms of

$$\hat{L}_{\sigma C} = d_\sigma^{(+)}(\hat{E}_C^{-1} - 1) + d_\sigma^{(-)}(\hat{E}_C^{+1} - 1) \tag{15.131}$$

and the raising and lowering operators defined as $\hat{E}_i^{\pm 1} \equiv \exp(\pm\partial_{Z_i})$ for $i = \mathrm{D}, \mathrm{C}$. We note that further transitions may be generated by crystal vibration and dielectric fluctuations (Utsumi et al., 2010; Golubev et al., 2011).

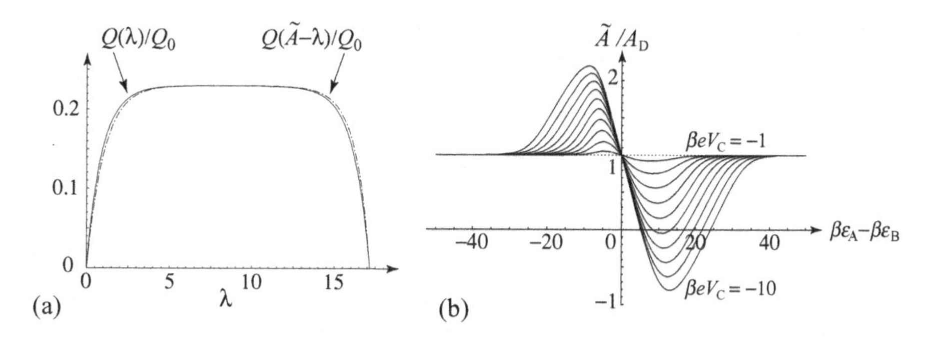

Figure 15.10 (a) The cumulant generating function $Q(\lambda) = Q(\lambda_D, 0; A_D, A_C)$ of the DQD current in the circuit of Figure 15.9(a) versus the counting parameter $\lambda = \lambda_D$ compared to the function transformed by the reflection $\lambda \to \tilde{A}_D - \lambda$ with the effective affinity $\tilde{A}_D \simeq 17.16$ (dotted-dashed line) for $A_D = 20$ and $A_C = 50$. The generating function is rescaled with Q_0, which is a constant proportional to the level density of the reservoirs. (b) The effective affinity $\tilde{A} = \tilde{A}_D$ rescaled by the actual affinity $A_D = 4$ versus the energy difference $\beta\varepsilon_A - \beta\varepsilon_B$ in the DQD for several values of the QPC affinity $A_C = \beta e V_C = -1, -2, \dots, -10$ and $\beta\varepsilon_A + \beta\varepsilon_B = 5$. Reprinted with permission from Bulnes Cuetara et al. (2013) © American Physical Society.

Now, we may introduce the following diagonal matrix composed with the Boltzmann factors of the DQD microstates in equilibrium with the reference reservoir $j = 2$,

$$\mathbf{M} = \begin{pmatrix} 1 & 0 & 0 \\ 0 & e^{-\beta(\varepsilon_+ - \mu_2)} & 0 \\ 0 & 0 & e^{-\beta(\varepsilon_- - \mu_2)} \end{pmatrix}. \tag{15.132}$$

Moreover, the operator (15.128) can be transformed according to equation (15.110) to include the counting parameters $\boldsymbol{\lambda} = (\lambda_D, \lambda_C)$. The resulting matrix $\mathbf{L}_{\boldsymbol{\lambda}}$ is satisfying the symmetry relation (15.111) with respect to the diagonal matrix (15.132) and the affinities defined as

$$A_D \equiv \beta\,(\mu_1 - \mu_2) \qquad \text{and} \qquad A_C \equiv \beta\,(\mu_3 - \mu_4) = \beta\,\Delta\mu_C, \tag{15.133}$$

which, respectively, drive the DQD and the QPC away from equilibrium. As a consequence, the cumulant generating function obeys the symmetry relation (15.112), so that the multivariate fluctuation relation (5.242) is satisfied for the two currents flowing in the DQD and the QPC (Golubev et al., 2011).

The issue is that the experiments can only perform the full counting statistics of the DQD current using the random trajectories of the QPC current (Utsumi et al., 2010; Küng et al., 2012, 2013). Therefore, the question arises as to whether the fluctuation relation holds for this single current. This question has been investigated, showing that the symmetry of the fluctuation relation is satisfied for the DQD current in the limit where the QPC current is much larger than the DQD current (Bulnes Cuetara et al., 2011, 2013). In this limit, the cumulant generating function for the DQD current, which is obtained by setting $\lambda_C = 0$, satisfies the approximate single-current symmetry relation

$$Q(\lambda_D, 0; A_D, A_C) \simeq Q(\tilde{A}_D - \lambda_D, 0; A_D, A_C). \tag{15.134}$$

However, this symmetry relation holds with respect to an effective affinity $\tilde{A}_D$, which differs in general from the actual one given by equation (15.133). As shown in Figure 15.10(a), the symmetry can be very closely satisfied, but the effective affinity $\tilde{A}_D \simeq 17.16$ is different from the actual affinity $A_D = 20$ because of the backaction of the QPC onto the DQD. The ratio between the effective and the actual affinities varies in particular with the energy difference $\varepsilon_A - \varepsilon_B$ between the DQD levels, as shown in Figure 15.10(b). The ratio $\tilde{A}_D/A_D$ can become negative, meaning that the DQD current can be reversed by the Coulomb drag effect due to the QCP (Bulnes Cuetara et al., 2013). This ratio also depends on the strength of the capacitive coupling between the two circuits (Bulnes Cuetara et al., 2011). Nevertheless, the effective affinity $\tilde{A}_D$ can be equal or close to the actual affinity A_D for $\varepsilon_A \simeq \varepsilon_B$ or for a large enough difference $|\varepsilon_A - \varepsilon_B|$. Accordingly, there exist conditions such that the single-current fluctuation relation $P(Z_D, t)/P(-Z_D, t) \sim \exp(A_D Z_D)$ holds with respect to the actual affinity $A_D = \beta(\mu_1 - \mu_2)$ within experimental accuracy (Küng et al., 2012, 2013).

15.8.5 Single-Electron Transistor

Single-electron transistors are composed of a conducting island or quantum dot connected to leads by three mesoscopic tunnel junctions (Ferry et al., 2009; Nazarov and Blanter, 2009), as shown in Figure 15.11. These tunnel junctions are characterized by their electric resistance or conductance $R_j = 1/G_j$, their capacitance C_j, and the voltage V_j applied to their lead with $j = \mathrm{L}, \mathrm{R}$, and G for gate. The conductive island can be large enough to be occupied by multiple electrons. Electron transport in such mesoscopic devices manifests Coulomb blockade effects.

According to classical electrodynamics, the voltage of the conductive island D occupied by $N \in \mathbb{Z}$ excess electrons is given by

$$V_D(N) = \frac{1}{C}(C_L V_L + C_R V_R + C_G V_G) - \frac{e}{C} N + V_p, \tag{15.135}$$

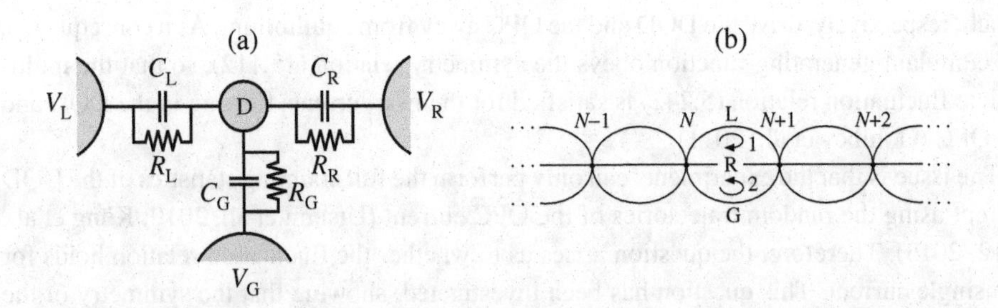

Figure 15.11 (a) Schematic representation of a single electron transistor. (b) Graph associated with the Markov process (Andrieux and Gaspard, 2006a).

where $C \equiv C_L + C_R + C_G$ is the total capacitance, e is the elementary charge, and V_p takes into account the possible misalignment of the Fermi level in the conductive island with respect to the leads. Since the occupancy of the multiple energy levels of the conductive island is described by a Fermi–Dirac distribution as for the reservoirs, the rates $W_j^{(\pm)}(N)$ for the transitions $N \to N \pm 1$ at the jth tunnel junction are also given by equation (15.119) with the corresponding conductances G_j and the energy changes

$$\Delta\varepsilon_j^{(\pm)}(N) = \mp e\left[V_D(N) - V_j\right] + \varepsilon_c, \qquad \text{where} \qquad \varepsilon_c = \frac{e^2}{2C} \tag{15.136}$$

is the capacitive charging energy (Amman et al., 1991).

The master equation for the time evolution of the probability distribution $P(N, \mathbf{Z}, t)$ with $\mathbf{Z} = (Z_L, Z_G)$ has the following form,

$$\begin{aligned}
\frac{d}{dt}P(N, Z_L, Z_G, t) = \Big[&\left(e^{-\partial_N} - 1\right) W_R^{(+)}(N) + \left(e^{+\partial_N} - 1\right) W_R^{(-)}(N) \\
&+ \left(e^{-\partial_N}\hat{E}_L^{-1} - 1\right) W_L^{(+)}(N) + \left(e^{+\partial_N}\hat{E}_L^{+1} - 1\right) W_L^{(-)}(N) \\
&+ \left(e^{-\partial_N}\hat{E}_G^{-1} - 1\right) W_G^{(+)}(N) \\
&+ \left(e^{+\partial_N}\hat{E}_G^{+1} - 1\right) W_G^{(-)}(N)\Big] P(N, Z_L, Z_G, t),
\end{aligned} \tag{15.137}$$

where $\hat{E}_j^{\pm 1} = \exp(\pm\partial_{Z_j})$ for $j = \text{L}, \text{G}$.

According to the network theory of Section 4.4.5, the two cycles shown in Figure 15.11(b) define the two following affinities,

$$A_j = \ln \frac{W_j^{(+)}(N)\, W_R^{(-)}(N+1)}{W_j^{(-)}(N+1)\, W_R^{(+)}(N)} = -\beta e(V_j - V_R) \qquad \text{for} \qquad j = \text{L}, \text{G}. \tag{15.138}$$

The modified operator (15.110) here obeys the symmetry relation (15.111) for the diagonal matrix $M(N, N') = M(N)\,\delta_{NN'}$ with the elements satisfying the recurrence relation $M(N) = M(N-1)\exp\left[\beta\Delta\varepsilon_R^{(-)}(N)\right]$, because the conditions $W_R^{(+)}(N-1)/W_R^{(-)}(N) = \exp\left[\beta\Delta\varepsilon_R^{(-)}(N)\right]$ hold for the transition rates with the reference reservoir $j = \text{R}$. Consequently, the cumulant generating function has the symmetry relation (15.112) of the fluctuation relation for the currents $\mathfrak{J}_j$ corresponding to the affinities (15.138) and the transport properties of Section 5.5 also hold (Andrieux and Gaspard, 2006a; Gu and Gaspard, 2020).

Single-electron transistors are often considered in the limit where the coupling to the gate is only capacitive with an infinite resistance $R_G = \infty$. In this limit, the gate current is equal to zero, i.e., $\mathfrak{J}_G = 0$, and the device can only be driven out of equilibrium by the affinity A_L.

15.9 Outlook

Further results have been obtained on the basis of fluctuation relations. In particular, the entropy production has been measured in a single-electron box at low temperature (Koski et al., 2013), as well as in an isolated spin-$\frac{1}{2}$ system with a nuclear magnetic resonance setup (Batalhõ et al., 2015). Fluctuation relations also hold for spintronics (López et al., 2012; Wang and Feldman, 2015). Similar considerations may be envisaged for bosonic and fermionic systems manifesting topological quantum effects (Tran et al., 2017).

These advances allow us to address the issue of energy dissipation and irreversibility in quantum-mechanical systems and to determine their linear and nonlinear transport properties in consistency with microreversibility.

Conclusion and Perspectives

Irreversible Phenomena and Statistical Mechanics

A great diversity of irreversible phenomena is observed in nature: diffusion, electric and heat conductions, viscosity, reactions, and their mutual coupling such as cross-diffusion, thermodiffusion, thermoelectricity, diffusiophoresis, thermophoresis, mechanochemical energy transduction, active ion transport, and many others. They are characterized by coefficients that should be determined by the microscopic motion of atoms and molecules composing matter and they contribute in general to entropy production. In order to bridge the gap between microdynamics and thermodynamics, statistical mechanics provides a unifying framework describing irreversible phenomena in a way that is consistent with the laws of nature, as presented in this book.

A key point is that the microscopic equations of motion are differential equations requiring initial conditions and boundary conditions to be solved. These conditions are free to take arbitrary values, which may be irrespective of the symmetries that the equations of motion are satisfying. Therefore, these symmetries can be broken by the solutions of the equations, although the equations remain themselves symmetric. This well-known mechanism in condensed matter physics also concerns microreversibility, which is the symmetry of the microscopic equations of motion under time reversal. Indeed, the time-reversal symmetry may be broken by the initial conditions or the boundary conditions, possibly combined with some time-dependent external forcing, which leads to symmetry breaking at the statistical level of description while microreversibility is always satisfied, as summarized in Table 2.2. Moreover, since interactions are local in nature, statistical correlations tend to evolve in time by spreading over more and more degrees of freedom and over larger and larger spatial volumes for trajectories issued from localized initial perturbations. In this way, the time asymmetry observed in irreversible phenomena and expressed by the second law of thermodynamics can be understood with the methods of statistical mechanics, establishing the unity of knowledge from the atomic scale up to the macroscopic realm and from quantum to classical systems.

With the local equilibrium approach, macroscopic hydrodynamics can be deduced from the microscopic Hamiltonian dynamics in multicomponent normal fluids, as well as the phases of matter with broken continuous symmetries such as crystals and liquid crystals.

In these phases, Green–Kubo formulas can be obtained for all the linear transport coefficients, including the viscosity coefficients, the electric and heat conductivities, the diffusion coefficients, the friction coefficients of the local order parameters, as well as the coefficients describing the coupling effects. In quantum statistical mechanics, the Green–Kubo formulas can also be obtained using functional time-reversal symmetry relations.

Markovian and non-Markovian master equations can be inferred from the microscopic dynamics for classical and quantum systems with slow and fast degrees of freedom or composed of weakly interacting parts. The interfacial nonequilibrium properties may also be described, notably, in the framework of fluctuating chemohydrodynamics.

Furthermore, transport can be rigorously studied in Hamiltonian dynamical models, allowing us to construct the diffusive modes of relaxation towards equilibrium from the Liouvillian phase-space dynamics, to establish Fourier's law in many-particle billiard models, and to show the importance of nonlinear transport coefficients in models for mechanothermal coupling.

Empowering Affinities

At the statistical level of description, thermodynamic equilibria correspond to statistical ensembles that are symmetric under time reversal in the space of microstates. Therefore, the principle of detailed balance holds at equilibrium, where every path and its time reversal have equal probabilities. This is no longer the case in systems driven away from equilibrium by the supply of free energy, where directionality and temporal ordering can manifest themselves.

A system can be driven out of equilibrium by time-dependent external forces, initial conditions, or boundary conditions. In particular, nonequilibrium boundary conditions can be maintained if the system is open and in contact with arbitrarily large energy or particle reservoirs at different temperatures or chemical potentials. The control parameters of the nonequilibrium constraints are the thermodynamic forces given by the thermal, chemical, and mechanical affinities, empowering the energy and particle fluxes flowing across nonequilibrium systems, as well as active motion in nonequilibrium matter. These global affinities extend to the boundaries of the system the local affinities that are introduced in nonequilibrium thermodynamics and in statistical mechanics with the local equilibrium approach. They thus play a central role in the formulation of the nonequilibrium properties. In particular, the entropy production rate is given by the sum of the global affinities multiplied by the mean currents for all the irreversible processes taking place in the system. In general, the mean currents can have nonlinear dependences on the affinities, which may cause current rectification in many nonequilibrium systems, including electric and thermal diodes, as well as chemical reactions.

Nonequilibrium Directionality and Temporal Ordering

In small systems, the currents of energy or particles fluctuate in time because of the thermal and molecular fluctuations due to the atomic structure of matter. The remarkable result is that these fluctuating currents and movements obey the fluctuation relations, which are valid from the vicinity of equilibrium to regimes arbitrarily far from equilibrium, and find their origin in microreversibility. They can be formulated in terms of either the probability distribution or the cumulant generating function for the full counting statistics of the currents. In this latter case, these relations have the universal form (5.91) or (5.177), which can be established for classical and quantum microdynamics, as well as for stochastic processes. With these relations, nonequilibrium directionality is expressed by the fact that the current fluctuations are more probable in the direction of the affinities than in the opposite direction, and that this imbalance exponentially increases with time. If the affinities are equal to zero, the system is at equilibrium and detailed balance is recovered. Fluctuation relations imply the nonnegativity of the entropy production rate, so that they are in agreement with the second law of thermodynamics, on the one hand, and microreversibility, on the other hand, allowing us to understand how these two aspects can be compatible with each other. Moreover, the multivariate fluctuation relation for the currents also implies time-reversal symmetry relations for the linear and nonlinear transport properties extending the Onsager–Casimir reciprocal relations and the fluctuation–dissipation relations from the linear to the nonlinear coefficients of nonequilibrium responses, as presented in Chapter 5. These properties are essential to describe the coupling between different processes because nonvanishing linear and nonlinear coupling coefficients enable energy transduction in nature.

In addition, the study of path probabilities shows that a fundamental relationship exists between the entropy production rate and time asymmetry in the temporal disorder of paths in nonequilibrium processes according to equation (6.103). The theorem of nonequilibrium temporal ordering expresses the prevalence of typical paths over their corresponding time reversal and, thus, the manifestation of dynamical order in nonequilibrium conditions. The analogy with other symmetry-breaking phenomena can be established in detail, as explained in Chapter 6.

Nonequilibrium Fluctuating Systems and Their Properties

The previous results can be shown to hold down to the nanoscale in several classes of fluctuating irreversible phenomena, including driven Brownian motion, effusion, flows in dilute and rarefied gases, transport by diffusion, ion transport, diodes, transistors, diffusion-influenced surface reactions, reactions, chemical clocks, enzymatic kinetics, copolymerization processes, molecular motors, chemically powered active particles, and transport in electronic mesoscopic devices, among other nonequilibrium processes. In open quantum systems composed of independent particles, the results can also be established within the scattering approach to the transport of bosons and fermions.

From statistical mechanics to thermodynamics, a clear distinction can thus be made between equilibrium and nonequilibrium properties. At equilibrium, the affinities, the mean currents, and the entropy production rate are equal to zero. No free-energy supply is needed. The principle of detailed balance is satisfied. The molecules and supramolecular assemblies form three-dimensional spatial structures after averaging over the thermal fluctuations. Instead, away from equilibrium, the affinities and mean currents take nonvanishing values and the entropy production rate is positive because of energy dissipation. Free-energy supply is required to sustain directionality and temporal ordering. As a consequence, the molecules and supramolecular assemblies perform four-dimensional spatiotemporal dynamics on average, which allow them to acquire functions, e.g., in biological cells (Gaspard, 2013b).

Perspectives

The results presented in this book shed new light, in particular, for extending our understanding of irreversible phenomena down to nanosystems in chemistry, biochemistry, and molecular biology. For this purpose, the atomic structure of large molecules such as enzymes and their interaction with some surrounding solution can be taken into account to investigate the mean movements and reactions that are induced by nonequilibrium concentrations for reactants and products. Similar methods can be envisaged for the collective dynamics of active particles and also large chemical reaction networks. These considerations also concern complex nonequilibrium processes such as nucleation (Lutsko, 2012), crystal growth (Ma et al., 2020), the dynamics of functionalized membranes or vesicles (Bachmann et al., 2016; Mognetti et al., 2019), and self-organization processes in general. Electron-transfer reactions and the quantum transport of electrons and ultracold atoms are also of great interest in this perspective.

Furthermore, the fluctuation relations provide new insights into the conception of methods for deducing Boltzmann's and other kinetic equations from microdynamics. Additionally, the local equilibrium approach to nonequilibrium statistical mechanics and the fluctuation relations are opening new roads for the systematic study of transport properties in the quantum phases of matter at low temperature, as well as in relativistic systems.

Appendix A

Complements on Thermodynamics

A.1 Thermodynamic Potentials

The first and second laws have for consequence the Gibbs relation (1.8), which can be generically written as $dE = \sum_{j=1}^{n} b_j dA_j$, where A_j and $b_j = \partial E/\partial A_j$ are thermodynamically conjugated variables. Energy thus depends on the independent variables $\{A_j\}_{j=1}^{n}$. In order to modify the set of independent variables, Legendre transforms can be performed; these exchange two thermodynamically conjugated variables. For instance, the new thermodynamic potential $X \equiv E - A_1 b_1$ has the Gibbs relation $dX = -A_1 db_1 + \sum_{j=2}^{n} b_j dA_j$, showing that the independent variables are now given by $(b_1, A_2, \ldots, A_n)$. In this way, 2^n different thermodynamic potentials may be defined. Table A.1 lists them for systems with one species of particles. Free energy is Helmholtz's free energy and free enthalpy is the Gibbs free energy.

A.2 Euler Relations in Homogeneous Systems

If the system is homogeneous, extensive variables such as energy, entropy, or particle numbers are proportional to the volume. Since all the independent variables $\{A_j\}_{j=1}^{n}$ of energy E are extensive, we have that $E(\{\lambda A_j\}_{j=1}^{n}) = \lambda E(\{A_j\}_{j=1}^{n})$ if the volume is multiplied by λ. Differentiating with respect to λ and taking $\lambda = 1$, we find Euler's relation $E = \sum_{j=1}^{n} b_j A_j$, which is only valid if the system is homogeneous. Euler's relation is given in Table A.1 for every listed thermodynamic potential. As a consequence, in systems with several particle species, the Gibbs free energy is given by $G = \sum_{k=1}^{c} \mu_k N_k$, where μ_k is the chemical potential per particle of species k and N_k is the number of these particles in the system. In addition, the grand thermodynamic potential is directly proportional to the pressure, $p = -J/V$. Although the thermodynamic potential K is vanishing in homogeneous systems (i.e., in bulk phases), it is nevertheless useful in heterogeneous systems where it collects the contributions of the interfaces and other heterogeneities. Another consequence of Euler's relations are the Gibbs–Duhem relations $\sum_{j=1}^{n} A_j db_j = 0$, obtained by differentiating Euler's relation and subtracting the corresponding Gibbs relation.

Dividing a quantity by the volume defines the associated density, which are intensive variables. In this way, we can define the energy density $e \equiv E/V$, the entropy density $s \equiv S/V$, the particle densities $n_k = N_k/V$, or the mass density $\rho \equiv M/V$ where

Table A.1. *Thermodynamic potentials for systems with one particle species.*

Potential	Variables	Definition	Gibbs Relation	Euler Relation
Energy	(S,V,N)	E by first law	$dE = TdS - pdV + \mu dN$	$E = TS - pV + \mu N$
Free energy	(T,V,N)	$F \equiv E - TS$	$dF = -SdT - pdV + \mu dN$	$F = -pV + \mu N$
Free enthalpy	(T,p,N)	$G \equiv F + pV$	$dG = -SdT + Vdp + \mu dN$	$G = \mu N$
Enthalpy	(S,p,N)	$H \equiv E + pV$	$dH = TdS + Vdp + \mu dN$	$H = TS + \mu N$
–	(S,V,μ)	$I \equiv E - \mu N$	$dI = TdS - pdV - Nd\mu$	$I = TS - pV$
Grand potential	(T,V,μ)	$J \equiv F - \mu N$	$dJ = -SdT - pdV - Nd\mu$	$J = -pV$
–	(T,p,μ)	$K \equiv G - \mu N$	$dK = -SdT + Vdp - Nd\mu$	$K = 0$
–	(S,p,μ)	$L \equiv H - \mu N$	$dL = TdS + Vdp - Nd\mu$	$L = TS$

$M \equiv \sum_{k=1}^{c} m_k N_k$ is the total mass expressed in terms of the masses $\{m_k\}_{k=1}^{c}$ of the particles. The energy density obeys the following Gibbs, Euler, and Gibbs–Duhem relations:

$$de = Tds + \sum_{k=1}^{c} \mu_k dn_k,$$

$$e = Ts - p + \sum_{k=1}^{c} \mu_k n_k, \qquad \text{and} \qquad s\,dT - dp + \sum_{k=1}^{c} n_k d\mu_k = 0. \qquad \text{(A.1)}$$

An alternative set of intensive variables can be defined using the total mass M instead of the volume (de Groot and Mazur, 1984), such as the energy per unit mass $\mathfrak{e} \equiv E/M$, the entropy per unit mass $\mathfrak{s} \equiv S/M$, and the volume per unit mass $\mathfrak{v} \equiv V/M = 1/\rho$. To characterize the composition, several alternative quantities may be used instead of the particle density (also called molar concentration), such as the mass concentration $\rho_k \equiv m_k n_k$ satisfying $\rho = \sum_{k=1}^{c} \rho_k$, the mass fraction $\mathfrak{y}_k \equiv \rho_k/\rho$, or the mole fraction $y_k \equiv n_k/\sum_{k=1}^{c} n_k$. We note that $\sum_{k=1}^{c} \mathfrak{y}_k = 1$ and $\sum_{k=1}^{c} y_k = 1$. For the intensive quantities per unit mass, the Gibbs, Euler, and Gibbs–Duhem relations read as

$$d\mathfrak{e} = Td\mathfrak{s} - p\,d\mathfrak{v} + \sum_{k=1}^{c} \tilde{\mu}_k d\mathfrak{y}_k,$$

$$\mathfrak{e} = T\mathfrak{s} - p\mathfrak{v} + \sum_{k=1}^{c} \tilde{\mu}_k \mathfrak{y}_k, \qquad \text{and} \qquad \mathfrak{s}\,dT - \mathfrak{v}\,dp + \sum_{k=1}^{c} \mathfrak{y}_k d\tilde{\mu}_k = 0, \qquad \text{(A.2)}$$

where $\tilde{\mu}_k \equiv \mu_k/m_k$ is the chemical potential per molecular mass of species k.

A.3 Equilibrium Properties of Materials

At equilibrium, various materials properties may be defined; these are listed in Table A.2 (Callen, 1985). Intensive variables characterizing heat capacity are defined per unit volume at constant volume and pressure, $c_\mathfrak{v} \equiv C_V/V$ and $c_p \equiv C_p/V$, or, similarly, per unit mass, $\mathfrak{c}_\mathfrak{v} \equiv C_V/M = c_\mathfrak{v}/\rho$ and $\mathfrak{c}_p \equiv C_p/M = c_p/\rho$. The adiabatic and isothermal bulk moduli are, respectively, defined by $K_S \equiv 1/\kappa_S$ and $K_T \equiv 1/\kappa_T$.

Table A.2. *Equilibrium properties of materials*

Property	Definition
Heat capacity at constant volume	$C_V \equiv \left(\frac{\partial E}{\partial T}\right)_{V,N} = T\left(\frac{\partial S}{\partial T}\right)_{V,N}$
Heat capacity at constant pressure	$C_p \equiv \left(\frac{\partial H}{\partial T}\right)_{p,N} = T\left(\frac{\partial S}{\partial T}\right)_{p,N}$
Coefficient of thermal expansion	$\alpha \equiv \frac{1}{V}\left(\frac{\partial V}{\partial T}\right)_{p,N}$
Coefficient of thermal pressure increase	$\varpi \equiv \frac{1}{p}\left(\frac{\partial p}{\partial T}\right)_{V,N}$
Adiabatic compressibility	$\kappa_S \equiv -\frac{1}{V}\left(\frac{\partial V}{\partial p}\right)_{S,N}$
Isothermal compressibility	$\kappa_T \equiv -\frac{1}{V}\left(\frac{\partial V}{\partial p}\right)_{T,N}$

These properties are not all independent since we have the relations,

$$T\alpha^2 = \rho\kappa_T(\mathfrak{c}_p - \mathfrak{c}_v) = \rho\mathfrak{c}_p(\kappa_T - \kappa_S), \qquad \frac{\mathfrak{c}_p}{\mathfrak{c}_v} = \frac{\kappa_T}{\kappa_S}, \qquad \varpi = \frac{\alpha}{\kappa_T p}, \tag{A.3}$$

and

$$\left(\frac{\partial \mu_k}{\partial T}\right)_{p,\{\mathfrak{n}_j\}_{j\neq k}} = \left(\frac{\partial \mu_k}{\partial T}\right)_{\rho,\{\mathfrak{n}_j\}_{j\neq k}} - \frac{\alpha}{\kappa_T n_k}, \tag{A.4}$$

which are obtained using the method of Jacobians (Landau and Lifshitz, 1980a).

A.4 Conditions for Thermodynamic Equilibrium

A.4.1 First Variation of the Entropy

For an isolated system composed of two parts A and B in contact with each other, the extensive variables are given by the sum of their values in the different parts, so that $E = E_A + E_B$, $S = S_A + S_B$, $V = V_A + V_B$, and $N_k = N_{kA} + N_{kB}$. As the total system is isolated, any change conserves the total energy and the same is assumed for the total volume and the total particle numbers, so that $\delta E_B = -\delta E_A$, $\delta V_B = -\delta V_A$, and $\delta N_{kB} = -\delta N_{kA}$. According to equation (1.12), the change in entropy $\delta S = \delta S_A + \delta S_B$, i.e., the first variation of entropy, is thus given by

$$\delta S = \left(\frac{1}{T_A} - \frac{1}{T_B}\right)\delta E_A + \left(\frac{p_A}{T_A} - \frac{p_B}{T_B}\right)\delta V_A - \sum_{k=1}^{c}\left(\frac{\mu_{kA}}{T_A} - \frac{\mu_{kB}}{T_B}\right)\delta N_{kA}. \tag{A.5}$$

Using the condition $\delta S = 0$ for the entropy to be extremal at equilibrium, we find

$$T_A = T_B, \qquad p_A = p_B, \qquad \text{and} \qquad \mu_{kA} = \mu_{kB} \qquad \text{for} \qquad k = 1, 2, \ldots, c, \tag{A.6}$$

which are, respectively, the conditions for thermal, mechanical, and chemical equilibrium. If the system forms a continuous medium, these conditions read

$$\nabla T = 0, \qquad \nabla p = 0, \qquad \text{and} \qquad \nabla \mu_k = 0, \tag{A.7}$$

in terms of the gradient ∇, expressing the uniformity of temperature, pressure, and chemical potentials at equilibrium.

A.4.2 Second Variation of the Entropy

At equilibrium, the entropy is not only extremal but also maximal, which implies that the necessary condition (A.7) should be supplemented by conditions of local stability for the equilibrium macrostate. In this respect, we need to calculate the second variation of the entropy. Continuing the expansion of the entropy up to the terms that are quadratic in the deviations, we obtain the following expression in every part (Landau and Lifshitz, 1980a; Callen, 1985; Kondepudi and Prigogine, 1998):

$$\delta^2 S = \frac{1}{2T}\left(-\delta T\,\delta S + \delta p\,\delta V - \sum_{k=1}^{c} \delta\mu_k\,\delta N_k\right). \tag{A.8}$$

Using the so-called Maxwell relations (Callen, 1985), we find

$$\delta^2 S = \frac{1}{2T}\left[-\frac{C_p}{T}\,\delta T^2 + 2V\alpha\,\delta T\,\delta p - V\kappa_T\,\delta p^2 - \sum_{k,l=1}^{c}\left(\frac{\partial \mu_k}{\partial N_l}\right)_{T,p}\delta N_k\,\delta N_l\right]. \tag{A.9}$$

The entropy is maximal if $\delta^2 S \leq 0$, which implies that

$$C_p > C_V > 0, \qquad \kappa_T > \kappa_S > 0, \qquad \left[\left(\frac{\partial \mu_k}{\partial N_l}\right)_{T,p}\right]_{k,l=1}^{c} > 0. \tag{A.10}$$

The extremal macrostate is stable if the entropy is globally maximal, but metastable if it is only locally maximal in the space of macrostates. Otherwise, the extremal macrostate is unstable and it does not exist as an equilibrium macrostate since it would be disrupted by the slightest fluctuations.

A.5 Hydrodynamic Equations in Eulerian and Lagrangian Forms

The Eulerian forms of hydrodynamic equations are given by the balance equations (1.16) with the decomposition (1.21) of the current density into its advective and other contributions for the quantities given in Table 1.1, supposing that there is no external force field. Their Lagrangian forms can thus be deduced by using the total time derivative along the streamlines of the fluid flow, which is defined as

$$\frac{dx}{dt} = \partial_t x + \mathbf{v}\cdot\nabla x, \tag{A.11}$$

where the densities x can be expressed in terms of the corresponding quantities per unit mass $\mathfrak{x} \equiv x/\rho$. In this way, the hydrodynamic equations for particle numbers, mass, linear momentum, and energy are obtained as follows in their respective Eulerian and Lagrangian forms:

$$\partial_t n_k + \nabla \cdot (n_k \mathbf{v} + \mathcal{J}_k) = \sigma_k, \qquad \rho \frac{d\mathfrak{y}_k}{dt} = -\nabla \cdot (m_k \mathcal{J}_k) + m_k \sigma_k, \tag{A.12}$$

$$\partial_t \rho + \nabla \cdot (\rho \mathbf{v}) = 0, \qquad \frac{d\rho}{dt} = -\rho \, \nabla \cdot \mathbf{v}, \tag{A.13}$$

$$\partial_t (\rho \mathbf{v}) + \nabla \cdot (\rho \mathbf{v}\mathbf{v} + \mathsf{P}) = 0, \qquad \rho \frac{d\mathbf{v}}{dt} = -\nabla \cdot \mathsf{P}, \tag{A.14}$$

$$\partial_t \epsilon + \nabla \cdot (\epsilon \mathbf{v} + \mathsf{P} \cdot \mathbf{v} + \mathcal{J}_q) = 0, \qquad \rho \frac{d\mathfrak{e}}{dt} = -\mathsf{P} : \nabla \mathbf{v} - \nabla \cdot \mathcal{J}_q, \tag{A.15}$$

where $\epsilon = (1/2)\rho \mathbf{v}^2 + e$ is the energy density, $\mathfrak{y}_k = m_k n_k / \rho$ is the mass fraction of species k, and $\mathfrak{e} = e/\rho$ the internal energy per unit mass (de Groot and Mazur, 1984; Balescu, 1975).

We note that the Lagrangian equation for the enthalpy per unit mass, $\mathfrak{h} \equiv \mathfrak{e} + p/\rho$, is given by

$$\rho \frac{d\mathfrak{h}}{dt} = \frac{dp}{dt} + p \nabla \cdot \mathbf{v} - \mathsf{P} : \nabla \mathbf{v} - \nabla \cdot \mathcal{J}_q, \tag{A.16}$$

as a consequence of equation (A.15) for energy.

A.6 Deduction of the Entropy Production in Normal Fluids

Using equation (A.13) for mass to transform equation (A.14) for linear momentum, we obtain the general form of Navier–Stokes equations

$$\rho \left(\partial_t \mathbf{v} + \mathbf{v} \cdot \nabla \mathbf{v} \right) = -\nabla \cdot \mathsf{P}. \tag{A.17}$$

Using equations (A.13) and (A.17), equation (A.15) for energy leads to the balance equation for the internal energy of density e,

$$\partial_t e + \nabla \cdot (e \mathbf{v} + \mathcal{J}_q) = -\mathsf{P} : \nabla \mathbf{v} \tag{A.18}$$

with the notation $\mathsf{P} : \nabla \mathbf{v} = \sum_{i,j=x,y,z} P_{ij} \partial_i v_j$. In equation (A.18), the source term represents the mechanical power density of the fluid flow.

Since local thermodynamic equilibrium is assumed in every fluid element, the differential of some density x can be replaced by the total time derivative (A.11) along the trajectories of the fluid elements. Accordingly, the Gibbs relation (1.22) is transformed into

$$T \left[\partial_t s + \nabla \cdot (s \mathbf{v}) \right] = \left[\partial_t e + \nabla \cdot (e \mathbf{v}) \right] - \sum_{k=1}^{c} \mu_k \left[\partial_t n_k + \nabla \cdot (n_k \mathbf{v}) \right] + p \, \nabla \cdot \mathbf{v}, \tag{A.19}$$

where the last term arises from Euler's relation in equation (A.1). Substituting the balance equation (A.18) for internal energy and those for the particle numbers in the left-hand side of equation (A.19), and dividing the result by the temperature T, we obtain

$$\partial_t s + \nabla \cdot (s\mathbf{v}) = -\frac{1}{T}\mathbf{\Pi} : \nabla\mathbf{v} - \frac{1}{T}\nabla \cdot \mathcal{J}_q + \sum_{k=1}^{c} \frac{\mu_k}{T}\nabla \cdot \mathcal{J}_k - \sum_{k=1}^{c} \frac{\mu_k}{T}\sigma_k \tag{A.20}$$

with the viscous pressure tensor $\mathbf{\Pi} = \mathbf{P} - p\,\mathbf{1}$, which is symmetric, i.e., $\mathbf{\Pi} = \mathbf{\Pi}^{\mathrm{T}}$. Using the following identity between any scalar field a and vectorial field $\mathcal{J}$, $a\nabla \cdot \mathcal{J} = \nabla \cdot (a\mathcal{J}) - \mathcal{J} \cdot \nabla a$, to transform the second and third terms with $\mathcal{J} = \mathcal{J}_q$ and $\mathcal{J} = \mathcal{J}_k$ and moving every divergence to the left-hand side, we find the balance equation for entropy,

$$\partial_t s + \nabla \cdot (s\mathbf{v} + \mathcal{J}_s) = -\frac{1}{T}\mathbf{\Pi} : \nabla\mathbf{v} + \mathcal{J}_q \cdot \nabla\frac{1}{T} - \sum_{k=1}^{c} \mathcal{J}_k \cdot \nabla\frac{\mu_k}{T} - \sum_{k=1}^{c}\sum_{r=1}^{m} \frac{\mu_k}{T}\nu_{kr}w_r \tag{A.21}$$

with the diffusive current density of entropy given by equation (1.23). Decomposing the viscous pressure tensor into its traceless part $\mathring{\mathbf{\Pi}} = \mathbf{\Pi} - \Pi\,\mathbf{1}$ with $\Pi = (\mathrm{tr}\,\mathbf{\Pi})/3$ and the other part $\Pi\,\mathbf{1}$, and carrying out the corresponding decomposition for the symmetric part of the velocity gradient, we get $-T^{-1}\,\mathbf{\Pi} : \nabla\mathbf{v} = \mathring{\mathbf{\Pi}} : \mathring{\mathbf{A}}_{\mathbf{g}} + \Pi\,\mathcal{A}_{\mathbf{g}}$ in terms of the traceless tensorial affinity $\mathring{\mathbf{A}}_{\mathbf{g}}$ and the scalar affinity $\mathcal{A}_{\mathbf{g}}$ defined in Table 1.2. Therefore, we obtain the entropy production rate density (1.24) with the affinities and current densities given in Table 1.2.

We note that the Gibbs relation (1.22) implies that

$$\left(\frac{\partial s}{\partial e}\right)_{\{n_k\}} = \frac{1}{T} \quad \text{and} \quad \left(\frac{\partial s}{\partial n_k}\right)_{e,\{n_j\}_{j\neq k}} = -\frac{\mu_k}{T}. \tag{A.22}$$

Hence, the local equilibrium assumption, according to which $s = s(e, \{n_k\})$, means that equation (1.22) holds with every differential dx replaced by either dx/dt, $\partial_t x$, or ∇x. Similar considerations apply to the Gibbs and Gibbs–Duhem relations in equation (A.1). For this reason, the expression of the entropy production rate density can be equivalently deduced using either total or partial derivatives with respect to time.

As a consequence of equation (A.21), the Lagrangian equation for the entropy per unit mass $\mathfrak{s} = s/\rho$ has the form,

$$\rho\frac{d\mathfrak{s}}{dt} = -\nabla \cdot \mathcal{J}_s + \sigma_s, \tag{A.23}$$

where σ_s is the entropy production rate density given by the right-hand side of equation (A.21) and, equivalently, by equation (1.24). If the irreversible processes are negligible (i.e., if $\mathcal{J}_s = 0$ and $\sigma_s = 0$), equation (A.23) reduces to $d\mathfrak{s}/dt = 0$, meaning that the entropy per unit mass remains constant along the streamlines.

A.7 The Heat Equation

The heat equation for the temperature can be deduced using the equation of state for the internal energy per unit mass $\mathfrak{e} = \mathfrak{e}\left(T, \rho, \{\mathfrak{y}_k\}_{k=1}^{c-1}\right)$, which depends on the temperature, the mass density, and the mass fractions of the solute species. Taking the time derivative of the internal energy per unit mass in a fluid volume element along the streamline of the flow, we have that

$$\frac{d\mathfrak{e}}{dt} = \left(\frac{\partial \mathfrak{e}}{\partial T}\right)_{\rho,\{\mathfrak{y}_k\}} \frac{dT}{dt} + \left(\frac{\partial \mathfrak{e}}{\partial \rho}\right)_{T,\{\mathfrak{y}_k\}} \frac{d\rho}{dt} + \sum_{k=1}^{c-1} \left(\frac{\partial \mathfrak{e}}{\partial \mathfrak{y}_k}\right)_{\rho,T,\{\mathfrak{y}_j\}_{j\neq k}} \frac{d\mathfrak{y}_k}{dt}. \tag{A.24}$$

The different coefficients can be determined using thermodynamics to get

$$\left(\frac{\partial \mathfrak{e}}{\partial T}\right)_{\rho,\{\mathfrak{y}_k\}} = T\left(\frac{\partial \mathfrak{s}}{\partial T}\right)_{\rho,\{\mathfrak{y}_k\}} = \mathfrak{c}_\mathfrak{v}, \tag{A.25}$$

$$\left(\frac{\partial \mathfrak{e}}{\partial \rho}\right)_{T,\{\mathfrak{y}_k\}} = \frac{1}{\rho^2}\left(p - T\frac{\partial p}{\partial T}\right)_{\rho,\{\mathfrak{y}_k\}}, \tag{A.26}$$

$$\left(\frac{\partial \mathfrak{e}}{\partial \mathfrak{y}_k}\right)_{\rho,T,\{\mathfrak{y}_j\}_{j\neq k}} = \frac{e_k}{m_k} \quad \text{with} \quad e_k \equiv \mu_k - T\left(\frac{\partial \mu_k}{\partial T}\right)_{\rho,\{\mathfrak{y}_j\}_{j\neq k}}, \tag{A.27}$$

where $\mathfrak{c}_\mathfrak{v}$ is the heat capacity per unit mass at constant volume (or constant density) and e_k is the energy per molecule of species k. Equations (A.26) and (A.27) are obtained by using the Helmholtz free energy per unit mass $\mathfrak{f}$, which is the thermodynamic potential depending on the relevant variables $\left(T, \rho, \{\mathfrak{y}_k\}_{k=1}^{c-1}\right)$. Its Gibbs relation is given by $d\mathfrak{f} = -\mathfrak{s}dT - pd(1/\rho) + \sum_{k=1}^{c-1}(\mu_k/m_k)d\mathfrak{y}_k$, so that $\mathfrak{e} = \mathfrak{f} - T\partial\mathfrak{f}/\partial T$, hence the relations (A.26) and (A.27).

Now, substituting the Lagrangian equations (A.12), (A.13), and (A.15) into equation (A.24), we obtain the following Lagrangian equation for the temperature of the moving fluid element:

$$\rho\mathfrak{c}_\mathfrak{v}\frac{dT}{dt} = -T\left(\frac{\partial p}{\partial T}\right)_{\rho,\{\mathfrak{y}_k\}} \nabla\cdot\mathbf{v} - \mathbf{\Pi} : \nabla\mathbf{v} - \nabla\cdot\mathcal{J}_q + \sum_{k=1}^{c-1} e_k\,\nabla\cdot\mathcal{J}_k - \sum_{k=1}^{c-1} e_k\,\sigma_k. \tag{A.28}$$

The last term can be written in the following form using $\sigma_k = \sum_{r=1}^{m} \nu_{kr}\,w_r$:

$$\sum_{k=1}^{c-1} e_k\,\sigma_k = \sum_{r=1}^{m} \Delta e_r\,w_r, \tag{A.29}$$

where $\Delta e_r \equiv \sum_{k=1}^{c-1} e_k \nu_{kr}$ is the energy of reaction r, such that $\Delta e_r > 0$ if the reaction stores chemical energy and $\Delta e_r < 0$ if it releases chemical energy. The Eulerian form of the heat equation is thus given by

$$c_v\,(\partial_t T + \mathbf{v}\cdot\nabla T) = -\frac{T\alpha}{\kappa_T}\nabla\cdot\mathbf{v} - \mathbf{\Pi} : \nabla\mathbf{v} - \nabla\cdot\mathcal{J}_q + \sum_{k=1}^{c-1} e_k\,\nabla\cdot\mathcal{J}_k - \sum_{r=1}^{m}\Delta e_r\,w_r, \tag{A.30}$$

where we used the relations $c_v = \rho\mathfrak{c}_\mathfrak{v}$ and $(\partial p/\partial T)_{\rho,\{\mathfrak{y}_k\}} = \alpha/\kappa_T$. The right-hand side gives the sources of heating in the fluid element. The first term describes heating due to compression and cooling due to expansion, the second heating due to viscosity, the third heating from heat conduction, the fourth from transport by diffusion, and the fifth from reactions.

Furthermore, the first term in the right-hand side of equation (A.28) can be expressed by using equation (A.13) in terms of the time derivative of the mass density. This latter can be supposed to depend on the temperature, the pressure, and the mass fractions: $\rho = \rho\left(T, p, \{\mathfrak{y}_k\}_{k=1}^{c-1}\right)$. According to equation (A.12), the heat equation becomes

$$c_p \frac{dT}{dt} = T\alpha \frac{dp}{dt} - \mathbf{\Pi} : \nabla\mathbf{v} - \nabla \cdot \mathcal{J}_q + \sum_{k=1}^{c-1} \mathfrak{h}_k \nabla \cdot \mathcal{J}_k - \sum_{r=1}^{m} \Delta\mathfrak{h}_r\, w_r, \tag{A.31}$$

with the heat capacity at constant pressure $c_p = c_\mathfrak{v} + T\alpha^2/\kappa_T$ (given by equation (A.3)), the enthalpy per molecule of species k

$$\mathfrak{h}_k \equiv \mu_k - T\left(\frac{\partial \mu_k}{\partial T}\right)_{p,\{\mathfrak{y}_j\}_{j\neq k}} = e_k + \frac{T\alpha}{\kappa_T n_k}, \tag{A.32}$$

and the enthalpy or heat of reaction r

$$\Delta\mathfrak{h}_r \equiv \sum_{k=1}^{c-1} \mathfrak{h}_k \nu_{kr} = \sum_{k=1}^{c-1} \mathfrak{h}_k \left(\nu_{kr}^{(-)} - \nu_{kr}^{(+)}\right) = \mathfrak{h}_r(\text{products}) - \mathfrak{h}_r(\text{reactants}), \tag{A.33}$$

which is positive if the reaction is endothermic and negative if it is exothermic (De Donder and Van Rysselberghe, 1936; Haase, 1969; Callen, 1985). We note that equation (A.31) can also be deduced from the Lagrangian equation (A.16) for the enthalpy per unit mass (Joulin and Vidal, 1998). If the process is isobaric, then $dp/dt = 0$, and the heat equation can thus be expressed in terms of the heat capacity at constant pressure and the enthalpies, instead of the heat capacity at constant volume and the energies (Landau and Lifshitz, 1987).

A.8 Deduction of the Hydrodynamic Modes in One-Component Fluids

In order to deduce the hydrodynamic modes from equations (1.44)–(1.46), the deviations of the temperature and the pressure around equilibrium are expanded into the deviations in the mass density and the entropy per unit mass according to

$$\delta T = \left(\frac{\partial T}{\partial \rho}\right)_\mathfrak{s} \delta\rho + \left(\frac{\partial T}{\partial \mathfrak{s}}\right)_\rho \delta\mathfrak{s} \quad \text{and} \quad \delta p = \left(\frac{\partial p}{\partial \rho}\right)_\mathfrak{s} \delta\rho + \left(\frac{\partial p}{\partial \mathfrak{s}}\right)_\rho \delta\mathfrak{s}. \tag{A.34}$$

Substituting these relations into equations (1.45) and (1.46), we obtain a closed set of linear partial differential equations for the five fields $(\delta\rho, \delta\mathfrak{s}, \delta\mathbf{v})$. Considering spatially periodic solutions $\delta\rho, \delta\mathfrak{s}, \delta\mathbf{v} \sim \exp(\imath q x + z_q t)$ propagating in the x-direction and with an exponential dependence on time, we find that the velocity components $(\delta v_y, \delta v_z)$ in the transverse y- and z-directions are decoupled from the deviations $\boldsymbol{\psi}_q = (\delta\rho, \delta\mathfrak{s}, \delta v_x)$. The transverse modes have the dispersion relation (1.47), while the dispersion relations of the three other modes are obtained by solving the eigenvalue problem $\mathsf{L} \cdot \boldsymbol{\psi}_q = z_q\, \boldsymbol{\psi}_q$ with the matrix

$$\mathsf{L} = \begin{bmatrix} 0 & 0 & -\imath\rho q \\ -\frac{\kappa}{\rho T}\left(\frac{\partial T}{\partial \rho}\right)_\mathfrak{s} q^2 & -\frac{\kappa}{\rho T}\left(\frac{\partial T}{\partial \mathfrak{s}}\right)_\rho q^2 & 0 \\ -\frac{\imath}{\rho}\left(\frac{\partial p}{\partial \rho}\right)_\mathfrak{s} q & -\frac{\imath}{\rho}\left(\frac{\partial p}{\partial \mathfrak{s}}\right)_\rho q & -\frac{1}{\rho}\left(\zeta + \frac{4}{3}\eta\right) q^2 \end{bmatrix}. \tag{A.35}$$

This problem can be solved by expanding the eigenvalues and the eigenvectors in powers of $(\imath q)$ up to second order (Balescu, 1975; Résibois and De Leener, 1977). The dispersion relations (1.48) and (1.49) are obtained using the heat capacities (1.51) and

$$\frac{1}{\mathfrak{c}_p} \equiv \frac{1}{T}\left(\frac{\partial T}{\partial \mathfrak{s}}\right)_p = \frac{1}{T}\frac{\partial(T,p)}{\partial(\mathfrak{s},p)} = \frac{1}{T}\frac{\frac{\partial(T,p)}{\partial(\mathfrak{s},\rho)}}{\frac{\partial(\mathfrak{s},p)}{\partial(\mathfrak{s},\rho)}} = \frac{1}{\mathfrak{c}_\mathfrak{v}} - \frac{\left(\frac{\partial T}{\partial \rho}\right)_\mathfrak{s}\left(\frac{\partial p}{\partial \mathfrak{s}}\right)_\rho}{T\left(\frac{\partial p}{\partial \rho}\right)_\mathfrak{s}} = \frac{1}{\mathfrak{c}_\mathfrak{v}} - \frac{T\alpha^2}{\rho\kappa_S \mathfrak{c}_p^2} \tag{A.36}$$

with equation (A.3) (Landau and Lifshitz, 1980a). We note that the heat capacities always satisfy the inequality $\mathfrak{c}_p \geq \mathfrak{c}_\mathfrak{v}$, so that the damping rate of the sound modes (1.49) is always nonnegative.

A.9 Interfacial Nonequilibrium Thermodynamics

A.9.1 Balance Equations in Heterogeneous Media

We consider a system with an interface between two immiscible bulk multicomponent phases. In such systems, nonequilibrium thermodynamics should be extended to include the contributions of the interfacial processes (Bedeaux et al., 1976; Bedeaux, 1986; Kjelstrup and Bedeaux, 2008).

The possibly moving interface is supposed to be located at the surface $f(\mathbf{r},t) = 0$. The interface divides the system into two continuous media, $+$ at $f(\mathbf{r},t) > 0$ and $-$ at $f(\mathbf{r},t) < 0$. These media may be fluids or solids, but at least one of them is a fluid. The vector normal to the interface is denoted $\mathbf{n} \equiv \nabla f(\mathbf{r},t)/\|\nabla f(\mathbf{r},t)\|$. It is convenient to introduce the Heaviside indicator functions of the two bulk phases $\theta^{\pm}(\mathbf{r},t) \equiv \theta\left[\pm f(\mathbf{r},t)\right]$, as well as a Dirac delta distribution located at the interface according to

$$\delta^{\rm s}(\mathbf{r},t) \equiv \|\nabla f(\mathbf{r},t)\|\, \delta\left[f(\mathbf{r},t)\right]. \tag{A.37}$$

Any density x can thus be decomposed as

$$x = x^{+}\theta^{+} + x^{\rm s}\delta^{\rm s} + x^{-}\theta^{-} \tag{A.38}$$

into the densities $x^{\pm}$ of the quantity X in the two bulk phases on both sides of the interface, and the corresponding excess surface density $x^{\rm s}$. As a consequence of these definitions, the bulk and surface densities obey the balance equations,

$$\partial_t x^{\pm} + \nabla\cdot\left(x^{\pm}\mathbf{v}^{\pm} + \mathcal{J}_x^{\pm}\right) = \sigma_x^{\pm}, \tag{A.39}$$

$$\partial_t x^{\rm s} + \nabla\cdot\left(x^{\rm s}\mathbf{v}^{\rm s} + \mathcal{J}_x^{\rm s}\right) = \sigma_x^{\rm s} - \mathbf{n}\cdot\left[x^{+}(\mathbf{v}^{+} - \mathbf{v}^{\rm s}) + \mathcal{J}_x^{+}\right] + \mathbf{n}\cdot\left[x^{-}(\mathbf{v}^{-} - \mathbf{v}^{\rm s}) + \mathcal{J}_x^{-}\right], \tag{A.40}$$

$$\mathbf{n}\cdot\mathcal{J}_x^{\rm s} = 0, \tag{A.41}$$

where $\mathbf{v}^{\pm}$ and $\mathbf{v}^{\rm s}$ are the bulk and interface velocities, $\mathcal{J}_x^{\pm}$ and $\mathcal{J}_x^{\rm s}$ are the bulk and surface diffusive current densities, and $\sigma_x^{\pm}$ and $\sigma_x^{\rm s}$ are the bulk and surface production densities. The quantities of interest in the bulk fluid phases are given in Table 1.1.

We notice that, if the bulk phases remain separated by the interface, the interface velocity is related to the bulk phase velocities according to $\mathbf{n}\cdot\mathbf{v}^{\pm}=\mathbf{n}\cdot\mathbf{v}^{\mathrm{s}}$. If this condition holds, the corresponding terms cancel each other in the right-hand side of equation (A.40) and the expressions of Bedeaux et al. (1976) are recovered.[1] However, this condition is not satisfied if the interface is a moving front of solidification or other phase transition, in which case we may have that $\mathbf{v}^{\pm}=0$ although $\mathbf{v}^{\mathrm{s}}\neq 0$, so that we need, in general, to keep the extra terms in equation (A.40) (Caroli et al., 1992).

A.9.2 Local Equilibrium at the Interface

The entropy balance equation can be established by assuming local equilibrium, implying the validity of the Gibbs and Euler relations in the bulk phases as well as at the interface. The relations (A.1) thus hold for the internal energy densities $e^{\pm}$, the temperatures $T^{\pm}$, the entropy densities $s^{\pm}$, the chemical potentials $\mu_k^{\pm}$ of species k, the particle densities $n_k^{\pm}$, and the hydrostatic pressures $p^{\pm}$ in the two bulk phases. At the interface, the analogous Gibbs and Euler relations are given by

$$de^{\mathrm{s}}=T^{\mathrm{s}}ds^{\mathrm{s}}+\sum_{k=1}^{c}\mu_k^{\mathrm{s}}\,d\Gamma_k\,,\qquad e^{\mathrm{s}}=T^{\mathrm{s}}s^{\mathrm{s}}+\gamma+\sum_{k=1}^{c}\mu_k^{\mathrm{s}}\,\Gamma_k\,,\tag{A.42}$$

where T^{s} is the surface temperature, μ_k^{s} the surface chemical potential of species k, $\gamma=-p^{\mathrm{s}}$ the surface tension defined as minus the hydrostatic surface pressure, and $\Gamma_k=n_k^{\mathrm{s}}$ the excess surface densities of species k. As a consequence, the entropy balance equations can be deduced in the form of equations (A.39)–(A.41) with $x=s$.

A.9.3 Contributions to the Entropy Production at an Interface

The total entropy production rate in a volume V is given by

$$\frac{d_{\mathrm{i}}S}{dt}=\int_V\sigma_s\,d^3r=\int_{V^+}\sigma_s^+\,d^3r+\int_{\text{interface}}\sigma_s^{\mathrm{s}}\,d\Sigma+\int_{V^-}\sigma_s^-\,d^3r\geq 0\tag{A.43}$$

in terms of all the contributions to the entropy production rate in the bulk phases and at the interface. In the bulk phases, the entropy production rate density has the standard bilinear form $\sigma_s^{\pm}=\sum_\alpha\mathcal{A}_\alpha^{\pm}\mathcal{J}_\alpha^{\pm}$ with the affinities and diffusive current densities of Table 1.2. At the interface, the entropy production rate can be expressed similarly in terms of $\sigma_s^{\mathrm{s}}=\sum_\alpha\mathcal{A}_\alpha^{\mathrm{s}}\mathcal{J}_\alpha^{\mathrm{s}}$ with the interfacial affinities and diffusive current densities of Table A.3, where

$$\tilde{\mathbf{\Pi}}=\mathbf{\Pi}+\frac{\rho}{2}(\mathbf{v}-\mathbf{v}^{\mathrm{s}})^2\,\mathbf{1},\quad\tilde{\mathcal{J}}_q=\mathcal{J}_q+h(\mathbf{v}-\mathbf{v}^{\mathrm{s}}),\quad\tilde{\mathcal{J}}_k=\mathcal{J}_k+n_k(\mathbf{v}-\mathbf{v}^{\mathrm{s}}),\tag{A.44}$$

and $h=Ts+g=Ts+\sum_k\mu_k n_k$ is the enthalpy density. If the condition $\mathbf{n}\cdot\mathbf{v}^{\pm}=\mathbf{n}\cdot\mathbf{v}^{\mathrm{s}}$ is satisfied, the contributions obtained by Bedeaux et al. (1976) with $\tilde{\mathbf{\Pi}}=\mathbf{\Pi}$, $\tilde{\mathcal{J}}_q=\mathcal{J}_q$, and $\tilde{\mathcal{J}}_k=\mathcal{J}_k$ are recovered.[1]

[1] This is the case, in particular, at a fluid–solid interface, as considered in Sections 10.7.1 and 12.3.1.

Table A.3. *The irreversible processes at an interface, their affinity $\mathcal{A}^{\rm s}_\alpha$, their associated diffusive current density $\mathcal{J}^{\rm s}_\alpha$, their space character, and the time-reversal parity of their affinity (Bedeaux et al., 1976; Bedeaux, 1986). Here, $(\nabla\mathbf{v}^{\rm s})^{\rm S} = (\nabla\mathbf{v}^{\rm s} + \nabla\mathbf{v}^{\rm s\,T})/2$ denotes the symmetrized gradient of the surface velocity, $\mathbf{1}_\parallel \equiv \mathbf{1} - \mathbf{nn}$, $\overset{\circ}{\mathbf{\Pi}}{}^{\rm s} = \mathbf{\Pi}^{\rm s} - \Pi^{\rm s}\mathbf{1}_\parallel$ is the traceless part of the viscous surface pressure tensor with $\Pi^{\rm s} = ({\rm tr}\,\mathbf{\Pi}^{\rm s})/2$, ν_{kr} the stoichiometric coefficient of species k in reaction r, and $\nabla_\parallel$ the tangential gradient.*

Irreversible Process	$\mathcal{A}^{\rm s}_\alpha$	$\mathcal{J}^{\rm s}_\alpha$	Space	Time
Interfacial shear viscosity	$\overset{\circ}{\mathbf{A}}{}^{\rm s}_{\mathbf{g}} = -\frac{1}{T^{\rm s}}\mathbf{1}_\parallel \cdot \left[(\nabla\mathbf{v}^{\rm s})^{\rm S} - \frac{1}{2}(\nabla\cdot\mathbf{v}^{\rm s})\mathbf{1}\right]\cdot\mathbf{1}_\parallel$	$\overset{\circ}{\mathbf{J}}{}^{\rm s}_{\mathbf{g}} = \overset{\circ}{\mathbf{\Pi}}{}^{\rm s}$	Tensor	Odd
Interfacial dilational viscosity	$\mathcal{A}^{\rm s}_{\mathbf{g}} = -\frac{1}{T^{\rm s}}\nabla\cdot\mathbf{v}^{\rm s}$	$\mathcal{J}^{\rm s}_{\mathbf{g}} = \Pi^{\rm s}$	Scalar	Odd
Interfacial reaction r	$\mathcal{A}^{\rm s}_r = -\frac{1}{T^{\rm s}}\sum_k \mu^{\rm s}_k \nu_{kr}$	$\mathcal{J}^{\rm s}_r = w^{\rm s}_r$	Scalar	Even
Heat conductivity inside the interface	$\boldsymbol{\mathcal{A}}^{\rm s}_{q\parallel} = \nabla_\parallel \frac{1}{T^{\rm s}}$	$\boldsymbol{\mathcal{J}}^{\rm s}_{q\parallel} = \boldsymbol{\mathcal{J}}^{\rm s}_q$	Vector	Even
Transport of species k inside the interface	$\boldsymbol{\mathcal{A}}^{\rm s}_{k\parallel} = -\nabla_\parallel \frac{\mu^{\rm s}_k}{T^{\rm s}}$	$\boldsymbol{\mathcal{J}}^{\rm s}_{k\parallel} = \boldsymbol{\mathcal{J}}^{\rm s}_k$	Vector	Even
Interfacial slippage	$\boldsymbol{\mathcal{A}}^{\rm s}_{\mathbf{v}\parallel} = -\frac{1}{T^{\rm s}}(\mathbf{v}^+ - \mathbf{v}^-)$	$\boldsymbol{\mathcal{J}}^{\rm s}_{\mathbf{v}\parallel} = \frac{1}{2}\mathbf{n}\cdot(\tilde{\mathbf{\Pi}}^+ + \tilde{\mathbf{\Pi}}^-)$	Vector	Odd
Interfacial displacement	$\boldsymbol{\mathcal{A}}^{\rm s}_{\mathbf{v}\top} = -\frac{1}{T^{\rm s}}(\frac{\mathbf{v}^+ + \mathbf{v}^-}{2} - \mathbf{v}^{\rm s})$	$\boldsymbol{\mathcal{J}}^{\rm s}_{\mathbf{v}\top} = \mathbf{n}\cdot(\tilde{\mathbf{\Pi}}^+ - \tilde{\mathbf{\Pi}}^-)$	Vector	Odd
Heat conductivity across the interface	$\mathcal{A}^{\rm s}_{q\perp} = \frac{1}{T^+} - \frac{1}{T^-}$	$\mathcal{J}^{\rm s}_{q\perp} = \frac{1}{2}\mathbf{n}\cdot\left(\tilde{\boldsymbol{\mathcal{J}}}^+_q + \tilde{\boldsymbol{\mathcal{J}}}^-_q\right)$	Scalar	Even
Heat conductivity to the interface	$\mathcal{A}^{\rm s}_{q\top} = \frac{1}{2}(\frac{1}{T^+} + \frac{1}{T^-}) - \frac{1}{T^{\rm s}}$	$\mathcal{J}^{\rm s}_{q\top} = \mathbf{n}\cdot\left(\tilde{\boldsymbol{\mathcal{J}}}^+_q - \tilde{\boldsymbol{\mathcal{J}}}^-_q\right)$	Scalar	Even
Transport of species k across the interface	$\mathcal{A}^{\rm s}_{k\perp} = -\frac{\mu^+_k}{T^+} + \frac{\mu^-_k}{T^-}$	$\mathcal{J}^{\rm s}_{k\perp} = \frac{1}{2}\mathbf{n}\cdot\left(\tilde{\boldsymbol{\mathcal{J}}}^+_k + \tilde{\boldsymbol{\mathcal{J}}}^-_k\right)$	Scalar	Even
Transport of species k to the interface	$\mathcal{A}^{\rm s}_{k\top} = -\frac{1}{2}(\frac{\mu^+_k}{T^+} + \frac{\mu^-_k}{T^-}) + \frac{\mu^{\rm s}_k}{T^{\rm s}}$	$\mathcal{J}^{\rm s}_{k\top} = \mathbf{n}\cdot\left(\tilde{\boldsymbol{\mathcal{J}}}^+_k - \tilde{\boldsymbol{\mathcal{J}}}^-_k\right)$	Scalar	Even

In the field of interfacial rheology, the coefficient $\eta^{\rm s}$ of interfacial shear viscosity can be introduced with the linear relation $\overset{\circ}{\mathbf{\Pi}}{}^{\rm s} = 2\,T^{\rm s}\eta^{\rm s}\,\overset{\circ}{\mathbf{A}}{}^{\rm s}_{\rm g}$ between the corresponding current density and affinity, while the coefficient $\zeta^{\rm s}$ of interfacial dilational viscosity is similarly defined with $\Pi^{\rm s} = T^{\rm s}\zeta^{\rm s}\mathcal{A}^{\rm s}_{\rm g}$ (Scriven, 1960; Edwards et al., 1991). In surface science, interfacial diffusion is described by the linear relation $\mathcal{J}^{\rm s}_{k\|} = \mathcal{L}^{\rm s}_{k,k}\mathcal{A}^{\rm s}_{k\|}$ between the current density and affinity associated with the transport of species k in the interface,[2] and surface reactions (including adsorption and desorption) by nonlinear relations between the interfacial reaction rates $w^{\rm s}_r$ and affinities $\mathcal{A}^{\rm s}_r$ (Kreuzer and Gortel, 1986; Garcia Cantú Ros et al., 2011). The interfacial processes of diffusion and reaction play essential roles in heterogeneous catalysis (McEwen et al., 2010a,b). Interfacial sliding friction is ruled by the linear relation $\mathcal{J}^{\rm s}_{{\rm v}\|} = \mathcal{L}^{\rm s}_{\rm v,v}\mathcal{A}^{\rm s}_{{\rm v}\|}$ between the current density and affinity associated with interfacial slippage. Diffusiophoresis results from the coupling between the current density of interfacial slippage $\mathcal{J}^{\rm s}_{{\rm v}\|}$ and the affinities $\mathcal{A}^{\rm s}_{k\|}$ for the transport of species k in the interface (Gaspard and Kapral, 2018b,c). Thermophoresis is caused by a similar coupling to the interfacial heat affinity $\mathcal{A}^{\rm s}_{q\|}$ (Gaspard and Kapral, 2019a). The Seebeck and Peltier thermoelectric effects at bimetallic junctions are similarly described (Kjelstrup and Bedeaux, 2008).

[2] As a consequence, Fick's law, $\mathcal{J}^{\rm s}_{k\|} = -\mathcal{D}^{\rm s}_k \nabla_{\|}\Gamma_k$, holds for surface diffusion with the coefficient $\mathcal{D}^{\rm s}_k = (\mathcal{L}^{\rm s}_{k,k}/T^{\rm s})(\partial\mu^{\rm s}_k/\partial\Gamma_k)_{T^{\rm s}}$.

Appendix B

Complements on Dynamical Systems Theory

B.1 Generalities

B.1.1 From Differential Equations to Flows

Dynamical systems are deterministic time evolution processes (Birkhoff, 1927). They can be defined locally in time in terms of ordinary or partial differential equations. These latter can be reduced to the former if the fields considered in spatially extended systems are decomposed into Fourier modes. Moreover, ordinary differential equations with an order larger than one can be written in the form of a set of coupled first-order differential equations by introducing supplementary variables to get

$$\frac{d\mathbf{X}}{dt} = \mathbf{V}(\mathbf{X}, t) \qquad \text{with} \qquad \mathbf{X} \in \mathcal{M} \subseteq \mathbb{R}^m, \tag{B.1}$$

where $\mathcal{M}$ is the phase space of dimension $m = \dim \mathcal{M}$. The m variables $\mathbf{X}$ are supposed to specify the state of the system. The time evolution is determined by the vector field $\mathbf{V}(\mathbf{X}, t)$ giving the speed of evolution at every point $\mathbf{X}$ of phase space and every instant of time t. This time dependence may be constant, periodic, quasiperiodic, or something else. If it is noisy, the process is not deterministic, but stochastic.

The vector field $\mathbf{V}(\mathbf{X}, t)$ is assumed to be smooth enough for the existence and uniqueness theorem to apply, so that the state at time t is given in terms of the initial conditions prepared at time t_0 according to $\mathbf{X}_t = \mathbf{\Phi}_{t_0}^t \mathbf{X}_{t_0}$, where the m-dimensional function $\mathbf{\Phi}_{t_0}^t$ is obtained by time integration. This function gives the trajectory or orbit $\mathbf{X}_t$ as a function of time and initial conditions. Since the initial conditions uniquely specify the trajectory at any time, this function satisfies $\mathbf{\Phi}_{t_0}^{t_2} = \mathbf{\Phi}_{t_1}^{t_2} \mathbf{\Phi}_{t_0}^{t_1}$ and also $\mathbf{\Phi}_t^t = \mathbf{1}$.

If the system is autonomous (i.e., its vector field is constant in time so that $\partial_t \mathbf{V} = 0$), the function $\mathbf{\Phi}_{t_0}^t$ only depends on the time difference $t - t_0$, thus defining the flow $\mathbf{\Phi}^{t-t_0}$, which forms the one-parameter continuous group obeying the properties (2.21).

In general, the vector field $\mathbf{V}(\mathbf{X}, t)$ is defined with nonlinear functions of the variables $\mathbf{X}$, so that complex dynamical behaviors such as chaotic dynamics may manifest themselves.

We note that dynamical systems called maps or mappings can be defined if time takes discrete values (Ott, 1993; Strogatz, 1994; Nicolis, 1995).

B.1.2 Linear Stability Analysis and Lyapunov Exponents

An important issue is to determine if some trajectory $\mathbf{X}_t$ is stable or unstable. In order to address this issue, the multidimensional ordinary differential equation (B.1) can be linearized in the vicinity of the trajectory $\mathbf{X}_t$. The infinitesimal deviations $\delta\mathbf{X} = \mathbf{X}'_t - \mathbf{X}_t$ with respect to the reference trajectory $\mathbf{X}_t$ satisfy the multidimensional linear equations

$$\frac{d}{dt}\,\delta\mathbf{X} = \frac{\partial\mathbf{V}}{\partial\mathbf{X}}(\mathbf{X}_t)\cdot\delta\mathbf{X}, \tag{B.2}$$

here assuming that the system is autonomous. Because of its linearity, its solutions can be expressed as

$$\delta\mathbf{X}_t = \mathsf{M}(t,\mathbf{X}_0)\cdot\delta\mathbf{X}_0, \qquad \text{where} \qquad \mathsf{M}(t,\mathbf{X}_0) \equiv \frac{\partial\mathbf{X}_t}{\partial\mathbf{X}_0} \tag{B.3}$$

is the $m\times m$ fundamental matrix giving the general solution of the linear equations (B.2) as a linear superposition, and $\mathbf{X}_t = \mathbf{\Phi}^t\mathbf{X}_0$. The reference trajectory is stable or unstable depending on the temporal behavior of the magnitude of the infinitesimal deviations. Denoting the transpose by the superscript T, the square of the deviations (B.3) can be written as

$$\|\delta\mathbf{X}_t\|^2 = \delta\mathbf{X}_0\cdot\mathsf{M}(t,\mathbf{X}_0)^{\mathrm{T}}\cdot\mathsf{M}(t,\mathbf{X}_0)\cdot\delta\mathbf{X}_0 = \sum_{j=1}^{m}\sigma_j(t,\mathbf{X}_0)\big[\mathbf{u}_j(t,\mathbf{X}_0)^{\mathrm{T}}\cdot\delta\mathbf{X}_0\big]^2, \tag{B.4}$$

where $\{\sigma_j\}_{j=1}^m$ and $\{\mathbf{u}_j\}_{j=1}^m$ are, respectively, the eigenvalues and orthonormal eigenvectors of the $m\times m$ real symmetric matrix $\mathsf{M}^{\mathrm{T}}\cdot\mathsf{M}$. In order to characterize the possible exponential increase or decrease in time of the infinitesimal deviations $\delta\mathbf{X}_t$, the *Lyapunov exponents* associated with the trajectory of initial conditions $\mathbf{X}_0$ are introduced as

$$\lambda_j(\mathbf{X}_0) \equiv \lim_{t\to\infty}\frac{1}{2t}\,\ln\sigma_j(t,\mathbf{X}_0). \tag{B.5}$$

Accordingly, the magnitude of the infinitesimal deviations would behave as $\|\delta\mathbf{X}_t\| \sim \exp(\lambda_j t)$ in the limit $t\to\infty$ if the initial deviations were oriented in the direction of the corresponding eigenvector, $\delta\mathbf{X}_0 \sim \mathbf{u}_j$. This definition is equivalent to that of equation (2.23). In this way, a spectrum of Lyapunov exponents $\{\lambda_j\}_{j=1}^m$ can be defined (Eckmann and Ruelle, 1985).

We note that, if the trajectory $\mathbf{X}_t$ does not converge to a stationary point $\mathbf{X}_{\mathrm{st}}$ such that $\mathbf{V}(\mathbf{X}_{\mathrm{st}}) = 0$, its speed $\mathbf{V}(\mathbf{X}_t)$ does not vanish and the Lyapunov exponent associated with the direction of the flow $\mathbf{u}_k \propto \mathbf{V}(\mathbf{X}_0)$ is equal to zero, i.e., $\lambda_k = 0$. Indeed, the infinitesimal deviations in the direction of the flow are given by $\delta\mathbf{X}_t = \mathbf{V}(\mathbf{X}_t)\,\delta t$, so that the associated Lyapunov exponent can be evaluated according to

$$\lambda_k = \lim_{t\to\infty}\frac{1}{t}\,\ln\frac{\|\delta\mathbf{X}_t\|}{\|\delta\mathbf{X}_0\|} = \lim_{t\to\infty}\frac{1}{t}\,\ln\frac{\|\mathbf{V}(\mathbf{X}_t)\|}{\|\mathbf{V}(\mathbf{X}_0)\|} = 0, \tag{B.6}$$

which is vanishing if the trajectory keeps away from fixed points. The Lyapunov exponents associated with the directions normal to the hypersurfaces defined by the constants of motion are also equal to zero.

Different definitions are adopted to characterize the stability or instability of a trajectory (Nicolis, 1995). The trajectory $\mathbf{X}_t$ is said to be stable in the sense of Lyapunov if the condition $\|\mathbf{X}'_t - \mathbf{X}_t\| \leq \varepsilon$ for all $t > 0$ implies that there exists some $\delta(\varepsilon) > 0$ such that $\|\mathbf{X}'_0 - \mathbf{X}_0\| \leq \delta$. The trajectory is said to be asymptotically stable if, moreover, $\lim_{t\to\infty} \|\mathbf{X}'_t - \mathbf{X}_t\| = 0$ for all initial conditions $\mathbf{X}'_0$ such that $\|\mathbf{X}'_0 - \mathbf{X}_0\| \leq \delta$. Otherwise, the trajectory is said to be unstable.

The trajectory is asymptotically stable if all the Lyapunov exponents are negative. If the trajectory is stable in the sense of Lyapunov, all its Lyapunov exponents are either negative or equal to zero. The trajectory is unstable if there is at least one positive Lyapunov exponent. In this latter case, the system manifests the property of sensitivity to initial conditions for the trajectory $\mathbf{X}_t$.

Now, the sum of all the Lyapunov exponents is related to the divergence of the vector field according to

$$\sum_{j=1}^{m} \lambda_j(\mathbf{X}_0) = \lim_{t\to\infty} \frac{1}{t} \int_0^t \operatorname{div} \mathbf{V}(\mathbf{\Phi}^\tau \mathbf{X}_0)\, d\tau, \tag{B.7}$$

where $\operatorname{div} \mathbf{V} = \operatorname{tr}(\partial \mathbf{V}/\partial \mathbf{X})$. Indeed, the infinitesimal phase-space volume $d^m X$ changes in time along the trajectory as

$$\frac{d^m X_t}{d^m X_0} = \left|\det \frac{\partial \mathbf{X}_t}{\partial \mathbf{X}_0}\right| = |\det \mathsf{M}(t, \mathbf{X}_0)| = \exp \int_0^t \operatorname{div} \mathbf{V}(\mathbf{\Phi}^\tau \mathbf{X}_0)\, d\tau, \tag{B.8}$$

while $|\det \mathsf{M}(t,\mathbf{X}_0)| = \prod_{j=1}^m \sigma_j(t,\mathbf{X}_0)^{1/2}$. Therefore, the phase-space volume contracts if $\int_0^t \operatorname{div} \mathbf{V}\, d\tau < 0$, so that the sum of Lyapunov exponents is negative. However, the phase-space volume is preserved if $\operatorname{div} \mathbf{V} = 0$, in which case the sum of Lyapunov exponents is equal to zero. This difference in the behavior of phase-space volumes constitutes a fundamental difference among the types of dynamical systems, as further discussed below.

B.1.3 Generalized Liouville Equation

As explained in Chapter 2, the time evolution of the probability density $p_t(\mathbf{X})$ to find the system in the state $\mathbf{X}$ at time t is ruled by the generalized Liouville equation (2.37), which is established by using the principle of local probability conservation in phase space (Nicolis, 1995). Since dynamical systems are deterministic, the probability density obeys the continuity equation $\partial_t p + \operatorname{div} \mathbf{J} = 0$ with the purely advective current density $\mathbf{J} = \mathbf{V} p$. Expanding the divergence, the generalized Liouville equation can be written in the form

$$\frac{dp}{dt} = \partial_t p + \mathbf{V} \cdot \operatorname{grad} p = -p \operatorname{div} \mathbf{V}. \tag{B.9}$$

Solving this equation in terms of the flow, $\mathbf{\Phi}^t$, shows that the probability density evolves in time according to

$$p_t(\mathbf{X}) = p_0\left(\mathbf{\Phi}^{-t}\mathbf{X}\right) \exp\left[-\int_0^t \operatorname{div} \mathbf{V}\left(\mathbf{\Phi}^{\tau-t}\mathbf{X}\right) d\tau\right]. \tag{B.10}$$

If the system preserves phase-space volumes (i.e., if $\text{div}\,\mathbf{V} = 0$), we recover the property (2.45), meaning that the probability density remains constant in time along the trajectory. However, if the phase-space volumes contract, the probability density increases in time.

B.1.4 Stationary Probability Distribution and Path Probabilities

An important issue is to determine the stationary probability density $p_{\text{st}}(\mathbf{X})$ of the generalized Liouville equation. For this purpose, ergodic theory can be considered, using the time average already introduced in equation (1.109) or (2.65) in order to obtain the stationary probability density (2.66), which is related to the ergodic property (2.68). The support of the stationary probability distribution is a subset $\mathcal{I}$ of phase space that is invariant under time evolution, i.e., such that $\mathcal{I} = \mathbf{\Phi}^t\mathcal{I}$ for all time t. We note that, if the system is ergodic, the Lyapunov exponents (B.5) take equal values for almost all trajectories with respect to the stationary probability distribution. Therefore, the spectrum of Lyapunov exponents provides a characterization of dynamical instability with respect to the so-defined stationary state.

On the one hand, ergodicity is discussed in Chapter 2 for Hamiltonian systems that are volume preserving, in which cases there is no possible point-like convergence of the probability density towards the stationary one. Volume-preserving systems may thus have invariant subsets extending over phase-space domains with positive phase-space volume.

On the other hand, for volume-contracting systems, some invariant subsets with zero phase-space volume, called *attractors*, may exist in phase space and they can attract all the trajectories coming from some phase-space domain with positive phase-space volume, called the basin of attraction (which contains the attractor). This is, in particular, the case for a fixed point that is asymptotically stable and attracting all the trajectories from initial conditions inside a ball of radius δ around it. For such a volume-contracting system, the probability density may undergo a point-like convergence towards the asymptotic stationary probability density that is concentrated on the fixed point in the form of the Dirac delta distribution, $p_{\text{st}}(\mathbf{X}) = \delta^m(\mathbf{X} - \mathbf{X}_{\text{st}})$. Such a phenomenon also manifests itself for other kinds of attractors such as attracting limit cycles (i.e., asymptotically stable periodic orbits), quasiperiodic attractors, or fractal attractors sustaining chaotic dynamics.

The properties of the stationary probability density may be related to the way phase-space volumes are locally contracted or stretched, as described by the Lyapunov exponents. In this regard, we may consider the time evolution for a ball of radius ε and centered on the point $\mathbf{X}_0$ in phase space, as schematically depicted in Figure B.1. During some finite time interval $t \in [0, T]$, the ball is stretched in the unstable directions associated with positive Lyapunov exponents and contracted in the stable directions with negative Lyapunov exponents. The question is to determine the stationary probability of all the initial conditions leading to orbits that remain within the distance ε of the reference orbit $\mathbf{X}_t$ during the time interval. These initial conditions belong to the following domain:

$$B_T(\mathbf{X}_0, \varepsilon) = \left\{\mathbf{X}'_0 \in \mathcal{M} : \left\|\mathbf{\Phi}^t\mathbf{X}'_0 - \mathbf{\Phi}^t\mathbf{X}_0\right\| < \varepsilon,\ \forall t \in [0, T]\right\}, \tag{B.11}$$

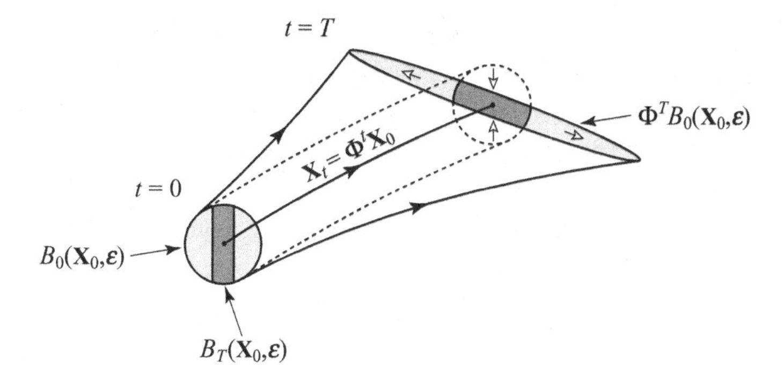

Figure B.1 Schematic representation in phase space of the ball $B_0(\mathbf{X}_0, \varepsilon)$ of radius ε and centered on the point $\mathbf{X}_0$ and its time evolution by the flow $\mathbf{\Phi}^t$ during the time interval $0 \leq t \leq T$. The ball undergoes stretching in the unstable directions, contraction in the stable directions, while the direction of the flow is neutral. The dashed lines delimit the cylinder of radius ε around the orbit $\mathbf{X}_t$ from the initial condition $\mathbf{X}_0$. The domain $B_T(\mathbf{X}_0, \varepsilon)$ contains all the initial conditions of orbits remaining within the distance ε of the orbit $\mathbf{X}_t$ during the whole time interval $0 \leq t \leq T$.

which corresponds to a tube around the reference trajectory $\mathbf{\Phi}^t\mathbf{X}_0$ with $0 \leq t \leq T$. If ε is small enough, the section of this tube at time t is an ellipsoid, because

$$\left\| \mathbf{\Phi}^t \mathbf{X}'_0 - \mathbf{\Phi}^t \mathbf{X}_0 \right\|^2 \simeq \left\| \frac{\partial \mathbf{X}_t}{\partial \mathbf{X}_0} \cdot \delta \mathbf{X}_0 \right\|^2 = \sum_{j=1}^{m} \sigma_j \left(\mathbf{u}_j^{\mathrm{T}} \cdot \delta \mathbf{X}_0 \right)^2 < \varepsilon^2, \tag{B.12}$$

for $\delta\mathbf{X}_0 = \mathbf{X}'_0 - \mathbf{X}_0$, according to equation (B.4). Therefore, the principal semiaxes of this ellipsoid extend over $\left|\mathbf{u}_j^{\mathrm{T}} \cdot \delta\mathbf{X}_0\right| < \varepsilon/\sqrt{\sigma_j}$. If this condition holds during the whole time interval $0 \leq t \leq T$, the domain is mainly determined by the directions of stretching, so that the stationary probability of the domain (B.11) behaves as

$$P_{\mathrm{st}}\left\{B_T(\mathbf{X}_0, \varepsilon)\right\} \sim \prod_{j:\, \sigma_j > 1} \frac{\varepsilon}{\sqrt{\sigma_j(T, \mathbf{X}_0)}} \sim \exp\left(- \sum_{\lambda_j > 0} \lambda_j T \right). \tag{B.13}$$

This probability is thus decreasing as the time interval T increases with a rate given by the sum of positive Lyapunov exponents. Since the domain (B.11) is the analogue of the cell $\mathcal{C}_\omega$ in the definition of path probability (6.1), the decay rate of the probability (B.13) gives the value of the so-called Kolmogorov–Sinai (KS) entropy per unit time, which is thus equal to the sum of positive Lyapunov exponents

$$h_{\mathrm{KS}} = \sum_{\lambda_j > 0} \lambda_j, \tag{B.14}$$

characterizing temporal disorder in the dynamics (Eckmann and Ruelle, 1985; Ott, 1993).

B.1.5 Hausdorff Dimension and Fractals

Any invariant subset $\mathcal{I}$ of phase space may be characterized by the Hausdorff dimension (Falconer, 1990). This dimension is evaluated by covering the subset $\mathcal{I}$ with small enough sets $\{\mathcal{U}_i\}$, such as balls. The diameter of these sets is denoted $\varepsilon_i = \sup\{\|\mathbf{X} - \mathbf{X}'\|; \forall \mathbf{X}, \mathbf{X}' \in \mathcal{U}_i\}$. All the small sets in the cover are supposed to have a diameter bounded by ε, $\varepsilon_i \leq \varepsilon$, which defines an ε-cover for the subset $\mathcal{I}$. For any nonnegative real number s, the following function is introduced:

$$\mathscr{H}_s(\mathcal{I}) \equiv \lim_{\varepsilon \to 0} \inf_{\{\mathcal{U}_i\}_\varepsilon} \sum_i \varepsilon_i^s , \tag{B.15}$$

where the infimum is taken over all the ε-covers $\{\mathcal{U}_i\}_\varepsilon$ of the subset $\mathcal{I}$. The *Hausdorff dimension* of the subset $\mathcal{I}$ is thus defined as

$$d_{\mathrm{H}}(\mathcal{I}) \equiv \inf\{s : \mathscr{H}_s(\mathcal{I}) = 0\} = \sup\{s : \mathscr{H}_s(\mathcal{I}) = \infty\}. \tag{B.16}$$

This dimension can be evaluated by requiring that

$$\sum_i \varepsilon_i^{d_{\mathrm{H}}} \sim 1 \tag{B.17}$$

in the limit where the diameters ε_i of the covering sets $\mathcal{U}_i$ are vanishing with $\varepsilon \to 0$. In this limit, their number is increasing to balance the reduction of their size.

A related concept is the so-called *box-counting dimension* defined according to

$$d_{\mathrm{B}}(\mathcal{I}) \equiv \lim_{\varepsilon \to 0} \frac{\ln N(\varepsilon;\mathcal{I})}{\ln(1/\varepsilon)} \tag{B.18}$$

by counting the number $N(\varepsilon;\mathcal{I})$ of small sets $\{\mathcal{U}_i\}$ of diameter ε needed to cover the subset $\mathcal{I}$. In general, the box-counting dimension is an upper bound on the Hausdorff dimension ($d_{\mathrm{H}} \leq d_{\mathrm{B}}$), but the equality holds for many regular enough subsets $\mathcal{I}$ (Falconer, 1990). Formula (B.18) is obtained from equation (B.17) by considering that all the diameters ε_i are equal to ε, in which case the sum reduces to the corresponding number $N(\varepsilon;\mathcal{I})$ of covering sets.

The Hausdorff and box-counting dimensions of points and lines are, respectively, equal to zero and one. There also exist subsets $\mathcal{I}$ with a noninteger Hausdorff dimension. This is often the case for subsets with a nontrivial self-similar structure, which were called *fractals* by Mandelbrot (1982).

If the invariant subset is equipped with a stationary probability distribution, the concepts of generalized dimension and multifractals have also been introduced (Halsey et al., 1986; Bessis et al., 1988).

B.2 Dissipative Dynamical Systems

Dissipative dynamical systems are encountered at the macroscale in hydrodynamics, chemical kinetics, or macroscopic mechanics, where the time evolution of macrovariables is strongly influenced by energy dissipation (Bergé et al., 1984; Scott, 1991; Ott, 1993; Strogatz, 1994; Nicolis, 1995). The macrovariables may be the amplitudes of the Fourier modes of hydrodynamic fields such as the fluid velocity, the temperature, or the concentrations of molecular species. If the system is uniform as in continuous-flow stirred tank reactors (introduced in Section 1.8.1), the macrovariables are the chemical concentrations themselves. Usually, dissipative systems are ruled by dynamical systems (B.1) satisfying

$$\frac{1}{t}\int_0^t \operatorname{div}\mathbf{V}\,d\tau < 0 \tag{B.19}$$

for large enough time t (Nicolis, 1995). Accordingly, the phase-space volumes contract and the sum of their Lyapunov exponents is negative, i.e., $\sum_{j=1}^{m}\lambda_j < 0$. Their trajectories often evolve towards attractors, as mentioned in Section 1.9.

For stationary attractors (i.e., asymptotically stable fixed points), all the Lyapunov exponents are negative, i.e., $\lambda_j < 0$ with $j = 1, 2, \ldots, m$.

For periodic attractors referred to as limit cycles, the Lyapunov exponent corresponding to the direction of the periodic orbit is equal to zero, while all the other Lyapunov exponents are negative.

Quasiperiodic attractors form tori in phase space. The dimension of the torus is equal to the number of independent frequencies of quasiperiodic motion. This dimension gives the number of zero Lyapunov exponents and all the others are negative.

Chaotic attractors such as the famous attractor of Lorenz (1963) have at least one positive Lyapunov exponent. Additionally, the Lyapunov exponent of the flow direction is equal to zero and others are negative. Chaotic attractors typically form fractals in phase space (Mandelbrot, 1982; Bergé et al., 1984; Ott, 1993; Nicolis, 1995).

B.3 Volume-Preserving Dynamical Systems

These systems satisfy the condition $\operatorname{div}\mathbf{V} = 0$ of volume preservation in phase space. There are several types of volume-preserving dynamical systems. The Hamiltonian systems preserve phase-space volumes since they obey Liouville's theorem. Additionally, the motion of a tracer particle advected in an incompressible fluid gives an example of non-Hamiltonian volume-preserving dynamics, because $d\mathbf{r}/dt = \mathbf{v}(\mathbf{r}, t)$ with $\mathbf{r}, \mathbf{v} \in \mathbb{R}^3$ and the incompressibility condition $\nabla \cdot \mathbf{v} = 0$ holds.

B.3.1 Hamiltonian Dynamical Systems

The Hamiltonian dynamical systems are introduced in Chapter 2. They find their origin in the semiclassical limit of the underlying quantum mechanics (Feynman and Hibbs, 1965). As a consequence, the classical trajectories are obtained by the action principle, also known as Hamilton's variational principle (Goldstein, 1950; Arnold, 1989). Accordingly, the action

$$W = \int [\mathbf{p}\cdot d\mathbf{q} - H(\mathbf{q}, \mathbf{p}, t)\,dt] \tag{B.20}$$

is defined in terms of the Hamiltonian function $H(\mathbf{q}, \mathbf{p}, t)$ in the phase space $\mathbf{X} = \mathbf{\Gamma} = (\mathbf{q}, \mathbf{p}) \in \mathbb{R}^{2f}$ for a system with f degrees of freedom and its first variation is vanishing, so that $\delta W = 0$, along the solutions of the Hamilton equations (2.9). The vector field of Hamiltonian systems can be expressed as

$$\frac{d\mathbf{\Gamma}}{dt} = \mathbf{V}(\mathbf{\Gamma},t) = \mathbf{\Sigma} \cdot \text{grad}_{\mathbb{R}^{2f}} H \qquad \text{with} \qquad \mathbf{\Sigma} = \begin{pmatrix} \mathbf{0} & \mathbf{1} \\ -\mathbf{1} & \mathbf{0} \end{pmatrix} \tag{B.21}$$

in terms of the $2f$-dimensional gradient of the Hamiltonian function $H(\mathbf{\Gamma},t)$ and the so-called fundamental symplectic matrix $\mathbf{\Sigma}$, where $\mathbf{1}$ denotes the $f \times f$ identity matrix. The phase-space dimension of Hamiltonian dynamical systems is thus always even, so $m = 2f$. The symplectic form (B.21) of the vector field implies that its divergence is vanishing, thus establishing the Liouville theorem (2.15). As a consequence of the symplectic structure of Hamiltonian dynamics, the $2f \times 2f$ matrix (B.3) of the linearized dynamics is symplectic, i.e., it satisfies the condition

$$\mathsf{M}^{\mathrm{T}} \cdot \mathbf{\Sigma} \cdot \mathsf{M} = \mathbf{\Sigma}. \tag{B.22}$$

This symmetry implies the pairing rule, according to which the Lyapunov exponents form pairs $\{\lambda_j, -\lambda_j\}$, each having the exact opposite value of the other. This remarkable property of the Lyapunov spectrum in Hamiltonian systems results from the symplectic symmetry such that, if the matrix $\mathsf{M}^{\mathrm{T}} \cdot \mathsf{M}$ has the eigenvalue σ_j and the associated eigenvector $\mathbf{u}_j$, then σ_j^{-1} is the eigenvalue corresponding to the eigenvector $\mathbf{\Sigma}^{\mathrm{T}} \cdot \mathbf{u}_j$. As a consequence of the pairing rule, the sum of Lyapunov exponents is equal to zero, which is consistent with the preservation of phase-space volumes by Liouville's theorem.

We note that Hamiltonian systems may be symmetric under time reversal if condition (2.28) is satisfied, or not if $H(\mathbf{\Theta\Gamma}) \neq H(\mathbf{\Gamma})$, which would be the case, for instance, in the presence of some given vector potential that is not reversed upon time reversal, i.e., $\mathbf{\Theta}(\mathbf{q}, \mathbf{p}) = (\mathbf{q}, -\mathbf{p})$.

B.3.2 Standard Map

In order to illustrate the specific properties of Hamiltonian dynamical systems, we may consider the so-called *standard map* (Chirikov, 1979; Lichtenberg and Lieberman, 1983). This map can be defined in terms of the periodically kicked one-dimensional system ruled by the Hamiltonian function

$$H(x,p,t) = \frac{p^2}{2m} + K \cos x \sum_{i=-\infty}^{+\infty} \delta(t - iT) \tag{B.23}$$

with mass $m = 1$ and period $T = 1$. If $x_i \in [0, 2\pi[$ denotes the position at time $t = iT$ and p_i the momentum for $(i-1)T < t < iT$, the time integration of Hamilton's equations gives the standard map

$$\phi \begin{cases} x_{i+1} = x_i + p_i + K \sin x_i, \\ p_{i+1} = p_i + K \sin x_i. \end{cases} \tag{B.24}$$

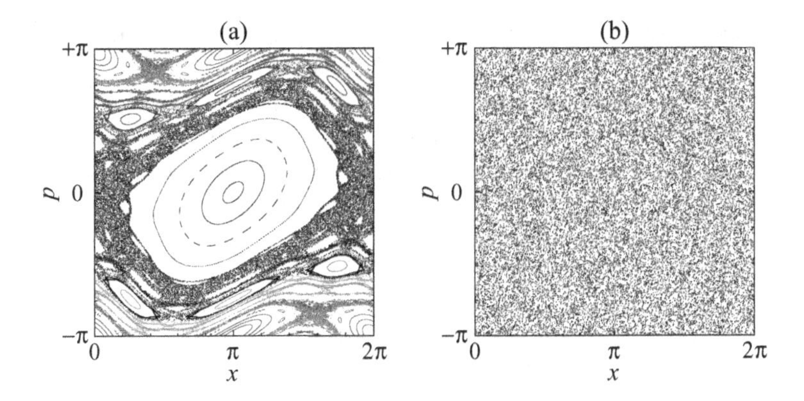

Figure B.2 Phase portraits of the standard map (B.24) for (a) $K = 1$; (b) $K = 10$.

This map satisfies the condition (2.29) of microreversibility,

$$\boldsymbol{\theta}\,\boldsymbol{\phi} = \boldsymbol{\phi}^{-1}\boldsymbol{\theta}, \qquad \text{where} \qquad \boldsymbol{\theta}(x,p) = (x-p,\,-p) \tag{B.25}$$

is the time-reversal transformation, which is an involution because $\boldsymbol{\theta}^2 = \mathbf{1}$ (MacKay, 1993).

The standard map can also be deduced from Hamilton's variational principle. Indeed, the time integration of the action (B.20) with the Hamiltonian function (B.23) and $x_t = x_i + p_{i+1}(t - iT)$ for $iT < t < (i+1)T$ leads to

$$W = \sum_{i=-\infty}^{+\infty} F(x_i, x_{i+1}) \quad \text{with} \quad F(x_i, x_{i+1}) = \frac{1}{2}(x_{i+1} - x_i)^2 - K\cos x_i . \tag{B.26}$$

Accordingly, the standard map is given by the canonical transformation

$$p_{i+1} = \frac{\partial F}{\partial x_{i+1}}, \qquad p_i = -\frac{\partial F}{\partial x_i}. \tag{B.27}$$

Therefore, the standard map has the symplectic structure, implying that the 2×2 matrix of the linearized dynamics

$$\mathbf{M}_i = \begin{pmatrix} \frac{\partial x_{i+1}}{\partial x_i} & \frac{\partial p_{i+1}}{\partial x_i} \\ \frac{\partial x_{i+1}}{\partial p_i} & \frac{\partial p_{i+1}}{\partial p_i} \end{pmatrix} = \begin{pmatrix} 1 + K\cos x_i & K\cos x_i \\ 1 & 1 \end{pmatrix} \tag{B.28}$$

satisfies the symplectic condition (B.22) with respect to the 2×2 fundamental symplectic matrix introduced in equation (B.21). Consequently, $|\det \mathbf{M}_i| = 1$, so that the standard map is area preserving. The uniform probability distribution $dx\,dp$, also called the Lebesgue measure, is thus stationary for this system.

Phase portraits are shown in Figure B.2 for the standard map. They are generated by trajectories starting from different initial conditions (x_0, p_0). For the parameter value $K = 1$, there exist many KAM islands (see Section 2.2.6) of regular quasiperiodic motion (Arnold and Avez, 1968; Ott, 1993), the largest one being centered around the point $(x, p) = (\pi, 0)$. Chaotic zones are characterized by uniform distributions of points. They are seen between the KAM islands in Figure B.2(a), for instance the one extending from the saddle point

$(x, p) = (0,0)$ to its image located at $(x, p) = (2\pi, 0)$ by the translational symmetry $x \to x + 2\pi$. Since the KAM islands and the chaotic zones are generated by distinct trajectories, they form different ergodic components. This is evidence that the standard map is not ergodic with respect to the uniform probability distribution, which can thus be decomposed into myriads of ergodic components. As the parameter K is increased, the chaotic zones become dominant in phase space. For instance, if $K = 10$, Figure B.2(b) shows that a single trajectory covers nearly the whole phase space. This does not necessarily exclude the possibility of tiny KAM islands, which are so small that they do not appear in the phase portrait. Nevertheless, here there is numerical evidence that the uniform probability distribution provides an excellent approximation for statistics carried out by time averaging.

B.3.3 Billiards and Hard Ball Gases

Billiards are dynamical systems where point particles or hard balls are moving in free flights between instantaneous elastic collisions either with each other or with the walls of the vessel containing them. The balls are disks if the physical space is two-dimensional, and spheres if the space is three-dimensional. In the limit where the diameter of the balls is equal to zero, these particles form an ideal gas and the elastic collisions only happen on the walls, so that the motion of a single particle is representative for the whole gas. In two dimensions, the vessel forms a billiard table, hence the name of these dynamical systems. We notice that there is no wall if the system is defined in a domain (such as a cube) with periodic boundary conditions.

We assume that the system is composed of N hard balls with radii $\{R_a\}_{a=1}^N$ and masses $\{m_a\}_{a=1}^N$. The map ruling their dynamics can be deduced from Hamilton's variational principle by considering the Hamiltonian function

$$H = \sum_{a=1}^{N} \frac{\mathbf{p}_a^2}{2m_a}, \tag{B.29}$$

supplemented by the conditions that the balls may not overlap:

$$\|\mathbf{r}_a - \mathbf{r}_b\| \geq R_a + R_b \qquad \text{for} \quad a, b = 1, 3, \ldots, N. \tag{B.30}$$

These conditions define hypersurfaces in configuration space, where the functions $f_{ab} = \|\mathbf{r}_a - \mathbf{r}_b\| - (R_a + R_b)$ are vanishing. The total energy $E = H$ is conserved because the Hamiltonian function (B.29) and the conditions (B.30) do not explicitly depend on time. Moreover, the total linear momentum $\mathbf{P} = \sum_{a=1}^N \mathbf{p}_a$ is also conserved for the dynamics with periodic boundary conditions because the conditions (B.30) are invariant under continuous spatial translations.

If we introduce the positions $\mathbf{q} = (\sqrt{m_a}\,\mathbf{r}_a)_{a=1}^N \in \mathbb{R}^{3N}$ and the canonically conjugated momenta $\mathbf{p} = (\mathbf{p}_a/\sqrt{m_a})_{a=1}^N \in \mathbb{R}^{3N}$ for the N hard spheres, the Hamiltonian function (B.29) becomes $H = \mathbf{p}^2/2$, so that the equations of motion are given by $\dot{\mathbf{q}} = \mathbf{p}$ and $\dot{\mathbf{p}} = 0$. Between the collisions, the momenta are thus conserved, so $\mathbf{p}_t = \mathbf{p}_{t_0}$, and the trajectories are the free flights $\mathbf{q}_t = \mathbf{q}_{t_0} + \mathbf{p}_{t_0}(t - t_0)$. Since the total energy $E = \mathbf{p}_t^2/2 = \dot{\mathbf{q}}_t^2/2$

is also conserved, the action (B.20) can be written as $W = \sqrt{E/2}\int d\mathscr{L}$, where $d\mathscr{L} = \|d\mathbf{q}\| = \sqrt{2E}\,dt$ is the element of length in the configuration space of variables $\mathbf{q} \in \mathbb{R}^{3N}$. Inside the energy shell $H = E$, Hamilton's variational principle $\delta W = 0$ thus leads to the vanishing of the first variation of the length: $\delta\mathscr{L} = 0$. Now, the constraints (B.30) can be taken into account by introducing Lagrange multipliers χ_i for every collision at the configuration $\mathbf{q}_i$ where $f(\mathbf{q}_i) = 0$ for some function f. The functional to be considered by the variational principle $\delta\mathscr{L} = 0$ is therefore given by

$$\mathscr{L} = \sum_{i=-\infty}^{+\infty} \left[\|\mathbf{q}_{i+1} - \mathbf{q}_i\| + \chi_i\, f(\mathbf{q}_i)\right]. \tag{B.31}$$

Taking the partial derivative with respect to $\mathbf{q}_i$, introducing the momenta after and before the ith collision as

$$\mathbf{p}_i^{(+)} \equiv \sqrt{2E}\,\frac{\mathbf{q}_{i+1} - \mathbf{q}_i}{\|\mathbf{q}_{i+1} - \mathbf{q}_i\|} \qquad \text{and} \qquad \mathbf{p}_i^{(-)} \equiv \sqrt{2E}\,\frac{\mathbf{q}_i - \mathbf{q}_{i-1}}{\|\mathbf{q}_i - \mathbf{q}_{i-1}\|}, \tag{B.32}$$

defining the unit vector $\mathbf{n}_i \equiv \|\partial_{\mathbf{q}} f(\mathbf{q}_i)\|^{-1}\partial_{\mathbf{q}} f(\mathbf{q}_i)$, which is normal to the hypersurface $f = 0$ at the impact point $\mathbf{q}_i$ in the $3N$-dimensional configuration space, and fixing the value of the Lagrange multiplier χ_i with the condition $\left(\mathbf{p}_i^{(+)}\right)^2 = \left(\mathbf{p}_i^{(-)}\right)^2$ of energy conservation in the elastic collision, we find the relation of specular reflection at every collision:

$$\mathbf{p}_i^{(+)} = \mathbf{p}_i^{(-)} - 2\left(\mathbf{n}_i \cdot \mathbf{p}_i^{(-)}\right)\mathbf{n}_i. \tag{B.33}$$

Between the collisions, the trajectories are the free flights: $\mathbf{q}_t = \mathbf{q}_i + \mathbf{p}_i^{(+)}(t - t_i)$ for $t_i < t < t_{i+1}$.

Going back to the original variables, we obtain

$$\partial_{\mathbf{q}} f = \left(\ldots, 0, \frac{1}{\sqrt{m_a}}\frac{\mathbf{r}_a - \mathbf{r}_b}{\|\mathbf{r}_a - \mathbf{r}_b\|}, 0, \ldots, 0, -\frac{1}{\sqrt{m_b}}\frac{\mathbf{r}_a - \mathbf{r}_b}{\|\mathbf{r}_a - \mathbf{r}_b\|}, 0, \ldots\right) \tag{B.34}$$

with $\|\mathbf{r}_a - \mathbf{r}_b\| = R_a + R_b$ upon collision between the balls a and b, so that $\left\|\partial_{\mathbf{q}} f\right\|^2 = m_a^{-1} + m_b^{-1}$. Substituting into equation (B.33) and denoting the positions $\mathbf{r}_a$ and the velocities $\mathbf{v}_a = \mathbf{p}_a/m_a$ of the particles after or before the collision at time t_i respectively as $\left[\mathbf{r}_a^{(\pm)}(t_i), \mathbf{v}_a^{(\pm)}(t_i)\right]_{a=1}^{N}$, we find that the binary collision between the ath and bth balls is ruled by

$$\mathbf{r}_a^{(+)} = \mathbf{r}_a^{(-)}, \qquad \mathbf{r}_b^{(+)} = \mathbf{r}_b^{(-)}, \tag{B.35}$$

$$\mathbf{v}_a^{(+)} = \mathbf{v}_a^{(-)} - 2\,\frac{m_b}{m_a + m_b}\left(\boldsymbol{\epsilon}_{ab} \cdot \mathbf{v}_{ab}^{(-)}\right)\boldsymbol{\epsilon}_{ab}, \tag{B.36}$$

$$\mathbf{v}_b^{(+)} = \mathbf{v}_b^{(-)} + 2\,\frac{m_a}{m_a + m_b}\left(\boldsymbol{\epsilon}_{ab} \cdot \mathbf{v}_{ab}^{(-)}\right)\boldsymbol{\epsilon}_{ab}, \tag{B.37}$$

where

$$\boldsymbol{\epsilon}_{ab} \equiv \frac{\mathbf{r}_a^{(\pm)} - \mathbf{r}_b^{(\pm)}}{R_a + R_b} \qquad \text{and} \qquad \mathbf{v}_{ab}^{(-)} \equiv \mathbf{v}_a^{(-)} - \mathbf{v}_b^{(-)} \tag{B.38}$$

are, respectively, the unit vector joining the centers of the colliding balls and the relative velocity vector, while $\mathbf{r}_c^{(+)} = \mathbf{r}_c^{(-)}$ and $\mathbf{v}_c^{(+)} = \mathbf{v}_c^{(-)}$ for the other balls $c \neq a, b$, which do not collide at time t_i (Gaspard and van Beijeren, 2002). Moreover, the free flight between two successive binary collisions is given by

$$\mathbf{r}_a^{(-)}(t_i) = \mathbf{r}_a^{(+)}(t_{i-1}) + (t_i - t_{i-1})\,\mathbf{v}_a^{(+)}(t_{i-1}), \tag{B.39}$$

$$\mathbf{v}_a^{(-)}(t_i) = \mathbf{v}_a^{(+)}(t_{i-1}). \tag{B.40}$$

We note that elastic collisions between hard balls are defocusing, so that hard ball gases have many positive Lyapunov exponents, as shown in Figure 2.2 (Dellago et al., 1996; Gaspard and van Beijeren, 2002).

The equations (B.35)–(B.40) reduce the continuous-time dynamics of the hard-ball gas to the discrete-time dynamics mapping collision to collision. In this regard, we may introduce the so-called Birkhoff coordinates that are intrinsic to the hypersurfaces where the elastic collisions happen in configuration space (Birkhoff, 1927; MacKay, 1993). At the ith collision, the positions can be decomposed as $\mathbf{q}_i = \left(\boldsymbol{\rho}_i, \rho_i^{(\mathrm{n})}\right)$ into the components $\boldsymbol{\rho}_i$ that are tangent to the hypersurface and the one $\rho_i^{(\mathrm{n})}$ that is normal, such that $\mathbf{n}_i \cdot \delta\boldsymbol{\rho}_i = 0$ and $\delta\rho_i^{(\mathrm{n})} = \mathbf{n}_i \cdot \delta\mathbf{q}_i$. The momenta that are canonically conjugated to the Birkhoff coordinates $\boldsymbol{\rho}_i$ are thus taken as $\boldsymbol{\pi}_i \equiv \mathbf{p}_i^{(-)} - \left(\mathbf{n}_i \cdot \mathbf{p}_i^{(-)}\right)\mathbf{n}_i$, satisfying $\mathbf{n}_i \cdot \boldsymbol{\pi}_i = 0$. The corresponding phase-space variables are given by $\mathbf{x}_i = (\boldsymbol{\rho}_i, \boldsymbol{\pi}_i) \in \mathbb{R}^{2f-2}$. The Birkhoff map $\mathbf{x}_{i+1} = \boldsymbol{\phi}(\mathbf{x}_i)$ is preserving the $(2f-2)$-dimensional volumes. The discrete times of the dynamics are given by the times of free flight between the elastic collisions. For two-dimensional billiards where a particle of mass m moves at the speed v in free flight between elastic collisions on the wall of the table, the Birkhoff coordinates are given by $(\rho_i, mv \sin\chi_i)$ where ρ_i is the arc of perimeter of the wall at impact and χ_i the angle between the incident velocity and the exterior direction normal to the wall at impact, so that the Birkhoff map is area preserving.

B.4 Escape-Rate Theory

In open Hamiltonian systems, trajectories may escape out of the phase-space domain where the particles interact (for instance because of absorbing boundary conditions). Considering a statistical ensemble of initial conditions inside this domain, the number of trajectories remaining therein after some time interval t will decrease. If the dynamical system is unstable enough, this decrease may be exponential for trajectories at given total energy in the long-time limit, $N_t \sim \exp(-\gamma t)$, defining the so-called escape rate γ. Accordingly, the probability to find the trajectory inside the domain is leaking at the rate γ, so that the escape rate is the leading Pollicott–Ruelle resonance in such open dynamical systems. In order to define a stationary probability distribution P_{st}, the left-hand side of equation (B.13) should, moreover, be divided by the factor $\exp(-\gamma t)$ for normalization purposes. This stationary probability distribution typically has a fractal for support in phase space (Ott, 1993; Gaspard, 1998).

If the system is chaotic inside the domain, the Kolmogorov–Sinai entropy per unit time associated with the so-defined stationary probability distribution is positive, but it is no longer equal to the sum of positive Lyapunov exponents, contrary to equation (B.14). Instead, the difference is related to the escape rate according to the following formula (Eckmann and Ruelle, 1985; Kantz and Grassberger, 1985):

$$\gamma = \sum_{\lambda_j > 0} \lambda_j - h_{\text{KS}}. \tag{B.41}$$

Additionally, we note that the Helfand moment (3.116) associated with some transport process undergoes a random walk with a diffusivity D giving the corresponding transport coefficient by the Einstein–Helfand formula (3.117). In this regard, the determination of the transport coefficient can be formulated as a first-passage problem of trajectories out of a phase-space domain defined by requiring that the Helfand moment $\delta G(t)$ satisfies the conditions $|\delta G(t)| \le \chi/2$. The escape rate out of this domain is given by $\gamma = D(\pi/\chi)^2$ for $\chi \to \infty$, as a consequence of the random walk of the Helfand moment in the interval $[-\chi/2, +\chi/2]$. Combining this with the formula (B.41), the transport coefficient can thus be determined as

$$D = \lim_{\chi \to \infty} \left(\frac{\chi}{\pi}\right)^2 \left(\sum_{\lambda_j > 0} \lambda_j - h_{\text{KS}}\right)_\chi \tag{B.42}$$

in terms of the Lyapunov exponents and the Kolmogorov–Sinai entropy per unit time, characterizing the microscopic dynamics, which is the basis of the escape-rate theory for transport (Gaspard and Nicolis, 1990; Gaspard, 1998; Dorfman, 1999; Viscardy and Gaspard, 2003; Gaspard, 2006; Klages, 2007).

To the extent that the Kolmogorov–Sinai entropy can be decomposed as $h_{\text{KS}} = \sum_{\lambda_j > 0} d_j \lambda_j$ in terms of the partial dimensions d_j associated with each unstable direction (Eckmann and Ruelle, 1985), we note that equation (13.20) linking the Hausdorff dimension of diffusion modes to the diffusion coefficient and the positive Lyapunov exponent in Lorentz gases is reminiscent of the escape-rate formula (B.42) (Gaspard et al., 2001; Gaspard, 2005, 2007b).

B.5 Non-Hamiltonian Time-Reversal Symmetric Dynamical Systems

There also exist dynamical systems that are time-reversal symmetric but non-Hamiltonian (Roberts and Quispel, 1992). Such systems are considered in the thermostatted-system approach where the nonequilibrium conditions are modeled by fictitious forces representing the interaction between the system and some hypothetical thermostat. At equilibrium, the modeling is such that these fictitious forces are switched off. Several kinds of thermostatted systems have been investigated (Evans et al., 1983; Nosé, 1984a,b; Hoover, 1985; Evans and Morriss, 1990). In the case of Gaussian thermostats, the fictitious forces are defined by imposing that the kinetic energy is exactly conserved instead of the total energy (Evans et al., 1983; Evans and Morriss, 1990). In the case of Nosé–Hoover thermostats (Nosé, 1984a,b; Hoover, 1985), an extra variable is introduced in such a way that the kinetic energy is conserved on average and the equilibrium canonical probability distribution is reproduced. These thermostatting forces do not depend on the many external degrees of freedom of real thermostats, so that the amount of dynamical randomness (or stochasticity) is determined by the internal degrees of freedom and the fictitious thermostatting forces do not include randomly fluctuating forces in such systems. In contrast to Hamiltonian systems, thermostatted systems do not preserve phase-space volumes even if they are symmetric under time reversal. Consequently, the phase-space volumes contract and the sum of all the Lyapunov exponents is negative in thermostatted systems where the nonequilibrium constraints are switched on. According to equation (B.7), the mean rate of phase-space contraction can thus be written as

$$\bar{\sigma} = - \lim_{t\to\infty} \frac{1}{t} \int_0^t \operatorname{div} \mathbf{V}\, d\tau = - \sum_{j=1}^{m} \lambda_j = \sum_{\lambda_j<0} |\lambda_j| - h_{\rm KS} \tag{B.43}$$

in terms of the KS entropy per unit time (B.14). In thermostatted systems, this contraction rate is interpreted as the entropy production rate (Evans and Morriss, 1990). Since $\bar{\sigma} > 0$ under nonequilibrium conditions, Liouville's theorem is not satisfied for thermostatted systems, in contrast to what is expected for a microscopic dynamics of quantum-mechanical origin.

In its latter form, equation (B.43) for the phase-space contraction rate can be compared with equation (B.41) for the escape rate. Both formulas have similar structures, although the sum extends over the positive Lyapunov exponents in equation (B.41) and the negative ones in equation (B.43). For many-particle systems, the sum of Lyapunov exponents and the Kolmogorov–Sinai entropy per unit time do not vanish at equilibrium in chaotic systems, where they have large values that are inversely proportional to the microscopic intercollisional time. In contrast, the escape rate and the contraction rate take significantly smaller values because they are inversely proportional to the macroscopic timescales of irreversible transport properties and they are equal to zero at equilibrium.

Appendix C

Complements on Statistical Mechanics

C.1 Equilibrium Statistical Ensembles

The equilibrium statistical ensembles mentioned in Chapter 2 are here described in detail. We suppose for simplicity that the system contains a single particle species for the microcanonical, canonical, grand canonical, and isobaric-isothermal ensembles, and two species for the semigrand canonical ensembles.

C.1.1 Microcanonical Ensemble

Isolated particle systems have given energy E, volume V, and particle number N. Their motion is ruled by the Hamiltonian function (2.2), where the external energy potential $u^{(\mathrm{ext})}(\mathbf{r})$ forms a square well fixing the volume V. If we consider a single trajectory starting from initial conditions $\mathbf{\Gamma}_0$, the time average and the assumption of ergodicity determine the stationary probability density (2.69) on the energy shell $E_0 = H(\mathbf{\Gamma}_0)$. However, in repeating the same experiment several times, the preparation of the initial conditions limits the accuracy of the initial total energy to some value ΔE. Therefore, the energy shell has some thickness ΔE. We could choose a Gaussian distribution in energy for the equilibrium density (2.76), but it is convenient to consider a uniform distribution of energy in the range $E < H(\mathbf{\Gamma}) < E + \Delta E$ by taking the *microcanonical probability distribution* of density

$$p_{\mathrm{eq}}(\mathbf{\Gamma}) = \frac{1}{h^{3N}\,\Omega(E,V,N;\Delta E)}\,\{\theta[E+\Delta E - H(\mathbf{\Gamma})] - \theta[E - H(\mathbf{\Gamma})]\}, \tag{C.1}$$

where $\theta(x)$ is the Heaviside function such that $\theta(x) = 1$ if $x > 0$ and zero otherwise. The normalization condition $\int_{\mathcal{M}} p_{\mathrm{eq}} d\mathbf{\Gamma} = 1$ is satisfied with

$$\Omega(E,V,N;\Delta E) = \frac{1}{h^{3N} N!}\int_{\mathbb{R}^{6N}} \{\theta[E+\Delta E - H(\mathbf{\Gamma})] - \theta[E - H(\mathbf{\Gamma})]\}\, d\mathbf{\Gamma}, \tag{C.2}$$

which is the number of quantum microstates inside the energy shell $E < H(\mathbf{\Gamma}) < E + \Delta E$, since every such microstate corresponds to the elementary quantal phase-space volume (2.7).

If $\Delta E \ll E$, the number (C.2) can be evaluated as $\Omega(E,V,N;\Delta E) \simeq \mathscr{D}(E,V,N)\Delta E$ in terms of the density of quantum microstates (2.8), because $d\theta(x)/dx = \delta(x)$. Under such circumstances, the probability density (C.1) can be rewritten in the form

$$\mathcal{P}_{\rm eq}(\mathbf{\Gamma}) \simeq \frac{\Delta E}{h^{3N}\Omega}\,\delta[E - H(\mathbf{\Gamma})], \quad \text{so that} \quad \langle A\rangle_{\rm eq} \simeq \frac{\int d\mathbf{\Gamma}\, A(\mathbf{\Gamma})\,\delta[E - H(\mathbf{\Gamma})]}{\int d\mathbf{\Gamma}\,\delta[E - H(\mathbf{\Gamma})]} \tag{C.3}$$

is the statistical average of some observable $A(\mathbf{\Gamma})$ over the microcanonical equilibrium probability distribution.

In this ensemble, the energy, volume, and particle number do not fluctuate, so that their standard deviations are equal to zero: $\sigma_E = 0$, $\sigma_V = 0$, and $\sigma_N = 0$.

According to Boltzmann (1896, 1898) and Planck (1914), the thermodynamic entropy of the macrostate described by the microcanonical statistical ensemble should be taken as

$$S(E,V,N) = k_{\rm B}\,\ln \Omega(E,V,N;\Delta E). \tag{C.4}$$

The entropy, as well as the energy, the volume, and the particle number, are extensive variables. However, the width ΔE is not supposed to be extensive, so that the dependence of the entropy on ΔE becomes negligible in the large-system limit.

The expression (C.4) for the entropy can be justified by considering the fluctuations of energy, particle number, or volume between two compartments A and B of the isolated system, separated, for instance, by a mobile wall with holes to allow energy and particle exchanges, as well as fluctuating volumes for both compartments. Since these three quantities are conserved in the isolated system, we have that $E_A + E_B = E$, $V_A + V_B = V$, and $N_A + N_B = N$. The probability density to find the compartment A with the energy E_A, the volume V_A, and the particle number N_A is given in the microcanonical ensemble by

$$\mathcal{P}_{\rm eq}(E_A,V_A,N_A) \simeq \frac{\mathscr{D}_A(E_A,V_A,N_A)\,\mathscr{D}_B(E-E_A,V-V_A,N-N_A)}{\sum_{N'_A}\int dE'_A dV'_A\,\mathscr{D}_A(E'_A,V'_A,N'_A)\,\mathscr{D}_B(E-E'_A,V-V'_A,N-N'_A)} \tag{C.5}$$

if the interactions between the particles have short or medium range, so that the contribution of the interface between both compartments can be neglected in large enough systems. If both compartments have similar sizes, the supposed relation (C.4) between the thermodynamic entropy and the number Ω of microstates has for consequence that the energy, volume, and particle number of the compartment A take their most probable values $(\bar{E}_A, \bar{V}_A, \bar{N}_A)$ if the conditions (A.6) of thermal, mechanical, and chemical equilibria are satisfied, which justifies the Boltzmann expression (C.4) for the entropy.

Now, if compartment B is assumed to be much larger than compartment A, we have that

$$E \simeq \bar{E}_B \gg \bar{E}_A, \qquad V \simeq \bar{V}_B \gg \bar{V}_A, \qquad \text{and} \qquad N \simeq \bar{N}_B \gg \bar{N}_A. \tag{C.6}$$

If these conditions hold, the probability density (C.5) becomes

$$\mathcal{P}_{\rm eq}(E_A,V_A,N_A) \simeq \frac{\mathscr{D}_A(E_A,V_A,N_A)\,{\rm e}^{-\beta E_A - \beta p V_A + \beta\mu N_A}}{\sum_{N'_A}\int dE'_A dV'_A\,\mathscr{D}_A(E'_A,V'_A,N'_A)\,{\rm e}^{-\beta E'_A - \beta p V'_A + \beta\mu N'_A}}, \tag{C.7}$$

where $\beta = (k_{\rm B}T)^{-1}$, while T, p, and μ are the temperature, pressure, and chemical potential given by equation (1.12) in compartment B, which plays the roles of thermostat, mechanostat, and chemostat. In this way, the Boltzmann energy distribution (as well as the distributions in volume and particle number) can be deduced by considering a small system inside a large isolated system in the microcanonical equilibrium statistical ensemble.

C.1.2 Canonical Ensemble

The canonical ensemble describes closed equilibrium systems, exchanging energy, but no particle with their environment. They are thus characterized by given values of their temperature T, volume V, and particle number N. As suggested by equation (C.5), the *canonical probability distribution* has the following density:

$$p_{\text{eq}}(\mathbf{\Gamma}) = \frac{1}{h^{3N} Z} \, \mathrm{e}^{-\beta H(\mathbf{\Gamma})} \qquad \text{with} \qquad Z(T, V, N) = \frac{1}{h^{3N} N!} \int_{\mathbb{R}^{6N}} \mathrm{e}^{-\beta H(\mathbf{\Gamma})} \, d\mathbf{\Gamma}, \quad \text{(C.8)}$$

where $\beta = (k_{\text{B}} T)^{-1}$ is the inverse temperature. The corresponding thermodynamic potential is the Helmholtz free energy

$$F(T, V, N) = -k_{\text{B}} T \, \ln Z(T, V, N). \quad \text{(C.9)}$$

The entropy, pressure, and chemical potential can thus be deduced from the Helmholtz free energy using the Gibbs relation $dF = -S dT - p dV + \mu dN$.

Since energy is exchanged between the system and its environment, energy fluctuates and its standard deviation is now positive and given by $\sigma_E = \sqrt{k_{\text{B}} T^2 C_V}$ in terms of the heat capacity at constant volume C_V.

For a nonrelativistic ideal gas of volume V, the partition function of the ideal gas is given by

$$Z_{\text{id}}(T, V, N) = \frac{1}{N!} \left(\frac{V}{\Lambda^3} \right)^N, \qquad \text{where} \qquad \Lambda \equiv \frac{h}{\sqrt{2\pi m k_{\text{B}} T}} \quad \text{(C.10)}$$

is the thermal de Broglie wavelength.

C.1.3 Grand Canonical Ensemble

If the system is open and exchanging energy and particles with its environment, its equilibrium macrostate is characterized by its temperature T, its volume V, and its chemical potential μ, so that its thermodynamic potential is the grand potential $J(T, V, \mu) \equiv F - \mu N = -pV$. Such an open system is described at equilibrium by the *grand canonical probability distribution* of density:

$$p_{\text{eq}}(\mathbf{\Gamma}_N, N) = \frac{1}{h^{3N} \Xi} \, \mathrm{e}^{-\beta H_N(\mathbf{\Gamma}_N) + \beta \mu N}, \quad \text{(C.11)}$$

where μ is the chemical potential, $\mathbf{\Gamma}_N$ denotes the set of positions and momenta for the N particles inside the system, and H_N is the corresponding Hamiltonian function. Here, the normalization condition takes into account the fact that the particle number is also a random variable,

$$\sum_{N=0}^{\infty} \int_{\mathcal{M}_N} p_{\text{eq}}(\mathbf{\Gamma}_N, N) \, d\mathbf{\Gamma}_N = \sum_{N=0}^{\infty} \frac{1}{N!} \int_{\mathbb{R}^{6N}} p_{\text{eq}}(\mathbf{\Gamma}_N, N) \, d\mathbf{\Gamma}_N = 1, \quad \text{(C.12)}$$

where $\mathcal{M}_N$ denotes the N-particle phase space defined by equation (2.4). The partition function can thus be expressed as

$$\Xi(T,V,\mu) = \sum_{N=0}^{\infty} e^{\beta\mu N}\, Z(T,V,N) \tag{C.13}$$

in terms of the partition function of the canonical ensemble (C.8) for the system with N particles. Therefore, the thermodynamic grand potential is given by

$$J(T,V,\mu) = -k_{\rm B}T\, \ln \Xi(T,V,\mu), \tag{C.14}$$

from which the different thermodynamic quantities can be deduced using the Gibbs relation $dJ = -SdT - pdV - \langle N\rangle d\mu$.

In the grand canonical ensemble, the particle number fluctuates with the standard deviation $\sigma_N = \sqrt{nk_{\rm B}T\kappa_T\langle N\rangle}$ where $n = \langle N\rangle/V$ is the particle density and $\kappa_T = (1/n)(\partial n/\partial p)_T$ is the isothermal compressibility. The variance of energy fluctuations is given by $\sigma_E^2 = k_{\rm B}T^2C_V + (\partial E/\partial N)_{T,V}^2\sigma_N^2$, which is higher than in the canonical ensemble because of particle exchanges in addition to those of energy. Moreover, the covariance between energy and particle number goes as $\langle\delta E\delta N\rangle \equiv \langle(E-\langle E\rangle)(N-\langle N\rangle)\rangle = nk_{\rm B}T\kappa_T\langle N\rangle(\partial E/\partial N)_{T,V}$.

The probability of finding N particles inside the system is obtained as

$$P_{\rm eq}(N) \equiv \int_{\mathcal{M}_N} p_{\rm eq}(\mathbf{\Gamma}_N, N)\, d\mathbf{\Gamma}_N = \frac{e^{\beta\mu N}\, Z(T,V,N)}{\Xi(T,V,\mu)}, \tag{C.15}$$

which is normalized according to $\sum_{N=0}^{\infty} P_{\rm eq}(N) = 1$. For an ideal gas of canonical partition function (C.10), we thus find the Poisson distribution

$$P_{\rm eq,\,id}(N) = \frac{1}{N!}\,\langle N\rangle^N\, e^{-\langle N\rangle}, \qquad \text{where} \qquad \langle N\rangle = e^{\beta\mu}\,\frac{V}{\Lambda^3} \tag{C.16}$$

is the mean value of the particle number at equilibrium.

C.1.4 Isobaric-Isothermal Ensemble

This statistical ensemble describes a system with given temperature T, pressure p, and particle number N, so that its energy and its volume fluctuate. This system can be envisaged as a system of N particles inside a cylinder with a fluctuating piston in contact with a mechanostat fixing the pressure p. The corresponding thermodynamic potential is the Gibbs free energy $G(T,p,N) \equiv F + pV = \mu N$, also called free enthalpy. This ensemble is described by the *isobaric-isothermal probability distribution* with the density,

$$p_{\rm eq}(\mathbf{\Gamma}, V) = \frac{1}{h^{3N}\upsilon\Upsilon}\, e^{-\beta H_N(\mathbf{\Gamma},V) - \beta pV}, \tag{C.17}$$

where the volume V is also a random variable, and υ is a constant having the units of a volume. If the piston had a one-dimensional motion with mass M, this volume would be taken as its cross-sectional area Σ multiplied by its thermal de Broglie wavelength, $\upsilon = \Sigma h/\sqrt{2\pi M k_{\rm B}T}$. In this case, the piston degree of freedom would add a negligible contribution to the Gibbs free energy if the system contains many other degrees of freedom.

Since the normalization condition has the form $\int dV \int_{\mathcal{M}} p_{\rm eq}(\mathbf{\Gamma}, V)\, d\mathbf{\Gamma} = 1$, the corresponding partition function is here given in terms of the partition function of the canonical ensemble (C.8) by

$$\Upsilon(T, p, N) = \frac{1}{\upsilon} \int dV\, {\rm e}^{-\beta p V}\, Z(T, V, N), \tag{C.18}$$

and the Gibbs free energy is related to this partition function according to

$$G(T, p, N) = -k_{\rm B}T\, \ln \Upsilon(T, p, N). \tag{C.19}$$

The thermodynamic quantities are deduced by the Gibbs relation $dG = -SdT + \langle V \rangle dp + \mu dN$.

In the isobaric-isothermal ensemble, the standard deviation of the volume is given by $\sigma_V = \sqrt{k_{\rm B}T\kappa_T \langle V \rangle}$, the variance of energy by $\sigma_E^2 = k_{\rm B}T^2 C_V + (\partial E/\partial V)_{T,N}^2 \sigma_V^2$, and the covariance between energy and volume by $\langle \delta E \delta V \rangle \equiv \langle (E - \langle E \rangle)(V - \langle V \rangle) \rangle = k_{\rm B}T\kappa_T \langle V \rangle (\partial E/\partial V)_{T,N}$.

C.1.5 Semigrand Canonical Ensembles

Further equilibrium statistical ensembles called semigrand canonical ensembles may be introduced in systems with several species of particles, for instance, to describe osmotic effects, in which cases the particle number is fixed for some species, but fluctuates for other species (Hill, 1956, 1960). Here, we consider systems with two particle species for simplicity.

Suppose that the system has a fixed volume V and is separated from its environment by a semipermeable membrane porous to the species $k = 1$, but impermeable to the other species $k = 2$. The environment plays the role of thermostat at the temperature T and chemostat at the chemical potential μ_1 for species $k = 1$. Therefore, the particle number fluctuates for species $k = 1$, although the particle number of species $k = 2$ is fixed at the value N_2. For this system, the thermodynamic potential is defined as $\tilde{J}(T, V, \mu_1, N_2) \equiv F(T, V, N_1, N_2) - \mu_1 N_1 = -pV + \mu_2 N_2$. The equilibrium statistics is here given by the *semigrand canonical probability distribution*,

$$p_{\rm eq}(\mathbf{\Gamma}_{N_1}, \mathbf{\Gamma}_{N_2}, N_1) = \frac{1}{h^{3N_1+3N_2}\, \tilde{\Xi}}\, {\rm e}^{-\beta H_{N_1}(\mathbf{\Gamma}_{N_1}, \mathbf{\Gamma}_{N_2}) + \beta \mu_1 N_1}, \tag{C.20}$$

where $\mathbf{\Gamma}_{N_k}$ is the set of positions and momenta for the N_k particles of species k, H_{N_1} is the Hamiltonian function when the system contains N_1 particles of species $k = 1$, and μ_1 is the corresponding chemical potential. The semigrand thermodynamic potential is defined as

$$\tilde{J}(T, V, \mu_1, N_2) = -k_{\rm B}T\, \ln \tilde{\Xi}(T, V, \mu_1, N_2), \tag{C.21}$$

giving the different thermodynamic quantities with the Gibbs relation $d\tilde{J} = -SdT - pdV - \langle N_1 \rangle d\mu_1 + \mu_2 dN_2$.

Another semigrand canonical ensemble can be introduced for a system separated from its environment by a semipermeable membrane porous to the species $k = 1$, but impermeable to the other species $k = 2$, and a mobile piston (Hill, 1960). In this case, the environment plays the role of the thermostat at the temperature T, chemostat at the chemical potential μ_1 for species $k = 1$, and mechanostat at the pressure p. Therefore, the system contains a fixed particle number N_2 for species $k = 2$, but its volume V and the particle number N_1 are fluctuating random variables. Here, the thermodynamic potential is defined as $\tilde{K}(T, p, \mu_1, N_2) \equiv G(T, p, N_1, N_2) - \mu_1 N_1 = \mu_2 N_2$, and the *semigrand canonical probability distribution* is taken as

$$p_{\text{eq}}(\mathbf{\Gamma}_{N_1}, \mathbf{\Gamma}_{N_2}, N_1, V) = \frac{1}{h^{3N_1+3N_2}\, \upsilon \tilde{\Upsilon}}\, \mathrm{e}^{-\beta H_{N_1}(\mathbf{\Gamma}_{N_1}, \mathbf{\Gamma}_{N_2}, V) - \beta p V + \beta \mu_1 N_1}, \tag{C.22}$$

where $\mathbf{\Gamma}_{N_k}$ is the set of positions and momenta for the N_k particles of species k, H_{N_1} is the Hamiltonian function when the system has the volume V and contains N_1 particles of species $k = 1$, μ_1 is the corresponding chemical potential, and υ is the same volume constant as defined for the isobaric-isothermal ensemble. The associated semigrand thermodynamic potential is given by

$$\tilde{K}(T, p, \mu_1, N_2) = -k_{\text{B}} T \, \ln \tilde{\Upsilon}(T, p, \mu_1, N_2), \tag{C.23}$$

to be used with the Gibbs relation $d\tilde{K} = -SdT + \langle V \rangle dp - \langle N_1 \rangle d\mu_1 + \mu_2 dN_2$.

In these semigrand canonical ensembles, the particle number has the standard deviations: $\sigma_{N_1} = \sqrt{k_{\text{B}} T (\partial \langle N_1 \rangle / \partial \mu_1)_{T,V,N_2}}$ and $\sigma_{N_2} = 0$.

C.2 Local Fluctuations at Equilibrium

A system at equilibrium has local fluctuations in energy and particle densities, which can be determined by considering a volume element of continuous medium. Since this element of given volume V is open to energy and particle exchanges with the rest of the system, the fluctuations can be computed in the grand canonical ensemble. We assume that the system is composed of a single particle species and described at equilibrium by the equations of state for the pressure $p(T, n)$ and the energy density $e(T, n)$, where T is the temperature and n the mean particle density. On spatial scales larger than the correlation length,[1] the densities are given by Gaussian white noise fields characterized by zero mean values $\langle \delta \hat{e}(\mathbf{r}) \rangle = 0$ and $\langle \delta \hat{n}(\mathbf{r}) \rangle = 0$, and the correlation functions,

$$\langle \delta \hat{e}(\mathbf{r})\, \delta \hat{e}(\mathbf{r}') \rangle = k_{\text{B}} T \left[T c_{\upsilon} + n \left(\frac{\partial n}{\partial p} \right)_T \left(\frac{\partial e}{\partial n} \right)_T^2 \right] \delta^3(\mathbf{r} - \mathbf{r}'), \tag{C.24}$$

$$\langle \delta \hat{e}(\mathbf{r})\, \delta \hat{n}(\mathbf{r}') \rangle = k_{\text{B}} T\, n \left(\frac{\partial n}{\partial p} \right)_T \left(\frac{\partial e}{\partial n} \right)_T \delta^3(\mathbf{r} - \mathbf{r}'), \tag{C.25}$$

$$\langle \delta \hat{n}(\mathbf{r})\, \delta \hat{n}(\mathbf{r}') \rangle = k_{\text{B}} T\, n \left(\frac{\partial n}{\partial p} \right)_T \delta^3(\mathbf{r} - \mathbf{r}'), \tag{C.26}$$

[1] The correlation length is the characteristic length, over which the correlation functions are exponentially decaying. In three-dimensional physical space, the correlation functions typically decay as $C(r) \sim r^{-1} \exp(-r/\ell_c)$ between two points separated by the distance r, which defines the correlation length ℓ_c. On scales $r \gg \ell_c$, the correlation functions can be approximated by Dirac delta distributions, as $C(r) = \gamma\, \delta^3(\mathbf{r})$ with the constant $\gamma = \int C(r)\, d^3 r$.

given in terms of the heat capacity at constant volume per unit volume $c_{\mathfrak{v}}$, the isothermal compressibility $\kappa_T = (1/n)(\partial n/\partial p)_T$, and $(\partial e/\partial n)_T = \mu - T\,(\partial\mu/\partial T)_n$.

If we introduce the heat excess density[2] as

$$\delta\hat{q}(\mathbf{r}) \equiv \delta\hat{e}(\mathbf{r}) - \left(\frac{\partial e}{\partial n}\right)_T \delta\hat{n}(\mathbf{r}), \tag{C.27}$$

the correlation functions simplify to

$$\langle\delta\hat{q}(\mathbf{r})\,\delta\hat{q}(\mathbf{r}')\rangle = k_{\rm B}T^2\,c_{\mathfrak{v}}\,\delta^3(\mathbf{r}-\mathbf{r}'), \tag{C.28}$$

$$\langle\delta\hat{q}(\mathbf{r})\,\delta\hat{n}(\mathbf{r}')\rangle = 0, \tag{C.29}$$

$$\langle\delta\hat{n}(\mathbf{r})\,\delta\hat{n}(\mathbf{r}')\rangle = k_{\rm B}T\,n^2\,\kappa_T\,\delta^3(\mathbf{r}-\mathbf{r}'). \tag{C.30}$$

Moreover, the density of linear momentum $\hat{\mathbf{g}}(\mathbf{r}) = \sum_{a=1}^{N} \mathbf{p}_a\,\delta^3(\mathbf{r}-\mathbf{r}_a)$ also has Gaussian fluctuations at equilibrium, which are characterized by zero mean values $\langle\delta\hat{\mathbf{g}}(\mathbf{r})\rangle = 0$ and the correlation functions

$$\langle\delta\hat{g}_i(\mathbf{r})\,\delta\hat{g}_j(\mathbf{r}')\rangle = \rho\,k_{\rm B}T\,\delta_{ij}\,\delta^3(\mathbf{r}-\mathbf{r}') \tag{C.31}$$

with $i, j = x, y, z$ and the mass density $\rho = mn$ (Landau and Lifshitz, 1980b). Accordingly, the components of the velocity field fluctuations, which are defined around equilibrium as $\delta\hat{\mathbf{v}} \equiv \delta\hat{\mathbf{g}}/\rho$, have the correlation functions

$$\langle\delta\hat{v}_i(\mathbf{r})\,\delta\hat{v}_j(\mathbf{r}')\rangle = \frac{k_{\rm B}T}{\rho}\,\delta_{ij}\,\delta^3(\mathbf{r}-\mathbf{r}'). \tag{C.32}$$

Similar results apply to a volume element at local equilibrium, moving in a continuous medium such as a fluid.

C.3 Dilute Solutions

We consider a solution containing N_c solvent particles and $\{N_k\}_{k=1}^{c-1}$ solute particles of species $k = 1, 2, \ldots, c-1$ in a volume V (Landau and Lifshitz, 1980a; Diu et al., 1989). The solution is dilute if the following condition holds:

$$N_c \gg \sum_{k=1}^{c-1} N_k \gg 1. \tag{C.33}$$

Since the solute particles are much less abundant than the solvent particles, the former are mainly interacting with the latter, while the mutual interaction between the solute particles is negligible. The Gibbs free energy of the solution can thus be decomposed as

$$G = N_c\,\mu_{c0}(T,p) + \Delta G \qquad \text{with} \qquad \Delta G = \Delta H - T\,\Delta S, \tag{C.34}$$

where $\mu_{c0}(T,p)$ is the chemical potential of pure solvent at temperature T and pressure p, while ΔH and ΔS are, respectively, the excess enthalpy and entropy due to the solute

[2] This quantity can be expressed as $\delta\hat{q} = c_{\mathfrak{v}}\delta T$ in terms of local excess temperature δT.

species. Since the solution is dilute, the main contribution to the excess enthalpy is proportional to the numbers of solute particles

$$\Delta H = \sum_{k=1}^{c-1} \psi_k(T, p)\, N_k \tag{C.35}$$

with some functions $\psi_k(T, p)$ depending on the interaction between every solute particle and the surrounding solvent particles. The excess entropy can be evaluated by counting all the possible microscopic configurations of the solute particles in the solvent. Since every solute particle can bind to a free solvent particle and the solute particles of each species are identical to each other, the number of possible configurations is given by

$$\Omega = \frac{N_c(N_c - 1)\cdots\left(N_c - \sum_{k=1}^{c-1} N_k + 1\right)}{N_1!\, N_2! \cdots N_{c-1}!} \simeq \prod_{k=1}^{c-1} \left(\frac{\mathrm{e}\, N_c}{N_k}\right)^{N_k}, \tag{C.36}$$

where the approximation is obtained with the condition (C.33) and Stirling's formula. Using the Boltzmann formula (C.4), the excess entropy is obtained as

$$\Delta S = k_{\mathrm{B}} \ln \Omega \simeq k_{\mathrm{B}} \sum_{k=1}^{c-1} N_k \ln \frac{\mathrm{e}\, N_c}{N_k}. \tag{C.37}$$

The Gibbs free energy for the dilute solution can thus be expressed as

$$G = N_c\, \mu_{c0}(T, p) + \sum_{k=1}^{c-1} N_k \left[\psi_k(T, p) + k_{\mathrm{B}} T \ln \frac{N_k}{\mathrm{e}\, N_c}\right]. \tag{C.38}$$

Accordingly, the chemical potentials of the solute and solvent species take the forms,

$$\mu_k = \left(\frac{\partial G}{\partial N_k}\right)_{T,p,\{N_l\}_{l \neq k,c}} = \psi_k(T, p) + k_{\mathrm{B}} T \ln \frac{N_k}{N_c} \qquad (k = 1, 2, \ldots, c-1), \tag{C.39}$$

$$\mu_c = \left(\frac{\partial G}{\partial N_c}\right)_{T,p,\{N_k\}_{k \neq c}} = \mu_{c0}(T, p) - k_{\mathrm{B}} T \sum_{k=1}^{c-1} \frac{N_k}{N_c}, \tag{C.40}$$

or equivalent expressions in terms of the densities $n_k = N_k/V$. In particular, equation (C.39) justifies the expression (1.36) for the chemical potential of solute species in a dilute solution. Furthermore, the entropy of the dilute solution is given by

$$S = -\left(\frac{\partial G}{\partial T}\right)_{p,\{N_k\}} = N_c\, s_{c0} + \sum_{k=1}^{c-1} N_k \left(s_k^0 - k_{\mathrm{B}} \ln \frac{N_k}{N^0}\right), \tag{C.41}$$

where $s_{c0} = -(\partial \mu_{c0}/\partial T)_p$ is the entropy per particle of the pure solvent and

$$s_k^0 = -\left(\frac{\partial \psi_k}{\partial T}\right)_p - k_{\mathrm{B}} \ln \frac{N^0}{\mathrm{e}\, N_c} \tag{C.42}$$

is the standard entropy per particle of the solute species k for $N^0 = n^0 V$ with the reference concentration n^0 equal to one mole per liter (Fermi, 1937).

We note that the expression (1.97) for the Gibbs free energy density is recovered with $n_{k,\mathrm{eq}} = n_c \exp(-\beta\psi_k)$ and $g_{\mathrm{eq}} = n_c\mu_{c0} - k_{\mathrm{B}}T \sum_{k=1}^{c-1} n_{k,\mathrm{eq}}$, where $\beta = (k_{\mathrm{B}}T)^{-1}$.

As a consequence of these considerations, if the dilute solution is separated from pure solvent by a membrane permeable to the solvent but impermeable to all the solute species, the dilute solution exerts on this semipermeable membrane an excess pressure called the osmotic pressure and given by $\Delta p = k_{\mathrm{B}}T \sum_{k=1}^{c-1} n_k$ (Landau and Lifshitz, 1980a; Diu et al., 1989).

Appendix D

Complements on Hydrodynamics

D.1 Hydrodynamics in Normal Fluids

D.1.1 Local Thermodynamic Relations

In the frame moving with the fluid element, the Gibbs relation in equation (3.61) gives

$$\beta = \frac{1}{k_{\rm B}}\left(\frac{\partial s}{\partial \epsilon_0}\right)_{\{n_k\}} \qquad \text{and} \qquad \beta\mu_{k0} = -\frac{1}{k_{\rm B}}\left(\frac{\partial s}{\partial n_k}\right)_{\epsilon_0,\{n_j\}_{j\neq k}}, \tag{D.1}$$

from which we deduce the Maxwell relations

$$\left(\frac{\partial \beta}{\partial n_k}\right)_{\epsilon_0,\{n_j\}_{j\neq k}} = \frac{1}{k_{\rm B}}\frac{\partial^2 s}{\partial \epsilon_0 \partial n_k} = -\left[\frac{\partial(\beta\mu_{k0})}{\partial \epsilon_0}\right]_{\{n_k\}}, \tag{D.2}$$

$$\left[\frac{\partial(\beta\mu_{k0})}{\partial n_l}\right]_{\epsilon_0,\{n_j\}_{j\neq k,l}} = -\frac{1}{k_{\rm B}}\frac{\partial^2 s}{\partial n_k \partial n_l} = \left[\frac{\partial(\beta\mu_{l0})}{\partial n_k}\right]_{\epsilon_0,\{n_j\}_{j\neq k,l}}. \tag{D.3}$$

In addition, the Gibbs–Duhem relation (3.65) for the Massieu density $\omega = \beta p$ implies that

$$\left(\frac{\partial p}{\partial \beta}\right)_{\{\beta\mu_{k0}\}} = -\frac{\epsilon_0 + p}{\beta} \qquad \text{and} \qquad \left[\frac{\partial p}{\partial(\beta\mu_{k0})}\right]_{\beta,\{\beta\mu_{j0}\}_{j\neq k}} = \frac{n_k}{\beta}. \tag{D.4}$$

D.1.2 Derivation of the Equations for the Conjugate Fields

Considering that β and $\beta\mu_{k0}$ are functions of the variables ϵ_0 and $\{n_k\}$, using the Maxwell relations (D.2), Equations (3.73) and (3.76) with equation (D.4) lead to

$$\frac{d\beta}{dt} = \left(\frac{\partial \beta}{\partial \epsilon_0}\right)_{\{n_k\}}\frac{d\epsilon_0}{dt} + \sum_k \left(\frac{\partial \beta}{\partial n_k}\right)_{\epsilon_0,\{n_j\}_{j\neq k}}\frac{dn_k}{dt} = C\,\nabla\cdot\mathbf{v} \tag{D.5}$$

with

$$\begin{aligned} C &= \beta\left\{\left(\frac{\partial p}{\partial \beta}\right)_{\{\beta\mu_{k0}\}}\left(\frac{\partial \beta}{\partial \epsilon_0}\right)_{\{n_k\}} - \sum_k\left[\frac{\partial p}{\partial(\beta\mu_{k0})}\right]_{\beta,\{\beta\mu_{j0}\}_{j\neq k}}\left(\frac{\partial \beta}{\partial n_k}\right)_{\epsilon_0,\{n_j\}_{j\neq k}}\right\} \\ &= \beta\left\{\left(\frac{\partial p}{\partial \beta}\right)_{\{\beta\mu_{k0}\}}\left(\frac{\partial \beta}{\partial \epsilon_0}\right)_{\{n_k\}} + \sum_k\left[\frac{\partial p}{\partial(\beta\mu_{k0})}\right]_{\beta,\{\beta\mu_{j0}\}_{j\neq k}}\left[\frac{\partial(\beta\mu_{k0})}{\partial \epsilon_0}\right]_{\{n_k\}}\right\} \\ &= \beta\left(\frac{\partial p}{\partial \epsilon_0}\right)_{\{n_k\}}, \end{aligned} \tag{D.6}$$

hence equation (3.77). Similarly, using the Maxwell relations (D.2) and (D.3), we have that

$$\frac{d(\beta\mu_{k0})}{dt} = \left[\frac{\partial(\beta\mu_{k0})}{\partial\epsilon_0}\right]_{\{n_k\}}\frac{d\epsilon_0}{dt} + \sum_l \left[\frac{\partial(\beta\mu_{k0})}{\partial n_l}\right]_{\epsilon_0,\{n_j\}_{j\neq l}}\frac{dn_l}{dt} = C_k\,\nabla\cdot\mathbf{v} \tag{D.7}$$

with $C_k = -\beta\,(\partial p/\partial n_k)_{\epsilon_0,\{n_j\}_{j\neq k}}$, and hence equation (3.78). Furthermore, using the relations (D.4), equation (3.75) becomes equation (3.79).

D.1.3 Derivation of the Dissipative Current Densities

Using the expressions (3.62)–(3.64) for the conjugate fields, the expansion $\nabla(\beta\mathbf{v}) = \mathbf{v}\nabla\beta + \beta\nabla\mathbf{v}$, the chemical potentials μ_{k0} in the frame moving with the fluid element, $\hat{\rho} = \sum_k m_k\hat{n}_k$, and $\hat{\mathbf{g}} = \sum_k m_k\hat{\jmath}_k$, the first and second series of terms in equation (3.87) are given by

$$\partial_\tau\chi\cdot\delta\hat{\mathbf{c}} = \partial_\tau\beta\left(\delta\hat{\epsilon} - \mathbf{v}\cdot\delta\hat{\mathbf{g}} + \frac{\mathbf{v}^2}{2}\delta\hat{\rho}\right) - \beta\,\partial_\tau\mathbf{v}\cdot\left(\delta\hat{\mathbf{g}} - \mathbf{v}\,\delta\hat{\rho}\right) - \sum_k \partial_\tau(\beta\mu_{k0})\,\delta\hat{n}_k \tag{D.8}$$

and

$$\nabla\chi : \delta\hat{\jmath} = \nabla\beta\cdot\left(\delta\hat{\jmath}_\epsilon - \mathbf{v}\cdot\delta\hat{\jmath}_{\mathbf{g}} + \frac{\mathbf{v}^2}{2}\delta\hat{\mathbf{g}}\right) - \beta\,\nabla\mathbf{v} : \left(\delta\hat{\jmath}_{\mathbf{g}} - \delta\hat{\mathbf{g}}\,\mathbf{v}\right) - \sum_k \nabla(\beta\mu_{k0})\cdot\delta\hat{\jmath}_k. \tag{D.9}$$

Next, equations (3.77)–(3.79) are used to replace the time partial derivatives by gradients in equation (D.8), which is added to equation (D.9) to get

$$\begin{aligned}
&\partial_\tau\chi\cdot\delta\hat{\mathbf{c}} + \nabla\chi : \delta\hat{\jmath} \\
&= \nabla\beta\cdot\left[\delta\hat{\jmath}_\epsilon - \mathbf{v}\cdot\delta\hat{\jmath}_{\mathbf{g}} + \frac{\mathbf{v}^2}{2}\,\delta\hat{\mathbf{g}} - \mathbf{v}\left(\delta\hat{\epsilon} - \mathbf{v}\cdot\delta\hat{\mathbf{g}} + \frac{\mathbf{v}^2}{2}\delta\hat{\rho}\right) - \frac{\epsilon_0 + p}{\rho}\left(\delta\hat{\mathbf{g}} - \mathbf{v}\,\delta\hat{\rho}\right)\right] \\
&\quad - \beta\,\nabla\mathbf{v} : \left[\delta\hat{\jmath}_{\mathbf{g}} - \delta\hat{\mathbf{g}}\,\mathbf{v} - \mathbf{v}\left(\delta\hat{\mathbf{g}} - \mathbf{v}\,\delta\hat{\rho}\right) - \left(\frac{\partial p}{\partial\epsilon_0}\right)_{\{n_k\}}\left(\delta\hat{\epsilon} - \mathbf{v}\cdot\delta\hat{\mathbf{g}} + \frac{\mathbf{v}^2}{2}\delta\hat{\rho}\right)\mathbf{1}\right. \\
&\qquad\qquad \left. - \sum_k \left(\frac{\partial p}{\partial n_k}\right)_{\epsilon_0,\{n_j\}_{j\neq k}} \delta\hat{n}_k\,\mathbf{1}\right] \\
&\quad - \sum_k \nabla(\beta\mu_{k0})\cdot\left[\delta\hat{\jmath}_k - \mathbf{v}\,\delta\hat{n}_k - \frac{n_k}{\rho}\left(\delta\hat{\mathbf{g}} - \mathbf{v}\,\delta\hat{\rho}\right)\right] \\
&\equiv \nabla\beta\cdot\delta\hat{\jmath}'_\epsilon - \beta\,\nabla\mathbf{v} : \delta\hat{\jmath}'_{\mathbf{g}} - \sum_k \nabla(\beta\mu_{k0})\cdot\delta\hat{\jmath}'_k.
\end{aligned} \tag{D.10}$$

Using equations (3.66)–(3.68) and neglecting the term $\hat{\mathbf{\Delta}}$ in equation (3.68) because it goes as the square of gradients, equation (3.90) is obtained by involving the quantities in the frame moving with the fluid element where $\mathbf{v} = 0$.

D.2 Microscopic Approach to Reactions in Fluids

For simplicity, we consider the reaction required for the formation of diatomic molecules A_2 in a fluid with monatomic species A at low enough temperature:

$$A + A + X \rightleftharpoons A_2 + X, \tag{D.11}$$

where X denotes either an atom A or a molecule A_2. We note that without a third particle the binding of two atoms to form a diatomic molecule would not be possible due to the requirement for energy conservation in binary collisions.

The Hamiltonian function ruling the motion of two atoms with mass m can always be decomposed into the kinetic energy of the center of mass and the energy of the relative motion of the two atoms according to

$$h_{ab} = \frac{\mathbf{p}_a^2}{2m} + \frac{\mathbf{p}_b^2}{2m} + u_{ab} = \frac{\mathbf{P}_{ab}^2}{2M} + \varepsilon_{ab} \qquad \text{with} \qquad \varepsilon_{ab} \equiv \frac{\mathbf{p}_{ab}^2}{2\mu} + u_{ab}, \tag{D.12}$$

where $M = 2m$ is the mass of the binary system, $\mathbf{P}_{ab} = \mathbf{p}_a + \mathbf{p}_b$ its momentum, $\mu = m/2$ the reduced mass, $\mathbf{p}_{ab} = (\mathbf{p}_a - \mathbf{p}_b)/2$ the momentum of relative motion, and $u_{ab} = u^{(2)}(r_{ab})$ the binary interaction energy expressed in terms of the relative distance $r_{ab} = \|\mathbf{r}_a - \mathbf{r}_b\|$. The interaction potential is assumed to vanish for arbitrarily large interatomic distance. In this case, the two atoms are bounded if the condition $\varepsilon_{ab} < 0$ is satisfied. The diatomic molecules are supposed to be located at the positions $\mathbf{R}_{ab} = (\mathbf{r}_a + \mathbf{r}_b)/2$ of their center of mass. Accordingly, the microscopic density of diatomic molecules can be defined as

$$\hat{n}_2(\mathbf{r};\mathbf{\Gamma}) \equiv \frac{1}{2}\sum_{a\neq b} \theta(-\varepsilon_{ab})\,\delta(\mathbf{r} - \mathbf{R}_{ab}). \tag{D.13}$$

Since the microscopic density of all the atoms in the system is always given by

$$\hat{n}(\mathbf{r};\mathbf{\Gamma}) \equiv \sum_a \delta(\mathbf{r} - \mathbf{r}_a), \tag{D.14}$$

the density of unbounded atoms can be defined according to $\hat{n}_1 \equiv \hat{n} - 2\hat{n}_2$.

Using the Hamilton equations of motion, the time evolution of these densities can be expressed as

$$\partial_t \hat{n}_1 + \nabla\cdot\hat{\jmath}_1 = -2\hat{\sigma} \qquad \text{and} \qquad \partial_t \hat{n}_2 + \nabla\cdot\hat{\jmath}_2 = \hat{\sigma} \tag{D.15}$$

with the microscopic rate density for the formation of diatomic molecules

$$\hat{\sigma} \equiv -\frac{1}{2}\sum_{a\neq b} \dot{\varepsilon}_{ab}\,\delta(-\varepsilon_{ab})\,\delta(\mathbf{r} - \mathbf{R}_{ab}), \tag{D.16}$$

the molecular current density

$$\hat{\jmath}_2 \equiv \frac{1}{2}\sum_{a\neq b} \theta(-\varepsilon_{ab})\,\dot{\mathbf{R}}_{ab}\,\delta(\mathbf{r} - \mathbf{R}_{ab}), \tag{D.17}$$

and the current density of unbounded atoms given by $\hat{\jmath}_1 \equiv \hat{\jmath} - 2\,\hat{\jmath}_2$ with $\hat{\jmath} \equiv \sum_a \dot{\mathbf{r}}_a \delta(\mathbf{r} - \mathbf{r}_a)$ due to the local conservation of all the atoms.

According to Hamilton's equations, the time derivative of the binary energy is given by

$$\dot{\varepsilon}_{ab} = \frac{1}{2m}\,(\mathbf{p}_a - \mathbf{p}_b)\cdot \sum_{c(\neq a,b)} (\mathbf{F}_{ac} - \mathbf{F}_{bc}), \tag{D.18}$$

where $c \neq a,b$ in the sum because of the action-reaction principle, $\mathbf{F}_{ab} = -\mathbf{F}_{ba}$, implying energy conservation within the corresponding binary system.

Taking statistical averages over an appropriate local equilibrium distribution leads to macroscopic diffusion–reaction equations. In particular, if the system is out of equilibrium with uniform atomic and molecular densities at the macroscale in its volume V, the net reaction rate density can be obtained as

$$w = \frac{1}{V}\int_V \langle\hat{\sigma}\rangle\, d\mathbf{r} = w_+ - w_-, \tag{D.19}$$

in terms of the forward and backward rate densities,

$$w_\pm \equiv \frac{1}{2V}\left\langle \sum_{a\neq b} \theta(\mp\dot{\varepsilon}_{ab})\,|\dot{\varepsilon}_{ab}|\,\delta(-\varepsilon_{ab})\right\rangle, \tag{D.20}$$

because molecules form if $\dot{\varepsilon}_{ab} < 0$, but dissociate if $\dot{\varepsilon}_{ab} > 0$. At equilibrium, the principle of detailed balance introduced in Section 2.5.2 holds, so that $w_+ = w_-$ and the net reaction rate density is vanishing. The rate densities (D.20) can be equivalently evaluated using the phase-space fluxes across the hypersurfaces defined by the conditions $\varepsilon_{ab} = 0$. In this way, the reaction rates can be calculated from the microdynamics of atoms.

Appendix E

Complements on Stochastic Processes

E.1 Central Limit Theorem

In probability theory, the mean value and the variance of successive random variables $\{X_i\}_{i=1}^{\infty}$ are, respectively, defined by

$$\langle X \rangle \equiv \lim_{k\to\infty} \frac{1}{k} S_k \quad \text{and} \quad \sigma^2 \equiv \lim_{k\to\infty} \frac{1}{k} \left\langle (S_k - k\,\langle X \rangle)^2 \right\rangle \quad \text{with} \quad S_k \equiv \sum_{i=1}^{k} X_i. \tag{E.1}$$

If they both exist, the central limit theorem says that the following probability converges to the error function,

$$\lim_{k\to\infty} P\left\{ \frac{S_k - k\,\langle X \rangle}{\sigma\,\sqrt{k}} < z \right\} = \frac{1}{\sqrt{2\pi}} \int_{-\infty}^{z} \exp\left(-y^2/2\right) dy, \tag{E.2}$$

meaning that the deviation of the sum S_k with respect to its mean value $k\langle X \rangle$ behaves as a Gaussian random variable in the asymptotic limit $k \to \infty$.

We note that the random variables X_i may be statistically independent or not. In the latter case, the variance is given as follows if the series is converging:

$$\sigma^2 = \sum_{k=-\infty}^{+\infty} C(k) \qquad \text{with} \qquad C(k) \equiv \langle (X_i - \langle X \rangle)(X_{i+k} - \langle X \rangle) \rangle. \tag{E.3}$$

The central limit theorem can be proved, in particular, with the method of characteristic functions (Feller, 1968, 1971). It can also be extended to integrals of random functions such as $S(t) = \int_0^t X(t')\,dt'$.

E.2 Large-Deviation Theory

Large-deviation theory goes beyond the central limit theorem (Ellis, 1985; Touchette, 2009). Although the central limit theorem captures the fluctuations in the close vicinity of the mean value, large-deviation theory allows us to characterize the fluctuations in the tails of their probability distributions in terms of all the (rescaled) statistical cumulants.

Large-deviation theory is based on two functions that are interconnected by the Legendre transform. On the one hand, the so-called *rate function* is defined as

$$I(\alpha) \equiv \lim_{k\to\infty} -\frac{1}{k} \ln P\left\{\frac{S_k}{k} \in (\alpha, \alpha + d\alpha)\right\} \qquad \text{for} \qquad S_k \equiv \sum_{i=1}^{k} X_i, \tag{E.4}$$

which determines how the probability of finding the sum around the value $S_k \simeq \alpha k$ decreases as k increases. On the other hand, the *cumulant generating function* can be defined as

$$J(\beta) \equiv \lim_{k\to\infty} \frac{1}{k} \ln\langle \exp(\beta\, S_k)\rangle. \tag{E.5}$$

All the statistical cumulants are obtained by taking successive derivatives of the generating function with respect to the parameter β and setting $\beta = 0$: $c_n = (d^n J/d\beta^n)(0)$ for $n = 0, 1, 2, 3, \ldots$. In particular, the first and second cumulants, i.e., the mean value and the variance, are given by

$$c_1 = \langle X\rangle = \frac{dJ}{d\beta}(0) \qquad \text{and} \qquad c_2 = \sigma^2 = \frac{d^2 J}{d\beta^2}(0). \tag{E.6}$$

Now, the probability that the sum takes the value $S_k \simeq \alpha k$ can be expressed for $k \to \infty$ in terms of the rate function (E.4) as

$$P\left\{\alpha k < S_k < (\alpha + d\alpha)\, k\right\} \simeq \mathcal{N}(k;\alpha)\, \mathrm{e}^{-I(\alpha)\, k}\, d\alpha, \tag{E.7}$$

where $\mathcal{N}(k;\alpha)$ is a normalization constant, which has a subexponential dependence on the variable k, i.e., such that $\lim_{k\to\infty} k^{-1} \ln \mathcal{N}(k;\alpha) = 0$. Replacing this result into the definition (E.5) and integrating over α with the method of steepest descent, the cumulant generating function is given by

$$J(\beta) = \alpha_\beta\, \beta - I(\alpha_\beta) \qquad \text{with} \qquad \frac{dI}{d\alpha}(\alpha_\beta) = \beta. \tag{E.8}$$

The result (E.8) shows that the cumulant generating function is given by the Legendre transform of the rate function: $J = \mathcal{L}e[I]$. Since this transform is an involution $\mathcal{L}e^2 = 1$, the rate function is recovered by the same Legendre transform $I = \mathcal{L}e[J]$:

$$I(\alpha) = \alpha\, \beta_\alpha - J(\beta_\alpha) \qquad \text{with} \qquad \frac{dJ}{d\beta}(\beta_\alpha) = \alpha. \tag{E.9}$$

If the rate function and the generating function are not differentiable, the Legendre transform can be extended into the Legendre–Fenchel transform to get

$$I(\alpha) = \sup_\beta\, [\alpha\, \beta - J(\beta)] \qquad \text{and} \qquad J(\beta) = \sup_\alpha\, [\alpha\, \beta - I(\alpha)]. \tag{E.10}$$

Supposing that these functions are differentiable, the cumulant generating function can be expanded around $\beta = 0$ as

$$J(\beta) = c_1\, \beta + \frac{c_2}{2!}\, \beta^2 + \frac{c_3}{3!}\, \beta^3 + \cdots \tag{E.11}$$

because $J(0) = 0$. Using the Legendre transform (E.9), the rate function is thus given by

$$I(\alpha) = \frac{(\alpha - c_1)^2}{2c_2} + O\left[(\alpha - c_1)^3\right]. \tag{E.12}$$

Substituting this expression into equation (E.7), the Gaussian probability distribution predicted by the central limit theorem is recovered, i.e.,

$$P(S_k) \sim \exp\left[-\frac{(S_k - c_1 k)^2}{2c_2 k}\right] \tag{E.13}$$

in the limit $k \to \infty$, by using only the first and second cumulants. The expansion of the generating function in terms of all the statistical cumulants (if they exist) or the whole rate function thus characterize more finely the asymptotic probability distribution of the sum than the central limit theorem, which is focused on the parabolic maximum of the distribution given by its first and second cumulants.

E.3 Standard Poisson Process

This continuous-time discrete-state Markovian process describes the counting of statistically independent particles crossing some area such as the window of some detector. This process is considered for counting α particles emitted from radioactive substances (Rutherford et al., 1910), electrons arriving and generating shot noise at the anode of vacuum tubes or other electronic devices (van Kampen, 1981; Gardiner, 2004), as well as similar processes for other particles.

Such processes are incremental, i.e., the number of particles can only increase one by one with time. If we suppose that their mean number increases linearly with time at the mean rate γ, the probability $P(n,t)$ that n particles have arrived by time t should obey the master equation

$$\frac{d}{dt} P(n,t) = \gamma\,[P(n-1,t) - P(n,t)] \qquad \text{for} \qquad n = 1,2,\ldots \tag{E.14}$$

and $dP(0,t)/dt = -\gamma P(0,t)$ for $n = 0$. This master equation is obtained by setting $W_+ = \gamma$ and $W_- = 0$ in equation (4.78), so that the standard Poisson process is a fully irreversible random drift with mean current $J = \gamma$ and diffusivity $D = \gamma/2$.

If the initial condition is $P(n, t = 0) = \delta_{n,0}$, the solution of these coupled ordinary differential equations is given by $P(n,t) = \mathrm{e}^{-\gamma t}(\gamma t)^n/(n!)$, which is a Poisson probability distribution with mean value $\langle n\rangle = \gamma t$ and variance $\mathrm{Var}(n) = \langle n^2\rangle - \langle n\rangle^2 = \gamma t$. The Poisson process is nonstationary since $\lim_{t\to\infty} P(n,t) = 0$ for any finite integer n in relation to the drift of the mean value $\langle n\rangle$ up to infinity.

The spectral density can be calculated using the MacDonald formula (4.16) with $Z(t) = n(t) - \langle n(t)\rangle$ such that $\langle Z(t)^2\rangle = \mathrm{Var}(n) = \gamma t$, giving $S_{nn}(\omega) = \gamma$ independently of the frequency ω. The standard Poisson process is thus a shot noise with $S_{nn}(\omega) = 2D = J$.

Remarkably, the superposition of two statistically independent Poisson processes is again a Poisson process. Indeed, if there are two independent sources of particles, the total number of particles arriving during the time interval $[0,t]$ is the sum of the numbers from both sources: $N = N_1 + N_2$. If the sources have emission rates γ_1 and γ_2, the numbers N_i have the statistically independent Poisson distributions

$$P_i(n,t) = \frac{a_i^n}{n!}\, \mathrm{e}^{-a_i} \qquad \text{with} \qquad a_i = \gamma_i t \qquad \text{for} \qquad i = 1,2. \tag{E.15}$$

Therefore, since $n_2 = n - n_1$, the probability distribution of their sum is given by

$$P(n,t) = \sum_{n_1=0}^{n} P_1(n_1,t)\, P_2(n-n_1,t) = \frac{(a_1+a_2)^n}{n!}\, \mathrm{e}^{-a_1-a_2}, \tag{E.16}$$

which is again a Poisson distribution, but with $\langle n \rangle = a_1 + a_1 = \langle n_1 \rangle + \langle n_2 \rangle$.

Accordingly, the sum of statistically independent Poisson processes $\mathcal{P}_i(a_i)$ is again a Poisson process, i.e., $\sum_i \mathcal{P}_i(a_i) = \mathcal{P}(\sum_i a_i)$. This result is used, for instance, in the expression (4.34) for Markov jump processes.

The stochasticity or temporal disorder of the standard Poisson process is characterized by the τ-entropy per unit time, $h(\tau) = \gamma \ln[\mathrm{e}/(\gamma\tau)] + O(\tau^2)$, for the stroboscopic observation of the process with sampling time τ (Gaspard and Wang, 1993). We note that this dynamical entropy increases without bound as $\tau \to 0$. The reason is that the random time intervals between the jumps is exponentially distributed and thus continuous. The τ-entropy per unit time is obtained by coarse graining the time axis of the exponential distribution $p(t) = \gamma \exp(-\gamma t)$ into time intervals $[t_j, t_j + \tau]$ with $t_j = j\tau$ and $j \in \mathbb{N}$, calculating the entropy and dividing by the mean time $\langle t \rangle = \gamma^{-1}$ between the jumps, giving $h(\tau) = -\gamma \sum_j P_j \ln P_j$ with $P_j \simeq p(t_j)\tau$. The degree of stochasticity of such a continuous-time process is thus much higher than for discrete-time processes such as Bernoulli or Markov chains.

E.4 Lower Bound on the Entropy Production Rate

For Markov jump processes ruled by the master equation (4.27), a lower bound on their entropy production rate (4.57) can be obtained using the log sum inequality

$$\sum_i a_i \ln \frac{a_i}{b_i} \geq \left(\sum_i a_i\right) \ln \frac{\sum_i a_i}{\sum_i b_i}, \tag{E.17}$$

holding for positive real numbers a_i and b_i (Cover and Thomas, 2006). Indeed, equation (4.57) can be written as $d_{\mathrm{i}}S/dt = k_{\mathrm{B}}\mathcal{R}$ with the rate,

$$\mathcal{R} \equiv \sum_{\omega,\omega'} \sum_{\rho} P(\omega',t)\, W\big(\omega' \overset{\rho}{\to} \omega\big) \ln \frac{P(\omega',t)\, W\big(\omega' \overset{\rho}{\to} \omega\big)}{P(\omega,t)\, W\big(\omega \overset{-\rho}{\to} \omega'\big)} \geq 0. \tag{E.18}$$

Lumping together the transition rates between the states ω and ω' for all the elementary transition mechanisms $\{\rho\}$ according to equation (4.26), the log sum inequality (E.17) can be used to find the lower bound $\mathcal{R} \geq \mathcal{R}_0$ with the quantity

$$\mathcal{R}_0 \equiv \sum_{\omega,\omega'} P(\omega',t)\, W(\omega' \to \omega) \ln \frac{P(\omega',t)\, W(\omega' \to \omega)}{P(\omega,t)\, W(\omega \to \omega')} \geq 0 \tag{E.19}$$

defined for the lumped transition rates $W(\omega \to \omega') = \sum_\rho W(\omega \overset{\rho}{\to} \omega')$. The probability distribution $P(\omega,t)$ is the same for both processes because the master equation (4.27) reduces to equation (4.25) with the lumped transition rates after summing over the elementary mechanisms $\{\rho\}$. However, the entropy production rate can be underevaluated by lumping together the transition rates.

As an example, we may consider the particle exchange process ruled by the master equation (4.91) with the transition rates (4.89). Summing over the random variable n counting the particle exchanges between the two reservoirs, the master equation for the probability distribution $\mathscr{P}(N,t) = \sum_{n\in\mathbb{Z}} P(N,n,t)$ reads

$$\frac{d}{dt}\mathscr{P}(N,t) = W_+(N-1)\,\mathscr{P}(N-1,t) + W_-(N+1)\,\mathscr{P}(N+1,t) - \big[W_+(N) + W_-(N)\big]\,\mathscr{P}(N,t) \tag{E.20}$$

with the lumped transition rates $W_\pm \equiv W_{\pm 1} + W_{\pm 2}$. The random variable $N(t)$ follows a reversible process of stationary distribution (4.94). Because of its reversibility, the rate (E.19) vanishes ($\mathcal{R}_0 = 0$) whether this process is at equilibrium or not. However, the full stochastic process for the joint random variables $N(t)$ and $n(t)$ is not reversible unless the system is at equilibrium and, consistently, the associated rate (E.18) giving the entropy production is positive out of equilibrium and only vanishing at equilibrium, where the affinity (4.92) is equal to zero.

Furthermore, in the limit where one of the rates $W(\omega \overset{\rho}{\to} \omega')$ vanishes although all the lumped rates $W(\omega \to \omega')$ are still positive, the entropy production rate given by equation (E.18) becomes infinite, so that the process is fully irreversible. In this other example, the rate (E.19) might give a finite positive value, drastically underevaluating the entropy production rate.

A similar lower bound can be obtained for the entropy production given by equation (6.103) in terms of the Kullback–Leibler divergence (6.101), if the probabilities of the paths (6.8) are summed over the sequences of elementary mechanisms $\boldsymbol{\rho} = \{\rho_1, \ldots, \rho_{k-1}\}$, giving the probabilities,

$$P_0(\boldsymbol{\omega}_0) = \sum_{\boldsymbol{\rho}} P(\boldsymbol{\omega}), \qquad \text{where} \qquad \boldsymbol{\omega}_0 = \omega_1 \to \omega_2 \to \cdots \to \omega_k \tag{E.21}$$

are the paths defined without specifying the sequence of elementary mechanisms followed during the process. According to the log sum inequality (E.17) with i denoting the sequence $\boldsymbol{\rho}$, the entropy production during the total time interval $t = k\tau$, given by

$$\Delta_{\rm i} S = \sum_{\boldsymbol{\omega}} P(\boldsymbol{\omega}) \ln \frac{P(\boldsymbol{\omega})}{P(\boldsymbol{\omega}^{\rm R})}, \tag{E.22}$$

has the lower bound

$$\Delta_{\rm i} S \geq \sum_{\boldsymbol{\omega}_0} \sum_{\boldsymbol{\rho}} P(\boldsymbol{\omega}) \ln \frac{\sum_{\boldsymbol{\rho}} P(\boldsymbol{\omega})}{\sum_{\boldsymbol{\rho}} P(\boldsymbol{\omega}^{\rm R})} = \sum_{\boldsymbol{\omega}_0} P_0(\boldsymbol{\omega}_0) \ln \frac{P_0(\boldsymbol{\omega}_0)}{P_0(\boldsymbol{\omega}_0^{\rm R})}, \tag{E.23}$$

because summing over the paths $\boldsymbol{\omega} = \omega_1 \overset{\rho_1}{\to} \omega_2 \overset{\rho_2}{\to} \cdots \overset{\rho_{k-1}}{\to} \omega_k$ is equivalent to summing over the paths $\boldsymbol{\omega}_0$ and the sequences $\boldsymbol{\rho}$.

E.5 The Martingale Property Underlying Entropy Production

A random process $\{X_i\}_{i=1}^{\infty}$ is said to be a *martingale* if, for all $l \geq k$, the property $\langle X_l | X_1, X_2, \ldots, X_k \rangle = X_k$ holds, where $\langle X|Y \rangle$ denotes the conditional expectation of X given that Y is known.

This property is satisfied by the random variable $\mathrm{e}^{-\Sigma(\boldsymbol{\omega})}$ defined by equation (6.19) in terms of the probability distributions $P(\boldsymbol{\omega})$ of the path $\boldsymbol{\omega}$ and $P\left(\boldsymbol{\omega}^{\mathrm{R}}\right)$ of its time reversal $\boldsymbol{\omega}^{\mathrm{R}}$. Indeed, we may consider the path $\boldsymbol{\omega}\boldsymbol{\omega}'$ obtained by the concatenation of the paths $\boldsymbol{\omega} = \omega_1 \cdots \omega_k$ and $\boldsymbol{\omega}' = \omega_{k+1} \cdots \omega_l$ with $l > k \geq 1$. First, we have that

$$P\left(\boldsymbol{\omega}^{\mathrm{R}}\right) = \sum_{\boldsymbol{\omega}'} P\left[\left(\boldsymbol{\omega}\boldsymbol{\omega}'\right)^{\mathrm{R}}\right], \tag{E.24}$$

because of the additive property (4.2). Next, using equation (6.19) in its left- and right-hand sides, we get

$$\mathrm{e}^{-\Sigma(\boldsymbol{\omega})} P(\boldsymbol{\omega}) = \sum_{\boldsymbol{\omega}'} \mathrm{e}^{-\Sigma(\boldsymbol{\omega}\boldsymbol{\omega}')} P(\boldsymbol{\omega}\boldsymbol{\omega}'). \tag{E.25}$$

Finally, dividing both sides by $P(\boldsymbol{\omega})$ and using the definition (4.7) for conditional probabilities, we find

$$\mathrm{e}^{-\Sigma(\boldsymbol{\omega})} = \sum_{\boldsymbol{\omega}'} \mathrm{e}^{-\Sigma(\boldsymbol{\omega}\boldsymbol{\omega}')} P(\boldsymbol{\omega}\boldsymbol{\omega}'|\boldsymbol{\omega}) = \left\langle \mathrm{e}^{-\Sigma(\boldsymbol{\omega}\boldsymbol{\omega}')} | \boldsymbol{\omega} \right\rangle, \tag{E.26}$$

which proves the martingale property (Neri et al., 2017).

E.6 Stochastic Integrals

There exist several iterative schemes equivalent to equation (4.115) for simulating the random trajectories of stochastic process ruled by the master equation (4.108) if we consider the intermediate points

$$\mathbf{x}_i^{(\alpha)} \equiv \mathbf{x}_i + \alpha\,(\mathbf{x}_{i+1} - \mathbf{x}_i), \quad t_i^{(\alpha)} \equiv t_i + \alpha\,(t_{i+1} - t_i), \quad \text{with} \quad 0 \leq \alpha \leq 1 \tag{E.27}$$

and $\mathbf{x}_i \equiv \mathbf{x}(t_i) \in \mathbb{R}^d$ for the discretization into time steps $\Delta t = t_{i+1} - t_i$. Defining the vector fields

$$\mathbf{A}^{(\alpha)} \equiv \mathbf{A} - \alpha \sum_{\mu=1}^{n} \left(\mathbf{B}_\mu \cdot \partial_{\mathbf{x}}\right) \mathbf{B}_\mu \qquad \text{and} \qquad \mathbf{B}_\mu^{(\alpha)} \equiv \mathbf{B}_\mu, \tag{E.28}$$

we have the following equivalent iterative schemes:

$$\mathbf{x}_{i+1} = \mathbf{x}_i + \mathbf{A}^{(\alpha)}\left(\mathbf{x}_i^{(\alpha)}, t_i^{(\alpha)}\right) \Delta t + \sum_{\mu=1}^{n} \mathbf{B}_\mu^{(\alpha)}\left(\mathbf{x}_i^{(\alpha)}, t_i^{(\alpha)}\right) G_{\mu i}\,\sqrt{\Delta t}. \tag{E.29}$$

First, equation (4.115) is recovered with the choice $\alpha = 0$. Next, equations (4.115) and (E.29) only differ by corrections of order $O(\Delta t^{3/2})$ after substituting into this latter the expressions (E.27), expanding in powers of $\Delta\mathbf{x} \equiv \mathbf{x}_{i+1} - \mathbf{x}_i$, and using the fact that $G_{\mu i} G_{\nu i} \simeq \langle G_{\mu i} G_{\nu i} \rangle = \delta_{\mu\nu}$. Consequently, the stochastic processes generated by the iterative schemes (E.29) are also ruled by the master equation (4.108) with the same vector field (4.102) and diffusivities (4.103) as for the Itô iterative scheme (4.115). The iterative scheme of Stratonovich is given by the choice $\alpha = 1/2$ corresponding to the midpoint discretization rule (Gardiner, 2004).

The issue is that Wiener's stochastic processes such as

$$dW_\mu(t) \equiv \chi_\mu(t)\, dt \tag{E.30}$$

are not differentiable functions of time, but instead behave as $\Delta W_\mu \propto \sqrt{\Delta t}$. Consequently, the squares of such quantities go as Δt and they contribute to the same order as the deterministic term. Accordingly, there are several possible definitions for stochastic integrals involving the products of the random variables (E.30) with some smooth enough functions $f[\mathbf{x}(t), t]$. As for Riemann integrals, time may be discretized into steps Δt, so that Wiener processes become independent identically distributed Gaussian random variables, $\Delta W_\mu(t_i) = G_{\mu i}\sqrt{\Delta t}$ with $\langle G_{\mu i}\rangle = 0$ and $\langle G_{\mu i} G_{\nu j}\rangle = \delta_{\mu\nu}\delta_{ij}$. Several types of stochastic integrals can be defined as

$$\int_{t_0}^{t} f[\mathbf{x}(t), t] \overset{\alpha}{\circ} dW_\mu(t) \equiv \lim_{k\to\infty} \sum_{i=1}^{k} f\left(\mathbf{x}_i^{(\alpha)}, t_i^{(\alpha)}\right) G_{\mu i}\sqrt{\Delta t}, \tag{E.31}$$

with the intermediate points (E.27) and $\Delta t = (t - t_0)/k$. Equivalence is obtained if the mean square between the left- and right-hand sides is vanishing in the limit $k \to \infty$ (Gardiner, 2004). The stochastic integrals by Itô and Stratonovich, respectively, correspond to the choices $\alpha = 0$ and $\alpha = 1/2$. Their equivalence is established by expanding the function f in powers of $\Delta\mathbf{x}$ and using $G_{\mu i} G_{\nu i} \simeq \delta_{\mu\nu}$, leading to

$$\int_{t_0}^{t} f \overset{\alpha}{\circ} dW_\mu = \int_{t_0}^{t} \left(f\, dW_\mu + \alpha\, \mathbf{B}_\mu \cdot \partial_{\mathbf{x}} f\, dt'\right), \tag{E.32}$$

where the first term in the right-hand side is a stochastic integral of Itô's type given by equation (E.32) with $\alpha = 0$ and the second term a standard integral. This extra term is consistent with the modification of the deterministic vector field $\mathbf{A}^{(\alpha)}$ in equation (E.28), if the intermediate points (E.27) with $\alpha \neq 0$ are used for integrating the stochastic differential equation. With a similar reasoning, we have that

$$\int_{t_0}^{t} f \overset{\alpha}{\circ} d\mathbf{x} \equiv \lim_{k\to\infty} \sum_{i=1}^{k} f\left(\mathbf{x}_i^{(\alpha)}, t_i^{(\alpha)}\right) (\mathbf{x}_{i+1} - \mathbf{x}_i) = \int_{t_0}^{t} \left(f\, d\mathbf{x} + 2\alpha\, \mathsf{D} \cdot \partial_{\mathbf{x}} f\, dt'\right) \tag{E.33}$$

in terms of the diffusivities (4.103). If the function f is replaced by the gradient $\partial_{\mathbf{x}} g$, the stochastic integral has the form,

$$\int_{t_0}^{t} \partial_{\mathbf{x}} g \overset{\alpha}{\circ} d\mathbf{x} = g(\mathbf{x}_t) - g(\mathbf{x}_{t_0}) + (2\alpha - 1) \int_{t_0}^{t} \mathsf{D} : \partial_{\mathbf{x}}^2 g\, dt', \tag{E.34}$$

so that the standard result of vector calculus is given by Stratonovich's integral with $\alpha = 1/2$ instead of Itô's integral with $\alpha = 0$.

As discussed in Section 4.5.3, the vector fields $\mathbf{B}_\mu$ can be taken as the d eigenvectors of the diffusivity matrix in equation (4.116), reducing the number of independent Gaussian white noises to $n = d$. Accordingly, the diffusivity matrix is given by

$$\mathsf{D} = \frac{1}{2}\,\mathsf{B}\cdot\mathsf{B}^{\mathrm{T}} \tag{E.35}$$

in terms of the $d \times d$ matrix $\mathsf{B} = (\mathbf{B}_\mu)$ composed with the d eigenvectors $\mathbf{B}_\mu$.

The path integrals constructed at the intermediate points (E.27) and corresponding to the iterative schemes (E.29) can be expressed as

$$p(\mathbf{x},t|\mathbf{x}_0,t_0) = \int \exp\left[-\int_{t_0}^{t} \mathcal{L}^{(\alpha)}(\mathbf{x},\dot{\mathbf{x}},t')\,dt'\right]\mathcal{D}\mathbf{x}(t), \tag{E.36}$$

in terms of the modified Lagrangian function (Itami and Sasa, 2017)

$$\mathcal{L}^{(\alpha)}(\mathbf{x},\dot{\mathbf{x}},t) = \frac{1}{4}\,\mathsf{D}^{-1} : \left(\dot{\mathbf{x}} - \mathbf{V}^{(\alpha)}\right)^2 + \Lambda^{(\alpha)}, \tag{E.37}$$

with the vector field

$$\mathbf{V}^{(\alpha)} \equiv \mathbf{A}^{(\alpha)} - \alpha\,\mathsf{B}\cdot(\partial_{\mathbf{x}}\cdot\mathsf{B}) = \mathbf{C} - 2\alpha\,\partial_{\mathbf{x}}\cdot\mathsf{D} = \mathbf{V} + (1-2\alpha)\,\partial_{\mathbf{x}}\cdot\mathsf{D}, \tag{E.38}$$

which has equivalent forms using $\mathbf{C} = \mathbf{A}$ or the drift velocity (4.112) and the diffusivity matrix (E.35); and with the following function of the variables $\mathbf{x}$:

$$\Lambda^{(\alpha)} \equiv \alpha\,\partial_{\mathbf{x}}\cdot\mathbf{A}^{(\alpha)} + \frac{\alpha^2}{2}\left[(\partial_{\mathbf{x}}\mathsf{B}) : (\partial_{\mathbf{x}}\mathsf{B})^{\mathrm{T}} - (\partial_{\mathbf{x}}\cdot\mathsf{B})^2\right]. \tag{E.39}$$

This result is proved by considering the decomposition (6.14) of the propagator into a multiple integral over the conditional probabilities to jump from $\mathbf{x} = \mathbf{x}_i$ to $\mathbf{x}' = \mathbf{x}_{i+1}$ over the time step $\tau = \Delta t$ according to the random iteration (E.29). This latter can be written as

$$\mathbf{x}' = \mathbf{x} + \mathbf{A}^{(\alpha)}\,\tau + \mathsf{B}\cdot\mathbf{G}\,\sqrt{\tau} \tag{E.40}$$

in terms of $\mathbf{A}^{(\alpha)} = \mathbf{A}^{(\alpha)}\left(\mathbf{x}_i^{(\alpha)}, t_i^{(\alpha)}\right)$, the matrix $\mathsf{B} = (\mathbf{B}_\mu) = (B_{l\mu})$ with the elements $B_{l\mu}\left(\mathbf{x}_i^{(\alpha)}, t_i^{(\alpha)}\right)$, and the d-dimensional vector $\mathbf{G} = (G_{\mu i})$. Supposing that the matrix B is invertible, so that the diffusivity matrix has full rank, the iteration (E.40) mapping the Gaussian variables $\mathbf{G}$ onto $\mathbf{x}'$ can be inverted to get

$$\mathbf{G} = \frac{1}{\sqrt{\tau}}\,\mathsf{B}^{-1}\cdot\left(\mathbf{x}' - \mathbf{x} - \mathbf{A}^{(\alpha)}\tau\right). \tag{E.41}$$

The probability distribution of the variables $\mathbf{x}'$ can thus be deduced from the one for $\mathbf{G}$ according to

$$p_\tau(\mathbf{x}'|\mathbf{x})\,d\mathbf{x}' = \mathrm{e}^{-\frac{1}{2}\mathbf{G}^{\mathrm{T}}\cdot\mathbf{G}}\,\frac{d\mathbf{G}}{\sqrt{(2\pi)^d}}, \tag{E.42}$$

using the change of variables from $\mathbf{G}$ to $\mathbf{x}'$. The corresponding Jacobian matrix is given by

$$\frac{\partial\mathbf{G}}{\partial\mathbf{x}'} = \frac{1}{\sqrt{\tau}}\,\mathsf{B}^{-1}\cdot(\mathsf{1} + \mathsf{M} + \mathsf{N}) \tag{E.43}$$

in terms of the matrices $\mathbf{1} = (\delta_{jk})$, $\mathbf{M} = (M_{jk})$, and $\mathbf{N} = (N_{jk})$ with the elements

$$M_{jk} = \alpha\, B_{jl}\, \frac{\partial\big(\mathbf{B}^{-1}\big)_{lm}}{\partial x_k} \left(x'_m - x_m - A_m^{(\alpha)}\,\tau\right), \qquad N_{jk} = -\alpha\, \frac{\partial A_j^{(\alpha)}}{\partial x_k}\,\tau, \tag{E.44}$$

where the partial derivatives are taken with respect to the arguments of the corresponding functions. The Jacobian determinant is thus obtained as

$$\det \frac{\partial \mathbf{G}}{\partial \mathbf{x}'} = \frac{1}{\sqrt{\tau^d}\,\det \mathbf{B}}\,\det\left(\mathbf{1} + \mathbf{M} + \mathbf{N}\right). \tag{E.45}$$

Since $\mathbf{M} = O\left(\tau^{1/2}\right)$ and $\mathbf{N} = O(\tau)$, we have that

$$\det\left(\mathbf{1} + \mathbf{M} + \mathbf{N}\right) = \exp\left[\operatorname{tr}\mathbf{M} + \operatorname{tr}\mathbf{N} - \frac{1}{2}\operatorname{tr}\mathbf{M}^2 + O\left(\tau^{3/2}\right)\right]. \tag{E.46}$$

The traces are evaluated as

$$\operatorname{tr}\mathbf{M} = M_{jj} = \alpha\, B_{jk}\, \frac{\partial\big(\mathbf{B}^{-1}\big)_{kl}}{\partial x_j} \left(x'_l - x_l - A_l^{(\alpha)}\,\tau\right), \tag{E.47}$$

$$\operatorname{tr}\mathbf{N} = N_{jj} = -\alpha\, \partial_{\mathbf{x}} \cdot \mathbf{A}^{(\alpha)}\,\tau, \tag{E.48}$$

$$\operatorname{tr}\mathbf{M}^2 = M_{jk}M_{kj} = \alpha^2\, B_{jl}\, \frac{\partial\big(\mathbf{B}^{-1}\big)_{lm}}{\partial x_k}\, B_{kp}\, \frac{\partial\big(\mathbf{B}^{-1}\big)_{pn}}{\partial x_j}\, B_{mq}\, B_{nq}\,\tau, \tag{E.49}$$

where we have used the fact that $\left(x'_m - x_m - A_m^{(\alpha)}\,\tau\right)\left(x'_n - x_n - A_n^{(\alpha)}\,\tau\right) \simeq 2D_{mn}\tau$ because of equations (E.41) and (E.35). These results can be substituted into equation (E.42) to get

$$p_\tau(\mathbf{x}'|\mathbf{x}) = \frac{1}{\sqrt{(4\pi\tau)^d\,\det\mathbf{D}}}\,\exp\Bigg[-\frac{1}{4\tau}\,\mathbf{D}^{-1} : \left(\mathbf{x}' - \mathbf{x} - \mathbf{A}^{(\alpha)}\,\tau - \boldsymbol{\Delta}\,\tau\right)^2$$
$$+ \frac{1}{4}\,\mathbf{D}^{-1} : \boldsymbol{\Delta}^2\,\tau + \operatorname{tr}\mathbf{N} - \frac{1}{2}\operatorname{tr}\mathbf{M}^2 + O\left(\tau^{3/2}\right)\Bigg] \tag{E.50}$$

with

$$\Delta_m = \alpha\, B_{jk}\, \frac{\partial\big(\mathbf{B}^{-1}\big)_{kl}}{\partial x_j}\, B_{ln}\, B_{mn} = -\alpha\, B_{mn}\, \frac{\partial B_{jn}}{\partial x_j}, \tag{E.51}$$

where the trace (E.47) has been incorporated inside the parentheses and the identity $\partial_j \mathbf{B}^{-1} \cdot \mathbf{B} = -\mathbf{B}^{-1} \cdot \partial_j \mathbf{B}$ has been used. In addition, we also get

$$\frac{1}{4}\,\mathbf{D}^{-1} : \boldsymbol{\Delta}^2\,\tau - \frac{1}{2}\operatorname{tr}\mathbf{M}^2 = \frac{\alpha^2}{2}\left(\frac{\partial B_{jl}}{\partial x_j}\frac{\partial B_{kl}}{\partial x_k} - \frac{\partial B_{jl}}{\partial x_k}\frac{\partial B_{kl}}{\partial x_j}\right)\tau. \tag{E.52}$$

In the limit $\tau \to 0$, we thus find the Lagrangian function (E.37) with the vector field (E.38) and the function (E.39) given by equation (E.48) for its first term and equation (E.52) for its second one.

In the case of one-dimensional processes ($d = 1$), equation (E.52) is equal to zero, so that the Lagrangian function (E.37) reduces to

$$\mathcal{L}^{(\alpha)}(x, \dot{x}, t) = \frac{1}{4D}\left(\dot{x} - V^{(\alpha)}\right)^2 + \alpha\,\partial_x A^{(\alpha)}, \tag{E.53}$$

with $V^{(\alpha)} = A^{(\alpha)} - \alpha B \partial_x B = V + (1 - 2\alpha)\partial_x D$ and $D = B^2/2$ (Langouche et al., 1979; Lau and Lubensky, 2007; Cugliandolo and Lecomte, 2017).

Stratonovich's midpoint discretization rule corresponds to the choice $\alpha = 1/2$, in which case $\mathbf{V}^{(\alpha)} = \mathbf{V}$. The Lagrangian function (6.16) corresponding to Itô's integration is recovered with $\alpha = 0$. The different discretization rules with $0 \leq \alpha \leq 1$ are meant to give equivalent physical results.

E.7 Ornstein–Uhlenbeck Process

This stochastic process, which has been studied in detail by Uhlenbeck and Ornstein (1930), can be defined in terms of the stochastic differential equation

$$\frac{dx}{dt} = -\frac{1}{\tau_{\mathrm{r}}} x + v_{\mathrm{fl}}(t) \tag{E.54}$$

with the Gaussian white noise such that $\langle v_{\mathrm{fl}}(t) \rangle = 0$ and $\langle v_{\mathrm{fl}}(t)\, v_{\mathrm{fl}}(t') \rangle = 2\mathcal{D}\, \delta(t - t')$, or, equivalently, using the Fokker–Planck equation

$$\partial_t p = \frac{1}{\tau_{\mathrm{r}}} \partial_x (x\, p) + \mathcal{D}\, \partial_x^2 p \equiv \hat{L} p, \tag{E.55}$$

where τ_{r} is the relaxation time of the mean value of the variable x and $\mathcal{D}$ is the diffusion coefficient (or diffusivity).

Taking the statistical average of equation (E.54), we get the time evolution $\langle x(t) \rangle = \langle x(0) \rangle \exp(-t/\tau_{\mathrm{r}})$, showing that this mean value undergoes relaxation to zero after transients over the timescale τ_{r}. The stationary solution of the Fokker–Planck equation (E.55) is the Gaussian distribution,

$$p_{\mathrm{st}}(x) = \frac{1}{\sqrt{2\pi \mathcal{D} \tau_{\mathrm{r}}}} \exp\left(-\frac{x^2}{2\mathcal{D}\tau_{\mathrm{r}}} \right) \tag{E.56}$$

with the variance $\sigma^2 = \langle x^2 \rangle_{\mathrm{st}} = \mathcal{D}\tau_{\mathrm{r}}$, so that the stochastic process is stationary. The autocorrelation function of the variable x and its spectral density (4.15) are, respectively, given by

$$\langle x(0)\, x(t) \rangle_{\mathrm{st}} = \mathcal{D}\tau_{\mathrm{r}} \exp(-|t|/\tau_{\mathrm{r}}) \qquad \text{and} \qquad \mathcal{S}_{xx}(\omega) = \frac{2\mathcal{D}}{\omega^2 + \tau_{\mathrm{r}}^{-2}}. \tag{E.57}$$

Furthermore, the conditional probability density to evolve from x_0 to x during the time t can be obtained by solving the Fokker–Planck equation (E.55) (Chandrasekhar, 1943; van Kampen, 1981), leading to

$$p_t(x|x_0) = \frac{1}{\sqrt{2\pi\sigma^2 \left(1 - \mathrm{e}^{-2t/\tau_{\mathrm{r}}}\right)}} \exp\left[-\frac{\left(x - \mathrm{e}^{-t/\tau_{\mathrm{r}}} x_0\right)^2}{2\sigma^2 \left(1 - \mathrm{e}^{-2t/\tau_{\mathrm{r}}}\right)} \right], \tag{E.58}$$

which converges to the stationary probability density (E.56) for $t \to \infty$.

We note that the Fokker–Planck operator $\hat{L}$ of equation (E.55) satisfies the reversibility condition (4.133) with the stationary density (E.56), so that the Ornstein–Uhlenbeck stochastic process is reversible.

The eigenvalues of the Fokker–Planck operator $\hat{L}$ are given by $z_m = -m/\tau_r$ for $m = 0, 1, 2, \ldots$. They are real, as expected for a reversible process. The associated right and left eigenfunctions such that $\hat{L}\psi_m = z_m \psi_m$ and $\hat{L}^\dagger \tilde{\psi}_m = z_m \tilde{\psi}_m$ can be expressed, respectively, as

$$\psi_m(x) = \frac{a}{\sqrt{\pi}\, 2^m m!} H_m(ax) \exp\left(-a^2 x^2\right) \qquad \text{and} \qquad \tilde{\psi}_m(x) = H_m(ax) \tag{E.59}$$

with $a = (2\mathcal{D}\tau_r)^{-1/2}$ in terms of the Hermite polynomials $H_m(\xi)$ (Risken, 1989; Gaspard et al., 1995; Abramowitz and Stegun, 1972). These eigenfunctions are biorthonormal: $\int_{-\infty}^{+\infty} \tilde{\psi}_m(x)\, \psi_n(x)\, dx = \delta_{mn}$. The right eigenfunction of the leading eigenvalue $z_0 = 0$ gives the stationary probability density (E.56): $\psi_0(x) = p_{\rm st}(x)$.

Examples for Ornstein–Uhlenbeck stochastic processes are given by the mechanical oscillator of the Langevin equation (4.195) in the overdamped limit $m = 0$, and the noisy RC electric circuit ruled by equation (4.194) in the overdamped limit $L = 0$. In both cases, the stationary state corresponds to thermodynamic equilibrium. However, the Ornstein–Uhlenbeck process may also describe nonequilibrium processes as discussed in Chapter 7.

The (ε, τ)-entropy per unit time of the Ornstein–Uhlenbeck process can be obtained as follows (Gaspard and Wang, 1993; Gaspard, 1998; Andrieux et al., 2008). Since the process is Markovian, the joint probability density that the continuous random variable $X(t)$ takes the values $\mathbf{x} = (x_1, x_2, \ldots, x_k)$ at the successive times $t_i = t_0 + i\tau$ with $1 \le i \le k$ can be factorized using the stationary and conditional probability densities (E.56) and (E.58) and expressed as the multivariate Gaussian distribution,

$$p(x_1, x_2, \ldots, x_k) = \prod_{i=2}^{k} p_\tau(x_i|x_{i-1})\, p_{\rm st}(x_1) = \frac{\exp\left(-\frac{1}{2}\mathbf{x}^{\rm T} \cdot \mathbf{C}_k^{-1} \cdot \mathbf{x}\right)}{(2\pi)^{\frac{k}{2}} (\det \mathbf{C}_k)^{\frac{1}{2}}}, \tag{E.60}$$

where the superscript T denotes the transpose and $(\mathbf{C}_k)_{ij} = \sigma^2 r^{|i-j|}$ with $r \equiv \exp(-\tau/\tau_r)$ for $1 \le i, j \le k$ is the correlation matrix. Its inverse is given by

$$\begin{aligned} \left(\mathbf{C}_k^{-1}\right)_{ij} &= \frac{1}{\sigma^2(1-r^2)} \left[-r\,\delta_{i-1,j} + (1+r^2)\,\delta_{ij} - r\,\delta_{i+1,j}\right] \quad \text{for} \quad 2 \le i \le k-1, \\ \left(\mathbf{C}_k^{-1}\right)_{1j} &= \left[\sigma^2(1-r^2)\right]^{-1} (\delta_{1j} - r\,\delta_{2j}), \text{ and } \left(\mathbf{C}_k^{-1}\right)_{kj} = \left[\sigma^2(1-r^2)\right]^{-1} (\delta_{kj} - r\,\delta_{k-1,j}). \end{aligned} \tag{E.61}$$

Its determinant has the value,

$$\det \mathbf{C}_k = \sigma^{2k} \left(1 - r^2\right)^{k-1}. \tag{E.62}$$

Now, the (ε, τ)-entropy per unit time is defined by

$$\begin{aligned} h(\varepsilon, \tau) = \lim_{k\to\infty} -\frac{1}{k\tau} \int_{-\infty}^{+\infty} dx_1 \cdots \int_{-\infty}^{+\infty} dx_k\, p(x_1, \ldots, x_k) \\ \times \ln \int_{-\varepsilon}^{+\varepsilon} d\xi_1 \cdots \int_{-\varepsilon}^{+\varepsilon} d\xi_k\, p(x_1 + \xi_1, \ldots, x_k + \xi_k), \end{aligned} \tag{E.63}$$

where the integrals in the logarithm extend over the tube of trajectories satisfying the conditions $|X(t_i) - x_i| < \varepsilon$ with $i = 1, 2, \ldots, k$, around the reference trajectory sampled at the successive positions $\{x_i\}_{i=1}^k$. Expanding in powers of the variables ξ_i and evaluating the integrals over $-\varepsilon < \xi_i < +\varepsilon$, the logarithm is obtained as

$$\ln \int_{-\varepsilon}^{+\varepsilon} d\xi_1 \cdots \int_{-\varepsilon}^{+\varepsilon} d\xi_k \; p(x_1 + \xi_1, \ldots, x_k + \xi_k)$$

$$= \ln \frac{(2\varepsilon)^k}{(2\pi)^{\frac{k}{2}} (\det \mathbf{C}_k)^{\frac{1}{2}}} - \frac{1}{2} \mathbf{x}^{\mathrm{T}} \cdot \mathbf{C}_k^{-1} \cdot \mathbf{x} + O(\varepsilon^2). \tag{E.64}$$

Carrying out the integrals over $-\infty < x_i < +\infty$ and using equation (E.62), the dynamical entropy (E.63) is found to be

$$h(\varepsilon, \tau) = \frac{1}{\tau} \ln \sqrt{\frac{\mathrm{e}\pi\sigma^2}{2\,\varepsilon^2} \left(1 - r^2\right)} + O\left(\varepsilon^2/\tau\right), \tag{E.65}$$

if the condition $\varepsilon \ll \sigma\sqrt{1 - r^2}$ holds.

E.8 Random Diffusion with Drift

We consider the random drift process with constant and uniform drift velocity and diffusivity tensor, so that the master equation (4.110)–(4.111) is given by

$$\partial_t\, p + \mathbf{V} \cdot \partial_{\mathbf{x}} p = \mathbf{D} : \partial_{\mathbf{x}}^2 p \tag{E.66}$$

with $\mathbf{x} \in \mathbb{R}^d$. Here, the spectrum is continuous with the eigenvalues $z_{\mathbf{q}} = -\imath \mathbf{V} \cdot \mathbf{q} - \mathbf{D} : \mathbf{q}^2$ associated with the eigenfunctions $\psi_{\mathbf{q}}(\mathbf{x}) = \exp(\imath \mathbf{q} \cdot \mathbf{x})$ with the wave vector $\mathbf{q} \in \mathbb{R}^d$. This process is not stationary.

If the process takes place on a one-dimensional ring of length L, the probability density $p(x,t)$ is defined for $0 \le x < L$ with the periodic boundary conditions $p(x,t) = p(x+L,t)$. Accordingly, the wave number takes the discrete values $q = 2\pi m/L$ with $m \in \mathbb{Z}$ and the eigenvalues are given by $z_m = -\imath V(2\pi m/L) - \mathcal{D}(2\pi m/L)^2$, so that the spectrum is discrete. If $\mathcal{D} > 0$, we have the property that $\operatorname{Re} z_m \le 0$ with $\operatorname{Im} z_m \ne 0$ as long as $V \ne 0$. This process is stationary.

The process is also stationary if $V < 0$ and $0 \le x < \infty$ with a reflecting boundary condition at $x = 0$, as for Brownian motion in the presence of gravity inside a semi-infinite fluid column. In this case, the stationary probability density is given by $p_{\mathrm{st}}(x) = (|V|/\mathcal{D}) \exp(-|V|x/\mathcal{D})$, which is also the eigenfunction $\psi_0 = p_{\mathrm{st}}$ associated with the zero eigenvalue $z_0 = 0$. The current density $\jmath = V\,p - \mathcal{D}\,\partial_x p$ is equal to zero for this stationary density, so that this stochastic process is reversible.

E.9 One-Dimensional Reversible Advection–Diffusion Processes

For processes with time-independent diffusivity and drift velocity in one dimension, the stationarity condition $\partial_t\, p_{\mathrm{st}} = 0$ implies that

$$\mathcal{D}(x)\, \frac{dp_{\mathrm{st}}}{dx} = V(x)\, p_{\mathrm{st}} - \jmath_{\mathrm{st}}, \tag{E.67}$$

where $j_{\rm st}$ is the uniform current density. The reversibility condition (4.134) is equivalent to supposing that this current density is equal to zero. Accordingly, the stationary probability density can be expressed as in equation (4.135) with the potential,

$$\Upsilon(x) = \Upsilon_0 - \int_0^x \frac{V(x')}{\mathcal{D}(x')}\, dx' . \tag{E.68}$$

If the drift velocity arises from a force field $F(x) = -\partial_x U$ deriving from some energy potential $U(x)$ according to $V(x) = \gamma^{-1}F(x) = -\gamma^{-1}\partial_x U(x)$ with the uniform friction coefficient γ related to the diffusion coefficient by the Einstein formula $\mathcal{D} = k_{\rm B}T/\gamma$, the stationary probability density is given by the Boltzmann distribution $p_{\rm st}(x) = \mathcal{N}\exp[-\beta U(x)]$ with the inverse temperature $\beta = (k_{\rm B}T)^{-1}$. In this case, the operator (4.137) and the function (4.138) are given by

$$\hat{K}\mathfrak{f} = \mathcal{D}\,\partial_x^2\mathfrak{f} - \Phi\,\mathfrak{f} \qquad \text{and} \qquad \Phi = \frac{1}{4}\,\beta^2\mathcal{D}\,(\partial_x U)^2 - \frac{1}{2}\,\beta\mathcal{D}\,\partial_x^2 U . \tag{E.69}$$

The spectrum is thus discrete if the energy potential increases as $U(x) \sim x^{2n}$ with $n \geq 1$ for $|x| \to \infty$, as it is in the case for the Ornstein–Uhlenbeck process (Risken, 1989; Gaspard et al., 1995).

E.10 Weak-Noise Limit Beyond the Quadratic Approximation

Starting from the master equation (4.32) with the same weak-noise assumptions as in Section 4.6, we note that

$$\mathrm{e}^{-\boldsymbol{\nu}_\rho\cdot\partial_{\mathbf{n}}} P = \mathrm{e}^{-\phi/\epsilon}\, \mathrm{e}^{\boldsymbol{\nu}_\rho\cdot\boldsymbol{\mathfrak{p}}}\,[1 + O(\epsilon)], \tag{E.70}$$

with $\mathbf{x} = \epsilon\mathbf{n}$, $P = \epsilon^d p$, equation (4.143), and the momenta $\boldsymbol{\mathfrak{p}} = \partial_{\mathbf{x}}\phi$. At leading order in ϵ, we thus obtain the Hamilton–Jacobi equation (4.147) with the Hamiltonian function,

$$\mathscr{H}(\mathbf{x}, \boldsymbol{\mathfrak{p}}) = \sum_\rho \left(\mathrm{e}^{\boldsymbol{\nu}_\rho\cdot\boldsymbol{\mathfrak{p}}} - 1\right) w_\rho(\mathbf{x}) \tag{E.71}$$

for the rates (4.140). In the quadratic approximation, the Freidlin–Wentzell Hamiltonian function (4.148) is recovered because of equation (4.141).

E.11 Examples of Non-Markovian Processes

Continuous-time random walks are known to be non-Markovian (Kenkre et al., 1973) and they may be described by generalized master equations such as

$$\frac{dP_\omega(t)}{dt} = \int_0^t \sum_{\omega'} [K_{\omega\omega'}(\tau)\, P_{\omega'}(t-\tau) - K_{\omega'\omega}(\tau)\, P_\omega(t-\tau)]\, d\tau \tag{E.72}$$

in terms of the memory kernels $K_{\omega\omega'}(\tau)$ (Kenkre et al., 1973).

A further example is Brownian motion including the effect of long-time tails due to hydrodynamics and described by the master equation (4.230) (Mazo, 2002) or by the generalized Langevin equation (10.146) (Kheifets et al., 2014).

Further examples are given by the semi-Markov processes, which are renewal processes defined with the probability densities $q_{\omega'\omega}(t) = P_{\omega'\omega}\, p_\omega(t)$ for making transitions $\omega \to \omega'$ in the time interval $[t, t + dt]$. These densities are supposed to be factorized into the time-independent transition probabilities $P_{\omega'\omega}$ and the functions $p_\omega(t) = \sum_{\omega'} q_{\omega'\omega}(t)$, which satisfy the normalization conditions $\sum_{\omega'} P_{\omega'\omega} = 1$ and $\int_0^\infty p_\omega(t)\, dt = 1$ (Goychuk, 2004; Andrieux and Gaspard, 2008c).

E.12 Rayleigh Gas

The Rayleigh gas or Rayleigh flight is a special type of Brownian motion that takes place in a rarefied gas instead of a fluid such as water (Rayleigh, 1891; Spohn, 1980, 1991). In this case, the mean free path of the gas molecules is larger than the radius of the spherical colloidal particle undergoing Brownian motion. The colloidal particle may be assumed to be a hard ball of mass M and radius R, and the gas molecules to be point-like and of lighter mass $m \ll M$, so that the gas molecules are much swifter than the colloidal particle. In the limit where the gas is ideal, the mean free path is infinite and every gas molecule arrives from large distances, most often performing a single elastic collision on the spherical particle before being scattered back to infinity. The elastic collisions are instantaneous and they happen at the successive times $\{t_i\}_{-\infty}^{+\infty}$. The force exerted by the gas on the colloidal particle has the form,

$$\mathbf{F}(t) = \sum_{i=-\infty}^{+\infty} \Delta\mathbf{p}_i\, \delta(t - t_i), \qquad \text{where} \qquad \Delta\mathbf{p}_i = 2m\, (\boldsymbol{\epsilon}_i \cdot \mathbf{v}_i)\, \boldsymbol{\epsilon}_i \tag{E.73}$$

is the transfer of momentum to the colloidal particle upon a collision with a molecule incoming with the velocity $\mathbf{v}_i$ according to equations (B.37) and (B.38), $\boldsymbol{\epsilon}_i$ being the unit vector oriented from the center of the colloidal particle to the point of impact on its surface. The velocities of the incoming molecules are oriented towards the center of the colloidal particle, so that $\boldsymbol{\epsilon}_i \cdot \mathbf{v}_i = -v_i \cos\theta_i < 0$, where $v_i = \|\mathbf{v}_i\|$ is the speed of the molecule and $0 \le \theta_i < \pi/2$. If the density of the gas is denoted n, the number of molecules arriving per unit time on the area $d\Sigma$, with a velocity $\mathbf{v}$ oriented in the direction (θ, φ) with respect to the unit vector $-\boldsymbol{\epsilon}$, and a speed ranging between v and $v + dv$ is given by the following frequency,

$$d\nu = n\, v \cos\theta\, d\Sigma\, \mathscr{P}_{\rm eq}(v)\, dv\, \frac{d\cos\theta\, d\varphi}{4\pi}, \quad \text{where} \quad \mathscr{P}_{\rm eq}(v) = 4\pi v^2 \frac{\exp\left(-\frac{mv^2}{2k_{\rm B}T}\right)}{(2\pi k_{\rm B}T/m)^{3/2}} \tag{E.74}$$

is the Maxwellian distribution of speeds such that $\int_0^\infty \mathscr{P}_{\rm eq}(v)\, dv = 1$, $\int d\Sigma = 4\pi R^2$, $0 \le \theta < \pi/2$, and $0 \le \varphi < 2\pi$. The arrival times of the molecules in every set (E.74) form independent Poisson processes. The total collision frequency thus has the value $\nu = \int d\nu = n\pi R^2 \langle v\rangle_{\rm eq}$ with the mean speed $\langle v\rangle_{\rm eq} = \sqrt{8k_{\rm B}T/(\pi m)}$, as expected.

Since the elastic collisions are instantaneous, the autocorrelation function of the random force (E.73) is given by $\langle \mathbf{F}(t) \cdot \mathbf{F}(t')\rangle_{\rm eq} = 6\,\gamma\, k_{\rm B}T\, \delta(t - t')$ with

$$\gamma = \frac{1}{6k_{\rm B}T} \lim_{\mathscr{T}\to\infty} \frac{1}{\mathscr{T}} \sum_{i:\, 0<t_i<\mathscr{T}} \left\langle \Delta\mathbf{p}_i^2 \right\rangle_{\rm eq} = \frac{1}{6k_{\rm B}T} \int \left\langle \Delta\mathbf{p}^2 \right\rangle_{\rm eq} d\nu. \tag{E.75}$$

Using the expression for $\Delta \mathbf{p}_i$ in equation (E.73), the friction coefficient of the Rayleigh gas is finally obtained as

$$\gamma = \frac{m^2}{3k_B T} n\pi R^2 \langle v^3 \rangle_{\rm eq} = \frac{8}{3} n R^2 \sqrt{2\pi m k_B T}, \tag{E.76}$$

because $\langle v^3 \rangle_{\rm eq} = (16/\sqrt{2\pi})(k_B T/m)^{3/2}$ (Spohn, 1991). The diffusion coefficient is here also given by $\mathcal{D} = k_B T/\gamma$.

An important difference with respect to Brownian motion in a fluid such as water is that the friction coefficient here scales as $\gamma \sim R^2$, which corresponds to the total cross section $\sigma_{\rm tot} = \pi R^2$ of the colloidal particle of radius R. In contrast, the friction coefficient scales as $\gamma \sim R$ according to Stokes' formula. Indeed, hydrodynamics manifests itself in the fluid if the mean free path is smaller than the radius of the colloidal particle, which instead leads to Stokes' friction coefficient.

E.13 Kac's Ring Model

This model was constructed by Kac (1956) for the purpose of explaining the role of probability concepts to understand how irreversibility arises in reversible systems. The model is composed of a ring with N lattice sites, each containing a black or white ball, as shown in Figure E.1(a). The ring rotates at the speed of one lattice site per time step. Moreover, there are $M \leq N$ markers or catalysts randomly distributed in front of the lattice sites. Their role is to change the color of the balls passing nearby. The fraction of markers with respect to the balls and lattice sites is equal to $\mu = M/N$. The configuration of the markers is given by $\{\epsilon_x\}_{x=1}^N$, where $\epsilon_x = -1$ if there is a marker in front of the xth lattice site, and $\epsilon_x = +1$ otherwise. The instantaneous state of the ball at the xth lattice site is given by $\eta_x(t) = \pm 1$, if the ball is, respectively, black or white at time $t \in \mathbb{Z}$.

The time evolution of the states is ruled by

$$\eta_x(t) = \epsilon_{x-1}\, \eta_{x-1}(t-1) = \epsilon_{x-1}\, \epsilon_{x-2} \cdots \epsilon_{x-t}\, \eta_{x-t}(0). \tag{E.77}$$

Let us denote the total numbers of black and white balls at time t by $N_{\rm b}(t)$ and $N_{\rm w}(t)$, respectively. Their sum is always equal to the number of lattice sites: $N_{\rm b}(t) + N_{\rm w}(t) = N$. Their difference is given by $\Delta N(t) \equiv N_{\rm b}(t) - N_{\rm w}(t) = \sum_{x=1}^N \eta_x(t)$.

We consider a statistical ensemble of Kac's rings with initial conditions such that $N_{\rm b}(0) - N_{\rm w}(0) = \Delta N(0)$. The ensemble average of the difference between the numbers of black and white balls is thus given by

$$\langle \Delta N(t) \rangle = \sum_{x=1}^{N} \langle \epsilon_{x-1}\, \epsilon_{x-2} \cdots \epsilon_{x-t} \rangle \, \langle \eta_{x-t}(0) \rangle = \langle \epsilon_t\, \epsilon_{t-1} \cdots \epsilon_1 \rangle \, \Delta N(0), \tag{E.78}$$

after changing the summation variable into $x' = x - t$. If the time t is not longer than the size N of the ring, the changes of color by the markers are independent of each other, but this is not the case for longer time, the dynamics being strictly periodic of period equal to $2N$. Since $\langle \epsilon_x \rangle = 1 - 2\mu$, the calculation of the ensemble average gives the following result (Kac, 1956; Dorfman, 1999):

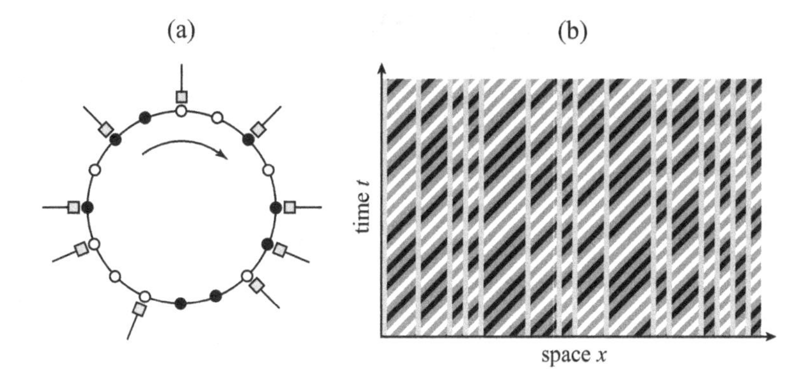

Figure E.1 (a) Schematic representation of Kac's ring model with black and white balls attached to the rotating ring and changing color upon meeting fixed markers (squares). (b) Example of a spacetime plot showing the time evolution of Kac's ring model. The vertical lines at fixed positions are the trajectories of the markers and the diagonal lines the trajectories of the balls moving with the ring and changing color.

$$\langle \Delta N(t) \rangle = \begin{cases} (1-2\mu)^t \, \Delta N(0) & \text{if} \quad 0 \le t \le N, \\ (1-2\mu)^{2N-t} \, \Delta N(0) & \text{if} \quad N < t \le 2N. \end{cases} \tag{E.79}$$

Therefore, there is exponential relaxation towards equilibrium configurations with equal probabilities to find black and white balls over the timescale $t \simeq \tau_{\rm r} = |\ln(1-2\mu)|^{-1}$. However, a Poincaré recurrence to the initial value happens at time $t = 2N$ because the system is strictly periodic. If the size N of the ring is very large, the period $2N$ can be much longer than the relaxation time $\tau_{\rm r}$, so that relaxation towards equilibrium with $\langle \Delta N \rangle_{\rm eq} = 0$ is actually observed over the physical timescale $t \simeq \tau_{\rm r} \ll N$.

Once the system is at equilibrium, it continues to manifest temporal disorder in the random transitions of the balls between the two colors, as long as half the period is not reached. This temporal disorder can be characterized by the mean decay rate (6.82) of the path probabilities for the system. The paths form spacetime plots, as illustrated in Figure E.1(b). If we consider the spacetime domain $X \ll T \ll N$, the spacetime plot is determined by the configuration of the markers $\{\epsilon_x\}_{x=1}^X$, the initial colors $\{\eta_x(0)\}_{x=1}^X$, and the colors of the balls ingoing the domain from its left-hand side at the times $1 \le t \le T$: $\{\eta_1(t)\}_{t=1}^T$. As long as $T \ll N$, these random variables are independent of each other and their probability distributions are given by $P(\epsilon_x = -1) = \mu$, $P(\epsilon_x = +1) = 1 - \mu$, and $P[\eta_x(t) = \pm 1] = 1/2$, if the system is at equilibrium. Consequently, the probability of a typical path can be evaluated as $P(\boldsymbol{\omega}) \sim \exp\{X[\mu \ln \mu + (1-\mu)\ln(1-\mu)] - (X+T)\ln 2\}$. Therefore, the entropy per unit time (6.82) is equal to $h = h_{\rm KS} = \ln 2$, which is positive and finite, thus defining a KS entropy per unit time (6.85) for the Kac ring model in the infinite-size limit $N \to \infty$. Here, dynamical randomness arises because new balls are incoming the domain of observation with random colors, as long as $T \ll N$. We note that this condition is the same as for observing exponential relaxation in this model.

Appendix F

Complements on Fluctuation Relations

F.1 Proof of the Nonlinear Response Properties at Higher Order

Here, the proof is given for the relation (5.155) of Section 5.5.6 (Barbier and Gaspard, 2018). The first step towards proving this relation is to expand both sides of the symmetry relation (5.154) in powers of both the counting parameters $\boldsymbol{\lambda}$ and the affinities $\mathbf{A}$ according to equation (5.151).

On the one hand, the left-hand side of equation (5.154) is given by

$$Q(-\boldsymbol{\lambda};\mathbf{A}) = \sum_{m=0}^{\infty}\sum_{n=0}^{\infty}\frac{(-1)^m}{m!\,n!}Q_{\alpha_1\cdots\alpha_m,\,\beta_1\cdots\beta_n}\lambda_{\alpha_1}\cdots\lambda_{\alpha_m}A_{\beta_1}\cdots A_{\beta_n}, \tag{F.1}$$

here using Einstein's convention of summations over the repeated indices $\alpha_1\cdots\alpha_m$ and $\beta_1\cdots\beta_n$.

On the other hand, the right-hand side of equation (5.154) can be written as

$$\hat{T}(\mathbf{A})\,Q(\boldsymbol{\lambda};\mathbf{A}) = \sum_{k=0}^{\infty}\frac{1}{k!}\left(\mathbf{A}\cdot\frac{\partial}{\partial\boldsymbol{\lambda}}\right)^k Q(\boldsymbol{\lambda};\mathbf{A}). \tag{F.2}$$

According to Newton's chain rule for the derivatives of products, we have that

$$\left(\mathbf{A}\cdot\frac{\partial}{\partial\boldsymbol{\lambda}}\right)\lambda_{\alpha_1}\cdots\lambda_{\alpha_m} = \sum_{j=1}^{m}\lambda_{\alpha_1}\cdots\lambda_{\alpha_{j-1}}A_{\alpha_j}\lambda_{\alpha_{j+1}}\cdots\lambda_{\alpha_m}, \tag{F.3}$$

so that the degree in the counting parameters $\boldsymbol{\lambda}$ is decreased by one, while the degree in the affinities $\mathbf{A}$ is increased by one. If we relabel the index of A_{α_j} into A_{β_j} in order to recover the notation chosen in equation (5.151), we see that the index α_j has been replaced by β_j. Therefore, the translation operator has the action of moving the indices β_j from the right to the left of the comma in the adopted notations. Iterating the action (F.3) of the translation operator (5.153), we thus obtain for any $k \geqslant 0$ that

$$\left(\mathbf{A}\cdot\frac{\partial}{\partial\boldsymbol{\lambda}}\right)^k Q(\boldsymbol{\lambda};\mathbf{A}) = \sum_{m=0}^{\infty}\sum_{n=k}^{\infty}\frac{1}{m!\,n!}Q^{\{k\}}_{\alpha_1\cdots\alpha_m,\,\beta_1\cdots\beta_n}\lambda_{\alpha_1}\cdots\lambda_{\alpha_m}A_{\beta_1}\cdots A_{\beta_n} \tag{F.4}$$

in terms of the quantities defined by

$$
\begin{aligned}
Q^{\{k\}}_{\alpha_1\cdots\alpha_m,\,\beta_1\cdots\beta_n} &\equiv \sum_{j=1}^{n} Q^{\{k-1\}}_{\alpha_1\cdots\alpha_m\beta_j,\,\beta_1\cdots\beta_{j-1}\beta_{j+1}\cdots\beta_n} \\
&= \sum_{j_1=1}^{n}\sum_{\substack{j_2=1\\ j_2\neq j_1}}^{n}\cdots\sum_{\substack{j_k=1\\ j_k\neq j_{k-1}}}^{n} Q_{\alpha_1\cdots\alpha_m\beta_{j_1}\cdots\beta_{j_k},\,(\cdot)}
\end{aligned}
\tag{F.5}
$$

for $k \geqslant 1$, with $Q^{\{0\}} \equiv Q$, and where $(\cdot)$ denotes the set of all indices β_k that are different from the indices $\beta_{j_1}\cdots\beta_{j_k}$, which are present to the left of the comma. Therefore, substituting the result (F.4) into (F.2) yields

$$
\hat{T}\,(\mathbf{A})\,Q\,(\boldsymbol{\lambda};\mathbf{A}) = \sum_{m=0}^{\infty}\frac{1}{m!}\sum_{k=0}^{\infty}\frac{1}{k!}\sum_{n=k}^{\infty}\frac{1}{n!}\,Q^{\{k\}}_{\alpha_1\cdots\alpha_m,\,\beta_1\cdots\beta_n}\lambda_{\alpha_1}\cdots\lambda_{\alpha_m}A_{\beta_1}\cdots A_{\beta_n}. \tag{F.6}
$$

Next, the two power series (F.1) and (F.6) are substituted into the symmetry relation (5.154) of the fluctuation relation for currents and the coefficients for the mth power of the counting parameters can be identified to get

$$
(-1)^m\sum_{n=0}^{\infty}\frac{1}{n!}\,Q_{\alpha_1\cdots\alpha_m,\,\beta_1\cdots\beta_n}A_{\beta_1}\cdots A_{\beta_n} = \sum_{k=0}^{\infty}\frac{1}{k!}\sum_{n=k}^{\infty}\frac{1}{n!}\,Q^{\{k\}}_{\alpha_1\cdots\alpha_m,\,\beta_1\cdots\beta_n}A_{\beta_1}\cdots A_{\beta_n}. \tag{F.7}
$$

In the right-hand side, the sums over k and n can be reordered to obtain

$$
(-1)^m\sum_{n=0}^{\infty}\frac{1}{n!}\,Q_{\alpha_1\cdots\alpha_m,\,\beta_1\cdots\beta_n}A_{\beta_1}\cdots A_{\beta_n} = \sum_{n=0}^{\infty}\frac{1}{n!}\sum_{j=0}^{n}\frac{1}{j!}\,Q^{\{j\}}_{\alpha_1\cdots\alpha_m,\,\beta_1\cdots\beta_n}A_{\beta_1}\cdots A_{\beta_n}, \tag{F.8}
$$

which is thus valid for any $m \geqslant 0$ and $\mathbf{A}$. Now, the coefficients for the nth power of the affinities can be identified on both sides of (F.8) to find

$$
Q_{\alpha_1\cdots\alpha_m,\,\beta_1\cdots\beta_n} = (-1)^m\sum_{j=0}^{n}\frac{1}{j!}\,Q^{\{j\}}_{\alpha_1\cdots\alpha_m,\,\beta_1\cdots\beta_n}, \tag{F.9}
$$

which holds for any $m,n \geqslant 0$.

Moreover, the quantity (F.5) with $k = j$ can be alternatively written as

$$
Q^{\{j\}}_{\alpha_1\cdots\alpha_m,\,\beta_1\cdots\beta_n} = j!\;Q^{(j)}_{\alpha_1\cdots\alpha_m,\,\beta_1\cdots\beta_n} \tag{F.10}
$$

in terms of the quantity $Q^{(j)}$ defined by equation (5.156). Indeed, because of the invariance of the coefficients (5.152) under the permutations of their indices on both sides of the comma, $Q^{\{j\}}$ can then be rewritten in a more convenient form by identifying the total number of terms $Q_{\alpha_1\cdots\alpha_m\beta_{l_1}\cdots\beta_{l_j},\,(\cdot)}$ with $l_1 < \cdots < l_j$ that occur in the right-hand side of (F.5) with $k = j$. The key point is that the latter involves sums over subscripts $k_1,\ldots,k_j$ that must all take different values. Therefore, counting all the $Q_{\alpha_1\cdots\alpha_m\beta_{l_1}\cdots\beta_{l_j},\,(\cdot)}$ such that

$l_1 < \cdots < l_j$ includes all the terms present in the right-hand side of (F.5). Furthermore, the total number of such terms $Q_{\alpha_1\cdots\alpha_m\beta_{l_1}\cdots\beta_{l_j},(\cdot)}$ can be readily seen to be nothing but the total number of permutations of the j distinct values $l_1,\ldots,l_j$, that is equal to $j!$. Therefore, the quantity $Q^{\{j\}}$ can be expressed by equation (F.10) in terms of the quantities (5.156).

Substituting (F.10) into (F.9) then yields the general relation

$$Q_{\alpha_1\cdots\alpha_m,\,\beta_1\cdots\beta_n} = (-1)^m \sum_{j=0}^{n} Q^{(j)}_{\alpha_1\cdots\alpha_m,\,\beta_1\cdots\beta_n} \qquad (\forall\, m,n \geqslant 0). \tag{F.11}$$

For m odd, the relations (5.155) are finally proved.

We note that the general relation (F.11) also gives

$$m \text{ even}: \qquad 0 = \sum_{j=1}^{n} Q^{(j)}_{\alpha_1\cdots\alpha_m,\,\beta_1\cdots\beta_n}. \tag{F.12}$$

However, these further relations can be deduced from the relations (5.155) for m odd (Barbier and Gaspard, 2018).

F.2 Consequences of Microreversibility in a Magnetizing Field

F.2.1 General Relations at Arbitrarily High Orders in a Magnetizing Field

With the substitution $\boldsymbol{\lambda} \to -\boldsymbol{\lambda}$, the symmetry relation (5.177) in the presence of the magnetizing field $\boldsymbol{\mathcal{H}}$ can be written in the following form

$$Q(-\boldsymbol{\lambda};\mathbf{A},\boldsymbol{\mathcal{H}}) = \hat{T}(\mathbf{A})\, Q(\boldsymbol{\lambda};\mathbf{A},-\boldsymbol{\mathcal{H}}) \tag{F.13}$$

in terms of the translation operator (5.153). Consequently, considerations similar to those of Section F.1 apply, leading to

$$Q_{\alpha_1\cdots\alpha_m,\,\beta_1\cdots\beta_n}(\boldsymbol{\mathcal{H}}) = (-1)^m \sum_{j=0}^{n} Q^{(j)}_{\alpha_1\cdots\alpha_m,\,\beta_1\cdots\beta_n}(-\boldsymbol{\mathcal{H}}) \tag{F.14}$$

with the quantities (5.156). These relations are the consequences of the fluctuation relation and they imply the results of Section 5.6.3 (Barbier and Gaspard, 2019).

Taking the symmetric and antisymmetric parts (5.189) of these relations gives

$$Q^{\mathrm{S}}_{\alpha_1\cdots\alpha_m,\,\beta_1\cdots\beta_n}(\boldsymbol{\mathcal{H}}) = (-1)^m \sum_{j=0}^{n} Q^{(j)\,\mathrm{S}}_{\alpha_1\cdots\alpha_m,\,\beta_1\cdots\beta_n}(\boldsymbol{\mathcal{H}}), \tag{F.15}$$

$$Q^{\mathrm{A}}_{\alpha_1\cdots\alpha_m,\,\beta_1\cdots\beta_n}(\boldsymbol{\mathcal{H}}) = (-1)^{m+1} \sum_{j=0}^{n} Q^{(j)\,\mathrm{A}}_{\alpha_1\cdots\alpha_m,\,\beta_1\cdots\beta_n}(\boldsymbol{\mathcal{H}}). \tag{F.16}$$

Again, the relations (F.15) for m even can be deduced from those with m odd. Furthermore, the relations (F.16) for m odd can be deduced from those with m even (Barbier and Gaspard, 2019). Therefore, microreversibility imposes constraints on the transport coefficients effectively reducing by half the amount of independent coefficients.

F.2.2 Consequences of Expanding the Hyperbolic Tangent

The relations (5.194) and (5.195), which are equivalent to the symmetry relation (5.177), generate further relations between the symmetric and antisymmetric parts of the transport coefficients in terms of the constant terms of Euler's polynomials. Indeed, the hyperbolic tangent function (5.161) can be expanded in Taylor series as

$$\tanh\frac{t}{2} = \sum_{j=1}^{\infty} c_{2j-1}\, t^{2j-1} \quad \text{with} \quad c_{2j-1} \equiv -\frac{e_0^{(2j-1)}}{(2j-1)!} = \frac{2}{(2j)!}\,(2^{2j}-1)\,B_{2j}, \tag{F.17}$$

where $e_0^{(n)} \equiv E_n(0)$ are the constant terms of the Euler polynomials $E_n(x)$, which are generated by

$$\frac{2}{\mathrm{e}^t + 1} = \sum_{n=0}^{\infty} \frac{1}{n!}\, e_0^{(n)}\, t^n, \tag{F.18}$$

and B_n are Bernoulli's numbers (Abramowitz and Stegun, 1972). In equation (F.18), the coefficients are given by $e_0^{(2j)} = 0$ for $j \geq 1$ and

$$e_0^{(0)} = 1, \quad e_0^{(1)} = -1/2, \quad e_0^{(3)} = 1/4, \quad e_0^{(5)} = -1/2, \ldots \tag{F.19}$$

Substituting the Taylor series (F.17) into equation (5.194) and expanding the symmetric part of the cumulant generating function in powers of the counting parameters $\boldsymbol{\lambda}$ and the affinities $\mathbf{A}$ is leading to

$$m \text{ odd}: \qquad Q^{\mathrm{S}}_{\alpha_1\cdots\alpha_m,\, \beta_1\cdots\beta_n}(\boldsymbol{\mathcal{H}}) = \sum_{j=1}^{\mathbb{E}\left(\frac{n+1}{2}\right)} e_0^{(2j-1)}\, Q^{(2j-1)\,\mathrm{S}}_{\alpha_1\cdots\alpha_m,\, \beta_1\cdots\beta_n}(\boldsymbol{\mathcal{H}}), \tag{F.20}$$

holding for any integer n, where $\mathbb{E}(x)$ denotes the integer part of the positive real number x (i.e., the natural number $k > 0$ such that $k \leq x < k+1$). Similarly, carrying out the substitution of the Taylor series (F.17) into equation (5.195) leads to

$$m \text{ even}: \qquad Q^{\mathrm{A}}_{\alpha_1\cdots\alpha_m,\, \beta_1\cdots\beta_n}(\boldsymbol{\mathcal{H}}) = \sum_{j=1}^{\mathbb{E}\left(\frac{n+1}{2}\right)} e_0^{(2j-1)}\, Q^{(2j-1)\,\mathrm{A}}_{\alpha_1\cdots\alpha_m,\, \beta_1\cdots\beta_n}(\boldsymbol{\mathcal{H}}), \tag{F.21}$$

for any integer n (Barbier and Gaspard, 2020b). These results show the existence of an alternating behavior for the symmetric and antisymmetric parts of the cumulants and their responses. In particular, equations (F.20) and (F.21) with $n = 0$ imply equation (5.201). The complementary nonvanishing parts of the cumulants obey relations involving the constant terms (F.19) of Euler's polynomials that are deduced from equations (F.20) and (F.21) with $n \geq 1$. The first few of these relations are equivalent to equations (5.179) and (5.183), and a combination of equations (5.186) and (5.187) (Barbier and Gaspard, 2020b).

The results (F.20) and (F.21) generalize, to systems with a nonzero magnetizing field, relations previously obtained in the absence of magnetizing field (Stratonovich, 1992;

Andrieux and Gaspard, 2007a; Andrieux, 2009). Indeed, when $\mathcal{H} = \mathbf{0}$ only (F.20) is non trivial, because (F.21) then gives $0 = 0$. In this case, expressions that are found in the cited references are recovered from (F.20) in view of (F.17).

F.3 Time-Reversal Symmetry of the Binary-Collision Operator

In this appendix, the time-reversal symmetry relation (9.92) is proved for the binary collision operator (9.68). Using equation (9.95), its adjoint is given by

$$\hat{L}^{\mathrm{C}\dagger}\Psi = \sum_{\lambda\mu\rho\sigma} W^{\mathrm{C}}_{\lambda\mu\rho\sigma}\, N_\rho\, N_\sigma \left(\hat{E}_\rho^{-1}\hat{E}_\sigma^{-1}\hat{E}_\lambda^{+1}\hat{E}_\mu^{+1} - 1\right)\Psi. \tag{F.22}$$

Using the time-reversal operator $\hat{\Theta}\,\Psi(\{N_\alpha\}) = \Psi(\{N_{\alpha^\Theta}\})$ and equation (9.95), the left-hand side of equation (9.92) is transformed into

$$\eta^{-1}\hat{\Theta}\,\hat{L}^{\mathrm{C}}\left(\hat{\Theta}\,\eta\,\Psi\right) = \sum_{\lambda\mu\rho\sigma} W^{\mathrm{C}}_{\lambda\mu\rho\sigma}\frac{\langle N_{\rho^\Theta}\rangle_{\mathrm{R}}\langle N_{\sigma^\Theta}\rangle_{\mathrm{R}}}{\langle N_{\lambda^\Theta}\rangle_{\mathrm{R}}\langle N_{\mu^\Theta}\rangle_{\mathrm{R}}}\, N_{\lambda^\Theta}\, N_{\mu^\Theta}\, \hat{E}_{\lambda^\Theta}^{-1}\hat{E}_{\mu^\Theta}^{-1}\hat{E}_{\rho^\Theta}^{+1}\hat{E}_{\sigma^\Theta}^{+1}\,\Psi$$
$$- \sum_{\lambda\mu\rho\sigma} W^{\mathrm{C}}_{\lambda\mu\rho\sigma}\, N_{\rho^\Theta}\, N_{\sigma^\Theta}\,\Psi. \tag{F.23}$$

Since summing over all the cells α or their time-reversal α^Θ gives the same result, we have that

$$\eta^{-1}\hat{\Theta}\,\hat{L}^{\mathrm{C}}\left(\hat{\Theta}\,\eta\,\Psi\right) = \sum_{\lambda\mu\rho\sigma} W^{\mathrm{C}}_{\rho^\Theta\sigma^\Theta\lambda^\Theta\mu^\Theta}\frac{\langle N_\lambda\rangle_{\mathrm{R}}\langle N_\mu\rangle_{\mathrm{R}}}{\langle N_\rho\rangle_{\mathrm{R}}\langle N_\sigma\rangle_{\mathrm{R}}}\, N_\rho\, N_\sigma\, \hat{E}_\rho^{-1}\hat{E}_\sigma^{-1}\hat{E}_\lambda^{+1}\hat{E}_\mu^{+1}\,\Psi$$
$$- \sum_{\lambda\mu\rho\sigma} W^{\mathrm{C}}_{\rho^\Theta\sigma^\Theta\lambda^\Theta\mu^\Theta}\, N_\lambda\, N_\mu\,\Psi. \tag{F.24}$$

By using the property of reciprocity (9.77) with respect to the equilibrium state of the right-hand reservoir to transform the first term, and equations (9.74)–(9.75) to transform the second term, we finally obtain

$$\eta^{-1}\hat{\Theta}\,\hat{L}^{\mathrm{C}}\left(\hat{\Theta}\,\eta\,\Psi\right) = \sum_{\lambda\mu\rho\sigma} W^{\mathrm{C}}_{\lambda\mu\rho\sigma}\, N_\rho\, N_\sigma \left(\hat{E}_\rho^{-1}\hat{E}_\sigma^{-1}\hat{E}_\lambda^{+1}\hat{E}_\mu^{+1} - 1\right)\Psi, \tag{F.25}$$

which is identical to the adjoint operator (F.22), thus proving the time-reversal symmetry relation (9.92) (Gaspard, 2013c).

F.4 Proof of the Finite-Time Fluctuation Relation for Diffusion

F.4.1 Markov Jump Process with Poisson Stationary Distribution

Here, the goal is to prove the finite-time fluctuation relation (10.56) for the Markov jump process (10.12) ruled by the master equation (10.13) for particle diffusion between two reservoirs with the transition rates

$$W_{+i}(\mathbf{N}) = k_{+i}\, N_i \qquad \text{and} \qquad W_{-i}(\mathbf{N}) = k_{-i}\, N_{i+1}, \tag{F.26}$$

which depend linearly on the particle numbers $\mathbf{N} = \{N_i\}_{i=1}^{l}$. In this case, the diffusion coefficient is assumed to be independent of the particle density in equations (10.16)–(10.17), so that the rate constants are given by

$$k_{\pm i} = \frac{\mathcal{D}}{\Delta x^2}\,\psi[\pm(u_{i+1} - u_i)], \quad \text{such that} \quad \frac{k_{+i}}{k_{-i}} = \mathrm{e}^{-\beta(u_{i+1}-u_i)}, \tag{F.27}$$

because of equation (10.19).

For such linear processes, the stationary solution of the reduced master equation (10.21) is the Poisson probability distribution given by

$$\mathscr{P}_{\rm st}(\mathbf{N}) = \prod_{i=1}^{l} \mathrm{e}^{-\langle N_i\rangle_{\rm st}} \frac{\langle N_i\rangle_{\rm st}^{N_i}}{N_i!} \tag{F.28}$$

in terms of the stationary mean particle numbers such that $(d/dt)\langle N_i\rangle_{\rm st} = 0$ for equation (10.33).

F.4.2 Full Moment Generating Function

Following Gardiner (2004), we introduce the full moment generating function

$$G(\mathbf{s}, z, t) \equiv \sum_{\mathbf{N},\mathcal{N}} z^{\mathcal{N}} \prod_{i=1}^{l} s_i^{N_i}\, P(\mathbf{N}, \mathcal{N}, t), \tag{F.29}$$

depending on the variables $\mathbf{s} = \{s_i\}_{i=1}^{l} \in \mathbb{R}^l$ and $z \equiv \exp(-\lambda)$. As a consequence of the master equation (10.13), this function obeys the partial differential equation of first order in the variables $\mathbf{s}$,

$$\partial_t G + (\mathsf{L}\cdot\mathbf{s} + \mathbf{f})\cdot\partial_{\mathbf{s}} G = (\mathbf{g}\cdot\mathbf{s} + h)\, G, \tag{F.30}$$

expressed in terms of the $l \times l$ matrix L with the nonvanishing elements $(\mathsf{L})_{i,i-1} = -k_{-(i-1)}$ for $2 \le i \le l$, $(\mathsf{L})_{i,i} = k_{-(i-1)} + k_{+i}$ for $1 \le i \le l$, and $(\mathsf{L})_{i,i+1} = -k_{+i}$ for $1 \le i \le l-1$; the vectors $(\mathbf{g})_i = k_{+0}\bar{N}_{\rm L} z\, \delta_{i,1} + k_{-l}\bar{N}_{\rm R}\, \delta_{i,l}$ and $(\mathbf{f})_i = -k_{-0}\, z^{-1}\delta_{i,1} - k_{+l}\,\delta_{i,l}$; and the scalar $h = -k_{+0}\bar{N}_{\rm L} - k_{-l}\bar{N}_{\rm R}$.

If $\mathbf{1}$ denotes the vector having all its elements equal to one, i.e., $(\mathbf{1})_i = 1$ for all $1 \le i \le l$, the matrix L has the property that $\mathbf{f}_0 = -\mathsf{L}\cdot\mathbf{1}$ with the vector $\mathbf{f}_0 = \mathbf{f}(z=1)$. We note that equation (10.33) for the mean particle numbers can be written in matrix form as

$$\langle\dot{\mathbf{N}}\rangle = -\mathsf{L}^{\rm T}\cdot\langle\mathbf{N}\rangle + \mathbf{g}_0, \tag{F.31}$$

where $\mathbf{g}_0 = \mathbf{g}(z=1)$. The corresponding stationary solution is thus given by $\langle\mathbf{N}\rangle_{\rm st} = \mathsf{L}^{\rm T-1}\cdot\mathbf{g}_0$. Furthermore, we have that $h = -\mathbf{1}\cdot\mathbf{g}_0$.

As a first-order partial differential equation, equation (F.30) can be solved using the method of characteristics (Gardiner, 2004). The equations for the characteristics are given by

$$\frac{d\mathbf{s}}{dt} = \mathsf{L}\cdot\mathbf{s} + \mathbf{f} \qquad \text{and} \qquad \frac{dG}{dt} = (\mathbf{g}\cdot\mathbf{s} + h)\, G, \tag{F.32}$$

which have the solutions,

$$\mathbf{s} = \mathrm{e}^{\mathbf{L}t} \cdot \left[\mathbf{s}_0 + \mathbf{L}^{-1} \cdot \left(\mathbf{I} - \mathrm{e}^{-\mathbf{L}t} \right) \cdot \mathbf{f} \right], \tag{F.33}$$

$$G = G_0 \exp\left[\mathbf{g} \cdot \mathbf{L}^{-1} \cdot \left(\mathbf{I} - \mathrm{e}^{-\mathbf{L}t} \right) \cdot \left(\mathbf{s} + \mathbf{L}^{-1} \cdot \mathbf{f} \right) + \left(h - \mathbf{g} \cdot \mathbf{L}^{-1} \cdot \mathbf{f} \right) t \right], \tag{F.34}$$

where $\mathbf{I}$ is the $l \times l$ unit matrix.

Now, the initial condition of the master equation (10.13) is taken as the stationary Poissonian distribution (F.28) multiplied by the Kronecker delta $\delta_{\mathcal{N},0}$, meaning that the counter is reset to zero, so $\mathcal{N} = 0$. The initial moment generating function (F.29) is thus given by $G_0(\mathbf{s}_0, z) = \exp[\langle \mathbf{N} \rangle_{\rm st} \cdot (\mathbf{s}_0 - \mathbf{1})]$. Substituting into equation (F.34) with $\mathbf{s}_0$ given by inverting equation (F.33), the solution of the partial differential equation (F.30) is thus equal to

$$G(\mathbf{s}, z, t) = \exp\left[\mathbf{g} \cdot \mathbf{L}^{-1} \cdot \left(\mathbf{I} - \mathrm{e}^{-\mathbf{L}t} \right) \cdot \left(\mathbf{s} + \mathbf{L}^{-1} \cdot \mathbf{f} \right) + \left(h - \mathbf{g} \cdot \mathbf{L}^{-1} \cdot \mathbf{f} \right) t \right]$$
$$\times \exp\left\{ \langle \mathbf{N} \rangle_{\rm st} \cdot \left[\mathrm{e}^{-\mathbf{L}t} \cdot \mathbf{s} - \mathbf{L}^{-1} \cdot \left(\mathbf{I} - \mathrm{e}^{-\mathbf{L}t} \right) \cdot \mathbf{f} - \mathbf{1} \right] \right\}. \tag{F.35}$$

F.4.3 Cumulant Generating Function for the Counting Statistics

The moment generating function of the random number $\mathcal{N}$ is given by

$$G\left(\mathbf{1}, z = \mathrm{e}^{-\lambda}, t \right) = \sum_{\mathcal{N}=-\infty}^{+\infty} \mathrm{e}^{-\lambda \mathcal{N}} \mathcal{P}_t(\mathcal{N}) = \sum_{\mathbf{N}, \mathcal{N}} \mathrm{e}^{-\lambda \mathcal{N}} P(\mathbf{N}, \mathcal{N}, t), \tag{F.36}$$

because of the definition (F.29), so that the time-dependent cumulant generating function (10.59) can be obtained as

$$Q_t(\lambda) \equiv -\frac{1}{t} \ln G\left(\mathbf{1}, z = \mathrm{e}^{-\lambda}, t \right). \tag{F.37}$$

Replacing therein the expression (F.35) for the full moment generating function, we find that

$$Q_t(\lambda) = \mathbf{g} \cdot \mathbf{L}^{-1} \cdot \mathbf{f} - \mathbf{g}_0 \cdot \mathbf{L}^{-1} \cdot \mathbf{f}_0 - \frac{1}{t} \left(\mathbf{g} - \mathbf{g}_0 \right) \cdot \mathbf{L}^{-1} \cdot \left(\mathbf{I} - \mathrm{e}^{-\mathbf{L}t} \right) \cdot \mathbf{L}^{-1} \cdot (\mathbf{f} - \mathbf{f}_0), \tag{F.38}$$

which can be written in the form (10.60) with the finite-time rates (10.58), having the asymptotic values

$$W_\infty^{(+)} = k_{+0} \, \bar{N}_{\rm L} \left(\mathbf{L}^{-1} \right)_{1l} k_{+l}, \qquad W_\infty^{(-)} = k_{-l} \, \bar{N}_{\rm R} \left(\mathbf{L}^{-1} \right)_{l1} k_{-0}, \tag{F.39}$$

and the function

$$\Upsilon(t) = k_{+0} \, k_{-0} \, \bar{N}_{\rm L} \, \mathbf{1}_{\rm L} \cdot \mathbf{L}^{-1} \cdot \left(\mathbf{I} - \mathrm{e}^{-\mathbf{L}t} \right) \cdot \mathbf{L}^{-1} \cdot \mathbf{1}_{\rm L}, \tag{F.40}$$

where the vector $\mathbf{1}_{\rm L} = (\delta_{i1})_{i=1}^{l}$ corresponds to the cell $i = 1$ in contact with the left-hand reservoir.

Accordingly, the moment generating function has the following expression:

$$G(\mathbf{1}, z, t) = \sum_{\mathcal{N}=-\infty}^{+\infty} z^{\mathcal{N}} \mathcal{P}_t(\mathcal{N}) = \exp\left[t \, W_t^{(+)} (z - 1) + t \, W_t^{(-)} (z^{-1} - 1) \right]. \tag{F.41}$$

Therefore, the probability distribution (10.55) is obtained by using the generating series of Bessel functions (Abramowitz and Stegun, 1972)

$$e^{x(q+q^{-1})/2} = \sum_{\mathcal{N}=-\infty}^{+\infty} q^{\mathcal{N}} I_{\mathcal{N}}(x) \qquad \text{for} \qquad q \neq 0 \tag{F.42}$$

with the variables (Andrieux and Gaspard, 2008f),

$$x = 2t\sqrt{W_t^{(+)} W_t^{(-)}} \qquad \text{and} \qquad q = z\sqrt{\frac{W_t^{(+)}}{W_t^{(-)}}}. \tag{F.43}$$

Finally, the finite-time fluctuation relation (10.56) is proved by $I_{\mathcal{N}}(x) = I_{-\mathcal{N}}(x)$.

F.4.4 The Case of Homogeneous Diffusion

In this case, the rate constants are equal to each other, so that $k_{\pm i} \equiv k$.

The asymptotic values (F.39) of the finite-time rates are thus given by

$$W_\infty^{(+)} = \frac{k}{l+1}\bar{N}_{\rm L} \qquad \text{and} \qquad W_\infty^{(-)} = \frac{k}{l+1}\bar{N}_{\rm R}, \tag{F.44}$$

which correspond to equation (10.65). The function (F.40) can be equivalently expressed as

$$\Upsilon(t) = \bar{N}_{\rm L}\,\mathbf{u}_0 \cdot (\mathbf{u}_0 - \mathbf{u}_t) \quad \text{with} \quad \mathbf{u}_t = e^{-\mathsf{L}t} \cdot \mathbf{u}_0 \quad \text{and} \quad \mathbf{u}_0 \equiv k\,\mathsf{L}^{-1} \cdot \mathbf{1}_{\rm L}. \tag{F.45}$$

The vector $\mathbf{u}_0$ is found to have the elements $(\mathbf{u}_0)_i = 1 - i/(l+1)$ for $1 \leq i \leq l$.

In the continuum limit where $\Delta x \to 0, l \to \infty$, and $l\Delta x = L$, the vector $\mathbf{u}_0$ corresponds to the function $u(x,0) = 1 - x/L$ defined on the interval $0 \leq x \leq L$, the vector $\mathbf{u}_t$ to the function $u(x,t)$ that is the solution of the deterministic diffusion equation with the boundary and initial conditions given in equation (10.67), and the function (F.45) becomes equation (10.66).

Similar considerations hold for the process of diffusion-influenced surface reaction presented in Section 10.4 (Gaspard and Kapral, 2018a).

F.5 Proof of the Finite-Time Fluctuation Relation for Reactions

The finite-time fluctuation relation (11.68) can be proved by methods similar to those used in Section F.4 (Andrieux and Gaspard, 2008f; Gu and Gaspard, 2020). For the Markov jump process ruled by the master equation (11.57), the full moment generating function can be introduced as (Gardiner, 2004),

$$G(s,\boldsymbol{\lambda},t) \equiv \sum_{X=0}^{\infty} s^X \sum_{\mathbf{n}} e^{-\boldsymbol{\lambda}\cdot\mathbf{n}} P(X,\mathbf{n},t), \tag{F.46}$$

which obeys the first-order partial differential equation given by

$$\partial_t G = (\mathcal{A}s - \mathcal{B})G + (C - \mathcal{D}s)\,\partial_s G, \tag{F.47}$$

where

$$\mathcal{A} = \Omega \left(\sum_{i=1}^{m-1} k_{+i}\, y_i\, \mathrm{e}^{-\lambda_i} + k_{+m}\, y_m \right), \qquad \mathcal{B} = \mathcal{A}(\boldsymbol{\lambda} = \mathbf{0}), \tag{F.48}$$

$$\mathcal{C} = \sum_{i=1}^{m-1} k_{-i}\, \mathrm{e}^{+\lambda_i} + k_{-m}, \qquad \mathcal{D} = \mathcal{C}(\boldsymbol{\lambda} = \mathbf{0}). \tag{F.49}$$

Equation (F.47) can be solved using the method of characteristics, for which we have

$$\frac{dG}{dt} = (\mathcal{A}s - \mathcal{B})G \qquad \text{and} \qquad \frac{ds}{dt} = -\mathcal{C} + \mathcal{D}s. \tag{F.50}$$

The initial condition is taken as the probability distribution $P(X, \boldsymbol{\lambda}, 0) = \mathscr{P}_{\rm st}(X)\, \delta_{\mathbf{n},\mathbf{0}}$ with the stationary Poisson distribution (11.59), so that $G(s, \boldsymbol{\lambda}, 0) = \mathrm{e}^{\langle X\rangle_{\rm st}(s-1)}$. The solution is given by

$$G(s, \boldsymbol{\lambda}, t) = \exp\left[t\, \frac{\mathcal{AC} - \mathcal{BD}}{\mathcal{D}} + \frac{\mathcal{B}}{\mathcal{D}}(s-1) - \frac{1}{\mathcal{D}^2}\left(1 - \mathrm{e}^{-\mathcal{D}t}\right)(\mathcal{A} - \mathcal{B})(\mathcal{C} - \mathcal{D}s)\right]. \tag{F.51}$$

The cumulant generating function at finite time is thus equal to

$$\mathcal{Q}_t(\boldsymbol{\lambda}) \equiv -\frac{1}{t} \ln G(1, \boldsymbol{\lambda}, t) = \frac{\mathcal{BD} - \mathcal{AC}}{\mathcal{D}} + \frac{1}{\mathcal{D}^2 t}\left(1 - \mathrm{e}^{-\mathcal{D}t}\right)(\mathcal{A} - \mathcal{B})(\mathcal{C} - \mathcal{D}), \tag{F.52}$$

which can be expressed in terms of the counting parameters and the rates (11.63) to get

$$\mathcal{Q}_t(\boldsymbol{\lambda}) = \sum_{i \neq j} W_{ij}\left(1 - \mathrm{e}^{-\lambda_i + \lambda_j}\right) + g(t) \sum_{i,j} W_{ij}\left(1 - \mathrm{e}^{-\lambda_i}\right)\left(1 - \mathrm{e}^{+\lambda_j}\right) \tag{F.53}$$

with the sums over $1 \leq i, j \leq m$, the convention $\lambda_m = 0$, and the function

$$g(t) \equiv \frac{1}{\mathcal{D}t}\left(1 - \mathrm{e}^{-\mathcal{D}t}\right). \tag{F.54}$$

This result should be compared with the moment generating function for a Poissonian process defined by

$$\mathcal{P}_t(\mathbf{n}) = \sum_{\mathbf{N}} \delta_{\mathbf{n}, \mathbf{n}(\mathbf{N})} \prod_{i \neq j} P(N_{ij}, t) \qquad \text{with} \qquad n_i(\mathbf{N}) \equiv \sum_{j=1}^{m} (N_{ij} - N_{ji}) \tag{F.55}$$

and the same Poisson distributions as in equation (5.251), but with time-dependent mean values $\langle N_{ij}\rangle_t$ to be determined by comparison with the previous calculation. Indeed, the finite-time cumulant generating function should here be given by

$$\mathcal{Q}_t(\boldsymbol{\lambda}) \equiv -\frac{1}{t} \ln G(1, \boldsymbol{\lambda}, t) = \frac{1}{t} \sum_{i \neq j} \langle N_{ij}\rangle_t \left(1 - \mathrm{e}^{-\lambda_i + \lambda_j}\right). \tag{F.56}$$

The comparison with equation (F.53) shows that the mean values of the Poisson distributions can be identified as

$$\langle N_{ij}\rangle_t = t\,(1-g)\,W_{ij} \quad \text{for} \quad i,\, j \neq m, \tag{F.57}$$

$$\langle N_{im}\rangle_t = t\,W_{im} + t\,g\sum_{j=1}^{m-1} W_{ij}, \quad \text{and} \quad \langle N_{mi}\rangle_t = t\,W_{mi} + t\,g\sum_{j=1}^{m-1} W_{ji}, \tag{F.58}$$

in terms of the function (F.54) and the rates (11.63). Since these latter can factorize as $W_{ij} = u_i v_j$ with $u_i \equiv k_{+i}\, y_i\, \Omega$ and $v_j \equiv k_{-j}/\mathcal{D}$, the mean values (F.57)–(F.58) satisfy the cyclic identities

$$\prod_{i=1}^{c}\langle N_{r_i r_{i+1}}\rangle_t = \prod_{i=1}^{c}\langle N_{r_{i+1} r_i}\rangle_t \qquad \text{with} \qquad r_{c+1} = r_1, \tag{F.59}$$

where $r_i \in \{1,2,\ldots,m\}$ and $c \geq 2$. As a consequence, we may introduce the time-dependent affinities,

$$A_{i,t} \equiv \ln\frac{\langle N_{im}\rangle_t}{\langle N_{mi}\rangle_t} = A_{i,\infty} + \ln\frac{1 + g(t)\sum_{j=1}^{m-1}\frac{k_{-j}}{k_{-m}}}{1 + g(t)\sum_{j=1}^{m-1}\frac{k_{+j}\, y_j}{k_{+m}\, y_m}}, \tag{F.60}$$

where $A_{i,\infty}$ denotes their asymptotic value given by equation (11.60). With these affinities, the finite-time cumulant generating function (F.56) obeys the symmetry

$$Q_t(\boldsymbol{\lambda}) = Q_t(\mathbf{A}_t - \boldsymbol{\lambda}). \tag{F.61}$$

Since the moment generating function $G(1,\boldsymbol{\lambda},t) = \mathrm{e}^{-Q_t(\boldsymbol{\lambda})t}$ uniquely characterizes the probability distribution (F.55), the finite-time fluctuation relation (11.68) is thus proved.

The timescale of convergence towards the asymptotic form (11.54) of the fluctuation relation can be obtained from the convergence of the time-dependent affinities (F.60) towards their asymptotic values according to the function (F.54).

References

Abramowitz, M., and Stegun, I. A. (eds). 1972. *Handbook of Mathematical Functions*. New York: Dover.

Agarwalla, B. K., Li, B., and Wang, J.-S. 2012. Full-counting statistics of heat transport in harmonic junctions: Transient, steady states, and fluctuation theorems. *Phys. Rev. E*, **85**, 051142.

Akhiezer, A. I., and Peletminskii, S. V. 1981. *Methods of Statistical Physics*. 1st ed. Oxford: Pergamon Press. Translated by M. Schukin.

Albano, A. M., Bedeaux, D., and Mazur, P. 1975. On the motion of a sphere with arbitrary slip in a viscous incompressible fluid. *Physica A*, **80**, 89–97.

Alberts, B., Bray, D., Johnson, A., Lewis, J., Raff, M., Roberts, K., and Walter, P. 1998. *Essential Cell Biology: An Introduction to the Molecular Biology of the Cell*. 1st ed. New York: Garland Science.

Alder, B. J., and Wainwright, T. E. 1970. Decay of the velocity autocorrelation function. *Phys. Rev. A*, **1**, 18–21.

Alder, B. J., Gass, D. M., and Wainwright, T. E. 1970. Studies in molecular dynamics. VIII. The transport coefficients for a hard-sphere fluid. *J. Chem. Phys.*, **53**, 3813–3826.

Allen, M. P., and Tildesley, D. J. 2017. *Computer Simulation of Liquids*. 2nd ed. Oxford: Oxford University Press.

Ambegaokar, V., and Halperin, B. I. 1969. Voltage due to thermal noise in the dc Josephson effect. *Phys. Rev. Lett.*, **22**, 1364–1366.

Amman, M., Wilkins, R., Ben-Jacob, E., Maker, P. D., and Jaklevic, R. C. 1991. Analytic solution for the current-voltage characteristic of two mesoscopic tunnel junctions coupled in series. *Phys. Rev. B*, **43**, 1146–1149.

Anderson, J. L. 1989. Colloid transport by interfacial forces. *Ann. Rev. Fluid Mech.*, **21**, 61–99.

Anderson, P. W. 1984. *Basic Notions of Condensed Matter Physics*. Menlo Park CA: Benjamin/Cummings.

Andrieux, D. 2009. *Nonequilibrium Statistical Thermodynamics at the Nanoscale: From Maxwell Demon to Biological Information Processing*. Saarbrücken: VDM Verlag. PhD thesis, Université libre de Bruxelles, Brussels.

Andrieux, D., and Gaspard, P. 2004. Fluctuation theorem and Onsager reciprocity relations. *J. Chem. Phys.*, **121**, 6167–6174.

Andrieux, D., and Gaspard, P. 2006a. Fluctuation theorem for transport in mesoscopic systems. *J. Stat. Mech.*, **2006**, P01011.

Andrieux, D., and Gaspard, P. 2006b. Fluctuation theorems and the nonequilibrium thermodynamics of molecular motors. *Phys. Rev. E*, **74**, 011906.

Andrieux, D., and Gaspard, P. 2007a. A fluctuation theorem for currents and non-linear response coefficients. *J. Stat. Mech.*, **2007**, P02006.

Andrieux, D., and Gaspard, P. 2007b. Fluctuation theorem for currents and Schnakenberg network theory. *J. Stat. Phys.*, **127**, 107–131.

Andrieux, D., and Gaspard, P. 2008a. Dynamical randomness, information, and Landauer's principle. *EPL*, **81**, 28004.

Andrieux, D., and Gaspard, P. 2008b. Fluctuation theorem and mesoscopic chemical clocks. *J. Chem. Phys.*, **128**, 154506.

Andrieux, D., and Gaspard, P. 2008c. The fluctuation theorem for currents in semi-Markov processes. *J. Stat. Mech.*, **2008**, P11007.

Andrieux, D., and Gaspard, P. 2008d. Nonequilibrium generation of information in copolymerization processes. *Proc. Natl. Acad. Sci.*, **105**, 9516–9521.

Andrieux, D., and Gaspard, P. 2008e. Quantum work relations and response theory. *Phys. Rev. Lett.*, **100**, 230404.

Andrieux, D., and Gaspard, P. 2008f. Temporal disorder and fluctuation theorem in chemical reactions. *Phys. Rev. E*, **77**, 031137.

Andrieux, D., and Gaspard, P. 2009. Stochastic approach and fluctuation theorem for ion transport. *J. Stat. Mech.*, **2009**, P02057.

Andrieux, D., and Gaspard, P. 2013. Information erasure in copolymers. *EPL*, **103**, 30004.

Andrieux, D., Gaspard, P., Ciliberto, S., Garnier, N., Joubaud, S., and Petrosyan, A. 2007. Entropy production and time asymmetry in nonequilibrium fluctuations. *Phys. Rev. Lett.*, **98**, 150601.

Andrieux, D., Gaspard, P., Ciliberto, S., Garnier, N., Joubaud, S., and Petrosyan, A. 2008. Thermodynamic time asymmetry in non-equilibrium fluctuations. *J. Stat. Mech.*, **2008**, P01002.

Andrieux, D., Gaspard, P., Monnai, T., and Tasaki, S. 2009. The fluctuation theorem for currents in open quantum systems. *New J. Phys.*, **11**, 043014. Erratum *ibid.* **11**, 109802.

Antczak, G., and Ehrlich, G. 2004. Long jump rates in surface diffusion: W on W(110). *Phys. Rev. Lett.*, **92**, 166105.

Aris, R. 1989. *Elementary Chemical Reactor Analysis*. Mineola: Dover.

Arnold, V. I. 1963. Proof of a theorem of A. N. Kolmogorov on the invariance of quasi-periodic motions under small perturbations of the Hamiltonian. *Russ. Math. Surv.*, **18**, 9–36.

Arnold, V. I. 1989. *Mathematical Methods of Classical Mechanics*. 2nd ed. New York: Springer.

Arnold, V. I., and Avez, A. 1968. *Ergodic Problems of Classical Mechanics*. New York: W. A. Benjamin.

Aron, C., Barci, D. G., Cugliandolo, L. F., Gonzalez Arenas, Z., and Lozano, G. S. 2016. Dynamical symmetries of Markov processes with multiplicative white noise. *J. Stat. Mech.*, **2016**, 053207.

Ashcroft, N. W., and Mermin, N. D. 1976. *Solid State Physics*. New York: Holt, Rinehart and Winston.

Astumian, R. D. 1997. Thermodynamics and kinetics of a Brownian motor. *Science*, **276**, 917–922.

Bachmann, S. J., Petitzon, M., and Mognetti, B. M. 2016. Bond formation kinetics affects self-assembly directed by ligand-receptor interactions. *Soft Matter*, **12**, 9585–9592.

Baiesi, M., Jacobs, T., Maes, C., and Skantzos, N. S. 2006. Fluctuation symmetries for work and heat. *Phys. Rev. E*, **74**, 021111.

Balakrishnan, V. 1979. Fluctuation-dissipation theorems from the generalized Langevin equation. *Pramana*, **12**, 301–315.

Balescu, R. 1975. *Equilibrium and Nonequilibrium Statistical Mechanics*. New York: Wiley.

Balian, R. 1991. *From Microphysics to Macrophysics*. Berlin: Springer.

Barato, A. C., and Chetrite, R. 2012. On the symmetry of current probability distributions in jump processes. *J. Phys. A: Math. Theor.*, **45**, 485002.

Barato, A. C., and Chetrite, R. 2015. A formal view on level 2.5 large deviations and fluctuation relations. *J. Stat. Phys.*, **160**, 1154–1172.

Barato, A. C., and Seifert, U. 2015. Thermodynamic uncertainty relation for biomolecular processes. *Phys. Rev. Lett.*, **114**, 158101.

Barato, A. C., Chetrite, R., Hinrichsen, H., and Mukamel, D. 2012. A Gallavotti-Cohen-Evans-Morriss like symmetry for a class of Markov jump processes. *J. Stat. Phys.*, **146**, 294–313.

Barbier, M., and Gaspard, P. 2018. Microreversibility, nonequilibrium current fluctuations, and response theory. *J. Phys. A: Math. Theor.*, **51**, 355001.

Barbier, M., and Gaspard, P. 2019. Microreversibility and nonequilibrium response theory in magnetic fields. *J. Phys. A: Math. Theor.*, **52**, 025003.

Barbier, M., and Gaspard, P. 2020a. Microreversibility and the statistics of currents in quantum transport. *Phys. Rev. E*, **102**, 022141.

Barbier, M., and Gaspard, P. 2020b. Microreversibility, nonequilibrium response, and Euler's polynomials. *J. Phys. A: Math. Theor.*, **53**, 145002.

Barrat, J.-L., and Bocquet, L. 1999. Large slip effect at a nonwetting fluid-solid interface. *Phys. Rev. Lett.*, **82**, 4671–4674.

Barroo, C., De Decker, Y., Visart de Bocarmé, T., and Gaspard, P. 2015. Fluctuating dynamics of nanoscale chemical oscillations: Theory and experiments. *J. Phys. Chem. Lett.*, **6**, 2189–2193.

Batalhõ, T. B., Souza, A. M., Sarthour, R. S., Oliveira, I. S., Paternostro, M., Lutz, E., and Serra, R. M. 2015. Irreversibility and the arrow of time in a quenched quantum system. *Phys. Rev. Lett.*, **115**, 190601.

Bechinger, C., Di Leonardo, R., Löwen, H., Reichhardt, C., Volpe, G., and Volpe, G. 2016. Active particles in complex and crowded environments. *Rev. Mod. Phys.*, **88**, 045006.

Bedeaux, D. 1986. Nonequilibrium thermodynamics and statistical physics of surfaces. *Adv. Chem. Phys.*, **64**, 47–109.

Bedeaux, D., and Mazur, P. 1974. Brownian motion and fluctuating hydrodynamics. *Physica A*, **76**, 247–258.

Bedeaux, D., Albano, A. M., and Mazur, P. 1976. Boundary conditions and non-equilibrium thermodynamics. *Physica A*, **82**, 438–462.

Bedeaux, D., Albano, A. M., and Mazur, P. 1977. Brownian motion and fluctuating hydrodynamics II; A fluctuation-dissipation theorem for the slip coefficient. *Physica A*, **88**, 574–582.

Benatti, F., Hudetz, T., and Knauf, A. 1998. Quantum chaos and dynamical entropy. *Commun. Math. Phys.*, **198**, 607–688.

Benenti, G., Lepri, S., and Livi, R. 2020. Anomalous heat transport in classical many-body systems: Overview and perspectives. *Frontiers Phys.*, **8**, 292.

Bennett, C. H. 1973. Logical reversibility of computation. *IBM J. Res. Dev.*, **17**, 525–532.

Bennett, C. H. 1979. Dissipation-error tradeoff in proofreading. *Biosystems*, **11**, 85–91.

Bennett, C. H. 1982. The thermodynamics of computation – A review. *Int. J. Theor. Phys.*, **21**, 905–940.

Benoist, T., Panati, A., and Pautrat, Y. 2020. Heat conservation and fluctuations between quantum reservoirs in the two-time measurement picture. *J. Stat. Phys.*, **178**, 893–925.
Berg-Sørensen, K., and Flyvbjerg, H. 2005. The colour of thermal noise in classical Brownian motion: A feasibility study of direct experimental observation. *New J. Phys.*, **7**, 38.
Bergé, P., Pomeau, Y., and Vidal, C. 1984. *Order Within Chaos*. New York: Wiley.
Bergquist, J. C., Hulet, R. G., Itano, W. M., and Wineland, D. J. 1986. Observation of quantum jumps in a single ion. *Phys. Rev. Lett.*, **57**, 1699–1702.
Bernard, W., and Callen, H. B. 1959. Irreversible thermodynamics of nonlinear processes and noise in driven systems. *Rev. Mod. Phys.*, **31**, 1017–1044.
Berne, B. J., and Pecora, R. 1976. *Dynamic Light Scattering*. New York: Wiley.
Berry, M. V., and Robbins, J. M. 1993. Chaotic classical and half-classical adiabatic reactions: Geometric magnetism and deterministic friction. *Proc. R. Soc. Lond. A*, **442**, 659–672.
Berry, R. S., Rice, S. A., and Ross, J. 1980. *Physical Chemistry*. New York: Wiley.
Bertini, L., De Sole, A., Gabrielli, D., Jona-Lasinio, G., and Landim, C. 2015. Macroscopic fluctuation theory. *Rev. Mod. Phys.*, **87**, 593–636.
Bérut, A., Arakelyan, A., Petrosyan, A., Ciliberto, S., Dillenschneider, R., and Lutz, E. 2012. Experimental verification of Landauer's principle linking information and thermodynamics. *Nature*, **483**, 187–189.
Bérut, A., Petrosyan, A., and Ciliberto, S. 2015. Information and thermodynamics: Experimental verification of Landauer's erasure principle. *J. Stat. Mech.*, **2015**, P06015.
Bessis, D., Paladin, G., Turchetti, G., and Vaienti, S. 1988. Generalized dimensions, entropies, and Liapunov exponents from the pressure function for strange sets. *J. Stat. Phys.*, **51**, 109–134.
Billingsley, P. 1978. *Ergodic Theory and Information*. Huntington, NY: Krieger Publishing Company.
Birkhoff, G. D. 1927. *Dynamical Systems*. New York: American Mathematical Society.
Birkhoff, G. D. 1931. Proof of the ergodic theorem. *Proc. Natl. Acad. Sci.*, **17**, 656–660.
Bixon, M., and Zwanzig, R. 1969. Boltzmann-Langevin equation and hydrodynamic fluctuations. *Phys. Rev.*, **187**, 267–272.
Blanter, Ya. M., and Büttiker, M. 2000. Shot noise in mesoscopic conductors. *Phys. Rep.*, **336**, 1–166.
Blickle, V., Speck, T., Helden, L., Seifert, U., and Bechinger, C. 2006. Thermodynamics of a colloidal particle in a time-dependent nonharmonic potential. *Phys. Rev. Lett.*, **96**, 070603.
Blokhuis, A., Lacoste, D., and Gaspard, P. 2018. Reaction kinetics in open reactors and serial transfers between closed reactors. *J. Chem. Phys.*, **148**, 144902.
Bochkov, G. N., and Kuzovlev, Yu. E. 1977. General theory of thermal fluctuations in nonlinear systems. *Sov. Phys. JETP*, **45**, 125–130.
Bochkov, G. N., and Kuzovlev, Yu. E. 1979. Fluctuation-dissipation relations for nonequilibrium processes in open systems. *Sov. Phys. JETP*, **49**, 543–551.
Bochkov, G. N., and Kuzovlev, Yu. E. 1981a. Nonlinear fluctuation-dissipation relations and stochastic models in nonequilibrium thermodynamics. I. Generalized fluctuation-dissipation theorem. *Physica A*, **106**, 443–479.
Bochkov, G. N., and Kuzovlev, Yu. E. 1981b. Nonlinear fluctuation-dissipation relations and stochastic models in nonequilibrium thermodynamics. II. Kinetic potential and variational principles for nonlinear irreversible processes. *Physica A*, **106**, 480–520.
Bocquet, L., and Barrat, J.-L. 1994. Hydrodynamic boundary conditions, correlation functions, and Kubo relations for confined fluids. *Phys. Rev. E*, **49**, 3079–3092.

Bogoliubov, N. N. 1946a. Kinetic Equations. *J. Phys. USSR*, **10**, 257–264.
Bogoliubov, N. N. 1946b. Kinetic Equations. *J. Phys. USSR*, **10**, 265–274.
Bollinger, J. J., and Wineland, D. J. 1990. Microplasmas. *Sci. Am.*, **262**(1), 114–120.
Boltzmann, L. 1871. On the thermal equilibrium between polyatomic gas molecules (in German). *Wiener Berichte*, **63**, 397–418.
Boltzmann, L. 1872. Further studies on the thermal equilibrium of gas molecules (in German). *Wiener Berichte*, **66**, 275–370.
Boltzmann, L. 1877. On the relationship between the second fundamental theorem of the mechanical theory of heat and probability calculations regarding the conditions of thermal equilibrium (in German). *Wiener Berichte*, **76**, 373–435.
Boltzmann, L. 1887. On the mechanical analogies for the second principle of thermodynamics (in German). *J. reine u. angew. Math.*, **100**, 201–212.
Boltzmann, L. 1896. *Lectures on Gas Theory I* (in German). Leipzig: Johann Ambrosius Barth.
Boltzmann, L. 1898. *Lectures on Gas Theory II* (in German). Leipzig: Johann Ambrosius Barth.
Boon, J. P., and Yip, S. 1980. *Molecular Hydrodynamics*. New York: McGraw-Hill.
Born, M., and Green, H. S. 1946. A general kinetic theory of liquids I: The molecular distribution functions. *Proc. R. Soc. Lond. A*, **188**, 10–18.
Bouchet, F. 2020. Is the Boltzmann equation reversible? A large deviation perspective on the irreversibility paradox. *J. Stat. Phys.*, **181**, 515–550.
Bowen, R. 1975. *Equilibrium States and the Ergodic Theory of Anosov Diffeomorphisms*. Berlin: Springer.
Brantut, J.-P., Grenier, C., Meineke, J., Stadler, D., Krinner, S., Kollath, C., Esslinger, T., and Georges, A. 2013. A thermoelectric heat engine with ultracold atoms. *Science*, **342**, 713–715.
Brenig, L., and Van den Broeck, C. 1980. Stochastic hydrodynamic theory for one-component systems. *Phys. Rev. A*, **21**, 1039–1048.
Brey, J. J., Zwanzig, R., and Dorfman, J. R. 1981. Nonlinear transport equations in statistical mechanics. *Physica A*, **109**, 425–444.
Briggs, M. E., Sengers, J. V., Francis, M. K., Gaspard, P., Gammon, R. W., Dorfman, J. R., and Calabrese, R. V. 2001. Tracking a colloidal particle for the measurement of dynamic entropies. *Physica A*, **296**, 42–59.
Brillouin, L. 1951. Maxwell's demon cannot operate: Information and entropy. *J. Appl. Phys.*, **22**, 334–337.
Brown, R. 1828. A brief account of microscopical observations, made in the months of June, July, and August 1827, on the particles contained in the pollen of plants; and on the general existence of active molecules in organic and inorganic bodies. *Edinb. New Phil. J.*, **5**, 358–371.
Brown, R., Ott, E., and Grebogi, C. 1987. Ergodic adiabatic invariants of chaotic systems. *Phys. Rev. Lett.*, **59**, 1173–1176.
Brun, R. 2009. *Introduction to Reactive Gas Dynamics*. Oxford: Oxford University Press.
Bulnes Cuetara, G., Esposito, M., and Gaspard, P. 2011. Fluctuation theorems for capacitively coupled electronic currents. *Phys. Rev. B*, **84**, 165114.
Bulnes Cuetara, G., Esposito, M., Schaller, G., and Gaspard, P. 2013. Effective fluctuation theorems for electron transport in a double quantum dot coupled to a quantum point contact. *Phys. Rev. B*, **88**, 115134.
Bunimovich, L. A. 1979. On the ergodic properties of nowhere dispersing billiards. *Commun. Math. Phys.*, **65**, 295–312.

Bunimovich, L. A., and Sinai, Ya. G. 1980a. Markov partitions for dispersed billiards. *Commun. Math. Phys.*, **78**, 247–280.

Bunimovich, L. A., and Sinai, Ya. G. 1980b. Statistical properties of Lorentz gas with periodic configuration of scatterers. *Commun. Math. Phys.*, **78**, 479–497.

Burgers, J. M. 1974. *The Nonlinear Diffusion Equation*. Dordrecht: D. Reidel.

Caldeira, A. O., and Leggett, A. J. 1983. Path integral approach to quantum Brownian motion. *Physica A*, **121**, 587–616.

Callen, H. B. 1985. *Thermodynamics and an Introduction to Thermostatistics*. 2nd ed. New York: Wiley.

Callen, H. B., and Welton, T. A. 1951. Irreversibility and generalized noise. *Phys. Rev.*, **83**, 34–40.

Callens, I., De Roeck, W., Jacobs, T., Maes, C., and Netočný, K. 2004. Quantum entropy production as a measure of irreversibility. *Physica D*, **187**, 383–391.

Calzetta, E. A., and Hu, B.-L. B. 2008. *Nonequilibrium Quantum Field Theory*. Cambridge, UK: Cambridge University Press.

Campbell, A. I., Ebbens, S. J., Illien, P., and Golestanian, R. 2019. Experimental observation of flow fields around Janus spheres. *Nat. Commun.*, **10**, 3952.

Campisi, M., Hänggi, P., and Talkner, P. 2011. Quantum fluctuation relations: Foundations and applications. *Rev. Mod. Phys.*, **83**, 771–791.

Carnot, S. 1824. *Reflections on the Motive Power of Fire and on Machines Fitted to Develop That Power* (in French). Paris: Bachelier.

Caroli, B., Caroli, C., and Roulet, B. 1992. Instabilities of planar solidification fronts, chap. 2. Pages 155–296 of: Godrèche, C. (ed.), *Solids Far From Equilibrium*. Cambridge, UK: Cambridge University Press.

Casimir, H. B. G. 1945. On Onsager's principle of microscopic reversibility. *Rev. Mod. Phys.*, **17**, 343–350.

Castiglione, P., Falcioni, M., Lesne, A., and Vulpiani, A. 2008. *Chaos and Coarse Graining in Statistical Mechanics*. Cambridge, UK: Cambridge University Press.

Cates, M. E., and Tailleur, J. 2015. Motility-induced phase separation. *Annu. Rev. Condens. Matter Phys.*, **6**, 219–244.

Cercignani, C. 1988. *The Boltzmann Equation and Its Applications*. New York: Springer.

Cercignani, C. 2000. *Rarefied Gas Dynamics*. Cambridge, UK: Cambridge University Press.

Chaikin, P. M., and Lubensky, T. C. 1995. *Principles of Condensed Matter Physics*. Cambridge, UK: Cambridge University Press.

Chandrasekhar, S. 1943. Stochastic problems in physics and astronomy. *Rev. Mod. Phys.*, **15**, 1–89.

Chang, C. W., Okawa, D., Majumdar, A., and Zettl, A. 2006. Solid-state thermal rectifier. *Science*, **314**, 1121–1124.

Chapman, S., and Cowling, T. G. 1960. *The Mathematical Theory of Non-Uniform Gases*. Cambridge, UK: Cambridge University Press.

Chernyak, V. Y., Chertkov, M., and Jarzynski, C. 2006. Path-integral analysis of fluctuation theorems for general Langevin processes. *J. Stat. Mech.*, **2006**, P08001.

Chetrite, R., and Gupta, S. 2011. Two refreshing views of fluctuation theorems through kinematics elements and exponential martingale. *J. Stat. Phys.*, **143**, 543–584.

Chetrite, R., and Touchette, H. 2015. Nonequilibrium Markov processes conditioned on large deviations. *Ann. Henri Poincaré*, **16**, 2005–2057.

Chirikov, B. V. 1979. A universal instability of many-dimensional oscillator systems. *Phys. Rep.*, **52**, 265–376.

Chowdhury, D. 2013. Stochastic mechano-chemical kinetics of molecular motors: A multidisciplinary enterprise from a physicist's perspective. *Phys. Rep.*, **529**, 1–197.

Ciliberto, S. 2017. Experiments in stochastic thermodynamics: Short history and perspectives. *Phys. Rev. X*, **7**, 021051.

Ciliberto, S., Gomez-Solano, R., and Petrosyan, A. 2013. Fluctuations, linear response, and currents in out-of-equilibrium Systems. *Annu. Rev. Condens. Matter Phys.*, **4**, 235–261.

Clausius, R. 1865. On several convenient forms for the fundamental equations of the mechanical theory of heat (in German). *Annalen der Physik und Chemie*, **125**, 353–400.

Clercx, H. J. H., and Schram, P. P. J. M. 1992. Brownian particles in shear flow and harmonic potentials: A study of long-time tails. *Phys. Rev. A*, **46**, 1942–1950.

Cleuren, B., Van den Broeck, C., and Kawai, R. 2006. Fluctuation theorem for the effusion of an ideal gas. *Phys. Rev. E*, **74**, 021117.

Cleuren, B., Willaert, K., Engel, A., and Van den Broeck, C. 2008. Fluctuation theorem for entropy production during effusion of a relativistic ideal gas. *Phys. Rev. E*, **77**, 022103.

Coddington, E. A., and Levinson, N. 1955. *Theory of Ordinary Differential Equations*. New York: McGraw-Hill.

Cohen-Tannoudji, C., Dupont-Roc, J., and Grynberg, G. 1992. *Atom-Photon Interactions: Basic Processes and Applications*. Weinheim: Wiley-VCH.

Coleman, B. D., and Fox, T. G. 1963a. General theory of stationary random sequences with applications to the tacticity of polymers. *J. Polym. Sci. A*, **1**, 3183–3197.

Coleman, B. D., and Fox, T. G. 1963b. Multistate mechanism for homogeneous ionic polymerization. I. The diastereosequence distribution. *J. Chem. Phys.*, **38**, 1065–1075.

Collin, D., Ritort, F., Jarzynski, C., Smith, S. B., Tinoco Jr., I., and Bustamante, C. 2005. Verification of the Crooks fluctuation theorem and recovery of RNA folding free energies. *Nature*, **437**, 231–234.

Colquhoun, D., and Sakmann, B. 1981. Fluctuations in the microsecond time range of the current through single acetylcholine receptor ion channels. *Nature*, **294**, 464–466.

Connes, A., Narnhofer, H., and Thirring, W. 1987. Dynamical entropy of C^* algebras and von Neumann algebras. *Commun. Math. Phys.*, **112**, 691–719.

Constantin, D., and Siwy, Z. S. 2007. Poisson–Nernst–Planck model of ion current rectification through a nanofuidic diode. *Phys. Rev. E*, **76**, 041202.

Cook, R. J., and Kimble, H. J. 1985. Possibility of direct observation of quantum jumps. *Phys. Rev. Lett.*, **54**, 1023–1026.

Cornfeld, I. P., Fomin, S. V., and Sinai, Ya. G. 1982. *Ergodic Theory*. New York: Springer.

Corrsin, S., and Lumley, J. 1956. On the equation of motion for a particle in turbulent fluid. *Appl. Sci. Res. A*, **6**, 114–116.

Coullet, P., and Iooss, G. 1990. Instabilities of one-dimensional cellular patterns. *Phys. Rev. Lett.*, **64**, 866–869.

Cover, T. M., and Thomas, J. A. 2006. *Elements of Information Theory*. 2nd ed. Hoboken, NJ: Wiley.

Crooks, G. E. 1998. Nonequilibrium measurements of free energy differences for microscopically reversible Markovian systems. *J. Stat. Phys.*, **90**, 1481–1487.

Crooks, G. E. 1999. Entropy production fluctuation theorem and the nonequilibrium work relation for free energy differences. *Phys. Rev. E*, **60**, 2721–2726.

Crooks, G. E. 2000. Path-ensemble averages in systems driven far from equilibrium. *Phys. Rev. E*, **61**, 2361–2366.

Cugliandolo, L. F., and Lecomte, V. 2017. Rules of calculus in the path integral representation of white noise Langevin equations: The Onsager–Machlup approach. *J. Phys. A: Math. Theor.*, **50**, 345001.

Curie, P. 1894. On symmetry in physical phenomena, symmetry on an electric field and a magnetic field (in French). *J. Phys. Théor. Appl.*, **3**, 393–415.

Curzon, F. L., and Ahlborn, B. 1975. Efficiency of a Carnot engine at maximum power output. *Am. J. Phys.*, **43**, 22–24.

Cvitanović, P., and Eckhardt, B. 1991. Periodic orbit expansions for classical smooth flows. *J. Phys. A: Math. Gen.*, **24**, L237–L241.

Datta, S. 1995. *Electronic Transport in Mesoscopic Systems*. Cambridge, UK: Cambridge University Press.

De Decker, Y. 2015. On the stochastic thermodynamics of reactive systems. *Physica A*, **428**, 178–193.

De Donder, T., and Van Rysselberghe, P. 1936. *Affinity*. Menlo Park, CA: Stanford University Press.

de Groot, S. R., and Mazur, P. 1984. *Nonequilibrium Thermodynamics*. New York: Dover.

de Schepper, I. M., and Cohen, E. G. D. 1980. Collective modes in fluids and neutron scattering. *Phys. Rev. A*, **22**, 287–289.

de Vega, I., and Alonso, D. 2017. Dynamics of non-Markovian open quantum systems. *Rev. Mod. Phys.*, **89**, 015001.

de Vega, I., Alonso, D., Gaspard, P., and Strunz, W. T. 2005. Non-Markovian stochastic Schrödinger equations in different temperature regimes: A study of the spin-boson model. *J. Chem. Phys.*, **122**, 124106.

Delbrück, M. 1940. Statistical fluctuations in autocatalytic reactions. *J. Chem. Phys.*, **8**, 120–124.

Dellago, Ch., Posch, H. A., and Hoover, W. G. 1996. Lyapunov instability in a system of hard disks in equilibrium and nonequilibrium steady states. *Phys. Rev. E*, **53**, 1485–1501.

Demaeyer, J., and Gaspard, P. 2009. Noise-induced escape from bifurcating attractors: Symplectic approach in the weak-noise limit. *Phys. Rev. E*, **80**, 031147.

Demaeyer, J., and Gaspard, P. 2013. A trace formula for activated escape in noisy maps. *J. Stat. Mech.*, **2013**, P10026.

Derrida, B. 2007. Non-equilibrium steady states: Fluctuations and large deviations of the density and of the current. *J. Stat. Mech.*, **2007**, P07023.

Desai, R. C., and Kapral, R. 1972. Translational hydrodynamics and light scattering from molecular fluids. *Phys. Rev. A*, **6**, 2377–2390.

DeVault, G. P., and McLennan, J. A. 1965. Statistical mechanics of viscoelasticity. *Phys. Rev.*, **137**, 724–730.

Dhar, A., Kundu, A., and Kundu, A. 2019. Anomalous heat transport in one dimensional systems: A description using non-local fractional-type diffusion Equation. *Frontiers Phys.*, **7**, 159.

Dirac, P. A. M. 1958. *The Principles of Quantum Mechanics*. 4th ed. Oxford: Clarendon Press.

Diu, B., Guthmann, C., Lederer, D., and Roulet, B. 1989. *Physique Statistique*. Paris: Hermann.

Dorfman, J. R. 1999. *An Introduction to Chaos in Nonequilibrium Statistical Mechanics*. Cambridge, UK: Cambridge University Press.

Dorfman, J. R., Kirkpatrick, T. R., and Sengers, J. V. 1994. Generic long-range correlations in molecular fluids. *Annu. Rev. Phys. Chem.*, **45**, 213–239.

Dorfman, J. R., Gaspard, P., and Gilbert, T. 2002. Entropy production of diffusion in spatially periodic deterministic systems. *Phys. Rev. E*, **66**, 026110.

Dorfman, J. R., van Beijeren, H., and Kirkpatrick, T. R. 2021. *Contemporary Kinetic Theory of Matter*. Cambridge, UK: Cambridge University Press.

Dufty, J. W. 1974. Gaussian model for fluctuation of a Brownian particle. *Phys. Fluids*, **17**, 328–333.

Dufty, J. W., Luo, K., and Wrighton, J. 2020. Generalized hydrodynamics revisited. *Phys. Rev. Res.*, **2**, 023036.

Duque-Zumajo, D., de la Torre, J. A., Camargo, D., and Español, P. 2019. Discrete hydrodynamics near solid walls: Non-Markovian effects and the slip boundary condition. *Phys. Rev. E*, **100**, 062133.

Ebers, J. J., and Moll, J. L. 1954. Large-signal behavior of junction transistors. *Proc. Inst. Radio Eng.*, **42**, 1761–1772.

Eckmann, J.-P., and Ruelle, D. 1985. Ergodic theory of chaos and strange attractors. *Rev. Mod. Phys.*, **57**, 617–656.

Eckmann, J.-P., Pillet, C.-A., and Rey-Bellet, L. 1999. Entropy production in nonlinear, thermally driven Hamiltonian systems. *J. Stat. Phys.*, **95**, 305–331.

Edwards, D. A., Brenner, H., and Wasan, D. T. 1991. *Interfacial Transport Processes and Rheology*. Boston: Butterworth-Heinemann.

Ehrenfest, P., and Ehrenfest, T. 1911. The conceptual foundations of the statistical approach in mechanics (in German). Tome IV, 2. Teil. Pages 3–90 of: Klein, F., and Müller, C. (eds), *Enzyklopädie der mathematischen Wissenschaften*. Leipzig: Teubner. English translation. 1990. New York: Dover.

Einstein, A. 1905. On the movement of small particles suspended in stationary liquids demanded by the molecular kinetic theory of heat (in German). *Ann. Physik*, **17**, 549–560.

Einstein, A. 1926. *Investigations on the Theory of the Brownian Movement*. London: Methuen & Co.

Ellis, R. S. 1985. *Entropy, Large Deviations, and Statistical Mechanics*. New York: Springer.

Epstein, I. R., and Pojman, J. A. 1998. *An Introduction to Nonlinear Chemical Dynamics*. New York & Oxford: Oxford University Press.

Ernst, M. H., and Cohen, E. G. D. 1981. Nonequilibrium fluctuations in μ space. *J. Stat. Phys.*, **25**, 153–180.

Ernst, M. H., and Dorfman, J. R. 1975. Nonanalytic dispersion relations for classical fluids. II. The general fluid. *J. Stat. Phys.*, **12**, 311–359.

Ernst, M. H., Hauge, E. H., and van Leeuwen, J. M. J. 1971. Asymptotic time behavior of correlation functions. I. Kinetic terms. *Phys. Rev. A*, **4**, 2055–2065.

Ernst, M. H., Dorfman, J. R., Nix, R., and Jacobs, D. 1995. Mean-field theory for Lyapunov exponents and Kolmogorov–Sinai entropy in Lorentz lattice gases. *Phys. Rev. Lett.*, **74**, 4416–4419.

Esposito, M., Harbola, U., and Mukamel, S. 2009. Nonequilibrium fluctuations, fluctuation theorems, and counting statistics in quantum systems. *Rev. Mod. Phys.*, **81**, 1665–1702.

Esposito, M., Kawai, R., Lindenberg, K., and Van den Broeck, C. 2010. Efficiency at maximum power of low-dissipation Carnot engines. *Phys. Rev. Lett.*, **105**, 150603.

Evans, D. J., and Morriss, G. P. 1990. *Statistical Mechanics of Nonequilibrium Liquids*. London: Academic Press.

Evans, D. J., and Searles, D. J. 2002. The fluctuation theorem. *Adv. Phys.*, **51**, 1529–1585.

Evans, D. J., Hoover, W. H., Failor, B. H., Moran, B., and Ladd, A. J. C. 1983. Nonequilibrium molecular dynamics via Gauss's principle of least constraint. *Phys. Rev. A*, **28**, 1016–1021.

Evans, D. J., Cohen, E. G. D., and Morriss, G. P. 1993. Probability of second law violations in shearing steady states. *Phys. Rev. Lett.*, **71**, 2401–2404.

Evans, R. 1979. The nature of liquid-vapour interface and other topics in the statistical mechanics of non-uniform, classical fluids. *Adv. Phys.*, **28**, 143–200.

Faggionato, A., and Di Pietro, D. 2011. Gallavotti–Cohen-type symmetry related to cycle decompositions for Markov chains and biochemical applications. *J. Stat. Phys.*, **143**, 11–32.

Falasco, G., Pfaller, R., Bregulla, A. P., Cichos, F., and Kroy, K. 2016. Exact symmetries in the velocity fluctuations of a hot Brownian swimmer. *Phys. Rev. E*, **94**, 030602(R).

Falconer, K. 1990. *Fractal Geometry*. Chichester: Wiley.

Feinberg, M. 2019. *Foundations of Chemical Reaction Network Theory*. Cham, Switzerland: Springer.

Felderhof, B. U. 1976. Force density induced on a sphere in linear hydrodynamics: II. Moving sphere, mixed boundary conditions. *Physica A*, **84**, 569–576.

Feller, W. 1968. *An Introduction to Probability Theory and Its Applications, Vol. I*. 3rd ed. New York: Wiley.

Feller, W. 1971. *An Introduction to Probability Theory and Its Applications, Vol. II*. 2nd ed. New York: Wiley.

Fermi, E. 1937. *Thermodynamics*. New York: Prentice Hall.

Ferry, D. K., Goodnick, S. M., and Bird, J. 2009. *Transport in Nanostructures*. 2nd ed. Cambridge, UK: Cambridge University Press.

Feshbach, H. 1962. A unified theory of nuclear reactions. II. *Ann. Phys.*, **19**, 287–313.

Feynman, R. P., and Hibbs, A. R. 1965. *Quantum Mechanics and Path Integrals*. New York: McGraw-Hill.

Fleming, P. D., and Cohen, C. 1976. Hydrodynamics of solids. *Phys. Rev. B*, **13**, 500–516.

Fogedby, H. C., and Imparato, A. 2012. Heat flows in chains driven by thermal noise. *J. Stat. Mech.*, **2012**, P04005.

Fogedby, H. C., and Jensen, M. H. 2005. Weak noise approach to the logistic map. *J. Stat. Phys.*, **121**, 759–778.

Fokker, A. D. 1914. The mean energy of rotating electrical dipoles in the radiation field (in German). *Ann. Physik*, **348**, 810–820.

Forster, D. 1975. *Hydrodynamic Fluctuations, Broken Symmetry, and Correlation Functions*. Reading, MA: Benjamin/Cummings.

Förster, H., and Büttiker, M. 2008. Fluctuation relations without microreversibility in nonlinear transport. *Phys. Rev. Lett.*, **101**, 136805.

Fowler, R. H. 1929. *Statistical Mechanics*. Cambridge, UK: Cambridge University Press.

Fox, R. F., and Uhlenbeck, G. E. 1970a. Contributions to non-equilibrium thermodynamics. I. Theory of hydrodynamic fluctuations. *Phys. Fluids*, **13**, 1893–1902.

Fox, R. F., and Uhlenbeck, G. E. 1970b. Contributions to non-equilibrium thermodynamics. II. Fluctuation theory for the Boltzmann equation. *Phys. Fluids*, **13**, 2881–2890.

Franosch, T., Grimm, M., Belushkin, M., Mor, F. M., Foffi, G., Forró, L., and Jeney, S. 2011. Resonances arising from hydrodynamic memory in Brownian motion. *Nature*, **478**, 85–88.

Freitas, N., Delvenne, J.-C., and Esposito, M. 2020. Stochastic and quantum thermodynamics of driven RLC networks. *Phys. Rev. X*, **10**, 031005.

Frenkel, D., and Smit, B. 2002. *Understanding Molecular Simulation*. 2nd ed. San Diego: Academic Press.

Fujisawa, T., Hayashi, T., Tomita, R., and Hirayama, Y. 2006. Bidirectional counting of single electrons. *Science*, **312**, 1634–1636.

Gallavotti, G. 1996. Extension of Onsager's reciprocity to large fields and the chaotic hypothesis. *Phys. Rev. Lett.*, **77**, 4334–4337.

Gallavotti, G., and Cohen, E. G. D. 1995. Dynamical ensembles in nonequilibrium statistical mechanics. *Phys. Rev. Lett.*, **74**, 2694–2697.

Gantmacher, F. R. 1959. *Applications of the Theory of Matrices*. New York: Interscience Publishers.

Garcia Cantú Ros, A., McEwen, J.-S., and Gaspard, P. 2011. Effect of ultrafast diffusion on adsorption, desorption, and reaction processes over heterogeneous surfaces. *Phys. Rev. E*, **83**, 021604.

Gardiner, C. W. 1979. A stochastic basis for isothermal equilibrium and nonequilibrium chemical thermodynamics. *J. Chem. Phys.*, **70**, 5778–5787.

Gardiner, C. W. 2004. *Handbook of Stochastic Methods for Physics, Chemistry and the Natural Sciences*. 3rd ed. Berlin: Springer.

Gardiner, C. W., and Zoller, P. 2000. *Quantum Noise*. Berlin: Springer.

Garnier, N., and Ciliberto, S. 2005. Nonequilibrium fluctuations in a resistor. *Phys. Rev. E*, **71**, 060101.

Gaspard, P. 1992. Dynamical chaos and many-body quantum systems. Pages 19–42 of: Cvitanović, P., Percival, I., and Wirzba, A. (eds), *Quantum Chaos – Quantum Measurement*. Dordrecht: Kluwer Academic Publishers.

Gaspard, P. 1994. Comment on dynamical randomness in quantum systems. *Prog. Theor. Phys. Suppl.*, **116**, 369–378.

Gaspard, P. 1997. Entropy production in open volume-preserving systems. *J. Stat. Phys.*, **88**, 1215–1240.

Gaspard, P. 1998. *Chaos, Scattering and Statistical Mechanics*. Cambridge, UK: Cambridge University Press.

Gaspard, P. 2000. Scattering, transport & stochasticity in quantum systems. Pages 425–456 of: Karkheck, J. (ed.), *Dynamics: Models and Kinetic Methods for Non-equilibrium Many Body Systems*. Dordrecht: Kluwer Academic Publishers.

Gaspard, P. 2001. Dynamical chaos and nonequilibrium statistical mechanics. *Int. J. Mod. Phys. B*, **15**, 209–235.

Gaspard, P. 2002a. The correlation time of mesoscopic chemical clocks. *J. Chem. Phys.*, **117**, 8905–8916.

Gaspard, P. 2002b. Trace formula for noisy flows. *J. Stat. Phys.*, **106**, 57–96.

Gaspard, P. 2003a. Lyapunov exponent of ion motion in microplasmas. *Phys. Rev. E*, **88**, 056209.

Gaspard, P. 2003b. Nonlinear dynamics and chaos in many-particle Hamiltonian systems. *Prog. Theor. Phys. Suppl.*, **150**, 64–80.

Gaspard, P. 2004a. Fluctuation theorem for nonequilibrium reactions. *J. Chem. Phys.*, **120**, 8898–8905.

Gaspard, P. 2004b. Time-reversed dynamical entropy and irreversibility in Markovian random processes. *J. Stat. Phys.*, **117**, 599–615. Erratum *ibid.* **126**, 1109 (2007).

Gaspard, P. 2005. Brownian motion, dynamical randomness and irreversibility. *New J. Phys.*, **7**, 77.

Gaspard, P. 2006. Hamiltonian dynamics, nanosystems, and nonequilibrium statistical mechanics. *Physica A*, **369**, 201–246.

Gaspard, P. 2007a. Temporal ordering of nonequilibrium fluctuations as a corollary of the second law of thermodynamics. *C. R. Physique*, **8**, 598–608.

Gaspard, P. 2007b. Time asymmetry in nonequilibrium statistical mechanics. *Adv. Chem. Phys.*, **135**, 83–133.

Gaspard, P. 2008. Thermodynamic time asymmetry and nonequilibrium statistical mechanics. Pages 67–87 of: Ishiwata, S., and Matsunaga, Y. (eds), *Physics of Self-Organization Systems*. New Jersey: World Scientific.

Gaspard, P. 2010. Nonequilibrium nanosystems. Pages 1–74 of: Radons, G., Rumpf, B., and Schuster, H. G. (eds), *Nonlinear Dynamics of Nanosystems*. Weinheim: Wiley-VCH.
Gaspard, P. 2012a. Broken $\mathbb{Z}_2$ symmetries and fluctuations in statistical mechanics. *Phys. Scr.*, **86**, 058504.
Gaspard, P. 2012b. Fluctuation relations for equilibrium states with broken discrete symmetries. *J. Stat. Mech.*, **2012**, P08021.
Gaspard, P. 2013a. Multivariate fluctuation relations for currents. *New J. Phys.*, **15**, 115014.
Gaspard, P. 2013b. Self-organization at the nanoscale in far-from-equilibrium surface reactions and copolymerizations. Pages 51–77 of: Mikhailov, A. S., and Ertl, G. (eds), *Engineering of Chemical Complexity*. Singapore: World Scientific.
Gaspard, P. 2013c. Time-reversal symmetry relation for nonequilibrium flows ruled by the fluctuating Boltzmann equation. *Physica A*, **392**, 639–655.
Gaspard, P. 2013d. Time-reversal symmetry relations for currents in quantum and stochastic nonequilibrium systems. Pages 213–257 of: Klages, R., Just, W., and Jarzynski, C. (eds), *Nonequilibrium Statistical Physics of Small Systems: Fluctuation Relations and Beyond*. Weinheim: Wiley-VCH.
Gaspard, P. 2013e. Time-reversal symmetry relations for fluctuating currents in nonequilibrium systems. *Acta Phys. Pol. B*, **44**, 815–845.
Gaspard, P. 2014. Random paths and current fluctuations in nonequilibrium statistical mechanics. *J. Math. Phys.*, **55**, 075208.
Gaspard, P. 2015a. Force-velocity relation for copolymerization processes. *New J. Phys.*, **17**, 045016.
Gaspard, P. 2015b. Scattering approach to the thermodynamics of quantum transport. *New J. Phys.*, **17**, 045001.
Gaspard, P. 2015c. Scattering theory and thermodynamics of quantum transport. *Ann. Phys. (Berlin)*, **527**, 663–683.
Gaspard, P. 2016a. Kinetics and thermodynamics of living copolymerization processes. *Phil. Trans. R. Soc. A*, **374**, 20160147.
Gaspard, P. 2016b. Template-directed copolymerization, random walks along disordered tracks, and fractals. *Phys. Rev. Lett.*, **117**, 238101.
Gaspard, P. 2020a. Microreversibility and driven Brownian motion with hydrodynamic long-time correlations. *Physica A*, **552**, 121823.
Gaspard, P. 2020b. Stochastic approach to entropy production in chemical chaos. *Chaos*, **30**, 113103.
Gaspard, P., and Andrieux, D. 2011a. From the multibaker map to the fluctuation theorem for currents. *Bussei Kenkyu*, **97**, 377–397.
Gaspard, P., and Andrieux, D. 2011b. Nonlinear transport effects in mass separation by effusion. *J. Stat. Mech.*, **2011**, P03024.
Gaspard, P., and Gerritsma, E. 2007. The stochastic chemomechanics of the F_1-ATPase molecular motor. *J. Theor. Biol.*, **247**, 672–686.
Gaspard, P., and Gilbert, T. 2008a. Heat conduction and Fourier's law by consecutive local mixing and thermalization. *Phys. Rev. Lett.*, **101**, 020601.
Gaspard, P., and Gilbert, T. 2008b. Heat conduction and Fourier's law in a class of many particle dispersing billiards. *New J. Phys.*, **10**, 103004.
Gaspard, P., and Gilbert, T. 2009. Heat transport in stochastic energy exchange models of locally confined hard sphere. *J. Stat. Mech.*, **2009**, P08020.
Gaspard, P., and Gilbert, T. 2017. Dynamical contribution to the heat conductivity in stochastic energy exchanges of locally confined gases. *J. Stat. Mech.*, **2017**, 043210.
Gaspard, P., and Kapral, R. 2017. Mechanochemical fluctuation theorem and thermodynamics of self-phoretic motors. *J. Chem. Phys.*, **147**, 211101.

Gaspard, P., and Kapral, R. 2018a. Finite-time fluctuation theorem for diffusion-influenced surface reactions. *J. Stat. Mech.*, **2018**, 083206.

Gaspard, P., and Kapral, R. 2018b. Fluctuating chemohydrodynamics and the stochastic motion of self-diffusiophoretic particles. *J. Chem. Phys.*, **148**, 134104.

Gaspard, P., and Kapral, R. 2018c. Nonequilibrium thermodynamics and boundary conditions for reaction and transport in heterogeneous media. *J. Chem. Phys.*, **148**, 194114.

Gaspard, P., and Kapral, R. 2019a. The stochastic motion of self-thermophoretic Janus particles. *J. Stat. Mech.*, **2019**, 074001.

Gaspard, P., and Kapral, R. 2019b. Thermodynamics and statistical mechanics of chemically powered synthetic nanomotors. *Adv. Phys.: X*, **4**, 1602480.

Gaspard, P., and Kapral, R. 2020. Active matter, microreversibility, and thermodynamics. *Research*, **2020**, 973923.

Gaspard, P., and Lutsko, J. 2004. Imploding shock wave in a fluid of hard-core particles. *Phys. Rev. E*, **70**, 026306.

Gaspard, P., and Nagaoka, M. 1999a. Non-Markovian stochastic Schrödinger equation. *J. Chem. Phys.*, **111**, 5676–5690.

Gaspard, P., and Nagaoka, M. 1999b. Slippage of initial conditions for the Redfield master equation. *J. Chem. Phys.*, **111**, 5668–5675.

Gaspard, P., and Nicolis, G. 1990. Transport properties, Lyapunov exponents, and entropy per unit time. *Phys. Rev. Lett.*, **65**, 1693–1696.

Gaspard, P., and van Beijeren, H. 2002. When do tracer particles dominate the Lyapunov spectrum? *J. Stat. Phys.*, **109**, 671–704.

Gaspard, P., and Wang, X.-J. 1988. Sporadicity: Between periodic and chaotic dynamical behaviors. *Proc. Natl. Acad. Sci.*, **85**, 4591–4595.

Gaspard, P., and Wang, X.-J. 1993. Noise, chaos, and (ε, τ)-entropy per unit time. *Phys. Rep.*, **235**, 291–343.

Gaspard, P., Nicolis, G., Provata, A., and Tasaki, S. 1995. Spectral signature of the pitchfork bifurcation: Liouville equation approach. *Phys. Rev. E*, **51**, 74–94.

Gaspard, P., Briggs, M. E., Francis, M. K., Sengers, J. V., Gammon, R. W., Dorfman, J. R., and Calabrese, R. V. 1998. Experimental evidence for microscopic chaos. *Nature*, **394**, 865–868.

Gaspard, P., Claus, I., Gilbert, T., and Dorfman, J. R. 2001. Fractality of the hydrodynamic modes of diffusion. *Phys. Rev. Lett.*, **86**, 1506–1509.

Gaspard, P., Grosfils, P., Huang, M.-J., and Kapral, R. 2018. Finite-time fluctuation theorem for diffusion-influenced surface reactions on spherical and Janus catalytic particles. *J. Stat. Mech.*, **2018**, 123206.

Ge, H., and Qian, H. 2009. Thermodynamic limit of a nonequilibrium steady state: Maxwell-type construction for a bistable biochemical system. *Phys. Rev. Lett.*, **103**, 148103.

Gerritsma, E., and Gaspard, P. 2010. Chemomechanical coupling and stochastic thermodynamics of the F_1-ATPase molecular motor with an applied external torque. *Biophys. Rev. Lett.*, **5**, 163–208.

Gibbs, J. W. 1902. *Elementary Principles in Statistical Mechanics*. New Haven, CT: Yale University Press.

Gilbert, T., Dorfman, J. R., and Gaspard, P. 2000. Entropy production, fractals, and relaxation to equilibrium. *Phys. Rev. Lett.*, **85**, 1606–1609.

Gillespie, D. T. 1976. A general method for numerically simulating the stochastic time evolution of coupled chemical reactions. *J. Comput. Phys.*, **22**, 403–434.

Gillespie, D. T. 1977. Exact stochastic simulation of coupled chemical reactions. *J. Phys. Chem.*, **81**, 2340–2361.

Gillespie, D. T. 2000. The chemical Langevin equation. *J. Chem. Phys.*, **113**, 297–306.
Gingrich, T. R., Horowitz, J. M., Perunov, N., and England, J. L. 2016. Dissipation bounds all steady-state current fluctuations. *Phys. Rev. Lett.*, **116**, 120601.
Glansdorff, P., and Prigogine, I. 1971. *Thermodynamics of Structure, Stability, and Fluctuations*. New York: Wiley.
Goldbeter, A. 1996. *Biochemical Oscillations and Cellular Rhythms*. Cambridge, UK: Cambridge University Press.
Goldenfeld, N. 1992. *Lectures on Phase Transitions and the Renormalization Group*. Reading, MA: Addison-Wesley.
Goldstein, H. 1950. *Classical Mechanics*. Reading, MA: Addison-Wesley.
Goldstone, J. 1961. Field theories with superconductor solutions. *Il Nuovo Cimento* (1955-1965), **19**, 154–164.
Golubev, D. S., Utsumi, Y., Marthaler, M., and Schön, G. 2011. Fluctuation theorem for a double quantum dot coupled to a point-contact electrometer. *Phys. Rev. B*, **84**, 075323.
Gombert, A. 2017. *Microreversibility and angular momentum transfer between a thermal system and a time-dependent field* (in French). Brussels: Université libre de Bruxelles. Master thesis in physics.
Gomez-Marin, A., Parrondo, J. M. R., and Van den Broeck, C. 2008. The "footprints" of irreversibility. *EPL*, **82**, 50002.
Gonze, D., Halloy, J., and Gaspard, P. 2002. Biochemical clocks and molecular noise: Theoretical study of robustness factors. *J. Chem. Phys.*, **116**, 10997–11010.
Goychuk, I. 2004. Quantum dynamics with non-Markovian fluctuating parameters. *Phys. Rev. E*, **70**, 016109.
Grad, H. 1958. Principles of the kinetic theory of gases. Pages 205–294 of: Flügge, S. (ed.), *Handbuch der Physik, vol. 12*. Berlin: Springer.
Green, H. S. 1952a. *The Molecular Theory of Fluids*. Amsterdam: North-Holland.
Green, M. S. 1952b. Markoff random processes and the statistical mechanics of time-dependent phenomena. *J. Chem. Phys.*, **20**, 1281–1295.
Green, M. S. 1954. Markoff random processes and the statistical mechanics of time-dependent phenomena. II. Irreversible processes in fluids. *J. Chem. Phys.*, **22**, 398–413.
Grosfils, P., Gaspard, P., and Visart de Bocarmé, T. 2015. The role of fluctuations in bistability and oscillations during the H_2+O_2 reaction on nanosized rhodium crystals. *J. Chem. Phys.*, **143**, 064705.
Gu, J., and Gaspard, P. 2018. Stochastic approach and fluctuation theorem for charge transport in diodes. *Phys. Rev. E*, **97**, 052138.
Gu, J., and Gaspard, P. 2019. Microreversibility, fluctuations, and nonlinear transport in transistors. *Phys. Rev. E*, **99**, 012137.
Gu, J., and Gaspard, P. 2020. Counting statistics and microreversibility in stochastic models of transistors. *J. Stat. Mech.*, **2020**, 103206.
Guldberg, C.M., and Waage, P. 1879. On chemical affinity (in German). *Erdmann's Journal für practische Chemie*, **127**, 69–114.
Gustavsson, S., Leturcq, R., Simovič, B., Schleser, R., Ihn, T., Studerus, P., Ensslin, K., Driscoll, D. C., and Gossard, A. C. 2006. Counting statistics of single electron transport in a quantum dot. *Phys. Rev. Lett.*, **96**, 076605.
Haase, R. 1969. *Thermodynamics of Irreversible Processes*. New York: Dover.
Haberland, H., Hippler, T., Donges, J., Kostko, O., Schmidt, M., and von Issendorff, B. 2005. Melting of sodium clusters: Where do the magic numbers come from? *Phys. Rev. Lett.*, **94**, 035701.

Haken, H. 1975. Cooperative phenomena in systems far from thermal equilibrium and in nonphysical systems. *Rev. Mod. Phys.*, **47**, 67–121.

Halsey, T. C., Jensen, M. H., Kadanoff, L. P., Procaccia, I., and Shraiman, B. I. 1986. Fractal measures and their singularities: The characterization of strange sets. *Phys. Rev. A*, **33**, 1141–1151.

Hänggi, P., Talkner, P., and Borkovec, M. 1990. Reaction-rate theory: Fifty years after Kramers. *Rev. Mod. Phys.*, **62**, 251–341.

Haraldsdóttir, H. S., and Fleming, R. M. T. 2016. Identification of conserved moieties in metabolic networks by graph theoretical analysis of atom transition networks. *PLoS Comput. Biol.*, **12**, e1004999.

Häring, J. M., Walz, C., Szamel, G., and Fuchs, M. 2015. Coarse-grained density and compressibility of nonideal crystals. *Phys. Rev. B*, **92**, 184103.

Harris, R. J., and Schütz, G. M. 2007. Fluctuation theorems for stochastic dynamics. *J. Stat. Mech.*, **2007**, P07020.

Hatano, T., and Sasa, S.-i. 2001. Steady state thermodynamics of Langevin systems. *Phys. Rev. Lett.*, **86**, 3463–3466.

Hauge, E. H., and Martin-Löf, A. 1973. Fluctuating hydrodynamics and Brownian motion. *J. Stat. Phys.*, **7**, 259–281.

Helfand, E. 1960. Transport coefficients from dissipation in a canonical ensemble. *Phys. Rev.*, **119**, 1–9.

Henss, A.-K., Sakong, S., Messer, P. K., Wiechers, J., Schuster, R., Lamb, D. C., Gross, A., and Wintterlin, J. 2019. Density fluctuations as door-opener for diffusion on crowded surfaces. *Science*, **363**, 715–718.

Hill, T. L. 1956. *Statistical Mechanics: Principles and Selected Applications*. New York: McGraw-Hill.

Hill, T. L. 1960. *An Introduction to Statistical Thermodynamics*. Reading, MA: Addison-Wesley.

Hill, T. L. 1989. *Free Energy Transduction and Biochemical Cycle Kinetics*. New York: Springer.

Hill, T. L., and Plesner, I. W. 1965. Studies in irreversible thermodynamics. II. A simple class of lattice models for open systems. *J. Chem. Phys.*, **43**, 267–285.

Hills, B. P. 1975. A generalized Langevin equation for the angular velocity of a spherical Brownian particle from fluctuating hydrodynamics. *Physica A*, **80**, 360–368.

Hirschfelder, J. O., Curtis, C. F., and Bird, R. B. 1954. *Molecular Theory of Gases and Liquids*. New York: Wiley.

Hoover, W. H. 1985. Canonical dynamics: Equilibrium phase-space distributions. *Phys. Rev. A*, **31**, 1695–1697.

Horn, F., and Jackson, R. 1972. General mass action kinetics. *Arch. Rational Mech. Anal.*, **47**, 81–116.

Horowitz, J. M. 2015. Diffusion approximations to the chemical master equation only have a consistent stochastic thermodynamics at chemical equilibrium. *J. Chem. Phys.*, **143**, 044111.

Huang, K. 1987. *Statistical Mechanics*. 2nd ed. New York: Wiley.

Huang, M.-J., Schofield, J., Gaspard, P., and Kapral, R. 2018. Dynamics of Janus motors with microscopically reversible kinetics. *J. Chem. Phys.*, **149**, 024904.

Huang, M.-J., Schofield, J., Gaspard, P., and Kapral, R. 2019. From single particle motion to collective dynamics in Janus motor systems. *J. Chem. Phys.*, **150**, 124110.

Hurtado, P. I., Pérez-Espigares, C. P., del Pozo, J. J., and Garrido, P. L. 2011. Symmetries in fluctuations far from equilibrium. *Proc. Natl. Acad. Sci.*, **108**, 7704–7709.

Ishizaki, A., and Fleming, G. R. 2009. Theoretical examination of quantum coherence in a photosynthetic system at physiological temperature. *Proc. Natl. Acad. Sci.*, **106**, 17255–17260.

Itami, M., and Sasa, S.-i. 2017. Universal form of stochastic evolution for slow variables in equilibrium systems. *J. Stat. Phys.*, **167**, 46–63.

Itoh, H., Takahashi, A., Adachi, K., Noji, H., Yasuda, R., Yoshida, M., and Kinosita, K. 2004. Mechanically driven ATP synthesis by F_1-ATPase. *Nature*, **427**, 465–468.

Ivanchenko, Yu. M., and Zil'berman, L. A. 1969. The Josephson effect in small tunnel contacts. *Soviet Phys. JETP*, **28**, 1272–1276.

Jackson, J. D. 1999. *Classical Electrodynamics*. 3rd ed. Hoboken, NJ: Wiley.

Jakšić, V., and Pillet, C.-A. 1996a. On a model for quantum friction. II. Fermi's golden rule and dynamics at positive temperature. *Commun. Math. Phys.*, **176**, 619–644.

Jakšić, V., and Pillet, C.-A. 1996b. On a model for quantum friction. III. Ergodic properties of the spin-boson system. *Commun. Math. Phys.*, **178**, 627–651.

Jarzynski, C. 1992. Diffusion equation for energy in ergodic adiabatic ensembles. *Phys. Rev. A*, **46**, 7498–7509.

Jarzynski, C. 1993. Multiple-time-scale approach to ergodic adiabatic systems: Another look. *Phys. Rev. Lett.*, **71**, 839–842.

Jarzynski, C. 1995. Thermalization of a Brownian particle via coupling to low-dimensional chaos. *Phys. Rev. Lett.*, **74**, 2937–2940.

Jarzynski, C. 1997. Nonequilibrium equality for free energy differences. *Phys. Rev. Lett.*, **78**, 2690–2693.

Jarzynski, C. 2000. Hamiltonian derivation of a detailed fluctuation theorem. *J. Stat. Phys.*, **98**, 77–102.

Jarzynski, C. 2006. Rare events and the convergence of exponentially averaged work values. *Phys. Rev. E*, **73**, 046105.

Jarzynski, C. 2011. Equalities and inequalities: Irreversibility and the second law of thermodynamics at the nanoscale. *Annu. Rev. Condens. Matter Phys.*, **2**, 329–351.

Jarzynski, C., and Wójcik, D. K. 2004. Classical and quantum fluctuation theorems for heat exchange. *Phys. Rev. Lett.*, **92**, 230602.

Jayannavar, A. M., and Kumar, N. 1981. Orbital diamagnetism of a charged Brownian particle undergoing a birth-death process. *J. Phys. A: Math. Gen.*, **14**, 1399–1405.

Jayannavar, A. M., and Sahoo, M. 2007. Charged particle in a magnetic field: Jarzynski equality. *Phys. Rev. E*, **75**, 032102.

Jiang, D.-Q., Qian, M., and Qian, M.-P. 2004. *Mathematical Theory of Nonequilibrium Steady States*. Berlin: Springer.

Joachain, C. J. 1975. *Quantum Collision Theory*. Amsterdam: North-Holland.

Johnson, J. B. 1928. Thermal agitation of electricity in conductors. *Phys. Rev.*, **32**, 97–109.

Joubaud, S., Garnier, N. B., and Ciliberto, S. 2007. Fluctuation theorems for harmonic oscillators. *J. Stat. Mech.*, **2007**, P09018.

Joubaud, S., Garnier, N. B., and Ciliberto, S. 2008. Fluctuations of the total entropy production in stochastic systems. *EPL*, **82**, 30007.

Joulin, G., and Vidal, P. 1998. An introduction to the instability of flames, shocks, and detonations. Pages 493–673 of: Godrèche, C., and Manneville, P. (eds), *Hydrodynamics and Nonlinear Instabilities*. Cambridge, UK: Cambridge University Press.

Jülicher, F., Ajdari, A., and Prost, J. 1997. Modeling molecular motors. *Rev. Mod. Phys.*, **69**, 1269–1282.

Jülicher, F., Grill, S. W., and Salbreux, G. 2018. Hydrodynamic theory of active matter. *Rep. Prog. Phys.*, **81**, 076601.

Kac, M. 1956. Some remarks on the use of probability in classical statistical mechanics. *Bull. Acad. Roy. Belg. (Cl. Sci.)*, **42**, 356–361.

Kadanoff, L. P., and Baym, G. 1962. *Quantum Statistical Mechanics*. New York: W. A. Benjamin.

Kadanoff, L. P., and Martin, P. C. 1963. Hydrodynamic equations and correlation functions. *Ann. Phys.*, **24**, 419–469.

Kantz, H., and Grassberger, P. 1985. Repellers, semi-attractors, and long-lived chaotic transients. *Physica D*, **17**, 75–86.

Kapral, R. 1972. Internal relaxation in chemically reacting fluids. *J. Chem. Phys.*, **56**, 1842–1847.

Kapral, R. 2013. Perspective: Nanomotors without moving parts that propel themselves in solution. *J. Chem. Phys.*, **138**, 020901.

Kardar, M., Parisi, G., and Zhang, Y. Z. 1986. Dynamic scaling of growing interfaces. *Phys. Rev. Lett.*, **56**, 889–892.

Kats, E. I., and Lebedev, V. V. 1994. *Fluctuational Effects in the Dynamics of Liquid Crystals*. New York: Springer.

Kavassalis, T. A., and Oppenheim, I. 1988. Derivation of the nonlinear hydrodynamic equations using multi-mode techniques. *Physica A*, **148**, 521–555.

Kawai, R., Parrondo, J. M. R., and Van den Broeck, C. 2007. Dissipation: The phase-space perspective. *Phys. Rev. Lett.*, **98**, 080602.

Kawasaki, K. 1971. Non-hydrodynamical behavior of two-dimensional fluids. *Phys. Lett. A*, **34**, 12–13.

Kenkre, V. M., Montroll, E. W., and Shlesinger, M. F. 1973. Generalized master equations for continuous-time random walks. *J. Stat. Phys.*, **9**, 45–50.

Kheifets, S., Simha, A., Melin, K., Li, T., and Raizen, M. G. 2014. Observation of Brownian motion in liquids at short times: Instantaneous velocity and memory loss. *Science*, **343**, 1493–1496.

Khinchin, A. Ya. 1932. To Birkhoff's solution to the ergodic problem (in German). *Math. Annalen*, **107**, 485–488.

Kinosita, K., Adachi, K., and Itoh, H. 2004. Rotation of F_1-ATPase: How an ATP-driven molecular machine may work. *Ann. Rev. Biophys. Biomol. Struct.*, **33**, 245–268.

Kirchhoff, G. R. 1847. On the solution of the equations to which one is led in the investigation of the linear distribution of galvanic currents (in German). *Poggendorff's Ann. Phys. Chem.*, **72**, 497–508.

Kirkwood, J. G. 1946. The statistical mechanical theory of transport processes I. General theory. *J. Chem. Phys.*, **14**, 180–201.

Kitamura, K., Tokunaga, M., Iwane, A. H., and Yanagida, T. 1999. A single myosin head moves along an actin filament with regular steps of 5.3 nanometers. *Nature*, **397**, 129–134.

Kjelstrup, S., and Bedeaux, D. 2008. *Non-Equilibrium Thermodynamics of Heterogeneous Systems*. New Jersey: World Scientific.

Klages, R. 2007. *Microscopic Chaos, Fractals and Transport in Nonequilibrium Statistical Mechanics*. Singapore: World Scientific.

Klich, I. 2003. Full counting statistics: An elementary derivation of Levitov's formula. Pages 397–402 of: Nazarov, Y. V. (ed.), *Quantum Noise in Mesoscopic Physics*. Dordrecht: Kluwer Academic Publishers.

Knauf, A. 1987. Ergodic and topological properties of Coulombic periodic potentials. *Commun. Math. Phys.*, **110**, 89–112.

Knudsen, M. 1909. Molecular flow of gases through orifices (and effusion). *Ann. Physik*, **28**, 999–1016.

Kolmogorov, A. N. 1954. On the conservation of quasi-periodic motions for a small change in the Hamiltonian function. *Dokl. Akad. Nauk*, **98**, 527–530.

Kolmogorov, A. N. 1956a. *Foundations of the Theory of Probability*. 2nd ed. New York: Chelsea Publishing Co.

Kolmogorov, A. N. 1956b. On the Shannon theory of information transmission in the case of continuous signals. *IRE Trans. Inform. Theory*, **1**, 102–108.

Kolmogorov, A. N. 1959. On entropy per unit time as a metric invariant of automorphisms. *Dokl. Akad. Nauk SSSR*, **124**, 754–755.

Kolomeisky, A. B., and Fisher, M. E. 2007. Molecular motors: A theorist's perspective. *Annu. Rev. Phys. Chem.*, **58**, 675–695.

Kondepudi, D., and Prigogine, I. 1998. *Modern Thermodynamics: From Heat Engines to Dissipative Structures*. Chichester, UK: Wiley.

Koopman, B. O. 1931. Hamiltonian systems and transformations in Hilbert space. *Proc. Natl. Acad. Sci.*, **17**, 315–318.

Koski, J. V., Sagawa, T., Saira, O.-P., Yoon, Y., Kutvonen, A., Solinas, P., Möttönen, M., Ala-Nissila, T., and Pekola, J. P. 2013. Distribution of entropy production in a single-electron box. *Nat. Phys.*, **9**, 644–648.

Kramers, H. A. 1940. Brownian motion in a field of force and the diffusion model of chemical reactions. *Physica*, **7**, 284–304.

Kreuzer, H. J., and Gortel, Z. W. 1986. *Physisorption Kinetics*. Berlin: Springer.

Krinner, S., Stadler, D., Husmann, D., Brantut, J.-P., and Esslinger, T. 2015. Observation of quantized conductance in neutral matter. *Nature*, **517**, 64–67.

Krinner, S., Esslinger, T., and Brantut, J.-P. 2017. Two-terminal transport measurements with cold atoms. *J. Phys.: Condens. Matter*, **29**, 343003.

Krylov, N. S. 1979. *Works on the Foundations of Statistical Physics*. Princeton, NJ: Princeton University Press.

Kubo, R. 1957. Statistical mechanical theory of irreversible processes. I. General theory and simple applications in magnetic and conduction problems. *J. Phys. Soc. Jpn.*, **12**, 570–586.

Küng, B., Rössler, C., Beck, M., Marthaler, M., Golubev, D. S., Utsumi, Y., Ihn, T., and Ensslin, K. 2012. Irreversibility on the level of single-electron tunneling. *Phys. Rev. X*, **2**, 011001.

Küng, B., Rössler, C., Beck, M., Marthaler, M., Golubev, D. S., Utsumi, Y., Ihn, T., and Ensslin, K. 2013. Test of the fluctuation theorem for single-electron transport. *J. Appl. Phys.*, **113**, 136507.

Kurchan, J. 1998. Fluctuation theorem for stochastic dynamics. *J. Phys. A: Math. Gen.*, **31**, 3719–3729.

Kurchan, J. 2000. A quantum fluctuation theorem. arXiv:cond-mat/0007360.

Kurchan, J. 2010. Six out of equilibrium lectures. In: Dauxois, T., Ruffo, S., and Cugliandolo, L. F. (eds), *Long-Range Interacting Systems* (Lecture Notes of the Les Houches Summer School, vol. 90). Oxford: Oxford University Press.

Kurtz, T. G. 1978. Strong approximation theorems for density dependent Markov chains. *Stoch. Proc. Appl.*, **6**, 223–240.

Lacoste, D., and Gaspard, P. 2014. Isometric fluctuation relations for equilibrium states with broken symmetry. *Phys. Rev. Lett.*, **113**, 240602.

Lacoste, D., and Gaspard, P. 2015. Fluctuation relations for equilibrium states with broken discrete or continuous symmetries. *J. Stat. Mech.*, **2015**, P11018.

Lacoste, D., and Mallick, K. 2009. Fluctuation theorem for the flashing ratchet model of molecular motors. *Phys. Rev. E*, **80**, 021923.

Lacoste, D., Lau, A. W. C., and Mallick, K. 2008. Fluctuation theorem and large deviation function for a solvable model of a molecular motors. *Phys. Rev. E*, **78**, 011915.
Landau, L. D., and Lifshitz, E. M. 1957. Hydrodynamic fluctuations. *JETP*, **5**, 512–513.
Landau, L. D., and Lifshitz, E. M. 1975. *Theory of Elasticity*. 2nd ed. Oxford: Pergamon Press.
Landau, L. D., and Lifshitz, E. M. 1976. *Mechanics*. 3rd ed. Oxford: Pergamon Press.
Landau, L. D., and Lifshitz, E. M. 1980a. *Statistical Physics, Part 1*. 3rd ed. Oxford: Pergamon Press.
Landau, L. D., and Lifshitz, E. M. 1980b. *Statistical Physics, Part 2*. Oxford: Pergamon Press.
Landau, L. D., and Lifshitz, E. M. 1981. *Physical Kinetics*. Oxford: Pergamon Press.
Landau, L. D., and Lifshitz, E. M. 1984. *Electrodynamics of Continuous Media*. 2nd ed. Oxford: Pergamon Press.
Landau, L. D., and Lifshitz, E. M. 1987. *Fluid Mechanics*. 2nd ed. Oxford: Pergamon Press.
Landauer, R. 1957. Spatial variation of currents and fields due to localized scatterers in metallic conduction. *IBM J. Res. Dev.*, **1**, 223–231.
Landauer, R. 1961. Irreversibility and heat generation in the computing process. *IBM J. Res. Dev.*, **5**, 183–191.
Langevin, P. 1908. On the theory of Brownian motion (in French). *C. R. Acad. Sci. Paris*, **146**, 530–532.
Langouche, F., Roekaerts, D., and Tirapegui, E. 1979. Functional integrals and the Fokker–Planck equation. *Il Nuovo Cimento B*, **53**, 135–159.
Lau, A. W. C., and Lubensky, T. C. 2007. State-dependent diffusion: Thermodynamic consistency and its path integral formulation. *Phys. Rev. E*, **76**, 011123.
Lebowitz, J. L., and Rubin, E. 1963. Dynamical study of Brownian motion. *Phys. Rev.*, **131**, 2381–2396.
Lebowitz, J. L., and Spohn, H. 1999. A Gallavotti–Cohen-type symmetry in the large deviation functional for stochastic dynamics. *J. Stat. Phys.*, **95**, 333–365.
Lecomte, V., Appert-Rolland, C., and van Wijland, F. 2005. Chaotic properties of systems with Markov dynamics. *Phys. Rev. Lett.*, **95**, 010601.
Lecomte, V., Appert-Rolland, C., and van Wijland, F. 2007. Thermodynamic formalism for systems with Markov dynamics. *J. Stat. Phys.*, **127**, 51–106.
Lee, H. K., Kwon, C., and Park, H. 2013. Fluctuation theorems and entropy production with odd-parity variables. *Phys. Rev. Lett.*, **110**, 050602.
Leggett, A. J. 2006. *Quantum Liquids*. Oxford: Oxford University Press.
Lesovik, G. B., and Sadovskyy, I. A. 2011. Scattering matrix approach to the description of quantum electron transport. *Phys. Usp.*, **54**, 1007–1059.
Levitov, L. S., and Lesovik, G. B. 1993. Charge distribution in quantum shot noise. *JETP Lett.*, **58**, 230–235.
Li, N., Ren, J., Wang, L., Zhang, G., Hänggi, P., and Li, B. 2012. Phononics: Manipulating heat flow with electronic analogs and beyond. *Rev. Mod. Phys.*, **84**, 1045–1066.
Liboff, R. L. 1990. *Kinetic Theory*. Englewood Cliffs, NJ: Prentice Hall.
Lichtenberg, A. J., and Lieberman, M. A. 1983. *Regular and Stochastic Motion*. New York: Springer.
Lide, D. R. (ed.). 2000. *CRC Handbook of Chemistry and Physics*. 81st ed. Boca Raton, FL: CRC Press.
Liebchen, B., Marenduzzo, D., Pagonabarraga, I., and Cates, M. E. 2015. Clustering and pattern formation in chemorepulsive active colloids. *Phys. Rev. Lett.*, **115**, 258301.
Lindblad, G. 1976. On the generators of quantum dynamical semigroups. *Commun. Math. Phys.*, **48**, 119–130.

Liu, S., Agarwalla, B. K., Wang, J.-S., and Li, B. 2013. Classical heat transport in anharmonic molecular junctions: Exact solutions. *Phys. Rev. E*, **87**, 022122.

Logan, J., and Kac, M. 1976. Fluctuations and the Boltzmann equation. I. *Phys. Rev. A*, **13**, 458–470.

López, R., Lim, J. S., and Sánchez, D. 2012. Fluctuation relations for spintronics. *Phys. Rev. Lett.*, **108**, 246603.

Lorentz, H. A. 1905. The motion of electrons in metallic bodies I. *Proc. R. Neth. Acad. Arts Sci. (KNAW)*, **7**, 438–453.

Lorentz, H. A. 1921. *Lessons on Theoretical Physics V, Kinetic Problems (1911-1912)* (in Dutch). Leiden: E. J. Brill.

Lorenz, E. N. 1963. Deterministic nonperiodic flow. *J. Atmos. Sci.*, **20**, 130–141.

Lovesey, S. W. 1980. *Condensed Matter Physics: Dynamic Correlations*. Reading, MA: Benjamin/Cummings.

Ludwig, G. 1962. An equation for description of fluctuation phenomena and turbulence in gases. *Physica*, **28**, 841–860.

Lugatio, L. A., and Lefever, R. 1987. Spatial Dissipative Structures in Passive Optical Systems. *Phys. Rev. Lett.*, **58**, 2209–2211.

Luo, J.-L., Van den Broeck, C., and Nicolis, G. 1984. Stability criteria and fluctuations around nonequilibrium states. *Z. Phys. B: Condens. Matter*, **56**, 165–170.

Lutsko, J. F. 2012. A dynamical theory of nucleation for colloids and macromolecules. *J. Chem. Phys.*, **136**, 034509.

Ma, W., Lutsko, J. F., Rimer, J. D., and Vekilov, P. G. 2020. Antagonistic cooperativity between crystal growth modifiers. *Nature*, **577**, 497–501.

Mabillard, J., and Gaspard, P. 2020. Microscopic approach to the macrodynamics of matter with broken symmetries. *J. Stat. Mech.*, **2020**, 103203.

Mabillard, J., and Gaspard, P. 2021. Nonequilibrium statistical mechanics of crystals. *J. Stat. Mech.*, **2021**, 063207.

MacDonald, D. K. C. 1948–1949. Spontaneous fluctuations. *Rep. Prog. Phys.*, **12**, 56–81.

Machlup, S., and Onsager, L. 1953. Fluctuations and irreversible processes. II. Systems with kinetic energy. *Phys. Rev.*, **91**, 1512–1515.

MacKay, R. S. 1993. *Renormalization in Area-Preserving Maps*. Singapore: World Scientific.

Maes, C. 1999. The fluctuation theorem as a Gibbs property. *J. Stat. Phys.*, **95**, 367–392.

Maes, C. 2020. Frenesy: Time-symmetric dynamical activity in nonequilibria. *Phys. Rep.*, **850**, 1–33.

Maes, C., and Netočný, K. 2003. Time-reversal and entropy. *J. Stat. Phys.*, **110**, 269–310.

Maes, C., and van Wieren, M. H. 2006. Time-symmetric fluctuations in nonequilibrium systems. *Phys. Rev. Lett.*, **96**, 240601.

Maes, C., Netočný, K., and Verschuere, M. 2003. Heat conduction networks. *J. Stat. Phys.*, **111**, 1219–1244.

Mandelbrot, B. B. 1982. *The Fractal Geometry of Nature*. San Francisco: Freeman and Co.

Maragakis, P., Ritort, F., Bustamante, C., Karplus, M., and Crooks, G. E. 2008. Bayesian estimates of free energies from nonequilibrium work data in the presence of instrument noise. *J. Chem. Phys.*, **129**, 024102.

Marchetti, M. C., Joanny, J. F., Ramaswamy, S., Liverpool, T. B., Prost, J., Rao, M., and Simha, R. A. 2013. Hydrodynamics of soft active matter. *Rev. Mod. Phys.*, **85**, 1143–1189.

Mareschal, M., and Kestemont, E. 1987. Experimental evidence for convective rolls in finite two-dimensional molecular models. *Nature*, **329**, 427–429.

Martin, P. C., Parodi, O., and Pershan, P. S. 1972. Unified hydrodynamic theory for crystals, liquid crystals, and normal fluids. *Phys. Rev. A*, **6**, 2401–2420.

Masters, A. J. 1998. Some notes on the dynamics of nematic liquid crystals. *Mol. Phys.*, **95**, 251–257.

Mátyás, L., and Gaspard, P. 2005. Entropy production in diffusion-reaction systems: The reactive random Lorentz gas. *Phys. Rev. E*, **71**, 036147.

Maxey, M. R., and Riley, J. J. 1983. Equation of motion for a small rigid sphere in a nonuniform flow. *Phys. Fluids*, **26**, 883–889.

Maxwell, J. C. 1860. Illustrations of the dynamical theory of gases. Part I. On the motions and collisions of perfectly elastic spheres. *Lond. Edinb. Dubl. Phil. Mag. J. Sci., 4th Series*, **19**, 19–32.

Maxwell, J. C. 1867. On the dynamical theory of gases. *Phil. Trans. R. Soc. London*, **157**, 49–88.

Maxwell, J. C. 1871. *Theory of Heat*. London: Longmans, Green, and Co.

Maxwell, J. C. 1879. On stresses in rarified gases arising from inequalities of temperature. *Phil. Trans. R. Soc. London*, **170**, 231–256.

Mazenko, G. F. 2006. *Nonequilibrium Statistical Mechanics*. Weinheim: Wiley-VCH.

Mazo, R. M. 2002. *Brownian Motion: Fluctuations, Dynamics and Applications*. Oxford: Clarendon Press.

Mazur, P. 1999. Mesoscopic nonequilibrium thermodynamics; irreversible processes and fluctuations. *Physica A*, **274**, 491–504.

Mazur, P., and Bedeaux, D. 1974. A generalization of Faxén's theorem to nonsteady motion of a sphere through an incompressible fluid in arbitrary flow. *Physica A*, **76**, 235–246.

McEwen, J.-S., Gaspard, P., Visart de Bocarmé, T., and Kruse, N. 2009. Nanometric chemical clocks. *Proc. Natl. Acad. Sci.*, **106**, 3006–3010.

McEwen, J.-S., Gaspard, P., Visart de Bocarmé, T., and Kruse, N. 2010a. Electric field induced oscillations in the catalytic water production on rhodium: A theoretical analysis. *Surf. Sci.*, **604**, 1353–1368.

McEwen, J.-S., Garcia Cantú Ros, A., Gaspard, P., Visart de Bocarmé, T., and Kruse, N. 2010b. Non-equilibrium surface pattern formation during catalytic reactions with nanoscale resolution: Investigations of the electric field influence. *Catalysis Today*, **154**, 75–84.

McLennan, J. A. 1960. Statistical mechanics of transport in fluids. *Phys. Fluids*, **3**, 493–502.

McLennan, J. A. 1961. Nonlinear effects in transport theory. *Phys. Fluids*, **4**, 1319–1324.

McLennan, J. A. 1963. The formal statistical theory of transport processes. *Adv. Chem. Phys.*, **5**, 261–317.

McQuarrie, D. A. 1967. Stochastic approach to chemical kinetics. *J. Appl. Prob.*, **4**, 413–478.

Mermin, N. D. 1968. Crystalline order in two dimensions. *Phys. Rev.*, **176**, 250–254.

Mermin, N. D., and Wagner, H. 1966. Absence of ferromagnetism or antiferromagnetism in one- or two-dimensional isotropic Heisenberg models. *Phys. Rev. Lett.*, **17**, 1133–1136.

Michaelis, L., and Menten, M. L. 1913. The kinetics of invertase action (in German). *Biochem. Z.*, **49**, 333–369.

Michal, G., and Schomburg, D. (eds). 2012. *Biochemical Pathways: An Atlas of Biochemistry and Molecular Biology*. 2nd ed. Hoboken, NJ: Wiley.

Miller, D. G. 1960. Thermodynamics of irreversible processes: The experimental verification of the Onsager reciprocal relations. *Chem. Rev.*, **60**, 15–37.

Min, W., Jiang, L., Yu, J., Kou, S. C., Qian, H., and Xie, X. S. 2005. Nonequilibrium steady state of a nanometric biochemical system: Determining the thermodynamic driving force from single enzyme turnover time traces. *Nano Lett.*, **5**, 2373–2378.

Moffitt, J. R., Chemla, Y. R., and Bustamante, C. 2010. Methods in statistical kinetics. *Methods Enzymol.*, **475**, 221–257.

Mognetti, B. M., Cicuta, P., and Di Michele, L. 2019. Programmable interactions with biomimetic DNA linkers at fluid membranes and interfaces. *Rep. Prog. Phys.*, **82**, 116601.

Moore, W. J. 1972. *Physical Chemistry*. 5th ed. London: Longman.

Mori, H. 1958. Statistical-mechanical theory of transport in fluids. *Phys. Rev.*, **112**, 1829–1842.

Mori, H. 1965. Transport, collective motion, and Brownian motion. *Prog. Theor. Phys.*, **33**, 423–455.

Moser, J. 1973. *Stable and Random Motions in Dynamical Systems*. Princeton, NJ: Princeton University Press.

Nakamura, S., Yamauchi, Y., Hashisaka, M., Chida, K., Kobayashi, K., Ono, T., Leturcq, R., Ensslin, K., Saito, K., Utsumi, Y., and Gossard, A. C. 2010. Nonequilibrium fluctuation relations in a quantum coherent conductor. *Phys. Rev. Lett.*, **104**, 080602.

Nakamura, S., Yamauchi, Y., Hashisaka, M., Chida, K., Kobayashi, K., Ono, T., Leturcq, R., Ensslin, K., Saito, K., Utsumi, Y., and Gossard, A. C. 2011. Fluctuation theorem and microreversibility in a quantum coherent conductor. *Phys. Rev. B*, **83**, 155431.

Nakano, H., and Sasa, S.-i. 2019. Statistical mechanical expressions of slip length. *J. Stat. Phys.*, **176**, 312–357.

Nambu, Y. 1960. Quasiparticles and gauge invariance in the theory of superconductivity. *Phys. Rev.*, **117**, 648–663.

Narnhofer, H., and Thirring, W. 1987. Dynamical entropy of quasifree automorphisms. *Lett. Math. Phys.*, **14**, 89–96.

Navier, C.-L. 1827. On the laws of motion of fluids (in French). *Mem. Acad. Sci. Inst. Fr.*, **6**, 389–440.

Nazarov, Yu. V., and Blanter, Ya. M. 2009. *Quantum Transport: Introduction to Nanoscience*. Cambridge, UK: Cambridge University Press.

Nelson, D. L., and Cox, M. M. 2017. *Lehninger Principles of Biochemistry*. 7th ed. New York: W. H. Freeman.

Neri, I. 2020. Second law of thermodynamics at stopping times. *Phys. Rev. Lett.*, **124**, 040601.

Neri, I., Roldán, E., and Jülicher, F. 2017. Statistics of infima and stopping times of entropy production and applications to active molecular processes. *Phys. Rev. X*, **7**, 011019.

Nicolis, G. 1972. Fluctuations around nonequilibrium states in open nonlinear systems. *J. Stat. Phys.*, **6**, 195–222.

Nicolis, G. 1979. Irreversible thermodynamics. *Rep. Prog. Phys.*, **42**, 225–268.

Nicolis, G. 1995. *Introduction to Nonlinear Science*. Cambridge, UK: Cambridge University Press.

Nicolis, G., and Malek Mansour, M. 1984. Onset of spatial correlations in nonequilibrium systems: A master-equation description. *Phys. Rev. A*, **29**, 2845–2853.

Nicolis, G., and Prigogine, I. 1971. Fluctuations in nonequilibrium systems. *Proc. Natl. Acad. Sci.*, **68**, 2102–2107.

Nicolis, G., and Prigogine, I. 1977. *Self-Organization in Nonequilibrium Systems*. New York: Wiley.

Noether, E. 1918. Invariant variation problems (in German). *Nachr. d. König. Gesellsch. d. Wiss. zu Göttingen, Math-phys. Klasse*, 235–257.

Noh, J. D., and Park, J.-M. 2012. Fluctuation relation for heat. *Phys. Rev. Lett.*, **108**, 240603.

Nonner, W., and Eisenberg, B. 1998. Ion permeation and glutamate residues linked by Poisson–Nernst–Planck theory in L-type calcium channels. *Biophys. J.*, **75**, 1287–1305.

Nosé, S. 1984a. A molecular dynamics method for simulations in the canonical ensemble. *Mol. Phys.*, **52**, 255–268.

Nosé, S. 1984b. A unified formulation of the constant temperature molecular dynamics methods. *J. Chem. Phys.*, **81**, 511–519.

Nyquist, H. 1928. Thermal agitation of electrical charges in conductors. *Phys. Rev.*, **32**, 110–113.

Onsager, L. 1931a. Reciprocal relations in irreversible processes I. *Phys. Rev.*, **37**, 405–426.

Onsager, L. 1931b. Reciprocal relations in irreversible processes II. *Phys. Rev.*, **38**, 2265–2279.

Onsager, L., and Machlup, S. 1953. Fluctuations and irreversible processes. *Phys. Rev.*, **91**, 1505–1512.

Onuki, A. 1978. On fluctuations in μ space. *J. Stat. Phys.*, **18**, 475–499.

Oono, Y., and Paniconi, M. 1998. Steady state thermodynamics. *Prog. Theor. Phys. Suppl.*, **130**, 29–44.

Oosawa, F., and Hayashi, S. 1986. The loose coupling mechanism in molecular machines of living cells. *Adv. Biophys.*, **22**, 151–183.

Oppenheim, I., and Levine, R. D. 1979. Nonlinear transport processes: Hydrodynamics. *Physica A*, **99**, 383–402.

Ortiz de Zárate, J. M., and Sengers, J. V. 2006. *Hydrodynamic Fluctuations in Fluids and Fluid Mixtures*. Amsterdam: Elsevier.

Ott, E. 1993. *Chaos in Dynamical Systems*. Cambridge, UK: Cambridge University Press.

Pathria, R. K. 1972. *Statistical Mechanics*. Oxford: Pergamon.

Paul, G. L., and Pusey, P. N. 1981. Observation of a long-time tail in Brownian motion. *J. Phys. A: Math. Gen.*, **14**, 3301–3327.

Pauli, W. 1928. On the H-theorem of entropy increase from the standpoint of the new quantum mechanics (in German). Pages 30–45 of: Debye, P. (ed.), *Probleme der Modernen Physik: Sommerfeld-Festschrift*. Leipzig: Hirzel.

Pauling, L. 1970. *General Chemistry*. 3rd ed. San Francisco: Freeman and Co.

Penrose, O., and Fife, P. C. 1990. Thermodynamically consistent models of phase-field type for the kinetics of phase transitions. *Physica D*, **43**, 44–62.

Perrin, J. 1910. *Brownian Movement and Molecular Reality*. London: Taylor and Francis.

Peterson, R. L. 1967. Formal theory of nonlinear response. *Rev. Mod. Phys.*, **39**, 69–77.

Piccirelli, R. A. 1968. Theory of the dynamics of simple fluids for large spatial gradients and long memory. *Phys. Rev.*, **175**, 77–98.

Pietzonka, P., Barato, A. C., and Seifert, U. 2016. Universal bounds on current fluctuations. *Phys. Rev. E*, **93**, 052145.

Pigolotti, S., Neri, I., Roldán, E., and Jülicher, F. 2017. Generic properties of stochastic entropy production. *Phys. Rev. Lett.*, **119**, 140604.

Planck, M. 1914. *The Theory of Heat Radiation*. 2nd ed. Philadelphia, PA: P. Blakiston's Son & Co.

Planck, M. 1917. On a theorem of statistical dynamics and its extension to quantum theory (in German). *Sitz. König. Preuss. Akad. Wiss.*, **24**, 324–341.

Polettini, M., and Esposito, M. 2014. Irreversible thermodynamics of open chemical networks. I. Emergent cycles and broken conservation laws. *J. Chem. Phys.*, **141**, 024117.

Pollack, G. L. 1969. Kapitza resistance. *Rev. Mod. Phys.*, **41**, 48–81.

Pollicott, M. 1985. On the rate of mixing of Axiom A flows. *Invent. Math.*, **81**, 413–426.

Pollicott, M. 1986. Meromorphic extensions of generalised zeta functions. *Invent. Math.*, **85**, 147–164.

Popkov, V., Schadschneider, A., Schmidt, J., and Schütz, G. M. 2015. Fibonacci family of dynamical universality classes. *Proc. Natl. Acad. Sci.*, **112**, 12645–12650.

Porporato, A., Rigby, J. R., and Daly, E. 2007. Irreversibility and fluctuation theorem in stationary time series. *Phys. Rev. Lett.*, **98**, 094101.

Pottier, N. 2009. *Nonequilibrium Statistical Physics: Linear Irreversible Processes*. Oxford: Oxford University Press.

Prähofer, M., and Spohn, H. 2004. Exact scaling functions for one-dimensional stationary KPZ growth. *J. Stat. Phys.*, **115**, 255–279.

Present, R. D. 1958. *Kinetic Theory of Gases*. New York: McGraw-Hill.

Pressé, S., Ghosh, K., Lee, J., and Dill, K. A. 2013. Principles of maximum entropy and maximum caliber in statistical physics. *Rev. Mod. Phys.*, **85**, 1115–1141.

Prigogine, I. 1949. The domain of validity of the thermodynamics of irreversible phenomena (in French). *Physica*, **15**, 272–284.

Prigogine, I. 1967. *Introduction to Thermodynamics of Irreversible Processes*. New York: Wiley.

Prigogine, I., and Lefever, R. 1968. Symmetry breaking instabilities in dissipative systems. II. *J. Chem. Phys.*, **48**, 1695–1700.

Prigogine, I., and Nicolis, G. 1967. Symmetry breaking instabilities in dissipative systems. *J. Chem. Phys.*, **46**, 3542–3550.

Probstein, R. F. 2003. *Physicochemical Hydrodynamics*. 2nd ed. Hoboken, NJ: Wiley.

Provata, A., Nicolis, C., and Nicolis, G. 2014. DNA viewed as an out-of-equilibrium structure. *Phys. Rev. E*, **89**, 052105.

Qian, H., and Beard, D. A. 2005. Thermodynamics of stoichiometric biochemical networks in living systems far from equilibrium. *Biophys. Chem.*, **114**, 213–220.

Qian, H., and Xie, X. S. 2006. Generalized Haldane equation and fluctuation theorem in the steady-state cycle kinetics of single enzymes. *Phys. Rev. E*, **74**, 010902.

Rao, R., and Esposito, M. 2016. Nonequilibrium thermodynamics of chemical reaction networks: Wisdom from stochastic thermodynamics. *Phys. Rev. X*, **6**, 041064.

Rao, R., and Esposito, M. 2018a. Conservation laws and work fluctuation relations in chemical reaction networks. *J. Chem. Phys.*, **149**, 245101.

Rao, R., and Esposito, M. 2018b. Conservation laws shape dissipation. *New J. Phys.*, **20**, 023007.

Rayleigh, Lord (J. W. Strutt). 1891. Dynamical problems in illustration of the theory of gases. *Lond. Edinb. Dubl. Phil. Mag. J. Sci., 5th Series*, **32**, 424–445.

Redfield, A. G. 1965. The theory of relaxation processes. *Adv. Magn. Opt. Reson.*, **1**, 1–32.

Reichl, L. E. 1998. *A Modern Course in Statistical Physics*. 2nd ed. New York: Wiley.

Reigh, S. Y., Huang, M.-J., Schofield, J., and Kapral, R. 2016. Microscopic and continuum descriptions of Janus motor fluid flow fields. *Phil. Trans. R. Soc. A*, **374**, 20160140.

Reitz, J. R., and Milford, F. J. 1967. *Foundations of Electromagnetic Theory*. 2nd ed. Reading, MA: Addison-Wesley.

Résibois, P., and De Leener, M. 1977. *Classical Kinetic Theory of Fluids*. New York: Wiley.

Résibois, P., and Lebowitz, J. L. 1965. Microscopic theory of Brownian motion in an oscillating field; Connection with macroscopic theory. *Phys. Rev.*, **139**, A1101–A1111.

Rice, S. A., and Zhao, M. 2000. *Optical Control of Molecular Dynamics*. New York: Wiley.

Risken, H. 1989. *The Fokker–Planck Equation*. 2nd ed. Berlin: Springer.

Roberts, J. A. G., and Quispel, G. R. W. 1992. Chaos and time-reversal symmetry: Order and chaos in reversible dynamical systems. *Phys. Rep.*, **216**, 63–177.
Robertson, B. 1966. Equations of motion in nonequilibrium statistical mechanics. *Phys. Rev.*, **144**, 151–161.
Robertson, B. 1967. Equations of motion in nonequilibrium statistical mechanics. II. Energy transport. *Phys. Rev.*, **160**, 175–183.
Robertson, B., Schofield, J., Gaspard, P., and Kapral, R. 2020. Molecular theory of Langevin dynamics for active self-diffusiophoretic colloids. *J. Chem. Phys.*, **153**, 124104.
Roldán, E. 2014. *Irreversibility and Dissipation in Microscopic Systems*. Cham: Springer.
Roldán, E., and Parrondo, J. M. R. 2010. Estimating dissipation from single stationary trajectories. *Phys. Rev. Lett.*, **105**, 150607.
Ronis, D., Kovac, J., and Oppenheim, I. 1977. Molecular hydrodynamics of inhomogeneous systems: The origin of slip boundary conditions. *Physica A*, **88**, 215–241.
Rowlinson, J. S., and Widom, B. 1989. *Molecular Theory of Capillarity*. Oxford: Clarendon Press.
Rubí, J. M., and Mazur, P. 2000. Nonequilibrium thermodynamics and hydrodynamic fluctuations. *Physica A*, **276**, 477–488.
Ruelle, D. 1978. *Thermodynamic Formalism*. Reading, MA: Addison-Wesley.
Ruelle, D. 1986a. Locating resonances for Axiom A dynamical systems. *J. Stat. Phys.*, **44**, 281–292.
Ruelle, D. 1986b. Resonances of chaotic dynamical systems. *Phys. Rev. Lett.*, **56**, 405–407.
Rutherford, E., Geiger, H., and Bateman, H. 1910. The probability variations in the distribution of α particles. *Lond. Edinb. Dubl. Phil. Mag. J. Sci.*, **20**, 698–707.
Sagawa, T., and Ueda, M. 2010. Generalized Jarzynski equality under nonequilibrium feedback control. *Phys. Rev. Lett.*, **104**, 090602.
Saha, A., and Jayannavar, A. M. 2008. Nonequilibrium work distributions for a trapped Brownian particle in a time-dependent magnetic field. *Phys. Rev. E*, **77**, 022105.
Saha, S., Golestanian, R., and Ramaswamy, S. 2014. Clusters, asters, and collective oscillations in chemotactic colloids. *Phys. Rev. E*, **89**, 062316.
Saito, K., and Dhar, A. 2011. Generating function formula of heat transfer in harmonic networks. *Phys. Rev. E*, **83**, 041121.
Saito, K., and Utsumi, Y. 2008. Symmetry in full counting statistics, fluctuation theorem, and relations among nonlinear transport coefficients in the presence of a magnetic field. *Phys. Rev. B*, **78**, 115429.
Sánchez, D., and Büttiker, M. 2004. Magnetic-field asymmetry of nonlinear mesoscopic transport. *Phys. Rev. Lett.*, **93**, 106802.
Sasa, S.-i. 2014. Derivation of hydrodynamics from the Hamiltonian description of particle systems. *Phys. Rev. Lett.*, **112**, 100602.
Sauter, Th., Neuhauser, W., Blatt, R., and Toschek, P. E. 1986. Observation of quantum jumps. *Phys. Rev. Lett.*, **57**, 1696–1698.
Schlögl, F. 1971. On thermodynamics near a steady state. *Z. Phys.*, **248**, 446–458.
Schlögl, F. 1972. Chemical reaction models for non-equilibrium phase transitions. *Z. Phys.*, **253**, 147–161.
Schnakenberg, J. 1976. Network theory of microscopic and macroscopic behavior of master equation systems. *Rev. Mod. Phys.*, **48**, 571–585.
Schnitzer, M. J., Visscher, K., and Block, S. M. 2000. Force production by single kinesin motors. *Nat. Cell Biol.*, **2**, 718–723.
Schöll, E. 2001. *Nonlinear Spatio-Temporal Dynamics and Chaos in Semiconductors*. Cambridge, UK: Cambridge University Press.
Scott, S. K. 1991. *Chemical Chaos*. Oxford: Clarendon Press.

Scriven, L. E. 1960. Dynamics of a fluid interface. *Chem. Eng. Sci.*, **12**, 98–108.
Sedra, A. S., and Smith, K. C. 2004. *Microelectronic Circuits*. 5th ed. New York: Oxford University Press.
Segel, I. H. 1975. *Enzyme Kinetics*. New York: Wiley.
Seifert, U. 2005a. Entropy production along a stochastic trajectory and an integral fluctuation theorem. *Phys. Rev. Lett.*, **95**, 040602.
Seifert, U. 2005b. Fluctuation theorem for a single enzyme or molecular motor. *Europhys. Lett.*, **70**, 36–41.
Seifert, U. 2011. Stochastic thermodynamics of single enzymes and molecular motors. *Eur. Phys. J. E*, **34**, 26.
Seifert, U. 2012. Stochastic thermodynamics, fluctuation theorems and molecular machines. *Rep. Prog. Phys.*, **75**, 126001.
Seitaridou, E., Inamdar, M. M., Phillips, R., Ghosh, K., and Dill, K. 2007. Measuring flux distributions for diffusion in the small-numbers limit. *J. Phys. Chem. B*, **111**, 2288–2292.
Sekimoto, K. 1997. Kinetic characterization of heat bath and the energetics of thermal ratchet models. *J. Phys. Soc. Jpn.*, **66**, 1234–1237.
Sekimoto, K. 1998. Langevin equation and thermodynamics. *Prog. Theor. Phys. Suppl.*, **130**, 17–27.
Sekimoto, K. 2010. *Stochastic Energetics*. Berlin: Springer.
Servantie, J., and Gaspard, P. 2003. Methods of calculation of a friction coefficient: Application to nanotubes. *Phys. Rev. Lett.*, **91**, 185503.
Servantie, J., and Gaspard, P. 2006a. Rotational dynamics and friction in double-walled carbon nanotubes. *Phys. Rev. Lett.*, **97**, 186106.
Servantie, J., and Gaspard, P. 2006b. Translational dynamics and friction in double-walled carbon nanotubes. *Phys. Rev. B*, **73**, 186106.
Sevick, E. M., Prabhakar, R., Williams, S. R., and Searles, D. J. 2008. Fluctuation theorems. *Annu. Rev. Phys. Chem.*, **59**, 603–633.
Shannon, C. E., and Weaver, W. 1949. *The Mathematical Theory of Communication*. Urbana, IL: The University of Illinois Press.
Shear, D. 1967. An analog of the Boltzmann H-theorem (a Liapunov function) for systems of coupled chemical reactions. *J. Theor. Biol.*, **16**, 212–228.
Shockley, W. 1949. The theory of *p-n* junctions in semiconductors and *p-n* junction transistors. *Bell Syst. Tech. J.*, **28**, 435–489.
Shockley, W., Sparks, M., and Teal, G. K. 1951. *p-n* junction transistors. *Phys. Rev.*, **83**, 151–162.
Siegert, A. J. F. 1949. On the approach to statistical equilibrium. *Phys. Rev.*, **76**, 1708–1714.
Sinai, Ya. G. 1959. On the notion of entropy of dynamical systems. *Dokl. Akad. Nauk SSSR*, **124**, 768–771.
Sinai, Ya. G. 1970. Dynamical systems with elastic reflections. *Russ. Math. Surv.*, **25**, 137–189.
Sköld, K., Rowe, J. M., Ostrowski, G., and Randolph, P. D. 1972. Coherent- and incoherent-scattering laws of liquid argon. *Phys. Rev. A*, **6**, 1107–1131.
Smeets, R. M. M., Keyser, U. F., Dekker, N. H., and Dekker, C. 2008. Noise in solid-state nanopores. *Proc. Natl. Acad. Sci.*, **105**, 417–421.
Sowa, Y., Rowe, A. D., Leake, M. C., Yakushi, T., Homma, M., Ishijima, A., and Berry, R. M. 2005. Direct observation of steps in rotation of the bacterial flagellar motor. *Nature*, **437**, 916–919.
Speck, T., and Seifert, U. 2005. Integral fluctuation theorem for the housekeeping heat. *J. Phys. A: Math. Gen.*, **38**, L581–L588.

Speck, T., Blickle, V., Bechinger, C., and Seifert, U. 2007. Distribution of entropy production for a colloidal particle in a nonequilibrium steady state. *EPL*, **79**, 30002.

Spinney, R. E., and Ford, I. J. 2012. Nonequilibrium thermodynamics of stochastic systems with odd and even variables. *Phys. Rev. Lett.*, **108**, 170603.

Spohn, H. 1980. Kinetic equations from Hamiltonian dynamics: Markovian limits. *Rev. Mod. Phys.*, **52**, 569–615.

Spohn, H. 1991. *Large Scale Dynamics of Interacting Particles*. Berlin: Springer.

Spohn, H. 2014. Nonlinear fluctuating hydrodynamics for anharmonic chains. *J. Stat. Phys.*, **154**, 1191–1227.

Srednicki, M. 1994. Chaos and quantum thermalization. *Phys. Rev. E*, **50**, 888–901.

Steckelmacher, W. 1986. Knudsen flow 75 years on. *Rep. Prog. Phys.*, **49**, 1083–1107.

Stratonovich, R. L. 1992. *Nonlinear nonequilibrium thermodynamics I*. Berlin: Springer.

Stratonovich, R. L. 1994. *Nonlinear nonequilibrium thermodynamics II. Advanced theory*. Berlin: Springer.

Strogatz, S. H. 1994. *Nonlinear Dynamics and Chaos*. Cambridge, MA: Perseus Books.

Strunz, W. T., Diósi, L., Gisin, N., and Yu, T. 1999. Quantum trajectories for Brownian motion. *Phys. Rev. Lett.*, **83**, 4909–4913.

Suárez, A., Silbey, R., and Oppenheim, I. 1992. Memory effects in the relaxation of quantum open systems. *J. Chem. Phys.*, **97**, 5101–5107.

Sutherland, W. 1905. A dynamical theory of diffusion for non-electrolytes and the molecular mass of albumin. *Phil. Mag.*, **9**, 781–785.

Svensson, E. C., Brockhouse, B. N., and Rowe, J. M. 1967. Crystal dynamics of copper. *Phys. Rev.*, **155**, 619–632.

Svoboda, K., Schmidt, C. F., Schnapp, B. J., and Block, S. M. 1993. Direct observation of kinesin stepping by optical trapping interferometry. *Nature*, **365**, 721–727.

Szamel, G. 1997. Statistical mechanics of dissipative transport in crystals. *J. Stat. Phys.*, **87**, 1067–1082.

Szamel, G., and Ernst, M. H. 1993. Slow modes in crystals: A method to study elastic constants. *Phys. Rev. B*, **48**, 112–118.

Szász, D. 1996. Boltzmann's ergodic hypothesis, a conjecture for centuries? *Studia Sci. Math. Hungarica*, **31**, 299–322.

Szilard, L. 1929. On the decrease of entropy in a thermodynamic system by the intervention of intelligent beings (in German). *Zeitschrift für Physik*, **53**, 840–856.

Tasaki, H. 2000. Jarzynski relations for quantum systems and some applications. arXiv:cond-mat/0009244 .

Tasaki, S., and Gaspard, P. 1995. Fick's law and fractality of nonequilibrium stationary states in a reversible multibaker map. *J. Stat. Phys.*, **81**, 935–987.

Thirring, W. 1983. *Quantum Mechanics of Large Systems*. New York: Springer.

Tobiska, J., and Nazarov, Yu. V. 2005. Inelastic interaction corrections and universal relations for full counting statistics in a quantum contact. *Phys. Rev. B*, **72**, 235328.

Tolman, R. C. 1938. *The Principles of Statistical Mechanics*. Oxford: Clarendon Press.

Touchette, H. 2009. The large deviation approach to statistical mechanics. *Phys. Rep.*, **478**, 1–70.

Toyabe, S., Watanabe-Nakayama, T., Okamoto, T., Kudo, S., and Muneyuki, E. 2011. Thermodynamic efficiency and mechanochemical coupling of F_1-ATPase. *Proc. Natl. Acad. Sci.*, **108**, 17951–17956.

Tran, D. T., Dauphin, A., Grushin, A. G., Zoller, P., and Goldman, N. 2017. Probing topology by "heating": Quantized circular dichroism in ultracold atoms. *Sci. Adv.*, **3**, e1701207.

Trepagnier, E. H., Jarzynski, C., Ritort, F., Crooks, G. E., Bustamante, C. J., and Liphardt, J. 2004. Experimental test of Hatano and Sasa's nonequilibrium steady-state equality. *Proc. Natl. Acad. Sci.*, **101**, 15038–15041.

Uehling, E. A., and Uhlenbeck, G. E. 1933. Transport phenomena in Einstein–Bose and Fermi–Dirac gases. I. *Phys. Rev.*, **43**, 552–561.

Uhlenbeck, G. E., and Ornstein, L. S. 1930. On the theory of the Brownian motion. *Phys. Rev.*, **36**, 823–841.

Utsumi, Y., and Saito, K. 2009. Fluctuation theorem in a quantum-dot Aharonov–Bohm interferometer. *Phys. Rev. B*, **79**, 235311.

Utsumi, Y., Golubev, D. S., Marthaler, M., Saito, K., Fujisawa, T., and Schön, G. 2010. Bidirectional single-electron counting and the fluctuation theorem. *Phys. Rev. B*, **81**, 125331.

Valadares, L. F., Tao, Y.-G., Zacharia, N. S., Kitaev, V., Galembeck, F., Kapral, R., and Ozin, G. A. 2010. Catalytic nanomotors: Self-propelled sphere dimers. *Small*, **6**, 565–572.

van Beijeren, H. 2012. Exact results for anomalous transport in one-dimensional Hamiltonian systems. *Phys. Rev. Lett.*, **108**, 180601.

van Beijeren, H., Dorfman, J. R., Posch, H. A., and Dellago, Ch. 1997. Kolmogorov–Sinai entropy for dilute gases in equilibrium. *Phys. Rev. E*, **56**, 5272–5277.

Van den Broeck, C. 2005. Thermodynamic efficiency at maximum power. *Phys. Rev. Lett.*, **95**, 190602.

Van den Broeck, C. 2013. Stochastic thermodynamics: A brief introduction. Pages 155–193 of: Bechinger, C., Sciortino, F., and Ziherl, P. (eds), *Physics of Complex Colloids, Proceedings of the International School of Physics "Enrico Fermi," vol. 184*. Amsterdam: IOS Press.

Van den Broeck, C., and Esposito, M. 2010a. Three detailed fluctuation theorems. *Phys. Rev. Lett.*, **104**, 090601.

Van den Broeck, C., and Esposito, M. 2010b. Three faces of the second law. I. Master equation formulation. *Phys. Rev. E*, **82**, 011143.

Van den Broeck, C., and Esposito, M. 2010c. Three faces of the second law. II. Fokker–Planck formulation. *Phys. Rev. E*, **82**, 011144.

Van den Broeck, C., and Esposito, M. 2015. Ensemble and trajectory thermodynamics: A brief introduction. *Physica A*, **418**, 6–16.

Van den Broeck, C., Meurs, P., and Kawai, R. 2005. From Maxwell demon to Brownian motor. *New J. Phys.*, **7**, 10.

van Hemmen, J. L. 1980. Dynamics and ergodicity of the infinite harmonic crystal. *Phys. Rep.*, **65**, 43–149.

Van Hove, L. 1954. Correlations in space and time and Born approximation scattering in systems of interacting particles. *Phys. Rev.*, **95**, 249–262.

van Kampen, N. G. 1974. Fluctuations in Boltzmann's equation. *Phys. Lett. A*, **50**, 237–238.

van Kampen, N. G. 1981. *Stochastic Processes in Physics and Chemistry*. Amsterdam: North-Holland.

van Zon, R., and Cohen, E. G. D. 2003a. Extension of the fluctuation theorem. *Phys. Rev. Lett.*, **91**, 110601.

van Zon, R., and Cohen, E. G. D. 2003b. Stationary and transient work-fluctuation theorems for a dragged Brownian particle. *Phys. Rev. E*, **67**, 046102.

van Zon, R., and Cohen, E. G. D. 2004. Extended heat-fluctuation theorems for a system with deterministic and stochastic forces. *Phys. Rev. E*, **69**, 056121.

van Zon, R., Ciliberto, S., and Cohen, E. G. D. 2004. Power and heat fluctuation theorems for electric circuits. *Phys. Rev. Lett.*, **92**, 130601.

Verley, G., Willaert, T., Van den Broeck, C., and Esposito, M. 2014. Universal theory of efficiency fluctuations. *Phys. Rev. E*, **90**, 052145.

Viscardy, S., and Gaspard, P. 2003. Viscosity in the escape-rate formalism. *Phys. Rev. E*, **68**, 041205.

Viscardy, S., Servantie, J., and Gaspard, P. 2007a. Transport and Helfand moments in the Lennard-Jones fluid. I. Shear viscosity. *J. Chem. Phys.*, **126**, 184512.

Viscardy, S., Servantie, J., and Gaspard, P. 2007b. Transport and Helfand moments in the Lennard-Jones fluid. II. Thermal conductivity. *J. Chem. Phys.*, **126**, 184513.

Vladimirsky, V., and Terletsky, Y. A. 1945. Hydrodynamic theory of translational Brownian motion. *Zh. Eksp. Theor. Fiz.*, **15**, 258–262.

von Smoluchowski, M. 1906. Kinetic theory of Brownian motion and suspension (in German). *Ann. Physik*, **21**, 756–780.

Wachtel, A., Rao, R., and Esposito, M. 2018. Thermodynamically consistent coarse graining of biocatalysts beyond Michaelis–Menten. *New J. Phys.*, **20**, 042002.

Wallace, D. C. 1987. On the role of density fluctuations in the entropy of a fluid. *J. Chem. Phys.*, **87**, 2282–2284.

Wallace, D. C. 1998. *Thermodynamics of Crystals*. Mineola, NY: Dover.

Walz, C., and Fuchs, M. 2010. Displacement field and elastic constants in nonideal crystals. *Phys. Rev. B*, **81**, 134110.

Wang, C., and Feldman, D. E. 2015. Fluctuation relations for spin currents. *Phys. Rev. B*, **92**, 064406.

Wang, G. M., Sevick, E. M., Mittag, E., Searles, D. J., and Evans, D. J. 2002. Experimental demonstration of violations of the second law of thermodynamics for small systems and short time scales. *Phys. Rev. Lett.*, **89**, 050601.

Wang, H., and Oster, G. 1998. Energy transduction in the F_1 motor of ATP synthase. *Nature*, **396**, 279–282.

Wang, S.-L., Sekerka, R. F., Wheeler, A. A., Murray, B. T., Coriell, S. R., Braun, R. J., and McFadden, G. B. 1993. Thermodynamically-consistent phase-field models for solidification. *Physica D*, **69**, 189–200.

Wegscheider, R. 1901. On simultaneous equilibria and the relationships between thermodynamics and reaction kinetics of homogeneous systems (in German). *Monatshefte für Chemie*, **32**, 849–906.

Weinberg, S. 1995. *The Quantum Theory of Fields, vol. I*. Cambridge, UK: Cambridge University Press.

Weinberg, S. 1996. *The Quantum Theory of Fields, vol. II*. Cambridge, UK: Cambridge University Press.

Weiss, G. 1975. Time reversibility of linear stochastic processes. *J. Appl. Prob.*, **12**, 831–836.

Wigner, E. P. 1932. On the quantum correction for thermodynamic equilibrium. *Phys. Rev.*, **40**, 749–760.

Wigner, E. P. 1963. Events, laws of nature, and invariance principles. *Nobel Lecture*, December 12. The Nobel Foundation, Stockholm, Sweden.

Wood, K., Van den Broeck, C., Kawai, R., and Lindenberg, K. 2007. Fluctuation theorem for entropy production during effusion of an ideal gas with momentum transfer. *Phys. Rev. E*, **75**, 061116.

Xiao, T. J., and Zhou, Y. 2018. Stochastic thermodynamics of mesoscopic electrochemical reactions. *Chin. J. Chem. Phys.*, **31**, 61–65.

Xie, S. 2001. Single-molecule approach to enzymology. *Single Mol.*, **2**, 229–236.

Yasuda, R., Noji, H., Yoshida, M., Kinosita Jr., K., and Itoh, H. 2001. Resolution of distinct rotational substeps by submillisecond kinetic analysis of F_1-ATPase. *Nature*, **410**, 898–904.

Yvon, J. 1935. *The Statistical Theory of Fluids and the Equation of State* (in French). Paris: Hermann & Cie.

Yvon, J. 1966. *Correlations and Entropy in Classical Statistical Mechanics* (in French). Paris: Dunod.

Zubarev, D. N. 1966. A statistical operator for non stationary processes. *Sov. Phys. Doklady*, **10**, 850–852.

Zwanzig, R. W. 1961. Statistical mechanics of irreversibility. Pages 106–141 of: *Lectures in Theoretical Physics, Vol. III*. New York: Interscience Publishers.

Zwanzig, R. 2001. *Nonequilibrium Statistical Mechanics*. Oxford: Oxford University Press.

Zwanzig, R., and Bixon, M. 1970. Hydrodynamic theory of the velocity correlation function. *Phys. Rev. A*, **2**, 2005–2012.

Index

Printed in the United States
by Baker & Taylor Publisher Services